Biology and Culture of Percid Fishes

Patrick Kestemont • Konrad Dabrowski
Robert C. Summerfelt
Editors

Biology and Culture of Percid Fishes

Principles and Practices

Springer

Editors
Patrick Kestemont
Research Unit in Environmental and Evolutionary Biology
University of Namur
Namur, Belgium

Konrad Dabrowski
School of Environment and Natural Resources
Ohio State University
Columbus, OH, USA

Robert C. Summerfelt
Department of Natural Resource Ecology and Management
Iowa State University of Science and Technology
Ames, IA, USA

ISBN 978-94-017-7226-6 ISBN 978-94-017-7227-3 (eBook)
DOI 10.1007/978-94-017-7227-3

Library of Congress Control Number: 2015947840

Springer Dordrecht Heidelberg New York London

Printed on acid-free paper

Springer Science+Business Media B.V. Dordrecht is part of Springer Science+Business Media (www.springer.com)

Foreword

Fish are consumed by their siblings, and some species of all other animals within the subphylum Vertebrata including humans. Fish provide humans high-quality and essential proteins as well as minerals. Some fishes of the finned variety, such as gadids, provide little fat (<1 %), while others, for example, salmonids and clupeids, have a high fat content containing a range of fat-soluble vitamins and essential fatty acids, all vital for human health. The demands of the growing human population outstrip the sustainability of wild fishes which have become increasingly depleted. Aquaculture production has increased in an effort to compensate for this depletion. Overfishing and aquaculture, however, have both come under scrutiny by environmentalists, although threats from aquaculture appear to be less than from the former.

The Percidae are an ancient northern hemisphere family of fishes containing about 266 species (mostly darters North American) and 11 genera. Several species are vulnerable and their conservation is of extreme concern. Others are economically important in recreational and commercial fishing. It is necessary to understand the biology of this family to conserve and manage their populations. In addition this knowledge can be used as a foundation for their culture. Four species of two genera, two in North America (yellow perch *Perca flavescens* and walleye *Sander vitreus*) and two in Europe (perch *Perca fluviatilis* and pike perch *Sander lucioperca*), have shown particular promise in aquaculture. This is probably aided by the similar biology of the two *Perca* and the two *Sander* species, which means that advances in aquaculture can be applied across all four. Much of the biology of percids is known, but as time moves on new techniques, which improve and develop our understanding of the basics, are applied in research. In addition there has been a significant advancement in percid aquaculture. So since my book on percids was published 15 years ago, there have been many developments, and the present tome edited by Patrick Kestemont, Robert Summerfelt, and Konrad Dabrowski is important in bringing us up to date in both the biology and culture of this group.

The book is extensive and comprehensive, particularly regarding the present state of aquaculture. It is divided into eight sections containing in total 35 chapters with topics including systematics, ecology, reproduction, early life stages, development, growth, metabolism, nutrition, behavior, husbandry, genetic modification, domestication, stress, immunology, diseases, health, commercial production, marketing, and economics. I was interested in reading about new molecular tools allowing the elucidation of evolutionary diversification of the percids. Other highlights include the development of out-of-season spawning in which the main environmental cues are temperature and photoperiod both of which can be programmed for intensive fish farming; the use of molecular analyses in determining the quality of ova, embryos, and larvae; the development of sensory systems and the gut; nutritional requirements and the function of lipids and fatty acids; biphasic growth; hormonal sexual reversal; selective breeding programs; domestication; the use of intensified recirculating aquaculture system(s) (RAS); intensive larviculture; bioenergetics modeling; and culture of percids for enhancement. I was surprised to learn that the country with the highest import and consumption of perch was Switzerland. The scarcity of information on artificially controlling reproduction by hormones and lack of knowledge on the immune system, although most diseases have been identified, indicate the need for future research. Percid fish farming, including the development of culture in different countries and defining the quality of fishes in the industry, has still a long way to go, but this book provides the information we have at hand at the present time. It will be of immense value to students and researchers in fish biology and fisheries and those working in the aquaculture industry.

Craig Consultancy
Dunscore
Dumfries
Scotland

John F. Craig

Preface

The Percidae family of fishes is a large, diverse, and economically important group of mostly freshwater fishes that includes 11 genera and 266 known species, belonging to the order Perciformes and the class Actinopterygii, and originally inhabiting the northern hemisphere. Several species of this family play important ecological roles in the functioning of lotic and lentic aquatic ecosystems as well as providing a valuable resource for both recreational and commercial fisheries. Considering the slow but continuous decline of percid capture fisheries, some of these species have been identified from the early 1970s as valuable candidates for aquaculture, and research efforts have been intensified from the 1990s, providing a huge amount of recent scientific and technical knowledge. In the past, some authors have dedicated a significant part of their time to compile the existing data on the biology and exploitation of several representatives of this diverse fish family, while some others assembled the knowledge on the aquaculture aspects. John Thorpe published in 1977 the first *Synopsis of Biological Data on the Perch Perca fluviatilis* (Linnaeus, 1758) *and Perca flavescens* (Mitchill, 1814) within the *Fisheries Synopsis* series produced by the Food and Agriculture Organization (FAO) of the United Nations. In 1979, Canadian fisheries biologists started to assemble the existing knowledge on walleye *Sander vitreus* biology and fisheries and produced the *Walleye Synopsis* (Colby et al., FAO Fisheries Synopsis 119). About a decade later, in 1987, John Craig published the first comprehensive monograph on percid fish biology (*The Biology of Perch and Related Fish*) and updated it in 2000 with the book *Percid Fishes: Systematics, Ecology, and Exploitation*, covering most taxonomical, ecological, and fisheries aspects of this family, but also including two new chapters, one on aquaculture (written by Patrick Kestemont and Charles Mélard) and one on North American darters (subfamily Etheostomatinae) (written by Larry Page). From an aquaculture perspective, a multi-contributor book dealing with the culture of walleye was edited by Robert Summerfelt in 1996 (*Walleye Culture Manual*), while, the same year, a special issue of the *Journal of Applied Ichthyology* (vol 12, Issue 3–4, 137–200) was published that focused on the *Recent Advances in the Aquaculture of Percid Fish* (Patrick Kestemont and Konrad Dabrowski as guest editors). More recently, the North Central Regional Aquaculture Center (NCRAC) of the USA promoted the compilation of information gleaned from scientific and extension reports in order to produce in 2006 the *Yellow Perch Perca flavescens Culture Guide* (edited by Hart, Garling and Malison). The American Fisheries Society supported the publication of the revised and updated walleye synopsis (*Biology, Management, and Culture of Walleye and Sauger*, edited by Bruce Barton in 2011). These publications represent the insight of fish cultural technicians and workers working in collaboration with many scientists and engineers on just a few economically important species of this family.

Biology and Culture of Percid Fishes: Principles and Practices intends to provide the reader a comprehensive scientific state of the art of percid biology and culture. All biological aspects relevant to the culture of different species belonging to the Percidae family are described in detail, including a complete critical peer review of the scientific literature complemented by unpublished data from different contributors.

This book represents the collaboration of 75 authors from 15 countries. It is composed of 35 chapters distributed into eight parts, dealing with systematics, ecology, and evolution of percids (2 chapters); reproductive biology (4 chapters); early life stages (development, metabolism, and husbandry) (5 chapters); juvenile and grow-out stages (growth, metabolism, behavior, and husbandry) (8 chapters); nutrition, feeds, and feeding practices (3 chapters); genetic improvement and domestication (5 chapters); stress, immunology, diseases, and health management (4 chapters); and commercial production, quality, marketing, and economics (4 chapters).

Chapter 1 (C.A. Stepien and A.E. Haponski) provides an updated knowledge on the taxonomy, distribution, and evolution of Percidae, using the results of the latest DNA sequence technology and morphological analyses to resolve the relationships of the family Percidae and its component genera and species. Newly assembled distribution maps are presented for the taxa, and their primary distinguishing morphological and life history is summarized. In Chap. 2, Z.S. Feiner and T.O. Höök review the basic information on environmental biology of the *Perca* and *Sander* genera, on which the majority of fisheries and aquaculture practices are focused. The authors discuss how individual- and population-level vital rates, including growth, foraging, reproduction, recruitment, and mortality, are mediated by biotic and abiotic environmental variables, with the aim to identify the major environmental drivers of biological variation in percids and thereby provide information for fisheries management practices for both wild and cultured percid populations.

In Part II, Chap. 3 (P. Fontaine et al.) describes with details the reproduction of the main cultured percid species, including the morpho-anatomical, histological, and physiological changes occurring during an annual reproductive cycle and the way to obtain out-of-season gonad maturation and spawning by photothermal regime regulation. In Chap. 4, written by D. Zarski and coauthors, endocrine regulation and reproductive protocols are reviewed, focusing on the final gamete maturation, spermiation, and ovulation processes which are the steps needed to be considered for successful reproduction. Characteristics of sperm morphology, physiology, and viability in Percidae are described in Chap. 5 (H. Alavi et al.) as well as the main factors affecting semen quality. This chapter also presents the methods for sperm short-term storage and cryopreservation. Part II is complemented by Chap. 6 (B. Schlaerlinger and D. Zarski) reviewing the current knowledge on ova characteristics and the proper embryonic development and larval metamorphosis for several percid species, allowing the identification of ova defects or developmental failures and the selection of relevant quality indicators for ova and larvae.

Part III focuses on the early life stages' development in Percidae and on the larval culture methods of *Perca* and *Sander* genera. In Chap. 7, M. Kamaszewski and T. Ostaszewska describe the development of sensory organs (olfactory placodes and epithelium, taste buds, eye and optic vesicles, inner ear, and neuromasts), starting during embryogenesis and evolving over the larval and juvenile stages, in relation with trophic and environmental preferences of the different percid species. Chapter 8 (N. Hamza et al.) examines the development and functionality of the digestive system, based on histological and enzymatic studies, focusing on the stomach, pancreatic, and intestinal digestive enzymes. The authors also show that the digestive structures and enzymes activities can be affected by the nature and the diet composition, providing some information about the nutritional requirements of percid larvae. In Chap. 9, P. Kestemont and coauthors present an overview of the different methods used to produce juveniles of the Eurasian perch and yellow perch, separating the production of fish in fertilized ponds, the fertilization in mesocosms and semi-intensive production systems, and the intensive production in tanks with supply of live prey gradually replaced by formulated feeds. For each system, the optimal husbandry conditions as well as the influence of main factors influencing the survival and growth of fish from larval to juvenile stages are described. Similarly, Chaps. 10 (S. Stenfeldt) and 11 (R.C. Summerfelt and J.A. Johnson) focus on the cultivated species of the *Sander* genus and review the intensive larviculture of pike perch (*Sander lucioperca*) and walleye, respectively. The description of husbandry aspects includes the embryonic development and egg incubation, the optimal biotic and abiotic rearing conditions, and the problems occurring in intensive larval culture system such as the non-inflation of gas bladder and occurrence of skeletal deformities.

Characteristics of cultivated percid species during grow-out stages, from juvenile to adult/market size fish, are described in Part IV. The current state of knowledge regarding fish skeletal muscle characteristics, factors affecting muscle growth, and proteomic-based research in teleost fish with emphasis on percids is reviewed in Chap. 12. The authors (K. Kwasek et al.) also compare skeletal muscle sarcoplasmic proteins/peptides between fast- and slow-growing yellow perch in order to identify the differences in expression of skeletal muscle proteins in fish exhibiting different growth capabilities and, ultimately, propose a selection tool for the production of larger percid fish. Bioenergetic aspects are examined in Chap. 13 (A. Alanärä and A. Strand) and Chap. 14 (C.P. Mandenjian). As percid fish species are relatively new in culture, there are no models available to estimate their energy requirement in intensive rearing systems, which in turn limits the opportunities to calculate the required daily feed allowance. In Chap. 13, authors put together data from the scientific literature to produce an alternative model for prediction of the daily growth and energy need of percid fish, with a special attention to Eurasian perch. This chapter also discusses how factors such as season and culture conditions influence the bioenergetic requirements and energy expenditures. As shown in Chap. 14, modeling can be used to identify the important factors determining the growth of percids in lakes, rivers, or seas. Bioenergetic modeling can also be applied to estimate the amount of food being annually consumed by the percid population or to quantify the effect of the difference in growth between sexes on contaminant accumulation in, for instance, walleye. Behavior is described in Chap. 15 by C. Magnhagen. This author highlights the importance of fish body size on trophic behavior and social interactions, but also the influence of the environment. Improving knowledge regarding the behavior of percids in the wild is suggested for obtaining the best result in culture. Chapters 16 (T. Polizar and coauthors) and 17 (I.A. Johnson and R.C. Summerfelt) provide updated information and recommendations regarding the main husbandry aspects affecting the culture of Eurasian perch and North American walleye, respectively. For both species optimal abiotic and biotic conditions are defined according to the rearing system, from the pond to recirculating aquaculture system(s) (RAS). Main advantages and constraints of these culture systems are presented, and opportunities for future research are discussed. In Chap. 18, Briland and coauthors analyze fish and plankton ecology in production ponds to provide a better understanding of the ecological and biological factors involved in optimal pond production of percid fingerlings for stocking. Factors affecting lifetime growth patterns of percids in natural populations are reviewed by M.D. Rennie and P.A. Venturelli in Chap. 19. The authors apply a biphasic growth model to describe the lifetime growth of Eurasian perch and pike perch populations and discuss the hypotheses for proximate mechanisms of female-biased sexual size dimorphism as well as the influence of fish density, food availability, predation, parasitism, and disease within an ecological context.

Part V deals with feeding and nutrition in percids. The influence of feed composition (especially the lipid and fatty acid type and content) on brood stock performances, with consequences for survival and growth of early life stages, is described in European percid species (Chap. 20, P. Kestemont et al.) and yellow perch (Chap. 21, K. Dabrowski et al.), respectively. Although larvae of some percid species still rely on live prey, significant progress has been achieved recently with the use of formulated diets as starting feed, namely, through an enrichment in phospholipids. In Chap. 22, F. Geay and P. Kestemont review the main abiotic and biotic factors influencing the feeding activity of juveniles and adults. The nutritional requirements of percid fish have been investigated, as well as the use of alternative oil sources, suggesting a high potential of these species to biosynthetize highly unsaturated fatty acids (HUFA) when fish oil is replaced by plant oil rich in polyunsaturated fatty acids (PUFA).

Intensive percid fish culture is rather new technology, and, up to date, despite a need to improve growth performances of these species, there have been very few attempts for genetic selection or designed domestication. However, different authors have investigated the potentialities of genetic manipulation and mass selection to improve the production of percid species in culture, as shown in Part VI. C. Rougeot (Chap. 23) presents an overview of the different methods used to produce triploids or to control the development of the phenotypic sex and the production of all-female populations, through hormonal treatment or gynogenesis. J. Held and coauthors (Chap. 24) report on the improved growth performances of interspecific hybrids (Eurasian perch and yellow perch, walleye and sauger). In Chap. 25, C.A. Stepien et al. focus on the comparative population structure and genetic diversity of yellow perch and walleye across North America, in relationship to historical patterns, habitat connectivity, dispersal ability, distributional abundances, and reproductive behavior. Such genetic characterization has been used to select yellow perch brood stock from different geographical regions of North America in order to enhance growth and evaluate the heritability of that growth (Chap. 26, F. W. Goetz et al.). The last chapter of Part VI written by R.J.W. Blonk and H. Komen (Chap. 27) describes the principles of selective breeding programs, the most commonly used selection methods and their implications for percids, as well as some insights into the optimization of breeding programs.

One general feature of the percid fishes used in aquaculture is their high sensitivity to stress under captive conditions, with negative consequences in terms of reduced immune resistance and, as a corollary, sensitivity to diseases and pathogens. The state of the art on these interrelated aspects is described in Part VII. As reported by S. Milla and coauthors in Chap. 28, accumulating evidence indicates that corticosteroids are strongly regulated in percids after exposure to stressors and play essential roles in the stress response. This chapter characterizes the corticosteroid synthesis and receptivity in percid fish, as well as their secondary and tertiary responses to stressors due to culture conditions. The possible reduction of stress sensitivity through progressive domestication is discussed in Chap. 29 by J. Douxfils et al. Comparing the responses of different captive generations of Eurasian perch to different kinds of stressors, the authors indicate that domestication positively influences fish tolerance to some stressors like chronic confinement but not to more severe ones such as hypoxia. The immune system is characterized by R. Mandiki and coauthors (Chap. 30), with a special emphasis on the immunocompetence changes in relation to environmental conditions as well as on the positive response of percid fish to some immunomodulatory compounds. In Chap. 31, H.D. Rodgers and N.B.D. Phelps describe the main infectious diseases associated with Eurasian perch and yellow perch, including viral and bacterial diseases, protozoan parasites, as well as noninfectious conditions (tail erosion, gill disease, and skeletal deformities) giving rise to significant livestock challenges and welfare problems.

The last section deals with the commercial production, quality as human food, marketing, and economics. It is by far the section for which the availability of data is the scarcest, indicating that the percid culture sector is still in its infancy as far as commercial production is concerned. Chapter 32 (S. Stenfeldt et al.) overviews the state of production in different countries, mostly European, involved in the production of Eurasian perch and pike perch. For each country, the main culture techniques are summarized, and production types are specified, according to the local or international markets. Aside from the production performances, usually expressed in terms of tonnage, the quality of the final product is now of prime importance, and the development of the percid fish industry calls for reflection on the concept and quantitative determination of quality. Chapter 33, written by M. Thomas et al., illustrates the complex picture of quality in percid fishes, based on nutritional, technological, sensory, and sanitary components, and identifies the main biotic and abiotic factors influencing this quality. The actual market of percid fishes is not easy to evaluate because it is largely considered, up to now, as a niche market, with large variations of consumer preferences according to the different European regions. The present picture of the European market of Eurasian perch is described in Chap. 34 (D. Toner), as well as perspectives regarding its evolution concurrently with the decline of captive fisheries supply and development of aquaculture production. The last chapter of this section (J.L. Overton et al.) aims to identify the main factors of success and constraints in percid fish aquaculture, including the positive aspects such as the control of reproductive cycle and out-of-season spawning or the better knowledge of feeding and nutritional requirements, but also the limitations for further upscaling of production, such as the slow growth, the need of domestication, or the lack of veterinary knowledge of percid diseases and prevention (vaccines, probiotics).

This book and others before it demonstrate that Percid culture has a promising future. The voluminous literature reported here demonstrates the existence of a sound foundation of basic and applied science as well as over 100 years of practical experience to move Percid cultural technology forward to produce fish for enhancement stocking as well as for commercial food fish markets. Yet, there is a critical need to see the advances in scientific knowledge integrated into production practices.

Namur, Belgium — Patrick Kestemont
Columbus, OH, USA — Konrad Dabrowski
Ames, IA, USA — Robert C. Summerfelt

Chapter Reviewers

This book has benefited from the contribution of many scientists, experts on the *Percidae* family, and specialists on topics covered by the different chapters, who accepted to serve as reviewers. The editors are grateful to the reviewers listed below:

Alami-Durante Hélène, France
Babiak Igor, Norway
Bégout Marie-Laure, France
Blonk Robbert, The Netherlands
Bower Paul R., USA
Brown Paul, USA
Cahu Chantal, France
Duncan Neil, Spain
Fontaine Pascal, France
Höök Tomas, USA
Jobling Malcolm, Norway
Kocovsky Patrick, USA
Komen Hans, The Netherlands
Magnhagen Carin, Sweden
Mélard Charles, Belgium
Milla Sylvain, France
Norberg Birgitta, Norway
Lund Ivar, Denmark
Portella Maria Cella, Brazil
Rennie Mickael, Canada
Roseman Edward, USA
Rougeot Carole, Belgium
Stejskal Vlastimil, Czech Republic
Stepien Carol, USA
Sturm Armin, United Kingdom
Tort Lluis, Spain
Vandeputte Marc, France
Witeska Malgorzata, Poland

Contents

Part I Systematics, Ecology, and Evolution of Percids

1 **Taxonomy, Distribution, and Evolution of the Percidae** 3
Carol A. Stepien and Amanda E. Haponski

2 **Environmental Biology of Percid Fishes** 61
Zachary S. Feiner and Tomas O. Höök

Part II Reproductive Biology

3 **Broodstock Management and Control of the Reproductive Cycle** 103
P. Fontaine, N. Wang, and B. Hermelink

4 **Artificial Reproduction of Percid Fishes** 123
D. Żarski, A. Horváth, J.A. Held, and D. Kucharczyk

5 **Sperm Morphology, Physiology, Motility, and Cryopreservation in Percidae** 163
Sayyed Mohammad Hadi Alavi, Andrzej Ciereszko, Azadeh Hatef, Jiří Křišťan, Boris Dzyuba, Sergei Boryshpolets, Marek Rodina, Jacky Cosson, and Otomar Linhart

6 **Evaluation and Improvements of Egg and Larval Quality in Percid Fishes** 193
B. Schaerlinger and D. Żarski

Part III Early Life Stages: Development, Metabolism, and Husbandry

7 **Development of the Sense Organs in Percid Fishes** 227
M. Kamaszewski and T. Ostaszewska

8 **Development and Functionality of the Digestive System in Percid Fishes Early Life Stages** 239
Neila Hamza, Teresa Ostaszewska, and Patrick Kestemont

9 **Culture Methods of Eurasian Perch and Yellow Perch Early Life Stages** 265
P. Kestemont, C. Mélard, J.A. Held, and K. Dabrowski

10 **Culture Methods of Pikeperch Early Life Stages** 295
Svend Steenfeldt

11 **Intensive Culture of Walleye from Egg Incubation to Juvenile** 313
Robert C. Summerfelt and J. Alan Johnson

Part IV Juvenile and Grow-Out Stages: Growth, Metabolism, Behavior and Husbandry

12 **Muscle Protein Characteristic and Its Association with Faster Growth in Percids and Other Teleosts** 339
Karolina Kwasek, Macdonald Wick, and Konrad Dabrowski

13 **The Energy Requirements of Percid Fish in Culture** 353
Anders Alanärä and Åsa Strand

14 **Bioenergetics Modeling of Percid Fishes** 369
Charles P. Madenjian

15 **Behaviour of Percid Fishes in the Wild and Its Relevance for Culture** 399
Carin Magnhagen

16 **Culture Methods of Eurasian Perch During Ongrowing** 417
Tomáš Policar, Azin Mohagheghi Samarin, and Charles Mélard

17 **Intensive Culture Methods of Walleye During Ongrowing** 437
J. Alan Johnson and Robert C. Summerfelt

18 **Large-Scale Production of Yellow Perch, Walleye, and Hybrid Walleye in Ponds** 469
Ruth D. Briland, Cathleen M. Doyle, and David A. Culver

19 **The Ecology of Lifetime Growth in Percid Fishes** 499
Michael D. Rennie and Paul A. Venturelli

Part V Nutrition, Feeds and Feeding Practices

20 **Nutritional Requirements and Feeding of Broodstock and Early Life Stages of Eurasian Perch and Pikeperch** 539
Patrick Kestemont and Emilie Henrotte

21 **Effects of Dietary Levels of PUFA Fed to Adult Yellow Perch on the Fatty Acid Composition of Eggs and Larvae Characteristics: New Research Directions** 565
Konrad Dabrowski, Jacques Rinchard, Sergiusz Czesny, and Malgorzata Korzeniowska

22 **Feeding and Nutrition of Percid Fishes During Ongrowing Stages** 587
Florian Geay and Patrick Kestemont

Part VI Genetic Improvement and Domestication

23 **Sex and Ploidy Manipulation in Percid Fishes** 625
Carole Rougeot

24 **Performance of Hybrid Percids** 635
James A. Held, Syaghalirwa N.M. Mandiki, Carole Rougeot, and Patrick Kestemont

25 **Comparative Genetic Diversity, Population Structure, and Adaptations of Walleye and Yellow Perch Across North America** 643
Carol A. Stepien, Osvaldo J. Sepulveda-Villet, and Amanda E. Haponski

26 **Production of Genetically Defined Perch Broodstocks and Their Selection for Fast Growth** 691
Frederick W. Goetz, Daniel R. Rosauer, Michael Grzybowski, Frederick P. Binkowski, and Brian S. Shepherd

27 **Genetic Improvement of Percids** 699
R.J.W. Blonk and J. Komen

Part VII Stress, Immunology, Diseases and Health Management

28 **Corticosteroids and the Stress Response in Percid Fish** 725
S. Milla, J. Douxfils, S.N.M. Mandiki, and M. Saroglia

29 **Domestication and Responses to Stress** 743
J. Douxfils, S.N.M. Mandiki, C. Mathieu, S. Milla, and P. Kestemont

30 **Immune Status and Immunomodulation in Percid Fish** 761
S.N.M. Mandiki, J. Douxfils, C. Mathieu, S. Milla, E. Henrotte, H. Jansen, and P. Kestemont

31 **Percid Fish Health and Disease** 799
H.D. Rodger and N.B.D. Phelps

Part VIII Commercial Production, Quality, Marketing and Economics

32 Current Status of Eurasian Percid Fishes Aquaculture 817
Svend Steenfeldt, Pascal Fontaine, Julia Lynne Overton, Tomáš Policar, Damien Toner, Bahram Falahatkar, Ákos Horváth, Ines Ben Khemis, Neila Hamza, and Mohammed Mhetli

33 Concept and Determinism of Quality in Percid Fishes 843
Marielle Thomas, Guillaume Mairesse, Jean-Noël Gardeur, and Jean Brun-Bellut

34 The Market for Eurasian Perch 865
Damien Toner

35 Commercial Production: Factors for Success and Limitations in European Percid Fish Culture 881
Julia L. Overton, Damien Toner, Tomáš Policar, and Dariusz Kucharczyk

Index 891

Part I
Systematics, Ecology, and Evolution of Percids

Chapter 1
Taxonomy, Distribution, and Evolution of the Percidae

Carol A. Stepien and Amanda E. Haponski

Abstract The family Percidae exclusively is native to freshwaters of the Northern Hemisphere, with just two of its genera divided between Eurasia and North America. Percidae comprises 11 genera and an estimated 266–275 species, reaching tremendous species richness in the North American darters. We provide an up-to-date account relating the results of the latest DNA sequence and morphological analyses to resolve the relationships of the family Percidae, including its component genera and species. We provide newly assembled distribution maps for the taxa, and summarize their primary distinguishing morphological characters and life history. For each genus, the latest phylogenetic tree of species relationships is shown and explained. We relate these findings to historic biogeography and contemporary distributions. Just recently, tremendous inroads have been made using new molecular tools and analyses that allow us to begin to understand the tremendous evolutionary diversification of the Percidae, as well as the landscape and climate factors that have shaped these patterns. This information may provide an important indication of the future responses of percid taxa to continued anthropogenic influences.

Keywords Percidae • Taxonomy • Evolution • Diversification • Biogeography

1.1 Introduction to the Family Percidae and Its Diversity

The fish family Percidae Rafinesque 1815 is classified as belonging to the Order Perciformes and the Class Actinopterygii. Percidae contains 11 genera, with an estimated 266–275 species recognized at present (lower limit from Eschmeyer's Catalog of Fishes; updated April 2014). These genera include: *Ammocrypta* Jordan 1877 (6 species), *Crystallaria* (Jordan and Gilbert in Jordan 1885) (2 species),

C.A. Stepien (✉) • A.E. Haponski
Great Lakes Genetics and Genomics Laboratory, Lake Erie Center and Department of Environmental Sciences, The University of Toledo,
6200 Bayshore Rd., Toledo, OH 43615, USA

Present Affiliation: Museum of Zoology and Department of Ecology and Evolutionary Biology, University of Michigan, 1109 Geddes Ave, Ann Arbor, MI 48109, USA
e-mail: carol.stepien@utoledo.edu

P. Kestemont et al. (eds.), *Biology and Culture of Percid Fishes*,
DOI 10.1007/978-94-017-7227-3_1

Etheostoma Rafinesque 1819 (~165–174 species; Near et al. 2011; Near and Keck 2013), *Gymnocephalus* Bloch 1793 (5 species), *Nothonotus* Putnam 1863 (~21 species), *Perca* Linnaeus 1758 (3 species), *Percarina* Nordmann 1840 (2 species), *Percina* (Haldeman 1842) (~52 species), *Romanichthys* Dumitrescu, Bănărescu and Stoica 1957 (1 species), *Sander* (Oken 1817) (5 species), and *Zingel* Cloquet 1817 (4 species). Percids are native to freshwaters in the Northern Hemisphere (Fig. 1.1), occurring in Eurasia (Fig. 1.1a) and/or North America (Fig. 1.1b), with a few species ranging into brackish waters.

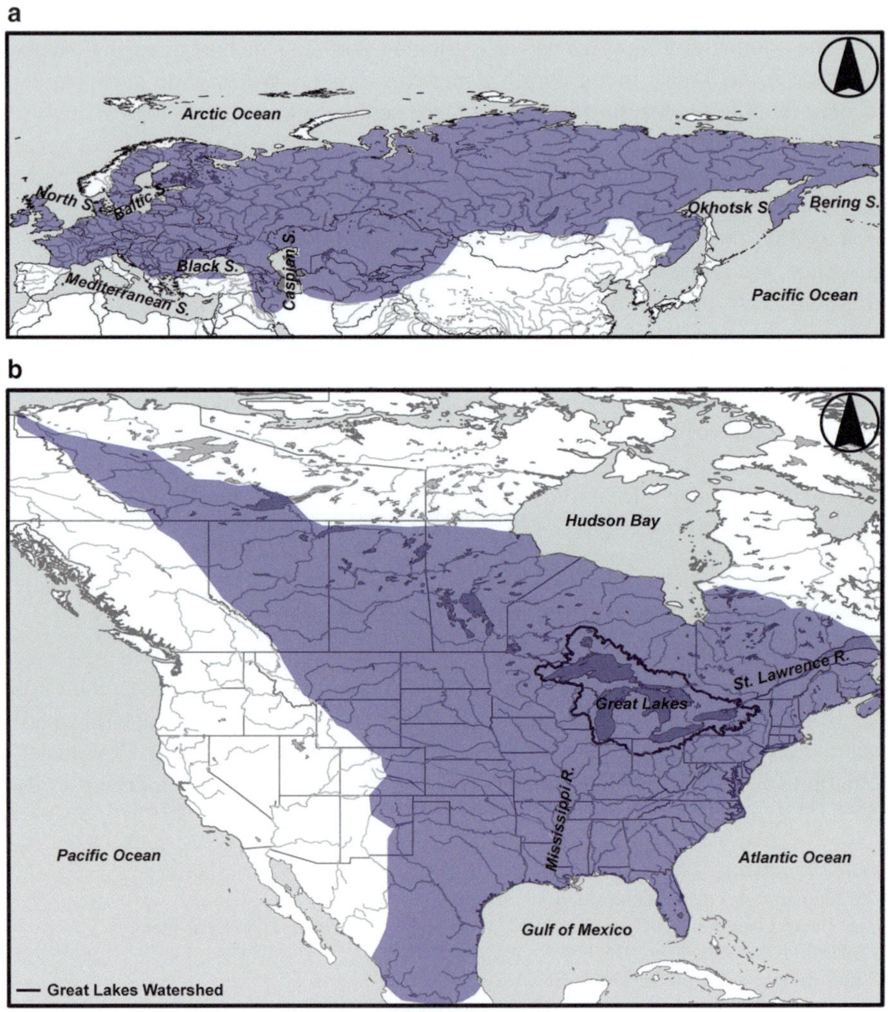

Fig. 1.1 Maps showing the native ranges of the family Percidae in (**a**) Eurasia and (**b**) North America, created using ArcGIS® software by Esri (Redlands, CA) from information in Collette and Bănărescu (1977), Craig (2000), and Maitland (2000)

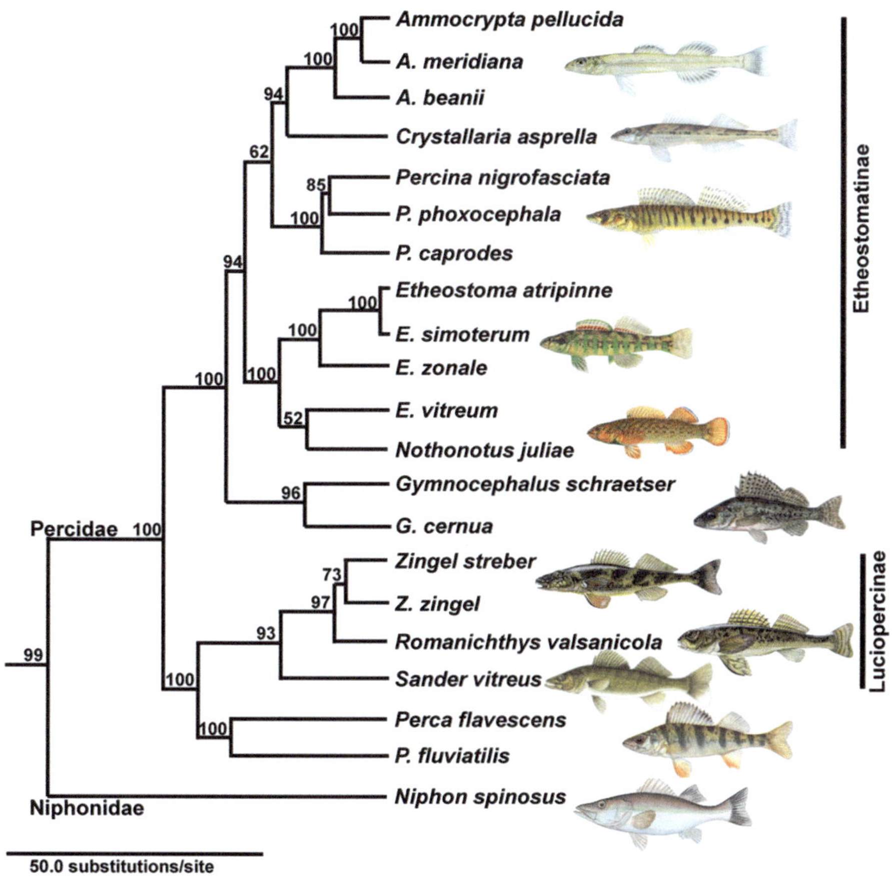

Fig. 1.2 Phylogenetic relationships of the family Percidae and its putative sister group the Niphonidae, modified from Betancur-R et al. (2013, 2014; http://www.deepfin.org/Classification_v2.htm). Tree was constructed by Betancur-R et al. (2013, 2014) from 21 DNA sequence regions distributed across the genome, including 19 nuclear exons, one nuclear intron, and mtDNA 16S rDNA sequences in RaXML (Stamatakis 2014; http://sco.h-its.org/exelixis/web/software/raxml/index.html). Specific model and analysis parameters are provided in Betancur-R et al. (2013, 2014). Above nodes = percentage support from 1000 bootstrap pseudo-replications; those with $\geq$50 % support are reported

The most recent and taxonomically complete DNA sequence phylogenetic analyses support a monophyletic Percidae, based on 19 nuclear exons, a nuclear intron, and the mtDNA 16S RNA gene by Betancur-R et al. (2013, 2014; Fig. 1.2), and on mitochondrial (mt) DNA sequences from the cytochrome oxidase *b* (cyt *b*) and the 12S RNA genes (Sloss et al. 2004; Billington and Sloss 2011; Fig. 1.3). Percids possess the following morphological characters: (1) ctenoid scales, (2) two dorsal fins (the first is spinous, and the second has soft rays), (3) thoracically-located pelvic fins with a single spine and five soft rays, (4) five to seven branchiostegal

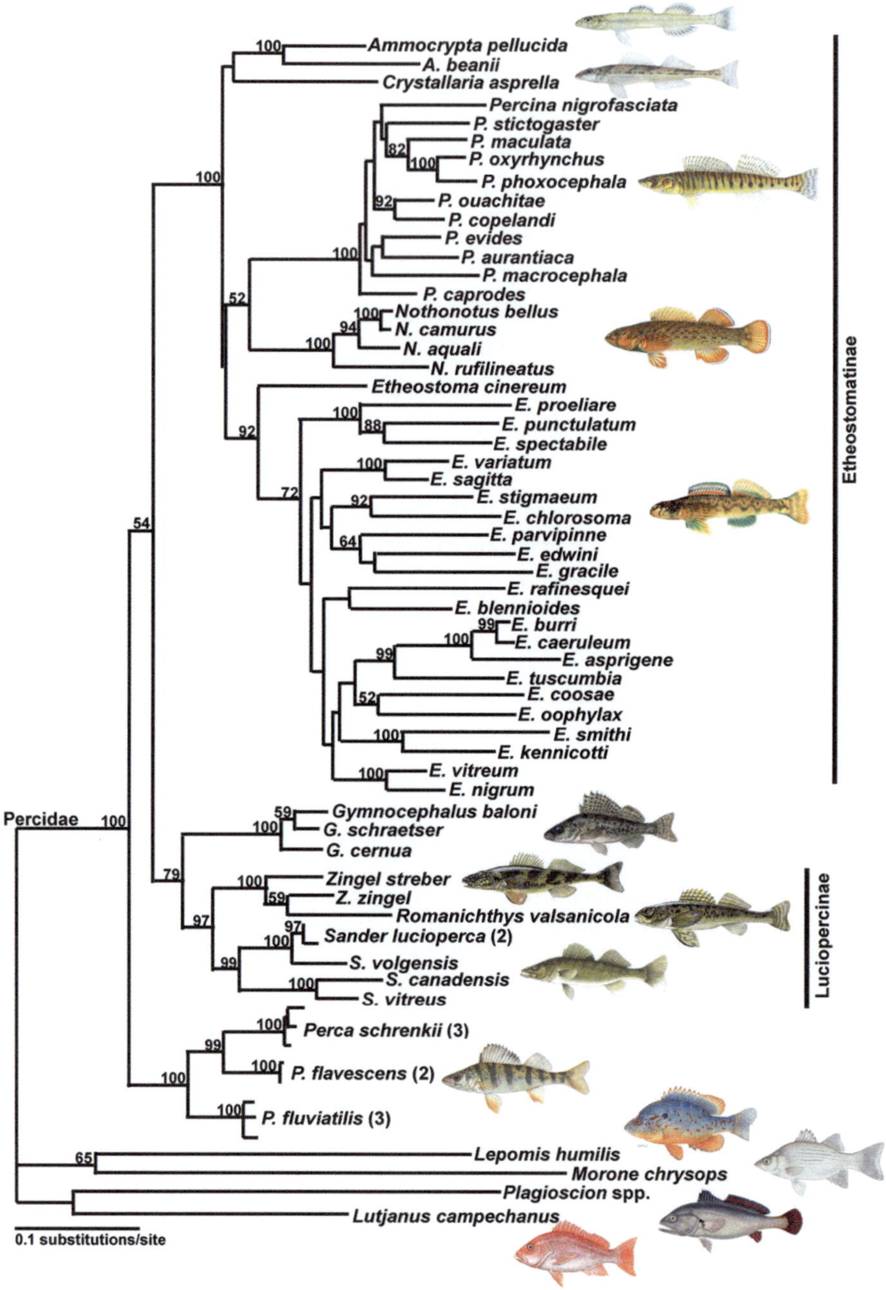

Fig. 1.3 Phylogenetic maximum likelihood tree constructed in PAUP* v.4.0b10 (Swofford 2002) showing relationships among members of the family Percidae based on combined mtDNA cytochrome *b* (cyt *b*) and 12S RNA gene sequence haplotypes, modified from Sloss et al. (2004; see for specific model analyses and parameters). Above nodes = percentage support from 500 bootstrap pseudo-replications ≥ 50 %. Numbers in parentheses indicate multiple haplotypes per species

rays, (5) spines on the cheeks and gill covers (opercula), (6) two narrow bands of numerous close-set teeth on the sides of the mouth (palatines), and (7) many have a heart-shaped plate of teeth on the roof of the mouth (vomer) (Craig 2000; Page 2000). Several derived osteological characters unite the Percidae, which were summarized by Bruner (2011) as including: lack of a lateral line tube on the supracleithrum, a single spine on the first proximal pterygiophore of the anal fin, two anal fin spines, 31–49 vertebrae, a neural arch fused to the centrum, and the presence of a single predorsal bone.

The Percidae family once was hypothesized to have originated in Eurasia, where there are presently fewer species than in North America (see Collette and Bănărescu 1977). Other studies have shown that the family instead originally was widely distributed across Eurasia and North America, at high northern latitudes during the Paleogene and Neogene periods (~66–5.3 million years ago (Mya)). Northerly percid populations then migrated southward during the end of the Neogene and beginning of the Quaternary period, during the Pliocene and Pleistocene epochs (~5.3–0.01 Mya), based on geographic distributions and fossil evidence (Cavender 1998). Two genera – *Perca* and *Sander* – are distributed across both Eurasia and North America, with separate species being native to each continent (Figs. 1.4 and 1.5). Recent phylogeographic and evolutionary studies (see Haponski and Stepien 2013 for *Sander*, and Stepien et al. 2015a for *Perca*) support continuous distributions of the common ancestors of each of the two genera across the North-Atlantic Land Bridge (NALB). Their respective species then diverged during the mid-Miocene disruption of the Land Bridge between the two continents. The native range of North American percids exclusively occurs east of the Rocky Mountains (Fig. 1.1b; Boschung and Mayden 2004), with the exception of the Mexican darter *Etheostoma pottsi* (Girard 1859), which is the sole darter to naturally occur in the Pacific drainage (Miller 2005). This general biogeographic distribution supports the NALB hypothesis.

Two species of Eurasian percids have been anthropogenically introduced to North America – the pikeperch *Sander lucioperca* (Linnaeus 1758) and the ruffe *Gymnocephalus cernua* (Linnaeus 1758). The pikeperch was stocked intentionally into Spiritwood Lake, North Dakota in 1989, in the hopes that it would become a valuable fishery. There since has been natural reproduction, but the population remains very small and does not support a fishery (Fuller 2014). The ruffe now is an established invasive species in the upper North American Great Lakes, after having been accidentally introduced in ~1986 via ballast water discharge from one or more oceanic vessels arriving from Europe (Pratt et al. 1992; Stepien et al. 1998, 2004, 2005). Its Great Lakes introduction was traced using genetic data to the European Elbe River population (Stepien et al. 2005), coinciding with increased trade between the U.S. and East Germany near the end of the "cold war".

The percid family is species rich in North America, where the darters are endemic and have radiated extensively, especially in the Central Highlands region. The darters include members of five genera – *Ammocrypta*, *Crystallaria*, *Etheostoma*, *Nothonotus*, and *Percina* – whose numbers total about 20 % of all North American freshwater fish species (summarized by Carlson and Wainwright 2010). The five darter genera are classified together as the subfamily Etheostomatinae Agassiz

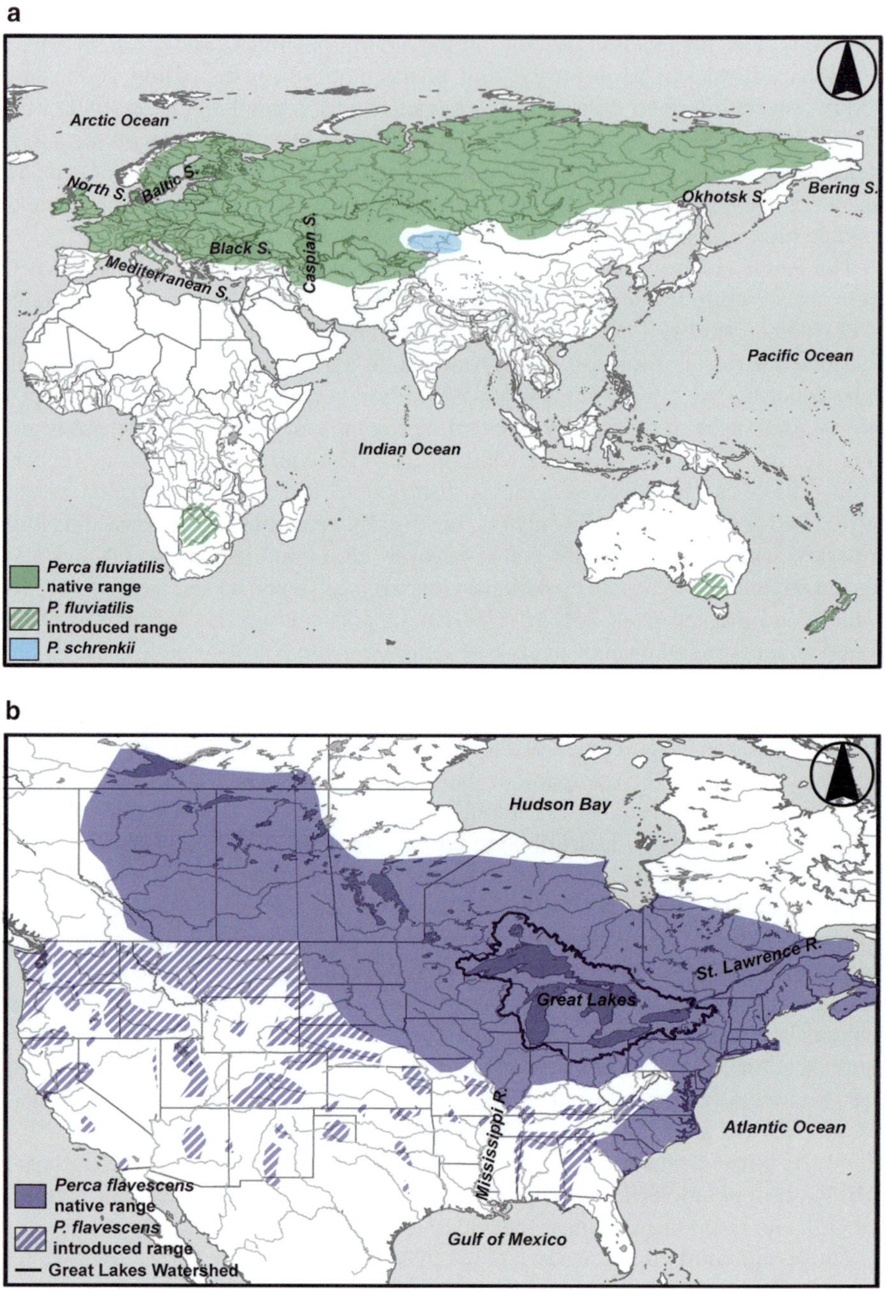

Fig. 1.4 Maps showing the native and introduced ranges of *Perca* spp. in (**a**) Eurasia and (**b**) North America, based on information from Collette and Bănărescu (1977), Craig (2000), Page and Burr (2011), and Fuller and Neilson (2012)

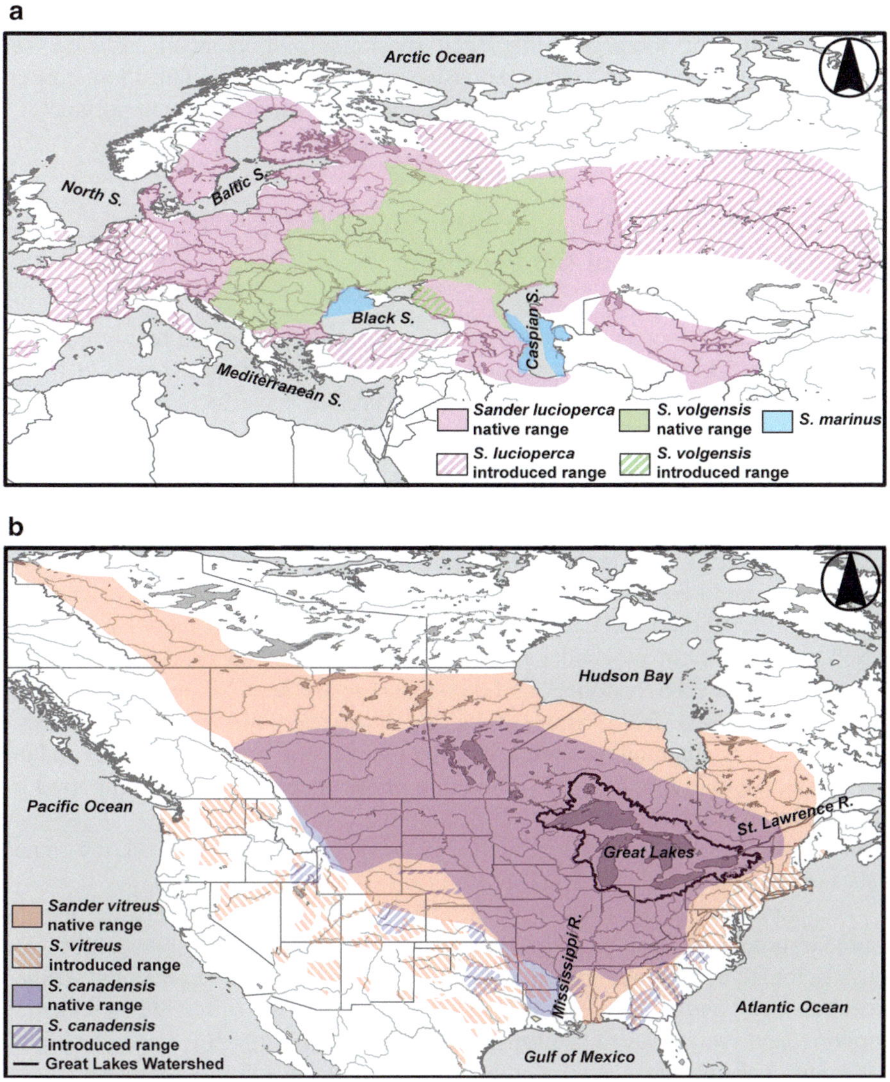

Fig. 1.5 Maps showing the native and introduced distributions of *Sander* spp. in: (**a**) Eurasia and (**b**) North America, based on information from Billington et al. (2011), the IUCN (International Union for Conservation of Nature and Natural Resources) Red List (http://www.iucnredlist.org), and Berg (1965). Note that range of *S. volgensis* is entirely contained within that of *S. lucioperca*

1850, whose monophyly is supported by nuclear and mtDNA sequence phylogenies (Figs. 1.2 and 1.3; Song et al. 1998; Sloss et al. 2004; Billington and Sloss 2011; Betancur-R et al. 2013, 2014). Some darter species are endemic to single streams, rendering them extremely vulnerable to extinction. Many of the darter species possess distinctive and beautiful color patterns, which especially are pronounced in reproductive males. It is believed that their sexual selection and mate choice have served as a driving force governing their speciation and abundant radiation (see Williams et al. 2013). Due to their extensive endemism and specialization, conservation of the darters and preservation of their habitats, merit special importance.

1.2 Evolutionary, Biogeographic, and Phylogenetic Diversification of the Percidae

Morphological characters and fossil dates have suggested that the family Percidae diverged ~66–58 Mya during the Paleocene Epoch (Collette and Bănărescu 1977). Hypothesized possible relationships for the Percidae have included an ancestor shared either with the widely distributed marine seabasses – the family Serranidae (Swainson 1839; Collette and Bănărescu 1977; Bruner 2011), or with the North American sunfishes – the family Centrarchidae (Bleeker 1859; Collette and Bănărescu 1977), or with members of the suborder Notothenioidei (e.g., the families Bovichtidae, Eleginopidae, Pseudaphritidae, and Nototheniidae (Chen et al. 2003; Dettai and Lecointre 2004, 2005) or the Bathydraconidae, Bempropidae, and Harpagiferidae (Smith and Craig 2007)). However, although the latter studies were based on a variety of molecular markers, including mitochondrial (mt) (12S and 16S) and nuclear (28S, histone 3, and rhodopsin) genes (see Chen et al. 2003; Dettai and Lecointre 2004, 2005, Smith and Craig 2007), they did not utilize the same taxa, precluding common resolution. A sister relationship between Percidae and the family Niphonidae Jordan 1923 (a Serranid offshoot) has been well-supported by consensus phylogenetic analysis (Fig. 1.2) of a variety of nuclear DNA gene sequences and the mt 16S RNA gene for 1591 taxa, which represented all ray-finned fishes (Betancur-R et al. 2013, 2014; see Fig. 1.2).

The phylogeny of Betancur-R et al. (2013, 2014; Fig. 1.2) indicated that the Percidae comprises five primary clades: the genus *Perca* (100 % support), the Luciopercinae Jordan and Evermann 1896 (93 % support) – containing *Zingel*, *Romanichthys*, and *Sander*, the genus *Gymnocephalus* (the ruffes; 96 % bootstrap support), and two clades of the Etheostomatinae (darters; 94 %) – one containing the genus *Etheostoma* + *Notothonus* (100 %) and the other comprising a less-supported clade (at just 62 % support) depicting *Percina* (100 %) as the sister group to *Crystallaria* (94 %) + *Ammocrypta* (100 %). In that tree (Fig. 1.2), a primary bifurcation of the Percidae occurred between *Perca* + Luciopercinae (100 % support) and *Gymnocephalus* + Etheostomatinae (100 %).

Phylogenetic analysis of two mtDNA gene regions (cytochrome *b* and 12S ribosomal DNA; Fig. 1.3) supported four primary groupings of the Percidae, including: *Perca* (100 % support), *Gymnocephalus* (100 %), Luciopercinae (*Zingel*, *Romanichthys*, and *Sander*; 97 %), and Etheostomatinae (*Ammocrypta*, *Crystallaria*, *Etheostoma*, *Nothonotus*, and *Percina*; 100 %) (Sloss et al. 2004; Billington and Sloss 2011). This phylogeny (Fig. 1.3) is quite similar to that of Betancur-R et al. (2013, 2014; Fig. 1.2). However, since closely related darter taxa have been shown by Near et al. (2011) to exhibit ~12.5 % interspecific introgression (i.e., species often possessed the mtDNA of another related species), correct interpretation of relationships based on mtDNA sequences alone likely is problematic. Thus, the nuclear DNA sequence phylogeny should be regarded as more reliable and robust.

In the mtDNA phylogeny (Fig. 1.3) by Sloss et al. (2004), *Perca* appeared as the sister taxon to all other Percidae, but support for monophyly of the remaining percids was weak (54 %). A major difference from the Betancur-R et al. (2013, 2014; Fig. 1.2) analysis is that the mtDNA tree (Fig. 1.3) depicted *Gymnocephalus* and the Luciopercinae as sister groups, again with relatively weak support (79 %). In both trees, the Etheostomatinae together are monophyletic. However, in the Betancur-R et al. (2013, 2014) tree, the genus *Etheostoma* is paraphyletic, with the genus *Notonotus* (however, only one of the latter species was included) (Fig. 1.2). This appears in contrast to the mtDNA tree (Sloss et al. 2004; Fig. 1.3), in which both genera – *Nothonotus* and *Etheostoma* – are monophyletic. However, only a single species of *Etheostoma* (*E. vitreum* (Cope 1870)) was analyzed in common between the two trees. Notably, the Sloss et al. (2004) mtDNA tree contains 23 species of *Etheostoma* and four *Nothonotus*, whereas the Betancur-R et al. (2013, 2014) phylogeny has just four *Etheostoma* and one *Nothonotus* species. The sister relationship between the genera *Crystallaria* and *Ammocrypta* is similar between the trees. In the Betancur-R et al. (2013, 2014) phylogeny, *Percina* is the sister group of *Crystallaria* + *Ammocrypta*, whereas in the Sloss et al. (2004) tree, *Percina* is the sister group of the *Nothonotus* clade (comprising *N. bellus* (Zorach 1968), *N. camurus* (Cope 1870), *N. aquali* (Williams and Etnier 1978), and *N. rufilineatuu* (Cope 1870)), and the *Crystallaria* + *Ammocrypta* clade forms the sister group to all other Etheostomatinae (albeit without bootstrap support).

The ancestral Percidae lineage that once was distributed across Eurasia and North America, which extended to the Atlantic coastlines of both continents later diverged, either following the final breakup of the Laurasian supercontinent 66–58 Mya (Briggs 1986; Wicander and Monroe 1993) or later via the closure and/or ecological inhospitality of continental land bridge connections (Tiffney and Manchester 2001; Milne 2006). Two historic land bridges variously allowed migration of taxa between the two continents: the Bering Land Bridge (BLB) across the North Pacific Ocean and the North Atlantic Land Bridge (NALB) across the North Atlantic Ocean. These each intermittently were available from the beginning of the Paleocene Epoch (~65 Mya), with the final subsidence of the NALB occurring during the late Miocene Epoch ~10 Mya (Tiffney 1985; Denk et al. 2011) and the BLB disappearing near the end of the Pleistocene Epoch ~0.01 Mya (Gladenkov et al. 2002). Recent fossil discoveries revealed two extinct *Sander* spp. in North America; one was dated to ~16.3–13.6 Mya in southern Saskatchewan, Canada (Murray and Divay 2011) and the other dated to ~5 Mya off Greenland (by Murray et al. 2009, who proposed the species as *S. teneri*). These have provided new data for interpreting the percid molecular phylogeny (Haponski and Stepien 2013; Stepien et al. 2015a; see Sect. 1.1.4.1). Date calibrations for the divergences between *Perca* (Fig. 1.6) and *Sander* (Fig. 1.7) species both match dates for disruption in the NALB during the late Miocene Epoch. The extent of their respective biogeographic distribution patterns (Fig. 1.4 for *Perca* and Fig. 1.5 for *Sander*), also indicate that the taxa diverged due to disruption of the NALB.

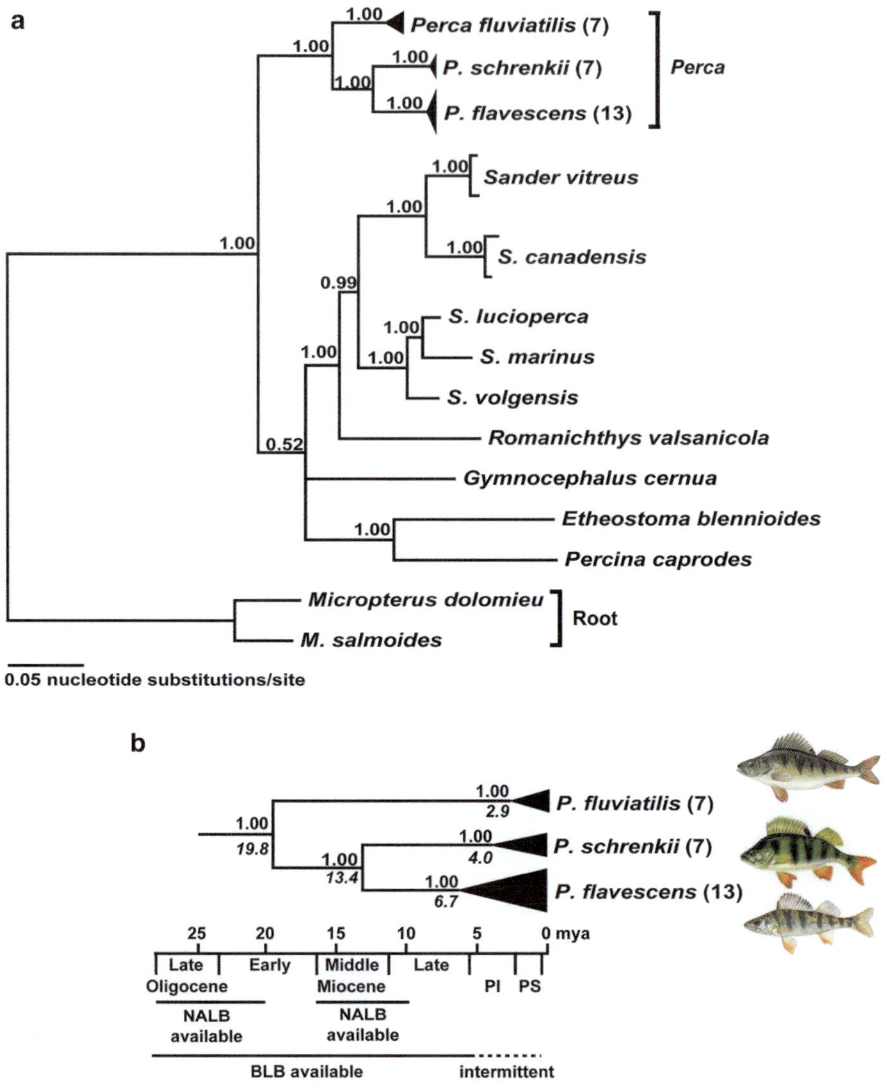

Fig. 1.6 (**a**) Phylogenetic tree (performed by us for this chapter and for Stepien et al. 2015a) from a partitioned Bayesian (MrBayes v.3.2.1; Ronquist and Huelsenbeck 2003, http://mrbayes.sourceforge.net/) analysis of the genus *Perca* and other percid relatives, based on a concatenated DNA sequence data set of three gene regions: mtDNA cytochrome *b* (cyt *b*) and cytochrome *c* oxidase I (COI), and nuclear recombination-activating gene intron 1 (RAG1), with posterior probabilities indicated on the tree. Tree was rooted to *Micropterus*, based on its hypothesized close relationship to Percidae, according to Song et al. (1998) and Sloss et al. (2004). Sequences were collected by the Stepien lab for all three *Perca* species and were compared by us to those deposited in GenBank (http://www.ncbi.nlm.nih.gov/genbank). Unique sequences alone were included in these analyses (for a complete list see Appendices 1 and 2). Corrected Akaike information criteria from jModeltest2 (Darriba et al. 2012; http://code.google.com/p/jmodeltest2/) were used to determine the most appropriate substitution models, which were: a transversional model with a gamma distribution

Following the older continental and land bridge divergences between the pairs of taxa, their respective distributions and population genetic patterns further were modified by loss and alterations of habitats during the Pleistocene glaciations ~2.6–0.01 Mya (Hewitt 1996, 2000; Bernatchez and Wilson 1998). The North American Laurentide Ice Sheet advanced further south than the Eurasian Scandinavian Ice Sheet (Hewitt 1996). However, North America had larger areas of glacial refugia due to its plains being oriented east–west in between north–south mountain ranges. In contrast, dispersal of Eurasian taxa was limited by boundaries of saline seas and mountain ranges that are oriented east–west (Hewitt 1996). Following the Pleistocene Ice Ages, most north temperate taxa extended their ranges northwards to recolonize old and new habitats, and these patterns now are accelerating due to anthropogenic climate change (Chu et al. 2005; Sharma et al. 2007). These biogeographic scenarios underlie the contemporary distributions of north temperate freshwater fishes, including the Percidae.

1.3 Genus *Perca*: Perch: Three Species: *P. fluviatilis*, *P. flavescens*, and *P. schrenkii*

Fossil record calibration data suggest that genus *Perca* originated ~19.8 Mya during the early Miocene Epoch (Fig. 1.6; new data original to the present study and also in Stepien et al. 2015a). *Perca* thus shares a similar biogeographic origin (Fig. 1.4)

Fig. 1.6 (continued) (0.236) for cyt *b*, a transitional model including invariable sites (0.645) and a gamma distribution (1.888) for COI, and a transitional model plus a gamma distribution (0.230) for RAG1. Bayesian analyses in MrBayes used a Metropolis-coupled Markov chain Monte Carlo (MC^3) approach and ran for 5,000,000 generations, with sampling every 100 generations. Four separate chains were run simultaneously per analysis, with two analyses run simultaneously. The burn-in period for the MC^3 was determined by plotting log likelihood values for each generation to identify when stationarity was reached. As burn-in, 25 % of the generations were discarded, along with the trees and parameter values sampled prior to the burn-in. A 50 % majority rule consensus tree was based on the remaining generations, whose branch support was determined from the posterior probability distribution (Holder and Lewis 2003) in MrBayes. (**b**) Time-calibrated phylogeny for *Perca* from BEAST (v.1.71; Drummond et al. 2012, http://beast.bio.ed.ac.uk/) analyses, which used the general time reversible nucleotide substitution model (Lanave et al. 1984) and a gamma distribution as identified by jModeltest2 for the individual gene trees. We used a relaxed molecular clock that assumed a lognormal distribution with the Yule speciation process (Gernhard 2008) as a tree prior. Two separate runs were conducted, each with a chain length of 50,000,000 generations and parameters were sampled every 100 generations. We used a single fossil calibration point: 12.0 Mya for the genus *Micropterus*, and three molecular calibration points from Haponski and Stepien (2013): 15.4 Mya for the origin of the North American *Sander*, 13.8 Mya for the Eurasian *Sander*, and 9.1 Mya for the divergence between *S. lucioperca* and *S. marinus*. Above branches = Bayesian posterior probabilities; Below (*italics*) = divergence estimates. Numbers in parentheses indicate multiple haplotypes per species. Dates for the availability of the North Atlantic Land Bridge (NALB) are from Tiffney (1985) and Denk et al. (2011), and for the Bering Land Bridge (BLB) from Gladenkov et al. (2002). *PI* Pliocene, *PS* Pleistocene. Note: We did not use the 26 Mya fossil calibration for *Perca* used in other trees, as that would "fix" the node that we were testing

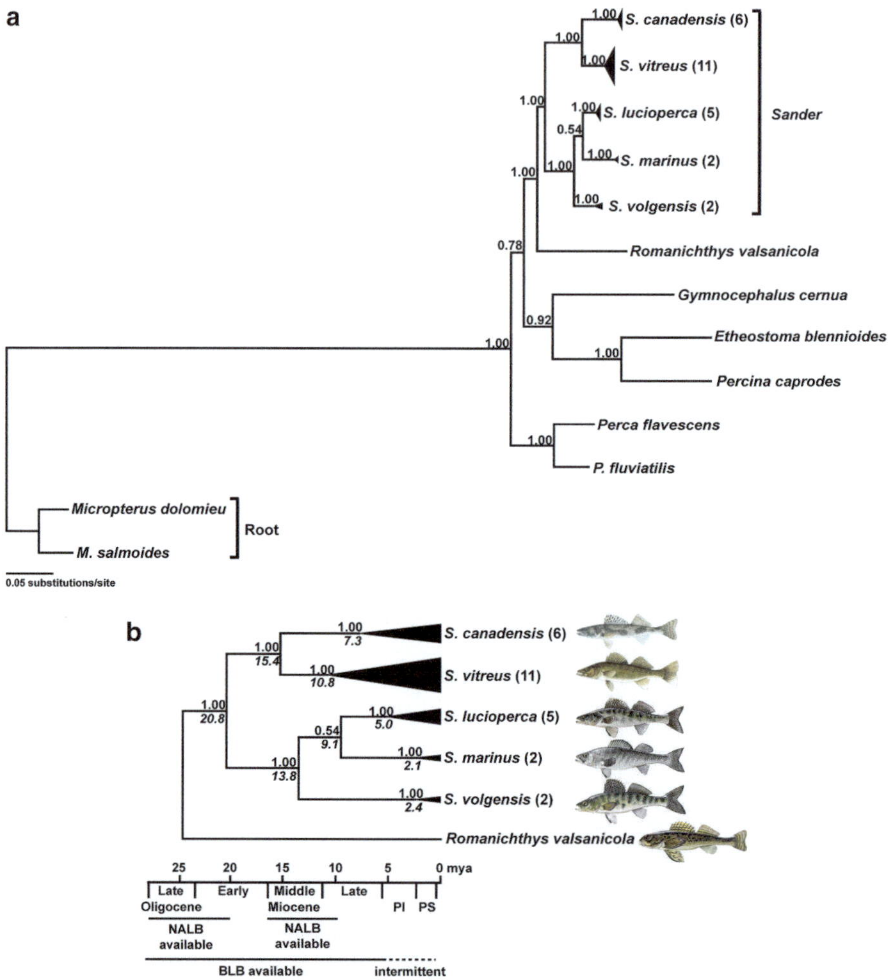

Fig. 1.7 (**a**) Phylogenetic tree of relationships within the genus *Sander* and to other percids from Haponski and Stepien (2013), using Bayesian analysis (MrBayes v.3.2.1; Ronquist and Huelsenbeck 2003, http://mrbayes.sourceforge.net/) of concatenated sequence data from six gene regions, showing posterior probabilities. Tree is rooted to *Micropterus*, based on its hypothesized close relationship to Percidae (Song et al. 1998; Sloss et al. 2004). (**b**) Time-calibrated phylogeny in BEAST (v.1.71; Drummond et al. 2012, http://beast.bio.ed.ac.uk/) for *Sander* and its sister taxon *Romanichthys* using two fossil calibration points: 26.0 Mya for the genus *Perca* and 12.0 Mya for the genus *Micropterus*. For specific model parameters and analysis settings, see Haponski and Stepien (2013). Above nodes = Bayesian posterior probabilities; Below (*italics*) = estimated divergences (Illustrations were used with permission from P. Maitland (Eurasian taxa) and J. Tomelleri (North American taxa)). Numbers in parentheses indicate multiple haplotypes per species. Dates for the North Atlantic Land Bridge (NALB) are from Tiffney (1985) and Denk et al. (2011), and Gladenkov et al. (2002) for the Bering Land Bridge (BLB) *PI* Pliocene, *PS* Pleistocene

and pattern with the percid genus *Sander* (see discussion below, and phylogeny by Haponski and Stepien (2013)). Phylogenies based on nuclear and/or mtDNA sequence data indicate that the genus *Perca* is clearly monophyletic and well-supported (with 1.0 Bayesian posterior probability and 100 % maximum likelihood bootstrap support on the tree shown in Fig. 1.6). Morphological characters that denote *Perca* include: a compressed body of moderate size, well-developed supraoccipital crest, presence of a predorsal bone, 7–8 branchiostegal rays, strongly serrated preopercle, well developed swimbladder, 33–43 vertebrae, large well developed anal spines, lateral line that ends prior to the caudal fin, absence of supplementary lateral lines on the caudal fin, and enlarged anteriormost interhaemal bone (Berg 1965; Collette and Bănărescu 1977; Bruner 2011).

The three *Perca* species have very significant fishery value and ecological importance as top piscivores in Eurasia and North America. Two *Perca* species are native to Eurasia – the European perch *P. fluviatilis* Linnaeus 1758 and the Balkhash perch *P. schrenkii* Kessler 1874, whereas the yellow perch *P. flavescens* (Mitchill 1814) is endemic to North America. Both the European perch and the yellow perch are widespread across much of their respective continents (Fig. 1.4), and support popular recreational and commercial fisheries. They each have been widely introduced for angling outside their native ranges. Perch also have been used as an important model in ecotoxicology studies (e.g., Couture et al. 2015).

All three species of *Perca* each are clearly monophyletic, having very high support comprising well-defined species (Fig. 1.6). The primary division in the genus divides the widely distributed *P. fluviatilis* lineage from a clade containing *P. schrenkii* and *P. flavescens*; the latter two species share a more recent common ancestry (Hai et al. 2008; Stepien et al. 2015a). *Perca schrenkii* and *P. flavescens* are sister species (nearest relatives), yet each are highly distinct, having diverged an estimated 13.4 Mya during the mid-Miocene Epoch (Fig. 1.6). Given that the native distribution of *P. flavescens* exclusively is east of the Rocky Mountains in North America and extends to the Atlantic Coast (Fig. 1.4b) and the distribution of *P. fluviatilis* extends northwest to the Atlantic Coast in Eurasia and is absent from far southeastern Asia (Fig. 1.4a), the distribution of their once-common ancestor most likely extended across the North Atlantic Land Bridge (NALB). This is concordant with findings by Wiley (1992) and Carney and Dick (2000). Moreover, a recent paper for the similarly-distributed pikeperch genus *Sander*, likewise pointed to historic connection over the NALB (Haponski and Stepien 2013). The NALB appears to have been disrupted during the estimated interval of 20–17 Mya, which matches this estimated time period for the *Perca* taxon divergence, determined from BEAST 1.71 analyses (Drummond et al. 2012) and fossil date calibrations shown in Fig. 1.6.

In contrast to their genetic divergence, the three *Perca* species appear very morphologically similar, with the exception of a few characters. Collette and Bănărescu (1977) found that the locations of the predorsal bone and the pterygiophore relative to the neural spines differ among them, with the predorsal bone of *P. flavescens* located between the first and second neural spines and its pterygiophore between the second and third. *Perca schrenkii* has a comparatively elongated body shape, wider spaces between its scales renders the scales appearing larger in size, the first dorsal spine is about half the length of the second, the posterior end of upper jaw is very wide and is usually larger than the width of the eye, the length of the jaw extends beyond the eye's mid-point, the origin of the dorsal fin is posterior to the pectoral fin base, and adults typically lack the dorsal bars characteristic of *P. fluviatilis* and *P. flavescens* (Berg 1965).

The European perch *P. fluviatilis* has the widest distribution of the three species of *Perca*, with its populations ranging across Eurasia from the United Kingdom and eastern France eastward to Siberia and Kolyma Russia (Collette and Bănărescu 1977; Craig 2000; Maitland 2000; Fig. 1.4a). The European perch has been widely introduced across Eurasia, especially into Spain, Italy, Albania, Siberia, and to other continents, such as Australia and Africa (Freyhof and Kottelat 2008). It occupies a wide variety and size of habitats from freshwater lakes and streams to estuarine lagoons, and is a prized commercial, recreational, and aquaculture species. In Europe, the most recent Pleistocene cold stages, especially the Weichselian glaciation 25,000–13,000 years ago (ya), shaped the habitats of many freshwater fish species – including *P. fluviatilis* (Nesbø et al. 1999). During the glaciations, *P. fluviatilis*, *S. lucioperca*, and other aquatic species migrated to waters south of the ice sheets, where their populations were concentrated in the restricted areas of glacial refugia (Hocutt and Wiley 1986).

The Balkhash perch *P. schrenkii* is native to Lakes Balkhash and Alakol in Kazakhstan, where it once supported abundant fisheries during the 1930s and 1940s. The Lake Balkash fishery has collapsed, but contemporary harvests remain in the Lake Alakol region (Sokolovsky et al. 2000). The species also occurs in associated catchments extending into China (Berg 1965). During the 1960s and 1970s, the Balkhash perch was introduced to various water bodies in Uzbekistan, Kazakhstan, and Middle Asia (Kamilov 1966; Nuriyev 1967; Pivnev 1985). It is classified in "The IUCN of Threatened Species" as being "data deficient" (Kottelat 1996; IUCN 2014; http://www.iucnredlist.org/).

The monotypic North American *P. flavescens* has an extensive native geographic distribution, which extends across much of the northeast and north central regions of North America, with a few isolated relict populations in the southeast (Fig. 1.4b; Sepulveda-Villet and Stepien 2012). It inhabits a diversity of lacustrine and fluvial habitats, ranging from large to small in geographic areas, with its most extensive habitats and greatest abundances occurring in the Laurentian Great Lakes – especially in Lake Erie (Scott and Crossman 1973; Hubbs and Lagler 2004) and the St. Lawrence River system (Bernatchez and Giroux 2012). Habitats in the northern areas of the yellow perch's range, as well as those of the other percids (and most other taxa), were reshaped by the Ice Age glaciations, leading to their present configuration about 12,000–4000 ya. These northerly populations are much younger than those in many areas that were unglaciated to the south; the latter likely have been more stable in terms of long-term genetic diversity and population sizes (see Sepulveda-Villet and Stepien 2012). Following the last glaciations, the formation of the present-day Great Lakes opened up a wide variety and expansive range of new habitats, leading to large populations. Foundings of Great Lakes populations from admixtures from at least three glacial refugia has manifested in appreciable genetic diversity (Stepien et al. 2015a).

1.4 Subfamily Luciopercinae: Three Genera: *Romanichthys*, *Zingel*, and *Sander*

Phylogenetic analyses based on nuclear and mtDNA genes strongly support monophyly of the subfamily Luciopercinae (Sloss et al. 2004; Billington and Sloss 2011; Betancur-R et al. 2013, 2014; Haponski and Stepien 2013; Figs. 1.2, 1.3, and 1.7), congruent with its morphological definition by Collette and Bănărescu (1977). Its three component genera are: *Romanichthys* (one species in Europe), *Zingel* (four species in Europe), and *Sander* (five species: three in Eurasia and two in North America). The Luciopercinae are united by the morphological characters of having weak anal spines, a lateral line that extends along the caudal fin – with supplementary lines above and below it, a non-serrated cleithrum on the pectoral girdle, and 41–50 vertebrae (Collette and Bănărescu 1977; Bruner 2011).

Romanichthys and *Zingel* are endemic to the Danube River watershed, which is surrounded by the mountain ranges of the Balkans, Carpathians, and the Alps. Analyses of nuclear and mitochondrial DNA sequences have indicated that *Romanichthys* and *Zingel* are sister genera, i.e., are more closely related to each other than to *Sander*, and form its sister group (see Sloss et al. 2004; Billington and Sloss 2011; Betancur-R et al. 2013, 2014; Haponski and Stepien 2013; Figs. 1.2, 1.3, and 1.7). Morphological differentiation supports this relationship, with *Romanichthys* and *Zingel* characterized by their small to moderate body size, well developed nasal flaps, absence of canine teeth, absence of the swimbladder, breeding tubercles in both sexes, fusiform or depressed body shape, weak and flexible anal fin spines, and 39–49 vertebrae (Collette and Bănărescu 1977; Bruner 2011). The *Romanichthys*/*Zingel* clade appears to diverged from *Sander* ~24.6 Mya during the late Oligocene Epoch, based on Bayesian phylogenetic analyses and fossil calibration points used by Haponski and Stepien (2013). Tectonic activity in the Periadriatic fault system of the southern Alps during the Oligocene Epoch (Viola et al. 2001) may have led to divergence of the *Romanichthys*/*Zingel* lineage from that of *Sander*.

1.4.1 Luciopercinae, Genus **Sander** *(=* **Stizostedion***): Pike-Perches: Five Species:* **S. lucioperca, S. volgensis, S. marinus, S. vitreus** *and* **S. canadensis**

The genus *Sander* originally was widely distributed across the Holarctic, extending from the Atlantic coasts to the central regions of both continents, as suggested by Cavender (1998) and shown in the map of Fig. 1.5. The name *Sander* (Oken 1817) replaced *Stizostedion* Rafinesque 1820, as recommended by Kottelat (1997) and then was accepted by the American Fisheries Society's Committee on Names of Fishes (see Nelson et al. 2003, 2004). The genus comprises five extant species, with three species in Eurasia: the pikeperch *S. lucioperca* (Linnaeus 1758), Volga pikeperch

S. volgensis (Gmelin 1789), and the extremely rare sea pikeperch *S. marinus* (Cuvier 1828) – and two species in North America: the sauger *S. canadensis* (Griffith and Smith 1834) and the walleye *S. vitreus* (Mitchill 1818). *Sander* species are key top piscivores in Eurasia and North America, and support economically valuable fisheries in both continents (Larsen and Berg 2006; Kuznetsov 2010; Schmalz et al. 2011). Species of *Sander* live in a wide variety of habitat types, including fast-flowing clear streams and slow turbid lakes (Collette and Bănărescu 1977; Billington et al. 2011).

Analyses of nuclear and/or mtDNA genes support monophyly of the genus *Sander*, which is the sister group of the *Romanichthys*/*Zingel* clade (Sloss et al. 2004; Billington and Sloss 2011; Betancur-R et al. 2013, 2014; Haponski and Stepien 2013; see Figs. 1.2, 1.3, and 1.7). Other molecular evolutionary analyses, which used a subset of the species and either were based on allozymes, mtDNA restriction fragment length polymorphisms (RFLPs; Billington et al. 1990, 1991), or mtDNA control region sequence data (Faber and Stepien 1998), similarly discerned *Sander* as monophyletic.

The body form of *Sander* is elongate and laterally compressed, ranging in size from ~450 mm total length for *S. volgensis* to ~1300 mm in *S. lucioperca*. The genus is characterized by: (1) canine teeth that are largest in *S. lucioperca* and absent in adult *S. volgensis*, whose juveniles possess them (Berg 1965), (2) narrow rows of teeth on the jaws, vomer, and palatines, (3) a strongly serrated preopercle, (4) a continuous lateral line extending from the head to the caudal fin, (5) accessory lateral lines on the upper and lower portions of the caudal fin, (6) strongly forked caudal fin, (7) lack of genital papilla, (8) 7–8 branchiostegal rays, (9) 12–13 anal fin rays (Berg 1965; Trautman 1981; Hubbs and Lagler 2004), and (10) a tapetum lucidum (reflective layer behind the retina), which aids in night vision (Moore 1944; Collette et al. 1977; Trautman 1981). Characters that distinguish among the species of *Sander* include: color, scale patterns, number of fin rays, and number of pyloric caeca (small blind sacs in the stomach that may aid in the breakdown of proteins (Moyle and Cech 2000), as detailed by Berg (1965); Trautman (1981), and Hubbs and Lagler (2004). For example, the cheeks of *S. volgensis* are fully scaled, *S. canadensis* possesses an intermediate amount, whereas scales are reduced or absent in *S. vitreus*, *S. lucioperca*, and *S. marinus* (Berg 1965; Trautman 1981; Hubbs and Lagler 2004).

The genus *Sander* diverged into two clades ~20.8 Mya during the Miocene Epoch (Fig. 1.7; Haponski and Stepien 2013), during the time of subsidence of the NALB (Denk et al. 2010). The lineages on the two continents then undertook separate evolutionary pathways. Contemporary Eurasian species appeared to originate ~13.8 Mya and the North American species ~15.4 Mya (Fig. 1.7).

1.4.1.1 Eurasian *Sander*: Zanders or Pike-Perches: Three Species: *S. lucioperca*, *S. volgensis*, and *S. marinus*

Of the three Eurasian *Sander* species, *S. lucioperca* has the largest geographic expanse (Fig. 1.5a; Berg 1965; Collette and Bănărescu 1977; Freyhof and Kottelat 2008), which extends from the Elbe River in Germany eastward to China and north

Fig. 1.8 Photograph of a sea pikeperch *Sander marinus* specimen from Baku Azerbaijan waters of the Caspian Sea, which now is archived at Stepien's Great Lakes Genetics/Genomics Laboratory (~350 mm TL, 5 pyloric caeca; genetically analyzed by Haponski and Stepien 2013)

into the Russian Federation, Sweden, and Finland. Its distribution includes the Aral, Azov, Baltic, Black, and Caspian Seas, where it ranges into brackish waters. The range of *S. volgensis* is smaller; it inhabits the Danube, Dnieper, and Don Rivers in the Black and Azov Sea basins and the Volga and Ural Rivers in the Caspian Sea basin, and its range is contained within that of *S. lucioperca*; i.e., they are sympatric in that region (Fig. 1.5a; Berg 1965; Collette and Bănărescu 1977; Freyhof 2011).

Until recently, little was known about the rare and enigmatic *S. marinus*, which has been reported to inhabit marine/estuarine waters of the Black and Caspian Sea basins (Fig. 1.5a; Berg 1965; Collette and Bănărescu 1977), yet Haponski and Stepien (2013) were unable to locate museum specimens or modern records from the Black Sea. The IUCN Red List has classified *S. marinus* as "data deficient" (World Conservation Monitoring Center 1996; IUCN 2014), and Maitland (2000) has listed it as "vulnerable". Two specimens of *S. marinus* from the Caspian Sea were obtained by Stepien (photograph in Fig. 1.8), which were genetically analyzed by Haponski and Stepien (2013); these results are included here.

The three Eurasian species diverged ~13.8 Mya into two lineages – with *S. volgensis* as the basal taxon, diverging ~13.8 Mya during the Mid-Miocene Epoch, during cooling temperatures (Fig. 1.7; Haponski and Stepien 2013). Congruent with the DNA sequence relationships, *S. volgensis* has retained the plesiomorphic life history condition of being a broadcast spawner that lacks parental care. This life history is shared with the North American taxa. In contrast, *S. lucioperca* and *S. marinus* share the derived life history characters of the males building nests and then guarding the eggs and fry (Guseva 1974; Collette et al. 1977; Craig 2000). This behavioral character unites the two species.

Sander lucioperca and *S. marinus* differentiated from each other ~9.1 Mya (Haponski and Stepien 2013), likely due to increases in salinity in the Ponto-Caspian Sea basin (including the Black and Caspian Seas), which then led to specialization of the *S. marinus* lineage in saline waters (see Reid and Orlova 2002). Today, *S. marinus* exclusively inhabits saline waters to ~13 ppt in today's Caspian Sea, where *S. lucioperca* is absent (restricted to rivers; S. Ibramihov, pers. commun.). *Sander lucioperca* appears to be adapted to a wide diversity of environmental temperatures, and is reported to briefly tolerate salinities to ~12 ppt (Craig 2000; Brown et al. 2001). The Ponto-Caspian basin is believed to have reached a higher salinity of ~17–20 ppt ~9 Mya (Reid and Orlova 2002), which may have isolated the ancestors of the two taxa.

Contemporary *S. lucioperca* genotypes differentiated from each other ~5.0 Mya, during which time the Ponto-Caspian basins experienced many geologic and climatic changes, including changes in sea levels and salinity (Reid and Orlova 2002), which likely isolated populations in different areas. Genotypes of *S. volgensis* and *S. marinus* originated during the Pleistocene Epoch ~2.4–2.1 Mya, when the Ponto-Caspian region experienced more fluctuations in water levels and salinities (Reid and Orlova 2002). *Sander lucioperca* contains greater genetic diversity than does *S. volgensis* across their respective ranges (Haponski and Stepien 2013). Post-glacial populations of *S. lucioperca* and *S. volgensis* likely re-colonized from a single small refugium near the Caspian Sea, where they experienced genetic bottlenecks and founder effects (Faber and Stepien 1998; Haponski and Stepien 2013). It is interesting that the rare and enigmatic *S. marinus* appears to have comparably greater diversity today than does the more abundant *S. volgensis* (Haponski and Stepien 2013). It may be that its genetic diversity was better protected in its geographic location from the population bottlenecks of the glaciations.

1.4.1.2 North American *Sander*: Two Species: *S. vitreus* and *S. canadensis*

Fossil evidence has dated the presence of extinct North American *Sander* to ~16–13 Mya in southern Saskatchewan Canada (Murray and Divay 2011). This timing corresponds to the divergence found between the North American taxa – the walleye *S. vitreus* and the sauger *S. canadensis* – estimated as ~15.4 Mya during the Miocene Epoch by Haponski and Stepien (2013; Fig. 1.7). The two North American species have widespread native distributions, with extensive overlap throughout most of the central region (Fig. 1.5b). *Sander vitreus* extends from the Mackenzie River in the Northwest Territories of Canada, south to the U.S. Gulf Coast, and northeastward to New Hampshire and Quebec (Billington et al. 2011; Page and Burr 2011). The range of *S. canadensis* includes the Mississippi River basin, Hudson Bay, the Great Lakes, and the St. Lawrence River drainage, ranging from Quebec to Alberta and south to Louisiana and Alabama (Billington et al. 2011; Page and Burr 2011). Morphologically, *S. vitreus* and *S. canadensis* are distinguishable by numbers of pyloric caeca (three in the former, five to eight in the latter) and their coloration and banding/barring patterns (Scott and Crossman 1973). The two species also significantly differ in body depths, head and upper and lower jaw lengths, head and interorbital widths, orbit diameters, and numbers of second dorsal and anal fin rays, as well as ratios of upper: lower jaw and orbit: interorbital lengths (Haponski 2013).

Today's genotypes of *S. vitreus* appear to date to ~10.8 Mya and *S. canadensis* to ~7.3 Mya during the Mid-Miocene Epoch (Haponski and Stepien 2013), when rapid cooling was beginning and the ice sheets were forming (Wolfe 1994; Hewitt 1996; Bruch et al. 2007). Genetic diversities of the North American species are greater than those in Eurasia, likely due to their larger and more consistent population sizes, and the greater number and geographic extent of their Pleistocene glacial refugia, suggesting fewer bottlenecks (Haponski and Stepien 2013).

Hubbs (1926) once described another possible North American *Sander* taxon, termed the "blue pike" – formerly designated as *S. vitreus* "*glaucus*" (Hubbs 1926; Trautman 1981) – whose range reportedly was restricted to Lakes Erie and Ontario. However, the "blue pike" intergraded with walleye in color and all morphological characters (Trautman 1981; Stepien and Faber 1998; Haponski and Stepien 2014) and, despite its common name, it was not a member of the pike family (Esocidae). The "blue pike" was reported to vary from walleye in having a steel grey-blue color, larger eyes that were located higher on the head, and a smaller interorbital distance (Trautman 1981; Hubbs and Lagler 2004). The latter two characters were recently re-assessed from paratype specimens by Haponski and Stepien (2014), who found slight differences but considerable overlap. Moreover, occasional steel-blue colored walleye variants regularly have been reported to occur in Lake Erie and other waters, including the Ohio River drainage (Scott and Crossman 1973; Trautman 1981, Haponski pers. obs.).

Additional confusion has resulted from the fact that some other walleye (as well as some yellow perch) in northern waters along the Canadian Shield possess turquoise-colored mucus (Yu et al. 2008; Schaefer et al. 2015); this color does not match the integumentary steel grey-blue color ascribed to the "blue pike" and typically rubs off. These turquoise mucus individuals are sympatric with "normal" yellow-colored individuals in the same waters, and the mucus pigment has been termed "sandercyanin" (Yu et al. 2008). In addition, the body shape, eye size, and interorbital distance of these walleye individuals also do not resemble those described for the "blue pike" (Stepien et al. 2015b). Nevertheless, some accounts in the literature and popular press have remained confused about these colors and whether they are related to the "blue pike" or display population genetic differentiation (e.g., Paradis and Magnan 2005; Yu et al. 2008; Laporte et al. 2011; see Haponski and Stepien 2014 for full account).

Haponski and Stepien (2014) evaluated the genetic and morphological characters of the historic paratypes of the "blue pike" designated by Hubbs (1926), along with historic and contemporary walleye, finding no differences that would merit taxonomic distinction. There was no statistical genetic difference between the historic "blue pike" and walleye samples and the morphological characters overlapped. There also was no resemblance of the historic or contemporary samples from Lakes Erie and Ontario to the Canadian shield populations. "Normal" yellow colored and turquoise mucus colored walleye from the same northern lake were genetically identical to one another. There thus are just five *Sander* spp., and neither the "blue pike" (*S. vitreus* "*glaucus*") nor the turquoise mucus *S. vitreus* merit taxonomic or population-level recognition.

1.4.2 *Luciopercinae, Genus* Zingel*: Zingels: Four Species:* Z. zingel, Z. asper, Z. balcanicus *and* Z. streber

The four species comprising the genus *Zingel* include: the zingel *Z. zingel* (Linneaus 1766), the asper (also known as the apron) *Z. asper* (Linnaeus 1758), Balkan streber *Z. balcanicus* (Karaman 1937), and the streber *Z. streber* (Siebold 1863); all are endemic to fast moving waters of the Danube River system in Europe (Fig. 1.9; Craig 2000). They are classified either as vulnerable or endangered (Maitland 2000; Laroche and Durand 2004). All are benthic, relatively sedentary, and feed at night on benthic invertebrates and small fishes (Craig 2000). Their shape is slender, stream-lined, and ventrally flattened, with two widely-separated dorsal fins and a ventral mouth. The genus is differentiated from the other Luciopercinae by the body shape, prominent snout, setiform teeth on the jaws, vomer, and palatines, a lateral line system on the lower portion of the head, seven infraorbitals, a moderately sized epioccipital process, tuberculate ball-shaped gill rakers on the first gill arch, and the ventral margins of the interopercle and subopercle being smooth (not-serrated; Berg 1965; Bruner 2011).

Zingel zingel reaches the largest size (300 mm) and lives in larger and deeper areas of the Danube and Dniester River systems. It has more spines (13–15 in the first dorsal fin) and soft rays (18–20 in the second) than the other two species (which share 8–9 and 12–13). It is believed to have always been relatively rare and is highly sensitive to pollution (Lelek 1987). *Zinger streber* is the most broadly distributed

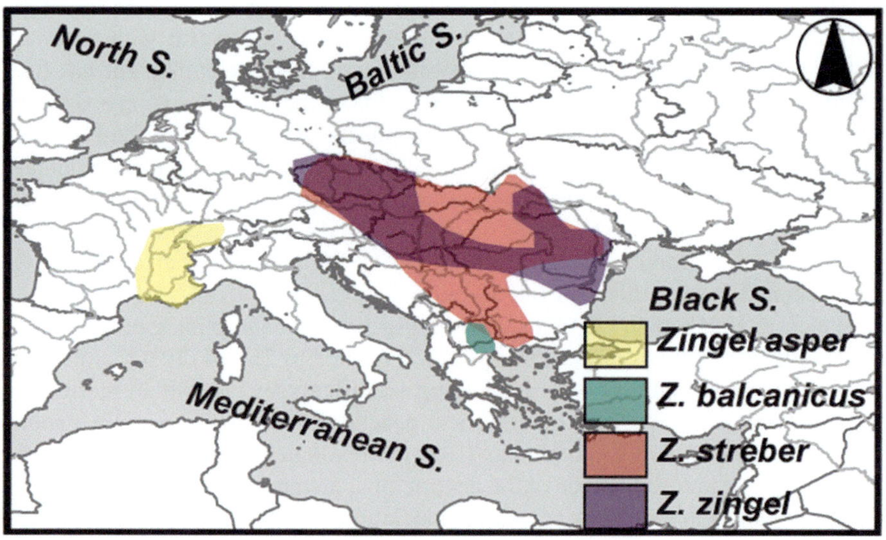

Fig. 1.9 Map showing the ranges of the four *Zingel* spp. in Europe, using information from Maitland (2000)

species; it lives in the Danube and Vardar Rivers and their associated tributaries, including faster flowing systems and higher headwater areas (Collette and Bănărescu 1977; Maitland 2000). Its caudal peduncle is longer and more slender than the other species (Craig 2000). *Zingel asper* is endemic to the Rhone River system and is classified as critically endangered; it now occupies <17 % of its original range (Laroche and Durand 2004). *Zingel balcanicus* has a restricted endemic range in the Vardar River of Macedonia and into Greece (Crivelli 2006; Economou et al. 2007). Very little is known about this species. There are just a few museum specimens, with the last collected in 1986 (Crivelli 2006); *Z. balcanicus* now may be extinct, with its pollution intolerance implicated (Georgijev 2004).

Phylogenetic relationships among *Zingel* species are unresolved, since prior published DNA studies lacked *Z. asper* and *Z. balcanicus* (e.g., Sloss et al. 2004; Betancur-R et al. 2013, 2014; Haponski and Stepien 2013). Earlier studies that included *Z. zingel* and *Z. streber* suggested that they appeared as sister species, comprising the sister group to *Romanichthys* (Fig. 1.10; Sloss et al. 2004; Billington and Sloss 2011). Here we mined genetic data for two mtDNA genes from the National Center for Biotechnology's GenBank (http://www.ncbi.nlm.nih.gov/), which indicated *Zingel* is not monophyletic, and may include *Romanichthys* (Fig. 1.10). The clade dates to an estimated 17.2 Mya. Our findings (Fig. 1.10) from mtDNA depict *Z. streber* as the sister taxon to *R. valsanicola* and the clade containing *Z. zingel* and *Z. asper*, with the latter two as sister species. Further analyses should investigate the relationships of all four *Zingel* spp. and *Romanichthys*.

Just a single study has examined the population genetic patterns of *Z. asper*, which analyzed five nuclear DNA microsatellite loci to discern high genetic diversity and significant genetic structuring among different tributaries in the Rhone River watershed (Laroche and Durand 2004). Samples within tributaries were undifferentiated, showing high gene flow (Laroche and Durand 2004); this pattern also characterizes species of the North American darters, which typically show high differentiation among systems but little within them (see Sect. 1.7.3).

1.4.3 Luciopercinae, Genus **Romanichthys***: One Species: The Asprete* **R. valsanicola**

The genus *Romanichthys* contains a single species – the asprete *R. valsanicola* Dumitrescu et al. 1957 – which is critically endangered and occupies a very restricted range (Fig. 1.11) in the upper reaches of the Arges, Vilsan, and Riul Rivers, which are tributaries of the Danube River in Romania (Maitland 2000). The species is believed to be extirpated from all but the Vilsan River (Craig 2000), with its disappearance attributed to the combination of deforestation, dams, herbicides, and pesticides (Lelek 1987). It lives in fast-flowing riverine waters and lacks a swim bladder. Prior phylogenetic analyses described the asprete as the sister taxon to

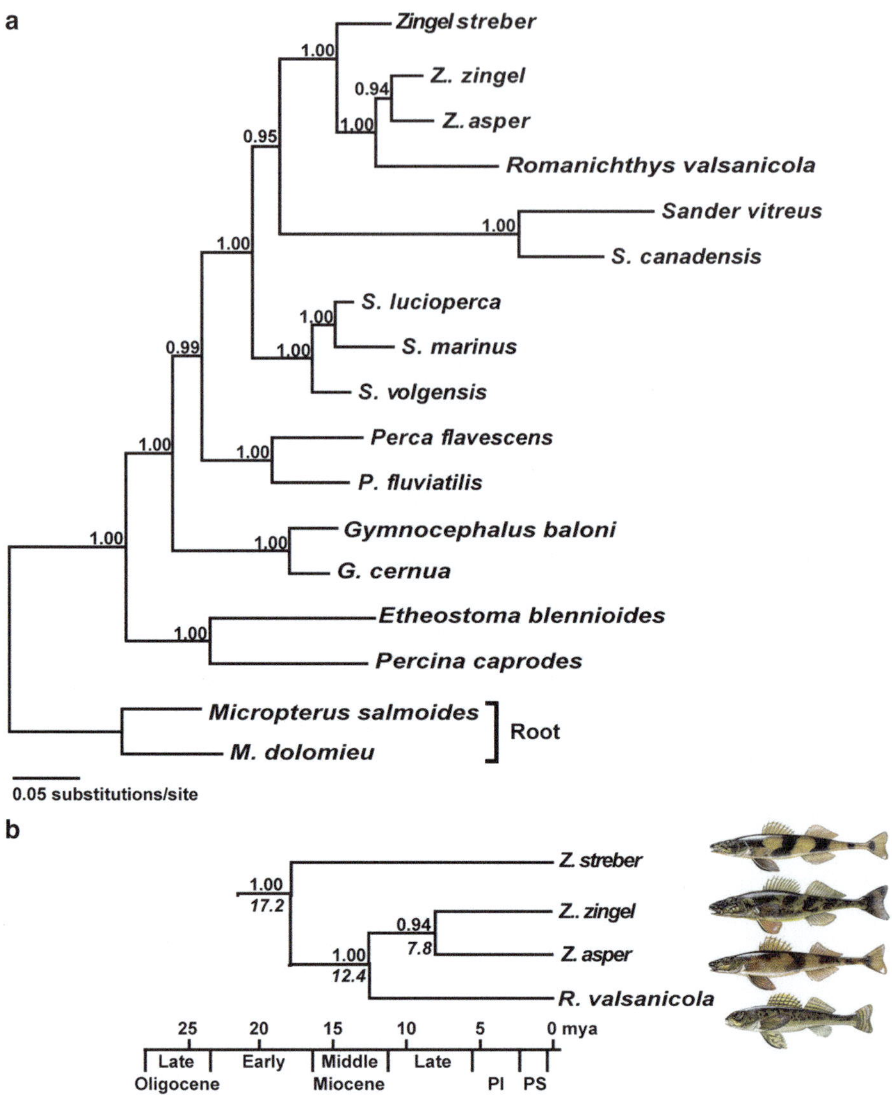

Fig. 1.10 (**a**) Phylogenetic tree of relationships within the genus *Zingel* in relation to other percids (performed by us for this chapter) using a partitioned Bayesian analysis (MrBayes v.3.2.1; Ronquist and Huelsenbeck 2003, http://mrbayes.sourceforge.net/) of concatenated sequences from two mtDNA gene regions – cytochrome *c* oxidase I (COI) and NADH dehydrogenase 2 (ND2). Sequences were collected from GenBank (http://www.ncbi.nlm.nih.gov/genbank) and those that were unique were included in the analyses (see Appendix 3a). Posterior probabilities are indicated. Tree is rooted to *Micropterus*, based on its hypothesized close relationship to Percidae (Song et al. 1998; Sloss et al. 2004).

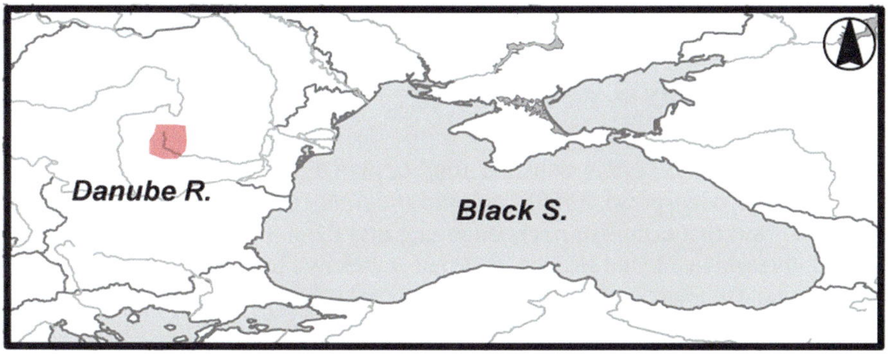

Fig. 1.11 Distribution of the rare and highly endangered *Romanichthys valsanicola* in Europe, based on information from Maitland (2000)

Zingel (Song et al. 1998; Sloss et al. 2004; Bruner 2011), however, mtDNA evidence (shown here) suggests that it is part of *Zingel* and may not be a valid genus (Fig. 1.10). This should be investigated further with nuclear genes. *Romanichthys valsanicola* is characterized by its two well-separated dorsal fins, with the second being larger than the first, dorsally-positioned eyes, tuberculate ball-shaped gill rakers on the first gill arch, and fine serrations on the ventral margins of the interopercle and subopercle (Craig 2000; Maitland 2000; Bruner 2011).

Fig. 1.10 (continued) Corrected Akaike information criteria from jModeltest2 (Darriba et al. 2012; http://code.google.com/p/jmodeltest2/) determined the most appropriate model of substitution for COI (transversional model including invariable sites (0.653) and gamma distribution (2.124)), and ND2 (transitional model plus invariable sites (0.348) and a gamma distribution (0.230)). Bayesian analyses in MrBayes used a Metropolis-coupled Markov chain Monte Carlo (MC^3) approach and ran for 5,000,000 generations, with sampling every 100 generations. Four separate chains were run simultaneously for each analysis, and two analyses ran simultaneously. The burn-in period for the MC^3 was determined by plotting log likelihood values for each generation to identify when stationarity was reached. As burn-in, 25 % of the generations were discarded, along with the trees and parameter values sampled prior to the burn-in. A 50 % majority rule consensus tree was based on the remaining generations, whose branch support was determined from the posterior probability distribution (Holder and Lewis 2003) in MrBayes. (**b**) Time-calibrated phylogeny in BEAST (v.1.71; Drummond et al. 2012, http://beast.bio.ed.ac.uk/) for *Zingel* and *Romanichthys* that used the general time reversible nucleotide substitution model (Lanave et al. 1984) and a gamma distribution as identified by jModeltest2 for the individual gene trees. BEAST analyses used a relaxed molecular clock that assumed a lognormal distribution with the Yule speciation process (Gernhard 2008) as a tree prior. Two separate runs were conducted, each with a chain length of 50,000,000 generations, and parameters sampled every 100 generations. We used two fossil calibration points: 26.0 Mya for the genus *Perca* and 12.0 Mya for the genus *Micropterus* and molecular calibration points of 15.4 for origin of North American *Sander*, 13.8 Mya for the Eurasian *Sander*, and 9.1 Mya between *S. lucioperca* and *S. marinus* from Haponski and Stepien (2013). Above nodes = Bayesian posterior probabilities; below (*italics*) = estimated divergences (Illustrations were used with permission from P. Maitland). *PI* Pliocene, *PS* Pleistocene

1.5 Genus *Gymnocephalus*: Ruffes: Four to Five Species: *G. cernua*, *G. acerina*, *G. baloni*, *G. schraetser*, and Possibly *G. ambriaelacus*

The genus *Gymnocephalus* contains four or five Eurasian species (Fig. 1.12), including: the widespread ruffe *G. cernua* (=*cernuus*), the Don ruffe *G. acerina* (Gmelin 1789) that occurs in rivers emptying into the northern Black Sea, Balon's ruffe *G. baloni* Holcík and Hensel 1974 that is endemic to fast-flowing areas of the Danube and Dnieper Rivers, striped ruffe *G. schraetser* (Linnaeus 1758) that is found in deeper waters of the Danube basin and estuaries leading into the Black Sea

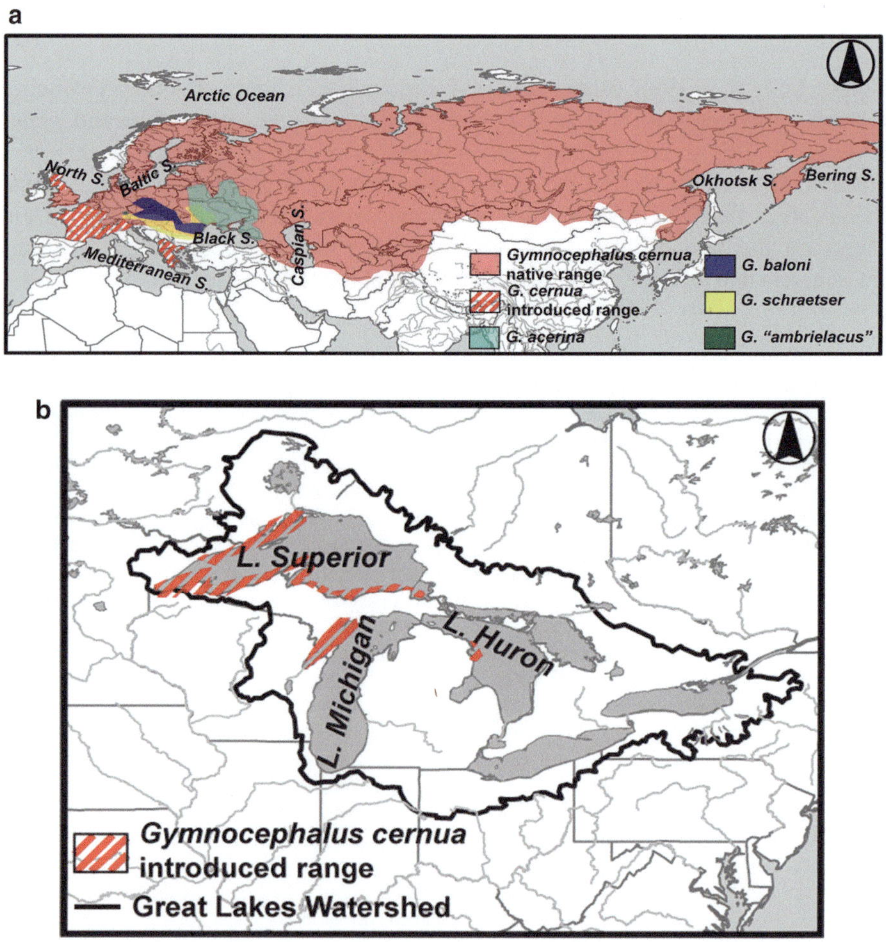

Fig. 1.12 Maps showing the native and introduced ranges of *Gymnocephalus* species in (**a**) Eurasia and (**b**) North America (solely introduced), based on information from Collette and Bănărescu (1977), Kålås (1995), Craig (2000), Maitland (2000), Walker and Yang (2009), and Fuller et al. (2014)

(Holcík and Hensel 1974; Shevtsova et al. 1986), and the proposed Ammersee ruffe *G. ambriaelacus* Geiger and Schliewen 2010 that was described as endemic to Lake Ammersee in southern Germany (Geiger and Schliewen 2010). *Gymnocephalus ambriaelacus* is critically endangered (Freyhof 2013), whereas the other species are listed as "Least Concern" in the IUCN Red List database (IUCN 2014). *Gymocephalus schraetser* significantly declined in some areas, including the Morava River, due to poor water quality in the 1940s and 1950s (Lelek 1987).

Morphological characters that unite *Gymnocephalus* spp. include partially joined dorsal fins, enlarged lateral line canals on the head, preorbital covering the maxillary bone, presence of setiform teeth, vomerine and palatine teeth few in number or absent, and three extrascapular bones in their lateral line system, with two of these as simple tubes (Berg 1965; Collette and Bănărescu 1977; Craig 2000; Bruner 2011). The genus is estimated here by us to date to about 13.4 Mya (Fig. 1.13).

Phylogenetic analyses of mtDNA and nuclear DNA indicate that *G. cernua* is the sister taxon of a clade containing *G. baloni* and *G. schraetser* (1.00/100 % support; Fig. 1.13; see Stepien et al. 1998, 2005; Sloss et al. 2004; Billington and Sloss 2011). *Gymnocephalus schraetser* and *G. baloni* are depicted on Fig. 1.13 as sharing a common ancestor about 8.0 Mya, according to our analyses here. Geiger and Schliewen (2010) proposed a new species as *G. ambriaelacus*, which likely is a population of *G. baloni*, to which it bears very close resemblance (since it differs by just a single nucleotide in a single gene); this relationship merits further investigation. Molecular genetic studies to date have lacked samples of *G. acerina*, which would provide valuable information.

The various species of *Gymnocephalus* differ in body shape, scale and fin ray counts, and some coloration features. *Gymnocephalus cernua* has 35–40 lateral line scales, 11–16 first dorsal spines, a single opercular spine, a blunt snout, and gray-green dorsal coloration with brownish speckles. *Gymnocephalus acerina* and *G. schraetser* possess a longer snout that is ~2–3× the length of the eye; the former has 50–55 lateral line scales and distinctive small darks spots along the sides of the body, whereas the latter has 55–62 lateral line scales and several dark longitudinal stripes down the sides of the body (Berg 1965; Maitland 2000). *Gymnocephalus baloni* is characterized by two opercular spines and several dark lateral bars (Maitland 2000). Lastly, *G. ambriaelacus* was described as having a larger eye diameter, more pectoral fin rays, and a steeper convex doral profile of the snout than *G. baloni* (Geiger and Schliewen 2010).

The widespread native range of the ruffe *G. cernua* includes southern England, northeastern France, the rivers entering the Baltic and White Seas, and most of central Europe and Siberia (Fig. Q; Holcik and Hensel 1974; Kovac 1998; Popova et al. 1998; Winfield et al. 1998). It was introduced to the St. Louis Harbor region of Lake Superior in the North American Great Lakes during the mid-1980s from ballast water arriving from an ocean-going vessel, and since has spread naturally and/or from other ballast water transport to some other areas of the upper Great Lakes (Busiahn 1997). *Gymnocephalus cernua* is tolerant of a wide variety of temperatures, salinities, water flow rates, and depths (Pratt et al. 1992; Ogle et al. 1995), rendering it an adaptable invader. Its population patterns are readily distinguishable across Eurasia by mtDNA differentiation (Stepien et al. 1998, 2004, 2005), as well as by

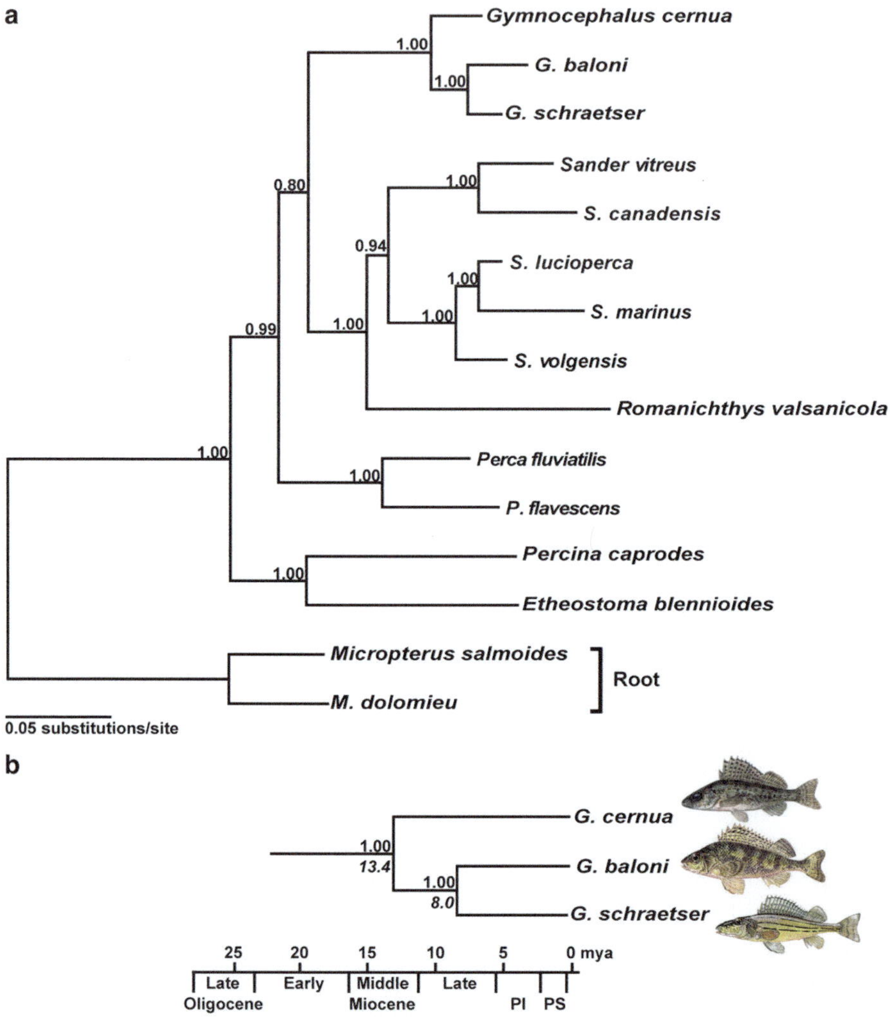

Fig. 1.13 (**a**) Phylogenetic tree (performed by us for this chapter) showing relationships within the genus *Gymnocephalus* and to other percids from a partitioned Bayesian analysis (MrBayes v.3.2.1; Ronquist and Huelsenbeck 2003, http://mrbayes.sourceforge.net/) of concatenated sequence data from two mitochondrial DNA regions – cytochrome (cyt) *b* and the control region – and the nuclear DNA lactate dehydrogenase intron 6 (LdhA6), with posterior probabilities indicated. Tree was rooted to *Micropterus*, based on its hypothesized close relationship to Percidae, according to Song et al. (1998) and Sloss et al. (2004). Sequences were collected from GenBank (http://www.ncbi.nlm.nih.gov/genbank) and those that were unique were included in the analyses (see Appendix 3b). Corrected Akaike information criteria from jModeltest2 (Darriba et al. 2012; http://code.google.com/p/jmodeltest2/) determined the most appropriate model of substitution, indicating the TPM2uf model with a gamma distribution (1.341) and invariable sites (0.548) for cyt *b*, a transversional model including a gamma distribution (0.316) for the control region, and a Kimura (1980) two

morphological characters (Stepien et al. 1998). Its Great Lakes colonization was genetically traced as matching a putative source population in or near the Elbe River, which drains into the North and Baltic seas (Stepien et al. 2004, 2005). Ecological adaptations that rendered *G. cernua* a successful and widespread colonizer in Eurasia following the retreating Pleistocene glaciers, likely predicated its later ecological successes in invading new habitats in Europe and the Great Lakes.

1.6 Genus *Percarina*: An Enigmatic Eurasian Taxon: One to Two Species: *P. demidoffi* and *P. maeotica*

Molecular phylogenetic studies have lacked *Percarina* material to date, circumventing the resolution of the relationship of this genus to others. Based on morphological characters, *Percarina* has been hypothesized to be closely related to *Perca* and *Gymnocephalus* (Collette and Bănărescu 1977; Craig 2000). *Percarina* possesses two widely separated dorsal fins, with 9–11 spines on the first dorsal (Craig 2000).

Two species are recognized: the common percarina *P. demidoffi* Nordman 1840 that is found in rivers draining into the northwestern Black Sea and shallow fresh and brackish waters within the Black Sea, and the Azov percarina *P. maeotica* Kuznetsov 1888, which is endemic to the Sea of Azov (see Fig. 1.14). Their distributions do not overlap. However, there is debate as to whether the genus comprises two species, a single species with two subspecies, or a single species. Kuznetsov (1888) described *P. maeotica* having distinctive scaled cheeks, an eye diameter shorter than the snout, and a dark spot at the snout tip. However, a recent morphological

Fig. 1.13 (continued) parameter model and a kappa distance (2.357) for LdhA6. Bayesian analyses in MrBayes used a Metropolis-coupled Markov chain Monte Carlo (MC^3) approach and ran for 5,000,000 generations, with sampling every 100 generations. Four separate chains were run simultaneously for each analysis, and two analyses ran simultaneously. The burn-in period for the MC^3 was determined by plotting log likelihood values for each generation to identify when stationarity was reached. As burn-in, 25 % of the generations were discarded, along with the trees and parameter values sampled prior to the burn-in. A 50 % majority rule consensus tree was based on the remaining generations, whose branch support was determined from the posterior probability distribution (Holder and Lewis 2003) in MrBayes. (**b**) Time-calibrated phylogeny in BEAST (v.1.71; Drummond et al. 2012, http://beast.bio.ed.ac.uk/) for *Gymnocephalus* that used the general time reversible nucleotide substitution model (Lanave et al. 1984) and a gamma distribution as identified by jModeltest2 for the individual gene trees. BEAST analyses used a relaxed molecular clock that assumed a lognormal distribution with the Yule speciation process (Gernhard 2008) as a tree prior. Two separate runs were conducted, each with a chain length of 50,000,000 generations, and parameters sampled every 100 generations. We used two fossil calibration points: 26.0 Mya for the genus *Perca* and 12.0 Mya for the genus *Micropterus*, and molecular calibration points of 15.4 Mya for the origin of the North American *Sander*, 13.8 Mya for the Eurasian *Sander*, and 9.1 Mya for between *S. lucioperca* and *S. marinus* from Haponski and Stepien (2013). Above nodes = Bayesian posterior probabilities; below (*italics*) = estimated divergences (Illustrations were used with permission from P. Maitland). *PI* Pliocene, *PS* Pleistocene

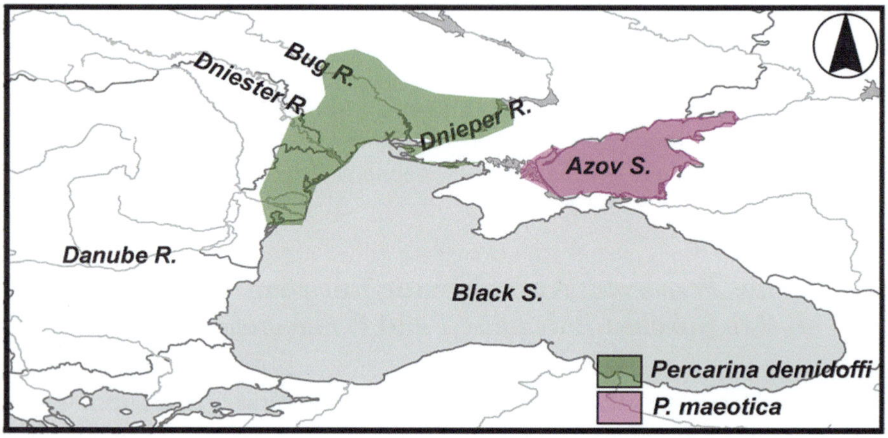

Fig. 1.14 Map showing the native ranges of *Percarina* spp. in Europe, based on information from Maitland (2000) and IUCN (2014)

analysis found that these and many other characters are not diagnostic between the two putative taxa (Vasil'eva 2006). *Percarina demidoffi* is listed as "near threatened" in the IUCN Red List (Freyhof 1996; IUCN 2014) and the European Union lists it as "endangered" in region 27, where increasing levels of salinity are impacting its breeding areas (Kottelat and Freyhof 2007). Molecular genetic data would be very valuable for resolving fundamental understanding of this group.

1.7 Subfamily Etheostomatinae: The Darters: Five Genera: *Ammocrypta*, *Crystallaria*, *Etheostoma*, *Nothonotus*, and *Percina*

Members of the subfamily Etheostomatinae – the darters – are endemic to North America (Fig. 1.15), inhabiting the Mississippi River basin and drainages of the Great Lakes, Hudson Bay, Atlantic Coast, Gulf of Mexico, and Mexican Pacific coast (Page 2000). They are small and benthic and typically "dart" from place to place. The males of most darter species have bright coloration patterns, especially during the reproductive period, which serve to attract females and function in species recognition. These color patterns constitute species-specific characters (see Boschung and Mayden 2004). Many darters spawn at specific sites and provide parental care of the eggs and fry. Diverse egg-deposition modes have evolved, including: egg burying (the hypothesized ancestral mode) – found in the genera *Ammocrypta*, *Crystallaria*, *Percina*, and some *Etheostoma*, egg attaching, and egg clustering (Page 2000). Specialized reproductive behavior – including the evolution of species-specific color patterns to attract mates – and high endemism have contributed to the large number of threatened and endangered darter species.

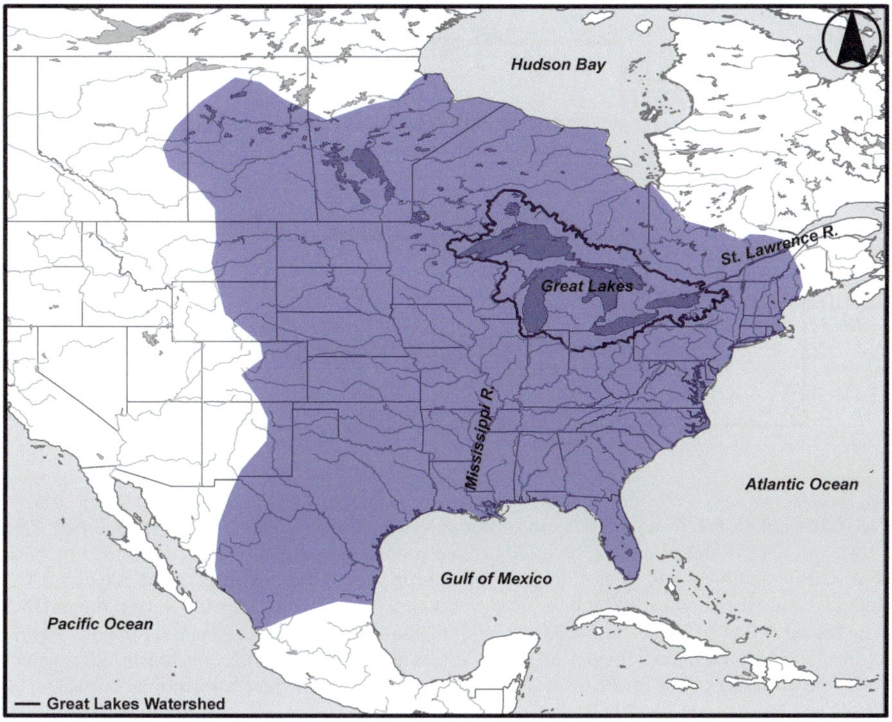

Fig. 1.15 Distribution of the Percidae subfamily Etheostomatinae, which is native to North America, based on information from Collette and Bănărescu (1977) and Craig (2000)

The darters constitute ~20 % of all North American freshwater fishes, and are second in species richness only to the minnows (Family Cyprinidae). Etheostomatinae includes ~248 species in five genera: *Ammocrypta* (6 species), *Crystallaria* (2 species), *Etheostoma* (~165–174 species), *Nothonotus* (~21 species), and *Percina* (~52 species). The molecular phylogenies of Figs. 1.2 and 1.3 show that Etheostomatinae is a monophyletic taxon. All Etheostomatinae have an absent or reduced swimbladder, and a reduced supraoccipital bone, as well as several other morphological distinctions (Page 2000). Phylogenetic analyses of mt and nuclear DNA sequences (Fig. 1.16) calibrated to the absolute age of centrarchid fossils estimate that Etheostomatinae differentiated ~38–31 Mya, as calculated by Near et al. (2011).

Species richness of Etheostomatinae is greatest in the Tennessee and Cumberland River systems and in tributaries in the Ozark and Ouachita Mountains; these areas are characterized by considerable spatial and topographic heterogeneity and long-term climatic stability (Page 2000). Individual species exhibit high degrees of morphological variability, which often differ among stream systems (Ayache and Near 2009). Some species are endemic to highly restricted ranges, including just a single stream system, rendering them extremely vulnerable to extinction (Page 2000).

Various genetic markers have revealed diverse relationships within the Etheostomatinae, hence resolution of their interrelationships has been very contro-

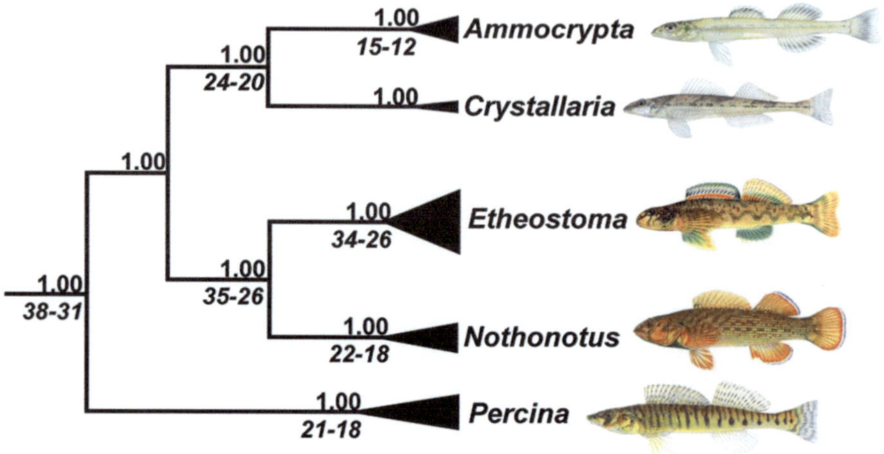

Fig. 1.16 Relationships among genera of the percid subfamily Etheostomatinae modified from Near et al. (2011) and Near and Keck (2013). Relationships originally were determined by Near et al. (2011) and Near and Keck (2013) using a partitioned Bayesian analysis (MrBayes v.3.1.2; Ronquist and Huelsenbeck 2003, http://mrbayes.sourceforge.net/) of sequences from the mtDNA cytochrome *b* (cyt *b*) gene and nuclear recombination-activating gene-1 (RAG1) exon 3 and S7 intron 1 and 13 protein coding nuclear genes. Values above nodes = posterior probabilities, values below are divergence time estimates from BEAST analyses for the genera performed by Near et al. (2011) and Near and Keck (2013)

versial. Many darter species have been found to share mtDNA haplotypes due to hybridization and mitochondrial replacement, confounding mtDNA phylogenies (Ray et al. 2008; Bossu and Near 2009; Keck and Near 2010). For example, the phylogenetic position of *Ammocrypta*, and the relationship of *Etheostoma zonale* to the snubnose darters have been disputed (e.g., Wood and Mayden 1997; Song et al. 1998; Sloss et al. 2004; Keck and Near 2008; Mendelson and Wong 2010). Smith et al. (2011) regarded *Percina* and *Nothonotus/Etheostoma* as sister groups, as further indicated by mtDNA data of Sloss et al. 2004 (see Fig. 1.3). However, many researchers have regarded *Percina* as comprising the sister group to all other darters (Kuehne and Barbour 1983; Page 1983); this scenario is supported by nuclear DNA phylogenies of Near et al. (2011) and Near and Keck (2013), as shown here in Fig. 1.16, and by morphological characters (Stephens et al. 2014). This phylogenetic consensus then groups the *Ammocrypta* + *Crystallaria* clade as the sister group to the *Etheostoma* + *Nothonotus* clade (Fig. 1.16).

1.7.1 Etheostomatinae, Genus Percina *~52 Species*

The genus *Percina* originated ~21–18 Mya (Near et al. 2011) and the latest and most comprehensive DNA evidence and morphological data indicate that it the sister group to all other darters (Fig. 1.16; Near et al. 2011; Stephens et al. 2014). This relationship was not supported by some molecular trees, as shown in Figs. 1.2 and 1.3, which were based on a relatively few number of taxa, and whose mtDNA relationships (Fig. 1.3) appeared confounded by the pronounced introgression of mtDNA (see account by Near et al. 2011). However, these molecular-based phylogenies all support monophyly of the genus (Figs. 1.2, 1.3, and 1.16).

Percina includes the largest-sized darters. Male *Percina* are characterized by unique ventral modified and enlarged scales that have pronounced serrated edges, which appear to function in providing tactile stimulation of the females during spawning (Page 2000). The females of some species also possess these modified scales. Members of the genus also have a complete and (almost) straight lateral line, ten preoperculomandibular pores, eight infraorbital pores, two anal spines, opaque color, and a moderate to deep body (Page 2000). The males typically become darker in color during the spawning season. *Percina* exhibits generalized reproductive behavior, spawning in stream riffles, where the females bury their eggs in sediment; they do not build nests and have no parental care. Most populations have relatively small numbers of individuals (Page 2000).

The genus *Percina* is widely distributed, ranging from as far south as the Gulf of Mexico and north to Hudson Bay (Fig. 1.17). Of the 52 *Percina* spp. the blackside

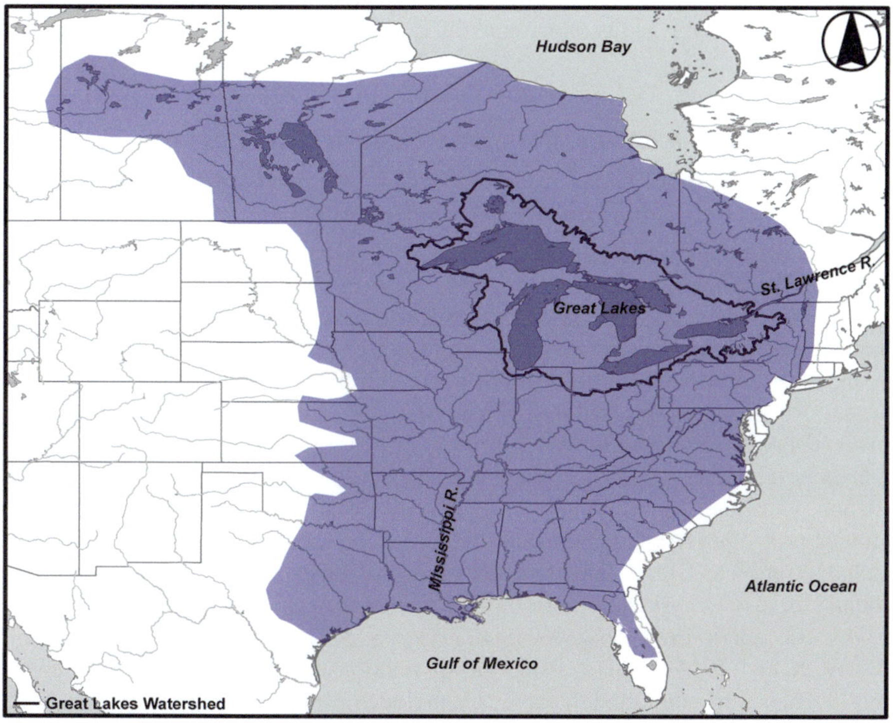

Fig. 1.17 Distribution of the genus *Percina* of the Percidae subfamily Etheostomatinae, based on information from Page and Burr (2011) and NatureServe Explorer (2014; http://explorer.natureserve.org/index.htm)

darter *P. maculata* (Girard 1859) and the logperch darter *P. caprodes* (Rafinesque 1818) have the widest distributions whereas others have restricted geographic ranges – such as the bluestripe darter *P. cymatotaenia* (Gilbert and Meek 1887 in Gilbert 1887) and the snail darter *P. tanasi* Etnier 1976 (Page and Burr 2011).

Relationships among *Percina* species have been well-resolved using the molecular characters of Near et al. (2011) and Near and Keck (2013), as illustrated in Fig. 1.18. Subgenera of *Percina* that were proposed on the basis of morphological characters and are supported by the molecular analyses include: *Cottogaster*, *Hypohomus*, *Odontopholis*, *Percina*, and *Swainia* (designated with *vertical bars* on the tree in Fig. 1.18). These clades also were supported by morphological characters, with the subgenus *Cottogaster* lacking a premaxillary frenum, *Percina* having a bulbous snout, *Odontopholis* possessing a caudal keel with strongly toothed scales and no modified midline scales along the belly, and *Swainia* having an elongated snout (summarized by Near 2002). A morphological hypothesis proposed that the freckled darter *P. lenticula* Richards and Knapp 1964 was the sister taxon of all other *Percina* (Page 1974). However, molecular data to date (Fig. 1.18) have indicated that *P. lenticula* is a more derived taxon (Near et al. 2011; Near and Keck 2013). In contrast, the Roanoke darter *P. roanoka* (Jordan and Jenkins 1889 in Jordan 1889) was once hypothesized to be derived (Page 1974, 1981; see Near 2002 for full account), but molecular data (Fig. 1.18) have indicated that it is one of the older members of the genus (Near et al. 2011).

1.7.2 Etheostomatinae, the Sister Genera **Ammocrypta** *(Six Species) and* **Crystallaria** *(Two Species)*

The genera *Ammocrypta* (six species) and *Crystallaria* (two species) are sister groups, whose members have elongate slender bodies that are translucent and possess a single anal spine (Simons 1992). They are the sole percids that have Spreitzer vertebrae, in which the first one to three anterior caudal vertebrae have open, unfused haemal arches, which allow the kidney to expand dorsally and posteriorly into the postabdominal vertebrae (Bruner 2004, 2011). The genus *Esox* (Esocidae) is the only other group of fishes that has this type of vertebrae. *Ammocrypta* and *Crystallaria* frequently bury themselves in the sand or gravel of streams with just the eyes visible. Their translucent color and burying behavior provide protection from predators (Williams 1975). Their morphological characters include complete lateral lines and uninterrupted head canals, and high meristic counts; these are considered plesiomorphic (ancestral) characters (Page 2000).

A close relationship between *Ammocrypta* and *Crystallaria* was proposed by Bailey et al. (1954), and a sister group relationship between the two genera (Fig. 1.16) is well-supported by the DNA phylogenies of Near et al. (2011) and Near and Keck (2013), AFLP (amplified fragment length polymorphism) results by Smith et al. (2011), and nuclear DNA sequences and morphological characters by Stephens et al. (2014). Smith et al. (2011) hypothesized that the *Ammocrypta*/*Crystallaria*

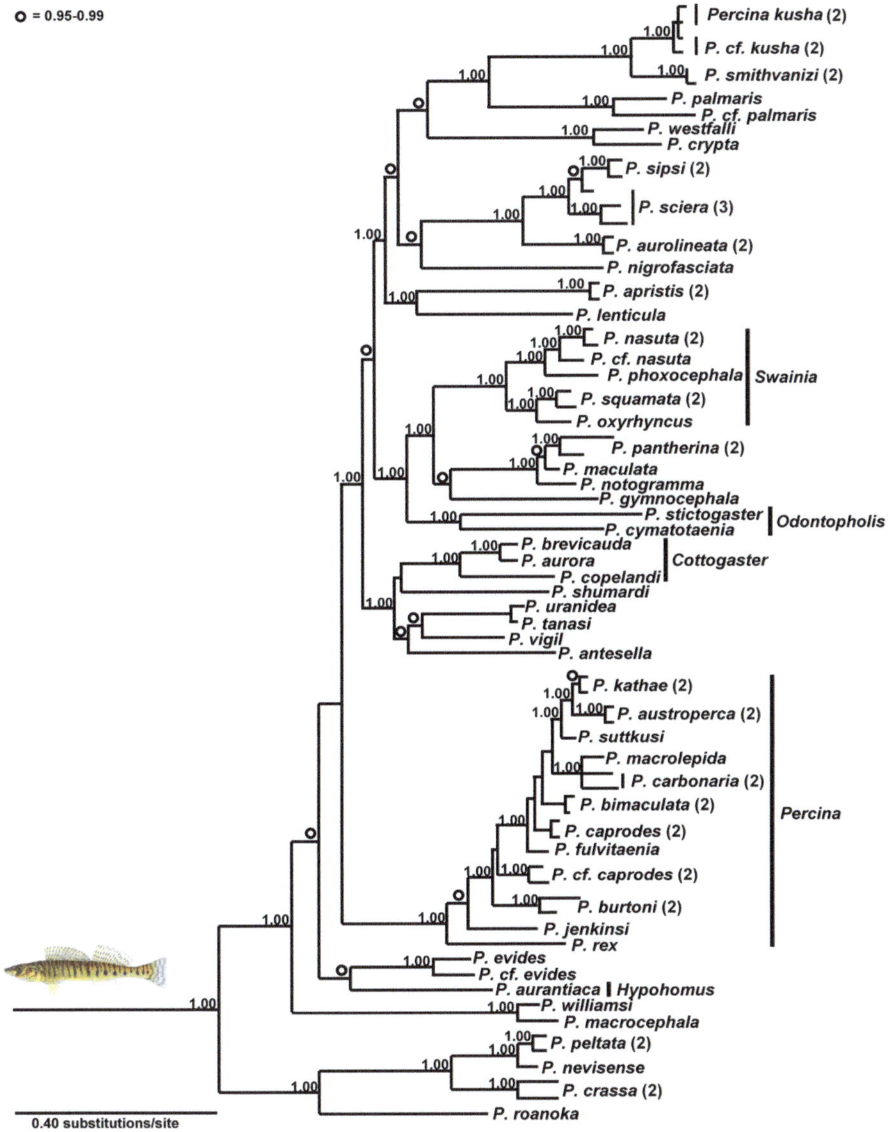

Fig. 1.18 Phylogenetic tree of relationships among members of the genus *Percina*, based on a partitioned Bayesian analysis (MrBayes v.3.1.2; Ronquist and Huelsenbeck 2003, http://mrbayes.sourceforge.net/) by Near et al. (2011) of concatenated sequences from the mtDNA cytochrome *b* (cyt *b*) gene, and the nuclear recombination-activating gene-1 (RAG1) exon 3 and S7 intron 1. Relationships are similar to those reported by Near and Keck (2013) using 13 protein coding nuvclear genes. Values on nodes = posterior probabilities (pp). *Circle* = range from 0.95 to 0.99 pp. Numbers in parentheses indicate multiple haplotypes per species. *Vertical bars* denote proposed subgenera, which are supported by this analysis

clade forms the sister group to all other darters and does not group with *Etheostoma*. However, AFLP results are an indirect approach for discerning the evolutionary history and patterns of characters, and their phylogenetic use should be regarded with extreme caution (summarized by Near and Keck 2012). We thus present the well-resolved DNA phylogeny shown in Fig. 1.16, based on Near et al. (2011) and Near and Keck (2013), which depicts *Ammocrypta/Crystallaria* as the sister group to the *Etheostoma/Nothonotus* clade. A study by Stephens et al. (2014) also recovered similar relationships among these genera using the nuclear recombination activating gene-1 (RAG-1) exon 3 and 77 morphological characters.

The six species of *Ammocrypta* Jordan 1878 commonly are known as the sand darters, which include the naked sand darter *A. beanii* Jordan 1878, Florida sand darter *A. bifascia* Williams 1975, western sand darter *A. clara* Jordan and Meek 1885, southern sand darter *A meridiana* Williams 1975, Eastern sand darter *A. pellucida* (Putnam 1863), and scaly sand darter *A. vivax* Hay 1882. *Ammocrypta* has a more extensive distribution than *Crystallaria*, with the former ranging from the Gulf of Mexico to the Great Lakes watershed (Fig. 1.19). *Ammocrypta* lacks a premaxillary frenum and palatine teeth, and has fewer dorsal spines (9–12) and rays (10–11) than does *Crystallaria* (Simons 1991, 1992). *Ammocrypta clara* is depicted as the sister species to all other members of the genus (Fig. 1.20).

The genus *Crystallaria* is differentiated from *Ammocrypta* and the other darters by several osteological characters, including a bifurcate supraoccipital crest and narrow premaxillary frenum (Simons 1991). *Crystallaria* also possesses greater numbers of dorsal spines, dorsal rays, and anal fin rays than does *Ammocrypta* (Simons 1991). There are just two species of *Crystallaria* (Jordan 1878) – the crystal darter *C. asprella* (Jordan 1878) and the more recently-described diamond darter *C. cincotta* Welsh and Wood 2008. The two species differ in gape size, number of dorsal spines and anal fin rays, scale numbers on the cheeks and opercle, and color (Welsh and Wood 2008).

All populations of both *Crystallaria* species have been shown to be genetically divergent, constituting independent evolutionary lineages that are highly imperiled (Wood and Raley 2000; Morrison et al. 2006). The crystal darter now is extirpated from much of its original range in the Mississippi River basin, east Gulf coast, and Mobile basin drainage (Fig. 1.19b; Stewart et al. 2005); it is listed as vulnerable due to "decline in area of occupancy, extent of occurrence, and/or quality of habitat" (NatureServe 2013a; IUCN 2014). The diamond darter once inhabited several drainages in the Ohio River basin, but its sole extant population occurs in the lower Elk River of West Virginia. It is listed as critically endangered (NatureServe 2013b; IUCN 2014).

1.7.3 Etheostomatinae, Genus Etheostoma*: ~165–174 Species*

The genus *Etheostoma* originated ~34–26 Mya and is the most species-rich of the darter genera, comprising ~165–174 species (Near et al. 2011; Near and Keck 2013). It is the largest genus of North American freshwater fishes and is broadly

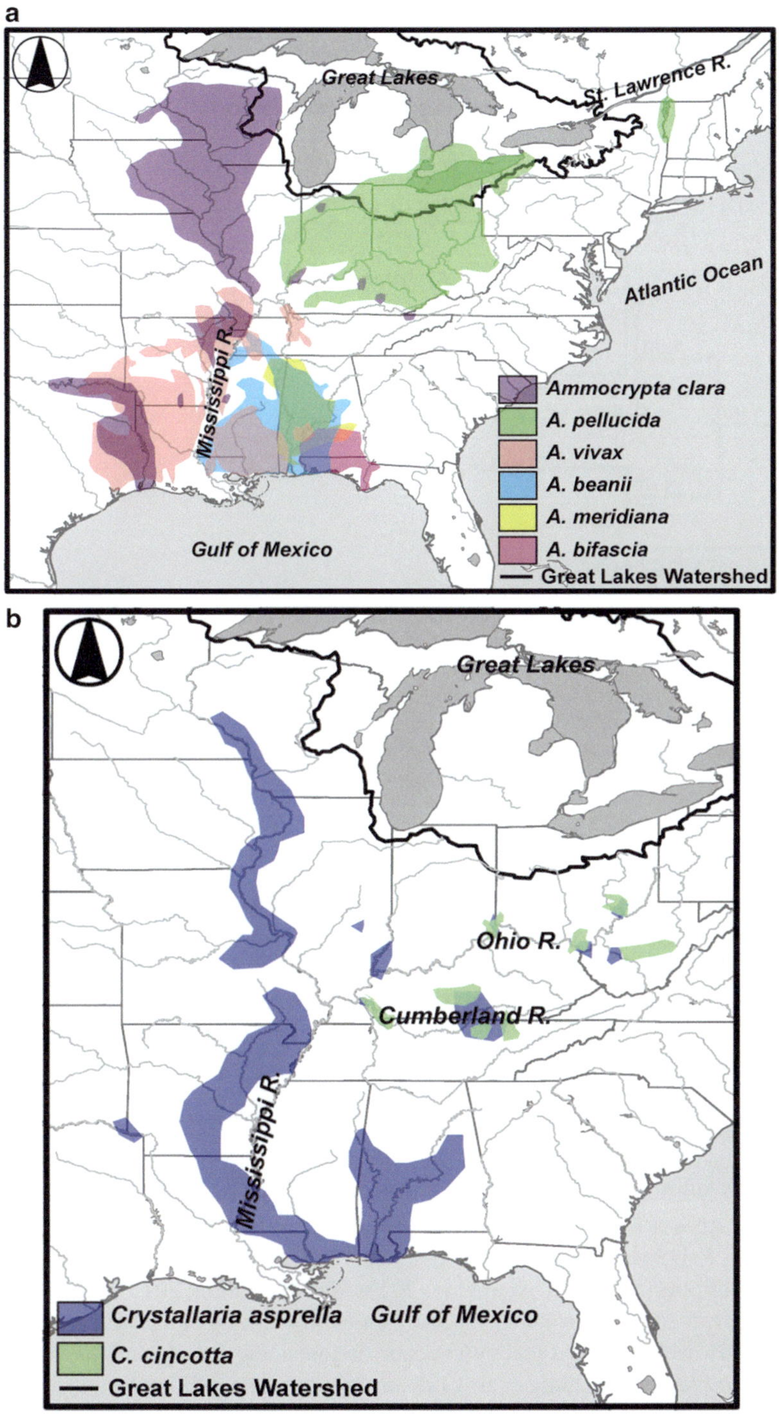

Fig. 1.19 Distributions for species belonging to the genera (**a**) *Ammocrypta* and (**b**) *Crystallaria* of the Percidae subfamily Etheostomatinae, based on information from Page and Burr (2011) and NatureServe Explorer (2014; http://explorer.natureserve.org/index.htm)

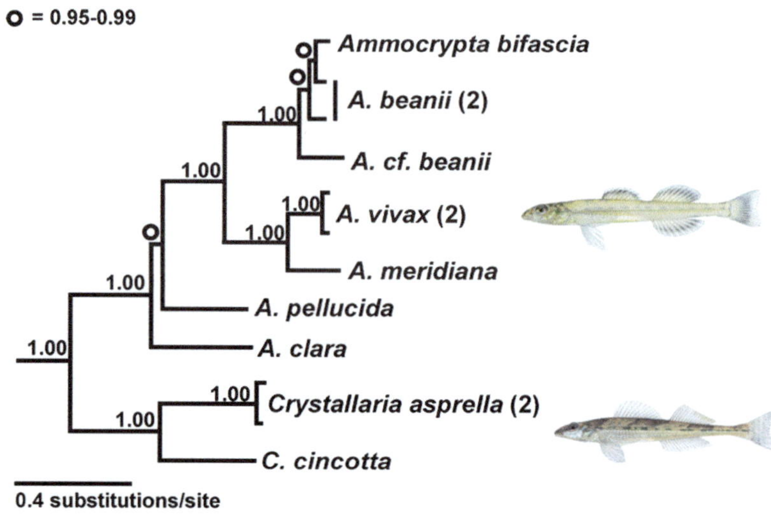

Fig. 1.20 Phylogenetic tree showing relationships among species of *Ammocrypta* and *Crystallaria*, based on a partitioned Bayesian analysis (MrBayes v.3.1.2; Ronquist and Huelsenbeck 2003, http://mrbayes.sourceforge.net/) by Near et al. (2011) using concatenated sequences from the mtDNA cytochrome *b* (cyt *b*) gene and nuclear recombination-activating gene-1 (RAG1) exon 3 and S7 intron 1. Relationships are similar to those reported by Near and Keck (2013) using 13 protein coding nuclear genes. Values on nodes = posterior probabilities (pp). *Circle* = range from 0.95 to 0.99 pp. Numbers in parentheses indicate multiple haplotypes per species

distributed from the Gulf of Mexico to Hudson Bay (Fig. 1.21). The genus has sometimes been regarded as a "catch-all" group, i.e., may not comprise a monophyletic group, as it lacks derived morphological characters (see Song et al. 1998). However, the most recent phylogenies by Near et al. (2011), Near and Keck (2013), and Stephens et al. (2014) depict it as monophyletic (see Fig. 1.16). Although the basal bifurcation of the tree shown in Fig. 1.22 lacks support, many of the internal relationships are well-supported.

Several subgenera have been proposed (e.g., Bailey and Gosline (1955), Page (1983)), however, the relationships among them and their species composition have been debated (see Near et al. 2011; Near and Keck 2013; Smith et al. 2014; Stephens et al. 2014), with most of the proposed subgenera un-supported by the most recent molecular trees (see Fig. 1.22). The exceptions are the subgenera *Doration*, *Catonotus*, *Psychromaster*, *Allohistium*, and *Fuscatelum*, which possess strong Bayesian support (Fig. 1.22; Near et al. 2011; Near and Keck 2013).

Some darter species have very narrow ranges; for example, the watercress darter *E. nuchale* Howell and Caldwell 1965 occurs in just a few springs and the vermillion darter *E. chermocki* Boschung et al. 1992 inhabits just a few kilometers of streambed (Boschung and Mayden 2004). Shapes and body features of darter species often display convergence according to type of habitats and feeding modes (see Page 2000). For example, mid-water species are larger in size and have fusiform shapes,

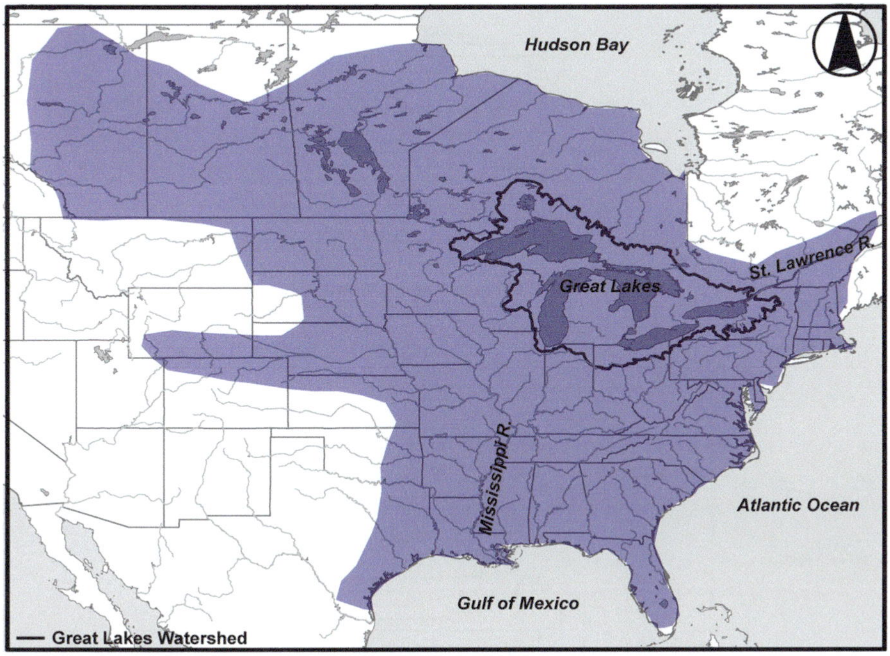

Fig. 1.21 Distribution of the genus *Etheostoma* from the Percidae subfamily Etheostomatinae, based on information from Page and Burr (2011) and NatureServe Explorer (2014; http://explorer.natureserve.org/index.htm)

smaller pectoral fins, pointed snouts, and terminal mouths. Benthic species are flattened to shift their gravity lower to reduce water resistance and have shorter snouts. Riffle-living species are more robust and have deeper caudal peduncles, and those living in sand have longer snouts to facilitate burrowing and foraging (see Page 2000). Green colors are common in benthic darters that live in vegetation, such as the greenside darter *E. blennioides* Rafinesque 1819 and the banded darter *E. zonale* (Cope 1868). Many of the benthic species have dark vertical bars, which serve to disrupt the pattern, camouflaging the darter, when viewed from above or the side by predators (see Page 2000).

In many *Etheostoma* species, males are distinctively brightly colored during the spawning season, with brightly-banded dorsal fins, whereas females are cryptic. Brightly colored males are particularly common in stream riffles. Their colors facilitate species and sex recognition. Males usually have larger fins than females, with the pelvic fins in some species (e.g., the least darter *E. microperca* Jordan and Gilbert 1888 in Jordan 1888) used to clap the back of the female during spawning. The males of many species have breeding tubercles, which are keratinized epidermal protuberances that are used for tactile stimulation of the females during spawning (see Page 2000). Females of a few species (e.g., the variegate darter *E. variatum* Kirkland 1840, and the glassy darter *E. vitreum* (Cope 1870)) and close relatives also develop these tubercles (see Collette 1965).

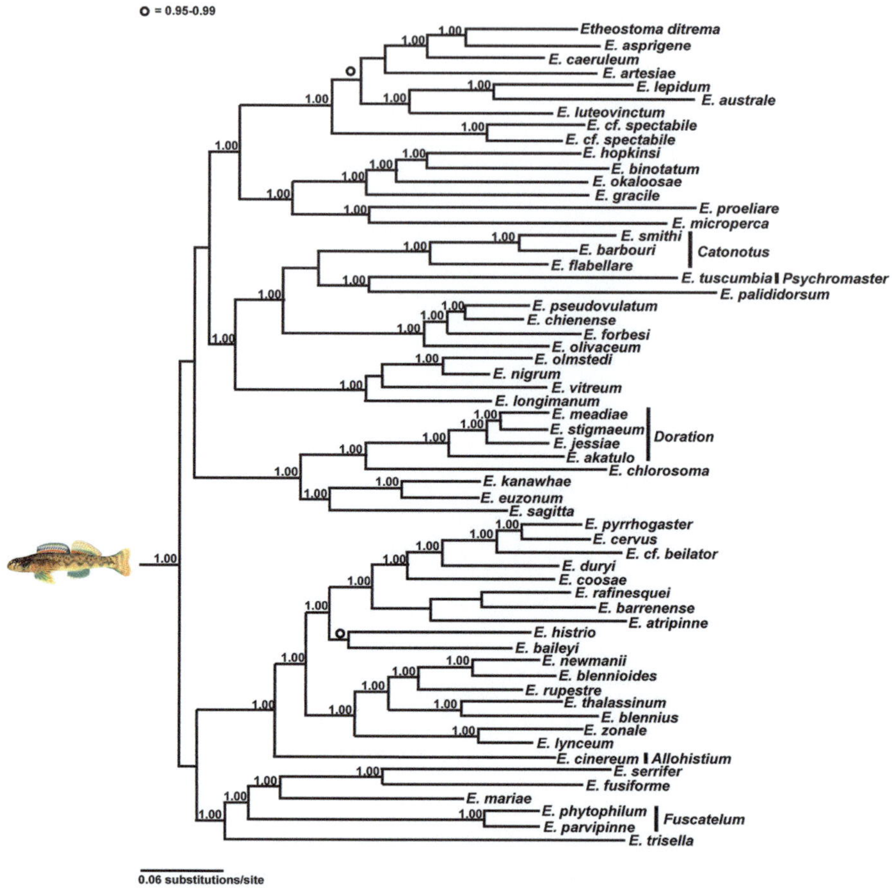

Fig. 1.22 Phylogenetic tree showing relationships among species of the genus *Etheostoma*, from Near and Keck (2013) using a partitioned Bayesian analysis (MrBayes v.3.1.2; Ronquist and Huelsenbeck 2003, http://mrbayes.sourceforge.net/) of 13 protein coding nuclear gene sequences. Relationships are similar to those reported by Near et al. (2011) for a Bayesian analysis of sequences from the mtDNA cytochrome *b* (cyt *b*) gene and nuclear recombination-activating gene-1 (RAG1) exon 3 and S7 intron 1. Values on nodes = posterior probabilities (pp). *Circle* = range from 0.95 to 0.99 pp. *Vertical bars* denote proposed subgenera from morphological hypotheses, which are supported by this analysis

Three spawning modes occur, including: burying, attaching, clumping, and clustering of eggs (see Page 2000). In burying, the fertilized eggs are buried over a wide area, providing protection from predators (e.g., rainbow darter *E. caeruleum* Storer 1845). Males of egg-clustering species select and defend a cavity in which one or more females lay eggs, and the male alone remains to guard the eggs. The eggs of egg-clustering species are adhesive and are deposited on the roofs of cavities, usually by successive females and guarded by a single male (e.g., fantail darter *E. flabellare* Rafinesque 1819).

Many species also exhibit considerable population-level differentiation across their respective ranges, which is opposed by some ongoing or recent historic gene flow. For example the widely-distributed rainbow darter *E. caeruleum* and greenside darter *E. blennioides* both show high divergences between populations in the Great Lakes versus those in the Ohio River drainage, dating to convergent separations during the Pleistocene glaciations an estimated 1.8–1.6 Mya (Haponski et al. 2009 for the rainbow darter; Haponski and Stepien 2008, and Stepien and Haponski 2010 for the greenside darter).

1.7.4 Etheostomatinae, Genus Nothonotus*: ~20 Species*

The genus *Nothonotus* is estimated to have diverged ~22–18 Mya (Near et al. 2011); this taxon formerly was regarded to be a subgenus of *Etheostoma. Nothonotus* is distributed across the Central Highlands, with most species occurring east of the Mississippi River (Fig. 1.23). Analyses by Near et al. (2011) using mt and nuclear DNA and by Near and Keck (2013) using nuclear genes alone concluded that *Nothonotus* constitutes a strongly supported clade separate from *Etheostoma*, and

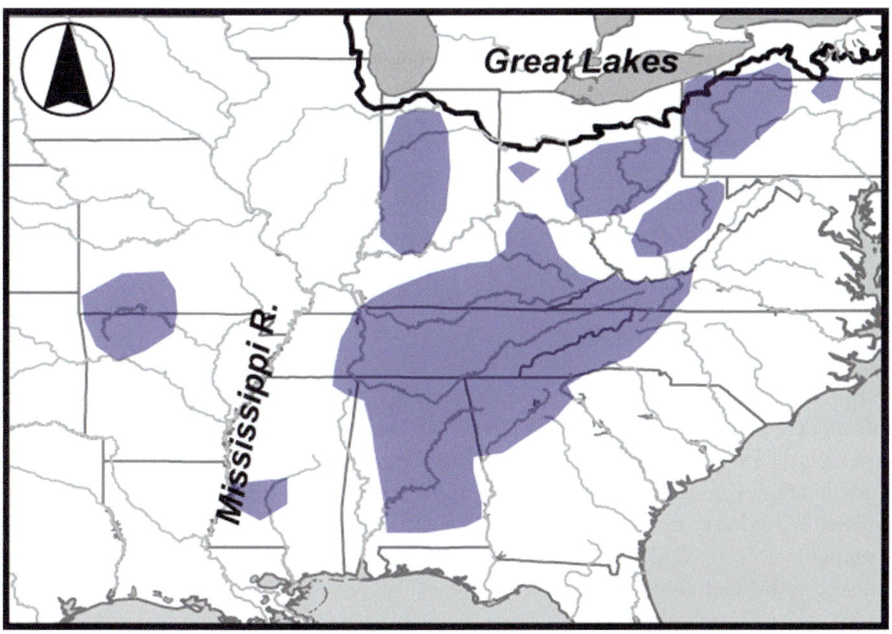

Fig. 1.23 Distribution of the genus *Nothonotus* from the Percidae subfamily Etheostomatinae, based on information from Page and Burr (2011) and NatureServe Explorer (2014; http://explorer.natureserve.org/index.htm)

recommended its elevation to generic status (Fig. 1.16). Stephens et al. (2014) also recovered *Nothonotus* and *Etheostoma* as separate monophyletic clades, based on nuclear DNA sequences and morphological characters.

Nothonotus possesses a deep slab sided body and a naked nape, distinguishing members of the genus from other darters (Etnier and Williams 1989; Wood 1996). It also has horizontal dark lines along the sides of the body, both above and below the lateral line. The reproductive mode of *Nothonotus* is egg-clumping (which is similar but simpler than egg-clustering of many *Etheostoma* darters), with the female laying eggs in crevices selected and guarded by the male, who then guards the eggs until hatching. Many of the species of *Nothonotus* are differentiated by distinctive male nuptial coloration patterns and scale counts.

Phylogenetic relationships among the 20 species of *Nothonotus* were analyzed by Near et al. (2011) and Near and Keck (2013), as shown in Fig. 1.24. All phylogenetic and morphological analyses have found that the yoke darter *N. juliae* Meek 1891 is the sister taxon to the remaining members of the genus (Fig. 1.23; Etnier and Williams 1989; Wood 1996; Keck and Near 2008; Near et al. 2011). Most proposed species were resolved as monophyletic by Near et al. (2011) and Near and Keck (2013), as shown on Fig. 1.23, with the exception of the bloodfin darter *N. sanguifluus* (Cope 1870), smallscale darter *N. microlepidus* (Raney and Zorach 1967), and spotted darter *N. maculatus* Kirtland 1840.

1.8 Anthropogenic Effects on Evolution of Percidae: Predictions for the Future

Many species of Percidae have been intentionally and unintentionally spread to areas outside of their native ranges. Notably, *Sander* and *Perca* spp. have been extensively stocked due to their fishery popularity. The walleye *Sander vitreus* has been the most widely introduced (Fig. 1.5b), including as far northeast as the St. Croix River in Maine, southeast to the Lower Oconee River in Georgia, west to rivers in Washington and Oregon that drain to the Pacific Ocean, and south to the Chattahoochee River, Casablanca Reservoir, and Guadalupe River in Texas (Billington et al. 2011; Fuller and Neilson 2012). The sauger *S. canadensis* has been introduced to the upper Savannah River in Georgia, Lake Texoma in Texas, the Apalachicola River in Florida, and the lower Bear River in Idaho (see Fig. 1.5b). In Eurasia (Fig. 1.5a), the pikeperch *S. lucioperca* has been transplanted to Spain, the United Kingdom, France, the Netherlands, western Germany, Denmark, Italy, Lithuania, Latvia, and Turkey (Larsen and Berg 2006). It also was intentionally stocked into Spiritwood Lake, North Dakota in 1989 with the misguided aspiration

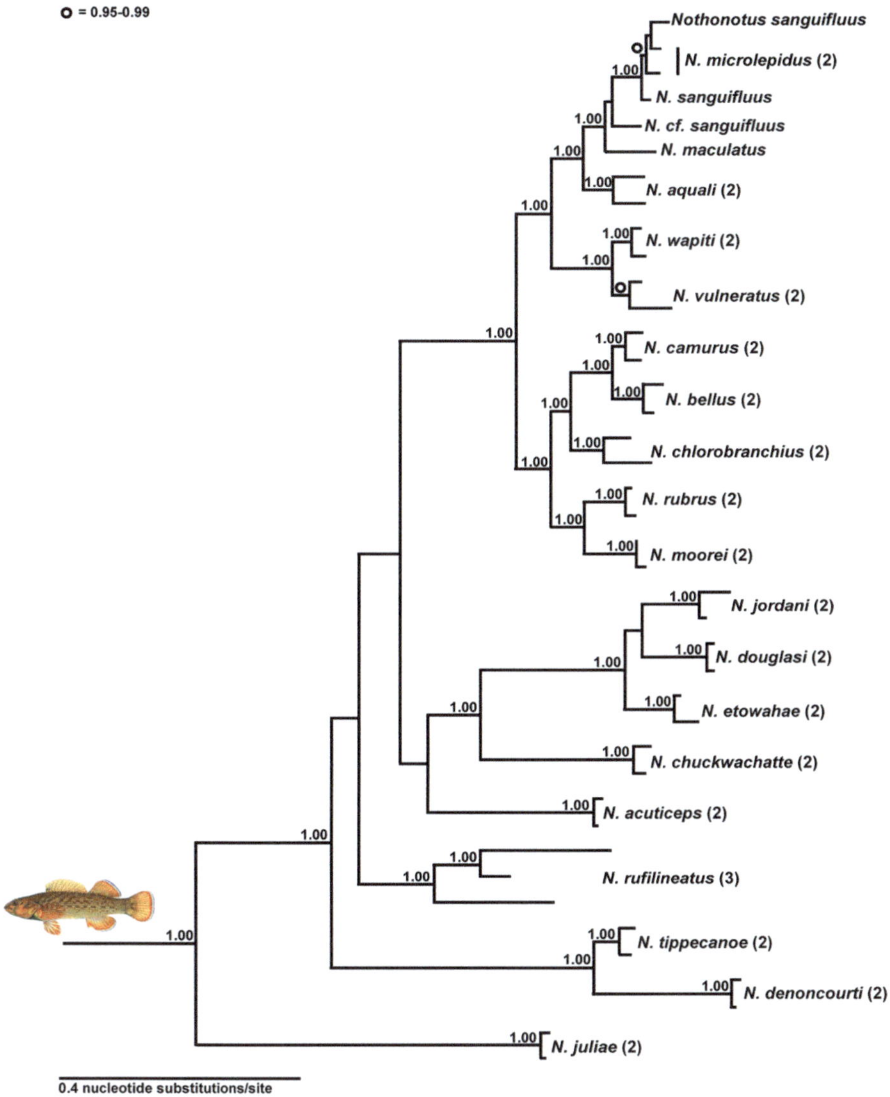

Fig. 1.24 Phylogenetic relationships among species of the genus *Nothonotus*, according to Near et al. (2011) using a partitioned Bayesian analysis (MrBayes v.3.1.2; Ronquist and Huelsenbeck 2003, http://mrbayes.sourceforge.net/) of sequences from the mtDNA cytochrome *b* (cyt *b*) gene and nuclear recombination-activating gene-1 (RAG1) exon 3 and S7 intron 1. Relationships are similar to those reported by Near and Keck (2013) from Bayesian analysis of sequences from 13 protein coding nuclear genes. Values on nodes = posterior probabilities. Numbers in parentheses indicate multiple haplotypes per species

that it would become a valuable fishery. There since has been natural reproduction, but the population remains very small and does not support a fishery (Fuller 2014). Due to high water conditions, there is considerable concern that *S. lucioperca* may have escaped into the James River, a tributary of the Mississippi River (Fuller 2014). This illustrates a very serious problem for native biodiversity, as it may introgress with native *S. vitreus* and/or *S. canadensis*, if the populations meet in the future. We caution that such translocations of species have no conservation or fishery merit, and must be discontinued.

The North American yellow perch *P. flavescens* has been intentionally released to other locations across the United States via bait bucket transfers or by fisheries agencies (summarized by Fuller and Neilson 2014). Stocked yellow perch can have damaging impacts to native food webs, having been shown to reduce prey abundance and diversity and outcompete native species. The European perch *P. fluviatilis* was introduced to Africa, Australia, and New Zealand due to its popularity as a food fish, and has been commercially cultured in Australia (Food and Agriculture Organization of the United Nations 2014).

The Eurasian ruffe *Gymnocephalus cernua* has been introduced into the Rhine River in France, northern Italy, Great Britain, and Greece in Europe via canals, shipping, and "bait bucket" transfers. In ~1986, the ruffe was accidentally introduced from ballast water discharge to Lake Superior's St. Louis Harbor in the Great Lakes (Simon and Vondruska 1991; Pratt et al. 1992). Genetic studies by Stepien et al. (1998, 2005) used mt and nuclear DNA sequence data to identify the Elbe River Germany as the likely source of the ruffe invasion, which dated to increased trade coinciding with the collapse of the former USSR. Following its initial introduction to the Great Lakes, the ruffe spread eastward along the southern shore of Lake Superior from 1986 to 1988 and to northern parts of Lakes Michigan (2002–2007) and Huron (1995) (Stepien et al. 1998, 2005; Fuller et al. 2014). The ruffe primarily forages on benthic prey items and has the potential to outcompete native species, such as yellow perch and European perch. Following introductions of ruffe, Scottish populations of *P. fluviatilis* and Russian populations of whitefish *Coregonus* declined due to egg consumption by ruffe (McLean 1993).

Many species of Percidae – including walleye, sauger, and yellow perch – have been stocked in areas within their native range with aims to re-establish or increase fishing abundance. This may lead to hybridization, especially in cases when two species meet that did not evolve together in that system (Billington and Sloss 2011). Stockings of saugeye (crosses of *S. vitreus* and *S. canadensis*) have been common in many areas of the central United States, which have interbred and introgressed (White and Schell 1995; White et al. 2005). White et al. (2005) found that 27 % of individuals identified as walleye in some stocked areas of the Ohio River actually were saugeye hybrids. Results may lead to competition between stocked and native individuals for food, resulting in loss of native population identity and biodiversity.

Global temperatures are predicted to increase over the next 50 years and likely will shift the distribution centers of most percid populations northward, extending their post-glacial expansion trajectory (Chu et al. 2005; Sepulveda-Villet and Stepien 2012). For example, Shuter and Post (1990) suggested that an increase of 4 °C would move the distributional extent of yellow perch north and, depending on lake morphometry and productivity, likely greatly affect survival, relative year-class strength, and ecosystem carrying capacity. Population genetic effects may include declines in variability, unique alleles, and local adaptations (see Sepulveda-Villet and Stepien 2012). Large available habitat expanses and high connectivity among waterways, in extensive regions such as the Great Lakes, may help to mitigate effects. However, these movements also may act to homogenize gene pools of distinctive spawning groups as they move northward and mix.

Populations inhabiting the lower latitudinal fringes of a species' native range likely house valuable genetic adaptations to warmer climates (Hampe and Petit 2005; Sepulveda-Villet and Stepien 2012). For example, the diverse South Atlantic coastal yellow perch populations may prove especially well-adapted to tolerating salinity fluctuations and increasing water temperatures, facilitating their northward coastal migration, if sea levels rise to eventually connect low-lying estuaries, which are currently isolated by barrier island and sandbar systems. Similarly, southerly walleye populations house unique genetic variation that may aid their survival in warmer climates (Stepien et al. 2009; Haponski and Stepien 2014). These may interbreed with existing populations in the north, given connection or transport opportunity. It is likely that some outlying and fringe populations will fail to adapt and experience increased isolation, and possible extirpation. It thus is possible that the adaptive potential of native populations may be either positively or negatively influenced by these alterations.

Climate change may alter perch growth rates (see Carlander 1977). For example, yellow perch were significantly affected by water level fluctuations of glacial lakes in North Dakota, with their lowest recorded abundances and body weights occurring during low water periods (Dembkowski et al. 2014), illustrating potential deleterious effects of increased evaporation and water losses that are linked to climate change. Temperature differences also may lead to genetic isolation of spawning groups through alteration of spawning times. For example, those spawning in warmer shallower waters may reproduce earlier than those in deeper cooler waters, leading to potential reproductive isolation (see European perch account by Bergek et al. 2010). Such findings highlight the significance of habitat and localized variations among their associated reproductive population groups. Genetic diversity and composition may thus be greatly influenced by the increasing pace of climate change.

Additionally the cumulative effects of multiple stressors on native species, including parasites, pathogens, and anthropogenic pollutants are a growing concern; these may negatively affect fish health and immune response of populations (Vidal-Martínez et al. 2010; Marcogliese and Pietrock 2011). Host-parasite coevolution and interactions will influence disease dynamics and survivability (Woolhouse et al. 2002; Stepien et al. 2015c). For example, the major histocompatibility complex

(MHC) class II receptor genes are highly variable in European perch (Michel et al. 2009; Oppelt and Behrmann-Godel 2012) and a long-term rise in temperature (over 35 years) appears to have resulted in extensive change in their genetic variability and allelic composition in an artificially heated Swedish lake (Björklund et al. 2015). Moreover, those results indicated an alteration of their parasitic community, implying long-term and extensive changes.

In conclusion, understanding the historical and present day factors that have shaped today's populations of percid species may aid their continued conservation in the face of future challenges. Evaluating their evolutionary diversification patterns stemming from post-glacial dispersal and adaptation in new environments, and the genetic and genomic reservoirs contained in isolated relict groups, may help us to predict the challenges faced by taxa during this era of rapid climate and habitat alterations.

Acknowledgements This is publication #2015–06 from the University of Toledo's Lake Erie Research Center. Support for our research reported here came from the National Science Foundation NSF GK-12 DGE#0742395, the USEPA #CR-83281401-0, NOAA Ohio Sea Grant R/LR-013, and USDA ARS 3655-31000-020-00D funded to CAS. We greatly appreciate percid samples supplied to us by many collectors, including: the USGS (Bruce Manny, Patrick Kocovsky, Wendylee Stott, Edward Roseman, Jeffrey Williamson), Ohio DNR (Alexander Ford, Ann Marie Gorman, Kevin Kayle, Cary Knight, Roger Knight, Matthew Turner, Jeffrey Tyson, Christopher Vandergoot), Michigan DNR (David Clapp, Robert Haas, Michael Thomas), New York Department of Environmental Conservation (Brian Beckwith, Donald Einhouse), Pennsylvania Fish and Boat Commission (Roger Kenyon), Ontario Ministry of Natural Resources (Christopher Wilson, Timothy Johnson), and from various colleagues including Colin Adams, Louis Bernatchez, Neil Billington, Vasily Boldyrev, John Clay Bruner, Mary Burnham-Curtis, Thomas Busiahn, Igor Grigorovich, Daniel Grul'a, Jeffrey Gunderson, Eric Hallerman, Anjie Hintz, Shaig Ibrahimov, Wolfgang Jansen, Douglas Jensen, Vladimir Kovac, Brandon Kulik, Yuriy Kvach, Ellen Marsden, Christine Mayer, Gregory Moyer, Andrew Naffziger, Matthew Neilson, Douglas Nelson, Camille Nesbø, Vadim Panov, Alex Parker, Webb Pearsall, Lars Rudstam, Mariusz Sapota, Wayne Schaefer, Jim Selgeby, Predrag Simonovic, George Spangler, David L. Stein, Roy Stein, Matthew White, James Williams, Ian Winfield, and Christopher Yoder. The manuscript benefitted from careful critical reviews by John C. Bruner, Guillermo Orti, and William Leo Smith.

Appendices

Appendix 1

Locations sampled for the three species of *Perca*, including latitude (lat) and longitude (long), number of individuals (N), number of haplotypes (N_H), and haplotypes recovered for mtDNA cytochrome *c* oxidase subunit 1 and cytochrome *b*, and for nuclear recombination activating gene 1.

					COI		Cyt *b*		RAG 1	
Species	Location	Lat.	Long.	N	N_H	Haplotype	N_H	Haplotype	N_H	Haplotype
P. flavescens	North America	–	–	12	6	–	11	–	4	–
	Chattahoochee R. AL USA	32.6047	85.0808	1	1	Pflacoi5	1	Pflacb2	2	Pflarag1,2
	Mobile Bay AL USA	30.7223	87.8745	1	1	Pflacoi3	1	Pflacb5	1	Pflarag3
	Morgan Cr. NC USA	35.5336	79.1931	2	2	Pflacoi2,4	2	Pflacb3,4	1	Pflarag1
	Muskegon L. MI USA	43.2324	86.2945	2	1	Pflacoi1	2	Pflacb6,7	2	Pflarag1,4
	Hudson R. NY USA	43.1453	73.4946	2	2	Pflacoi1,6	2	Pflacb1,8	1	Pflarag1
	L. Erie OH USA	41.8060	81.4180	2	1	Pflacoi1	2	Pflacb1,11	1	Pflarag1
	Bush R. MD USA	37.2519	122.0506	2	1	Pflacoi2	2	Pflacb9,10	1	Pflarag1
P. fluviatilis	Europe	–	–	8	4	–	6	–	4	–
	Volga R. Vinovka RUS	48.8708	44.6601	2	1	Pflucoi3	2	Pflucb2,3	2	Pflurag1,2
	Volga R. Stupino RUS	48.3103	45.7973	1	1	Pflucoi1	0	–	0	–
	Dojran L. Doirani GRC	41.1200	22.4500	1	1	Pflucoi2	1	Pflucb4	1	Pflurag3
	Danube R. Vienna AUT	48.1383	16.2211	2	1	Pflucoi1	2	Pflucb1,5	1	Pflurag4
	Lake Baikal Ust-Barguzin RUS	53.3000	108.0000	2	1	Pflucoi4	1	Pflucb6	2	Pflurag1,2
P. schrenkii	Europe	–	–	8	3	–	5	–	3	–
	Alakol L. KAZ	46.1000	81.3500	4	3	Psccoi1,2,3	4	Psccb1,2,3,4	2	Pscrag1,2

Appendix 2

Comparison of *Perca* sequences from GenBank to those found in this study. NR=not recovered by us.

Gene/region	Species	Our haplotype	GenBank accession #
mtDNA COI	*P. flavescens*	Pflacoi1	KC819884 (Haponski and Stepien 2013), EU524238, EU524240-43 (Hubert et al. 2008), JX516851, JX516904, JX516930, JX516933, JX516938, JX516951, JX516977, JX516993, JX517139 (April et al. 2013)
		Pflacoi2	EU524239, EU524244-45 (Hubert et al. 2008), HQ557132 (April et al. unpub.), JX516793, JX516848, JX517095, JX517165 (April et al. 2013)
		NR	JX516865 (April et al. 2013)
	P. fluviatilis	Pflucoi1	KC819887 (Haponski and Stepien 2013), EF609432 (Ward and Holmes 2007), HQ600747-49 (Triantafyllidis et al. 2011), HQ960502-04, HQ960529, HQ960564, HQ960575, HQ960578-79, HQ960620, HQ960725-27, HQ960979-80, HQ961086-88 (iBOL unpub.), JN027875-76 (April et al. 2011)
		Pflucoi2	HQ600750-52 (Triantafyllidis et al. 2011)
		NR	HQ960464, HQ960510, HQ960557, HQ960863, HQ960865, HQ960925, HQ960944 (iBOL unpub.)
		NR	JN027871-74 (April et al. 2011)
		NR	HQ960866 (iBOL unpub.); KJ128570-71 (Zuccon and Noren unpub.)
		NR	HQ960653, HQ960670 (iBOL unpub.)
		NR	HQ600745 (Triantafyllidis et al. 2011)
		NR	HQ960864 (iBOL unpub.)
		NR	HQ600746 (Triantafyllidis et al. 2011)
	P. schrenkii	NR	JN027877 (April et al. 2011)

(continued)

Gene/region	Species	Our haplotype	GenBank accession #
mtDNA cyt *b*	*P. flavescens*	NR	AF045357 (Song et al. 1998), AF546115 (Moyer et al. unpub.)
		NR	AY374280 (Sloss et al. 2004)
		NR	EU348833 (Hai et al. unpub.)
		NR	EU348834 (Hai et al. unpub.)
		NR	EU348835 (Hai et al. unpub.)
		NR	EU348836 (Hai et al. unpub.)
		NR	EU348837 (Hai et al. unpub.)
		NR	EU348838 (Hai et al. unpub.)
	P. fluviatilis	Pflucbh1	KC819836 (Haponski and Stepien 2013), FJ788392-93, FJ788400-04, FJ788406-10 (Kalous et al. unpub.)
		NR	AF546116 (Moyer et al. unpub.), AY374281 (Sloss et al. 2004)
		NR	AF045358 (Song et al. 1998)
		NR	AF546117 (Moyer et al. unpub.)
		NR	AY929376 (Noren unpub.)
		NR	EU348839 (Hai et al. unpub.)
		NR	EU348840 (Hai et al. unpub.)
		NR	EU348841 (Hai et al. unpub.)
		NR	EU348842 (Hai et al. unpub.)
		NR	EU348843 (Hai et al. unpub.)
		NR	EU348844 (Hai et al. unpub.)
		NR	EU348845 (Hai et al. unpub.)
		NR	EU348846 (Hai et al. unpub.)
		NR	FJ788389 (Kalous et al. unpub.)
		NR	FJ788391 (Kalous et al. unpub.)
		NR	FJ788405 (Kalous et al. unpub.)
		NR	FJ788411 (Kalous et al. unpub.)
	P. schrenkii	NR	EU348848 (Hai et al. 2008)
		NR	AF546118 (Moyer et al. unpub.)
		NR	EU348847 (Hai et al. unpub.)
		NR	EU348849 (Hai et al. 2008)
		NR	EU348852 (Hai et al. 2008)
		NR	EU348851 (Hai et al. 2008)
		NR	AF546120 (Moyer et al. unpub.)
		NR	AF546119 (Moyer et al. unpub.)
		NR	EU348850 (Hai et al. unpub.)
Nuclear RAG 1	*P. flavescens*	Pflarag11	KC819906 (Haponski and Stepien 2013), AY308768 (Holcroft 2004), FJ381301 (Bossu and Near 2009), JX189840 (Wainwright et al. 2012)
		NR	AY308768 (Holcroft 2004)
	P. fluviatilis	Pflurag11	KC819904 (Haponski and Stepien 2013)

Appendix 3

GenBank sequences used to build phylogenetic trees for a. *Zingel* spp. and b. *Gymnocephalus* spp. Sequences include mtDNA cytochrome *c* oxidase I (COI) and NADH dehydrogenase 2 (ND2) for *Zingel* spp. and mtDNA control region and cytochrome (cyt) *b* and the nuclear DNA lactate dehydrogenase intron 6 (LdhA6) for *Gymnocephalus*.

a. *Zingel* spp.

Species	COI	ND2
Z. asper	KJ554652 (Geiger et al. 2014)	JQ088647 (Lang et al. unpub.)
Z. streber	JN028454 (April et al. 2011)	JQ088648 (Lang et al. unpub.)
	JN028452 (April et al. 2011)	
	JN028453 (April et al. 2011)	
Z. zingel	KC819882 (Haponski and Stepien 2013)	JQ088649 (Lang et al. unpub.)

b. *Gymnocephalus* spp.

Species	Control region	Cyt *b*	LdhA6
G. baloni	AF025360 (Faber and Stepien 1997)	AY374279 (Sloss et al. 2004)	AY034782 (Stepien et al. 2005)
G. cernua	AF025355-56 (Faber and Stepien 1997)	AF045356 (Song et al. 1998)	AY034781 (Stepien et al. 2005)
	AF025357-58 (Faber and Stepien 1997)	AF386598 (Song et al. 1998)	
	KC819863 (Haponski and Stepien 2013)	KC819833 (Haponski and Stepien 2013)	
	U90620 (Faber and Stepien 1997)		
G. schraetser	AF025361 (Faber and Stepien 1997)	AF546114 (Moyer et al. unpub.)	AY034783 (Stepien et al. 2005)
	AF025362 (Faber and Stepien 1997)	HM049946 (Matschiner et al. 2011)	

References

Agassiz L (1850) Lake superior: its physical character, vegetation, and animals, compared with those of other and similar regions. Gould, Kendall, and Lincoln, Boston

April J, Mayden RL, Hanner RH, Bernatchez L (2011) Genetic calibration of species diversity among North America's freshwater fishes. Proc Natl Acad Sci U S A 108:10602–10607

April J, Hanner R, Mayden R, Bernatchez L (2013) Metabolic rate and climatic fluctuations shape continental wide pattern of genetic divergence and biodiversity in fishes. PloSOne 8:e70296

Ayache NC, Near TJ (2009) The utility of morphological data in resolving phylogenetic relationships of darters as exemplified with *Etheostoma* (Teleostei: Percidae). Bull Peabody Nat Hist Mus 50:327–346

Bailey RM, Gosline WA (1955) Variation and systematic significance of vertebral counts in the American fishes of the family Percidae. Misc Pub Mus Zool Univ Mich 93:1–52

Bailey RM, Winn HE, Smith CL (1954) Fishes from the Escambia river, Alabama and Florida, with ecologic and taxonomic notes. Proc Acad Nat Sci Phila 106:109–164

Berg LS (1965) Freshwater fishes of the U.S.S.R. and adjacent countries. Academy of Sciences of the U.S.S.R. Zoological Institute, Moscow-Leningrad

Bergek S, Sundblad G, Björklund M (2010) Population differentiation in perch *Perca fluviatilis*: environmental effects on gene flow? J Fish Biol 76:1159–1172

Bernatchez L, Giroux M (2012) Les poisons d'eau douce du Quebec et leur repartition dans l'Est du Canada. Broquet, Saint-Constant

Bernatchez L, Wilson CC (1998) Comparative phylogeography of nearctic and palearctic fishes. Mol Ecol 7:431–452

Betancur-R R, Broughton RE, Wiley EO, Carpenter K, Lopez JA, Li Ch, Holcroft NI, Arcila D, Sanciangco M, Cureton JC, Zhang F, Buser T, Campbell MA, Ballesteros JA, Roa-Varon A, Willis S, Borden WC, Rowley T, Reneau PC, Hough DJ, Lu G, Grande T, Arratia G, Orti G (2013) The tree of life and a new classification of bony fishes. PLOS Curr 5: ecurrents.tol.53b a26640df0ccaee75bb165c8c26288

Betancur-R R, Wiley EO, Bailly N, Miya M, Lecointre G, Orti G (2014) New and revised classification of bony fishes version 2. http://www.deepfin.org/Classification_v2.htm

Billington N, Sloss BL (2011) Molecular systematics of *Sander*, and hybridization between walleye and sauger. In: Barton BA (ed) Biology, management, and culture of walleye and sauger. American Fisheries Society, Bethesda, pp 85–104

Billington N, Hebert PDN, Ward RD (1990) Allozyme and mitochondrial DNA variation among three species of *Stizostedion* (Percidae): phylogenetic and zoogeographic implications. Can J Fish Aquat Sci 47:1093–1102

Billington N, Danzmann RG, Hebert PDN, Wards RD (1991) Phylogenetic relationships among four members of *Stizostedion* (Percidae) determined by mitochondrial DNA and allozyme analyses. J Fish Biol 39:251–258

Billington N, Wilson C, Sloss BL (2011) Distribution and population genetics of walleye and sauger. In: Barton BA (ed) Biology, management, and culture of walleye and sauger. American Fisheries Society, Bethesda, pp 105–132

Björklund M, Aho T, Behrmann-Godel J (2015) Change in temperature regime over 35 years causes selection on MHC class II genes in the Eurasian perch (*Perca fluviatilis* L.). Ecol Evol 5:1440–1455

Bleeker P (1859) Enumeratio specierum piscirum hucusque in archipelago Indico observataum, adjectis habitationibus citationibusque, ubi descriptions earum recentiores reperiuntur, nec non speciebus Musei Bleekeriani Bengalensibus, Japonicis, Capensibus Tasmanicisque. Acta Societatis Regiae Scientiarum Indo-Neerlandicae v. 6 i–xxxvi 6:1–276

Bloch ME (1793) Naturgeschichte der Auslandischen Fische, vol 7. Konigl Akad, Berlin

Boschung HT, Mayden RL (2004) Fishes of Alabama. Smithsonian Books, Washington, DC

Boschung HT, Mayden RL, Tomelleri JR (1992) *Etheostoma chermocki*, a new species of darter (Teleostei: Percidae) from the Black Warrior River drainage of Alabama. Bull Ala Mus Nat Hist 13:11–20

Bossu CM, Near TJ (2009) Gene trees reveal repeated instances of mitochondrial DNA introgression in orangethroat darters (Percidae: *Etheostoma*). Syst Biol 58:114–129

Briggs JC (1986) Introduction to the zoogeography of north American fishes. In: Hocutt CH, Wiley EO (eds) The zoogeography of north American freshwater fishes. Wiley, New York, pp 1–16

Brown JA, Moore WM, Quabius ES (2001) Physiological effects of saline waters on zander. J Fish Biol 59:1544–1555

Bruch AA, Uhl D, Mosbrugger V (2007) Micoene climate in Europe – patterns and evolution a first synthesis of NECLIME. Palaeogeogr Palaeoclimatol 253:1–7

Bruner JC (2004) "Spreitzer" vertebrae, a unique character found only in *Ammocrypta* (crystal and sand darters). In: Barry TP, Malison JA (eds) Proceedings of PERCIS III, the Third International Percid Fish Symposium. University of Wisconsin-Madison, Wisconsin Sea Grant, pp 57–58

Bruner JC (2011) A phylogenetic analysis of Percidae using osteology. In: Barton BA (ed) Biology, management, and culture of walleye and sauger. American Fisheries Society, Bethesda, pp 5–84
Busiahn TR (1997) Ruffe control: a case study of an aquatic nuisance species control program. In: D'Itri FM (ed) Zebra mussels and aquatic nuisance species. Ann Arbor Press, Ann Arbor, pp 69–86
Carlander KD (1977) Handbook of freshwater fishery biology, vol 3: Life history data on Ichthyopercid and Percid fishes of the United States and Canada. Iowa State University Press, Ames
Carlson RL, Wainwright PC (2010) The ecological morphology of darter fishes (Percidae: Etheostomatinae). Biol J Linn Soc 100:30–45
Carney JP, Dick TA (2000) The historical ecology of yellow perch (*Perca flavescens* [Mitchill]) and their parasites. J Biogr 27:1337–1347
Cavender TM (1998) Development of the north American tertiary freshwater fish fauna with a look at parallel trends found in the European record. Ital J Zool 65:149–161
Chen W, Bonillo C, Lecointre G (2003) Repeatability of clades as a criterion of reliability: a case study for molecular phylogeny of Acanthomorpha (Teleostei) with larger number of taxa. Mol Phylogenet Evol 26:262–288
Chu C, Mandrak NE, Minns CK (2005) Potential impacts of climate change on the distributions of several common and rare freshwater fishes in Canada. Divers Distrib 11:299–310
Cloquet H (1817) Cingle. In: Dictionnaire des sciences naturelles de Levrault 9:240–241
Collette BB (1965) Systematic significance of breeding tubercles in fishes of the family Percidae. Proc US Natl Mus 177:567–614
Collette BB, Bănărescu P (1977) Systematics and zoogeography of the fishes of the family Percidae. J Fish Res Board Can 34:1450–1463
Collette BB, Ali MA, Hokanson KEF, Nagiec M, Smirnov SA, Thorpe JE, Weatherly AH, Willemsen J (1977) Biology of the percids. J Fish Res Board Can 34:1891–1899
Cope ED (1868) On the distribution of fresh-water fishes in the Allegheny region of southwestern Virginia. J Acad Nat Sci Phil 6:207–247
Cope ED (1870) A partial synopsis of the fishes of the fresh waters of north Carolina. Proc Am Phil Soc 11:448–495
Couture P, Pyle G, Campbell PGC, Hontela A (2015) Chapter 10: Using *Perca* as biomonitors in ecotoxicological studies. In: Couture P, Pyle G (eds) Biology of perch. CRC Press, Boca Raton (in press)
Craig JF (2000) Percid fishes systematics, ecology, and exploitation. Blackwell Science, Oxford
Crivelli AJ (2006) *Zingel balcanicus*. In: IUCN 2014. IUCN red list of threatened species. Version 2014.2. Available at: http://www.iucnredlist.org/details/60334/0
Cuvier, G. L C. F D. 1828. Tableau historique des progrès de l'ichtyologie, depuis son origine jusqu'à nos jours, p. 1–270,. In: Histoire naturelle des poissons. Vol. 1. G. Cuvier and A. Valenciennes. Levrault, Paris.
Darriba D, Taboada GL, Doallo R, Posada D (2012) jModelTest 2: more models, new heuristics and parallel computing. Nat Method 9:772, Available at http://code.google.com/p/jmodeltest2/
Dembkowski DJ, Chipps SR, Blackwell BG (2014) Response of walleye and yellow perch to water-level fluctuations in glacial lakes. Fish Manag Ecol 21:89–95
Denk T, Grimsson F, Zetter R (2010) Episodic migration of oaks to Iceland: evidence for a North Atlantic "Land Bridge" in the latest Miocene. Am J Bot 97:276–287
Denk T, Grimsson F, Zetter R, Simonarson LA (2011) The biogeographic history of Iceland – the North Atlantic land bridge revisited. Top Geobiol 35:647–668
Dettaï A, Lecointre G (2004) In search of notothenoid (Teleostei) relatives. Antarc Sci 16:1–15
Dettaï A, Lecointre G (2005) Further support for the clades obtained by multiple molecular phylogenies in the acanthomorph bush. CR Biol 328:674–689
Drummond AJ, Suchard MA, Xie D, Rambaut A (2012) Bayesian phylogenetics with BEAUti and the BEAST 1.7. Mol Biol Evol 29:1969–1973, Available at http://beast.bio.ed.ac.uk/Main_Page#Citing_BEAST

Dumitrescu M, Bănărescu PM, Stoica N (1957) *Romanichthys valsanicola* nov. gen. nov. sp. (Pisces, Percidae). Trav Mus Hist Natur Travaux "Grigore Antipa" 1:225–244
Economou AN, Giakoumi S, Vardakas L, Barbieri R, Stoumboudi M, Zogaris S (2007) The freshwater ichthyofauna of Greece – an update based on a hydrographic basin survey. Med Mar Sci 8:91–166
Etnier DA (1976) *Percina (Imostoma) tanasi*, a new percid fish from the Little Tennessee River, Tennessee. P Biol Soc Wash 88:469–488
Etnier DA, Williams JD (1989) *Etheostoma (Nothonotus) wapiti* (Osteichthyes: Percidae), a new darter from the southern bend of the Tennessee River system in Alabama and Tennessee. Proc Biol Soc Wash 102:987–1000
Faber JE, Stepien CA (1997) The utility of mitochondrial DNA control region sequences for analyzing phylogenetic relationships among populations, species, and genera of the Percidae. In: Kocher TD, Stepien CA (eds) Molecular systematics of fishes. Academic Press, London, pp 129–144
Faber JE, Stepien CA (1998) Tandemly repeated sequences in the mitochondrial DNA control region and phylogeography of the pike–perches *Stizostedion*. Mol Phylogenet Evol 10:310–322
Food and Agriculture Organization of the United Nations (2014) *Perca fluviatilis* species fact sheet. Fisheries and Aquaculture Department. http://www.fao.org/fishery/species/2298/en
Freyhof J (1996) *Percarina demidoffi*. In: IUCN 2014. IUCN red list of threatened species. Version 2014.2. Available at http://www.iucnredlist.org/details/40710/0
Freyhof J (2011) *Sander volgensis*. In: IUCN 2011. IUCN red list of threatened species. Version 2011.2. Available at http://www.iucnredlist.org/apps/redlist/details/20862/0
Freyhof J (2013) *Gymnocephalus ambriaelacus*. In: IUCN 2014. IUCN red list of threatened species. Version 2014.2. Available at http://www.iucnredlist.org/details/188654/0
Freyhof J, Kottelat M (2008) *Sander lucioperca*. In: IUCN 2011. IUCN red list of threatened species. Version 2011.2. Available at http://www.iucnredlist.org/apps/redlist/details/20860/0
Fuller P (2014) *Sander lucioperca*. USGS Nonindigenous aquatic species database. Gainesville. http://nas.er.usgs.gov/queries/factsheet.aspx?SpeciesID=830
Fuller P, Neilson M (2012) *Sander vitreus*. USGS nonindigenous aquatic species database, Gainesville. http://nas.er.usgs.gov/queries/FactSheet.aspx?speciesID=831
Fuller P, Neilson M (2014) *Perca flavescens*. USGS nonindigenous aquatic species database, Gainesville. http://nas.er.usgs.gov/queries/factsheet.aspx?SpeciesID=820
Fuller P, Jacobs PG, Larson J, Makled TH, Fusaro A (2014) *Gymnocephalus cernua*. USGS nonindigenous aquatic species database, Gainesville. http://nas.er.usgs.gov/queries/factsheet.aspx?SpeciesID=7
Geiger MF, Schliewen UK (2010) *Gymnocephalus ambriaeacus*, a new species of ruffe from Lake Ammersee, southern Germany. Spixiana 33:119–137
Geiger MF, Herder F, Monaghan MT, Almada V, Barbieri R, Bariche M, Berrebi P, Bohlen J, Casal-Lopez M, Delmastro GB, Denys GP, Dettai A, Doadrio I, Kalogianni E, Karst H, Kottelat M, Kovacic M, Laporte M, Lorenzoni M, Marcic Z, Ozulug M, Perdices A, Perea S, Persat H, Porcelotti S, Puzzi C, Robalo J, Sanda R, Schneider M, Slechtova V, Stoumboudi M, Walter S, Freyhof J (2014) Spatial heterogeneity in the Mediterranean biodiversity hotspot affects barcoding accuracy of its freshwater fishes. Mol Ecol Resour 14:1210–1221
Georgijev SB (2004) Recent knowledge of the endemicity of the Republic of Macedonia's ichthyofauna. Croa J Fish 62:43–58
Gernhard T (2008) New analytical results for speciation times in neutral models. Bull Math Biol 70:1082–1097
Gilbert CH, Meek SE (1887) *Etheostoma cymatoaenia*. In: Gilbert CH (ed) Descriptions of new and little known etheostomids. Proc US Nat Mus 10:47–64, p 51
Girard CF (1859) Ichthyological notices. Proc Acad Natl Sci Phila 11:56–68
Gladenkov AY, Oleinik AE, Marincovich L, Barinov KB (2002) A refined age for the earliest opening of Bering strait. Palaeogeogr Palaeoclimatol 183:321–328

Gmelin JF (1789) Caroli a Linné. Systema naturae pe regna tria naturae, secundum classes, ordines, genera, species, cum characteribus, differentiis, synonymis, locis. Tomus Primus. Editio decimo tertia, aucta, reformata

Griffith E, Smith CH (1834) The class pisces, arranged by the Baron Cuvier, with supplementary additions, by Edward Griffith, F.R.S., and Lieut.–Col. Charles Hamilton Smith, F.R., L.S.S. In: The Animal Kingdom (Cuvier, G.L.C.F.D.). Whittaker and Co., London, pp 1–62

Guseva TV (1974) The ecology of spawning and embryonic development of the sea zander, *Lucioperca marina*. Izvestiya SSSR 2:87–91

Hai S, Li J, Feng J, Mu L (2008) Systematics of *Perca* based on multivariate morphometrics and mitochondrial cytochrome *b* gene variation analysis. Zool Res 29:113–120

Haldeman SS (1842) Description of two new species of the genus *Perca*, from the Susquehanna river. J Acad Nat Sci Phila 8:330

Hampe A, Petit RJ (2005) Conserving biodiversity under climate change: the rear edge matters. Ecol Lett 8:461–467

Haponski AE (2013) Evolutionary, biogeographic, and population genetic patterns of walleye and other *Sander*: relationships across continents, corridors, and spawning sites. Ph.D. Dissertation, University of Toledo

Haponski AE, Stepien CA (2008) Molecular, morphological, and biogeographic resolution of cryptic taxa in the greenside darter *Etheostoma blennioides* complex. Mol Phylogenet Evol 49:69–83

Haponski AE, Stepien CA (2013) Phylogenetic and biogeographic relationships of the *Sander* pikeperches (Perciformes: Percidae): patterns across north America and Eurasia. Biol J Linn Soc 110:156–179

Haponski AE, Stepien CA (2014) A population genetic window into the past and future of the walleye *Sander vitreus*: relation to historic walleye and the extinct "blue pike" *S. v.* "*glaucus*". BMC Evol Biol 14:133

Haponski AE, Bollin TL, Jedlicka MA, Stepien CA (2009) Landscape genetic patterns of the rainbow darter *Etheostoma caeruleum*: a catchment analysis of mitochondrial DNA sequences and nuclear microsatellites. J Fish Biol 75:2244–2268

Hay OP (1882) On a collection of fishes from the lower Mississippi Valley. Bull US Fish Commun 2:57–75

Hewitt GM (1996) Some genetic consequences of ice ages, and their role in divergence and speciation. Biol J Linn Soc 58:247–276

Hewitt GM (2000) The genetic legacy of the quaternary ice ages. Nature 405:907–913

Hocutt CH, Wiley EO (1986) The zoogeography of north American freshwater fishes. Wiley, New York

Holčík J, Hensel K (1974) A new species of *Gymnocephalus* (Pisces: Percidae) from the Danube, with remarks on the genus. Copeia 1974:481–486

Holcroft NI (2004) A molecular test of alternative hypotheses of tetraodontiform (Acanthomorpha: Tetraodontiformes) sister group relationships using data from the RAG1 gene. Mol Phylogenet Evol 32:749–760

Holder M, Lewis PO (2003) Phylogeny estimation: traditional and Bayesian approaches. Nat Rev Genet 4:275–284

Howell WM, Caldwell RD (1965) *Etheostoma* (*Oligocephalus*) *nuchale*, a new darter from a limestone spring in Alabama. Tul Stud Zool 12:101–108

Hubbs CL (1926) A check–list of the fishes of the great lakes and tributary waters, with nomenclatorial notes and analytical keys. University of Michigan Museum of Zoology Miscellaneous Publication 15, Ann Arbor

Hubbs CL, Lagler KF (2004) Fishes of the Great Lakes region. (Smith GR, revised). University of Michigan, Ann Arbor

Hubert N, Zhang J, April J, Bernatchez L, Hanner R, Holm E, Mandrak NE, Taylor E, Burridge M, Watkinson D, Curry A, Bentzen P (2008) Identifying Canadian freshwater fishes through DNA barcodes. PLoSOne 3:E2490

IUCN (International Union for Conservation of Nature) (2014) Red list. Available at http://www.iucnredlist.org/search. Accessed 24 Nov 2014

Jordan DS (1877) Contributions to North American ichthyology; based primarily on the collections of the United States National Museum. II. A-Notes on Cottidae, Etheostomatidae, Percidae, Centrarchidae, Aphododeridae, Dorysomatidae, and Cyprinidae, with revisions of the genera and descriptions of new or little known species. Bull US Natl Mus 10:5–68

Jordan DS (1878) A catalogue of the fishes of Illinois. Bull Ill Stat Lab Natl Hist 1:37–70

Jordan DS (1885) A catalogue of the fishes known to inhabit the waters of North America, north of the Tropic of Cancer, with notes on species discovered in 1883 and 1884. Ann Rep Commun Fish Fish 1885(1887):789–973

Jordan DS (1888) A manual of vertebrate animals of the northern United States, including the district north and east of the Ozark mountains, south of the Laurentian hills, north of Virginia, and east of the Missouri River; inclusive of marine species, 5th edn. Jansen, McClurg, and Co, Chicago

Jordan DS (1889) Descriptions of fourteen species of fresh-water fishes collected by the U.S. Fish Commission in the summer of 188. Proc US Nat Mus 11:351–362

Jordan DS (1923) A classification of fishes including families and genera as far as known, vol 3, Stanford University publications, University series, Biological sciences. Stanford University, Stanford

Jordan DS, Evermann BW (1896) The fishes of North and Middle America: a descriptive catalogue of the species of fish-like vertebrates found in the waters of North America, north of the Isthmus of Panama. Bull US Natl Mus 47:1240

Jordan DS, Gilbert CH (1888) *Etheostoma microperca*. In: Jordan DS (ed) A manual of vertebrate animals of the northern United States, including the district north and east of the Ozark mountains, south of the Laurentian hills, north of Virginia, and east of the Missouri River; inclusive of marine species, 5th edn. AC McClurg and Company, Chicago, p 134

Jordan DS, Jenkins OP (1889) *Etheostoma roanoka*. In: Jordan DS (ed) Descriptions of fourteen species of fresh-water fishes collected by the U.S. Fish Commission in the summer of 1888. Proc US Nat Mus 11:351–362, p 358

Jordan DS, Meek SE (1885) List of fishes collected in Iowa and Missouri in August, 1884, with descriptions of three new species. Proc US Natl Mus 8:1–17

Kålås S (1995) The ecology of ruffe, *Gymnocephalus cernuus* (Pisces:Percidae) introduced to Mildevatn, western Norway. Environ Biol Fishes 42:219–232

Kamilov G (1966) New data on the composition of ichthyofauna of reservoirs in Uzbekistan. In: Biological background of commercial fishery in reservoirs of Middle Asia and Kazakhstan. Kainar, Alma-Ata

Karaman S (1937) Beitrag zur Kenntnis der Süβwasserfische Jugoslaviens. Glasnik Skopskog Naucnog Drustva. Bull Soc Sci Skoplje 17:55–64

Keck BP, Near TJ (2008) Assessing phylogenetic resolution among mitochondrial, nuclear, and morphological datasets in *Nothonotus* darters (Teleostei: Percidae). Mol Phylogenet Evol 46:708–720

Keck BP, Near TJ (2010) Geographic and temporal aspects of mitochondrial replacement in *Nothonotus* darters (Teleostei: Percidae: Etheostomatinae). Evolution 64:1410–1428

Kessler KT (1874) Pisces. In: Fedtschensko's Expedition to Turkestan. Zoogeographical Researches. Izvestiia Imperatorskago Obschchestva Liubitelei Estestvozanii, Antropologii i Etnografii 11:1–63

Kimura M (1980) A simple method for estimating evolutionary rates of base substitutions through comparative studies of nucleotide sequences. J Mol Evol 16:111–120

Kirtland JP (1840) Descriptions of four new species of fishes. Bost J Natl Hist 3:273–277

Kottelat M (1996) *Perca schrenkii*. In: IUCN 2014. IUCN red list of threatened species. Version 2014.2. Available at http://www.iucnredlist.org/details/40709/0

Kottelat M (1997) European freshwater fishes. Biologia 52:1–271

Kottelat M, Freyhof J (2007) Handbook of European freshwater dishes. Publications Kottelat, Cornol

Kovac V (1998) Biology of Eurasian ruffe from Slovakia and adjacent central European countries. J Great Lakes Res 24:205–216

Kuehne RA, Barbour RW (1983) The American darters. University Press of Kentucky, Lexington

Kuznetsov ID (1888) *Percarina* (Nordm.) and *Benthophilus* (Eichw.) of the Sea of Azov. Trudy St.-Peterburgskogo Obscestva Estestvoispytatelej = Travaux de la Societe des Naturalistes de St. Petersbourg 19:189–212

Kuznetsov VA (2010) Growth, size–age structure of catches and reproduction of the Volga zander *Sander volgensis* (Percidae) in the upper part of the Volga stretch of the Kuibyshev Reservoir. J Ichthyol 50:772–777

Lanave C, Preparata G, Saccone C, Serio G (1984) A new method for calculating evolutionary substitution rates. J Mol Evol 20:86–93

Laporte M, Magnan P, Angers B (2011) Genetic differentiation between the blue and the yellow phenotypes of walleye (*Sander vitreus*): an example of parallel evolution. Ecoscience 18:124–129

Laroche J, Durand JD (2004) Genetic structure of fragmented populations of a threatened endemic percid of the Rhone river: *Zingel asper*. Heredity 92:329–334

Larsen LK, Berg S (2006) NOBANIS – invasive alien species fact sheet – *Sander lucioperca* – from: online database of the north European and Baltic network on invasive alien species – NOBANIS. www.nobanis.org

Lelek A (1987) The freshwater fishes of Europe: threatened fishes of Europe. AULA-Verlag, Wiesbaden

Linnaeus C (1758) Tomus I. Systema naturae pe regna tria naturae, secundum classes, ordines, genera, species, cum characteribus, differentiis, synonymis, locis. Editio decimo reformata. Holmiae. Laurentii Salvii

Linnaeus C (1766) Systema naturae sive regna tria naturae, secundum classes, ordines, genera, species, cum characteribus, differntiis, synonymis, locis, 12th edn. Laurentii Salvii, Holmiae. 1:1–532

Maitland PS (2000) Freshwater fish of Britain and Europe. Hamlyn, London

Marcogliese DJ, Pietrock M (2011) Combined effects of parasites and contaminants on animal health: parasites do matter. Trends Parasitol 27:123–130

Matschiner M, Hanel R, Salzburger W (2011) On the origin and trigger of the notothenioid adaptive radiation. PLoSOne 6:E18911

McLean M (1993) Ruffe (*Gymnocephalus cernuus*) fact sheet. Minnesota Sea Grant Program, Great Lakes Sea Grant Network

Meek SE (1891) Report of explorations made in Missouri and Arkansas during 1889, with an account of the fishes observed in each of the river basins examined. Bull US Fish Commun 9:113–141

Mendelson TC, Wong MK (2010) AFLP phylogeny of the snubnose darters and allies (Percidae: *Etheostoma*) provides resolution across multiple levels of divergence. Mol Phylogenet Evol 57:1253–1259

Michel C, Bernatchez L, Behrmann-Godel J (2009) Diversity and evolution of MHII β genes in a non-model percid species – the Eurasian perch (*Perca fluviatilis* L.). Mol Immunol 46:3399–3410

Miller RR (2005) Freshwater fishes of Mexico. University of Chicago Press, Chicago

Milne RI (2006) Northern hemisphere plant disjunctions: a window on tertiary land bridges and climate change? Ann Bot Lond 98:465–472

Mitchill SL (1814) Report in part of Samuel L. Mitchill, M.D., on the fishes of New York. New York

Mitchill SL (1818) Memoir on ichythology. The fishes of New York, described and arranged. Am Mon Mag Crit Rev 6:321–328

Moore GA (1944) The retinae of two North American teleosts, with special reference to their tapeta lucida. J Comp Neurol 80:369–379

Morrison CL, Lemarie DP, Wood RM, King TL (2006) Phylogeographic analyses suggest multiple lineages of *Crystallaria asprella* (Percidae: Etheostominae). Conserv Genet 7:129–147

Moyle PB, Cech JJ (2000) Fishes. An introduction to ichthyology, 4th edn. Prentice Hall, Upper Saddle

Murray AM, Divay JD (2011) First evidence of percids (Teleostei: Perciformes) in the Miocene of north America. Can J Earth Sci 48:1419–1424
Murray AM, Cumbaa SL, Harington CR, Smith GR, Rybczynski N (2009) Early Pliocene fish remains from Arctic Canada support a pre-Pleistocene dispersal of percids (Teleostei: Perciformes). Can J Earth Sci 46:557–570
NatureServe (2013a) *Crystallaria asprella*. In: IUCN 2014. IUCN red list of threatened species. Version 2014.2. Available at http://www.iucnredlist.org/details/5786/0
NatureServe (2013b) *Crystallaria cincotta*. In: IUCN 2014. IUCN red list of threatened species. Version 2014.2. Available at http://www.iucnredlist.org/details/202435/0
NatureServe Explorer (2014) An online encyclopedia of life. http://www.natureserve.org/conservation-tools/data-maps-tools/natureserve-explore
Near TJ (2002) Phylogenetic relationships of *Percina* (Percidae: Etheostomatinae). Copeia 2002:1–14
Near TJ, Keck BP (2012) AFLPs do not support deep phylogenetic relationships among darters (Teleostei: Percidae: Etheostomatinae). Heredity 108:647–648
Near TJ, Keck BP (2013) Free from mitochondrial DNA: nuclear genes and the inference of species trees among closely related darter lineages (Teleostei: Percidae: Etheostomatinae). Mol Phylogenet Evol 66:868–876
Near TJ, Bossu CM, Bradburd GS, Carlson RL, Harrington RC, Hollingsworth PR, Keck BP, Etnier DA (2011) Phylogeny and temporal diversification of darters (Percidae: Etheostomatinae). Syst Biol 60:565–595
Nelson JS, Crossman EJ, Espinosa-Perez H, Findley LT, Gilbert CR, Lea RN, Williams JD (2003) The "names of fishes" list, including recommended changes in fish names: Chinook salmon for chinook salmon, and *Sander* to replace *Stizostedion* for the sauger and walleye. Fisheries 28:38–39
Nelson JS, Crossman EJ, Espinosa-Perez H, Findley LT, Gilbert CR, Lea RN, Williams JD (2004) Common and scientific names of fishes from the United States, Canada, and Mexico, vol 9, 6th edn, Special publication. American Fisheries Society, Bethesda
Nesbø CL, Fossheim T, Vollestad LA, Jakobsen KS (1999) Genetic divergence and phylogeographic relationships among European perch (*Perca fluviatilis*) populations reflect glacial refugia and postglacial colonization. Mol Ecol 8:1387–1404
Nordmann A (1840) Prodrome de l'ichthyologie Pontique. In: Voyage dans la Russie meridionale et la Crimee, par la Hongrie, la Valachie et la Moldavie execute en 1857, sous la direction de M. Anatole de Demidoff, vol III. pp 353–549
Nuriyev H (1967) Occurrence of Balkhash perch in Kattakurgan reservoir (pool of Zeravshan River). In: Biological background of commercial fishery in reservoirs of Middle Asia and Kazakhstan (summaries of reports), Balkhash, pp 208–209
Ogle DH, Selgeby JH, Newman RM, Henry MG (1995) Diet and feeding periodicity of ruffe in the St. Louis River Estuary, Lake superior. Trans Am Fish Soc 124:356–369
Oken L (1817) V. Kl. Fische. Isis von Oken 8:1779–1782
Oppelt C, Behrmann-Godel J (2012) Genotyping MHC classIIB in non-model species by reference strand-mediated conformational analysis (RSCA). Conserv Genet Resour 4:841–844
Page LM (1974) The subgenera of *Percina*. Copeia 1974:66–86
Page LM (1981) The genera and subgenera of darters (Percidae, Etheostomatini). Occ Pap Mus Natl Hist Univ Kansas 90:1–69
Page LM (1983) Handbook of darters. TFH Publications, Neptune City
Page LM (2000) Etheostomatinae. In: Craig JF (ed) Percid fishes systematics, ecology, and exploitation. Blackwell Science, Oxford, pp 225–253
Page LM, Burr BM (2011) Field guide to the freshwater fishes of north America, 2nd edn. Houghton Mifflin Harcourt, Boston
Paradis Y, Magnan P (2005) Phenotypic variation of walleye, *Sander vitreus*, in Canadian Shield lakes: new insights on percid polymorphism. Environ Biol Fish 73:357–366
Pivnev I (1985) Fish species in pool of Chu and Talas rivers. Ilim, Frunze
Popova OA, Reshetnikov YS, Kiyashko VI, Dgebuadze YY, Mikheev VN (1998) Ruffe from the former USSR: variability within the largest part of its natural range. J Great Lakes Res 24:263–284

Pratt DM, Blust WH, Selgeby JH (1992) Ruffe, *Gymnocephalus cernuus*: newly introduced in North America. Can J Fish Aquat Sci 49:1616–1618

Putnam FW (1863) List of the fishes sent by the museum to different institutions, in exchange for other specimens, with annotations. Bull Mus Comp Zool 1:2–16

Rafinesque CS (1815) Analyse de la nature, ou tableau de l'univers et des corps organizes, Palerme

Rafinesque CS (1818) Second decade of new North American fishes. Am Mon Mag Crit Rev 2:204–206

Rafinesque CS (1819) Prodrome de 70 nouveau genres d'animaux découverts dans l'interieur des États–Unis d'Amerique durant l'année 1818. J Phys Chim Hist Nat Arts, tom 88:417–429

Rafinesque CS (1820) Ichthyologia ohiensis, or Natural history of the fishes inhabiting the river Ohio and its tributary streams, preceded by a physical description of the Ohio and its branches. WG Hunt, Lexington

Raney EC, Zorach T (1967) *Etheostoma microlepidum*, a new percid fish of the subgenus *Nothonotus* from the Cumberland and Tennessee River systems. Am Midl Nat 77:93–103

Ray JM, Lang NJ, Wood RM, Mayden RL (2008) History repeated: recent and historical mitochondrial introgression between the current darter *Etheostoma uniporum* and rainbow darter *Etheostoma caeruleum* (Teleostei: Percidae). J Fish Biol 72:418–434

Reid DF, Orlova MI (2002) Geological and evolutionary underpinnings for the success of Ponto–Caspian species invasions in the Baltic Sea and North American Great Lakes. Can J Fish Aquat Sci 59:1144–1158

Richards WJ, Knapp LW (1964) *Percina lenticula*, a new percid fish, with a redescription of the subgenus *Hadropterus*. Copeia 1964:690–701

Ronquist F, Huelsenbeck JP (2003) MrBayes 3: Bayesian phylogenetic inference under mixed models. Bioinformatics 19:1572–1574. http://mrbayes.sourceforge.net/. (v3.1.2, 2005)

Schaefer WF, Schmitz MH, Blazer VS, Ehlinger TJ, Berges JA (2015) Localization and seasonal variation of blue pigment (sandercyanin) in walleye (*Sander vitreus*). Can J Fish Aquat Sci 72:1–9

Schmalz PJ, Fayram AH, Isermann DA, Newman SP, Edwards CJ (2011) Harvest and exploitation. In: Barton BA (ed) Biology, management, and culture of walleye and sauger. American Fisheries Society, Bethesda, pp 375–402

Scott WB, Crossman EJ (1973) Freshwater fishes of Canada. J Fish Res Board Can 184:1–196

Sepulveda–Villet OJ, Stepien CA (2012) Waterscape genetics of the yellow perch (*Perca flavescens*): patterns across large connected ecosystems and isolated relict populations. Mol Ecol 21:5795–5826

Sharma S, Jackson DA, Minns CK, Shuter BJ (2007) Will northern fish populations be in hot water because of climate change? Glob Change Biol 13:2052–2064

Shevtsova TM, Kunitskii TM, Ryhevskii DF, Serov VK, Pluta MV (1986) *Gymnocephalus baloni* novyi vid ikhtiofauny USSR (*Gymnocephalus baloni*- new species of fish of the USSR fish fauna). Dokl Akad Nauk Belorusskoi SSR 30:954–955

Shuter BJ, Post JR (1990) Climate, population viability, and the zoogeography of temperate fishes. Trans Am Fish Soc 119:314–336

Siebold CTE (1863) Die Susswasserfische von Mitteleuropa. Leipzig. i–viii:1–430

Simon TP, Vondruska JT (1991) Larval identification of the ruffe, *Gymnocephalus cernuus* (Linnaeus) (Percidae: Percini), in the St. Louis River Estuary, Lake Superior drainage basin, Minnesota. Can J Zool 69:436–441

Simons AM (1991) Phylogenetic relationships of the crystal darter, *Crystallaria asprella* (Teleostei: Percidae). Copeia 1991:927–936

Simons AM (1992) Phylogenetic relationships of the *Boleosoma* species group (Percidae: *Etheostoma*). In: Mayden RL (ed) Systematics, historical ecology, and North American freshwater fishes. Stanford University Press, Stanford, pp 268–292

Sloss BL, Billington N, Burr BM (2004) A molecular phylogeny of the Percidae (Teleostei, Perciformes) based on mitochondrial DNA sequence. Mol Phylogenet Evol 32:545–562

Smith WL, Craig MT (2007) Casting the percomorph net widely: the importance of broad taxonomic sampling in the search for the placement of serranid and percid fishes. Copeia 2007:35–55

Smith TA, Mendelson TC, Page LM (2011) AFLPs support deep relationships among darters (Percidae: Etheostomatinae) consistent with morphological hypotheses. Heredity 107:579–588

Smith TA, Ciccotto PJ, Mendelson TC, Page LM (2014) Dense taxon sampling using classification of darters (Percidae: *Etheostoma*). Copeia 2014:257–268
Sokolovsky V, Galouschak S, Skakun V (2000) Current condition of Balkhash perch *Perca schrenki* (Percidae) in lakes of Alakol system. Probl Ichthyol 40:228–234
Song CB, Near TJ, Page LM (1998) Phylogenetic relations among Percid fishes as inferred from mitochondrial cytochrome *b* DNA sequence data. Mol Phylogenet Evol 10:343–353
Stamatakis A (2014) RAxML version 8: a tool for phylogenetic analysis and post-analysis of large phylogenies. Bioinformatics 30:1312–1313
Stephens RR, Johnson KA, Grady MP (2014) Phylogenetic placement of the extinct *Etheostoma sellare* and other darters with a compilation of morphological character state polarizations for darters. Copeia 2014:540–555
Stepien CA, Faber JE (1998) Population genetic structure, phylogeography, and spawning philopatry in walleye (*Stizostedion vitreum*) from mtDNA control region sequences. Mol Ecol 7:1757–1769
Stepien CA, Haponski AE (2010) Systematics of the greenside darter *Etheostoma blennioides* complex: consensus from nuclear and mitochondrial DNA sequences. Mol Phyl Evol 57:434–447
Stepien CA, Dillon AK, Chandler MD (1998) Genetic identity, phylogeography, and systematics of ruffe *Gymnocephalus* in the North American Great Lakes and Eurasia. J Great Lakes Res 24:361–378
Stepien CA, Ford AM, Dillon-Klika AK (2004) Risk analysis and genetic identity of the Eurasian source population for the ruffe (*Gymnocephalus cernuus*) invasion in the Great Lakes. In: Barry TP, Malison JA (eds) Proceedings of Percis III, the 3rd International Symposium on Percid Fishes. University of Wisconsin Sea Grant, Madison, pp 91–92
Stepien CA, Brown JE, Neilson ME, Tumeo MA (2005) Genetic diversity of invasive species in the Great Lakes versus their Eurasian source populations: insights for risk analysis. Risk Anal 25:1043–1060
Stepien CA, Murphy DJ, Lohner RN, Sepulveda–Villet OJ, Haponski AE (2009) Signatures of vicariance, postglacial dispersal, and spawning philopatry: population genetics and biogeography of the walleye *Sander vitreus*. Mol Ecol 18:3411–3428
Stepien CA, Behrmann-Godel J, Bernatchez L (2015a) Evolutionary relationships, population genetics, and adaptations of perch (Perca). In: Couture P, Moyer G (eds) Biology of perch. CRC Press (in press), pp 7–46, Chapter 2
Stepien CA, Sepulveda-Villet OJ, Haponski AE (2015b) Comparative genetic diversity, population structure, and adaptations of walleye and yellow perch across North America. In: Kestlemont P (ed) Biology and culture of percid fishes – principles and practices. Springer (this book)
Stepien CA, Pierce LR, Leaman D, Niner M, Shepherd B (2015c) Genetic diversification of an emerging pathogen: a decade of mutation in a novel fish viral hemorrhagic septicemia (VHS) substrain in the Laurentian Great Lakes PLoS One (accepted)
Stewart JG, Barko VA, Henry DB, Herzog DP, Ridings JW, Kelley AF, Wallace JE (2005) New records of the crystal darter (*Crystallaria asprella*) in the middle Mississippi river. Am Midl Nat 154:471–473
Storer DH (1845) Description of a hitherto undescribed species of fish. Proc Boston Soc Natl Hist 2:47–48
Swainson W (1839) The natural history and classification of fishes, amphibians, and reptiles, or monocardian animals, vol 2. Spottiswoode and Company, London
Swofford DL (2002) PAUP*. Phylogenetic analysis using parsimony (*and other methods) 4.0. Sinauer Associates, Sunderland
Tiffney BH (1985) The Eocene North Atlantic land bridge: its importance in tertiary and modern phytogeography of the northern hemisphere. J Arnold Arbor 66:243–273
Tiffney BH, Manchester SR (2001) The use of geological and paleontological evidence in evaluating plant phylogeographic hypotheses in the northern hemisphere tertiary. Int J Plant Sci 162:S3–S17
Trautman MB (1981) The fishes of Ohio. Ohio State University Press, Columbus
Triantafyllidis A, Bobori D, Koliamitra C, Gbandi E, Mpanti M, Petriki O, Karaiskou N (2011) DNA barcoding analysis of fish species diversity in four north Greek lakes. Mitochondr DNA 22:37–42

Vasil'eva ED (2006) Variation of the external characters and taxonomic relationships of Azov and Black-Sea populations of *Percarina demidoffii* (Percidae). J Ichthyol 46:292–300
Vidal-Martinez V, Pech D, Sures B, Purucker ST, Poulin R (2010) Can parasites really reveal environmental impact? Trends Parasitol 26:44–51
Viola G, Mancktelow NS, Seward D (2001) Late Oligocene–Neogene evolution of Europe–Adria collision: new structural and geochronological evidence from the Giudicarie fault system (Italian Eastern Alps). Tectonics 20:999–1020
Wainwright PC, Smith WL, Price SA, Tang KL, Sparks JS, Ferry LA, Kuhn KL, Eytan RI, Near TJ (2012) The evolution of pharyngognathy: a phylogenetic and functional appraisal of the pharyngeal jaw key innovation in labroid fishes and beyond. Syst Biol 61:1001–1027
Walker KF, Yang HZ (2009) Fish and fisheries in western China. In: Petr T (ed) Fish and fisheries at higher altitudes: Asia. FAO Fisheries technical paper no. 385, Rome. Available at: http://www.fao.org/docrep/003/x2614e/x2614e12.htm#P5_0
Ward RD, Holmes BH (2007) An analysis of nucleotide and amino acid variability in the barcode region of cytochrome *c* oxidase I (cox1) in fishes. Mol Ecol Notes 7:899–907
Welsh SA, Wood RM (2008) *Crystallaria cincotta*, a new species of darter (Teleostei: Percidae) from the Elk River of the Ohio River drainage, west Virginia. Zootaxa 1680:62–68
White MM, Schell S (1995) An evaluation of the genetic integrity of Ohio River walleye and sauger stocks. In: Schramm HL, Piper RG (eds.) Uses an defects of cultured fishes in aquatic ecosystems. American Fisheries Society Symposium 15, Bethesda, pp 52–60
White MM, Kassler TW, Philipp DP, Schell SA (2005) A genetic assessment of Ohio River walleyes. Trans Am Fish Soc 134:661–675
Wicander R, Monroe JS (1993) Historical geology: evolution of the earth and life through time, 2nd edn. West Publishing Company, St. Paul
Wiley EO (1992) Phylogenetic relationships of the Percidae (Teleostei: Perciformes): a preliminary hypothesis. In: Mayden RL (ed) Systematics, historical ecology, and North American freshwater fishes. Stanford University Press, Stanford, pp 247–267
Williams JD (1975) Systematics of the percid fishes of the subgenus *Ammocrypta*, genus *Ammocrypta*, with descriptions of two new species. Bull Ala Mus Natl Hist 1:1056
Williams JD, Etnier DA (1978) *Etheostoma aquali*, a new percid fish (subgenus *Nothonotus*) from the Duck and Buffalo Rivers, Tennessee. Proc Biol Soc Wash 91:463–471
Williams TH, Gumm JM, Mendelson TC (2013) Sexual selection acting on a speciation trait in darters (Percidae: *Etheostoma*). Behav Ecol 24:1407–1414
Winfield IJ, Rosch R, Appelberg M, Kinnerback A, Rask M (1998) Recent introductions of the ruffe (*Gymnocephalus cernuus*) to *Coregonus* and *Perca* lakes in Europe and an analysis of their natural distributions in Sweden and Finland. J Great Lakes Res 24:235–248
Wolfe JA (1994) Tertiary climate changes at middle latitudes of western North America. Palaeogeogr Palaeoclimatol 108:195–205
Wood RM (1996) Phylogenetic systematics of the darter subgenus *Nothonotus* (Teleostei: Percidae). Copeia 1996:300–318
Wood RM, Mayden RL (1997) Phylogenetic relationships among selected darter subgenera (Teleostei: Percidae) as inferred from analysis of allozymes. Copeia 1997:265–274
Wood RM, Raley ME (2000) Cytochrome *b* sequence variation in the crystal darter *Crystallaria asprella* (Actinopterygii: Percidae). Copeia 2000:20–26
Woolhouse MEJ, Webster JP, Domingo E, Charlesworth B, Levin BR (2002) Biological and biomedical implications of the co-evolution of pathogens and their hosts. Nat Genet 32:569–577
World Conservation Monitoring Center (1996) *Sander marinus*. In: IUCN 2014. IUCN red list of threatened species. Version 2014.2. Available at http://www.iucnredlist.org/details/20861/0
Yu C, Ferraro D, Ramaswamy S, Schmitz MH, Schaefer WF, Gibson DT (2008) Purification and properties of sandercyanin, a blue protein secreted in the mucus of blue forms of walleye, *Sander vitreus*. Environ Biol Fish 82:51–58
Zorach T (1968) *Etheostoma bellum*, a new darter of the subgenus *Nothonotus* from the green river system, Kentucky and Tennessee. Copeia 1968:474–482

Chapter 2
Environmental Biology of Percid Fishes

Zachary S. Feiner and Tomas O. Höök

Abstract The large percids, including *Perca* and *Sander* species, are economically and ecologically important species that inhabit large temperature regions of the Northern Hemisphere. In this chapter, we provide an overview of the environmental biology of the *Perca* (including yellow perch *P. flavescens* and Eurasian perch *P. fluviatilis*) and *Sander* (including walleye *S. vitreus*, pikeperch *S. lucioperca*, and sauger *S. canadensis*) genera, on which the majority of fisheries and aquaculture practices are focused. Through a comprehensive literature review, we discuss how individual- and population-level vital rates, including growth, foraging, reproduction, recruitment, and mortality, are mediated by biotic (e.g., density dependence, resource availability) and abiotic (e.g., temperature, light) environmental variables. As fisheries exploitation is a major source of size-selective mortality in many percid populations, we also examine the potential impacts of fishing mortality on both population metrics and individual vital rates, and identify several research areas that require further investigation. Through this review we aim to identify the major environmental drivers of variation in percid vital rates and thereby inform management practices for both wild and cultured percid populations.

Keywords *Perca* • *Sander* • Fisheries • Exploitation • Ecology

2.1 Introduction

The large percids (principally *Perca* and *Sander*) constitute ecologically and economically important species throughout large temperate regions of the Northern Hemisphere. Percids have been characterized as cool to warm water fish species, persisting in habitats ranging from warm, highly eutrophied waters to cold, oligotrophic systems. Percids exhibit high levels of phenotypic plasticity and inhabit a broad range of environmental conditions throughout their geographic ranges. While the feeding ecology of these species varies during ontogeny, generally progressing

Z.S. Feiner • T.O. Höök (✉)
Department of Forestry and Natural Resources, Illinois-Indiana Sea Grant,
Purdue University, 195 Marsteller St., West Lafayette, IN 47907, USA
e-mail: thook@purdue.edu

P. Kestemont et al. (eds.), *Biology and Culture of Percid Fishes*,
DOI 10.1007/978-94-017-7227-3_2

from zooplanktivory (e.g., Pycha and Smith 1955) to primarily benthivory and piscivory (e.g., Paradis et al. 2006), percids are opportunistic, euryphagous predators consuming differential prey bases across systems. In many systems, they serve as top predators, exerting influential top-down control on lower trophic levels (Guzzo et al. 2013). Simultaneously, percids may also provide important forage for piscivorous fishes, birds (DeBruyne et al. 2013), and mammals and support a variety of recreational, commercial, and subsistence fisheries.

We review percid environmental biology by considering key processes regulating population abundance and biomass, including individual growth, reproduction, recruitment, and mortality. We focus on both perch, including North American yellow perch (*Perca flavescens*) and Eurasian perch (*P. fluviatilis*), and *Sander*, including North American walleye (*Sander vitreus*), sauger (*S. canadensis*), and Eurasian pikeperch or zander (*S. lucioperca*). However, given our familiarity with lentic North American systems and the multitude of studies on yellow perch and walleye, our examples are likely biased towards these species.

2.2 Individual Growth and Foraging Behavior

Individual growth is one of the most important facets of percid biology, as it may influence mortality, maturation, future growth and reproductive potential, and foraging success of individuals (Madenjian et al. 1996; Quist et al. 2003a; Galarowicz and Wahl 2005; Heibo et al. 2005). Fish growth and foraging behavior can also determine potential yields in aquaculture, commercial, and recreational harvest (Machiels and Wijsman 1996; Isermann et al. 2005). Percids exhibit substantial geographical and sex-related variation in growth both within and among species. *Perca* species tend to grow more slowly and to smaller maximum sizes than *Sander* species (Mehner et al. 1996; Graeb et al. 2005; Ljunggren and Sandström 2007). Growth rates of percids are sexually dimorphic, as females grow more quickly and to larger sizes than males prior to and following maturation, potentially due to higher growth efficiency in females (Mooij et al. 1999; Henderson et al. 2003; Mandiki et al. 2004a; Heibo et al. 2005; Lappalainen et al. 2005; Headly and Lauer 2008; Venturelli et al. 2009; Sect. 19.3.2). Percids also exhibit latitudinal variation in growth rates fairly typical of many fish species; northern populations tend to grow more slowly and live longer than southern populations (Quist et al. 2003a; Sects. 19.3.1 and 19.4.1). Much of this variation between populations and species may be due to genetic differences driven by adaptation to local environmental conditions (Wang and Eckmann 1994; Galarowicz and Wahl 2003; Mandiki et al. 2004b; Zhao et al. 2007; Wang et al. 2009a). In terms of foraging behavior, percid species follow similar ontogenetic patterns in diet throughout life, shifting from zooplankton to benthivory to piscivory as they increase in size (e.g., Graeb et al. 2005; Roswell et al. 2014). However, *Sander* shift to piscivory earlier and are seemingly obligate piscivores, meaning cannibalism can become a regulatory factor on population size in some systems (Frankiewicz et al. 1999). *Perca*, in contrast, exhibit a more generalist foraging strategy (Knight et al. 1984; Mehner et al. 1996; Campbell 1998; Tyson and Knight 2001; Graeb et al. 2005). Intraspecific variation in growth and foraging behavior is strongly dependent on a suite of environmental variables, including temperature, light, dissolved oxygen, prey availability, and harvest rates. Here, we focus on the responses of percid growth and foraging to these environmental variables, while addressing the implications of changes in growth with respect to aquaculture and fisheries management.

2.2.1 Temperature

Temperature is arguably the most important environmental factor affecting percid growth (Chap. 19) and behavior (Chap. 15). Percids have been described as temperate mesotherms, capable of tolerating a wide range distribution of temperatures (Hokanson 1977; Kitchell et al. 1977; Frisk et al. 2012). Across species and populations, the thermal environment an individual experiences has been consistently and strongly linked to growth rates, often through the use of the growing degree days metric (GDD; Colby and Nepszy 1981; Power and Van Den Heuvel 1999; Zhao et al. 2007; Venturelli et al. 2009; Chezik et al. 2014; Sect. 19.4.1). Thermal optima in fish are dependent on feeding rates (Kitchell et al. 1977), reproductive status (Hokanson 1977), possible pathogen infection (Hockett and Mundahl 1989), and other variables. Among percid adults, pikeperch generally have the warmest thermal optima for growth (27–30 °C; Hokanson 1977; Marshall 1977; Wang et al. 2009b), followed by yellow and Eurasian perch (22–28 °C; Hokanson 1977; Kitchell et al. 1977; Karås and Thoresson 1992; Tidwell et al. 1999; Strand et al. 2011), and walleye (21–23 °C; Huh et al. 1976; Hokanson 1977; Kitchell et al. 1977; Hokanson and Koenst 1986; Kocovsky and Carline 2001; Quist et al. 2003a). There is, however, significant ontogenetic variation in optimal temperatures within species, and juveniles tend to have higher thermal optima (pikeperch, 26–30 °C; perch, 24–29 °C; walleye, 24–28 °C) than adults (Kitchell et al. 1977; Hokanson 1977; Kocovsky and Carline 2001; Rónyai and Csengeri 2008; Frisk et al. 2012). Larval percids tend to exhibit lower thermal optima, ranging from 24 to 29 °C in pikeperch (Lehtonen et al. 1996) to 20–22 °C in perch and walleye (Hokanson 1977; Wang and Eckmann 1994; Jonas and Wahl 1998), although optimal temperatures from 15 to 25 °C have been observed depending on species and habitat (Hokanson 1977; Wang and Eckmann 1994; Lehtonen et al. 1996; Jonas and Wahl 1998). Larval growth is strongly positively influenced by temperature up to thermal optima (Hoxmeier et al. 2006; Irwin et al. 2009). The response of juvenile and adult percid growth to temperature can differ between cool and warm systems (Kitchell et al. 1977; Power and Van Den Heuvel 1999). In cooler systems that do not surpass thermal optima, growth is often positively related to temperature (Colby and Nepszy 1981; Lappalainen et al. 2005; Zhao et al. 2007; Venturelli et al. 2009; Wang et al. 2009b), while the inverse is true in warm systems where high summer temperatures may limit growth (Mehner and Wieser 1994; Hartman and Margraf 1992; Kershner et al.

1999; Tidwell et al. 1999; Quist et al. 2003a; Ložys 2004; Hoxmeier et al. 2006). Due to these mesothermic temperature preferences, adult growth rates may also vary seasonally within intermediate thermal systems, increasing in early summer, slowing in the warmest months, and increasing again as temperatures cool in fall (Kitchell et al. 1977; Mehner et al. 1996; Kocovsky and Carline 2001). There is also evidence of adaptive variation in the thermal tolerances of percids. Southern walleye populations exhibit the highest growth in warmer temperatures (25 °C) than northern populations (22 °C) (Galarowicz and Wahl 2003; Chapter, Fig. 19.5). Thus, the effects of temperature on percid growth are largely non-linear, and may partially depend on an interaction of genetic adaptive variation and developmental stage.

The direct effects of temperature on percid growth can be traced back to bioenergetic responses in metabolic rate and consumption. Conceptually, energy available to fish growth can be described as the difference between consumption and the sum of respiration, metabolism, egestion, and excretion, with each of these rates responding to changes in temperature to determine the potential for individual growth (Kitchell et al. 1977; Karås and Thoresson 1992). In percids, increasing temperatures tend to increase excretion, respiration, and standard metabolic rates assuming a fixed level of consumed energy (Kitchell et al. 1977; Cai and Summerfelt 1992; Karås and Thoresson 1992; Galarowicz and Wahl 2003; Frisk et al. 2012). These responses result in less energy available for growth for percids experiencing increasingly elevated temperatures.

Temperature also drives development, behavior, activity, and habitat selection in percids. In pikeperch, larval development is optimized at 24–29 °C, while temperatures above 20 °C promote feeding activity in walleye and perch (Wang and Eckmann 1994; Lehtonen et al. 1996). Temperatures below 14–15 °C appear to decrease larval walleye feeding activity (Koenst and Smith 1976). In addition, foraging activity, exhibited in traits such as attack rate, swimming performance, and consumption, have all been positively related to temperature in larval percids, while searching and handling time is negatively related (Bergman 1987; Johnston and Mathias 1994; Jonas and Wahl 1998). At particularly high temperatures, juvenile and adult percids tend to have opposite behavioral responses. Adult percids will attempt to compensate for decreased energy efficiency at high temperatures by decreasing activity levels. However, this decrease in activity may not be sufficient to completely compensate for increased metabolic rates and may also decrease consumption through reduced foraging activity, thereby reducing individual growth (Mehner and Wieser 1994; Kocovsky and Carline 2001; Quist et al. 2002). If alternative habitats are available, many percids will migrate to areas with more preferable temperatures (Kocovsky and Carline 2001; Quist et al. 2002), moving either laterally (e.g., among lake basins or lagoon and sea habitats; Kershner et al. 1999; Ložys 2004) or vertically (e.g., from nearshore or epilimnetic areas to the meta- and hypolimnion in stratified lakes; Lester et al. 2004; Jones et al. 2006).

2.2.2 Light and Vision

Sander and *Perca* species exhibit unique adaptations and habitat preferences with regard to light intensity, visibility, and photoperiod that differentially influence their foraging behavior and growth. Much of this variation is driven by differences in eye structure among species. *Sander* species possess the *tapetum lucidum*, a highly reflective layer of the retina that greatly increases the sensitivity of the eye to light, enabling the use of dimmer habitats, while *Perca* species lack such a structure (Ali et al. 1977). As a result, the species strongly diverge in photic habitat preferences. Larvae of all percids begin life exhibiting positive phototaxis, but *Sander* larvae switch to negative phototaxis as their eyes develop, at approximately 8 weeks of age (Schumann 1963; Bulkowski and Meade 1983; Luchiari et al. 2006; Wang and Appenzeller 2006). Juvenile and adult walleye and pikeperch prefer low light conditions, exhibit nocturnal and crepuscular activity peaks, and feed more efficiently in turbid, low light, and nighttime conditions (Scherer 1971; Ali et al. 1977; Mathias and Li 1982; Lester et al. 2004; Luchiari et al. 2006). In contrast, juvenile and adult yellow and Eurasian perch prefer high light conditions and exhibit diurnal feeding behavior, although they may also be most active during crepuscular periods to balance foraging success with predation risk (Huh et al. 1976; Ali et al. 1977; Huusko et al. 1996; Gardeur et al. 2007; Sects. 15.2 and 15.3). Photoperiod also differentially affects growth of percids. In experiments examining the relative importance of temperature and photoperiod to percid growth, the growth of *Perca* species was highly sensitive to photoperiod, decreasing when individuals are exposed to extended or unnatural lighting conditions, while the growth of *Sander* species was driven more by temperature (Huh et al. 1976; Gardeur et al. 2007; Shewmon et al. 2007).

Differences in eye structure may also partially explain variation in the response of *Perca* and *Sander* species to eutrophication and turbidity. Growth, predation success, and consumption are strongly negatively affected by increasing turbidity in Eurasian and yellow perch while their foraging activity increases, potentially in an attempt to increase the probability of encountering prey or in response to decreased predation risk in high turbidity environments (Radke and Gaupisch 2005; Ljunggren and Sandström 2007; Irwin et al. 2009; Reichert et al. 2010; Wellington et al. 2010). The contrast of prey to the environment is also extremely important, especially in larval *Perca* species, where high contrast environments promote larval survival and growth and low light or contrast may delay onset of feeding, slow growth, and increase mortality (Hinshaw 1985; Tamazouzt et al. 2000). Interestingly, there is a difference between responses to sedimentary and phytoplanktonic turbidity in *Perca*. Phytoplankton has a much stronger negative effect on *Perca* growth and foraging than sediment, which may imply that algal blooms are particularly detrimental to larval and juvenile growth (Radke and Gaupisch 2005; Wellington et al. 2010). *Sander* species, on the other hand, do not appear to be affected by turbidity until it reaches extremely high levels, up to an order of magnitude higher than levels that affect *Perca* (Vandenbyllaardt et al. 1991; Ljunggren and Sandström

2007). Even then, their growth is not as strongly affected, although their prey selection may change toward larger items (e.g., large zooplankton in juvenile pikeperch), which may affect the ontogenetic switch to piscivory (Vandenbyllaardt et al. 1991; Ljunggren and Sandström 2007; Zingel and Paaver 2010). In the wild, *Sander* species may even prefer more eutrophic, stained, or turbid systems than *Perca* species, as pikeperch have been observed to forage more efficiently in eutrophic systems with higher levels of water color (Lehtonen et al. 1996; Keskinen and Marjomäki 2003).

2.2.3 Dissolved Oxygen

Percid species appear to be fairly tolerant of low dissolved oxygen (DO) concentrations (Petrosky and Magnuson 1973; Suthers and Gee 1986). In terms of direct effects of DO on growth, percids exhibit ontogenetic differences in DO tolerance. Small individuals can tolerate lower DO than larger fish, which may play an important role in predator-prey dynamics, potentially allowing small fish to use low DO habitats as refugia from larger predators (Robb and Abrahams 2003). In general, percids appear to suffer negative effects on growth when DO concentrations decline below 2 mg/L (Suthers and Gee 1986; Arend et al. 2011; Brandt et al. 2011; Roberts et al. 2011), although feeding rates and consumption may begin to decline at DO levels as high as 3.5 mg/L without significant changes in growth (Carlson et al. 1980).

Most effects of DO are observed as sub-lethal effects on percid habitat selection and behavior (Chap. 15). Percids have been shown to avoid areas below 1.5–3 mg/L DO (Suthers and Gee 1986; Roberts et al. 2009, 2012). In lakes with hypolimnetic hypoxia, fish move upward to the metalimnion, thereby reducing benthivory and increasing zooplanktivory, which may lead to reduced growth rates (Roberts et al. 2009; Arend et al. 2011; Brandt et al. 2011). However, diet and hydroacoustic observations in Lake Erie, North America, have revealed that yellow perch make foraging "dives" through the hypolimnetic hypoxic zone in summer and therefore are able to consume benthic invertebrates (Roberts et al. 2012). Interestingly, the potential effects of these short-term exposures to hypoxic habitats appear to be minimal (Roberts et al. 2009, 2011). In walleye, DO may moderate thermo- and phototactic responses, forcing fish to reduce their residence time in or entirely abandon preferred habitats when DO falls below 2–3 mg/L (Scherer 1971; Brandt et al. 2011). Because preferred temperatures and light levels for walleye often spatially overlap with hypoxic zones in lakes, they may be forced into habitats above thermal and photic optima. However, hypoxia may not be entirely negative for large predators such as walleye, as many forage fish species may also aggregate at the metalimnion due to tradeoffs between thermal and oxygen preferences, thereby condensing prey and increasing walleye consumption (Brandt et al. 2011). Thus, while hypoxia may be an important moderator of behavior and habitat selection, its effects on percid biology may be mixed.

2.2.4 Ontogeny and Prey Availability

Sander and *Perca* species follow fairly similar trends in their diet ontogeny, but differ substantially in the timing of ontogenetic diet shifts (especially to piscivory) and foraging behavior (Sect. 15.2). *Sander* species are specialist piscivores with large gape sizes and canine teeth, actively selecting preferred prey types (Campbell 1998; Graeb et al. 2005). In contrast, *Perca* species lack canine teeth, exhibit slower increases in gape size with growth, and exhibit more generalist, passive prey selection behavior dependent on prey availability and foraging efficiency (Knight et al. 1984; Mehner et al. 1996; Graeb et al. 2005). *Perca* species also tend to exhibit higher foraging and attack rates than *Sander* species (Graeb et al. 2005; Ljunggren and Sandström 2007).

Perca and *Sander* larvae begin to feed roughly 4–5 days post hatch, at about 9 mm in length for walleye and 6 mm for pikeperch, yellow perch, and Eurasian perch (Johnston and Mathias 1994; Wang and Eckmann 1994; Peterka et al. 2002; Fulford et al. 2006). Newly hatched larvae begin feeding on small planktonic items such as algae, rotifers, molluscan veligers, and copepod nauplii, eventually graduating to larger prey items like adult copepods, *Ceriodaphnia*, and larger daphnid species as they increase in length to around 20–35 mm (Fox and Flowers 1990; Wang and Eckmann 1994; Specziár and Biró 2003; Graeb et al. 2004; Specziár 2005; Galarowicz et al. 2006; Withers 2012). Depending on availability, percid juveniles larger than 35–40 mm will subsequently switch to larger benthic prey, consisting of organisms such as chironomid larvae, mayfly nymphs, amphipods, and *Mysis* (Fox and Flowers 1990; Johnston and Mathias 1994; Specziár and Biró 2003; Galarowicz and Wahl 2005; Specziár 2005; Roswell et al. 2013). The ontogenies of *Perca* and *Sander* diverge at this point; the faster growing walleye and pikeperch will switch to almost exclusive piscivory at sizes from 35 to 100 mm depending on the abundance and size of available fish prey, preferring fish from one fifth to one tenth their own length (Turesson et al. 2002; Persson and Brönmark 2002; Specziár and Biró 2003; Galarowicz and Wahl 2005; Graeb et al. 2005; Specziár 2005; Galarowicz et al. 2006; Hoxmeier et al. 2006), while yellow and Eurasian perch continue to eat a mixture of zooplankton and benthic invertebrates (Diehl 1993; Schaeffer et al. 2000; Svänback and Eklöv 2001; Tyson and Knight 2001; Roswell et al. 2013). Some *Perca* populations eventually switch to piscivory at sizes larger than 150 mm, often around 3–5 years of age, eating young-of-year cyprinid and clupeid species when abundant (Knight et al. 1984; Tyson and Knight 2001; Svanbäck and Eklöv 2001; Headly and Lauer 2008). However, *Perca* species are especially plastic in their dietary preferences, and there are examples of *Perca* populations greatly delaying ontogenetic shifts or never switching to benthivory or piscivory throughout life (e.g., Quevedo et al. 2009; Roswell et al. 2014).

Prey size and availability have important implications for growth throughout ontogeny. Percids exhibit a type II functional response to increasing prey densities (Hoxmeier et al. 2004, 2006; Galarowicz and Wahl 2005), and growth rates of diverse ages of percids have been positively related to prey density (Hartman and

Margraf 1992; Diehl 1993; Wang and Eckmann 1994; Hoxmeier et al. 2004; but see Hoxmeier et al. 2006). However, growth is driven not simply by prey density, but also the availability and capture efficiency of appropriately sized and energy-rich prey throughout ontogeny. Early in life, the lack of large zooplankton, such as *Daphnia*, may restrict growth and delay switching to either benthic invertebrates or fish (Mills et al. 1989; Graeb et al. 2004; Fulford et al. 2006; Lappalainen et al. 2005; Vinni et al. 2009). The absence of large benthic invertebrates during the juvenile stage may negatively impact growth and condition, which could delay the final switch to piscivory (Hayward and Margraf 1987; Diehl 1993; van Densen et al. 1996; Schaeffer et al. 2000; Tyson and Knight 2001; Vinni et al. 2009). Finally, the switch to piscivory appears to be largely driven by the size and density of forage fish, creating a potential growth bottleneck if small forage fish (e.g., young of year clupeids or cyprinids) are not available (Knight et al. 1984; Hartman and Margraf 1992; Persson and Brönmark 2002; Sherwood et al. 2002; Kolar et al. 2003; Keskinen and Marjomäki 2004; Galarowicz and Wahl 2005; Headly and Lauer 2008; Sect. 15.2.1). This heavy reliance of growth on consumption based upon prey size and availability likely drives the strong compensatory density dependent growth effects commonly observed in yellow perch (e.g., Post et al. 1997; Headly and Lauer 2008; Irwin et al. 2009; Ivan et al. 2011), Eurasian perch (e.g., Diehl 1993; Byström and Garcia-Berthou 1999; Heibo et al. 2005), walleye (e.g., Knight et al. 1984; Muth and Wolfert 1986; Fox and Flowers 1990; Schueller et al. 2005; Venturelli et al. 2009), and pikeperch (e.g., Lehtonen et al. 1996; Lappalainen et al. 2005; Sect. 19.4.3).

Prey densities may also affect population behavior and size distributions. Eurasian perch and yellow perch exhibit rapid, plastic niche partitioning in lakes based on foraging success, diverging into littoral and pelagic sub-populations with distinct habitats, morphologies, diet preferences, and growth patterns (Post et al. 1997; Svanbäck and Eklöv 2001, 2006; Quevedo et al. 2009; Roswell et al. 2013; Sects. 15.2 and 15.5). Varied growth rates and limited forage fish may result in populations with bimodal size distributions, with faster growing individuals able to switch to piscivory and reap the associated energetic benefits of fish prey, while slower growing individuals continue to eat invertebrates and suffer reduced growth (Post and Evans 1989; van Densen et al. 1996; Post et al. 1997; Svanbäck and Eklöv 2001; Specziár 2005; Johnston et al. 2012a). In extreme cases, lack of appropriately-sized prey may even result in stunted populations (Heath and Roff 1996; Heibo et al. 2005; Persson and Brönmark 2002; Vinni et al. 2009).

At an individual level, growth is strongly related to prey consumption rate (Kitchell et al. 1977; Karås and Thoresson 1992), and the implications of food limitation or starvation are severe. Larvae may be able to resist starvation for 11–19 days depending on water temperature (Jonas and Wahl 1998; Olin et al. 2012; Andree et al. 2014), but their energy densities decline rapidly in the first 5 days without food, reaching a "point of no return" after 6–8 days where their activity levels, prey capture and handling efficiencies, and predation avoidance abilities decline precipitously (Jonas and Wahl 1998). Juveniles can withstand starvation for a longer time period (16–21 days), but ultimately follow the same patterns of decreasing activity and increasing predation susceptibility as larvae (Jonas and Wahl 1998). Percids may attempt to compensate for lack of food by decreasing respiration rates and activity (Cai and Summerfelt 1992; Mehner and Wieser 1994) and increasing consumption when food is available (Mattila et al. 2009). However, these measures are often not sufficient to avoid decreases in growth (Mehner and Wieser 1994; Mattila et al. 2009). As a result, prey availability and consumption can have wide-ranging effects at both the population and individual levels.

2.2.5 Exploitation

All of the percid species discussed in this chapter are highly important economic resources, and many populations are subject to high rates of fishing mortality that may affect both population size structures and individual growth patterns. Size selective harvest has the potential to reduce mean age and size in percid populations through selective removal of older, larger individuals, as seen in walleye (Colby and Baccante 1996) and yellow perch (Heyer et al. 2001; Lauer et al. 2008; Collingsworth and Marschall 2011a). Reductions in abundance may also release the population from intraspecific competition and other density-dependent effects, leading to faster growth (Paukert and Willis 2001; Lauer et al. 2008). Alternatively, size selective fishing that removes the largest individuals may also remove the fastest growing fish, leaving individuals that suffer from poor growth, thus reducing the average individual growth potential in the population (Machiels and Wijsman 1996). This can lead to differential mortality rates between growth trajectories; in one Eurasian perch population, the smallest fish in a given year class suffered highest mortality from ages 0 to 1 from natural causes, while mortality of larger fish from the same year class peaked from ages 2 to 4 due to fishing (Machiels and Wijsman 1996). Changing population growth patterns and size structure may also affect the potential value of the stock, either by reducing commercial yields (as seen in Lake Michigan yellow perch; Marsden and Robillard 2004; Lauer et al. 2008) or decreasing popularity to recreational anglers, who generally prefer relatively fewer but larger fish over many small ones (Isermann et al. 2005). Because of the economic and intrinsic value of these fisheries and the strong influence growth may have on their production, a clear understanding of the potential ramifications of fishing pressure is needed to properly manage and maintain percid populations.

2.3 Reproductive Ecology

Life history traits of large percid species, namely maturation, gonad development, fecundity, spawning behavior, and offspring traits, exhibit a wide range of variation both within (e.g., Colby and Nepszy 1981) and across percid species (e.g., Marshall 1977). In general, all *Perca* and *Sander* species are iteroparous, periodic spawners that typically spawn in spring (Johnston and Leggett 2002; Lappalainen et al. 2003; Heibo et al. 2005). Each species exhibits a capital spawning strategy, with gonad development taking place throughout the fall and winter months prior to spawning (Malison et al. 1994; Henderson et al. 2000). Within species, variation in life history traits appears to be driven by a combination of plastic responses to environmental conditions (Jansen 1996; Johnston et al. 2012a) and adaptation to local environmental conditions (Hokanson 1977; Wang et al. 2012; Collingsworth and Marschall 2011a). Because variation in reproductive characteristics can have strong effects on population traits such as individual and population growth, mortality, and recruitment (Lester et al. 2000; Collingsworth and Marschall 2011a), understanding the environmental mechanisms controlling observed reproductive life history trait variation can provide vital information for both management of wild populations and improved aquaculture practices.

In this section, we present an overview of the general reproductive ecology and life history patterns exhibited by *Perca* and *Sander* species. We also describe environmental factors, including temperature, light, dissolved oxygen, spawning habitat, and food resources, that appear to drive variation in many aspects of percid reproductive biology. In addition, we examine recent evidence that maternal effects (i.e., correlations between the traits of an individual female and her offspring) may be an important factor in determining the size and fitness of percid eggs and larvae. Finally, because all of our focal species support highly valuable recreational and commercial fisheries world-wide (Baccante and Reid 1988; Lappalainen et al. 2003; Lauer et al. 2008; Vainikka et al. 2012) we discuss the potential implications of fishing exploitation with regard to variation in percid reproductive life history traits.

2.3.1 Maturation

Maturation, i.e., the initial development of gonads in preparation for spawning and subsequent allocation of energy to gonadal development and spawning behavior, can affect future growth, reproductive potential, and mortality (Hayes and Taylor 1990; Bronte et al. 1993; Lester et al. 2000; Roff et al. 2006). In general, percid maturation schedules are typical of a periodic life history strategy (Winemiller and Rose 1992). Maturation is sexually dimorphic among species, wherein males mature at ages 1 or 2 and females mature at ages 3–4 (Muth and Wolfert 1986; Diana and Salz 1990; Jansen 1996; Henderson and Morgan 2002; Lappalainen et al. 2003; Wang et al. 2009c). However, all species exhibit broad variation in maturation schedules among populations, with estimates of age at maturity ranging from as early as age 1 (Muth and Wolfert 1986; Houthuijzen et al. 1993; Madenjian et al. 1996) to as late as age 10 (Colby and Nepszy 1981; Raikova-Petrova and Živkov 1998; Lester et al. 2000; Lappalainen et al. 2003). Length at maturation also varies considerably depending on sex and species (Lester et al. 2000; Heibo et al. 2005; Purchase et al. 2005). Across several populations, males tend to mature around 80–150 and 300–350 mm TL in *Perca* and *Sander* (Diana and Salz 1990; Purchase et al. 2005; Wang et al. 2009c; Venturelli et al. 2010a), while females mature between 150–200 and 350–450 mm TL (Diana and Salz 1990; Heibo et al. 2005; Wang et al. 2009c; Venturelli et al. 2010a), respectively.

Variation of percid maturation schedules is related to differences in growth and mortality rates, especially in juveniles (Colby and Nepszy 1981; Lappalainen et al. 2003; Heibo et al. 2005). Populations with faster individual growth rates tend to mature at younger ages than populations with slower growing individuals (Madenjian et al. 1996; Heibo et al. 2005; Schueller et al. 2005). Individual growth and mortality rates are strongly linked with maturation of percids (Chap. 19); therefore, any environmental variable that affects growth or mortality rates (Sects. 2.2, 2.5, and 19.4) may also impact maturation. For instance, temperature has a negative effect on age at maturation in many populations, mainly through its positive effects on growth (Raikova-Petrova and Živkov 1998; Venturelli et al. 2009). This trend contributes to latitudinal variation in maturation schedules among percid populations, where colder, slower growing northern populations tend to mature at older ages and larger sizes and warmer, faster growing southern populations tend toward maturation at younger ages and smaller sizes (Lappalainen et al. 2003; Heibo et al. 2005). Variation in annual temperatures may also influence short-term patterns in maturation schedules, and Venturelli et al. (2009) found that maturation in walleye was better described as a function of growing degree days (GDD) than either age or length. Percids may require an extended cooling period less than 10–12 °C during winter for proper gonad development (discussed below); extension of this period may allow a larger proportion of smaller females to develop their gonads and mature than during a relatively warm year (Hokanson 1977; Hermelink et al. 2011).

In addition to temperature, maturation may also be dependent on fish condition, often driven by food availability and population density. Maturation of female walleyes has been found to be positively related to body lipid stores, especially the initial maturation of females age 2–5 in the transition period from juveniles to adults (Henderson et al. 2000; Henderson and Morgan 2002). This is thought to be due to tradeoffs in present and future reproductive success, where small females in poor condition would delay maturation in favor of increasing size, and thus reproductive output, the following year (Henderson and Morgan 2002). There are also several examples of compensatory density-dependent shifts in maturation schedules in percids following changes in population abundance (Schueller et al. 2005; Sect. 19.4.3). For example, following rehabilitation of the walleye fishery in Lake Erie, population abundance increased, leading to decreased condition and growth, which resulted in increased age and length at maturity (Muth and Wolfert 1986). In the opposite case, abundant food resources in a newly freshened reservoir led to rapid growth and early maturity in Eurasian perch (Houthuijzen et al. 1993). Therefore, availability of food resources, mediated through intraspecific competition or changes in the environment, may have important effects on percid maturation schedules.

In many percid populations it remains unclear whether observed variation in maturation schedules is primarily reflective of phenotypic plasticity (Purchase et al. 2005) or evidence of adaptation to local environmental regimes (Wang et al. 2009b; Collingsworth and Marschall 2011a). Evidence from the literature suggests both processes may play a role in shaping maturation schedules of percids. Johnston et al. (2012a) suggested that older age and larger size at maturation in walleye than sauger may allow for more plasticity in those traits, which would enable walleye to respond to short-term changes in the environment. However, Wang et al. (2009b) found significant adaptive variation in the maturation schedules of Laurentian Great Lakes walleye populations, suggesting each had adapted to their local environment and incumbent selection pressures. Unfortunately, quantitative assessment of plastic and adaptive variation in maturation is lacking for most percid species, despite the wide range of literature documenting differences in these traits (e.g., Lappalainen et al. 2003 for pikeperch). Disentangling plastic and adaptive variation can be difficult, but may provide important information into the underlying differences among species and populations and can also indicate important changes in the selection pressures on populations, whether through fisheries exploitation (e.g., Barot et al. 2004) or other ecosystem changes (e.g., Wang et al. 2008).

2.3.2 Fecundity

Fecundity varies significantly among percid species due to differences in body and egg size. Pikeperch, with their small eggs but relatively large body size, have by far the highest fecundity, with estimates ranging from 24,000 to over 2.5 million eggs per female (Lehtonen et al. 1996; Kosior and Wandzel 2001). Pikeperch are followed by walleye, which produce 24,000–614,000 eggs per female (Wolfert 1969; Serns 1982; Muth and Ickes 1993; Colby and Baccante 1996). Finally, yellow and Eurasian perch have been estimated to produce 1,910–157,594 eggs per female (Jansen 1996; Lauer et al. 2005; Pedicillo et al. 2008). Much of this observed variation is due to the positive correlations between fecundity and maternal length, age, and, most strongly, mass (Serns 1982; Muth and Ickes 1993; Jansen 1996; Lauer et al. 2005; Johnston et al. 2012a, b). Large, old females can produce orders of magnitude more eggs than small, young females, drastically increasing their reproductive output in a given year (Venturelli et al. 2009; Collingsworth and Marschall 2011b).

Variation in fecundity has been linked to the thermal environment, photoperiod, and food supply (Colby and Nepszy 1981). Compensatory density dependent responses in walleye fecundity have been attributed to changes in food supply, fish growth, and condition through intraspecific competition (Colby and Nepszy 1981; Baccante and Reid 1988; Rose et al. 1999; Moles et al. 2008; Collingsworth and Marschall 2011b; Johnston et al. 2012a). Differences in fecundity among populations may also be related to adaptation of egg size to local environments (Johnston and Leggett 2002); given equal reproductive investment, populations adapted to produce smaller eggs would be assumed to exhibit higher fecundity. This trend was observed across several northern Ontario lakes in walleye, where fecundity increased with increasing lake-specific GDD (>5 °C), potentially through adaptive mechanisms to decrease egg size (and thus increase fecundity) in more benign environments (Baccante and Colby 1996).

2.3.3 Gonad Development and Egg Size

All percids can be described as capital spawners, meaning gonad development takes place throughout the year preceding spawning in spring (Malison et al. 1994; Henderson et al. 2000). *Perca* and *Sander* species follow very similar trajectories of gonad development following a short, 1–3 month quiescent period after spawning (Hokanson 1977; Ciereszko et al. 1997; Henderson et al. 2000). Male gonads generally start to develop in fall and are fully developed by the mid-winter months, able to sustain spermiation for up to 6 months until spawning in early spring (Malison and Held 1995; Shewmon et al. 2007). In females, ovary growth begins from August to October, and continues throughout the winter (Diana and Salz 1990; Malison et al. 1994; Malison and Held 1995; Ciereszko et al. 1997; Henderson et al. 2000; Lappalainen et al. 2003). Vitellogenesis begins in oocytes in November and continues until just before spawning (Hokanson 1977; Malison et al. 1994; Henderson et al. 2000; Hermelink et al. 2011). Final hydration and maturation of the oocytes occurs just before spawning (Malison et al. 1994; Sandström et al. 1995). It is also at this time that the gelatinous, ribbon-like skein forms around the eggs of *Perca* species, linking them in a single long strand (Sandström et al. 1995, 1997). Oocyte number and energy content are generally determined by December or January; although the ovary will roughly double in size between January and spawning, additional gains are of little energetic value (Sandström et al. 1995, 1997; Henderson et al. 2000). Oocyte size is determined by January in *Perca* (Sandström et al. 1997; Henderson et al. 2000) but may continue to develop until just before spawning in *Sander* (Malison et al. 1994). Among percids, walleyes produce the largest eggs (1.4–2.1 mm in diameter; Wolfert 1969; Muth and Ickes 1993; Wang et al. 2012), followed by *Perca* spp. (0.94–2.0 mm; Ciereszko et al. 1997; Lauer et al. 2005; Pedicillo et al. 2008) and pikeperch (0.5–1.5 mm; Marshall 1977; Lappalainen et al. 2003).

In mature percids, the annual onset and quality of gonad development is driven by both endogenous mechanisms and exogenous inputs, the most important of which are temperature and photoperiod (Ciereszko et al. 1997). Percids require an extended winter period of 3–5 months with temperatures below 10 °C (perch and walleye) or 15 °C (pikeperch) for proper gonad development to occur (Hokanson 1977; Sandström et al. 1997; Hermelink et al. 2011). Winters that are either characterized by a short cold period (i.e., less than 3 months) or warm (i.e., mean temperatures above 15 °C) can lead to suboptimal ovary development, decreased egg size, and atresia of oocytes (Ciereszko et al. 1997; Sandström et al. 1997; Migaud et al. 2002; Shewmon et al. 2007; Hermelink et al. 2011). In *Perca* species, warm winter and spring temperatures may also lead to poor development of the skein, which can break down and cause high egg mortality during incubation (Sandström et al. 1997). Thus, poor thermal conditions may significantly affect reproductive success in some populations of percids (e.g., Schlumberger and Proteau 1996).

Photoperiod appears to control the onset of gonad development in percids (Malison et al. 1994; Migaud et al. 2004), as a natural decline in photoperiod has initiated gonad development in perch even without a concomitant decrease in temperature (Wang et al. 2006; Shewmon et al. 2007). In experimental treatments, photoperiod also plays a role in proper gonad development, as perch exposed to unnaturally long or constant photoperiod regimes either failed to mature or exhibited low gonadosomatic indices and poor oocyte quality (Migaud et al. 2004, 2006; Wang et al. 2006). Thus, it appears that an interaction of photoperiod and temperature is vitally important to the proper maturation and reproduction of percid species, where photoperiod initiates the process and optimal temperatures are needed for later development and maturation.

Maternal size, condition, and adaptation to local environments also appear to play a role in gonad development and egg size in percids. Many studies have linked increases in egg size and quality to increases in maternal length, mass, and age (Johnston and Leggett 2002; Lauer et al. 2005; Johnston et al. 2012b; Olin et al. 2012; Andree et al. 2014). Larger, older females tend to exhibit increased GSI and invest relatively more in reproduction than smaller or younger females (Diana and Salz 1990; Heyer et al. 2001; Olin et al. 2012). Female body condition, driven through acquisition of resources throughout the summer and fall preceding ovary development, may also determine the amount of energy available to invest in reproduction (Johnston and Leggett 2002; Lappalainen et al. 2003; Wang et al. 2006; Collingsworth and Marschall 2011b; Johnston et al. 2012b). Therefore, food intake and lipid availability in females may be an important factor determining egg size and quality, with potential ramifications for recruitment the following year (Madenjian et al. 1996; Tyson and Knight 2001).

Interestingly, these maternal effects appear to vary among populations, and there does not appear to be a single optimum egg size for any species (Johnston and Leggett 2002; Collingsworth and Marschall 2011b). Rather, the relative strength of maternal effects and egg size appear to be highly conserved within populations and highly varied among populations (Wang and Eckmann 1994; Venturelli et al. 2010b; Wang et al. 2012). For instance, Moodie et al. (1989) showed that larval survival in walleye was related to relative larval size differences within populations rather than size variation among populations (i.e., the smallest larvae from a single population suffered high mortality rates, but larval size differences among populations did not explain patterns in larval survival). System productivity often has a negative relationship with egg size, likely because more productive systems offer more food resources, reducing the strength of size-dependent mortality on larvae (Johnston and Leggett 2002; Heibo et al. 2005; Wang et al. 2012). Egg size in walleye and Eurasian perch has also been shown to decrease with increasing latitude, potentially because longer lived northern populations invest less in annual reproduction. However, maternal effects appear to be strongest at the northern and southern edges of a species' distribution, where environmental conditions are usually marginal and contribute a greater benefit to the production of relatively larger offspring by larger females (Johnston and Leggett 2002; Heibo et al. 2005). Thus, some of the variation in maternal effects, reproductive investment, and egg size in percids may be largely determined through adaptive mechanisms responding to long term environmental conditions to optimize reproductive success.

2.3.4 Spawning Behavior

Percids exhibit a strikingly wide array of spawning strategies, behaviors, and habitats among species (Sect. 15.6). A common theme among percid species is that spring spawning bouts are prompted primarily by temperature and photoperiod (Hokanson 1977; Ciereszko et al. 1997; Rinchard et al. 2011), although the optimum temperatures range widely among species. Walleye spawning takes place at temperatures from 2 to 15 °C, with peak spawning often occurring at temperatures between 4 and 10 °C (Hokanson 1977; Roseman et al. 1996; Ivan et al. 2010; Martin et al. 2012). Walleye undergo spawning migrations to rivers, streams, and reefs within lakes, and exhibit high levels of spawning site fidelity, which may be strong enough to cause genetic differences among stocks (Rinchard et al. 2011; Stepien et al. 2012). Males are usually first to the spawning areas, followed initially by older females and later by younger females, which may prolong the spawning season in populations with extended age structures (Casselman et al. 2006; Venturelli et al. 2010a). Slightly adhesive, demersal eggs are broadcast onto preferred substrates at depths less than 5 m, usually gravel and cobble in river beds and exposed shorelines and reefs in lakes (Roseman et al. 1996; Rinchard et al. 2011; Raabe and Bozek 2012). Fertilization is external, and fertilization success in the wild may be fairly low (Heidinger et al. 1997), potentially because walleye sperm activity decreases by up to 70 % in the 30–40 s following activation, which is a trait common among all percids (Casselman et al. 2006). While pikeperch are similar to walleye in that they undergo spawning migrations and exhibit strong site fidelity (Lappalainen et al. 2003), the remainder of their spawning behaviors differ significantly from those of their congener (Sect. 15.6). Pikeperch prefer warmer temperatures but have highly plastic thermal preferences, spawning between 4.5 and 26 °C throughout their range (Hokanson 1977), although preferred temperatures range from 8 to 18 °C (Lappalainen et al. 2003; Hermelink et al. 2011). Males appear first on the spawning grounds and will excavate nests in a variety of substrates, including mud, vegetation, sand, and gravel (Hokanson 1977; Lappalainen et al. 2003). Nesting sites can be as deep as 16 m, taking into account preferences for cool temperatures and low light intensity (Schlumberger and Proteau 1996). Females deposit a clutch of highly adhesive eggs in the nest and leave soon after. Males will aggressively defend the clutch until hatching and swim up of fry (Schlumberger and Proteau 1996; Hermelink et al. 2011). The additional parental care provided by pikeperch appears to yield higher fertilization rates and egg survival than found in walleye (Marshall 1977).

The *Perca* species exhibit yet another set of spawning traits that dissociates them from either pikeperch or walleye. Like those species, spawning in perch is triggered by increases in temperature and photoperiod (Migaud et al. 2002; Collingsworth and Marschall 2011b). Spawning occurs at temperatures intermediate to those of the *Sander* species, ranging from 4 to 19 °C with an optimum around 8–10 °C (Forney 1971; Collingsworth and Marschall 2011b). Perch in warmer areas spawn earlier in the year and may experience a protracted spawning period compared to those in cooler systems (Sandström et al. 1997). Alternatively, very cold temperatures (2–3 °C) have also been observed to suspend spawning in perch (Sandström et al. 1997). Spawning aggregations of perch concentrate on preferred substrates, namely dead vegetation, coarse woody debris, and large cobble (Robillard and Marsden 2001; Pedicillo et al. 2008; Čech et al. 2009; Sect. 15.6). Spawning depth can range from 2 to 20 m, and is influenced by a number of factors that may influence egg survival, including temperature, dissolved oxygen, wind, and ultraviolet radiation (Huff et al. 2004; Williamson et al. 1997; Čech et al. 2009). In a unique adaptation, both yellow and Eurasian perch lay their eggs as part of a gelatinous, accordion-like skein that may reach up to 2 m in length (Scott and Crossman 1973). The skein is draped over woody debris and vegetation, and is believed to deter predators, hold the egg mass in place, promote fertilization, and aid in oxygenating the eggs (Treasurer 1983; Newsome and Tompkins 1985; Reyes et al. 1992).

2.3.5 *Incubation and Offspring Survival*

Despite the striking diversity in their spawning strategies and behaviors, perch, pikeperch, and walleye eggs exhibit similar trends in their responses to temperature, light, and dissolved oxygen. Optimal temperatures for incubation of perch and walleye eggs range from 9 to 20 °C (Koenst and Smith 1976; Hokanson 1977; Guma'a 1978; Wang and Eckmann 1994; Huff et al. 2004), while pikeperch incubation is optimized at 11.5–20 °C (Muntyan 1977; Lappalainen et al. 2003). Percid eggs appear fairly resistant to fluctuations in temperature during incubation, although very large perturbations (20 °C or more) have been shown to decrease hatching success and swim up rates in walleye (Schneider et al. 2002). Warm temperatures may also promote bacterial and fungal infection of eggs and cause perch skeins to deteriorate, which can greatly increase egg mortality (Sandström et al. 1997). On average, percid eggs will hatch in 9–13 days at preferred temperatures, although hatch times are negatively related to temperature and may vary from 3–4 days at warmer temperatures (15–21 °C) to 19–35 days at colder temperatures (7–11 °C) (Oseid and Smith 1971; Malison et al. 1994; Huff et al. 2004). Incubation of eggs at or near optimal temperatures has also been shown to result in increased larval length at hatch, which may increase larval survival (Wang and Eckmann 1994). Extension of the incubation period by low temperatures can leave eggs more vulnerable to mortality through wave displacement and predation (Johnson 1961; Clady 1976; Roseman et al. 1996; Lappalainen et al. 2003; Huff et al. 2004; Ivan et al. 2010). Percid eggs are also relatively tolerant to low dissolved oxygen (DO) conditions. However, higher DO may shorten the incubation period and increase larval size at hatch, which may promote early life survival (Oseid and Smith 1971). Average larval length at hatch varies among species. Walleye larvae are among the largest of any percid species at hatch, ranging from 6 to 8.6 mm (Houde and Forney 1970; Marshall 1977). Yellow and Eurasian perch produce intermediately sized larvae (4.6–6.0 mm; Thorpe 1977; Weber et al. 2011), and pikeperch larvae are smallest at hatch (4.5–5.5 mm; Lappalainen et al. 2003).

Egg and larval mortality can be extremely high (up to 80 %) in the first 5 days after fertilization (Moodie et al. 1989; Latif et al. 1999), and parental effects may contribute to variability in egg and larval survival following spawning (Wang and Eckmann 1994; Andree et al. 2015). In walleye, Moodie et al. (1989) found that smaller females produced smaller eggs and higher proportions of deformed larvae that failed to survive. In addition, larval size at hatch and survival have each been shown to increase with maternal age, ova lipid provisioning, and fatty acid content of eggs (Wiegand et al. 2004, 2007; Johnston et al. 2012b). For yellow perch, Heyer et al. (2001) found that larger females produced shorter larvae with larger yolk sacs that could potentially lead to improved survival during the pelagic stage. Similarly, in Eurasian perch Olin et al. (2012) demonstrated that larger females produced larger, heavier larvae that better survived starvation. In contrast, Andree et al. (2014) found negative relationships between maternal size and age and larval survival in yellow perch during the first 5 days post-hatch using older and larger yellow perch females than Olin et al. (2012), but no significant relationship during days 6–14, which suggests that maternal effects may vary with both maternal and larval ontogeny. Finally, the added parental care provided by pikeperch may increase egg and larval survival relative to that of walleye, despite their relatively small eggs and larvae (Marshall 1977; Lappalainen et al. 2003). Therefore, parental effects may not only influence egg size (as discussed above) but also larval survival, which can potentially affect recruitment of these species (Roseman et al. 1996; Venturelli et al. 2009).

2.3.6 Fishery-Dependent Selection

As described above, percid species support highly valuable commercial and recreational fisheries throughout their respective ranges (Lehtonen et al. 1996; Lester et al. 2000; Lappalainen et al. 2003; Lauer et al. 2008). Therefore, fishing mortality may very well be an evolutionarily recent and important factor influencing life history trait expression in these species. Harvest of percids has been shown to decrease population abundance in many areas (e.g., Marsden and Robillard 2004; Lauer et al. 2008), which may lead to increased growth and fecundity due to release from intraspecific competition for resources (Baccante and Reid 1988; Colby and Baccante 1996). However, size selective harvest can also cause truncation of size and age distributions, skewed sex ratios, and reduction of the spawning stock, which in turn may result in a shortened spawning period (due to the lack of old, large females) and increase variability in recruitment (Colby and Nepszy 1981; Lauer et al. 2008; Collingsworth and Marschall 2011b). Harvest of older, larger individuals coupled with faster growth can also drive maturation schedules toward maturity at younger ages through both phenotypically plastic responses to growth and contemporary evolution in response to harvest (Muth and Wolfert 1986; Jansen 1996; Heibo et al. 2005; Sharpe and Hendry 2009; Wang et al. 2009c). Finally, maternal effects, coupled with maturation at younger ages and smaller sizes, can result in the production of fewer eggs of poorer quality, reducing offspring fitness and further depressing recruitment (Heyer et al. 2001; Venturelli et al. 2009; Olin et al. 2012). Thus, exploitation can have major effects on the life histories of percid populations, which in turn may determine sensitivity to fisheries exploitation (Lester et al. 2000). Because reproductive characteristics of percids are highly variable, as we have demonstrated in this section, it is crucial that information on specific populations of interest are collected in order to determine proper management actions and gauge population-wide responses to changing conditions (Purchase et al. 2005).

2.4 Early Life Survival and Recruitment

2.4.1 Patterns of Percid Early Life Survival and Recruitment

For most fish populations, recruitment success is the key determinant of future population size. While the precise definition of recruitment will vary across fish stocks depending upon life histories, available data, and stage-specific survival rates, we generally define recruitment as annual population-level reproductive success, i.e., the number of young fish which survive through early life to contribute to adult or harvestable components of a population. For many fish populations, annual recruitment success may be set during very early life and the number of individuals which survive through a very early life stage (e.g., larval stage) may be strongly correlated with the number of individuals which ultimately recruit to the adult population. In such cases, biotic and abiotic factors affecting survival through these very early life stages will determine recruitment success. In contrast, recruitment of other fish populations may not be set until later in life, and factors affecting survival of later life stages may have stronger influence on recruitment.

Similar to most fish species, percid populations display high inter-annual recruitment variation. In fact, year-class strength of several percid populations exhibit particularly strong *boom-and-bust* patterns. As mentioned above, percids may be characterized as periodic species based on Winemiller and Rose's (1992) life-history continuum (i.e., they produce relatively large clutches of small to medium sized offspring; Sect. 2.2). Moreover, excepting nest-guarding behavior in pikeperch, they provide low parental care for offspring, although active selection of spawning location and packaging of eggs in skeins by perch could be viewed as pre-hatching parental care. In turn, these reproductive traits result in large clutches of larvae with small to medium sized yolk sacs and low survival rates. Due to their high abundance, small changes in survival rates of larval percids have the potential to lead to large responses in the number of individuals that survive to subsequent life-stages and ultimately recruit to adult populations.

Houde (1994) compared recruitment patterns of marine and freshwater fishes and suggested that recruitment of marine fishes is often set during early larval stages, while recruitment of fish in small freshwater systems is set during later juvenile stages. In general, marine larvae are small, making them susceptible to drift and spatio-temporally variable physical, chemical, and lower trophic level processes which affect their survival through both density-dependent and -independent processes. In comparison, freshwater larvae tend to be larger and their recruitment may be more influenced by biological interactions and density-dependent processes acting on later life stages. Percids inhabit a diversity of environments, from small ponds and streams to huge lacustrine, brackish, and riverine systems. Thus, not surprisingly, the age when recruitment is set varies among percid populations. While recruitment success of many percid populations appears to be dependent on processes affecting early life stages (e.g., Anderson et al. 1998; Kjellman et al. 2003), other populations respond to later life processes (e.g., Hansen et al. 1998; Ivan et al. 2011). The age when recruitment is set may even vary among species within a single system. Ivan et al. (2011) examined when recruitment was set for yellow perch and walleye in Saginaw Bay, Lake Huron, USA (1970–2008) and found that the catch of age-0 walleye in the fall was strongly correlated to the catch of age-1 walleye in the following year and the catch of age-2 walleye 2 years later; indicating that walleye recruitment was set by fall of age-0. However, Ivan et al. (2011) also demonstrated that in Saginaw Bay catch of age-0 yellow perch in the fall was not related to catch of yellow perch 2 years later. Rather, catch of age-1 yellow perch in the fall was strongly related to catch of age-2 yellow perch the following year, indicating that yellow perch recruitment was set by fall of age-1.

Across most fish species, survival during early life is strongly related to size and hence feeding opportunities and growth rates. While some studies have pointed out mechanisms (e.g., Marshall et al. 2010) and specific cases (e.g., Litvak and Leggett 1992; Dibattista et al. 2007) whereby the *bigger-is-better* axiom may not hold, as a general rule larger larval fish are less susceptible to both starvation and predation mortality (Miller et al. 1988). Obviously, high feeding rates increase growth rates and decrease the likelihood of starvation mortality. Moreover, as fish increase in size (a) swimming speeds and gape sizes tend to increase, allowing for consumption of additional larger and more active prey and (b) mass-specific metabolic rates decrease, allowing a greater proportion of consumed energy to be directed towards growth, while minimizing energy depletion and starvation risk. For example, as larval yellow perch increase in size their swimming speeds increase (Houde 1969) and they expand their diets (e.g., Fulford et al. 2006). Similarly, Graham and Sprules (1992) found that the size of zooplankton prey consumed by larval walleye increased with size, and Johnston and Mathias (1994) demonstrated that prey consumption increased exponentially with larval walleye length. In addition, metabolic energy utilization decreases as size of larval walleye increases (Madon and Culver 1993), suggesting that larger larval walleye can tolerate a longer period without food compared to smaller larvae. Jonas and Wahl (1998) estimated that larval walleye reach a "point-of-no-return" after a 6 day starvation period and that after 5–6 days of starvation larval walleye were less successful at capturing prey and more vulnerable to predators, and Letcher et al. (1996) found that larger larval yellow perch can survive longer without feeding than smaller larval perch. Thus, increased growth of young fish tends to facilitate further growth and minimizes the probability of starvation. Further, as swimming speeds increase, larger larval fish are able to both actively avoid predators and overcome predators' gape limitations. Thereby, risk of predation mortality also generally decreases as larval fish increase in size. Consistent with these expectations, Brandt et al. (1987) found that alewife (*Alosa pseudoharengus*) predators in Lake Ontario selectively fed on smaller larval yellow perch, and Brooking et al. (1998) revealed that larger walleye larvae were better able to escape alewife predation. Collectively, these studies demonstrate that generalized cross-taxa positive linkages among larval fish size, feeding success, swimming speeds, growth, and survival hold for larval percids.

The effects of growth and size on survival rates may extend beyond very early life stages and also influence survival during later young-of-year and yearling life stages. Similar to larval stages, as later stage juvenile fish increase in size they may be better able to actively avoid large piscivores, and their larger body size coupled with development of larger, more rigid spines may deter potential gape-limited predators. Post and Evans (1989) experimentally demonstrated that the risk of predation for young-of-year yellow perch may decrease with total length. Similarly, Roswell et al. (2014) found that during July–November piscivorous walleye in Saginaw Bay, Lake Huron selectively consumed small young-of-year yellow perch. However, predation risk may also indirectly influence growth and activity, and therefore mortality, rates of juvenile percids. In yellow perch, Rennie et al. (2010) observed positive relationships between an index of predation and metrics of growth such as maximum size, specific growth rates, and growth efficiencies. However, these effects traded off with negative relationships between predation and specific consumption and activity rates, suggesting predation-induced shifts in behavior can modify percid life history patterns (Sect. 15.3).

In addition to predation mortality during the growing season, size may also mediate overwinter mortality of juvenile fishes. Winter represents a time of resource scarcity, and young percids, both *Perca* (Griffiths and Kirkwood 1995; Huss et al. 2008; Heerman et al. 2009; Pothoven et al. 2014; Roswell et al. 2014) and *Sander* (Post and Evans 1989; Rennert et al. 2005; Pothoven et al. 2014) species, may lose significant energy stores over winter, which could leave them vulnerable to starvation mortality. Such mortality may be size dependent; larger fish are less likely to experience starvation during winter, as their larger size allows for greater energy storage and lower mass-specific metabolic rates. Several authors have indicated that size-dependent overwinter mortality could be important for young percids (e.g., Post and Evans 1989; Fitzgerald et al. 2006; Heerman et al. 2009) and poor growth before winter or very long, severe winter conditions could lead to poor population-level recruitment (e.g., Post and Evans 1989). In contrast, other studies have revealed that size-dependent overwinter mortality may be minimal for young percids (Johnson and Evans 1991; Pratt and Fox 2002). Thus, in many cases young percids may obtain sufficient size and energy content prior to winter or may be able to readily feed during winter, such that starvation mortality is minimal. However, even if young fish do not directly starve to death during winter, the threat of starvation may lead to risky foraging behavior, increasing predation mortality of young fish and compromised physiological condition, potentially enhancing prevalence of disease-related mortality (Garvey et al. 2004; Sect. 4.4.5). Shuter and Post (1990) suggested that overwinter mortality may strongly mediate the northerly latitudinal distributions of yellow perch in Canada; in more northerly systems the growing season may be too short for young perch to reach a sufficient size to survive winter. Further, they speculated that with climate warming this interaction between growing season and overwinter survival could be relaxed, allowing yellow perch populations to persist in more northern lakes (Shuter and Post 1990). In short, the importance of overwinter mortality is likely system-specific and has the strongest influence in colder, less productive systems. In addition, overwinter mortality is more likely to affect perch rather than sander species; because, despite their physiological similarities, perch tend to hatch later and obtain a much smaller size by winter relative to sander (Pothoven et al. 2014; Sects. 2.1 and 2.2).

2.4.2 Influence of Environmental Variables on Percid Recruitment

Annual percid year-class success appears to be influenced by both abiotic and biotic factors, with recruitment of some populations responding most strongly to abiotic conditions (e.g., temperatures, water levels, structural habitats), some populations controlled by biotic interactions (e.g., prey availability, predation pressure, competition, intra-population compensatory density-dependent effects) and some influenced by the interaction of abiotic and biotic processes. The breadth of factors controlling percid recruitment partially reflects the diversity of systems occupied by percids and the premise that abiotic factors may have strong effects on year-strength in large systems, while biotic interactions dominate in small systems. However, various studies of percid early life survival have also demonstrated the strong influence of biotic interactions in large systems (e.g., Brandt et al. 1987; Ljunggren et al. 2010; Redman et al. 2011) and the effect of abiotic factors in small systems (e.g., Kallemeyn 1987; Pope et al. 1996; Quist et al. 2003b).

Abiotic influences: Myers (1998) suggested that annual recruitment success of fish populations at the edge of their latitudinal ranges are most likely to be affected by inter-annual variation in weather and climatic conditions. For example, recruitment rates of some percid populations towards the northern extreme of their range are positively associated with growing season (spring, summer, and autumn) water temperatures (e.g., Tolonen et al. 2003). However, such positive associations between growing season temperatures and year-class strength are also evident for percid populations located far south of such northern distributional extremes (e.g., Koonce et al. 1977; Kallemeyn 1987; Sarvala and Helminen 1996; Nyberg et al. 2001; Pitlo 2002; Wysujack et al. 2002; Kjellman et al. 2003; Quist et al. 2003b, 2004; Paxton et al. 2004; Redman et al. 2011). While most studies demonstrating a positive association between growing season temperatures and year-class strength are not able to robustly evaluate the mechanisms underlying such relationships, growing season temperatures likely act both directly and indirectly. For example, water temperatures may directly (a) influence the production of eggs by adults, (b) development rates of eggs once deposited onto substrates, and (c) affect growth rates of both larvae and later stage young percids. Further, water temperatures may act indirectly by (a) affecting production of prey and (b) perhaps more importantly, by influencing the phenology of young percids and their potential prey, predators, and competitors (see below). Rapid warming in the spring may in particular favor strong year-classes as such conditions may (a) allow for rapid egg development and short duration of the egg stage when storm events could dislodge eggs from suitable substrates and (b) maximize the likelihood that percid larvae will emerge just prior to spring peaks in zooplankton prey (Busch et al. 1975; Pitlo 2002). In contrast, associations between year-class strength and late summer and fall temperatures (e.g., Henderson and Nepszy 1988; Sarvala and Helminen 1996) likely reflect the positive effect of water temperatures on young-of-year growth, allowing young fish to grow large enough to avoid size-selective predators and minimize size-based overwinter mortality.

Climatic influences on percid recruitment extend beyond the effects of water temperature. Beard et al. (2003) found a strong year effect on walleye recruitment success across lakes in Wisconsin, USA, likely reflecting some regional climatic process. Similarly, Schupp (2002) demonstrated that walleye recruitment in Minnesota, USA, lakes was negatively influenced by the 1991 Mt. Pinatubo eruption in the Philippines, which may have influenced both water temperatures and light penetration, thereby affecting system productivities. Pope et al. (1996) found that in a South Dakota, USA, lake, yellow perch year-class strength was greatest during years with high precipitation and fairly stable conditions (low wind and minimal daily temperature deviation). Precipitation patterns structure stream discharge and flow dynamics and may influence percids spawning in fluvial systems. Moreover, discharge of riverine waters may strongly influence environmental conditions in receiving basins. High seasonal precipitation may lead to large sediment and nutrient-rich discharge plumes. Such plumes may support phytoplankton production and lead to high densities of zooplankton prey (thereby facilitating growth of larval percids), while simultaneously providing a poor visibility environment, where young percids may be less conspicuous to visual predators (Reichert et al. 2010).

Water currents may rapidly transport larval fish into or out of favorable nursery habitats, and thereby affect growth and survival. As such, annual differences in regional water circulation and local water currents have the potential to shape recruitment patterns. The role of water currents in structuring recruitment of marine fishes was initially suggested by Hjort (1914) and has subsequently been evaluated for a plethora of marine fish species. In contrast, the role of water currents in influencing recruitment success of freshwater fishes has received relatively limited attention. However, several relatively recent studies of walleye and yellow perch in the Laurentian Great Lakes have demonstrated the potential for water currents to affect percid early life survival in large freshwater systems (e.g., Dettmers et al. 2005; Beletsky et al. 2007; Zhao et al. 2009). Both eggs and larval percids are susceptible to transport by water currents (e.g., Roseman et al. 2001, 2005; Höök et al. 2006). Intense storm events and resulting strong water currents have the potential to push eggs off of suitable incubation reefs, leading to very low recruitment success during some years (Roseman et al. 2001; Zhao et al. 2009). Similarly, transport of larval percids to suitable nearshore or offshore nursery areas has been linked to annual recruitment success of both yellow perch and walleye (Dettmers et al. 2005; Beletsky et al. 2007; Zhao et al. 2009).

Biotic influences: In addition to abiotic effects, young percid survival and recruitment success may be influenced by a variety of biotic factors. However, it is difficult to draw generalities regarding biotic controls on percid recruitment, as biotic controls (a) have been identified in both small (e.g., Nielsen 1980; Quist et al. 2004) and large (e.g., Ljunggren et al. 2010; Forsythe et al. 2012) systems, (b) have been proposed to affect both larval (Brandt et al. 1987; Kjellman et al. 2003) and later juvenile life-stages (Nielsen 1980; Ritchie and Colby 1988; Hartman and Margraf 1993), and (c) may include a variety of biotic interactions, such as, predation (Hartman and Margraf 1993; Quist et al. 2003b), prey availability (Ritchie and Colby 1988; Ljunggren et al. 2010), disease (Paxton et al. 2004), and inter-specific competition (Quist et al. 2004; Forsythe et al. 2012). Further, biotic controls on young percid survival may be mediated by abiotic processes (e.g., turbidity affecting vulnerability to predation or foraging success; water currents influencing overlap with prey) and a variety of biotic processes may interactively affect early life survival.

Intra-population density-dependence is a common line of evidence for biotic controls on percid recruitment. Several researchers have developed Ricker-type stock-recruitment models to describe annual variation of percid year-class strength (e.g., Craig and Kipling 1983; Paxton et al. 2004; Forsythe et al. 2012), and such models are based on the assumption of compensatory density-dependent effects (i.e., reduced per capita recruitment success with increasing spawning stock size). While such models do not specify the mechanisms of compensatory density-dependent controls, percids may be limited by spawning habitat, leading to reduced per capita reproductive success at high spawner densities, or high densities of young percids may compete for limited prey or lead to concentrated cannibalism. Similar to most ecological systems, studies to identify density-dependent effects for percids have focused on compensatory effects (Sect. 19.4.3). Nonetheless, while less prominent, depensatory effects may also influence recruitment success. Forney (1971) suggested that survival of young yellow perch in Oneida Lake, New York, USA was influenced by depensatory density-dependence. Specifically, Forney (1971) found that during the 1960s large numbers of young yellow perch were able to swamp predation pressure by walleye, thereby leading to higher per capita survival for young yellow perch. However, when subsequently examining this pattern over a longer time series (40 years), Irwin et al. (2009) found that depensatory controls were replaced by compensatory controls on yellow perch survival and growth.

Young percids are selective foragers, initially selecting small-bodied zooplankton prey and then transitioning to feed on larger zooplankton, benthic invertebrates and fish (Sect. 2.1). Thus, their early life growth and survival may be dependent on availability of suitable prey for specific ontogenetic stanza. If larval hatching and exogenous feeding does not overlap spatially and temporally with available small prey, their survival may be compromised. Such overlap may be strongly mediated by physical processes, e.g., spring warming affecting temporal overlap and water currents affecting spatial overlap. Moreover, consumption of zooplankton or other preferred prey by competitors may lead to dramatic reductions in small-bodied prey and lead to very low percid recruitment success. For example, Ljunggren et al. (2010) found that in coastal areas of the Baltic Sea, intense planktivory by sprat (*Sprattus sprattus*) led to severe reductions in zooplankton densities and almost complete recruitment failure of Eurasian perch. Similarly, Quist et al. (2004) suggested that in cool years walleye hatch later and grow more slowly in Kansas, USA, reservoirs and therefore compete directly with abundant gizzard shad (*Dorosoma cepedianum*) larvae for zooplankton prey. In contrast, early hatching and rapid growth in warm years allowed walleye to gain a size advantage and prey directly on gizzard shad larvae, rather than compete with them (Quist et al. 2004).

Both larval and juvenile percid survival may be strongly controlled by predation by piscivores, and studies have demonstrated the possible influence of predators targeting young percids during both larval and later life stages. For example, Brandt et al. (1987) highlighted the potential for alewife to consume huge numbers of larval yellow perch in Lake Ontario, North America, and Quist et al. (2003b) pointed to the importance of predation by white crappie on walleye in reservoirs. In contrast, predation by walleye on post-larval young-of-year yellow perch may strongly limit year-class strength in Lake Erie and Saginaw Bay, Lake Huron (Hartman and Margraf 1993; Fielder and Thomas 2006; Roswell 2011). Given the potential high abundance and local concentration of larval percids, we suggest that intense, short-term predation on larval percids has the potential to strongly compromise cohort success. However, such an effect may be under-appreciated because larval percids are difficult to document in stomachs of potential piscivores due to rapid digestion rates.

The effect of predation on recruitment success may be mediated by other biotic interactions. As described above, predation mortality may be strongly size-selective (e.g., Post and Evans 1989; Roswell et al. 2014). Therefore, high prey availability and conditions that favor rapid growth by young percids should minimize predation mortality. Moreover, potential predators on young percids do not rely solely on percids as prey. For example, Ritchie and Colby (1988) found that alternating years of high production of burrowing mayflies, *Hexagenia*, corresponded to walleye recruitment success, leading to the hypothesis that *Hexagenia* served as a preferred prey for potential walleye predators. In addition, Fitzgerald et al. (2006) found that overwinter survival of young yellow perch was inversely related to abundances of alternative prey in gizzard shad and white perch (*Morone americana*). Thus, the presence of alternative prey may serve as important buffers for predation and thereby enhance young percid survival.

In many cases, researchers have documented negative associations between percid year-class strength and abundance of potential predators without demonstrating the exact mechanism underlying such negative associations. For example, Redman et al. (2011) and Forsythe et al. (2012) demonstrated negative associations between alewife densities and yellow perch year-class strength in southern Lake Michigan, USA. Fielder et al. (2007) demonstrated a strong negative effect of alewives on walleye recruitment in Saginaw Bay, Lake Huron; after the alewife population crashed, walleye recruitment increased dramatically. Mercado-Silva et al. (2007) suggested a negative association by rainbow smelt (*Osmerus mordax*) on young walleye in lakes in Wisconsin, USA, may limit walleye recruitment. While alewife and rainbow smelt are known to consume larval percids, they also consume zooplankton and small benthic invertebrates. Thus, such negative associations may develop through both predatory and competitive effects.

2.5 Adult Mortality

As compared to larval and juvenile fish, adult fish experience much lower mortality rates. Similar to many other moderate- to long-lived fish species, percids display highly variable mortality rates across populations and over time within individual systems. This variability is likely related to the broad set of natural and anthropogenic-induced phenomena that can contribute to mortality of adult percids.

Mortality rates of fishes are often divided into mortality from fisheries harvest and natural mortality. The latter can encapsulate a variety of phenomena, including mortality related to starvation, predation, disease, parasitism, severe temperatures, low oxygen, and responses to biotic and abiotic chemicals. A number of studies suggest that adult mortality rates are size-dependent; e.g., through a cross-taxa analysis of marine fish, Pauly (1980) demonstrated that natural mortality declined with asymptotic length. Similarly, Lorenzen (1996) analyzed annual instantaneous natural mortality rates (M) across a broad set of marine, freshwater, and cultured fish species and demonstrated that natural mortality is size-dependent: ln(M) scales negatively and linearly with ln(mean mass), such that $\ln(M) = \ln(a) + b \times \ln(\text{mass})$ where b represents a dimensionless scaling parameter. After accounting for mass, Lorenzen's (1996) analysis suggested that mortality rates did not differ significantly among natural environments (lakes, rivers, oceans); however, natural mortality rates were significantly higher in natural systems than in cultured ponds, cages, and tanks. Moreover, Lorenzen (1996) found that the mass-mortality scaling parameter was significantly more negative in culture systems than in natural systems. Lorenzen (1996) attributed this latter finding to the presence of predation in natural systems and argued that non-predation natural mortality must scale more negatively with body size than predation mortality. Specific to percids, Lorenzen (1996) (a) estimated a mass-mortality scaling parameter (b) of roughly-0.4 (which is more negative than the −0.29 value estimated across species and natural systems) and (b) estimated a significantly higher mortality at unit mass compared to natural populations of other similarly-sized species (see Fig. 4 in Lorenzen 1996).

While a variety of phenomena may contribute to individual mortality events, various studies have attempted to describe broad, across-system variation of intra-specific mortality rates in fishes. Blanck and Lamouroux (2007) examined such broad patterns in life-history traits (including longevity and mortality) of European freshwater fishes, and found that latitude had a stronger effect on life-history traits than habitat type: specifically, higher latitude populations tended to live longer (i.e., lower adult mortality). While Blanck and Lamouroux (2007) included pikeperch (n = 38 populations) in their analysis, they did not appear to find a significant effect of latitude for this percid species. In contrast, Heibo et al. (2005) explored how life-history traits of Eurasian perch varied across latitude and found the expected patterns: mortality decreased with latitude and life-span increased with latitude.

A number of specific natural sources of mortality have been identified for adult percids, despite their relatively low mortality rates. They are susceptible to a variety of diseases and parasites (e.g., Szalai and Dick 1991; Kane-Sutton et al. 2010; Jensen et al. 2011), and disease outbreaks have led to large die-offs of certain populations, e.g., a large die-off of yellow perch in Lake Erie, North America, was associated with viral hemorrhagic septicemia virus (Kane-Sutton et al. 2010). In addition, percids are susceptible to a large number of both piscine and terrestrial predators. Szalai and Dick (1991) estimated a natural mortality rate for yellow perch in Dauphin Lake, Manitoba, Canada as 61 %, with predation by northern pike (*Esox lucius*) accounting for 19 % and 45 % for females and males, respectively. Over the past three to four decades, double-crested cormorants (*Phalacrocorax auritus*) have expanded dramatically throughout North America and high local abundances of these piscivorous birds have been blamed for percid population declines in many systems (e.g., Burnett et al. 2002; VanDeValk et al. 2002; Rudstam et al. 2004; Fielder 2010).

Percids can persist in a fairly broad range of temperatures (Sect. 2.1). Lethal temperatures are dependent on acclimation temperature, but according to Hokanson (1977) the ultimate upper incipient lethal temperature for percids is between 28 and 35 °C. Similarly, percids can tolerate fairly low dissolved oxygen levels. For example, Roberts et al. (2011, 2012) demonstrated that yellow perch can survive and even grow at dissolved oxygen concentrations as low as 2 mg/l. And, percids can tolerate a relatively broad range of pH. Walleye may tolerate a range of 6.0–9.0, while yellow perch may persist in water with pH as low as 4.2 (Schneider et al. 2002). Moreover, Heibo and Vøllestad (2002) did not find a difference in mortality rates of European perch in highly acidic (pH $\leq$5.0) versus non-acidic (pH $\geq$6.0) lakes.

While non-fisheries-related mortality is often grouped into a *natural mortality* category, several non-fisheries sources of mortality may be heavily mediated by anthropogenic activities (i.e., they may not truly be *natural* phenomena). For example, humans have contributed to the spread of several pathogens, including ones that target percids and various intentional and unintentional introduced species prey upon and/or parasitize percids. Further, anthropogenic chemical contamination, thermal pollution, and both point-source and non-point-source loading of nutrients can cause oxygen concentration and thermal conditions leading to lethality.

Mortality through fisheries harvest can greatly exceed natural mortality. In fact, collapses of some percid populations have been partially attributed to fisheries harvest (e.g., Marsden and Robillard 2004). In many systems, percids are simultaneously targeted by both commercial and recreational fisheries. While commercial fisheries have historically been blamed for crashes of many fish stocks, declines of some percid populations are undoubtedly related to intense recreational harvest

(Post et al. 2002). Moreover, while direct fisheries harvest may be a major component of total mortality, indirect effects of fisheries practices may also be responsible for high mortality. For instance, by-catch of percids by fisheries targeting other species may lead to high percid mortality (e.g., MacMillan and Roth 2012). Additionally, both recreational and commercial fishing practices which capture percids and then release them back into the environment may nonetheless result in high indirect mortality, as capture and handling may lead to various physiological costs and compromise the short-term and long-term survival of adult percids (e.g., Hyvarinen et al. 2008).

2.6 Conclusions

As we have reviewed in this chapter, percid species represent a suite of highly plastic species capable of thriving in systems ranging from small streams to large lakes and bays. As such, the mechanisms controlling the important vital rates discussed in this chapter (i.e., growth, reproduction, recruitment, and mortality) likely vary both spatially and temporally across different species, populations, and environments. In many systems, percid populations have strong influences on local ecosystem dynamics through predatory and competitive interactions with other species while also supporting valuable recreational and commercial fisheries harvest. Because of their economic and ecological importance, a large body of research, partially reviewed here, exists on many aspects of percid biology and ecology. However, important gaps remain in many areas of percid biology, including, for example, the roles of biotic and abiotic factors in influencing recruitment of populations, the importance of harvest rates on the life history ecology of percid species, and the respective contribution of different sources of mortality to adult longevity and population abundance. Some of these questions remain unanswered partially because percids are such a diverse and widespread group and local environmental conditions, adaptation, and plasticity all likely play a role in population- and individual-level variation. Thus, future research should focus on the level of and causes of observed variation in population traits, and may lead to a clearer understanding of percid biology and its responses to local environmental phenomena. Such information should aid in developing management practices and promoting the persistence of percid populations throughout the globe.

References

Ali MA, Ryder RA, Anctil M (1977) Photoreceptors and visual pigments as related to behavioral responses and preferred habitats of perches (*Perca* spp.) and pikeperches (*Stizostedion* spp.). J Fish Res Board Can 34:1475–1480

Anderson MR, Fisher SJ, Willis DW (1998) Relationship between larval and juvenile yellow perch abundance in eastern South Dakota glacial lakes. N Am J Fish Manag 18:989–991

Andree SR, Feiner ZS, Bledsoe JW, Cragun AM, Höök TO (2015) Ontogenetic variability of maternal effects in an iteroparous fish. Ecol Freshw Fish. doi:10.1111/eff.12153
Arend KK, Beletsky D, DePinto JV, Ludsin SA, Roberts JJ, Rucinski DK, Scavia D, Schwab DJ, Höök TO (2011) Seasonal and interannual effects of hypoxia on fish habitat quality in central Lake Erie. Freshw Biol 56:366–383
Baccante DA, Colby PJ (1996) Harvest, density, and reproductive characteristics of North American walleye populations. Ann Zool Fenn 33:601–615
Baccante DA, Reid DM (1988) Fecundity changes in two exploited walleye populations. N Am J Fish Manag 8:199–209
Barot S, Heino M, O'Brien L, Dieckmann U (2004) Long-term trend in the maturation reaction norm of two cod stocks. Ecol Appl 14:1257–1271
Beard TD Jr, Hansen MJ, Carpenter SR (2003) Development of a regional stock-recruitment model for understanding factors affecting walleye recruitment in northern Wisconsin lakes. Trans Am Fish Soc 132:382–391
Beletsky D, Mason DM, Schwab DJ, Rutherford ES, Janssen J, Clapp DF, Dettmers JM (2007) Biophysical model of larval yellow perch advection and settlement in Lake Michigan. J Great Lakes Res 33:842–866
Bergman E (1987) Temperature-dependent differences in foraging ability of two percids, *Perca fluviatilis* and *Gymnocephalus cernuus*. Environ Biol Fish 19:45–53
Blanck A, Lamouroux N (2007) Large-scale intraspecific variation in life-history traits of European freshwater fish. J Biogeogr 34:862–875
Brandt SB, Mason DM, MacNeill DB, Coates T, Gannon JE (1987) Predation by alewives on larvae of yellow perch in Lake Ontario. Trans Am Fish Soc 116:641–645
Brandt SB, Costantini M, Kolesar S, Ludsin SA, Mason DM, Rae CM, Zhang H (2011) Does hypoxia reduce habitat quality for Lake Erie walleye (*Sander vitreus*)? A bioenergetics perspective. Can J Fish Aquat Sci 68:857–879
Bronte CR, Selgeby JH, Swedberg DV (1993) Dynamics of a yellow perch population in western Lake Superior. N Am J Fish Manag 13:511–523
Brooking TE, Rudstam LG, Olson MH, VanDeValk AJ (1998) Size-dependent alewife predation on larval walleyes in laboratory experiments. N Am J Fish Manag 18:960–965
Bulkowski L, Meade JW (1983) Changes in phototaxis during early development of walleye. Trans Am Fish Soc 112:445–447
Burnett JAD, Ringler NH, Lantry BF, Johnson JH (2002) Double-crested cormorant predation on yellow perch in the eastern basin of Lake Ontario. J Great Lakes Res 28:202–211
Busch W-DN, Scholl RL, Hartman WL (1975) Environmental factors affecting the strength of walleye (*Stizostedion vitreum vitreum*) year-classes in western Lake Erie, 1960–70. J Fish Res Board Can 32:1733–1743
Byström P, Garcia-Berthou E (1999) Density dependent growth and size-specific competitive interactions in young fish. Oikos 86:217–232
Cai Y, Summerfelt RC (1992) Effects of temperature and size on oxygen consumption and ammonia excretion by walleye. Aquaculture 104:127–138
Campbell EA (1998) Predation by small walleyes on yellow perch: effects of prey size distribution. Trans Am Fish Soc 127:588–597
Carlson AR, Blocher J, Herman LJ (1980) Growth and survival of channel catfish and yellow perch exposed to lowered constant and diurnally fluctuating dissolved oxygen concentrations. Prog Fish Cult 42:73–78
Casselman SJ, Schulte-Hostedde AI, Montgomerie R (2006) Sperm quality influences male fertilization success in walleye (*Sander vitreus*). Can J Fish Aquat Sci 63:2119–2125
Čech M, Peterka J, Říha M, Jůza T, Kubečka J (2009) Distribution of egg strands of perch (*Perca fluviatilis* L.) with respect to depth and spawning substrate. Hydrobiologia 630:105–114
Chezik KA, Lester NP, Venturelli PA (2014) Fish growth and degree-days I: selecting a base temperature for a within-population study. Can J Fish Aquat Sci 71:47–55
Ciereszko RE, Dabrowski K, Ciereszko A, Ebeling J, Ottobre JS (1997) Effects of temperature and photoperiod on reproduction of female yellow perch *Perca flavescens*: plasma concentrations

of steroid hormones, spontaneous and induced ovulation, and quality of eggs. J World Aquacult Soc 28:344–356
Clady MD (1976) Influence of temperature and wind on the survival of early states of yellow perch, *Perca flavescens*. J Fish Res Board Can 33:1887–1893
Colby PJ, Baccante DA (1996) Dynamics of an experimentally exploited walleye population: sustainable yield estimate. Ann Zool Fenn 33:589–599
Colby PJ, Nepszy SJ (1981) Variation among stocks of walleye (*Stizostedion vitreum vitreum*): management implications. Can J Fish Aquat Res 38:1814–1831
Collingsworth PD, Marschall EA (2011a) Spatial and temporal patterns in maternal energetic traits of yellow perch (*Perca flavescens*) in Lake Erie. Freshw Biol 56:2500–2513
Collingsworth PD, Marschall EA (2011b) Identifying relationships between catches of spawning condition yellow perch and environmental variables in the western basin of Lake Erie. Trans Am Fish Soc 140:31–36
Craig JF, Kipling C (1983) Reproduction effort versus the environment; case histories of Windermere perch, *Perca fluviatilis* L., and pike, *Esox lucius* L. J Fish Biol 22:713–727
DeBruyne RL, Coleman JTH, Jackson JR, Rudstam LG, VanDeValk AJ (2013) Analysis of prey selection by double-crested cormorants: a 15-year diet study in Oneida Lake, New York. Trans Am Fish Soc 142:430–446
Dettmers JM, Janssen J, Pientka B, Fulford RS, Jude DJ (2005) Evidence across multiple scales for offshore transport of yellow perch (*Perca flavescens*) larvae in Lake Michigan. Can J Fish Aquat Sci 62:2683–2693
Diana JS, Salz R (1990) Energy storage, growth, and maturation of yellow perch from different locations in Saginaw Bay, Michigan. Trans Am Fish Soc 119:976–984
Dibattista JD, Feldheim KA, Gruber SH, Hendry AP (2007) When bigger is not better: selection against large size, high condition and fast growth in juvenile lemon sharks. J Evol Biol 20:201–212
Diehl S (1993) Effects of habitat structure on resource availability, diet and growth of benthivorous perch, *Perca fluviatilis*. Oikos 61:403–414
Fielder DG (2010) Response of yellow perch in Les Cheneaux Islands, Lake Huron to declining numbers of double-crested cormorants stemming from control activities. J Great Lakes Res 36:207–214
Fielder DG, Thomas MV (2006) Fish population dynamics of Saginaw Bay, Lake Huron 1998–2004. Michigan department of natural resources, fisheries research report 2083, Ann Arbor
Fielder DG, Schaeffer JS, Thomas MV (2007) Environmental and ecological conditions surrounding the production of large year classes of walleye (*Sander vitreus*) in Saginaw Bay, Lake Huron. J Great Lakes Res 33(sp1):118–132
Fitzgerald DG, Forney JL, Rudstam LG, Irwin BJ, VanDeValk AJ (2006) Gizzard shad put a freeze on winter mortality of age-0 yellow perch but not white perch. Ecol Appl 16:1487–1501
Forney JL (1971) Development of dominant year classes in a yellow perch population. Trans Am Fish Soc 100:739–749
Forsythe PS, Doll JC, Lauer TE (2012) Abiotic and biotic correlates of yellow perch recruitment to age-2 in southern Lake Michigan, 1984–2007. Fish Manag Ecol 19:389–399
Fox MG, Flowers DD (1990) Effect of fish density on growth, survival, and food consumption by juvenile walleyes in rearing ponds. Trans Am Fish Soc 119:112–121
Frankiewicz P, Dabrowski K, Martyniak A, Zalewski M (1999) Cannibalism as a regulatory force of pike perch, *Stizostedion lucioperca* (L.), population dynamics in the lowland Sulejou reservoir (Central Poland). Hydrobiologia 408(409):47–55
Frisk M, Skov PV, Steffensen JF (2012) Thermal optimum for pikeperch (*Sander lucioperca*) and the use of ventilation frequency as a predictor of metabolic rate. Aquaculture 324–325:151–157
Fulford RS, Rice JA, Miller TJ, Binkowski FP, Dettmers JM, Belonger B (2006) Foraging selectivity by larval yellow perch (*Perca flavescens*): implications for understanding recruitment in small and large lakes. Can J Fish Aquat Sci 63:28–42

Galarowicz TL, Wahl DH (2003) Differences in growth, consumption, and metabolism among walleyes from different latitudes. Trans Am Fish Soc 132:425–437
Galarowicz TL, Wahl DH (2005) Foraging by a young-of-the-year piscivore: the role of predator size, prey type, and density. Can J Fish Aquat Sci 62:2330–2342
Galarowicz TL, Adams JA, Wahl DH (2006) The influence of prey availability on ontogenetic diet shifts of a juvenile piscivore. Can J Fish Aquat Sci 63:1722–1733
Gardeur J-N, Mathis N, Kobilinksy A, Brun-Bellut J (2007) Simultaneous effects of nutritional and environmental factors on growth and flesh quality of *Perca fluviatilis* using a fractional factorial design study. Aquaculture 273:50–63
Garvey JE, Ostrand KG, Wahl DH (2004) Energetics, predation, and ration affect size-dependent growth of fish during winter. Ecology 85:2860–2871
Graeb BDS, Dettmers JM, Wahl DH, Cáceres CE (2004) Fish size and prey availability affect growth, survival, prey selection, and foraging behavior of larval yellow perch. Trans Am Fish Soc 133:504–514
Graeb BDS, Galarowicz T, Wahl DH, Dettmers JM, Simpson MJ (2005) Foraging behavior, morphology, and life history variation determine the ontogeny of piscivory in two closely related predators. Can J Fish Aquat Sci 62:2010–2020
Graham DM, Sprules WG (1992) Size and species selection of zooplankton by larval and juvenile walleye (*Stizostedion vitreum vitreum*) in Oneida Lake, New York. Can J Zool 70:2059–2067
Griffiths D, Kirkwood RC (1995) Seasonal variation in growth, mortality and fat stores of roach and perch in Lough Neagh, Northern Ireland. J Fish Biol 47:537–554
Guma'a SA (1978) The effects of temperature on the development and mortality of eggs of perch, *Perca fluviatilis*. Freshw Biol 8:221–227
Guzzo MM, Haffner GD, Legler ND, Rush SA, Fisk AT (2013) Fifty years later: trophic ecology and niche overlap of a native and non-indigenous fish species in the western basin of Lake Erie. Biol Invasions 15:1695–1711
Hansen MJ, Bozek MA, Newby JR, Newman SP, Staggs MD (1998) Factors affecting recruitment of walleyes in Escanaba Lake, Wisconsin, 1958–1996. N Am J Fish Manag 18:764–774
Hartman KJ, Margraf FJ (1992) Effects of prey and predator abundances on prey consumption and growth of walleyes in western Lake Erie. Trans Am Fish Soc 121:245–260
Hartman KJ, Margraf FJ (1993) Evidence of predatory control of yellow perch (*Perca flavescens*) recruitment in Lake Erie, U.S.A. J Fish Biol 43:109–119
Hayes DB, Taylor WW (1990) Reproductive strategy in yellow perch (*Perca flavescens*): effects of diet ontogeny, mortality, and survival costs. Can J Fish Aquat Sci 47:921–927
Hayward RS, Margraf FJ (1987) Eutrophication effects on prey size and food available to yellow perch in Lake Erie. Trans Am Fish Soc 116:210–223
Headly HC, Lauer TE (2008) Density-dependent growth of yellow perch in southern Lake Michigan, 1984–2004. N Am J Fish Manag 28:57–69
Heath DD, Roff DA (1996) The role of trophic bottlenecks in stunting: a field test of an allocation model of growth and reproduction in yellow perch, *Perca flavescens*. Environ Biol Fish 45:53–63
Heerman L, Eriksson L-O, Magnhagen C, Borcherding J (2009) Size-dependent energy storage and winter mortality of perch. Ecol Freshw Fish 18:560–571
Heibo E, Vøllestad LA (2002) Life-history variation in perch (*Perca fluviatilis*) in five neighbouring Norwegian lakes. Ecol Freshw Fish 11:270–280
Heibo E, Magnhagen C, Vøllestad LA (2005) Latitudinal variation in life-history traits in Eurasian perch. Ecology 86:3377–3386
Heidinger RC, Brooks RC, Leitner D, Soderstrom I (1997) Prediction of walleye egg and embryo survival at two stages of development. Prog Fish Cult 59:64–67
Henderson BA, Morgan GE (2002) Maturation of walleye by age, size, and surplus energy. J Fish Biol 61:999–1011
Henderson BA, Nepszy SJ (1988) Recruitment of yellow perch (*Perca flavescens*) affected by stock size and water temperature in Lakes Erie and St. Clair, 1965–1985. J Great Lakes Res 14:205–215

Henderson BA, Trivedi T, Collins N (2000) Annual cycle of energy allocation to growth and reproduction of yellow perch. J Fish Biol 57:122–133
Henderson BA, Collins N, Morgan GE, Vaillancourt A (2003) Sexual size dimorphism of walleye (*Stizostedion vitreum vitreum*). Can J Fish Aquat Sci 60:1345–1352
Hermelink B, Wuertz S, Trubiroha A, Rennert B, Kloas W, Schulz C (2011) Influence of temperature on puberty and maturation of pikeperch, *Sander lucioperca*. Gen Comp Endocrinol 172:282–292
Heyer CJ, Miller TJ, Binkowski FP, Caldarone EM, Rice JA (2001) Maternal effects as a recruitment mechanism in Lake Michigan yellow perch (*Perca flavescens*). Can J Fish Aquat Sci 58:1477–1487
Hinshaw JM (1985) Effects of illumination and prey contrast on survival and growth of larval yellow perch *Perca flavescens*. Trans Am Fish Soc 114:540–545
Hjort J (1914) Fluctuations in the great fisheries of northern Europe reviewed in the light of biological research. Rapp p-v réun – Cons int explor mer 20:1–228
Hockett CT, Mundahl ND (1989) Effects of black spot disease on thermal tolerances and condition factors of three cyprinid fishes. J Freshw Ecol 5:67–72
Hokanson KEF (1977) Temperature requirements of some percids and adaptations to the seasonal temperature cycle. J Fish Res Board Can 34:1524–1550
Hokanson KEF, Koenst WM (1986) Revised estimate of growth requirements and lethal temperature limits of juvenile walleyes. Prog Fish Cult 48:90–94
Höök TO, McCormick MJ, Rutherford ES, Mason D, Carter GS (2006) Short-term water mass movements in Lake Michigan: implications for larval fish transport. J Great Lakes Res 32:728–737
Houde ED (1969) Sustained swimming ability of larvae of walleye (*Stizostidion vitreum vitreum*) and yellow perch (*Perca flavescens*). J Fish Res Board Can 26:1647–1659
Houde (1994) Differences between marine and freshwater fish larvae: implications for recruitment. ICES J Mar Sci 51:91–97
Houde ED, Forney JL (1970) Effects of water currents on distribution of walleye larvae in Oneida Lake, New York. J Fish Res Board Can 27:445–456
Houthuijzen RP, Backx JJGM, Buuse AD (1993) Exceptionally rapid growth and early maturation of perch in a freshwater lake recently converted from an estuary. J Fish Biol 43:320–324
Hoxmeier RJH, Wahl DH, Hooe ML, Pierce CL (2004) Growth and survival of larval walleyes in response to prey availability. Trans Am Fish Soc 133:45–54
Hoxmeier RJH, Wahl DH, Brooks RC, Heidinger RC (2006) Growth and survival of age-0 walleye (*Sander vitreus*): interactions among walleye size, prey availability, predation, and abiotic factors. Can J Fish Aquat Sci 63:2173–2182
Huff DD, Grad G, Williamson CE (2004) Environmental constraints on spawning depth of yellow perch: the roles of low temperature and high solar ultraviolet radiation. Trans Am Fish Soc 133:718–726
Huh HT, Calbert HE, Stuiber DA (1976) Effects of temperature and light on growth of yellow perch and walleye using formulated feed. Trans Am Fish Soc 105:254–258
Huss M, Strand BPA, Eriksson L, Persson L (2008) Influence of growth history on the accumulation of energy reserves and winter mortality in young fish. Can J Fish Aquat Sci 65:2149–2156
Huusko A, Vuorimies O, Sutela T (1996) Temperature- and light-mediated predation by perch on vendace larvae. J Fish Biol 49:441–457
Hyvarinen P, Leppaniemi V, Johansson K, Korhonen P, Suuronen P (2008) Stress and survival of small pike-perch *Sander lucioperca* (L.) after trawling and chilling. J Fish Biol 72:2677–2688
Irwin BJ, Rudstam LG, Jackson JR, VanDeValk AJ, Forney JL, Fitzgerald DG (2009) Depensatory mortality, density-dependent growth, and delayed compensation: disentangling the interplay of mortality, growth, and density during early life stages of yellow perch. Trans Am Fish Soc 138:99–110

Isermann DA, Willis DW, Lucchesi DO, Blackwell BG (2005) Seasonal harvest, exploitation, size selectivity, and catch preferences associated with winter yellow perch anglers on South Dakota lakes. N Am J Fish Manag 25:827–840

Ivan LN, Rutherford ES, Riesing C, Tyler JA (2010) Density, production, and survival of walleye (*Sander vitreus*) eggs in the Muskegon River, Michigan. J Great Lakes Res 36:328–328

Ivan LN, Höök TO, Thomas MV, Fielder DG (2011) Long-term and interannual dynamics of walleye and yellow perch in Saginaw Bay, Lake Huron. Trans Am Fish Soc 140:1078–1092

Jansen WA (1996) Plasticity in maturity and fecundity of yellow perch, *Perca flavescens* (Mitchill): comparisons of stunted and normal-growing populations. Ann Zool Fenn 33:403–415

Jensen BB, Holopainen R, Tapiovaara H, Ariel E (2011) Susceptibility of pike-perch *Sander lucioperca* to a panel of ranavirus isolates. Aquaculture 313:24–30

Johnson FH (1961) Walleye egg survival during incubation on several types of bottom in Lake Winnibigoshish, Minnesota, and connecting waters. Trans Am Fish Soc 90:312–322

Johnson TB, Evans DO (1991) Behaviour, energetics, and associated mortality of young-of-the-year white perch (*Morone americana*) and yellow perch (*Perca flavescens*) under simulated winter conditions. Can J Fish Aquat Sci 48:672–680

Johnston TA, Leggett WC (2002) Maternal and environmental gradients in the egg size of an iteroparous fish. Ecology 83:1777–1791

Johnston TA, Mathias JA (1994) The effects of temperature on feeding in zooplanktivorous walleye, *Stizostedion vitreum*, larvae. Environ Biol Fish 40:189–198

Johnston TA, Lysack W, Leggett WC (2012a) Abundance, growth, and life history characteristics of sympatric walleye (*Sander vitreus*) and sauger (*Sander Canadensis*) in Lake Winnipeg, Manitoba. J Great Lakes Res 38:35–46

Johnston TA, Wong DM-M, Moles MD, Wiegand MD, Casselman JM, Leggett WC (2012b) Reproductive allocation in exploited lake whitefish (*Coregonus clupeaformis*) and walleye (*Sander vitreus*) populations. Fish Res 125–126:225–234

Jonas JL, Wahl DH (1998) Relative importance of direct and indirect effects of starvation for young walleyes. Trans Am Fish Soc 127:192–205

Jones ML, Shuter BJ, Zhao Y, Stockwell JD (2006) Forecasting effects of climate change on Great Lakes fisheries: models that link habitat supply to population dynamics can help. Can J Fish Aquat Sci 63:457–468

Kallemeyn LW (1987) Correlations of regulated lake levels and climatic factors with abundance of young-of-the-year walleye and yellow perch in four lakes in Voyageurs National Park. N Am J Fish Manag 7:513–521

Kane-Sutton M, Kinter B, Dennis PM, Koonce JF (2010) Viral hemorrhagic septicemia virus infection in yellow perch, Perca flavescens, in Lake Erie. J Great Lakes Res 36:37–43

Karås P, Thoresson G (1992) An application of a bioenergetics model to Eurasian perch (*Perca fluviatilis* L.). J Fish Biol 41:217–230

Kershner MW, Schael DM, Knight RL, Stein RA, Marschall EA (1999) Modeling sources of variation for growth and predatory demand of Lake Erie (*Stizostedion vitreum*), 1986–1995. Can J Fish Aquat Sci 56:527–538

Keskinen T, Marjomäki TJ (2003) Growth of pikeperch in relation to lake characteristics: total phosphorous, water colour, lake area and depth. J Fish Biol 63:1274–1282

Keskinen T, Marjomäki TJ (2004) Diet and prey size spectrum of pikeperch in lakes in central Finland. J Fish Biol 65:1147–1153

Kitchell JF, Stewart DJ, Weininger D (1977) Applications of a bioenergetics model to yellow perch (*Perca flavescens*) and walleye (*Stizostedion vitreum vitreum*). J Fish Res Board Can 34:1922–1935

Kjellman J, Lappalainen J, Urho L, Hudd R (2003) Early determination of perch and pikeperch recruitment in the northern Baltic Sea. Hydrobiologia 495:181–191

Knight RL, Margraf FJ, Carline RF (1984) Piscivory by walleyes and yellow perch in western Lake Erie. Trans Am Fish Soc 113:677–693

Kocovsky PM, Carline RF (2001) Influence of extreme temperatures on consumption and condition of walleyes in Pymatuning Sanctuary, Pennsylvania. N Am J Fish Manag 21:198–207

Koenst WM, Smith LL Jr (1976) Thermal requirements of the early life history stages of walleye, *Stizostedion vitreum vitreum*, and sauger, *Stizostedion canadense*. J Fish Res Board Can 33:1130–1138
Kolar CS, Wahl DH, Hooe ML (2003) Piscivory in juvenile walleyes: relative importance of prey species, timing of spawning prey fish, and density on growth and survival. Trans Am Fish Soc 132:679–690
Koonce JF, Bagenal TB, Carline RF, Hokanson KEF, Nagięć M (1977) J Fish Res Board Can 34:1900–1909
Kosior M, Wendzel T (2001) Comparison of fecundity of pikeperch (*Stizostedion lucioperca* [L.]) in three lagoons in the southern Baltic Sea. Bull Sea Fish Inst 3(154):3–28
Lappalainen J, Dörner H, Wysujack K (2003) Reproduction biology of pikeperch (*Sander lucioperca* (L.)) – a review. Ecol Freshw Fish 12:95–106
Lappalainen J, Malinen T, Rahikainen M, Vinni M, Nyberg K, Ruuhijärvi J, Salminen M (2005) Temperature dependent growth and yield of pikeperch, *Sander lucioperca*, in Finnish lakes. Fish Manag Ecol 12:27–35
Latif MA, Bodally RA, Johnston TA, Fudge RJP (1999) Critical stage in developing walleye eggs. N Am J Aquacult 61:34–37
Lauer TE, Shroyer SM, Kilpatrick JM, McComish TS, Allen PJ (2005) Yellow perch length-fecundity and length-egg size relationships in Indiana waters of Lake Michigan. N Am J Fish Manag 25:791–796
Lauer TE, Doll JC, Allen PJ, Breidert B, Palla J (2008) Changes in yellow perch length frequencies and sex ratios following closure of the commercial fishery and reduction in sport bag limits in southern Lake Michigan. Fish Manag Ecol 15:39–47
Lehtonen H, Hansson S, Winkler H (1996) Biology and exploitation of pikeperch, *Stizostedion lucioperca* (L.), in the Baltic Sea area. Ann Zool Fenn 33:525–535
Lester NP, Shuter BJ, Kushneriuk RS, Marshall TR (2000) Life history variation in Ontario walleye populations: implications for safe rates of fishing. Percid Community Synthesis, Population and Yield Characteristics Working Group. Ontario Ministry of Natural Resources, Peterborough
Lester NP, Dextrase AJ, Kushneriuk RS, Lawson MR, Ryan PA (2004) Light and temperature: key factors affecting walleye abundance and production. Trans Am Fish Soc 133:588–605
Letcher BH, Rice JA, Crowder LB, Binkowski FP (1996) Size-dependent effects of continuous and intermittent feeding on starvation time and mass loss in starving yellow perch larvae and juveniles. Trans Am Fish Soc 125:14–26
Litvak MK, Leggett WC (1992) Age and size-selective predation on larval fishes: the bigger-is-better hypothesis revisited. Mar Ecol Prog Ser 81:13–24
Ljunggren L, Sandström A (2007) Influence of visual conditions on foraging and growth of juvenile fishes with dissimilar sensory physiology. J Fish Biol 70:1319–1334
Ljunggren L, Sandström A, Bergström U, Mattila J, Lappalainen A, Johansson G, Sundblad G, Casini M, Kaljuste O, Eriksson BK (2010) Recruitment failure of coastal predatory fish in the Baltic Sea coincident with an offshore ecosystem regime shift. ICES J Mar Sci 67:1587–1595
Lorenzen K (1996) The relationship between body weight and natural mortality in juvenile and adult fish: a comparison of natural ecosystems and aquaculture. J Fish Biol 49:627–647
Ložys L (2004) The growth of pikeperch (*Sander lucioperca* L.) and perch (*Perca fluviatilis* L.) under different water temperature and salinity conditions in the Curonian Lagoon and Lithuanian coastal waters of the Baltic Sea. Hydrobiologia 514:105–113
Luchiari AC, de Morais Freire FA, Koskela J, Pirhonen J (2006) Light intensity preference of juvenile pikeperch *Sander lucioperca* (L.). Aquacult Res 37:1572–1577
Machiels MAM, Wijsman J (1996) Size-selective mortality in an exploited perch population and the reconstruction of potential growth. Ann Zool Fenn 33:397–401
MacMillan EP, Roth BM (2012) By-catch in the Saginaw Bay, Lake Huron commercial trap net fishery. J Great Lakes Res 38:353–361
Madenjian CP, Tyson JT, Knight RL, Kershner MW, Hansen MJ (1996) First-year growth, recruitment, and maturity of walleyes in western Lake Erie. Trans Am Fish Soc 125:821–830

Madon SP, Culver DA (1993) Bioenergetics model for larval and juvenile walleyes: an in situ approach with experimental ponds. Trans Am Fish Soc 122:797–813
Malison JA, Held JA (1995) Reproduction and spawning in walleye (*Stizostedion vitreum*). J Appl Ichthyol 12:153–156
Malison JA, Procarione LS, Barry TP, Kapuscinski AR, Kayes TB (1994) Endocrine and gonadal changes during the annual reproductive cycle of the freshwater teleost, *Stizostedion vitreum*. Fish Physiol Biochem 13:473–484
Mandiki SNM, Houbart M, Babiak I, Vandeloise E, Gardeur JN, Kestemont P (2004a) Are sex steroids involved in the sexual growth dimorphism in Eurasian perch juveniles? Physiol Behav 80:603–609
Mandiki SNM, Blanchard G, Mélard C, Koskela J, Kucharczyk D, Fontaine P, Kestemont P (2004b) Effects of geographic origin on growth and food intake in Eurasian perch (*Perca fluviatilis* L.) juveniles under intensive culture conditions. Aquaculture 229:117–128
Marsden JE, Robillard SR (2004) Decline of yellow perch in southwestern Lake Michigan, 1987–1997. N Am J Fish Manag 24:952–966
Marshall TR (1977) Morphological, physiological, and ethological differences between walleye (*Stizostedion vitreum vitreum*) and pikeperch (*S. lucioperca*). J Fish Res Board Can 34:1515–1523
Marshall DJ, Heppell SS, Munch SB, Warner RR (2010) The relationship between maternal phenotype and offspring quality: do older mothers really produce the best offspring? Ecology 91:2862–2873
Martin DR, Powell LA, Pope KL (2012) Habitat selection by adult walleye during spawning season in irrigation reservoirs: a patch occupancy modeling approach. Environ Biol Fish 93:589–598
Mathias JA, Li S (1982) Feeding habits of walleye larvae and juveniles: comparative laboratory and field studies. Trans Am Fish Soc 111:722–735
Mattila J, Koskela J, Pirhonen J (2009) The effect of the length of repeated feed deprivation between single meals on compensatory growth of pikeperch *Sander lucioperca*. Aquaculture 296:65–70
Mehner T, Wieser W (1994) Effects of temperature on allocation of metabolic energy in perch (*Perca fluviatilis*) fed submaximal rations. J Fish Biol 45:1079–1086
Mehner T, Heinz S, Bauer D, Herbst R, Voigt H, Benndorf J (1996) Intraguild predation and cannibalism in age-0 perch (*Perca fluviatilis*) and age-0 zander (*Stizostedion lucioperca*): interactions with zooplankton succession, prey fish availability and temperature. Ann Zool Fenn 33:353–361
Mercado-Silva N, Sass GG, Roth BM, Gilbert S, Vander Zanden MJ (2007) Impact of rainbow smelt (*Osmerus mordax*) invasion on walleye (*Sander vitreus*) recruitment in Wisconsin lakes. Can J Fish Aquat Sci 64:1543–1550
Migaud H, Fontaine P, Sulistyo I, Kestemont P, Gardeur J-N (2002) Induction of out-of-season spawning in Eurasian perch *Perca fluviatilis*: effects of rates of cooling and cooling durations on female gametogenesis and spawning. Aquaculture 3–4:253–267
Migaud H, Fontaine P, Kestemont P, Wang N, Brun-Bellut J (2004) Influence of photoperiod on the onset of gonadogenesis in Eurasian perch *Perca fluviatilis*. Aquaculture 241:561–574
Migaud H, Wang N, Gardeur JN, Fontaine P (2006) Influence of photoperiod on reproductive performances in Eurasian perch *Perca fluviatilis*. Aquaculture 252:385–393
Miller TJ, Crowder LB, Rice JA, Marschall EB (1988) Larval size and recruitment mechanisms in fishes: toward a conceptual framework. Can J Fish Aquat Sci 45:1657–1670
Mills EL, Sherman R, Robson DS (1989) Effect of zooplankton abundance and body size on growth of age-0 yellow perch (*Perca flavescens*) in Oneida Lake, New York, 1975–1986. Can J Fish Aquat Sci 46:880–886
Moles MD, Johnston TA, Robinson BW, Leggett WC, Casselman JM (2008) Is gonadal investment in walleye (*Sander vitreus*) dependent on body lipid reserves? A multipopulation comparative analysis. Can J Fish Aquat Sci 65:600–614

Moodie GEE, Loadman NL, Wiegand MD, Mathias JA (1989) Influence of egg characteristics on survival, growth, and feeding in larval walleye (*Stizostedion vitreum*). Can J Fish Aquat Sci 46:516–521
Mooij WM, Van Rooij JM, Wijnhoven S (1999) Analysis and comparison of fish growth from small samples of length-at-age data: detection of sexual dimorphism in Eurasian perch as an example. Trans Am Fish Soc 128:483–490
Muntyan SP (1977) Effect of constant incubation temperatures on hatching and the morphological characteristics of pike-perch embryos. In: Metabolism and biochemistry of fishes. Nauka Press, Moscow, pp 214–221
Muth KM, Ickes BS (1993) Fecundity of walleyes in western Lake Erie, 1966 and 1990–91. J Great Lakes Res 19:715–719
Muth KM, Wolfert DR (1986) Changes in growth and maturity of walleyes associated with stock rehabilitation in western Lake Erie, 1964–1983. N Am J Fish Manag 6:1968–1983
Myers RA (1998) When do environment – recruitment correlations work? Rev Fish Biol Fish 8:285–305
Newsome GE, Tompkins J (1985) Yellow perch egg masses deter predators. Can J Zool 63:2882–2884
Nielsen LA (1980) Effect of walleye (*Stizostedion vitreum vitreum*) predation on juvenile mortality and recruitment of yellow perch (*Perca flavescens*) in Oneida Lake, New York. Can J Fish Aquat Sci 37:11–19
Nyberg P, Bergstand E, Degerman E, Enderlein O (2001) Recruitment of pelagic fish in an unstable climate: studies of Sweden's four largest lakes. Ambio 30:559–564
Olin M, Jutila J, Lehtonen H, Vinni M, Ruuhijärvi J, Estlander S, Rask M, Kuparinen A, Lappalainen J (2012) Importance of maternal size on the reproductive success of perch, *Perca fluviatilis*, in small forest lakes: implications for fisheries management. Fish Manag Ecol 19:363–374
Oseid DM, Smith LL Jr (1971) Survival and hatching of walleye eggs at various dissolved oxygen levels. Progress Fish Cult 33:81–85
Paradis Y, Bertolo A, Perot A, Débas S, Magnan P (2006) Do benthivory and piscivory result in similar growth in walleye? J Fish Biol 69:1317–1329
Paukert CP, Willis DW (2001) Comparison of exploited and unexploited yellow perch *Perca flavescens* (Mitchill) populations in Nebraska Sandhill lakes. Fish Manag Ecol 8:533–542
Pauly D (1980) On the interrelationships between natural mortality, growth parameters, and mean environmental temperature in 175 fish stocks. J Cons Int Explor Mer 39:175–192
Paxton CGM, Winfield IJ, Fletcher JM, George DG, Hewitt DP (2004) Biotic and abiotic influences on the recruitment of male perch in Windmere, U.K. J Fish Biol 65:1622–1642
Pedicillo G, Merulli F, Carosi A, Viali P, Lorenzoni M (2008) The use of artificial spawning substrates as media to support the reproduction of Eurasian perch in Lake Piediluco. Hydrobiologia 609:219–223
Persson A, Brönmark C (2002) Foraging capacity and resource synchronization in an ontogenetic diet switcher, pikeperch (*Stizostedion lucioperca*). Ecology 83:3014–3022
Peterka J, Matěna J, Lipka J (2002) The diet and growth of larval and juvenile pikeperch (*Stizostedion lucioperca* (L.)): a comparative study of fishponds and a reservoir. Aquac Int 11:337–348
Petrosky BR, Magnuson JJ (1973) Behavioral responses of northern pike, yellow perch and bluegill to oxygen concentrations under simulated winterkill conditions. Copeia 1973:124–133
Pitlo J Jr (2002) Effects of environmental factors on walleye and sauger recruitment in Pool 13, Upper Mississippi River. N Am J Fish Manag 22:1021–1031
Pope KL, Willis DW, Lucchesi DO (1996) Differential relations of age-0 black crappie and yellow perch to climatological variables in a natural lake. J Freshw Ecol 11:345–350
Post JR, Evans DO (1989) Size-dependent overwinter mortality of young-of-the-year yellow perch (*Perca flavescens*): laboratory, in situ enclosure, and field experiments. Can J Fish Aquat Sci 46:1958–1968

Post JR, Johannes MRS, McQueen DJ (1997) Evidence of density-dependent cohort splitting in age-0 yellow perch, (*Perca flavescens*): potential behavioural mechanisms and population-level consequences. Can J Fish Aquat Sci 54:867–875

Post JR, Sullivan M, Cox S, Lester NP, Walters CJ, Parkinson EA, Paul AJ, Jackson L, Shuter BJ (2002) Canada's recreational fisheries: the invisible collapse? Fisheries 27(1):6–17

Pothoven SA, Höök TO, Roswell CR (2014) Energy content of young yellow perch and walleye in Saginaw Bay. J Great Lakes Res 40(Suppl 1):133–138

Power M, Van Den Heuvel MR (1999) Age-0 yellow perch growth and its relationship to temperature. Trans Am Fish Soc 128:687–700

Pratt TC, Fox MG (2002) Influence of predation risk on the overwinter mortality and energetic relationships of young-of-year walleyes. Trans Am Fish Soc 131:885–898

Purchase CF, Collins NC, Shuter BJ (2005) Sensitivity of maximum sustainable harvest rates to intra-specific life history variability of lake trout (*Salvelinus namaycush*) and walleye (*Sander vitreus*). Fish Res 72:141–148

Pycha RL, Smith LL Jr (1955) Early life history of the yellow perch, *Perca flavescens* (Mitchell), in the Red Lakes, Minnesota. Trans Am Fish Soc 84:249–260

Quevedo M, Svanbäck R, Eklöv P (2009) Intrapopulation niche partitioning in a generalist predator limits food web connectivity. Ecology 90:2263–2274

Quist MC, Guy CS, Bernot RJ, Stephen JL (2002) Seasonal variation in condition, growth and food habits of walleye in a Great Plains reservoir and simulated effects of an altered thermal regime. J Fish Biol 61:1329–1344

Quist MC, Guy CS, Schultz RD, Stephen JL (2003a) Latitudinal comparisons of walleye growth in North America and factors influencing growth of walleyes in Kansas reservoirs. N Am J Fish Manag 23:677–692

Quist MC, Guy CS, Stephen JL (2003b) Recruitment dynamics of walleyes (*Stizostedion vitreum*) in Kansas reservoirs: generalities with natural systems and effects of a centrarchid predator. Can J Fish Aquat Sci 60:830–839

Quist MC, Guy CS, Bernot RJ, Stephen JL (2004) Factors related to growth and survival of larval walleyes: implications for recruitment in a southern Great Plains reservoir. Fish Res 67:215–225

Raabe JK, Bozek MA (2012) Quantity, structure, and habitat selection of natural spawning reefs by walleyes in a north temperate lake: a multiscale analysis. Trans Am Fish Soc 141:1097–1108

Radke RJ, Gaupisch A (2005) Effects of phytoplankton-induced turbidity on predation success of piscivorous Eurasian perch (*Perca fluviatilis*): possible implications for fish community structure in lakes. Naturwissenschaften 92:91–94

Raikova-Petrova G, Živkov M (1998) Maturity, spawning and sex ratio of pike perch, *Stizostedion lucioperca* (L.), in two Bulgarian reservoirs as compared to other European habitats. J Appl Ichthyol 14:31–35

Redman RA, Czesny SJ, Dettmers JM, Weber MJ, Makauskas D (2011) Old tales in a recent context: current perspective on yellow perch recruitment in Lake Michigan. Trans Am Fish Soc 140:1277–1289

Reichert JM, Fryer BJ, Pangle KL, Johnson TB, Tyson JT, Drelich AB, Ludsin SA (2010) River-plume use during the pelagic larval stage benefits recruitment of a lentic fish. Can J Fish Aquat Sci 67:987–1004

Rennert B, Wirth M, Günther S, Shulz C (2005) Effect of feeding under-year zander (*Sander lucioperca*) on size, body mass and body composition before and after wintering. J Appl Ichthyol 21:429–432

Rennie MD, Purchase CF, Shuter BJ, Collins NC, Abrams PA, Morgan GE (2010) Prey life-history and bioenergetics responses across a predation gradient. J Fish Biol 77:1230–1251

Reyes MRD, Arzbach H-H, Braum E (1992) *In situ* development of perch eggs, *Perca fluviatilis* L. (Pisces, Percidae) in a small eutrophic lake, Lake Plusee, Holstein, Germany. Int Rev Gesamten Hydrobiol Hydrogr 77:467–481

Rinchard J, Ware K, Dabrowski K, Van Tassell JJ, Marschall EA, Stein RA (2011) Egg thiamine concentration affects embryo survival in Lake Erie walleye. Environ Biol Fish 90:53–60
Ritchie BJ, Colby PJ (1988) Even-odd year differences in walleye year-class strength related to mayfly production. N Am J Fish Manag 8:210–215
Robb T, Abrahams MV (2003) Variation in tolerance to hypoxia in a predator and prey species: an ecological advantage of being small? J Fish Biol 62:1067–1081
Roberts JJ, Höök TO, Ludsin SA, Pothoven SA, Vanderploeg HA, Brandt SB (2009) Effects of hypolimnetic hypoxia on foraging and distributions of Lake Erie yellow perch. J Exp Mar Biol Ecol 381:S132–S142
Roberts JJ, Brandt SB, Fanslow D, Ludsin SA, Pothoven SA, Scavia D, Höök TO (2011) Effects of hypoxia on consumption, growth, and RNA:DNA ratios of young yellow perch. Trans Am Fish Soc 40:1574–1586
Roberts JJ, Grecay PA, Ludsin SA, Pothoven SA, Vanderploeg HA, Höök TO (2012) Evidence of hypoxic foraging forays by yellow perch (*Perca flavescens*) and potential consequences for prey consumption. Freshw Biol 57:922–937
Robillard SE, Marsden JE (2001) Spawning substrate preferences of yellow perch in Lake Michigan. N Am J Fish Manage 21:208–215
Roff DA, Heibo E, Vøllestad LA (2006) The importance of growth and mortality costs in the evolution of the optimal life history. J Evol Biol 19:1920–1930
Rónyai A, Csengeri I (2008) Effect of feeding regime and temperature on growing results of pikeperch (*Sander lucioperca* L.). Aquacult Res 39:820–827
Rose K, Rutherford ES, McDermot DS, Forney JL, Mills EL (1999) Individual-based model of yellow perch and walleye populations in Oneida Lake. Ecol Monogr 69:127–154
Roseman EF, Taylor WW, Hayes DB, Haas RC, Knight RL, Paxton KO (1996) Walleye egg deposition and survival on reefs in western Lake Erie (USA). Ann Zool Fenn 33:341–351
Roseman EF, Taylor WW, Hayes DB, Knight RL, Haas RC (2001) Removal of walleye eggs from reefs in western Lake Erie by a catastrophic storm. Trans Am Fish Soc 130:341–346
Roseman EF, Taylor WW, Hayes DB, Tyson JT, Haas RC (2005) Spatial patterns emphasize the importance of coastal zones as nursery areas for larval walleye in western Lake Erie. J Great Lakes Res 31(Suppl 1):28–44
Roswell CR (2011) Growth, condition, and diets of age-0 Saginaw Bay yellow perch, implications for recruitment. MS thesis, Purdue University, West Lafayette
Roswell CR, Pothoven SA, Höök TO (2013) Spatio-temporal, ontogenetic and interindividual variation of age-0 diets in a population of yellow perch. Ecol Freshw Fish 22:479–493
Roswell CR, Pothoven SA, Höök TO (2014) Patterns of age-0 yellow perch growth, diets, and mortality in Saginaw Bay, Lake Huron. J Great Lakes Res 40(Suppl 1):123–132
Rudstam LG, VanDeValk AJ, Adams CM, Coleman JTH, Forney JL, Richmond ME (2004) Cormorant predation and the population dynamics of walleye and yellow perch in Oneida Lake. Ecol Appl 14:149–163
Sandström O, Neuman E, Thoresson G (1995) Effects of temperature on life history variables in perch. J Fish Biol 47:652–670
Sandström O, Abrahamsson I, Andersson J, Vetemaa M (1997) Temperature effects on spawning and egg development in Eurasian perch. J Fish Biol 51:1015–1024
Sarvala J, Helminen H (1996) Year-class fluctuations of perch (*Perca fluviatilis*) in Lake Pyhäjärvi, Southwest Finland. Ann Zool Fenn 33:389–396
Schaeffer JS, Diana JS, Haas RC (2000) Effects of long-term changes in the benthic community on yellow perch in Saginaw Bay, Lake Huron. J Great Lakes Res 26:340–351
Scherer E (1971) Effects of oxygen depletion and of carbon dioxide buildup on the photic behavior of the walleye (*Stizostedion vitreum vitreum*). J Fish Res Board Can 28:1303–1307
Schlumberger O, Proteau J-P (1996) Reproduction of pike-perch (*Stizostedion lucioperca*) in captivity. J Appl Ichthyol 12:149–152
Schneider JC, Copeland J, Wolgamood M (2002) Tolerance of incubating walleye eggs to temperature fluctuation. N Am J Aquac 64:75–78

Schueller AM, Hansen MJ, Newman SP, Edwards CJ (2005) Density dependence of walleye maturity and fecundity in Big Crooked Lake, Wisconsin, 1997–2003. N Am J Fish Manag 25:841–847
Schumann GO (1963) Artificial light to attract young perch: a new method of augmenting the food supply of predaceous fish fry in hatcheries. Prog Fish Cult 25:171–174
Schupp DH (2002) What does Mt. Pinatubo have to do with walleyes? N Am J Fish Manag 22:1014–1020
Scott WB, Crossman EJ (1973) Freshwater fishes of Canada. Fish Res Board Can Bull 184:1–966
Serns SL (1982) Walleye fecundity, potential egg deposition, and survival from egg to fall young-of-year in Escanaba Lake, Wisconsin, 1979–1981. N Am J Fish Manag 2:388–394
Sharpe DMT, Hendry AP (2009) Life history change in commercially exploited fish stocks: an analysis of trends across studies. Evol Appl 2:260–275
Sherwood GD, Pazzia I, Moeser A, Hontela A, Rasmussen JB (2002) Shifting gears: enzymatic evidence for the energetic advantage of switching diet in wild-living fish. Can J Fish Aquat Sci 59:229–241
Shewmon LN, Godwin JR, Murashige RS, Daniels HV, Losordo TM (2007) Environmental manipulation of growth and sexual maturation in yellow perch, *Perca flavescens*. J World Aquac Soc 38:383–394
Shuter BJ, Post JR (1990) Climate, population viability, and the zoogeography of temperate fishes. Trans Am Fish Soc 119:314–336
Specziár A (2005) First year ontogenetic diet patterns in two coexisting *Sander* species, *S. lucioperca* and *S. volgensis* in Lake Balaton. Hydrobiologia 549:115–130
Specziár A, Bíró P (2003) Population structure and feeding characteristics of Volga pikeperch *Sander volgensis* (Pisces, Percidae), in Lake Balaton. Hydrobiologia 506–509:503–510
Stepien CA, Banda JA, Murphy DM, Haponski AE (2012) Temporal and spatial genetic consistency of walleye spawning groups. Trans Am Fish Soc 141:660–672
Strand Å, Magnhagen C, Alanärä A (2011) Growth and energy expenditures of Eurasian perch *Perca fluviatilis* (Linneaus) in different temperatures and of different body sizes. J Aquac Res Dev 2:114
Suthers IM, Gee JH (1986) Role of hypoxia in limiting diel spring and summer distribution of juvenile yellow perch (*Perca flavescens*) in a prairie marsh. Can J Fish Aquat Sci 43:1562–1570
Svanbäck R, Eklöv P (2001) Effects of habitat and food resources on morphology and ontogenetic growth trajectories in perch. Oecologia 131:61–70
Svanbäck R, Eklöv P (2006) Genetic variation and phenotypic plasticity: causes of morphological and dietary variation in Eurasian perch. Evol Ecol Res 8:37–49
Szalai AJ, Dick TA (1991) Role of predation and parasitism in growth and mortality of yellow perch in Dauphin Lake, Manitoba. Trans Am Fish Soc 120:739–751
Tamazouzt L, Chatain B, Fontaine P (2000) Tank wall colour and light level affect growth and survival of Eurasian perch larvae (*Perca fluviatilis* L.). Aquaculture 182:85–90
Thorpe JE (1977) Morphology, physiology, behavior, and ecology of *Perca fluviatilis* L. and *P. flavescens* Mitchill. J Fish Res Board Can 34:1504–1514
Tidwell JH, Coyle SD, Evans J, Weibel C, McKinney J, Dodson K, Jones H (1999) Effect of culture temperature on growth, survival, and biochemical composition of yellow perch *Perca flavescens*. J World Aquac Soc 30:324–330
Tolonen A, Lappalainen J, Pulliainen E (2003) Seasonal and year class strength variations of perch near the northern limits of its distribution range. J Fish Biol 63:176–186
Treasurer JW (1983) Estimates of egg and viable embryo production in a lacustrine perch, *Perca fluviatilis*. Environ Biol Fish 8:3–16
Turesson H, Persson A, Brönmark C (2002) Prey size selection in piscivorous pikeperch (*Stizostedion lucioperca*) includes active prey choice. Ecol Freshw Fish 11:223–233
Tyson JT, Knight RL (2001) Response of yellow perch to changes in the benthic invertebrate community of western Lake Erie. Trans Am Fish Soc 130:766–782
Vainikka A, Koskimäki J, Niemela PT, Kortet R (2012) Composition of the Eurasian perch (*Perca fluviatilis*) catches in ice fishing: does capture order predict body size? Fish Res 115–116:24–30

van Densen WLT, Ligtvoet W, Roozen RWM (1996) Intra-cohort variation in the individual size of juvenile pikeperch, *Stizostedion lucioperca*, and perch, *Perca fluviatilis*, in relation to the size spectrum of their food items. Ann Zool Fenn 33:495–506

Vandenbyllaardt L, Ward FJ, Braekevelt CR, McIntyre DB (1991) Relationships between turbidity, piscivory, and development of the retina in juvenile walleyes. Trans Am Fish Soc 120:382–390

VanDeValk AJ, Adams CM, Rudstam LG, Forney JL, Brooking TE, Gerken MA, Young BP, Hooper JT (2002) Comparison of angler and cormorant harvest of walleye and yellow perch in Oneida Lake, New York. Trans Am Fish Soc 131:27–39

Venturelli PA, Shuter BJ, Murphy CA (2009) Evidence for harvest-induced maternal influences on the reproductive rates of fish populations. Proc R Soc B 276:919–924

Venturelli PA, Lester NP, Marshall TR, Shuter BJ (2010a) Consistent patterns of maturity and density-dependent growth among populations of walleye (*Sander vitreus*): application of the growing degree-day metric. Can J Fish Aquat Sci 67:1057–1067

Venturelli PA, Murphy CA, Shuter BJ, Johnston TA, Boag PT, Casselman JM, Montgomerie R, Wiegand MD, Leggett WC (2010b) Maternal influences on population dynamics: evidence from an exploited freshwater fish. Ecology 91:2003–2012

Vinni M, Lappalainen J, Malinen T, Lehtonen H (2009) Stunted growth of pikeperch *Sander lucioperca* in Lake Sahajärvi, Finland. J Fish Biol 74:967–972

Wang N, Appenzeller A (2006) Abundance, depth distribution, diet composition and growth of perch (*Perca fluviatilis*) and burbot (*Lota lota*) larvae and juveniles in the pelagic of Lake Constance. Ecol Freshw Fish 7:176–183

Wang N, Eckmann R (1994) Effects of temperature and food density on egg development, larval survival and growth of perch (*Perca fluviatilis* L.). Aquaculture 122:323–333

Wang N, Gardeur J-N, Henrotte E, Marie M, Kestemont P, Fontaine P (2006) Determinism of the induction of the reproductive cycle in female Eurasian perch, *Perca fluviatilis*: identification of environmental cues and permissive factors. Aquaculture 261:706–714

Wang H-Y, Höök TO, Ebener MP, Mohr LC, Schneeberger PJ (2008) Spatial and temporal variation of maturation schedules of lake whitefish (*Coregonus clupeaformis*) in the Great Lakes. Can J Fish Aquat Sci 65:2157–2169

Wang H-P, Li L, Wallat G, Brown B, Yao H, Gao Z, Tiu L, O'Bryant P, Rapp D, MacDonald R (2009a) Evaluation of relative growth performance and genotype by environment effects for cross-bred yellow perch families reared in communal ponds using DNA parentage analyses. Aquacult Res 40:1363–1373

Wang H-Y, Cook HA, Einhouse DW, Fielder DG, Kayle KA, Rudstam LG, Höök TO (2009b) Maturation schedules in the Great Lakes region: comparison of maturation indices and evaluation of sampling-induced biases. N Am J Fish Manag 29:1540–1554

Wang N, Xu X, Kestemont P (2009c) Effect of temperature and feeding frequency on growth performances, feed efficiency and body composition of pikeperch juveniles (*Sander lucioperca*). Aquaculture 289:70–73

Wang H-Y, Einhouse DW, Fielder DG, Rudstam LG, Vandergoot C, VanDeValk AJ, Zorn T, Höök TO (2012) Maternal and stock effects on egg-size variation among walleye *Sander vitreus* stocks from the Great Lakes region. J Great Lakes Res 38:477–489

Weber MJ, Dettmers JM, Wahl DH (2011) Growth and survival of age-0 yellow perch across habitats in southwestern Lake Michigan: early life history in a large freshwater environment. Trans Am Fish Soc 140:1172–1185

Wellington CG, Mayer CM, Bossenbroek JM, Stroh NA (2010) Effects of turbidity and prey density on the foraging success of age 0 year yellow perch *Perca flavescens*. J Fish Biol 76:1729–1741

Wiegand MD, Johnston TA, Martin J, Leggett WC (2004) Variation in neutral and polar lipid compositions of ova in ten reproductively isolated populations of walleye (*Sander vitreus*). Can J Fish Aquat Sci 61:110–121

Wiegand MD, Johnston TA, Leggett WC, Watchorn KE, Ballevona AJ, Porteous LR, Casselman JM (2007) Contrasting strategies of ova lipid provisioning in relation to maternal characteristics in three walleye (*Sander vitreus*) populations. Can J Fish Aquat Sci 64:700–712

Williamson CE, Metgar SL, Lovera PA, Moeller RE (1997) Solar ultraviolet radiation and the spawning habitat of yellow perch, *Perca flavescens*. Ecol Appl 7:1017–1023
Winemiller KO, Rose KA (1992) Patterns of life-history diversification in North American fishes: implications for population regulation. Can J Fish Aquat Sci 49:2196–2218
Withers J (2012) Potential invasive species impacts on prey availability and foraging by larval yellow perch and alewife in the near-shore zone of southeastern Lake Michigan. MS thesis, Purdue University, West Lafayette
Wolfert DR (1969) Maturity and fecundity of walleye from the eastern and western basins of Lake Erie. J Fish Res Board Can 26:1877–1888
Wysujack K, Kasprzak P, Laude U, Mehner T (2002) Management of a pikeperch stock in a long-term biomanipulated stratified lake: efficient predation vs. low recruitment. Hydrobiologia 479:169–180
Zhao Y, Schuter BJ, Jackson DA (2007) Life history variation parallels phylogeographics patterns in North American walleye (*Sander vitreus*) populations. Can J Fish Aquat Sci 65:198–211
Zhao Y, Jones ML, Shuter BJ, Roseman EF (2009) A biophysical model of Lake Erie walleye (*Sander vitreus*) explains interannual variations in recruitment. Can J Fish Aquat Sci 66:114–125
Zingel P, Paaver T (2010) Effects of turbidity on feeding of the young-of-the-year pikeperch (*Sander lucioperca*) in fishponds. Aquacult Res 41:89–197

Part II
Reproductive Biology

Chapter 3
Broodstock Management and Control of the Reproductive Cycle

P. Fontaine, N. Wang, and B. Hermelink

Abstract The annual reproductive cycles of percid fish have been clearly characterized for males and females in various natural habitats mainly from North-America or Eurasia, including morpho-anatomical, histological and physiological studies. From this strong basal knowledge, numerous experimental approaches were conducted to understand the environmental and hormonal controls of these reproductive cycles, and to develop artificial programmes in order to obtain delayed or out-of season spawning. It was demonstrated that photoperiod and temperature variations were the major environmental cues. Now efficient photo-thermal programmes exist and have been used by SMEs in intensive fish farming conditions (water recirculating systems).

Keywords Percids • Spawning • Reproductive cycle • Hormonal control • Maturity

3.1 Introduction

Over the last two decades, percid fish culture has progressively appeared as a real way of diversification for the freshwater inland aquaculture sector in temperate areas (Kestemont and Mélard 2000; Fontaine 2004). Consequently, domestication of percid species has been and still is progressing (Teletchea and Fontaine 2014) and the first commercial farms have been recently built in Europe, mainly based on water recirculating aquaculture system(s) (RAS) (Dalsgaard et al. 2013). To support this development focusing on human consumption markets, secured hatchery and nursery productions must be expanded in order to highly increase the availability of

---

P. Fontaine (✉) • N. Wang
Unité de Recherche Animal et Fonctionnalités des Produits Animaux, Université de Lorraine, USC INRA 340, Vandoeuvre-lès-Nancy F-54505, France
e-mail: p.fontaine@univ-lorraine.fr

B. Hermelink
Leibniz-Institute of Freshwater Ecology and Inland Fisheries, Mueggelseedamm 310, 12587 Berlin, Germany

P. Kestemont et al. (eds.), *Biology and Culture of Percid Fishes*,
DOI 10.1007/978-94-017-7227-3_3

juveniles for on-growing farms (particularly by obtaining several reproductions per year). To achieve this objective, a precise knowledge of the natural reproductive cycle (part 1 of this chapter) and its environmental control (part 2) is required to obtain out-of-season spawning and high quality gametes and larvae (Bromage and Roberts 1995). Strict control of the reproductive cycle is a key factor in the domestication of new species, like percid fishes, and the future implementation of breeding programmes for further improvement of cultured percid performances (growth, stress or disease resistance, etc.).

3.2 Natural Reproductive Cycle of Percid Fishes

3.2.1 Gonad Ontogenesis, Sexual Maturity and Gonad Morpho-anatomy

Percid oogenesis and spermatogenesis can start in fish of very small size, i.e. between 3 and 8 cm in length (Malison et al. 1986; Zakes and Demska-Zakes 1996; Lappalainen et al. 2003). This timing is age and size dependent and differs between males and females. In pikeperch *Sander lucioperca*, the first onset of oogenesis in females takes place at a size of approximately 8 cm total length (TL), while males seem to start spermatogenesis later at a larger body size of approximately 12 cm TL (Demska-Zakes and Zakes 1995; Zakes and Demska-Zakes 1996; Lappalainen et al. 2003). A similar pattern can be observed in walleye *Sander vitreus*, a species closely related to pikeperch. In this species, the first differentiation of oogonia occurs in fish from 7.5 cm TL upward, while first spermatogenesis does not begin before fish reach 15 cm TL (Malison et al. 1990).

The age at the onset of sexual maturity (puberty) varies greatly depending on sex and geographical origin (latitude). With respect to the effect of sex, males are mature at a younger age and at a smaller size than females (Alm 1954; Kukuradze 1974; Craig 1977; Papageorgiou 1977; Raikova-Petrova and Zivkov 1998; Henderson and Morgan 2002; Lappalainen et al. 2003). For example, male pikeperch become mature earlier than females at an age of (1) 2–3 years when they have grown up to more than 21 cm, while females do not usually mature before they have reached their third year of age and sizes of more than 28 cm (Raikova-Petrova and Zivkov 1998; Lappalainen et al. 2003). The effect of geographical origin (fish location) is linked to environmental conditions and food availability over the year. More favourable environmental and nutritional conditions allow higher growth and consequently earlier sexual maturity (Colby and Nepszy 1981; Houthuijzen et al. 1993). Maturity is reached at younger ages and smaller sizes in southern populations as compared to northern populations (Raikova-Petrova and Zivkov 1998; Lappalainen et al. 2003). In the southern Baltic Sea, pikeperch mature earlier (3–4 year old) whereas, due to lower growth rates at lower temperatures, pikeperch of the northern Baltic mature generally later (4–5 year old)

(Lehtonen et al. 1996). In percid fishes, as observed in yellow perch (Malison et al. 1986), body size is the most important factor in the control of sexual maturity. All these data result from studies of wild or pond cultured populations subjected to natural climate variations (temperature, photoperiod). Recently, it has been shown that sexual maturity is achieved much earlier when fish are reared under RAS conditions and maintained under constant photo-thermal conditions (20–22 °C, L:D 16:8) (Ben Ammar et al. 2012). Eurasian perch puberty can be achieved within 11–14 months post-hatching in both sexes.

Regarding the anatomy of the reproductive system, the ovaries of percid fishes are initially paired. However, a particularity exists in *Perca* species in which the ovaries fuse together during gonadogenesis to form a single ovarian mass, whereas two paired ovaries are conserved in *Sander* species (Figs. 3.1 and 3.2) (Craig 2000).

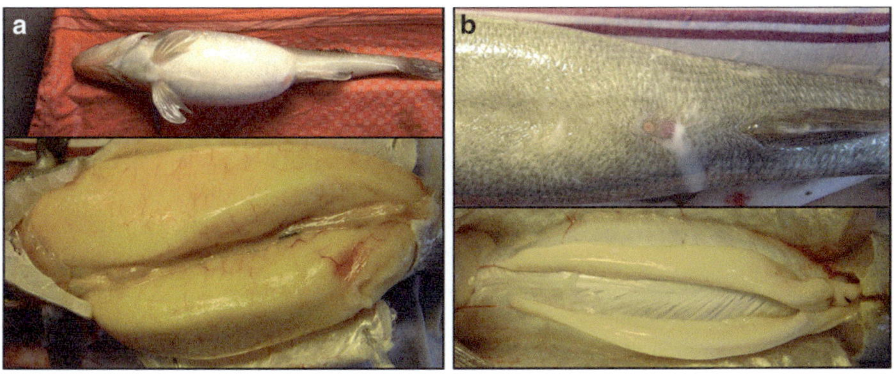

Fig. 3.1 Fully mature female pikeperch and corresponding ovaries (**a**) and fully mature male pikeperch and corresponding testicles (**b**)

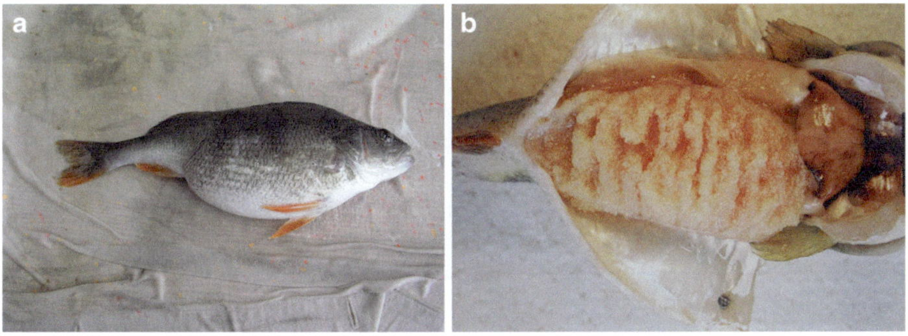

Fig. 3.2 Fully mature female Eurasian perch (**a**) and ovarian mass organized in one unique ribbon (**b**)

3.2.2 *Male Reproductive Cycle*

3.2.2.1 Testicle Development

In wild populations of perch, walleye and pikeperch caught in natural habitats (lakes, ponds or rivers) of northern temperate or Mediterranean areas, it is observed that males display low gonado-somatic index GSI ([gonad weight/body weight] × 100, %) during the summer period (Turner 1919; Le Cren 1951; Craig 1974; Treasurer and Holliday 1981; Tanasichuk and Mackay 1989; Hayes and Taylor 1994; Malison et al. 1994; Sulistyo et al. 2000; M'Hetli et al. 2011). During this period, male GSI is below 1 % and only spermatogonia are found in the testicles (Craig 1974; Sulistyo et al. 2000). Depending on the species and geographical location, multiplication and differentiation of spermatogonia into advanced spermatogenic cells begin over the period late August–December. Consequently, male GSI highly and rapidly increases during this period and often reaches a peak: 0.5–1 % in pikeperch (Uysal et al. 2006; M'Hetli et al. 2011), 2.5 % in walleye (Malison et al. 1994) and 6–10 % in Eurasian perch (Turner 1919; Le Cren 1951; Craig 1974; Treasurer and Holliday 1981; Hayes and Taylor 1994; Sulistyo et al. 2000). At the end of the autumnal development, spermatogenesis is actually completed. Testes are filled with spermatozoa (>95 % of germ cells) and males are often able to express fluid milt (Malison et al. 1994; Sulistyo et al. 2000). Then male GSI remains relatively stable from October to November until spawning in spring (Tanasichuk and Mackay 1989; Dabrowski et al. 1994; Dabrowki and Ciereszko 1996; Henderson et al. 1996). Sometimes, a slight decrease or increase in GSI was reported prior to the spawning season in both Eurasian and yellow perch (GSI=4–6 %) (Craig 1974; Hayes and Taylor 1994; Sulistyo et al. 2000), or in walleye (3.2 %) (Malison et al. 1994).

3.2.2.2 Testicular Steroidogenesis

Plasma levels of steroid hormones show significant variations during male gonad maturation (Schulz et al. 2010). Androgens (Testosterone, T; 11-ketotestosterone, 11KT) were identified as major sex steroids involved in spermatogenesis and spermiation in fish (Billard et al. 1990; Borg 1994). Androgens are effective in supporting either the whole process of spermatogenesis, or at least some steps such as spermatogonia multiplication, spermatocyte formation/maturation and spermiation (Schulz et al. 2010). For instance, when water temperatures decrease to 10–12 °C in autumn, spermatogenesis in male pikeperch leads to rising plasma levels of the androgens testosterone (T) and 11-ketotestosterone (11-KT) (Hermelink et al. 2011, 2013).

Serum T levels vary according to variations in GSI, with very low levels (often undetectable) in summer (<0.5 ng mL^{-1}), followed by a high increase (first peak) in November-December (1.6 ng mL^{-1} in walleye, 12.3 ng mL^{-1} in Eurasian perch), a decrease in winter (January-February) and finally a second peak (2.8 ng mL^{-1} in walleye and 6.8 ng mL^{-1} in Eurasian perch) prior to spawning in March-April

(Malison et al. 1994; Sulistyo et al. 2000). The first increase is thought to regulate male germ cell differentiation and the pre-spawning peak may stimulate secondary sexual behaviours, increase the pituitary GTH levels or serve as a precursor for the production of other steroids like 11-KT (Crim et al. 1981; Kobayashi et al. 1989). As for 11-KT, plasma levels in males remain low during the summer period and the beginning of autumn (<10 ng mL^{-1} in walleye and <0.5 ng mL^{-1} in Eurasian perch) (Malison et al. 1994; Sulistyo et al. 2000). Then, a high increase is observed concomitantly with the T plasma level increase in yellow perch in autumn (Dabrowski et al. 1996) or later in January-March as measured in walleye (Malison et al. 1994) and Eurasian perch (Sulistyo et al. 2000). Consequently, higher 11-KT levels are measured in winter: 1.2–1.5 ng mL^{-1} in yellow perch, 4.5–5 ng mL^{-1} in Eurasian perch and 40 ng mL^{-1} in walleye. Levels of 11-KT remain high until spawning and may have a major role in maintaining spermatozoa viability during the prolonged period of sperm storage within the testes. Levels of both T and 11-KT decrease sharply at the onset of spawning. Such a rapid drop in androgens may reflect a shift in the steroidogenic pathway from C19 androgens to C21 progestogen production at the onset of spermiation to regulate the process of spermiation and control male spawning behaviour (Malison et al. 1994).

3.2.3 Female Reproductive Cycle

3.2.3.1 Ovary Development

According to many studies (Malservisi and Magnin 1968; Treasurer and Holliday 1981; Jamet and Desmolles 1994; Malison et al. 1994; Dabrowski et al. 1996; Sulistyo et al. 1998; Lappalainen et al. 2003), percid ovarian development is group synchronous meaning that at most two distinct oocyte developmental stages can be seen within the ovary during sexual resting and maturation (Wallace and Selman 1981; Nagahama 1983). One population consists of a heterogeneous group of small oocytes, and the other is made up of a group of larger oocytes developing synchronously the following spawning season (Malison and Held 1996a). As for GSI and oocyte diameter (OD) variations over an annual reproductive cycle, similar patterns are observed in all percid fishes including five main stages: (1) very low values over the post-spawning period and summer (quiescent or sexual rest period), (2) a rapid increase in autumn (vitellogenesis), (3) a slow or reduced increase in winter, (4) a second high GSI increase before spawning (oocyte hydratation) and (5) a sharp decrease after spawning. In walleye (Minnesota, USA), GSI and OD increase rapidly from October to November (up to 7.6 % and 1,000 μm, respectively) and reach a maximum just prior to spawning (15 %, 1,500 μm) (Malison et al. 1994). Higher GSIs (18–22 %) were recorded by Henderson and Nepszy (1993) in Lake Erie. In pikeperch (Poulet 2004; Uysal et al. 2006), Eurasian perch (Craig 1974; Sulistyo et al. 1998; Noaksson et al. 2004) and yellow perch (Dabrowski et al. 1994; Hayes and Taylor 1994; Dabrowki and Ciereszko 1996; Henderson et al. 2000), female

GSI increases steadily from the beginning of autumn until the spawning period. For example, in Eurasian perch, the OD increases from 200 μm in August to 750 μm in December (Treasurer and Holliday 1981; Sulistyo et al. 1998). However, the GSI increase is slower during the winter period (Sulistyo et al. 1998; Poulet 2004). Prior to spawning, maximum female GSIs are 20–31 % for Eurasian and yellow perch (Treasurer and Holliday 1981; Heidinger and Kayes 1986; Jamet and Desmolles 1994; Dabrowki and Ciereszko 1996; Sulistyo et al. 1998) and 8–10 % for pikeperch (Poulet 2004). Consequently, in percid fishes, especially in female perch, the investment in reproduction is very high and much higher than in males. Moreover, with increasing age, more and more energy is required for ovary maturation. Fecundity and gonad size correlate with somatic and gonad energy density (Henderson et al. 2000). For example, the energy invested in eggs by Eurasian perch females can be superior to 86 % of the energy stored by the soma in 1 year, while the energy invested in the testes is at most 10 % (Craig 1977).

3.2.3.2 Ovarian Steroidogenesis

In females, increases in LH and FSH activate the synthesis of 17β-estradiol (E2) and its precursor hormone T catalysed by aromatase (Hermelink et al. 2011, 2013). Oogenesis correlates with changes in serum levels of sex steroids (17β-estradiol-, E_2; testosterone, T) (Lubzens et al. 2010). E2 is secreted by the ovaries and released into the bloodstream where it is transported to the liver, bound to the sex hormone binding globulin (SHBG). In response to the increasing E2 blood concentrations, vitellogenin is synthesised in the liver. Vitellogenin in turn is also released in the bloodstream and is taken up by the oocytes by receptor-mediated endocytosis (Tao et al. 1996). In the oocytes vitellogenin is cleaved into the final egg-yolk proteins (lipovitellins and phosvitins) and deposited in the yolk granules (Wallace 1978). The process of vitellogenesis lasts until rising water temperatures induce final maturation. At the end of the vitellogenesis process, the oocytes are filled up with a few large lipid droplets and the nucleus is located centripetally. Final maturation is characterised by germinal vesicle breakdown (GVBD). GVBD is regulated by a maturation-inducing steroid (MIS). In salmonids and many other species 17,20β-P is this MIS, while 17,20β,21-trihydroxy-4-pregnen-3-one is also often found in perciforms (Young et al. 2005; Nagahama and Yamashita 2008; Lubzens et al. 2010). Pankhurst et al. (1986), Barry et al. (1995), Fontaine et al. (2003), and Migaud et al. (2003a) reported that 17,20β-P is the potential MIS in walleye and Eurasian perch. After GVBD the ripe eggs can be ovulated and fertilized.

As far as the natural female reproductive cycle is concerned, serum E_2 levels in percid females increase rapidly from basal values in early autumn (September-October, <0.1 ng mL^{-1}) to peak values in late autumn (late October-November, 3–4 ng mL^{-1}), which coincides with the period of maximum ovarian growth (Malison et al. 1994; Sulistyo et al. 1998; Noaksson et al. 2004). Subsequently, in walleye, E_2 levels decrease steadily from December through spawning (Malison et al. 1994), whereas in Eurasian perch they remain high until spawning (Sulistyo et al. 1998; Noaksson et al. 2004). In this species, a peak (4 ng mL^{-1}) is often measured in April during the peri-ovulatory period (Sulistyo et al. 1998; Migaud et al. 2003a). In yellow perch, a study dealing with in vitro production of ovarian steroids shows peaks of E_2 production in early autumn (October) and during the peri-ovulatory period (Dabrowski et al. 1996) and confirms data obtained from in vivo studies. E_2 stimulates the synthesis of vitellogenin (De Vlaming et al. 1977, 1980; Wallace and Selman 1981). In walleye, vitellogenesis is completed by early winter, whereas an active vitellogenesis is maintained until spawning in Eurasian perch. In walleye (Malison et al. 1994), serum T levels exhibit a bimodal pattern with two peaks in November (1.6 ng mL^{-1}) and April, just prior to spawning (3.3 ng mL^{-1}). This second peak could be related to final oocyte maturation and ovulation (Young et al. 1983). The simultaneous E_2 level decrease and T level increase from November to the spawning period suggest a decrease in aromatase activity. In Eurasian perch, the plasma T level remains very low until November, then increases dramatically in December (15–20 ng mL^{-1}) and stabilizes until the peri-ovulatory period (Sulistyo et al. 1998). A similar plasma T profile was observed by Noaksson et al. (2004) in a Swedish lake (Lake Djursjön). Such high plasma T values in perch females during the peri-ovulatory period have been also observed by Dabrowski et al. (1996) and Migaud et al. (2003a).

3.2.4 Reproductive Strategy

Mainly based on eco-physiological studies of wild populations from various water bodies, the previous parts of this chapter suggest that all percid fishes display similar reproductive cycles. A recent analysis of 29 reproductive traits in 65 temperate freshwater fish species (including 5 percid fishes from Eurasia and North-America) demonstrates that all percid fishes are clustered in the “early-spring spawners” group (Teletchea et al. 2009a). The (three) main phases of the percid reproductive cycle have been specified by Wang et al. (2010): (1) induction of gametogenesis by decreasing both temperature and photoperiod, (2) vitellogenesis during the autumnal chilling period and (3) synchronization of final stages (oocyte maturation, spawning) by increasing temperatures. Thus the reproductive cycle of percid fishes is controlled by annual variations in both photoperiod and temperature. Its environmental control is therefore more complex than in other species like salmonids. As photoperiod is the most accurate environmental zeitgeber in temperate areas (Migaud et al. 2010), this factor is strictly implicated into the control of the percid reproductive cycle. Numerous studies suggest that temperature changes also play their part, especially during winter and spring periods (Hokanson 1977; Schlumberger and Proteau 1996; Sandström et al. 1997; Gillet and Dubois 2007). For example, induction and intensity of pikeperch spawning are controlled by variations of water temperature (Lappalainen et al. 2003).

Despite many similarities between percid reproductive cycles, high ecological plasticity related to both their huge biogeographical distribution (Thorpe 1977) and

Table 3.1 Intraspecific variations of spawning period and temperature in Eurasian perch populations from the North hemisphere

Location and country	Spawning period	Spawning temperature (°C)	Authors
Mirgenbach reservoir, France[a]	Late February–mid April	12–14	Flesch 1994
Lake Agios Vasilios, Greece	Mid March–early April	8	Papageorgiou 1977
Slapton Ley, England	Mid March–early May	8–14	Craig 1974
Ivan'kovo reservoir, Russia	April–May	7–15	Makarova 1973
Forsmark and Oskarshamn basins, Sweden[a]	Early April–late June	7–24	Sandström et al. 1997
Loch Kinord and Davan, Scotland	Late April–mid May	9–11	Treasurer 1983
Lake Zürich, Switzerland	Late April–early June	8–15	Zeh et al. 1989
Lake Geneva, France	Late April–early June	8–16	Gillet and Dubois 1995
La Gombe, Belgium	Late April–mid June	8.5–13.5	Dalimier et al. 1982
Loch Leven, Scotland	Late April–mid June	9–11.5	Jones 1982
Windermere, England	Mid May–mid June	9–18	Guma'a 1978
Lake Saarlampi, Finland	Mid may–late May	12–14	Urho 1996

[a]Water bodies receiving cooling water discharged by nuclear power stations

habitat diversity can be noted for numerous reproductive traits for each species. For example, the broodstock characteristics at maturity for pikeperch (Lappalainen et al. 2003) or spawning time and temperature for Eurasian perch (see Table 3.1) are highly variable. In fact, over the northern hemisphere, Eurasian perch populations spawn between February and June over a 7–20 °C temperature range (Sandström et al. 1997; Gillet and Dubois 2007). Furthermore, within a same area or ecosystem, significant variations can be observed. Indeed, in the southern Baltic Sea pikeperch spawn in late April–early May and in late May–June in the northern Baltic (Baltic Proper, Gulf of Finland) due to a water temperature rise occurring later in the season (Lehtonen et al. 1996; Lappalainen et al. 2003).

In addition, ovarian development is not fully synchronous within a same population, explaining the long spawning period (Henderson et al. 2000; Migaud et al. 2003b). Similar variations are also observed for yellow perch (Scott and Crossman 1973; Hokanson 1977; Heidinger and Kayes 1986) and walleye (Malison and Held 1996b). The geographical effect is often related to marked variations in fish reproductive performance. For example, based on a study of 75 Eurasian perch populations, Heibo et al. (2005) reported a negative correlation between fecundity and latitude. Likewise, the spawning performance of percid fishes is also affected by the condition of the females, presumably reflecting the net acquisition of energy in the preceding growing season (Henderson and Nepszy 1993; Henderson et al. 1996; Henderson and Morgan 2002).

Finally, population structure or water quality could also affect the reproductive performance. In pikeperch a greater range of age classes participating to spawning increases the duration of the spawning season (Lappalainen et al. 2003). The oldest pikeperch initiate the spawning season while the youngest and smallest end it. Interestingly, freshwater pikeperch of the same latitude spawn earlier than the individuals living in brackish water. Habitat parameters within the population location also influence fecundity. Pikeperch which have spent their whole lifetime in brackish water show higher fecundity than freshwater ones (Lehtonen et al. 1996).

3.2.5 Conclusion

All percid fishes have similar reproductive cycles and belong to a homogenous group (early spring spawners). In the North hemisphere, after a short quiescent post-spawning period (May-June), gonadal recrudescence and oocyte growth in females occur over a decreasing photoperiod and water temperature period in fall. After a long wintering period, final oocyte maturation, ovulation and spawning are observed in spring (mainly over April-May) when water temperature and photoperiod increase, confirming the importance of these two environmental factors in the control of percid reproductive cycle. Finally, according to the diversity of the habitat and the geographical location, high ecological plasticity is recorded for major reproductive traits like female fecundity. It suggests that numerous factors (environment, nutrition and population) can modulate the reproductive performance.

3.3 Environmental Control of the Reproductive Cycle

3.3.1 Hormonal Regulation of the Reproductive Cycle

In fish, especially in temperate areas, photoperiod and temperature are the only two environmental signals that can provide a consistent timing message for broodstock (Craig 2000; Migaud et al. 2010). According to these authors, photoperiod variations are transduced by a photoneuroendocrine system (retina, suprachiasmatic nucleus and pineal gland) which releases the hormone melatonin exclusively at night. The duration of the melatonin secretion varies according to night length. Melatonin mediates the transduction of photoperiodic information to the hypothalamus-pituitary-gonad (HPG) axis (Amano et al. 2000). Like in other teleosts, the reproduction of male and female percid fishes is regulated by photo-thermal variations and the hypothalamus-pituitary-gonad (HPG) axis (via hormones). The hypothalamus acts as an interface between the nervous system and the endocrine system, incorporating internal (e.g. nutritional conditions) as well as external (e.g. photoperiod, temperature) triggers. Key regulators of the HPG axis are the pituitary gonadotropins, the luteinizing hormone (LH) and the follicle-stimulating hormone (FSH) and at the level of the gonads, the sex steroids (Zohar et al. 2010). The knowledge on neuroendocrine regulation of fish reproduction has increased tremendously over the past 10 years (Migaud et al. 2010). The brain-pituitary-gonad (BPG) cascade has been well described, notably the kisspeptin control of gonadotropin-releasing hormone (GnRH) synthesis. Synthesis of the sex steroids is stimulated by the gonadotrophins. This is crucially important for reproductive processes like the spermatogenesis in males or the vitellogenesis in females and the final maturation in both (Lubzens et al. 2010; Schulz et al. 2010).

Consequently, over the last two decades, some studies were conducted to control percid reproductive cycles by regulating photoperiod and temperature for to either (1) extending the natural spawning period (advanced or postponed spawning) or (2) obtaining out-of-season spawning (often in combination with hormonal injection to induce ovulation and spawning in females and sperm production in males). In the first case, broodstock was first maintained under natural conditions in ponds or floating cages (i.e. natural induction of the reproductive cycle), then subjected to artificial conditions during the final part of their reproduction cycle. In the second case, broodstock was exposed to artificial conditions during the whole reproduction cycle. In that case, broodstock is disconnected from natural cycles and their manipulation starts with fish at a sexual resting stage.

3.3.2 *Environmental Control of the Reproductive Cycle*

3.3.2.1 Extending the Natural Spawning Season by Advancing Spawning Time

Due to a longer duration of the vitellogenesis process in perch females, causing spawning to start earlier seems to be an effective option for pikeperch and walleye, but not a reliable one for perch. In walleye, gonadal maturation is essentially completed by mid-winter (several months before spawning) suggesting the possibility to induce walleye to spawn earlier (i.e. from January to March) using relatively simple environmental/hormonal manipulations (Malison et al. 1994). A few years later, using wild broodstock caught in lakes or reservoirs in autumn and then overwintered in ponds, Malison et al. (1998) showed that walleye can be induced to spawn up to 10 weeks (late January) prior to the normal spawning season by using relatively simple environmental (temperature increase from 2 to 10 °C over a 1-week period, 12L:12D) and hormonal treatments (human chorionic gonadotropin, hCG; des-Gly^{10} [D-Ala^{6}] LHRH-ethylamide, LHRHa). They reported that the eggs are of similar quality (% of live eggs at 6 days post fertilization) to those from walleye spawning during the normal reproductive season. In pikeperch, using a very similar method (broodstock under natural photo-thermal conditions until late December, then gradual increase in temperature and photoperiod up to 12 °C and 14L:10D, respectively, over a 1-month period and injection of human chorionic gonadotropin hCG), similar results were obtained by Zakes and Szczepkowski (2004) or Zakes (2007) with spawning advanced by 3 months (early February) compared to the natural spawning season. Similar advanced pikeperch spawning (2 months before the natural spawning season) can be achieved without hormonal injection, but only with temperature and photoperiod manipulations. In their study, Müller-Belecke and Zienert (2008) accelerated maturation and spawning of mature (adult) pikeperch at 15 °C and 16 h light after a cooling period below 10 °C combined with a photoperiod of 8 light (L):16 dark (D). However, these authors did not succeed in obtaining postponed spawning with viable larvae.

In yellow perch, Kayes and Calbert (1979) did not succeed to advance spawning in wild breeders caught in February by warming up the water temperature precociously (up to 12 °C). In their experiment, spawning occurred at the same time as in the corresponding wild population. In a similar trial, using yellow perch broodstock caught in September, an artificial photo-thermal programme (photoperiod and temperature decrease down to L:D 10:14 and 7.5 °C from September to October, followed by stable conditions and finally photoperiod and temperature increase up to L:D 15:9 and 13 °C in January) allowed advanced spawning to be obtained (1 month earlier as compared to the natural spawning occurring in April in the control group) (Dabrowski et al. 1996). However, lower maximum GSI (14 % vs 22 %) and viability of eggs (36 % vs 81 %) were recorded. These authors concluded that condensing the photo-thermal cycle, in expense of the vitellogenic phase, is not an effective method to induce advanced spawning in female yellow perch. According to them, photo-thermal manipulation during the period of active ova growth and vitellogenesis, as well as during the post vitellogenic phase of ovarian development, does not lead to accelerated and/or normal spawning in yellow perch. In males, patterns of changes in GSI, milt weight and sperm concentrations in groups with one out of the two factors shifted (temperature, photoperiod) were not significantly different from those of the control group (ambient temperature and photoperiod). It thus seems that variation of only one factor is enough to drive spermatogenesis until the spawning period. Males and females yellow perch respond differently to such photoperiod and temperature variations.

3.3.2.2 Obtaining Out-of Season Spawning

In percid fish, photoperiod and/or temperature manipulations can be used to delay the onset of the reproductive cycle and to avoid negative effects of undesirable reproductive cycles during the on-growing phase (negative interactions with growth). It was shown that the application of continuous light (L:D 24:0) or constant light based on a long day length (L:D 14:10, L:D 16:8 or L:D 17:7) inhibits the onset of puberty in Eurasian perch (Migaud et al. 2003b, 2004a; Abdulfatah et al. 2011), yellow perch (Shewmon et al. 2007) or pikeperch (Ben Ammar et al. 2015), even if water temperature is decreased. In Eurasian perch a photoperiod increase is also effective in delaying the onset of puberty (Fontaine et al. 2006). Generally, simulation of summer conditions delays gonad development and puberty (Dabrowski et al. 1996). The inhibition of the female reproductive cycle by constant photo-thermal conditions may be related to lower sex steroid levels and to an inhibition of ovarian regulation by gonadotropins, probably stopping gonadogenesis before the vitellogenesis stage (Milla et al. 2009).

Under farm conditions (RAS), out-of-season spawning based on a complete control of the reproductive cycle, from its initial induction to spawning, is currently used for Eurasian perch or pikeperch production in Europe (between 4 and 12 spawnings per year according to SMEs). Management of the reproductive cycle is based only on temperature and photoperiod manipulations, combined or not with hormonal treatments for spawning synchronization (see Chap. 4). In percid fishes, the reproductive cycle is induced by decreasing both temperature and photoperiod, then a chilling period to allow vitellogenesis and finally increasing temperature so as to synchronize the final stages of the maturational process (Wang et al. 2010). A reliable photo-thermal protocol has been finalized for Eurasian perch, which allows 100 % gravid females, spawning and spermiating males to be obtained at each out-of-season spawning period after 11 months of photo-thermal manipulations (Fig. 3.3, Abdulfatah 2010). To achieve such performance, the main critical steps that need to be followed are:

- The application of a photoperiod decrease (high amplitude, 4 or 8 h) followed by a gradual temperature decrease 1 month later (Wang et al. 2006; Abdulfatah et al. 2011, 2013),
- A long wintering period with a low temperature (6 °C) and constant short day length (L:D 8:16) (Abdulfatah 2010; Abdulfatah et al. 2013) and
- A water temperature increase up to 12–14 °C (Abdulfatah et al. 2012).

Furthermore, some recommendations should be mentioned. An initial low amplitude photoperiod decrease (1 h) is too close to the threshold of sensitivity below which females do not perceive photoperiod variations (Abdulfatah et al. 2011). Even if the water temperature remains constant at 22 °C, a photoperiod decrease from summer conditions (L:D 16:8) to winter conditions (L:D 8:16) induces the

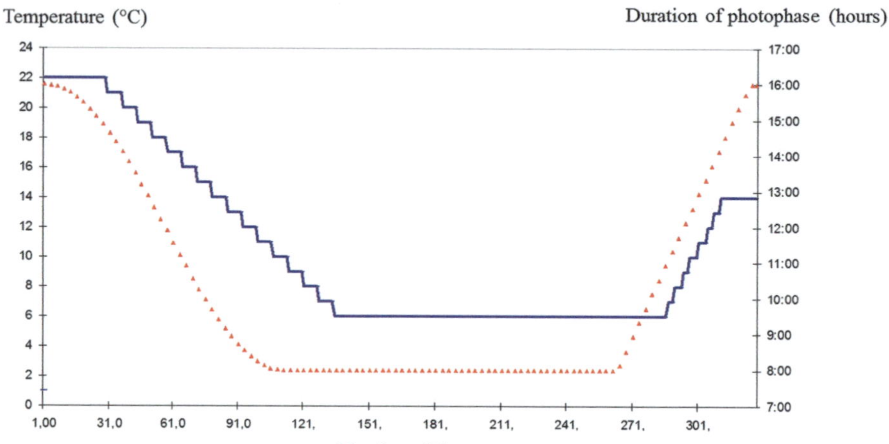

Fig. 3.3 Photo-thermal protocol applied under farm conditions to obtain out-of-season spawning in Eurasian perch (Abdulfatah 2010). *Blue line*: variations of temperature, *red line*: variations of photophase duration

reproductive cycle in females (Abdulfatah et al. 2013). Similar results were observed in yellow perch (Shewmon et al. 2007).

Recently, research has also been examining how to induce out-of-season spawning in pikeperch. Taking into account the results of Müller-Belecke and Zienert (2008) and Hermelink et al. (2011, 2013) set up an experiment about the induction of puberty in virgin pikeperch of 2 years old (mean body weight: 350 g) by different temperature regimes under a constant photoperiod (L:D 12:12). In their first study, broodstock was subjected to five different cooling regimes (from 23 °C to 6 °C, 9 °C, 12 °C, 15 °C or 23 °C over a 30-day period) and then fish were reared at these 5 temperatures for 5 months. After 4 months (including the initial cooling period), female pikeperch reared at 12 °C showed a significant increase in GSI and plasma E2 concentration and 90 % of the females were in mid-vitellogenesis. Later, significant oogenesis development was also observed at 9 and 15 °C, whereas at 6 and 23 °C in nearly all females stagnation of oogenesis was recorded. They concluded first that a moderate temperature (12 °C) is most efficient and that high (23 °C, no thermal change) and low (6 °C) temperatures prevent gonadal maturation, and second that temperature is the prime factor in the induction of puberty in pikeperch (Hermelink et al. 2011). In the framework of a second study (Hermelink et al. 2013), gametogenesis was induced by a 14-day cooling period (23 °C down to 12 °C) followed by subsequent rearing at 12 °C for 12 weeks at a fixed L:D 12:12 photoperiod, and finally breeders (250 g) were kept for 12 more weeks at 12 °C, 14 °C, 16 °C or 18 °C. It was shown that post pubertal temperatures about 14 °C promoted complete ripening in pikeperch of both sexes within 8 weeks, whereas higher temperatures constrained full gonadal maturation. Furthermore, with reference to the optimal temperature for the induction of puberty, slightly higher temperatures about 14 °C are optimal for post pubertal maturation.

All these studies indicate that the induction of out-of-season reproduction in percid fishes (based on a complete control of the reproductive cycle) requires using RAS systems equipped with cooling units allowing temperature to be accurately controlled. Pikeperch seem to be less dependent on photoperiod variations than Eurasian perch. Nevertheless, it was recently demonstrated that photoperiod conditions modify pikeperch spawning performances such as spawning timing, so that fish under L:D 24:0 spawned earlier than those of other photoperiods (Sara Poursaeid et al. 2012). As it was demonstrated in Eurasian perch that photoperiod conditions are determinant during the initial induction phase and in the control of the final gamete quality (Migaud et al. 2004b, 2006) further studies will be necessary to specify the role of photoperiod in pikeperch out-of-season spawning.

3.4 Regulation of Reproductive Performance

The regulation of fish reproductive performance is multifactorial, with many environmental, nutritional or populational (related to broodstock characteristics) factors affecting them (Bromage and Roberts 1995). In percid fishes, few studies have examined this issue. Within the environmental factors, the effect of light has already been considered. It was shown that photoperiod affects gametogenesis, spawning time, spawning rate, egg quality and broodstock mortality (Migaud et al. 2006; Abdulfatah 2010; Sara Poursaeid et al. 2012). More particularly, Migaud et al. (2006) demonstrated that daily light variations are important in the control of Eurasian perch spawning. Likewise, Wang (2006) observed that male Eurasian perch sperm quality is affected by light intensity in interaction with their initial nutritional status. Males with low initial reserves (perivisceral adipose tissue) subjected to low light intensity (100 vs 1,000 lx) develop higher GSI with sperm in larger amount and with higher spermatozoa concentrations. Obviously, temperature is another key factor to obtain normal gametogenesis and high spawning quality. The wintering temperature must be kept low, under a threshold that seems to be different for Eurasian perch and pikeperch (Abdulfatah 2010; Hermelink et al. 2011, 2013). For example, in Eurasian perch, a wintering temperature of 14 °C is associated with advanced spawning with "white eggs" (no larvae) (Abdulfatah 2010). For this species, the wintering temperature must be below 14 °C in accordance with what Hokanson (1977) suggested for yellow perch (below 10 °C). Handling also appears as an important factor regulating percid reproductive performance. In Eurasian perch (Wang et al. 2006), handling combined with temperature modulated the broodstock response. One hundred per cent of the unhandled females held at 14 °C after the initial cooling period (9 weeks) were at the vitellogenesis stage with low plasma levels of cortisol (9 ng mL^{-1}) and high levels of E2 (1.6 ng mL^{-1}). In comparison, handled fish held at 6 °C exhibited a lower rate of females at the vitellogenesis stage (40–73 %) associated with higher plasma levels of cortisol (40–90 ng mL^{-1}) and lower levels of E2 (0.6–1.1 ng mL^{-1}). When a handling stress is applied later, for example during the spawning season, it can result in a delay or lack of spawning as observed in pikeperch (Sara Poursaeid et al. 2012). Broodstock diet was also shown to be another major issue for percid reproduction. Concerning nutritional factors, the initial status of broodstock is important while the initial fat storage of male pikeperch influences the quality of sperm, in terms of spermatozoa concentration and motility, and timing of spermiation. Fish with higher initial fat storage (fat index: 1.11 % vs 0.58 %) spermiated earlier during the spawning season than broodstock with a low level of fat (Teletchea et al. 2009b). Indeed, Wang et al. (2009) reported that the dry feed they had used was not totally adequate for percid reproduction. Henrotte et al. (2010a) have proved that the optimal dietary ratio for DHA/EPA/ARA must be 3/2/2 for Eurasian perch broodstock. These authors also showed that the dietary n-3/n-6 ratio affects the lipid composition of male perch semen but not the indicators of sperm quality (Henrotte et al. 2010b and Chap. 22).

3.5 Conclusions

Valuable results obtained over the last two decades have provided a real knowledge of percid broodstock management and the control of their reproductive cycles, especially to extend the natural spawning season (advanced spawning) and to induce out-of-season spawning with a complete control of the reproductive cycle and production of viable larvae. Reliable protocols are now available. However, further research is needed to secure the reproductive performance and to reduce the cost of juvenile production. Priority should be given to (1) the identification of the upper threshold for the wintering period, and (2) a better knowledge of the multifactorial regulation of reproductive performance.

References

Abdulfatah A (2010) Etude du déterminisme environnemental du cycle de reproduction de la perche commune (*Perca fluviatilis*). Thèse de l'I.N.P.L., Nancy, 182 p. (in French)

Abdulfatah A, Fontaine P, Kestemont P, Gardeur JN, Marie M (2011) Effects of photothermal kinetic and amplitude of photoperiod decrease on the induction of the reproduction cycle in female Eurasian perch *Perca fluviatilis*. Aquaculture 322–323:169–176

Abdulfatah A, Fontaine P, Marie M (2012) Effect of final temperature and photoperiod kinetics on female Eurasian perch spawning induction. AQUA 2012: global aquaculture – securing our future, 1–5 Sept, Prague

Abdulfatah A, Fontaine P, Kestemont P, Milla S, Marie M (2013) Effects of the thermal threshold and the timing of temperature reduction on the initiation of reproduction cycle in female of Eurasian perch *Perca fluviatilis*. Aquaculture 376–379:90–96

Alm G (1954) Maturity, mortality and growth of Perch, *Perca fluviatilis* L, grown in ponds. Rep Inst Freshw Res Drottningholm 35:11–20

Amano M, Iigo M, Ikuta K, Kitamura S, Yamada H, Yamamori K (2000) Roles of melatonin in gonadal maturation of underyearling precocious male Masu salmon. Gen Comp Endocrinol 120:190–197

Barry TP, Malison JA, Lapp AF, Procarione LS (1995) Effects of selected hormones and male cohorts on final oocyte maturation, ovulation, and steroid production in walleye (*Stizostedion vitreum*). Aquaculture 138:331–347

Ben Ammar I, Le Doré Y, Iuretig A, Teletchea F, Schaerlinger B, Milla S, Fontaine P (2012) Age effect on puberty onset of Eurasian perch reared under RAS conditions. AQUA 2012, global aquaculture – securing our future, 1–5 Sept, Prague

Ben Ammar I, Teletchea F, Milla S, Ndiaye WN, Ledoré Y, Missaoui H, Fontaine P (2015) Continuous lighting inhibits the onset of reproductive cycle pikeperch males and females. Fish Physiol Biochem 41(2):345–356

Billard R, Le Gac F, Loir M (1990) Hormonal control of sperm production in teleost fish. In: Epple A, Scanes CG, Stetson MH (eds) Progress in comparative endocrinology. Willey-Liss, New York, pp 329–335

Borg B (1994) Androgens in teleost fishes. Comp Biochem Physiol C 109:219–245

Bromage NR, Roberts RJ (1995) Broodstock management and egg and larval quality. Blackwell Science, Oxford, 424 p

Colby PJ, Nepszy SJ (1981) Variations among stocks of walleye (*Stizostedion vitreum vitreum*): management implications. Can J Fish Aquat Sci 38:1814–1831

Craig JF (1974) Population dynamics of perch, *Perca fluviatilis* L. in Slapton Ley, Devon. Freshwat Biol 4:433–444

Craig JF (1977) The body composition of adult perch, *Perca fluviatilis*, in Windermere, with reference to seasonal changes and reproduction. J Anim Ecol 46:617–632

Craig JF (2000) Percid fishes: systematics, ecology and exploitation. Blackwell Science, Oxford

Crim LW, Peter RE, Billard R (1981) Onset of gonadotropic hormone accumulation in the immature trout pituitary gland in response to estrogen of aromatizable androgen steroid hormones. Gen Comp Endocrinol 37:192–196

Dabrowki K, Ciereszko A (1996) The dynamics of gonad growth and ascorbate status in yellow perch, *Perca flavescens* (Mitchill). Aquac Res 27:539–542

Dabrowski K, Ciereszko A, Ramseyer L, Culver D, Kestemont P (1994) Effects of hormonal treatment on induced spermiation and ovulation in the yellow perch (*Perca flavescens*). Aquaculture 120:171–180

Dabrowski K, Ciereszko RE, Ciereszko A, Toth GP, Christ SA, El-Saidy D, Ottobre JS (1996) Reproductive physiology of yellow perch (*Perca flavescens*): environmental and endocrinological cues. J Appl Ichthyol 12:139–148

Dalimier N, Philippart JC, Voss J (1982) Etude éco-éthologique de la reproduction de la perche (*Perca fluviatilis* L.): observations en plongée dans une carrière inondée. Cah Ethol Appl 2:37–50 (in French)

Dalsgaard J, Lund I, Thorarinsdottir R, Drengstig A, Arvonen K, Pedersen P (2013) Farming different species in RAS in Nordic countries: current status and future perspectives. Aquac Eng 53:2–13

De Vlaming VL, Vodicnik MJ, Bauer G, Murphy T, Evans D (1977) Estradiol-17β effects on lipid and carbohydrate metabolism and on the induction of a yolk precursor in goldfish, *Carassius auratus*. Life Sci 20:1945–1952

De Vlaming VL, Wiley HS, Delahunty G, Wallace RA (1980) Goldfish (*Carassius auratus*) vitellogenin: induction, isolation, properties and relationship to yolk proteins. Comp Biochem Physiol 67B:613–623

Demska-Zakes K, Zakes Z (1995) Sex differentiation in pikeperch, *Stizostedion lucioperca* (L.). Archiwum Rybactawa Polskiego 3:51–57

Flesch A (1994) Biologie de la perche (*Perca fluviatilis*) dans le reservoir du Mirgenbach (Cattenom, Moselle). PhD thesis, University of Metz, 241 p. (in French)

Fontaine P (2004) The intensive culture of perch and pikeperch. In: Abstract book of the fourth workshop of the PROFET network, Inland fisheries and freshwater aquaculture, Budapest, p 17, 20–21 Feb 2004

Fontaine P, Migaud H, Mandiki R, Gardeur JN, Kestemont P, Fostier A (2003) Involvement of sex steroids in final stages of oogenesis in Eurasian perch, *Perca fluviatilis*. Fish Physiol Biochem 28:331–332

Fontaine P, Pereira C, Wang N, Marie M (2006) Influence of pre-inductive photoperiod variations on Eurasian perch *Perca fluviatilis* broodstock response to an inductive photothermal program. Aquaculture 255:410–416

Gillet C, Dubois JP (1995) A survey of the spawning of perch (*Perca fluviatilis*), pike (*Esox lucius*), and roach (*Rutilus rutilus*), using artificial spawning substrates in lakes. Hydrobiologia 300–301:409–415

Gillet C, Dubois JP (2007) Effect of water temperature and size of females on the timing of spawning of perch *Perca fluviatilis* L. in Lake Geneva from 1984 to 2003. J Fish Biol 70:1001–1014

Guma'a SA (1978) The effects of temperature on the development and mortality of eggs of perch, *Perca fluviatilis*. Freshw Biol 8:221–227

Hayes DB, Taylor WW (1994) Changes in the composition of somatic and gonadal tissues of yellow perch following white sucker removal. Trans Am Fish Soc 123:204–216

Heibo E, Magnhagen C, Vøllestad LA (2005) Latitudinal variation in life-history traits in Eurasian perch. Ecology 86(12):3377–3386

Heidinger RC, Kayes TB (1986) Chapter 7: Yellow perch. In: Stickney RR (ed) Culture of non salmonid freshwater fishes. CRC Press, Boca Raton, pp 103–113

Henderson BA, Morgan GE (2002) Maturation of walleye by age, size and surplus energy. J Fish Biol 61:999–1011

Henderson BA, Nepszy SJ (1993) Reproductive tactics of walleye (*Stizostedion vitreum*) in Lake Erie. Can J Fish Aquat Sci 51:986–997

Henderson BA, Wong JL, Nepszy SJ (1996) Reproduction of walleye in Lake Erie: allocation of energy. Can J Fish Aquat Sci 53:127–133

Henderson BA, Trivedi T, Collins N (2000) Annual cycle of energy allocation to growth and reproduction of yellow perch. J Fish Biol 57:122–133

Henrotte E, Mandiki R, Agbohessi TP, Vandecan M, Mélard C, Kestemont P (2010a) Egg and larval quality, and egg fatty acid composition of Eurasian perch breeders (*Perca fluviatilis*) fed different dietary DHA/EPA/AA ratios. Aquac Res 41:51–63

Henrotte E, Kaspar V, Rodina M, Psenicka M, Linhart O, Kestemont P (2010b) Dietary n-3/n-6 ratio affects the biochemical composition of Eurasian perch (*Perca fluviatilis*) semen but not indicators of sperm quality. Aquac Res 41:31–38

Hermelink B, Wuertz S, Trubiroha A, Rennert B, Kloas W, Schulz C (2011) Influence of temperature on puberty and maturation of pikeperch, *Sander lucioperca*. Gen Comp Endocrinol 172:282–292

Hermelink B, Wuertz S, Rennert B, Kloas W, Schulz C (2013) Temperature control of pikeperch (*Sander lucioperca*) maturation in recirculating aquaculture systems – induction of puberty and course of gametogenesis. Aquaculture 400–401:36–45

Hokanson KEF (1977) Temperature requirements of some percids and adaptations to the seasonal temperature cycle. J Fish Res Board Can 34:1524–1550

Houthuijzen RP, Backx JJGM, Buijse AD (1993) Exceptionally rapid growth and early maturation of perch in a freshwater lake recently converted from an estuary. J Fish Biol 43:320–324

Jamet JL, Desmolles F (1994) Growth, reproduction and condition of roach (*Rutilus rutilus* L.), perch (*Perca fluviatilis* L.) and ruffe (*Gymnocephalus cernuus* L.) in eutrophic Lake Aydat (France). Int Rev Gesamten Hydrobiol 79:305–322

Jones DH (1982) The spawning of perch (*Perca fluviatilis* L.) in Loch Leven, Kinross, Scotland. Fish Manag 13(4):139–151

Kayes TG, Calbert HE (1979) Effects of photoperiod and temperature on the spawning of yellow perch (*Perca flavescens*). Proc World Maric Soc 10:306–316

Kestemont P, Mélard C (2000) Chapter 11: Aquaculture. In: Craig JF (ed) Percid fishes: systematics, ecology and exploitation, vol 3, Fish and aquatic resources series. Blackwell Science, Oxford, pp 191–224

Kobayashi M, Aida K, Hanyu I (1989) Induction of gonadotropin surge by steroid hormone implantation in ovariectomized and sexually regressed female goldfish. Gen Comp Endocrinol 73:469–476

Kukuradze AM (1974) A description of the spawning stock and reproduction of pikeperch (*Lucioperca lucioperca*) of the Danube delta and waters in the Danubian region. J Ichthyol 14:385–392

Lappalainen J, Dörner H, Wysujack K (2003) Reproductive biology of pikeperch (*Sander lucioperca* (L.)) – a review. Ecol Freshw Fish 12:95–106

Le Cren ED (1951) The length-weight relationship and seasonal cycle in gonad weight and condition in the perch (*Perca fluviatilis*). J Anim Ecol 20:201–2019

Lehtonen H, Hansson S, Winkler H (1996) Biology and exploitation of pikeperch, *Stizostedion lucioperca* (L.), in the Baltic Sea area. Ann Zool Fenn 33:525–535

Lubzens E, Young G, Bobe J, Cerdà J (2010) Oogenesis in teleosts: how fish eggs are formed. Gen Comp Endocrinol 165:367–489

M'Hetli M, Ben Khemis I, Hamza N, Turki B, Turki O (2011) Allometric growth and reproductive biology traits of pikeperch *Sander lucioperca* at the southern edge of its range. J Fish Biol 78:567–579

Makarova NP (1973) Seasonal changes in some of the physiological characteristics of the perch (*Perca fluviatilis* L.) of Ivan'kovo Reservoir. J Ichthyol 13:742–752

Malison JA, Held JA (1996a) Chapter 2: Reproductive biology and spawning. In: Summerfelt RC (ed) Walleye culture manual, vol 101, NCRAC culture series. North Central Regional Aquaculture Center Publications Office, Iowa State University, Ames, pp 11–18

Malison JA, Held JA (1996b) Reproductive and spawning in walleye (*Stizostedion vitreum*). J Appl Ichthyol 12(3–4):153–156

Malison JA, Kayes TB, Best CD, Amundson CH, Wentworth B (1986) Sexual differentiation and use of hormones to control sex in yellow perch (*Perca flavescens*). Can J Fish Aquat Sci 43(1):26–35

Malison JA, Kayes TB, Held JA, Amundson CH (1990) Comparative survival, growth, and reproductive development of juvenile walleye and sauger and their hybrids reared under intensive culture conditions. Progr Fish Cult 52:73–82

Malison JA, Procarione LS, Barry TP, Kapuscinski AR, Kayes TB (1994) Endocrine and gonadal changes during the annual reproductive cycle of the freshwater teleost, *Stizostedion vitreum*. Fish Physiol Biochem 13(6):473–484

Malison JA, Procarione LS, Kayes TB, Hansen JF, Held JA (1998) Induction of out-of-season spawning in walleye (*Stizostedion vitreum*). Aquaculture 163:151–161

Malservisi A, Magnin E (1968) Changements cycliques annuels se produisant dans les ovaires de *Perca fluviatilis flavescens* (Mitchill) de la region de Montréal. Natur Can 95:929–945 (in French)

Migaud H, Mandiki R, Gardeur JN, Fostier A, Kestemont P, Fontaine P (2003a) Synthesis of sex steroids in final oocyte maturation and induced ovulation in female Eurasian perch, *Perca fluviatilis*. Aquat Living Resour 16(4):380–388

Migaud H, Mandiki R, Gardeur JN, Kestemont P, Bromage N, Fontaine P (2003b) Influence of photoperiod regimes on the Eurasian perch gonadogenesis, spawning and eggs and larvae quality. Fish Physiol Biochem 28:395–397

Migaud H, Fontaine P, Kestemont P, Wang N, Brun-Bellut J (2004a) Influence of photoperiod on the onset of gonadogenesis in Eurasian perch *Perca fluviatilis*. Aquaculture 241:561–574

Migaud H, Gardeur JN, Kestemont P, Fontaine P (2004b) Off-season spawning of Eurasian perch *Perca fluviatilis*. Aquac Int 12:87–102

Migaud H, Wang N, Gardeur JN, Fontaine P (2006) Influence of photoperiod on reproductive performances in Eurasian perch *Perca fluviatilis*. Aquaculture 252:385–393

Migaud H, Davie A, Taylor JF (2010) Current knowledge on the photoneuroendocrine regulation of reproduction in temperate fish species. J Fish Biol 76:27–68

Milla S, Mandiki SNM, Hubermont P, Rougeot C, Mélard C, Kestemont P (2009) Ovarian steroidogenesis inhibition by constant photothermal conditions is caused by a lack of gonadotropin stimulation in Eurasian perch. Gen Comp Endocrinol 163:242–250

Müller-Belecke A, Zienert S (2008) Out-of-season spawning of pikeperch (*Sander lucioperca* L.) without the need of hormonal treatments. Aquac Res 39:1279–1285

Nagahama Y (1983) The functional morphology of teleost gonads. In: Hoar WS, Randall DJ, Donaldson EM (eds) Endocrine tissues and hormones, vol 9, Fish physiology, part A. Academic, New York, pp 223–275

Nagahama Y, Yamashita M (2008) Regulation of oocyte maturation in fish. Develop Growth Differ 50:195–219

Noaksson E, Gustavsson B, Linderoth M, Zebühr Y, Broman D, Balk L (2004) Gonad development and plasma steroid profiles by HRGC/HRMS during one reproductive cycle in reference and leachate-exposed female perch (*Perca fluviatilis*). Toxicol Appl Pharmacol 195:247–261

Pankhurst NW, van der Kraak G, Peter RE (1986) Effects of human chorionic gonadotropin, Des-Gly10 (D-Ala^{6}) LHRH-ethylamide and pimozide on oocyte final maturation, ovulation and levels of plasma sex steroids in the walleye (*Stizostedion vitreum*). Fish Physiol Biochem 1:45–54

Papageorgiou NK (1977) Fecundity and reproduction of perch (*Perca fluviatilis* L.) in Lake Agios Vasilios, Greece. Freshw Biol 7:559–565

Poulet N (2004) Le sandre (*Sander lucioperca* (L.)): biologie, comportement et dynamique des populations en Camargue (Bouches du Rhône, France). Thèse Université de Toulouse III: 286 p. (in French)

Raikova-Petrova G, Zivkov M (1998) Maturity, spawning and sex ratio of pikeperch, *Stizostedion lucioperca* (L.) in two Bulgarian reservoirs as compared to other European habitats. J Appl Ichthyol 14:31–35

Sandström O, Abrahamsson I, Andersson J, Vetemaa M (1997) Temperature effects on spawning and egg development in Eurasian perch. J Fish Biol 51:1015–1024

Sara Poursaeid S, Falahatkar B, Takami GA, Efatpanah I (2012) Effects of different photoperiods and handling stress on spawning and reproductive performance of pikeperch *Sander lucioperca*. Anim Reprod Sci 132:213–222

Schlumberger O, Proteau JP (1996) Reproduction of pike-perch (*Stizostedion lucioperca*) in captivity. J Appl Ichthyol 12:149–152

Schulz RW, França LR, Lareyre JJ, Le Gac F, Chiarini-Garcia H, Nobrega RH, Miura T (2010) Spermatogenesis in fish. Gen Comp Endocrinol 165:390–411

Scott WB, Crossman EJ (1973) Freshwater fishes of Canada. J Fish Res Board Can 194:1–966

Shewmon LN, Godwin JR, Murashige RS, Daniels HV (2007) Environmental manipulation of growth and sexual maturation in yellow perch, Perca flavescens. J World Aquacult Soc 38(3):383–394

Sulistyo I, Rinchard J, Fontaine P, Gardeur JN, Capdeville B, Kestemont P (1998) Reproductive cycle and plasma levels of sex steroids in female Eurasian perch *Perca fluviatilis*. Aquat Living Resour 11(2):101–110

Sulistyo I, Fontaine P, Rinchard J, Gardeur JN, Migaud H, Capdeville B, Kestemont P (2000) Reproductive cycle and plasma levels of sex steroids in male Eurasian perch *Perca fluviatilis*. Aquat Living Resour 13(2):80–106

Tanasichuk RW, MacKay WC (1989) Quantitative and qualitative characteristics of somatic and gonadal growth of yellow perch (*Perca flavescens*) from Lac Ste. Anne, Alberta. Can J Fish Aquat Sci 46:989–994

Tao Y, Berlinsky DL, Sullivan CV (1996) Characterization of a vitellogenin receptor in white perch (*Morone americana*). Biol Reprod 55:646–656

Teletchea F, Fontaine P (2014) Levels of domestication in fish: implications for the sustainable future in aquaculture. Fish Fish 15:181–195

Teletchea F, Fostier A, Kamler E, Gardeur JN, Le Bail PY, Jalabert B, Fontaine P (2009a) Comparative analysis of reproductive traits in 65 freshwater fish species: application to the domestication of new species. Rev Fish Biol Fish 19:403–430

Teletchea F, Gardeur JN, Psenicka M, Kaspar V, Le Doré Y, Linhart O, Fontaine P (2009b) Effects of four factors on the quality of male reproductive cycle in pikeperch *Sander lucioperca*. Aquaculture 291:217–223

Thorpe JE (1977) Morphology, physiology, behaviour and ecology of *Perca fluviatilis* L. and *Perca flavescens* Mitchill. J Fish Res Board Can 34:1504–1514

Treasurer JW (1983) Estimate of egg and viable embryo production in a lacustrine perch, *Perca fluviatilis*. Environ Biol Fish 8(1):3–16

Treasurer JW, Holliday FGT (1981) Some aspects of the reproductive biology of perch *Perca fluviatilis* L. A- histology description of the reproductive cycle. J Fish Biol 18:359–376

Turner CL (1919) The seasonal cycle in the spermary of the perch. J Morphol 32:681–711

Urho L (1996) Habitat shifts of perch larvae as survival strategy. Ann Zool Fenn 33(3–4):329–340

Uysal K, Yerlikaya A, Aksoylar MY, Yöntem M, Ulupinar M (2006) Variations in fatty acids composition of pikeperch (*Sander lucioperca*) liver with respect to gonad maturation. Ecol Freshw Fish 15:441–445

Wallace RA (1978) Oocyte growth in non mammalian vertebrates. In: Jones RE (ed) The vertebrate ovary. Plenum Press, New York, pp 469–502

Wallace RA, Selman K (1981) Cellular and dynamic aspects of oocyte growth in teleosts. Am Zool 21:325–343

Wang N (2006) Déterminisme multifactoriel de la qualité du cycle de reproduction chez la perche commune *Perca fluviatilis*. Approche multifactorielle. Thèse de l'Université Henri Poincaré, Nancy 1, 140 p. (in French)

Wang N, Gardeur JN, Henrotte E, Marie M, Kestemont P, Fontaine P (2006) Determinism of the induction of the reproductive cycle in female Eurasian Perch, *Perca fluviatilis*: effects of environmental cues and modulating factors. Aquaculture 261:706–714

Wang N, Mandiki SNM, Henrotte E, Bouyahia AG, Mairesse G, Rougeot C, Mélard C, Kestemont P (2009) Effect of partial or total replacement of forage fish by a dry diet on the quality of reproduction in pikeperch, *Sander lucioperca*. Aquac Res 40:376–383

Wang N, Teletchea F, Kestemont P, Milla S, Fontaine P (2010) Photothermal control of the reproductive cycle in temperate fishes. Rev Aquac 2:209–222

Young G, Kagawa H, Nagahama Y (1983) Evidence for a decrease in aromatase activity in the ovarian granulosa cells of amago salmon (*Oncorhynchus rhodurus*) associated with final oocyte maturation. Biol Reprod 29:310–315

Young G, Lokman PM, Kusakabe M, Nakamura I, Goetz FW (2005) Gonadal steroidogenesis in teleost fish. In: Hew C (ed) Molecular aspects of fish and marine biology, vol 2. In: Sherwood N, Melamed P (eds) Hormones and their receptors in fish reproduction. World Scientific Press, Singapore, pp 155–223

Zakes Z (2007) Out-of-season spawning of cultured pikeperch [*Sander lucioperca* (L.)]. Aquac Res 38:1419–1427

Zakes Z, Demska-Zakes K (1996) Effects of diets on growth and reproductive development of juvenile pikeperch, *Stizostedion lucioperca* (L.), reared under intensive culture conditions. Aquac Res 27:841–845

Zakes Z, Szczepkowski M (2004) Induction of out-of-season spawning of pikeperch, *Sander lucioperca* (L.). Aquac Int 12:11–18

Zeh M, Ritter E, Ribi G (1989) Spawning and egg development of *Perca fluviatilis* in Lake Zürich. Aquat Sci 51(2):100–107

Zohar Y, Munoz-Cueto JA, Elizur A, Kah O (2010) Neuroendocrinology of reproduction in teleost fish. Gen Comp Endocrinol 165:438–455

Chapter 4
Artificial Reproduction of Percid Fishes

D. Żarski, A. Horváth, J.A. Held, and D. Kucharczyk

Abstract Artificial reproduction, being a specific human intervention in the process of reproduction, is a key step in aquaculture of percid fishes. This group of fish, exhibits specific traits, considered amenable to artificial reproductive protocols. For example, this is the only extensively studied group of freshwater teleosts where application of human chorionic gonadotropin (hCG) and gonadoliberine analogues (GnRHa) alone, promotes final oocyte maturation (FOM) and spawning without any other hormonal therapy, whereas in other species (cyprinids, catfishes or salmonids) anti-dopaminergic treatment is also needed. Another characteristic trait is that percid females can release their eggs spontaneously in the tank, regardless of the presence of males. This makes artificial spawning of these fish relatively difficult. In the present chapter endocrine regulation as well as reproductive protocols applied to this group of fish are reviewed extensively, however, the focus of this review is on the final gamete maturation, spermiation and ovulation processes are the steps considered from artificial reproduction perspectives. The published data revealed that scientific activity was focused mainly on the problem of synchronization of ovulation and the effectiveness of different hormonal therapies. This evolved into the development of several specific protocols and methods (e.g. percid-specific preovulatory maturational stages of oocytes), which allowed improvement of that in these species. It was also established, that hCG or GnRHa applied alone are the most effective spawning agents in wild or pond-reared percids. However, there is still a considerable lack of data considering the effectiveness of these protocols in controlled reproduction of domesticated broodstocks. Apart from that, there are many other aspects to be investigated. Such as hormonal regulation of final gamete maturation and spawning, verification of some reproductive protocols as possible

D. Żarski (✉) • D. Kucharczyk
Department of Lake and River Fisheries,
University of Warmia and Mazury, Olsztyn, Poland
e-mail: danielzarski@interia.pl

A. Horváth
Department of Aquaculture, Szent István University, Gödöllő, Hungary

J.A. Held
University of Wisconsin-Extension Madison, Madison, WI, USA
e-mail: jaheld@wisc.edu

P. Kestemont et al. (eds.), *Biology and Culture of Percid Fishes*,
DOI 10.1007/978-94-017-7227-3_4

gamete quality determinants and gamete management protocols (prior to and following fertilization), which were relatively scarcely studied.

Keywords Hormonal regulation • Fertilization • Spermiation • Ovulation • Percids

4.1 Introduction

Spawning, either natural or artificial, is a key step in fish lifecycle. In the wild the act of spawning closes the development of the organism into adult specimen and in this way ensures the survival of the species. In captivity spawning is a necessary element securing constant production. All the commercially important percid fishes exhibit some differences considering the reproductive features, where the structure of the eggs laid (in the form of 'ribbon' or as a batch of single eggs) is the most distinctive one (for details see Chap. 2). However, there are also many similarities allowing consideration of these species as a one specific group representing comparable pattern of gonadal development as well as gametes maturation and spawning (see also Chaps. 2 and 3).

Artificial reproduction may be defined as a human intervention in the process of reproduction (Woynarovich and Horváth 1980). It takes into account a number of techniques aimed at producing high quality offspring suitable for further culture process or restocking. Generally, as considering artificial reproduction human intervention was concentrated on the final phases of the maturation of spawners (final oocyte maturation, ovulation and spermiation) and the methods of control over these processes as well as gametes management and other procedures (in vitro fertilization, egg treatment, incubation) up to hatching. This concerns, among others, manipulation of environmental conditions, hormonal therapies, gametes handling and other protocols which were proved to be crucial for successful reproduction.

From an aquaculture perspective, successful reproduction is one of the crucial steps in the culture process. Its effectiveness directly affects the production capacity and in this way may influence the production profitability. In all cultivated percids the reproduction of captive broodstock is feasible. Nevertheless, the artificial reproduction of wild or pond reared fish (which are exposed to natural food source and natural photo-thermal conditions) is still a major concern. Therefore, this chapter extensively described the procedures of artificial reproduction of wild (or pond reared) fish as well the one of breeders held in intensive conditions.

4.2 Endocrine Regulation of Spermiation

Endocrine regulation of sexual maturation and spermatogenesis in male percids is generally similar to that of other teleosts. As in most animal species spermatogenesis of fish is also controlled by the hypothalamic-pituitary-gonad axis. The gonadotropin releasing hormone (GnRH) produced in the hypothalamus stimulates the release of gonadotropins (GtH) of the pituitary which in turn stimulate the steroid (androgen) production in the testis (Vizziano et al. 2008). While the function of androgens has been studied to some extent in percids, very little information is available on the male-specific actions of either GnRH or GtH-s in these species. GnRH is a decapeptide, a neurohormone that is mainly localized in the anteroventral preoptic region of the telencephalon (Peter 1983) which does not seem to exert male- or female-specific action. Its production is controlled by an antagonist, the dopamine, which is a unique feature in teleosts. Inhibition of GnRH action by dopamine is not observed in all teleosts, and particularly little information is available regarding this process in males. Fish gonadotropins are present in two forms characterized as either GtH I and GtH II being homologous with mammalian FSH and LH (Vizziano et al. 2008) or as FSH and LH proper (Schulz et al. 2010). Unlike in mammals where LH regulates sex steroid release by Leydig cells and FSH is responsible for the regulation of Sertoli cell activities in males, in fish the roles of the two forms are not clearly separated. Receptors of FSH display a preference for FSH, however, they can be activated by LH, too, although at higher concentrations. LH receptors seem to be more LH-specific in fish where cross-activation by FSH is possible only in non-physiologically high concentrations (Bogerd et al. 2005). Nevertheless, FSH is known to stimulate steroidogenesis in fish, suggesting that FSH receptors are present in the Leydig cells. The expression of FSH and LH receptor mRNA in Leydig cells was demonstrated for the first time in fish (and vertebrates) in the African catfish (*Clarias gariepinus*) by García-López et al. (2009). Generally, FSH concentrations are high during the onset of spermatogenesis (proliferation of spermatogonia, meiosis and early spermiogenesis) while LH levels start to increase close to spawning (spermiogenesis and spermiation) what was described extensively in Chap. 3.

The unique study on the gonadotropins of percid males has shown an increased expression of FSHβ and LHβ (the hormone specific β-subunit of gonadotropins) mRNA in pikeperch (*Sander lucioperca*) during early and mid-spermatogenesis compared to pre-spermatogenesis (Hermelink et al. 2011). The increase of LHβ mRNA expression was more pronounced than that of the FSHβ mRNA which corresponds with the findings in other teleost species. The main steroid hormones found in the testis include estrogens such as 17β-estradiol (E2), androgens such as testosterone (T) and 11-ketotestosterone (11-KT) and progestins such as 17α,20β-dihydroxy-4-pregnen-3-one (abbreviated as DHP or 17,20β-P) and 17α,20β,21-trihydroxy-4-pregnen-3-one (20β-S) (Schulz et al. 2010). The major sites of sex steroid synthesis in the testis are the Leydig cells that can be found in the interstitium between individual tubules or lobules. These are large polygonal cells characterized by extensive agranular endoplasmatic reticulum and mitochondria with tubular christae. These cells appear before or during sex differentiation depending on the species. These cells are believed to be homologous with the steroid-producing Leydig cells of higher vertebrates (Billard et al. 1982).

The function of estrogens in the testis is still largely unclear but it is suggested that it plays a role in spermatogonial renewal and multiplication (Vizziano et al. 2008). They have also been reported to play a major role in gene expression in the testis (Schulz et al. 2010). In percid males E2 levels were rarely studied. As a notable exception, E2 concentration in males of yellow perch (*Perca flavescens*) was found to be low throughout the year with a five to tenfold increase in November–December to 50–100 pg/ml while decreasing just before the spawning season (Ciereszko et al. 1998). In contrast, in Eurasian perch (*Perca fluviatilis*) males E2 concentrations were found to be high during the spawning season (probably sustaining spermiation) and decreased significantly after the end of spawning (Rougeot et al. 2004). This corresponds with the findings of Hermelink et al. (2011) who reported that E2 levels remained at high and stable levels during pre-, early and mid-spermatogenesis in pikeperch.

Androgen receptors are expressed in the testicular somatic cells (primarily Sertoli and interstitial cells) but not in the germ cells (Schulz et al. 2010). They effectively support the entire process of spermatogenesis from spermatogonial multiplication to spermiation. Although both can be found in females, too, levels of 11-KT were found to be ten times higher in pikeperch males than in females (Teletchea et al. 2009a). Seasonal changes in plasma T and 11-KT levels have been relatively well described in percids and follow the same general pattern of decreasing abruptly following spawning and beginning to increase roughly 4–5 months before the subsequent spawning (described in greater details in Chap. 3).

Progestins have been shown to be present in elevated concentrations throughout the spermiation process in many fish species (Schulz et al. 2010). They play a particularly important role in sperm maturation and release, increasing pH of sperm ducts and probably exert their action through a membrane receptor of spermatozoa (Vizziano et al. 2008). In percid males the presence and concentrations of progestings are rarely measured. Interestingly, in the ruffe (*Gymnocephalus cernuus*) metabolites of 20β-S were found to act as pheromones that stimulated swimming activity and social interactions in conspecific males. When males were exposed to the urine of females injected with 20β-S they started swimming more actively and inspecting each other. In contrast, direct exposure to either progestins or to the urine of females injected with DHP caused no noticeable change in the behavior of males (Sorensen et al. 2004). In the pikeperch, concentrations of DHP stayed stable and low (below 100 pg/ml) during pre-, early and mid-spermatogenesis, suggesting that they only increase during active spermiation (as they do in other teleosts) which was not studied, yet (Hermelink et al. 2011).

4.3 Endocrine Regulation of Final Oocyte Maturation (FOM) and Ovulation

In teleosts process of oocyte maturation and ovulation is preceded by the oocyte growth (vitellogenesis) which involves incorporation of yolk proteins and lipids into oocytes (Tyler and Sumpter 1996; Devlin and Nagahama 2002; Patino and Sullivan 2002). During the vitellogenesis process oocytes are arrested in the first meiotic prophase (Nagahama 1994). The process is up-regulated by hypothalamic-pituitary-gonadal axis, where the major role plays follicle-stimulating hormone (FSH). These processes are extensively described in Chap. 3.

After the complete vitellogenesis and just before ovulation the final oocyte maturation (FOM) process begins. During the FOM, the first meiotic division is completed and the second is started, which proceeds until the metaphase II stage (Nagahama and Yagashita 2008). The process is accompanied by a number of morphological changes involving, for example, migration and decomposition of the germinal vesicle, formation of yolk and coalescence of oil droplets into one large droplet (Migaud et al. 2003; Żarski et al. 2011a, 2012a, c). In percids the most characteristic feature during FOM was the oil droplet coalescence process. It was found that the small oil droplets coalesce into one big oil droplet, what was accompanied together with the GV migration (Żarski et al. 2011a, 2012a, c). During this phase the most important role plays second gonadotropic hormone which is luteinising hormone (LH) secreted by the pituitary gland during FOM (e.g., Kagawa et al. 1998; Patino et al. 2001; Patino and Sulivan 2002). However, Fontaine et al. (2003) suggest also that in the whole process E_2 and T may have been involved. Basically, FOM is triggered by the LH which stimulates theca cell layers of the ovarian follicle for production of 17α-hydroxyprogesterone which traverses the basal lamina and is then converted to the maturation inducing steroid (MIS) by the granulosa cells of the ovarian follicle (e.g., Nagahama 1994; Nagahama and Yamashita 2008). The MIS was proven to bind with the specific receptors of oocyte plasma membrane in postvitellogenic oocytes of teleosts (Nagahama and Yamashita 2008). In the case of freshwater percids, like in most fish species, the role of MIS the most probably plays 17α,20β-dihydroxy-4-pregnen-3-one (DHP) (Goetz and Theofan 1979; Goetz et al. 1989; Barry et al. 1995; Fontaine et al. 2003; Nagahama and Yamashita 2008; see also Chap. 3). The MIS promote the formation and activation of the maturation promoting factor (MPF) which is the complex of cdc2 kinase and cyclin-B (for details see: Yamashita et al. 1992; Nagahama and Yamashita 2008). The MPF is the final inducer of the oocyte maturation process (Nagahama and Yamashita 2008). The data on the MPF in freshwater percids are still missing, however already published data suggest that the pathways and mechanisms of this process is very similar to those described for other teleosts or even amphibians (extensively revised by Nagahama and Yamashita 2008).

In some of the fish species, such as cyprinids, maturation and ovulation are also controlled by the dopamine (DA), which acts as a LH inhibitory factor (Peter and Yu 1997; Mylonas et al. 2010). It is well reflected in the cases when GnRH analogues are administered for induction of maturation and ovulation *in vivo*, since elevated level of gonadotropins (as a result of GnRH administration) in the blood stream affect production of DA and in this way decreases effectiveness of such hormonal therapy (e.g., Peter and Yu 1997; Mylonas et al. 2010). In consequence, administration of GnRH alone does not affect maturation and ovulation. In such cases dopamine antagonist (e.g., metoclopramide, pimozide) administration is needed to affect ovulation (Krejszeff et al. 2009, 2010; Targońska et al. 2010; Kujawa et al. 2011). However, in the case of percids such additional treatment is not needed and administration of GnRH alone affects ovulation successfully (e.g.,

Kouril et al. 1997; Kucharczyk et al. 1998; Ronyai and Lengyel 2010; Żarski et al. 2013a). It may be then concluded that inhibitory effect of DA does not occur in the freshwater percids or it is very weak. Nevertheless, as considering endocrine regulation of FOM and ovulation of those species it seems that this aspect may be omitted.

Ovulation process includes expulsion of the matured oocyte from the ovarian follicles (Goetz 1983; Goetz and Garczynski 1997). Generally, this process is up-regulated by the LH secretion what affects consequently production of MIS (as described above and in Chap. 3). However, it is still not clear which factor directly affect process of ovulation. It was found that DHP plays an essential role in this process in yellow perch (Goetz et al. 1989; Goetz and Garczynski 1997). For the act of follicular rupture during ovulation the most probably proteinases are responsible (Nagahama and Yamashita 2008). However, the biochemical pathways between DHP (17,20β-P) and activation of proteinases remains unknown, although the arachidonic acid and prostaglandins were reported to be involved in this process (Bradley and Goetz 1994; Goetz and Garczynski 1997; Patino et al. 2003). It was reported, that indomethacin, a prostaglandin endoperoxide synthase inhibitor, has been responsible for inhibition of ovulation in yellow perch. It may be then suggested that cyclooxygenase metabolites are responsible in some way in regulation of ovulation (Goetz and Theofan 1979; Goetz and Garczynski 1997).

4.4 Artificial Spawning

4.4.1 Spontaneous Captive Spawning

One of the easiest methods of fish artificial spawning is so called 'semi controlled' ('semi artificial') spawning in tanks, small ponds or cages placed in the pond or lake (Steffens et al. 1996; Kucharczyk et al. 2007; Zakęś and Demska-Zakęś 2009). This method was widely used by inexperienced farmers, who wanted to produce larvae of those species. In general, the spontaneous spawning, regardless if performed in ponds, cages or tanks, is a simple method where spawners of both sexes are kept in a closed environment allowing them spawning without strict human control and stimulated only by the environmental conditions (e.g. Kucharczyk et al. 2007; Policar et al. 2008). The big advantage of this method is that fish are not disturbed by human, what limits the stress and they spawn whenever they are really ready to reproduce. The big disadvantages of such reproduction method are, very often, low fertilization rate, no possibility of controlled fertilization (for example specific crossbreeding, genome manipulation etc.) and limited possibilities to predict the moment of spawning. It may be improved by the application of hormonal treatment (e.g. Kucharczyk et al. 2000, 2007), but it still remains rather difficult to control the reproduction, although fish will spawn more synchronously.

In the case of yellow and Eurasian perch the spontaneous spawning was reported to be used many times (Hinshaw 2006; Ronyai and Lengyel 2010; Żarski et al.

2011a). This is caused by the fact that these species lay eggs in the form of ribbon (described in details in Chap. 2) which may be then easily removed and incubated in a regular hatchery devices. In this case, the fertilization rate should not be a problem if fish are kept in a small tank where spermiating male is able to produce a 'sperm cloud' which allows the spermatozoa surround and fertilize the ribbon effectively (Kucharczyk et al. 2000). However, there are no possibilities to perform selective breeding and/or other specific techniques such as fertilization of eggs with cryopreserved sperm or genome manipulation.

Application of this method of reproduction in the case of walleye and pikeperch is more complicated due to the fact that these species lay eggs as a batch of hundred thousands of sticky eggs strongly attached to any kind of substrate. Therefore, it is necessary to provide the nests in the spawning area. The nests are usually made of different natural (e.g. juniper branches, conifer branches, roots, sedges, alder, rice grass, wood wool, Fig. 4.1) or synthetic materials (Wojda et al. 1994; Skrzypczak et al. 1998; Kucharczyk et al. 2007). After the spawning the eggs together with the substrate were moved to the hatchery and incubated (Fig. 4.2). The incubation of the eggs attached to the substrate is quite problematic. The unfertilized or not developing eggs cannot be removed what usually leading to the fungal infection significantly reducing the incubation effectiveness. The substrate itself, when natural materials are used, is also a very good medium for fungal infection. Therefore, some improvement may be achieved by using the artificial substrate, although the problem of providing good water exchange around all the eggs still exists. Of course, this method of propagation may also be applied with or without hormonal treatment of spawners. However, this method is still much less effective than in the case of

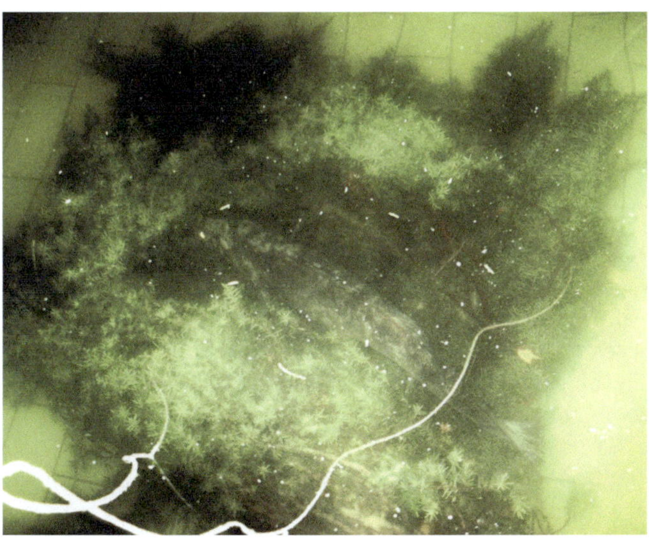

Fig. 4.1 A male pikeperch above the nest made of juniper branches with the eggs laid on it (Photo: D. Żarski)

Fig. 4.2 A juniper branch with the eggs of pikeperch laid on it. The abundance of white eggs indicates a low fertility and the whitish cloudy areas the beginning of fungal infection (Photo: D. Żarski)

yellow and Eurasian perch, since fertilization rate, and consequently survival rate, are usually much lower than the methods involving controlled fertilization.

Actually, the captive spontaneous spawning is the method which may be recommended to inexperienced farmers who has limited knowledge and skills in controlled reproduction. However, it has to be stressed that this method may generate low and unpredictable spawning effectiveness. While this method may be successfully applied in commercial farming of yellow and Eurasian perch, commercial aquaculture of pikeperch and walleye should not rely solely on this kind of propagation method.

4.4.2 Controlled Reproduction

Controlled reproduction involves the procedures of stimulation of spermiation and ovulation in already matured fish, gametes collection and in vitro fertilization. For culture purposes it is becoming a more and more required way for obtaining eggs and larvae. The biggest advantage of this method of propagation is the possibility of selective breeding of specimens with required pheno- and genotypes and to having control over the entire procedure of reproduction (Thorgaard 1995; Zohar and Mylonas 2001; Mylonas et al. 2010). It includes, among others, timing of fertilization, duration of incubation and timing of hatching. The latter, in the case of percids, may be crucial for larviculture effectiveness where size heterogeneity of larvae, caused by the different hatching times, may increase intensity of cannibalism and, in consequence, production outcome (e.g. Baras et al. 2003; Kestemont et al. 2003; Kooten et al. 2010), as extensively described in Chaps. 9, 10 and 11

4.4.2.1 Sex Recognition

In all percids there is no clear sexual dimorphism. It was already reported that during the culture process some sex-dependent growth heterogeneity may be observed (for more details see Chap. 2). However, fish at the same size beyond the spawning season usually looks very similar and it is very hard to distinguish males from females. Only just before or during the spawning season the fish gender may be usually recognized as the males are spermiating after gentle pressure of their abdomen and females display distended abdominal part of body. However, it may still happen (especially prior to spawning season) that the sex be indistinguishable. In that case catheterization (Fig. 4.3) may be a useful method for gender verification (Ross 1984).

4.4.2.2 Determination of Maturity Stage of Females

The effectiveness of hormonal stimulation as well as latency time following hormonal treatment in females are strictly dependent on the maturity stage (Brzuska 1979, 1988). Until recently, in freshwater aquaculture, the germinal

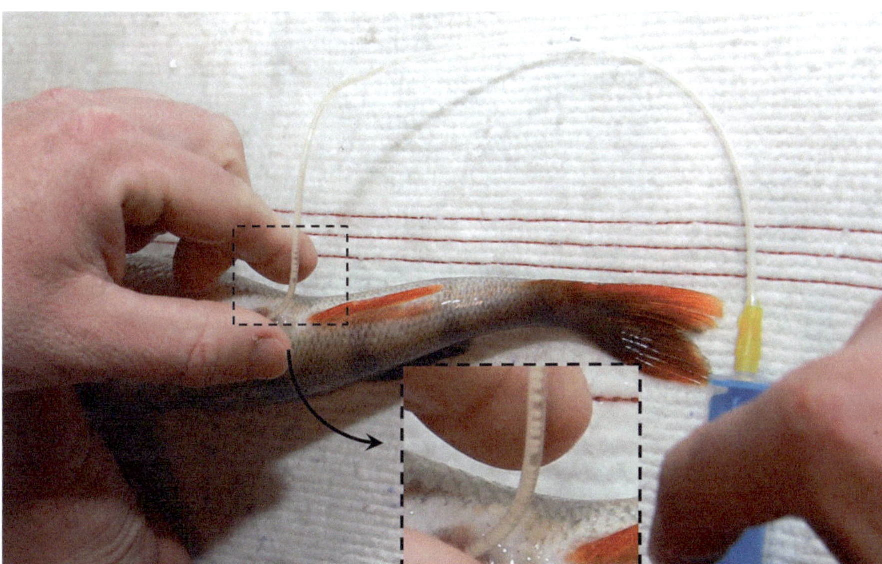

Fig. 4.3 Catheterization of the Eurasian perch female. Close to the genital pore oocytes may be seen in the lumen of the catheter (Photo: S. Krejszeff)

vesicle (GV) position in preovulatory oocytes was taken into consideration for determining the maturation stage of females. In effect four stages were distinguished: central position of GV (stage 1), early migration of GV (stage 2), late migration of GV (stage 3), peripheral position of GV or oocytes after the GV breakdown, i.e. GVBD (stage 4). However, the application of this method in freshwater percids in many cases created many ambiguities and consequently revealed weakness of this classification in this group of fish (Żarski et al. 2011a, 2012a). The largest ambiguity concerned the last stage (stage 4), where some authors to this stage qualified oocytes with the GV at a peripheral position (Kucharczyk et al. 1996, 1998, 2001; Targońska et al. 2014; Żarski et al. 2011b) and others oocytes which already undergone GVBD (Barry et al. 1995; Malison et al. 1998; Ronyai 2007; Zakęś and Demska-Zakęś 2009). Therefore, a verification of the final oocyte maturation (FOM, as described earlier in Sect. 4.3) process was made in order to distinguish a percid-specific features of preovulatory oocytes. Actually, six preovulatory oocyte maturation stages were distinguished, first in Eurasian perch (Żarski et al. 2011a):

- Stage I: the GV was situated in the oocyte centre, oil droplets were poorly visible (Figs. 4.4a and 4.5a);
- Stage II: beginning of GV migration, beginning of coagulation of clearly visible oil droplets (Figs. 4.4b and 4.5b);
- Stage III: migrating of GV (above half of the oocyte diameter), oil droplets were clearly visible (Figs. 4.4c and 4.5c);
- Stage IV: the GV is at the oocyte periphery, a large forming oil droplet was clearly visible; the droplet diameter was greater than the GV diameter and it reached the size of about one third of the oocyte diameter; also smaller droplets were visible (Figs. 4.4d and 4.5d);
- Stage V: the GV is situated at the oocyte edge, clearly visible one large (size of about half the oocyte diameter) oil droplet (Figs. 4.4e and 4.5e);
- Stage VI: oocyte samples taken for analysis were macroscopically transparent; no visible GV after they were placed in Serra's solution (following GVBD), oocytes at the pre-ovulation stage (Figs. 4.4f and 4.5f, g);

It may be found, that for the classification, not only the position of germinal vesicle but also the oil droplets coalescence rate were considered. This phenomenon, next investigated histologically (Żarski et al. 2012c, Fig. 4.5), seems to be very specific to percid fishes as this classification was later successfully transferred also to pikeperch (Żarski et al. 2012a). Therefore, it may be suggested that this classification may be used (with some minor modifications) in all percids allowing higher synchronization and more accurate prediction of ovulation.

4.4.2.3 Stimulation of Spermiation

Regarding hormonal induction of spermiation in percids, little information is available that is specific to this taxon. Percids are known to release an ample amount of sperm during the reproductive season. For example, non-treated Eurasian perch was

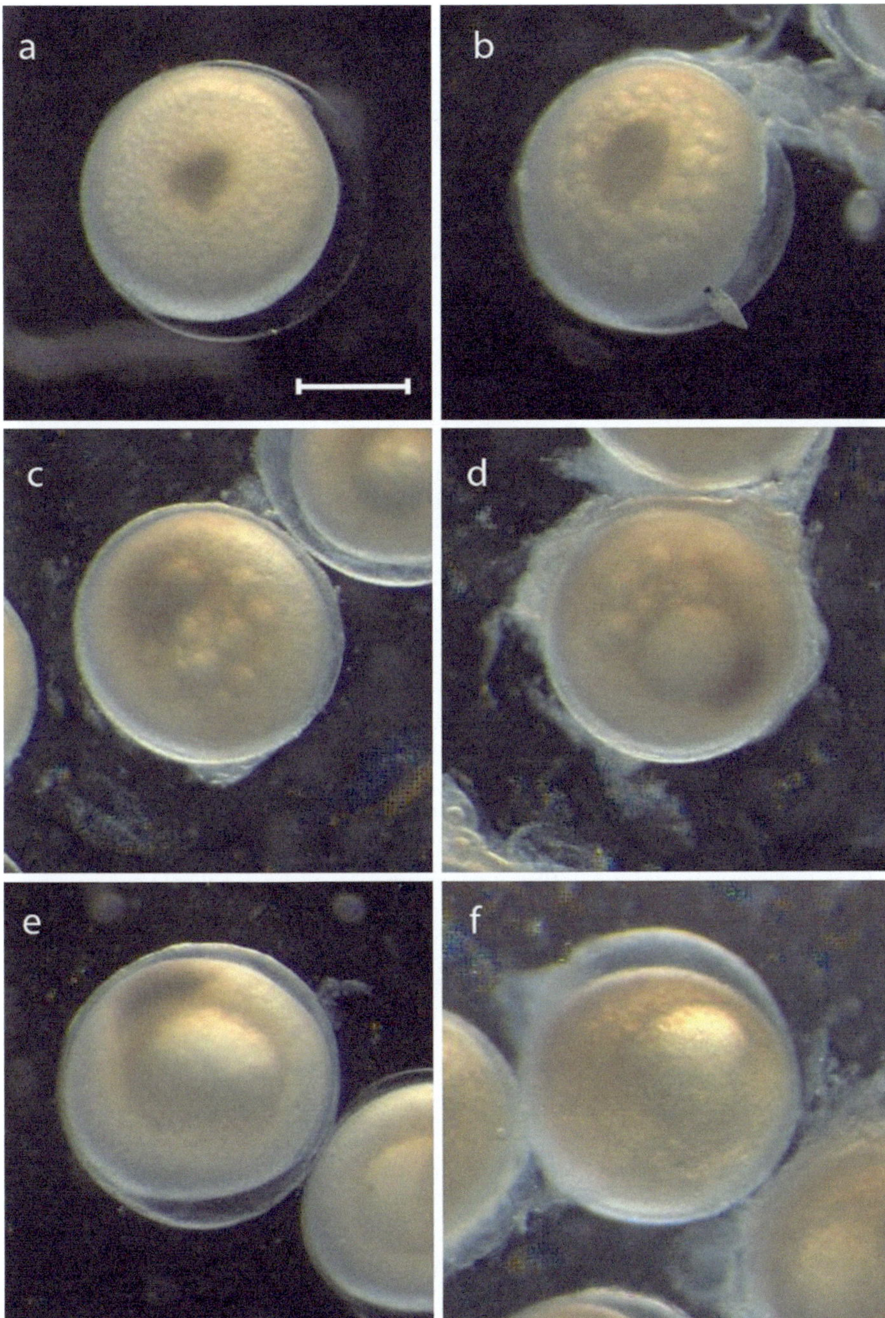

Fig. 4.4 Oocyte maturation stages for percids on the example of Eurasian perch (according to Żarski et al. 2011a); (**a**) stage I, (**b**) stage II, (**c**) stage III, (**d**) stage IV, (**e**) stage V, (**f**) stage VI; for details see the text; bar: 0.5 mm

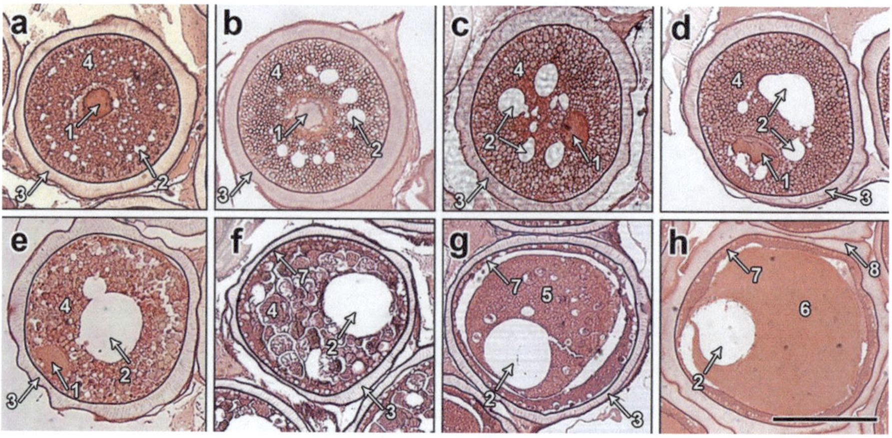

Fig. 4.5 Histological cross-sections of oocytes and ovulated eggs of Eurasian perch at different maturational stages (according to Żarski et al. 2012c): (**a**) stage I, (**b**) stage II, (**c**) stage III, (**d**) stage IV, (**e**) stage V, (**f**) early phase of stage VI, (**g**) late phase (after proteolysis) of stage VI, (**h**) stage VII (ovulated eggs). *1* germinal vesicle, *2* oil droplets, *3 zona radiata externa*, *4* yolk vesicles and globules, *5* homogenous (after proteolysis) yolk before hydration, *6* yolk after hydration, *7* cortical alveoli, *8 zona radiata externa* without visible microvillosities. *Bars* represent 0.5 mm

found to release approximately 13–20 mL of sperm per kg of body weight during the spawning season (Kucharczyk et al. 1996, 1998; Alavi et al. 2010). Therefore hormonal induction of males is only a supplementary measure to enhance milt production. Spermiation can be induced by a change in environmental conditions, such as temperature increase during transfer from one habitat to another as it was described for the walleye (Dąbrowski et al. 2000). Human chorionic gonadotropin (hCG), carp pituitary extract (CPE) or a combination of the two were used to stimulate spermiation in the pikeperch and Volga pikeperch (*Sander volgensis*) but none of them were applied male-specifically (Bokor et al. 2007, 2008). Żarski et al. (2013a) reported, that neither gonadotropins (i.e. hCG, CPE or combination of hCG and pregnant mare serum gonadotropin) nor neurohormone (gonadoliberine analog) affected spermiation rate and spermatozoa motility in pikeperch. In the yellow perch monthly injections of luteinizing hormone releasing hormone analog (LHRHa) starting in January increased production of milt in February and March (Dabrowski et al. 1994). Males of the Eurasian perch produced significantly higher volumes of semen following injection with either 2 mg per kg of body weight of carp pituitary extract or 0.5 pellet of the synthetic GnRH analog Ovopel as compared to the untreated control (Kucharczyk et al. 2001). Also combined injection of FSH (25 μg kg^{-1}) and LH (25 μg kg^{-1}) administered together with a dopamine antagonist (metoclopramide at a dose of 2.5 mg kg^{-1}) significantly increased sperm production in this species (Kucharczyk et al. 1996). During out-of-season spawning trials of Eurasian perch, smaller fish (individual weight: 30–70 g) displayed a higher percentage of spermiating individuals than did larger fish (200–400 g) following injection with various doses of Ovopel, however, it was unclear whether this was an effect of hormonal treatment or that of the pre-spawning chilling period as these conditions were not investigated independently (Szczerbowski et al. 2009). Thus, administration of hormonal products to percid males can enhance spermiation parameters but it is not an inevitable procedure. However, it is very important to note that the sperm volume and quality may be dependent on the timing of its collection, with the highest parameters being recorded during the spawning season (Alavi et al. 2010). However, there is still missing data about the effectiveness of different hormonal treatment protocols and the latency time following hormonal treatment on the quantity and quality of sperm in percids, what has already been proven to have huge impact on the effectiveness of stimulation of spermiation in other fish species (Król et al. 2009; Cejko et al. 2012, 2013).

4.4.2.4 Stimulation of Ovulation

There are numerous differences in the methods and effectiveness of artificial reproduction of cultured and wild females of percid fishes. In the cultured stocks the most important aspect is to provide proper feeding regime and promote vitellogenesis with the use of proper photothermal manipulations prior to spawning (see Chaps. 3, 20 and 21 for details). These factors directly affect gamete maturation process and consequently eggs quality (Ciereszko et al. 1997, 1998; Sulistyo et al. 1998; Henrotte et al. 2008, 2010; Abdulfatah et al. 2011). Induction of FOM and ovulation in cultured females is possible with the application of photothermal manipulations only, even out-of the spawning season (Müller-Bellecke and Zienert 2008). So, there is no need for any hormonal therapy to obtain the eggs in contrast to wild fish, where hormonal stimulation is usually required. In the case of cultured stocks reproductive procedures are adjusted to the local conditions (within particular fish farm) and in accordance with specific production cycle.

Stimulation of FOM and ovulation under the controlled conditions is conducted mainly by the photo-thermal manipulations (Kolkovski and Dąbrowski 1998; Zohar and Mylonas 2001; Müller-Bellecke and Zienert 2008) and hormonal treatment (Brzuska 2005; Krejszeff et al. 2009, 2010; Mylonas et al. 2010). Additionally, some authors suggest that pheromonal stimulation can play a significant role in these processes (Stacey 2003; Żarski 2012). In the case of percids, among the environmental cues, induction of FOM (which includes morphological changes in oocytes after the completed vitellogenesis as well as germinal vesicle [GV] migration and its breakdown [GVBD]; for more details see Sects. 4.3 and 4.4.2.2) as well as ovulation are mainly controlled by the temperature (Dąbrowski et al. 1996). Photoperiod in those processes plays a minor role (Kayes and Calbert 1979; Dąbrowski et al. 1996; Ciereszko et al. 1997). Thus, in the practice of artificially induced reproduction more attention is paid to the thermal manipulations (Dąbrowski et al. 1994; Kucharczyk et al. 2007; Żarski et al. 2013a). Basically, in all percids similar thermal conditions are applied (Tables 4.1, 4.2, 4.3 and 4.4). The optimal temperature for artificial reproduction ranges between 10 and 15 °C (Kayes and Calbert 1979; Dąbrowski et al. 1994;

Table 4.1 Hormonal preparations, general reproductive methods applied and results of artificial reproduction of Eurasian perch *Perca fluviatilis*

Spawning agent	The dose (per kg of body weight)			Dopamine antagonist		Interval between injections (h)	Method of injection	Temperature (°C)[f]	Photoperiod	Maturity stage prior to hormonal stimulation	Spawning method	Ovulation rate (%)	Latency time (h)[g]	Embryonic survival (%)[h]	Reference
	I injection	II injection	III injection	Preparation	Dose (per kg of body weight)[h]										
hCG	200 IU	500 IU	5000 IU	–	–	24	IM	10–11	nd	2–3 [S4]	CS	50	96–128	84.1	1
	500 IU	–	–	–	–	–	IP	14	14L:10D	1 [S6]	TS	75	120–140	63	2
	500 IU	–	–	–	–	–	IP	12	14L:10D	1 [S6]	CS	67	110	53	2
	500 IU	–	–	–	–	–	IP	12	14L:10D	2 [S6]	CS	63	92	61.7	2
	500 IU	–	–	–	–	–	IP	12	14L:10D	3 [S6]	CS	78	68	76.2	2
	500 IU	–	–	–	–	–	IP	12	14L:10D	4 [S6]	CS	100	49	77.2	2
	500 IU	–	–	–	–	–	IP	12	14L:10D	5 [S6]	CS	100	29	81.2	2
	500 IU	–	–	–	–	–	IP	12	14L:10D	6 [S6]	CS	100	18	86.5	2
	500 IU[d]	–	–	–	–	–	IP	14	12L:12D	1 [S4]	CS	100	108	87.9	3
hCG + CPE	500 IU (hCG) + 0.4 mg (CPE)	3.6 mg CPE	–	–	–	24	IM/IP	10–11	nd	2–3 [S4]	CS	100	38–40	60.3	1
	200 IU hCG	500 IU (hCG) + 0.4 mg (CPE)	3.6 mg CPE	–	–	24	IM/IP	10–11	nd	2–3 [S4]	CS	100	62–64	72.1	1
	1000 IU (hCG) + 0.4 mg (CPE)	3.6 mg CPE	–	–	–	24	IM	10–11	14L:10D	2–3 [S4]	CS	92	38–64	53.3	4

CPE	0.4 mg	3.6 mg	–	–	–	24	IP	10–11	nd	2–3 [S4]	CS	100	38–40	61.9	1
	5 mg	–	–	–	–	–	IM	17–19	12L:12D	nd	TS	80	93	26	5
	4 mg[d]	–	–	–	–	–	IP	14	12L:12D	1 [S4]	CS	18	139	33.9	3
FSH + LH	75 IU (FSH) + 75 IU (LH)	–	–	PIM	5 mg	–	IM	14	14L:10D	2–3 [S4]	CS	50	16–22	55.5	6
	75 IU (FSH) + 75 IU (LH)	–	–	PIM	10 mg	–	IM	14	14L:10D	2–3 [S4]	CS	75	21.5–22	61.0	6
	75 IU (FSH) + 75 IU (LH)	–	–	MTC	5 mg	–	IM	14	14L:10D	2–3 [S4]	CS	100	20–21.5	66.8	6
	75 IU (FSH) + 75 IU (LH)	–	–	MTC	10 mg	–	IM	14	14L:10D	2–3 [S4]	CS	100	16–18	75.4	6
	75 IU (FSH) + 75 IU (LH)	–	–	–	–	–	IM	14	14L:10D	2–3 [S4]	CS	50	22–24	52.1	6

(continued)

Table 4.1 (continued)

Spawning agent	The dose (per kg of body weight)			Dopamine antagonist		Interval between injections (h)	Method of injection	Temperature (°C)[f]	Photoperiod	Maturity stage prior to hormonal stimulation	Spawning method	Ovulation rate (%)	Latency time (h)[g]	Embryonic survival (%)[h]	Reference
	I injection	II injection	III injection	Preparation	Dose (per kg of body weight)[h]										
mGnRHa	1 μg	–	–	–	–	–	IM	13–15	12L:12D	nd	TS	–	–	–	7
	5 μg	–	–	–	–	–	IM	13–15	12L:12D	nd	TS	11	94	98.5 (9)	7
	25 μg	–	–	–	–	–	IM	13–15	12L:12D	nd	TS	28	103	98.6 (9)	7
	125 μg	–	–	–	–	–	IM	13–15	12L:12D	nd	TS	78	115	94.1 (9)	7
	40 μg[d]	–	–	MTC	20 mg	–	IP	14	12L:12D	1 [S4]	CS	55	139	35.8	3
	0.2 μg[a]	20 μg	–	MTC	0.1 mg + 10 mg	24	IM	10–11	14L:10D	2–3 [S4]	CS	92	42–68	53.4	4
	0.2 μg[a, b]	20 μg	–	MTC	0.1 mg + 10 mg	24	IP	14	14L:10D	nd	CS	0	–	–	8
	40 μg[a, b]	–	–	MTC	20 mg	–	IP	14	14L:10D	nd	CS	0	–	–	8
	0.2 μg[a, c]	20 μg	–	MTC	0.1 mg + 10 mg	24	IP	14	14L:10D	nd	CS	95	144–168	36	8
	40 μg[a, c]	–	–	MTC	20 mg	–	IP	14	14L:10D	nd	CS	100	96–120	62	8
AN–GnRHa	40 μg	–	–			–	IM	17–19	12L:12D	nd	TS	100	79	29	5
	40 μg	–	–	MTC	10 mg	–	IM	17–19	12L:12D	nd	TS	60	83	25	5
sGnRHa	10 μg[d, e]	–	–	DOM	5 mg	–	IP	14	12L:12D	1 [S4]	CS	100	129	74.6	3
MTC	–	–	–	MTC	11 mg	–	IM	17–19	12L:12D	nd	TS	0	–	–	5

NaCl	+	+	–	–	–	24	nd	10–12	nd	2–3 [S4]	CS	25	72–120	34.1	1
	+	–	–	–	–	–	IM	14	14L:10D	2–3 [S4]	CS	25	22	44.1	6
	+	+	–	–	–	24	IM	10–11	14L:10D	2–3 [S4]	CS	17	64–68	7.5	4
	+	–	–	–	–	–	IM	13–15	12L:12D	nd	TS	–	–	–	7
	+	–	–	–	–	–	IP	14	14L:10D	1 [S6]	TS	–	–	–	2
	+	–	–	–	–	–	IM	17–19	12L:12D	nd	TS	100	108	43	5
	+[d]	–	–	–	–	–	IP	14	12L:12D	1 [S4]	CS	0	–	–	3
	+[b, c]	–	–	–	–	–	IP	14	14L:10D	nd	CS	0	–	–	8

References: 1. Kucharczyk et al. 1996, 2. Żarski et al. 2011a, 3. Targońska et al. 2014, 4. Kucharczyk et al. 2001, 5. Ronyai and Lengyel 2010, 6. Kucharczyk et al. 1998, 7. Kouril et al. 1997, 8. Szczerbowski et al. 2009

nd no data, *mGnRHa* mamalian analogue of gonadoliberine, *sGnRHa* salmon gonadoliberine analogue, *AN-GnRH* azagly-nafaraelin, *CPE* carp pituitary extract, *hCG* human chorionic gonadotropin, *MTC* metoclopramide, *PIM* pimozide, *DOM* domperidone, *S4* according to 4-stage classification of oocyte maturation stages, *S6* according to 6-stage classification of oocyte maturation stages (according to Żarski et al. 2011a), *TS* tank spawning, *CS* controlled spawning (eggs stripped manually, artificial fertilization etc.), *IP* injection applied intraperitoneally, *IM* injection applied intramuscularly (at the dorsal part)

[a]Analogue contained in the commercial preparation Ovopel (complex with metoclopramide)

[b]Spawning performed out-of season with 60 day-long period of temperature below 10 °C

[c]Spawning performed out-of season with 90 day-long period of temperature below 10 °C

[d]Out-of season

[e]Analogue contained in the commercial preparation Ovaprim (complex with domperidone)

[f]Temperature regime following initial injection

[g]Latency time between first injection and ovulation

[h]Embryonic survival determined at the eyed-egg stage

[i]Unknown moment of embryonic survival determination,

(+) fish were treated with saline (placebo group)

Table 4.2 Hormonal preparations, general reproductive methods applied and results of artificial reproduction of yellow perch, *Perca flavescens*

Spawning agent	The dose (per kg of body weight)		Dopamine antagonist dose (per kg of body weight)[h]	Interval between injections (h)	Method of injection	Temperature (°C)[j]	Photoperiod	Spawning method	Ovulation rate (%)	Latency time (h)[k]	Embryonic survival (%)	Reference
	I injection	II injection										
CPE	1 mg	–	–	–	nd	14	nd	TS	100	96–120	58.6–85	1[f, g]
hCG	230 IU	–	–	–	nd	14	nd	nd	100	96–120	nd	1[f, g]
mGnRHa	10 μg[a]	100 μg	10 mg[i]	48	IP	13	12L:12D	CS	50	nd[n]	nd	2
	10 μg[b]	100 μg	10 mg[i]	48	IP	13	12L:12D	CS	50	96–144	nd	2
	10 μg[c]	100 μg	10 mg[i]	48	IP	13	12L:12D	CS	100	96–144	82.7[m]	2
	100 μg[c]	–	10 mg	–	IP	13	12L:12D	CS	100	96–145	84.4[m]	2
	300 μg[c]	–	10 mg	–	IP	13	12L:12D	CS	100	96–146	78.3[m]	2
	10 μg[d]	100 μg	10 mg[i]	48	IP	12–13	15L:9D	CS	100	0–72[l]	48.8[m]	3
	10 μg[e]	100 μg	10 mg[i]	48	IP	12–13	15L:9D	CS	100	0–120[l]	25.3[m]	3
	10 μg[d]	100 μg	10 mg[i]	48	IP	12–13	10L:14D	CS	70	192–288	76.2[m]	3
	10 μg[e]	100 μg	10 mg[i]	48	IP	12–13	10L:14D	CS	80	144–240	87.1[m]	3

NaCl	+[a, b]	+	–	48	IP	13	12L:12D	CS	0	–	–	2
	+[c]	+	–	48	IP	13	12L:12D	CS	20	–	96.4[m]	2

References: 1. Kayes 1977 (following Craig 2000; Dąbrowski et al. 1996), 2. Dąbrowski et al. 1994, 3. Ciereszko et al. 1997

nd no data, '+' fish were treated with saline (placebo group) at a volume of 0.05 mL per fish, *mGnRHa* mamalian analogue of gonadoliberine, *hCG* human chorionic gonadotropin, *CPE* carp pituitary extract, *TS* tank spawning, *CS* controlled spawning (eggs stripped manually, artificial fertilization etc.), *IP* injection applied intraperitoneally

[a]Spawning performed in February

[b]Spawning performed in April

[c]Spawning performed in May

[d]For spawning 2-year old pond reared fish were used

[e]For spawning 3-year old pond reared fish were used

[f]Data taken from Craig 2000

[g]Data taken from Dąbrowski 1996

[h]As a dopamine antagonist pimozide was used

[i]Applied only with the first injection

[j]Temperature regime following initial injection

[k]Latency time between first injection and ovulation

[l]Spontaneous ovulation was recorded just before and just after first injection

[m]Embryonic survival determined at the eyed-egg stage

[n]Ovulation occurred within 3 weeks following injection

Table 4.3 Hormonal preparations, general reproductive methods applied and results of artificial reproduction of pikeperch, *Sander lucioperca*

Spawning agent	The dose (per kg of body weight)		Dopamine antagonist dose (per kg of body weight) [d]	Interval between injections (h)	Method of injection	Temperature (°C)[f]	Photoperiod	Maturity stage prior to hormonal stimulation	Spawning method	Ovulation rate (%)	Latency time (h)[g]	Embryonic survival (%)	Reference
	I injection	II injection											
hCG	500 IU	–	–	–	IP	13	12L:12D	3–4 [S6]	CS	100	96–120	88.3[h]	1
	500 IU	–	–	–	IP	15	12L:12D	3–4 [S6]	CS	100	72–96	84.4[h]	1
	500 IU	–	–	–	IP	12	nd	2 [S6]	CS	75	78–98	71.3[h]	2
	500 IU	–	–	–	IP	12	nd	3 [S6]	CS	100	57–78	73.3[h]	2
	500 IU	–	–	–	IP	12	nd	4 [S6]	CS	100	48–58	77[h]	2
	500 IU	–	–	–	IP	12	nd	5 [S6]	CS	83	32–49	76.5[h]	2
	500 IU	–	–	–	IP	12	nd	6 [S6]	CS	80	5–30	79.5[h]	2
	200 IU	200 IU	–	24	IP	14.5	nd	1–2 [S4]	CS	83.3	75	72.4[i]	3
	200 IU	500 IU	–	24	IP	14.5	nd	1–2 [S4]	CS	100	77	68[i]	3
	600 IU	–	–	–	IP	nd	nd	2 [S4]	CS/TS	100	60	nd	4
	150 IU	500 IU	–	48	IP	nd	nd	2 [S4]	CS/TS	100	133	nd	4
	250 IU	–	–	–	IM	15	14L:10D	2–4 [S4]	CS	71	85	70.9[k]	5
	500 IU	–	–	–	IM	16	14L:10D	2–4 [S4]	CS	100	78.1	84.2[k]	5
	750 IU	–	–	–	IM	17	14L:10D	2–4 [S4]	CS	100	78.6	86.8[k]	5
	1000 IU	–	–	–	IM	18	14L:10D	2–4 [S4]	CS	83	88	52.5[k]	5
PG–600	500 IU	–	–	–	IP	13	12L:12D	3–4 [S6]	CS	83	96–120	80.3[h]	1
	500 IU	–	–	–	IP	15	12L:12D	3–4 [S6]	CS	83	72–96	78.3[h]	1
hCG + CPE	200 IU[b]	3 mg	–	24	IM	15.5–16.7	12L:12D	nd	CS/TS	100	94	91[j]	6
CPE	4 mg	–	–	–	IP	13	12L:12D	3–4 [S6]	CS	67	72–96	78.2[h]	1
	4 mg	–	–	–	IP	15	12L:12D	3–4 [S6]	CS	67	48–72	74.9[h]	1
	3 mg[b]	3 mg	–	24	IM	15.5–16.6	12L:12D	nd	CS/TS	100	96	93[j]	6
	3 mg[b, c]	3 mg	–	24	IM	15.5–16.6	12L:12D	nd	CS/TS	94	81.5	86.5[j]	6

mGnRHa	5 µg	10 µg	2.5+5 mg	24	IP	14.5	nd	1–2 [S4]	CS	0	–	–	3
	5 µg	20 µg	2.5 mg+10 mg	24	IP	14.5	nd	1–2 [S4]	CS	50	94	3.2[i]	3
	1 µg	–	–	–	IM	19	14L:10D	2–4 [S4]	CS	86	89.3	52.3[k]	5
	2.5 µg	–	–	–	IM	20	14L:10D	2–4 [S4]	CS	71	84.42	65.5[k]	5
	5 µg	–	–	–	IM	21	14L:10D	2–4 [S4]	CS	71	83.9	51.1[k]	5
	10 µg	–	–	–	IM	22	14L:10D	2–4 [S4]	CS	71	79.4	52.2[k]	5
	25 µg	–	–	–	IM	23	14L:10D	2–4 [S4]	CS	100	93	60.5[k]	5
	50 µg	–	–	–	IM	24	14L:10D	2–4 [S4]	CS	86	86.5	22.1[k]	5
	40 µg[a]	–	20 mg	–	IP	13	12L:12D	3–4 [S6]	CS	67	96–120	80.3[h]	1
	40 µg[a]	–	20 mg	–	IP	15	12L:12D	3–4 [S6]	CS	67	72–96	72.3[h]	1
	2 µg[a, b, c]	2 µg	1+1 mg	24	IM	15.5–16.7	12L:12D	nd	CS/TS	85	102	72[j]	6
O–GnRHa	5 µg[b]	15 µg	–	24	IM	15.5–16.5	12L:12D	nd	CS/TS	100	110	43[j]	6
	5 µg[b]	15 µg	10 mg[e]	24	IM	15.5–16.5	12L:12D	nd	CS/TS	50	117	56[j]	6
NaCl	+	–	–	–	IP	13	12L:12D	3–4 [S6]	CS	17	144	–	1
	+	–	–	–	IP	15	12L:12D	3–4 [S6]	CS	–	–	–	1
	+	+	–	24	IP	14.5	nd	1–2 [S4]	CS	50	100	70.5[i]	3
	+	–	–	–	IM	25	14L:10D	2–4 [S4]	CS	0	–	–	5

References: 1. Żarski et al. 2013a, 2. Żarski et al. 2012a, 3. Zakęś and Demska-Zakęś 2005, 4. Korbuly et al. 2010, 5. Kristan et al. 2013, 6. Ronyai 2007

nd no data, '+' fish were treated with saline (placebo group), *mGnRHa* mamalian analogue of gonadoliberine, *hCG* human chorionic gonadotropin, *PG-600* mixture of hCG and PMSG (pregnant mare serum gonadotropin) in the proportion of 1:2, *O-GnRH* ovurelin, *CPE* carp pituitary extract, *S4* according to 4-stage classification of oocyte maturation stages, *S6* according to 6-stage classification of oocyte maturation stages (according to Żarski et al. 2012a), *TS* tank spawning, *CS* controlled spawning (eggs stripped manually, artificial fertilization etc.), *IP* injection applied intraperitoneally, *IM* injection applied intramuscularly (at the dorsal part)

[a]Analogue contained in the commercial preparation Ovopel (complex with metoclopramide)

[b]Spawning performed out-of season

[c]Average of the two treatments

[d]Metoclopramide was used as a dopamine antagonist

[e]Applied with first injection only

[f]Temperature regime following initial injection

[g]Latency time between first injection and ovulation

[h]Embryonic survival determined 72 h following fertilization

[i]Embryonic survival determined at the eyed-egg stage

[j]Embryonic survival determined 16–18 h following fertilization

[k]Embryonic survival determined at hatching

Table 4.4 Hormonal preparations, general reproductive methods applied and results of artificial reproduction of walleye, *Sander vitreum*

Spawning agent	The dose (per kg of body weight)		Interval between injections (h)	Method of injection	Temperature (°C)[b]	Photo-period	Maturity stage prior to hormonal stimulation[g]	Spawning method	Ovulation rate (%)	Latency time (h)[c]	Embryonic survival (%)	Reference
	I. Injection	II. Injection										
CPE	13,4 mg	13,4 mg	72	IP	10	N	nd	ST	100	nd	nd	1
	3,1 mg	2 mg	72	IP	10	N	nd	ST	67	nd	nd	1
	13,1 mg	–	–	IP	10	N	nd	ST	50	nd	nd	1
hCG	500 IU	–	–	IM	9–10	nd	3	CS	40	96–120	92[e]	2
	500 IU	–	–	IM	9–10	nd	2	CS	25	96	nd	2
	150 IU[a]	500 IU	35	nd	nd	nd	nd	CS	50	155–179	nd	3
	150 IU	500 IU	48	IM	10	12L:12D	1	CS	60	120	41,4[f]	4
	150 IU	500 IU	48	IM	10	12L:12D	2	CS	100	120	71,7[f]	4
	220 IU	–	–	IP	10–14	N	nd	ST	69	24–120+	nd	5
	440 IU	–	–	IP	10–15	N	nd	ST	57	0–95[d]	nd	5
	660 IU	–	–	IP	10–16	N	nd	ST	74	0–120+[d]	nd	5
	880 IU	–	–	IP	10–17	N	nd	ST	68	0–120+[d]	nd	5
	1100 IU	–	–	IP	10–18	N	nd	ST	78	0–120+[d]	nd	5
	1320 IU	–	–	IP	10–19	N	nd	ST	75	24–119	nd	5
	1540 IU	–	–	IP	10–20	N	nd	ST	70	0–120+[d]	nd	5
hCG+ DHP	150 IU	2 mg	48	IM	10	12L:12D	1	CS	40	48–120	1,4[f]	4
	150 IU	2 mg	48	IM	10	12L:12D	2	CS	100	120–144	0[f]	4
DHP	100 µg	–	–	IM	9–10	nd	2	CS	0	–	–	2
mGnRHa	100 µg	–	–	IM	9–10	nd	3	CS	60	72–120	90[e]	2
	100 µg	–	–	IM	9–10	nd	2	CS	0	–	–	2
	35 µg	100 µg	48	IM	10	12L:12D	1	CS	20	120	62,8[f]	4
	35 µg	100 µg	48	IM	10	12L:12D	2	CS	100	120	30,6[f]	4

NaCl	+	–	–	IM	9–10	nd	2	CS	0	–	–	2
	+	–	–	IM	9–10	nd	3	CS	0	–	–	2
	+	+	48	IM	10	12L:12D	1	CS	0	–	–	4
	+	+	48	IM	10	12L:12D	2	CS	0	–	–	4
	+	–	–	IP	10–13	N	nd	ST	40	24–119	nd	5
RS	2 mL	–	–	IP	11	N	nd	ST	0	nd	nd	1

References: 1. Lessman 1978, 2. Barry et al. 1995, 3. Dąbrowski et al. 2000, 4. Malison et al. 1998, 5. Hearn 1980

nd no data, '+' fish were treated with saline (placebo group), *mGnRHa* mamalian analogue of gonadoliberine, *hCG* human chorionic gonadotropin, *CPE* carp pituitary extract, *DHP* 17a,20b-di-hydroxy-4-pregnen-3-one, *RS* control group received 'Ringer's' solution (for details see Lessman 1978), *CS* controlled spawning (eggs stripped manually, artificial fertilization etc.), *ST* eggs were stripped manually and no fertilization was made, *IP* injection applied intraperitoneally, *IM* injection applied intramuscularly (at the dorsal part), *N* photoperiod was natural (coincident with the daily cycle)

[a]Spawning performed out-of season

[b]Temperature regime following initial injection

[c]Latency time between first injection and ovulation

[d]Some females ovulated at the day of the injection, i.e. up to 23 h post injection

[e]Embryonic survival determined 2 days following fertilization

[f]Embryonic survival determined at the eyed-egg stage

[g]Maturation stage was determined on the basis of four-stage classification

Kucharczyk et al. 1996, 1998; Kouril et al. 1997; Malison et al. 1998; Zakęś and Demska-Zakęś 2009; Żarski et al. 2011b, 2013a; Křišťan et al. 2012a). More recently it was shown that thermal manipulations could be a very useful tool in synchronization of ovulation in pikeperch (Żarski et al. 2013a). However, in order to evaluate possible reaction to such manipulation, and consequently apply it successfully in artificial reproduction, more studies are required to include species- and population-specific differences.

In the practice of hormonal stimulation two hormonal therapies are generally used. First, is to stimulate releasing of endogenous gonadotropins (GtH) from pituitary gland with the application of preparations containing gonadoliberine (GnRH, for more details see Sects. 4.2 and 4.3). The second one is to inject fish with GtH containing preparations (e.g. carp pituitary extract [CPE], human chorionic gonadotropin [hCG]) directly affecting Leydig cells in the testis or follicle cells in ovaries. This results in production of maturation inducing steroid (MIS) promoting final maturation of gametes and, in consequence, spermiation or ovulation (for more details see Sects. 4.2 and 4.3).

In the case of wild females, the necessity of hormonal stimulation during the spawning season mainly depends on the maturity stage of females (Kucharczyk et al. 2007). Fish at the latest maturation stage VI (after the GVBD; according to the classification of Żarski et al. 2011a, 2012a) would most likely ovulate without the hormonal treatment. However, the moment of ovulation is then very hard to predict. Fish at less advanced maturational stages very often may not ovulate (Kucharczyk et al. 2001; Żarski et al. 2011a, 2013a) or ovulation is extended up to few weeks and its effectiveness is usually very low (Dąbrowski et al. 1994; Kucharczyk et al. 1996, 1998; Ronyai and Lengyel 2010). Reproductive effectiveness in such cases may also depend on other factors such as temperature (Dąbrowski et al. 1994) or possible pheromonal stimulation which was proven to promote GVBD in pikeperch (Barry et al. 1995) or FOM in Eurasian perch (Żarski 2012). Such differences may also stem from handling stress affecting reproductive effectiveness negatively (Schreck et al. 2001; Wang et al. 2006). It is then almost not possible to predict the timing of ovulation and therefore perform *in vitro* fertilization. Also, the effectiveness of artificial reproduction of wild fish may be dependent on the time of fish capture. Kestemont et al. (1999) reported lower egg quality in late spawning season (when fish are the most advanced in maturation and hormonal treatment may be not necessary) compared with those obtained earlier during the spawning period. That is why many authors have focused on the development of hormonal treatment protocols allowing higher synchronization and predictability of ovulation in females of percids.

In percids the most commonly applied hormonal preparations are human Chorionic Gonadotropin (hCG) and GnRH analogues (GnRHa) while carp pituitary extract (CPH) was used less frequently (see Tables 4.1, 4.2, 4.3 and 4.4). Although, application of GnRHa in the case of percids was reported to be effective without dopamine antagonist (DA) injection, contrarily to some other fish species such as the cyprinids (Kucharczyk et al. 2005, 2008), in many cases administration of DA was reported (see Tables 4.1, 4.2, 4.3 and 4.4). Interestingly, in some cases such preparations (GnRHa+DA) were reported to affect ovulation rate and egg quality negatively (Zakęś and Demska-Zakęś 2009), whereas GnRHa administered alone revealed usually satisfied results (Kouril et al. 1997; Schlumberger and Proteau 1996; Ronyai and Lengyel 2010). However, it was reported in Walleye (Barry et al. 1995; Malison et al. 1998) and pikeperch (Żarski et al. 2013a) that the effectiveness of GnRHa could be also dependent on the maturity stage of the females where reproductive success was higher when more advanced in maturation fish were spawned.

Induction of ovulation is very often dependent on the dose of the hormone applied. In percids the most commonly applied preparation, which is hCG, was proved to be effective within a relatively wide range of doses, from 200 to 5700 IU per kg of females body weight. The most commonly applied dose ranged between 400 and 500 IU kg^{-1} (Tables 4.1, 4.2, 4.3 and 4.4).

The most effective doses of GnRHa (administered with 10 mg kg^{-1} of pimozide which is a DA) in yellow perch ranged between 110 (administered in two doses of 10 and 100 μg kg^{-1} respectively; pimozide was administered with the first dose) and 300 μg kg^{-1} (Dąbrowski et al. 1994). Similarly, injection of Eurasian perch with 125 μg kg^{-1} resulted in higher spawning and fertilization rate as compared to groups treated with 25 μg kg^{-1} (Kouril et al. 1997). In walleye a single GnRHa dose of 100 μg kg^{-1} was effective to cause GVBD (Barry et al. 1995). The effective dose of mammalian GnRHa (administered together with DA – metoclopramide) in Eurasian perch was 22 μg kg^{-1} (Kucharczyk et al. 2001) and in pikeperch 40 μg kg^{-1} (Żarski et al. 2013a). Malison et al. (1998) reported that doubled injection of GnRHa (35 and 100 μg kg^{-1}) caused ovulation only in walleye females with already migrating GV, as compared to the fish with oocyte exhibiting GV in the central position (less advanced in maturation). Such differences in effectiveness of GnRHa application could stem from difference in both fish maturation stages (Żarski et al. 2013a) and activeness of GnRHa forms (amino acid sequence), as already reported in other fish species (Podhorec and Kouril 2009; Targońska et al. 2010).

Application of CPH was very often performed in two doses, with the first one (priming dose) constituting 10–50 % of the second (resolving) one. In general, the average applied total doses ranged between 3 and 5 mg kg^{-1} (Tables 4.1, 4.2, 4.3 and 4.4). However, in pikeperch (Müller et al. 2004) and walleye up to 15 mg kg^{-1} and 26.8 mg kg^{-1} (in two equal doses) (Lessman 1978) were used, respectively. Even such high doses seemed to be not harmful to the fish. Nevertheless, the data on the application of this spawning agent are quite limited, as compared to the other hormonal preparations.

As mentioned above, when considering reproduction of wild females of percids, the procedures which may be undertaken (photothermal manipulation, hormonal and/or pheromonal stimulation) are strictly dependent on the maturational stages of females. And the biggest problem is that wild females during the spawning season, even caught from the same population and at the same time, are usually at different maturity stages (from I to VI as described by Żarski et al. 2011a, 2012a). It is worth mentioning that stimulation with hCG at a maturation stage I and II (according to the classification given by Żarski et al. 2011a, 2012b, c) resulted in lower ovulation and fertilization rates compared with the fish hormonally stimulated at stages III–VI. Thus, it may be recommended to reproduce wild percids females at a stage III or later.

4.4.2.5 Gamete Collection

Collection of gametes in percids does not create any difficulties as it is in sturgeon females where special surgical techniques are necessary for obtaining the eggs from the genital pore (Mims et al. 2004; Pourasadi et al. 2009) or in catfish males where testes have to be dissected to extract a sufficient amount of sperm (Legendre et al. 1996; Bokor et al. 2010). For the artificial spawning purposes collection of sperm and eggs in percids is proceeded through simple stripping (gentle massaging of the abdomen parts of the fish body) (Rinchard et al. 2011; Kucharczyk et al. 2007; Zakęś and Demska-Zakęś 2009; Żarski et al. 2011a, 2012a).

In males there is a need for taking care to avoid urine contamination of stripped semen when the sperm is collected into the syringes (or any other container) because spermatozoa might have been activated by the urine (Bokor et al. 2007). This can lead to low motility rate of spermatozoa, even after short period of storage (Satterfield and Flickinger 1995a) and, consequently, affect fertilization rate negatively. A possible method is to use a catheter for sperm collection (Bokor et al. 2007; Grozea et al. 2008; Korbuly et al. 2010). It was previously reported that such method allows avoiding urine contamination in salmonids (Glogowski et al. 2000). Another way is to strip the sperm shortly before using it for fertilization (Kucharczyk et al. 2007). More details are provided in the Chap. 5.

Collection of eggs is very simple when fish is taken for stripping shortly after ovulation (Fig. 4.6). However, it exist huge differences among percids regarding the type of ovulated eggs. Eggs of pikeperch and walleye are released as a batch of separated eggs, whereas Eurasian and yellow perch eggs are included within a large, cylindrical gelatinous strand called ribbon (Probst et al. 2009; Formicki et al. 2009, see also Chap. 2). This creates some differences in the in vitro fertilization procedures. Eggs obtained from few pikeperch or walleye females are usually mixed in one bowl and are fertilized according to the "dry" method which involves mixing of eggs and sperm before activation with activating solution. In the case of both perch species it is recommended to fertilize the eggs from each female separately in order to avoid covering one egg ribbon by another, as already reported in Eurasian perch by Żarski et al. (2012d).

One of the biggest obstacles in percids reproduction is to get the 'dry' eggs. This stems from the fact that females are able to release eggs in the tank spontaneously (Żarski et al. 2011a, 2012a). In order to get the dry eggs a frequent control of ovulation, sometimes even every 1 h (Křišťan et al. 2013), is necessary. This, in

Fig. 4.6 Eggs stripping from Eurasian perch (on the *left*) and pikeperch (on the *right*). In the case of Eurasian perch eggs in the form of 'ribbon' may be noticed (Photo: S. Krejszeff)

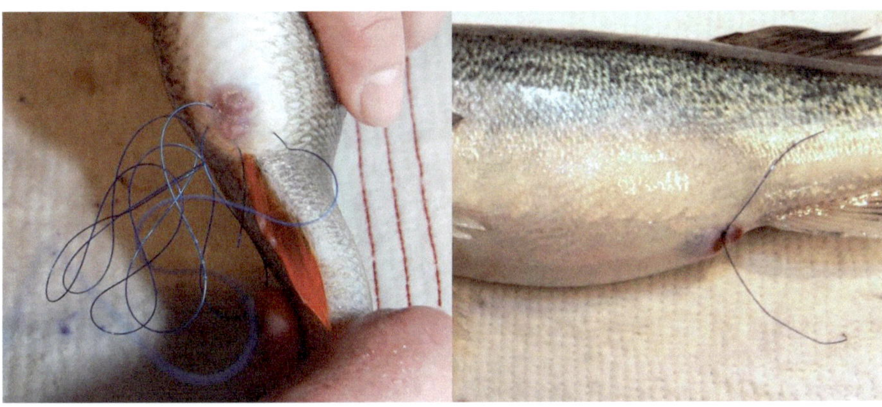

Fig. 4.7 A genital pore of Eurasian perch female being sewn up with the sterile surgical thread (on the *left*) and female pikeperch with sewn up genital pore (on the *right*) (Photo: S. Krejszeff)

turn, causes huge stress to the fish what can presumably affect the egg quality and health status of the female (e.g. Lessman 1978). Therefore, in some cases in Eurasian perch the genital pore of the fish was sewn up to prevent the spontaneous release of eggs in the tank (Kucharczyk et al. 1996, 1998, 2001). This method, with the high attention paid to the sterility of the procedure performed, was also successfully applied to pikeperch (Żarski, unpublished data, Fig. 4.7). Recently, it was found that the eggs of pikeperch may be 'stored' in the ovaries of the female for at least 12 h following ovulation at a temperature of 15 °C without negative effect on eggs viability (Mohagheghi Samarin, Blecha, Policar, unpublished data). Although, the 'operation' of the sewing up the genital pore of the fish may be quite laborious, this method is very promising alternative when the 'dry' eggs are absolutely essential from the breeding point of view.

4.4.2.6 *In Vitro* Fertilization

The *in vitro* fertilization is very important part of finfish controlled reproduction. This process involves exposure of the gametes (eggs and sperm) in a medium (so called 'activating solution' [AS]) which is able to activate sperm and eggs at the same time (Żarski et al. 2015b). After the contact with water (or any other activating solution) eggs acquiring the ability to be fertilized (Coward et al. 2002; Minin and Ozerova 2008). This ability retains for a certain period, which for many fish species is unknown. Żarski et al. (2012d) found, that eggs of Eurasian perch eggs remain active up to 2.5 min post-activation with hatchery water. However, the sperm after activation was motile only up to 37 s. And the spermatozoa motility rate was highly decreasing in time. Alavi et al. (2007) reported that only 7 % of spermatozoa are motile after 45 s post activation, whereas almost 92 % motility is recorded after 15 s post activation. Taking into account that eggs of cyprinids (Żarski et al. 2015b, 2014) and pikeperch (Żarski et al. 2013b) remain active for about 1 min following activation, eggs of Eurasian perch retain the fertilizability for a relatively long period of time. Interestingly, it was reported, that application of the 'dry method' for fertilization in this species (eggs and sperm mixed first and then exposed to AS; e.g. Dąbrowski et al. 1994) affected lower fertilization rate in comparison to the method where eggs were activated first and after 15 s sperm was added (Żarski et al. 2012d). This phenomenon was the most probably related with the fact that eggs within the ribbon upon activation are covered one by another. This enables to penetrate all the eggs by spermatozoa, which very suddenly lose their motility (see Chap. 5) and ability to successfully fertilize the eggs. Therefore for procedure of *in vitro* fertilization in Eurasian perch it may be recommended to activate the eggs first and next, after 15–30 s, add the sufficient amount of sperm (Fig. 4.8). This should allow spreading the whole ribbon within the activating solution and exposing all the eggs to the active spermatozoa, when added. Moore (2003) also reported that repeated addition of sperm to the eggs (at 0, 30 and 60 s post-activation of eggs) improved fertilization effectiveness in walleye, although mechanism is unclear since there is no data on the duration of eggs and sperm activation. Considering the fact that eggs of pikeperch lose their fertilizability very suddenly (within first minute following activation; Żarski et al. 2013b), it may be suggested that the observations made by Moore (2003) in walleye stemmed from other reasons. Nevertheless, in the view of the recently published data and considering the high variability in the fertilization rate usually observed in percid fishes there is still need for development of novel techniques of eggs fertilization (Satterfield and Flickinger 1995b; Rinchard et al. 2005; Zakęś and Demska-Zakęś 2009; Żarski et al. 2011a, 2012d).

For the purposes of *in vitro* fertilization as an activating solution (AS) the most commonly 'hatchery water' was used. However, composition of such AS is not standardized and parameters (e.g. osmolality, ionic composition) are different for different hatcheries. That is why more often other AS-s are used instead of hatchery water, which are very often precisely formulated and standardized. Different

Fig. 4.8 A procedure of *in vitro* fertilization of eggs of Eurasian perch (according to Żarski et al. 2012d). Eggs are first activated with 'hatchery water' in order to spread the egg-ribbon in the water (in order to avoid of covering the eggs one by another) (on the *left*) and after 15–30 s sperm is added (on the *right*) (Photo: S. Krejszeff)

activating solutions have been tested only in Eurasian perch so far. It was proven that application of the Woynarovich solution (3 g of urea and 4 g of NaCl in 1 L of distilled water; Woynarovich and Woynarovich 1980) may improve the fertilization as compared to hatchery water. It stems from the fact that in Woynarovich solution eggs retain the ability to be fertilized for over 3 min while motility of spermatozoa was observed up to 89 s post activation (Żarski et al. 2012d). Prolonged motility of percids spermatozoa after the application of specifically formulated AS-s was reported earlier (for details see Chap. 5). However, the possibility of application of those AS-s for *in vitro* fertilization remains unknown. Especially, when it was already reported that AS which successfully activated sperm motility was not suitable for fertilization (Saad and Billard 1987; Żarski et al. 2014). Considering the fact that application of AS-s other than hatchery water (such as Woynarovich solution in Eurasian perch) may bring huge benefits for the fish farmers increasing fertilization effectiveness, more work is needed aiming at the development of the activating solutions specifically dedicated to the particular species.

One of the most important aspect of *in vitro* fertilization is to use proper amount of sperm per particular amount of eggs, what is called in the literature as a "sperm:egg ratio" (e.g. Linhart et al. 2006). Rinchard et al. (2005) reported, that maximization of fertilizing ability in walleye may be secured with the minimal sperm:egg ratio of 25,000: 1. On the other hand, Casselman et al. (2006) reported that only 5000 of spermatozoa per egg affected very high (over 90 %) survival rate of embryos in this species. Such contrary results may stem from the fact that different methods for estimation of sperm concentration were used (spermatocrit method by Casselman et al. (2006) and counting chambers method by Rinchard et al. (2005)). Most recently, Křišťan et al. (2012b) found that 100,000 spermatozoa per egg may be considered as a secure sperm:egg ratio in pikeperch. Mean pikeperch sperm concentration may be considered at a level of 20×10^9 spermatozoa per mL (see e.g., Zakęś and Demska-Zakęś 2009; Bokor et al. 2008). Whereas, mean number of eggs may be considered at a level of 1000 eggs g^{-1}, because recorded number of "dry eggs" (before activation) per gram ranged approximately between 800 and 1300 (e.g., Bokor et al. 2008, Żarski, unpublished data). Considering the above mentioned it may be recommended to use 0.5 mL of very good quality sperm for each 100 g of eggs in pikeperch and in walleye. Until now, no published data about the effect of sperm:egg ratio on fertilization effectiveness in Eurasian and yellow perch are available.

4.4.2.7 Eggs Treatment, Incubation and Hatching

After *in vitro* fertilization eggs need to be incubated under optimal conditions. In Eurasian and yellow perch incubation of eggs is usually performed without any problem as far as oxygenated water flow is provided washing all the eggs within the whole ribbon. For that purpose small floating cages or flow-through chambers have been successfully used (Kucharczyk et al. 1996; Żarski et al. 2011b, Fig. 4.9). However, pikeperch and walleye eggs are getting to be sticky after their contact with water. Thus, prior to incubation in regular hatchery devices (such as Weiss or Zuger jars), an unsticking procedure must be applied. For that purpose talc with sodium chloride (Schlumpberger and Schmidt 1980; Steffens et al. 1996; Kucharczyk et al. 2007), enzymes (Krise et al. 1986, 1988; Zakęś and Demska-Zakęś 2009) and

Fig. 4.9 An incubation of the Eurasian perch egg-ribbons on the wooden frame covered with a net (mesh size 0.5 mm) from the bottom (on the *left*) and in the floating plastic 'cages' where bottom was replaced with the net (on the *right*) (Photo: S. Krejszeff)

tannic acid (sometimes preceded by immersion in Woynarovich solution; Bokor et al. 2008) have been used (Demska-Zakęś et al. 2005; Rinchard et al. 2005, 2011; Kucharczyk et al. 2007). In percid fishes the latter compound is the most commonly used method, due to, among others, the low cost of its application. However, opinions about its application for commercial purposes were very often inconsistent (Demska-Zakęś et al. 2005; Kucharczyk et al. 2007; Zakęś and Demska-Zakęś 2009). A procedure involving 2–5 min baths in a tannic acid solution a few min following fertilization, with concentrations ranging between 0.4 and 1.5 g L^{-1}, was usually applied (Czesny et al. 2005; Demska-Zakęś et al. 2005; Rinchard et al. 2005, 2011; Kucharczyk et al. 2007). Demska-Zakęś et al. (2005) suggested that this procedure affect low survival rate of embryos or eggs did not lose its adhesive properties. However, the most recent studies indicated that the problem of low effectiveness of this method was related with the time of application of tannic acid following eggs activation (Żarski et al. 2015a). It was proven that in the eggs of pikeperch following activation an extreme chorion deformation, caused by the cortical reaction, occurred (see also Chap. 6). This deformation lasted up to 10 min post activation (Żarski et al. 2012b), whereas tannic acid was usually applied shortly after activation (up to 5 min post activation) (Czesny and Dąbrowski 1998; Czesny et al. 2005; Demska-Zakęś et al. 2005; Rinchard et al. 2005, 2011; Kucharczyk et al. 2007). According to the findings of Żarski et al. 2015a the best moment of application of tannic acid (at a dose of 0.75 g L^{-1}) was 30 min post fertilization coinciding with the end of the egg swelling process (Żarski et al. 2012b, 2015a). Such procedure allowed applying tannic acid for removing the pikeperch egg stickiness in commercial hatcheries successfully (Żarski, unpublished data, Fig. 4.10).

Generally, all percids have similar thermal requirements as considering egg incubation. Median thermal tolerance limits (on the base of normal hatched larvae following incubation at a constant temperature from fertilization to hatching) for walleye, Eurasian and yellow perch is actually almost the same and ranging from 6.0–6.8 °C to 19.2–20.9 °C. Only pikeperch eggs require slightly higher thermal regimes ranging from 9.0–10.0 °C to 21.5–24.0 °C (after Hokanson 1977). Koenst and Smith (1976) reported that optimal temperature for incubation of walleye eggs ranged between 9 and 15 °C. For pikeperch eggs incubation temperatures between 12 and 16 °C were recommended (Kokurewicz 1969; Kucharczyk et al. 2007; Zakęś and Demska-Zakęś 2009). Saat and Veersalu (1996) reported that optimal temperatures for Eurasian perch ranged between 8 and 18 °C, while for yellow perch Hinshaw (2006) suggested temperatures from 5 to 15 °C. However, on the base of the literature data Teletchea et al. (2009b) indicated that for walleye, Eurasian and yellow perch thermal regime for incubation should be fixed at 12.5 °C, and 14 °C for pikeperch. At such temperatures incubation of Eurasian and yellow perch lasted 165 degree-days, while eggs of walleye and pikeperch had to be incubated for 140 and 115 degree-days, respectively (Teletchea et al. 2009b).

Hatching of percids larvae is usually an asynchronous event. In some cases it may last even several days among the same egg batch or egg-ribbon. It is important to emphasize that even very low quality larvae (highly malformed) were observed to hatch spontaneously (Żarski et al. 2011b, see also Chap. 6 for more details). It

Fig. 4.10 A commercial-size incubation of eggs treated with tannic acid (0.75 g L^{-1}, 30 min following eggs activation) according to the method described by Żarski et al. 2015a (Photo: D. Żarski)

was already reported that early hatched Eurasian perch larvae were characterized by a higher survival rate than those hatched 2 days later. Also, it was proven that joint rearing of larvae hatched at different times may affect survival and growth rate of the larvae significantly (Kestemont et al. 2003, see also Chaps. 9, 10 and 11). Although it is not clear whether the larvae hatching earlier or later are of different biological properties, a special attention must be given to the extension of hatching time, and possible separation of early and late hatched larvae can be recommended before the initial larval rearing stage.

4.5 Conclusions

Based on published data, it may be concluded that percids represent a quite unique group of fish exhibiting specific traits from the perspective of artificial reproduction. For example, one of the most characteristic features is that in this group of fish, hormonal stimulation does not have to be supported by the anti-dopaminergic treatment, which is necessary in other freshwater fishes (cyprinids, catfishes and salmonids). Despite the efforts undertaken in the development of protocols allowing full control over reproduction (mostly synchronization of ovulation and hormonal therapy) in percids, there are still many aspects to be investigated. There is a considerable lack of data regarding hormonal regulation of the final gamete maturation process as well as stress-related effects on these processes. This knowledge may help to understand the variability of gamete quality, still widely observed in these species. Also, there are many ambiguities related to gamete management, both prior to and following fertilization (e.g. short-term storage, removing the adhesiveness of eggs), which may facilitate and thus improve this important step of artificial reproduction. Therefore, future research on the artificial reproduction of percids should employ modern scientific tools (e.g. proteomics, transcriptomics, etc.) possibly allowing to answer many unanswered questions.

References

Abdulfatah A, Fontaine P, Kestemont P, Gardeur JN, Marie M (2011) Effects of photothermal kinetics and amplitude of photoperiod decrease on the induction of the reproduction cycle in female Eurasian perch *Perca fluviatilis*. Aquaculture 322:169–176

Alavi SMH, Rodina M, Policar T, Kozak P, Pšenicka M, Linhart O (2007) Semen of *Perca fluviatilis* L.: sperm volume and density, seminal plasma indices and effects of dilution ratio, ions and osmolality on sperm motility. Theriogenology 68:276–283

Alavi SMH, Rodina M, Hatef A, Stejskal V, Policar T, Hamačkova J, Linhart O (2010) Sperm motility and monthly variations of semen characteristics in *Perca fluviatilis* (Teleostei: Percidae). Czech J Anim Sci 55:174–182

Baras E, Kestemont P, Mélard C (2003) Effect of stocking density on the dynamics of cannibalism in sibling larvae of *Perca fluviatilis* under controlled conditions. Aquaculture 219:241–255

Barry TP, Malison JA, Lapp AF, Procarione LS (1995) Effects of selected hormones and male cohorts onfinal oocyte maturation, ovulation, and steroid production in walleye (*Stizostedion vitreum*). Aquaculture 138:331–347

Billard R, Fostier A, Weil C, Breton B (1982) Endocrine control of spermatogenesis in teleost fish. Can J Fish Aquat Sci 39:65–79

Bogerd J, Granneman JCM, Schulz RW, Vischer HF (2005) Fish FSH receptors bind LH: how to make the human FSH receptor to be more fishy? Gen Comp Endocrinol 142:34–43

Bokor Z, Müller T, Bercsényi M, Horváth L, Urbányi B, Horváth A (2007) Cryopreservation of sperm of two European percid species, the pikeperch (*Sander lucioperca*) and the Volga pikeperch (*S. volgensis*). Acta Biol Hung 58(2):199–207

Bokor Z, Horvath A, Horvath L, Urbanyi B (2008) Cryopreservation of pike perch sperm in hatchery conditions. Isr J Aquacult-Bamidgeh 60(3):166–169

Bokor Z, Urbanyi B, Horvath L, Horvath A (2010) Commercial-scale cryopreservation of wels catfish (*Silurus glanis*) semen. Aquac Res 41:1549–1551

Bradley JA, Goetz FW (1994) The inhibitory effect of indomethacin, nordihydroguaiaretic acid, and pyrrolidinedithiocarbamate on ovulation and prostaglandin synthesis in yellow perch (*Perca flavens*) follicle incubates. Prostaglandins 48:11–20

Brzuska E (1979) The in vivo method of estimating the stages of oocytes maturation in carp *Cyprinus carpio* L. Acta Hydrobiol 21:423–433

Brzuska E (1988) Further investigation of in vivo estimation of the maturation stages of carp (*Cyprinus carpio* L.) females. Acta Hydrobiol 30:421–428

Brzuska E (2005) Artificial spawning of carp (cyprinus carpio L.): differences between females of polish strain 6 and Hungarian strain W treated with carp pituitary homogenate, ovopel or dagin. Aquac Res 36:1015–1025

Casselman SJ, Schulte-Hostedde AI, Montgomerie R (2006) Sperm quality influences male fertilization success in walleye (*Sander vitreus*). Can J Fish Aquat Sci 63:2119–2125

Cejko BI, Kowalski RK, Żarski D, Dryl K, Targońska K, Chwaluczyk R, Kucharczyk D, Glogowski J (2012) The influence of the length of time after hormonal treatment with [(d-ala 6, pro 9 NEt)-

mGnRH + metoclopramide] i.e. ovopel on barbel *Barbus barbus* (L.) milt quality and quantity indicators. J Appl Ichthyol 28:249–253

Cejko BI, Żarski D, Krejszeff S, Kucharczyk D, Kowalski RK (2013) Effect of hormonal stimulation of the Crucian Carp *Carassius carassius* (L.) on milt volume, number of sperm and its motility. Isr J Aquacult Bamidgeh (65: 912.)

Ciereszko RE, Dąbrowski K, Ciereszko A (1997) Effects of temperature and photoperiod on reproduction of female yellow perch *Perca flavescens*: plasma concentrations of steroid hormones, spontaneous and induced ovulation, and quality of eggs. J World Aquacult Soc 28:344–356

Ciereszko RE, Dabrowski K, Ciereszko A, Ottobre JS (1998) Plasma concentrations of steroid hormones in male yellow perch, *Perca flavescens*: the effect of age and photothermal manipulation. Environ Biol Fish 51:97–105

Coward K, Bromage NR, Hibbitt O, Parrington J (2002) Gamete physiology, fertilization and egg activation in teleost fish. Rev Fish Biol Fish 12:33–58

Craig JF (2000) Percid fishes systematics. Ecology and exploitation. Blackwell Science, Dunscore

Czesny S, Dabrowski K (1998) The effect of egg fatty acid concentrations on embryo viability in wild and domesticated walleye (*Stizostedion vitreum*). Aquat Living Resour 11:371–378

Czesny S, Rinchard J, Dabrowski K (2005) Intrapopulation variation in egg lipid and fatty acid composition and embryo viability in a naturally spawning walleye population from an inland reservoir. N Am J Fish Manag 25:122–129

Dąbrowski K, Ciereszko A, Ramseyer L, Culver D, Kestemont P (1994) Effects of hormonal treatment on induced spermiation and ovulation in the yellow perch (*Perca flavescens*). Aquaculture 120(1–2):171–180

Dąbrowski K, Ciereszko RE, Ciereszko A, Toth GP, Christ SA, El-Saidy D, Ottobre JS (1996) Reproductive physiology of yellow perch (*Perca flavescens*): environmental and endocrinological cues. J Appl Ichthyol 12:139–148

Dąbrowski K, Czesny S, Kolkovski S, Lynch WE Jr (2000) Intensive culture of walleye larvae produced out of season and during regular season spaning. N Am J Aquac 62:219–224

Demska-Zakes K, Zakes Z, Roszuk J (2005) The use of tannic acid to remove adhesiveness from pikeperch, *Sander lucioperca*, eggs. Aquac Res 36:1458–1464

Devlin RH, Nagahama Y (2002) Sex determination and sex differentiation in fish: an overview of genetic, physiological, and environmental influences. Aquaculture 208:191–364

Fontaine P, Migaud H, Mandiki R, Gardeur JN, Kestemont P, Fostier A (2003) Involvement of sex steroids in final stages of oogenesis in Eurasian perch, *Perca fluviatilis*. Fish Physiol Biochem 28:331–332

Formicki K, Smaruj I, Szulc J, Winnicki A (2009) Microtubular network of the gelatinous egg envelope within the egg ribbon of European perch, *Perca fluviatilis* L. Acta Ichthyol Piscat 39:147–151

Garcìa-López Á, Bogerd J, Granneman JCM, van Dijk W, Trant JM, Taranger GL, Schulz RW (2009) Leydig cells express follicle-stimulating hormone receptors in African catfish. Endocrinology 150:357–365

Glogowski J, Kwasnik M, Piros B, Dabrowski K, Goryczko K, Dobosz S, Kuzminski H, Ciereszko A (2000) Characterization of rainbow trout milt collected with a catheter: semen parameters and cryopreservation success. Aquac Res 31:289–296

Goetz FW (1983) Hormonal control of oocyte final maturation and ovulation in fishes. In: Hoar WS, Randall DJ, Donaldson EM (eds) Fish physiology, vol IXB. Academic, New York, pp 117–170

Goetz FW, Garczynski M (1997) The ovarian regulation of ovulation in teleost fish. Fish Physiol Biochem 17:33–38

Goetz FW, Theofan G (1979) In vitro stimulation of germinal vesicle breakdown and ovulation of yellow perch (*Perca flavescens*) oocytes. Effects of 17a-hydroxy-20ß-dihydroprogesterone and prostaglandins. Gen Comp Endocrinol 37:273–285

Goetz FW, Duman P, Berndtson A, Janowsky EG (1989) The role of prostaglandins in the control of ovulation in yellow perch, *Perca flavescens*. Fish Physiol Biochem 7(1–6):163–168

Grozea A, Telea A, Korbuly B (2008) Preliminary study regarding the efficiency of different hormones on pikeperch spermination. Faculty of animal science and biotechnologies. Timisoara 41(2):59–64

Hearn MC (1980) Ovulation of pond-reared walleyes in response to various injection levels of human chorionic gonadotropin. Prog Fish Cult 42:228–230

Henrotte É, Overton JL, Kestemont P (2008) Effects of dietary n-3 and n-6 fatty acids levels on egg and larval quality of eurasian perch. Cybium 32(2 Suppl):271–272

Henrotte E, Mandiki RSNM et al (2010) Egg and larval quality, and egg fatty acid composition of eurasian perch breeders (*Perca fluviatilis*) fed different dietary DHA/EPA/AA ratios. Aquac Res 41(9):53–61

Hermelink B, Wuertz S, Trubiroha A, Rennert B, Kloas W, Schulz C (2011) Influence of temperature on puberty and maturation of pikeperch, *Sander lucioperca*. Gen Comp Endocrinol 172:282–292

Hinshaw JM (2006) Species profile: yellow perch. Southern Regional Aquaculture Center publication no 7204. North Carolina State University. Stoneville, MS, USA

Hokanson KEF (1977) Temperature requirements of some percids and adaptations to the seasonal temperature cycle. J Fish Res Board Can 34:1524–1550

Kagawa H, Tanaka H, Okuzawa K, Kobayashi M (1998) GTH II but not GTH I induces final maturation and the development of maturational competence of oocytes of red seabream in vitro. Gen Comp Endocrinol 112:80–88

Kayes T (1977) Reproductive biology and artificial propagation methods for adult perch. In: Soderberg RW (ed) Perch fingerling production for aquaculture, Advisory Report #421. University of Wisconsin Sea Grant Program, Madison, pp 6–21

Kayes TB, Calhert HE (1979) Effect of photoperiod and temperature on the spawning of yellow perch (*Perca flavescens*). Proc World Maricult Soc 10:306–316

Kestemont P, Cooremans J, Abi-Ayad A, Mélard C (1999) Cathepsin L in eggs and larvae of perch perca fluviatilis: variations with developmental stage and spawning period. Fish Physiol Biochem 21(1):59–64

Kestemont P, Jourdan S, Houbart M, Me'lard C, Paspatis M, Fontaine P, Cuvier A, Kentouri M, Baras E (2003) Size heterogeneity, cannibalism and competition in cultured predatory fish larvae: biotic and abiotic influences. Aquaculture 227:333–356

Koenst WM, Smith LL Jr (1976) Thermal requirements of the early life history stages of walleye, *Stizostedion vitreum vitreum*, and sauger, *Stizostedion canadense*. J Fish Res Board Can 33:1130–1138

Kokurewicz B (1969) The influence of temperature on the embryonic development of the perches, *Perca fluviatilis* L. and *Lucioperca lucioperca* (L.). Zool Pol 19:47–67

Kolkovski S, Dąbrowski K (1998) Off-season spawning of yellow perch. Prog Fish Cult 60:133–136

Kooten T, Andersson J, Byström P, Persson L, Roos AM (2010) Size at hatching determines population dynamics and response to harvesting in cannibalistic fish. Can J Fish Aquat Sci 67:401–416

Korbuly B, Grozea A, Cean A, Bănăţean-Dunea I, Păcală N (2010) Spawning latency period in hormonal induced reproduction of pikeperch (*Sander lucioperca*). SPASB 43:32–35

Kouril J, Linhart O, Relot P (1997) Induced spawning of perch by means of a GnRH analogue. Aquac Int 5:375–377

Krejszeff S, Targońska K, Żarski D, Kucharczyk D (2009) Domestication affects spawning of the ide (*Leuciscus idus*)-preliminary study. Aquaculture 295:145–147

Krejszeff S, Targońska K, Żarski D, Kucharczyk D (2010) Artificial reproduction of two different spawn-forms of the chub. Reprod Biol 10:67–74

Krise WF (1988) Optimum protease exposure time for removing adhesiveness of walleye eggs. Prog Fish Cult 50:126–127

Krise WF, Bulkowski- Cummings L, Shellman AD, Kraus KA, Gould RW (1986) Increased walleye egg hatch and larval survival after protease treatment of eggs. Prog Fish Cult 48:95–100

Křišťan J, Stejskal V, Policar T (2012a) Comparison of reproduction characteristics and broodstock mortality in farmed and wild eurasian perch (*Perca fluviatilis* L.) females during spawning season under controlled conditions. Turk J Fish Aquat Sci 12:191–197

Křišťan J, Żarski D, Palińska K, Krejszeff S, Nowosad J, Alavi SMH, Policar T, Kucharczyk D (2012b) The effect of sperm to egg ratio on fertilization success in artificial insemination of pike-perch *Sander lucioperca* (L.). Domestication in Finfish Aquaculture, Department of Lake and River Fisheries of University of Warmia and Mazury, Olsztyn, 23–25 Oct, p 70.

Křišťan J, Alavi SMH, Stejskal V, Policar T (2013) Hormonal induction of ovulation in pikeperch (*Sander lucioperca* L.) using human chorionic gonadotropin (hCG) and mammalian GnRH analogue. Aquac Int 21:811–818

Król J, Kowalski RK, Hliwa P, Dietrich GJ, Stabiński R, Ciereszko A (2009) The effects of commercial preparations containing two different GnRH analogues and dopamine antagonists on spermiation and sperm characteristics in the European smelt *Osmerus eperlanus* (L.). Aquaculture 286:328–331

Kucharczyk D, Kujawa R, Mamcarz A, Skrzypczak A, Wyszomirska E (1996) Induced spawning in perch. *Perca fluviatilis* L. using carp pituitary extract and HCG. Aquac Res 27(11):847–852

Kucharczyk D, Kujawa R, Mamcarz A, Skrzypczak A, Wyszomirska E (1998) Induced spawning in perch, *Perca fluviatilis* L., using FSH + LH with pimozide or metoclopramide. Aquac Res 29:131–136

Kucharczyk D, Mamcarz A, Kujawa R, Wyszomirska E, Skrzypczak A (2000) Spontaneous and artificial spawning of perch (*Perca fluviatilis* L.) as studied on individuals collected from two populations. Folia Univ Agric Stetin 214(27):135–146

Kucharczyk D, Szczerbowski A, Łuczynski MJ, Kujawa R, Mamcarz A, Wyszomirska E, Szabó T, Ratajski S (2001) Artificial spawning of Eurasian perch, *Perca fluviatilis* L. Using ovopel. Arch Pol Fish 9(1):39–49

Kucharczyk D, Kujawa R, Mamcarz A, Targońska-Dietrich K, Wyszomirska E, Glogowski J, Babiak I, Szabó T (2005) Induced spawning in bream (*Abramis brama* L.) using pellets containing GnRH. Czech J Anim Sci 50:89–95

Kucharczyk D, Kestemont P, Mamcarz A (2007) Artificial reproduction of pikeperch. Mercurius, Olsztyn, 80 p

Kucharczyk D, Targońska K, Hliwa P, Gomułka P, Kwiatkowski M, Krejszeff S, Perkowski J (2008) Reproductive parameters of common carp (*Cyprinus carpio* L) spawners during natural season and out-of-season spawning. Reprod Biol 8:285–289

Kujawa R, Kucharczyk D, Mamcarz A, Zarski D, Targońska K (2011) Artificial spawning of common tench *Tinca tinca* (Linnaeus, 1758), obtained from wild and domestic stocks. Aquac Int 19:513–521

Legendre M, Linhart O, Billard R (1996) Spawning and management of gametes, fertilized eggs and embryos in Siluroidei. Aquat Living Resour 9:59–80

Lessman CA (1978) Effects of gonadotropin mixtures and two steroids on inducing ovulation in walleyes. Prog Fish-Cult 40:3–5

Linhart O, Rodina M, Kocour M, Gela D (2006) Insemination, fertilization and gamete management in tench, *Tinca tinca* (L.). Aquac Int 14:61–73

Malison JA, Procarione LS, Kayes TB, Hansen JF, Held JA (1998) Industion of out-of-season spawning in walleye (*Stizostedion vitreum*). Aquaculture 163:151–161

Migaud H, Mandiki R, Gardeur JN, Kestemont P, Bromage N, Fontaine P (2003) Influence of photoperiod regimes on the eurasian perch gonad genesis and spawning. Fish Physiol Biochem 28:395–397

Mims SD, Onders RJ, Gomelsky B, Shelton WL (2004) Effectiveness of the minimally invasive surgical technique (MIST) for removal of ovulated eggs from first-time and second-time mist-spawned paddlefish. N Am J Aquac 66:70–72

Minin AA, Ozerova SG (2008) Spontaneous activation of fish eggs is abolished by protease inhibitors. Russ J Dev Biol 38:293–296

Moore AA (2003) Manipulation of fertilization procedures to improve hatchery walleye egg fertility and survival. N Am J Aquac 65:56–59

Müller T, Taller J, Nyitrai G, Kucska B, Cernak I, Bercsenyi M (2004) Hybrid of pikeperch, *Sander lucioperca* L. and Volga perch, *S. viologens* (Gamelin). Aquac Res 35(9):915–916
Müller-Bellecke A, Zienert S (2008) Out-of-season spawning of pike perch (*Sander lucioperca* L.) without the need for hormonal treatments. Aquac Res 39:1279–1285
Mylonas CC, Fostier A, Zanuy S (2010) Broodstock management and hormonal manipulations of fish reproduction. Gen Comp Endocrinol 165:516–534
Nagahama Y (1994) Endocrine regulation of gametogenesis in fish. Int J Dev Biol 38:217–229
Nagahama Y, Yamashita M (2008) Regulation of oocyte maturation in fish. Dev Growth Differ 50:195–219
Patino R, Sullivan CV (2002) Ovarian follicle growth, maturation, and ovulation in teleost fish. Fish Physiol Biochem 26:57–70
Patino R, Yoshizaki G, Thomas P, Kagawa H (2001) Gonadotropic control of ovarian follicle maturation: the two-stage concept and its mechanisms. Comp Biochem Physiol B129:427–439
Patino R, Yoshizaki G, Bolamba D, Thomas P (2003) Role of arachidonic acid and protein kinase c during maturation-inducing hormone-dependent meiotic resumption and ovulation in ovarian follicles of atlantic croaker. Biol Rep 68:516–523
Peter RE (1983) The brain and neurohormones in teleost reproduction. In: Randall D, Hoar WS, Donaldson E (eds) Reproduction endocrine tissues and hormones, vol 9, Part A of fish physiology. Academic, New York, pp 97–135
Peter RE, Yu KL (1997) Neuroendocrine regulation of ovulation in fishes: basic and applied aspects. Rev Fish Biol Fish 7:173–197
Podhorec P, Kouril J (2009) Induction of final oocyte maturation in Cyprinidae fish by hypothalamic factors: a review. Vet Med 54:97–110
Policar T, Toner D, Alavi SMH, Linhart O (2008) Chapter 4: Reproduction and spawning. In: Toner D, Rougeot C (eds) Farming of Eurasian Perch volume 1–Juvenile Production, Aquaculture explained special publication no 24, Dun Laoghaire
Pourasadi M, Falahatkar B, Takami GA (2009) Minimally invasive surgical technique for egg collection from the Persian sturgeon, *Acipenser persicus*. Aquac Int 17:317–321
Probst WN, Stoll S, Hofmann H, Fischer P, Eckmann R (2009) Spawning site selection by eurasian perch (*Perca fluviatilis* L.) in relation to temperature and wave exposure. Ecol Freshw Fish 18(1):1–7
Rinchard J, Dabrowski K, Tassell JJ, Stein RA (2005) Optimization of fertilization success in *Sander vitreus* is influenced by the sperm : egg ratio and ova storage. J Fish Biol 67:1157–1161
Rinchard J, WareK DK, Tassell JJ, Marschall EA, Stein RA (2011) Egg thiamine concentration affects embryo survival in Lake Erie walleye. Environ Biol Fish 90:53–60
Rónyai A (2007) Induced out-of-season and seasonal tank spawning and stripping of pike perch (*Sander lucioperca* L.). Aquac Int 38:1144–1151
Rónyai A, Lengyel SA (2010) Effects of hormonal treatments on induced tank spawning of eurasian perch (*Perca fluviatilis* L.). Aquac Int 41(9):345–347
Ross RM (1984) Catheterization: a non-harmful method of sex identification for sexually monomorphic fishes. Prog Fish-Cult 46:151–152
Rougeot C, Nicayenzi F, Mandiki SNM, Rurangwa E, Kestemont P, Mélard C (2004) Comparative study of the reproductive characteristics of XY male and hormonally sex-reversed XX male eurasian perch, *Perca fluviatilis*. Theriogenology 62:790–800
Saad A, Billard R (1987) Composition et emploi d'un dilueur d'insemination chez la carpe, *Cyprinus carpio*. Aquaculture 66:329–345
Saat T, Veersalu A (1996) The rate of early development in perch *Perca fluviatilis* L. and ruffe *Gymnocephalus cernuus* (L.) at different temperatures. Ann Bot Fenn 33:693–698
Satterfield JR Jr, Flickinger SA (1995a) Factors influencing storage potential of preserved walleye semen. Prog Fish-Cult 57:175–181
Satterfield JR Jr, Flickinger SA (1995b) Field collection and short-term storage of walleye semen. Prog Fish-Cult 57:182–187
Schlumberger O, Proteau JP (1996) Reproduction of pikeperch (*Stizostedion lucioperca*) in captivity. J Appl Ichthyol 12:149–152

Schlumpberger W, Schmidt K (1980) Vorläufiger Stand der Technologie zur Aufzucht von vorgestreckten Zander (*Stizostedion lucioperca*). Z Binnenfisch DDR 27:284–286
Schreck CB, Contreras-Sanchez W, Fitzpatrick MS (2001) Effects of stress on fish reproduction, gamete quality, and progeny. Aquaculture 197:3–24
Schulz RW, Renato de França L, Lareyre J-J, LeGac F, Chiarini-Garcia C, Nobrega RE, Miura T (2010) Spermatogenesis in fish. Gen Comp Endocrinol 165:390–411
Skrzypczak A, Kucharczyk D, Mamcarz A, Kujawa R, Furgala-Selezniow G (1998) A new breeding technique of pikeperch (*Stizostedion lucioperca*) in net cages. Czech J Anim Sci 43:398–398
Sorensen P, Murphy C, Loomis K, Maniak P, Thomas P (2004) Evidence that 4-pregnen-17,20beta,21-triol-3-one functions as a maturation-inducing hormone and pheromonal precursor in the percid fish, Gymnocephalus cernuus. Gen Comp Endocrinol 139(1):1–11
Stacey N (2003) Hormones, pheromones and reproductive behavior. Fish Physiol Biochem 28:229–235
Steffens W, Geldhauser F, Gerstner P, Hilge V (1996) German experiences in the propagation and rearing of fingerlink pikeperch (*Stizostedion lucioperca*). Ann Zool Fenn 33:627–634
Sulistyo I, Rinchard J, Fontaine P, Gardeur J, Capdeville B, Kestemont P (1998) Reproductive cycle and plasma levels of sex steroids in female eurasian perch *Perca fluviatilis*. Aquat Living Resour 11(2):101–110
Szczerbowski A, Kucharczyk D, Mamcarz A, Łuczyński M, Targońska K, Kujawa R (2009) Artificial off-season spawning of eurasian perch *Perca fluviatilis* L. Arch Pol Fish 17(2):95–98
Targońska K, Kucharczyk D, Kujawa R, Mamcarz A, Żarski D (2010) Controlled reproduction of asp, *Aspius aspius* (L.) using luteinizing hormone releasing hormone (LHRH) analogues with dopamine inhibitors. Aquaculture 306:407–410
Targońska K, Szczerbowski A, Żarski D, Łuczyński M, Szkudlarek M, Gomułka P, Kucharczyk D (2014) Comparison of different spawning agents in artificial out-of-season spawning of Eurasian perch, *Perca fluviatilis* L. Aquac Res 45:765–767
Teletchea F, Gardeur J-N, Psenicka M, Kaspar V, Le Dore Y, Linhart O, Fontaine P (2009a) Effects of four factors on the quality of male reproductive cycle in pikeperch *Sander lucioperca*. Aquaculture 291(3–4):217–223
Teletchea F, Gardeur JN, Kamler E, Fontaine P (2009b) The relationship of oocyte diameter and incubation temperature to incubation time in temperate freshwater fish species. J Fish Biol 74:652–668
Thorgaard GH (1995) Biotechnological approaches to broodstock management. In: Bromage NR, Roberts RJ (eds) Broodstock management and egg and larval quality. Blackwell Science, Oxford, pp 76–93
Tyler CR, Sumpter JP (1996) Oocyte growth and development in teleosts. Rev Fish Biol Fish 6:287–318
Vizziano D, Fostier A, Loir M, LeGac F (2008) Testis development, its hormonal regulation and spermiation induction in teleost fish. In: Alavi SMH, Cosson JJ, Coward K, Rafiee G (eds) Fish spermatology. Alpha Science, Oxford, pp 103–139
Wang N, Gardeur JN, Henrotte E, Marie M, Kestemont P, Fontaine P (2006) Determinism of the induction of the reproductive cycle in female Eurasian perch, *Perca fluviatilis*: identification of environmental cues and permissive factors. Aquaculture 261:706–714
Wojda R, Sliwinski J, Ciesla M (1994) Results of studies on natural spawning in pikeperch, *Stizostedion lucioperca* (L.). Rocz Nauk PZW 7:71–80 (In Polish with English summary)
Woynarovich E, Horváth L (1980) The artificial propagation of warmwater finfishes – a manual for extension. FAO Fisheries Technical Paper, Rome, Italy. (201), 183 p
Woynarovich E, Woynarovich A (1980) Modified technology for elimination of stickiness of common carp *Cyprinus carpio* eggs. Aquacult Hung 2:19–21
Yamashita M, Fukada S, Yoshikuni M, Bulet P, Hirai T, Yamaguchi A, Lou YH, Zhao Z, Nagahama Y (1992) Purification and characterization of maturation-promoting factor in fish. Dev Biol 149:8–15

Zakęś Z, Demska-Zakęś K (2005) Artificial spawning of pikeperch (*Sander lucioperca* (L.)) stimulated with human chorionic gonadotropin (hCG) and mammalian GnRH analogue with a dopamine inhibitor. Arch Pol Fish 13:63–75

Zakęś Z, Demska–Zakęś K (2009) Controlled reproduction of pikeperch *Sander lucioperca* (L.): a review. Arch Pol Fish 17:153–170

Żarski D (2012) First evidence of pheromonal stimulation of maturation in Eurasian perch, *Perca fluviatilis* L., females. TrJFAS 12:771–776

Żarski D, Bokor Z, Kotrik L, Urbányi B, Horváth A, Targońska K, Krejszeff S, Palińska K, Kucharczyk D (2011a) A new classification of a preovulatory oocyte maturation stage suitable for the synchronization of ovulation in controlled reproduction of Eurasian perch, *Perca fluviatilis* L. Reprod Biol 3:194–209

Żarski D, Palińska K, Targońska K, Bokor Z, Kotrik L, Krejszeff S, Kupren K, Horvath A, Urbanyi B, Kucharczyk D (2011b) Oocyte quality indicators in Eurasian perch, Perca fluviatilis L., during reproduction under controlled conditions. Aquaculture 313:84–91

Żarski D, Kucharczyk D, Targońska K, Palińska K, Kupren K, Fontaine P, Kestemont P (2012a) A new classification of pre-ovulatory oocyte maturation stages in pikeperch, *Sander lucioperca* (L.), and its application during artificial reproduction. Aquac Res 43:713–721

Żarski D, Krejszeff S, Palińska K, Targońska K, Kupren K, Fontaine P, Kestemont P, Kucharczyk D (2012b) Cortical reaction as an egg quality indicator in artificial reproduction of pikeperch, *Sander lucioperca*. Reprod Fertil Dev 24:843–850

Żarski D, Krejszeff S, Horváth Á, Bokor Z, Palińska K, Szentes K, Łuczyńska J, Targońska K, Kupren K, Urbányi B, Kucharczyk D (2012c) Dynamics of composition and morphology in oocytes of Eurasian perch, *Perca fluviatilis* L., during induced spawning. Aquaculture 364–365:103–110

Żarski D, Horvath A, Kotrik L, Targońska K, Palińska K, Krejszeff S, Bokor Z, Urbanyi B, Kucharczyk D (2012d) Effect of different activating solutions on the fertilization ability of Eurasian perch, *Perca fluviatilis* L., eggs. J Appl Ichthyol 28:967–972

Żarski D, Targońska K, Kaszubowski R, Kestemont P, Fontaine P, Krejszeff S, Kupren K, Kucharczyk D (2013a) Effect of different commercial spawning agents and thermal regime on the effectiveness of pikeperch, *Sander lucioperca* (L.), reproduction under controlled conditions. Aquac Int 21:819–828

Żarski D, Křišťan J, Palińska-Żarska K, Krejszeff S, Policar T, Kucharczyk D (2013b) The period of pikeperch, *Sander lucioperca* (L.), eggs ability to fertilization and its dependence on egg quality. Diversification in Inland Finfish Aquaculture II (DIFA II), 24–26 Sept, University of South Bohemia in České Budějovice, Faculty of Fisheries and Protection of Waters, Vodnany

Żarski D, Horváth A, Bernáth G, Palińska-Żarska K, Krejszeff S, Müller T, Kucharczyk D (2014) Application of different activating solutions to in vitro fertilization of crucian carp, *Carassius carassius* (L.), eggs. Aquac Int 22:173–184

Żarski D, Targońska K, Palińska K, Kucharczyk D, Krejszeff S, Kupren K, Fontaine P, Kestemont P (2015a) The application of tannic acid to the elimination of egg stickiness at varied moments of the egg swelling process in pikeperch, *Sander lucioperca* (L.). Aquac Res 46:324–334

Żarski D, Cejko BI, Krejszeff S, Palińska-Zarska K, Horváth Á, Sarosiek B, Judycka S, Kowalski RK, Łaczyńska B, Kucharczyk D (2015b) The effect of osmolality on egg fertilization in common carp, Cyprinus carpio Linnaeus, 1758. J Appl Ichthyol 31:159–163

Zohar Y, Mylonas CC (2001) Endocrine manipulations of spawning in cultured fish: from hormones to genes. Aquaculture 197:99–136

Chapter 5
Sperm Morphology, Physiology, Motility, and Cryopreservation in Percidae

Sayyed Mohammad Hadi Alavi, Andrzej Ciereszko, Azadeh Hatef, Jiří Křišťan, Boris Dzyuba, Sergei Boryshpolets, Marek Rodina, Jacky Cosson, and Otomar Linhart

Abstract In percids, the spermatozoon is acrosomeless and asymmetrical in shape. The head of the spermatozoon is spherical and contains the genomic material. Mitochondria and proximal and distal centrioles are located in the midpiece of the spermatozoon. The flagellum consists of an axoneme with a "9+2" microtubule structure surrounded by a plasma membrane. The length of spermatozoa flagella is between 30 and 35 μm. The volume of sperm and spermatozoa concentration highly differs among species and individuals. Seminal plasma is composed of both mineral and organic compounds and has osmolality about 300 mOsmol kg^{-1} to maintain the spermatozoa in the quiescent state. A hypo-osmotic shock is required to trigger initiation of spermatozoa motility after discharge into an aquatic environment. The duration of sperm motility lasts from several seconds to a few minutes, however sperm

S.M.H. Alavi (✉)
South Bohemian Research Center of Aquaculture and Biodiversity of Hydrocenoses, Faculty of Fisheries and Protection of Waters, University of South Bohemia in České Budějovice, Vodňany 389 25, Czech Republic

Department of Veterinary Sciences, Faculty of Agrobiology, Food and Natural Resources, Czech University of Life Sciences, 165 21 Prague 6, Czech Republic
e-mail: alavi@af.czu.cz; alavi@frov.jcu.cz; smhalavi1@gmail.com

A. Ciereszko
Institute of Animal Reproduction and Food Research, Polish Academy of Sciences, Olsztyn 10-747, Poland

A. Hatef
South Bohemian Research Center of Aquaculture and Biodiversity of Hydrocenoses, Faculty of Fisheries and Protection of Waters, University of South Bohemia in České Budějovice, Vodňany 389 25, Czech Republic

Veterinary Biomedical Sciences Department, Western College of Veterinary Medicine, University of Saskatchewan, Saskatoon, Saskatchewan S7N 5B4, Canada

J. Křišťan • B. Dzyuba • S. Boryshpolets • M. Rodina • J. Cosson • O. Linhart (✉)
South Bohemian Research Center of Aquaculture and Biodiversity of Hydrocenoses, Faculty of Fisheries and Protection of Waters, University of South Bohemia in České Budějovice, Vodňany 389 25, Czech Republic
e-mail: linhart@frov.jcu.cz

P. Kestemont et al. (eds.), *Biology and Culture of Percid Fishes*,
DOI 10.1007/978-94-017-7227-3_5

motility kinetics (percentage of motile spermatozoa, spermatozoa velocity and beating frequency of flagella) rapidly decrease after initiation of sperm motility due to rapid depletion of energy source required for the axonemal beating. The environmental osmolality, pH and ionic concentrations affect sperm motility kinetics. The highest percentage of spermatozoa motility and spermatozoa velocity are observed in an activation medium with osmolality of 100 mOSmol kg^{-1}. There are various factors that affect semen quality in male broodfish including photoperiod and seasonal regimes, nutrition and antinutritional factors, rearing condition and age and size of broodfish. For short-term storage, it is essential to dilute semen in an ionic extender (300 mOsmol kg^{-1}) with or without antibiotics. Methanol (6–10 %) and dimethyl sulfoxide (10 %) can be used as cryoprotectant for sperm cryopreservation.

Keywords Percidae • Sperm • Motility • Cryopreservation • Semen quality

5.1 Introduction

In commercial fish culture, effective broodfish management is a prerequisite for successful artificial breeding (Alavi et al. 2008a). The management of male broodfish is highly family-specific and depends upon gonad morphology, the neuroendocrine and endocrine regulation of spermatogenesis and spermatozoa maturation, and the physiology and biochemistry of sperm (Alavi et al. 2008a, 2012; Ciereszko 2008). This chapter summarizes methods of sperm collection, handling of sperm in the hatchery, short-term storage and cryopreservation of sperm in Percidae, and reviews studies of physiology and biochemistry of seminal plasma, sperm motility, and energetics, as well as factors influencing sperm quality.

5.2 Semen Collection and Stripping

Prior to stripping, broodfish should be anesthetized for 3–10 min. Tricaine methanesulfonate (99–132 mg L^{-1}), 2-phenoxyethanol solution (0.2–0.3 mL L^{-1}), clove oil, *Syzygium aromaticum* (approximately 0.5 mL L^{-1}), and propiscin (1.5–2 mL L^{-1} water) have been used for walleye, Eurasian perch, and pikeperch (Satterfield and Flickinger 1995b; Król et al. 2006; Bokor et al. 2007; Kazun and Siwicki 2001; Teletchea et al. 2009). The anesthetized fish should be rinsed in freshwater to remove residual anesthetic and wiped dry to avoid water contamination of semen during stripping. Semen is collected by applying gentle pressure on the abdomen, anterior to the anal pore. Initially released drops should generally be discarded, to avoid contamination by urine spontaneously released from the bladder (Alavi et al. 2007). The container of stripped semen should be placed on ice immediately. The broodfish may be left to recover in freshwater with a prophylactic treatment of 5 mg L^{-1} furacin and 0.5 % NaCl (Satterfield and Flickinger 1995b).

5.3 Spermatozoa Morphology and Fine Structure

Of 159 species in nine genera of Percidae (Lahnsteiner and Patzner 2008), spermatozoa morphology and ultrastructure have been studied only for Eurasian perch, *Perca fluviatilis* (Lahnsteiner et al. 1995; Hatef et al. 2011) and pikeperch, *Sander lucioperca* (Lahnsteiner and Mansour 2004; Křišťan et al. 2014) (Fig. 5.1). In both species, the spermatozoon is identified as an acrosomeless aquasperm and differentiated into head, small midpiece, and flagellum (Fig. 5.1Aa, Ba). The spermatozoon is asymmetrical, with the flagellum inserting mediolaterally into the head (Fig. 5.1Ab, Bb). The head of the spermatozoon is spherical and contains the genomic material. Head

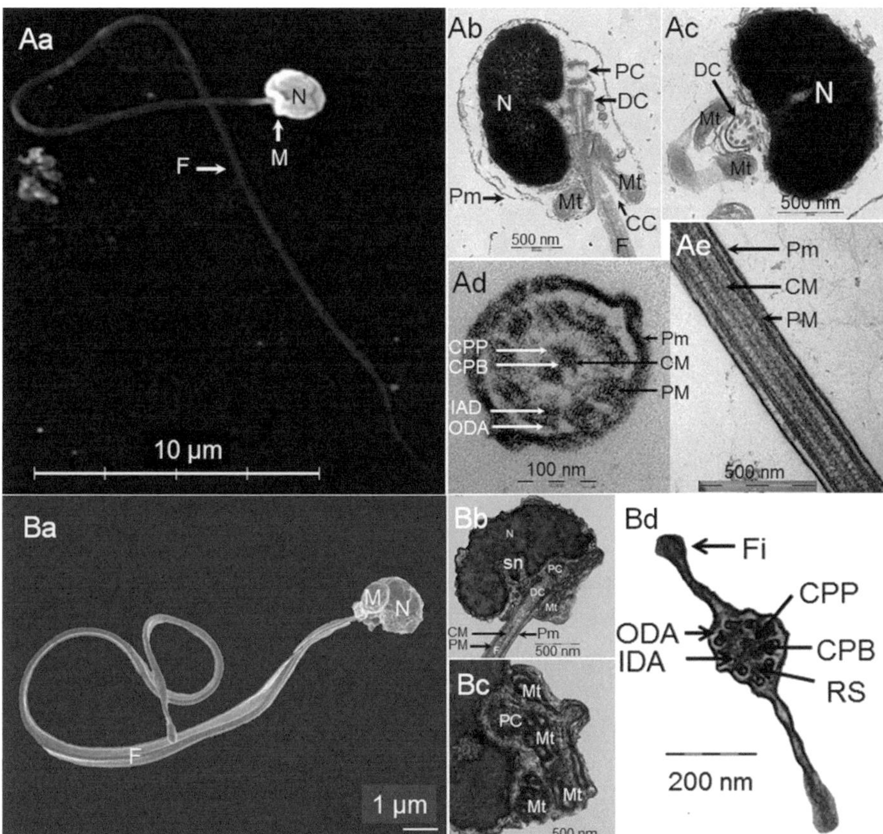

Fig. 5.1 Spermatozoon morphology and ultrastructure of (**a**) Eurasian perch, *Perca fluviatilis* and (**b**) pikeperch, *Sander lucioperca* (Hatef et al. 2011; Křišťan et al. 2012b). Figures Aa and Ba SEM of spermatozoon composed of head, midpiece and flagellum (JSM 6300 or JSM 7401-F, JEOL Ltd., Tokyo, Japan). Figures Ab–Ae and Bb–Bd are TEMs of longitudinal (**b**, **e**) or cross (**c**, **d**) sections of the mid-piece (**b**, **c**) and the flagellum (**d**, **e**) (JEOL 1010, JEOL Ltd., Tokyo, Japan). *Ax* axoneme, *CC* cytoplasmic channel, *CM* central microtubules, *CPB* central pair bridge, *CPP* central pair projections, *DC* distal centriole, *F* flagellum, *Fi* fin-like structure, *IDA* inner dynein arm, *M* mid-piece, *Mt* mitochondria, *N* nucleus, *ODA* outer dynein arm, *PC* proximal centriole, *Pm* plasma membrane, *PM* peripheral microtubules, *RS* radial spokes, *sn* satellite nuclear notches

dimensions are greater in Eurasian perch (length: 1.9 ± 0.2 μm and width: 1.8 ± 0.1 μm) than in pikeperch (length 1.6–1.8 ± 0.2 μm, width 1.3–1.6 ± 0.1 μm). In Eurasian perch, a small cylindrical midpiece (diameter: 0.8 ± 0.1 μm) is located at the base of the head (Fig. 5.1Aa). In pikeperch, the midpiece is integrated into the head region or lateral to the head (Fig. 5.1Ba). The nucleus of pikeperch contains two invaginations, the satellite nuclear notches (Fig. 5.1Bb). Proximal and distal centrioles are present in the midpiece, each consisting of nine triplets of microtubules (Fig. 5.1Ab, Ac, Bb, and Bc). In pikeperch, the proximal centriole is inclined relative to the distal centriole at 110° compared to 90° in Eurasian perch. The distal centriole serves as a basal body for the formation of the flagellum. Lahnsteiner et al. (1995) have reported the existence of a mitochondrion surrounding the cytoplasmic channel and forming an incomplete ring in Eurasian perch, while Hatef et al. (2011) reported two mitochondria (Fig. 5.1Ab, Ac). In pikeperch, two to four mitochondria are irregularly arranged in the midpiece region (Fig. 5.1Bb, Bc). In both species, a cytoplasmic channel (diameter 0.4 ± 0.1 μm) is clearly visible around the midpiece and flagellum (Fig. 5.1Ab, Bb). The length of the flagellum in Eurasian perch (30–35 or 25.3–32.5 μm) and pikeperch (length: 33.2 ± 0.9 μm) is similar and consists of an axoneme surrounded by a plasma membrane (Lahnsteiner et al. 1995; Wirtz and Steinmann 2006; Křišťan et al. 2014) (Fig. 5.1Ae, Bb). The axoneme exhibits typical eukaryotic microtubule-based organization with a central pair of single microtubules surrounded by nine peripheral doublets of microtubules (Fig. 5.1Ad, Bd). Outer and inner dynein arms, radial spokes, and other ultrastructural features of the axoneme are observed in spermatozoa of both species. Well-developed lateral fin-like projections are observed in pikeperch spermatozoa (Fig. 5.1Ba, Bd). Although Lahnsteiner et al. (1995) reported similar structures in Eurasian perch spermatozoa, but micrographs obtained from cross sections of flagella of Eurasian perch spermatozoa do not reveal such structures (Hatef et al. 2011) (Fig. 5.1Aa, Ad).

5.4 Semen Production

Mature Eurasian perch males of 89–117 g body mass produce 0.6–6.8 mL of semen (Alavi et al. 2007). Sperm dry weight (expressed as percent of wet weight of semen) in walleye (*Sander vitreus*) weighing 967–2747 g is reported to be between 5.3 % and 28.2 %, with a mean value of 20.8 % (Gregory 1970). The average volume of semen produced by walleyes in first, second, and third daily strippings is 3.1, 1.3, and 0.5 mL, respectively (Satterfield and Flickinger 1995b). Brown and Moore (1996) reported average values of 3.6 mL of semen per walleye male, but provided no information regarding broodfish body mass. Pikeperch weighing 778–1284 g produce 0.4–1.1 mL of semen (Korbuly et al. 2009). Most studies show high inter-individual variations in production of semen in Percidae.

Semen in Percidae is characterized by a high spermatozoa concentration, reaching 19–128 × 10^9 cells mL^{-1} in Eurasian perch, 40–75 × 10^9 cells mL^{-1} in yellow perch (*Perca flavescens*), and 21–69 × 10^9 cells mL^{-1} in walleye (Table 5.1). Reported spermatozoa concentration in other percid species is much lower, for

Table 5.1 Ionic contents, pH and osmolality of the seminal plasma in Percidae

Fish	Sodium (mM)	Potassium (mM)	Chloride (mM)	Calcium (mM)	pH	Osmolality (mOsmol.kg^{-1})	Spermatozoa concentration (× 10^9 cells per mL)	Reference
Eurasian perch	96.3–172.1 (124.0 ± 21.7)	7.9–11.6 (10.2 ± 1.1)		0.4–1.3 (0.7 ± 0.3)	8.1–8.4 (8.3 ± 0.1)	253–353 (283.9 ± 37.2)		Lahnsteiner et al. (1995)[b]
							(76.2 ± 3.2)	Piironen and Hyvärinen (1983)[a]
							28–42 (32)	Rougeot et al. (2004)
							29.1–32.4	Król et al. (2006)
							18·8–127·5 (53·6)	Wirtz and Steinmann (2006)
	112.5–138.3 (131.0 ± 2.2)	7.7–15.0 (10.7 ± 0.6)	88.0–115.0 (106.8 ± 2.4)	1.7–2.8 (2.4 ± 0.1)		255–334 (298.1 ± 5.1)	(29.2 ± 3.2)	Alavi et al. (2007)[a]
							(37.8 ± 6.3)[1]	Rodina et al. (2008)[c]
							(45.3 ± 5.4)[2]	
						(298 ± 12)		Boryshpolets et al. (2009)[b]
						(423.6 ± 19.1)[1]	(59.2 ± 2.2)[1]	Alavi et al. (2010)[a, d]
						(523.6 ± 42.1)[2]	(66.5 ± 3.1)[2]	
						(373.0 ± 57.7)[3]	(36.0 ± 4.6)[3]	
						(292.0 ± 9.1)[4]	(45.5 ± 2.9)[4]	
						300–306	54.3–68.6	Hatef et al. (2010)
							(47.0 ± 8.9)	Nynca et al. (2010)[b]
						293.6–299.0	46.1–58.9	Henrotte et al. (2010)
						(305.8 ± 1.7)	(61.5 ± 2.7)	Hatef et al. (2011)[a]
							(1.10 ± 0.65)	Lahnsteiner (2010)

(continued)

Table 5.1 (continued)

Fish	Sodium (mM)	Potassium (mM)	Chloride (mM)	Calcium (mM)	pH	Osmolality (mOsmol.kg^{-1})	Spermatozoa concentration (× 10^9 cells per mL)	Reference
Pikeperch						221–287		Kowalski et al. (2003)
						178–227	4.3–5.3	Cejko et al. (2008)
							7.0–8.3	Jarmołowicz et al. (2010)
							(14.6 ± 7.7)	Nynca et al. (2010)[b]
							8.1–20.6	Křišťan et al. (2014)
Volga pikeperch							(8.4 ± 5.4)	Nynca et al. (2010)[b]
Hybrid (pikeperch×Volga pikeperch)							(13.2 ± 2.6)	Nynca et al. (2010)[b]
Walleye	3.8–2.4	1.0–1.6	4.3–4.7	0.02–0.03				Gregory (1970)[e]
					8.5		24–59 (38.6)	Brown and Moore (1996)
							21–50	Rinchard et al. (2005)
Yellow perch							(41.6 ± 3.5)	Ciereszko and Dabrowski (1993)[b]
	120	10			8.3	316.7		Dabrowski et al. (1996)
							40–75	Ciereszko et al. (1998)

[a]Data in brackets are mean ± standard error of mean
[b]Data in brackets are mean ± standard deviation
[c]Values are for stripped sperm of normal males (1) and testicular sperm of neomales (2)
[d]Sperm samples were collected on November 29, 2005 (1), January 10, 2006 (2), February 21, 2006 (3) and April 7, 2006 (4)
[e]Values are measured in g L^{-1}

example 4.3–20.6 × 10^9 cells mL^{-1} in pikeperch, 8.4 × 10^9 cells mL^{-1} in Volga pikeperch, and 13.2 × 10^9 cells mL^{-1} in hybrids of pikeperch and Volga pikeperch (Table 5.1). This may be due to the existence of a urinary bladder at the urogenital pore in pikeperch, which readily allows contamination of semen by urine during stripping (Křišťan et al. 2014).

High spermatozoa concentration in semen of percids implies a relatively low volume of seminal plasma, and spermatozoa volume can represent up to 95 % of the total volume of semen (Piironen and Hyvärinen 1983). Spermatocrit (the proportion of white packed material relative to the total volume of semen following centrifugation in a capillary tube) is reported to be 66–70 % in Eurasian perch (Hatef et al. 2010, 2011), and 36–94 % in walleye (Gregory 1970).

Wide variation in semen volume and spermatozoa concentration reported in different studies may be related to broodfish rearing conditions, methods of spawning induction, duration of broodfish participation in spawning, and duration and time of spermiation within the reproductive season (Alavi et al. 2008a; Ciereszko 2008).

5.5 Seminal Plasma Composition

Seminal plasma comprises both mineral and organic compounds produced by the testicular main duct and the sperm duct (Alavi et al. 2008a; Ciereszko 2008). It plays a key role in protecting and maintaining vital spermatozoa functions including viability, motility, and fertilizing ability for a prolonged period. Maintenance of a quiescent state, supply of adequate levels of nutrients, and protection against damage by microbes or xenobiotics are key elements for integrity of spermatozoa function (Wojtczak et al. 2005; Dietrich et al. 2011).

5.5.1 Mineral Composition of Seminal Plasma

Sodium ions predominate in the seminal plasma; their concentration is tenfold that of potassium ions (Table 5.1). Potassium ion concentration is five to tenfold that of calcium ions (Table 5.1). The osmolality of seminal plasma is similar among percids, with measured values of approximately 300 mOsmol kg^{-1} in mid-spawning season (Table 5.1). It is assumed that seminal plasma osmolality is a major factor in maintaining spermatozoa in a quiescent state during storage in the reproductive system (see Sect. 5.6 in this chapter). The slight inter-species differences in seminal plasma osmolality among percids may be related to the amount of ions and inorganic compounds (Lahnsteiner et al. 1995; Ciereszko 2008).

5.5.2 *Organic Composition of Seminal Plasma*

5.5.2.1 Organic Substances with Low Molecular Weight

Cholesterol and monosaccharides (glucose, fructose, and galactose) are among numerous organic substances identified in the seminal plasma of Eurasian perch that are potentially important for steroid hormone biosynthesis and the energy of spermatozoa (Piironen and Hyvärinen 1983; Lahnsteiner et al. 1995; Henrotte et al. 2010). Piironen and Hyvärinen (1983) identified glycerol and citric acid in perch seminal plasma and linked the former with the lipolytic activity of testes and the latter with chelating of divalent ions, which have importance for keeping spermatozoa in a quiescent state (Ciereszko et al. 2000).

The amino acids alanine (2.1 mM), arginine (7.8 mM), asparagine (1.1 mM), cysteine (0.6 mM), glutamic acid (1.0 mM), isoleucine (1.1 mM), lysine (0.02 mM), methionine (1.6 mM), phenylalanine (0.3 mM), tryptophan (0.1 mM), and valine (0.5 mM) have recently been detected in the seminal plasma of Eurasian perch (Lahnsteiner 2010). Composition of free amino acids in the seminal plasma of Eurasian perch compared to other fish species highlights species-specific differences (Lahnsteiner 2009, 2010). It has been shown also that 48 h in vitro incubation of sperm in the presence of 2.5 mM of asparagine, lysine, methionine, or valine increases Eurasian perch spermatozoa motility and velocity (Lahnsteiner 2010). This is strong evidence of supporting roles of amino acids in spermatozoa viability. Further studies are required to investigate the physiological effects of specific amino acids on spermatozoa function (Lahnsteiner 2009, 2010).

Seminal plasma contains numerous low and high molecular weight antioxidative substances to protect spermatozoa. Uric acid has been identified in the seminal plasma of yellow perch (Ciereszko et al. 1999) and cysteine and glutathione in the semen of Eurasian perch and pikeperch (Stejskal et al. 2008). Lahnsteiner and Mansour (2010) identified ascorbic acid, glutathione, methionine, and uric acid and suggested a physiological role in the antioxidant system of Eurasian perch semen. Their studies show that uric acid is the most abundant, possibly being a major antioxidant in semen, as it improves sperm motility and membrane integrity and decreases sperm lipid peroxidation.

5.5.2.2 Proteins and Enzymes

Concentration of proteins in seminal plasma of Eurasian perch shows higher values than in other percids (Table 5.2). High protein concentration in seminal plasma correlates with high spermatozoa concentration in these species (Nynca et al. 2010).

Similar to other teleosts (Dabrowski and Ciereszko 1994), seminal plasma of percids is characterized by a species-specific system of proteins with anti-trypsin activity. Anti-trypsin activity in Eurasian perch is one of the highest among percids (Table 5.2). Multiple forms of anti-trypsin activity are observed in the seminal plasma of percids; at least five compared to one to three in other species (Dabrowski

Table 5.2 Protein concentration and anti-trypsin activity in seminal plasma of Percidae

Fish	Protein concentration (g.L^{-1})	Anti-trypsin activity (U.L^{-1})	Reference
Eurasian perch	2.3–5.6 (3.5 ± 0.6)		Lahnsteiner et al. (1995)
	3.9–6.1		Król et al. (2006)
	(8.0 ± 5.6)	(1406.4 ± 648.0)	Nynca et al. (2010)
		265–724	Król et al. (2011)
Pikeperch	1.0–2.4		Cejko et al. (2008)
	(1.8 ± 0.1)	(416.4 ± 123.3)	Nynca et al. (2010)
Volga pikeperch	(1.2 ± 0.2)	(178.4 ± 3.5)	Nynca et al. (2010)
Hybrid (pikeperch × Volga pikeperch)	(1.8 ± 0.4)	(340.3 ± 32.8)	Nynca et al. (2010)
Yellow perch	3.2		Dabrowski and Ciereszko (1994)

Data in brackets are mean ± standard deviation

and Ciereszko 1994; Ciereszko et al. 2000; Król et al. 2006; Nynca et al. 2010). The physiological importance of multiple forms of anti-trypsin activity is largely unknown, but it is species-specific and could potentially be used for the identification of hybrids of Percidae. The anti-trypsin activity profile in the seminal plasma shows similarities to that in the blood plasma. This suggests that inhibitors in seminal plasma can originate from blood. It is assumed that these inhibitors play key roles in protection of spermatozoa and reproductive tissue from proteolytic attack and/or in the regulation of spermatogenesis and spermatozoa motility (Ciereszko 2008). Potential targets for anti-trypsin activity in percid fish are proteolytic enzymes and proteolytic activity (Lahnsteiner et al. 1995; Kowalski et al. 2003, 2004; Król et al. 2011).

A significant correlation is shown between protein concentration and anti-trypsin activity in percids, including Eurasian perch and Volga pikeperch (Nynca et al. 2010). This association suggests that serine protease inhibitors are major proteins in the seminal plasma of percid fish and/or that their synthesis is regulated by the same mechanism as for other seminal plasma proteins.

Aspartate aminotransferase (AspAT) is present in seminal plasma of yellow perch, and its activity is about 0.25 % that seen in spermatozoa (Dabrowski at al. 1996). It can potentially be used as an indicator of sperm quality, since leakage of AspAT from spermatozoa into seminal plasma due to sperm damage can be measured.

Lactic dehydrogenase (LDH) has been identified in seminal plasma of Eurasian perch (Lahnsteiner et al. 1995). The activity of sperm LDH has been employed to demonstrate the cytotoxic effect of gossypol on yellow perch spermatozoa (Ciereszko and Dabrowski 2000).

Major oxidant defensive enzymes are also identified in Eurasian perch seminal plasma, including catalase, glutathione reductase, methionine reductase, peroxidase, and superoxide dismutase (Lahnsteiner and Mansour 2010). Treating semen with catalase results in substantial improvement of spermatozoa motility and membrane integrity.

Other enzymes described in seminal plasma include proteases, alkaline and acid phosphatases, ATPase, β-glucoronidase, butyryl cholinesterase, and glucose-6-phosphate dehydrogenase (Lahnsteiner et al. 1995; Kowalski et al. 2003, 2004).

5.5.2.3 Lysozyme Activity and Immunoglobulins

Lahnsteiner and Radner (2010) demonstrated the presence of the lysozyme N-acetylmuramide glycanohydrolase and immunoglobulins in seminal plasma of Eurasian perch (1–2 U L^{-1}). Both proteins are important components of the defense system and are related to quality of fish sperm.

5.6 Spermatozoa Motility

Similar to most other freshwater and marine fish, spermatozoa in percids are immotile in the sperm duct due to osmolality of the seminal plasma (Lahnsteiner et al. 1995; Alavi et al. 2007). This contrasts with the mechanism of spermatozoa quiescence in Salmonidae and Acipenseridae, in which K^+ is the main inhibitor of sperm activation in the seminal plasma (Cosson 2010; Alavi et al. 2012).

Initiation of spermatozoa motility in percids is triggered at discharge into hypo-osmotic freshwater (Lahnsteiner et al. 1995; Alavi et al. 2007; Boryshpolets et al. 2009). Studies report greater than 80 % spermatozoa motility in Eurasian perch (Lahnsteiner et al. 1995; Alavi et al. 2007, 2010; Henrotte et al. 2010; Lahnsteiner 2011), walleye (Satterfield and Flickinger 1995a; Bergeron et al. 2002; Casselman et al. 2006), and pikeperch (Křišťan et al. 2014). The initial spermatozoa motility rate does not differ among percids and cannot be used as a species indicator. Potential for motility in fish sperm is a hormone-dependent mechanism regulated by the hypothalamus-pituitary-gonad axis leading to increase of intracellular pH and cAMP in spermatozoa during maturation of spermatozoa in the sperm duct (Miura et al. 1992). In percids, as in other fish species (Alavi et al. 2009, 2012), waves propagate along the flagellum starting immediately upon initiation of sperm motility, while later in the motility phase waves are restricted to the region proximal to the head and becoming absent at the end of the motility period (Alavi et al. 2010) (Fig. 5.2). After initiation of motility, the percent of motile spermatozoa, velocity, and flagellar beat frequency decrease in Eurasian perch (Fig. 5.3) and pikeperch (Fig. 5.4). Decrease in these parameters is chiefly due to depletion of intracellular ATP required for axonemal beating (Fig. 5.3). Duration of spermatozoa motility in percids ranges from a few seconds to a few minutes and is reflected in inter-individual differences within a species and inter-specific differences among Percidae. In Eurasian perch, most studies report duration lasting less than 120 s after activation in freshwater, distilled water, or saline medium (Fig. 5.3) (Lahnsteiner et al. 1995; Alavi et al. 2007, 2010; Rodina et al. 2008). In walleye, the spermatozoa motility period is less than 120 s after activation in freshwater (Satterfield and Flickinger 1995a; Casselman et al. 2006; Green and Kelly 2008). In pikeperch, the duration of

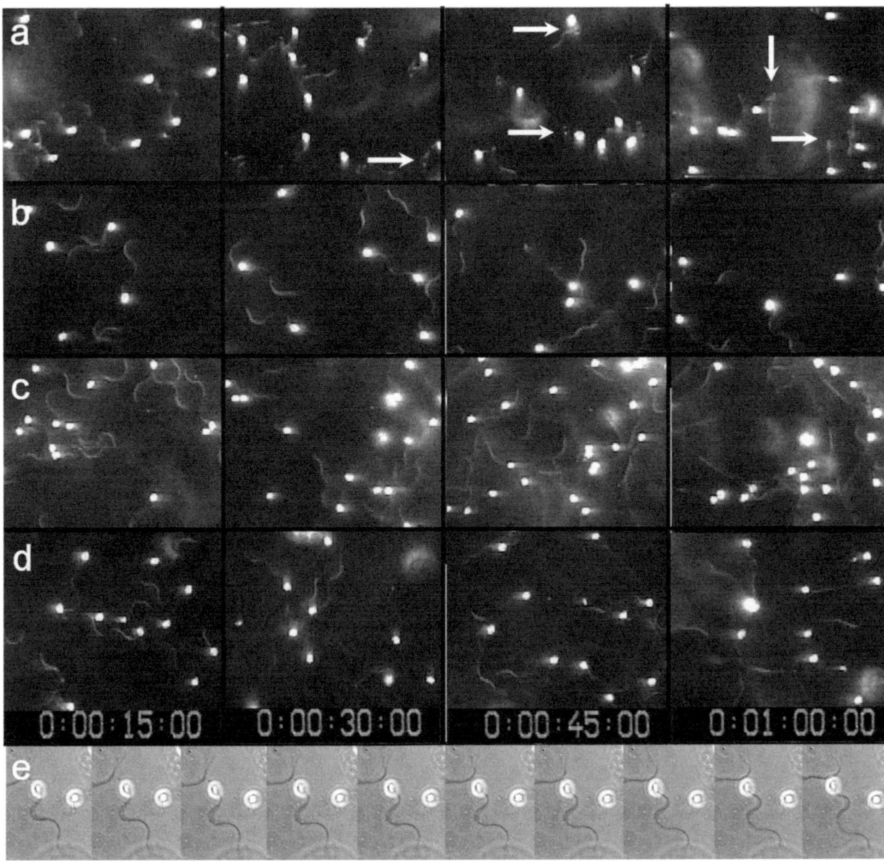

Fig. 5.2 Dark field stroboscope illuminated flagellum images of Eurasian perch (*Perca fluviatilis*) spermatozoa at 15, 30, 45 and 60 s post-activation in (**a**) buffered distilled water, 20 mM Tris, pH 8.0, (**b**) NaCl 50 mM, Tris 20 mM, pH 8.0 (**c**) sucrose 100 mM, Tris 20 mM, pH 8.0 and (**d**) NaCl 50 mM, $CaCl_2$ 1 mM, Tris 20 mM, pH 8.0. Semen was firstly diluted in NaCl 200 mM, $NaHCO_3$ 2.38 mM, pH 7.5 (osmolality 380 mOsmol kg^{-1}) and motility was observed under (for details see Alavi et al. 2007). Beating of flagella were observed either in ionic or non-ionic activation media. (*Arrows*), Damage to flagella including blebs and loops. (In motile spermatozoa, flagellar wave are seen, while immotile spermatozoa can be recognized by straight non-moving flagellum.) (**e**) Eurasian perch spermatozoa beating 20 s post-activation in tap water (22 °C). Time interval between each image is 4 ms. The flagellum propagates 3 bends or 1.5 sine wave (High-speed video, Olympus i-speed TR 1000 frames/s)

sperm motility is reported as 5–9 min (Korbuly et al. 2009) or 1–2 min (Křišťan et al. 2014) after activation in freshwater (Fig. 5.4). Recently, Lahnsteiner (2011) reported duration of sperm motility in Eurasian perch longer than 2 h following activation in a medium composed of 75 mM NaCl, 2 mM KCl, 1 mM $MgSO_4$, 1 mM $CaCl_2$, 20 mM Tris, pH 8.0 (210 mOsmol kg^{-1}). The observed differences in spermatozoa velocity and duration of motility among species may be related to initial ATP content. Duration of sperm motility shows a positive relationship to spermatozoa ATP content. Green and Kelly (2008) reported a longer period of

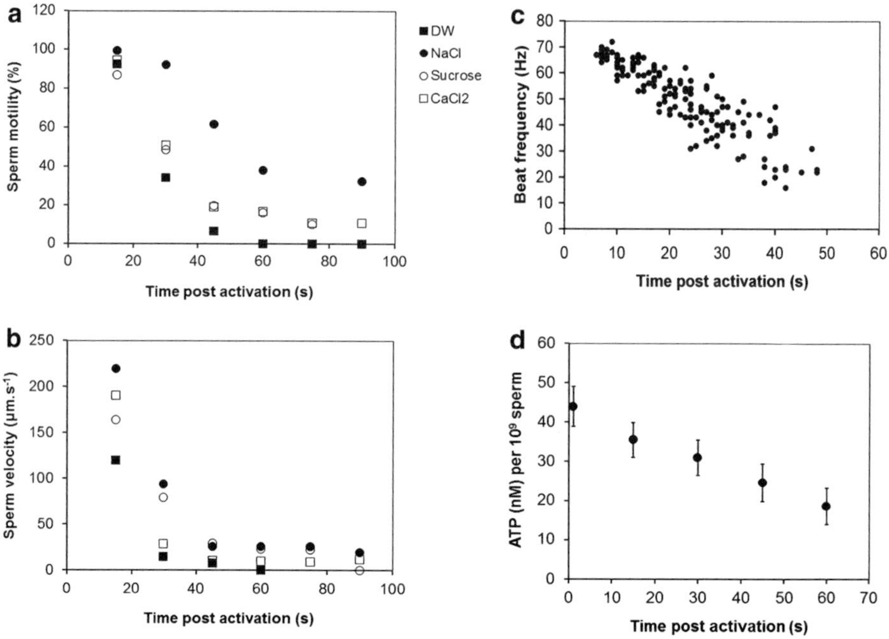

Fig. 5.3 Changes in (**a**) spermatozoa motility, (**b**) velocity, (**c**) beating frequency of flagellum and (**d**) ATP contents in Eurasian perch (*Perca fluviatilis*) with time post-activation in buffered distilled water (*DW*), NaCl 50 mM, Tris 20 mM (NaCl), sucrose 100 mM, Tris 20 mM (sucrose) and NaCl 50 mM, $CaCl_2$ 1 mM, Tris 20 mM ($CaCl_2$), pH 8.0–8.5 (Alavi et al. 2007, 2010; Hatef et al. 2010, 2011). Semen was diluted in NaCl 200 mM, $NaHCO_3$ 2.38 mM, pH 7.5 (osmolality 380 mOsmol kg^{-1}). Spermatozoa motility and velocity are higher after activation in ionic or non-ionic activation media with 100–150 mOsmol kg^{-1} compared to those of buffered distilled water. Adding Ca^{2+} to the activation medium increase spermatozoa velocity

motility in walleye spermatozoa containing a greater amount of ATP. Initial spermatozoa velocity also differs among percids. Values are reported as 122 µm s^{-1} in Eurasian perch (Lahnsteiner et al. 1995; Alavi et al. 2007, 2010), 94–120 µm s^{-1} in walleye (Casselman et al. 2006), and 158–165 µm s^{-1} (Křišťan et al. 2014). Spermatozoa velocity depends primarily on axonemal beating (flagellar beating frequency and wave parameters) and the morphology of the spermatozoa such as head size as well as the presence of a fin-like structure along the flagellum, which enhances flagellar beating (Alavi et al. 2009; Cosson 2010; Gillies et al. 2012).

5.7 Mechanism of Initiation of Spermatozoa Motility

Numerous factors contribute to initiation of spermatozoa motility in freshwater fish. The effects of pH, ions, and osmolality are critical to understanding the triggers of activation following sperm release from the genital pore into freshwater (Alavi and

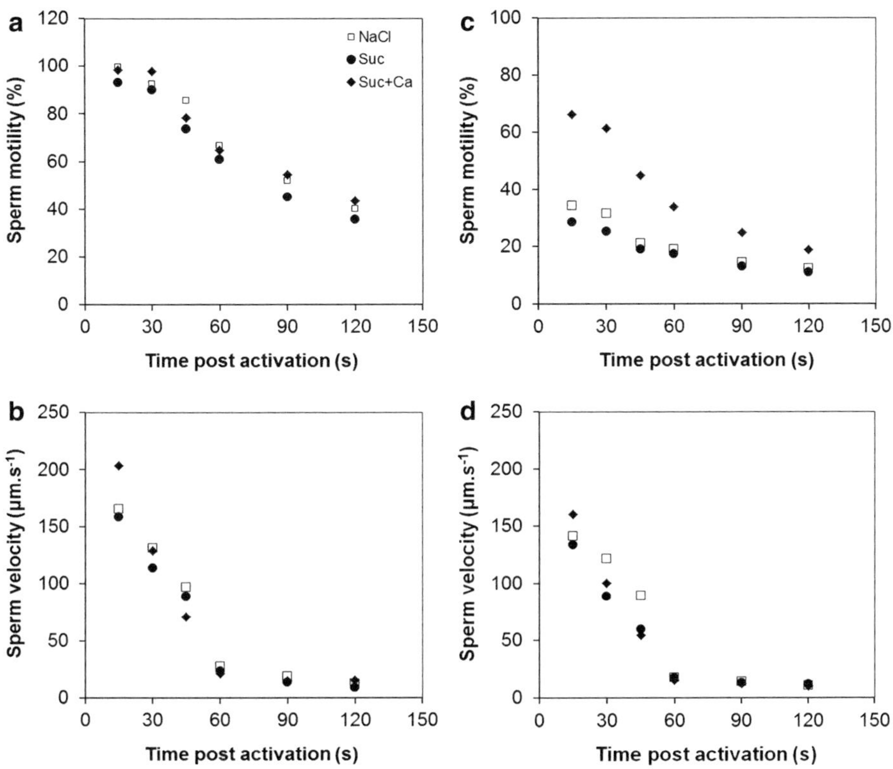

Fig. 5.4 Changes in (**a**, **c**) sperm motility (**b**, **d**) velocity in pikeperch (*Sander lucioperca*) post-activation in 45 mM NaCl, 5 mM KCl, 20 mM Tris, pH 8.5 (NaCl), 100 mM sucrose, 20 mM Tris, pH 8.5 (Suc) or 100 mM sucrose, 1 mM $CaCl_2$, 20 mM Tris, pH 8.5 (Suc+Ca). (**a**, **b**) fresh spermatozoa; (**c**, **d**) spermatozoa incubated 24 h on ice. Addition of Ca^{2+} to the activation medium increase motility and velocity of incubated spermatozoa (**c**, **d**). No pre-dilution was performed in these observations

Cosson 2005, 2006; Morisawa 2008) and can provide valuable information for development of effective sperm immobilization and activation solutions that can be used for development of protocols for short-term storage, cryopreservation, and artificial insemination at fish farms (Ciereszko et al. 2000).

5.7.1 Effects of pH on Sperm Activation

Motility of Eurasian perch spermatozoa has been demonstrated to be induced upon dilution in a buffered NaCl solution (100 mOsmol kg^{-1}) at pH 5.5–9.0 (Lahnsteiner et al. 1995; Hatef et al. 2011). The percent motile spermatozoa at pH 7.5 is not significantly different from that at pH 5.5, but decreases have been demonstrated at pH higher than 9.0. Lahnsteiner et al. (1995) reported lower spermatozoa velocity after activation in an acidic (121 μm s^{-1} at pH 6.5) or alkaline condition (119 μm s^{-1} at pH 9.0) compared to that of pH 7.5 (174 μm s^{-1}). Therefore, pH is not the key factor regulating spermatozoa motility in seminal plasma or after release into the aquatic environment. Seminal plasma pH has been reported as 8.3 in Eurasian perch (Lahnsteiner et al. 1995).

5.7.2 *Effects of Ions and Osmolality on Initiation of Spermatozoa Motility*

The initiation of spermatozoa motility appears to be ion-independent, as motility can be activated in both ionic (NaCl or KCl) and non-ionic (sucrose or glucose) solution (Fig. 5.5) (Lahnsteiner et al. 1995; Alavi et al. 2007, 2008b). Spermatozoa motility is suppressed with dilution in 150–175 mM NaCl or KCl (300–350 mOsmol kg^{-1}) (Fig. 5.5a), and sucrose or glucose at 300 mM prevents spermatozoa motility in Eurasian perch (Fig. 5.5b), demonstrating that osmolality is the principal inhibitor of spermatozoa motility in the seminal plasma of percids. Upon dilution of sperm in NaCl, KCl, or sucrose-based solution, the initial motility rate is usually not different from that in distilled water, but decreases at osmolalities above 200 mOsmol kg^{-1}. The highest spermatozoa velocity is reported at 100 mOsmol kg^{-1} (Lahnsteiner et al. 1995; Alavi et al. 2007, 2008b). Therefore, media containing NaCl or KCl or sucrose at 100 mOsmol kg^{-1} and media of 300–350 mOsmol kg^{-1} are suggested as activation and immobilizing media for percid sperm, respectively (Fig. 5.6).

Spermatozoa velocity in Eurasian perch and pikeperch increases when calcium (Ca^{2+}) is added to the activation medium, but the percent motile spermatozoa does not differ (Alavi et al. 2007; Křišťan et al. 2014) (Figs. 5.3 and 5.4). This is evidence of involvement of Ca^{2+} in axonemal beating, and further studies using specific Ca^{2+} channel blockers to investigate the role of Ca^{2+} in the initiation of spermatozoa motility in percids is essential.

In extremely low osmolality activation media, such as buffered distilled water, blebs appear along the flagellum that prevent correct and efficient wave propagation (Alavi et al. 2007) (Fig. 5.2). Later in the motile period, the tip of the flagellum becomes curled into a loop, shortening its length and affecting sperm velocity (Fig. 5.2). Similar damage has been observed in other freshwater fish and is probably the primary reason for the rapid decrease in sperm motility and velocity after sperm activation under very low osmolality conditions (Alavi et al. 2009).

5.8 Factors Affecting Semen Quality

As in other teleosts (Ciereszko 2008; Alavi et al. 2008a), numerous factors influence semen quality in percids (Ciereszko 2008). Season, photothermal regime, hormonal stimulation, nutrition, rearing conditions, status of males, age, and semen contamination have been studied in percids.

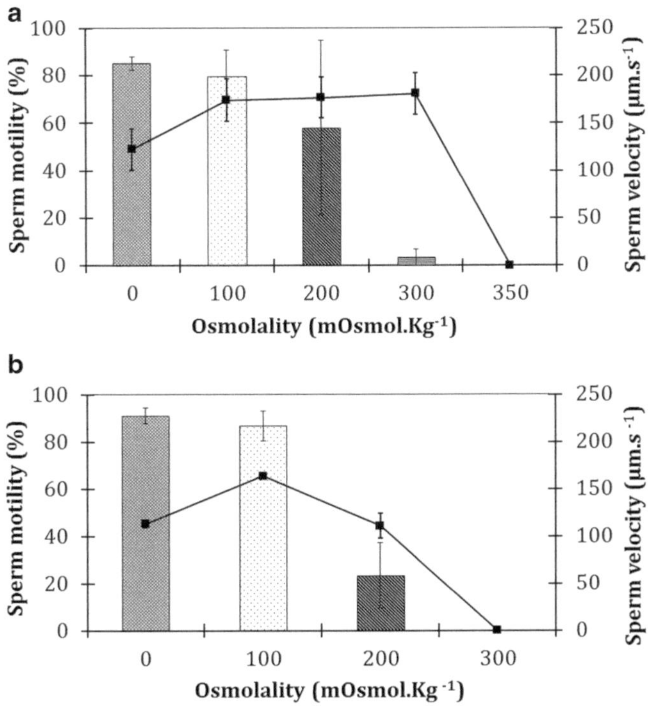

Fig. 5.5 Effects of osmolality on spermatozoa motility and velocity in Eurasian perch (*Perca fluviatilis*) after activation in (**a**) NaCl or (**b**) sucrose based activation media. Osmolality higher than 300–350 suppresses initiation of spermatozoa motility

5.8.1 Season

Reproduction of temperate fish, including percids, is clearly affected by the season (Wang et al. 2010). In yellow perch, semen volume increases from January to April (Dabrowski et al. 1994). Before the spawning season (November-January), the semen of Eurasian perch is characterized by high spermatozoa concentration (59–66 × 10^9 cells mL^{-1}) as compared to 36–45 × 10^9 cells mL^{-1} in February and April (Alavi et al. 2010). This is associated with decrease in seminal plasma osmolality from a range of 423–523 mOsmol kg^{-1} in November and January to 373–292 mOsmol kg^{-1} in February and April (Alavi et al. 2010). These changes are likely related to the semen hydration phenomenon, a physiological characteristic at the final stage of testicular development that leads to increase in semen volume (Ciereszko 2008). Along with physiological changes with spermatozoa maturation, spermatozoa motility increases from the beginning to the middle of the spawning season, and subsequently decreases at the end of the spawning season, likely due to aging of spermatozoa (Dabrowski et al. 1996; Król et al. 2006; Alavi et al. 2010).

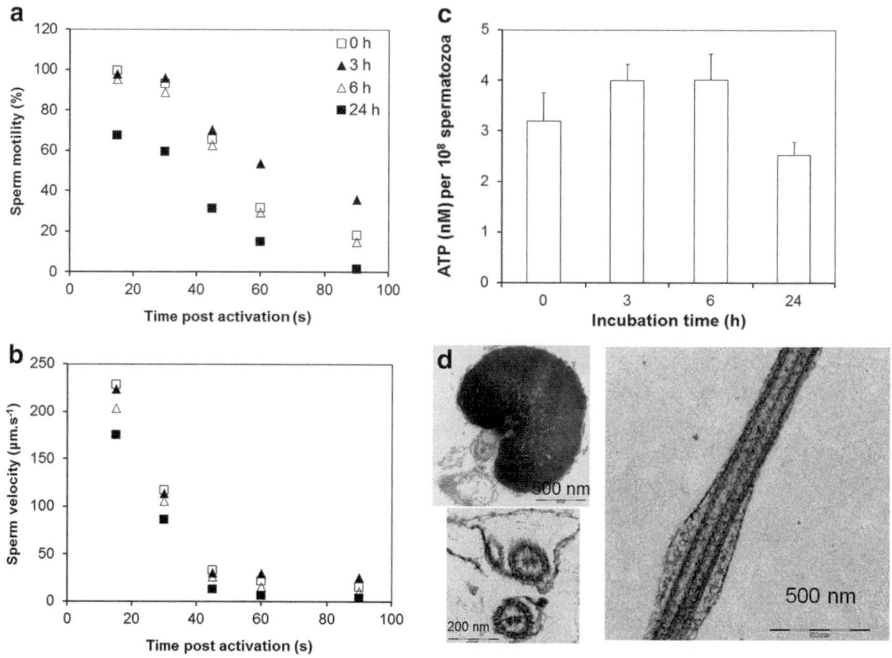

Fig. 5.6 Changes in (**a**) sperm motility, (**b**) velocity, (**c**) ATP contents and (**d**) ultrastructure in Eurasian perch (*Perca fluviatilis*) following 24 h incubation in NaCl 200 mM, $NaHCO_3$ 2.38 mM, pH 7.5 (osmolality 380 mOsmol kg^{-1}) (Hatef et al. 2011). Spermatozoa motility and velocity activated in 50 mM NaCl, 20 mM Tris, pH 8.5 (110 mOsmol kg^{-1}) decrease following incubation which are associated with decrease in ATP content and damage to flagella and mitochondria

5.8.2 Photothermal Regimes

Photoperiod and water temperature are basic environmental factors regulating reproduction of temperate fish. Phase-shifted photothermal cycles are frequently used to regulate time of sexual maturation and spawning in yellow perch (Dabrowski et al. 1996; Ciereszko et al. 1998) and Eurasian perch (Migaud et al. 2004, 2006). These studies showed that the shortened photothermal cycle accelerates testicular development, while continuous light inhibits it. Wang et al. (2008) determined that light intensity increases semen production.

5.8.3 Hormone Stimulation

As in most finfish with importance in aquaculture (Alavi et al. 2008a), hormone treatment has been applied to stimulate release of semen and/or synchronize spawning in percids. LHRHa, carp pituitary extract, or human chorionic gonadotropin (hCG), and a combination of follicle stimulating hormone (FSH), luteinizing hormone (LH), and metoclopramide have been evaluated for yellow perch, Eurasian perch, and pikeperch, respectively (Dabrowski et al. 1996; Kucharczyk et al. 1996, 1998; Zakęś 2007; Zakęś and Demska-Zakęś 2009). However, the success of spermiation induction may be associated with reproductive season. For example, more effective stimulation by LHRHa has been reported in yellow perch in February and March compared to April (Dabrowski et al. 1996).

5.8.4 Nutrition and Antinutritional Factors

Most information concerning effects of nutrition on reproductive performance of males involves essential fatty acids and antioxidants. Lee and Dabrowski (2004) studied effects of supplemental ascorbic acid and α-tocopherol and demonstrated potential transfer of both vitamins from to the testis of yellow perch, resulting in an increase in semen quality. Henrotte et al. (2010) showed that the composition of spermatozoa fatty acids is highly correlated with dietary fatty acid composition. However, no effects have been detected on semen quality. Considering the function of unsaturated fatty acids in metabolism and cryo-resistance of spermatozoa, further studies are required to investigate whether supplemented fatty acids can improve spermatozoa quality in percids. Wang et al. (2008) determined that the initial nutritional status of fish (together with light intensity) is important to increase semen quality in Eurasian perch.

Plant-derived ingredients of fish meal contain numerous anti-nutritional substances, such as gossypol, which is present in the cotton plant and has been shown to have male contraceptive activity. In vitro studies have demonstrated that gossypol inhibits spermatozoa motility and LDH activity in yellow perch, leading to decrease in fertilization ability (Ciereszko and Dabrowski 2000). These results suggest a potential for reproductive impairment in fish when cottonseed-containing diets or organic fertilizers are used in perch aquaculture.

5.8.5 Rearing Conditions

Cejko et al. (2008) reported higher spermatozoa motility in pikeperch maintained in tanks in photothermal controlled conditions compared to those in ponds or cages. The photothermal regime has been shown to highly influence sperm quality in percids (Król et al. 2006).

5.8.6 Status and Age of Broodfish

Three status-dependent mating tactics have been identified for Eurasian perch according to body size: sneaking (small males), group spawning (midsized males) and dominant spawning (large males) (Wirtz and Steinmann 2006). In this classification, larger males produce spermatozoa superior to that of smaller males. Korbuly et al. (2009) also observed that large healthy pikeperch males generally produce more semen than smaller specimens. Therefore, size of broodfish could be considered a biological indicator in percids, since the number of spermatozoa and their swimming velocity are critical indicators of fertilization success in walleye (Casselman et al. 2006). Ciereszko et al. (1998) found 3-year-old males to have higher spermatozoa concentration compared to 2-year-old males, and that these differences could be influenced by the photothermal regime.

5.8.7 Semen Contamination

Contamination of semen by urine is a serious threat to spermatozoa in percids. Studies show low initial spermatozoa motility in percids, particularly in pikeperch (Bokor et al. 2007; Cejko et al. 2008). This can result from spontaneous contamination of semen by urine during stripping, altering the osmolality of the seminal plasma. Semen contamination by urine not only influences spermatozoa motility but can result in decrease of spermatozoa fertilizing ability and affect success of cryopreservation (Linhart et al. 2003).

5.9 Short-Term Storage of Semen

Preserved semen can be useful in aquaculture to overcome problems such as shortage of males when females are ripe and can eliminate the need for transport of adult fish. For instance, walleye males are usually in short supply at the end of the spawning season, as their sexual maturation period is shifted 7–14 days in advance of females (Satterfield and Flickinger 1995b). Development of protocols for short-term storage is critical to ensuring a reliable source of semen and also provides a convenient means of increasing genetic diversity in hatcheries (Cloud et al. 1990). Essential steps in short-term storage of semen are to dilute semen in an extender as soon as possible after stripping, to store diluted semen in an oxygen-rich environment with the semen layer not exceeding 2–5 mm in thickness, and to maintain a temperature of 1–5 °C (Satterfield and Flickinger 1995a). Use of an extender provides ion concentrations and osmotic pressure at levels isosmotic to the seminal plasma, preventing initiation of spermatozoa motility (Rodina et al. 2004), protecting spermatozoa from osmotic damage and contaminants such as urine, and maintaining an ATP source required for flagellar beating as well as fertilizing ability (Linhart et al. 2003). At low temperatures, spermatozoa metabolism is reduced (Rurangwa et al. 2004). However, prolonged cool storage conditions can greatly affect the quality of semen, since anaerobic conditions and associated potential microbial contamination (possibly prevented by addition of antibiotics) may reduce spermatozoa motility and viability (Wayman 2003).

For walleye semen, three immobilizing media have been introduced as extenders (Moore 1987; Brown and Moore 1996; Malison and Held 1996; Satterfield and Flickinger 1995a): (a) $CaCl_2$ 0.117 g, $MgCl_2$ 0.134 g, Na_2HPO_4 0.236 g, KCl 1.872 g, NaCl 6.578 g, D-glucose 10.0 g, citric acid monohydrate 0.10 g, KOH (1.27 g/100 mL) 10 mL and bicine (5.3 g bicine/100 mL) 20 mL, pH 9.0; (b) NaCl 8.760 g, 5000 units of penicillin and 5 mg streptomycin per mL of 0.9 % NaCl 10 mL, pH 7.6; and (c) $CaCl_2$ 0.103 g, $MgCl_2$ 0.220 g, Na_2HCO_3 0.235 g, KCl 2.558 g, NaCl 5.780 g, pyruvate 6.0 g, citric acid monohydrate 0.1 g, KOH (1.27 g/100 mL) 10 mL, pen-strep 10 mL and HEPES 2.380 g, pH 9.0. Following 14 days incubation, more than 50 % spermatozoa were motile for 50–60 s in all three extenders. Different semen to extender ratios have been also evaluated, with no differences observed in fertilization rate of fresh semen and that preserved in each of the three extenders following 2–10 days storage with addition of antibiotics. Further studies have shown that antibiotics might not be necessary, since the semen contains natural antibiotics (Brown and Moore 1996).

Hatef et al. (2011) reported decrease in Eurasian perch spermatozoa motility from 99 % to 67 % and decrease in spermatozoa velocity from 228 to 175 μm s^{-1} following 24 h incubation of semen in an extender composed of 180 mM NaCl, 2.68 mM KCl, 1.36 mM $CaCl_2$, and 2.38 mM $NaHCO_3$, adjusted to pH 8.5 (340 mOsmol. Kg^{-1}) at ratio 1:50 (sperm:extender). Neither spermatozoa motility nor velocity decreased following 3 and 6 h incubation compared to fresh semen. Decrease in spermatozoa motility might be caused by damage to the plasma membrane, a key element for receiving hypo-osmotic signals required for activation of spermatozoa. Further analyses showed that the decrease in spermatozoa velocity might be related to depletion of ATP content as a source of energy for axonemal dynein ATPase activity, which has been also associated with damage to the fine structure of sperm, particularly mitochondria. Further studies are required to investigate fertilizing ability of spermatozoa in the above-mentioned formulated extender during short-term storage.

In pikeperch, Korbuly et al. (2009) studied incubation of semen using various immobilizing media and observed more than 50 % of spermatozoa were motile for over 120 s when the semen was diluted in Ringer's solution (309 mOsmol. kg^{-1}) or phosphate buffered saline (295 mOsmol kg^{-1}) after 3 h incubation compared to fresh semen. Non-diluted semen, stripped manually from pikeperch, can be stored for up to 48 h at 4 °C (Křišťan et al. 2014). During period of storage, both spermatozoa motility and velocity decrease (Fig. 5.5), which is associated with the decrease in ATP content and damage to spermatozoa morphology and fine structure. In pikeperch, short-term semen storage has been investigated using two extenders: (a) 180 mM NaCl, 2.68 mM KCl, 1.36 mM $CaCl_2$, 2.38 mM $NaHCO_3$, pH 8.0 (343 mOsmol kg^{-1}) and (b) 180 mM NaCl, 2.68 mM KCl, 2.38 mM $NaHCO_3$, pH 8.0 (380 mOsmol kg^{-1}). Results showed higher spermatozoa motility and velocity in diluted sperm compared to non-diluted sperm. Both non-diluted semen and semen diluted in an extender exhibited higher spermatozoa motility and velocity during the period of storage when 1 mM Ca^{2+} was added to the activation medium (Křišťan et al. 2014).

5.10 Sperm Cryopreservation

Cryopreservation is a method for long-term storage of viable spermatozoa in a frozen state in liquid nitrogen (−196 °C). Sperm cryopreservation provides biological advantages to conserve threatened fish, to reduce genetic diversity, and to increase effectiveness of artificial reproduction. Cryopreservation may provide valuable spermatozoa for late-ripening females, for crossbreeding fish from different populations, and for crossbreeding strains that mature at different times. Cryopreservation success is highly variable and depends on factors including biological characteristics of broodfish and physico-chemical conditions of cryopreservation: composition of cryoprotective extender, sperm to extender ratio, and freezing and thawing rates. Since sperm physiology, and particularly seminal plasma composition, varies widely among species, cryopreservation methods should be designed and/or evaluated for each species (Ciereszko et al. 2000; Kopeika and Kopeika 2008).

According to available data (Table 5.3), the first successful percid semen cryopreservation was achieved by Moore (1987) in walleye and by Ciereszko et al. (1993) in yellow perch. In both cases, multi-component cryoprotective extenders and a two-step freezing process in which sperm was initially cooled on the surface of dry ice were used (Table 5.3). Simplified extenders, i.e. ion-, sucrose-, or glucose-based solutions, have been used by Bergeron et al. (2002) and by Bokor et al. (2007, 2008) using pellet-freezing and floating-frame methods, respectively. Results showed higher numbers of frozen-thawed spermatozoa with a floating-frame method following dilution of semen in simplified extenders. Several studies of semen of various percids showed no significant difference between dimethyl sulfoxide (DMSO) and methanol (MeOH) as cryoprotectants (Table 5.3). These studies also indicated that cryoprotectants could be applied in the concentration range 6–10 % without significant effects on fertilizing ability of spermatozoa after thawing.

Procedures for cryopreservation of percid sperm can be summarized as follows: (1) sperm uncontaminated with urine is required, (2) sperm should be diluted 1:1–15 with a simplified ionic- or organic-based solution or extenders, (3) 6–10 % MeOH or 10 % DMSO can be used as cryoprotectant, (4) semen suspension in cryoprotective medium should be drawn into 0.25 or 0.5 mL straws prior to freezing, and (5) straws should be placed on a 3 cm thick floating styrofoam frame in liquid nitrogen for 3–10 min prior to deep freezing. Most frequently used thawing conditions are to place straws in a 40 °C water bath for 13 s; however, the thawing condition also depends on the volume of straws.

Table 5.3 Sperm cryopreservation in Percidae

Species	Extender[a]	Cryoprotectant (concentration)	Straw (mL)	Freezing method	Thawing condition	Sperm motility (%)	Fertilization rate[b] (%)	Reference
Yellow perch	Control						53.4	Ciereszko et al. (1993)
	125 mM sucrose, 6.5 mM glutathione, 100 mM $KHCO_3$ (1:3 and 1:7)	DMSO 8 % and egg yolk 10 %	0.1 Pellet	Pellet-freezing	Pellets were thawed in 3 mL of physiological saline at room temperature	n.d.	28.7 ± 27.2	
Yellow perch	Control					>90	61.1	
	450 mM sucrose	DMSO 15 %	0.05 pellet	Pellet-freezing	Pellets were thawed in 5 mL of 0.5 % NaCl at 21 °C	n.d.	42.5–47.2 (3×10^6)	Glogowski et al. (1999)
	450 mM sucrose	DMSO 15 % and egg yolk 10 %	0.05 Pellet			n.d.		
	450 mM sucrose	DMA 15 %	0.05 Pellet			n.d.		
	450 mM sucrose,	DMA 15 % and egg yolk 10 %	0.05 pellet			n.d.		
Pikeperch	Control					63	61	Bokor et al. (2008)
	350 mM glucose, 30 mM Tris, pH 8.0 (1:1)	MeOH 10 %	0.5	Frame-floating	Straws were thawed in 40 °C water bath for 13 s	40	47 (3.4×10^5)	
							55 (1.1×10^5)	
							87 (0.7×10^5)	
Pikeperch	Control					50	51	Bokor et al. (2007)

(continued)

Table 5.3 (continued)

Species	Extender[a]	Cryoprotectant (concentration)	Straw (mL)	Freezing method	Thawing condition	Sperm motility (%)	Fertilization rate[b] (%)	Reference
	350 mM glucose, 30 mM Tris, pH 8.0 (1:9)	DMSO or MeOH 10 %	0.5	Frame-floating	Straws were thawed in 40 °C water bath for 13 s	18 (MeOH) 28 (DMSO)	26 (MeOH) 43 (DMSO)	
	200 mM KCl, 30 mM Tris, pH 8.0 (1:9)	DMSO or MeOH 10 %				17 (MeOH) 11 (DMSO)	33 (MeOH) 38 (DMSO)	
	300 mM sucrose, 30 mM Tris, pH 8.0 (1:9)	DMSO or MeOH 10 %				27 (MeOH) 23 (DMSO)	35 (MeOH) 35 (DMSO)	
Volga Pikeperch	Control					60		
	350 mM glucose, 30 mM Tris, pH 8.0 (1:1)	MeOH 10 %	0.5	Frame-floating	Straws were thawed in 40 °C water bath for 13 s	30 %	42–60 (Hatching rate)	
Walleye	Values in $g.L^{-1}$: 0.117 $CaCl_2$, 0.0134 $MgCl_2$, 0.236 Na_2HPO_4, 1.872 KCl, 6.578 NaCl, 10 glucose, 0.1 citric acid, 0.127 KOH, 1.06 bicine, pH 9.0, 4 $mg.mL^{-1}$ BSA, 7.5 $mg.ml^{-1}$ soy protein (1:2)	DMSO 7 %		Pellet freezing	Straws were thawed in 32 or 21 °C water bath	n.d.	82–97	Moore (1987)
						n.d.	66–73	
Walleye	Control					80	82	

	Values in $g.L^{-1}$: 0.117 $CaCl_2$, 0.0134 $MgCl_2$, 0.236 Na_2HPO_4, 1.872 KCl, 6.578 NaCl, 10 glucose, 0.1 citric acid, 0.127 KOH, 1.06 bicine, pH 9.0, 4 $mg.mL^{-1}$ BSA, 7,5 $mg.ml^{-1}$ soy protein (1:5, 1:9, 1:15)	DMSO, 7 %	0.25	Frame-floating	Straws were thawed in 4 °C water bath for 30 s	16 (1:5)	28–59	Bergeron et al. (2002) and Moore (1987)
						37 (1:9)	(5.2×10^5)	
						46 (1:15)		
	Values in $mg.mL^{-1}$: 10.0 KCl, 3.6 NaCl, 0.09 $MgCl_2$, 0.2 Na_2HCO_3 (1:5, 1:9, 1:15)	DMSO 10 %				12 (1:5)	n.d.	
						10 (1:9)		
						7 (1:15)		
	300 mM glucose	DMSO 10 %				9 (1:5)	n.d.	
						6 (1:9)		
						2 (1:15)		
Eurasian perch (normal male)	Control						49.3 (12×10^5)	Rodina et al. (2008)
	300 mM glucose (1:6)	DMSO 10 %	0.5	Frame-floating	Straws were thawed in 40 °C water bath for 8 s	n.d.	29.1 (12×10^5)	
							37.2 (2.4×10^5)	
Eurasian perch (neomale male)	Control						42.5 (12×10^5)	Rodina et al. (2008)
		DMSO 10 %	0.5		Straws were thawed in 40 °C water bath for 8 s	n.d.	7.3 (12×10^5)	
							6.6 (2.4×10^5)	

(continued)

Table 5.3 (continued)

Species	Extender[a]	Cryoprotectant (concentration)	Straw (mL)	Freezing method	Thawing condition	Sperm motility (%)	Fertilization rate[b] (%)	Reference
Eurasian perch	Control							Dzyuba et al. (2008)
	300 mM glucose (1:6)	MeOH 4–10 %	0.5	Frame-floating	Straws were thawed in 40 °C water bath for 8 s	20 %	65–70 %	

Pellet freezing method, pellets of sperm suspension were kept on dry ice for 5 min and then transferred to liquid nitrogen
Frame-floating method, straws were kept on styrofoam frame floating 3 cm above liquid nitrogen for 3 min and then transferred to it
n.d. Not determined, *DMA* Dimethylacetamide, *DMSO* Dimethyl sulfoxide, *MeOH* Methanol
[a]Sperm to extender ratio is shown in parentheses
[b]Number of spermatozoa per egg is shown in parentheses

5.11 Conclusions

Although spermatozoa morphology seems to be similar within Percidae, studies of species belonging to this family will provide valuable information on inter-species differences in the ultrastructure, which is associated with sperm motility. In this context, the position of proximal and distal centrioles, number of mitochondria, flagellum length, and the presence of a fin-like structure are important.

The composition of seminal plasma in Percidae is well documented. Since seminal plasma is critical to spermatozoa viability and to protect against damage caused by microbes, xenobiotics, and oxidative stress, future investigations are required to characterize the function of seminal plasma compounds that contribute to protection against sperm aging in the reproductive system, adverse effects of endocrine disrupting chemicals, and effects of nutrients and anti-oxidants. It is also important to determine whether seminal plasma can be considered as an extender for cryopreservation and/or for in vitro maturation of spermatozoa.

Sperm motility in Percidae is induced by a hypo-osmotic signal that probably mediates Ca^{2+}-dependent axonemal beating. In this context, it is important to understand the role of specific Ca^{2+} channels in the sperm membrane and Ca^{2+}-dependent proteins in the molecular structure of the axoneme. Available literature shows that media with 100–200 and 300 mOsmol kg^{-1} can be considered potent activating and immobilizing solutions, respectively. Further studies on the effects of externally supplemented ATP, Ca^{2+}, and anti-oxidants will provide valuable information for development of activation and immobilizing media for short-term storage and cryopreservation. ATP content has been investigated in the sperm of several percid species, but studies of other percids can identify inter-species differences in duration of sperm motility and velocity corresponding to ATP content and/or flagellum length.

This literature review shows that semen quality can vary in response to numerous factors and often reflects variability in environmental and handling conditions as well as individual and social conditions. Among environmental factors, season seems to be the most significant, and phase-shifted photothermal cycles can be developed to regulate sexual maturation and spawning season. In aquaculture, maintenance and nutrition can affect the quality of semen. Contamination is a serious threat to semen quality. Expanded knowledge of factors contributing to sperm quality is necessary to allow improvement of the control of reproduction.

Further studies are required to define the effects of parameters involved in sperm quality on success of cryopreservation. In this context, it is important to study how fatty acids, vitamins, and anti-oxidants in the diet may influence sperm cryo-resistance.

Acknowledgments This study was funded by the Grant Agency of the Czech Republic (523/09/1793, P503/12/1834, P503/13/34049P, P502/11/0090, P502/12/1973); Grant Agency of the Czech Academy of Science (IAA 608030801); Ministry of Agriculture (NAZV QI101C033), Ministry of Education, Youth, and Sport (CENAKVA CZ.1.05/2.1.00/01.0024, CZ.1.07/2.3.00/20.0024, LO1205); Grant Agency of the University of South Bohemia (114/2013/Z); and funding to the Institute of Animal Reproduction and Food Research, Olsztyn, Poland.

References

Alavi SMH, Cosson J (2005) Sperm motility in fishes: (I) effects of pH and temperature. Cell Biol Int 29:101–110

Alavi SMH, Cosson J (2006) Sperm motility in fishes: (II) Effects of ions and osmotic pressure. Cell Biol Int 30:1–14

Alavi SMH, Rodina M, Policar T, Kozak P, Psenicka M, Linhart O (2007) Semen of *Perca fluviatilis* L. Sperm volume and density, seminal plasma indices and effects of dilution ratio, ions and osmolality on sperm motility. Theriogenology 68:276–283

Alavi SMH, Linhart O, Coward K, Rodina M (2008a) Fish spermatology: implication for aquaculture management. In: Alavi SMH, Cosson JJ, Coward K, Rafiee R (eds) Fish spermatology. Alpha Science, Oxford, pp 397–460

Alavi SMH, Rodina M, Policar T, Cosson J, Kozak P, Psenicak M, Linhart O (2008b) Physiology and behavior of stripped and testicular sperm in *Perca fluviatilis* L. 1758. Cybium 32:162–163

Alavi SMH, Rodina M, Viveiros ATM, Cosson J, Gela D, Boryshpolets S, Linhart O (2009) Effects of osmolality on sperm morphology, motility and flagellar wave parameters in Northern pike (*Esox lucius* L.). Theriogenology 72:32–43

Alavi SMH, Rodina M, Hatef A, Stejskal V, Policar T, Hamackova J, Linhart O (2010) Sperm motility and monthly variations of semen characteristics in *Perca fluviatilis* (Teleostei: Percidae). Czech J Anim Sci 55:174–182

Alavi SMH, Hatef A, Pšenička M, Kašpar V, Boryshpolets S, Dzyuba B, Cosson J, Bondarenko V, Rodina M, Gela D, Linhart O (2012) Sperm biology and control of reproduction in sturgeon: (II) sperm morphology, acrosome reaction, motility and cryopreservation. Rev Fish Biol Fish 22:861–886

Bergeron A, Vandenberg G, Proulx D, Bailey JL (2002) Comparison of extenders, dilution ratios and theophylline addition on the function of cryopreserved walleye semen. Theriogenology 57:1061–1071

Bokor Z, Müller T, Bercsenyi M, Horvath L, Urbanyi B, Horvath A (2007) Cryopreservation of sperm of two european percid species, the pikeperch (*Sander lucioperca*) and the volga pikeperch (*S. -volgensis*). Acta Biol Hung 58:199–207

Bokor Z, Horvath A, Horvath L, Urbanyi B (2008) Cryopreservation of pike perch sperm in hatchery conditions. Isr J Aquacult Bamidgeh 60:166–169

Boryshpolets S, Dzyuba B, Stejskal V, Linhart O (2009) Dynamics of ATP and movement in Eurasian perch (*Perca fluviatilis* L.) sperm in conditions of decreasing osmolality. Theriogenology 72:851–859

Brown GG, Moore AA (1996) Comparative storage methods and fertility studies of walleye semen. In: Summerfelt RC (ed) Walleye culture manual. North Central Regional Aquaculture Center publications series 101. Iowa State University, Ames, pp 45–49

Casselman SJ, Schulte-Hostedde AI, Montgomerie R (2006) Sperm quality influences male fertilization success in walleye (*Sander vitreus).* Can J Fish Aquat Sci 63:2119–2125

Cejko BI, Glogowski J, Kowalski RK, Kucharczyk D, Targońska K (2008) Description of pikeperch, *Sander lucioperca* (L.), semen obtained from males held under different rearing conditions. Arch Pol Fish 16:93–100

Ciereszko A (2008) Chemical composition of seminal plasma and its physiological relationship with sperm motility, fertilizing capacity, and cryopreservation success in fish. In: Alavi SMH, Cosson JJ, Coward K, Rafiee R (eds) Fish spermatology. Alpha Science, Oxford, pp 215–240

Ciereszko A, Dabrowski K (1993) Estimation of sperm concentration of rainbow trout, whitefish and yellow perch by spectrophotometric technique. Aquaculture 109:367–373

Ciereszko A, Dabrowski K (2000) *In vitro* effect of gossypol acetate on yellow perch (*Perca flavescens*) spermatozoa. Aquat Toxicol 49:181–187

Ciereszko A, Ramseyer L, Dabrowski K (1993) Cryopreservation of yellow perch semen. Prog Fish Cult 55:261–264

Ciereszko RE, Dabrowski K, Ciereszko A, Ottobre JS (1998) Plasma concentrations of steroid hormones in male yellow perch, *Perca flavescens*: the effect of age and photothermal manipulation. Environ Biol Fish 51:97–105

Ciereszko A, Dabrowski K, Kucharczyk D, Dobosz S, Goryczko K, Glogowski J (1999) The presence of uric acid, an antioxidative substance, in fish seminal plasma. Fish Physiol Biochem 21:313–315

Ciereszko A, Glogowski J, Dabrowski K (2000) Biochemical characteristics of seminal plasma and spermatozoa of freshwater fish and the relation to semen biology, quality and cryopreservation. In: Tiersch TR, Mazik PM (eds) Cryopreservation in aquatic species. World Aquaculture Society, Baton Rouge, pp 20–48

Cloud JG, Miller WH, Levanduski MJ (1990) Cryopreservation of sperm as a means to store salmonid germ plasm and to transfer genes from wild fish to hatchery populations. Prog Fish Cult 52:51–53

Cosson J (2010) Frenetic activation of fish spermatozoa flagella entails short-term motility, portending their precocious decadence. J Fish Biol 76:240–279

Dabrowski K, Ciereszko A (1994) Proteinase inhibitor(s) in seminal plasma of teleost fish. J Fish Biol 45:801–809

Dabrowski K, Ciereszko A, Ramseyer L, Culver D, Kestemont P (1994) Effects of hormonal treatment on induced spermiation and ovulation of the yellow perch (*Perca flavescens*). Aquaculture 120:171–180

Dabrowski K, Ciereszko RE, Ciereszko A, Toth GP, Christ SA, El-Saidy D, Ottobre JS (1996) Reproductive physiology of yellow perch (*Perca flavescens*): environmental and endocrinological cues. J Appl Ichthyol 12:136–148

Dietrich MA, Dietrich GJ, Hliwa P, Ciereszko A (2011) Carp transferrin can protect spermatozoa against toxic effects of cadmium ions. Comp Biochem Physiol 153C:422–429

Dzyuba BB, Boryshpolets S, Stejskal V, Linhart O (2008) Cryopreservation of European perch (*Perca fluvitilis* L.) sperm with using methanol as a cryoprotectant. Biophy Living Cell 9:48 (in Russian)

Gillies EA, Bondarenko V, Cosson J, Pacey AA (2012) Fins improve the swimming performance of fish sperm: a hydrodynamic analysis of the Siberian sturgeon *Acipenser baerii*. Cytoskeleton 70:85–100

Glogowski J, Ciereszko A, Dabrowski K (1999) Cryopreservation of muskellunge and yellow perch semen. N Am J Aquac 61:258–262

Green CC, Kelly AM (2008) Effect of the exogenous soyabean phyto-oestrogen genistein on sperm quality, ATP content and fertilization rates in channel catfish *Ictalurus punctatus* (Rafinesque) and walleye *Sander vitreus* (Mitchill). J Fish Biol 72:2485–2499

Gregory RW (1970) Physical and chemical properties of walleye sperm and seminal plasma. T Am Fish Soc 99:518–525

Hatef A, Alavi SMH, Linhartova Z, Rodina M, Policar T, Linhart O (2010) *In vitro* effects of Bisphenol A on sperm motility characteristics in *Perca fluviatilis* L. (Percidae; Teleostei). J Appl Ichthyol 26:696–701

Hatef A, Alavi SMH, Butts IAE, Policar T, Linhart O (2011) The mechanisms of action of mercury on sperm morphology, Adenosine-5′-triphosphate content and motility in *Perca fluviatilis* (Percidae; Teleostei). Environ Toxicol Chem 30:905–914

Henrotte E, Kaspar V, Rodina M, Psenicka M, Linhart O, Kestemont P (2010) Dietary n-3/n-6 ratio affects the biochemical composition of Eurasian perch (*Perca fluviatilis*) semen but not indicators of sperm quality. Aquac Res 41:e31–e38

Jarmołowicz S, Demska-Zakeś K, Kowalski R, Cejko B, Glogowski J, Zakeś Z (2010) Impact of dibutyl phthalate and benzyl butyl phthalate on motility parameters of sperm from the European pikeperch *Sander lucioperca* (L.). Arch Pol Fish 18:149–156

Kazun K, Siwicki AK (2001) Propiscin – a safe new anaesthetic for fish. Arch Pol Fish 9:183–190

Kopeika EF, Kopeika J (2008) Variability of sperm quality after cryopreservation in fish. In: Alavi SMH, Cosson JJ, Coward K, Rafiee R (eds) Fish spermatology. Alpha Science, Oxford, pp 347–396

Korbuly B, Grozea A, Cean A, Banatean-Dunea I, Pacala N, Valean A (2009) Milt dilution effectiveness on pikeperch *Sander lucioperca* sperm and DNA inactivation. Zooteh Biotehnol 42:65–70

Kowalski R, Glogowski J, Kucharczyk D, Goryczko K, Dobosz S, Ciereszko A (2003) Proteolytic activity and electrophoretic profiles of proteases from seminal plasma of teleosts. J Fish Biol 63:1008–1019

Kowalski R, Glogowski J, Kucharczyk D, Mak M, Dobosz S, Zakes Z, Ciereszko A (2004) Characterization of gelatinolytic activity in seminal plasma of some teleost fish. Aquac Int 12:57–68

Król J, Glogowski J, Demska-Zakeś K, Hliwa P (2006) Quality of semen and histological analysis of testes in Eurasian perch *Perca fluviatilis* L. during a spawning period. Czech J Anim Sci 51:220–226

Król J, Kowalski R, Demska-Zakeś K, Hliwa P, Glogowski J (2011) Proteolytic and anti-proteolytic activity in the seminal plasma of Eurasian perch (*Perca fluviatilis* L.) during the spawning period. Czech J Anim Sci 56:390–397

Křišťan J, Hatef A, Alavi SMH, Policar T (2014) Sperm morphology, ultrastructure, and motility in pikeperch Sander lucioperca (Percidae, Teleostei) associated with various activation media. Czech J Life Sci 59:1–10

Kucharczyk D, Kujawa R, Mamcarz A, Skrzypczak A, Wyszomirska E (1996) Induced spawning in perch, *Perca fluviatilis* L. using carp pituitary extract and HCG. Aquac Res 27:847–852

Kucharczyk D, Kujawa R, Mamcarz A, Skrzypczak A, Wyszomirska E (1998) Induced spawning in perch, *Perca fluviatilis* L. using FSH + LH with pimozide or metoclopramide. Aquac Res 29:131–136

Lahnsteiner F (2009) The role of free amino acids in semen of rainbow trout (*Oncorhynchus mykiss*) and carp (*Cyprinus carpio*). J Fish Biol 75:816–833

Lahnsteiner F (2010) A comparative study on the composition and importance of free amino acids in semen of gilthead sea bream, *Sparus aurata*, and perch, *Perca fluviatilis*. Fish Physiol Biochem 36:1297–1305

Lahnsteiner F (2011) Spermatozoa of the teleost fish *Perca fluviatilis* (perch) have the ability to swim for more than two hours in saline solutions. Aquaculture 314:221–224

Lahnsteiner F, Mansour N (2004) Sperm fine structure of the pikeperch, *Sander lucioperca* (Percidae, Teleostei). J Submicrosc Cytol Pathol 36:309–312

Lahnsteiner F, Mansour N (2010) A comparative study on antioxidant systems in semen of species of the Percidae, Salmonidae, Cyprinidae, and Lotidae for improving semen storage techniques. Aquaculture 307:130–140

Lahnsteiner F, Patzner RA (2008) Sperm morphology and ultrastructure in fish. In: Alavi SMH, Cosson JJ, Coward K, Rafiee R (eds) Fish spermatology. Alpha Science, Oxford, pp 1–61

Lahnsteiner F, Berger B, Weismann T, Patzner R (1995) Fine structure and motility of spermatozoa and composition of the seminal plasma in the perch. J Fish Biol 47:492–508

Lee KJ, Dabrowski K (2004) Long-term effects and interactions of dietary vitamins C and E on growth and reproduction of yellow perch, *Perca flavescens*. Aquaculture 230:377–389

Linhart O, Rodina M, Bastl J, Cosson J (2003) Urinary bladder, ionic composition of seminal fluid and urine with characterization of sperm motility in tench (*Tinca tinca* L.). J Appl Ichthyol 19:177–181

Malison JA, Held JA (1996). Reproductive biology and spawning. In: Summerfelt RC (ed) Walleye culture manual. North Central Regional Aquaculture Center publications series 101. Iowa State University, Ames, pp 11–18

Migaud H, Fontaine P, Kestemont P, Wang N, Brun-Bellut J (2004) Influence of photoperiod on the onset of gonadogenesis in Eurasian perch *Perca fluviatilis*. Aquaculture 241:561–574

Migaud H, Wang N, Gardeur J-N, Fontaine P (2006) Influence of photoperiod on reproductive performances in Eurasian perch *Perca fluviatilis*. Aquaculture 252:385–393

Miura T, Yamauchi K, Takahashi H, Nagahama Y (1992) The role of hormones in the acquisition of sperm motility in salmonid fish. J Exp Zool 261:359–363

Moore AA (1987) Short-term storage and cryopreservation of walleye semen. Prog Fish-Cult 49:40–43

Morisawa M (2008) Adaptation and strategy for fertilization in the sperm of teleosts fish. J Appl Ichthyol 24:362–370

Nynca J, Horvath A, Dietrich MA, Muller T, Karol H, Urbanyi B, Kotrik L, Ciereszko A (2010) Serine proteinase inhibitors in the seminal plasma of percid fish. J Appl Ichthyol 26:742–745

Piironen J, Hyvärinen H (1983) Composition of the milt of some teleost fishes. J Fish Biol 22:351–361

Rinchard J, Dabrowski K, Van Tassell JJ, Stein RA (2005) Optimization of fertilization success in *Sander vitreus* is influenced by the sperm:egg ratio and ova storage. J Fish Biol 67:1157–1161

Rodina M, Cosson J, Gela D, Linhart O (2004) Kurokura solution as immobilizing medium for spermatozoa of tench (*Tinca tinca* L.). Aquac Int 12:119–131

Rodina M, Policar T, Linhart O, Rougeot C (2008) Sperm motility and fertilizing ability of frozen spermatozoa of males (XY) and neomales (XX) of perch (*Perca fluviatilis*). J Appl Ichthyol 24:438–442

Rougeot C, Nicayenzi F, Mandiki SNM, Rurangwa E, Kestemont P, Melard C (2004) Comparative study of the reproductive characteristics of XY male and hormonally sex-reversed XX male Eurasian perch, *Perca fluviatilis*. Theriogenology 62:790–800

Rurangwa E, Kime DE, Ollevier F, Nash JP (2004) Measurement of sperm motility and factors affecting sperm quality in cultured fish. Aquaculture 234:1–28

Satterfield JR Jr, Flickinger SA (1995a) Factors influencing storage potential of preserved walleye semen. Prog Fish Cult 57:175–181

Satterfield JR Jr, Flickinger SA (1995b) Field collection and short-term storage of walleye semen. Prog Fish Cult 57:182–187

Stejskal K, Svobodova Z, Fabrik I, Adam V, Beklova M, Rodina M, Kizek R (2008) Content of cysteine, reduced and oxidized glutathione in spermatozoa of representatives of Acipenseriformes (*Acipenser baerii* and *A. ruthenus*) as well as teleosts (*Perca fluviatilis* and *Sander lucioperca*). J Appl Ichthyol 24:519–521

Teletchea F, Gardeur JN, Psenicka M, Kaspar V, Le Doré Y, Linhart O, Fontaine P (2009) Effects of four factors on the quality of male reproductive cycle in pikeperch, *Sander lucioperca*. Aquaculture 291:217–223

Wang N, Rodina M, Gardeur JN, Vuillard JT, Policar T, Henrotte E, Mandiki S, Kestemont P, Linhart O, Fontaine P (2008) Determinism of the quality of reproduction in male Eurasian perch, *Perca fluviatilis*: a multifactorial study. Cybium 32:192–193

Wang N, Teletchea F, Kestemont P, Milla S, Fontaine P (2010) Photothermal control of the reproductive cycle in temperate fishes. Rev Aquac 2:209–222

Wayman WR (2003) From gamete collection to database development: development of a model cryopreserved germplasm respiratory for aquatic species with emphasis on sturgeon. Dissertation, Louisiana State University

Wirtz S, Steinmann P (2006) Sperm characteristics in perch *Perca fluviatilis* L. J Fish Biol 68:1896–1902

Wojtczak M, Dietrich GJ, Ciereszko A (2005) Transferrin and antiproteases are major proteins of common carp seminal plasma. Fish Shellfish Immunol 19:387–391

Zakęś Z (2007) Out-of-season spawning of cultured pikeperch (*Sander lucioperca* L.). Aquac Res 38:1419–1427

Zakęś Z, Demska-Zakęś K (2009) Controlled reproduction of pikeperch *Sander lucioperca* (L.): a review. Arch Pol Fish 17:153–170

Chapter 6
Evaluation and Improvements of Egg and Larval Quality in Percid Fishes

B. Schaerlinger and D. Żarski

Abstract For a sustainable breeding process, the optimization of recirculating aquaculture system(s) (RAS) fish rearing conditions and the control of out-of-season reproduction, it is important (i) to define intrinsic and extrinsic factors regulating the fish life cycle and (ii) to have a good knowledge of what makes a good ovum, embryo or larva. In the present chapter, we first describe the current knowledge on ova characteristics and the proper embryonic and larval development progress for several percid species. Indeed, it is important to well define the correct sequence of events in order to better characterize potential impairments. The characterization of ova defects or developmental failures (mortality or abnormalities occurrence) may allow the definition of different categories/levels of quality. This quality scale could help scientists and fish breeders to choose the most relevant quality indicators depending on their technical or scientific problem. Indeed, indicators could be either predictive, to assess ova quality, or studied after fertilization to determine embryonic and/or larval abilities to develop properly and reach key steps in addition to the ova quality. However, the possible indicators allowing precise determination of the egg and larvae quality are actually scarce. Some morphological parameters in most cases allow indicating, with high probability, high (e.g. intensiveness of cortical reaction in pikeperch) or low (e.g. oil droplets fragmentation in ovulated eggs of Eurasian perch and pikeperch) egg quality rather than quantifying the real quality. Moreover, there is still no clear molecular predictors of the egg and larvae quality due to the too few or ambiguous data obtained in the field. On the base of the most recent studies it seems that in many cases molecular analyses are one of the most promising methods possibly allowing an estimation of the ova, embryos and larvae quality. But the research activities in this field are still in progress.

Keywords Percids • Egg • Embryo • Larva • Quality

B. Schaerlinger (✉)
Unité Recherche Animal et Fonctionnalités des Produits Animaux, Université de Lorraine, 2 avenue de la Forêt de Haye, B.P. 172, 54506 Vandoeuvre-lès-Nancy, France
e-mail: Berenice.Schaerlinger@univ-lorraine.fr

D. Żarski
Department of Lake and River Fisheries, University of Warmia and Mazury in Olsztyn, Oczapowskiego 5, pok. 325, 10-957 Olsztyn, Poland

P. Kestemont et al. (eds.), *Biology and Culture of Percid Fishes*,
DOI 10.1007/978-94-017-7227-3_6

6.1 Introduction

As a whole, fish reproduction in captive conditions often leads to various levels of success. Indeed, the living conditions of breeders influencing gametogenesis on the one hand, and incubation and larval rearing conditions on the other hand can lead to reproduction failures. Assessing reproduction quality indicators represents an important field of research in aquaculture. Egg quality has recently been defined as its ability to be fertilized and develops into a healthy embryo and larva (Bobe and Labbe 2010; Migaud et al. 2013). One can add that the embryo and larva quality can be defined as their abilities to properly reach several key steps during the development. In order to evaluate properly ova, embryonic and larval quality, two questions need to be addressed. First, what is considered as good, medium or low quality? Second, what is the most relevant way to evaluate reproduction success, and if possible in a predictive manner? To answer the first question, it appears necessary to establish criteria including mortality stages, deformities occurrence or larval resistance. These criteria could help to define more properly some quality levels/categories. Indeed, in the literature, each study defined their own criteria and making a careful comparison become impossible.

This chapter mainly aims to present methods currently available to predict, assess and improve reproductive performance success in percid species, but also briefly describes the developmental steps of the embryos and larvae allowing wider overview on the problem of the egg quality. The level of knowledge on this aspect is various depending on the studied species. However, in every case there are many variables needed to be improved in order to obtain good quality offspring. Here we first aimed at defining the morphological and biochemical composition of good quality ova and the proper sequence of the embryonic and larval development. This knowledge is necessary to properly define failures that can occur during oogenesis leading to poor quality ova and/or developmental process. Finally, the last part of the chapter presents several categories of factors that could affect reproductive success.

6.2 Ova Structure and Biochemical Composition

6.2.1 Ova Structure

The fish egg is the final product of the entire oogenesis process (Tyler and Sumpter 1996). In general, the ovulated egg consists of an external layer called the chorion (*zona radiata*) and an internal layer called the vitelline membrane directly surrounding the egg cell (see e.g., Cotelli et al. 1988; Riehl and Patzner 1998; Quagio-Grassiotto and Guimaraes 2003; Mansour et al. 2009). Just beneath the plasma membrane, cortical alveoli (granules) are usually located, which play a very important role upon the activation (Hart and Yu 1980; Lee et al. 1999). In percids, the surface of the chorion is coated with structures forming a sticky (walleye and pikeperch) or thick jelly-like (Eurasian and yellow perch) layer (Riehl and Patzner 1998;

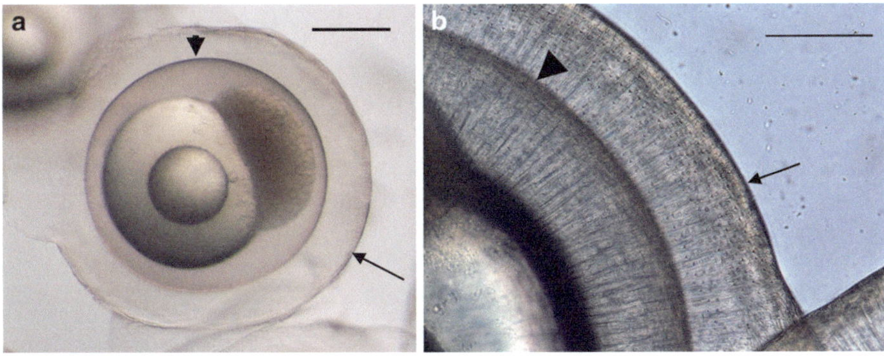

Fig. 6.1 Structure of the Eurasian perch *zona radiata* (**a**) Embryo at that the blastula stage (**b**) higher magnification to show the structure of the *zona radiata externa* (Pictures of D. Zarski (**a**) and M. Alix (**b**)). *Arrows* present the *zona radiata externa* and *arrowhead* the emplacement of the chorion. *Scale bars* correspond to 500 µm in (**a**) and 200 µm in (**b**)

Formicki et al. 2009) which is also called the *zona radiata externa* (ZRE, Fig. 6.1). Actually, only the ZRE creates the differences in the morphology of the eggs among the freshwater percids. Morphologically, egg cells of those species (without the ZRE) look very similar. The eggs are fully transparent with one, clearly-visible oil droplet (with a diameter of about 0.6 mm) within the egg cell and the main volume of the non-activated egg consists of yolk mass (Mansueti 1964; Żarski et al. 2011a, 2012a, b). The average diameter of such eggs ranged between 1.0 and 1.3 mm in pikeperch (Demska-Zakęś et al. 2005; Żarski et al. 2012b), between 0.86 and 0.91 mm (without ZRE) in Eurasian perch (Sulistyo et al. 1998; Żarski et al. 2012c) and approximately 1.2 mm in yellow perch (Mansueti 1964).

Eggs of walleye and pikeperch are deposited (ovulated) as a batch of single eggs, similarly to other freshwater fish species, whereas eggs of Eurasian and yellow perch are situated within the cylindrical, gelatinous strand which was in the literature named as ribbon (e.g., Probst et al. 2009; Formicki et al. 2009). This structure is unique among the freshwater fish and is usually deposited in one piece with a length of up to 5.5 m (Korzelecka et al. 1998; Formicki et al. 2009). There is a relationship between the size of females and the one of egg ribbons that has been established within a size of breeders ranging from 150 to 350 mm (Dubois et al. 1996). Eggs within the ribbon are connected by a very thick ZRE, which may occupy from 25 % to 30 % of the entire egg diameter (Mansueti 1964).

After activation, the eggs, due to the cortical reaction (where the content of cortical alveoli is released between the vitelline membrane and the chorion forming a perivitelline space), start to swell due to the infusion of water into the perivitelline space through the osmotic gradient, caused by content of cortical alveoli (for details see e.g., Coward et al. 2002; Minin and Ozerova 2008). This process leads to form the final volume of the egg, and was also described as water hardening (e.g., Mansueti 1964; Czesny et al. 2005a; Demska-Zakęś et al. 2005). In the case of walleye and pikeperch, after the contact with water (Fig. 6.2), the eggs become

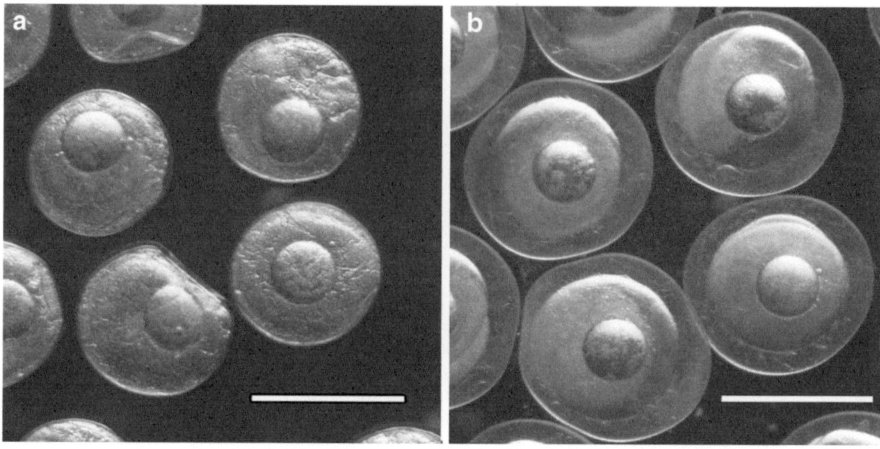

Fig. 6.2 Eggs of pikeperch at the moment of contact with water (**a**) and after water hardening (60 min post activation) (**b**). The *scale bar* represents 1 mm (Pictures of D. Zarski)

sticky (e.g., Schlumpberger and Schmidt 1980; Rinchard et al. 2005; Demska-Zakęś et al. 2005; Kucharczyk et al. 2007), whereas in perch, large water acquisition by the gelatinous layer was observed (Formicki et al. 2009). After the swelling process, percid eggs look the same, except the formed perivitelline space (and thus increased diameter). After water hardening, diameter of the walleye eggs ranges between 1.6 and 2.0 mm (Serns 1982; Czesny et al. 2005a) and for pikeperch between 1.26 and 1.44 (Demska-Zakęś et al. 2005; Żarski et al. 2012b). In the case of yellow perch, the internal (without ZRE) egg diameter ranges approximately between 1.6 and 1.8 mm (Mansueti 1964) and in Eurasian perch between 1.30 and 1.43 mm (Żarski et al. 2011b). The ZRE constitutes up to 50 % of the total egg volume (Mansueti 1964; Korzelecka et al. 1998).

6.2.2 Molecular Composition of the Ova

Except for gas and few molecules, fish embryos in the egg don't have any exchange of material with the outside. Therefore, the process of incorporation of molecules in the oocyte during the oogenesis is extremely important since its correctness ensures the proper development and survival of the offspring. On the contrary, bad incorporation/expression of particular molecules can have a huge negative impact during the embryonic and larval development (mortality, abnormalities occurrence depending on the cellular function which is affected). These molecules are involved in three main cellular functions: (i) the energetic reserves for the embryo (glycogen, lipids, free amino acids and Vitellogenins), (ii) the cellular process and the structure in the embryos (proteins, free amino acids, lipids, transcripts and several ions), and (iii) molecules necessary to keep the osmotic balance in the egg (water, free amino acids).

In percids, the biochemical composition of ova has been studied in some targeted species. No large scale studies have been performed to investigate their transcriptome, proteome, glycogen or free amino acid composition. Most studies focused on lipids composition of ova. The most abundant lipids are neutral lipids (wax ester, triacylglycerols, cholesterol and free fatty acids (around 85 % of the total lipid content)) while the polar lipids (phosphoglycerlipids, sphingolipids) correspond to around 15 % of the total lipid content (Henrotte et al. 2010). Moreover the most abundant fatty acids are n-3 polyunsaturated fatty acids (PUFA) and monounsaturated fatty acids (MUFA) while the saturated fatty acids (SFA) and the n-6 PUFA are less represented in *P. fluviatilis* (Blanchard et al. 2005; Żarski et al. 2012c; Henrotte et al. 2010). Moreover, it has been shown that the most represented fatty acids are not only docohexaenoic acid (DHA) but also 16:1 and 18:1 fatty acids. These last types of lipids have never been studied in ova so far. On the contrary, eicosapentaenoic acid (EPA) and arachidonic acid (AA) were poorly represented in eggs of wild Eurasian perch (Żarski et al. 2012c). In the walleye n-3 PUFA is the most abundant fatty acid in ova followed by SFA and MUFA while n-6 PUFA are poorly represented (Moodie et al. 1989). In the Eurasian perch, it has been shown that the diet of females can affect the fatty acid proportions in the ova and may be linked to the reproductive success (Henrotte et al. 2010). Percid ova are characterized by the presence of an oil globule in addition to the yolk. Moodie et al. (1989) have shown in the walleye that while the yolk is composed by lipoproteins and polar lipids (e.g. phospahtidylcholine and phosphatidylethanolamine), the oil globule are composed by neutral lipids, mainly triacylglycérols (TAG). In addition, the fatty acid composition of each compartment is different as the yolk is mainly composed by SFA and n-3 PUFA. The neutral lipids of the oil droplet are composed by n-3 PUFA and MUFA. It has been suggested that the polar lipids of the yolk are used before the neutral lipids of the oil globule during the embryonic development, mostly as energetic reserves (Wiegand 1996).

Molecules necessary to keep the osmotic balance in the egg are usually important for marine species which eggs undergo a huge hydration just prior to ovulation. This mechanism has poorly been described in freshwater fish species such as percids. A recent study investigated the hydration level during *P. fluviatilis* oocyte maturation and showed an increase of around 4 % of water intake during this process. Up to now, this data corresponds to the fewer hydration process ever described (Żarski et al. 2012c). Moreover, recent studies indicated that some populations of Eurasian perch *Perca fluviatilis* could spawn in brackish or even salty water (Tibblin et al. 2012; Skovring et al. 2013), suggesting an adaptation of perch eggs to various salinity conditions. Same data have previously been observed in ruffe *Gymnocephalus cernua* (Vetemaa and Saat 1996). It could be interesting to investigate more carefully the potential hydration process in freshwater fish as percids.

6.3 Embryonic and Larval Development

6.3.1 Embryonic Development

Fish embryonic development is exposed to several extrinsic and intrinsic factors that can affect the speed of the embryogenesis and consequently leads to asynchronous development within and between spawn. Among these factors the temperature of incubation is probably the most important and several studies take into account this parameter by using the degree day parameter (number of days elapsed since the fertilization at a given temperature (°C)). However, this classification has mainly two important drawbacks. First, it doesn't take into account other parameters that could modulate fish embryogenesis such as the water flow, oxygenation or the water composition, and second it doesn't allow any easy comparison of the embryonic development between fish species. In the current scientific context requiring more careful studies of embryonic stages (corresponding to the mortality or deformities apparition stages) or the comparison of the embryonic development of several species, it becomes thus more accurate to characterize fish embryogenesis through morphological criteria.

Fish embryogenesis elapses from the ova activation (described in the Sect. 6.2.1) to the first oral feeding which are easily observable and stable stages. The choice of the first oral feeding as the end of embryogenesis instead of the hatching period is done because hatching may occur at different embryonic stages depending on the fish species. Moreover, within a species, the total length of the embryos at hatching may be different depending on the environmental conditions and, to some extent, the size of the female being positively correlated with the size of the egg. It suggests that the hatching stage is not fixed even within a spawn. This has been shown in *Perca fluviatilis* for which embryos can hatch with size ranging from 4 to 9 mm (Konstatinov 1957) and with different eye pigmentations (M. Alix and B. Schaerlinger personal observations). Similar data have been observed in *Perca flavescens* (Mansueti 1964). On the other hand, the first oral feeding corresponds to the end of the period in the course of which the embryo depends only on its own reserves. Once the animals begin to eat, they enter into the larval stages (see Sect. 6.3.2) until they have the same morphology than the adults (juvenile stages). The embryonic development can be divided into five main steps (cell cleavage, gastrulation, organogenesis, hatching and the free embryo period). Even if several differences can be observed among fish species, these main periods of embryonic development chronologically occur in the same manner.

Up to now, only few works carefully examined percids embryonic staging. As those studies were conducted with different aims, they pointed on diverse stages of the embryonic development. Most of them have been performed on North American species as the fantail (*Etheostoma flabellare*), rainbow (*Etheostoma caeruleum*) and banded darters (*Etheostoma zonale*) (Cooper 1979; Paine 1984; Paine and Balon 1984a; Mendelson et al. 2006), the American yellow perch (*Perca flavescens*) (Mansueti 1964), the logperch (*Percina caprodes*) (Paine and Balon 1984b; Cooper

1978) and the walleye (*Stizostedion vitreum*) (Mc Elmann and Balon 1979). These fish represent *Etheostoma*, *Perca*, *Percina* and *Sander* genders, respectively. Others have been done on European species as the ruffe (*Gymnocephalus cernuus*), Balon's ruffe (*Gymnocephalus baloni*) and the yellow pope (*Gymnocephalus schraester*) representing the *Gymnocephalus* gender (Kovac 1992; Kovac 1993a, b, 1994) and the Danube streber (*Zingel streber*) for the *Zingel* gender (Kovac 2000). The reproduction traits of several of these species are given in the Table 6.1. As a whole, the total duration of the embryonic development of those percid species is highly variable from activation to the first feeding (Figs. 6.3 and 6.4). A compared analysis of each step allows the determination of common features and differences at each step.

6.3.1.1 Cell Cleavage Stages

After the entry of the spermatozoa in the ova, the yolk, that was previously uniformly distributed, begins to segregate in the vegetative pole while the cytoplasm segregates in the animal pole (Fig. 6.3). It corresponds to the first embryonic cell called the zygote period. Thereafter begins a series of cell cleavages at the animal pole to obtain a multicellular embryo called blastula. In percids, the duration of this period can last from 12 h (for the streber) to 30 h (for the yellow perch) (Fig. 6.4, Table 6.2) suggesting that the regulation of the speed of cell division is different between species. This phenomenon is interesting in regard to the fact that both species are those presenting the longest embryonic development among percids (Mansueti 1964; Kovac 2000) (Table 6.2). The cell cleavage period can be subdivided into two phases: first cell divisions are synchronous while in the second phase, corresponding to the beginning of the blastula period, the cell divisions become asynchronous. Finally, the cell cleavage is characterized by the midblastula transition (MBT) that corresponds to a switching of genes expression from the maternal genes to the zygotic genes (expression of the embryonic genes). The timing of the MBT has not been determined in any studied percids species but it may correspond to the middle end of the cleavage period such as in *Brachydanio rerio* (Kimmel et al. 1994).

6.3.1.2 Gastrulation Process

The gastrulation phase corresponds to a series of cell migrations called epiboly (Fig. 6.3), and allowing the cell distribution around the vitelline reserves. During that step, the cells migrate from the animal pole to the vegetative pole. The staging is indicated as a percentage of epiboly corresponding to the proportion of yolk surrounded. One key step corresponds to the germ ring step. During that step, the epiboly stops and an involution of the unique cell layer occurs to form two cell layers: (i) the epiblast (future ectoderm mainly precursor of the epiderm and the nervous system), (ii) the hypoblast (precursor of the mesoderm that will lead to the circulation system, muscles, bones and most of the internal organs and the endoderm, precursor

Table 6.1 Summary of the breeders and reproduction conditions of representative percid fish

Scientific name	Common name	Mean adult size (cm)	Size at sexual maturity (females)	Age of sexual maturity (females)	Geographical origin	Spawning substratum	Reproductive season	Egg batches	References
Etheostoma caeruleum	Rainbow darter	5.3	ND	ND	North America	Burried in the substrate	Spring (april–may)	ND	Fishbase
Gymnocephalus schraetser	Yellow pope	15	12 cm	2 years	Europe (east)	Fixed on stones	Late spring	ND	Fishbase
Perca flavescens	Yellow perch	19	16 cm	2–4 years	North America	Fixed on plants or stones	Early spring	One batch	Fishbase; Schneider (1984)
Percina caprodes	Logperch	12.5	ND	2 years	North America	Burried in the substrate	Late winter to late spring	Several batches	Fishbase; Hubbs (1985)
Stizostedion vitreum	Walleye	54	36 cm	2 years	North America	Demersal	Spring (april-june)	One batch	Fishbase
Zingel streber	Danube streber	12	ND	ND	Europe (east)	Burried in the substrate	Early spring	ND	Fishbase

ND non determined

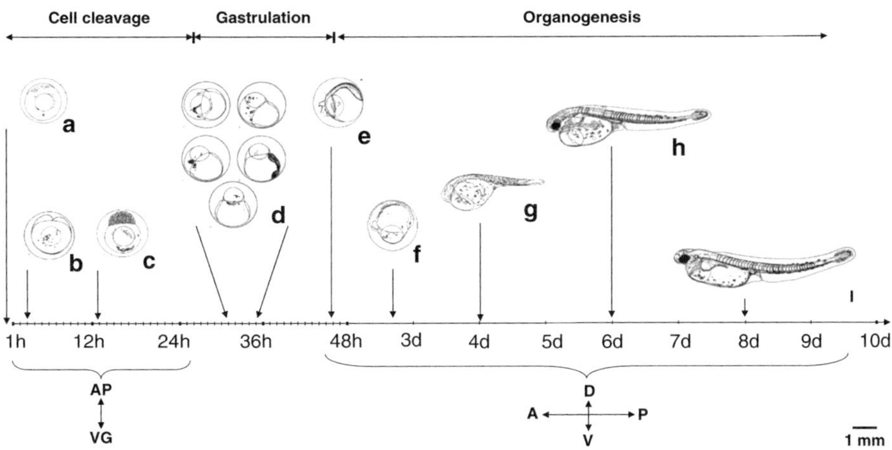

Fig. 6.3 Embryonic developmental table of the walleye (*S. vitreus*). (**a**) ova before activation; (**b**, **c**) cell cleavage; (**b**) 2 cells step; (**c**) blastula step; (**d**) gastrulation steps, the orientation of the embryos varies while the cell movements process; (**e–i**) organogenesis; (**e**) beginning of the antero-posterior axis formation; (**f**) somites formation; (**g**) the tail is detached from the yolk, the somitogenesis proceeds to elongate the embryo, the eyes are developed and the otic vesicle is visible; (**h**) circulation system development (**i**) embryo at hatching. *AP* animal pole, *VP* Vegetative pole, *A* anterior, *D* dorsal, *P* posterior, *V* ventral (Drawings from Mc Elman and Balon 1979)

of the digestive system and its derivatives). Moreover, rapidly a cell accumulation appears on one side of the embryo called the embryonic shield. It marks the dorsal part of the embryos and is the first manifestation of the dorso-ventral axis.

As for the cell cleavage, the gastrulation total duration is diverse among percids from 6 h for the rainbow darter to 23 h for the streber (Paine and Balon 1984a; Kovac 2000) (Fig. 6.4, Table 6.2). The germ ring and embryonic shield formation occurs early at 25 % of epiboly in *Etheostoma* species, the timing seems to be important as premature or delayed involution may lead to lethality (Mendelson et al. 2006). This has not been described in any other percid, probably because the authors didn't pay attention on that particular point.

In several percid species, due to the presence of the oil droplet and the cellular migration, the gastrulation may be accompanied by a rotation of the embryos (Fig. 6.3). Indeed, while the first part of the epiboly occurs in the horizontal plane, from the stage corresponding to the 50 % of epiboly, the embryo rotates by 90° so the oil droplet protrudes and the embryonic shield takes a different position. This has been described in the walleye (Mc Elmann and Balon 1979), the northern logperch (Paine and Balon 1984b) but not in darters (Paine and Balon 1984a; Mendelson et al. 2006) or in the *Gymnocephalus* gender (Kovac 1994) (Fig. 6.4). For other percids, the gastrulation observation was not described clearly enough to know whether this rotation occurs or not. It may be due to a difference of cell migration speed between the dorsal and ventral part of the embryos as proposed in other fish species embryos (Kimmel et al. 1994). As a consequence, the blastopore doesn't

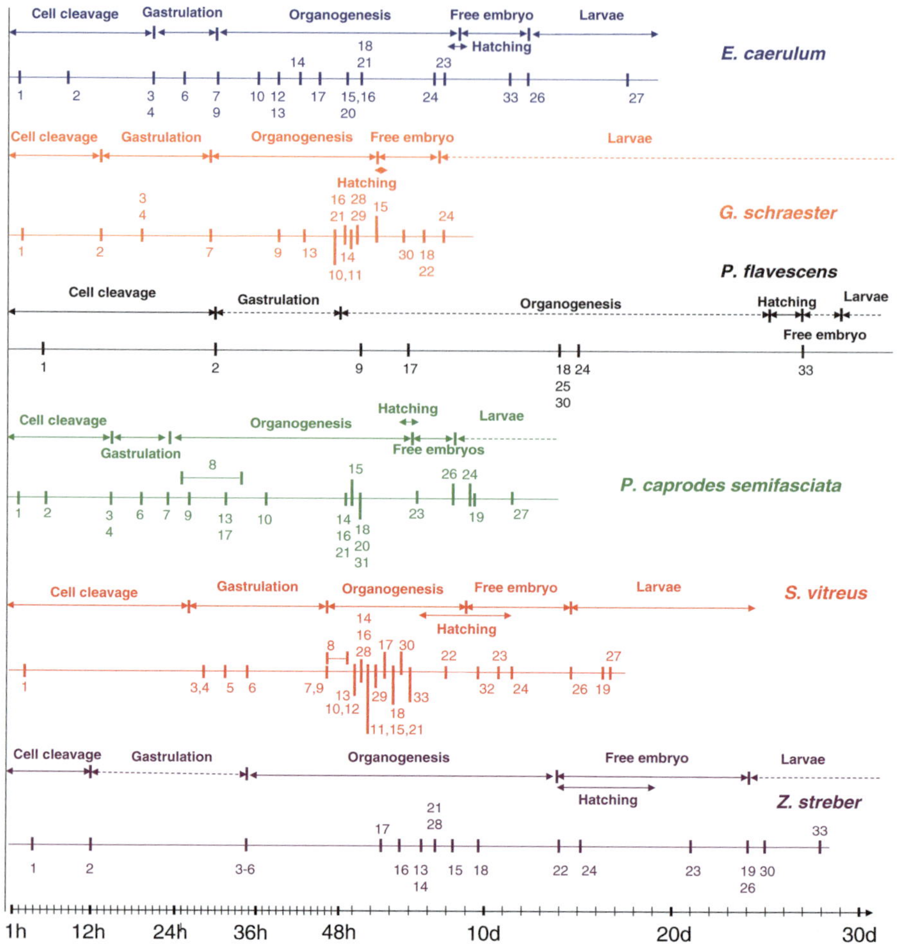

1 First division completed
2 Morula
3 Germ ring apparition
4 Embryonic shield
5 50 % epiboly
6 Rotation complete
7 Tail bud closure
8 Translocation
9 Optic vesicles
10 Otic vesicles
11 Otoliths apparition
12 Olfactory vesicles
13 Kupfer vesicle
14 Melanophores on the yolk
15 Melanophores on the body
16 Body contraction
17 Tail free
18 Eye pigmentation
19 Swimbladder inflation
20 Hatching glands
21 Heart beating
22 Mouth formation
23 First teeth formation
24 Mouth opening
25 Gills formation
26 First oral feeding
27 Yolk completely used
28 Median finfold apparent
29 Ventral finfold apparent
30 Pectoral fin apparent
31 Caudal fin apparent
32 Pectoral fins movement
33 Rays in the caudal fin

Fig. 6.4 Comparison of the embryonic development of representative species of each percid gender *blue* E. *caerulum* (Paine 1984; Cooper 1979; Medelson et al. 2006; Paine and Balon 1984) (20 °C), *orange* G. *schraester* (Kovac 1992) (16–20 °C), *black* P. *flavescens* (Mansueti 1964) (6.7 °C, mean temperature at spawning), *green* P. *caprodes semifasciata* (Paine and Balon 1984; Cooper 1978) (20 °C), *red* S. *vitreus* (McElman and Balon 1979) (9–14 °C) and *purple* Z. *streber* (Kovac 2000) (12–17 °C). The time scale of each step (cell cleavage, gastrulation, organogenesis, hatching, free embryo and larval) are represented in bold lines when the stage is properly documented in term of time or dotted lines when the stage is not well determined. The time scale

close at the vegetative pole but rather ventrally in these fish. Surprisingly, species undergoing early germ ring and embryonic shield apparition are not those presenting the early rotation while the two processes may probably be linked. One can thus conclude that data describing the gastrulation in percids are not precise enough because each author focused on diverse staging and didn't pay enough attention on others to properly understand these processes and thus make a general rule.

6.3.1.3 Organogenesis

This step can last from 5 days for *Percina caprodes* to 23 days for *Perca flavescens* (Mansueti 1964; Paine and Balon 1984b) (Fig. 6.4, Table 6.1). During this step, tissues will begin to differentiate and most of organs will progressively appear. This process continues during the free embryos and larval stages until the animal become a juvenile.

After the complete covering of the yolk at the end of the gastrulation, embryos begin to elongate and grow. Both antero-posterior and dorso-ventral axes become obvious. In addition, the Kupfer vesicles appears. It is a transient spherical ciliated organ specific to teleosts fish which seems to be involved in the left-right determination of the heart, brain and gut (Essner et al. 2005). The organogenesis subdivision can be staged thanks to the number of somites (repetitive units appearing sequentially from the anterior part of the trunk to the end of the tail). Moreover, the first cell differentiation occur leading to the organogenesis with the apparition of the neural plate giving rise to the future central nervous system. Interestingly, for the walleye and the rainbow darter the first step leading to the formation of the neural plate occurs before the blastopore closure thus showing that the first organogenesis manifestation is earlier in comparison to other fish species. As during the gastrulation, an embryonic rotation is observed in several species (walleye, logperch). This step is called translocation and is characterized by the decrease of the distance between the head of the embryo and the oil globule through embryonic shifts (Mc Elmann and Balon 1979). This phenomenon has not been observed in other percid species.

As a whole, the chronology of the organ differentiation and development is the same in every species. However, some organs finish their development during the hatching period, free embryos and larval stages for some species while for others, their development is complete before hatching. One example is the gaping of the mouth that takes place early during organogenesis in *Perca flavescens* while it takes place later in other species (Mansueti 1964) (Fig. 6.4). Several organs easily observable with a binocular loupe begin to develop (e.g. the optic, otic and olfactory precursors, the heart beating, the melanophores apparition in the yolk and later on the

Fig. 6.4 (continued) is presented as hours and days elapsed from the activation without taking into account the temperature. General features commonly studied in papers are presented below the time scale for each species as indicated in the legend. The larval stages are not complete due to the lack of information for some species

Table 6.2 Summary of the developmental characteristics of representative percid fish

Scientific name	Common name	Embryonic incubation temperature (°C)	Larval incubation temperature (°C)	Egg size (mm)	Total embryonic duration
Etheostoma caeruleum	Rainbow darter	20	20	1.9–2.2	Around 12 days
Gymnocephalus schraetser	Yellow pope	16–20	16–20	1.22	Around 7.5 days
Perca flavescens	Yellow perch	10–22	10–22	1.6	29 days
Percina caprodes	Logperch	20	20	1.23	Around 8 days
Stizostedion vitreum	Walleye	15	15	1.85	14.5 days
Zingel streber	Danube streber	12–17	15–17	1.94	Around 24 days

Common name	Cell cleavage duration (h)	Gastrulation duration (h)	Organogenesis duration	Size at hatching (mm)	Hatching duration	Free embryos phase duration
Rainbow darter	20	6	Around 7.5 days	8	Few hours	Around 3 days
Yellow pope	13	16	Around 3 days	3.7–4.4	18 h	Around 3.5 days
Yellow perch	30	18	Around 23 days	4–6.6	Around 2 days	3 days
Logperch	15	8	Around 5 days	6.2	Few hours	Around 2 days
Walleye	26	20	Around 7 days	6.8–7.3	Around 5 days	5.5 days
Danube streber	12	23	Around 12 days 12 h	6.4–7.6	Around 4 days	10 days

Common name	Larval size (first feeding) (mm)	Larval developement duration	Juvenile size	Beginning of juvenile stage	References
Rainbow darter	9.6	Around 7 days	9.6–10.6 mm	19 days	Fishbase; Paine and Balon (1984a)
Yellow pope	5.2–6.8	Around 32 days	12–12.9 mm	39 days	Fishbase; Kovac (1992)
Yellow perch	7	ND	14 mm	ND	Fishbase; Teletchea et al. (2009); Mansuetti (1964)
Logperch	6.2	ND	ND	ND	Fishbase; Paine and Balon (1984b)
Walleye	9–9.7	Around 10 days	8.9–9.8 mm	25 days	Fishbase; McElman and Balon (1979)
Danube streber	9.2	32 days	14·7 mm	56 days	Fishbase, Kovac (2000)

ND non determined

embryonic body and the eyes pigmentation (eyed-stage)). The ability of embryo to reach these steps should be important to follow and thus predict the apparition of deformities or mortality occurrence. Indeed, up to now, the most studied developmental rates are checked at 72 h post-fertilization or eyed stages (with few other steps as shown in the Sect. 6.3.3.1.1) in works studying the reproductive performance quality. Investigating parameters cited above may allow the identification of developmental failures within this interval in order to be more precise.

6.3.1.4 The Hatching Period

During the organogenesis, several granules called the hatching glands appear. They contain hatching enzymes that are released before hatching and weakens the chorion and/or the *zonula radiata externa* (Kimmel et al. 1994). Firstly, the development duration within the envelope is highly variable among species. For example, the comparison between *Zingel streber* and *Perca flavescens* shows that while the total duration of the embryonic development (from activation to the first feeding) is quite the same (around 23–29 days), the yellow perch hatches after a long period in the envelope (around 23 days), while the streber development in the envelope is very short (around 12 days) (Mansueti 1964; Kovac 2000) (Fig. 6.4, Table 6.1). By comparing the embryonic morphology of both species, there are mainly differences of the yolk sac utilization. Indeed, authors of both studies didn't pay attention to the same criteria to describe embryos. As proposed by Mansueti (1964), this difference can mainly be due to the spawning mode. The yellow perch eggs are surrounded by a thick jelly coat (as for the Eurasian perch *Perca fluviatilis*) protecting the embryos from the outside aggression and thus allowing a longer period of development in the envelope. In contrast, other percid species eggs are solely protected by a tough chorion and are either buried in the substrate or dermersal (Table 6.1). In those cases the developmental phase within the envelope is shorter, potentially avoiding predatory behaviors from other species. From a developmental point of view, the staging of embryos is diverse between species mostly thanks to the skeletal, branchial, intestinal and fins developments. For example, while some individuals hatch with a quite well developed mouth with teeth or fins allowing movements in every direction, others acquire these abilities during the free embryonic stages (Fig. 6.4).

Interestingly, depending on the species, the hatching duration lasts from several hours to 5 days for the same spawn suggesting that even within a species the developmental stages can be variable. For most of percids, authors observed differences in the embryos total length at hatching and sometimes differences in the circulation or skeletal systems advancement. In *Perca fluviatilis*, Konstantinov characterized four developmental stages at hatching (Konstantinov 1957) that differ in term of developmental advancement from those of *Perca flavescens* (Mansueti 1964). Criteria to distinguish each stage are mainly the yolk absorption level and the total length. However, other criteria as the eye development should be established in the future to better characterize these stages (M. Alix and B. Schaerlinger, personal observation). This difference of hatching development may confer an advantage for a given species because the dispersion of the size of free embryos allows a better adaptation to the environment. Indeed, embryos of different sizes can, on the one hand, easily avoid predators and, on the other hand, take advantage of preys from diverse species and sizes. So, even if several embryos die during this phase, others may pass through this step, allowing thus the recruitment success of the species whatever the environmental conditions.

6.3.1.5 Free Embryos Stages

Mainly this step allows the preparation of embryos to ensure an efficient transition to the larval stages (Fig. 6.3). Indeed, the free embryos phase corresponds to a period of final development of the intestinal, branchial, fins and circulatory systems which become progressively active (Fig. 6.4). The vitelline reserves have greatly reduced and the circulation system begins to be deviated from the yolk to the gut and branchial systems allowing a new way of oxygenation and alimentation of the body. Several studies shows in percid species a high rate of mortality at that stage that can mainly be due to an impairment of this circulation deviation. Moreover, the embryos continue to grow except for the logperch (Paine and Balon 1984b).

6.3.2 Larval Development

The onset of the larval development is controversial as some scientists rather use the hatching period, others the first oral feeding or the yolk resorption (Teletchea and Fontaine 2011). As explained above, we choose to determine the first oral feeding as the onset of the larval development. Indeed, hatching and yolk resorption can be dependent upon spawn characteristics or incubation conditions. They thus can't correspond to a developmental step. On the contrary, the first oral feeding elapses for a small period of the development and may probably correspond to a specific stage of maturation of the intestine.

This part of the chapter used mainly data from *Etheostoma*, *Percina*, *Gymnocephalus*, *Sander* and *Zingel* genders (Paine and Balon 1984a, b; McElmann and Balon 1979; Kovac 1993b, 2000). However as for the embryonic stages, the larval development of each species has not been studied with the same purposes and don't focused on the same characteristics. The larval development can be divided into two large steps, the finfold phase and the finformed phases (Kovac 2000). Each of these stages could be further cut into several steps but we choose to present only the most important characteristics of each phase.

6.3.2.1 The Finfold Phase

This stage begins with the first feeding, but the larvae still need yolk supply, principally lipids from the oil globule. In the walleye, it has been shown that the oil globule fastly decreases during this period (McElman and Balon 1979). In the meantime, the stomach and the intestine develop and replace the yolk in the ventral space.

Moreover, this phenomenon is accompanied by the final reduction of the vitelline circulation and a definitive switch toward the intestine circulation (Paine and Balon 1984b; Kovac 2000). Once the oil globule is entirely used, larvae depend exclusively upon exogenous food. As for the free embryonic stage, this period of mixed feeding may allow larvae to adapt their digestive physiology to the exogenous food. The time elapsing from the first oral feeding to the first defecation can be quite long, suggesting a progressive starter of the digestive function (e.g. in the walleye the first oral feeding takes place 15 days post fertilization in the walleye and the first defecation was observed 2 days later which is quiet long (McElman and Balon 1979)). Moreover, temperate fish larvae need to have a developmental synchronization with the preys availability in the nature. The transition period could be involved in this synchronization. Another important characteristic of this stage is the filling of the swim bladder allowing the control of buoyancy. This step is usually concomitant with the onset of oral feeding and it corresponds to an important key step during the larval development. Indeed, yellow perch larvae with uninflated swim bladders display lower survival rate than those with an inflated one due to a lack of preying efficiency (Czesny et al. 2005b). The respiration function is mediated by gills that are fully protected by the operculum and larvae display regular mouth movements demonstrating active breathing in every species. The finfold stage is also the time for the beginning of cartilage and bone formation. It begins with a progressive chondrification and ossification of the head and the body axis with the formation of vertebrae (Paine and Balon 1984b; Kovac 2000; McElman and Balon 1979). Moreover first teeth begin to appear. Finally, the body pigmentation increases in the head and along the ventral side of the body axis.

6.3.2.2 The Finformed Phase

While during the first larval stage animals are mainly surrounded by the finfold, the second step of the development is characterized by the differentiation of fins. The finfold degenerates all around the larvae except in the part corresponding to the emplacement of fins in adults (anal, caudal, dorsal and pelvic). Moreover, rays begin to differentiate in every fins before undergoing an ossification (Paine and Balon 1984b; Kovac 2000; McElman and Balon 1979). Finally, the body pigmentation is almost complete and larvae can present colour diversity. At the end of the step, larvae look like adults but sexually immature and thus become juveniles.

6.4 Evaluation of Egg and Larval Quality

6.4.1 What Are the Egg Quality Indicators?

Prediction of the egg quality in aquaculture is one of the most important and challenging step of controlled reproduction. Objective indicators allow not only facilitating the commercial fish production (by choosing only the highest quality eggs for further steps of reproduction) but also more efficient and faster development of the reproductive protocols through the research activity. Some indicators can be predictive (measured before fertilization) and characterize principally ova quality while other are evaluated after the fertilization (reproductive performance) and characterize ova quality in addition to embryonic and larval quality. Predictive indicators are often species specific and it can thus be difficult to find relevant ones for each species. However, assessing reproductive performance is very relevant but not predictive leading sometimes to long and expensive studies. On another level, indicators can be divided into morphological (Brooks et al. 1997) or biochemical indicators. Morphological are easy to evaluate but have to be determined for every species. Biochemical indicators can need access to specific equipments but have the advantage to lean on general molecular mechanisms the deregulation of which would affect the developmental success of every fish species. Moreover, even if the genome of a fish species is unknown (which is the case in most percid species), biochemical assessment of the quality could be performed with few technical adjustments. Finally, molecular indicators may mainly be investigated on ova and would thus be predictive. Performing the characterization of developmental failure could probably be important to define levels of quality which in turn would help to choose the most relevant quality indicator. This approach could thus help scientist and fish breeders to improve their assessment of reproductive success and finally their breeding practices.

6.4.2 *Description of Developmental Defects*

6.4.2.1 Occurrence of Mortality During Embryonic and Larval Stages

Reproduction failures are often due to lethality occurrence among embryos and larvae. Sometimes, when rearing conditions are well established in captive conditions, the survival rate is higher than in the nature as shown for the walleye (Ivan et al. 2010). However, when the rearing conditions are not well controlled, mortality occurrence is very high in hatcheries. Lethality stages have poorly been studied in fish and almost no work has been performed on percids. However, this question needs to be addressed as the stage and the phenotype of mortality could reveal clues to understand rearing conditions problems. As a whole in the Eurasian perch several steps of mortality has been observed. The first one takes place during the first 24 h probably corresponding to cleavage defects and/or zygotic genome activation (MBT) impairment (M. Alix and B. Schaerlinger unpublished data). Moreover other lethality stages have been observed by the time of hatching or first feeding. It may be interesting to better characterize mortality occurrences in order to define general phenotypes and thus improving rearing conditions to avoid lethality. As proposed in the Sect. 6.3.1.3, assessing the proper developmental advancement by checking several specific stages within the interval between the fertilization and the eyed-stage should help to define other key steps of the embryonic susceptibility.

6.4.2.2 Occurrence of Developmental Abnormalities

Developmental abnormalities are a very unwanted element of aquaculture production. The high rate of abnormal embryos and larvae leads to a reduced number of fry which could be intended for further culture purposes. Abnormal development in fish embryos and, consequently, freshly-hatched larvae, may be caused by several factors. Among the environmental factors, temperature and salinity have been observed to affect embryonic development negatively when they reach values exceeding the optimal range for a particular species (e.g., Bermudes and Ritar 1999; Haddy and Pankhurst 2000; Kupren et al. 2011). In addition, it was proven that improper broodstock management (including diet composition) and/or reproductive procedure (including photothermal manipulations and hormonal treatment) were responsible for developmental abnormalities in embryos and freshly-hatched larvae (e.g., Kjørsvik et al. 1990; Aegerter and Jalabert 2004; Bonnet et al. 2007; Palińska et al. 2011; Żarski et al. 2011a). Additionally, toxic substances, including heavy metals (e.g., Von Westernhagen et al. 1988; Jezierska et al. 2000; Ługowska 2007; Ługowska and Kubik 2011), present in the water during incubation as well as in the natural environment of the spawners (Black et al. 1988; Cameron et al. 1992; Von Westernhagen et al. 1988; Singh et al. 2008) were found to affect embryonic development and induce larval deformations.

In the case of freshwater percids, data on deformations in newly-hatched free embryos are very limited. It has recently been reported that the procedure of artificial reproduction may affect deformations in freshly-hatched free embryos, where the most common scoliosis and lordosis of different parts of the spine, yolk sac malformation, yolk sac oedema, gape jaw and cardiac oedema were found (Żarski et al. 2011a). However, in other fish species, many other deformations have been found, such as kyphosis, axial curvature in the abdominal and caudal region, severe spine curved-in axial and caudal region, C-shaped larva, pigment-deficient eye, deformed skull, body shortened (for details see: Jezierska et al. 2000; Ługowska 2007; Palińska et al. 2011) (Figs. 6.5 and 6.6). Each particular deformation may occur alone or together with other deformations (Jezierska et al. 2000; Ługowska 2007; Palińska et al. 2011; Żarski et al. 2012a). In Eurasian perch, deformation occurrence and frequency were found to be strictly related with the quality of eggs and ranged between 6.36 % and 86.14 % in the highest and lowest egg quality, respectively (Żarski et al. 2011a). However, such a high deformation rate was probably a result of an artificial reproduction protocol, while deformation in perch spawned naturally in the natural environment did not exceed 1.78 % (Treasurer 1983).

It has already been reported that larval deformations in pikeperch may be affected by the weaning protocol (from live to compound diet) and feeding regime. In such cases, the retraction of upper and lower jaws as well as scoliosis were described (Kestemont et al. 2007). Kowalska et al. (2006) reported mainly lordosis, whereas Hamza et al. (2008) recorded mainly kyphosis and lordosis. Authors very rarely

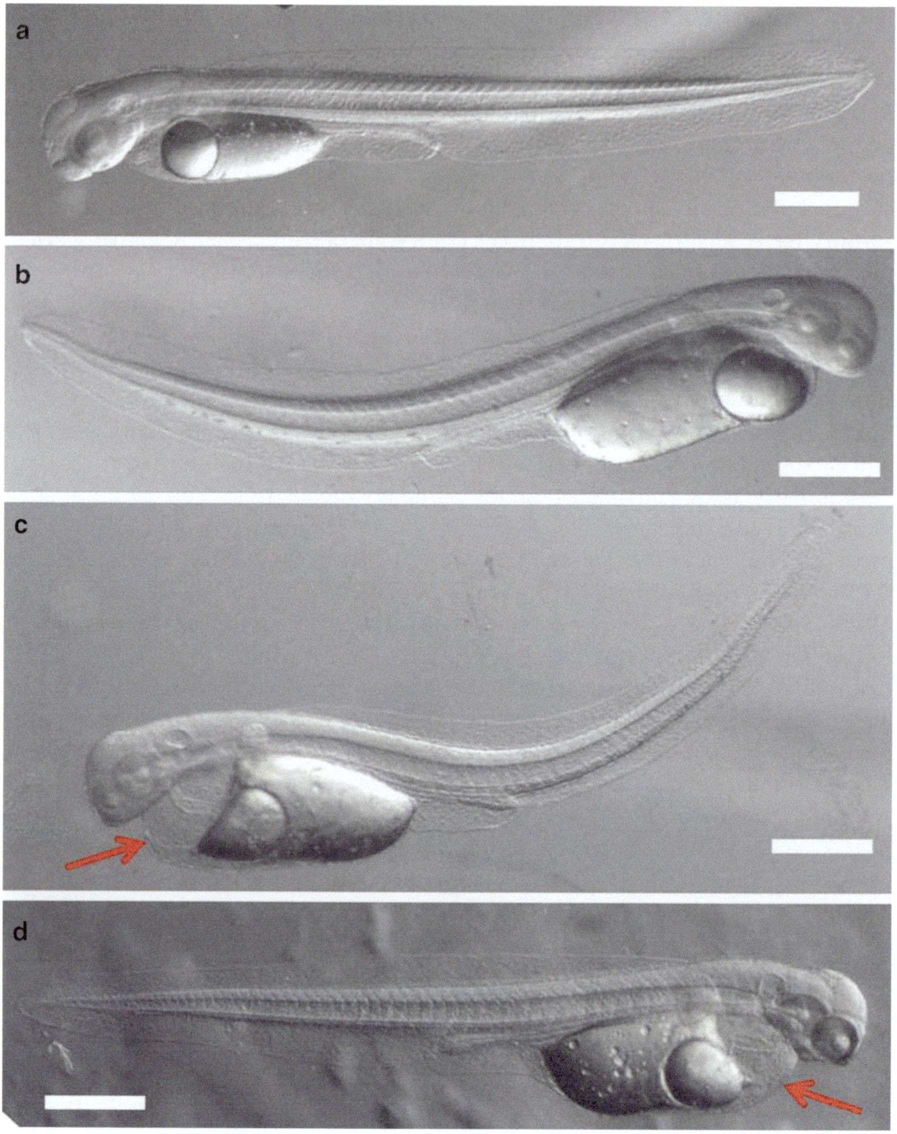

Fig. 6.5 Pikeperch, *Sander lucioperca*, larvae with normal shape (**a**) and with the most common morphological deformations (**b-d**). (**b**) spinal lordosis, (**c**) spinal lordosis and cardiac oedema, (**d**) cardiac oedema. *Arrows* indicate cardiac oedema. The *Scale bar* represents 500 μm (Pictures of D. Zarski)

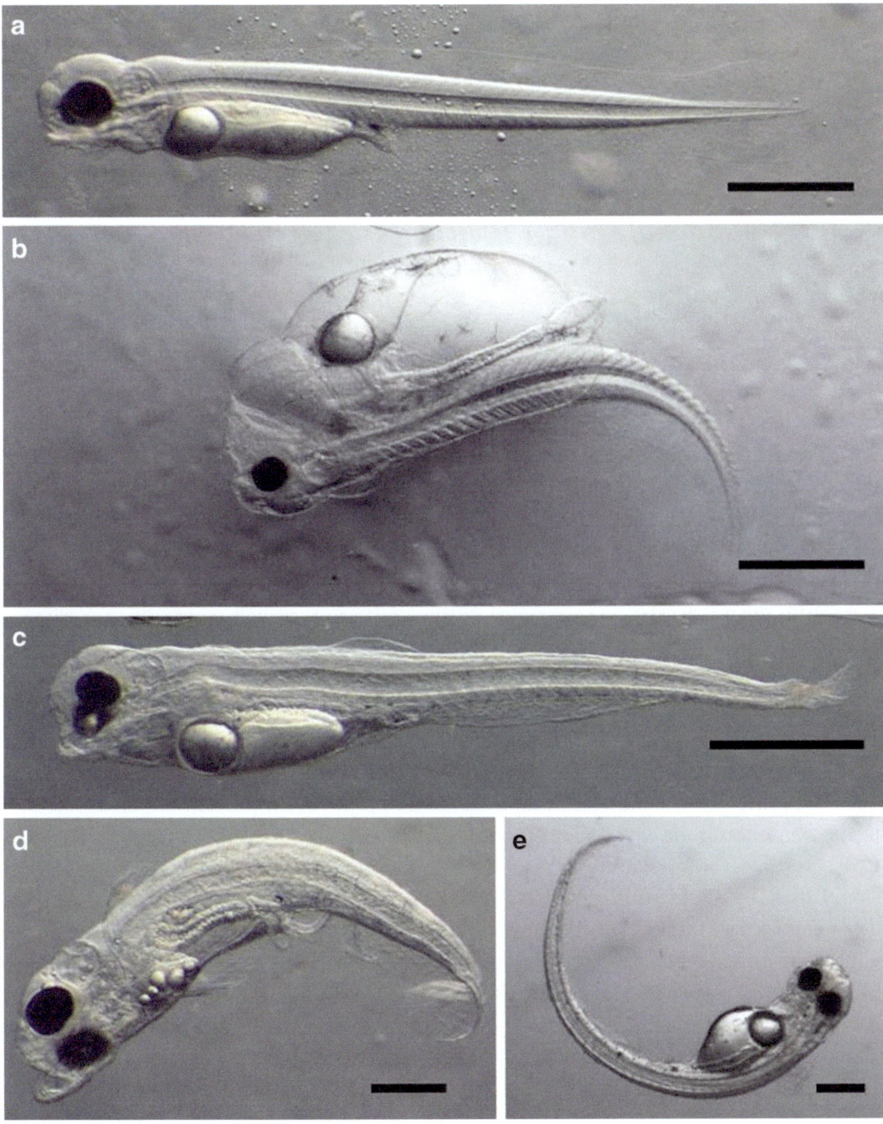

Fig. 6.6 Eurasian perch, *Perca fluviatilis*, with normal shape (**a**) and different type of deformations (**b–e**). (**b**) larva with several deformations (cardiac and yolk sac oedema, lordosis, jaw malformation), (**c**) larva with jaw malformation and tail deterioration, (**d**) larva with several oil droplets in the 'yolk-sac area', (**e**) 'c-shaped' larva. The *scale bar* represents 500 μm (Pictures of D. Zarski)

noted scoliosis or incomplete jaw development, as compared to the data described by Kestemont et al. (2007). The frequency of total deformations reported by those authors ranged between 15.5 % and 35.5 % (Kestemont et al. 2007), 9.5–16.5 % (Kowalska et al. 2006) and 7.5–12.5 % (Hamza et al. 2008).

6.4.3 Definition of Quality Indicators

6.4.3.1 Egg Quality

6.4.3.1.1 Morphological Indicators

In freshwater percids, the most common practice of egg quality evaluation was the determination of embryo survival rate shortly after fertilization (Migaud et al. 2004, 2006; Żarski et al. 2011a), at the eyed-egg stage (Czesny and Dąbrowski 1998; Kucharczyk et al. 1998; Czesny et al. 2005a; Żarski et al. 2011b) or at hatching (Migaud et al. 2006). Egg diameter has also been reported to be correlated with embryo viability in the walleye (Malison et al. 1998). A simple general morphological examination was also very helpful in the evaluation of fish egg quality. Partially or totally opaque eggs or visible internal damage of the yolk structure were also usually reliable indicators of poor quality (e.g. Treasurer 1983; Pavlov and Emel'yanova 2008; Teletchea and Fontaine 2011).

In the case of Eurasian and yellow perch, a very simple method of indication of low egg quality was egg-ribbon fragmentation (Dąbrowski et al. 1994; Overton et al. 2008). Recently, it was reported that fragmentation of oil droplets in ovulated eggs of the Eurasian perch may be a useful tool for its quality evaluation, with more than one big oil droplet in an ovulated egg proven to be an indicator of decreased quality (Żarski et al. 2011a). However, this method may be used only when "dry" eggs (before contact with water) are evaluated, while many lipid droplets may coalesce into one large droplet after the contact of eggs with water. A similar morphological indicator of oil droplets in ovulated eggs was certified in pikeperch (Fig. 6.7). In the egg batches where oil droplet was found to be fragmented (n = 7) the embryonic survival rate (at 72 h post fertilization) and hatching rate did not exceed 49 % and 20 %, respectively. Among the hatched larvae over 94 % of different types of deformities were recorded (D. Żarski unpublished data). However, this phenomenon has only been observed sporadically in this species. A much more reliable egg quality indicator of pikeperch eggs was the observation and determination of egg deformation (extreme chorion deformation) rate shortly (3–5 min) after egg activation in the water. This method is based on observations of the cortical reaction which, during the first few minutes, occurs very violently in eggs of this species and is an indicator of high egg quality. This phenomenon was clear and easily noticeable (Żarski et al. 2012b). For the walleye, the turbidity of ovarian fluid (measured spectrophotometrically) caused by the increased amount of protein after short-term storage (8 h) of the eggs was negatively correlated with egg quality at the end of the spawning season (Dietrich et al. 2012). However, according

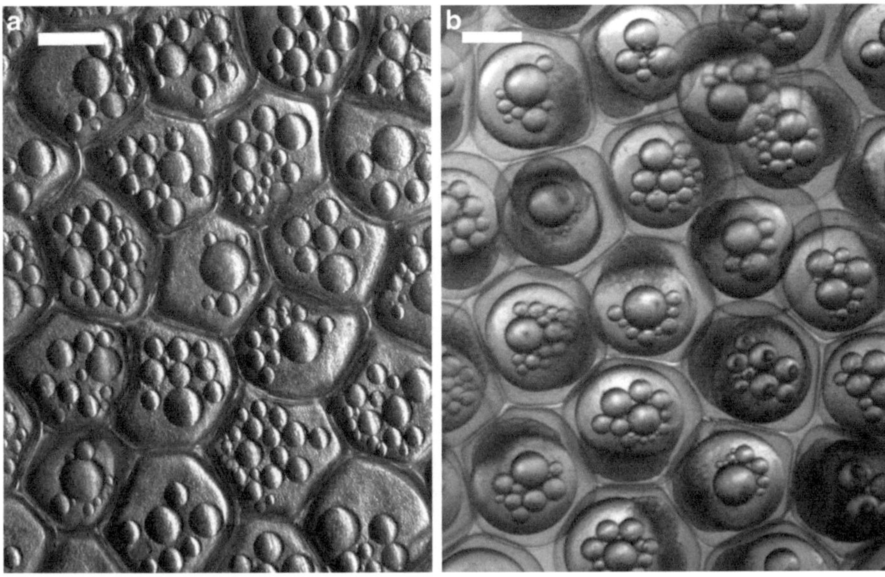

Fig. 6.7 Eggs of pikeperch, *Sander lucioperca*, with fragmented oil droplets. Eggs before activation (**a**) and during the early development (**b**). The *scale bar* represents 500 μm (Pictures of D. Zarski)

to the authors, this method seems to be an indicator of the changing proportion of over-ripened and/or broken eggs rather than egg quality itself. Schrader and Schrader (1922) reported that abnormal blastomere morphology (during the first cell divisions shortly after fertilization) were very good indicators of low egg quality in the walleye. Similar observations were made for other fish species, such as Atlantic cod, *Gadus morhua* L. (Hansen and Puvanendran 2010). The data on other non-biochemical egg quality indicators in freshwater percids is very scarce and more work is still needed in this field.

6.4.3.1.2 Biochemical Indicators

Few works investigated potential biochemical indicators of egg quality that are mainly predictive. Some studies focused on the relationship between the fatty acid composition of the Eurasian perch eggs and their reproductive performance. However, the comparison of these data shows that conclusions cannot be generalized. Indeed, while Henrotte et al. (2008) show that the n-3/n-6 ratio above 2.5 in the egg could be an indicator of low quality gametes, data obtained by Blanchard et al. (2005) (ratio ranging from 2.9 to 3.5 in wild populations) contradict this hypothesis. In the same manner, the DHA/EPA ratio within the egg doesn't seem to be a reproducible egg quality indicator. Indeed, it can range from 3.5 (Abi-Ayab et al. 2000) to 8.4 (Henrotte et al. 2010) for good quality eggs. These data suggest that even if the yolk composition is crucial for the proper development of the embryo, other parameters should be studied in the meantime to predict ova quality. One of these parameters could be the yolk utilization during the embryogenesis. Another study performed on the same species showed that high activity of the Cathepsin L (enzyme involved in the Vitellogenin proteolysis during the embryonic and larval development) just before hatching is negatively linked to the hatching rate or the embryonic resistance to osmotic stress (Kestemont et al. 1999).

The proteomic profile of *Perca fluviatilis* ova has been investigated on spawn of various qualities that have been estimated thanks to the study of reproductive performance (Castets et al. 2012). Data showed that proteins involved in cell protection against stress, protein degradation, Vitellogenins or cell metabolism display various expression level depending on the ova quality. They thus could be promising indicators of ova quality.

Addressing the question of the molecular composition of the ova to predict its quality is important because it will give clues to find good predictive indicators that could be used for several species. Indeed, the affected cellular functions are, in general, conserved among fish species and data obtained in one percid species could be applied to others. This research should first be done at a large scale (e.g. transcriptomic, proteomic) but it is also important to confirm results with smaller scale techniques (e.g. western blotting, RT-PCR…) to choose the better indicators. Moreover, it appears that assessing gametes quality through molecular techniques would rather need to investigate the level/ratio of diverse molecules and the combination of all these data should give an indication of the quality rather than investigating only one molecule expression. This field of research deserves to be carefully studied.

6.4.3.2 Free Embryo and Larval Quality

Evaluation of egg quality on the basis of the survival rate of embryos (e.g. to the eyed-egg stage) was not always a reliable indicator of the effectiveness of reproductive procedures or induction of maturation processes, while the development of embryos in eggs of low quality could be observed even up to the hatching stage (e.g. Żarski et al. 2011a), leading to the need for an evaluation of larval quality.

Evaluation of the larval quality in fish was mostly based on the determination of survival rate at particular life stages (Dąbrowski et al. 2000). Initially, the ability of free embryos to hatch was indicated as a reliable indicator of its quality (e.g., Bobe and Labbe 2010) as it was suspected that mainly properly-developed larvae and/or larvae with appropriate energy content were able to get out of the egg shell. However, it was proven that even embryos with developmental malformations were able to hatch successfully (Ługowska and Sarnowski 2011; Żarski et al. 2012a). Next, the deformation rate of freshly-hatched embryos was used as an indicator of quality. But hatched embryos without evident deformities could still be of variable quality, which may be reflected in a failure of inflation of the swim bladder or the start of exogenous feeding (which were also considered to be quality indicators in percids) (Dąbrowski et al. 2000; Żarski et al. 2011a). Recently, a test of salinity stress was developed and applied for evaluation of percid hatched embryos quality. This method indicated their ability to perform osmoregulation and energy mobilization to counteract stress (Kestemont et al. 2007). This method is very simple to use and seems to give a reliable indication of their quality (Kestemont et al. 1999; Henrotte et al. 2010) for both research and aquacultural purposes.

6.5 Improvement of Egg and Larval Quality

Reproduction success depends upon breeders rearing conditions on the one hand, and egg and larval incubation environment on the other hand. The control of these conditions needs to be understood for every species in order to improve zootechnical practices and reproduction success in captive breeding. This part will mainly focus on data obtained for *Perca* and *Sander* genders.

6.5.1 Broodstock Management

Fish life cycle depends upon two main categories of factors: (i) determining factors and (ii) modulating factors (Wang et al. 2010). Determining factors seem to be involved in triggering key steps of the gametogenesis (e.g. vitellogenesis, ovulation). Identifying and controlling determining factors will allow the establishment of environmental programs to perform out-of-season reproduction. However, even if determining factors are well controlled, the reproduction success remains unstable in recirculating aquaculture system(s) (RAS). Modulating factors are able to do fine-tuning of reproduction conditions. They correspond to parameters as the stress of manipulations, nutrition, intrinsic characteristics of the fish population or the individual breeders or hormonal injections. Thus, while determining factors could be seen as a switch for fish reproduction, modulating factors further act as dimmer switch to progressively improve or damage the quality of reproductive performance.

6.5.1.1 Temperature and Photoperiod

In the nature, temperate fish, among which most of percids belong, are submitted to important amplitudes of temperature and/or photoperiod variations (Taranger 2010; Wang et al. 2010). For Eurasian perch *Perca fluviatilis*, it has been shown that female's gametogenesis steps follow these variations suggesting that they could play the role of determining factors of egg quality (Sulistyo et al. 1998). Further works confirmed this hypothesis and allowed the establishment of a photo-thermal program capable of controlling Eurasian perch reproduction cycle in RAS conditions although the reproductive performance remain various (Migaud et al. 2004, 2006; Wang et al. 2006; Fontaine et al. 2006; Abdlufatah et al. 2011, 2013). However, the yellow perch (*Perca flavescens*) gametogenesis is principally controlled by the temperature (Dabrowski et al. 1996). For the pikeperch (*Sander lucioperca*) only few works investigated the effect of both parameters. Although, there is very limited data available on the effect of these variables on the spawning effectiveness in percids, it is already well established that photo-thermal manipulations are among the most important factors affecting gametes quality in domesticated broodstock of these species. This creates the necessity for more intensive studies in this specific field where more attention should be paid not only for the spawning performance in general, but more specifically to gametes quality.

6.5.1.2 Nutrition

Breeders' nutrition influences greatly their fecundity and the gametes quality. Indeed, it has been shown in several species that the fecundity seem to be directly linked to the energetic value of their regime (e.g. forage fish vs. pellets) (Cerda et al. 1994; Bromage et al. 1992). Moreover, as previously explained, fish eggs contain large amount of nutrients among which lipids and amino acids. These are either taken from female's intrinsic reserves or directly from the food during the reproduction season. In percids, very few works focused on the improvement of the nutritional contribution for the reproduction success. A preliminary study performed on the pikeperch showed that commercial food choose in the study leads to fewer reproduction success in comparison to forage fish or mixed nutrition (Wang et al. 2009). This is mainly due to diverse lipid compositions between pellets and forage fish. Another study showed that a diet composed of 3/2/2 ratio of respectively DHA/EPA/AA led to the Eurasian perch eggs and larvae of good quality (Henrotte et al. 2010).

6.5.1.3 Hormonal Treatments

In order to synchronize fish spawning by triggering ovulation of females in the same time or provoke spawning earlier than the regular spawning season, hormonal injections are often performed. Those treatments include injections of pituitary extracts, human chorionic gonadotropin (hCG), luteinizing hormone-releasing hormone analog (LHRHa), pregnant mare serum gonadotropin (PMSG) or Gonadotropin-releasing hormone (GnRH). Several studies showed that pikeperch or Eurasian perch spawning can be induced by diverse hormonal stimulations (Żarski et al. 2011a; Zakęś 2007). Depending on the nature and the doses of hormones, reproductive performances can sometimes be altered, although wild fish (which has completed the gametes maturation process in the wild or in the wild-like environment [i.e. earthen ponds]) were examined. For example, pond-reared pikeperch hatching rate was more negatively affected by the application GnRH analogous in comparison to hCG injections (Krist'an et al. 2013). Moreover, those treatments could induce some stress in fish as it was reported in the pikeperch (Falahatkar and Poursaeid 2014). As considering the effect of hormonal treatment on the spawning effectiveness more details may be found in the Chap. 4.

6.5.1.4 Other Factors

Several other factors are able to modulate reproduction success. They can be divided into three main categories: (i) environmental (e.g. temperature, light intensity), (ii) populational (e.g. domestication level, females weight) and (iii) nutritional (e.g. feed composition, feeding rate). Up to now, almost no studies investigated the impact of these factors on percid reproduction success. Moreover, it is commonly known that fish are susceptible to stressors, such as inevitable manipulations, however, this aspect was hardly studied in the percids. And the recent study proved in pikeperch that handling affects the fertilization and hatching rates, or even may inhibit the spawning (Sarameh et al. 2012). Therefore, the stress should be also considered from the perspective of the modulating factors affecting gametes quality. This creates the need to consider the stress level as a factor affecting egg quality and thus as an important research priority in this field.

6.5.2 Egg and Larval Manipulation and Incubation

As for breeders, egg and larval incubation conditions are of first importance for the proper control of fish life cycle. It includes physicochemical parameters of the water as the temperature, salinity, oxygen and pH, handling or incubation conditions (zoug bottles, tray-type incubators, etc.), the nature and the quality of substratum and the food supply for larvae. A detailed description of egg incubation optimal conditions is given in the Chap. 4. Some works investigated the effect of embryonic and larval rearing conditions on their quality. The optimal temperature of incubation has largely been documented in *Perca fluviatilis* and *Perca flavescens* with various data. Indeed, incubation temperature ranges can be 12–20 °C (Wang and Eckmann 1994), 8–16 °C (Swift 1965), 12–16 °C (Kokurewicz 1969) or 10–16 °C (Guma'a 1978) for *P. fluviatilis* and 10–18 °C (Hokanson and Kleiner 1974) for *P. flavescens*. Moreover, for the Eurasian perch, the survival rate observed at 22 °C was 75.7 %, 7.5 %, 2.3 % and 0 % respectively in Wang and Eckmann (1994), Hokanson and Kleiner (1974), Swift (1965) and Guma'a (1978) studies. This difference can be explained by fish population adaptation to various spawning temperatures depending to their natural living conditions (Wang and Eckmann 1994). For example, populations used by Swift and Guma'a studies had been taken in Windermere (NW of England) with an incubation temperature ranging from 9 to 18 °C. On the other hand, Wang fish population from the Lake Constance (at the border between

Germany, Austria and Switzerland, in the Alps) with temperature ranging from 12 to 18 °C. This hypothesis has been confirmed with northern European *P. fluviatilis* populations (Sandstrom et al. 1997). Moreover, there were some differences of susceptibility of developmental stages to the temperature (Wang and Eckmann 1994). As shown for the temperature, it is important to optimize egg and larval incubation conditions not only for every species but also depending on the fish populations used in the fisheries.

6.6 Conclusion and Prospects

In conclusion, it appears that in most of percid fish species the reproduction practices still need to be more carefully studied. In some species as Eurasian perch, out-of-season reproduction of the RAS-reared broodstock is possible but the reproduction success still needs to be improved since the egg and larvae quality are still in many cases variable, although the same photo-thermal and feeding regimes are applied. In other species as the pikeperch, hatchery practices are not well established. It may be presumed that with clear egg and larvae quality indicators the development of aquaculture of this species will progress much faster. Especially, that in every case that it is important (i) to define determining and modulating factors regulating fish life cycle (ii) to properly describe the developmental process of each species to determine the proper timing to reach key steps (iii) to have a large overview of developmental failures (mortality, deformities) that can occur in the studied species (this knowledge could help to define some level of developmental impairment and thus categories of ova, embryos or larval qualities) and (iv) to choose relevant quality indicators specific to the studied species and depending upon the stage that need to be checked (ova, embryo, larva). Current methods to evaluate egg and larval quality are still infrequent. All of them are morphological and can either indicate high or low probability of the egg development but fail to predict the nature of the developmental impairment. Biochemical indicators seem promising but further work need to be done. Once these questions answered (either successively or in parallel), it could help to improve rearing practices of fish species and may help, in the future, to define a methodology to study new candidates for the diversification of aquaculture.

References

Abdulfatah A, Fontaine P, Kestemont P, Gardeur JN, Marie M (2011) Effects of photothermal kinetics and amplitude of photoperiod decrease on the induction of the reproduction cycle in female Eurasian perch *Perca fluviatilis*. Aquaculture 322–323:169–176

Abdulfatah A, Fontaine P, Kestemont P, Milla S, Marie M (2013) Effects of the thermal threshold and the timing of temperature reduction on the initiation and course of oocyte development in cultured female of Eurasian perch *Perca fluviatilis*. Aquaculture 376–379:90–96

Abi-Ayad S-ME-A, Kestemont P, Mélard C (2000) Dynamics of total lipids and fatty acids during embryogenesis and larval development of Eurasian perch (*Perca fluviatilis*). Fish Physiol Biochem 23:233–243

Aegerter S, Jalabert B (2004) Effects of post-ovulatory oocyte ageing and temperature on egg quality and on the occurrence of triploid fry in rainbow trout, *Oncorhynchus mykiss*. Aquaculture 231:59–71

Bermudes M, Ritar AJ (1999) Effects of temperature on the embryonic development of the striped trumpeter (*Latris lineata* Bloch and Schneider, 1801). Aquaculture 176:245–255

Black DE, Phelps DK, Lapan RL (1988) The effect of inherited contamination on egg and larval winter flounder, *Pseudopleuronectes americanus*. Mar Environ Res 25:45–62

Blanchard G, Druart X, Kestemont P (2005) Lipid content and fatty acid composition of target tissues in wild *Perca fluviatilis* females in relation to hepatic status and gonad maturation. J Fish Biol 66:73–85

Bobe J, Labbé C (2010) Egg and sperm quality in fish. Gen Comp Endocrinol 165:535–548

Bonnet E, Fostier A, Bobe J (2007) Characterization of rainbow trout egg quality: a case study using four different breeding protocols, with emphasis on the incidence of embryonic malformations. Theriogenology 67:786–794

Bromage N, Jones J, Randall C, Thrush M, Davies B, Springate J, Duston J, Barker G (1992) Broodstock management, fecundity, egg quality and the timing of egg production in the rainbow trout (*Oncarhynchus mykiss*). Aquaculture 100:141–166

Brooks S, Tyler CR, Sumpter JP (1997) Egg quality in fish: what makes a good egg? Rev Fish Biol Fish 7:387–416

Cameron P, Berg J, Dethlefsen V, Von Westernhagen H (1992) Developmental defects in pelagic embryos of several flatfish species in the Southern North sea. Neth J Sea Res 29:239–256

Castets MD, Schaerlinger B, Silvestre F, Gardeur JN, Dieu M, Corbier C, Kestemont P, Fontaine P (2012) Combined analysis of *Perca fluviatilis* reproductive performance and oocyte proteomic profile. Theriogenology 78:432–442

Cerda J, Carillo M, Zanuy S, Ramos J, Delahiguera M (1994) Influence of nutritional composition of diet on sea bass. Dicentrarchus labrax L. reproductive performance on egg and larval quality. Aquaculture 128(3–4):345–361

Cooper JE (1978) Eggs and larvae of the logperch, Percina caprodes (Rafinesque). Am Midland Nat 2:257–269

Cooper JE (1979) Description of eggs and larvae of fantail (Etheostoma flabellare) and rainbow (E. caerleum) darters from Lake Erie Tributaries. Trans Am Fish Soc 108:46–56

Cotelli F, Andronico F, Brivio M, Lamia CL (1988) Structure and composition of the fish egg chorion (Carassius auratus). J Ultrastruct Mol Struct Res 99:70–78

Coward K, Bromage NR, Hibbitt O, Parrington J (2002) Gamete physiology, fertilization and egg activation in teleost fish. Rev Fish Biol Fish 12:33–58

Czesny S, Dąbrowski K (1998) The effect of egg fatty acid concentration on embryo viability in wild and domesticated walleye (*Stizostedion lucioperca*). Aquat Living Resour 11:371–378

Czesny S, Rinchard J, Dabrowski K (2005a) Intrapopulation variation in egg lipid and fatty acid composition and embryo viability in a naturally spawning walleye population from an inland reservoir. N Am J Fish Manag 25:122–129

Czesny S, Graeb BD, Dettmers JM (2005b) Ecological consequences of swim bladder noninflation. Trans Am Fish Soc 134:1011–1020

Dabrowski K, Ciereszko A, Ramseyer L, Culver D, Kestemont P (1994) Effects of hormonal treatment on induced spermiation and ovulation in the yellow perch (*Perca flavescens*). Aquaculture 120:171–180

Dabrowski K, Ciereszko RE, Ciereszko A, Toth GP, Christ SA, El-Saidy D, Ottobre JS (1996) Reproductive physiology of yellow perch (*Perca flavescens*): environmental and endocrinological cues. J Appl Ichthyol 12:139–148

Dabrowski K, Czesny S, Kolkovski S, Lynch WE Jr, Bajer P, Culver DA (2000) Intensive culture of walleye larvae produced out of season and during regular season spawning. N Am J Aquacult 62:219–224

Demska-Zakęś K, Zakęś Z, Roszuk J (2005) The use of tannic acid to remove adhesiveness from pikeperch, *Sander lucioperca*, eggs. Aquacult Res 36:1458–1464

Dietrich MA, Dabrowski K, Arslan M, Ware K, Van Tassell J (2012) Quantifying quality attributes of walleye eggs prior to fertilization – impact of time of 2 ovulation and gametes storage. J Great Lakes Res 38:403–584

Dubois JP, Gillet C, Bonnet S, Chevalier-Weber Y (1996) Correlation between the size of mature female perch (*Perca fluviatilis* L.) and the width of their egg strands in Lake Geneva. Ann Zool Fennici 33:417–420

Essner JJ, Amack JD, Nyholm MK, Harris EB, Yost HJ (2005) Kupffer's vesicle is a ciliated organ of asymmetry in the zebrafish embryo that initiates left-right development of the brain, heart and gut. Development 132:1247–1260

Falahatkar B, Poursaeid S (2014) Effects of hormonal manipulation on stress responses in male and female broodstocks of pikeperch *Sander lucioperca*. Aqua Int 30(12):1102–1109

Fishbase website: http://www.fishbase.org

Fontaine P, Pereira C, Wang N, Marie M (2006) Influence of pre-inductive photoperiod variations on Eurasian perch *Perca fluviatilis* broodstock response to an inductive photothermal program. Aquaculture 255:410–416

Formicki K, Smaruj I, Szulc J, Winnicki A (2009) Microtubular network of the gelatinous egg envelope within the egg ribbon of European perch, *Perca fluviatilis* L. Acta Ichthyol Piscat 39:147–151

Guma'a SA (1978) The effects of temperature on the development and mortality of eggs of perch, *Perca fluviatilis*. Freshw Biol 8:221–227

Haddy JA, Pankhurst NW (2000) The effects of salinity on reproductive development, plasma steroid levels, fertilisation and egg survival in black bream *Acanthopagrus butcheri*. Aquaculture 188:115–131

Hamza N, Mhetli M, Khemis IB, Cahu C, Kestemont P (2008) Effect of dietary phospholipid levels on performance, enzyme activities and fatty acid composition of pikeperch (*Sander lucioperca*) larvae. Aquaculture 275:274–282

Hansen JH, Puvanendran V (2010) Fertilization success and blastomere morphology as predictors of egg and juvenile quality for domesticated Atlantic cod, *Gadus mor-hua*, broodstock. Aquacult Res 41:1791–1798

Hart NH, Yu S-F (1980) Cortical granule exocytosis and cell surface reorganization in eggs of *Brachydanio*. J Exp Zool 213:137–159

Henrotte E, Overton J, Kestemont P (2008) Effects of dietary n-3 and n-6 fatty acids levels on egg and larval quality of Eurasian perch. Cybium 32:271–272

Henrotte E, Mandiki RSNM et al (2010) Egg and larval quality, and egg fatty acid composition of eurasian perch breeders (*Perca fluviatilis*) fed different dietary DHA/EPA/AA ratios. Aquac Res 41(9):53–61

Hokanson KEF, Kleiner CF (1974) Effects of constant and rising temperatures on survival and developmental rates of embryonic and larval yellow perch, *Perca flavescens* (Mitchill). In: The early life history of fish, Springer-Verlag, Berlin, pp 437–448

Hubbs CL (1985) Darter reproductive seasons. Copeia 1:56–68

Ivan LN, Rutherford ES, Riseng C, Tyler JA (2010) Density, production, and survival of walleye (Sander vitreus) eggs in the Muskegon River, Michigan. J Great Lakes Res 36:328–337

Jezierska B, Lugowska K, Witewska M, Sarnowski P (2000) Malformations of newly hatched common carp larvae. EJPAU 3:#2

Kestemont P, Cooremans J, Abi-Ayad A, Mélard C (1999) Cathepsin L in eggs and larvae of perch perca fluviatilis: variations with developmental stage and spawning period. Fish Physiol Biochem 21(1):59–64

Kestemont P, Xueliang X, Hamza N, Maboudou J, Toko II (2007) Effect of weaning age and diet on pikeperch larviculture. Aquaculture 264:197–204

Kimmel CB, Ballard WW, Kimmel SR, Ullmann B, Schilling TF (1994) Stages of embryonic development of the zebrafish. Dev Dyn 203:253–310

Kjørsvik E, Mangor-Jensen A, Holmefjord I, Blaxter JHS, Southward AJ (1990) Egg quality in fishes. Advan Mar Biol 26:71–113

Kokurewicz B (1969) The influence of temperature on the embryonic development of the perches: *Perca fluviatilis* L. and *Lucioperca lucioperca* (L.). Zool Pol 19:47–67

Konstantinov KG (1957) Comparative analysis of the morphology and biology of perch, pikeperch and Volga pike-perch during several developmental stages. Tr Inst Morfol Zhovoth Akad Nauk SSSR 16:181–236

Korzelecka A, Bonisławska M, Winnicki A (1998) Structure, size and spatial distribution of perch (*Perca fluviatilis* L.) egg components during incubation. EJPAU 1:#05

Kovac V (1992) Early development of the yellow pope, *Gymnocephalus schraester*. Fol Zool 41(4):365–377

Kovac V (1993a) Early development of the Balon's ruff, *Gymnocephalus baloni* Holcik and Hensel, 1974. Fol Zool 42(4):349–360

Kovac V (1993b) Early development of ruff, *Gymnocephalus cernuus*. Fol Zool 42(3):269–280

Kovac V (1994) Early ontogeny of three Gymnocephalus species (Pisces: Percidae) reflections on the evolution of the genus. Environ Biol Fish 40:241–253

Kovac V (2000) Early development of *Zingel streber*. J Fish Biol 57:1381–1403

Kowalska A, Zakęś Z, Demska-Zakęś K (2006) The impact of feeding on the results of rearing larval pikeperch, *Sander lucioperca* (L.) with regard to the development of the digestive tract. EJPAU Fish 9(2)

Krist'an J, Alavi SMH, Stejskal V, Policar T (2013) Hormonal induction of ovulation in pikeperch (*Sander lucioperca* L.) using human chorionic gonadotropin (hCG) and mammalian GnRH analogue. Aquac Int 21:811–818

Kucharczyk D, Kujawa R, Mamcarz A, Skrzypczak A (1998) Induced spawning in perch, *Perca fluviatilis* L., using FSH+LH with pimozide or metoclopramide. Aquac Int 29:131–136

Kucharczyk D, Kestemont P, Mamcarz A (2007) Artificial reproduction of pikeperch. Mercurius, Olsztyn, p 80

Kupren K, Żarski D, Krejszeff S, Kucharczyk D, Targońska K (2011) Effect of stocking density on growth, survival and development of asp *Aspius aspius* (L.), ide *Leuciscus idus* (L.) and chub *Leuciscus cephalus* (L.) larvae during initial rearing under laboratory conditions. Ital J Anim Sci 10:178–184

Lee KW, Webb SE, Miller AL (1999) A wave of free cytosolic calcium traverses zebrafish eggs on activation. Dev Biol 214:168–180

Ługowska K (2007) The effect of cadmium and cadmium/copper mixture during the embryonic development on deformation of common carp larvae. Electron J Ichthyol 2:46–60

Ługowska K, Kubik J (2011) Malformations of barbel (*Barbus barbus*) larvae induced by copper and cadmium. Ochr Środ Zas Nat 48:170–178

Ługowska K, Sarnowski P (2011) Heads or tails – fish hatching. Acta Ichthyol Piscat 41:13–17

Malison JA, Procarione LS, Kayes TB, Hansen JF, Held JA (1998) Induction of out-of-season spawning in walleye *Stizostedion vitreum*. Aquaculture 163:151–161

Mansour N, Lahnsteiner F, Patzner RA (2009) Physiological and biochemical investigations on egg stickiness in common carp. Anim Reprod Sci 114:256–268

Mansueti AJ (1964) Early development of the yellow perch, *Perca flavescens*. Chesap Sci 5:46–66

McElman JF, Balon EK (1979) Early ontogeny of walleye, *Stizostedion vitreum*, with steps of saltatory development. Environ Biol Fish 4:309–348

Mendelson TC, Imhoff VE, Iovine MC (2006) Analysis of early embryogenesis in rainbow and banded darters (Percidae: Etheostoma) reveals asymmetric postmating barrier. Environ Biol Fish 76:351–360

Migaud H, Gardeur J, Kestemont P, Fontaine P (2004) Off-season spawning of eurasian perch *Perca fluviatilis*. Aquac Int 12:87–102

Migaud H, Wang N, Gardeur J, Fontaine P (2006) Influence of photoperiod on reproductive performances in eurasian perch *Perca fluviatilis*. Aquaculture 252:385–393

Migaud H, Bell G, Cabrita E, McAndrew B, Davie A, Bobe J, Herraez MP, Carrillo M (2013) Gamete quality and broodstock management in temperate fish. Rev Aquacult 5:S194–S223
Minin AA, Ozerova SG (2008) Spontaneous activation of fish eggs is abolished by protease inhibitors. Russ J Dev Biol 38:293–296
Moodie GEE, Loadman N, Wiegand MD, Mathias JA (1989) Influence of egg characteristics on survival, growth and feeding in larval walleye (*Stizostedion vitreum*). Can J Fish Aquat Sci 46:516–521
Overton J, Paulsen H, Kucharczyk D, Szczerbowski A (2008) Bornholm Salmon hatchery: control of out-of-season spawning of Eurasian perch. In: Fontaine P, Kestemont P, Teletchea F, Wang N (eds) Percid fish culture, from research to production. Presses Universitaires de Namur, pp 40–43
Paine MD (1984) Ecological and evolutionary consequences of early ontogenies of darters (Etheostomatini). Environ Biol Fish 11(2):97–106
Paine MD, Balon EK (1984a) Early development of the rainbow darter, *Etheostoma caeruleum*, according to the theory of saltatory ontogeny. Environ Biol Fish 11(4):277–299
Paine MD, Balon EK (1984b) Early development of the northern logperch, *Percina caprodes semifasciata*, according to the theory of saltatory ontogeny. Environ Biol Fish 11(3):173–190
Palińska K, Kucharczyk D, Żarski D, Krejszeff S, Targońska K, Kupren K (2011) Deformation of newly hatched larvae in ide *Leuciscus idus* L. In: Mediterranean aquaculture 2020, Aquaculture Europe 2011 conference, 18–21 Oct, Rhodes, pp 817–818
Pavlov DA, Emel'yanova NG (2008) Morphological criteria of egg quality in marine fishes: activation and cleavage of eggs of *Zebrasoma scopas* (Acanthuridae). J Ichthyol 48:533–548
Probst WN, Stoll S, Hofmann H, Fischer P, Eckmann R (2009) Spawning site selection by eurasian perch (*Perca fluviatilis* L.) in relation to temperature and wave exposure. Ecol Freshw Fish 18:1–7
Quagio-Grassiotto I, Guimarães ACD (2003) Follicular epithelium, theca and egg envelope formation in *Serrasalmus spilopleura* (Teleostei, Characiformes, Characidae). Acta Zool (Stockholm) 84:121–129
Riehl R, Patzner RA (1998) Mini-review: the modes of eggs attachment in teleost fishes. Ital J Zool 65:415–420
Rinchard J, Dabrowski K, Tassell JJ, Stein RA (2005) Optimization of fertilization success in *Sander vitreus* is influenced by the sperm: egg ratio and ova storage. J Fish Biol 67:1157–1161
Sandstrom O, Abrahamsson I, Andersson J, Vetemaa M (1997) Temperature effects on spawning and egg development in Eurasian perch. J Fish Biol 51:1015–1024
Sarameh S, Falahatkarb B, Takamic GA, Efatpanahd I (2012) Effects of different photoperiods and handling stress on spawning and reproductive performance of pikeperch *Sander lucioperca*. Anim Reprod Sci 132:213–222
Schlumpberger W, Schmidt K (1980) Vorläufiger Stand der Technologie zur Aufzucht von vorgestreckten Zander (*Stizostedion lucioperca*). Z Binnenfisch DDR 27:284–286
Schneider JC (1984) Yellow perch maturity and fecundity as a function of age and growth. Michigan department of natural resources- Fisheries division- Fisheries Research Report. N°1915
Serns SL (1982) Walleye fecundity, potential egg deposition, and survival from egg to fall young-of-year in Escanaba Lake, Wisconsin, 1979–1981. N Am J Fish Manag 4:388–394
Shrader F, Shrader SP (1922) Mortality in pike-perch eggs in hatcheries. Appendix V to the report of the VS Commissioner of Fisheries for 1922. BF Doc No 926, Washington, 11 p
Singh PB, Sahu V, Singh V, Nigam SK, Singh HK (2008) Sperm motility in the fishes of pesticide exposed and from polluted rivers of Gomti and Ganga of north India. Food Chem Toxicol 46:3764–3769
Sulistyo I, Rinchard J, Fontaine P, Gardeur J, Capdeville B, Kestemont P (1998) Reproductive cycle and plasma levels of sex steroids in female eurasian perch *Perca fluviatilis*. Aquat Living Resour 11:101–110
Skovring M, Christensen EAF, Carl H, Jacobsen L, Moller PR (2013) Marine spawning sites of Perca fluviatilis revealed by oviduct-inserted acoustic transmitters. Aquat Biol 19:201–206

Swift DR (1965) Effect of temperature on mortality and rate of development of the eggs of the pike (*Esox lucius* L.) and the perch (*Perca fluciatilis* L.). Nature 206:528

Taranger GL, Carrillo M, Schulz RW, Fontaine P, Zanuy S, Felip A, Weltzien FA, Dufour S, Karlsen O, Norberg B, Andersson E, Hausen T (2010) Control of puberty in farmed fish. Gen Comp Endocrinol 165(3):483–515

Teletchea F, Fontaine P (2011) Particularities of early life stages in temperate freshwater fish species: comparisons with marine species and implications for aquaculture practices. Aquac Res 42:630–654

Teletchea F, Gardeur JN, Kamler E, Fontaine P (2009) The relationship of oocyte diameter and incubation temperature to incubation time in temperate freshwater fish species. J Fish Biol 74:652–668

Tibblin P, Koch-Schmidt P, Larsson P, Stenroth P (2012) Effects of salinity on growth and mortality of migratory and resident forms of Eurasian perch in the Baltic Sea. Ecol Freshw Fish 21:200–206

Treasurer JW (1983) Estimates of egg and viable embryo production in a lacustrine perch, *Perca fluviatilis*. Env Biol Fish 8:3–16

Tyler CR, Sumpter JP (1996) Oocyte growth and development in teleosts. Rev Fish Biol Fish 6:287–318

Vetemaa M, Saat T (1996) Effects of salinity on the development of fresh-water and brackish-water ruffe *Gymnocephalus cernuus* (L.) embryos. Ann Zool Fenn 33:687–691

Von Westernhagen H, Hoar WS, Randall DJ (1988) 4 sublethal effects of pollutants on fish eggs and larvae. Fish Physiol 11:253–346

Wang N, Eckmann R (1994) Effects of temperature and food density on egg development, larval survival and growth of perch (*Perca fluvia tilis* L.). Aquaculture 122:323–333

Wang N, Gardeur JN, Henrotte E, Marie M, Kestemont P, Fontaine P (2006) Influence of pre-inductive photoperiod variations on Eurasian perch *Perca fluviatilis* broodstock response to an inductive photothermal program. Aquaculture 261:706–714

Wang N, Mandiki SNM, Henrotte E, Bouyahia AG, Mairesse G, Rougeot C, Melard C, Kestemont P (2009) Effect of partial or total replacement of forage fish by a dry diet on the quality of reproduction in pikeperch, *Sander lucioperca*. Aquac Res 40:376–383

Wang N, Teletchea F, Kestemont P, Milla S, Fontaine P (2010) Photothermal control of the reproductive cycle in temperate fishes. Rev Aquac 2:209–222

Wiegand MD (1996) Composition, accumulation and utilization of yolk lipids in teleost fish. Rev Fish Biol Fisher 6:259–286

Zakęś Z (2007) Out-of-season spawning of cultured pikeperch [*Sander lucioperca* (L.)]. Aquac Res 38:1419–1427

Żarski D, Palińska K, Targońska K, Bokor Z, Kotrik L, Krejszeff S, Kupren K, Horvath A, Urbanyi B, Kucharczyk D (2011a) Oocyte quality indicators in Eurasian perch, *Perca fluviatilis* L., during reproduction under controlled conditions. Aquaculture 313:84–91

Żarski D, Bokor Z, Kotrik L, Urbányi B, Horváth A, Targońska K, Krejszeff S, Palińska K, Kucharczyk D (2011b) A new classification of a preovulatory oocyte maturation stage suitable for the synchronization of ovulation in controlled reproduction of Eurasian perch, *Perca fluviatilis* L. Reprod Biol 3:194–209

Żarski D, Kucharczyk D, Targońska K, Palińska K, Kupren K, Fontaine P, Kestemont P (2012a) A new classification of pre-ovulatory oocyte maturation stages in pikeperch, Sander lucioperca (L.), and its application during artificial reproduction. Aquacult Res 43:713–721

Żarski D, Krejszeff S, Palinska K, Targońska K, Kupren K, Fontaine P, Kestemont P, Kucharczyk D (2012b) Cortical reaction as an egg quality indicator in artificial reproduction of pikeperch, *Sander lucioperca*. Reprod Fertil Dev 24:843–850

Żarski D, Krejszeff S, Horváth A, Bokor Z, Palińska K, Szentes K, Łuczyńska J, Targońska K, Kupren K, Urbányi B, Kucharczyk D (2012c) Dynamics of composition and morphology in oocytes of Eurasian perch, Perca fluviatilis L., during induced spawning. Aquaculture 364–365:103–110

Part III
Early Life Stages: Development, Metabolism, and Husbandry

Chapter 7
Development of the Sense Organs in Percid Fishes

M. Kamaszewski and T. Ostaszewska

Abstract Morphology and arrangement of various receptors in Teleostei indicate trophic and environmental preferences of different species. In the Percidae family, the sensory system begins to develop during embryogenesis and evolves over the larval and juvenile stages. The olfactory placodes develop between 26 (*Gymnocephalus* sp.) and 78 (walleye, *Sander vitreus*) hours post fertilization (hpf). However, the olfactory epithelium becomes fully developed at the age of about 1 month, as in the pikeperch (*Sander lucioperca*). During the ontogeny of fish, taste buds develop later than the olfactory system. In pike-perch juveniles, the first taste buds appear 13 days post hatching, primarily in the mouth and the gills, but later on they become visible in other parts of the body. During embryonic development of percids, the eye is the first sense organ to form. *Gymnocephalus* sp. show well developed eyes already after 26 hpf, while in walleye embryos the optic vesicles are fully formed after 70 hpf. Fish maintain body balance thanks to their inner ear (labyrinth), while the movement and vibration in the surrounding water is detected by canal neuromasts of the lateral line and superficial neuromasts of the skin. In the embryos of different percid species auditory vesicles appear after 26–70 h of development, while the lateral line – after 103 h.

Keywords Sensory organs • Ontogeny • Olfactory placode • Eye • Neuromasts

7.1 Introduction

Sense organs of fish detect environmental stimuli and provide information about any changes of ambient conditions. Various stimuli received by sensory cells are transformed into the membrane action potentials and directed via neurons to the appropriate brain regions. Taking into consideration type of stimulus and physiological properties sensory cells of sense organs can be divided into: mechanoreceptors

M. Kamaszewski (✉) • T. Ostaszewska
Division of Ichthyobiology and Fisheries, Faculty of Animal Science,
Warsaw University of Life Sciences, Ciszewskiego 8, 02-786 Warsaw, Poland
e-mail: maciej_kamaszewski@sggw.pl

P. Kestemont et al. (eds.), *Biology and Culture of Percid Fishes*,
DOI 10.1007/978-94-017-7227-3_7

(detecting mechanical stimuli), thermoreceptors (detecting thermal stimuli), chemoreceptors (detecting chemical stimuli), photoreceptors (detecting electromagnetic waves), and electroreceptors (detecting electric stimuli) (Kilarski 2012). Morphology and arrangement of various receptors in Teleostei indicate trophic and environmental preferences of different fish species.

7.2 The Olfactory Organ

Fish detect ambient chemical changes using olfactory cells, the bipolar primary neurons. Olfactory organ develops as first chemosensory organ during fish ontogenesis. Degree of olfactory organ development depends on environmental conditions, while the surface area of olfactory epithelium indicates olfactory acuity (Kuciel et al. 2013). Fish olfactory organ provides invaluable information not only from proximate vicinity but also from distant environment, which is inaccessible for other senses. Olfactory sense involves three factors: the source, signal and receiver. Fish are able to detect directional signals concerning direct threat or water contamination (Tierney et al. 2010). Formation and morphology of olfactory organs of vertebrates have been extensively studied in some taxa (Zeiske et al. 2003). The olfactory epithelium of vertebrates develops from a pair of olfactory placodes present in anterior head region. Placodes are visible as thickened ectoderm area and develop during early embryogenesis (Zeiske et al. 2003). This structure, apparently homogenous and consisting of identical ectodermic cells, detaches from the neural plate in its anterior region (Whitlock 2004; Camacho et al. 2010). Before development of olfactory placodes, the cells of presumptive olfactory epithelium are covered by epidermis and differentiate into neurons. Their axons penetrate across basal lamina and grow towards the brain where they reach olfactory bulb (Hansen and Zeiske 1993; Camacho et al. 2010). In fish, similarly as in other vertebrates, usually a pair of olfactory organs occurs located in rostral part of snout. The olfactory epithelium lines the olfactory rosette. During the embryonic development, olfactory sensory epithelium (OE) develops from the thickened ectoderm in front of the embryonic neural tube. Various types of olfactory sensor cells differentiate already during early ontogenesis.

Olfactory epithelium in various fish species may differ but always consists of three types of cells: receptor cells (neurons), supporting cells and basal cells (Hara 1994). Fish belonging to Perciformes order show high variability of olfactory organ structure. First perciform fishes appeared in early Cenozoic, and now about 7,800 species belonging to 150 families are known (Hanse and Zielinski 2005). In Perciform fishes olfactory organ may be present as olfactory rosette, e.g. in Eurasian perch *Perca fluviatilis* (Døving et al. 1977), as large olfactory chamber lined with olfactosensory epithelium, e.g. in *Neogobius melanostomus* (Belanger et al. 2003), or as a channel with islets of olfactosensory epithelium, e.g. in *Periophthalmus barbarus* (Kuciel et al. 2011). Despite extensive studies of olfactory organs in some Perciform species, their detailed morphology in various groups belonging to this order is still poorly known (Kuciel et al. 2013). Craig (2000) described the olfactory organs in yellow perch (*Perca flavescens*) as oval or rosette shaped and nearly fill two cartilaginous chambers located in the front part of the head. Olfactory chamber of fish is covered with lining epithelium, and cylindrical olfactory epithelium occurs only at the very bottom of it. Lining of the olfactory chamber together with sensory epithelium is folded forming a rosette. The most commonly observed type of rosette shows a single longitudinal fold with transversal folds at both sides, so called olfactory lamellae. Eurasian perch (*Perca fluviatilis*) has olfactory rosettes with about 20 lamellae, kinociliated cells are present mainly in the sensory epithelium, but may also be present in the non-sensory epithelium (Cox 2008). In Perciform fishes, similarly as in Acipenseriforms, Cypriniforms, and Salmonids, the cells of olfactory epithelium are rich in mitochondria and show well developed tubular reticulum. The presence of chloride cells, i.e. cells with a clearly seen ion-transporting specialization was also observed (Ruzhinskaya et al. 2001). According to Hansen and Zielinski (2005) and Hara (2011a), the olfactory epithelium contains three different types of olfactory sensory neurons (OSNs): ciliated, microvillous and crypt cells. Basal cells facilitate growth and regeneration of the olfactory epithelium. Supporting cells and ciliated non-sensory cells are scattered among the OSNs. Olfactory epithelium may also contain goblet cells which are always present in non-sensory epithelium of the nasal cavity.

Olfactory placodes appear very early during embryonic development of Percid fish. Kovac (1994) reported in fishes of genus *Gymnocephalus* olfactory placodes already in 26 h post fertilization. During embryonic development of walleye (*Sander vitreus*) at 15 °C olfactory placodes develop between the 70 and 78 h post fertilization (hpf) (McElman and Balon 1979). In pikeperch (*Sander lucioperca*) embryos developing at 18 °C, Jara and Szymoniewski (1953) observed the olfactory placodes already in 48.5 h. At hatching olfactory pits of pikeperch larvae are visible in anterior part of head (Fig. 7.1a). In 5 days post-hatching (dph) the ciliate cells are distinct in the sensory epithelium (Fig. 7.1b). At the age of about 1 month olfactory epithelium is fully developed (Fig. 7.1c).

The sensitivity of fish olfaction is odorant dependent. Generally, fish are able to detect the presence of substances dissolved in water at concentrations ranging down to the parts per billion (10^{-9}) or trillion (10^{-12}) (Belanger et al. 2006). These substances are various natural chemicals such as bile salt, amino acids, sex steroids or prostaglandins (Tierney et al. 2010; Hara 2011b).

7.3 The Taste Buds

In vertebrates, the taste buds are derived from the endoderm, as opposed to other receptor cells. During the ontogeny of fish, these chemoreceptors develop later than the cells of the olfactory system (Hara 2011c). In fish, the taste buds are usually situated in the oral cavity, larynx, palate, oesophagus, gills and skin, but they also may appear on the lips, barbels and fins (Kasumyan and Døving 2003; Hara 2011c).

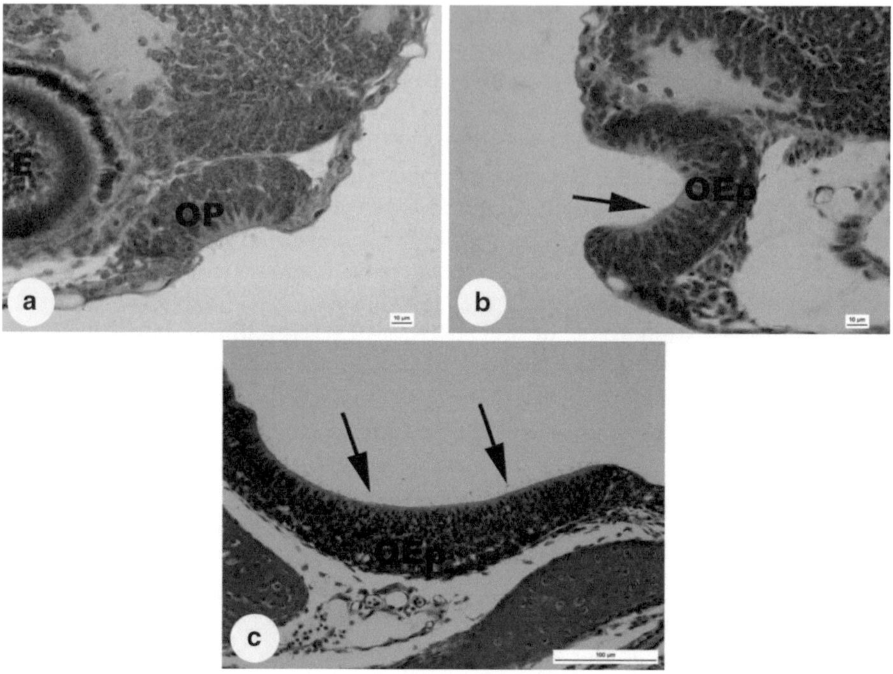

Fig. 7.1 (**a**) Longitudinal section of pike-perch head at hatching. Olfactory pits (*OP*) of pikeperch larvae are visible in anterior part of head (*E-eye*). Scale bar 10 μm. (**b**) Longitudinal section of pike-perch olfactory pits at 5 days post hatching. The ciliate cells (*arrow*) are visible in the olfactory epithelium (*OEp*). Scale bar 10 μm. (**c**) Longitudinal section of pike-perch olfactory pits at 30 days post hatching. The ciliate cells (*arrows*) are visible in the fully developed olfactory epithelium (*OEp*). Scale bar 100 μm

Taste bud distribution indicates the ecological niche and trophic preferences of the given fish species.

Taste buds are incorporated in the epithelium or lie on the dermal papilla. They consist of three types of cells: (1) those ending with a rod-shaped apical protrusion or cilia, (2) those ending with microvilli (also known as supporting cells) and (3) basal cells. The type-1 and (probably) type-2 cells are receptor cells. Type-3 cells are located on the bottom of the taste buds and are connected to both types of receptor cells by desmosomes. Due to the high cytoplasmatic serotonin content and the synaptic connections to receptor cells and nerve fibers, basal cells are probably responsible for the modulation of gustatory activity (Hara 2011c). According to Hara (1994), electrophysiological thresholds for the stimulatory amino acids range from 10^{-10} M for the channel catfish to 10^{-6} M for the puffer. The highest stimulation of receptors can be observed few seconds after stimulation.

In developing pike-perch juveniles, the first taste buds appear 13 days post hatching (Ostaszewska 2005). Primarily taste buds appear in the mouth and the gills; later on they become visible in other parts of the body. The number of chemoreceptor cells increases with age (Fig. 7.2).

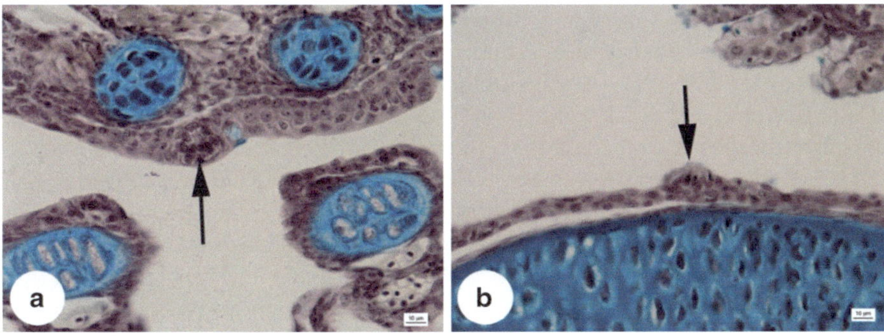

Fig. 7.2 (**a**) Longitudinal section of pike-perch head, 13 days post hatching. (**b**) Longitudinal section of pike-perch head, 30 days post hatching. *Arrow* – taste buds. Scale bars 10 μm

7.4 The Eye

Visual telereceptor is an organ sensitive to electromagnetic waves of various lengths. Vision is the most important sense in most larval fish. It plays a key role in feeding, spatial orientation, and predator avoidance (Rodriguez and Gisbert 2001).

Retina of teleost fishes, similarly as in other vertebrates, is unified with the optic nerve, from both, developmental and functional point of view. It consists of two types of photosensitive structures: rods and three types of cones (Kilarski 2007). The rods are photosensitive receptors enabling vision in black and white at low light intensity. In contrast, cones are photosensitive receptors enabling color vision in bright light. Therefore nocturnal fishes, e.g. burbot *Lota lota* show higher density of rods compared to cones.

The eye develops from the second brain vesicle (diencephalon) that invaginates towards the skin forming the visual cup and then develops into retina and optic nerve, while the lens develops from skin epithelium. Retina and lens form as first parts of the eye. Choroid and sclera develop from the cells that detached from presumptive meninges and migrated along the optic nerve onto the existing retina and lens. Vitreous body, another constituent of fish eye is probably derived from retinal cells. The lens in teleost fishes is almost spherical and hard, unable to change the shape (Grodziński 1981). Vertebrate eyes continue to grow postembryonically, and growth involves rather extension of already existent nerve tissue than neurogenesis (Fernald 1985). In teleosts and amphibians the number of retinal cells may increase with increase of the eye and whole body size (Fernald 1985).

The wall of the eye consists of three layers: external (sclera or fibrous tunic), internal (choroid) and retina. The sclera is a hard and adaptable layer being a skeleton of the eye. Its inner surface is lined with strongly pigmented choroid being a kind of *camera obscura* that shields the retina from light. Retina itself is a proper photosensitive layer, which is adjacent to choroid and iris (Sattari et al. 2012). The optic system of the eye includes cornea, which does not play any important role in fish vision, pupil, lens, and vitreous body (Grodziński 1981).

During larval development of Siberian sturgeon (*Acipenser baerii*) first retinal layer was observed on the 3 dph as unilayered pigmented epithelium (Ostaszewska and Dabrowski 2009). In 5–6 days-old fish a crystalline lens was found consisting of multilayered cuboidal epithelium and nucleated fiber cells. The lens of acipenserid fishes is connected to the cornea, which consists of connective tissue and lamellar cells. The retina of Acipenseridae shows seven layers: (1) the pigment epithelium organized as a single layer of cuboidal cells, (2) the outer nuclear layer formed by two types of photoreceptor cells, rods and single cones, (3) the outer plexiform layer organized as a thin reticular tissue, (4) the inner nuclear layer (with horizontal, amacrine and bipolar cells, (5) the inner plexiform layer consisting of reticular tissue, (6) the ganglion cell layer composed of a row of spherical cells, (7) the nerve fiber layer, which consists of the nerve fibers leading to the optic tract (Rodriguez and Gisbert 2001). In 7-days old sturgeons the eye was already completely developed (Ostaszewska and Dabrowski 2009). During embryonic development of percid fishes the eye forms as a first sense organ. According to Kovac (1994), fish of genus *Gymnocephalus* showed developed eyes already in 26 hpf, while McElman and Balon (1979) reported that development of optic vesicles in walleye embryos started in 29 hpf, and in 70 hpf the vesicles were fully developed. At the same time also development of the lens took place (McElman and Balon 1979). Similarly, Jara and Szymoniewski (1953) reported the presence of optic germ in pike-perch embryos incubated at 18 °C already on the 43.5 hpf, while the lens was observed at 73.5 dph. At hatching the eyes of pikeperch larvae were pigmented and distinctly visible, and differentiation of retinal layers took place, and was completed at 7 dph (Fig. 7.3).

Due to different biology of percid fishes, the differences in they eye morphology and function are also observed. In sauger (*Sander canadensis*) the *tapetum lucidum* of the retina is better developed than in walleye (*S. vitreum*). According to Ali and Anctil (1977), retina of sauger is more photosensitive than in walleye. Structure of retina in both species is similar but in walleye retinal pigment is more densely distributed, while there is less reflecting material. These differences are related to different environment inhabited by both species and differences in their behavior. Walleye lives in both, transparent and turbid waters but preys mainly at dawn and dusk, while sauger prefers extremely turbid waters and preys all day. The data show that retina of sauger is better adapted to darker environment. Adaptation includes more reflecting material and better arrangement of retinal cells, uniform distribution of tapetum within entire retina, and lower relative abundance of cones in retina and in other areas (Ali and Anctil 1977).

The retinas of darter fish (*Etheostoma*, *Percidae*) have rod photoreceptors and both single and double cones. The λ_{max} range for rod cells is 513–535 nm, for cones in the middle-wavelength sensitive class 504–543 nm, while for cones in the long-wavelength sensitive class 582–628 nm (Gumm et al. 2012). According to Loew et al. (1993), the action spectrum of yellow perch photoreceptors shows three peaks: one between 640 and 700 nm, one between 490 and 525 nm, and one in the near-ultraviolet range, between 360 and 400 nm. Only yellow perch juveniles have near-ultraviolet-sensitive photoreceptors, adults have not.

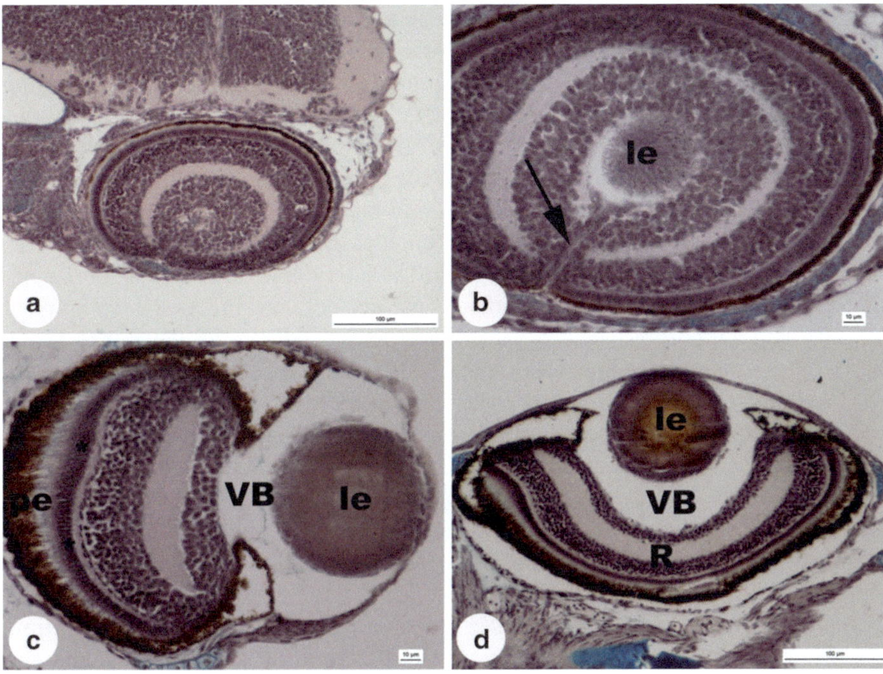

Fig. 7.3 (**a**) Longitudinal section of pike-perch eye at day of hatching. Scale bars 10 μm. (**b**) Longitudinal section of pikeperch eye at 2 dph. Lens (*le*) and optic nerve (*arrow*) are visible. Scale bars 10 μm. (**c**) Cross section of pikeperch eye at 7 dph. Retinal layers are completely differentiated. Lens (*le*), vitreous body (*VB*), rods and cones layer (*stars*) and pigmented epithelium (*pe*) are visible. Scale bars 10 μm. (**d**) Cross section of pikeperch eye at 13 dph. Lens (*le*), vitreous body (*VB*), and retina (*R*) are visible. Scale bars 100 μm

7.5 The Inner Ear and Lateral Line

Body balance in fish is maintained due to inner ear (labyrinth), while canal neuromasts of lateral line and superficial neuromasts of skin detect movement and vibration in the surrounding water (Montgomey et al. 1997). The lateral line and the organ of body balance and hearing develop early, initially as a common placode adjacent to the medulla oblongata (Grodziński 1981). The formation of the otic placode is the earliest morphologically visible event in inner ear development. The placode is a single layer of ectoderm situated over the hindbrain, and gives rise to mechanoreceptors (Ohyama et al. 2007). On the other hand, development of otic neuromasts is related to the formation of lateral line. The area of the otic placode caudal to the otic vesicles is the presumptive posterior lateral line placode (Miyake et al. 1997).

The labyrinth (inner ear) of percid species and other teleost fishes is a paired organ situated in the skull cavity, and not separated from the brain with bone tissue. It consists of three semicircular canals, and relative sensory areas, and three otolithic organs: sacculum, lagena and utriculus. Utriculus and semicircular canals are present in most teleost fishes as the vestibular organ and are responsible for maintaining of body position (Platt 1983), while sacculum and lagena are related to hearing (Kawamura 1984; Salem and Zaghloul 2001). At the base of each of three semicircular canals there are so called ampullae with sensory epithelium. Inner surface of canals is lined with unilayered epithelium with thickened areas called maculae. The cells of sensory epithelium: the hair cells and basement cells in area of ampullar cristae are covered with gelatinous matter (cupula). Labyrinth contains calcium carbonate structures – otoliths (Kilarski 2007). The lateral line organ is a set of mechanoreceptors detecting water movements and its cellular functional component (sensory and supporting cells) is called neuromast and is fully developed at the onset of exogenous feeding (Northcutt 2003). Two types of neuromasts were described: superficial – in skin and canal neuromast of the lateral line (Kilarski 2007). In percid fishes structure of this organ is diverse. The canal neuromasts of ruffe are large and their diameter reaches 3 mm, while superficial neuromasts of this species are small, of diameter no more than 50 μm (Kilarski 2007). Pikeperch have 139–198 superficial neuromasts and 66 canal neuromasts (the ratio 2.5:1), while in perch 86–121 superficial neuromasts are accompanied by 64 canal neuromasts (ratio 1.6:1) (Grodziński 1981).

Auditory vesicles and otolith and/or otoconia primordia were for the first time found in perch embryos on the 6 day post fertilization (at 15–17 °C) (Salem and Zaghloul 2001). According to these authors, the saccular macula overlain by a small round otolith was established by 7 days after fertilization and the stato-acoustic ganglion which contains the neuronal precursors of the macular epithelium was also seen at this stage. Saccular macula started to differentiate on the 12 dpf and was well developed already on the 1 dph. It consists of two types of epithelium: sensory and transitional (Salem and Zaghloul 2001). The inner ear develops as follows: sacculus forms on the 7 dpf, utriculus – on the 10 dpf, while semicircular canals were observed in perch on the day of hatching (Salem and Zaghloul 2001).

In other species of percids auditory vesicles developed in similar way. In the embryos of *Gymnocephalus* auditory vesicles appeared after 26 h of development (Kovac 1994), while the lateral line – after 103 h. Also in other percid species auditory placodes develop early – on the 70 h in walleye (McElman and Balon 1979), and on the 73.5 h in pikeperch (Jara and Szymoniewski 1953). According to Jara and Szymoniewski (1953) and McElman and Balon (1979), the otoliths in these two species were observed on the 92.5 h (in pikeperch), and 162 h (in walleye). On the day of hatching of pikeperch larvae sensory epithelium of the labyrinth was fully developed, while on the 2 dph three semicircular canals were visible (Fig. 7.4).

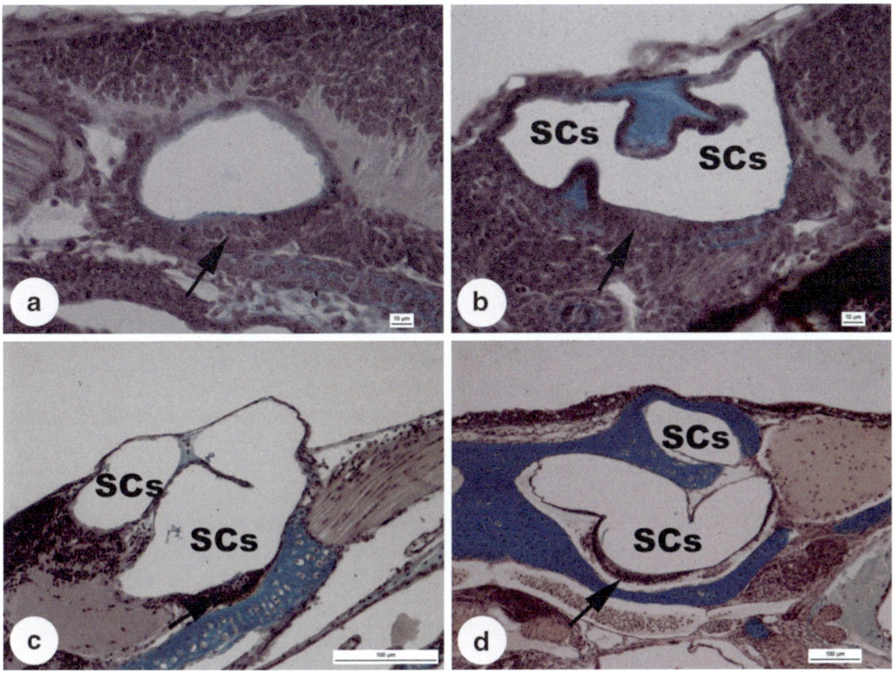

Fig. 7.4 Longitudinal section of pikeperch labyrinth: (**a**) at day of hatching; (**b**) at 2 dph; (**c**) 11 dph and (**d**) 30 dph. *SCs* semicircular canals (lateral at figure (**b**), anterior and lateral at figure (**c**) and lateral and posterior at figure (**d**)). *Arrow* – sensory epithelium Scale bars 10 μm (at figure **a**, **b**) and 100 μm (at figure **c**, **d**)

References

Ali MA, Anctil M (1977) Retinal structure and function in the walleye (*Stizostedion vitreum vitreum*) and sauger (*S. canadense*). J Fish Res Board Can 34:1467–1474

Belanger RM, Smith CM, Corkum LD, Zielinski BS (2003) Morphology and histochemistry of the peripheral olfactory organ in the round goby, *Neogobius melanostomus* (Teleostei: Gobiidae). J Morphol 257:62–71

Belanger RM, Corkum LD, Li W, Zielinski BS (2006) Olfactory sensory input increases gill ventilation in male round gobies (*Neogobius melanostomus*) during exposure to steroids. Comp Biochem Physiol A 144:196–202

Camacho S, Ostos-Garrido MV, Domezain A, Carmona (2010) Study of the olfactory epithelium in the developing sturgeon. Characterization of the crypt cells. Chem Senses 35:147–156

Cox JPL (2008) Hydrodynamic aspects of fish olfaction. J R Soc Interface 5:575–593

Craig JF (2000) Morphology and anatomy. In: Percid fishes: systematics, ecology and exploitation. Blackwell Science, Oxford

Døving KB, Dubois-Dauphin M, Holley A, Jourdan F (1977) Functional anatomy of the olfactory organ of fish and the ciliary mechanism of water transport. Acta Zool (Stockholm) 58:245–255

Fernald RD (1985) Growth of the teleost eye: novel solutions to complex constraints. Environ Biol Fishes 13:113–123

Grodziński Z (1981) Anatomia i embriologia ryb. Państwowe Wydawnictwo Rolnicze i Lene, Warsaw. In Polish
Gumm JM, Loew ER, Mendelson TC (2012) Differences in spectral sensitivity within and among species of darters (genus *Etheostoma*). Vision Res 55:19–23
Hansen A, Zeiske E (1993) Development of the olfactory organ in the zebrafish, *Brachydanio rerio*. J Comp Neurol 333:289–300
Hansen A, Zielinski BS (2005) Diversity in the olfactory epithelium of bony fishes: development, lamellar arrangement, sensory neuron cell types and transduction components. J Neurocytol 34:183–208
Hara TJ (1994) The diversity of chemical stimulation in fish olfaction and gustation. Rev Fish Biol Fish 4:1–35
Hara TJ (2011a) Morphology of the olfactory (smell) system in fishes. In: Farrell AP (ed) Encyclopedia of fish physiology: from genome to environment. Academic, San Diego, pp 194–207
Hara TJ (2011b) Neurophysiology of olfaction. In: Farrell AP (ed) Encyclopedia of fish physiology: from genome to environment. Academic, San Diego, pp 208–217
Hara TJ (2011c) Morphology of the gustatory (Taste) system in fishes. In: Farrell AP (ed) Encyclopedia of fish physiology: from genome to environment. Academic, San Diego, pp 187–193
Jara Z, Szymoniewski R (1953) Rozwój ikry sandacza (*Stizostedion lucioperca lucioperca* Cuv.). Roczniki Nauk Rolniczych Polish Agricult Ann 67:143–160. In Polish with English abstract
Kasumyan AO, Døving KB (2003) Taste preferences in fish. Fish Fish 4:289–347
Kawamura G (1984) The sense organs and behavior of milkfish fry in relation to collection techniques. In: Juario JV, Ferraris RP, Beitez LV (eds) Advance in milkfish biology and culture. Island Publication House, Manilla, pp 69–85
Kilarski W (2007) Zarys anatomii i histologii ryb doskonałokostnych. Wydawnictwo Instytutu Rybactwa ródlądowego, Olsztyn, In Polish
Kilarski W (2012) Anatomia ryb. Powszechne Wydawnictwo Rolnicze i Lene, Poznań, In Polish
Kováč C (1994) Early ontogeny of three *Gymnocephalus* species (Pisces: Percidae): reflections on the evolution of the genus. Environ Biol Fish 40:241–253
Kuciel M, Zuwała K, Jakubowski M (2011) A new type of fish olfactory organ structure in *Periophthalmus barbarus* (Oxudercinae). Acta Zool (Stockholm) 92:276–280
Kuciel M, Zuwała K, Satapoomin U (2013) Comparative morphology (SEM) of the peripheral olfactory organ in the Oxudercinae subfamily (Gobiidae, Perciformes). Zool Anzeiger J Comp Zool 252:424–430
Loew ER, McFarland WN, Mills EL, Hunter D (1993) A chromatic action spectrum for planktonic predation by juvenile yellow perch, *Perca flavescens*. Can J Zool 71:384–386
McElman JF, Balon EK (1979) Early ontogeny of walleye, *Stizostedion vitreum*, with steps of saltatory development. Environ Biol Fish 4:309–348
Miyake T, von Herbing IH, Hall BK (1997) Neural ectoderm, neural crest, and placodes: contribution of the otic placode to the ectodermal lining of the embryonic opercular cavity in Atlantic cod (Teleostei). J Morphol 231:231–252
Montgomery JC, Baker CF, Carton AG (1997) The lateral line can mediate rheotaxis in fish. Nature 389:960–963
Northcutt RG (2003) Development of the lateral line system in the channel catfish. In: Browman HI, Skiftesvik AB (eds) The Big Fish Bang. Proceedings of the 26th annual larval fish conference. Institute of Marine Research, Bergen
Ohyama T, Groves AK, Martin K (2007) The first steps towards hearing: mechanisms of otic placode induction. Int J Dev Biol 51:463–472
Ostaszewska T (2005) Developmental changes of digestive system structures in pike-perch (*Sander lucioperca* L.). Electron J Ichthyol 2:65–78
Ostaszewska T, Dabrowski K (2009) Early development of Acipenseriformes (Chondrostei: Actinopterygii). In: Kunz YV, Luer CA, Kapoor BG (eds) Development of non-teleost fishes. Science, Enfield, pp 171–230

Platt CT (1983) The peripheral vestibular system in fishes. In: Northcutt RG, Davis RE (eds) Fish neurobiology and behavior. University of Michigan Press, Ann Arbor, pp 89–100
Rodríguez A, Gisbert E (2001) Morphogenesis of the eye of Siberian sturgeon. J Fish Biol 59:1427–1429
Ruzhinskaya NN, Gdovskii PA, Devitsina GV (2001) Chloride cells, a constituent of the fish olfactory epithelium. J Evol Biochem Physiol 37:89–94
Salem MA, Zaghloul MS (2001) Structure and development of the saccular sensory epithelium in relation to otolith growth in the perch *Perca fluviatilis* (Teleostei). In: Proceedings of the first international conference (Egyptian British Biological Society, EBB Soc). Egypt J Biol 3:1–12
Sattari A, Asli M, Mansoori FS, Kheirandish R, Yavari H (2012) Histological study of middle layer of rabbit fish eye (*Siganus javus*). Asian Pac J Trop Biomed 2:1086–1089
Tierney KB, Baldwin DH, Hara TJ, Ross PS, Scholz NL, Kennedy CJ (2010) Olfactory toxicity in fishes. Aquat Toxicol 96:2–26
Whitlock KE (2004) A new model for olfactory placode development. Brain Behav Evol 64:126–140
Zeiske E, Kasumyan A, Bartsch P, Hansen A (2003) Early development of the olfactory organ in sturgeons of the genus Acipenser: a comparative and electron microscopic study. Anat Embryol 206:357–372

Chapter 8
Development and Functionality of the Digestive System in Percid Fishes Early Life Stages

Neila Hamza, Teresa Ostaszewska, and Patrick Kestemont

Abstract In percid fishes, the development of digestive structures and activities is quite similar to that of other carnivorous species (sea bass and sea bream). In most species at hatching, the digestive tract is a simple tube consisting of undifferentiated cells. The mouth and anus are closed and the esophagus is not connected with the intestine. Liver and pancreas are undifferentiated. Digestive enzymatic activities (pancreas, intestine) are detected shortly after hatching. According to histological and enzymatic studies, important changes occur around mouth opening (fifth to seventh dph). The pancreas shows exocrine activity and the liver becomes functional with adipogenic and glycogenic functions. The primary stomach develops in pikeperch and even earlier in Eurasian perch. Pancreatic (trypsin, amylase) and intestinal (leucine-alanine peptidase, Alkaline phosphatase, aminopeptidase N) enzyme activities increase at mouth opening. Leucine-alanine peptidase (cytosolic enzyme) activity declines after mouth opening concurrently with the strong increase of the brush border membrane enzymes (Alkaline phosphatase, leucine-aminopeptidase N) activity indicating the development of the brush border membrane of intestinal enterocytes. The stomach development occurs between 15th and 20th dph in pikeperch and between 21th and 35th in Eurasian perch. Pepsin activity is detected only on day 29 in pikeperch as well as in Eurasian perch larvae and is concurrent with the development of gastric glands. The development of brush border membrane of the intestinal enterocytes and the gastric activity indicate that larvae acquire an adult mode of digestion. The digestive structures and activities can be affected by the

N. Hamza (✉)
Institut National des Sciences et Technologies de la Mer,
28 rue du 2 mars 1934, Salammbô 2025, Tunisie
e-mail: hamza.neila@gmail.com

T. Ostaszewska
Division of Ichthyobiology and Fisheries, Faculty of Animal Science,
Warsaw University of Life Sciences, Ciszewskiego 8, 02-786 Warsaw, Poland
e-mail: teresa_ostaszewska@sggw.pl

P. Kestemont
Research Unit in Environmental and Evolutionary Biology, University of Namur,
Rue de Bruxelles 61, 5000 Namur, Belgium

P. Kestemont et al. (eds.), *Biology and Culture of Percid Fishes*,
DOI 10.1007/978-94-017-7227-3_8

nature and the diet composition. This aspect is also discussed in this chapter allowing an approach of the nutritional requirements of percid larvae.

Keywords Digestive system • Enzymatic activities • Stomach • Pancreas • Intestine

8.1 Introduction

In the past, percids were commonly reared in fertilized ponds, as complementary carnivorous species in ponds mainly devoted to the production of cyprinids. In consequence, this fish generally fed on available zooplankton during larval stage. Protozoans, rotifers and young copepods are considered as important feed sources at this stage. Due to temperature and trophic conditions, suitable periods of larval production in ponds are, however, rather limited and usually restricted to spring. In order to secure mass production of juveniles all the year round and reduce the variability in the larval rearing success and juvenile yields obtained from pond system, more intensive percid larval culture has been developed during the last decades, first for North-American percid species (mainly walleye) and more recently for European ones (Eurasian perch and pikeperch) (see Chaps. 9 and 10 for more details). The growing of percid fish larvae in intensive indoor conditions have several advantages but, as feed availability fully depends on the quantity and type of feed supplied by the farmer, the feeding and nutritional requirements of the larvae must be well understood. Acceptance and efficiency of feed provided largely depend on the development of the digestive tract and associated organs (liver, pancreas, pyloric caeca, etc.).

Compared with most carnivorous fish, hatching size (total length = 5.4 mm, body weight = 0.7–0.8 mg) and mouth gape (0.36 mm) of percid fish like Eurasian perch or pikeperch are relatively small and thus constitute a limiting factor in initial feeding with live organisms (Kestemont et al. 1996). As many other fish species, percid larvae have a rudimentary digestive tract at first feeding and are not "equipped" with a functional stomach (Mani-Ponset et al. 1994; Vlavonou 1996; Ostaszewska 2005), which leads to a reduced ability to digest complex dietary proteins (Tonheim et al. 2004; Ronnestad et al. 2007). In many fish species including percids, the anatomy of the larval digestive system is quite different from that of juveniles and adults (Beccaria et al. 1991; Zambonino and Cahu 1994). Thus, a better knowledge of the nutrition physiology during early fish ontogeny is essential for the understanding of larval nutritional needs and for suggesting adequate feeding scheme.

8.2 Passage from Endotrophy to Exotrophy

At mouth opening (around 95° days), pikeperch larvae have a substantial vitelline vesicle under the anterior area of the gut. This vesicle includes the oil globule in front and the vitellus at the back (Mani-Ponset et al. 1994). These two parts, surrounded by periblast, are isolated from the digestive tract but connected with liver by vitelline veins. Yolk reserves cover the energetic expenses during the embryonic stage. At the start of exogenous feeding, most of the yolk sac is utilized. The resorption of vitellus begins on 2 days post-hatching (dph) in both Eurasian perch (Vlavonou 1996) and pikeperch (Hamza et al. 2007). Its total resorption occurs around 400° days and corresponds to the end of endo-exotrophic period.

Mani-Ponset et al. (1994) distinguished three phases in the post-embryonic development of pikeperch:

- An endotrophic phase when the pre-larva lives on its yolk reserves (0–95° days);
- An endo-exotrophic phase characterized by the depletion of yolk reserves and the beginning of feeding (95–400° days);
- And an exotrophic phase when exogenous feeding is the only nutrient source (after 400° days)

8.3 Ontogeny of Digestive Organs

Development of the alimentary tract is similar in pikeperch (*Sander lucioperca*), perch (*Perca fluviatilis*), and streber (*Zingel streber*) (Mani-Ponset et al. 1994; Kestemont et al. 1996; Kováč 2000; Ostaszewska 2002). However, there are some differences concerning the time of differentiation of various structures, and thus, also the moment of functional ability in food utilization by the digestive tract.

In most species of percid fish at hatching, the digestive tract is a simple tube consisting of cells which are not differentiated. The mouth and anus is closed and the esophagus is not connected with the intestine (Fig. 8.1a, b).

The total body length of pikeperch at hatching is about 5–6 mm. The body of newly hatched pikeperch larvae is transparent and surrounded by a fin fold. The larvae display oval yolk sac divided into two sections: oil globule in the anterior part, and yolk in the posterior part. Both oil globule and yolk are enclosed within the yolk syncytial layer (YSL). Two syncytial zones are visible: one surrounding the oil globule, and the other surrounding the yolk. On the basis of ultrastructural YSL

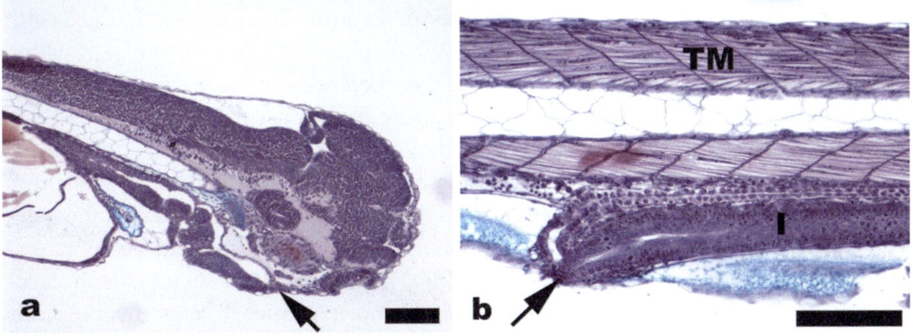

Fig. 8.1 (**a**) Longitudinal section of pikeperch head at hatching. The closed mouth (*arrow*). Scale bar 100 μm. (**b**) Longitudinal section of the posterior body part. Primary digestive tract (*I*) at hatching. Abdominal muscles (*TM*), anus closed (*arrow*). Scale bar 100 μm

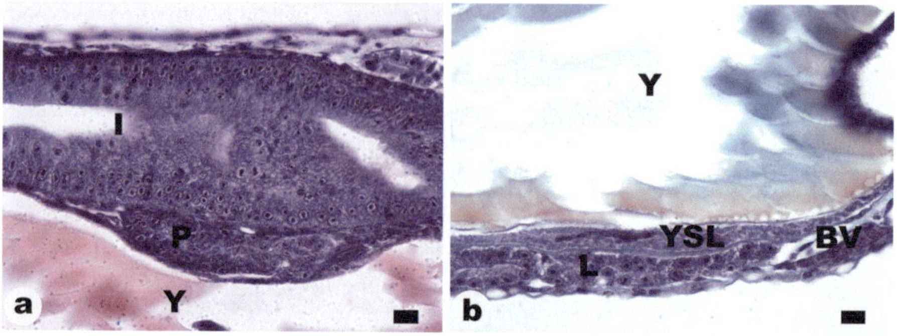

Fig. 8.2 (**a**) Longitudinal section of pikeperch of differentiated pancreas (*P*) cells are located over the yolk sac (*Y*). Intestine (*I*). (**b**) Longitudinal section of pikeperch of differentiated liver cells (*L*) adjacent to the blood vessels (*BV*) and syncytium (*YSL*) at 2 dph. Yolk sac (*Y*). Scale bars 10 μm

studies it was found that the spare substances of yolk sac are synthesized in periblast and overload in the circulatory (Mani-Ponset et al. 1994; Ostaszewska 2002). No blood vessel networks are observed on the yolk sac.

The mouth and pharynx of pikeperch are lined with cubic epithelial cells of irregular shape. Over the development time, these cells gradually flatten, and on the third day they transform into the multilayered squamous epithelium. The esophagus is undifferentiated. At hatching, the intestinal lumen is lined with irregular cubic cells that transform the second day into the unilayered cylindrical epithelium (Fig. 8.2a) (Ostaszewska 2005). On the first day post hatching the gut of Eurasian perch is extended and some folds are present (Kestemont et al. 1996). The liver and pancreas of pikeperch are located closed to the yolk sac, as undifferentiated cells (Fig. 8.2a, b). The undifferentiated liver and pancreas are separated from the yolk sac by syncytial layer. The cells of liver adjoin blood vessels and the syncytium (Fig. 8.2b) (Ostaszewska 2005). At hatching, the liver of Eurasian perch overhangs the anterior intestine while the pancreas is not distinguishable. The pancreas becomes visible under the stomach on day 2 post hatching (dph) (Kestemont et al. 1996).

On the third day post hatching (TL = 5 mm) the intestine length and lumen of pikeperch increase, particularly in the anterior section. The intestine is lined with slightly folded unilayered mucosa of cylindrical epithelial cells, the nuclei of which is situated in the basal region (Fig. 8.3a). The hepatocytes with centrally located nucleus and distinct nucleolus increase, resulting in liver growth (Fig. 8.3b). At this stage, no glycogen storage is observed (Ostaszewska 2005).

In pikeperch considerable changes occur between the fifth and seventh dph (TL = 6 mm). The mouth opens, and so does the esophagus which connects the anterior intestine (Fig. 8.4). The primary stomach develops between esophagus and anterior intestine (Fig. 8.4) (Ostaszewska 2005). The “stomach section” is discernible because, contrary to the entire esophagus and intestine, it lacks mucous cells producing acidic mucus. They are absent from the future stomach region, and pyloric sphincter. The stomach of Eurasian perch starts differentiating earlier. At the

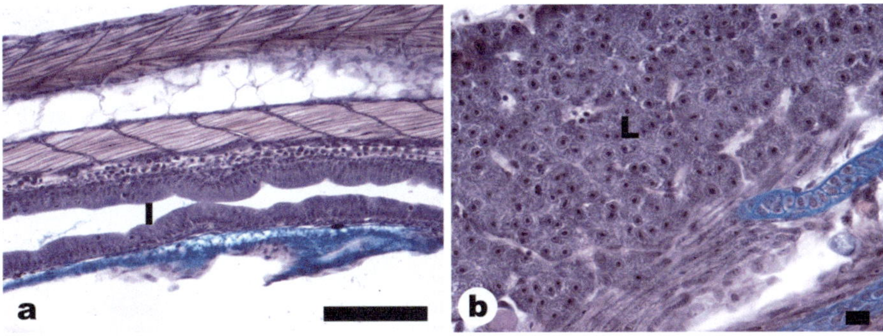

Fig. 8.3 (**a**) Longitudinal section of pikeperch of the primary intestine (*I*) lined with single layer of columnar epithelium of variable cell height at 3 dph. Scale bar 100 μm. (**b**) Longitudinal section of differentiated liver cells (*L*) at 3 dph. Scale bar 10 μm

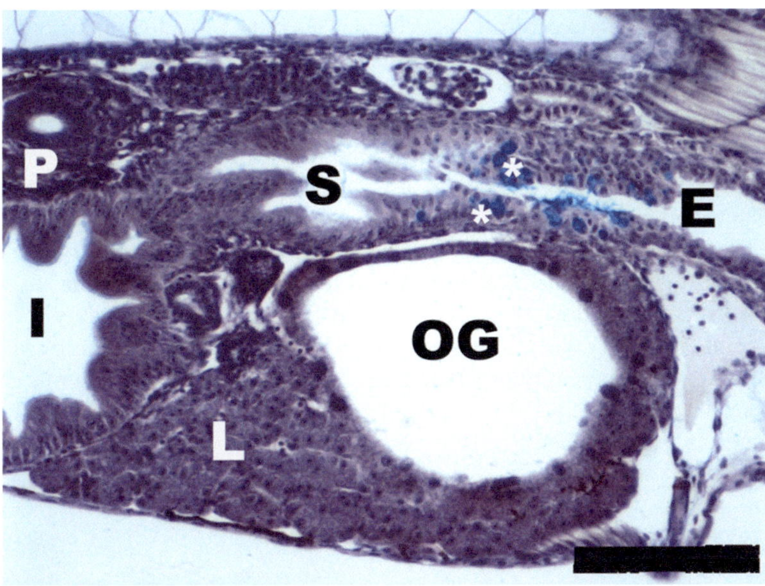

Fig. 8.4 Longitudinal section of pikeperch. Connection between the esophagus (*E*) and anterior intestine (*I*) at 5 dph. Liver (*L*), pancreas (*P*), primary stomach (*S*) oil globule (*OG*) mucous cells (*). Scale bar 100 μm

second dph a large part of the stomach inner surface is smooth and bordered by a mucus layer (Kestemont et al. 1996).

The mucous cells develop in the esophagus of pikeperch larvae after the mouth opening, simultaneously with the onset of a mixed endo-exogenous feeding (Fig. 8.4) (Ostaszewska 2005). At the same time mucous cells appear also in *Solea senegalensis* (Kaup) (Sarasquete et al. 1996), and in *Solea solea* (L.) (Boulhic and Gabaudan 1992), while they develop later in other species such as *Sparus aurata* L.

and *Scophthalmus maximus* L. (Sarasquete et al. 1995). According to Gisbert et al. (1999), esophageal mucous cells develop 2 days before the beginning of active feeding in *Acipenser baeri*, similarly to *Aspius aspius* L. (Ostaszewska and Wegiel 2002). The mucous cells of pikeperch posterior esophagus secrete mainly acidic carbohydrate compounds, while those in the pharynx and anterior esophagus secrete neutral glycoproteins. The number and size of mucous cells increase with fish age. Similar pattern of mixed secretion (of acidic and neutral mucins) was observed in the larvae of *Sparus aurata* (Domeneghini et al. 1998), *Acipenser baeri* (Gisbert et al. 1999), and *Melanogrammus aeglefinus* (Hamlin et al. 2000). According to Boulhic and Gabaudan (1992), the esophagus of *Solea solea* larvae secretes exclusively acidic glycoproteins.

In pikeperch, further increase in liver volume takes place between the fifth and seventh dph, and hepatic blood vessels filled with blood cells become visible (Ostaszewska 2005). Glycogen storage is observed (Fig. 8.5a). Similarly, in Eurasian perch at the sixth day post hatching the vascularization of the liver is secured by numerous porte vein and centro-lobular vein. This indicates a liver development in which adipogenic and glycogenic functions are evidenced (Kestemont et al. 1996). Similar observations concerning the activity of this gland were reported by Boulhic and Gabaudan (1992) in *Solea solea* L. organogenesis. According to Diaz and Connes (1991), glycogen appears at the same time as hepatocytes differentiation, and its storage starts when the animal still relies on maternal nutrition (mammals) or yolk nutrients (birds, fish).

The pancreas of pikeperch is situated above the liver, and shows exocrine activity. On the fifth day post hatching, first proenzyme granules appear in the basophilic pancreatic cells. Two days later, the number of granules considerably increases (Fig. 8.6) and a large Langerhans islet appears (Fig. 8.5b). At the beginning of endo-exogenous feeding, increased pancreatic activity indicates an important role of pancreatic secretory products before the stomach development (Ribeiro et al. 1999; Zambonino Infante and Cahu 2001).

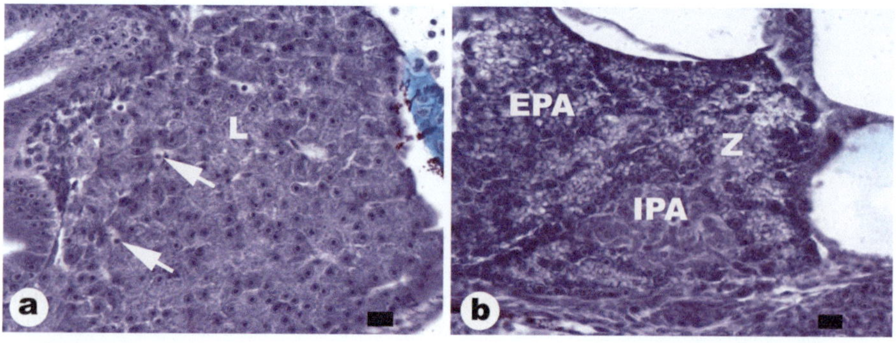

Fig. 8.5 Longitudinal section of pikeperch of liver (*L*) and pancreas at 6 dph. (**a**) Blood vessels (*arrows*) among the hepatocytes. Glycogen PAS-positive areas. (**b**) Exocrine part of pancreas (*EPA*) and Langerhans islet (*IPA*). Proenzyme (*Z*). Scale bars 10 μm

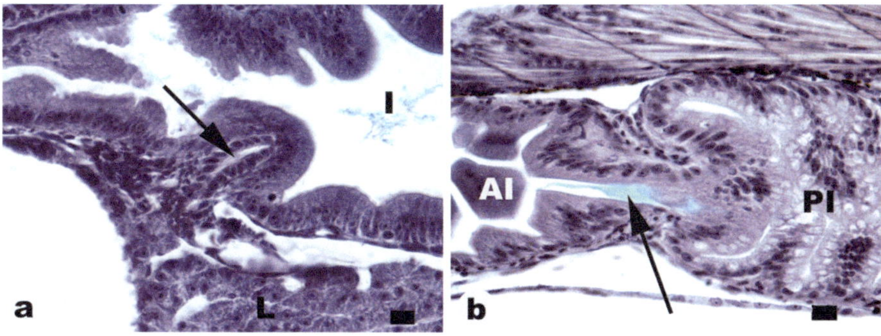

Fig. 8.6 (**a**) Longitudinal section of pikeperch of anterior intestine (*I*) at 7 dph. Connection (*arrow*) between the bile duct with intestine (*I*). Liver (*L*). (**b**) Intestinal valve (*arrow*) between the anterior (*AI*) and posterior intestine (*PI*) at 7 dph. Scale bars 10 μm

From the very beginning of mixed, endo-exogenous feeding (six to seven dph), a bile duct is observed in pikeperch, connecting the liver with the intestine. It opens into the anterior intestine section (Fig. 8.6a). Both, the gall bladder and the bile duct mucosa consist of unilayered cubic epithelium. The gall bladder is situated between the liver and pancreas.

From the beginning of mixed feeding (six to seven dph), the intestine of pikeperch is divided into two sections: anterior and posterior, separated by a valve (Fig. 8.6b). In Eurasian perch larvae, on the first dph, the intestinal valve appears between the median and posterior intestine (Kestemont et al. 1996). According to Pedersen and Hjelmeland (1988), the intestinal valve plays an important role at early larval stage, preventing enzyme escape from the intestine.

The yolk sac of pikeperch content considerably decreases between the 7th and 15th dph. The esophagus becomes longer and the secretory activity and number of mucous cells increase. The end of endogenous feeding of pikeperch larvae reared at 20 °C takes place at the same time as macroscopic yolk sac resorption, on day 6 post hatching. The remaining traces of yolk nutrients visible under the microscope are absorbed until day 15 (Ostaszewska 2005). At the same time the yolk of Eurasian perch reared at 21 °C almost disappears (Kestemont et al. 1996). Histological observations revealed that *Petenia splendida* larvae retained endogenous yolk reserves until 24 dph (Treviño et al. 2011).

In enterocyte cytoplasm of pikeperch, a small number of light vacuoles are observed on the ninth dph (Fig. 8.7a). In the anterior part, they disappear after 1–2 days, while, in the posterior part, the supranuclear regions contain vacuoles with acidophilic granules (Fig. 8.7b). These supranuclear vacuoles do not disappear before stomach development. Similar supranuclear vacuoles containing acidophilic granules visible in the cytoplasm of pikeperch posterior intestine were found also in larvae of other fish species: *Gadus morhua*; *Dicentrarchus labrax*; *Pleuronectes ferrunginea*; *Pagrus pagrus* (Dabrowski and Culver 1991; Kjørsvik et al. 1991; Deplano et al. 1991; Baglole et al. 1998; Darias et al. 2007). These results suggest that the vacuoles are the result of pinocytotic absorption of protein macromolecules

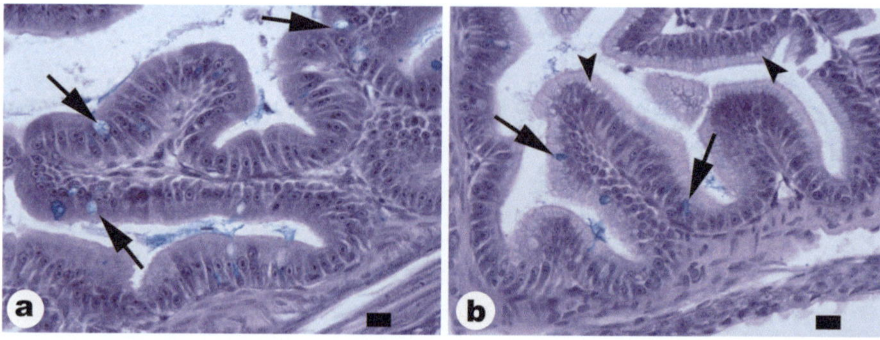

Fig. 8.7 Longitudinal section of pikeperch of anterior (a) and posterior (b) intestine. Small vacuoles with acidophilic granules (*arrowhead*) in the supranuclear regions of the enterocyte at 9 dph. Mucous cells (*arrows*). Scale bars 10 μm

from the intestine. This process is typical for fish larvae showing very low secretion of digestive enzymes, and lacking the gastric protease-producing glands (Govoni et al. 1986). Watanabe (1984) suggests that acidophilic granules found in this part of epithelium indicate active intracellular digestion necessary for protein assimilation, before the stomach development. The presence of supranuclear vacuoles indicates lipid storage (De Silva and Anderson 1995; Fontagné et al. 1998; Crespo et al. 2001). Lipids are hydrolyzed in the intestine to fatty acids and monoglycerides, and then absorbed. Then they are resynthesized in the smooth endoplasmic reticulum, and stored as fat droplets in enterocytes (De Silva and Anderson 1995).

The intestine of pikeperch grows as the fish ages: the number and size of mucosal folds as well as the length of brush border increase. In the epithelium of anterior and central intestine mucous cells appear, and their number increase with fish age. In the hindmost section of the intestine, a short anal channel develops, lined with cubic epithelium without mucous cells.

Over the entire endo-exogenous feeding period of pikeperch, the liver and pancreas gradually increase. At the seventh dph, lipid storage begins in the hepatocytes. Exocrine activity of pancreas significantly decreases from the nine to tenth dph, and lower number of proenzyme granules is observed. On the 13th dph, first tooth germs and taste buds appear in the mouth (Fig. 8.8).

In pikeperch, the stomach development occurs between the 15th and 20th dph (TL = 8–15 mm), with the appearance of gastric glands (Fig. 8.9a) and pyloric sphincter (Fig. 8.9b). Also the intestine elongates and develops first loop (Fig. 8.9b). Around day 20 post-hatching the stomach is morphologically developed (Fig. 8.10).

Pikeperch juvenile stomach is Y-shaped and consists of the cardia, pyloric stomach, and blind sac (Fig. 8.10). These parts are identified according to the terminology developed by Harder (1975), based on the differences in histological and histochemical properties. Fish species displaying the Y-shaped stomach with the blind sac belong, among others, to the *Engraulidae*, *Gadidae*, *Clupeidae*, and *Percidae* families (Harder 1975).

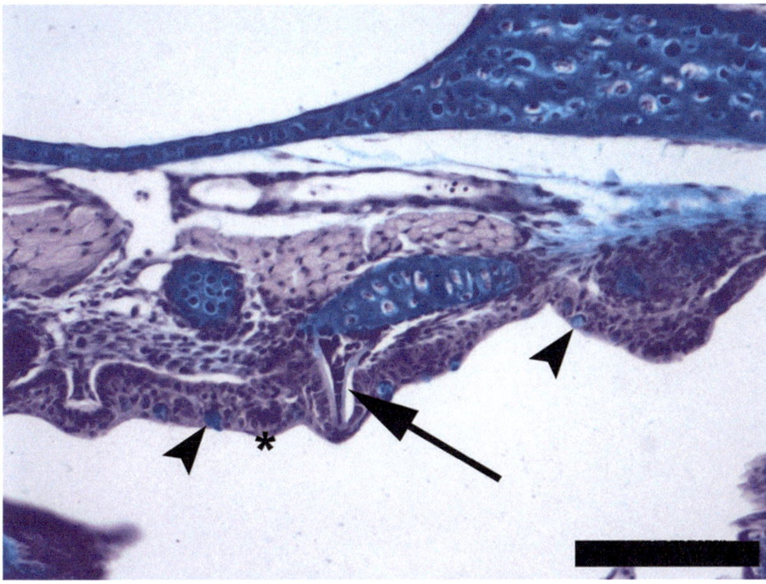

Fig. 8.8 Longitudinal section of buccopharynx of pikeperch. Buccopharyngeal epithelium with scattered mucous cells (*arrowheads*), taste buds (*) and teeth (*arrow*). Scale bar 100 μm

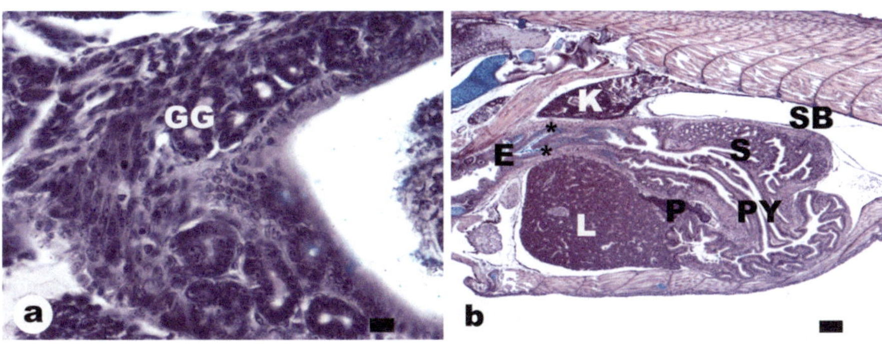

Fig. 8.9 Longitudinal section of pikeperch at 16 dph. (**a**) The beginning of gastric gland (*GG*) development. Scale bar 10 μm. (**b**) Stomach (*S*) developed from esophagus extension. Mucous cells (*) of esophagus (*E*). Liver (*L*), pyloric sphincter (*PY*), pancreas (*P*), swim bladder (*SB*), kidney (*K*). Scale bars 100 μm

The mucosa of both, cardia and blind sac consist of unilayered cylindrical epithelium, mucosal *lamina propria*, and muscle layer. The mucosa of cardia and blind sac contain long, tubular and divided gastric glands situated in the *lamina propria* and surrounded by a loose connective tissue. The gland bottoms consist of polygonal secretory cells of round nuclei and cytoplasm containing acidophilic granules visible after staining with hematoxylin and eosin. The glands open to the mucosa fold crypts. The glands of the cardiac section are considerably longer, comparing to those in the blind sac (Fig. 8.10).

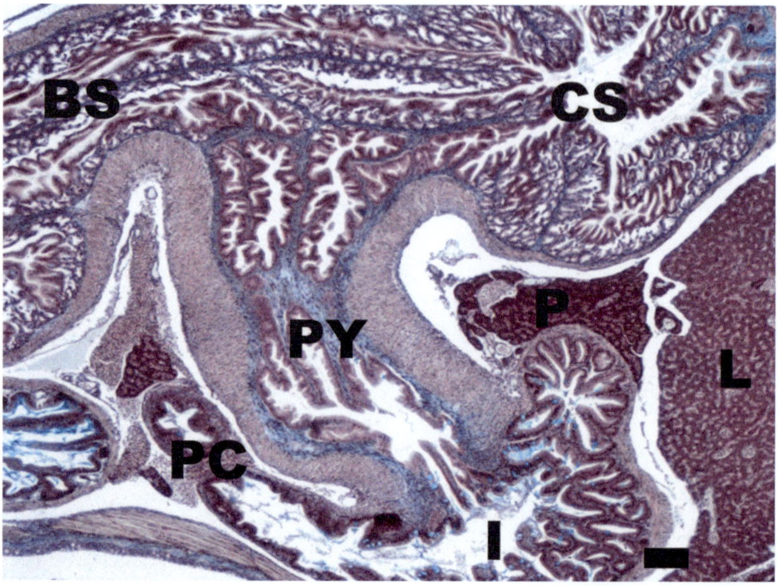

Fig. 8.10 Longitudinal section of pikeperch of digestive tract. Fully morphologically developed stomach at 30 dph. Liver (*L*), pancreas (*P*), intestine (*I*), pyloric caeca (*PC*), pyloric sphincter (*PY*), cardia stomach (*CS*) and blind sac (*BS*). Scale bar 100 μm

The appearance of gastric glands in pikeperch larvae occurs quite early during digestive tract ontogeny comparing with other fish species such as *Solea solea*, *Pagrus pagrus* (Boulhic and Gabaudan 1992; Darias et al. 2007) but later than in *Acipenser baeri* (Gisbert et al. 1999). They appears in both cardia and blind sac. In *Paralabrax maculatofasciatus* larvae they are found only in the anterior part of the stomach (Peña et al. 2003). However, the development of gastric glands is not necessarily accompanied by the onset of stomach activity. In *Coregonus lavaretus* larvae, pepsin activity and acidity were observed a long time after the development of gastric glands (Mähr et al. 1983). Similarity, pikeperch gastric glands contain eosinophilic particles indicating pepsinogen presence, thus a digestive activity of the stomach 10 days after the development of the gastric glands. The presence of neutral mucins in the gastric epithelium also suggests activity of gastric glands. Neutral mucins protect the epithelium against self-digestion by hydrochloric acid, and digestive enzymes produced by the gastric glands (Gisbert et al. 1999). Secretion of neutral mucins was observed in the cardia of adult *Sparus aurata* (Domeneghini et al. 1998) and larvae of *Acipenser baeri* and *Paralichthys californicus* (Gisbert et al. 1999, 2004).

Between the 20th and 30th dph (TL = 15–30 mm), mouth cavity and pharynx of juveniles are lined with the multilayered squamous epithelium, with numerous taste cells. Teeth appear in the upper and lower pharyngeal region. In the endmost section of esophagus, near the stomach, the multilayered squamous epithelium is replaced by multilayered cubic cells which are present also in the stomach wall adjacent to the esophagus. Among cubic cells, numerous mucous cells secrete acidic carboxyl and sulfate mucins.

The pyloric stomach mucosa consists of unilayered cylindrical epithelium, *lamina propria* and muscle layer. The long mucosal folds are developed as fanlike divisions (Fig. 8.10). The apical cytoplasm of cylindrical cells of cardia, pyloric stomach and blind sac are PAS-positive which indicates the presence of neutral compounds. The gastric glands contain eosinophilic particles indicating pepsinogen secretion between the 25th and 30th dph. The intestine, including the pyloric caeca, begins with the pyloric sphincter, and its anterior section is curved downwards. The pyloric caeca of Eurasian perch are observed at the same time as those of pikeperch (Kestemont et al. 1996). Craig (1987) reported that the genus Perca have three pyloric caeca. Pyloric caeca of pikeperch are different from the intestinal ones by higher amount of submucosa, and shorter folds (Fig. 8.10). Pyloric caeca lumen cross-section is starlike. Development of pyloric caeca is the last of a series of important morphological changes occurring in the digestive tract, suggesting that fish attained the juvenile stage (Bisbal and Bengtson 1995). According to Balon (1975), metamorphosis is completed and fish reach the juvenile stage when they show all fins distinct and well developed. This definition is based on external morphological features, and does not take into consideration internal changes, which do not always occur simultaneously. According to Stroband and Dabrowski (1981), the juvenile stage begins when the gastric glands develop, stomach displays digestive activity, and pyloric caeca appear.

In the older pikeperch, from day 11 post-hatching, mucous cells appear among the enterocytes of the anterior and posterior intestines. Similar cells were found by Domeneghini et al. (1998) in *Sparus aurata* L., Darias et al. (2007) in *Pagrus pagrus* and identified as basic secretory cells. They synthesize neutral and acid glycoconjugates. According to the same authors, carbohydrate compounds are the main component of intestinal mucus in vertebrates. Grau et al. (1992) reported that neutral mucous compounds of the intestine participate in enzymatic food digestion, formation of food mass, and absorption. In fish and in mammals, intestinal mucus plays an important protective role (Domeneghini et al. 1998). In pikeperch juveniles, after metamorphosis, mucous cells are very abundant in the entire intestine, but their number is different in various sections. They are the most numerous in the posterior intestine. According to Domeneghini et al. (1998), high density of mucous cells in the colon is essential for easy defecation. The mucus produced by the fish mucous cells plays the same role as in mammals (Scocco et al. 1998), protecting the mucosa of the digestive tract.

In pikeperch juveniles the U-shaped liver is centrally located in the peritoneal cavity, ventrally from the esophagus, and anterior-ventrally from the stomach (Fig. 8.11). The hepatocytes contain numerous light vacuoles indicating lipid storage, and dark PAS-positive glycogen granules. The gall bladder is situated between the liver and anterior intestine with which it is connected with the bile duct (Fig. 8.11).

As reported by Kestemont et al. (1996) and Ostaszewska (2005), the pancreas of Eurasian perch and pikeperch are dispersed within the mesentery of the anterior stomach section, among the pyloric caeca, and bile duct (Fig. 8.10). In pikeperch juvenile exocrine pancreatic tissue is present around the hepatic portal veins. The Langerhans islets are dispersed within entire gland. Similarly as in the other

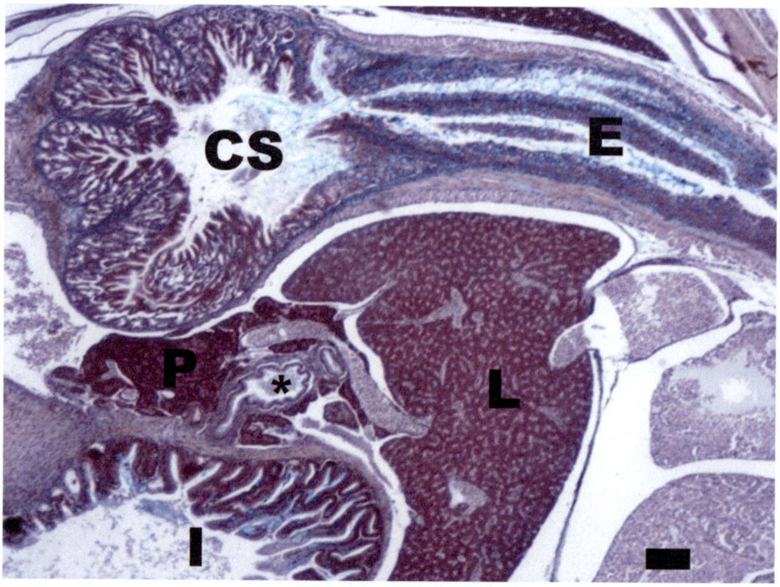

Fig. 8.11 Longitudinal section of pikeperch of esophagus (*E*), anterior stomach part (*CS*), and gall bladder (*) at 30 dph. Pancreas (*P*), intestine (*I*), liver (*L*). Scale bar 100 μm

vertebrates, liver and pancreas of pikeperch are of endodermal origin. Hepatocytes proliferate very quickly during the development, and due to morphogenetic translocation, already in the fifth day of life surround the intestine.

The pikeperch digestive system is developed before the complete metamorphosis, since the alimentary tract is distinctly divided into esophagus, anterior and posterior intestine, and food particles are assimilated. Also liver and pancreas are active, and gall bladder store bile. There are no anatomical differences between larval and juvenile stages, except for the presence of gastric gland and pyloric caeca as well as intestine length. Thus, literature data (Ruuhijärvi et al. 1991; Schlumberger and Proteau 1991; Mani-Ponset et al. 1994; Ostaszewska 2002) suggest that despite such an advanced digestive system activity, feeding of pikeperch with artificial feeds is difficult before fish attain the juvenile stage.

8.4 Ontogeny of Digestive Enzymes

Available data on digestive enzymes at larval stage reported in this chapter concern essentially Eurasian perch and pikeperch. The ontogeny of digestive enzymes in these larvae have essentially been described by Kestemont et al. (1996), Cuvier-Péres and Kestemont (2002), and Hamza et al. (2007) and their pattern and values range are quite similar. The following description will focus essentially on pikeperch larvae. The onset of digestive functions, associated with morphological transformations, follows a sequential chronology in developing fish like that in developing mammals

(Cahu and Zambonino Infante 2001). The digestive tract of fish larvae is not achieved at hatching but undergoes major developmental changes over several weeks (Cousin and Baudin Laurencin 1985; Boulhic and Gabaudan 1992). In percid fishes the development of digestive structures and activities seems similar to that of other carnivorous species. Indeed, according to Mani-Ponset et al. (1994), the development of the digestive tract of pikeperch seems chronologically similar to that of European sea bass *Dicentrarchus labrax* and sea bream *Sparus aurata* larvae.

Digestive enzyme activities, particularly pancreatic and intestinal ones, display a pattern characteristic of developing animals. Their activities are already detectable at hatching and are independent from exogenous feeding since the larvae do not feed before three (Eurasian perch) to five dph (pikeperch). Profiles of pancreatic and intestinal enzymes are depicted below for pikeperch larvae reared at 19–20 °C.

8.4.1 Pancreatic Enzymes

Pancreatic enzyme activities (trypsin and amylase) are detected shortly after hatching in Eurasian perch and pikeperch larvae (Cuvier-Péres and Kestemont 2002; Hamza et al. 2007). From hatching to mouth opening, trypsin and amylase activities increase (in Eurasian perch and pikeperch) concurrently with the exocrine activity of the pancreas signaled by first proenzyme granules in exocrine cells (Mani-Ponset et al. 1994; Ostaszewska 2005). Their relative decrease (from day 15 to 30) is due to protein deposition and secretion of other enzymes (Cahu and Zambonino Infante 1994). It does not reflect a decline of trypsin or amylase synthesis as segmental activity increases with larval development (Cuvier-Péres and Kestemont 2002).

- Trypsin is a *proteolytic* enzyme secreted by pancreas acini. It is detected from hatching at a level of 30 mU mg protein^{-1} and reaches a peak for pikeperch larvae fed with live prey (50 mU mg protein^{-1} (Fig. 8.12a) compared with larvae fed with an artificial diet (Hamza et al. 2007).
- Amylase is a carbohydrase also secreted by pancreas. Also detected from hatching (0.001 U mg protein^{-1} in pikeperch) its activity reaches fivefolds the hatching value at mouth opening (Fig. 8.12b). Amylase is significantly affected by age but not by the nature of the diet (live prey or artificial diet) (Hamza et al. 2007). Its activity is relatively low in carnivorous species like percid fishes compared with the activity reported in omnivorous species like lipped grey mullet (Zouiten et al. 2008).

8.4.2 Intestinal Enzymes

- Leucine-alanine peptidase (Leu-ala) is a peptidase present in the cytosol of the enterocytes. Its activity increases after hatching and reaches a peak at mouth opening (800 U mg protein^{-1} in pikeperch). This high activity reveals the development of intracellular digestion and is concurrent with the appearing of acidophilic granules observed on day 9 in the enterocytes (Fig. 8.7). A further decrease

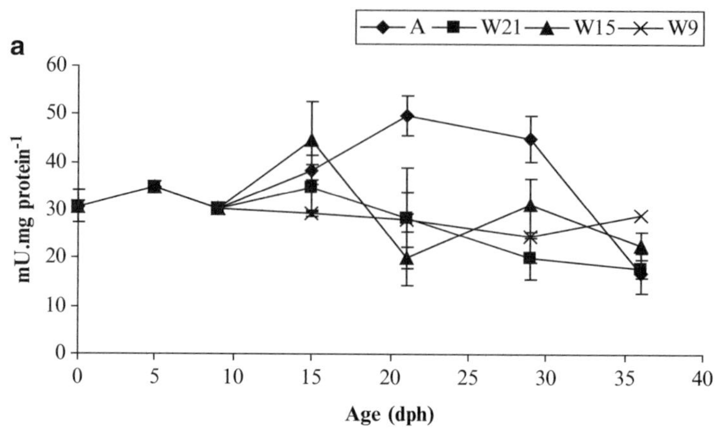

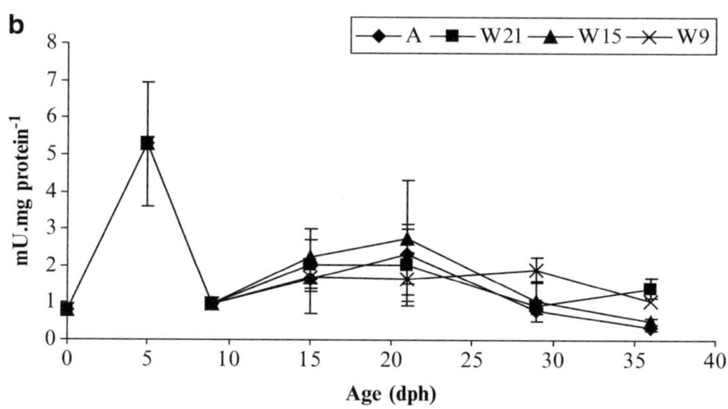

Fig. 8.12 Specific activities of trypsin (**a**) and amylase (**b**) in pikeperch larvae fed *Artemia* (A) and weaned on day 9, 15, 21 (W9, 15, 21). Data are means ± SD (n=4) (Modified from Hamza et al. 2007)

of leu-ala activity (Cuvier-Péres and Kestemont 2002; Hamza et al. 2007) indicates a lowering in intracellular digestion (Fig. 8.13a) relayed by digestion at the level of brush border membrane of the enterocytes.

- Alkaline phosphatase (AP) is an esterase whose activity is induced by phosphorylated substrates such as phospholipids and phospholipoproteins. Its increase from hatching to 21 dph in pikeperch (Fig. 8.13b) traduces the development of enterocytes brush border membrane and extracellular digestion (Zambonino Infante and Cahu 2001).
- Leucine-Aminopeptidase N (AN) is one of the intestinal brush border enzymes and hydrolyzes peptides to amino acids in the final process of protein digestion (Ugolev and De Laey 1973). It increases from hatching to mouth opening (around 40 mU mg protein^{-1}) then decreases and finally reaches a second peak on 29 dph (35 mU mg protein^{-1}) in pikeperch larvae (Fig. 8.13c). These two enzymes are often cited as indicator of brush border membrane development.

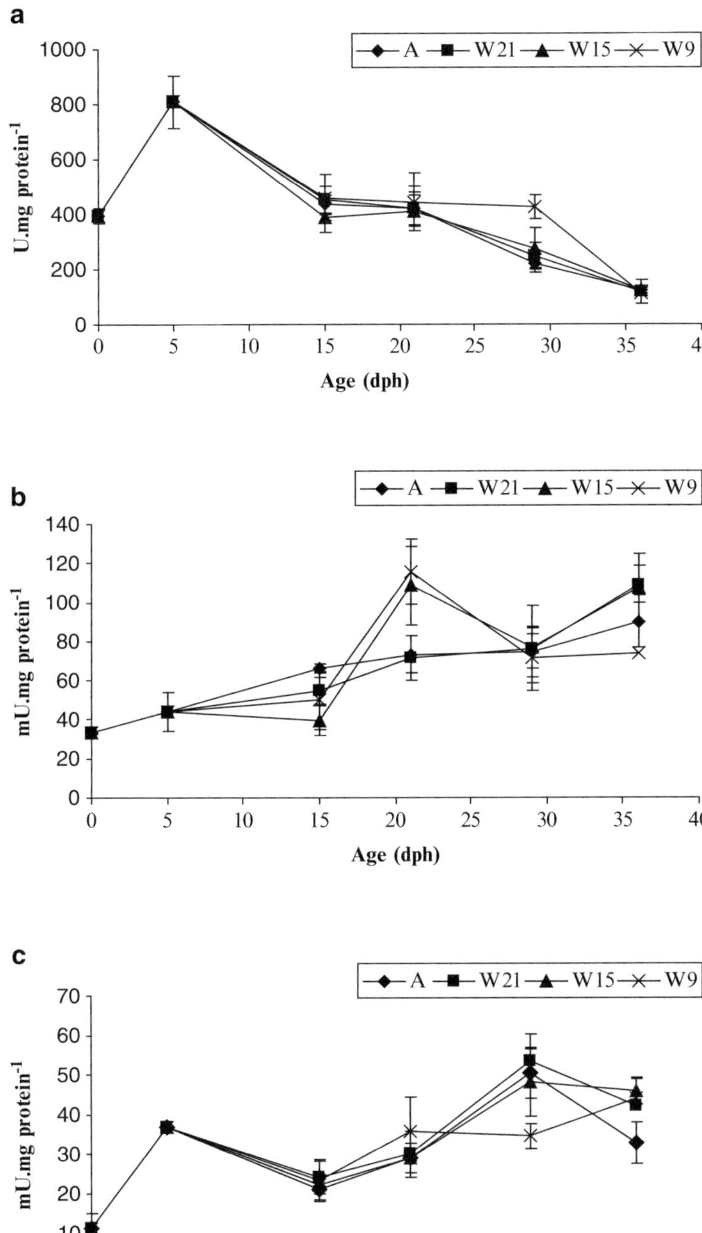

Fig. 8.13 Specific activities of intestinal enzymes leucine-alanine peptidase (**a**), alkaline phosphatase (**b**) and leucine-aminopeptidase N (**c**) in pikeperch larvae fed *Artemia* (A) and weaned on day 9, 15, 21 dph (W9, 15, 21). Data are means ± SD (n=4) (Modified from Hamza et al. 2007)

The strong increase of brush border membrane enzymes (leucine-aminopeptidase N, alkaline phosphatase, maltase and γ-glutamyl transpeptidase) activity occurres concurrently with the decrease of cytosolic enzyme activity (Fig. 8.13). This phenomenon occurs between 15 and 30 dph in Eurasian perch (Cuvier-Péres and Kestemont 2002) and pikeperch (Hamza et al. 2007) and reflects the development of the brush border membrane of intestinal enterocytes. Then, the larvae acquire an adult mode of digestion (Cahu and Zambonino Infante 1994).

8.4.3 Gastric Enzyme

- Pepsin is a proteolytic enzyme secreted by the cells of gastric glands. The stomach development occurs between 15 and 20 dph in pikeperch (Hamza et al. 2007; Ostaszewska et al. 2005) and between 21 and 35 dph in Eurasian perch (Vlavonou 1996). It is fully achieved around day 26 dph in Eurasian perch larvae according to Domeneghini et al. (2008). Pepsin activity is detected for the first time on day 29 in pikeperch larvae (Ostaszewska 2005) and reaches 55–112 mU mg protein^{-1} (Hamza et al. 2007). This activity is concurrent with the gastric glands development in pikeperch (Vlavonou 1996; Ostaszewska et al. 2005; Hamza et al. 2007) as well as in Eurasian perch (Cuvier-Péres and Kestemont 2002). The functionality of the stomach (pepsin secretion) is synchronized with the metamorphosis and indicates that larvae acquire an adult mode of gastric digestion (Zambonino and Cahu 1994; Ronnestad et al. 2007). This profile has been shown in several species like turbot (*Psetta maxima*) and European sea bass (Cousin and Baudin Laurencin 1985).

8.5 Effect of the Diet on Digestive Structures and Enzymes

The enzyme activities and digestive structures development can be modified by the nature (live prey versus artificial diet) and/or biochemical composition of the diet. The importance of these factors has been demonstrated in fish larvae like European sea bass (Zambonino and Cahu 1994; Cahu and Zambonino 1995; Cahu et al. 1998, 2003a) sea bream *Sparus aurata* (Silva et al. 2010; Gisbert et al. 2012) as well as common carp (Escaffre et al. 1997; Gisbert et al. 2012) and percid species (Kestemont et al. 1996, 2001; Ostaszewska et al. 2005; Ostaszewska and Boruta 2006; Hamza et al. 2007, 2008).

Moreover, several authors have demonstrated that the moment of weaning and dietary nutrients like protein hydrolysates, lipids or phospholipids (PL) can affect the digestive maturation process and enzyme activities (Zambonino and Cahu 1999; Buchet et al. 2000; Cahu et al. 2003a; Wold et al. 2007; Hamza et al. 2008, 2012).

8.5.1 Influence of Live Prey Versus Artificial Diet

A study comparing growth, digestive structures and enzymes during ontogeny of pikeperch larvae fed live (*Artemia*) or artificial diets at different ages showed that larvae precociously weaned (day 9) show a lower growth (9 mg) than larvae fed with *Artemia* (80 mg) or weaned on day 21 (100 mg on day 36). Moreover, enzymatic assays and histological observations in digestive tract evidenced that, maturation processes of enterocytes can be impaired or delayed by an inadequate diet or a precocious weaning (Hamza et al. 2007). Indeed, in larvae weaned on day 9, number and height of enterocytes are strongly reduced and the epithelium appears atrophied compared with the larvae fed live preys or weaned later (Fig. 8.14).

The modulation of enzyme activity by dietary factors varies with the type of enzyme. In pikeperch larvae as in another carnivorous species, the level of amylase specific activity is weak, and is not significantly affected by the nature of the diet (live prey or artificial diet). This was shown by Hamza et al. (2007) in an experiment based on pikeperch larvae fed live preys or weaned at different ages. The general enzymatic pattern is similar to the one of other carnivorous species (Fig. 8.12b). In fish larvae, it has been shown that trypsin is modulated by the dietary protein content (Cahu et al. 2004). In percid fishes, larvae fed live preys usually exhibit a higher trypsin activity than those fed artificial diet, as shown in Fig. 8.12a (Hamza et al. 2007).

The adaptation to dietary changes was clearly observed for intestinal enzyme activities in pikeperch larvae. In an experiment comparing the enzymatic activities of pikeperch larvae fed live (*Artemia*) or artificial diets at different ages, leu-ala activity remains significantly higher for larvae weaned on day 9 compared with those fed live preys or weaned later. Higher AP activity is observed after weaning, especially in larvae weaned in first days. The increase of AN in all groups except in larvae precociously weaned (day 9) may reveal a perturbation in secretory process and/or a delay in the maturation of the intestine (Fig. 8.13b, c).

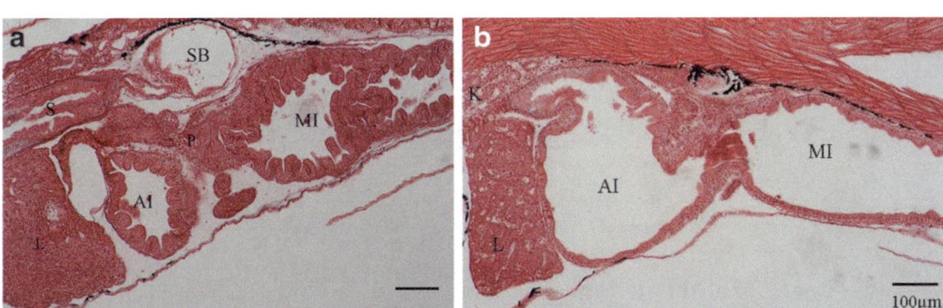

Fig. 8.14 Sagittal section of pikeperch larva (15 dph) fed *Artemia* (**a**) and weaned on day 9 (**b**). GX100. *AI* anterior intestine, *K* kidney, *L* liver, *MI* median intestine, *P* pancreas, *S* stomach (here gastric area), *SB* swimbladder (Modified from Hamza et al. 2007)

8.5.2 Influence of a Diet Containing Casein

Some studies revealed that incorporation of protein hydrolysate in diets for fish larvae stimulates secretion of pancreatic and membranous intestinal enzymes, and improves fish survival and growth (Cahu et al. 1999; Carvalho et al. 2004). Ostaszewska et al. (2005) showed that pikeperch larvae fed formulated diets based on casein or casein hydrolysates show very low growth (around 50 mg on day 35) compared with larvae fed with *Artemia* or commercial diets (200 mg on day 35) and anomalies in digestive system: lower and less numerous intestinal folds (Fig. 8.15a), retarded development of gastric glands (Fig. 8.15b) and smaller hepatocytes in the liver (Fig. 8.15c).

These morphological changes indicated that the diets did not fulfill the fish requirements during their development. On the other hand, fish fed commercial diets and live *Artemia* nauplii show no anomalies in intestinal folds (Fig. 8.15d), stomach (Fig. 8.15e) and liver (Fig. 8.15f) (Ostaszewska et al. 2005). Also, the digestive enzyme (amylase, trypsin and lipase) activity is higher in fish fed commercial diets comparing with fish fed casein hydrolysates diet (Kamaszewski et al. 2010).

8.5.3 Influence of the Phospholipids

Several studies showed that certain nutrients can affect the digestive ontogeny, including both the digestive enzymes and structures. It is well known that PL play a major role in the structure and function of cellular membranes (Tocher 2003), in the intestinal absorption and transport of long chain fatty acids (Fontagné et al. 2000) as well as in the skeletal development (Cahu et al. 2003b). In this way, several authors showed that larvae fed with high dietary levels of lipids or phospholipids (PL) display higher digestive activities and a more developed brush border membrane and hepatocytes (Fontagné et al. 1998; Zambonino Infante and Cahu 1999; Buchet et al. 2000; Cahu et al. 2003a; Gisbert et al. 2005; Wold et al. 2007).

In pikeperch larvae, an increase in dietary PL level from 1.4 % to 9.5 % led to 50 % enhancement in final weight: 160 mg to 240 mg respectively on day 34 (Hamza et al. 2008). Moreover, a dietary level of 9.5 % PL induces a higher specific activity of AP and AN than 1.4 and 4.7 % PL suggesting better development and earlier maturation of digestive structures (Fig. 8.16).

8.6 Genetic Determinism

In fish larvae, digestive enzyme synthesis is age dependent and genetically programmed (Segner et al. 1993). In this way, several studies demonstrated that pancreatic enzymes like trypsin, lipase, phospholipase or amylase are regulated at a transcriptional level during larval development (Péres et al. 1998; Cahu and

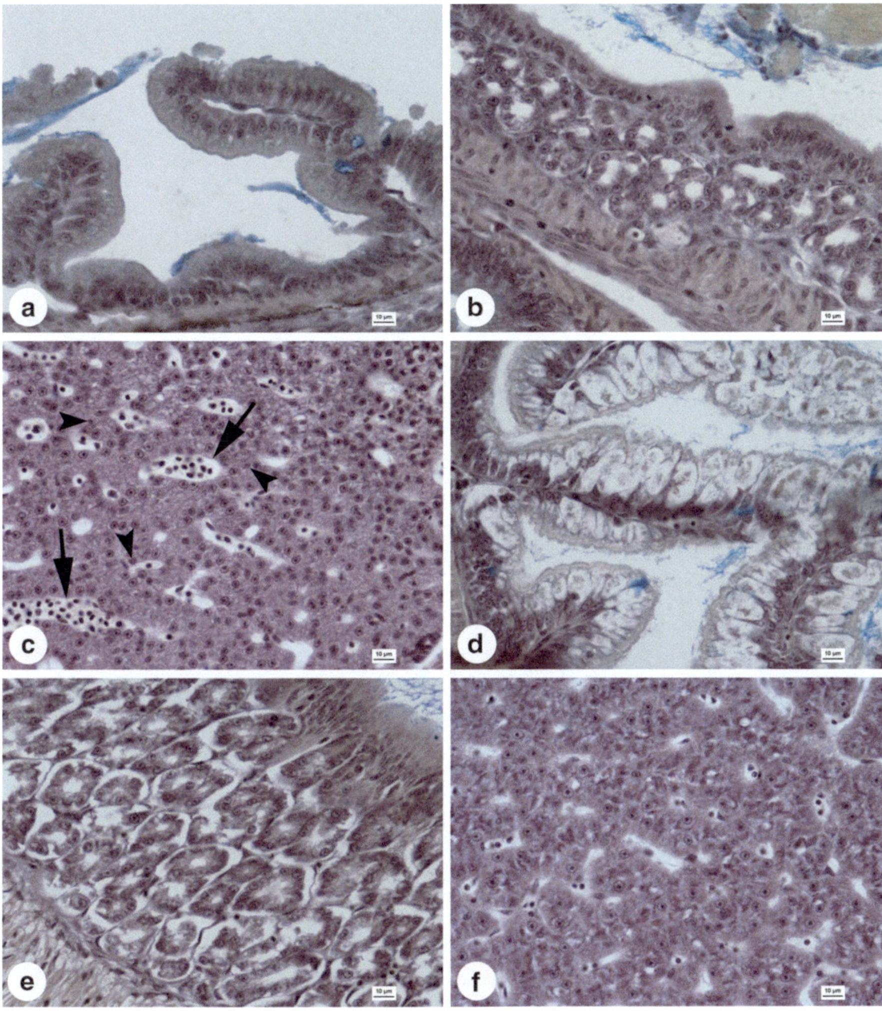

Fig. 8.15 Longitudinal section of pikeperch fed diet based on casein: posterior intestine epithelium (**a**), the gastric glands (**b**) and liver (**c**). Dispersed groups of necrotic hepatocytes (*arrows head*), dilated blood vessels (*arrows*) in liver. The longitudinal section of pikeperch fed *Artemia* nauplii: posterior intestine epithelium (**d**), the gastric glands (**e**) and liver (**f**). Scale bars 10 μm

Zambonino Infante 2001; Kurokawa et al. 2002; Cahu et al. 2003a). Eurasian perch and pikeperch display strong pancreatic and intestinal enzyme activities before mouth opening (Cuvier-Péres and Kestemont 2002; Hamza et al. 2007). Thus, this activity is not promoted by food intake and it can be assumed that this increase is genetically programmed.

Recent studies have investigated the effect of exogenous nutrition on the genes involved in morphogenesis. Villeneuve et al. (2006) showed that the genes involved in morphogenesis of the European sea bass can be modulated by nutrient levels like

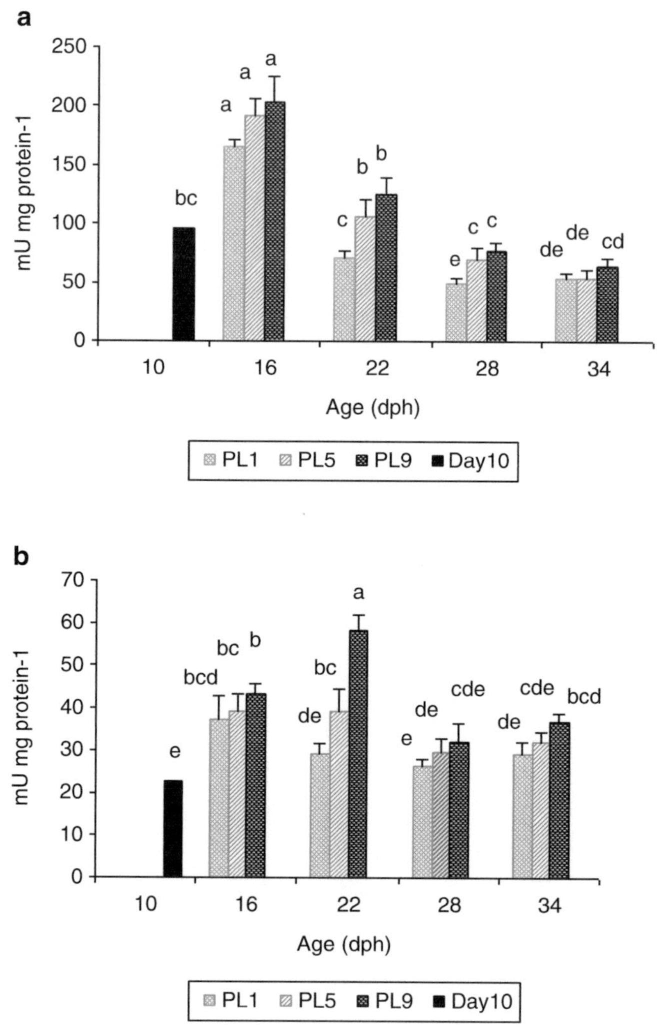

Fig. 8.16 Specific activity of alkaline phosphatase (**a**) and leucine-aminopeptidase N (**b**) during larval development of pikeperch fed with different dietary phospholipid levels (PL1, PL5, PL9 = 1.4 %, 4.7 % and 9.5 % of PL, respectively. Data are means ± SD (n = 4). Bars with different superscript letters are significantly different (*P < 0.05*; two-way ANOVA followed by Tukey HSD) (Modified from Hamza et al. 2008)

vitamin A or polyunsaturated fatty acids of phospholipids between 8 and 13 dph. In the same way, Darias et al. (2011) evidenced an effect of dietary ascorbic acid on the expression of the genes involved in acid ascorbic and calcium absorption as well as those implicated in skeletogenesis and bone mineralization. Concerning percid species, few studies treated the interactions between diet and metabolism at the cellular level. A recent study investigated the effect of nutrients on metabolic pathways in pikeperch larvae. Using proteomic, Hamza et al. (2010) evidenced that dietary PL level affected protein expression profile in the liver of 34 days old larvae. The differentially expressed proteins were involved in several cellular processes, and especially in glycolysis and gluconeogenesis. Methionine metabolism is also affected as well as structure and stress status.

8.7 Conclusions

This chapter aimed to resume the knowledge concerning the development and functionality of digestive structures during the ontogenesis of percid larvae (recap in Fig. 8.17). The difficulties of fish larvae to accept and digest artificial diets have often been attributed to their immature digestive system at hatching and low enzymatic capacities or to the diet composition (Lauff and Hofer 1984; Person-Le Ruyet et al. 1989; Cahu and Zambonino 2001; Ronnestad et al. 2007). As it has been demonstrated by Cahu and Zambonino (2001) for marine species, studies on digestive physiology of percid larvae showed that the digestive enzymes are present very

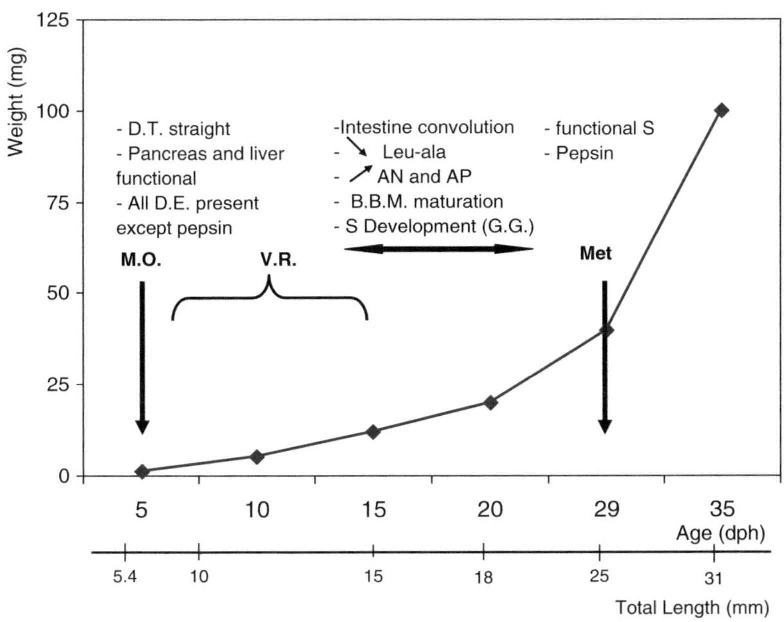

B.B.M.: brush border membrane (enterocytes); D.T.: digestive tract; D.E.: digestive enzymes; G.G.: gastric glands; Met: metamorphosis; M.O.: mouth opening; S: Stomach; V.R.: Vitellus resorption

N.B. : Weight and length data obtained with larvae fed *Artemia* (Hamza et al. 2007; Ostaszewska, 2005)

Fig. 8.17 Main events in morphological and functional development of digestive tract during pikeperch ontogenesis

early and their activities are high relatively to their weight (Kestemont et al. 1996; Diaz et al. 1997; Cuvier-Péres and Kestemont 2002; Hamza et al. 2007, 2008). Diaz et al. (1997) evidenced the absorption of lipids in the first stage of pikeperch development which implies suitable enzyme machinery from initiation of exogenous feeding. In the same way, Ostaszewska (2005) and Ostaszewska et al. (2005) showed an important role of pancreatic secretory products at the beginning of endo-exogenous feeding.

So, as it has been shown in recent studies, enzymatic capacities of the larvae do not seem to be a limiting factor (Lazo et al. 2000; Cahu and Zambonino Infante 2001; Kolkovski 2001) to digest microparticulate diets and to ensure proper growth. In this way, Hamza et al. (2007, 2008) showed that digestive enzyme activities in pikeperch larvae can be modulated by the biochemical composition of the diet. Thus, the success of feeding and better performances seems to depend more of the diet suitability to the nutritional requirements of the larvae. A better knowledge about larval digestive capacities and better approach of nutritional requirements in recent years allowed the improvement of formulation and manufacturing of artificial diets as well as weaning success (Ostaszewska et al. 2005; Hamza et al. 2007, 2008, 2012; Kestemont et al. 2007).

References

Baglole CJ, Goff GP, Wright GM (1998) Distribution and ontogeny of digestive enzymes in larval yellowtail and winter flounder. J Fish Biol 53:767–784

Balon EK (1975) Terminology of intervals in fish development. J Fish Res Board Can 32:1663–1670

Beccaria C, Diaz JP, Connes R, Chatain B (1991) Organogenesis of the exocrine pancreas in the sea bass, *Dicentrarchus labrax* L., reared extensively and intensively. Aquaculture 99:339–354

Bisbal GA, Bengtson DA (1995) Development of digestive tract in larval summer flounder. J Fish Biol 47:277–291

Boulhic M, Gabaudan J (1992) Histological study of the organogenesis of the digestive system and swim bladder of the dover sole, *Solea solea* (Linnaeus 1758). Aquaculture 102:373–396

Buchet V, Zambonino Infante JL, Cahu CL (2000) Effect of lipid level in a compound diet on the development of red drum (*Sciaenops ocellatus*) larvae. Aquaculture 184:339–347

Cahu CL, Zambonino JL (1995) Maturation of the pancreatic and intestinal digestive functions in sea bass (*Dicentrarchus labrax*): effect of weaning with different protein sources. Fish Physiol Biochem 14:431–437

Cahu CL, Zambonino Infante JL (1994) Early weaning of sea bass (*Dicentrarchus labrax*) larvae with a compound diet: effect on digestive enzymes. Comp Biochem Physiol A 109:213–222

Cahu C, Zambonino Infante JL (2001) Substitution of live food by formulated diets in marine fish larvae. Aquaculture 200:161–180

Cahu C, Zambonino Infante J, Escaffre AM, Bergot P, Kaushik S (1998) Preliminary results on sea bass (*Dicentrarchus labrax*) larvae rearing with compound diet from first feeding. Comparison with carp (*Cyprinus carpio*) larvae. Aquaculture 169:1–7

Cahu C, Zambonino Infante J, Quazuguel P, Le Gall MM (1999) Protein hydrolysate vs fish meal in compound diets for 10-day old sea bass *Dicentrarchus labrax* larvae. Aquaculture 171:109–119

Cahu CL, Zambonino Infante JL, Barbosa V (2003a) Effect of dietary phospholipid level and phospholipid: neutral lipid value on the development of sea bass (*Dicentrarchus labrax*) larvae fed a compound diet. Br J Nutr 90:21–28

Cahu C, Zambonino Infante J, Takeuchi T (2003b) Nutritional components affecting skeletal development in fish larvae. Aquaculture 227:245–258

Cahu C, Ronnestad I, Grangier V, Zambonino Infante JL (2004) Expression and activities of pancreatic enzymes in developing sea bass larvae (Dicentrarchus labrax) in relation to intact and hydrolyzed dietary protein; involvement of cholecystokinin. Aquaculture 238:295–308

Carvalho AP, Sa R, Oliva-Teles A, Bergot P (2004) Solubility and peptide profile affect the utilization of dietary protein by common carp (*Cyprinus carpio*) during early larval stages. Aquaculture 234:319–333

Cousin JCB, Baudin Laurencin F (1985) Morphogenèse de l'appareil digestif et de la vessie gazeuse du turbot, *Scophtalmus maximus*. Aquaculture 47:305–319

Craig JF (1987) The biology of perch and related fish. Croom Helm, Beckenham, 333 pp

Crespo S, Marín de Mateo M, Santamaria CA, Sala R, Grau A, Pastor E (2001) Histopathological observations during larval rearing of common dentex *Dentex dentex* L. (Sparidae). Aquaculture 192:121–132

Cuvier-Péres A, Kestemont P (2002) Development of some digestive enzymes in Eurasian perch larvae *Perca fluviatilis*. Fish Physiol Biochem 24:279–285

Dabrowski K, Culver D (1991) The physiology of larval fish; digestive tract and formulation of starter diets. Aquac Mag Mar/Apr:49–61

Darias MJ, Ortiz-Delgado JB, Sarasquete C, Martínez-Rodríguez G, Yúfera M (2007) Larval organogenesis of *Pagrus pagrus* L., 1758 with special attention to the digestive system development. Histol Histopathol 22:753–768

Darias MJ, Mazurais D, Koumoundouros G, Le Gall MM, Huelvan C, Desbruyeres E, Quazuguel P, Cahu CL, Zambonino-Infante JL (2011) Imbalanced dietary ascorbic acid alters molecular pathways involved in skeletogenesis of developing European sea bass (*Dicentrarchus labrax*). Comp Biochem Physiol A 159:46–55

De Silva SS, Anderson TA (1995) Fish nutrition in aquaculture. Chapman and Hall, London, 340 pp

Deplano M, Diaz JP, Connes R, Kentouri-Divanach M, Cavalier F (1991) Appearance of lipid absorption capacities in larvae of the sea bass *Dicentrarchus labrax* during transition to the exotrophic phase. Mar Biol 108:361–371

Diaz JP, Connes R (1991) Development of liver of the sea bass, *Dicentrarchus labrax* L. (Teleost, fish): II. Hepatocyte differentiation. Biol Struct Morphol 3:57–65

Diaz JP, Mani-Ponset L, Guyot E, Connes R (1997) Biliary lipid secretion during early post-embryonic development in three fiches of aquacultural interest: sea bass, *Dicentrarchus labrax* L., sea bream, *Sparus aurata* L., and pike-perch, *Stizostedion lucioperca* (L). J Exp Zool 277:365–370

Domeneghini C, Pannelli Straini R, Veggetti A (1998) Gut glycoconjugates *in Sparus aurata* L. (Pisces, Teleostei). A comparative histochemical study in larval and adult ages. Histol Histopathol 13:359–372

Domeneghini C, Vasconi M, Di Giancamillo A, Bosi G, Saroglia M (2008) Gut development in *Perca fluviatilis*: a micro-anatomical study. Vet Res Commun 32:S143–S145

Escaffre AM, Zambonino Infante JL, Cahu CL, Mambrini M, Bergot P, Kaushik SJ (1997) Nutritional value of soy protein concentrate for larvae of common carp (*Cyprinus carpio*) based on growth performance and digestive enzyme activities. Aquaculture 153:63–80

Fontagné S, Geurden I, Escaffre AM, Bergot P (1998) Histological changes induced by dietary phospholipids in intestine and liver of common carp (*Cyprinus carpio* L.) larvae. Aquaculture 161:213–223

Fontagné S, Burtaire L, Corraze G, Bergot P (2000) Effects of dietary medium-chain triacylglycerols (tricaprylin and tricaproin) and phospholipid supply on survival, growth and lipid metabolism in common carp (*Cyprinus carpio* L.) larvae. Aquaculture 190:289–303

Gisbert E, Sarasquete MC, Williot P, Castelló-Orvay F (1999) Histochemistry of the development of the digestive system of Siberian sturgeon during early ontogeny. J Fish Biol 55:596–616
Gisbert E, Piedrahita RH, Conklin DE (2004) Ontogenetic development of the digestive system in California halibut (*Paralichthys californicus*) with notes on feeding practices. Aquaculture 232:455–470
Gisbert E, Villeneuve L, Zambonino-Infante JL, Quazuguel P, Cahu CL (2005) Dietary phospholipids are more efficient than neutral lipids for long-chain polyunsaturated fatty acid supply in European sea bass *Dicentrarchus labrax* larval development. Lipids 40:1–11
Gisbert E, Skalli A, Fernández I, Kotzamanis Y, Zambonino-Infante JL, Fabregat R (2012) Protein hydrolysates from yeast and pig blood as alternative raw materials in microdiets for gilthead sea bream (*Sparus aurata*) larvae. Aquaculture 338–341:96–104
Govoni JJ, Boehlert GW, Watanabe Y (1986) The physiology of digestion in fish larvae. Environ Biol Fish 16:59–77
Grau A, Crespo S, Sarasquete MC, Gonzalez de Canales ML (1992) The digestive tract of the amberjack *Seriola dumerili*, Risso: a light and scanning electron microscopy study. J Fish Biol 41:287–303
Hamlin HJ, Hunt von Herbing I, Kling LJ (2000) Histological and morphological evaluations of digestive tract and associated organs of haddock throughout post-hatching ontogeny. J Fish Biol 57:716–732
Hamza N, Mhetli M, Kestemont P (2007) Effects of weaning age and diets on ontogeny of digestive activities and structures of pikeperch (*Sander lucioperca*) larvae. Fish Physiol Biochem 33:121–133
Hamza N, Mhetli M, Ben Khemis I, Cahu C, Kestemont P (2008) Effect of dietary phospholipid levels on performance, enzyme activities and fatty acid composition of pikeperch (*Sander lucioperca*) larvae. Aquaculture 275:274–282
Hamza N, Silvestre F, Mhetli M, Ben Khemis I, Dieu M, Raes M, Cahu C, Kestemont P (2010) Differential protein expression profile in the liver of pikeperch (*Sander lucioperca*) larvae fed with increasing levels of phospholipids. Comp Biochem Physiol D 5:130–137
Hamza N, Kestemont P, Ben Khemis I, Mhetli M, Cahu C (2012) Effect of different sources and levels of dietary phospholipids on performances and fatty acid composition of pikeperch (*Sander lucioperca*) larvae. Aquac Nutr 18:249–257
Harder W (1975) Anatomy of fishes. E. Schweizerbart'sche Verlagsbuchhandlung (Nägele u. Obermiller) Stuttgart. 612 pp
Kamaszewski M, Napora-Rutkowski Ł, Ostaszewska T (2010) The effect of feeding on activity of digestive enzymes and morphological changes in pike-perch (*Sander lucioperca*) liver and pancreas. Isr J Aquac 62:225–236
Kestemont P, Mélard C, Fiogbé E, Vlavonou R, Masson G (1996) Nutritional and animal husbandry aspects of rearing early life stages of Eurasian perch *Perca fluviatilis*. J Appl Ichthyol 12:157–165
Kestemont P, Vandeloise E, Mélard C, Fontaine P, Brown PB (2001) Growth and nutritional status of Eurasian perch *Perca fluviatilis* fed graded levels of dietary lipids with or without ethoxyquin. Aquaculture 203:85–99
Kestemont P, Xueliang X, Hamza N, Maboudou J, Imorou Toko I (2007) Effect of weaning age and diet on pikeperch larviculture. Aquaculture 264:197–204
Kjørsvik E, van der Meeren T, Kryvi H, Arnfinnson J, Kvenseth PG (1991) Early development of the digestive tract of cod larvae, *Gadus morhua* L., during start-feeding and starvation. J Fish Biol 38:1–15
Kolkovski S (2001) Digestive enzymes in fish larvae and juveniles- implications and applications to formulated diets. Aquaculture 200:181–201
Kováč V (2000) Early development of *Zingel streber*. J Fish Biol 57:1381–1403
Kurokawa T, Suzuki T, Ohta H, Kagawa H, Tanaka H, Unuma T (2002) Expression of pancreatic enzyme genes during the early larval stage of Japanese eel *Anguilla japonica*. Fish Sci 68:736–744

Lauff M, Hofer R (1984) Proteolytic enzymes in fish development and the importance of dietary enzymes. Aquaculture 37:335–346

Lazo JP, Holt GJ, Arnold CR (2000) Ontogeny of pancreatic enzymes in larval red drum *Sciaenops ocellatus*. Aquac Nutr 6:183–192

Mähr K, Grabner M, Hofer R, Moser H (1983) Histological and physiological development of the stomach of *Coregonus* sp. Arch Hydrobiol 98:344–353

Mani-Ponset L, Diaz JP, Schlumberger O, Connes R (1994) Development of yolk complex, liver and anterior intestine in pike-perch larvae, *Stizostedion lucioperca* (Percidae), according to the first diet during rearing. Aquat Living Resour 7:191–202

Ostaszewska T (2002) The morphological and histological development of digestive tract and swim bladder in early organogenesis of pike-perch larvae (*Stizostedion lucioperca* L.) in different rearing environments. Treatises and monographs. SGGW, Warsaw (in Polish)

Ostaszewska T (2005) Developmental changes of digestive system structures in pike-perch (*Sander lucioperca* L.). Electron J Ichthyol 2:65–78

Ostaszewska T, Boruta A (2006) The effect of diet on fatty acid composition and liver histology of pikeperch (*Sander lucioperca* (L.)) larvae. Arch Pol Fish 14:53–66

Ostaszewska T, Wegiel M (2002) Differentiation of alimentary tract during organogenesis in larval asp (*Aspius aspius* L.). Acta Sci Pol Piscaria 1:23–33

Ostaszewska T, Dabrowski K, Czuminska K, Olech W, Olejniczak M (2005) Rearing of pike-perch larvae using formulated diets- first success with starter feeds. Aquac Res 36:1167–1176

Pedersen BH, Hjelmeland K (1988) Fate of trypsin and assimilation efficiency in larval herring (*Clupea harengus*) following digestion of copepods. Mar Biol 97:467–476

Peña R, Dumas S, Villalejo-Fuerte M, Ortiz-Galindo JL (2003) Ontogenetic development of the digestive tract in reared spotted sand bass *Paralabrax maculatofasciatus* larvae. Aquaculture 219:633–644

Péres A, Zambonino Infante JL, Cahu C (1998) Dietary regulation of activities and mRNA levels of trypsin and amylase in sea bass (*Dicentrarchus labrax*) larvae. Fish Physiol Biochem 19:145–152

Person-Le Ruyet J, Samain JF, Daniel JY (1989) Evolution de l'activité de la trypsine et de l'amylase au cours du développement chez la larve de bar (*Dicentrarchus labrax*). Effet de l'âge au sevrage. Océanis 15:465–480

Ribeiro L, Zambonino Infante JL, Cahu C, Dinis MT (1999) Development of digestive enzymes in larvae *Solea senegalensis*, Kaup 1858. Aquaculture 179:465–473

Rønnestad I, Kamisaka Y, Conceição LEC, Morais S, Tonheim SK (2007) Digestive physiology of marine fish larvae: hormonal control and processing capacity for proteins, peptides and amino acids. Aquaculture 268:82–97

Ruuhijärvi J, Virtanen E, Salminen M, Muynda M (1991) The growth and survival of pike – perch, *Stizostedion lucioperca* L., larvae fed on formulated feeds. In: Lavens P et al (eds) Larvi "91. European Aquaculture Society, Special Publication No. 15, Gent, pp 154–156

Sarasquete MC, Polo A, Yúfera M (1995) Histology and histochemistry of the development of digestive system of larval gilthead seabream, *Sparus aurata* L. Aquaculture 130:79–92

Sarasquete MC, González Canales ML, Arellano JM, Muñoz-Cueto JA, Riberio L, Dinis MT (1996) Histochemical aspects of the yolk sac and digestive tract of larvae of Senegal sole, *Solea senegalensis* (Kaup, 1858). Histol Histopathol 11:881–888

Schlumberger O, Proteau JP (1991) Production de juvéniles de sandre (*Stizostedion lucioperca*). Aqua Rev 36:25–28

Scocco P, Accili D, Menghi G, Ceccarelli P (1998) Unusual glycoconjugates in the oesophagus of a tilapine polyhybrid. J Fish Biol 53:39–48

Segner H, Rösh R, Verreth J, Witt U (1993) Larval nutritional physiology: studies with *Clarias gariepinus*, *Coregonus lavaretus* and *Scophtalmus maximus*. J World Aquac Soc 24:121–134

Silva FCP, Nicoli JR, Zambonino-Infante JL, Le Gall MM, Kaushik S, Gatesoupe FJ (2010) Influence of partial substitution of dietary fish meal on the activity of digestive enzymes in the intestinal brush border membrane of gilthead sea bream, *Sparus aurata* and goldfish, *Carassius auratus*. Aquaculture 306:233–237

Stroband HWJ, Dabrowski KR (1981) Morphological and physiological aspects of digestive system and feeding in fresh-water fish larvae. In: Fontain M (ed) Nutrition des Poissons. Centre National de la Recherche Scientifique, Paris, pp 355–374

Tocher DR (2003) Metabolism and functions of lipids and fatty acids in teleost fish. Rev Fish Sci 11:107–184

Tonheim SK, Espe M, Raae AJ, Darias MJ, Rønnestad I (2004) In vivo incorporation of [U]-^{14}C-amino acids: an alternative protein labelling procedure for use in examining larval digestive physiology. Aquaculture 235:553–567

Treviño L, Alvarez-González CA, Perales-García N, Arévalo-Galán L, Uscanga-Martínez A, Márquez-Couturier G, Fernández I, Gisbert EA (2011) A histological study of the organogenesis of the digestive system in bay snook *Petenia splendida* Günther, 1862 from hatching to the juvenile stage. J Appl Ichthyol 27:73–82

Ugolev AM, De Laey P (1973) Membrane digestion. A concept of enzymic hydrolysis on cell membranes. Biochem Biophys Acta 300:105–128

Villeneuve LAN, Gisbert E, Moriceau J, Cahu CL, Zambonino Infante JL (2006) Intake of high levels of vitamin A and polyunsaturated fatty acids during different developmental periods modifies the expression of morphogenesis genes in European sea bass (*Dicentrarchus labrax*). Br J Nutr 95:677–687

Vlavonou RS (1996) Elevage expérimental de la perche *Perca fluviatilis* L., Développement larvaire et croissance. Dissertation, University of Metz

Watanabe Y (1984) Morphological and functional changes in rectal epithelial cells of pond smelt during postembryonic development. Bull Jpn Soc Sci Fish 50:805–814

Wold P-A, Hoehne-Reitan K, Cahu C, Zambonino Infante J, Rainuzzo J, Kjørsvik E (2007) Phospholipids vs. neutral lipids: effects on digestive enzymes in Atlantic cod (*Gadus morhua*) larvae. Aquaculture 272:502–513

Zambonino Infante JL, Cahu CL (1994) Influence of diet on pepsin and some pancreatic enzymes in sea bass (*Dicentrarchus labrax*) larvae. Comp Biochem Physiol A 109:209–212

Zambonino Infante JL, Cahu L (1999) High dietary lipid levels enhance digestive tract maturation and improve *Dicentrarchus labrax* larval development. J Nutr 129:1195–1200

Zambonino Infante JL, Cahu CL (2001) Ontogeny of the gastrointestinal tract of marine fish larvae. Comp Biochem Physiol C 130:477–487

Zouiten D, Ben Khemis I, Besbes R, Cahu C (2008) Ontogeny of the digestive tract of thick lipped grey mullet (*Chelon labrosus*) larvae reared in “mesocosms”. Aquaculture 279:166–172

Chapter 9
Culture Methods of Eurasian Perch and Yellow Perch Early Life Stages

P. Kestemont, C. Mélard, J.A. Held, and K. Dabrowski

Abstract For carnivorous species producing relatively small larvae, as Eurasian perch and yellow perch, the successful rearing of early life stages is still a matter of concern, even if significant improvements have been achieved during these last two decades. This chapter presents an overview of the different methods used to produce juveniles of these two species: (i) production of fish in fertilized ponds, with fingerling habituation to artificial feed before or after pond harvest, (ii) fertilization of mesocosms and semi-intensive production up to 45 days old, (iii) intensive production in tanks with supply of live prey progressively replaced by artificial feed. For each system, the optimal husbandry conditions as well as the influence of main factors (stocking density, temperature, growth heterogeneity and management of cannibalism, non-inflation of swim bladder,…) influencing the survival and growth of fish from larval to juvenile stages are described.

Keywords Perch • Larval rearing • Cannibalism • Swim bladder

9.1 Introduction

Several reviews dealing with the larval rearing methods of Eurasian perch (Kestemont et al. 1996, 2008; Kestemont and Mélard 2000) and yellow perch (Hart et al. 2006) have been produced during the last two decades. The culture methods

P. Kestemont (✉)
Research Unit in Environmental and Evolutionary Biology, University of Namur, Rue de Bruxelles 61, 5000 Namur, Belgium
e-mail: patrick.kestemont@unamur.be

C. Mélard
Aquaculture Research and Education Center (CEFRA), University of Liège, Chemin de la Justice, 10, 4500 Tihange, Belgium

J.A. Held
University of Wisconsin-Extension Madison, Madison, WI, USA

K. Dabrowski
School of Environment and Natural Resources, Ohio State University, 2021 Coffey Rd, 43210 Columbus, OH, USA

P. Kestemont et al. (eds.), *Biology and Culture of Percid Fishes*,
DOI 10.1007/978-94-017-7227-3_9

are rather diversified, from extensive production in ponds in Central Europe, usually targeting the polyculture of different freshwater species, mainly cyprinids, to more complex perch monoculture systems such as the tandem pond/tank rearing widely used in the north-central region of the USA, or the fully intensive larval rearing in recirculating systems as performed in several European countries.

This chapter describes the different culture methods of Eurasian perch and yellow perch early life stages and, for each method, discusses the main driving factors to be considered for optimizing the rearing performances.

9.2 Pond Culture of Yellow Perch and Eurasian Perch

In Europe, pond production of Eurasian perch is practiced in several countries or regions such as the North of France, Ireland, Czech Republic, Hungary and Poland. In some operations, particularly in central and Eastern Europe, pond culture is not limited to juvenile production, but continues up to market size fish, while, in other facilities, larvae are stocked in ponds up to the juvenile stage, and then harvested and subsequently trained in tanks (Kestemont et al. 2008). However, mass production of Eurasian perch early life stages is mainly performed under intensive tank conditions, allowing for year-round production of juveniles. Alternatively, most yellow perch fingerlings destined for commercial food-fish operations are produced using a tandem pond-tank protocol (Hartleb et al. 2012). In this method, fertilized ponds are employed to rear yellow perch from hatch to the juvenile stage (18–30 mm total length, TL), then the fingerlings are harvested from the ponds and placed into tanks for habituation to formulated feeds (Heidinger and Kayes 1986). Once the habituation process is completed, the fingerlings are returned to ponds or stocked into tanks for grow out to market size. This tandem pond-tank culture is also successfully applied in Ireland and Czech Republic with Eurasian perch.

Production ponds for both Eurasian and yellow perch fingerlings range from 0.1 to 0.8 ha. Although ponds smaller than 0.1 ha can be used to culture perch fingerlings, constructions costs for a commercial scale operation based on such ponds may be economically unfeasible. Fingerling ponds larger than 0.8 ha can be productive but they can be unwieldy to manage and harvest. Ponds can be lined or unlined, drainable or undrainable. Unlined (earthen bottom) ponds are preferred due to limited construction costs and the capability to provide a substrate for zooplankton eggs, thus ensuring continuing populations of forage. In some applications lined ponds are necessary, however, plastic liners are expensive and lined ponds usually require inoculations of phyto- and zooplankton to produce an adequate food web for concentrated fingerling production. Fingerling production ponds are generally 1.5–4 m deep with steeply sloped sides to eliminate shallow littoral zones that encourage the growth of rooted macrophytes.

Drainable ponds provide easy access to fingerlings at harvest and can be dried during the off-season to help control feral fish populations, aquatic nuisance species, parasites and disease vectors. The bottoms of drainable ponds are sloped

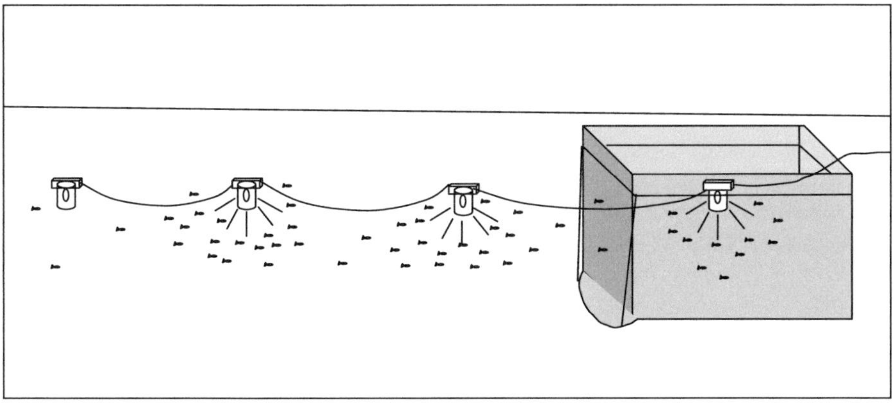

Fig. 9.1 Using lights to trap small yellow perch fingerlings from a pond at night

toward the drain to promote a rapid discharge of water during the final stages of drawdown and flush the fingerlings from the pond. Undrainable ponds require seines or traps for harvest and eradication of all remaining fish in the pond prior to use the following season. Undrainable ponds also tend to support populations of snails that can act as intermediate hosts for digenetic trematodes.

One method of removing small fingerlings from undrainable ponds uses lights at night to attract the highly photopositive fish into a box net (Fig. 9.1). A series of lights suspended under floats is distributed in the pond so that the final light is located in the box. Initially all lights are turned on and schools of fish are attracted. Over the course of approximately 30 min the lights are sequentially extinguished until only the light in the trap is lit. The fish follow the lights into the trap and the open side of the box is pulled shut, capturing the fingerlings. This technique is limited to small (<35 mm TL) fingerlings; as the perch grow beyond 35 mm TL they become less phototactic. While not as efficient as seining, the light harvest technique can be effective for the partial harvest of ponds or in situations where seines are not practical, and is a very low stress method of collection.

9.2.1 Pond Fertilization

The most critical aspect to the successful pond production of fingerlings is fertilization. Pond waters are supplemented with organic materials or plant nutrients to stimulate the development of decomposer or phytoplankton populations at the food web base and eventually provide prey for the developing fish. Fertilization protocols vary widely between yellow perch culturists but generally fall into three categories: organic, inorganic or a combination of the two.

Organic fertilizers are any material of direct or indirect plant origin that is subject to microbial decomposition (Soderberg 2012). Commonly used organic fertilizers include dried or fresh plant material such as hay or straw, animal manures, seed residues such as soybean or cottonseed meal, and yeast. Some protocols call for a terrestrial crop to be planted in the dried pond bottom that later decomposes when the pond is flooded.

In general, organic fertilizers support a food web primarily based on heterotrophic bacteria and fungi that zooplankton directly feed on (Tice et al. 1996; Morris and Mischke 1999). The release of nitrogen and phosphorus from heterotrophic decomposition can stimulate secondary production of phytoplankton that is also fed upon by zooplankton (Hartleb et al. 2012). Decomposition of the organic material can lead to dissolved oxygen depletion, particularly at the bottom of the pond. At low dissolved oxygen levels (<1 ppm) aerobic bacteria, whose activity results in nontoxic end products can be limited by the lack of oxygen and replaced by anaerobic decomposers whose activity results in hydrogen sulfide that is highly toxic to fish. Furthermore, an abundance of aerobic decomposition can have a negative impact on the overall dissolved oxygen levels of the pond. Accordingly, aeration, water circulation and oxygen monitoring are important management activities when using organic fertilization protocols.

Boyd (1977) noted that because of their slow release of nutrients, organic materials are not as good as inorganic fertilizers at creating the large, fast algal blooms that are needed during a short larval production period. Hartleb et al. (2012) observed that the slow release of nutrients by organic materials makes it difficult to respond to environmental changes that may impact phytoplankton populations. Additionally, the delayed release of nutrients (particularly phosphorus) from organics applied late in the fingerling production cycle can stimulate the formation of filamentous algae, which is problematic during pond harvest. The observations of Boyd and Hartleb et al. are particularly important to yellow perch fingerling culture because the pond phase of production is relatively short in duration (5–7 weeks) and yellow perch fingerlings are usually harvested at a small size for habituation to formulated feed.

Inorganic fertilizers are materials high in soluble amount of nitrogen, phosphorus and potassium, or combinations of such plant nutrients; although potassium is generally not considered limiting in aquatic culture (Soderberg 2012). Common inorganic fertilizers include urea, ammonium phosphate, phosphoric acid, and a variety of complex agricultural fertilizers such as triple superphosphate. Inorganic fertilizers support an autotrophic food web by supplying nutrients that increase the abundance of algae for zooplankton consumption. While algae require numerous macro- and micronutrients to thrive, they are most commonly limited by inadequate levels of nitrogen and phosphorus (Boyd 2012). Carbon availability is generally not thought of as a limiting factor in aquatic ecosystems (Boyd 2012). Agricultural lime (calcite or dolomite), however is employed in some fertilization protocols to increase the alkalinity and hardness of acidic waters and thereby increase the effectiveness of the phosphorus additions. Since solar energy input is critical to photosynthesis, a lack of sunshine can also limit algal production (Neal and Kröger 2012).

Inorganic fertilization regimens are designed to support sustained zooplankton and benthic populations, and are used in combination with varying fry stocking densities to produce desired fingerling quantity and quality (Hartleb et al. 2012). A considerable amount of research has been conducted to develop inorganic fertilization protocols for the production of walleye fingerlings (see Tice et al. 1996; Culver 1996; Jacob and Culver 2010). These methods are generally applicable to yellow perch fingerling production (Soderberg 1977).

Variables such as timing the initiation and frequency of fertilizer application, desired nutrient levels and relative amounts of nitrogen and phosphorus all have implications for the successful culture of yellow perch fingerlings. Because of their small size at hatch and correspondingly small mouth gape, providing the appropriate organisms for first feeding is critically important to success culturing yellow perch fingerlings. Published reports on larval yellow perch feeding indicate that rotifers, copepod nauplii and adult copepods are important initial food sources (Siefert 1972; Bremigan et al. 2003; Fulford et al. 2006). Hartleb (2003) found larval diet selection included benthic insects, ostracods and green algae, although the presence of green algae in the diet coincided with poor growth and high mortality.

The time interval between pond filling with the associated initial application of fertilizer and fry stocking should be managed to assure an abundance of small prey for the larval fish. This interval is generally 3–10 days but will vary with a number of factors including water source, water temperature and fertility of the pond substrate. Poor results will be experienced if fry stocking occurs after rotifer and copepod nauplii populations decline in favor of larger copepods and cladocerans.

Culver (1996) recommended a nitrogen to phosphorus (N:P) ratio of 20:1 by weight with no more than 30 μg of P L^{-1}. This ratio favors the production of small algae that is most easily consumed by crustacean zooplankton; the low concentrations of P limit the production of large blue-green and green filamentous species. Culver (1996) also suggested weekly additions of inorganics based on water quality analysis to maintain the N:P balance, however water quality monitoring at the Lake Mills State Fish Hatchery (Lake Mills, WI) indicated that nitrogen was virtually consumed within 3 days. A protocol employing twice-weekly applications was adopted to maintain the proper N:P ratio in walleye fingerling ponds (Robert Fahey Wisconsin Department of Natural Resources, pers. com.). After approximately 3 weeks, phosphorus begins to recycle from decaying plant material and in many cases phosphorus additions are discontinued to maintain N:P ratios and limit the growth of filamentous algae (Hartleb et al. 2012).

Table 9.1 compares the species distribution and relative abundance of zooplankton sampled from yellow perch fingerling production ponds fertilized with either organic or inorganic regimens. The organic protocol called for an initial application of soybean meal at a rate of 336.3 kg ha^{-1} with weekly additions of 112.1 kg ha^{-1}. The inorganic protocol employed initial applications of 35.9 kg ha^{-1} urea, 9.4 kg ha^{-1} triple superphosphate and 197.3 kg ha^{-1} agricultural lime with weekly additions of 23.5 kg ha^{-1} urea, 4.0 kg ha^{-1} triple superphosphate and 89.7 kg ha^{-1} agricultural lime. As can be seen from the data the inorganic protocol resulted in roughly 3.5 times the number of small prey (rotifers and juvenile stages of copepods) and 1.5 times the number of large prey (adult copepods and cladocerans as the organic treatment.

Of course, this is not the whole story. Availability of appropriate sized prey for the growing fingerlings is critical to success. Figure 9.2 describes the zooplankton populations of the two treatments at each of the sampling points. The inorganic

Table 9.1 Relative abundance of zooplankton in yellow perch fingerling production ponds using organic and inorganic fertilization regimens. The data represents the mean organisms L^{-1} and S.E.M. of four ponds per treatment sampled six times at weekly intervals from fry introduction to harvest

	Organic fertilization	Inorganic fertilization
Rotifera		
Asplanchna spp.	20.0 ± 6.5	51.0 ± 37.7
Brachionus spp.	7.3 ± 3.0	80.8 ± 43.7
Filinia terminalis	1.1 ± 0.4	18.0 ± 8.6
Keratella spp.	340.3 ± 121.7	1414.5 ± 575.1
Polyarthra spp.	21.4 ± 11.4	46.5 ± 19.5
Synchaeta spp.	9.5 ± 2.9	43.6 ± 29.0
Misc.Rotifers	2.0 ± 1.1	0.6 ± 0.3
Total	401.6	1655.0
Copepoda (Juvenile)		
Copepodites	6.2 ± 3.6	16.1 ± 8.2
Nauplii	137.8 ± 42.6	205.7 ± 57.0
Total	144.0	221.8
Copepoda (Adult)		
Calanoid Copepods	1.1 ± 0.6	0.6 ± 0.4
Cyclopoid Copepods	5.9 ± 3.4	26.6 ± 13.2
Harpactoid Copepods	0 ± 0	0 ± 0
Total	7.0	27.2
Cladocera		
Bosmina longirostris	4.4 ± 1.8	6.9 ± 3.0
Chydoridae	5.3 ± 1.4	1.7 ± 1.0
Daphnia spp.	14.4 ± 4.5	12.2 ± 6.6
Misc Cladocerans	4.2 ± 1.4	1.5 ± 0.5
Total	28.3	22.3

protocol resulted in considerably higher numbers of small prey during the first 1–3 weeks of culture and slightly higher numbers of large prey during the final weeks of culture. The wide variability between ponds and the limited number of replicates makes it difficult to draw meaningful conclusions using standard statistical methods, however the trends, especially for the small prey, appear obvious. To the producer the only meaningful measure of success is the number of fish produced. In this study, the ponds using the inorganic protocol averaged approximately 437,000 ± 127,000 fingerlings ha^{-1} and the organic 310,000 ± 80,000 fingerlings ha^{-1}. Survival averaged 70 % and 49 % for inorganic and organic ponds, respectively. All ponds were stocked at a rate of a 673,000 fry ha^{-1} and ranged in size from 0.04 to 0.57 ha.

Combined fertilization protocols rely on both organic and inorganic fertilizers. This method provides for the development of multiple webs as well as the ability to respond to environmental changes (Hartleb et al. 2012). Combined fertilization techniques are the same as inorganic methods with the exception of initial and supplemental applications of organics. The organics support the heterotrophs and

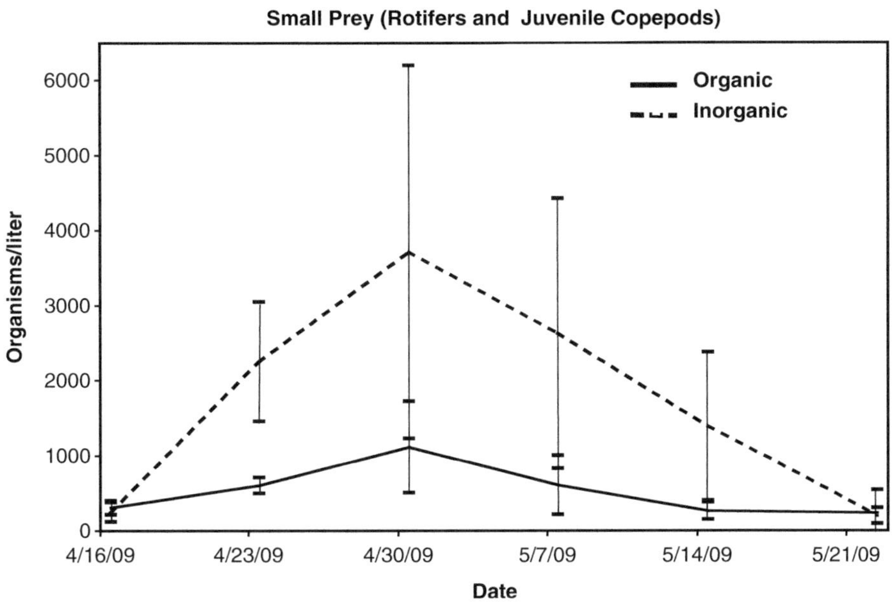

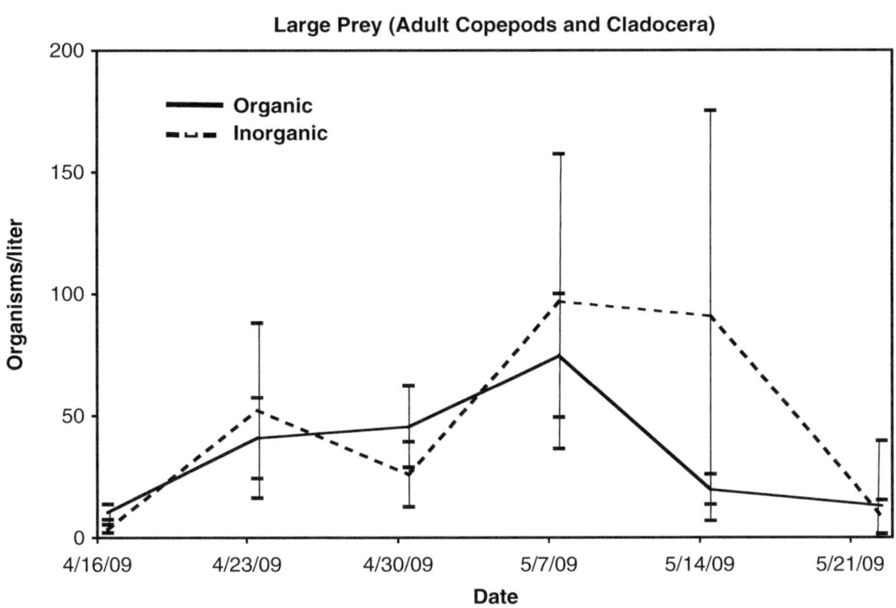

Fig. 9.2 Abundance of small (*top* graph) and large (*bottom* graph) prey over the course of the culture period in yellow perch fingerling production ponds fertilized using organic and inorganic fertilization protocols. The data represents the mean ± S.E.M. of n = 4 ponds/treatment

the secondary release of nitrogen and phosphorus from decomposition is accounted for in the water quality monitoring of the inorganic protocol. While the combined method results in the use of less chemical fertilizers, its advantage may be limited to years in which a lack of sunshine results in poor algae production (Hartleb et al. 2012) and therefore a greater reliance on the heterotrophic food web.

9.2.2 *Larval Stocking Density, Survival and Growth of Eurasian Perch in Ponds*

Depending on climatic conditions prevailing in European countries, pond stocking of newly hatched larvae is usually undertaken from late March to late April. Alternatively stocking can be accomplished with fertilized eggs at the eyed stage approximately 1–2 days prior to hatching. In this case eyed eggs are placed in small floating cages (0.4 × 0.4 × 0.4 m) with small mesh sizes (1.5 mm) retaining the eggs but allowing the passage of hatched larvae. Fertilized ribbons can also be draped over supporting branches laid throughout the pond (Kestemont et al. 2008). Pond stocking rates for Eurasian perch are highly variable ranging from 10,000 to 60,000 eyed eggs per 100 m^2.

Several studies have shown that the production of juveniles in ponds rarely exceeds 1,000–5,000 juveniles per 100 m^2 at a mean body weight of 0.5–1.5 g after 2 months of culture regardless of initial larval stocking density. This represents a survival rate of 5–20 %. Ponds should be stocked with egg ribbons or larvae that are close in age to reduce size heterogeneity and the potential impact of cannibalism caused by early-hatched fish (more details on cannibalism in perch are provided in Sect. 9.3.3). High stocking densities induce a rapid depletion of natural resources and, consequently, the emergence of cannibalistic behavior. A regular assessment of zooplankton populations allows for identification of harvest times whereby fingerlings are harvested before complete zooplankton population declines, resulting in fewer instances of in-pond cannibalism (Kestemont et al. 2008). In Ireland, the practice is to remove Eurasian perch juveniles after about 8 weeks (mid to late June), at a body weight of <0.5 g, for feed training in tanks. At this stage, fish display shoaling behavior, and the shoals usually contain fish of the same size, limiting the need for initial grading when transferred into tanks.

9.2.3 *Yellow Perch and Eurasian Perch Fingerling Habituation*

Habituation of yellow perch fingerlings to formulated feeds and intensive culture conditions is an important step in the production protocol. In this process, yellow perch fingerlings are harvested from production ponds, concentrated in tanks and trained to accept formulated feed. Once trained, the fingerlings can be stocked into ponds or tanks for grow out to market size.

Pond stocking rates for yellow perch fry range from 375,000 to 1,500,000 fish ha^{-1} depending on production goals. Stocking rates determine the length of time fingerlings can remain in the pond before food requirements of the fish exceed available zooplankton populations. Lower stocking rates allow a longer pond phase and therefore larger fingerlings. Larger (>35 mm TL) fingerlings, however, are not necessarily the most desirable subjects for habituation to formulated feed. Malison and Held (1992) found no difference in the habituation success of yellow perch when comparing well-conformed fish averaging 16.9, 32.5, and 42.6 mm TL. Held et al. (1998) demonstrated that a combination of high fry stocking densities (1,000,000 fry ha^{-1}) and early pond harvest, when fish reach 17–25 mm TL, resulted in production as high as 494,193 fish ha^{-1} of which 50–70 % can be habituated to formulated feed. The physical condition and body conformation of harvested fingerlings are important to habituation success. Predictably, efforts at habituating overly stressed or starving emaciated fingerlings regardless of length results in very poor success.

Another important consideration in deciding what size fingerling to harvest is the duration of the habituation period, that is, the number of days it takes for all fish in the cohort to be actively feeding on the formulated diet. Most commercial yellow perch fingerling operations are limited by tank space to undertake the habituation process. Far more perch fingerlings can be produced in ponds than can be accommodated in tanks for habituation at any one time. Accordingly, a short habituation period allows fish to be advanced through the process more rapidly, resulting in more feed-trained fish over the course of the season.

Recording daily mortalities in the habituation tanks will help to define the end of the habituation period. Since the fish in each cohort are virtually the same size, those not accepting the formulated diet will starve and die within days of each other; a graph of daily mortalities will show a well-defined peak (Fig. 9.3). Malison and Held (1992) found that yellow perch harvested at 16.9 mm TL required 13 days to complete the habituation process, yellow perch at 32.5 mm TL required 32 days and fish at 42.6 mm TL required 47 days (Fig. 9.4).

Recent modifications of the habituation protocol developed by researchers at the University of Wisconsin Aquaculture Program (Lake Mills, WI) have resulted in habituation success rates of 80–90 % and are currently in practice by several yellow perch culturists. The use of freeze-dried krill as an attractant in combination with the formulated diet has demonstrated noticeably improved initial acceptance of the food. Yellow perch fingerlings respond to the krill within hours of harvest rather than days as previously experienced with formulated feed alone. Malison and Held (1992) found that subdued internal tank lighting improved the habituation success of yellow perch fingerlings. Acting on these results, commercial culturists have employed internal lighting or dim overhead lighting during habituation to reduce the agitated activity of perch fingerlings responding to shadows and movement near the tanks (David Northey Coolwater Farms LLC, pers. com.).

Other modifications of the habituation protocol employed by culturists at Coolwater Farms (Cambridge, WI) include partial (seine) harvest of fingerling production ponds and the use of micro ponds. Coolwater Farms culturists remove a portion of the fingerling population from production ponds at the smallest practical

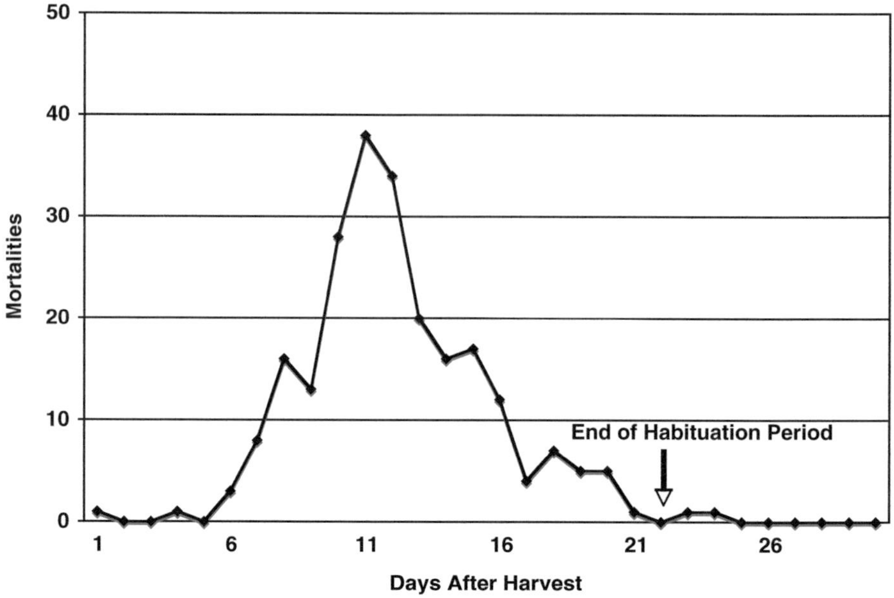

Fig. 9.3 Typical mortality curve generated from in-tank habituation of yellow perch fingerlings harvested from ponds at 21 mmTL. The data represents the number of mortalities removed from the tank each day

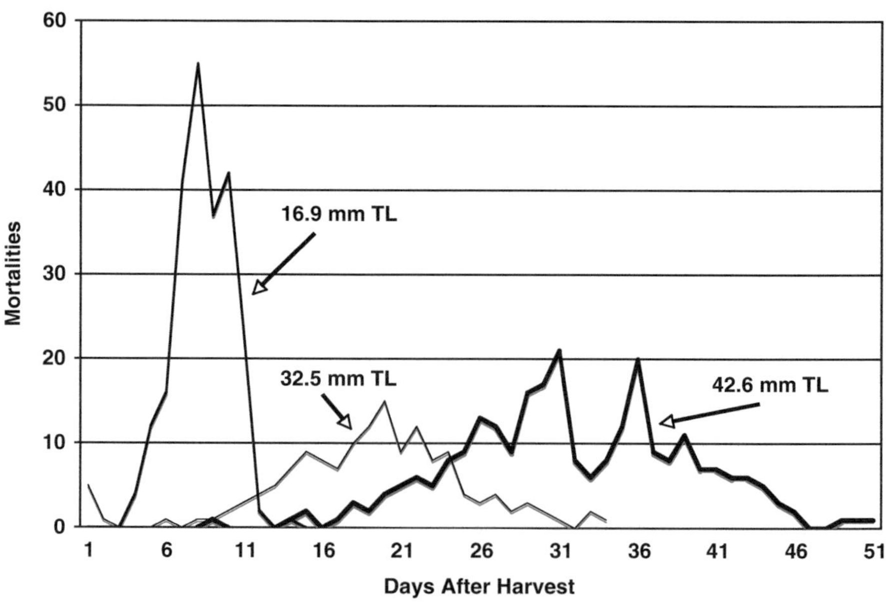

Fig. 9.4 Mortality curves during in-tank habituation for yellow perch fingerlings harvested from pond at 16.9, 32.5, and 42.6 mmTL (Data from Malison and Held (1992))

size for tank habituation (~20 mm TL) thereby reducing the predation pressure on zooplankton populations in the ponds. At this time in-pond feeding is initiated to supplement natural productivity and expose the remaining fingerlings to formulated feed. The harvested fingerlings are habituated in tanks for 4–7 days and then transferred to very small (12 m^2) micro ponds for a further 7 days of habituation and conditioning to outdoor conditions. Finally the fish are released to the grow out pond. Using this protocol, limited tank space can be reallocated to a new harvest of juveniles and the semi-habituated fingerlings can experience a transition from indoor to outdoor culture. A significant number (approx. 5–15 %) of habituated fingerlings will revert to natural forage when released to grow out ponds without this transition. At the end of the first season of growth, these foragers can be easily identified by their small size and vivid coloration.

Habituation of yellow perch fingerlings can also be undertaken in ponds, initially using lights at night to attract the highly photopositive fingerlings to vibrating feeders that distribute diet at frequent intervals. Ultimately, feeding can be extended to daylight hours with fish being attracted by the vibration of the feeders. In-pond techniques eliminate the need for indoor tank facilities and the considerable husbandry associated with tank habituation. Additionally, in-pond habituation allows for earlier exposure of formulated feed to fingerlings that are too small and delicate to harvest and an extended transition from natural to formulated diet. In this protocol, production as high as 615,000-trained fingerlings ha^{-1} has been documented (Malison et al. 1994). Subsequent (unpublished) trials of this technique have indicated that harvest and size grading of the habituated fingerlings was necessary to maintain a uniform size cohort and avoid social dominance and potential cannibalism (Fig. 9.5).

Fig. 9.5 Typical size distribution of yellow perch after 6 weeks of in-pond habituation

9.3 Semi-intensive Culture in Mesocosms

The mesocosm (green water) system is one of the optimal culture systems for the production of weaned perch (*Perca fluviatilis*) fry under semi-intensive conditions (Mélard et al. 1996; Kestemont et al. 1996). At the end of the larval rearing period, survival ranges from 20 % to 40 % depending on culture conditions (initial stocking density, temperature, rotifer abundance) and the intensity of cannibalism. No deformities were recorded in Eurasian perch reared in this semi-intensive fertilized system. Good survival rates and fry quality probably result from the initial access to live preys: rotifers, protozoa and algae. Algae have been considered to be beneficial to larval fish during the first feeding, even in carnivorous fish like perch (Nash et al. 1974; Dabrowski 1984).

9.3.1 Rearing Method

A large set of experiments aimed at determining optimal rearing methods and culture conditions are summarized hereafter. Larval rearing facilities consisting of rectangular 10 m^2 (5 m^3) out-door concrete tanks were supplied with river water, or better with de-saturated pathogen-free well water during a 44-days rearing period. Water temperatures ranged from 17 to 23 °C. Air diffusers placed in each tank maintained the oxygen level around 8 ppm. The tanks were fertilized with an initial single input of 1.5 kg of chicken manure pellets put in a basket that stimulated the development of phytoplankton then zooplankton (rotifers) (Fig. 9.6). Chicken manure pellets were removed after 4–5 days. During the 7-day fertilization period, the temperature was maintained at 24–25 °C (heating system, green-house) to accelerate and increase the plankton production. During this period no water renewal was applied. Since young perch are unlikely to survive handling at the larval stage, ready-to-hatch eyed perch eggs were stocked on trays in the tank at day 8 after tank fertilization, when the availability of zooplankton is at its maximum (density up to 2,500 rotifers L^{-1}). Five days after hatching, the tanks were partly (80 %) shaded to prevent any further development of filamentous algae that may trap larvae. During the first days, water renewal rate was null to prevent the flushing of algae and rotifers. Tanks were then progressively transformed to flow through systems with water renewal rates from 1 to 2 $m^3 h^{-1}$ depending on stocking density and biomass increase.

Fish were fed with *Artemia* nauplii 2 days after hatching as a complementary food to rotifers. Depending on initial larval stocking density, rotifers were completely consumed by Eurasian perch larvae 2–4 days post-hatching. The weaning period extended from the 7th to the 30th day of feeding during which *Artemia* nauplii were progressively replaced by formulated larval dry diet (54 % protein content), then by a standard fry trout diet (50 % protein content). Food distribution was exclusively diurnal using automatic feeders for formulated diets and manual distribution (four times daily) for *Artemia* nauplii. Feed introduction took place in the

Fig. 9.6 Mesocosms used for the semi-intensive (*green water*) rearing of Eurasian perch larvae, after fertilization with chicken manure

non-shaded area of the tank. The use of high quality larval formulated feed could probably reduce or even suppress the use of *Artemia* but to date this substitution has not been tested with Eurasian perch. Fish were fed close to the maximum food ration determined from the following equation:

$$R(\%\text{body weight}) = 45.902\,W^{-0.265}, W = \text{body weight}(\text{mg}).$$

At the end of the 44-day rearing period, fish were collected and graded for the first time into two categories, based on an obviously bimodal size and weight distribution (Fig. 9.7). Fast growing fish were ranked as potential cannibals.

9.3.2 Optimal Temperature and Stocking Densities

At stocking densities ranging from 400 to 4,000 larvae m^{-2}, growth is negatively correlated with stocking density at 20 °C, but positively at 23 °C (Fig. 9.8). Growth rate of fish is clearly enhanced at 23 °C. However, survival rate is lower (mean = 18.7 %) at 23 °C than at 20 °C (mean = 29.2 %) (Fig. 9.9).

Regardless of temperature, coefficient of weight variation and proportion of cannibals are reduced at high stocking densities (Fig. 9.9): the number of cannibals when plotted against stocking density describes a semi-logarithmic relationship (Fig. 9.10). Cannibals represent ±2 % of the final population at initial stocking den-

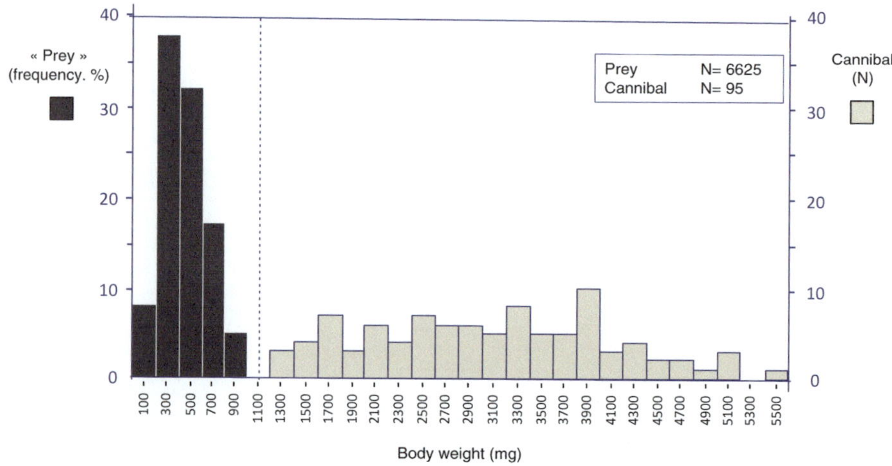

Fig. 9.7 Typical bimodal weight distribution at the end of the rearing period (44 days) (From Mélard et al. (1996))

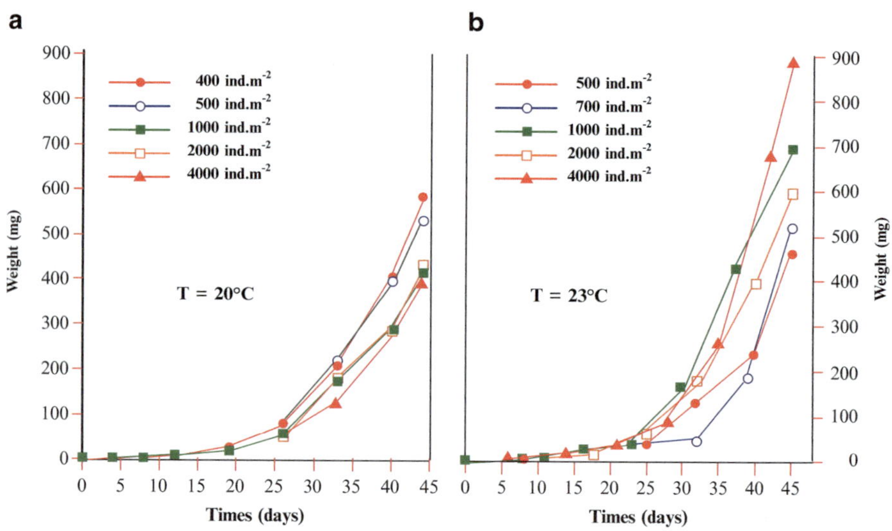

Fig. 9.8 Effects of stocking density and temperature on the growth of perch larvae in semi-intensive system (From Kestemont et al. (1996))

sity of 4,000 larvae m^{-2} and ±4 % at 500 larvae m^{-2}. This model clearly recommends increasing stocking density as much as possible to reduce the impact of cannibalism and attain the optimum growth-survival combinations. Still, it should be emphasized that higher stocking densities up to 4,000 larvae m^{-2} would probably generate physicochemical conditions beyond the tolerance range of Eurasian perch larvae. Similarly, the quantity of live natural planktonic prey in the rearing system is lim-

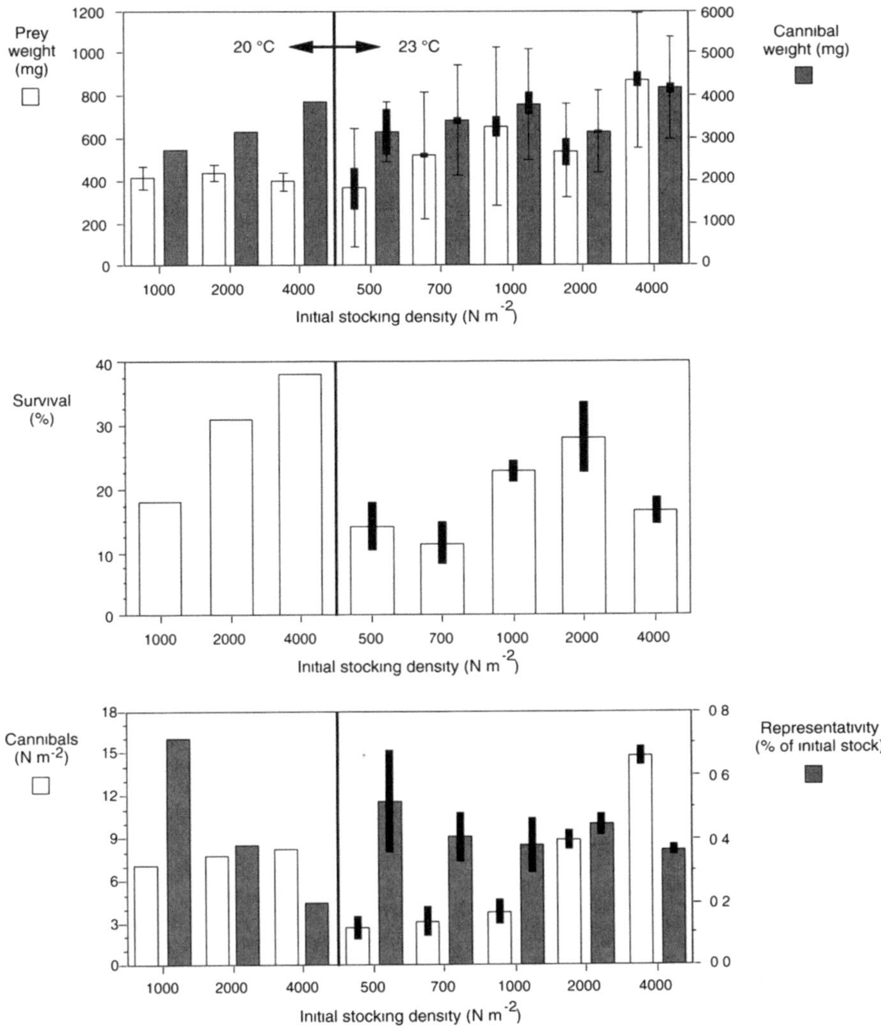

Fig. 9.9 Effect of stocking density and temperature on growth, survival and emergence of cannibalism in perch larvae reared in semi-intensive system during 45 days (From Mélard et al. (1996))

ited, resulting in a possible shortage of prey/larvae at initial densities higher than 4,000 larvae m^{-2}.

Increased initial density of rotifers (expressed as the number of rotifers $fish^{-1}$) improves the fish survival rate and markedly reduces cannibalism but negatively affects the growth rate of Eurasian perch larvae. The early weaning of perch reared under a restricted ration of rotifers but trained to ingest formulated dry food earlier probably explains this inverse relationship between rotifer availability and growth of larvae.

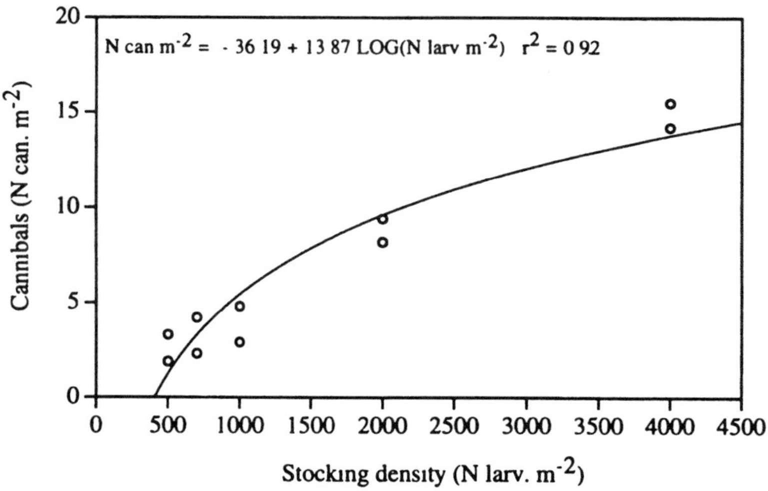

Fig. 9.10 Relationship between initial stocking density and cannibal frequency in perch larvae reared in semi-intensive system during 45 days (temperature 23 °C) (From Kestemont et al. (1996))

Table 9.2 Cannibalism, growth and survival rate of perch larvae reared during 44 days in mesocosm (10 m^2) tank at 17, 20 and 23 °C. Initial weight: 0.9 mg. Initial stocking density: 4,000 larvae m^{-2} (From Mélard et al. (1996b))

Temperature (°C)	17	20	23
Mean final weight (mg, ± SD)	195 ± 49.3	336 ± 107.3	607 ± 212.3
Growth (mg $fish^{-1}$ d^{-1})	4.4	7.6	13.8
Specific growth rate (% d^{-1})	12.22	13.46	14.80
Coefficient of variation (%)	33	40	42
Survival (%)	32.4	13.9	6.3
Cannibal fish (final N m^{-2})	0.7	6.3	8.6
Cannibal fish (% final population)	0.05	1.30	3.50

Rearing at lower temperature also limits the emergence of cannibalism during the larval stage: the ratio of cannibals decreases from 3.50 % of the final population at 23 °C to 0.05 % at 17 °C (Table 9.2). Consequently, survival is significantly higher at 17 °C but growth rate is 70 % lower than at 23 °C (Table 9.2). More frequent pathologies, mainly parasites (Grignard et al. 1996), also induce higher mortality at 23 °C.

9.4 Intensive Culture in Recirculating Systems

As stressed by Kestemont and Mélard (2000), intensive culture of fry has many advantages (e.g. fairly stable culture conditions, rearing of fry produced by off-season spawning, more predictable production of juveniles and easier control of cannibalism) compared with extensive production in ponds or semi-intensive methods in mesocosms. However, there are several critical factors affecting the success of perch intensive culture including: poor knowledge of fish nutritional requirements, clinging and non-feeding behaviors, non-inflation of the gas bladder and cannibalism. The small mouth gape of larvae at hatching, particularly in yellow perch, and their dependence on live feed organisms have also been considered as limiting factors in intensive fry culture. Despite highly variable and sometimes unsatisfactory survival rates, a new generation of high quality artificial feeds, formulated for marine fish species, has made the problem of starter feeding a concern of low priority. Co-feeding (combining rotifers and/or *Artemia* nauplii as live preys and dry feed, either concurrently or successively) is still largely used in commercial Eurasian and yellow perch hatcheries, but feeding newly-hatched larvae with artificial dry feed is also feasible. Feeding and nutritional aspects of Eurasian and yellow perch early life stages are not developed in this chapter as more details are provided in Chaps. 20 and 21.

The system developed at the Ohio State University (Wojno et al. 2012) for rearing yellow perch larvae-juveniles (10–14 days) includes conical tanks with upwelling water and sprinklers designed to break surface tension (Fig. 9.11). Tanks are continuously provided with algal suspension (Instant Algae Premium 3,000, Reed Mariculture, Campbell, CA) to create turbidity of 10–14 NTU. Larvae are stocked at the density of 20–100 per L and feeding is initiated 2–3 days after hatching at 20–25 °C. The first

Fig. 9.11 Conical tanks (70 L) with central outlet, sprinkler for water tension break up and continuous supply of algae suspension to achieve turbidity of 10–14 NTU

food constitutes live rotifers (3–4 days) followed by 1–2 day transition to live brine shrimp. In the successful rearing conditions, yellow perch larvae were provided with 30 rotifers mL^{-1} as early as 2 days after hatching. This was followed with *Artemia* feeding (4 nauplii mL^{-1}) 1–2 days after receiving rotifers. After 10–13 days in the intensive culture system, provided exclusively with live food, larvae were weaned to dry feeds. The second phase of experiments was carried out in larger tanks to wean from live food to dry-formulated feeds. Larvae cultured under 24 h light conditions had significantly decreased swim bladder inflation (39.6 ± 8.3 %) compared to those that had 12L:12D photoperiod (60.7 ± 14.5 %). The 24 h light, however, had no effect on the survival (59 % versus 54 %) and growth (12.0 ± 1.3 versus 11.5 ± 1.5 mm body length). The experiment showed no effect of the sprinkler angle on swim bladder inflation or the survival. No effect was shown when sprinklers alone or together with oil-absorbing pads were used. Body lengths after 13 days of rearing in intensive conditions were influenced by the origin of broodstock fish and amounted to 11.5 ± 1.7, 10.2 ± 0.4 and 9.0 ± 0.5 mm for progenies from three different females. Furthermore, this study suggested that survival of larvae can also be affected by the origin of females: wild (20 %) or domesticated (80 %).

9.4.1 Size Heterogeneity and Cannibalism

As reviewed by Kestemont et al. (2003), fish of both marine and freshwater origins generally produce sibling eggs and newly hatched larvae of very similar size, with a coefficient of variation in weight (calculated as 100 $SD_{weight}/mean_{weight}$) ranging between 7.5 % in salmonids to 15–20 % in European seabass. Coefficients of variation in weight of Eurasian perch and yellow perch are within this range (15–18 % and 8–10 %, respectively, Kestemont et al. 1996; Kestemont and Mélard 2000). As in other percomorphs, a substantial proportion of percid fish larvae (e.g. 30–40 % in Eurasian perch) are unable to prey on *Artemia* nauplii, inducing an early mortality in starved larvae. Beyond day 11, cannibalism contributes to eliminating the few remaining individuals weakened by starvation (Baras et al. 2003). Braband (1995) reported that Eurasian perch consumed sibling tail first or head first. Cuff (1980) described the type I cannibalism as a prey capture tail first. The prey is swallowed progressively up to the head, which is eventually discarded. As size heterogeneity develops, cannibals switch to complete (type II) cannibalism, capture the prey head first and swallow them completely. According to Baras et al. (2003), type I cannibalism (which has a relatively low impact on overall survival) is exerted by Eurasian perch from day 11 onwards, and ceases around day 16–18 post-hatching, while the appearance of type II cannibalism (inducing up to 50 % of total mortality) is determined by gape-size limitations and occurs from days 12 to 14 post-hatching. Type II cannibalism can be mitigated by size-sorting or other methods of limiting size heterogeneity.

Huston and DeAngelis (1987) suggested that changes in size distribution results from the interaction of four primary factors related to characteristics of individuals composing the population: initial sizes, distribution of growth rates due to differences among individuals, size and time dependence of each individual's growth rate and mortality that may affect size classes differently. In an attempt to classify the biological mechanisms producing bimodality within monospecific populations, Kestemont et al. (2003) adapted this four-class ranking proposed by Huston and DeAngelis (1987) by classifying all mechanisms as inherent (i.e. having a strong genetic component) or imposed (requiring certain biotic or abiotic conditions to be manifested). Both inherent (e.g. genetic variation, changes in morpho-physiology of larvae) and imposed mechanisms (temporal and spatial heterogeneity, competition, density-dependent or independent mortality), may or may not require interactions between individuals to be expressed.

Consequences of size heterogeneity at the early life stages are usually more dramatic than at the juvenile or adult stages, principally because larvae have a large mouth relative to body size, and they can exert cannibalism on prey (here their siblings) that are just slightly smaller (Baras 1998). Cannibalism generally requires a size difference between the potential cannibal and the victim, and is thus facilitated by size heterogeneity (Claessen et al. 2000), but it also affects size heterogeneity, since the smallest fish are consumed by the largest ones. As stressed by many authors, cannibalism in wild as well as culture environments can thus be viewed as a cause or consequence of size heterogeneity (DeAngelis et al. 1979; Hecht and Appelbaum 1988; Baras 1998).

The respective importance of several intrinsic and extrinsic factors behind growth depensation (i.e. the increase of size heterogeneity over time) and size heterogeneity was investigated within a large multi-factorial study conducted in the framework of a European project to determine the biological bases for minimizing fish competition and cannibalism in Eurasian perch larvae and post-larvae (Kestemont et al. 2003). The larval stage was defined as the period from hatching to weaning on dry feed, while the post-larvae referred to the post-weaning period (from 50 mg to several hundred mg), assuming that the fish have acquired an adult mode of digestion (see Chap. 8 for complete description of digestive system during early life stages) . The effects of different variables, grouped into three main categories (population, abiotic environment and feeding variables) were analyzed in terms of final size variation and growth, survival and cannibalism. Table 9.3 summarizes the main significant outputs of this study.

Among the population variables, hatching time and stocking density significantly impacted on the population dynamics of Eurasian perch. Within a cohort, earlier hatch resulted in higher survival and faster growth. This was observed both in mixed groups (larvae hatched on D1 + D2 were labeled in alizarin red solution to be distinguished from larvae hatched later) and when fish hatching at different times were reared separately. Obviously, fish hatching earlier have greater competitive capabilities; they gain access to food earlier, grow faster and can eventually exhibit cannibalistic behavior on their younger siblings. Different hatching times within a single batch of eggs can also reflect different quality of eggs, and consequently different quality of larvae (Kestemont et al. 2003). The effects of stocking density on rearing performance differed between perch larvae and post-larvae. At the larval stage, the slowest growth and higher rates of

Table 9.3 Population, environment and feeding variables significantly affecting the final size variation, growth[a], survival and/or cannibalism in Eurasian perch larvae and post-larvae reared in tanks (Data compiled from (Kestemont et al. 2000, 2003; Baras et al. 2003))

Variables	Stage	Tested levels	Initial mass (mg)	Final mass (mg)	CV_{weight} (%)	Survival (%)	Cannibalism (%)
Population variables							
Hatching time	Larvae	D1+D2[b]	0.8	84.7[a]	30.9	62.2[a]	10.7
		D3[b]		59.1[b]	53.5	39.0[b]	3.97
		D1–D3[b]		59.4[b]	46.9	30.0[b]	6.56
Initial size heterogeneity	Larvae	Standard[c]	8.60	121	29.7	47.3	8.70
		High[c]		155	40.0	56.0	11.9
		Controlled[c]		144	40.3	47.2	20.6
	Post-larvae	Standard[c]	84.4	442	35.3	54.5	2.2[a]
		High[c]		500	38.1	36.2	19.8[b]
		Controlled[c]		280	35.9	42.9	16.4[b]
Stocking density	Larvae	10 fish L^{-1}	1.04	7.3[a]	46.2	14.7	54.6[a]
		32 fish L^{-1}		9.7[b]	43.1	18.9	29.6[b]
		100 fish L^{-1}		11.2[c]	36.1	20.1	35.7[a,b]
	Post-larvae	1 fish L^{-1}	188	816[a]	27.2	97.3[a]	0[a]
		3.2 fish L^{-1}		932[b]	31.4	95.8[a]	0.3[a]
		10 fish L^{-1}		865[a,b]	32.5	75.3[b]	21.1[b]
Environmental variables							
Day length	Larvae	LD 8 :16	1.04	21.7[a]	43.4[a]	39.6[a]	19.8
		LD 12:12		68.3[b]	34.8[a,b]	45[a,b]	18.7
		LD16:8		62.2[b]	24.0[c]	49[a,b]	21.8
		LD 24:0		67.8[b]	25.0[b,c]	56.2[b]	25.0
	Post-larvae	LD 8:16	83.8	290	28	93.2	10.2[a]
		LD 12:12		296	24.3	87.0	7.7[a]
		LD16:8		313	29.5	90.7	4.0[a]
		LD 24:0		280	24.7	88	0.4[b]
Light intensity	Larvae	5 Lux	1.04	58.7[a]	45.6	44.4	27.5
		30 Lux		90.0[a,b]	48.3	47.7	39.6
		90 Lux		88.5[a,b]	38.1	53.3	19.9
		400 Lux		112[b]	36.6	52.7	25.1
	Post-larvae	5 Lux	45.9	408[a]	30.3	73.3	22.3
		30 Lux		324[a,b]	39.1	74.8	11.4
		90 Lux		290[b,c]	30.5	67.8	15.5
		400 Lux		219[c]	34.2	69.9	19.2
Feeding variables							
Feeding level (% biomass day^{-1})	Larvae	1	8.6	12[a]	33	30[a]	5.8
		5		15[a]	42	78[b]	6.2
		10		39[b]	39	78[b]	10
		20		52[c]	55	67[b]	8

Values with different superscripts within a same column are significantly different ($p < 0.05$)

[a]Mass values refer to non-cannibal fish

[b]D1 + D2 = larvae hatched during the first 2 days of the hatching period; D3 = larvae hatched on day 3; D1–D3 = larvae hatched on days 1–3

[c]Standard size heterogeneity = sibling fish reared in the same tank from hatching under standard conditions, high = sibling fish initially reared at two different stocking densities to create size differences, then mixed altogether; controlled = sibling fish with regular size-sorting (*i.e.* removal of both large and small individuals)

cannibalism were observed at low stocking density and can be attributed to several factors, such as the development of dominance hierarchies, growth depensation and territoriality. Baras et al. (2003) described a shoaling behavior of larvae held at intermediate and high stocking density while fish reared at low density were dispersed. At high stocking density the number of available territories is proportionally limited, as is the number of cannibals, explaining why the cannibalism was less intense at the highest density. The number of fish classified as cannibals at the end of the experiment was proportional to stocking density, but the proportion of cannibals relative to the initial number of fish was negatively correlated to stocking density (Baras et al. 2003). Similar observations were made by Mélard et al. (1996) in large-scale experiments conducted in mesocosms (see also Sect. 9.3). At the post-larval stage, unlike the larval stage, Eurasian perch were negatively affected by high stocking density, which induced slow growth and high rates of cannibalism. A density-dependent variation in food availability was suggested to explain these results; all fish reared at low and intermediate stocking density having access to food, while not at the highest density. Surprisingly, initial size heterogeneity had very little influence on final size heterogeneity and cannibalism in both larvae and post larvae of Eurasian perch while this factor is frequently invoked as a key to cannibalism and growth depensation in many other species, including the walleye *Sander vitreum* (Loadman et al. 1986). In Eurasian perch, the frequent removal of larger, supposedly cannibalistic individuals did not improve survival or growth, probably because their removal promoted the rapid establishment of new dominance hierarchies and resulted in more frequent agonistic interactions between fish. Therefore, the usefulness of size-sorting practices at the larval stage is highly questionable, except for the initial sorting based on hatching time. The same statement was made by Babiak et al. (2004) when investigating the effects of initial weight and its variation on quantitative characteristics of juvenile cohorts in Eurasian perch larvae. Despite considerable differences in initial CV of different Eurasian perch progenies (CV: 219 % versus 34.5 % and 56 % at 7 mg initial body weight), intermediate and final CVs of weight were stable at 30–40 % (for fish of 2–5 g final body weight) regardless of progenies and growth characteristics (Babiak et al. 2004).

Rearing temperature is probably among the main environmental variables susceptible to influence the size heterogeneity and, consequently, the cannibalism rate, as it governs the fish growth rate, and thus may induce growth depensation. Since a temperature of 20–23 °C is usually recommended as optimal for the intensive larval culture of Eurasian perch, there are few data available on the effects of low or high temperature on growth, size heterogeneity and cannibalism in fish reared in recirculating systems. Kestemont and Mélard (2000) reported that temperature significantly influences the specific growth rate (SGR) of Eurasian perch larvae (from 16.4 to 118 mg body mass), with a maximum of 22.0 % day^{-1} at 23 °C when fish were fed high daily rations (35 % body mass per day) while, at 14 and 26 °C, the SGR did not exceed 11 and 15 % day^{-1}, respectively (see also Sect. 9.3).

As shown in Table 3.1, survival, growth and size homogeneity among larvae were improved in Eurasian perch when day length was increased. Percid fish larvae are visual feeders, exhibiting a stronger feeding behavior in light than in darkness. Moreover, the increased and well distributed number of meals under long day length or continuous lighting might have enhanced larval food access, reducing growth depensation and, as a corollary, the incidence of cannibalism. An alternative explanation may be that potential cannibals can forage over longer periods of time thus reducing their tendency for cannibalistic activity (Kestemont et al. 2003). It is likely that increased day length impacts the behavior of both potential cannibals and potential prey. At the post-larval stage, day length did not significantly influence the survival of Eurasian perch but the proportion of mortality due to cannibalism was significantly reduced when day length was increased from LD 8:16 to LD 24:0.

Feed supply greatly affects the establishment of dominance hierarchies, growth depensation and cannibalism (Paller and Lewis 1987; Hecht and Pienaar 1993). Even in species rarely displaying cannibalistic behavior such as the goldfish *Carassius auratus*, cannibalism occurs frequently among larvae when they are poorly fed or starved (Kestemont 1995). However in Eurasian perch, cannibalism was higher at an intermediate ration when compared to low or high rations. If high feeding level has limited the impact of cannibalism, low ration has impaired the growth rate of all fish, postponing the cannibalism until some fish reached the minimum size to exert their cannibalistic behavior.

9.4.2 Clinging, Non-feeding Behavior and Non-inflation of the Gas Bladder

In nature, the presence of suspended matter (phytoplankton, clay, etc.) is responsible for rather high turbidity, limiting the light penetration into water. In tanks, water turbidity is low (usually below 5 NTU), light passes through the water and is reflected from tank surfaces. The term "clinging behavior" has been initially used by Summerfelt (1996) for describing the attraction of larval walleyes to these tank surfaces. This behavior is usually associated with the refusal of feed intake, regardless of the type of feed supplied (live prey or artificial feed). Clinging and associated non-feeding behaviors are also quite common in Eurasian perch and yellow perch when newly-hatched larvae are stocked in tanks of small sizes, while these behaviors are rarely observed in large tanks. Once feed intake is initiated, the larvae continue to ingest feed regardless of tank size, and the clinging and non-feeding behaviors disappear. The addition of clay to make the water turbid (up to 50 NTU) and to increase light diffusion has been recommended for walleye larvae by Bristow and Summerfelt (1994). Algae suspension was successfully applied to yellow perch and in this case the turbidity was in the range of 10–15 NTU during the first 10–12 days of rearing (Kwasek et al. 2013). Fish were then transferred to tanks without turbidity for further rearing. The combination of light intensity and tank color also modifies the efficiency of food perception and the feeding behavior of perch larvae. Hinshaw (1986) suggested that yellow perch larvae were attracted by the white walls and ceased feeding while dark wall tanks facilitated prey detection by improving the contrast between the food and the walls. The use of grey walls and a diffuse light source is thus appropriate (Hinshaw 1985). Similar results were obtained for European perch where black tank walls resulted in improved body weight and condition factor in comparison to groups reared in tanks with gray walls (Tamazouzt 1995; Jentoft et al. 2006). However, results of these experiments are of limited value as all treatments averaged 75 % mortality. Tamazouzt et al. (2000) recommended the use of white or light colored walls for the larval culture of Eurasian perch under a maximum light source of 800 lux at the water surface in the case of fish being reared with inert prey (frozen zooplankton and formulated dry feeds). Kestemont et al. (2003, 2008) reported the highest growth and survival rates with a light intensity of 90–400 lux and a long day length (LD 16:8 or 24:0) if a continuous feed supply is ensured. Most recent data suggest that yellow perch tolerate a wide range of light intensity, whereas prey density (25 vs. 150 zooplankton L^{-1}) has a significant impact on growth and survival (Martin et al. 2012). Eurasian perch can feed to certain extent in complete darkness using lateral line perception if food density is high (Dabrowski 1982). A less explored area of visual feeding in perch larvae is the spectral characteristic of the light employed for culture (Loew et al. 1993). Yellow perch juveniles (length 25–32 mm) have the highest relative quantal sensitivity at 360 nm, near ultraviolet wavelength, although the number of UV-sensitive cones in their retina, like in other teleosts, decreases dramatically in comparison to larval stages. Visual perception based on near ultraviolet and UV-A and UV-B vision (320–360 nm) must also take into account detrimental effect of UV-A light intensity as demonstrated in yellow perch larvae subjected to this short wavelength (Boily et al. 2011).

Non-inflation of gas bladder (NIGB) has been described in the wild in European lakes (Egloff 1996) as well in N. American Lake Michgan (Czesny et al. 2005). The latter authors demonstrated that the frequency of non-inflation in yellow perch was between 5 % and 25 %. The affected fish were in the size range of 10–15 mm. only those with filled swim bladders survive to a size larger than 20 mm. This is not the case in intensive systems with available food where individuals with uninflated swim bladder were observed.

Eurasian perch and yellow perch are species that are much more sensitive to an increase surface tension than walleye and pikeperch (Colesante et al. 1986; Barrows et al. 1988; Summerfelt 1996), most likely due to their smaller size at the time of swim bladder inflation (6–8 mm total body length). As recommended for these latter species, spray flow, avoiding the accumulation of an oil film at the water surface, can be used successfully to reduce the risk of NGIB and, consequently, the rate of skeletal deformities (pre-haemal lordosis and/or scoliosis). Equipment such as surface skimmers and degassers may also be employed but were not specifically tested with walleye or yellow perch. As described for pikeperch and walleye in Chap. 10 and 11, fish without inflated gas bladders will survive in aquaculture systems but more energy will be devoted to swimming activities, and this re-allocation of energy will be done at the detriment of growth (Ostaszewska 2005).

Furthermore, Jacquemond (2004) described the phenomenon of late swim bladder inflation (up to 66 % of the fish population) in Eurasian perch and argued that fish recovered completely at the size of 68–72 mm body length. We tested this assumption in yellow perch (Fig. 9.12) using the same method of separation of inflated and non-inflated swim bladder fish with anesthesia. We validated our method with X-ray analysis at the time of separation. We found no evidence of late swim bladder inflation in yellow perch and although fish continue to grow, the growth rate of non-inflated perch was inferior in comparison to control and severe skeletal deformities were observed.

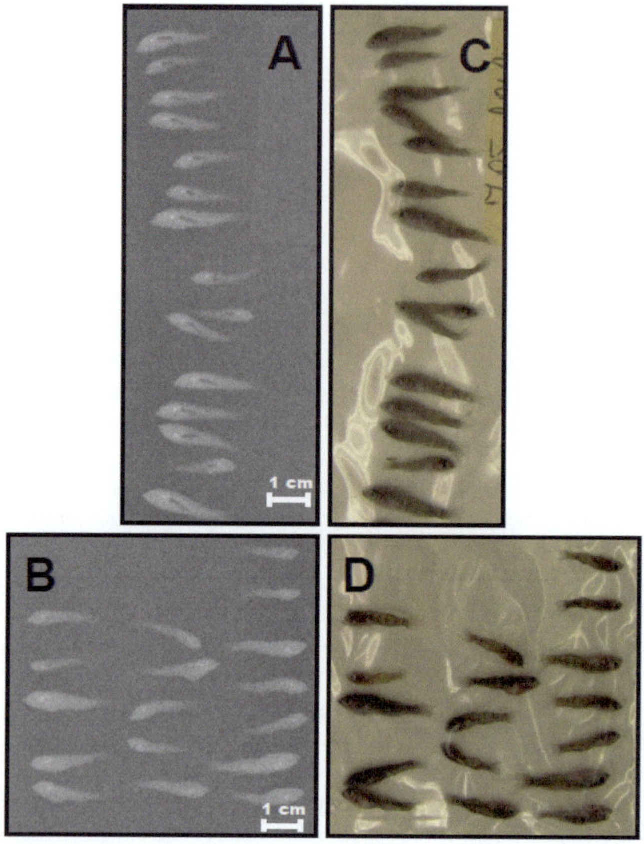

Fig. 9.12 X-ray and direct images of larval and juvenile yellow perch (0.01–0.1 g) with inflated (**a**, **c**) and un-inflated (**b**, **d**) swim bladder

9.4.3 Culture Tanks and Larval Rearing Conditions in Semi-commercial Conditions

Intensive culture of Eurasian and yellow perch larvae is usually performed in recirculating aquaculture system(s) (RAS) in order to efficiently control water quality and to avoid or reduce the presence of pathogens. Larval rearing tanks are of various shapes and materials, squared or cylindro-conical, or made of cages (with 250–300 μm mesh size) placed in large concrete tanks. Regardless of shape, they must be rather large, ranging from 300 L to several m^3, to avoid the clinging and non-feeding behaviors described above. As said before, dark walls or increased turbidity are recommended for facilitating the detection of feed and avoidance of cannibalism by the juveniles of 12–15 mm. At hatching, Eurasian perch and yellow perch are rather fragile and sensitive to handling. Transfer of perch at the eyed egg stage is thus recommended to reduce the mortalities induced by manipulation. Eggs can be suspended on trays about 1 day before hatching, initially at a temperature close to that of incubation (17–18 °C). Within 1 day temperature is raised to 20 °C, and then maintained constant between 20 and 23 °C, depending on strains, throughout the larval rearing phase, up to the time of transfer of juveniles to grow-out tanks (Kestemont et al. 1996, 2008).

Initial stocking density ranges between 20 and 50 larvae L^{-1}. Higher densities, up to 100 newly hatched larvae L^{-1} have been successful, but fish density must be reduced after the weaning phase, as the fish grow (Kestemont et al. 2008).

Feeding starts 2–3 days post-hatching (dph), usually with small strains of *Artemia* nauplii (350–380 μm) during few days and then with regular size *Artemia* nauplii (420–450 μm). During the first week, a daily ration of 500–1,000 nauplii per larvae is provided, distributed very frequently by hand or with a peristaltic pump during the daylight period. From the end of the first week to the end of week 3, the daily ration, calculated on the dry matter basis of *Artemia*, can be reduced from 35 % to 10 % of fish biomass. On day 21 post-hatching, when fish have a mean body mass of 50 mg, they can be trained to accept dry feed, by replacing progressively the live prey with a high quality compound feed (300–500 μm) within 4 days (Kestemont and Mélard 2000; Kestemont et al. 2008). Recently, fish feed companies have developed high quality dry diets enriched in phospholipids as unique starter feed, supporting satisfactory survival and growth rates, together with low skeletal deformities, similar to the results obtained with live feed. A detailed description of feeding and nutrition requirements of Eurasian perch and yellow perch is provided in Chaps. 20 and 21, respectively.

9.4.4 Production Data

In Eurasian perch, larval stocking densities ranging from 20 to 100 fish L^{-1} have been tested under commercial conditions in a hatchery using a RAS technology until day 49 post-hatching. Fish were maintained in triplicate in 500-L squared polyester tanks at a temperature of 21–22 °C and LD 12:12 (50 lux at water surface). They were fed eight times a day with *Artemia* nauplii from day 3 to day 21, followed by a progressive habituation to dry feed (co-feeding) during 10 days, and then exclusively dry diet. Growth curves were rather similar in all treatments (SGR = 13–14 % d^{-1}), and fish body weight reached between 0.59 and 0.71 g. Final biomass was significantly influenced by the initial stocking density, ranging from 5.8 kg m^{-3} at the low initial density to 35.4 kg m^{-3} at the highest one (Fontaine et al. 2001). Similar results with a final biomass of 35 kg m^{-3} at the end of the nursery phase are reported by different French and Swiss producers of Eurasian perch (P. Fontaine and T. Janssen, pers. com.). Higher stocking density, up to 150 fish L^{-1}, is practiced in Ireland where fish are weaned to dry diet from day 16 onwards (D. Toner, pers. com.). Depending on fish farms, survival rates of Eurasian perch vary between hatch and the end of the nursery cycle, but usually range between 15 % and 35 % within 3–4 months, from larval stage to 5 g juvenile. In Irish nursery farms, a survival rate of up to 50 % is a consistent result, but some 25 % of these fish are culled during ongrowing because of slow growth (D. Toner, pers. com.).

9.5 Conclusions

While significant progress in the rearing of early life stages of Eurasian perch and yellow perch has been achieved during the last two decades, the cost and availability of feed trained fingerlings remains a considerable bottleneck in the production protocol for the perch food fish industry. In yellow perch the tandem pond/tank is the most widely used method for producing weaned juveniles in the USA. Large numbers of juveniles can be produced by a rather simple method, with lowered incidence of nutritional deformities. The use of this method however, requires both pond and tank facilities neither of which are employed to their maximum potential. Development of efficient and reliable intensive fry culture protocols as well as year-round availability of yellow perch eggs will have a significant impact on the expansion of the industry. In Eurasian perch, the semi-intensive rearing in small green water tanks (mesocosms) combined with daily supply of live food and dry diet also constitutes a rather simple and cheap alternative to the intensive production in tank for the production of weaned juveniles. The logistics of producing and supplying adequate levels of live prey on a commercial scale however represents additional demands on labour and facilities for success of the overall protocol. The recent availability of high quality commercial diets that were initially formulated for marine fish larvae but provided good results in percid fish species may limit the need for *Artemia* nauplii as starter feed and could simplify the rearing procedure. Intensive rearing of percid larvae in RAS has been extensively investigated in experimental conditions and optimal husbandry conditions are now available for both species, at least at the laboratory scale. Applications of these techniques on a commercial scale will be required to determine their efficiency and productivity. Mortalities, largely caused by cannibalism and skeletal deformities, still limit the reliability of this system, but significant improvements have been achieved by controlling the physical environment, the fish stocking density and the feeding scheme. Commercial production of Eurasian perch larvae is now operational throughout the year in several European countries and some initiatives have started recently in USA.

References

Babiak I, Mandiki SNM, Ratsinjomanana K, Kestemont P (2004) Initial weight and its variation in post-larval Eurasian perch affect quantitative characteristics of juvenile cohorts under controlled conditions. Aquaculture 243:263–276

Baras E (1998) Bases biologiques du cannibalisme chez les poissons. Cah Ethol 18:53–98

Baras E, Kestemont P, Mélard C (2003) Effect of stocking density on the dynamics of cannibalism in sibling larvae of *Perca fluviatilis* under controlled conditions. Aquaculture 219:241–255

Barrows FT, Lellis WA, Nickum JG (1988) Intensive culture of larval walleye with dry or formulated feed: note on swim bladder inflation. Prog Fish Cult 50:160–166

Boily V, Bertolo A, Magnan P, Martinoli M-G, Therien H-M (2011) The effect of UVR irradiance and spectral composition on yellow perch (*Perca flavescens*) larvae survival. Aquat Sci 73:345–354

Boyd CE (1977) Practical aspects of chemistry in pond aquaculture. Prog Fish Cult 59:85–93

Boyd CE (2012) Nutrient cycling. In: Mischke CC (ed) Aquaculture pond fertilization: impacts of nutrient input on production. Wiley-Blackwell, Oxford, pp 3–21
Braband A (1995) Intracohort cannibalism among larval stages of perch (*Perca fluviatilis*). Ecol Freshw Fish 4:70–76
Bremigan MT, Dettmers JM, Mahan AL (2003) Zooplankton selectivity by larval yellow perch in Green Bay, Lake Michigan. J Great Lakes Res 29:501–510
Bristow BT, Summerfelt RC (1994) Performance of larval walleye cultured intensively in clear and turbid water. J World Aquac Soc 25:454–464
Claessen D, De Roos AM, Persson L (2000) Dwarfs and giants: cannibalism and competition in size-structured populations. Am Nat 155:219–237
Colesante RT, Youmans NB, Ziolkoski B (1986) Intensive culture of walleye fry with live food and formulated diets. Prog Fish Cult 45:126–127
Cuff WR (1980) Behavioral aspects of cannibalism in larval walleye, *Stizostedion vitreum*. Can J Zool 58:1504–1507
Culver DA (1996) Fertilization procedures for pond culture of walleye and saugeye. In: Summerfelt RC (ed) Walleye culture manual, vol 101, NCRAC culture series. North Central Regional Aquaculture Center Publications Office, Iowa State University, Ames, pp 115–122
Czesny SJ, Graeb BDS, Dettmers JM (2005) Ecological consequences of swim bladder non inflation for larval yellow perch. Trans Am Fish Soc 134:1011–1020
Dabrowski KR (1982) The influence of light intensity on feeding of fish larvae and fry. II. *Rutilus rutilus* (L.) and *Perca fluviatilis* (L.). Zool J Physiol 86:353–360
Dabrowski K (1984) The feeding of fish larvae. Present state of the art and perspectives. Reprod Nutr Dev 24:807–833
DeAngelis DL, Cox DK, Coutant CC (1979) Cannibalism and size dispersal in young-of-the-year largemouth bass: experiment and model. Ecol Model 8:133–148
Egloff M (1996) Failure of swim bladder inflation of perch, *Perca fluviatilis* L. found in natural populations. Aquat Sci 58:15–23
Fontaine P, Kestemont P, Mélard, Verstraete W (2001) New microbial technologies applied to the rearing of Eurasian perch (*Perca fluviatilis*) in recirculating aquaculture system. Final report FAIR CT98-9241 (1/1/99-31/12/00). Nancy, 149p
Fulford RS, Rice JA, Miller TJ, Binkowski FP, Dettmers JM, Belonger B (2006) Foraging selectivity by larval yellow perch (*Perca flavescens*): implications for understanding recruitment in small and large lakes. Can J Fish Aquat Sci 63:28–42
Grignard JC, Melard C, Kestemont P (1996) A preliminary study of parasites and diseases of perch in an intensive culture system. J Appl Ichthyol 12(3–4):195–199
Hart SD, Garling DL, Malison JA (eds) (2006) Yellow perch (*Perca flavesc*ens) culture guide, vol 103, NCRAC culture series. Iowa State University, Ames
Hartleb CF (2003) Food chain dynamics and diets of larval and post-larval yellow perch in culture ponds. In: Barry TP, Malison JA (eds) Proceedings of Percis III: the third international percid fish symposium. University of Wisconsin Sea Grant Institute, Madison
Hartleb CF, Johnson JA, Held JA (2012) Walleye and yellow perch pond fertilization. In: Mischke CC (ed) Pond fertilization: impacts of nutrient input on aquaculture production. Wiley-Blackwell, Oxford, pp 147–161
Hecht T, Appelbaum S (1988) Observations on intraspecific aggression and coeval sibling cannibalism by larva and juvenile *Clarias gariepinus* (Clariidae: Pisces) under controlled conditions. J Zool Lond 214:21–44
Hecht T, Pienaar AG (1993) A review of cannibalism and its implications in fish larviculture. J World Aquac Soc 24:246–261
Heidinger RC, Kayes TB (1986) Yellow perch. In: Stickney RR (ed) Culture of nonsalmonid freshwater fishes. CRC Press, Boca Raton, pp 103–113
Held JA, Malison JA, Kuczynski TE (1998) Techniques for the commercial production of feed-trained yellow perch (*Perca flavescens*) fingerlings. World Aquaculture Society book of abstracts 29th annual meeting of the World Aquaculture Society, Las Vegas

Hinshaw JM (1985) Effects of illumination and prey contrast on survival and growth of larval yellow perch *Perca flavescens*. Trans Am Fish Soc 114:540–545

Hinshaw JM (1986) Factors affecting feeding, survival and growth of larval and early juvenile yellow perch (*Perca flavescens* Mitchill). PhD thesis, North Carolina State University, Raleigh, 80 p

Huston MA, DeAngelis DL (1987) Size bimodality in monospecific populations: a critical review of potential mechanisms. Am Nat 129:678–707

Jacob AP, Culver DA (2010) Experimental evaluation of the impacts of reduced inorganic phosphorus fertilization rates on juvenile saugeye production. Aquaculture 304:22–33

Jacquemond F (2004) Separated breeding of perch fingerlings (*Perca fluviatilis* L.) with and without initial inflated swim bladder: comparison of swim bladder development, skeleton conformation and growth performance. Aquaculture 239:261–273

Jentoft AH, Oxnevad S, Aastevit A, Andersen Q (2006) Effects of tank wall color and up-welling water flow on growth and survival of Eurasian perch larvae (*Perca fluviatilis*). J World Aquac Soc 37:313–317

Kestemont P (1995) Influence of feed supply, temperature and body size on the growth of goldfish *Carassius auratus* larvae. Aquaculture 136:341–349

Kestemont P, Mélard C (2000) Chapter 11: Aquaculture. In: Craig JF (ed) Percid fishes. Systematics, ecology and exploitation. Blackwell Science, Oxford, pp 191–224

Kestemont P, Mélard C, Fiogbé E, Vlavonou R, Masson G (1996) Nutritional and animal husbandry aspects of rearing early life stages of Eurasian perch *Perca fluviatilis*. J Appl Ichthyol 12:157–165

Kestemont P, Cuvier A, Houbart M, Mélard C, Baras E, Fontaine P, Jourdan S, Gardeur J-N, Anthouard M, Boujard T, Kentouri M, Pasapatis M, Ducarme C (2000) An integrated study of interindividual competition and its relationships with feeding physiology and behavior in cultured predatory fish. Fair CT96-1572 final report, 118 p

Kestemont P, Jourdan S, Houbart M, Mélard C, Paspatis M, Fontaine P, Cuvier A, Kentouri M, Baras E (2003) Size heterogeneity, cannibalism and competition in cultured predatory fish larvae: biotic and abiotic influences. Aquaculture 227:333–356

Kestemont P, Rougeot C, Musil J, Toner D (2008) Chapter 5: Larval and juvenile production. In: Toner D, Rougeot C (eds) Farming of Eurasian perch, vol 1. Juvenile production. Aquaculture explained, n 24. Irish Sea Fisheries Board, Dublin, pp 30–41

Kwasek K, Wojno M, Niass F, Grayson J, Dabrowski K (2013) The art of raising yellow perch in captivity. World Aqaculture Society, book of abstract, Nashville, 21–25 Feb

Loadman NL, Moodie GEE, Mathias JA (1986) Significance of cannibalism in larval walleye (*Stizostedion vitreum*). Can J Fish Aquat Sci 43:613–618

Loew ER, McFarland WN, Mills EL, Hunter D (1993) A chromatic action spectrum for planktonic predation by juvenile yellow perch, *Perca flavescens*. Can J Zool 71:384–386

Malison JA, Held JA (1992) Effects of fish size at harvest, initial stocking density and tank lighting conditions on the habituation of pond-reared yellow perch (*Perca flavescens*) to intensive culture conditions. Aquaculture 104:67–88

Malison JA, Northey DL, Held JA, Kuczynski TE (1994) Habituation of yellow perch (*Perca flavescens*) fingerlings to formulated feeds in ponds using lights and vibrating feeders. World Aquaculture Society book of abstracts 25th annual meeting of the World Aquaculture Society, New Orleans

Martin BT, Wahl DH, Czesny SJ (2012) Effects of light intensity, prey density, and ontogeny on foraging success and prey selection of larval yellow perch (*Perca flavescens*). Ecol Freshw Fish 21:588–596

Mélard C, Baras E, Mary L, Kestemont P (1996) Relationships between stocking density, growth, cannibalism and survival rate in intensively cultured larvae and juveniles of perch (*Perca fluviatilis*). Ann Zool Fenn 33:643–651

Morris JE, Mischke CC (1999) Plankton management for fish culture ponds, vol 114, North Central Regional Aquaculture Center (NCRAC) technical bulletin series. North Central Regional Aquaculture Center Publications Office, Iowa State University, Ames

Nash CE, Kuo CM, McConnel SC (1974) Operational procedure for rearing larvae of grey mullet (Mugil cephalus L.). Aquaculture 3:15–24

Neal JW, Kröger C (2012) Sportfish pond fertilization. In: Mischke CC (ed) Pond fertilization: impacts of nutrient input on aquaculture production. Wiley-Blackwell, Oxford, pp 245–258

Ostaszewska T (2005) Developmental changes of digestive system structures in pikeperch (*Sander lucioperca* L.). Electron J Ichthyol 2:65–78

Paller MH, Lewis WM (1987) Effects of diet on growth depensation and cannibalism among intensively cultured larval striped bass. Prog Fish Cult 49:270–275

Siefert RE (1972) First food of larval yellow perch, white sucker, bluegill, emerald shiner, and rainbow smelt. Trans Am Fish Soc 101:219–225

Soderberg RW (1977) Perch fingerling production for aquaculture. University of Wisconsin Sea Grant College Program, advisory report #421, Madison

Soderberg RW (2012) Organic and inorganic fertilization. In: Mischke CC (ed) Pond fertilization: impacts of nutrient input on aquaculture production. Wiley-Blackwell, Oxford, pp 33–45

Summerfelt RC (1996) Intensive culture of walleye fry. In: Summerfelt RC (ed) Walleye culture manual, vol 101, NCRAC culture series. North Central Regional Aquaculture Center Publications Office, Iowa State University, Ames, pp 161–185

Tamazouzt L (1995) L'alimentation artificielle de la perche *Perca fluviatilis* en milieu confinés (eau recycle, cage flottante). Thèse de doctorat, Université H. Poincaré, Nancy, 128 p

Tamazouzt L, Chatain B, Fontaine P (2000) Tank wall color and light level affect growth and survival of Eurasian perch larvae (*Perca fluviatilis* L.). Aquaculture 182:85–90

Tice BJ, Soderberg RW, Kirby JM, Marcinko MT (1996) Growth and survival of walleyes reared in ponds fertilized with organic or inorganic materials. Prog Fish Cult 58:135–139

Wojno M, Kwasek K, Dabrowski K, Wick M (2012) Yellow perch rearing in intensive conditions –update for 2011. Lake Erie-Inland waters annual review, Ohio Division of Wildlife, Columbus, 26–27 Jan

Chapter 10
Culture Methods of Pikeperch Early Life Stages

Svend Steenfeldt

Abstract Throughout the last decade, intensive rearing of pike perch fry have developed from small research based setups to full commercial scale operations with capacities to support the fry requirements of large scale highly intensified recirculating aquaculture system(s) (RAS) for ongrowing of the species. The methodology has to a large extent been transferred from the knowledge and prior research in marine larval rearing, using live feeds and recirculation technology. Specific adaptations to pikeperch have included feeding strategies that takes into account that pikeperch larvae are reared in fresh water, and the fact that pikeperch are highly cannibalistic already at the pre weaning stage.

Keywords Sander • Weaning • Cannibalism • Larval rearing • Recirculating system

10.1 Introduction

Intensive rearing of pikeperch has reached the commercial stage in several European countries including Denmark and Holland. The coming challenges include improvements in the reliability of fry production and refinements in the procedures leading to reduced production prices of fry, and improved competitiveness of the operators.

Production of fingerlings is the primary bottleneck in pikeperch culture. A critical phase in fingerling production is the early stage of exogenous feeding, where larval digestive capabilities are limited, and only live feeds can be utilized by the larvae (Nyina-wamwiza et al. 2005). These challenges have been overcome in a number of species and methodologies have been transferred between species. The majority of work on development of techniques for production and use of live feed has been conducted on marine species (Støttrup and McEvoy 2003).

S. Steenfeldt (✉)
BioMar A/S, The North Sea Science Park, Willemoesvej 2, 9850 Hirtshals, Denmark
e-mail: sjs@biomar.dk

P. Kestemont et al. (eds.), *Biology and Culture of Percid Fishes*,
DOI 10.1007/978-94-017-7227-3_10

Pikeperch is a freshwater fish, but from a larval rearing perspective, pikeperch in many ways resembles marine species. It produces high numbers of small eggs (Lehtonen et al. 1996) and consequently, it has small first feeding larvae compared with other fresh water fish e.g. salmonids. At an early age pikeperch in natural environments become piscivorous (Hilge and Steffens 1996). In intensive larviculture of pikeperch, cannibalism may cause severe losses if the husbandry requirements and/or nutritional needs are not better defined. Much effort in intensive pikeperch hatcheries is allocated to prevent cannibalism that seems irreversible if established in a batch of fry (Steenfeldt et al. 2010).

10.2 Live Feed Production

10.2.1 Artemia

Production of live feed is a resource demanding task. Much effort must be put into production of a stable quantity and quality of live feed for the larvae (Støttrup and McEvoy 2003).

In commercial intensive pikeperch larval rearing, *Artemia* with an optional supply of rotifers and algae is used for start feeding and larval rearing till the time of weaning (Rønfeldt and Nielsen 2010).

Natural populations of *Artemia* are found in over 300 places around the globe. The environmental conditions vary considerably. *Artemia* are found in waters with temperatures ranging from 6 to 35 °C and a wide range of salinities. *Artemia* produces two types of eggs. During environmentally beneficial conditions *Artemia* produces thin-shelled eggs. These eggs will hatch within short time. During more hostile conditions, with high salinities or low oxygen levels, resting eggs are developed. These eggs can survive for years, are harvested and later used in the aquaculture sector (Lavens and Sorgeloos 1996).

The *Artemia* cysts are hatched in sea water or alternatively fresh water supplemented with 3 % salt e.g. NaCl or sea water derived salt at 27–30 °C. Oxygen consumption is high and sufficient supply is often secured through use of pure oxygen. A harvesting device with a 150 μm mesh is used to concentrate and rinse the cysts after hatching (Lavens and Sorgeloos 1996).

After hatching pikeperch larvae utilize energy from their yolk sac, before shifting to exogenous feeding. Mouth opening happens at 95 Day degrees equivalent to 5 days at 18 °C (Mani-Ponset et al. 1994). This is a critical stage in the rearing process. Fish larval size at time of first feeding or rather the size of its mouth and diameter of esophagus determines the prey size that it will be able to ingest (Busch 1996).

Larvae should be presented with feed that fulfills their nutritional demands and are available in sufficient quantities to ensure, that ingestion rates are maintained at high enough levels to cover assimilation for retention and catabolism.

Rearing of pikeperch larvae on inert diets have been tried with some success (Ostaszewska et al. 2005), but live feed is still the preferred feed used in rearing of pikeperch.

Pikeperch larvae are able to ingest Artemia from first feeding (Steenfeldt et al. 2010). Selected small strains of newly hatched *Artemia* will be ingested by the first feeding larvae. When fish larval mouth gape has increased, larger instar two to three stage Artemia can be used. To increase the nutritional value of the latter Artemia, emulsified preparations of essential nutrients including HUFAs and vitamins are marketed. These contain self-dispersing selections of marine oils, vitamins and carotenoids. Enrichment emulsions are suspended in the water of hatched *Artemia* for 6–20 h before *Artemia* are harvested. The oil globules are readily ingested by *Artemia* and thereby boost their nutritional composition (Moretti et al. 2004). Specially prepared emulsions may be prepared to investigate dietary needs of fish larvae (Lund and Steenfeldt 2011).

10.2.2 Rotifers

Supplementation of the *Artemia* diet with a starter diet of rotifers is practiced at some commercial pikeperch hatcheries (Rønfeldt and Nielsen 2010). Rotifers are produced in tanks typically with a volume of 0.5–2 m^3, a size that facilitates handling of the rotifers and cleaning of the production systems. The rotifers are fed three to five times daily with baker's yeast and algae or marine oil. It is possible to enrich rotifers by use of algae. The high content of the fatty acid eicosapentaeoic acid (EPA) in e.g. *Nanocholropsis occulata* and docosahexaenoic acid DHA in e.g. *Isochrysis galbana* have made them well suited as live food for rotifers in marine fish hatcheries (Støttrup and McEvoy 2003; Divanach and Kentouri 2000). However, culture of algae is labor intensive (Tapie and Bernard 1988). Alternative enrichment products as the ones mentioned for *Artemia* have been developed. The advantages of these are that they provide a short-term boosting of essential fatty acids and vitamins immediately before the rotifers are transferred to the fish larval rearing tanks (Lavens and Sorgeloos 1996). During the short term boosting the rotifers will fill their guts with oil globules (i.e., bioencapsulate) and serve as vectors allocating the oils to the fish larvae.

10.3 Intensive Larval Rearing

The methodology of intensive rearing of marine fish fry that is practiced in Western Europe has been applied to pikeperch with success. Investment and running costs are high and must be balanced with high productivity to be economical sustainable.

10.3.1 Tank Design

Tanks used for larval rearing in commercial pikeperch farms are typically cylindrical conical (Szkudlarek and Zakes 2007; Steenfeldt et al. 2010). In experimental research, aquaria are frequently used (Molnár et al. 2004). A comparison between cylindrical versus cuboidal tanks concluded, that cylindrical tanks provided a better rearing environment, mainly due to the differences in water flow patterns generated by the two systems (Moore et al. 1994b).

Commercial scale intensive larval rearing in Denmark is typically carried out in cylindrical tanks of 3.5 m^3 and a depth of approximately 2 m (Fig. 10.1).

10.3.2 Aeration

Aeration in cylindrical conical tanks is administered by placing an aquarium type air diffuser in the tip of the cone. This creates a water flow where dead zones with stagnant water will not form (Moretti et al. 2005). The flow generated by the gentle aeration lifts the central column of water vertically up through the center of the tank.

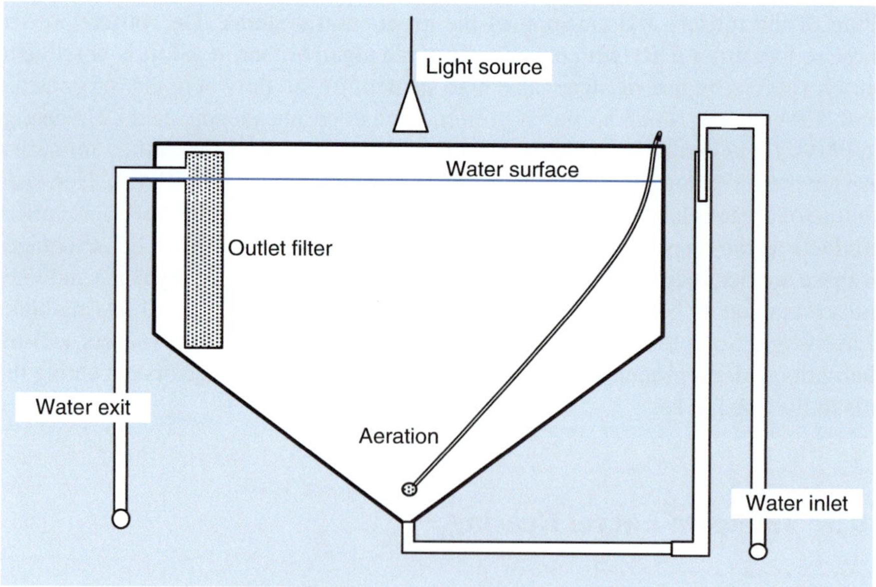

Fig. 10.1 Cylindric-conical larval rearing tank with inlet through the *bottom* and outlet filter through cylindrical submerged filter in the left side of the tank

At the surface the water moves horizontally to the tank edge before being forced down along the tank side towards the cone tip. Besides oxygenation the aeration device also homogenizes the concentration of prey and fish larvae in the tank (Barahona-Fernandes 1978).

10.3.3 Water Exchange

Water inlet to the tank is in some systems through the tip of the cone (Steenfeldt et al. 2010), or through a pipe protruding down into the water from the surface (Lekang 2007b). In rearing of larval fish, currents in the water must be kept low (Kolkovski et al. 2004). A water exchange rate of 25–50 % per hour is typically used; increasing as the larvae grow and more feed is administered (Steenfeldt et al. 2010).

Tank outlet or drain has two functions: Removal of wastes and maintaining water level in the tank while withholding the reared organisms in the tank (Lekang 2007b). In larval rearing tanks, the mesh size of the outlet screen must let feed organisms pass through the filter, while avoiding that larvae are flushed out. When stocking pikeperch yolk sac larva in Danish systems, a mesh size of 400–500 μm is used (Rønfeldt and Nielsen 2010; Lund and Steenfeldt 2011). The surface area of the filter must be large enough to ensure that the current passing the filter is sufficiently low, to avoid larvae being trapped (impinged) on the mesh. Often cylindrical outlet filters are used i.e. pipes that are perforated and covered with mesh (Moretti et al. 2005).

10.3.4 Water Currents

Water currents in the tanks generated by water exchange affects the stocked fish. Currents velocities less than 10 cm/s are recommended for sea bass (*Dicentrarchus labrax*) and sea bream (*Sparus aurata*) (Divanach et al. 1997a). This is accomplished by letting the water in through multiple inlets, or by using large diameter pipes to reduce water pressure to a minimum. Low water exchange in larval rearing tanks i.e. 10–33 %/h (Büke 2002; Lund and Steenfeldt 2011) helps avoiding currents from forming. Skeletal deformities may be a consequence of high currents in the rearing tanks (Divanach et al. 1997a). Causative mechanism likely includes forced high swimming activity (Kihara et al. 2002).

Removing settled solids are traditionally performed by manually siphoning flocculating material from the bottom once or twice daily (Steenfeldt et al. 2010). The vertical sides of cylindrical conical tanks will not accumulate material settling in the tanks. This material will instead build up on the bottom surface, even though this is sloping towards the tip of the cone with an angle of typically 45°. Mechanical automated sweepers are being used in the private sector, consequently reducing labor requirements.

10.3.5 Recirculating Aquaculture System(s) (RAS)

In Europe, intensive larval rearing systems in commercial culture are exclusively recirculating aquaculture system(s) (RAS). These systems are preferred by commercial enterprises because of reduced water consumption and high level of control of rearing environment. Because intake of new water to the system is less than 5 % daily, purification of this in accordance to environmental impact legislations are reduced when compared to flow through systems (Steenfeldt et al. 2010).

Intensive rearing systems for pikeperch are in many ways similar to the systems used in rearing of marine larvae. They are typically based on recirculation of the water using mechanical and biological treatment of the water. Recirculation reduces energy consumption related to temperature conditioning of the water, and has advantages in relation to maintaining water quality with limited fluctuations (Blancheton 2000).

In a RAS larval rearing unit the water leaves the fish tanks and passes a mechanical filter where particulate matters i.e. faeces and uneaten feed are removed. The water then passes a biological filter where ammonia is oxidized to nitrite and nitrate before it reaches a trickling filter where carbon dioxide and other gasses in excess are removed and oxygen content is returned to close to saturation. Before the water returns to the fish it may pass a UV irradiation unit for disinfection (Fig. 10.2).

The chemical parameters of water quality should follow the general guidelines for larvae rearing systems including marine larval rearing systems. Supplementation of the inlet water with pure oxygen may be relevant if stoking density is high and water flow low. In larviculture of walleye 112.7 % oxygen saturation did not enhance gas bladder inflation after a 23-days rearing period compared with larvae cultured at ambient oxygen concentrations (Summerfelt 1991).

When feeding pikeperch larvae with *Artemia* having a protein content of 31 % (Helland et al. 2003), 5 % of the feed is nitrogen and the part of this not allocated to growth will be lost to the water surrounding the fish, mainly as ammonia (NH3) and urea (Poppe 1990). Ammonia is toxic to fish at concentrations as low as 0.002 mg/l, but will be in equilibrium with the ammonium ion (NH_4^+). The equilibrium strongly

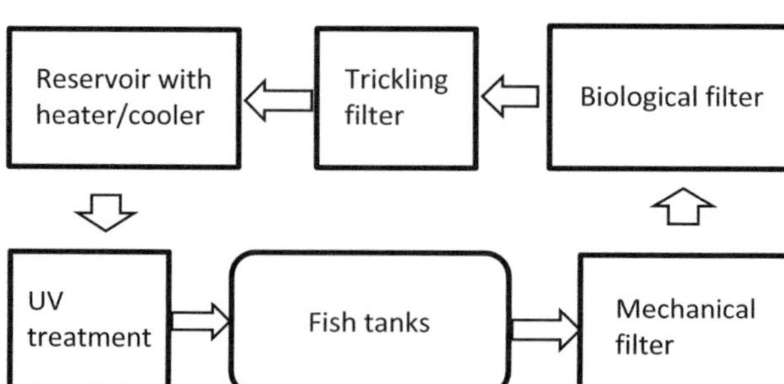

Fig. 10.2 Schematic drawing of water flow in a RAS system, sequencing the main units of holding and processing the water

depends on pH. Below pH 9 the equilibrium is shifted towards the ammonia ion and at pH 7 the shift towards the ammonia ion is almost 100 %. Nitrification where ammonia is oxidized to nitrite and nitrate is a fundamental part of a RAS system. Nitrification is optimal at pH around 8–9 (Lekang 2007a). The derived products nitrite and nitrate are much less toxic to fish. Rearing of salmon in constant concentrations of 84–99 mg/l nitrate-nitrogen had no adverse effects. Nitrite toxicity in freshwater though is reported at concentrations 50–100 times lower than in seawater; caused by the nitrite having an affinity for the Cl^- uptake system of the fish (Kroupova et al. 2005).

The rearing environment in recirculating aquaculture system(s) (RAS) is more stable than in flow through systems. Parameters such as temperature, pH, salinity etc. can be adjusted to ideal levels of the target species. Although bacterial load of recirculation systems can be high, the bacteria may have probiotic effects.

It is well known that the RAS microbial flora includes both chemoautotrophic (e.g. nitrifiers) and heterotrophic bacteria, that actively consume oxygen and organic matter. Such bacterial communities harbor species that are obligate or facultative pathogens and that may cause disease in fish (Michaud et al. 2009). However, the main part of the heterotrophic bacterial compartment is constituted by neutral microbes, contributing to maintain a good microbial water quality by occupying niches and prevent proliferation of harmful species (Michaud et al. 2009; Attramadal 2011).

10.3.6 Feeding Strategy

Using rotifers is believed to reduce the difference in time between early and late first feeding larvae in a batch (Rønfeldt and Nielsen 2010). The procedure used was 2 days of feeding with rotifers followed by 1 day of co-feeding with *Artemia*, where after only *Artemia* was used.

Feeding pikeperch larvae with live feed normally used in marine aquaculture i.e. rotifers and *Artemia* is not an ideal situation. *Brachionus plicatilis* and *Artemia* originates from saline ecosystems, and will not survive for more than a few hours in the fresh water environment in a pikeperch rearing facility. Use of freshwater rotifers is a possibility, but there is no commercially available alternative live prey organism to *Artemia*.

A commonly adapted feeding strategy is pulse feeding where live feed organisms are added to the tanks a number of times daily. The alternative continuous feeding seems not to increase growth or survival in larval fish, presumably due to their inherited behavioral and physiological adaptation to conditions of varying prey availability (Rabe and Brown 2000).

Overfeeding will increase number of uneaten organisms accumulating on the bottom, consequently deteriorating water quality. Cost of live feed is an economic burden and feed loss is minimized by the industry. The nutritional value of live feed is highest at the time of feeding (Lavens and Sorgeloos 1996). Maximizing the quantity of feed ingested at this time will result in the highest possible nutritional quality of the feed ingested and can be increased by introducing a starvation period before feeding.

Optimized pulse feeding is a balance between addition of high quality newly enriched live feed to the tank and high ingestion rates by the larvae followed by a period where the prey is continuously flushed from the tanks by the water exchange until the prey density is low in the tanks and larval ingestion is reduced and appetite increases before next feeding. Administering live feed two to four times daily seems enough to cover the demands of pikeperch larvae (Rønfeldt and Nielsen 2010) and a period of darkness reduces larval activities (Batty 1987) and ensures high ingestion rates when feed is introduced in the morning before light is turned on (Lavens and Sorgeloos 1996).

By distributing the prey organisms from the reservoir tanks to the larval rearing tanks by pipelines, manpower costs can be reduced. If the reservoir tanks can be cooled to e.g. 5 °C, the detrimental effect of the nutritional quality of the prey will be small (Lavens and Sorgeloos 1996).

Maintaining a high level of hygiene in the rearing system is a balance between how frequent deposited material is removed from the tank and how often the larvae are disturbed by this procedure. Survival of *Artemia* in fresh water is minutes to hours. Consequently significant numbers of *Artemia* will die in the tanks before being ingested by larvae. Dead Artemia accumulate on the tank bottom and if not removed bacteria and fungi will proliferate. Consequently tanks need to be physically cleaned daily. This is in most facilities carried out by siphoning up from the tank bottom by suction devices powered by gravity (Moretti et al. 2005). Manual tank cleaning is labor intensive and automated systems for tank cleaning has been implemented at commercial farms. Automated systems are designed as brushes moving along the bottom. This has the advantage of being able to run for long time periods at slow speeds, thereby not disturbing the fish. Even with automated systems, material is resuspended in the water column and cleaning remains a balance between removal of material and minimizing resuspension of debris.

10.3.7 Stocking Density

Before stocking, the concentration of larvae in the transport unit can be quantified by subsampling a number of known volumes followed by manually counting the number of larvae in each. Transfer of a specific number of larvae to the larval rearing unit is now possible by volumetric transfer (Steenfeldt et al. 2011).

In commercial aquaculture, economic considerations necessitate culture regimes that may induce stress responses in the reared organisms. Various stressors may induce stress responses (Costas et al. 2008). High stocking densities are known to elevate levels of plasma cortisol in larger sized fish (Pickering and Potinger 1989).

Experimental evidence of an optimal stocking density of pikeperch larvae in intensive rearing units have generally found increased survival and growth rates with reduced densities (Szkudlarek and Zakes 2007). This has been counterbalanced though, by the increased number of larvae stocked at higher densities up to the highest tested density of 100 larvae per liter. The same overall conclusion was made by (Moore et al. 1994a) when comparing the densities of 20 and 60 larvae per liter. Laboratory experiments in very small units (1–32 ml) did result in a positive correlation between mortality and reduced density (Tagawa et al. 2004). The use of very small water volumes may have influenced these results though. High densities reduce production costs if larval quality and survival is not affected. In crucian carp (*Carassius carassius* L.) best results were obtained using 50 larvae/l. Effects of high stocking densities were limited though at densities of 200–600 larvae/l (Zarsky et al. 2011). In Asian seabass (*Lates calcarifer*) stocking densities of 20–80 larvae/l were correlated to survival ($R^2=0.89$) but growth was not correlated to stocking density ($R^2=0.08$), (Salama 2007).

Commercial scale hatcheries producing marine fish e.g. sea bass (*Dicentrarchus labrax*) stock around 100 larvae per liter (Büke 2002) which is comparable to the density of larvae stocked in commercial intensive production of pikeperch. Stocking density varies with species with the highest stocking densities at up to 140 larvae/l in sea bass and sea bream, 30 larvae/ l in turbot and <10 larvae/l for halibut (Shields 2001).

10.3.8 Water Currents and Turbulence

Sea bass (*Dicentrarchus labrax*) develop higher ratios of skeletal deformities in currents higher than 10 cm s^{-1} (Divanach et al. 1997b). General recommendations for aeration in 2–12 m^3 tanks for sea bass and sea bream larvae are below 60 l min^{-1} (Moretti et al. 2005). Increasing aeration to above 600 ml/min in 250 l rearing tanks had negative effects on survival on turbot larvae (Person-Le-Ruyet 1989). Effects of high aeration levels in larval rearing tanks were attributed to direct physical effects on turbot larvae rather than indirect effects on prey availability (Gaignon et al. 1998).

When exposing pikeperch in 54-L tanks to turbulent conditions by administering 800 ml air min^{-1} through an air stone at the bottom of the tank, larval growth decreased and gas bladder inflation was delayed and never reached the almost 100 % inflation rate that characterized the control group (Rønfeldt and Nielsen 2010). A low water exchange rate is necessary to avoid feed being flushed from the tanks and to ensure that larvae are not trapped on the outlet screen of the tank.

10.3.9 Surface Film

Removal of surface film is necessary to reduce bacterial buildup in the tanks and to create conditions that facilitate filling of the gas bladder (Summerfelt 1996; Moore et al. 1994b; Barrows et al. 1993). Excessive buildup of a surface film will also act as substrate for bacteria and fungi. Removal of surface film can be done manually, by skimming the whole surface at regular intervals. Surface skimmers that aggregate the surface film and trap it in a reduced area of the surface are used in European sea bass and gilthead sea bream culture (Moretti et al. 2004). Surface skimmers are floating frames that use air that blows along the surface through a hole in the frame. The airflow draws the surface film through the hole where it is collected inside the trap. The trapped surface film must be removed regularly either manually, or by automated suction devices. An alternative method used in walleye and pikeperch culture is the surface spray method (Summerfelt 1996; Summerfelt et al. 2011). Use of this method implies that the surface film and its associated microorganisms are forced down into the water column.

The advantage of this system is that no manual removal of surface film is needed. A comparison between effects of surface skimmers and surface sprays on gas bladder inflation in pikeperch concluded that the surface spray gave slightly higher gas bladder inflation rates than surface skimmers i.e. 96 % versus 100 % (Rønfeldt and Nielsen 2010) whereas the control group with no surface film removal resulted in 62 % gas bladder inflation. In an experiment on walleye surface spray removal gave 62.3 % gas bladder inflation compared with 41.7 % inflation using surface skimmers and 21.7 % in a control group (Boggs and Summerfelt 2003).

A surface film will develop in tanks when feeding live prey. Especially when using enriched prey, an oily surface film quickly develops on the water surface (Moretti et al. 2005). The surface film will not pass through the outlet filters and must be removed to avoid bacterial build up in the tank environment.

10.3.10 Gas Bladder Inflation

Removal of the surface film is especially important during the period where the larvae fill their gas bladder. This is a very critical phase in the larval rearing period (Summerfelt 1996; Rønfeldt and Nielsen 2010).

Gas bladder inflation in pikeperch reared at around 18–20 °C takes place 7–14 days post hatch (Rønfeldt and Nielsen 2010).

Pikeperch is a physoclist fish i.e. the opening between the gas bladder and the digestive tract is only open during the few days when the gas bladder is initially filled. Filling of the gas bladder is an active process where the fish larvae swims towards the water surface and gulps an air bubble by protruding its jaws through the water surface (Summerfelt 1996). If the water surface is covered by a film of oily substance, the larvae will not be able to gulp the air (Chatain and Ounais-Guschemann 1990). The pneumatic opening between the gas bladder and the digestive tract closes after a few days, and the fish will have lost the ability to fill and benefit from the gas bladder later in life. Fish without gas bladders will be negatively buoyant and will orient the body in a non-horizontal position in the water, constantly compensating for the negative buoyancy by jerking forward movements of the body. The excess energy used for this constant anti gravitational swimming activity, causes these fish to allocate less energy to somatic growth and they will become smaller than fish with filled gas bladders (Demska-Zakes et al. 2003). Fish filling their gas bladders in tanks where surface film has not been removed adequately will risk transferring bacteria from the surface film to the gas bladder, leading to gas bladder inflammation or aerocystitis (Ostaszewska 2003).

In walleye gas bladder noninflation was concluded to be a result of ingestion of bacteria and organic debris into the gas bladder 6–11 days posthatch. At 12 days post hatch the pyloric sphincter developed and separated common bile duct the intestine from the pneumatic duct in the stomach (Summerfelt 1996).

Fish without inflated gas bladders will survive relatively well in aquaculture systems but will allocate more energy to swimming activity and thus show reduced growth (Ostaszewska 2003). If no effort is done to remove these, they will form part of the standing stock during juvenile rearing and ongrowing. Within the group of fish with non-inflated gas bladders high frequencies of deformed fish will be present. The typical deformity observed in these fish is pre-haemal lordosis where the mid-posterior part of the vertebrae bends upwards. This is caused by the excessive swimming to compensate for negative buoyancy that builds up muscle that leads to an uneven load on the spine (Steenfeldt et al. 2010). Fish with pre-haemal lordosis do not present themselves well on the market. By discarding the fish at an early stage the economic burden of growing fish that are to be discarded is avoided (Steenfeldt et al. 2010). The earliest convenient time is when fish are moved from the hatchery to the weaning facility. The fish are anesthetized in a bucket using e.g. MS-222 or ethylene glycol mono phenyl ether. Fish with non-inflated gas bladders will sink to the bottom, and the ones with inflated gas bladders will float on the surface. Fish with non-inflated gas bladders can then easily be discarded and the remaining moved to the weaning system.

10.3.11 Turbidity

Turbidity is traditionally reduced if possible, by mechanical filtration and adsorbtion to biofilters or by water renewal. In intensive larval rearing systems, abioseston is minimized to reduce the availability of substrate to pathogenic bacteria.

Pikeperch in nature thrive in turbid water though (Jepsen et al. 1999). In a number of species, increased turbidity has positive effects on larval performance (Utne 1997, Vogel and Beauchamp 1999; Robertis et al., 2003). Increased turbidity likely reduces interactions among individuals but may have varying effects on different species, depending on their search volume and foraging mode (Utne 1997).

Positive effects of turbidity on larval feeding have been related to the modification of light and contrast conditions and reduced interactions between individual larvae (Meager and Utne-Palm 2008). Reduced foraging search volume in turbid water may reduce prey ingestion rates in turbid conditions (Vogel and Beauchamp 1999). The risk of being preyed upon may be increased in clear water though (Robertis et al. 2003). The multiple factors affected by turbidity advocates for species specific intermediate optimal turbidities (Meager and Utne-Palm 2008).

Effects of turbidity on larval walleye are well studied (Summerfelt 1996), but effects on larval pikeperch are less investigated. Wild pikeperch thrive in turbid water (Jepsen et al. 1999). When adding clay to the rearing water, (Rønfeldt and Nielsen 2010) found earlier gas bladder inflation and earlier onset of first feeding in tanks with turbid water. Larval rearing of pikeperch is often carried out in clear water despite the knowledge that both larval pikeperch and walleye seems to perform better in turbid water (Bristow and Summerfelt 1994; Bristow et al. 1996).

In pikeperch larvae in turbid water (59.7 ± 7,5 NTU), 21 ± 8 % had inflated gas bladder on day 8 post hatching, compared to 0 % larvae reared in clear water (>1 NTU) ($P \leq 0.001$). The effect was alto significantly different at day 11, ($P \leq 0.001$) but not on day 9 and 10 where gas bladder inflation only tended to be higher in larvae reared in turbid water (Rønfeldt and Nielsen 2010).

In the same experiment, a higher percentage of larvae in turbid water had started feeding on day 6 and 7 post hatching (30 ± 10 and 100 ± 0 respectively) compared to larvae reared in clear water (10 ± 10 and 45.5 ± 5.7 respectively). Length of the larvae reared in turbid water was significantly higher on day 13 ($P = 0.044$) and 15 ($P = 0.001$) and tended to be higher in the remaining days from day 8 to 14. Survival was not affected significantly by turbidity.

Rearing larvae in green water has the combined effects of increased turbidity, plus the effects of possible positive nutritional effects of microalgae if assimilated by the fish larvae. The algae may serve as a water quality stabilizer (Houde 1975). The positive nutritional effects of microalgae can be caused by the fish assimilating microalgae directly or by zooplankton feeding on microalgae and thus increase their nutritional value to the fish (Houde 1975). Juvenile pikeperch reared in turbid water ponds selected larger invertebrate prey species to a fish size where piscivorous feeding was expected (Zingel and Paaver 2010).

The effect of green water technology on pikeperch was tested by addition of chlorella algal paste to the rearing water (Rønfeldt and Nielsen 2010). The effects were not as clear as the effects of adding clay. This may have been due to the lower turbidity reached (16.7 ± 3.5 NTU) compared to the clay caused turbidity reaching 59.7 ± 7.5 NTU.

10.3.12 Light Intensity

Light intensity is a very species specific factor. In rearing of European sea bass 500 lux at the surface is recommended. In sea bream 3000–5000 lux is needed (Moretti et al. 2004). Pikeperch larvae are reared at lower light intensities e.g. 50 lux (Lund and Steenfeldt 2011; Steenfeldt et al. 2011) and 140 lux (Rønfeldt and Nielsen 2010). Higher light intensities at the surface will aggregate the larvae in the lower part of the tank. Introduction of a daily dark period might be beneficial if larvae will not approach the brightly illuminated water surface, especially during the period of gas bladder inflation (Trotter et al. 2003). A dark period can be part of a feeding strategy where feeding are omitted during the dark period. An 8–12 h dark period is typically used in intensive pikeperch larval rearing. Continuous feeding seems not to have any positive effects.

10.3.13 Quantification of Mortality

When larval tanks are siphoned, quantification of mortality is to some extent possible (Lund and Steenfeldt 2011). Larger outbreaks of mortalities will be clearly visible in the siphoned material (Steenfeldt et al. 2011). By counting the numbers of dead fish the loss of fish from each tank can be estimated. Exact quantification of mortality is difficult though. At temperatures of 18 °C bacterial activity is significant and dead larva quickly disintegrates and become difficult to quantify.

10.4 Weaning

The weaning facility basically consists of ordinary fish rearing tanks of PE, Fiber glass or concrete. Water flow must be sufficient to maintain good rearing conditions in the tank i.e. change of one tank volume/hour (Steenfeldt et al. 2010). Increasing the water exchange rate to three times/hour reduced survival and growth at a commercial pikeperch farm (Steenfeldt et al. 2010).

10.4.1 System Design

Tank shape and water in- and outlet vary between facilities and attention must be paid to observation of larval behavior in the tanks. Currents in the tanks must be sufficiently low to allow larvae to maintain position in the tank and outlet screens must be observed to ensure that no live larvae are trapped on the screen.

Weaning facilities for intensive rearing of pikeperch fry in Western Europe are recirculation systems much resembling systems used for rearing of marine larvae. Outlet water from the tanks passes mechanical, biological and trickling filters before returning to the fish tanks. Temperature and water quality monitoring and manipulation equipment are implemented and often UV sterilization units will reduce bacterial load in the systems (Moretti et al. 2004).

The larvae are moved to a dedicated weaning facility prior to weaning. At this stage the larvae are easily irreversibly damaged by physical handling (Lund and Steenfeldt 2011). Consequently transfer are best carried out without trapping them on a landing net, but by reducing the volume of water in the larval rearing tanks and transferring the larvae with the remaining water to the weaning facility.

Temperature changes between the larval rearing unit and the weaning facility should be minimized and the fish starved a day before transfer (Moretti et al. 2004).

10.4.2 Weaning Strategy

In pikeperch attempts to wean larvae at 9 days post hatch resulted in significantly reduced growth when compared with weaning at day 15 post hatch (Hamza et al. 2007). Kestemont et al. (2007) found weaning of pikeperch on 19 days post hatch to have highest survival and growth but also highest cannibalism when compared with weaning at 12 or 26 days post hatch (Steenfeldt et al. 2010) found no significant differences in growth between pikeperch larvae weaned on day 16 or larvae fed *Artemia* till day 29. Larvae weaned on day 7 were significantly smaller though, already on day 16 post hatch and remained smaller till day 29. In commercial intensive rearing of pikeperch, weaning as early as possible is targeted to reduce the costs of feeding *Artemia* when compared with feeding dry feeds (Steenfeldt et al. 2010). The industry aims at initiating weaning on day 14–17 post hatch.

During weaning the quantity of *Artemia* is gradually reduced while the quantity of dry diets is increased (Kestemont et al. 2007) or remaining constantly high to facilitate larval ingestion of the inert diet.

Tanks must be cleaned daily to remove feed loss and mortality outbreaks ameliorated promptly to reduce losses to i.e. bacterial infections.

The dietary requirements of pikeperch larvae are not as high as in many marine larval species with respect to HUFA levels (Lund and Steenfeldt 2011). Use of high quality diets with pre-hydrolized diets (Kolkovski 2001) may improve larval performance and increase larval quality. A test of four commercially available weaning diets i.e. Nippai, EWOS Aglonorse No 1, Weanex 500 (Dana feed) and Gemmawean (Skretting) were tested on pikeperch in 2007 (Steenfeldt and Lund 2008). The experiment was conducted in a triplicate setup in 350 L-tanks measuring 1 by 1 m. Co-feeding with *Artemia* lasted 5 days with a gradual delay of time of feeding *Artemia* until on day 6 where no *Artemia* was added. After 20 days fish fed Dana feed and EWOS were larger ($\alpha=0.05$) than fish fed Skretting or Nippai. Fish fed Weanex 500 had higher survival than the remaining treatments ($\alpha=0.05$). Average survival of all fish during weaning was 79.6 %. There was a tendency towards lowest mortality caused by cannibalism in the fish fed Nippai. This may have been an effect of the lowest growth registered for this feed and lowest number of fish registered with a weight of above 1.1 g at the end of the trial. A similar relation between high growth and low cannibalism was observed by (Kestemont et al. 2007).

10.5 Cannibalism

Cannibalism is often categorized as type I and type II. In type I cannibalism the prey is only partly ingested whereas in type II the whole prey is swallowed (Kestemont et al. 2003). The intensity of cannibalism is generally determined by the ratio between predator gape size and prey size (Smith and Reay 1991). A high predator to prey gape size ratio is usually required in type II cannibalism (Baras and Jobling 2002; van Damme et al. 1989; Hecht and Appelbaum 1988).

Cannibalism types 1 and 2 also make quantification difficult since parts of or whole larvae may be ingested by cannibalistic individuals in the tanks.

Cannibalism is a known problem in rearing of a number of fish larval species (Hecht and Pienaar 1993). Both genetic and behavioral factors seem to cause cannibalism. In European seabass and Eurasian perch initial stocking density and initial size heterogeneity influenced cannibalism in both species whereas a number of other factors gave ambiguous results (Kestemont et al. 2003).

In pikeperch high mortality due to cannibalism may reduce the number of larvae produced significantly (Steenfeldt et al. 2011). The cause of cannibalism in pikeperch seems not to be different times of hatching, (Steenfeldt et al. 2011) but seems to be related to high growth rates during larval life (Kestemont et al. 2007).

In walleye survival in turbid water was significantly higher than in clear water (Bristow and Summerfelt 1994). The higher mortality in the clear water may have been caused by cannibalism since the more uniform distribution of fish larvae in turbid water has been reported to reduce cannibalism in walleye larvae (Loadman et al. 1989). In another study on walleye cannibalism was not affected by turbidity though (Rieger and Summerfelt 1997). Consequently the mechanisms associated to cannibalistic behavior and turbidity is not clear.

By staining otoliths of pikeperch larvae early in life it is possible at a later time to estimate individual larval growth prior to the time of capture. Estimation of prior growth of pikeperch larvae revealed that cannibals that were among the largest of fish in a batch at day 35 post hatch also had been among the largest fish on day 14. Non cannibalistic individuals with only *Artemia* in their guts of the same size on day 35 had also been among the largest fish in the batch on day 14 (Steenfeldt et al. 2010). It was concluded that cannibals are among the largest fish from early larval life (day 14 post hatch), but that non cannibals of the same size at day 14 post hatch can grow at similar rates till day 35 post hatch. Size of prey fish in cannibal stomachs could also be estimated based on otolith size, and were found to be 65.4 ± 6.7 % of cannibal length. Since cannibals are significantly larger than their prey, grading of early stage fish may help ameliorate cannibalism.

References

Attramadal KJK (2011) Water treatment as an approach to increase microbial control in the culture of cold water marine larvae. Dissertation, NTNU

Barahona-Fernandes MH (1978) Effect of aeration on the survival and growth of sea bass (Dicentrarchus labrax L.) larvae: a preliminary study. Aquaculture 14:67–74

Baras E, Jobling M (2002) Dynamics of intracohort cannibalism in cultured fish. Aquac Res 33:461–479

Barrows FT, Zitzow RE, Kindschi GA (1993) Effects of surface water spray, diet, and phase feeding on swim bladder inflation, survival, and cost of production of intensively reared larval walleyes. Prog Fish Cult 55:224–228

Batty RS (1987) Effect of light intensity and food-searching of larval herring, *Clupea harengus*: a laboratory study. Mar Biol 94:323–327

Blancheton JP (2000) Developments in recirculation systems for mediterranean fish species. Aquac Eng 22:17–31

Boggs CT, Summerfelt RC (2003) Enhancing gas bladder inflation in larval walleye: comparison of two methods for removing an oily film from the water surface of culture tanks, pp 19–20, 20–24 July 2003

Bristow BT, Summerfelt RC (1994) Performance of larval walleye cultured intensively in clear and turbid water. J World Aquacult Soc 25:454–464

Bristow BT, Summerfelt RC, Clayton RD (1996) Comparative performance of intensively cultured larval walleye in clear, turbid and colored water. Prog Fish Cult 58:1–10

Büke E (2002) Sea bass (*Dicentrarchus labrax* L., 1781) seed production. Turk J Fish Aquat Sci 2:61–70

Busch A (1996) Transition from endogenous to exogenous nutrition: larval size parameters determining the start of external feeding and size of prey ingested by Ruegen spring herring *Clupea harengus*. Mar Ecol Prog Ser 130:39–46

Chatain B, Ounais-Guschemann N (1990) Improved rate of initial swim bladder inflation in intensively reared *Sparus auratus*. Aquaculture 84:345–353

Costas B, Aragão C, Mancera JM, Dinis MT, Conceição LEC (2008) High stocking density induces crowding stress and affects amino acid metabolism in Senegalese sole *Solea senegalensis* (Kaup 1858) juveniles. Aquac Res 39:1–9

Demska-Zakes K, Kowalska A, Zakes Z (2003) The development of the swimbladder of pike perch *Sander lucioperca* reared in intensive culture. Arch Pol Fish 11:45–55

De Robertis A, Ryer CH, Veloza A, Brodeur RD (2003) Differential effects of turbidity on prey consumption of piscivorous and planktivorous fish. Can J Fish Aquat Sci 6012:1517–1526

Divanach P, Kentouri M (2000) Hatchery techniques for specific diversification in Mediterranean finfish larviculture. Rome, Anonymous

Divanach P, Papandroulakis N, Anastasiadis P, Koumoundouros G, Kentouri M (1997a) Effect of water currents on the development of skeletal deformities in sea bass (*Dicentrarchus labrax*L.) with functional swimbladder during postlarval and nursery phase. Aquaculture 156:145–155

Divanach P, Papandroulakis N, Anastasiadis P, Koumoundouros G, Kentouri M (1997b) Effect of water currents on the development of skeletal deformities in sea bass (Dicentrarchus labrax L.) with functional swimbladder during postlarval and nursery phase. Aquaculture 156:145–155

Gaignon J L, Petton B, Quemener L (1998) Hydrodynamic effects in larvae rearing tanks on survival and growth of turbot (*Psetta maxima*). Bull Fr Pesche Piscic 350-351:303–323

Hamza N, M'Hetli M, Kestemont P (2007) Effects of weaning age and diets on ontogeny of digestive activities and structures of pikeperch (*Sander lucioperca*) larvae. Fish Physiol Biochem 33:121–133

Hecht T, Appelbaum S (1988) Observations on intraspecific aggression and coeval sibling cannibalism by larval and juvenile *Clarias gariepinus* (*Clariidae*: *Pisces*) under controlled conditions. J Zool 214:21–44

Hecht T, Pienaar AG (1993) A review of cannibalism and its implications in fish larviculture. J World Aquacult Soc 24:246–261

Helland S, Terjesen BF, Berg L (2003) Free amino acid and protein content in the planktonic copepod Temora longicornis compared to *Artemia franciscana*. Aquaculture 215:213–228

Hilge V, Steffens W (1996) Aquaculture of fry and fingerlings of pike-perch (Stizostedion lucioperca L.) – a short review. J Appl Ichthyol 12:167–170

Houde ED (1975) Effects of stocking density and food density on survival, growth and yield of laboratoryreared larvae of seabream *Archosargus rhomboidalis* (L.) (Sparidae). J Fish Biol 7:115–127

Jepsen N, Koed A, ØKland F (1999) The movements of pikeperch in a shallow reservoir. J Fish Biol 54:1083–1093

Kestemont P, Jourdan S, Houbart M, Mélard C, Paspatis M, Fontaine P, Cuvier A, Kentouri M, Baras E (2003) Size heterogeneity, cannibalism and competition in cultured predatory fish larvae: biotic and abiotic influences. Aquaculture 227:333–356

Kestemont P, Xueliang X, Hamza N, Maboudou J, Imorou Toko I (2007) Effect of weaning age and diet on pikeperch larviculture. Aquaculture 264:197–204

Kihara M, Ogata S, Kawano N, Ryoichi Yamaguchi R (2002) Lordosis induction in juvenile red sea bream, Pagrus major, by high swimming activity. Aquaculture 212:149–158

Kolkovski S (2001) Digestive enzymes in fish larvae and juveniles—implications and applications to formulated diets. Aquaculture 200:181–201

Kolkovski S, Curnow J, King J (2004) Intensive rearing system for fish larvae research I. Marine fish larval rearing system. Aquac Eng 31:295–308

Kroupova H, Machova J, Svobodova Z (2005) Nitrite influence on fish: a review. Vet Med Czech 50:461–471

Lavens P, Sorgeloos P (eds) (1996) Manual on the production and use of live food for aquaculture. FAO, Rome

Lehtonen H, Hansson S, Winkler H (1996) Biology and exploitation of pikeperch, *Stizostedion lucioperca* (L.), in the Baltic Sea area. Ann Zool Fenn 33:525–535

Lekang O (2007a) 17. Internal transport and size grading. In: Anonymous aquaculture engineering. Blackwell, Oxford, pp 227–254

Lekang O (ed) (2007b) Aquaculture engineering. Blackwell, Oxford

Loadman NL, Mathias JA, Moodie GEE (1989) Method for the intensive culture of walleye. Prog Fish Cult 51:1–9

Lund I, Steenfeldt S (2011) The effects of dietary long-chain essential fatty acids on growth and stress tolerance in pikeperch larvae (*Sander lucioperca* L.). Aquac Nutr 17:191–199

Meager JJ, Utne-Palm AC (2008) Effect of turbidity on habitat preference of juvenile atlantic cod, gadhus morhua. Environ Biol Fish 81:149–155

Mani-Ponset L, Diaz JP, Schlurnberger O, Connes R (1994) Development of yolk complex, liver and anterior intestine in pike-perch larvae, *Stizostedion lucioperca* (Percidae), according to the first diet during rearing. Aquat Living Resour 7:191–202

Michaud L, Lo Giudice A, Troussellier M, Smedile F, Bruni V, Blancheton JP (2009) Phylogenetic characterization of the heterotrophic bacterial communities inhabiting a marine recirculating system. J Appl Microbiol 107:1935–1956

Molnár T, Hancz C, Bódis M, Müller T, Bercsényi M, Horn P (2004) The effect of initial stocking density on growth and survival of pike-perch fingerlings reared under intensive conditions. Aquac Int 12:181–189

Moore A, Prange MA, Bristow BT, Summerfelt RC (1994a) Influence of stocking densities on walleye fry viability in experimental and production tanks. Prog Fish Cult 56:194–201

Moore A, Prange MA, Summerfelt RC, Bushman RP (1994b) Evaluation of tank shape and a surface spray for intensive culture of larval walleyes fed formulated feed. Prog Fish Cult 56:100–110

Moretti A, Fernandez-Criado FD, Vetillart R (2004) Manual on hatchery production of seabass and gilthead seabream. FAO, Rome

Moretti A, Fernandez-Criado MP, Vetillart R (2005) Manual on hatchery production of seabass and gilthead seabream. FAO, Rome

Nyina-wamwiza L, Xu X, Blanchard G, Kestemont P (2005) Effect of dietary protein, lipid and carbohydrate ratio on growth, feed efficiency and body composition of pikeperch *Sander lucioperca* fingerlings. Aquacult Res 36:486–492

Ostaszewska T (2003) Histopathological changes during pikeperch Sander lucioperca swim bladder development. 20–24 July 2003, pp 111–112

Ostaszewska T, Dabrowski K, Czumińska K, Olech W, Olejniczak M (2005) Rearing of pike-perch larvae using formulated diets – first success with starter feeds. Aquacult Res 36:1167–1176

Person-Le Ruyet J (1989) The hatchery rearing of turbot larvae (Scopthalmus maximus, L.). Cuad Area Cienc Mar 3:57–91

Pickering AD, Potinger TG (1989) Stress responses and disease resistance in salmonid fish: effects of chronic elevation of plasma cortisol. Fish Physiol Biochem 7:253–258

Poppe TT (ed) (1990) Fiskehelse (fish health). John Grieg forlag AS, Stavanger

Rabe J, Brown JA (2000) A pulse feeding strategy for rearing larval fish: an experiment with yellowtail flounder. Aquaculture 191:289–302

Rieger PW, Summerfelt RC (1997) The influence of turbidity on larval walleye, Stizostedion vitreum, behavior and development in tank culture. Aquaculture 159:19–32
Rønfeldt JL, Nielsen JW (2010) Filling of gas bladder, growth and the survival in pikeperch larvae (Sander lucioperca) in intensive aquaculture (in Danish). MSc thesis, University of Copenhagen 350–351:303–323
Salama AJ (2007) Effects of stocking density on fry survival and growth of asian sea bass (*Lates calcarifer*). JKAU Mar Sci 18:53–61
Shields RJ (2001) Larviculture of marine finfish in Europe. Aquaculture 200:55–88
Smith C, Reay P (1991) Cannibalism in teleost fish. Rev Fish Biol Fish 1:41–64
Steenfeldt S, Lund I (2008) Development of production methodology for intensive rearing of pikeperch fry. Denmark, Anonymous, Hirtshals. (in Danish)
Steenfeldt S, Vestergaard M, Overton JL, Paulsen H, Larsen VJ, Henriksen NH (2010) Development of intensive rearing of pikeperch in Denmark. Denmark, Anonymous, Hirtshals. (in Danish)
Steenfeldt S, Lund I, Höglund E (2011) Is batch variability in hatching time related to size heterogeneity and cannibalism in pikeperch (*Sander lucioperca*)? Aquac Res 42:727–732
Støttrup JG, McEvoy LA (eds) (2003) Live feeds in aquaculture. Blackwell, Oxford
Summerfelt RC (1996) Intensive culture of walleye fry. In: Summerfelt RC (ed) Walleye culture manual. NRAC culture series, vol 101 edn. North central regional aquaculture center publications office, Iowa state university, Ames, pp 161–185
Summerfelt RC, Johnson JA, Clouse CP (2011) Culture of walleye, sauger, and hybrid walleye. In: Barton B (ed) Biology, management, and culture of walleye, sauger, and hybrid walleye. American Fisheries Society Special Publication, Bethesda, pp 451–570
Szkudlarek M, Zakes Z (2007) Effects of stocking density on survival and growth performance of pikeperch, Sander lucioperca L., under controlled conditions. Aquac Int 15:67–81
Tagawa M, Kaji T, Kinoshita M, Tanaka M (2004) Effects of stocking density and addition of proteins on larval survival in Japanese flounder, *Paralichthys olivaceus*. Aquaculture 230:517
Tapie P, Bernard A (1988) Microalgae production: technical and economic evaluations. Biotechnol Bioeng 32:873–885
Trotter AJ, Battaglene SC, Pankhurst PM (2003) Effects of photoperiod and light intensity on initial swim bladder inflation, growth and post-inflation viability in cultured striped trumpeter (*Latris lineata*) larvae. Aquaculture 224:141–158
Utne ACW (1997) The effect of turbidity and illumination on the reaction distance and search time of a marine planktivore (Gobiusculus flavescens). J Fish Biol 50:926–938
van Damme P, Appelbaum S, Hecht T (1989) Sibling cannibalism in Koi carp, *Cyprinus carpio* L., larvae and juveniles reared under controlled conditions. J Fish Biol 34:855–863
Vogel JL, Beauchamp DA (1999) Effects of light, prey size, and turbidity on reaction distances of lake trout *Salvelinus namaycush* to salmonid prey. Can J Fish Aquat Sci 56:1293–1297
Zarsky D, Targonska K, Krejszeff S, Kwiatkowski M, Kupren K, Kucharczyk D (2011) Influence of stocking density and type of feed on the rearing of crucian carp, *Carassius carassius* (L.) larvae under controlled conditions. Aquac Int 19:1105–1117
Zingel P, Paaver T (2010) Effects of turbidity on feeding of the young-of-the-year pikeperch (*Sander lucioperca*) in fishponds. Aquacult Res 41:189–197

Chapter 11
Intensive Culture of Walleye from Egg Incubation to Juvenile

Robert C. Summerfelt and J. Alan Johnson

Abstract This chapter describes early life stages and reviews intensive larviculture of Walleye (*Sander vitreus*) from hatch to 35 days posthatch (dph). Embryonic development, egg incubation and chemotherapy for eggs, as well as details of gas bladder inflation and methodology to overcome non-inflation of the gas bladder (NGB) are illustrated with photographs from microvideography and artwork. Husbandry includes description of stocking density, as well as environmental features (light, tank color, turbid water, surface spray, and tank hygiene) needed to overcome the problem of clinging behavior and NGB. Consideration is given to use of live and manufactured feeds as well as feeding rate and frequency. The problem of deformities as well as occurrence and treatment of disease and are given appropriate attention. The chapter demonstrates a science-based, production-scale protocol for Walleye fry culture that can achieve 60–70 % survival from hatch to 35 days post hatch (dph). The chapter supports the viewpoint that intensive larviculture offers a practical alternative to pond-culture for production of feed-trained juvenile Walleyes.

Keywords Walleye • Incubation • Larviculture • Husbandry • Juveniles

11.1 Introduction

This essay on intensive fry culture is an abridged retrospective of an extensive review (Summerfelt et al. 2011) included in Biology, Management, and Culture of Walleye and Sauger (Barton 2011). That review was comprehensive in subject matter, documented with 437 references, and it included many topics presented

R.C. Summerfelt (✉)
Department of Natural Resource Ecology and Management, Iowa State University of Science and Technology, Ames, IA 50011-3221, USA
e-mail: rsummerf@gmail.com

J.A. Johnson
Rathbun Fish Culture Research Facility, Iowa Department of Natural Resources, 15053 Hatchery Place, Moravia, IA 52571, USA

P. Kestemont et al. (eds.), *Biology and Culture of Percid Fishes*,
DOI 10.1007/978-94-017-7227-3_11

in other chapters of this book. Therefore, we have reduced the details and scope of this review, but incorporated new sources relevant to the present topic. Pond culture of Walleye from larva to juvenile is described in Chap. 18 by Briland et al. The present chapter focuses on intensive (i.e., tank) culture of Walleye from hatch to 35 dph. We have not endeavored to systematically compare biology and husbandry of Walleye with that of pikeperch because that species is reviewed by Steenfeldt (Chap. 10).

11.2 Life Stages from Hatch to Juvenile

The interval from fertilization to hatch includes a short early phase of cleavage with conspicuous cell division, and the post-cleavage embryonic phase involving transition from blastula to the blastoderm over-growth (engulfing) of the yolk; gastrula; formation of germ layers and organs (organogenesis), until a little fish with dark eye spots (eyed-egg stage) and a beating heart can be seen sitting on top of the massive yolk (McElman and Balon 1979). At 15 °C cleavage lasts from 2.5 to 48 h after fertilization and the embryonic interval 11 days (165 daily temperature units, DTU). A critical embryological stage of early organogenesis occurs in 50–100 h (21–42 DTU, which is the sum of the mean daily temperatures above 0 °C) after gastrulation when the germinal layers are transformed into various organs of the embryo (Latif et al. 1999).

The common perspective is that the embryo stage terminates at hatch, but McElman and Balon (1979) regard newly hatched fish with a yolk sac a free embryo (eleutheroembryo, eleuthero- "free"); an interval (~4 days duration) when the embryo transitions from yolk sac nutrition (endogenous feeding) to the start of external feeding. The free embryo is equivalent to the prolarval (yolk sac stage) of Li and Mathias (1982). Hubbs (1943) and Nelson (1968) recognized only two larval stages, the prolarval and postlarval stages. We have followed the description of three larval stages by Li and Mathias (1982) that are collectively called fry by hatchery personnel in the U.S. (Fig. 11.1). The prolarva and postlarva I stages are each about 5 days duration at 18–20 °C.

11.2.1 Prolarva

The interval with the yolk sac, thus the term sac fry, starts at hatch and lasts until the yolk sac is absorbed. The first prolarvae observed with open mouths were 3 days old (47 DTU), but all fish we examined at 4 days (64 DTU) had open mouths (Phillips and Summerfelt 1999). In that study, a few 4 days old (64 DTU) larva were feeding but all 7-days-old postlarvae I (113 DTU) were feeding. Prolarvae at the end of this stage are 8–9 mm long.

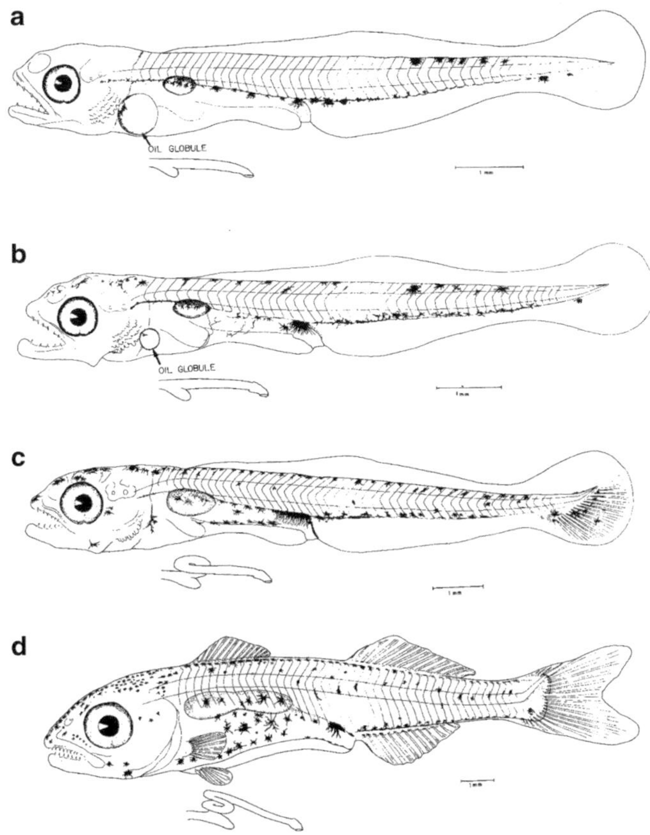

Fig. 11.1 The larval stages (**a**–**c**) to early (scaleless) juvenile (**d**): (**a**) Late (~ 5 dph) prolarva without yolk sac, but a large oil globule and inflated gas bladder; (**b**) Post-larva I (with oil globule), (**c**) Post-larva II (without oil globule), development of fin rays on caudal fin, and first loop of intestine that cuts off secretions from the gall bladder from reaching the ostium of the pneumatic duct. Juvenile (**d**) has lost median fin-fold, developed paired fins and fin rays, and has more pigmentation. The bar represents 1 mm (From Li and Mathias 1982, with permission of the American Fisheries Society)

11.2.2 Postlarva I

The postlarva I stage immediately follows yolk sac absorption but the stage is recognized by presence of the oil globule. This stage is an interval of mixed nutrition when the larval fish start feeding (236 DTU, McElman and Balon 1979), but before the oil globule is fully absorbed. The oil globule provides a nutritional buffer from starvation (Li and Mathias 1982). The larvae are stronger swimmers and exogenous feeding begins including the onset of cannibalism. This is also a critical time because it is the interval when the lava inflates their gas bladder (Marty et al. 1995). Postlarva I ends when the oil globule is adsorbed. Phillips and Summerfelt (1999) reported that Walleye measured 10.0 mm at 6 days (81 DTU) when the yolk sac disappeared.

Gas bladder inflation is a critical developmental event because fish that fail to inflate their gas bladder are unable to maintain their position in the water column without excessive expenditure of energy (Summerfelt 2013). It is doubtful that fish without an inflated gas bladder will survive in nature, but in tank culture, some do live to juvenile and older but they have much slower growth. Phillips and Summerfelt (1999) observed the first fish with an inflated gas bladder at 5-days old (96 DTU), but all 7-days-old postlarvae I (113 DTU) had inflated gas bladders; the interval for gas bladder inflation is from 5 to 12 dph.

First filling of the gas bladder requires that the larva penetrate the water surface with its head to gulp air (Fig. 11.2; Rieger and Summerfelt 1998). Because the air bubble that is swallowed is too large to pass from the foregut

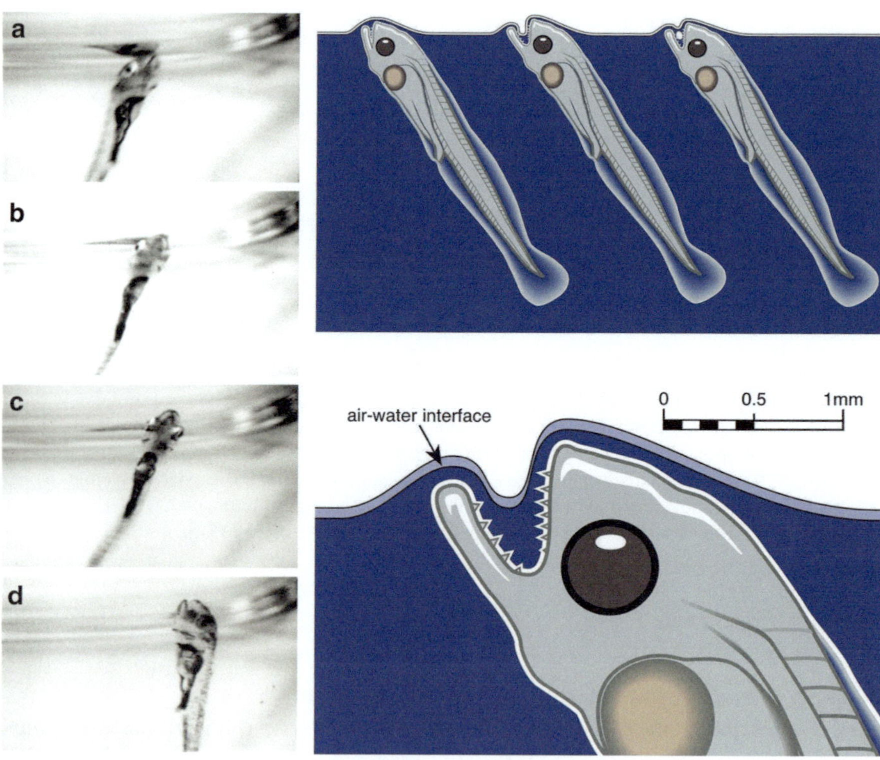

Fig. 11.2 Photographic (*left*) sequence from microvideography (From Rieger and Summerfelt 1998) of Walleye larva (prolarva I) penetrating the water surface to gulp air: (**a**) larva approach surface; (**b**) pushes against surface film with rapid movements of the caudal fin; (**c**) larva at instant of penetration; (**d**) larva following penetration, tail movement stopped and the larva was momentarily suspended by the surface tension for 0.5 s. Artistic representation (*right*) of the larvae penetrating water surface to gulp air: third fish in *upper right* has gulped an air bubble, and *lower right* illustrates surface film folding into the fish's mouth

through the small diameter of the pneumatic duct (25–45 μm), the bubble must first be emulsified to smaller size by action of surfactants in gallbladder secretions that discharge near the ostium of the pneumatic duct (Marty et al. 1995). The interval for gas bladder inflation ends after 12 dph when closure of the pyloric sphincter prevents surfactants from the gall bladder to discharge into the foregut (Marty et al. 1995).

11.2.3 Postlarva II

The postlarva II stage begins after oil globule disappearance and lasts about 8 days when the larvae have grown to ~16–19 mm (Li and Mathias 1982). The first loop of the intestine is formed (Li and Mathias 1982), and the adult complement of pyloric caeca (3) are present (Nelson 1968); fin rays are developing on the caudal fin; the median fin fold is diminishing, and pigmentation is occurring.

11.2.4 Juvenile

For convenience, young fish ≥20 mm are considered juvenile. The fin fold is gone, the gas bladder has elongated, paired fins are obvious, and fin rays and spines have appeared. Scale development begins in the peduncle region but full-body coverage is slow. Walleye sampled from a 1.8-ha hatchery pond near Gunton, Manitoba had no scales at 22 dph and 20.9 mm, 66.7 % of fish were scaled at 29 days, 93.3 % at 36 dph at 34.0 mm; all fish >52 mm and 45 dph were scaled (Glenn and Mathias 1985). Full coverage with scale is important for fish husbandry because fish are more likely to suffer electrolyte loss (chloride) from handling stress before they are completely scaled.

11.3 Egg Incubation

11.3.1 Incubation Interval

The number of days from water hardening to hatch is the incubation interval. For individual fish, it is noted that eggs from the same female hatch over several days, with peak hatch somewhere between the first and last. Reports by hatchery personnel of the incubation interval represent the number of days at the hatchery when there are eggs incubating, but they represent eggs of many females that were spawned over many days. Also, the hatching interval at the same hatchery will be extended when the spawning season is interrupted by abrupt changes in spring weather. The incubation interval at the same hatchery may be constant for several years (10 days in Oklahoma 2005–2008) or show substantial annual variation; e.g., at the Milford Hatchery, in for 23 years (1985–2008, missing 1995), the interval from first to last hatch ranged from 13 to 23 days at a mean temperature of 13.4 °C (Summerfelt et al. 2011). There is evidence that climate change (warming) is resulting in earlier spawning (Schneider et al. 2010).

Heidinger et al. (1997) noted most mortality (83 %) during egg incubation occurred before 6 h post-fertilization at 3 DTU (°C), which occurs when the embryo is a sensitive blastula stage as the cells are forming germinal layers. A broader sensitive interval would be mortality during the first 5 days of egg incubation when more than 80 % of the total mortality that occurs (Latif et al. 1999, 2001).

Hatcheries that use lake water without artificial heating often start egg incubation at about 7 °C, followed by a slow water temperature increase reaching 13–15 °C when hatching occurs. Nelson et al. (1965) report hatching occurred over a 4 day interval, starting at 14 days at 12.8 °C (179 DTU). At the Oneida Lake Hatchery, New York, hatching begins as early as 131 DTU (°C) but usually peaks at 204 DTU (°C), and the incubation interval is typically 21 days, with a range of 15–30 days (Colesante 1996). Harvey and Hood (1996) reported an incubation interval of 21 days using water at a constant temperature of 11 °C (231 DTU); however, 27 days to hatch (range, 22–31 days) and 265 DTU (°C) was the average for a 7-year data set for eggs of Vermillion Lake (Minnesota) Walleye spawning in the Pike River (Thompson 1996).

Genetics of the broodstock, maternal nutrition, water quality, and other factors affect survival during egg incubation to eyed-egg stage, but temperature relationship has been most often examined. Smith and Koenst (1975) observed that the acclimation temperature of mature females affected the rate of development and percentage of eggs that hatched of eggs. The highest percent hatch (circa 85 %) was for eggs incubated at 6 °C that came from females acclimated to 6 °C, and no fish hatched from eggs incubated at 21 °C from females acclimated to that same temperature.

Although it is reasonable to assume a detrimental effect of diurnal temperature fluctuations, a summary of several studies show that Walleye eggs are tolerant to 24-h temperature changes as great as ±19 °C (Allbaught and Manz 1964; Zitzow 1991). In fact, Schneider et al. (2002) reported that short term temperature reduction as much as −9.3 or an increase of +20.2 °C had no effect on eye-up rate, although a substantial reduction in swim-up (hatch rate) occurred when eggs were subjected to extreme (±21.1 °C) cold and heat; yet, they concluded that "temperature fluctuations great enough to directly affect incubating Walleye eggs are unlikely to occur in hatcheries or on natural spawning grounds."

After the sensitive interval has passed, especially by eye-up, the incubation interval can be extended using colder water to address special circumstances at the hatchery. In one case, the incubation interval was intentionally extended to 42 days (Paddock 1996) because the ponds where the fry were to be stocked were first used for northern pike *Esox lucious*. In another, the incubation interval was extended to allow time for disease testing of the parental fish in order to maintain disease-free certification of the stocked fish (Moodie and Mathias 1996).

11.3.2 Incubation Jars

In North America, Walleye eggs are typically incubated in a "McDonald" jar, although commonly used, the name is historically a misnomer (Summerfelt et al. 2011). Originally made of glass, modern commercial versions of the jar are manufactured of a clear plastic (Fig. 11.3). The size varies slightly depending on the source, but a common size has a volume of ~6.0 L (height 45.8 cm, diameter 15.9 cm). The key feature of this jar is that the water goes down a tube (pipe) to a parabolic bottom creating an upwelling that gently rolls and suspends the eggs. Several smaller variations of hatching jars have described for research applications where large numbers of eggs are not needed.

At hatch, the prolarvae swim-up and flow out of the jar to an open trough that receives larvae from many incubation jars, often from long rows of jars with double shelf (Fig. 11.3). The trough empties into a common collection tank ("catch tank"). The drain from the catch tank is covered with a brass screen 680 uμ (e.g., Thompson 1996), 50-mesh stainless steel (Colesante 1996), or nylon screening of 670 μm (Rathbun Fish Hatchery) to prevent the fry from escaping.

A portable, low-flow incubator was developed for use in remote locations because it requires only 0.5 L/min for an incubator with a 10.4-L capacity (Zitzow 1991; Greiff 1996). They are considered beneficial for use at times when heating cold groundwater is needed (Gallinat 1996), or when water quality of ground (well) water has high concentrations of manganese, ammonia, or iron. The low-flow incubator has been used to extend the incubation interval of Walleye eggs with "no apparent detrimental effects on the developing embryos" (Zitzow 1991).

Fig. 11.3 Incubation jar (*line drawing*), a double row lineup of jars with incubating Walleye eggs, and a close-up of jar showing layer (*double bar*) of dead eggs that will be removed with siphon

11.3.3 Stocking Density and Water Exchange

Stocking densities of Walleye eggs (volume of eggs per jar) varies among hatcheries reporting data in the Walleye Culture Manual: e.g., 2.8 L Colesante (1996) to 4.7 L (Thompson 1996). Eggs require a constant supply of water for respiration and to prevent clumping without excessive turbulence. Commonly, water flow rate is subjectively judged by a flow rate that ensures a continuous but gentle movement within the jar (Harvey and Hood 1996). Colesante (1996) reported a water flow rate of 5.7 L min^{-1} for up to 3 days after water hardening, and then 3.8 L min^{-1} for the duration of incubation. Copeland and Wolgamood (1996) used 1 L min^{-1} without rolling the eggs.

11.3.4 Egg Treatments for Fungus and Viruses

The eggs of Walleye and most artificially propagated fishes (e.g., especially salmonids) need treatment during incubation to prevent fungus overgrowth of the eggs. In the U.S., the choice is usually between a Food and Drug Administration Center for Veterinary Medicine (FDACVM) registered formulation of formalin or hydrogen peroxide. Formalin is used at a concentration of 1000–1667 μL L^{-1} (ppm) of flow for 15-min exposure, and hydrogen peroxide is used at 500 mg L^{-1} for 15-min exposure in a flow-through treatment to prevent fungus on incubating Walleye eggs.

Substantial attention has been given to determining safe and effective measures for disinfection of Walleye eggs of the viral hemorrhagic septicemia virus (VHSV) after U.S. Department of Agriculture included Walleye on federal order restricting the interstate transport of live fish because a new virulent viral strain (VHSV Ivb) of viral hemorrhagic septicemia virus (VHSV) was discovered in the Great Lakes (Phelps et al. 2012). Thereafter, several studies were undertaken to determine effective and safe methods for disinfecting Walleye eggs with compounds containing iodine:

- Iodophor treatment of 100 mg/L for 30-min post fertilization (during water hardening) was effective at eliminating VHSV tested by viral isolation in cell culture (Tuttle-Lau et al. 2010).
- Treatment of water-hardened eggs with 100-ppm iodine for 30–45 min was done without detriment to swim up rate or fry and fingerling survival (Getchell 2007).
- After thoroughly rinsing fertilized eggs that were treated with tannic acid (400 mg L^{-1}) for 2 min 30 s to remove stickiness, an iodophor treatment of 100 mg L^{-1} for 30 min "sufficiently eliminated VHSV as detected by viral isolation (Groocock et al. 2012). Viral isolation, however, requires 28–30 days compared with 2–4 days for reverse transcription polymerase chain reaction (rRT-PCR) assay, but detection of a particular sequence of viral RNA does not prove viable virus (Phelps et al. 2012).
- Exposure of incubating eggs to 380 mg L^{-1} of iodine for 30 or 180 min decreased embryo survival, but treatments up to 200 mg L^{-1} for 15-min did not cause post-fertilization mortality of embryos or harm viability of Walleye larvae (Dabrowski et al. 2009).

Because there is higher toxicity to eggs when the treatment is applied during water hardening compared with treatments of eggs after water hardening, treatments are 50 mg L^{-1} for 30-min during water hardening, and 100 mg L^{-1} for 10-min during egg incubation.

11.3.5 Removal of Dead Eggs

Removal of dead eggs reduces the opportunity for fungus to infect the jar and cause additional egg mortality. Dead eggs appear white and are more opaque than viable eggs. Because dead eggs are slightly more buoyant than live eggs, the gentle rolling of eggs in the incubation jar brings the dead eggs to the surface within several days after fertilization (Fig. 11.3). Eggs are siphoned as soon as 3 days after placement in jars and is repeated as necessary until just before egg hatching. A clear plexiglas tube with a length of silicon rubber tubing attached and a bucket are the basic equipment needed for this task. Water flow to the jar does not need to be turned off during this process, and an inflow is necessary to prevent draining the jar of all water. Once the siphon is running, the tip is held 2–5 cm above the layer of dead eggs and the suction will lift the eggs into the siphon; the distance can be varied to increase or decrease the speed of egg removal. Movement of the siphon is slow and methodic to prevent stirring up the live eggs in the dead eggs. Nearly all the dead eggs at the surface can be removed by this method. Removed eggs should be examined to ensure that siphoning process has not removed an excessive number of live eggs. Some dead eggs may be mixed deeper in the jar and may be removed on later days.

The separation of live and dead eggs into two distinct layers is not always 100 % and siphoning must be done carefully, but because the process inevitably results in removal of some live eggs, one hatchery uses a second jar to place the eggs after siphoning where the siphoning process is repeated to recapture some of the live eggs. Generally, however, the number of eggs incubated at the hatchery is in excess of their needs, and the loss of some live eggs in clearing the jars of the dead eggs is accepted.

11.3.6 Enumeration of Eggs and Determining Hatching Success

An estimation of hatching success is typically calculated from volume measurements immediately after eggs are placed in the jar and after eye-up, but not at hatch. The incubation jars are marked in 0.5 L increments to facilitate measurement. Water flow is turned off for a brief period, generally less than a minute, to allow eggs to settle so that volume can be extrapolated from the marks on the jar. Once the measurement is taken, water flow is restored. Egg counts per unit volume and the measured jar volume then determine the number of eggs. The difference between initial number and number after eye-up is used to calculate the percent at hatch.

11.4 Husbandry

Although at this time, intensive culture is uncommon at most public and private hatcheries, a considerable body of science-based, applied research now provides information essential for cultural technology to overcome the major biological constraints needed to support production-scale (i.e., commercial-scale) application. The critical problems of noninflation of the gas bladder (NGB), clinging behavior, cannibalism, and lack of suitable larval feed have been resolved to provide opportunity for successful, large-scale production of fingerlings.

11.4.1 Environmental Aspects

Size of aquaria or tanks used in laboratory-scale research on intensive larviculture of Walleye has varied over two orders of magnitude. Li and Mathias (1982),notable among the first systematic work on Walleye larviculture, used 0.3-L fingerbowls, 200 L glass aquaria, and 10 L polyethylene washtubs. They found that even with constant stocking density, "the effect of culture volume was difficult to interpret," but as they said a minimum of 20 L was essential for at least 30 % survival. In our research studies, tank volume ranged from 150 to 279 L, but these two sizes were not compared, however, Moore et al. (1994b) compared Walleye growth and survival to 21-dph in 278-L and 679-L tanks, but the results were inconsistent. There was significantly higher growth and survival for fish cultured in the 278-L tanks in one trial, but lower in another trial, and not significantly different in another trial. They regarded the 679-L tank a production-scale size (Fig. 11.4).

The largest operating production-scale culture system for culture of larval Walleye is that of the Oneida Fish Culture Station, New York State Department of Environmental Conservation, Constantia, NY (Colesante (1996), where rectangular, fiberglass tanks with a volume of 4.2 m^3, are used. In Canada, Moodie and Mathias (1996) described an experimental, commercial-scale (2.6 m^3) rectangular trough for larval culture of Walleye, but it apparently it has not been used in commercial culture.

Larval Walleye exhibit a strong, positive phototactic response to light for the first 30–40 dph. In clear water, both direct or reflected light from tank surfaces is such a strong attraction for Walleye that they cling or seem to adhere to tank surfaces; they will cling to the tank sidewall, to a shiny metal standpipe screen, or to a metal screen on the sidewall of the tank, especially when light from outside of the tank that passes through the screen. Clinging behavior is detrimental to fry survival because they stop feeding and seem more vulnerable to sibling cannibalism.

The first effort to overcome the clinging problem was to paint the tank surfaces a flat black color; gray or black tank walls are preferred to tank to white, yellow or green walls. Colesante (1996) stated that darkened sides of the rearing tanks minimizes clinging behavior and he reported that high-intensity lights (680 lux at the

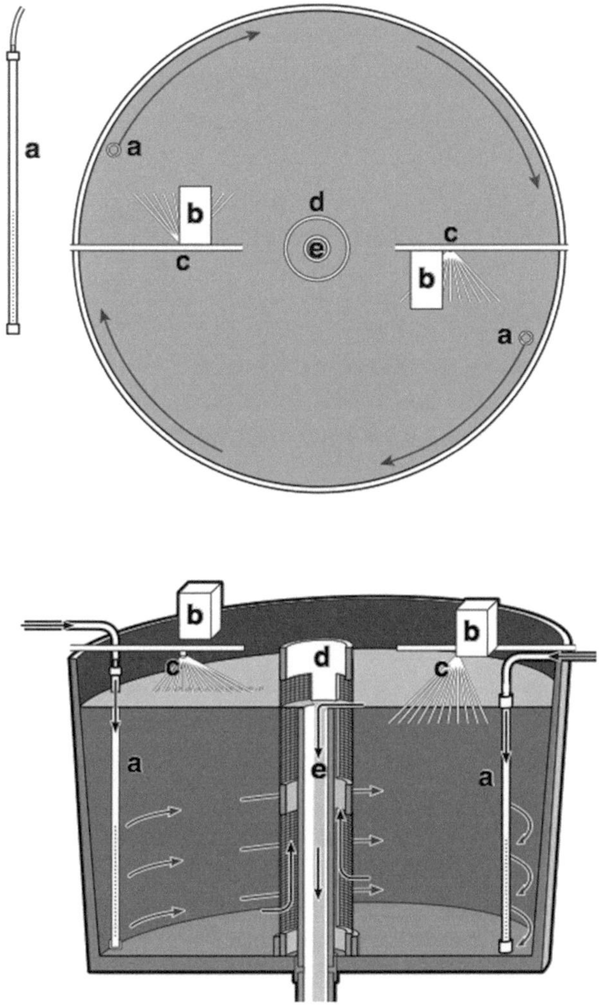

Fig. 11.4 Features of a 680-L, production-scale fiberglass tank for larviculture of Walleye: (**a**) Inlet pipe with vertical row of 3.2 mm holes; (**b**) fish feeder; (**c**) surface spray (see Fig. 11.5 for details); (**d**) removable screen surrounding standpipe; (**e**) standpipe

water surface) help achieve uniform distribution and attract the fry to the surface to aid in the process of gas bladder inflation.

Although we recommend black tank surfaces, the most effective method to disperse fry and eliminate clinging behavior in tank culture of Walleye is to use turbid water (Fig. 11.5). Turbidity makes the water more opaque because it causes light to be scattered and absorbed rather than transmitted in straight lines. In the first intentional use of turbid water for larviculture of fish, Bristow and Summerfelt (1994) reported statistically significant higher survival, and total length and

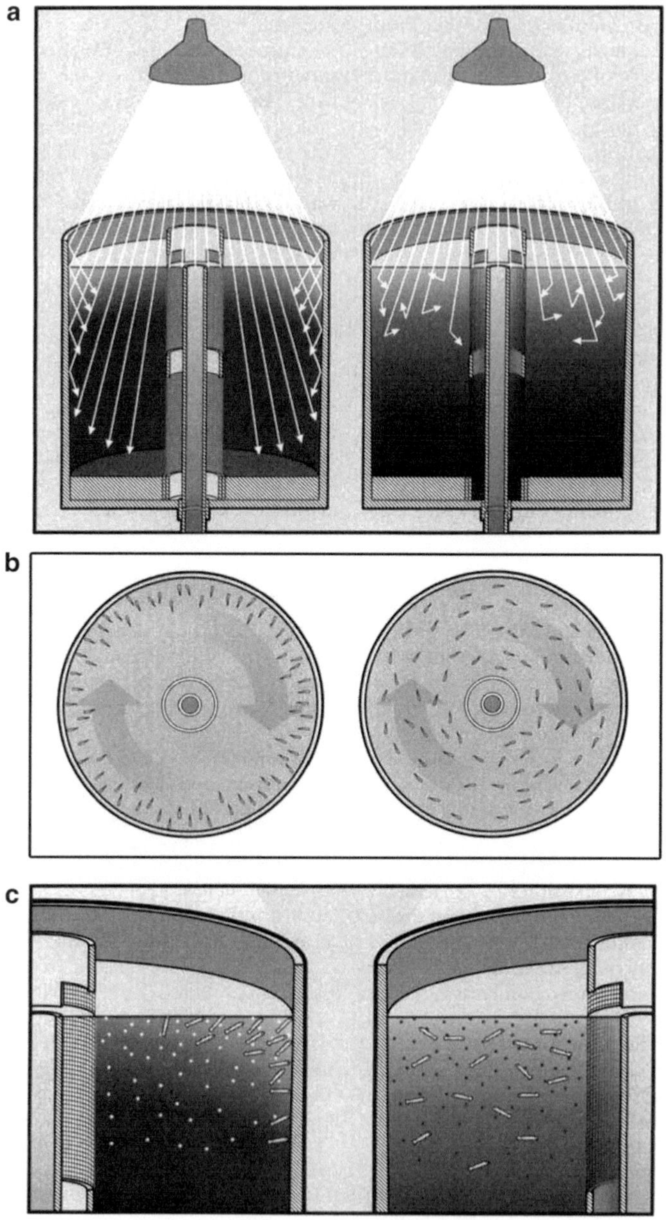

Fig. 11.5 Comparison of Walleye distribution in tanks with clear (*left side* of each panel) and in turbid water (*right side*): (**a**) Overhead light in clear water passes from surface to bottom, and it reflects from tank walls, but light in turbid water is dispersed and does not reach to bottom of the tank; (**b**), (**c**) Larval Walleye have strong positive phototaxis, which attracts them to both direct and reflected light from tank surfaces, whereas in turbid water they are dispersed (Revised from Bristow and Summerfelt 1994)

weight at 21–30 dph of fish reared in turbid (16.1–49.7 NTU) water compared with fish reared in clear water (NTU 0.3). The larvae in the turbid water were closer to the water surface and they were more even dispersed (Fig. 11.2). Turbid-water culture reduces clinging behavior, larvae begin feeding sooner, and survival and growth are substantially improved (Bristow and Summerfelt 1994, 1996; Rieger and Summerfelt 1997).

Artificial turbidity is achieved by addition of small volume of clay slurry to each tank every 20-min or by adding the clay slurry to water supply to the tanks; each method has its merits depending on the physical arrangements of the culture system, but the latter produces a more uniform level of turbidity. Because there is a highly significant rectilinear correlation between turbidity, which is expressed nephelometric turbidity units (NTU), and mg L^{-1} of fine clay, a specific NTU level can be established by addition of a corresponding quantity of clay (Phillips and Summerfelt 2003).

A turbidity as low as 25 NTU is beneficial (Bristow and Summerfelt 1996), but performance was better at a turbidity of $\geq$50 NTU (Phillips and Summerfelt 2003). Literature reviewed by Rønfeldt and Nielsen (2010) includes reports of enhanced survival, growth and GBI in several species of fish in turbid water culture using clay or microalgae; pikeperch (*Sander lucioperca*) larvae were fed live food (rotifers then *Artemia* nauplii) in black, 56 L conical tanks in clear water, turbid water using clay (59.7 NTU), or turbid green water (16.7 NTU, Chlorella). They reported earlier GBI and feeding (food in gut) for larvae raised in turbid water compared with larvae reared in either water made turbid with algae or clear water. Survival and larval length were greater for larvae cultured in turbid water when the study ended on 15 dph. Overall, larval performance in water made turbid with clay was better than in green water, but performance was better in green water than clear water.

Tanks for larviculture of fish require a removable screen that has a mesh small enough to retain the fish. In a round tank, the screen is usually around a center standpipe, on the sidewall of a round tank with a sidewall drain design, or located at the end of a rectangular tank (i.e. raceway). The screen needs to be removable for daily cleaning with pressurized water to remove waste feed, oil, and biofilm. The standard standpipe screen used in circular tanks has an open bottom, but when it is replaced after cleaning some of the fry may be trapped in the annulus (space) between the standpipe and the screen. A modification of the standpipe screen has been described that prevents the fry from entrapment in the space between screen and standpipe (Clayton and Summerfelt 2011). That screen type is used in our university research facility and at the RFCRF.

A small mesh size of a stainless steel (SS) or synthetic material (nylon) is essential to retain the larva. A "50-mesh" SS screen was used by Colesante (1996) in a raceway larviculture system. Barrows et al. (1988) used a SS screen with 600 μm mesh. Kindschi and Barrows (1991) used nylon screen and recommended a mesh not larger than 710 μm for the first 21-days posthatch. At the RFCRF, the 710 μm screen is replaced with a 1-mm screen at 9–10 dph. Further increases in screen opening are recommended but they have not been evaluated, however, as the fish grow and feeding rate is increased, larger mesh would allow more material to pass through the screen, thus, improving water quality.

A nylon screen with mesh width of 710 μm (0.71 mm) with an open area of 44 % is used in our research facilities, however, we spray-painted the screen black to reduce lipophilic character of the screen (Boggs 1994) and reduce light reflection of the light color of the unpainted nylon. A smaller mesh may be required to retain smaller fry that are produced by advancing the spawning date 2 months and the fry of sauger and hybrid Walleye, which are smaller than Walleye.

Waste removal from the tank can be enhanced by increasing the mesh size of the screen as the fish grow. In our facilities, a second set of nylon screens that have a 1-mm mesh with 58 % open area are used from 22-dph to the end of the larviculture interval. The use of critical swimming speed has been proposed to calculate acceptable drainage screen areas and through-screen velocities for larval rearing tanks (Peterson and Harmon 2001).

Larvae must penetrate the water surface to gulp air for first filling of their gas bladder but the larvae may be prevented from this essential act by presence of an oil film that increases surface tension (Reiger and Summerfelt 1998). Until this problem was resolved, non-inflation of the gas bladder (NGB) was a dominating obstacle to larviculture of many physoclistous marine and freshwater fishes (Czesny et al. 2005; Summerfelt 2013). Tanks for larviculture must be equipped with a method remove oil from the surface of the culture tank during the critical period when the larvae must inflate their gas bladder. Walleye, other percids, and nearly all spiny-rayed fishes are physoclists, meaning that air gulped at the water surface is able to passes through the pneumatic duct for only a short interval after yolk sac absorption, beyond that interval, it is generally assumed that GBI cannot occur because the pneumatic duct disappears. Gas bladder inflation (GBI) for larval begins at 6 days posthatch (post-larval I) and takes place from 6th to 12th day posthatch. Marty et al. (1995) demonstrated that it was not disappearance of the pneumatic duct, rather the opportunity for inflation of the gas bladder ended when the pyloric sphincter developed which prevented surfactants in the bile secretions from emulsifying air bubbles in the foregut to a size small enough to pass through the small diameter of the pneumatic duct to the gas bladder; thus differentiation of the foregut prevents inflation after 12 dph, not disappearance of the pneumatic duct.

Without effective measures to prevent oil contamination and a means to eliminate oily water surface, NGB occurs in a high percentage of the larvae, causing high mortality of the larvae and lordosis, a distinctive skeletal defect associated with NGB (Kitajima et al. 1981, 1991). Oil on the surface will not pass through the screens used in fry culture tanks (Boggs 1994) but may be removed from the surface with a strong surface spray (Fig. 11.6) that impacts the water at a right angle to the plane of the surface with sufficient intensity to homogenize the oil droplets to a size that will pass through the standpipe screen (Moore et al. 1994a), or collected in surface trap (Chatain and Ounais-Guschemann 1990) or air jet, oil skimmer (Lim 1993).

Gas bladder inflation (GBI) was typically poor (<25 %) in the larviculture of many species of fish until it was discovered that a spray of water to the surface would enhance gas bladder inflation (Barrows et al. 1993; Moore et al. 1994a). With an effective spray, nearly 100 % of juveniles will have inflated gas bladder (Clayton and Summerfelt 2010). The spray removes the oil film and cleans the surface of feed

Fig. 11.6 Surface spray in 150-L tank tank used in Walleye larviculture research (*left*), and schematic representation of spray homogenizing oil film by the strong impact on the surface (Authors illustrations)

and debris. In circular tanks with a circular flow pattern, the water passes under the spray head with each revolution of the water mass. In order to homogenize the oil to facilitate passage out of the tank, it is important that the spray impact the water surface with enough force to produce a slight depression in the water under the spray. Moore et al. (1994a) observed slightly lower GBI in 679-L (88.7, 96.3, and 88.0 %) than 278-L tanks (99.5, 99.5, and 100.0 %) in three trials for fry stocked at 20 L^{-1}, where the larger tanks had one spray per 4778 cm^2 compared with one spray per 5941^2 of tank surface for the smaller tanks. Because even larger tanks may be used in commercial applications, it is important that the sprays be sufficient in number and intensity.

A steady inflow water is required to add oxygen and remove dissolved and suspended wastes. For comparative purposes, it is helpful to consider the inflow per hour (Q, L h^{-1}) relative to tank volume (V_T). This is expressed as the number of water turnovers per hour (i.e., the exchange rate, R), where $R = Q/V_T$. An R value of 1 means that the total volume of inflow in 1 h (L $min^{-1} \times 60$-min) equals the water capacity of the tank, doubling the inflow velocity will double R, and so forth. At a fixed oxygen content of the inflowing water, the exchange rate is the major factor determining carrying capacity of intensive, flow-through aquaculture systems (Westers 2001). Moodie and Mathias (1996) expressed the exchanges per day rather than per hour.

The current in the culture tanks is proportional to the velocity and the direction of the inflow from the openings in the vertical inflow standpipe. It is important that the current from inflow to the drain does not overwhelm the swimming ability of the larvae. Because of the hydrodynamic burden of a bulky yolk sac, newly hatched larvae are poor swimmers, and cannot maintain a level body orientation. They are quickly exhausted when forced to maintain position, and they will sink when not swimming (Li and Mathias 1982). A fast current has negative effect on the ability of the larvae to maintain contact with and penetrate the surface film for inflation of the GB. In other species, lordosis was present in larvae forced to swim against a current of at least 20 cm/s (Chatain 1994). Larval movement to a light-colored plastic or shiny metallic screen is a function of their attraction to light and an inability to swim against the current, which means that they will "go with the flow," and impinge on the outlet screen. Although their swimming ability increases with development, the velocity of the outflow should not exceed their swimming ability, even after yolk sac absorption, larval fish have to generate extreme tail beat frequencies to generate length-specific swimming speeds needed to escape predation (Bainbridge 1958), or to being pulled along with current to the drain where they will be impinged on the outlet screen.

The application of basic fish swimming capacity and development to larviculture means that feeding should not begin for the first 3 days posthatch as the larvae have not opened their mouth and yolk provides an endogenous source of nutrition. Moreover, in the first few days metabolism is low because water temperatures are low, similar to that of the incubation jars (10–14 °C), and only gradually increased to 18–20 °C by 30-days posthatch. Because feeding by prolarva is not necessary, the oxygen demand of water is minimal. The initial R values (per hour) at production scale range from 0.26 per hour (Moore et al. 1994a) to 0.44-0-0.54 (Çolesante 1996). Using the tank volumes and flow rates given in the article, we calculated that Colesante (1996) increased R from about 0.5 to 0.89 to 1.35 2 days preceeding and during the interval of GBI, which is after yolk sac absorption. Moodie and Mathias (1996) used an exchange rate of 2.5 per day (~0.1 per hour) during the interval of gas bladder inflation (GBI) to reduce accumulation of surface oil and films that inhibit GBI, yet GBI in their system was only 16–23 % because they did not have a surface spray or oil trap to homogenize (surface spray) or to capture the oil (oil trap).

After feeding begins, the value of R should be gradually increased as needed to maintain oxygen content of the water close to saturation level for the temperature, and to remove ammonia, carbon dioxide and solid wastes. At our university laboratory, using small experimental tanks (150 L), the standard protocol was to have an R value of 0.5 for prolarva, 0.75 for postlarva I, and 1 by 21-days posthatch. At the RFCRF, R values begin at 0.44 until 13 dph when R is increased to 0.65. At 23 dph the flow is increased to an R of 1.1 and raised again on 32 dph to 1.75 until harvest. Tanks must be monitored regularly to ensure adequate oxygen levels are maintained.

High water quality for larviculture is essential and it should be frequently monitored. Water from surface sources (e.g., eutrophic pond or lake) may have oxygen super-saturation during daylight and high carbon dioxide at night. Water from wells will invariably have low concentration of oxygen, high carbon dioxide, and often soluble iron and manganese that precipitate on tank surfaces and the screens. It should not have more than 105 % total gas pressure or a delta P of more than 10. The water should be degassed through a column with packed media. Larvae should not be exposed to prolonged hyperoxic conditions because unlike mammals, the ventilation rate of fish is determine by the oxygen content of the blood rather than CO_2.

Thus, hyperoxia decreases ventilation rate, resulting in a rise in CO_2 tension (P_{CO2}) in the blood, a condition called hypercapnia, and a corresponding decrease in blood pH (Takeda 1990; Wedemeyer 1996).

Compressors should not be used to aerate water destined for use in intensive culture of fry because they will contaminate the air with oil from the pistons. Oil-contaminated air that is bubbled through water will transfer the oil to the water, which will rise to the surface and interfere with gas bladder inflation. Before use of turbid water, some hatcheries placed an air line around the center standpipe to keep fry from being impinged on the screen. A column of rising air bubbles in a fry culture tanks may cause undesirable turbulence, and fast rising air bubbles will even throw fry out of the water where they will stick to the side walls above the water line.

Light intensity of 100–700 lx (lux) is acceptable, but higher intensity is desirable through the larval stages. Colesante (1996) used 680 lux at the water surface; and Moore et al. (1994a) recommended 500 lux. Moodie and Mathias (1996) maintained a temperature of 20 °C throughout the 30-days fry rearing interval. A sudden increase of 5 °C for 24-h when the larvae were 5 days posthatch may stimulate fry to accept the manufactured feed (Summerfelt 1996). Johnson and Rudacille (2010) maintained water temperature at 18.5 °C and lux levels at 75–100 lux, but the RFCRF now uses 100 lux and a temperature of 20 °C as their protocol.

11.4.2 Biological Aspects

A stocking density that yields the highest survival from prolarval to the juvenile stage has been examined several times. In small, glass culture chambers (30–70 L), survival through the 5 day prolarval stage was unaffected by density as high as 100 fish L^{-1} but survival during the postlarva I and II stages decreased substantially at densities greater than one fish L^{-1} (Li and Mathias 1982). Practicality, however, requires finding the highest stocking density that optimizes survival, growth or yield for a production-scale system. In series of density trials by Moore et al. (1994b) survival in 278-L cylindrical tanks to 21-dph did not differ significantly among density treatments of 20, 30, and 40 fry L^{-1}, but in another trial, survival was greater at a stocking density of 20 L^{-1} than at 50 or 60 L^{-1}, however, in spite of lower survival, the number of fry produced (yield) was 216 % greater in the tanks stocked at 60 L^{-1} than at 20 L^{-1}. Growth rate was not affected by density in any trial. Survival differences between 278-L and 679-L tanks stocked at 20 larvae L^{-1} sizes were inconsistent, but the highest survival was an average of 74.9 % to 21 dph in the three 679-L tanks. Cannibalism was 7.5 % of the number stocked in the small tanks but less than 3.5 % in the larger tanks.

Peterson et al. (1997) obtained survival to 23 dph of 33.7 % at 40 fry L^{-1} and 36.1 % at 60 fry L^{-1} in 1200 L tanks; total length (mm) of juveniles were 17.9 mm at 40 fry L^{-1} and 17.1 mm at 60 fry L^{-1}. At the end of the 23 days interval, the fish were transferred to hatchery ponds for addition 19 days, which is a reversal to the traditional tandem pond to tank culture system (Summerfelt et al. 2011).

Current stocking regimen at Rathbun Fish Culture Research Facility (RFCRF) is 30 fry L^{-1}. Although higher stocking density would result in higher yield per tank, with improvements in survival, a lower stocking density is required to prevent overcrowding of the larviculture tanks when the goal is to produce a larger, fully scaled fingerling that will tolerate handling with low mortality during the transfer to raceways for further growout. In 2012, with a stocking density of 30 prolarvae and a culture interval to 35 dph, survival was 71 % and the juveniles had a mean weight of 0.67 g. The final density was 14.9 g L^{-1}, substantially higher than the maximum density previously achieved in the same system (9.7 g L^{-1}).

Both formulated feed and live feeds are used for larviculture of Walleye. The feeding protocol at the Oneida Fish Culture Station, Constantia, NY (Colesante 1996) is to start Walleye fry on brine shrimp nauplii, first at 800–1000 nauplii $fish^{-1}$ day^{-1}, then increased on day 30 to 1300 nauplii $fish^{-1}$ day^{-1} when feeding formulated diet is started. The mixed feeding continued through 44 dph when feeding brine shrimp was stopped, then the juveniles were fed another 2–3 weeks on only formulated feed before they are made available for enhancement stocking.

Feeding rates for formulated feeds are usually expressed as g 1000^{-1} fry, or in terms of particle density in the culture tank (feed particles L^{-1}) irrespective of fish density. An example of the latter, 100 particles L^{-1} for dry diets (Moodie and Mathias 1996) or 100–200 L^{-1} daphnids (1.2 mm) for postlarvae I and II (Li and Mathias 1982). Colesante (1996), however, based feeding rates on fish density; i.e., brine shrimp at 800–1000 nauplii $fish^{-1}$ L^{-1}. Our experience has been to feed larval Walleye commercial, microparticulate diets in terms of g of feed 1000^{-1} fry, the rate progressively increased with increasing fish size. Feeding rates have been reduced from that reported by Summerfelt (1996), Johnson et al. (2008), and Summerfelt et al. (2011). At the RFCRF, we start with 3–5 g 1000^{-1} prolarvae, then progressively increase the daily rate to 7 g 1000^{-1} fish when they are 23 mm TL juveniles, then 10 % bodyweight thereafter until 35 dph and fish are ~45 mm TL, a size when they are moved from the larviculture to grow-out tanks.

In small aquaria (30–200-L), cannibalism among sibling Walleye occurred only during the interval from 6 to 16 dph postlarval I to juvenile, but it disappeared by the juvenile stage (Li and Mathias 1982). Peterson et al. (1997) also observed exceptionally high cannibalism during a 4 days interval 8–11 dph that accounted for at least 50 % of all mortality during the 23-days culture interval. Their opinion supported that of Loadman et al. (1986) that most mortality ascribed to cannibalism was not from direct consumption but morbidity resulting from injuries from unsuccessful attacks. Stocked at the same N L^{-1}, cannibalism was significantly lower in larger than smaller tanks (Moore et al. 1994b).

The intensity of sibling cannibalism is also related to genetics and age differences of the larvae which determine the onset of feeding. A highly significant correlation was obtained between DTU to initiation of feeding and onset of cannibalism among larval Walleye that were progeny of feral broodstock collected from Ohio, Kansas, Iowa, Wisconsin, Minnesota, and North Dakota (Bristow and Summerfelt 2003). In that study, cannibalism of a stock from a northern Minnesota lake was 4.5 % compared with 1.2–1.5 % for the other five stocks (Bristow 1993).

The mouth of prolarvae is not open until they are 3 dph, although the rate of development is affected by temperature, age affects onset of first feeding. Fish differing in age by 1 day may produce bimodal size groups within the population that may substantially increase the problem of cannibalism. Difference in size among siblings is a critical factor in occurrence and intensity of cannibalism (Change and Liao 2003). A substantial reduction in cannibalism may be expected when tanks are stocked only with fish that hatched within a short interval. In our experimental studies, this interval has been as short as 1 h, but for production-scale culture the objective should be to use only fish that hatched within 24-h.

Appropriate feed size, frequency, and feeding rate are important factors to reduce the severity of cannibalism. Feed should be dispensed in small doses at frequent intervals of 5 min. In small (280 L) tanks, one feeding location has been sufficient because some feed particles break the water surface and descend immediately, and some particles travel on the surface in the water current until the surface spray breaks the surface tension and the particles then descend. Large, 679-L, tanks were provided two feeders compared to one feeder on a 280-L tank (Moore et al. 1994b). As fry grow, fecal casts may be distinguished from feed particles, thus if waste collected from fry tanks are all feces, then feed rates should be increased to allow some waste.

Columnaris disease and bacterial gill disease are more commonly observed during fry culture than other diseases. Columnaris disease (*Flavobacterium columnare*, formerly *Flexibacter columnaris*) can be caused by stress and mechanical injury (Hussain and Summerfelt 1991) and therefore cannibalism may cause columnaris in larvae. Mortality caused by columnaris may be controlled by static bath treatment with diquat dibromide (Reward®) administered under the supervision of an investigational new animal drug exemption (INAD; AADAP 2011). Larvae with bacterial gill disease have been treated with Chloramine-T to control mortality using INAD 9321. Both drugs have been applied to tanks of larvae at the RFCRF in 1-h static bath treatments on consecutive days as allowed by the INAD.

Deformities—There are many genetic, environmental, and nutritional influences affecting occurrence of deformities (also called malformations) in fish. Although the topic is too complex to be reviewed in detail, certainly, nutrition is a major factor affecting skeletal malformations such as scoliosis, lordosis, bending opercle, and twisted jaw (Cahu et al. 2003). Johnson et al. (2008) reported incidence of cataract, small pupil, short operculum, cranial deformities, lordosis and scoliosis in 28 dph Walleye with a significant difference in occurrence between two commercial larval diets. Deformities of young juvenile Walleye were illustrated by Summerfelt et al. (2011).

Lordosis is a distinctive skeletal defect often associated with NGB (Kitajima et al. 1981, 1991). In other species, lordosis was present in fish forced to swim against a current of at least 20 cm/s, while only 20–30 % of fish reared in static waters were affected (Chatain 1994).

Environmental contaminants in the water supply also cause deformities; e.g., Latif et al. (2001) cited several publications that have demonstrated a relationship between acute developmental abnormalities in fishes at high concentrations of total Hg or methylmercury (MeHg), which is a major environmental problem in northern lakes of the United States and Canada that prompt fish consumption advisory for

Walleye and other fishes (Lavigne et al. 2010). Thus, it is expected that accumulation of toxins in spawning female Walleye would show a relationship to the incidence of deformities; "even at sublethal concentration of methylmercury (MeHg), developing Walleye eggs suffered mortality during the transformation of the germinal layers into various organs of the embryo (i.e., at the onset of gastrulation and early organogenesis)," a process described as a critical period in Walleye egg development (Latif et al. 1999).

The growth rate, as is true of survival, during larviculture is influenced by many variables such as genetics of the stock, the environment—especially use of turbid water—stocking density, diet, feeding rate, and other factors. A summation and analysis of such information are beyond the scope of this review. At the RFCRF, larval Walleye from hatch to 24.5 dph when they were fed Fry Feed Kyowa was 0.73 mm d^{-1}, but 0.88–1.03 mm d^{-1} to 34–37 dph when fed Otohime diet. In 2012, using the Otohime feed, at a mean temperature of 19.7 °C, fry grew to 43.1 mm and 0.68 g at 35 dph.

Research scale larvaculture tanks (275 L) stocked with an initial density of 30 fry L^{-1} have produced as many as 6270 fry per tank (76 % survival) or 22.8 fish L^{-1}. Harvest density at 35 dph has reached as high as14.9 g L^{-1} with fish 0.67 g and 43.5 mm. Densities this high require close monitoring of oxygen levels and flow rates to maintain suitable water quality, and adequate feed rates and feeder calibration to prevent cannibalism and poor growth.

11.5 Conclusion

Research findings and personal experience convince us that intensive larviculture now offers a practical alternative to pond-culture for production of feed-trained juvenile Walleye. The analysis of technical reports demonstrate a science-based, production-scale protocol for Walleye fry culture that can achieve 60–70 % survival from hatch to 35 days post hatch (dph). On-growing feed-trained fingerlings to food-size is described by Johnson and Summerfelt in Chap. 17.

References

Allbaugh CA, Manz JV (1964) Preliminary study of the effects of temperature fluctuations on developing walleye eggs and fry. Prog Fish Cult 26:175–180

Bainbridge R (1958) The speed of swimming fish as related to size and to the frequency and amplitude of the tail beat. J Exp Biol 35:109–133

Barrows FT, Lellis WA, Nickum JG (1988) Intensive culture of larval walleye with dry or formulated feed: note on swim bladder inflation. Prog Fish Cult 50:160–166

Barrows FT, Zitzow RE, Kindschi GA (1993) Effects of surface water spray, diet, and phase feeding on swim bladder inflation, survival, and cost of production of intensively reared larval walleyes. Prog Fish Cult 55:224–228

Barton BA (ed) (2011) Biology, management, and culture of walleye and sauger. American Fisheries Society, Bethesda

Boggs CT (1994) Origin and characterization of oil films from larval walleye culture tanks. MS thesis, Iowa State University, Ames

Bristow BT (1993) Comparison of larval walleye stocks in intensive culture. Master's thesis, Iowa State University, Ames

Bristow BT (1996) The extensive–intensive production of advanced fingerling walleyes at the Spirit Lake State Fish Hatchery. In: Summerfelt RC (ed) Walleye culture manual. NCRAC Culture Series 101, North Central Regional Aquaculture Center Publications Office. Iowa State University, Ames, pp 209–212

Bristow BT, Summerfelt RC (1994) Performance of larval walleye cultured intensively in clear and turbid water. J World Aquac Soc 25:454–464

Bristow BT, Summerfelt RC (2003) Variation among stocks of walleye (*Sander vitreus*): length at hatch, and larval development. In: Barry TP, Malison JA (eds) Proceedings of Percis III: the third international Percid fish symposium. University of Wisconsin Sea Grant Institute, Madison, pp 21–22

Bristow BT, Summerfelt RC (1996) Comparative performance of intensively cultured larval walleye in clear, turbid, and colored water. Prog Fish Cult 58:1–10

Cahu C, Infante J, Takeuchi T (2003) Nutritional components affecting skeletal development in fish larvae. Aquaculture 227:254–258

Change EY-Y, Liao IC (2003) Sibling cannibalism of young red drum, *Sciaenops ocellatus*, in relation to size disparity and metabolic rates. Environ Biol Fish 68:407–415

Chatain B (1994) Abnormal swim bladder development and lordosis in sea bass (*Dicentarchus labrax*) and sea bream (*Sparus auratus*). Aquaculture 119:371–379

Chatain B, Qunais-Guschemann N (1990) Improved rate of initial gas bladder inflation in intensively reared *Sparus auratus*. Aquaculture 84:345–353

Clayton RD, Summerfelt RC (2010) Gas bladder inflation in walleye fry cultured in turbid water with and without a surface spray. N Am J Aquac 72:338–342

Clayton RD, Summerfelt RC (2011) A standpipe screen design to prevent fry loss during tank cleaning. N Am J Aquac 73:104–106

Colesante RT (1996) Intensive culture of walleye using brine shrimp and formulated diets. In: Summerfelt RC (ed) Walleye culture manual, vol 101, NCRAC culture series. North Central Regional Aquaculture Center Publications Office, Iowa State University, Ames, pp 191–194

Copeland JA, Wolgamood MW (1996) Walleye spawning in Michigan. In: Summerfelt RC (ed) Walleye culture manual, vol 101, NCRAC culture series. North Central Regional Aquaculture Center Publications Office, Iowa State University, Ames, pp 21–23

Czesny S, Rinchard J, Dabrowski K (2005) Intrapopulation variation in egg lipid and fatty acid composition and embryo viability in a naturally spawning walleye population from an inland reservoir. N Am J Fish Manag 25:122–129

Dabrowski K, Ware K, Jaroszewska M, Kwasek K (2009) Evaluation of walleye embryo survival and larval viability after iodine treatment. N Am J Aquac 71:122–129

FDACMV (Federal Drug Administration Center for Veterinary Medicine) (2013) Approved drugs for use in aquaculture. Available: http://www.fda.gov/AnimalVeterinary/DevelopmentApprovalProcess/Aquaculture/ucm132954.htm. Jan 2013

Gallinat MP (1996) Stripping walleye eggs collected from speared fish and incubation of eggs using Big Redd incubators. In: Summerfelt RC (ed) Walleye culture manual, vol 101, NCRAC culture series. North Central Regional Aquaculture Center Publications Office, Iowa State University, Ames, pp 37–40

Getchell, R. 2007. Egg disinfection may keep viruses at bay. Fish Farming News 2007(4):8–9

Glenn CL, Mathias JA (1985) Circuli development on body scales of young pond-reared walleye (*Stizostedion vitreum*). Can J Zool 63:912–915

Greiff E (1996) Stripping, fertilizing, and incubating walleye eggs with Big Redd incubators. In: Summerfelt RC (ed) Walleye culture manual, vol 101, NCRAC culture series. North Central Regional Aquaculture Center Publications Office, Iowa State University, Ames, pp 31–36

Groocock GH, Getchell RG, Cornwell ER, Frattinii SA, Wooster GH, Bowser PR, LaPan SR (2012) Iodophor disinfection of walleye eggs exposed to viral hemorrhagic septicemia virus Type IVb. North Am J Aquacult 75:25–33

Harvey JE, Hood SE (1996) Collection and spawning of walleye broodfish and incubation of fertilized eggs. In: Summerfelt RC (ed) Walleye culture manual, vol 101, NCRAC culture series. North Central Regional Aquaculture Center Publications Office, Iowa State University, Ames, pp 25–28

Heidinger RC, Brooks RC, Leitner D, Soderstrom I (1997) Prediction of walleye egg and embryo survival at two stages of development. Progress Fish Cult 59:64–67

Hubbs CL (1943) Terminology of early stages of fishes. Copeia 1943:260

Hussain M, Summerfelt RC (1991) The role of mechanical injury in an experimental transmission of *Flexibacter columnaris* to fingerling walleye. J Iowa Acad Sci 98:93–98

INAD AADA (2011) Fact sheet: diquat INAD 10-969. Aquatic Animal Drug Approval Partnership (AADP) Program. http://www.fws.gov/fisheries/aadap/diquat.htm. Accessed June 2015

Johnson JA, Rudacille JB (2010) Intensive culture of walleye converted to a dry diet as small fingerlings. Iowa Department of Natural Resources, Federal Aid in Sport Fish Restoration, Completion report, project F-160-R, Des Moines

Johnson JA, Summerfelt RC, Clayton RD (2008) Evaluation of replacement feeds for Fry Feed Kyowa for larval walleyes. N Am J Aquac 70:445–451

Kindschi GA, Barrows FT (1991) Optimal screen mesh size for restraining walleye fry. Prog Fish Cult 53:53–55

Kitajima C, Tsukashima Y, Fujita S, Watanable T, Yone Y (1981) Relationship between uninflated swim bladders and lorodic deformity in hatchery-reared red sea bream *Pagurus major*. Bull Jpn Soc Sci Fish 47:1289–1294

Kitajima C, Watanabe T, Tsukashima Y, Fujita S (1991) Lordotic deformation and abnormal development of swim bladder in some hatchery-bred marine physoclistious fishes. In: Lavens P, Sorgeloos P, Jaspers PE, Ollevier FP (eds) LARVI '91 – fish & crustacean larviculture symposium. European Aquaculture Society, Special Publication, no 15, Gent, p 316

Latif MA, Bodaly RA, Johnston TA, Fudge RJP (1999) Critical stage in developing walleye eggs. N Am J Aquac 61:34–37

Latif MA, Bodaly RA, Johnston TA, Fudge RJP (2001) Effects of environmental and maternally derived methylmercury on the embryonic and larval stages of walleye (*Stizostedion vitreum*). Environ Pollut 111:139–148

Lavigne M, Lucotte M, Paquet S (2010) Relationship between mercury concentration and growth rates for walleyes, northern pike, and lake trout from Quebec lakes. N Am J Fish Manag 30:1221–1237

Li S, Mathias JA (1982) Causes of high mortality among cultured larval walleyes. Trans Am Fish Soc 111:710–721

Lim LC (1993) Larviculture of the greasy grouper, *Epinephelus tauvina* and the brown-marbled grouper, *E fuscoguttatus* in Singapore. J World Aquac Soc 24:262–273

Loadman NL, Moodie GEE, Mathias JA (1986) Significance of cannibalism in larval walleye (*Stizostedion vitreum*). Can J Fish Aquat Sci 43:613–618

Marty GD, Hinton DE, Summerfelt RC (1995) Histopathology of swimbladder noninflation in walleye (*Stizostedion vitreum*) larvae: role of development and inflammation. Aquaculture 138:35–48

McElman JF, Balon EK (1979) Early ontogeny of walleye, *Stizostedion vitreum*, with steps of saltatory development. Environ Biol Fish 4:300–348

Moodie GEE, Mathias JA (1996) Intensive culture of larval walleye on formulated food. In: Summerfelt RC (ed) Walleye culture manual, vol 101, NCRAC culture series. North Central Regional Aquaculture Center Publications Office, Iowa State University, Ames, pp 187–190

Moore A, Prange MA, Summerfelt RC, Bushman RP (1994a) Evaluation of tank shape and a surface spray for rearing larval walleye on formulated feed. Prog Fish Cult 56:100–110

Moore A, Prange MA, Bristow BT, Summerfelt RC (1994b) Influence of stocking densities on walleye fry viability in experimental and production tanks. Prog Fish Cult 56:194–201

Nelson WR (1968) Embryo and larval characteristics of sauger, walleye, and their reciprocal hybrids. Trans Am Fish Soc 97:167–174

Nelson WR, Hines NR, Beckman LG (1965) Artificial propagation of saugers and hybridization with walleyes. Prog Fish Cult 27:216–218

Paddock D (1996) Incubation of walleye eggs at Garrison Dam National Fish Hatchery. In: Summerfelt RC (ed) Walleye culture manual, vol 101, NCRAC culture series. North Central Regional Aquaculture Center Publications Office, Iowa State University, Ames, pp 29–30

Peterson RH, Harmon P (2001) Swimming ability of pre-feeding striped bass larvae. Aquac Int 9:361–366

Peterson DL, Carline RF, Wilson TA, Hendricks ML (1997) Production-scale methods for intensive culture of walleye fry. Prog Fish Cult 59:14–19

Phelps NBD, Patnayak DP, Jiang Y, Goyal SM (2012) The use of a one-step real-time reverse transcription polymerase chain reaction (rRT-PCR) for the surveillance of viral hemorrhagic septicemia virus (VHSV) in Minnesota. J Aquat Anim Health 24:238–243

Phillips TA, Summerfelt RC (1999) Gill development in larval walleye. Trans Am Fish Soc 128:162–168

Phillips TA, Summerfelt RC (2003) An estimate of optimal turbidity to maximize survival and growth rate of larval walleye in intensive culture. In: Barry TP, Malison JA (eds) Proceedings of Percis III: the third international Percid fish symposium. University Wisconsin Sea Grant Instititue, Madison, pp 51–52

Rieger PW, Summerfelt RC (1997) The influence of turbidity on larval walleye, *Stizostedion vitreum*, behavior and development in tank culture. Aquaculture 159:19–32

Rieger PW, Summerfelt RC (1998) Microvideography of gas bladder inflation in larval walleye. J Fish Biol 53:93–99

Rønfeldt JL, Nielsen JW (2010) Effect of turbidity on rearing of pikeperch larvae (*Sander lucioperca* L.). In: Biologispeciale udarbejdet, Detnaturvidenskabelige Fakultet Københavns Universitet, pp 19–27

Schneider JC, Copeland J, Wolgamood M (2002) Tolerance of incubating walleye eggs to temperature fluctuations. N Am J Aquac 64:75–78

Schneider KN, Newman RM, Card V, Weisberg S, Pereira DL (2010) Timing of walleye spawning as an indicator of climate change. Trans Am Fish Soc 139:1198–1210

Smith LL Jr, Koenst WM (1975) Temperature effects on eggs and fry of percoid fishes. U.S. Environmental Protection Agency. Ecol Res Ser EPA-660/3–75–017

Summerfelt RC (2013) Gas bladder inflation in larval fish aquaculture. In: Qin JG (ed) Larval fish aquaculture. Nova Science, Hauppauge, pp 125–142

Summerfelt RC, Johnson JA, Clouse CP (2011) Culture of walleye, sauger, and hybrid walleye. In: Barton B (ed) Biology, management, and culture of walleye, sauger, and hybrid walleye. American Fisheries Society Special Publication, Bethesda, pp 451–570

Takeda T (1990) Ventilation, cardiac output and blood respiratory parameters in the carp, *Cyprinus carpio*, during hyperoxia. Respir Physiol 81:227–239

Thompson D (1996) Stripping, fertilizing, and incubating walleye eggs at a Minnesota hatchery. In: Summerfelt RC (ed) Walleye culture manual, vol 101, NCRAC culture series. North Central Regional Aquaculture Center Publications Office, Iowa State University, Ames, pp 41–44

Tuttle-Lau MT, KA Phillips, Gaikowski MP (2010) Evaluation of the efficacy of iodophor disinfection of walleye and northern pike eggs to eliminate viral hemorrhagic septicemia virus. U.S. Geological Survey Fact Sheet 2009–3107, 4 p

Wedemeyer GA (1996) Physiology of fish in intensive culture systems. Chapman & Hall, New York, 250 pp

Westers H (2001) Production. In: Wedemeyer GA (ed) Fish hatchery management, 2nd edn. American Fisheries Society, Bethesda, pp 31–89

Zitzow RE (1991) Extended incubation of walleye eggs with low-flow incubators. Prog Fish Cult 53:188–189

Part IV
Juvenile and Grow-Out Stages: Growth, Metabolism, Behavior and Husbandry

Chapter 12
Muscle Protein Characteristic and Its Association with Faster Growth in Percids and Other Teleosts

Karolina Kwasek, Macdonald Wick, and Konrad Dabrowski

Abstract This chapter provides a brief review of the current state of knowledge regarding fish skeletal muscle characteristics, factors affecting muscle growth, and proteomic based research in teleost fish with emphasis on percids. Part of the review includes a study that investigated genetic outcome that underlie the growth potential of muscle of yellow perch *Perca flavescens*. More specifically, it compared skeletal muscle sarcoplasmic proteins/peptides between fast- and slow-growing yellow perch in order to identify the differences in expression of skeletal muscle proteins in fish exhibiting different growth capabilities. Briefly, the study identified bands that presented different staining intensities between fast- and slow-growing fish by using 1D electrophoresis. It also demonstrated muscle metabolic enzymes identified by protein sequencing using nano-LC/MS/MS. The results of the present work contribute to the identification of genetic traits that affect the growth superiority in fish in controlled conditions. Therefore it could become a tool for selection of breeders with the potential for increased protein accretion associated with rapid muscle growth, and hence, the production of larger fish.

Keywords Yellow perch • Skeletal muscle • Myogenesis • Growth hormone • Proteomics

12.1 Introduction

One of the main goals of aquaculture is to increase fish production by selecting for the growth of fast skeletal muscle. One barrier to accomplishing this is a lack of knowledge of the genes controlling and participating in the cellular and molecular

---

K. Kwasek (✉) • K. Dabrowski
School of Environment and Natural Resources, Ohio State University, 2021 Coffey Rd, Columbus, OH 43210, USA
e-mail: karolina.kwasek@gmail.com

M. Wick
Animal Sciences, The Ohio State University, 2027 Coffey Rd, Columbus, OH 43210, USA

P. Kestemont et al. (eds.), *Biology and Culture of Percid Fishes*,
DOI 10.1007/978-94-017-7227-3_12

mechanisms responsible for muscle growth and development in fish species. Determining the gene expression events giving rise to the proteins underlying muscle growth in fish species will provide valuable information for brood stock selection strategies and improved muscle yield. Currently there is a distinct lack of information on events during the growth of fish muscle. A discussion of recent proteomic based research will be presented.

12.2 Anatomy

The skeletal muscle of teleosts consists of long sheets of muscles (myotomes) extending on both sides of fish body, where connective tissue is in the form of fine membranes (myocommata) separating the muscles into segments (Fig. 12.1, Kiessling et al. 2006). Unlike other vertebrates, except poultry, percids and teleosts partition their muscles into separate layers, based on fiber type. The predominant muscle, by mass, resides in the deep inner layer consisting of white fast twitch Type IIb glycolytic/anaerobic fibers. The second most predominant muscle type consists of red slow twitch Type I aerobic fibers with least predominant being intermediate fast twitch Type IIa oxidative/glycolytic (Fig. 12.2). The unique characteristics of each fiber type are listed in Table 12.1.

12.3 Myogenesis

Embryonic muscle development is unique in teleosts (Fig. 12.3). During embryogenesis uncommitted mesodermal stem cells give rise to a population of cells that become committed to the myogenic pathway. These cells proliferate. At one point

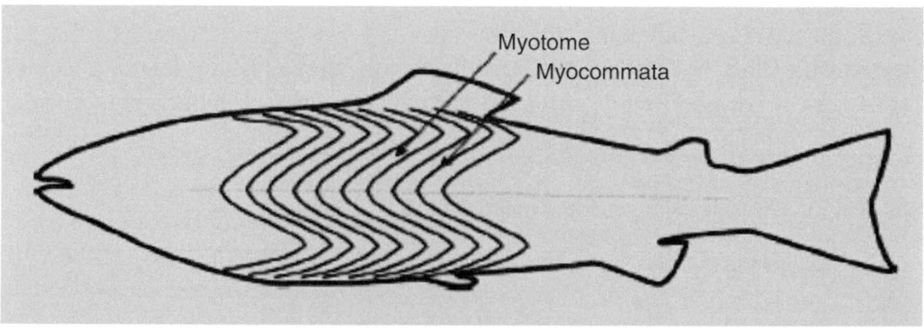

Fig. 12.1 The collocation of muscle fibers (myotomes) and connective tissue (myocommata) in teleost fish skeletal muscle (Adapted from Kiessling et al 2006)

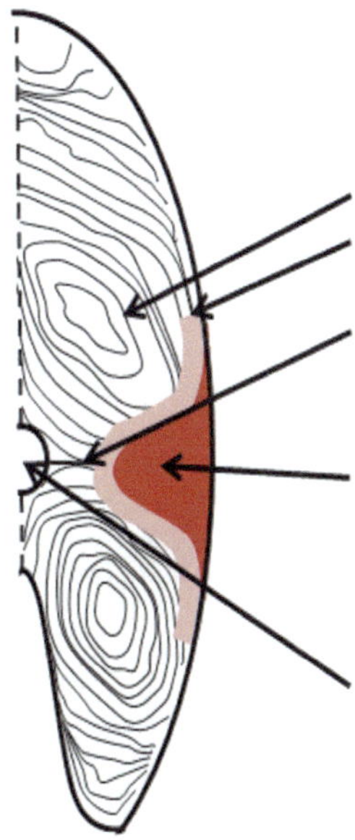

Fig. 12.2 Generalized diagram of a cross section indicating the structural components of the muscle fiber types that make up the muscle anatomy of teleosts

Table 12.1 Skeletal muscle fiber type characteristics

Fiber type/characteristic	Type I	Type IIa	Type IIb/X
Color	Red	Pink	White
Contraction speed	Slow	Intermediate	Fast
Resistance to fatigue	High	Intermediate	Slow
Force production	Low	Intermediate	High
Mitochondria density	High	High	Low
Capillary density	High	Intermediate	Low
Energy source	Triglycerides	Glycogen	Glycogen
		Creatine phosphate	Creatine phosphate
Myoglobin content	High	Intermediate	Low
Motor neuron size	Low	Intermediate	High

one sub-population fuses to become multinucleated myotubes or myofibers. The second sub-population remains quiescent and resides under the basal lamina of the myofiber to be used in further myogenic events including hyperplasia, hypertrophy and muscle repair.

Postnatal fish muscle growth is unique compared to birds and mammals because it is determined by not only hypertrophy which is the increase in the size of existing muscle fibers, but also by hyperplasia which is an increase in the number of muscle fibers through fiber recruitment. Both of these processes continue throughout the

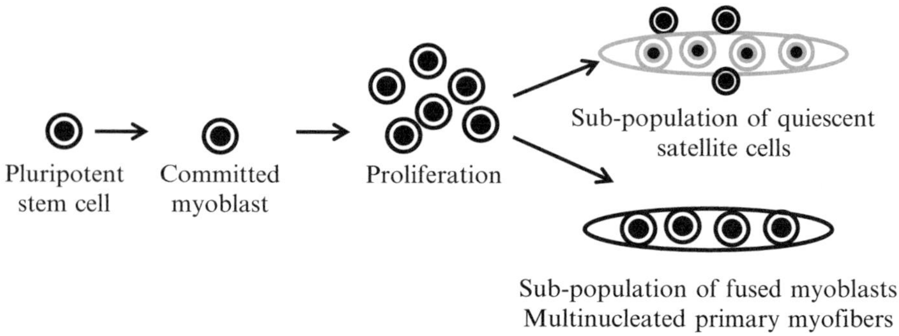

Fig. 12.3 Generalized schematic of myogenesis from an uncommitted mesodermal stem cell population resulting in two populations. One sub-population which fuses into a multinucleated myotube or myofiber. The second sub-population remains quiescent under the basal lamina of the myofiber to be used in further myogenic events including hyperplasia, hypertrophy and muscle repair

whole life cycle of a fish (Johnston et al. 2009). In a representative of Perciformes fishes, angelfish (*Pterophylum scalare*) muscle fiber diameter increased (hypertrophy) sixfold whereas the number of new muscle fibers did not change significantly during embryonic development (from eyed embryo to free-swimming juvenile) (Kacperczyk et al. 2011). However, Ostaszewska et al. (2008) analyzed contribution of hyperplasia and hypertrophy in the early life of pike-perch (*Sander lucioperca*) and concluded that hypertrophy of white muscle fibers in larval pike-perch was impacted by diet, age and size. For instance in pike-perch larvae the hypertrophy was responsible for 5.4–21 % of muscle growth. In juvenile pike-perch (26–40 day old) this proportion increased further to 28–56 % in comparison to hyperplasia.

Ayala et al. (2013) reported effect of photoperiod on muscle structure differentiation in larval fish, shi drum (*Umbrina cirrosa*). The authors concluded that in white muscle growth hyperplasia was higher in 24 h light regime, whereas hypertrophy was higher in 16 h light: 8 h dark regime.

Molecular mechanisms of hyperplastic muscle growth are proposed to reside in the differences in satellite cell proliferation (Koumans et al. 1993; Rowlerson et al. 1995), as well as the involvement of undefined subpopulations of myogenic cells. Myosin isoform expression has also been shown to be associated with different muscle growth patterns (Karasinski 1993; Rowlerson et al. 1985; Cole and Johnston 2001). Johnston (1999) distinguished three phases of myogenesis (muscle growth) in teleosts.

In the initial stage of myogenesis undifferentiated myoblasts fuse to become multinucleated myotubes (Konigsberg 1963). These myoblasts are derived from a population of progenitor cells (Kacperczyk et al. 2009). The latter authors described migration of progenitor cells into the myotomes and becoming satellite cells that will participate in muscle growth through hyperplasia. The population consists of at least two sub-populations. One sub-population differentiates during embryogenesis where they fuse to give rise to muscle cells. The other sub-population, becomes quiescent and is termed as satellite cells used for muscle repair and growth, later in life. Myogenic satellite cells have been isolated from a variety of fish species (Powell et al. 1989; Greenlee et al. 1995; Venkateswaran et al. 1995; Burton et al. 2000). During the second phase distinct germinal zones of myoblast proliferation are observed at the dorsal and ventral summits of the myotomes. Finally, during the third phase myoblasts on the surface of embryonic muscle fibers are activated.

12.4 Factors Affecting Muscle Growth

The growth of a muscle is the result of a balance between protein accumulation (anabolism) and degradation (catabolism). The physiology of muscle, including muscle mass, is the result of dynamic interactions between the environmental factors and varying expression of genes that affect its development, growth, and structure (Fig. 12.3). The environmental factors that influence muscle growth have been extensively studied and include photoperiod (Johnston et al. 2003), temperature (Wilkes et al. 2001), oxygen, pH (Johnston 2006) as well as feed availability (Bureau et al. 2006; Ostaszewska et al. 2008). The genetic factors, many of which still remain unknown, include, for example, age at sexual maturation of fish which is related to the body size since it reduces the somatic growth of muscle of the fish at the onset of maturation (Martyniuk et al. 2003; Wringe et al. 2010).

A greater growth rate reduces production costs by decreasing the time of fish culture to reach market weight, and therefore, fast muscle growth is the major quantitative trait in genetic breeding programs in aquaculture (Sánchez-Molano et al. 2013). Consequently, the selection for faster growing fish contributes not only to more time efficient production but also results in greater feed consumption and more efficient feed utilization (Thodesen et al. 1999). For that reason, in order to select broodfish giving rise to individuals with faster muscle growth the gene products underlying the mechanisms of more rapid muscle growth need to be identified. Because fast skeletal muscle fibers constitute the main edible part of fish, understanding the factors that affect muscle cellularity (fiber number, size, etc.) as well as expression of muscle proteins, is crucial for producing fish of desired size and quality. Nevertheless, breeding programs using genetically improved stocks are scarce and equal only 2 % of aquaculture production, with the exception for Atlantic salmon *Salmo salar* and tilapia *Oreochromis niloticus*.

A controlled environment favors very different traits in fish than nature does (Saikkonen et al. 2011). Consequently, the important traits for genetic selection vary depending on animal status: wild or domesticated. The most favorable traits in wild animals are frequently related to maintenance and reproduction. The traits for genetic selection of domesticated fish, desired and influenced by humans, are mostly associated with adaptation of fish to new conditions, resistance to diseases, successful reproduction, feed efficiency, and mostly fast growth. Meshcheryakova et al. (2004) reported a decrease in the activity of proteins involved in the metabolism of carbohydrates and increased production of lactate in yellow perch skeletal muscle from humus enriched lake versus control. The quality of yellow perch fillets has also been found to exhibit several differences between farmed and wild fish in terms of fat, fatty acid profiles (arachidonic acid) and protein content (Gonzalez et al. 2006). In nature, fish with larger muscle mass are more competitive and less vulnerable to predation (Meekan and Fortier 1996). However, they may be more vulnerable to fishing because of their greater appetite and consequently greater foraging-related activity that could increase their exposure to fishing gear (Biro and Post 2008). Under controlled conditions, conversely, a faster growth rate reduces production costs by decreasing the time of fish culture. Therefore, fast muscle growth is the major focus for production enhancement (Sánchez-Molano et al. 2013).

Faster muscle growth performance of fish has been attributed to greater appetite which enhances the capacity of the fish to ingest food (Valente et al. 2001). Valente et al. (1998) showed differences in feed consumption between two rainbow trout *Oncorhynchus mykiss* strains that contributed to higher weight gain of the fast-growing group. In winter, Atlantic salmon displayed diminished appetite resulting in reduced rate of muscle growth due to lower temperatures and shorter photoperiod (Johnston et al. 2003). Mambrini et al. (2004) reported that brown trout *Salmo trutta* selected for growth demonstrated superior growth performance when fed ad libitum, meaning, when their appetite was satisfied. However, the physiological basis and genetic determinants for such increased feed intake regulation in fish, which can often be the result of additional factors, remain unresolved.

12.5 Proteomic Analysis of Skeletal Muscle Proteins

The aquaculture industry relies on sustainable fish production, which can be enhanced by the selection for faster muscle growth of economically viable fish species. The commercially important myotomal muscle is the largest tissue fraction in the majority of fish species (de Almeida et al. 2010). Understanding the molecular mechanisms responsible for fish muscle growth is critical to developing biomarkers for further development of the industry. Fish white muscle, which comprises most of the fish musculature (~70 %), is composed of fast-contracting muscle fibers using the glycolytic metabolic pathways in fast swimming during, for instance, predation or escape. Using proteomics analysis Reddish et al. (2008) identified 20 proteins/peptides associated with body mass and body length in yellow perch. These results provide insight into gene products both positively and negatively associated with muscle growth in yellow perch. These investigators identified five of the seven glycolytic enzymes associated with fish mass and size. This is consistent with the notion that increasing the flow of glucose through the glycolytic pathway might be associated with the increase of the size of fast skeletal muscle, the predominant contributor to fish growth. Finally, the structural/contraction proteins actin and myosin also need to increase in order to increase the mass of fast skeletal muscle. This is reasonable because these proteins constitute greater than 40 % of the total protein of skeletal muscle.

Table 12.2 Gene products associated with fast and slow growing of yellow perch

Proteins associated with fast growing yellow perch	Proteins associated with slow growing yellow perch
Alpha actinin	SERCA
Glycogen phosphorylase	Alpha actinin
Phosphoglucose isomerase	Phosphoglucose isomerase
ATP synthase	Creatine kinase
Enolase	GAPDH
Creatine kinase	
GAPDH	
Lactate dehydrogenase	
Adenylate kinase	

Kwasek (2012) compared skeletal muscle sarcoplasmic proteins/peptides between fast- and slow-growing yellow perch in order to identify the differences in expression of skeletal muscle proteins in fish exhibiting different growth capabilities by using a combination of 1D electrophoresis followed by image and statistical analysis. The study identified 58 bands out of which 18 bands had different staining intensities between fast- and slow-growing yellow perch ($P<0.05$) (Table 12.2).

Nine proteins/peptides were identified as predominantly metabolic enzymes, including: phosphoglucoseisomerase 2, enolase (alpha and beta), GAPDH, glycogen phosphorylase (muscle isoform), lactate dehydrogenase chain A, F1 ATP synthase beta subunit, adenylate kinase isoenzyme 1, and creatine kinase muscle isoform 2. All of these proteins with the exception for phosphoglucoseisomerase 2 were found in bands that showed increased expression in fast-growing fish compared to slow-growing fish suggesting possible association of these proteins with faster growth of yellow perch when expressed at higher levels. Nagakawa and Nagayama (1989) indicated that aldolase is one of the major sarcoplasmic proteins in muscle of a number of fish species. It has also been suggested that the higher expression rates of aldolase as well as other enzymes involved in glycolytic pathways (enolase, phosphoglucose isomerase, GAPDH) might reflect increased ATP turnover to support the rapid growth. The reduced expression in slow-growers could alternatively reflect a way of saving energy by suppressing production of the same proteins (Gomez-Requeni et al. 2011). These results are consistant with those of Reddish et al. (2008) that reported metabolic enzymes, such as aldolase, pyruvate kinase, and enolase as well as structural and contractile proteins (myosin heavy and light chains, actin, alpha-actinin, etc.) associated with yellow perch growth. Rescan et al. (2007) suggested that fish that appear to grow slower might have up-regulated proteolysis pathways, such as lysosomal system including cathepsins and the ubiquitin-proteasome system as shown in rainbow trout during a fasting period. Moreover, it has been shown that slower muscle growth rate in fish might be associated with higher expression of the nucleoside diphosphate kinase enzyme, mainly related to the regulation of growth, development, and differentiation (Gomez-Requeni et al. 2011). Some studies on rainbow trout indicated differential expression of metabolic and structural proteins, including troponin-T-1, related, possibly, to the more rapid growth of rainbow trout receiving higher levels of dietary nucleotides (Keyvanshokooh and Tahmasebi-Kohyani 2012). Additionally, nutrients such as lysine were shown to down-regulate muscle proteins and up-regulate proteins affecting fasting, energy deficit, growth arrest, and apoptosis, when withdrawn from zebrafish diet (Gomez-Requeni et al. 2011). In yellow perch proteomic fingerprinting revealed 13 electrophoretically resolved protein/peptide bands from the skeletal muscle affected by molecular form of dietary lysine supplement, free amino acid or dipeptide (lysine-glycine), suggesting that skeletal muscle protein expression can be altered by dietary treatments in the species (Fig. 12.4, Kwasek et al. 2012). In another percid, walleye *Sander vitreus*, the expression of myosin heavy chain isoforms in white muscle was shown to be altered immediately by factors affecting fish growth such as fasting (Dhillon et al. 2009). In addition, analysis of extracts of white skeletal muscle of walleye found six phenotypes of malate dehydrogenase isozymes (Clayton et al. 1971).

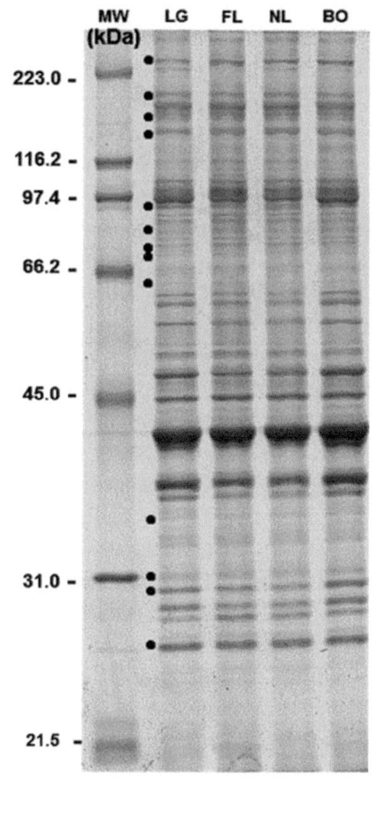

Fig. 12.4 Representative 10 % SDS-PAGE gel of yellow perch skeletal muscle sarcoplasmic fraction. The dots direct bands that had different ($P<0.05$) staining intensities between dietary treatments [wheat gluten based diets supplemented with lysine-glycine dipeptide (LG), free lysine (FL), no lysine supplementation (NL) and commercial diet (BO)]. MW is the molecular weight marker (Adopted from Kwasek et al 2012)

12.6 Growth Hormone System Proteins Affecting Muscle Growth

Growth hormone which in vertebrates, including teleost fish, is produced by the pituitary gland, plays an important physiological role in protein, carbohydrate, and lipid metabolism, as well as skeletal muscle growth (Reinecke et al. 2005). Roberts et al. (2004) observed elevated growth hormone levels in yellow perch treated with estradiol-17β in the feed for 1.5 month. Jentoft et al. (2005) injected juvenile yellow perch with bovine growth hormone, estradiol-17β or both but no growth enhancement was reported. Malison et al. (1985) instead reported that estradiol-17β promoted yellow perch growth but only after a certain size or after reaching maturation status. Surprisingly, no effect of estradiol on feed intake and growth was reported in Eurasian perch *Perca fluviatilis* (Mandiki et al. 2004). In percids and other teleost fish, the growth hormone induces muscle growth by modifying the expression of several genes: atrophy, myostatin (negative regulators), insulin-like growth factor (IGF) system as well as myogenic regulatory factors (MRFs) (positive regulators) (Fig. 12.5, Fuentes et al. 2013). Growth hormone stimulates somatic growth mainly via the IGF pathway in which it arouses insu-

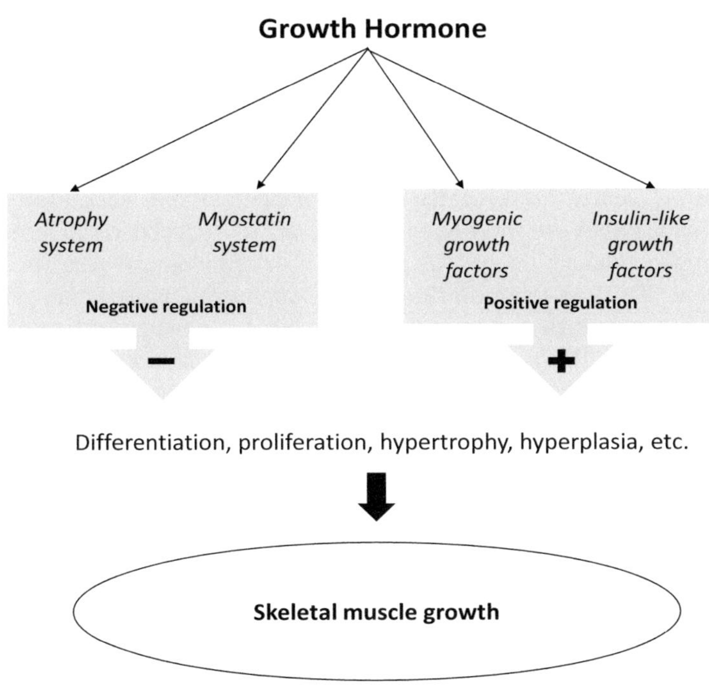

Fig. 12.5 Regulation of skeletal muscle growth by growth hormone. The growth hormone affects expression of genes involved in atrophy, myostatin, MRFs and IGFs systems which ultimately affect skeletal muscle outputs (Fuentes et al. 2013 modified)

lin-like growth factor-1 (IGF-1) synthesis in the liver and its transport to target cells which ultimately influences growth through the stimulation of amino acid and lipid metabolism (Roberts et al. 2004). Insulin-like growth factors, such as IGF-I, are major anabolic agents responsible for tissue growth in vertebrates and any changes in their expression can partly account for changes in the growth rate (Wood et al. 2005). However, the mechanism by which IGF affects muscle mass in percids remains uncertain. Proteins of the IGF system (i.e. IGF-I, IGF-II, growth hormone, IGF-binding proteins (IGFBP) or IGF-receptors) when highly expressed can stimulate muscle accretion and thus, increase fish growth. IGF-I regulates protein accretion and degradation via Akt-mTor signaling pathway. Whereas, synthesis of IGF-II required for muscle cell differentiation is regulated by mTOR and amino acid availability (Johnston 2006). Hevrøy et al. (2007) for instance showed that lysine intake induces growth activation in fast muscle in Atlantic salmon through increased mRNA levels of IGF-II. Bower et al. (2008) reported that transition to fast growth in Atlantic salmon skeletal muscle involved the local up-regulation of IGF-I, IGFBP-5.2, and IGFBP-4, as well as the down-regulation of IGFBP-2.1. In addition, increased IGF-II mRNA levels were shown in fast-growing muscle of channel catfish *Ictalurus punctatus* compared to slow-growing fish (Peterson et al. 2008). In perch, treatments with Southern black bream *Acanthopagrus butcheri* or bovine growth hormones did not affect the level of mRNA transcript copies of IGF-1 gene (Jentoft et al. 2005).

Negative regulators of muscle growth include myostatin, a member of the transforming growth factor-β superfamily expressed mainly in the skeletal muscle. Myostatin is an important negative regulator of muscle growth. In fact, in mammals, including humans, mutations of myostatin gene result in double-muscling phenotype (Clop et al. 2006; Mosher et al. 2007; Schuelke et al. 2004). In sea bass *Dicentrarchus labrax*, however, the inhibition of myostatin gene expression by double-stranded RNA and short-hairpin interfering RNA did not result in significant muscle growth although the mRNA copy number of myostatin gene was reduced (Terova et al. 2013). In zebrafish *Danio rerio* suppression of myostatin gene led to not only 45 % larger fish compared to controls but it also increased levels of MRFs (Lee et al. 2009). MRFs regulate skeletal muscle hyperplasia and hypertrophy and they include transcripts of the following genes: MyoD, myogenin, Mrf4, and Myf5. MyoD and Myf5 are mainly involved in the activation of myoblasts and their proliferation as well as determination of myogenic lineage. Whereas, myogenin and Myf4 are involved in myoblast differentiation (myotubes fusion) (Alves-Costa et al. 2013). In gilthead seabream *Sparus aurata* cultured myocytes (muscle cells) growth hormone and IGFs increased expression of MRFs involved in muscle cell differentiation and proliferation (Jiménez-Amilburu et al. 2013). In flounder *Paralichthys olivaceus* the expressions of Myf5, MyoD, and myogenin at the early life stages are characterized by gradual surge followed by drop to very low levels. In the adults, the expression of MyoD, Myf5, and myogenin in muscle is higher compared to other tissues, indicating the importance of these MRFs in muscle growth (Zhang et al. 2010). Currently, no available information exists on expression patterns of myostatin or MRFs in percids suggesting possible area for exploitation.

12.7 Conclusions

Wild fish catches need to be avoided in order to create sustainable aquaculture able to deliver fish as one of the main protein sources. The identification of genetic traits that contribute to growth superiority in fish in controlled conditions could become a tool for selection of broodfish with the potential for increased protein accretion associated with rapid muscle growth, and hence, the production of larger fish. However, culturing fish for food requires in-depth knowledge about muscle morphology, development, and composition, many aspects of skeletal muscle growth which still require further investigation particularly in percids fish.

References

Alves-Costa FA, Barbosa CM, Aguiar RCM, Mareco EA, Dal-Pai Silva M (2013) Differential expression of myogenic regulatory factor genes in the skeletal muscles of tambaqui *Colossoma macropomum* (Cuvier 1818) from Amazonian Black and clear water. J Genom Article ID 465727, 9 p

Ayala MD, Abella E, Arizcun M, Garcia-Alcazar A, Navarro F, Blanco A, Lopez-Albors OM (2013) Muscle development and body growth in larvae and early post-larvae of shi drum, *Umbrina cirrosa* L., reared under different larval photoperiod: muscle structural and ultrastructural study. Fish Physiol Biochem 39:807–827

Biro PA, Post JR (2008) Rapid depletion of genotypes with fast growth and bold personality traits from harvested fish populations. Proc Natl Acad Sci U S A 105:2919–2922

Bower NI, Li X, Taylor R, Johnston IA (2008) Switching to fast growth: the insulin-like growth factor (IGF) system in skeletal muscle of Atlantic salmon. J Exp Biol 211:3859–3870

Bureau DP, Hua K, Cho CY (2006) Effect of feeding level on growth and nutrient deposition in rainbow trout (*Oncorhynchus mykiss* Walbaum) growing from 150 to 600 g. Aquac Res 37:1090–1098

Burton NM, Byrne KM, Vierck JL, Dodson MV (2000) Methods for satellite cell culture under a variety of conditions. Methods Cell Sci 22:51–61

Clayton JW, Tretiak DN, Koyman AH (1971) Genetics of multiple malate dehydrogenase isoenzyme of skeletal muscle of walleye (*Stizostedion vitreum vitreurn*). J Fish Res Board Can 28:1005–1008

Clop A, Marcq F, Takeda H, Pirottin D, Tordoir X, Bibé B, Bouix J, Caiment F, Elsen JM, Eychenne F, Larzul C, Laville E, Meish F, Milenkovic D, Tobin J, Charlier C, Georges M (2006) A mutation creating a potential illegitimate microRNA target site in the myostatin gene affects muscularity in sheep. Nat Genet 38:813–818

Cole NJ, Johnston IA (2001) Plasticity of myosin heavy chain expression with temperature acclimation is gradually acquired during ontogeny in the common carp (*Cyprinus carpio* L.). J Comp Phys B 171:21–326

de Almeida FLA, Pessotti NS, Pinhal D, Padovani CR, Leitao NDJ, Carvalho RF, Martins C, Portella MC, Pai-Silva MD (2010) Quantitative expression of myogenic regulatory factors MyoD and myogenin in pacu (*Piaractus mesopotamicus*) skeletal muscle during growth. Micron 41:997–1004

Dhillon RS, Esbaugh AJ, Wang YS, Tufts BL (2009) Characterization and expression of a myosin heavy–chain isoform in juvenile walleye *Sander vitreus*. J Fish Biol 75:1048–1062

Fuentes EN, Valdés JA, Molina A, Björnsson BT (2013) Regulation of skeletal muscle growth in fish by the growth hormone–insulin-like growth factor system. Gen Comp Endocrinol 192:136–148

Gomez-Requeni P, Vareille M, Kousoulaki K, Jordal A-EO, Conceicao LEC, Ronnestad I (2011) Whole body proteome response to a dietary lysine imbalance in zebrafish *Danio rerio*. Comp Biochem Physiol 6:178–186

Gonzalez S, Flick GJ, O'Keefe SF, Duncan SE, McLean E, Craig SR (2006) Composition of farmed and wild yellow perch (Perca flavescens). J Food Compos Anal 19:720–726

Greenlee AR, Dodson MV, Yablonka-Reuveni Z, Kersten CA, Cloud JG (1995) In vitro differentiation of myoblasts from skeletal muscle of rainbow trout. J Fish Biol 46:731–747

Hevrøy EM, El-Mowafi A, Taylor RG, Olsvik PA, Norberg B, Espe M (2007) Lysine intake affects gene expression of anabolic hormones in Atlantic salmon, Salmo salar. Gen Comp Endocrinol 152:39–46

Jentoft S, Topp N, Seeliger M, Malison JA, Barry TP, Held JA, Roberts S, Goetz F (2005) Lack of growth enhancement by exogenous growth hormone treatment in yellow perch (Perca flavescens) in four separate experiments. Aquaculture 250:471–479

Jiménez-Amilburu V, Salmerón C, Codina M, Navarro I, Capilla E, Gutiérrez J (2013) Insulin-like growth factors effects on the expression of myogenic regulatory factors in gilthead sea bream muscle cells. Gen Comp Endocrinol 188:151–158

Johnston IA (1999) Muscle development and growth: potential implications for flesh quality in fish. Aquaculture 177:99–115

Johnston IA (2006) Review: environment and plasticity of myogenesis in teleost fish. J Exp Biol 209:2249–2264

Johnston IA, Manthri S, Smart A, Campbell P, Nickell D, Alderson R (2003) Plasticity of muscle fibre number in seawater stages of Atlantic salmon in response to photoperiod manipulation. J Exp Biol 206:3425–3435

Johnston IA, Lee HT, Macqueen DJ, Paranthaman K, Kawashima C, Anwar A, Kinghorn JR, Dalmay T (2009) Embryonic temperature affects muscle fibre recruitment in adult zebrafish: genome-wide changes in gene and microRNA expression associated with the transition from hyperplastic to hypertrophic growth phenotypes. J Exp Biol 12:1781–1793

Kacperczyk A, Jagla T, Daczewska M (2009) Pax-3 and pax-7 label muscle progenitor cells during myotomal myogenesis in *Coregonus lavaretus* (Teleostei: *Coregonidae*). Anat Histol Embryol 38:411–418

Kacperczyk A, Jedrzejowska I, Daczewska M (2011) Differentiation and growth of myotomal muscles in a non-model tropical fish pterophyllum scalare (Teleostei: *Cichlidae*). Anat Histol Embryol 40:411–418

Karasinski J (1993) Diversity of native myosin and myosin heavy chain in fish skeletal muscles. Comp Biochem Physiol B Biochem Mol Biol 106:1041–1047

Kiessling A, Ruohonen K· Bjørnevik M (2006) Muscle fibre growth and quality in fish. Arch Tierz, Dummerstorf 49 Special Issue, 137–146

Konigsberg IR (1963) Clonal analysis of myogenesis. Science 140:1273–1284

Koumans JTM, Akster HA, Booms RGH, Osse JWM (1993) Influence of fish size on proliferation and differentiation of cultured myosatellite cells of white axial muscle of carp, (*Cyprinus carpio* L.). Differentiation 53:1–6

Kwasek K (2012) The nutritional and genetic effects on body growth, reproduction and molecular mechanisms responsible for muscle growth in yellow perch *Perca flavescens*. Unpublished doctoral dissertation, The Ohio State University

Kwasek K, Wojno M, Terova G, Ostaszewska T, Wick M, Dabrowski K (2012) The effect of the dipeptide, Lys-Gly, supplement on growth, muscle proteins and PEPT1 gene expression in juvenile yellow perch. Rev Fish Biol Fish 22:797–812

Lee CY, Hu SY, Gong HY, Chen MH, Lu JK, Wu JL (2009) Suppression of myostatin with vector-based RNA interference causes a double-muscle effect in transgenic zebrafish. Biochem Biophys Res Commun 387:766–771

Malison JA, Best CD, Kayes TB, Amundson CH (1985) Hormonal growth promotion and evidence for a size-related difference in response to estradio 17β in yellow perch (*Perca flavescens*) can. J Fish Aquat Sci 42:1627–1633

Mambrini M, Medale F, Sanchez MP, Recalde B, Chevassus B, Labbe L, Quillet E, Boujard T (2004) Selection for growth in brown trout increases feed intake capacity without affecting maintenance and growth requirements. J Anim Sci 82:2865–2875
Mandiki SNM, Houbart M, Babiak I, Vandeloise E, Gardeur JN, Kestemont P (2004) Are sex steroids involved in the sexual growth dimorphism in Eurasian perch juveniles? Physiol Behav 80:603–609
Martyniuk CJ, Perry JML, Mogahadam HK, Ferguson MM, Danzmann RG (2003) The genetic architecture of correlations among growth-related traits and male age at maturation in rainbow trout. J Fish Biol 63:746–764
Meekan MG, Fortier L (1996) Selection for fast growth during the larval life of Atlantic cod *Gadus morhua* on the Scotian Shelf. Mar Ecol Prog Ser 137:25–37
Meshcheryakova OV, Gruzdev AI, Nemova NN (2004) Comparative evaluation of carbohydrate metabolism in perch (Perca fluviatilis L) from water bodies with different contents of humic acids. Biol Bull 31:15–20
Mosher DS, Quignon P, Bustamante CD, Sutter NB, Mellersh CS, Parker HG, Ostrander EA (2007) A mutation in the myostatin gene increases muscle mass and enhances racing performance in heterozygote dogs. PLoS Genet 3:779–786
Nagakawa T, Nagayama F (1989) Enzymatic properties of fish muscle aldolase. Comp Biochem Physiol 92:405–410
Ostaszewska T, Dabrowski K, Wegner A, Krawiec M (2008) The effects of feeding on muscle growth dynamics and the proliferation of myogenic progenitor cells during pike perch development (*Sander lucioperca*). J World Aquac Soc 39:184–195
Peterson BC, Small BC, Waldbieser GC, Bosworth BG (2008) Endocrine responses of fast- and slow-growing families of channel catfish. N Am J Aquac 70:240–250
Powell R, Dodson M, Cloud J (1989) Cultivation and differentiation of satellite cells from skeletal muscle of the rainbow trout *Salmo gairdneri*. J Exp Zool 250:333–338
Reddish JM, St-Pierre N, Nichols A, Green-Church C, Wick M (2008) Proteomic analysis of proteins associated with body mass and length in yellow perch, *Perca flavescens*. Proteomics 8:2333–2343
Reinecke M, Björnsson BT, Walton W, Dickhoff WW, McCormick SD, Navarro I, Power DM, Gutiérrez J (2005) Growth hormone and insulin-like growth factors in fish: where we are and where to go. Gen Comp Endocrinol 142:20–24
Rescan Y-P, Montfort J, Ralliere C, Le Cam A, Esquerre D, Hugot K (2007) Dynamic gene expression in fish muscle during recovery growth induced by a fasting-refeeding schedule. BMC Genomics 8:438
Roberts S, Barry T, Malison J, Goetz F (2004) Production of a recombinantly derived growth hormone antibody and the characterization of growth hormone levels in yellow perch. Aquaculture 232:591–602
Rowlerson A, Scapolo P, Mascarello F, Carpene E, Veggetti A (1985) Comparative study of myosins present in the lateral muscle of some fish: species variations in myosin isoforms and their distribution in red, pink and white muscle. J Muscle Res Cell Motil 6:601
Rowlerson A, Mascarello F, Radaelli G, Veggetti A (1995) Differentiation and growth of muscle in the fish, *Sparus aurat*a (L): II. Hyperplastic and hypertrophic growth of lateral muscle from hatching to adult. J Muscle Res Cell Motil 16:223–236
Saikkonen A, Kekalainen J, Piironen J (2011) Rapid growth of atlantic salmon juveniles in captivity may indicate poor performance in nature. Biol Conserv 144:2320–2327
Sánchez-Molano E, Cerna A, Toro MA, Bouza C, Hermida M, Pardo BG, Terova G, Rimoldi S, Bernardini G, Saroglia M (2013) Inhibition of myostatin gene expression in skeletal muscle of fish by in vivo electrically mediated dsRNA and shRNAi delivery. Mol Biotechnol 54:673–684
Schuelke M, Wagner KR, Stolz LE, Hübner C, Riebel T, Komen W, Braun T, Tobin JF, Lee SJ (2004) Myostatin mutation associated with gross muscle hypertrophy in a child. N Engl J Med 350:2682–2688
Terova G, Rimoldi S, Bernardini G, Saroglia M (2013) Inhibition of myostatin gene expression in skeletal muscle of fish by in vivo electrically mediated dsRNA and shRNAi delivery. Mol Biotechnol 54:673–84

Thodesen J, Grisdale-Helland B, Helland S, Gierde B (1999) Feed intake, growth and feed utilization of offspring from wild and selected Atlantic salmon (*Salmo solar*). Aquaculture 180:237–246

Valente LMP, Fauconneau B, Gomes EFS (1998) Voluntary feed intake, feed and nutrient utilisation in slow and fast growing rainbow trout strains. Aquat Living Resour 11:93–99

Valente LMP, Fauconneau B, Gomes EFS, Boujard T (2001) Feed intake and growth of fast and slow growing strains of rainbow trout *Oncorhynchus mykiss* fed by automatic feeders or by self-feeders. Aquaculture 195:121–131

Venkateswaran V, Brackett EJ, Vierck J, Cloud JG, Dodson MV (1995) Substratum is an important determinant in growth factor regulation of trout-derived satellite cells. Basic Appl Myol 5:297–304

Wilkes D, Xie SQ, Stickland NC, Alami-Durante H, Kentouri M, Sterioti A, Koumoundouros G, Fauconneau B, Goldspink G (2001) Temperature and myogenic factor transcript levels during early development determines muscle growth potential in rainbow trout (*Oncorhynchus mykiss*) and sea bass (*Dicentrarchus Labrax*). J Exp Biol 204:2763–2771

Wood AW, Duan C, Bern HA (2005) Insulin-like growth factor signaling in fish. Int Rev Cytol 243:215–285

Wringe BF, Devlin RH, Ferguson MM, Moghadam HK, Sakhrani D, Danzmann RG (2010) Growth-related quantitative trait loci in domestic and wild rainbow trout (*Oncorhynchus mykiss*). BMC Genet 11:63

Zhang Y, Tan X, Xu P, Sun W, Xu Y, Zhang P (2010) Quantitative comparison of the expression of myogenic regulatory factors in flounder (*Paralichthys olivaceus*) embryos and adult tissues. Chin J Oceanol Limnol 28:248–253

Chapter 13
The Energy Requirements of Percid Fish in Culture

Anders Alanärä and Åsa Strand

Abstract In commercial aquaculture, knowledge about and means for predicting growth rates, feed intake and energy requirements of the farmed animal in different conditions is essential for the viability of the enterprise. As percid fish species are relatively new in culture, there are no models available to estimate the energy requirement of the cultured fish, which in turn limits the opportunities to calculate the required daily feed allowance. Classical bioenergy budgets are often used to describe energy intake in relation to different energy expenditures of fish by quantifying steps where energy expenditures occur. However, in commercial aquaculture the objective is to optimize the output (growth) in relation to the energy intake, e.g. where energy expenditures occur is less important. In this chapter, we put together data from the scientific literature to produce an alternative model for prediction of the daily growth and energy need of percid fish in general and Eurasian perch (*Perca fluviatilis* L.) in particular. A practice for calculating the daily feed allowance is presented where local rearing conditions can be taken into account. This makes the model applicable to commercial enterprises and may improve feed management, fish growth and thus economics of the fish farms. This chapter also discusses how factors such as season and culture conditions influence the energy requirements and energy expenditures of the percid fish.

Keywords Eurasian perch • *Perca fluviatilis* • Energetics • Growth • Feed intake

---

A. Alanärä (✉)
Department of Wildlife, Fish, and Environmental Studies,
Swedish University of Agricultural Sciences, Umeå, Sweden
e-mail: Anders.Alanara@slu.se

Å. Strand
Department of Biological and Environmental Sciences,
University of Gothenburg, Gothenburg, Sweden

P. Kestemont et al. (eds.), *Biology and Culture of Percid Fishes*,
DOI 10.1007/978-94-017-7227-3_13

13.1 Introduction

There is a long tradition to utilize energetic principles for calculation of the daily feed allowances of farm animals (MacEwan 1945; Bull and Carroll 1946; Lloyd et al. 1978). For livestock, the extent of nutritional knowledge is well developed, while in contrast, in aquaculture feed rations are based on growth rate estimates rather than the actual energy requirements of the cultured fish. As the energy expenditures of fish vary with local conditions, many culture operations end up with poor feed efficiency. Feed remains as one of the largest costs in semi-intensive and intensive aquaculture (Riepe 1997; Cho and Bureau 1998; Dunning et al. 1998), therefore inefficient feed management will have a negative impact on farm economics. Overfeeding fish also leads to an increased environmental load, increased cost for water reconditioning and reduced welfare of the fish. It is thus necessary to optimize feeding to improve the economic and environmental sustainability of aquaculture.

In this chapter we will present an alternative scheme for constructing species specific models on energy requirements and growth of fish in culture, using the Eurasian perch (*Perca fluviatilis*) as an example species. These models may be used for production planning and daily feed allowance purposes in commercial rearing situations.

13.2 The Energy Requirements of Percids

Traditionally the energy requirements of fish has been estimated by constructing complete energy budgets, balancing energy intake against energy expenditures such as faecal production, nitrogen excretion, metabolism and growth (Brett and Groves 1979; Brafield 1985; Jobling 1994; De Silva and Anderson 1995). Despite improvements in methodology, this approach is often associated with several potential sources of error (Jobling 1983, 1994; Brafield 1985; Talbot 1985) and many of the developed energy budgets have been proven to be inaccurate when tested (Cui and Wootton 1989; Ney 1993; Cui and Xie 2000). Alanärä et al. (2001) have developed an alternative model for estimation of the daily energy requirements and calculation of feed budgets for fish in culture. The model is based on two major components; (A) the daily growth increment (T*Wi*, $g \cdot d^{-1}$) of the fish and (B) the amount of digestible energy needed (DEN, kJ DE) to obtain one unit of biomass gain (g) of the fish.

Component A is retrieved by constructing a species specific growth model. The most commonly used estimate of fish growth is the specific growth rate (SGR; Ricker 1979). SGR is expressed mathematically as:

$$\mathrm{SGR} = \left(\ln W_2 - \ln W_1\right) / \Delta t \cdot 100 \qquad (13.1)$$

where W_2 is the final weight (g), W_1 is the initial weight (g) and Δt is the number of days between weightings. The form of the equation assumes that fish weight increases exponentially. However, this assumption is only valid for young fish cultured for short periods of time, and consequently, SGR is not suitable for reporting growth of large fish or longer culture periods (Hopkins 1992). Furthermore, as an organism increases in size, the rate of its metabolic activities slows down (Brett 1979; Brett and Groves 1979; Jobling 1994) and as a result, the relative growth rate will decrease. Any growth increment (in real terms) is also smaller for a large individual. Thus, SGR will decrease as the size of the fish increases (Brett 1979; Iwama and Tautz 1981; Jobling 1983, 1994). Moreover, as fish are ectothermic animals, the ambient water temperature will affect metabolic rates of the fish, with increasing metabolic rates at increasing temperatures (Brett and Groves 1979; Jobling 1994, 1997; De Silva and Anderson 1995). Consequently, at high temperatures the relative growth rate will be higher than at low temperatures, and SGR will therefore increase with increasing temperature (Brett 1979; Jobling 1994; Wootton 1998). The temperature and size dependence of SGR make data collection for model construction very time consuming and labour demanding, and as a consequence, only a few models describing SGR for fish in culture are available (Alanärä et al. 2001). Strand et al. (2011a, b) studied the effect of temperature (8–27 °C) and body size (20–180 g) on SGR in Eurasian perch. As predicted, SGR was highly affected by both variables. The optimal temperature for growth was found to be around 23 °C, which is similar to other studies on Eurasian perch (Mélard et al. 1996; Kestemont et al. 2003).

To reduce the problem of body size and temperature, the thermal unit growth coefficient (TGC) was developed by Iwama and Tautz (1981) and later modified by Cho (1990). TGC is expressed mathematically as:

$$\mathrm{TGC} = \left(W_2^{(1/3)} - W_1^{(1/3)}\right) / \left(T \cdot \Delta t\right) \cdot 1000 \qquad (13.2)$$

where T is the water temperature (°C). Instead of using the logarithm of the fish weight ($\ln W$) for calculating growth rate as SGR does, TGC uses a power function ($W^{(1/3)}$). This mathematical adjustment provides a better fit of the growth coefficient to the actual growth pattern of the fish (Cho 1992). Thus, due to the power function and the inclusion of temperature, TGC is thought to be less affected by body size of the fish (Kaushik 1995, 1998; Bureau et al. 2000) and temperature (Azevedo et al. 1998; Cho and Bureau 1998; Bureau et al. 2000; Bailey and Alanärä 2006) than SGR. In addition, the TGC coefficient predicts growth over time quite accurately (Bureau et al. 2000). Consequently, in contrast to the complex SGR models, TGC data collected for fish of a given size at one temperature may ideally be used to predict the weight increment of fish at other sizes and temperatures. In a number of experiments on Eurasian perch, Strand et al. (2011a, b) showed that TGC responded in a similar way as SGR to both temperature and body size. TGC, however, was more or less unaffected by temperature within the range of 17–23 °C (Fig. 13.1). The relationship between body size (W; 20–180 g) and TGC within this range can be expressed as (Strand et al. 2011b):

$$\mathrm{TGC} = 0.373 + 8.024 / W \qquad (13.3)$$

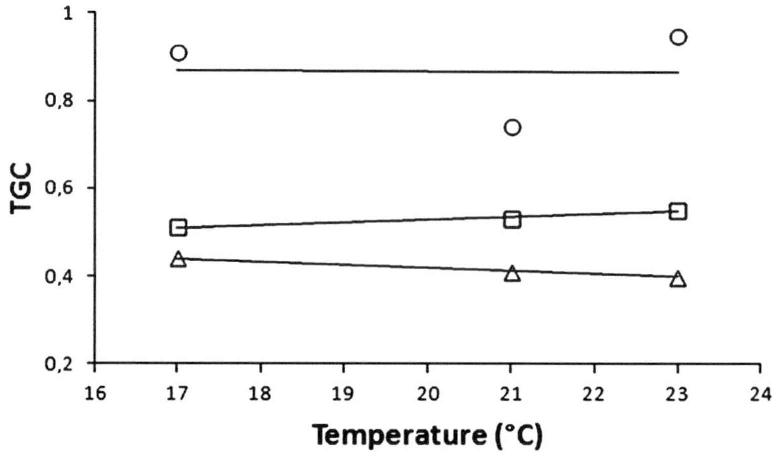

Fig. 13.1 The relationship between temperature and TGC for Eurasian perch of different body sizes (*circles* 20 g, *squares* 35 g and *triangles* 50 g) (Data from Strand et al. 2011b)

Component B is retrieved by quantifying the amount of digestible energy needed to produce one unit of weight gain. According to Cho et al. (1982), the principles of bioenergetics were applied to fish already in 1914 by Ege and Krogh (1914) and several years later by Ivlev (1939), and Cui and Xie (2000) states that the first bioenergetic model for fish was developed by Kitchell et al. (1974). Models based on similar principles had been proposed earlier by other researchers (Ursin 1967; Kerr 1971), but the model developed by Kitchell et al. (1974), is probably the most influential (Cui and Xie 2000). The simplest form of a bioenergetic model can be derived from the basics of bioenergetics: "*any change in body weight results from the difference between what enters the body and what leaves it*" (Jobling 1997). Growth can thus be expressed as the net energy gain obtained when all energy expenditures are subtracted from the total energy ingested. To provide a more detailed view of the energy budget, the energy expenditures can be divided into smaller units and the energy budget can then be expressed as:

$$I = F + E + M + G \qquad (13.4)$$

where I is the energy content of the ingested feed, F is the energy lost in faeces, E is the energy lost to excretion, M is the energy lost in metabolism and G is the energy retained as growth and gonad production (Brett and Groves 1979; Jobling 1994; De Silva and Anderson 1995).

As can be expected, several problems arise when the different units of the energy budget are to be measured. So far it has not been possible to determine all the components in the energy budget simultaneously, and often one or more of the major units have been estimated "by difference" to produce a balanced budget (Jobling 1994). Different experimental procedures will also produce differing results (Talbot 1985). Furthermore, bioenergetic studies of fish have largely been theoretical and performed in laboratories (Knights 1985), and the experiments also often impose unnatural or unrealistic feeding regimes and living conditions on the fish which exposes the fish to both acute and chronic stress (Talbot 1985). Despite improvements in methodology, the bioenergetic approach is thus often associated with measurement errors (Brafield 1985) and several of the developed energy budgets prove to deliver inaccurate results when tested (Cui and Wootton 1989; Ney 1993; Cui and Xie 2000). Thus, the experimental approaches used to develop energy budgets tend to produce results that are rarely transferable to aquaculture (Alanärä et al. 2001).

In order to construct an energy requirement model that is useful in culture situations we do not need to estimate all pathways of energy losses, but rather to measure the amount of energy digested and the energy allocated in terms of growth. This assumes that energy losses for excretion (E) and metabolism (M) are more or less constant when the fish are held in a specific rearing environment. In addition, by using the value on digestible energy content of the feed, the energy losses in faeces (F) is accounted for. The digestible energy needed to obtain one unit of biomass gain (DEN; kJ DE·g^{-1}) of fish in culture is calculated as:

$$\mathrm{DEN} = \left(FI \cdot DE\right) / \left(W_2 - W_1\right) \tag{13.5}$$

where *FI* is the feed intake (g) and *DE* is the digestible energy content of the feed (kJ·g^{-1}).

The big advantage with the model developed by Alanärä et al. (2001) is that since the values for the different energy expenditures need not be quantified, estimates of DEN can be made when the fish are being raised under experimental conditions similar to those in commercial culture. Due to differences in rearing environment, however, values on DEN may differ. As will be discussed later in this chapter, the energy expenditures in fish farms are often related to stress caused by either suboptimal rearing environments or handling of the fish. Ideally, farm specific DEN values should be estimated.

At and below optimal temperatures for growth (i.e. 23 °C), DEN of Eurasian perch have been found not to be affected by temperature (Fig. 13.2, Strand et al. 2011a, b). This is in accordance with data presented by Bailey and Alanärä (2006), where DEN of salmonid species like rainbow trout (*Oncorhynchus mykiss*), Atlantic salmon (*Salmo salar*) and Arctic charr (*Salvelinus alpinus*) was shown to be unaffected by temperatures below the optimal. As standard metabolic rate of fish increases with increasing temperature (Brett and Groves 1979; Jobling 1994, 1997; De Silva and Anderson 1995) so does the energy expenditures of the fish, and hence a connection between temperature and DEN should exist. However, at normal rearing temperatures the effect of metabolic costs on the overall energy budget is negligible. At high temperatures though, metabolic costs increases dramatically. Strand et al. (2011a) showed an exponential increase in energy expenditures in Eurasian perch when temperature exceeded the optimal growing temperature (Fig. 13.2). A similar exponential effect has been shown for salmonids (Bailey and Alanärä 2006).

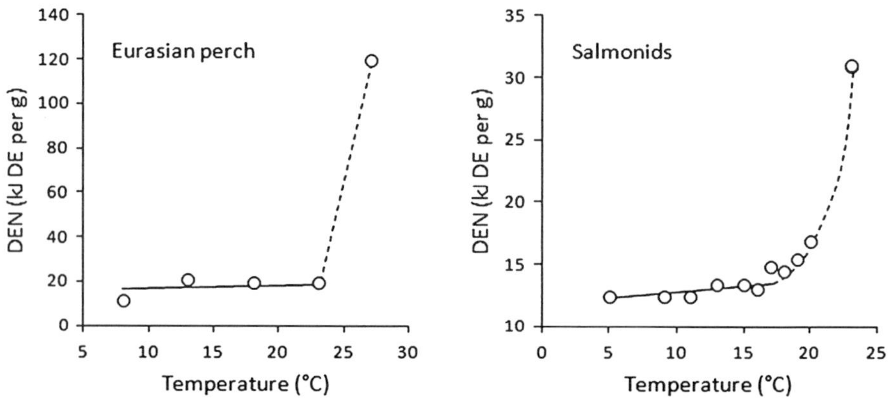

Fig. 13.2 Data showing the exponential increase in digestible energy need as temperature exceeds the optimal one for growth in Eurasian perch and salmonids (Atlantic salmon and Arctic charr) (Data from Bailey and Alanärä 2006 and Strand et al. 2011a)

Furthermore, the DEN value of Eurasian perch has been found to increase with increasing body size (Strand et al. 2011b). The same relationship have been demonstrated for different salmonid and flatfish species (Bailey and Alanärä 2006), as well as for cod (*Gadus morhua* L.; Björnsson et al. 2001). One explanation for this may be the allometric changes in the ratio of lipid, protein, and water storage that occurs with increasing size of the fish. Thus, as fish grow, the ratio of these macronutrients stored in the tissue changes (Jobling 2001). The storage of high-energy molecules of lipid is more "costly", in terms of the energy ingested, than muscle (protein) growth. In addition, 1 g of lipid deposition leads to a weight increase of 1 g, whereas deposition of 1 g protein is associated with an additional deposition of glycogen and 3–4 g of water. Fish that store more lipids should therefore require more energy and gain less weight. This can be seen as a higher DEN value. The relationship between body size and DEN for Eurasian perch can be expressed as (Strand et al. 2011b):

$$\mathrm{DEN} = 6.422 + 3.407 \cdot \ln W \tag{13.6}$$

A literature search on the "Web of Science" for feeding trials data on Eurasian perch and yellow perch (*Perca flavescens*) gave 30 articles in total. Out of that, 12 contained data that could be used to calculate DEN (Table 13.1). In comparison with the DEN model presented by Strand et al. (2011b) most values are higher than what the model predicts (Fig. 13.3). Eurasian perch have been shown to be rather stress sensitive when held in rearing facilities (see discussion below). Within the study of Strand et al. (2011b), care was taken not to disturb and stress the fish during the experiments. The higher energy requirements reported in other studies may thus be the result of higher energy expenditures for fish kept in sub-optimal experimental conditions.

The differences in DEN between perch and salmonid species seen in Fig. 13.3 may be related to the higher energetic costs of life in warm water. Brett and Groves (1979) compared values on standard metabolism for tropical, temperate

Table 13.1 Feed efficiency (FE) and digestible energy need (DEN) for Eurasian perch and yellow perch at different temperatures and body sizes. Values are retrieved based on feed intake data, digestible energy content of the feed and weight gain. Energy values of macronutrients (23.7, 36.3 and 17.2 kJ·g^{-1} for protein, fat and carbohydrates, respectively) were obtained from Brett and Groves (1979). Apparent digestibility coefficients (ADC) used was 0.87, 0.90 and 0.65 for protein, fat and carbohydrates, respectively. These data were used to calculate the digestible energy content of the feed in each experiment (DE)

Author	Species	Temperature	Weight	FE	DE (MJ DE/kg)	DEN
Xu et al. 2001	Eurasian perch	23	59	0.80	17.8	22.2
Jourdan et al. 2000	Eurasian perch	23	14	0.80	17.1	21.3
Blanchard et al. 2008	Eurasian perch	23	75	0.88	18.1	20.6
Fiogbe and Kestemont 2003	Eurasian perch	23	3	0.54	18.1	33.5
Juell and Lekang 2001	Eurasian perch	18	41	1.14	16.9	14.9
Mandiki et al. 2004	Eurasian perch	22	8	0.76	18.1	23.8
Mandiki et al. 2004	Eurasian perch	22	47	0.66	18.4	27.9
Mandiki et al. 2004	Eurasian perch	24	11	0.70	18.1	25.8
Mandiki et al. 2004	Eurasian perch	24	38	0.68	18.4	27.1
Kestemont et al. 2001	Eurasian perch	23	36	0.64	18.6	29.1
Mathis et al. 2003	Eurasian perch	23	84	0.93	18.8	20.2
Twibell and Brown 2000	Yellow perch	20	25	0.69	12.8	18.5
Twibell et al. 2001	Yellow perch	21	45	0.60	12.8	21.3
Gould et al. 2003	Yellow perch	22	497	0.48	17.8	37.5
Hart et al. 2010	Yellow perch	22	44	0.65	18.1	27.8

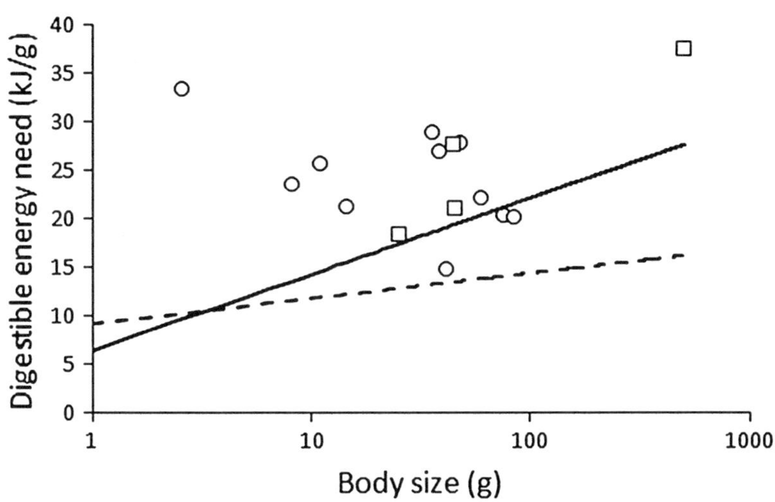

Fig. 13.3 Digestible energy need (DEN) at different body sizes. *Circles* are previously published data for Eurasian perch and *squares* data for yellow perch. The *solid line* represents the model developed for Eurasian perch by Strand et al. (2011b) and the *hatched line* represents the model for salmonids ($\mathrm{DEN} = 9.22 + 1.12 * \ln \mathrm{W}$) developed by Bailey and Alanärä (2006)

and polar fishes against temperature, and concluded that warm-water adapted species operate at a higher metabolic maintenance level in accordance with the higher water temperatures. The tropical species incur an energy expenditure that is about 70 % higher than that for temperate species. The amount of energy needed to produce one unit of weight increase in Eurasian perch is 50–70 % higher than for salmonid species (compare values in Fig. 13.3). Thus, differences in DEN between warm-water species and cold-water species are similar to that Brett and Grove (1979) found for energy expenditures. The practical implication of this is that the feed requirements for producing one unit of fish gain will be higher for percids compared to salmonids. This will consequently influence production costs. Domestication processes and selective breeding may significantly reduce DEN of cultured fish. Thodesen et al. (1999) compared the growth and feed efficiency of a selected Atlantic salmon strain (five generations) against the offspring of the founder population over a full commercial rearing cycle. The results showed a 25 % improvement of DEN for the selected stock, indicating a possible space for development in percid fish.

By combining data on daily growth (component A) and DEN (component B), a model expressing the theoretical energy requirements (TER; kJ·day^{-1}) can be developed. The model is written as:

$$\mathrm{TER} = \mathrm{T}Wi \cdot \mathrm{DEN} \tag{13.7}$$

where T*Wi* is the theoretical weight increment per day (g·day^{-1}). Figures on theoretical weight increment for Eurasian perch can be obtained by using the TGC model (Eq. 13.3) or any other growth model. By using the TGC model, the expected weight of the fish (W_2, g) after a period of time can be calculated as:

$$W_2 = \left(W_1^{(1/3)} + \left(\mathrm{TGC}/1000 \cdot T \cdot D\right)\right)^3 \tag{13.8}$$

where *W1* is the initial weight, *TGC* the growth coefficient (value from Eq. 13.3), *T* is temperature and *D* is days (*T* and *D* forms the sum of temperature). T*Wi* is then obtained by calculating the weight of the fish after 1 day of growth and subtracting the initial weight. Thus, by combining the theoretical daily weight gain and the DEN values, a model describing the daily theoretical energy requirement (TER, kJ·d^{-1}) at different temperatures and for fish of different sizes can be expressed. For Eurasian perch the models looks like this (Strand et al. 2011b):

$$\mathrm{TER} = 0.039 \cdot T \cdot W^{0.614 - 0.014/T} \tag{13.9}$$

An example of a theoretical energy requirement chart based on Eq. 13.9 is demonstrated in Table 13.2. This model by Strand et al. (2011b) corresponds well with data on feed requirements obtained by Mélard et al. (1996) and Fiogbé and Kestemont (2003) for Eurasian perch in culture conditions.

Table 13.2 TER (kJ·day^{-1}) values for Eurasian perch (*Perca fluviatilis* L.) reared at different temperatures and of different sizes (based on Eq. 13.9)

		Size of fish (g)				
		20	50	100	150	200
Temperature °C	16	3.9	6.9	10.5	13.5	16.1
	18	4.4	7.7	11.8	15.2	18.1
	20	4.9	8.6	13.1	16.9	20.1
	22	5.4	9.5	14.5	18.5	22.1
	24	5.9	10.3	15.8	20.2	24.1

13.3 Daily Feed Allowance

Values on daily feed allowance given by feed companies, feeding system companies or others generally lack the ability to allow adjustments to be made in relation to local rearing conditions. By using previous growth records or theoretical models (Eqs. 13.3 and 13.8), the first component in the feed budget model (component A; the daily weight increment), can be estimated. To obtain farm specific data of the daily weight increment is of high importance as these may vary considerably between rearing facilities (sites), strains and different times of the year. Growth rate data is, however, reasonable easy to collect by regular weightings, thus this should not constitute a problem in a commercial culture situation. By using Eq. 13.6 or estimates of DEN based on own measurements, the second component (B) in the energy requirement model is achieved. Values of DEN on the other hand are more difficult to obtain in practical rearing situations as accurate measures on feed intake are required. The value of DEN is probably more robust than growth rate data, and is mainly affected by poor rearing conditions and a stressful environment. Fish farmers should be able to rely on expert models for DEN, i.e. like Eq. 13.6 for use in the feed budget model. Once both components are known, farmers can then create their own feed budget model or feeding chart. For evaluation of the daily feed allowance of the fish (FA, g·day^{-1}), the following calculation is made:

$$\mathrm{FA} = \mathrm{TER} \cdot \mathrm{n} / DE \tag{13.10}$$

where n is the number of fish in the rearing unit and DE is the digestible energy content of the feed (kJ/kg). The feed allowance model was tested by Bailey and Alanärä (2001) with good results on hatchery-reared rainbow trout. In Table 13.3, an example is given on how the daily feed allowance for a group of Eurasian perch can be calculated based on the methodology described in this chapter.

Table 13.3 Example on how the daily feed allowance can be calculated for a group of Eurasian perch based on the methodology described in this chapter

Data		
Fish size		50 g
Temperature		22 °C
Feed, digestible energy content		18 MJ per kg (or kJ per g)
Number of fish		3,000
Model	**Calculation**	**Result**
$TGC = 0.373 + 8.024 / W$	$TGC = 0.373 + 8.024 / 50$	0.53
$W_2 = \left(W^{(1/3)} + (TGC / 1000 \cdot T \cdot D)\right)^3$	$W_2 = \left(50^{(1/3)} + (0.53 / 1000 \cdot 22 \cdot 1)\right)^3$	50.48 g
$TWi = W2 - W1$	$TWi = 50.48 - 50$	0.48 g/d
$DEN = 6.422 + 3.407 \cdot \ln W$	$DEN = 6.422 + 3.407 \cdot \ln 50$	19.8 kJ/g
$TER = TWi \cdot DEN$	$TER = 0.48 \cdot 19.8$	9.5 kJ/d
$TER_{mod} = 0.039 \cdot T \cdot W^{0.614 - 0.014/T}$	$TER_{mod} = 0.039 \cdot 22 \cdot 50^{0.614 - 0.014/22}$	9.5 kJ/d
$FA = TER \cdot n / DE$	$FA = 9.5 \cdot 3000 / 18$	1578 g

13.4 Seasonal Variations in Energy Requirements and Growth

Energy requirement and growth of percid fish may vary considerably at different times of the year, which in turn will affect the daily feed allowance. Staffan et al. (2005) demonstrated an increase in feed intake and growth in hatchery reared Eurasian perch during spring; despite constant water temperatures and day length (Fig. 13.4). In addition, Strand et al. (2007a) showed a corresponding decrease in feed intake and growth between September and November in Eurasian perch held in constant environmental conditions (Fig. 13.4). Similar seasonal patterns have also been noted in free living Eurasian perch (Griffiths and Kirkwood 1995). The seasonal variation in feed intake has been most widely studied in salmonid species such as Atlantic salmon (Thorpe 1994), Arctic charr (Sæther et al. 1996), and chinook salmon (*Oncorhynchus tshawytscha*; Clarke and Blackburn 1994). Similar to Eurasian perch, these species typically increase feeding in spring and experience a depression in feed intake during the autumn (Smith et al. 1993; Tveiten et al. 1996). These seasonal variations in appetite and growth despite constant environmental cues such as temperature and day length are referred to as an endogenous rhythm (Eriksson and Alanärä 1992).

A physiological explanation for reduced feeding during the autumn is the level of fat depots. It has been proposed that fish may reduce feeding once they have acquired sufficient energy reserves to survive the winter (Tveiten et al. 1996). Furthermore, there may be an inverse relationship between body fat content and feed intake, which would partly explain the large increase in feed intake after a long winter, when energy reserves are depleted (Metcalfe and Thorpe 1992; Jobling and

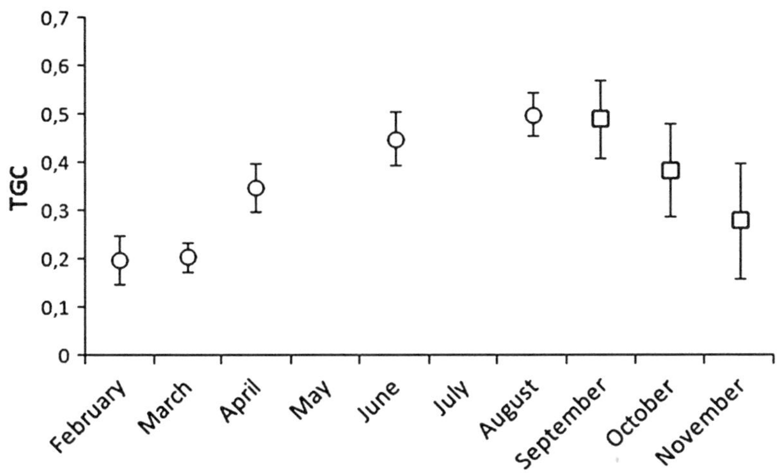

Fig. 13.4 The growth rate (TGC) of Eurasian perch at different times of the year. Fish of the size 20–30 g were held in similar rearing conditions at 17 °C and constant day length (LD 18:6) (*Circles* represent data from Staffan et al. (2005) and *squares* data from Strand et al. (2007a))

Miglavs 1993; Shearer et al. 1997; Silverstein et al. 1999). If this explanation is valid also for the seasonal variation in feed intake and growth of percid fish needs to be studied further.

There is no evidence that this seasonal variation influence the digestible energy need of fish (Strand et al. 2007a, b, 2011a). It is more likely that it solely work on appetite and thereby growth. As there is strong evidence for large seasonal variations in the feeding and growth of Eurasian perch (Karås 1990; Staffan et al. 2005; Strand et al. 2007a, b, 2011a), any growth model used to predict the daily weight increase should include a seasonal factor. By doing so, the daily feed rations are adjusted so that the peak in growth during summer and the depression in autumn are accounted for.

13.5 Culture Conditions Affecting Energy Requirements and Growth

Fish held in culture are sensitive to disturbance, and feeding activity and growth may be severely reduced by exposure of the fish to stressful events such as cleaning of tanks (Head and Malison 2000; Kestemont and Baras 2001), inappropriate feeding regimes (Brännäs et al. 2001) or rearing environments (Malison and Held 1992; Brännäs et al. 2001; Papoutsoglou et al. 2000, 2005), handling (Acerete et al. 2004; Jentoft et al. 2005) and social dominance hierarchies (Brännäs et al. 2001). Culture management should therefore aim to optimize the farming environment to maximise growth and welfare of the fish.

Percid fish have been found to be disturbed by ordinary farming procedures such as handling (Acerete et al. 2004; Jentoft et al. 2005), cleaning of tanks and shadows created by human activities near tanks (Acerete et al. 2004; Jentoft et al. 2005). Strand et al. (2007a) subjected juvenile Eurasian perch to daily disturbance either by creating shadows over the tanks three times daily (moderate disturbance), or by cleaning the tanks with a brush once daily in addition to creation of shadows (severe disturbance). Both types of disturbance caused significantly lower feed intake and growth rate (up to approximately 50 %) for groups with disturbed fish compared to control groups. This is in accordance with data calculated from Jentoft et al. (2005) where disturbance of Eurasian perch and rainbow trout reduced weight increase of the fish by 46 and 27 %, respectively. In the study performed by Strand et al. (2007a), disturbed fish also demonstrated up to 40 % higher energy expenditures than the undisturbed fish. It was formerly believed that the reduced feed intake was the reason for the lower growth rate demonstrated by a stressed organism (Pickering 1993; Jobling 1994). However, the data presented by Strand et al. (2007a) show that fish exposed to disturbances also experienced increased energy expenditures compared to undisturbed fish. This is further supported by results obtained for other species such as rainbow trout (Barton and Schreck 1987) and largemouth bass (*Micropterus salmoides;* Rice 1990). Brief disturbance caused these species to increase metabolic rate by 25 % and 20 % for rainbow trout and largemouth bass, respectively. The 20 % reduction in metabolic rate for largemouth bass was also calculated to reduce weight increase by about 40 %.

One way to reduce stress in culture is to keep the fish in a suitable culture environment. Fish may demonstrate a preference for a specific background colour, probably to decrease their conspicuousness (Bradner and McRobert 2001), and in general, dark tank colours are preferred by most species (Brännäs et al. 2001). However, Strand et al. (2007b) found no effect of either tank colour (black, grey and white) or light intensity on energy expenditures of juvenile Eurasian perch. However, a clear difference in body colour of the fish was noted, with dark, almost black, perch coming from the black tanks and very pale perch coming from the white tanks. This is similar to the findings of other perch studies (Parker 1948; Mairesse et al. 2005). The lack of effect of tank colour on energy expenditures of the fish thus indicates that the capacity of perch to change body colour in accordance with its background may reduce the problem of conspicuousness and thus reduce a potential source of stress for the fish. This is in agreement with results obtained by Staffan (2004), who performed an experiment in which perch juveniles could move freely between two tanks of different colours, but did not show general preferences for any specific colour.

Tank colour, however, does affect feed intake and growth rate of perch kept at low light intensities, with reduced efficiency in tanks with darker walls (Strand et al. 2007b). The higher feed intake and corresponding higher growth rates in light, compared to in dark tanks, are suggested to be an effect of higher visibility of feed in light tanks, resulting from higher contrast between the feed and the tank's background. At high light intensity, however, the effect of tank colour was reduced and feed intake and growth rates were similar for all groups (Strand et al. 2007b). The importance of a high contrast between the food object and the background has been previously demonstrated in studies on the effect of turbidity on feeding success in fish (Fiksen et al. 1998; Utne-Palm 1999).

References

Acerete L, Balasch JC, Espinnosa E, Josa A, Tort L (2004) Physical responses in Eurasian perch (*Perca fluviatilis* L.) subjected to stress by transport and handling. Aquaculture 237:167–178

Alanärä A, Kadri S, Paspatis M (2001) Feeding management. In: Houlihan D, Jobling M, Boujard T (eds) Food intake in fish. Blackwell Science, Oxford

Azevedo PA, Cho CY, Leeson S, Bureau DP (1998) Effects of feeding level and water temperature on growth, nutrient and energy utilization and waste outputs of rainbow trout (*Oncorhynchus mykiss*). Aquat Living Resour 11:227–238

Bailey J, Alanärä A (2001) A test of a feed budget model for rainbow trout, *Oncorhynchus mykiss* (Walbaum). Aquac Res 32:465–469

Bailey J, Alanärä A (2006) Digestible energy need (DEN) of selected farmed fish species. Aquaculture 251:438–455

Barton BA, Schreck CB (1987) Metabolic cost of acute physical stress in juvenile steelhead. Trans Am Fish Soc 116:257–263

Björnsson B, Steinarsson A, Oddgeirsson M (2001) Optimal temperature for growth and feed conversion of immature cod (*Gadus morhua* L.). J Mar Sci 58:29–38

Blanchard G, Gardeur JN, Mathis N, Brun-Bellut J, Kestemont P (2008) Ultrastructural features of hepatocytes in cultured Eurasian perch (*Perca fluviatilis* L.) as affected by nutritional and husbandry conditions. Br J Nut 100:317–331

Bradner J, McRobert SP (2001) Background colouration influences body colour segregation in mollies. J Fish Biol 59:673–681

Brafield AE (1985) Laboratory studies of energy budgets. In: Tytler P, Calow P (eds) Fish energetics: new perspectives. Croom Helm, Sydney

Brännäs E, Alanärä A, Magnhagen C (2001) The social behaviour of fish. In: Keeling LJ, Gonyou HW (eds) Social behaviour in farm animals. CABI, New York

Brett JR (1979) Environmental factors and growth. In: Hoar WS, Randall DJ, Brett JR (eds) Fish physiology, vol VIII. Academic, London

Brett JR, Groves TDD (1979) Physiological energetics. In: Hoar WS, Randall DJ, Brett JR (eds) Fish physiology, vol VIII. Academic, London

Bull S, Carroll WE (1946) Principles of feeding farm animals. The Interstate Printers and Publishers, Danville

Bureau DP, Azevedo PA, Tapia-Salazar M, Cuzon G (2000) Pattern and cost of growth and nutrient deposition in fish and shrimp: potential implications and applications. In: Cruz-Suárez LE, Ricque-Marie D, Tapia-Salazar M, Olvera-Novoa MA, Civera-Cerecedo R (eds) Avances en Nutrición Acuícola V. Fish Nutrition Research Laboratory, Mérida

Cho CY (1990) Fish nutrition, feeds, and feeding: with special emphasis on salmonid aquaculture. Food Rev Inter 6:333–357

Cho CY (1992) Feeding systems for rainbow trout and other salmonids with reference to current estimates of energy and protein requirements. Aquaculture 100:107–123

Cho CY, Bureau DP (1998) Development of bioenergetic models and the Fish-PrFEQ software to estimate production, feeding ration and waste output in aquaculture. Aquat Living Resour 11:199–210

Cho CY, Slinger SJ, Bayley HS (1982) Bioenergetics of salmonid fishes: energy intake, expenditure and productivity. Comp Biochem Physiol 73B:25–41

Clarke WC, Blackburn J (1994) Effect of growth on early sexual maturation in stream-type chinook salmon (*Oncorhynchus tshawytscha*). Aquaculture 121:95–103

Cui Y, Wootton RJ (1989) Bioenergetics of growth of a cyprinid, *Phoxinus phoxinus* (L.): development and testing of a growth model. J Fish Biol 34:47–64
Cui Y, Xie S (2000) Modelling growth in fish. In: Theodorou MK, France J (eds) Feeding systems and feed evaluation models. CAB International, Wallingford
De Silva SS, Anderson TA (1995) Energetics. In: De Silva SS, Anderson TA (eds) Fish nutrition in aquaculture. St Edmundsbury Press, Suffolk
Dunning RD, Losordo TM, Hobbs AO (1998) The economics of recirculating tank systems: a spreadsheet for individual analysis. SRAC Publication No. 456. Southern Regional Aquaculture Center, Stoneville, 7 pp
Ege R, Krogh A (1914) On the relation between the temperature and the respiratory exchange in fishes. Int Rev Hydrobiol 1:48–55
Eriksson L-O, Alanärä A (1992) Timing of feeding behavior in salmonids. In: Thorpe JE, Huntingford FA (eds) The importance of feeding behavior for the efficient culture of salmonid fishes. World Aquaculture Society, Baton Rouge
Fiksen O, Utne ACW, Aksnes DL, Eiane K, Helvik JV, Sundby S (1998) Modelling the influence of light, turbulence and ontogeny on ingestion rates in larval cod and herring. Fish Ocean 7:355–363
Fiogbé ED, Kestemont P (2003) Optimum daily ration for Eurasian perch Perca fluviatilis L. reared at its optimum growing temperature. Aquaculture 216:243–252
Gould NL, Glover MM, Davidson LD, Brown PB (2003) Dietary flavor additives influence consumption of feeds by yellow perch *Perca flavescens*. J World Aquac Soc 34:412–417
Griffiths D, Kirkwood RC (1995) Seasonal variation in growth, mortality and fat stores of roach and perch in Lough Neagh, Northern Ireland. J Fish Biol 47:537–554
Hart SD, Brown BJ, Gould NL, Robar ML, Witt EM, Brown PB (2010) Predicting the optimal dietary essential amino acid profile for growth of juvenile yellow perch with whole body amino acid concentrations. Aquac Nut 16:248–253
Head AB, Malison JA (2000) Effects of lightning spectrum and disturbance level on the growth and stress response of yellow perch *Perca flavescens*. J World Aquac Soc 31:73–80
Hopkins KD (1992) Reporting fish growth: a review of the basics. J World Aquac Soc 23:173–179
Ivlev VS (1939) Energy balance in the carp. Zoolicheskii Zhurnal 18:303–318
Iwama GK, Tautz AF (1981) A simple growth model for salmonids in hatcheries. Can J Fish Aquat Sci 38:649–656
Jentoft S, Aastveit AH, Torjesen PA, Andersen Ø (2005) Effects of stress on growth, cortisol and glucose levels in non-domesticated Eurasian perch (*Perca fluviatilis*) and domesticated rainbow trout (*Oncorhynchus mykiss*). Comp Biochem Physiol A: Mol Integr Physiol 141:353–358
Jobling M (1983) Growth studies with fish-overcoming the problem of size variation. J Fish Biol 22:153–157
Jobling M (1994) Fish bioenergetics. Chapman & Hall, London
Jobling M (1997) Temperature and growth: modulation of growth rate via temperature change. In: Wood CM, McDonald DG (eds) Global warming: implications for freshwater and marine fish. Cambridge University Press, Ontario
Jobling M (2001) Nutrient partitioning and the influence of feed composition on body composition. In: Houlihan D, Jobling M, Boujard T (eds) Food intake in fish. Blackwell Science, Oxford
Jobling M, Miglavs I (1993) The size of lipid depots – a factor contributing to the control of food intake in Arctic charr, *Salvelinus alpinus*? J Fish Biol 43:487–489
Jourdan S, Fontaine P, Boujard T, Vandeloise E, Gardeur JN, Anthouard M, Kestemont P (2000) Influence of daylength on growth, heterogeneity, gonad development, sexual steroid and thyroid levels, and N and P budgets in *Perca fluviatilis*. Aquaculture 186:253–265
Juell J-E, Lekang OI (2001) The effect of feed supply rate on growth of juvenile perch (*Perca fluviatilis*). Aquac Res 32:459–464

Karås P (1990) Seasonal changes in growth and standard metabolic rate of juvenile perch, *Perca fluviatilis* L. J Fish Biol 37:913–920
Kaushik SJ (1995) Nutrient requirements, supply and utilization in the context of carp culture. Aquaculture 129:225–241
Kaushik SJ (1998) Nutritional bioenergetics and estimation of waste production in non-salmonids. Aquat Liv Res 11:211–217
Kerr SR (1971) A simulation model of lake trout growth. J Fish Res Board Can 28:815–819
Kestemont P, Baras E (2001) Environmental factors and feed intake: mechanisms and interactions. In: Houlihan D, Jobling M, Boujard T (eds) Food intake in fish. Blackwell Science, Oxford
Kestemont P, Vandeloise E, Mèlard C, Fontaine P, Brown PB (2001) Growth and nutritional status of Eurasian perch *Perca fluviatilis* fed graded levels of dietary lipids with or without added ethoxyquin. Aquaculture 203:85–99
Kestemont P, Jourdan S, Houbart M, Mélard C, Paspatis M et al (2003) Size heterogeneity, cannibalism and competition in cultured predatory fish larvae: biotic and abiotic influences. Aquaculture 227:333–356
Kitchell JF, Koonge JF, O'Neill RV, Shugart HH, Magnuson JJ, Booth RS (1974) Model of fish biomass dynamics. Trans Am Fish Soc 103:786–798
Knights B (1985) Energetics and fish farming. In: Tytler P, Calow P (eds) Fish energetics: new perspectives. Croom Helm, Sydney
Lloyd LE, McDonald BE, Crampton EW (1978) Fundamentals of nutrition, 2nd edn. Freeman, San Francisco
MacEwan G (1945) The feeding of farm animals. The Hunter-Rose Co. Limited, Toronto
Mairesse G, Thomas M, Gardeur J-N, Brun-Bellut J (2005) Appearance and technological characteristics in wild and reared Eurasian perch, *Perca fluviatilis* (L.). Aquaculture 246:295–311
Malison JA, Held JA (1992) Effects of fish size at harvest, initial stocking density and tank lightning conditions on the habituation of pond-reared yellow perch (*Perca flavescens*) to intensive culture conditions. Aquaculture 104:67–78
Mandiki SNM, Blanchard G, Mélard C, Koskela J, Kucharczyk D, Fontaine P, Kestemont P (2004) Effects of geographic origin on growth and food intake in Eurasian perch (*Perca fluviatilis* L.) juveniles under intensive culture conditions. Aquaculture 229:117–128
Mathis N, Feidt C, Brun-Bellut J (2003) Influence of protein/energy ratio on carcass quality during the growing period of Eurasian perch (*Perca fluviatilis*). Aquaculture 217:453–464
Mélard C, Kestemont P, Grignard JC (1996) Intensive culture of juvenile and adult Eurasian perch (*P. fluviatilis*): effect of major biotic and abiotic factors on growth. J App Ichthyol 12:175–180
Metcalfe NB, Thorpe JE (1992) Anorexia and defended energy levels in over-wintering juvenile salmon. J Anim Ecol 61:175–181
Ney JJ (1993) Bioenergetics modelling today: growing pains on the cutting edge. Trans Am Fish Soc 122:736–748
Papoutsoglou SE, Mylonakis G, Miliou H, Karakatsouli NP, Chadio S (2000) The effect of background colour on growth performances and physiological responses of scaled carp (*Cyprinus carpio* L.) reared in a closed circulated system. Aquac Eng 22:309–318
Papoutsoglou SE, Karakatsouli N, Chiras G (2005) Dietary L-tryptophan and tank colour effects on growth performance of rainbow trout (*Oncorhynchus mykiss*) juveniles reared in a recirculating system. Aquac Eng 32:277–284
Parker GH (1948) Colour change in vertebrates. In: Parker GH (ed) Animal colour changes and their neurohumours. Cambridge University Press, Cambridge
Pickering AD (1993) Husbandry and stress. In: Muir J, Roberts RJ (eds) Recent advances in aquaculture. Blackwell Scientific Publications, Oxford
Rice JA (1990) Bioenergetics modelling approaches to evolution of stress in fish. In: Adams SM (ed) Biological indicators of stress in fish. American Fisheries Society, Bethesda
Ricker WE (1979) Growth rates and models. In: Hoar WS, Randall DJ, Brett JR (eds) Fish physiology, vol VIII. Academic, London

Riepe JR (1997) Enterprise budgets for yellow perch production in cages and ponds in the north central region, 1994/95, vol 111, Technical bulletin series. Iowa State University, Ames

Sæther B-S, Johnsen HK, Jobling M (1996) Seasonal changes in food consumption and growth of Arctic charr exposed to either simulated natural or a 12:12 LD photoperiod at constant water temperature. J Fish Biol 48:1113–1122

Shearer KD, Silverstein JT, Plisetskaya EM (1997) Role of adiposity in food intake control of juvenile chinook salmon (*Oncorhynchus tshawytscha*). Comp Biochem Physiol 118:1209–1215

Silverstein JT, Shearer KD, Dickhoff WW, Plisetskaya EM (1999) Regulation of nutrient intake and energy balance in salmon. Aquaculture 177:161–169

Smith IP, Metcalfe NB, Huntingford FA, Kadri S (1993) Daily and seasonal patterns in the feeding behaviour of Atlantic salmon (*Salmo salar*) in a sea cage. Aquaculture 117:165–178

Staffan F (2004) Food competition and its relation to aquaculture in juvenile *Perca fluviatilis*. Dissertation, Swedish University of Agricultural Sciences

Staffan F, Magnhagen C, Alanärä A (2005) Individual feeding success of juvenile perch is consistent over time in aquaria and under farming conditions. J Fish Biol 66:798–809

Strand Å, Alanärä A, Staffan F, Magnhagen C (2007a) Effects of tank colour and light intensity on feed intake, growth rate and energy expenditure of juvenile Eurasian perch, *Perca fluviatilis* L. Aquaculture 272:312–318

Strand Å, Magnhagen C, Alanärä A (2007b) Effects of repeated disturbances on feed intake, growth rates and energy expenditures of juvenile perch, *Perca fluviatilis*. Aquaculture 265:163–168

Strand Å, Magnhagen C, Alanärä A (2011a) Growth and energy expenditures of Eurasian perch *Perca fluviatilis* (Linnaeus) in different temperatures and of different body sizes. J Aquac Res Dev 2:114

Strand Å, Overton JL, Alanärä A (2011b) The energy requirements of perch (*Perca fluviatilis* L.) in intensive culture. Aquac Nut. doi:10.1111/j.1365-2095.2010.00838.x

Talbot C (1985) Laboratory methods in fish feeding and nutritional studies. In: Tytler P, Calow P (eds) Fish energetics: new perspectives. Croom Helm, Sydney

Thodesen J, Grisdale-Helland B, Helland SJ, Gjerde B (1999) Feed intake, growth and feed utilization of offspring from wild and selected Atlantic salmon (*Salmo salar*). Aquaculture 180:237–246

Thorpe JE (1994) Reproductive strategies in Atlantic salmon, *Salmo salar* L. Aquac Fish Manag 25:77–87

Tveiten H, Johnsen HK, Jobling M (1996) Influence of maturity status on the annual cycles of feeding and growth in Arctic charr reared at constant temperature. J Fish Biol 48:910–924

Twibell DG, Brown PB (2000) Dietary choline requirement of juvenile yellow perch (*Perca flavescens*). J Nut 130:95–99

Twibell DG, Watkins BA, Brown PB (2001) Dietary conjugated linoleic acids and lipid source alter fatty acid composition of juvenile yellow perch, *Perca flavescens*. J Nut 131:2322–2328

Ursin E (1967) A mathematical model of some aspects of fish growth, respiration and mortality. J Fish Res Bd Can 24:2355–2453

Utne-Palm AC (1999) The effect of prey mobility, prey contrast, turbidity and spectral composition on the reaction distance of *Gobiusculus flavescens* to its planktonic prey. J Fish Biol 54:1244

Wootton RJ (1998) Ecology of teleost fishes, vol 24, 2nd edn, Fish and fisheries. Kluwer, Dordrecht

Xu XL, Fontaine P, Mélard C, Kestemont P (2001) Effects of dietary fat levels on growth, feed efficiency and biochemical compositions of Eurasian perch *Perca fluviatilis*. Aquac Int 9:437–449

Chapter 14
Bioenergetics Modeling of Percid Fishes

Charles P. Madenjian

Abstract A bioenergetics model for a percid fish represents a quantitative description of the fish's energy budget. Bioenergetics modeling can be used to identify the important factors determining growth of percids in lakes, rivers, or seas. For example, bioenergetics modeling applied to yellow perch (*Perca flavescens*) in the western and central basins of Lake Erie revealed that the slower growth in the western basin was attributable to limitations in suitably sized prey in western Lake Erie, rather than differences in water temperature between the two basins. Bioenergetics modeling can also be applied to a percid population to estimate the amount of food being annually consumed by the percid population. For example, bioenergetics modeling applied to the walleye (*Sander vitreus*) population in Lake Erie has provided fishery managers valuable insights into changes in the population's predatory demand over time. In addition, bioenergetics modeling has been used to quantify the effect of the difference in growth between the sexes on contaminant accumulation in walleye. Field and laboratory evaluations of percid bioenergetics model performance have documented a systematic bias, such that the models overestimate consumption at low feeding rates but underestimate consumption at high feeding rates. However, more recent studies have shown that this systematic bias was due, at least in part, to an error in the energy budget balancing algorithm used in the computer software. Future research work is needed to more thoroughly assess the field and laboratory performance of percid bioenergetics models and to quantify differences in activity and standard metabolic rate between the sexes of mature percids.

Keywords Walleye • Yellow perch • Bioenergetics • Modeling • Energy budget

C.P. Madenjian (✉)
U. S. Geological Survey, Great Lakes Science Center, 1451 Green Road,
Ann Arbor, MI 48105, USA
e-mail: cmadenjian@usgs.gov

P. Kestemont et al. (eds.), *Biology and Culture of Percid Fishes*,
DOI 10.1007/978-94-017-7227-3_14

14.1 Introduction

The intent of this book chapter is to review and characterize the current state of knowledge of bioenergetics modeling of percid fishes, and then suggest future directions for research in percid bioenergetics modeling. To accomplish this goal, I first provide a brief description of bioenergetics models for percid fishes, including a general equation for the energy budget, a brief description of the bioenergetics model components, a list of inputs to the bioenergetics models, and a list of bioenergetics models developed specifically for percids. Next, studies in which percid bioenergetics modeling has been used to investigate various aspects of percid biology and population dynamics are discussed. In addition, I provide an up-to-date synthesis on the evaluation of percid bioenergetics model performance. With this synthesis, I review the studies on assessment of percid bioenergetics model performance, interpret the results, and then draw conclusions with regard to the reliability of percid bioenergetics models. Finally, I make recommendations for future research directions for bioenergetics modeling of percid fishes.

14.2 Description of Percid Bioenergetics Models

14.2.1 Background and General Form of Models

As a consequence of the law of conservation of energy, the amount of energy ingested by a fish should equal the sum of the total energy expenditure by the fish and the fish's growth (Kitchell et al. 1977; Hewett and Johnson 1987, 1992; Hanson et al. 1997). Most of the various energy expenditures by a fish can be measured independently in the laboratory. Nonetheless, these laboratory determinations of energy expenditures may require a substantial amount of effort (Stewart 1980; Stewart et al. 1983). Typically, fish growth is more easily measured than the various energy expenditures or food consumption. Thus, fish growth is usually the most easily determined component of all of the components in the fish's energy budget. Growth trajectories can be developed for a fish population within a given aquatic ecosystem by sampling fish from the aquatic ecosystem, determining the weights and ages of the sampled fish, and then estimating a mean weight for the various ages (Ricker 1975; Hilborn and Walters 1992).

A direct measurement of the amount of food consumed by a fish in the field (i.e., in lakes, rivers, and oceans) is not easily accomplished (Hewett and Johnson 1987, 1992), because this technique typically involves examining and weighing stomach contents of fish caught in the field and then applying knowledge of gastric evacuation rates to estimate food consumption rate (Elliott and Persson 1978; Eggers 1979; Adams et al. 1982). Problems associated with this technique include: (1) the amount of food found in a fish's stomach can be highly variable within a given fish population, and therefore a considerable amount of sampling is needed to obtain a representative estimate of food consumption rate, and (2) food consumption rates typically exhibit a high degree of variability over time. Moreover, use of gastric evacuation rates to estimate food consumption rate complicates the estimation procedure because gastric evacuation rate can depend on water temperature, fish size, and potentially other factors as well (Hewett et al. 1991). Consequently, the estimate of gastric evacuation rate used to derive the food consumption rate could potentially be biased for certain combinations of water temperature and fish size. Given the above-mentioned problems, Kitchell et al. (1977) and Hewett and Johnson (1987, 1992) proposed application of fish bioenergetics models as an alternative approach toward estimation of food consumption by fish in the field.

Although the more common use of fish bioenergetics models is to estimate food consumption for an observed amount of growth in the field, fish bioenergetics models may also be used to predict fish growth for a given rate of food consumption. Predicting fish growth for a given rate of food consumption has provided insights into the importance of various factors for determining fish growth (Hayward and Margraf 1987), as well as into the effects of climate change on fish growth (Hill and Magnuson 1990).

A fish's energy budget can be depicted as:

$$G = C - R - F - U - S \tag{14.1}$$

where G = growth, C = consumption, R = respiration, F = egestion, U = excretion, and S = release of gametes (typically at spawning). Energy intake for the fish is via food consumption (C) (Fig. 14.1). Energy is lost via respiration (R), which is defined as the metabolic process by which an organism assimilates oxygen and releases carbon dioxide and other products of oxidation. Energy is also lost via egestion (F), which is the expulsion of feces, and via excretion (U), which is the discharge of urine (Fig. 14.1). When a fish matures, energy is also lost via release of milt or eggs

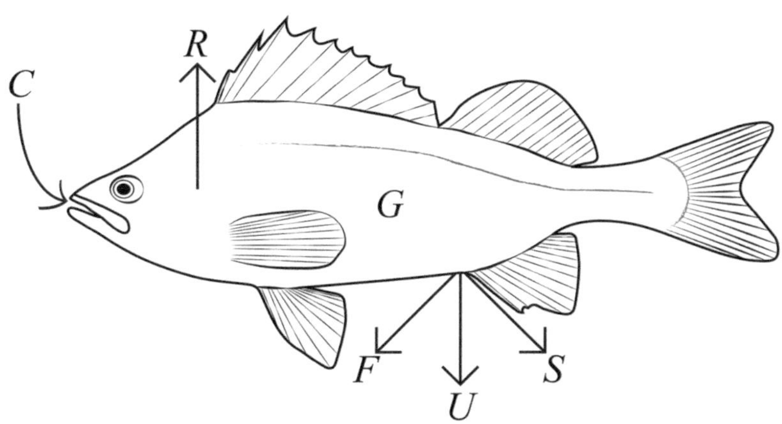

Fig. 14.1 Diagram illustrating the energy budget for a fish. *C* consumption, *G* growth, *R* respiration, *F* egestion, *U* excretion, *S* release of gametes at spawning

at the time of spawning (S). Fish growth (G) is equal to the energy intake via consumption minus all of the energy losses (R, F, U, and S) (Fig. 14.1). Equation 14.1 can be rewritten, using algebra, to isolate C on the left-hand side of the equation:

$$C = G + R + F + U + S \tag{14.2}$$

where all of the terms are defined as in Eq. 14.1. Consumption is equal to the sum of the energy incorporated into fish growth and the various energy expenditure components.

Foundations for the development of fish bioenergetics models were laid by earlier ecological energetics research work, including studies by Winberg (1960) and Phillipson (1966). Substantial progress toward development of fish bioenergetics models was made as a result of the International Biological Program (IBP) during 1964–1974, with the IBP effort culminating in the development of a bioenergetics model for bluegill (*Lepomis macrochirus*) (Kitchell et al. 1974). Additional insights into physiological energetics of fishes have been provided by other fish physiologists, notably Brett and Groves (1979).

Eventually, dynamic energy budget (DEB) theory, as proposed by Kooijman (1993, 2010), may become applicable to bioenergetics modeling of percid fishes, but much work will be needed to bridge DEB theory with traditional bioenergetics models for percids. According to DEB theory, simple quantitative rules for organization of metabolism of individual organisms can be understood from basic first principles. The DEB theory approach is abstract, the DEB models are parameter-sparse, and the DEB theory approach is generally applicable to all organisms, as well as to energy and mass fluxes in any biologically relevant environment (Kooijman 1993, 2010). However, the state variables in the DEB models are not directly measurable, which makes evaluations of such models problematic (Nisbet et al. 2012). Further, observable fluxes such as respiration rate are linear combinations of individually unobservable fluxes in DEB models. In contrast, traditional fish bioenergetics models are typically parameter-rich and specific to a fish species, but the performance of such models can be directly evaluated (Ney 1993). Nisbet et al. (2012) attempted to bridge DEB models with traditional bioenergetics models for Pacific bluefin tuna (*Thunnus orientalis*) and Pacific salmon (*Oncorhynchus* spp.), but further research will be required to link DEB theory with traditional bioenergetics models for percid fishes.

14.2.2 Model Components

For the percid fishes, standard respiration (SR), which is defined as the respiration rate of a fish with routine activity, is modeled as a function of both fish weight and water temperature (Kitchell et al. 1977). Respiration is modeled as:

$$R = ACT \cdot SR \tag{14.3}$$

where ACT = the activity multiplier. For percid fishes, egestion and excretion are typically modeled as functions of fish weight, water temperature, and feeding level (Kitchell et al. 1977). To simulate the release of gametes at spawning, mature fish lose a percentage of its body weight on the day of spawning during the computer simulation run. A 1-day time step is typically used in fish bioenergetics model simulations. Maximum consumption (C_{max}) is modeled as a function of fish weight and water temperature (Kitchell et al. 1977). The proportion of maximum consumption, known as the *P-value*, is typically assumed to be constant during a model simulation run. The *P-value* is an index of a fish's actual feeding rate relative to the maximum feeding rate possible for the fish's weight and water temperature experienced by the fish. More details on the percid bioenergetics model components are provided by Kitchell et al. (1977), Hewett and Johnson (1987, 1992), Karås and Thoresson (1992), and Hanson et al. (1997).

14.2.3 Model Inputs

The following four sets of input information are needed to apply fish bioenergetics models:

- Water temperature regime experienced by the fish
- Diet composition of the fish
- Energy density of the fish's prey
- Energy density of the fish.

Energy densities of both the fish and its prey are needed to calculate the amount of energy contained in a given weight of the fish and its prey. Because the bioenergetics model is based on an energy balance (see Eqs. 14.1 and 14.2), energy densities of both the fish and its prey are required inputs for model application. Typically, analysis of stomach contents of fish is used to determine the fish's diet composition (Hewett and Johnson 1987, 1992; Hanson et al. 1997). Alternatively, diet composition of the fish can be reconstructed based on stable isotope signatures for carbon and nitrogen (and possibly sulfur as well) in the fish and its prey, although certain assumptions are involved in this type of diet composition determination (Peterson and Fry 1987; Ciancio et al. 2010).

The two different modes of operation for fish bioenergetics models are:

- Input fish growth to estimate consumption
- Input consumption rate to estimate fish growth.

As previously mentioned, in most cases, the objective of the bioenergetics modeling is to estimate consumption of food by the fish, given a known amount of growth by the fish. However, in some cases, the objective of the bioenergetics modeling may be to predict fish growth, given a known food consumption rate for the fish.

Fish bioenergetics models can be used to estimate consumption by a single fish or to estimate consumption by a fish population (Hewett and Johnson 1987, 1992;

Hanson et al. 1997). Information on population size and population mortality rate is required for application of the fish bioenergetics model to a fish population. Regardless of whether the bioenergetics model is applied to a single fish or a fish population, the growth trajectory for an average fish within the fish population is typically used to estimate consumption given a known amount of growth (Hewett and Johnson 1987, 1992). For the typical application of the fish bioenergetics model to a fish population, population size is multiplied by the daily food consumption for an average-sized fish to estimate food consumption by the fish population on each day of the simulation.

14.2.4 Bioenergetics Models Developed Specifically for Percids

Kitchell et al. (1977) developed bioenergetics models for yellow perch (*Perca flavescens*) and walleye (*Sander vitreus*) (Table 14.1). Their yellow perch and walleye bioenergetics models differed only by the optimum and maximum temperatures used to model maximum consumption and respiration. Optimum and maximum temperatures for walleye were lower than those for yellow perch, reflecting the walleye's preference for lower water temperature compared with yellow perch (Kitchell et al. 1977). The respiration component of both models was partially based on measurements of standard respiration rate of Eurasian perch (*Perca fluviatilis*) by Solomon and Brafield (1972). The activity multiplier *ACT* was assigned a value of 1.0 for both models (Kitchell et al. 1977). The Kitchell et al. (1977) bioenergetics models were very slightly modified in preparing them for a computer software

Table 14.1 List of bioenergetics models developed specifically for percids

Species	Lifestages	Comments	Reference(s)
Yellow perch	Larger (>2 g) juveniles through adults	–	Kitchell et al. (1977), Hewett and Johnson (1987)
Walleye	Larger (>2 g) juveniles through adults	–	Kitchell et al. (1977), Hewett and Johnson (1987)
Eurasian perch	Larger (>2 g) juveniles through adults	–	Karås and Thoresson (1992)
Zander	Larger (>2 g) juveniles through adults	–	Keskinen et al. (2008)
Yellow perch	Larvae through smaller (<2 g) juveniles	Partial model (respiration component)	Post (1990)
Walleye	Larvae through smaller (<2 g) juveniles	–	Madon and Culver (1993)

package of fish bioenergetics models (Hewett and Johnson 1987), and the model formulations from Hewett and Johnson (1987) are typically referred to as the Kitchell et al. (1977) bioenergetics models for yellow perch and walleye. The suite of fish bioenergetics models included in the Hewett and Johnson (1987) computer software package are referred to as Wisconsin fish bioenergetics models because these models were developed by researchers associated with the University of Wisconsin Center for Limnology in Madison, Wisconsin (USA). Thus, the bioenergetics models for yellow perch and walleye appearing in the Hewett and Johnson (1987) software package are synonymous with the Wisconsin bioenergetics models for yellow perch and walleye, respectively. The computer software package by Hewett and Johnson (1987) was updated in 1992 (Hewett and Johnson 1992) and in 1997 (Hanson et al. 1997), but in both cases the bioenergetics models for yellow perch and walleye remained unchanged from the model formulations appearing in Hewett and Johnson (1987).

Karås and Thoresson (1992) made substantial modifications to the Kitchell et al. (1977) yellow perch bioenergetics model, and proposed that their modified model was a more accurate depiction of Eurasian perch bioenergetics than the Kitchell et al. (1977) yellow perch bioenergetics model (Table 14.1). Consequently, Karås and Thoresson (1992) recommended that their model be used for Eurasian perch rather than the Kitchell et al. (1977) yellow perch bioenergetics model and suggested that their model may also be a more accurate depiction of yellow perch bioenergetics than the Kitchell et al. (1977) model. Modifications made by Karås and Thoresson (1992) primarily involved revising the relationships between maximum consumption and water temperature and between respiration and water temperature. Karås and Thoresson (1992) argued that the Kitchell et al. (1977) yellow perch model was developed using a mixture of data from Eurasian perch, yellow perch, and walleye, and therefore more accurate predictions for Eurasian perch would be expected from their model, based solely on Eurasian perch data, rather than from the Kitchell et al. (1977) yellow perch model.

Keskinen et al. (2008) made substantial modifications to the Wisconsin walleye bioenergetics model to develop a bioenergetics model for zander (*Sander lucioperca*), a close relative of the walleye (Table 14.1). Keskinen et al. (2008) revised the relationships between maximum consumption and water temperature and respiration and water temperature. A slight modification to the egestion and excretion components was also made. These researchers concluded that their modified model was a more accurate depiction of zander bioenergetics than the Kitchell et al. (1977) model.

Because application of the Hewett and Johnson (1987) bioenergetics models for yellow perch and walleye to larvae and small (<2 g wet weight) juveniles appeared to be yielding inaccurate estimates of food consumption, new models for larvae and small juveniles were fully developed for walleye by Madon and Culver (1993) and partly developed for yellow perch by Post (1990) (Table 14.1). In constructing these new models, these researchers used some of the same parameters used by Kitchell et al. (1977), but also attempted to measure respiration of larval and small juvenile fish. In contrast, Solomon and Brafield (1972) measured standard respiration of larger juvenile and adult Eurasian perch. Thus, both Post (1990) and Madon and Culver (1993) developed a new submodel for the respiration component of the bioenergetics model. In addition, Madon and Culver (1993) estimated maximum consumption for larval and small juvenile walleye. Again, in contrast, all of the data used by Kitchell et al. (1977) in developing the yellow perch and walleye bioenergetics models were for larger juvenile and adult fish. Post (1990) and Madon and Culver (1993) concluded that the Hewett and Johnson (1987) bioenergetics models for yellow perch and walleye, respectively, were inappropriate for larval fish and small juveniles.

14.3 Applications of Percid Bioenergetics Models

14.3.1 Overview

In this section, I present examples of applications of percid bioenergetics models used to address issues and problems in fish ecology and fishery management. These applications can be grouped into the following four categories: (1) understanding percid growth and consumption in ecosystems, (2) quantifying the trophic link between predator and prey populations, (3) quantifying the role of percids in cycling nutrients within ecosystems, and (4) understanding contaminant accumulation in percids. My intention was not to cover all of the percid bioenergetics model applications that have been conducted, but rather to select a few applications from each of the categories to illustrate the usefulness of the bioenergetics model approach. For each of the examples presented in this section except the Kraft (1992) application to larval yellow perch, the percid bioenergetics models were applied to larger juvenile and adult fish. Further, in all applications except the Kraft (1992) application to larval yellow perch, either the yellow perch bioenergetics model or walleye bioenergetics model, as presented in Hewett and Johnson (1987), was used. For larval yellow perch, Kraft (1992) applied the yellow perch bioenergetics model developed by Post (1990).

14.3.2 Understanding Percid Growth and Consumption in Ecosystems

To explain the difference in yellow perch growth between the western basin and central basin of Lake Erie, Hayward and Margraf (1987) applied the Wisconsin yellow perch bioenergetics model (Table 14.2). Two hypotheses were considered in the explaining the faster growth observed in the central basin: (1) the water temperature regime in the central basin was more favorable for yellow perch growth than that in the western basin, and (2) food availability was greater in the central basin. These researchers fit the bioenergetics model to observed yellow perch growth in the western basin to estimate the proportions of maximum consumption (*P-values*) for

Table 14.2 Examples of percid bioenergetics model applications

Population(s)	Application	Reference
Yellow perch, Lake Erie	Wisconsin yellow perch bioenergetics model was applied to determine the relative effects of water temperature and food availability on yellow perch growth in the western and central basins of Lake Erie	Hayward and Margraf (1987)
Walleye, Lake Erie	Wisconsin walleye bioenergetics model was applied to determine the effects of walleye movement, and water temperature regimes associated with these movements, on walleye growth and food consumption in Lake Erie	Kershner et al. (1999)
Yellow perch, Lake Erie	Wisconsin yellow perch bioenergetics model was applied to forecast the effects of climate change on yellow perch growth and food consumption in Lake Erie	Hill and Magnuson (1990)
Yellow perch, Quebec lakes	Wisconsin yellow perch bioenergetics model was applied to estimate the activity multiplier *ACT* for yellow perch populations in each of the 12 lakes	Boisclair and Leggett (1989)
Walleye, Lake Erie	Wisconsin walleye bioenergetics model was applied to map the potential for walleye growth in central Lake Erie in response to hypoxia	Brandt et al. (2011)
Walleye, Lake Erie	Wisconsin walleye bioenergetics model was coupled with walleye population size and mortality estimates to estimate annual prey fish consumption by the walleye population in western Lake Erie, 1986–1988	Hartman and Margraf (1992)
Walleye, Lake Erie	Wisconsin walleye bioenergetics model was coupled with walleye population size and mortality estimates to estimate annual prey fish consumption by the walleye population in Lake Erie, 1986–1995	Kershner et al. (1999)
Walleye, Lake Erie	Wisconsin walleye bioenergetics model was coupled with walleye population size and mortality estimates to estimate annual consumption of age-0 yellow perch by the walleye population in western Lake Erie	Hartman and Margraf (1993)
Walleye, Lake Mendota	Wisconsin walleye bioenergetics model was coupled with a walleye population model to estimate annual consumption of prey fish by the walleye population in Lake Mendota	Johnson et al. (1992)
Walleye, Sparkling Lake	Wisconsin walleye bioenergetics model was coupled with walleye population size and mortality estimates to estimate annual consumption of prey fish by the walleye population in Sparkling Lake	Lyons and Magnuson (1987)
Yellow perch, Lake Memphremagog	Post (1990) model and Wisconsin yellow perch bioenergetics model were applied to larval yellow perch population and age-2 and older yellow perch population, respectively, to determine nitrogen and phosphorus excretion rates by these two populations in Lake Memphremagog	Kraft (1992)
Walleye, South Manistique Lake	Wisconsin walleye bioenergetics model was used to determine the contribution of the growth dilution effect toward the difference in PCB concentrations between the sexes of mature walleyes from South Manistique Lake	Madenjian et al. (2009)

combinations of weeks during the growing season and yellow perch age. Then, they inputted these *P-values* into the bioenergetics model, but used the central basin water temperature regime, to simulate yellow perch growth under the scenario of western basin food availability coupled with central basin water temperatures. The result was very little or no improvement in yellow perch growth from observed growth in the western basin. Hayward and Margraf (1987) also weighed stomach contents of yellow perch and used a gastric evacuation model to estimate daily food consumption. Results clearly showed that daily food consumption was substantially higher in the central basin compared with the western basin. Hayward and Margraf (1987) concluded that the slower growth of western basin yellow perch was due to the paucity of larger-sized benthic invertebrates in the western basin relative to the central basin rather than the water temperature regime in the central basin being more favorable for yellow perch growth.

Kershner et al. (1999) explored the effects of water temperature on walleye growth in Lake Erie using the Wisconsin walleye bioenergetics model. These researchers fixed the *P-value* to 0.4, and then simulated walleye growth under both a western basin water temperature regime and a central basin water temperature regime. Although walleye subjected to the western basin water temperatures consumed more food than walleye subjected to central basin water temperatures, walleye growth was slower in the western basin than in the central basin because summer water temperatures in the western basin tended to exceed the optimal temperature for walleye food consumption to a greater degree than in the central basin, leading to greater limitation of summer growth of walleye in the western basin. Kershner et al. (1999) also simulated the growth of walleye moving from the western basin to the central basin in June and then returning to the western basin in September, and they found that walleye growth and consumption under this temperature regime exceeded those under either the western basin temperature regime or the central basin temperature regime. Further, Kershner et al. (1999) noted that walleye tagging studies conducted in Lake Erie indicated movement of a portion of the population from the western basin to the central basin in early June, and they proposed that this movement could be explained by the higher growth rates achieved by this migratory behavior.

Bioenergetics modeling is especially appropriate for predicting the effects of climate change on growth and food consumption by fish, because the predicted change in the water temperature regime due to climate change can be directly inputted into the bioenergetics model to forecast changes in growth and consumption under climate change scenarios. Hill and Magnuson (1990) used this approach to predict changes in growth and food consumption by yearling yellow perch in western Lake Erie under climate change (Table 14.2). Water temperatures were predicted under the scenario of a doubling of the present-day carbon dioxide concentration in the Earth's atmosphere. The two water temperature regimes for the future that were constructed by these researchers included one allowing for thermoregulation by the yellow perch and the other not allowing for thermoregulation. Present-day growth of yellow perch in Lake Erie was attained by assigning the *P-value* equal to 0.45. Assuming the *P-value* was equal to 0.45 in the future, yellow perch growth increased by 30 % but decreased by 35 % from the present-day level under the thermoregulation and no-thermoregulation climate change scenarios, respectively. If the mass of food consumed in the future was equal to the present-day consumption, then yellow perch growth decreased by 30 % and 65 % from the present-day level under the thermoregulation and no-thermoregulation climate change scenarios, respectively.

Boisclair and Leggett (1989) used the Wisconsin yellow perch bioenergetics model to estimate the activity multiplier *ACT* for 28 combinations of age class and population, with yellow perch populations from 12 different lakes being sampled. These researchers caught yellow perch from all 12 lakes, weighed the stomach contents, and applied a gastric evacuation model to estimate daily food consumption. Then, for each of the 28 combinations, they adjusted the value of *ACT*, while inputting the estimated food consumption into the bioenergetics model, until predicted yellow perch growth matched observed yellow perch growth. Estimates of *ACT* ranged from 0.1 to 2.9, with an average value of 1.8. Boisclair and Leggett (1989) concluded that yellow perch activity varied substantially across age classes and populations, and that assigning *ACT* a value of 1 when applying the bioenergetics model may yield inaccurate estimates of food consumption. Hewett et al. (1991) commented that the high variability in these *ACT* estimates was partly attributable to the impreciseness of the gastric evacuation technique. Moreover, Hewett et al. (1991) suspected that the gastric evacuation technique yielded inaccurate estimates of daily food consumption because yellow perch weight was not factored into the calculation of evacuation rate, with the result being overestimation of daily food consumption. In turn, overestimation of food consumption led to overestimation of *ACT*. In addition, energy density of yellow perch was not determined in any of the 12 lakes. Rather, for all 28 combinations of age class and population, yellow perch energy density was assumed to be equal to a value obtained from the literature, and yellow perch energy density was assumed to remain constant over time. Hewett et al. (1991) contended that these gross assumptions about yellow perch energy density further contributed to the inaccuracy of the *ACT* estimates.

Brandt et al. (2011) applied the Wisconsin walleye bioenergetics model to Lake Erie walleye to map the potential for walleye growth in central Lake Erie in response to hypoxia (Table 14.2). These researchers used the bioenergetics model to estimate the growth rate potential (GRP) for age-3 walleye in grids along transects through the central basin of Lake Erie, based on hydroacoustic estimates of prey fish density, water temperature, light intensity, and dissolved oxygen (DO) concentration. A feeding model was developed to estimate food consumption rate for the walleye as a function of prey fish density and light intensity. The estimated food consumption rate was then adjusted by a DO concentration factor, which ranged from 0 to 1 in value. The DO concentration factor approached a value of 1 as DO concentration increased above 4 mg O_2/l, and approached a value of 0 as DO concentration decreased below 2 mg O_2/l. This adjusted feeding rate was then inputted into the walleye bioenergetics model, along with water temperature and walleye weight, to estimate the walleye daily growth rate, and GRP was assigned the value of the estimated walleye daily growth rate. GRP was then mapped for each of the grids in each transect. Results indicated that a hypoxic event during September 2005 caused a slight (<2 %) decrease in the availability of high quality habitat, defined as GRP > 0 $g \cdot g^{-1} \cdot day^{-1}$, for walleye in central Lake Erie. Nevertheless, mean GRP was substantially higher in September (during the hypoxic event) than during August (pre-hypoxia) or October (post-hypoxia), and this higher GRP was attributed to the concentration of prey within favorable temperature, DO, and light conditions in response to the hypoxia. Brandt et al. (2011) concluded that their results do not suggest that hypoxia is negatively influencing walleye through reduced habitat quality.

14.3.3 Quantifying the Trophic Link Between Percid Populations and Their Prey

Bioenergetics modeling can be used to estimate annual consumption of a particular species of prey by a predator population, thereby enabling the quantification of the trophic link between predator and prey populations. This type of bioenergetics model application has value in both basic and applied science contexts. From a basic science standpoint, quantifying this trophic link allows the ecologist to assess the role of the predator in structuring the ecosystem (Carpenter et al. 1985). From an applied science perspective, construction of a food web model, using bioenergetics modeling, can eventually be valuable in making decisions on the management of important populations within the ecosystem (Pauly et al. 2000). Further, quantification of the trophic link between piscivore and prey fish populations can be directly used by fishery managers in deciding stocking rates for piscivores. For example, bioenergetics modeling used to estimate annual consumption of alewives (*Alosa pseudoharengus*) by salmonines in Lakes Michigan and Ontario has played a role in guiding fishery management decisions to reduce Chinook salmon (*Oncorhynchus tshawytscha*) stocking rates since the 1980s (Stewart et al. 1981; Jones et al. 1993; Hansen et al. 1993; Madenjian 2011a). Reductions in stocking rates were aimed at balancing predator demand with supply of prey fish.

Hartman and Margraf (1992) applied the Wisconsin walleye bioenergetics model to the walleye population in western Lake Erie to estimate that the walleye population was annually consuming between 84,000 and 94,000 metric tons of prey fish during years 1986–1988. In addition, these researchers used the bioenergetics model to predict the reduction in prey fish consumption with 5 %, 10 %, and 20 % increases in the fishing mortality rates for walleye in western Lake Erie. Results from these bioenergetics model simulations indicated that increases in walleye fishing mortality had only modest effects on annual prey fish consumption, as annual prey fish consumption decreased by only 3 %, 7 %, and 15 % corresponding to the 5 %, 10 %, and 20 % increases, respectively, in walleye fishing mortality. Kershner et al. (1999) applied the Wisconsin walleye bioenergetics model to the Lake Erie adult (age-2 and older) walleye population to estimate that lakewide annual food consumption by the adult walleye population ranged from roughly 100,000 to 500,000 metric tons during years 1986–1995 (Table 14.2). These researchers concluded that the predatory demand can vary substantially with walleye population size and age structure. Hartman and Margraf (1993) applied the Wisconsin walleye bioenergetics model to the young (age-2 and younger) walleye population in western Lake Erie to estimate annual consumption of age-0 yellow perch by the young walleye population during 1988. Modeling results indicated that young walleyes ate 6.82 billion age-0 yellow perch in western Lake Erie during June and July of 1988. Estimates of the number of age-0 yellow perch in western Lake Erie in early June 1988 ranged from 7.6 to 24.0 billion, and therefore young walleyes ate between 28 % and 90 % of the age-0 yellow perch in western Lake Erie during June–July 1988. Hartman and Margraf (1993) concluded that walleye predation on age-0 yellow perch could be an important determinant of yellow perch year-class strength in Lake Erie during some years.

Johnson et al. (1992) coupled the Wisconsin walleye bioenergetics model with a walleye population model to estimate the annual consumption of prey fish by the walleye population Lake Mendota, Wisconsin (USA) (Table 14.2). In their simulations, these researchers varied walleye stocking rates and harvest regulations on walleye to determine effects of such variations on prey fish consumption. Results indicated that annual consumption of age-0 and age-1 yellow perch by the Lake Mendota walleye population increased as the legal minimum size for harvesting a walleye increased. Further, at a walleye stocking rate equivalent to 8,000 yearlings per year combined with a 381-mm minimum size limit for harvest of a walleye, simulation results indicated annual consumption of yellow perch by the walleye population would be sufficient to reduce yellow perch recruitment in Lake Mendota in most years. Lyons and Magnuson (1987) applied the Wisconsin walleye bioenergetics model to the walleye population in Sparkling Lake, a small lake in northern Wisconsin (USA), to estimate annual consumption of age-0 yellow perch, darters, and minnows during years 1982–1983. According to the modeling results, in 1982, when age-0 yellow perch were scarce, walleye predation accounted for practically all of the adult darter annual mortality (estimated at 88 %) and 75 % of the adult minnow annual mortality (estimated at 54 %). In 1983, when age-0 yellow perch were abundant, walleye predation accounted for 80 % of the adult darter annual mortality (estimated at 71 %), but only 35 % of the adult minnow annual mortality (estimated at 68 %). Lyons and Magnuson (1987) concluded that an abundant age-0 year-class of yellow perch buffers darters and minnows from predation by walleyes in Sparkling Lake, and that walleye predation had a much greater impact on the darter population than on the minnow population.

14.3.4 Quantifying the Role of Percids in Cycling Nutrients Within Ecosystems

Wisconsin fish bioenergetics models are well suited to estimate excretion of nutrients by fish, because excretion is a specified component of these models (Kraft 1992). Nitrogen and phosphorus are essential nutrients for growth of phytoplankton and other aquatic plants, and fish excrete both of these nutrients in their urine.

Phosphorus has been identified as the limiting nutrient for primary production in freshwater ecosystems, whereas nitrogen is regarded as the limiting nutrient in many marine ecosystems. Fish bioenergetics modeling can be used to estimate the amounts of phosphorus and nitrogen excreted by a population of fish within an aquatic ecosystem. In some cases, fish bioenergetics modeling has shown that certain fish populations play an important role in cycling nutrients within some aquatic ecosystems, whereas results for other aquatic ecosystems have indicated that certain fish populations have little effect on nutrient cycling within the ecosystem. For example, Kraft (1993), using the Wisconsin alewife bioenergetics model, estimated that the alewife population in Lake Michigan during the mid-1970s regenerated phosphorus via egestion and excretion at the same rate as the zooplankton population in Lake Michigan. Thus, he concluded that alewives played an important role in phosphorus recycling within the Lake Michigan ecosystem during the mid-1970s. In contrast, Bunnell et al. (2005), using a round goby (*Neogobius melanostomus*) bioenergetics model, estimated that the round goby population in central Lake Erie excreted less than 1 % of the phosphorus needed by the benthic community for primary production. These researchers concluded that the round goby invasion of central Lake Erie had very little influence on phosphorus cycling within the central Lake Erie ecosystem.

Kraft (1992) applied the Post (1990) bioenergetics model and the Wisconsin yellow perch bioenergetics model to the larval yellow perch population and age-2 and older yellow perch population, respectively, in Lake Memphremagog (Vermont, USA and Quebec, Canada) to determine the nitrogen (N) and phosphorus (P) excretion rates for both populations in the lake during June through October (Table 14.2). He estimated that volumetric phosphorus excretion rate by the age-0 yellow perch population in the lake peaked at 0.3 $\mu g\ P \cdot l^{-1} \cdot day^{-1}$, which was nearly an order of magnitude higher than the peak rate for the age-2 and older yellow perch population. Kraft (1992) concluded that: (1) the age-0 yellow perch population in Lake Memphremagog played a more significant role in limnetic nutrient cycling than the adult yellow perch population, (2) the ratio of excreted N to P by yellow perch in the lake could vary fourfold during the course of the summer, and (3) the age-0 yellow perch population in the lake could serve as an important phosphorus sink.

14.3.5 Understanding Contaminant Accumulation in Percids

Bioenergetics modeling has played a key role in identifying the most important factors regulating contaminant accumulation in fish in aquatic ecosystems (Madenjian 2011a). At one time, direct uptake of polychlorinated biphenyls (PCBs) from the water into fish via the gills was believed to be the major pathway of PCBs into fish living in lakes, rivers, and oceans. However, Weininger (1978) applied the Wisconsin lake trout (*Salvelinus namaycush*) bioenergetics model to Lake Michigan lake trout to show that nearly all of the PCBs in an adult lake trout from Lake Michigan entered the lake trout via food consumption. The Weininger (1978) model for PCB accumulation in lake trout allowed for direct uptake of PCBs from the water, using the best available information on the kinetics of direct uptake of PCBs from the water and the best available estimate of the PCB concentration in Lake Michigan water, yet the modeling results indicated that food consumption was responsible for nearly all of the PCB body burden in Lake Michigan lake trout. Since the time of the Weininger (1978) application, application of new technology for measuring PCB concentrations in water has revealed that the PCB concentration in Lake Michigan water had been overestimated, and therefore the importance of direct uptake of PCBs from water was even less than that assessed by Weininger (Madenjian et al. 1993).

Bioenergetics modeling has been useful in quantifying the effect of the difference in growth rates between the sexes on PCB accumulation in walleye (Madenjian et al. 1998, 2009). Female walleyes grow considerably faster than male walleyes, with mean weight at age for mature females ranging from 20 % to 50 % higher than that for mature males. All other factors being equal, faster-growing fish have lower contaminant concentrations than slower-growing fish, and this phenomenon is referred to as the growth dilution effect. Madenjian et al. (2009) applied the Wisconsin walleye bioenergetics model to walleye from South Manistique Lake (Michigan, USA) to estimate food consumption by both male and female walleye. Respiration rates and activity are assumed to not differ between the sexes in the Wisconsin fish bioenergetics models (Hewett and Johnson 1987, 1992). Nevertheless, these bioenergetics models can be used to determine the effect of differences in growth between the sexes on PCB accumulation in fish (Madenjian 2011b). Bioenergetics modeling results indicated that the difference in growth rate between the sexes could account for mature male walleyes being 12 % higher in PCB concentration than mature female walleyes from South Manistique Lake. Interestingly, mature male walleyes were actually 34 % higher in PCB concentration than mature female walleyes. Madenjian et al. (2009) attributed this 34 % higher PCB concentration in the mature males to not only a higher growth rate in the females, but also to a higher rate of energy expenditure in the males stemming from higher activity in males and a higher resting metabolic rate in the males. This characteristic of a higher energy expenditure rate in males may be common to most fish populations around the world (Madenjian 2011b).

14.4 Evaluations of Percid Bioenergetics Model Performance

14.4.1 Overview

Fish bioenergetics models can be evaluated in both the field and laboratory (Chipps and Wahl 2008), and advantages and disadvantages are associated with either approach (Madenjian et al. 2010). To thoroughly test the performance of a fish bioenergetics model, use of both approaches is recommended (Madenjian et al. 2000). Unless otherwise specified, the evaluations of percid bioenergetics model performance discussed in this section pertain to larger juvenile and adult fish.

The traditional approach for evaluating a fish bioenergetics model in the field involves capturing fish in the field, weighing the stomach contents, and applying a gastric evacuation rate model to estimate daily food consumption. Use of a gastric evacuation model could result in biased estimates of food consumption rate for certain combinations of water temperature and fish size; please refer to Sects. 14.2.1 and 14.3.2 for more details. One advantage of the field approach is that the energy budget, including activity, for the fish in its natural environment is tested. An alternative method for evaluating fish bioenergetics models in the field is to use PCBs as a tracer for food consumption (Madenjian et al. 2000). This method involves determining net trophic transfer efficiency (γ), which is the efficiency with which the fish retains PCBs, in the laboratory, and estimating γ in the field using the fish bioenergetics model. The ratio of the laboratory estimate of γ to the field estimate of γ furnishes an index of the accuracy of the bioenergetics model's estimates of food consumption in the field. Disadvantages of this method are discussed by Madenjian et al. (2000). Because γ may depend on feeding rate, the feeding rates selected in the laboratory should bracket the feeding rate in the field (Madenjian et al. 2000). Another alternative to evaluating fish bioenergetics models in the field is to use mercury as a tracer of food consumption (Trudel et al. 2000; Keskinen et al. 2008). One caveat associated with this mercury technique is that estimates of food consumption have been shown to be sensitive to the estimate of the rate of elimination of mercury from the fish's body, and previously derived estimates of mercury elimination rates for fish have been found to be inaccurate (Van Walleghem et al. 2007; Madenjian et al. 2012a).

Laboratory evaluation of fish bioenergetics models is typically accomplished by recording food consumption and growth by fish in laboratory tanks, and then assessing the accuracy of the fish bioenergetics model predictions of food consumption and growth (Whitledge and Hayward 1997). The main advantage of this approach is that food consumption and growth can be directly measured in the laboratory. Disadvantages include: (1) activity in the laboratory tanks may be substantially less than the activity in the field, (2) high feeding rates attained by fish in the field may not be easily attained by fish confined to laboratory tanks, and (3) artifacts of fish confined to laboratory tanks may confound interpretation of the model evaluation results in some cases (Madenjian et al. 2004, 2012b).

14.4.2 Field Evaluations

Schaeffer et al. (1999) evaluated the field performance of both the Wisconsin yellow perch bioenergetics model and the Karås-Thoresson bioenergetics model applied to yellow perch in Saginaw Bay (Lake Huron, USA) (Table 14.3). Yellow perch were captured in Saginaw Bay, stomach contents were weighed, and the gastric evacuation rate model by Persson (1979) was applied to estimate daily food consumption. Results indicated that the Wisconsin yellow perch bioenergetics model predictions of daily ration were lower than the daily ration estimated by application of gastric

Table 14.3 List of studies evaluating performance of percid bioenergetics models

Model	Type of evaluation	Reference
Yellow perch (Kitchell et al. 1977; Hewett and Johnson 1987)	Field evaluation	Schaeffer et al. (1999)
Eurasian perch (Karås and Thoresson 1992)	Field evaluation	Schaeffer et al. (1999)
Walleye (Kitchell et al. 1977; Hewett and Johnson 1987)	Field evaluation	Lantry et al. (2008)
Zander (Keskinen et al. 2008)	Field evaluation	Keskinen et al. (2008)
Yellow perch (Post 1990)	Field evaluation	Worischka and Mehner (1998)
Walleye (Madon and Culver 1993)	Field evaluation	Worischka and Mehner (1998)
Yellow perch (Kitchell et al. 1977; Hewett and Johnson 1987)	Laboratory evaluation	Bajer et al. (2003)
Eurasian perch (Karås and Thoresson 1992)	Laboratory evaluation	Bajer et al. (2003)
Walleye (Kitchell et al. 1977; Hewett and Johnson 1987)	Laboratory evaluation	Madenjian et al. (2010)
Walleye (Kitchell et al. 1977; Hewett and Johnson 1987)	Laboratory evaluation	Madenjian and Wang (2013)
Zander (Keskinen et al. 2008)	Laboratory evaluation	Keskinen et al. (2008)

evacuation rates to weight of stomach contents from yellow perch caught in the field when water temperatures exceeded 22 °C, such that the model predictions of cumulative consumption over the growing season were 25–50 % lower than the field estimates. Karås-Thoresson model predictions were in good agreement with the field estimates over the full range of water temperatures (from 10 to 26 °C) for ages 1–3 yellow perch, but the Karås-Thoresson model predictions were lower than the field estimates for age-4 yellow perch. Schaeffer et al. (1999) concluded that the Karås-Thorreson model produced better fits to the field estimates than the Wisconsin yellow perch bioenergetics model. However, these researchers acknowledged that the field estimates of daily ration may be in error, especially at water temperatures above 22 °C. Persson (1979) measured rates of gastric evacuation of Eurasian perch in the laboratory for water temperatures ranging from 4 to 22 °C, and observed an exponential increase over this range. But, he cautioned potential users of his gastric evacuation rate model that gastric evacuation rate would likely level off at temperatures exceeding 22 °C. Nonetheless, Schaeffer et al. (1999) extrapolated the exponential curve developed by Persson (1979) to water temperatures up to 26 °C to estimate gastric evacuation rates at these higher temperatures. Persson (1979) also cautioned that gastric evacuation rate may also vary with Eurasian perch size, but he neither measured gastric evacuation rate over a wide range of Eurasian perch sizes nor included Eurasian perch size in his gastric evacuation rate model. All of the bioenergetics model predictions in the Schaeffer et al. (1999) study were generated using the Hewett and Johnson (1992) software package.

Using an approach similar to that of Schaeffer et al. (1999), Lantry et al. (2008) compared the field estimates of daily ration with the Wisconsin walleye bioenergetics model predictions of daily ration for walleye caught in Oneida Lake, New York (USA) (Table 14.3). The Forney (1977) model, in which gastric evacuation rate was a function of prey weight, was used to estimate gastric evacuation rates of the walleye. Neither water temperature nor walleye size was included in the model. The model was developed based on data for water temperatures ranging from 16.8 to 23.5 °C. Predictions of daily ration from the Wisconsin bioenergetics model tended to be higher than the field estimate when the field daily ration was less than 3.7 % of the walleye weight, whereas bioenergetics model predictions tended to be lower than the field estimate when the field daily ration was greater than 3.7 % of the walleye weight. Lantry et al. (2008) noted that their results mimicked the pattern observed by Bajer et al. (2003), who performed a laboratory evaluation of the Wisconsin yellow perch bioenergetics model and the Karås-Thoresson bioenergetics model. Bajer et al. (2003) concluded that a systematic bias was apparent in both models, whereby the models overestimate food consumption when feeding rate is low and underestimate food consumption when feeding rate is high. Apparently, the bioenergetics model predictions were generated using the Hanson et al. (1997) software package for both the Lantry et al. (2008) study and Bajer et al. (2003) study.

Keskinen et al. (2008) used mercury as a tracer of food consumption to evaluate the performance of their zander bioenergetics model in two Finnish lakes (Table 14.3). Initially, these researchers evaluated their zander bioenergetics model in the laboratory and noted that the model underestimated consumption at high feeding rates. Consequently, they developed a correction factor to apply to the model predictions of consumption, and applied their model, including the correction factor, to zander in both lakes. Mercury concentrations in skinless fillets of zander and in whole prey fish, including European smelt (*Osmerus eperlanus*) and Eurasian perch, were determined. Predictions of zander mercury concentration, based on their corrected zander bioenergetics model, matched the observed mercury concentrations of zander in both lakes, and Keskinen et al. (2008) concluded that their corrected zander bioenergetics model was providing reasonably accurate estimates of food consumption by zander in the field. Because *ACT* was assigned a value of 1 in their zander bioenergetics model, Keskinen et al. (2008) also concluded that zander exhibited routine activity in the field. Energy densities of zander and their prey in the two Finnish lakes were not determined by these researchers. Rather, energy densities from the literature were used in their bioenergetics modeling. Apparently, bioenergetics modeling predictions were generated using the Hanson et al. (1997) software package in the Keskinen et al. (2008) study.

Worischka and Mehner (1998) compared field estimates of daily ration with predictions of food consumption from the larval fish bioenergetics models for yellow perch and walleye developed by Post (1990) and Madon and Culver (1993), respectively (Table 14.3). These researchers captured larval Eurasian perch, a close relative of yellow perch, and zander, a close relative of the walleye, from a man-made lake in Germany, weighed the stomach contents, and then used gastric evacuation rate models by Jobling (1981) and Elliott and Persson (1978) to estimate daily food consumption. Fish larvae were sampled every 30 min over the course of a 24-h period. Worischka and Mehner (1998) concluded that larval fish models were accurate when *ACT* was assigned a value of 1, rather than the values of 4.4 and 3 estimated by Post (1990) and Madon and Culver (1993) for larval yellow perch and larval walleye, respectively. Both Post (1990) and Madon and Culver (1993) sampled fish larvae every 3 h over a 24-h period to estimate daily food ration. Worischka and Mehner (1998) showed that sampling fish larvae every 3 h can lead to substantial overestimation of gastric evacuation rate, which in turn would lead to overestimation of both food consumption and *ACT*. Worischka and Mehner (1998) also pointed out that their conclusion of *ACT* being equal to 1 agreed with the conclusion reached by Karjalainen et al. (1997), who had also evaluated the larval yellow perch bioenergetics model in an earlier study. The Hewett and Johnson (1992) software package was used to generate bioenergetics model predictions in both the Worischka and Mehner (1998) study and the Karjalainen et al. (1997) study.

14.4.3 Laboratory Evaluations

As previously mentioned, Bajer et al. (2003) evaluated the performance of the Wisconsin yellow perch bioenergetics model and the Karås-Thoresson bioenergetics model in the laboratory (Table 14.3). Yellow perch were fed meal worms (*Tenebrio molitor*) ad libitum during 125-day and 144-day experiments. Average weight of the yellow perch at the start of the experiments ranged from 19 to 35 g, and the average water temperature during the experiments was 21 °C. These researchers concluded that both bioenergetics models showed systematic bias in their predictions of daily food consumption, whereby the models overestimated daily food consumption when daily feeding rate was less than 1.5 % of the yellow perch body weight but underestimated daily food consumption when daily feeding rate was greater than 1.5 % of the yellow perch body weight. Bajer et al. (2003, 2004) contended that this systematic bias was common to most fish bioenergetics models, and suggested that this systematic bias may due to inaccurate submodels for egestion and excretion and to overestimation of respiration rate at low feeding levels. Bajer et al. (2003) also noted that cumulative consumption over the course of the entire experiment was better predicted by the Wisconsin yellow perch bioenergetics model than the Karås-Thoresson model. Apparently, all of the bioenergetics model predictions in the Bajer et al. (2003) study were generated using the Hanson et al. (1997) computer software package.

Madenjian et al. (2010) evaluated the performance of the Wisconsin walleye bioenergetics model in the laboratory (Table 14.3). Walleye were fed rainbow smelt (*Osmerus mordax*) in four laboratory tanks during a 126-day experiment. Daily feeding rates ranged from 1.4 % to 1.8 % of walleye body weight, and water temperatures ranged from 12.7 to 15.7 °C, and averaged 14.7 °C, during the course of the experiment. Average walleye weights at the start and end of the experiment were 153 and 209 g, respectively. Madenjian et al. (2010) concluded that the Wisconsin walleye bioenergetics model significantly underestimated food consumption when daily feeding rate exceeded 1.5 % of walleye body weight. These researchers suggested that this underestimation of food consumption at the higher feeding rates could have been due to laboratory artifacts of fish confined to tanks; specifically, respiration rate may have been artificially inflated for fish feeding at the higher rates. Madenjian et al. (2010) also concluded that the Wisconsin walleye bioenergetics model significantly overestimated walleye growth when daily feeding rate exceeded 1.5 % of walleye body weight. All bioenergetics model predictions in the Madenjian et al. (2010) study were generated using the Hanson et al. (1997) computer software package.

Because Madenjian et al. (2012b) discovered an error in the algorithm used to balance the fish's energy budget in the Hanson et al. (1997) computer software package, Madenjian and Wang (2013) reevaluated the Wisconsin walleye bioenergetics model using the same data from the Madenjian et al. (2010) laboratory evaluation and the correct energy budget balancing algorithm. Results of this reevaluation indicated that the Wisconsin walleye bioenergetics model predicted food consumption and growth by walleye without any detectable bias. In other words, when the correct algorithm for balancing the energy budget was used, all detectable bias in bioenergetics model predictions disappeared. Similarly, Madenjian et al. (2012b, 2013) showed that nearly all of the previously documented detectable bias in the predictions of the Wisconsin lake trout bioenergetics model disappeared when the correct energy budget balancing algorithm was used. Moreover, Madenjian et al. (2012b, 2013) and Madenjian and Wang (2013) suggested that the systematic bias in fish bioenergetics model predictions described by Bajer et al. (2004) was a consequence of using the incorrect energy budget balancing algorithm. Both the Hewett and Johnson (1987) computer software package and the Hewett and Johnson (1992) computer software package contain the correct algorithm for balancing the fish's energy budget.

As previously mentioned, Keskinen et al. (2008) evaluated the performance of their zander bioenergetics model in the laboratory (Table 14.3). Zander were allowed to feed ad libitum on Eurasian perch in laboratory tanks during a 28-day experiment. Zander were kept at 10, 14, 18, and 22 °C. Initial weights of the zander ranged from 85 to 185 g. Energy densities of the zander and the Eurasian perch used in this laboratory experiment were not determined, rather both zander energy density and Eurasian perch energy density were assigned a value of 4,186 J/g, a value taken from the literature, for application of the bioenergetics models. Keskinen et al. (2008) applied both their zander bioenergetics model and the Wisconsin walleye bioenergetics to the data generated from the laboratory experiment. Laboratory performance of their zander bioenergetics model was quite similar to that of the Wisconsin walleye bioenergetics model, with both models underestimating food consumption at the higher feeding rates. Keskinen et al. (2008) then developed a correction factor for their zander bioenergetics model predictions of consumption, and evaluated the corrected zander bioenergetics model in the field. Again, apparently the Hanson et al. (1997) software package was used to generate bioenergetics model predictions in the Keskinen et al. (2008) study.

14.4.4 Conclusions

Considering the findings from all of these bioenergetics model evaluations, results were inconclusive as to whether: (1) either the Wisconsin yellow perch bioenergetics model or the Karås-Thoresson model exhibited a systematic bias in their predictions of food consumption by yellow perch such that consumption was overestimated when feeding rate was low and consumption was underestimated when feeding rate was high, and (2) the Karås-Thoresson model outperformed the Wisconsin yellow perch bioenergetics model. In the Schaeffer et al. (1999) study, gut evacuation rate was likely overestimated at water temperatures exceeding 22 °C, and perhaps for the larger-sized yellow perch as well. If so, we would expect that the reported bias in the Wisconsin yellow perch bioenergetics model predictions of food consumption by yellow perch was an artifact of poor performance of the gut evacuation rate model at relatively high water temperatures and large yellow perch sizes. The systematic bias described by Bajer et al. (2003) was likely just a consequence of the incorrect energy budget balancing algorithm. Based on the Madenjian et al. (2012b, 2013) and Madenjian and Wang (2013) results, most of this systematic bias would be expected to disappear after correcting the energy budget balancing algorithm. If both of these expectations prove to be correct, then the Wisconsin yellow perch bioenergetics model would likely be more accurate than the Karås-Thoresson model. According to the Schaeffer et al. (1999) study results, Karås-Thoresson model predictions of consumption matched the field estimates at the higher feeding levels, which coincided with water temperatures over 22 °C, for three of four age classes. If gastric evacuation rate was indeed overestimated at water temperatures over 22 °C, then the Karås-Thoresson model was overestimating food consumption at the higher feeding rates for ages 1–3 yellow perch. Further, using the correct gastric evacuation rate for water temperatures over 22 °C and for very large yellow perch would likely show that the Karås-Thoresson model was also overestimating food consumption at the higher feeding rates for age-4 yellow perch as well.

Findings from all of these bioenergetics model evaluations were inconclusive as to whether the Wisconsin walleye bioenergetics model exhibited systematic bias in its predictions of food consumption by walleye such that the model overestimated food consumption at low feeding rates and underestimated food consumption at high feeding rates. Again, based on the Madenjian et al. (2012b, 2013) and Madenjian and Wang (2013) results, most of the systematic bias described by Lantry et al. (2008) would be expected to disappear with use of the correct energy budget balancing algorithm. No bias was detected in the predictions of Wisconsin walleye bioenergetics model applied to walleye in laboratory tanks at an average water temperature of about 15 °C and daily feeding rates of between 1.4 % and 1.8 % of walleye body weight (Madenjian and Wang 2013).

In addition, findings from all of these bioenergetics model evaluations were inconclusive as to whether either the Wisconsin walleye bioenergetics model or the zander bioenergetics model were actually underestimating food consumption by zander feeding at relatively high rates in the laboratory. Keskinen et al. (2008) did not measure the energy densities of any of the zander and Eurasian perch used in their laboratory experiment. Rather energy densities of all of the zander and Eurasian perch used in the experiment were assumed to be equal to 4,186 J/g. Yet, accurate determinations of the energy densities of both predator and prey, including changes in predator energy density over time, are critical for the accurate assessment of bioenergetics model performance (Madenjian et al. 2013). Zander feeding at relatively high rates in the Keskinen et al. (2008) experiment likely had energy densities exceeding the assumed value and likely showed a greater increase in energy density over the course of the experiment than zander feeding at relatively low rates. Thus, food consumption by the zander feeding at a relatively high rate was probably underestimated by the bioenergetics models. Not only does laboratory performance of the fish bioenergetics depend on use of the correct energy budget balancing algorithm, but accuracy of the laboratory evaluation also hinges on accurate tracking of both predator and prey energy densities over the course of the laboratory experiment (Madenjian et al. 2013). Both the Keskinen et al. (2008) zander bioenergetics model, without any kind of correction factor, and the Wisconsin walleye bioenergetics model may be capable of predicting food consumption by zander in the laboratory without any detectable bias, provided that accurate estimates of the energy densities of both zander and its food are used in the model applications. Further, based on the findings from the bioenergetics model evaluations, the Keskinen et al. (2008) zander bioenergetics model cannot be concluded, with any certainty, to be a better predictor of food consumption by zander than the Wisconsin walleye bioenergetics model or vice versa.

The modified bioenergetics models developed by Worischka and Mehner (1998) for yellow perch (or Eurasian perch) larvae and walleye (or zander) larvae appeared to be more accurate than the models by Post (1990) for yellow perch larvae and Madon and Culver (1993) for walleye larvae, respectively. Worischka and Mehner (1998) sampled fish larvae at 30-min intervals, rather than using the 3-h interval regimen used by Post (1990) and Madon and Culver (1993), to estimate gastric evacuation rates of the fish larvae. This more frequent sampling by Worischka and Mehner (1998) revealed that gastric evacuation rate is overestimated using the 3-h interval sampling. Consequently, *ACT* was overestimated by Post (1990) and Madon and Culver (1993). Based on these results, *ACT* should be assigned a value of 1 in bioenergetics models for percid larvae.

14.5 Future Research Directions in Percid Bioenergetics Modeling

Even though the Wisconsin yellow perch and walleye bioenergetics models developed by Kitchell et al. (1977) are nearly 40 years old, and a substantial amount of knowledge about percid bioenergetics modeling has been gained since 1977, more research work is needed to fully resolve some of the issues that have emerged from the evaluations of percid bioenergetics model performance. Gastric evacuation rate models need to be developed over wider ranges of water temperature and percid size. Percid bioenergetics models will likely need to be reevaluated in the field, especially if the new gastric evacuation rate models differ substantially from the old ones. New laboratory and field evaluations of percid bioenergetics models will be needed. Additional reevaluations in the laboratory and field will be needed just to correct for use of the incorrect energy budget balancing algorithm. Unless otherwise specified, the future research directions for percid bioenergetics modeling discussed in this section pertain to larger juvenile and adult fish.

Certainly, yellow perch and Eurasian perch gastric evacuation rates need to be measured at water temperatures greater than 22 °C to determine whether gastric evacuation rate begins to level off at these higher temperatures. If leveling off occurs, then these new data would need to combined with those of Persson (1979) to build a new gastric evacuation rate model for perch across a wider range of water temperatures. Further, it still needs to be established that gastric evacuation rates for Eurasian perch are practically identical to those for yellow perch. In addition, gastric evacuation rate should be measured in yellow perch and Eurasian perch considerably larger than the sizes used by Persson (1979), and at a range of water temperatures from 4 to 30 °C, to determine whether perch size is a determinant of gastric evacuation rate. If perch size is found to affect gastric evacuation rate, then a new model for perch gastric evacuation rate as a function of both water temperature and perch size should be constructed. Additional measurements of walleye evacuation rates would also be desirable. Specifically, whether walleye evacuation rate continues to be independent of walleye size and water temperature at walleye sizes and water temperatures outside the ranges used by Forney (1977) needs to be investigated. Moreover, it still needs to be established that gastric evacuation rates for walleye are practically identical to those for zander.

Reevaluations and additional evaluations of the perch (yellow perch and Eurasian perch) bioenergetics models will definitely be needed. Will reevaluation of the Wisconsin yellow perch bioenergetics model and the Karås-Thoresson model using new estimates of field ration, based on a new perch gastric evacuation rate model, but using all of the other data from Schaeffer et al. (1999) for Saginaw Bay yellow perch, reveal reasonably accurate estimates of consumption by the Wisconsin yellow perch bioenergetics model but overestimation of consumption by the Karås-Thoresson model at higher feeding rates? This question will be ripe for answering once the new perch gastric evacuation rate model has been developed. Will reevaluation of the laboratory data generated by Bajer et al. (2003) using the correct energy budget balancing algorithm show that the previously described systematic bias in the Wisconsin yellow perch bioenergetics model predictions of consumption disappears, but the Karås-Thorsson model overestimates consumption at higher feeding rates? This question can be answered by using the correct energy budget balancing algorithm. Additional laboratory evaluations of the perch bioenergetics models are needed to investigate model reliability at water temperatures greater than and less than 21 °C and at perch sizes considerably greater than 40 g. I would also recommend field evaluations of the perch bioenergetics models using PCBs as a tracer of food consumption. Finally, it still needs to be established that yellow perch bioenergetics and Eurasian perch bioenergetics are interchangeable.

Reevaluations and additional evaluations of the Wisconsin walleye bioenergetics models will be needed. The Lantry et al. (2008) data set should be reevaluated using the correct energy budget balancing algorithm to determine whether the previously described systematic bias in the Wisconsin walleye bioenergetics model predictions of field ration disappears. Additional laboratory evaluations of the Wisconsin walleye bioenergetics model will be needed to investigate the reliability of model predictions at temperatures greater than and less than 15 °C and at walleye sizes considerably greater than 250 g. Field evaluations of the Wisconsin walleye bioenergetics model using PCBs as a tracer of food consumption would also be useful in further assessing the field performance of the model. Care should be taken to avoid "hot spot" effects, as described by Madenjian et al. (1998, 2009), confounding the interpretation of the field evaluation results. In addition, the maximum consumption (C_{max}) function in the Wisconsin walleye bioenergetics model may need further refinement. The C_{max} function was originally developed based on data from Swenson and Smith (1973), who fed fathead minnows (*Pimephales promelas*) ad libitum to walleye of various sizes and at several different water temperatures in the laboratory for about 5 days. Perhaps a more accurate estimate of the sustained maximum consumption rate would be attained by extending this type of laboratory experiment for a period of 80 days or more (Stewart et al. 1983).

To more accurately assess the Keskinen et al. (2008) zander bioenergetics model performance, additional laboratory and field evaluations would be instructive. For both approaches to yield highly accurate assessments of bioenergetics model performance, energy densities of both zander and its prey must be accurately tracked over time. Repeating these evaluations, but making the necessary energy density determinations, will allow for more accurate assessment. Perhaps the zander model yields unbiased predictions of consumption and growth by zander without any kind of correction factor. Or perhaps the Wisconsin walleye bioenergetics model yields unbiased predictions of consumption by zander, whereas the Keskinen et al. (2008) zander bioenergetics model slightly overestimates consumption. These issues can be resolved by repeating the evaluations. For the field evaluation, use of new estimates of mercury elimination rates would also contribute to a more accurate assessment of model performance, because the mercury elimination rates used in the Trudel et al. (2000) approach have been found to be overestimates (Van Walleghem et al. 2007; Madenjian et al. 2012a). Moreover, determinations of mercury concentration in the zander on a whole-fish basis, rather than in fillets, would improve the accuracy of the evaluation, because mercury concentration in the fillet may substantially differ from the whole-fish mercury concentration in some cases (Becker and Bigham 1995). Finally, researchers will eventually need to address the question of whether walleye bioenergetics and zander bioenergetics are interchangeable.

For certain research questions, differences in the energy budget of percids between the sexes must be taken into account. To answer these types of questions, standard respiration (*SR*) and activity (*ACT*) for both male and female percids will need to be determined. Madenjian (2011b) proposed that *SR* does not significantly differ between the sexes for juvenile fish, but *SR* of females becomes lower than that of males as the fish approach maturity, and *SR* of mature females remains lower than that of mature males throughout the adult years. Laboratory determinations of *SR* for both males and females are needed to test this hypothesis. Of course, *SR* should be determined over a range of sizes of the percids, over a range of water temperatures, and at different times of the year. In addition, laboratory determinations of percid respiration rate at various swimming speeds for both males and females may eventually be useful in further refining percid bioenergetics models. Field evaluations using PCBs as a tracer of food consumption could be used to estimate *ACT* in both mature males and mature females, provided *SR* is known for both mature males and mature females. Alternatively, estimation of field rations for both males and females based on gastric evacuation rate models, similar to the approach used by Boisclair and Leggett (1989), could be used to estimate *ACT* for both males and females. For this approach, accurate estimates of gastric evacuation rates are paramount to obtaining accurate estimates of *ACT* for both sexes.

Additional evaluations of the larval fish bioenergetics models for perch and walleye (and zander) would be valuable in corroborating the conclusions by Wörischka and Mehner (1998) that *ACT* is equal to 1 for both perch and zander (or walleye) larvae. Further, it still needs to be established that the bioenergetics of larval yellow perch and the bioenergetics of larval Eurasian perch are interchangeable, and that the bioenergetics of larval walleye and the bioenergetics of larval zander are interchangeable.

Bioenergetics of percids other than yellow perch, Eurasian perch, walleye, and zander has yet to be investigated. Are the bioenergetics models for the four above-mentioned species appropriate for other species of percids? If not, which modifications to these models would make them applicable to other percid species? These questions remain to be answered.

Percid bioenergetics models can be applied in ways other than those already documented. For example, the Wisconsin walleye bioenergetics model could be used to model accumulation of PCBs and polybrominated diphenyl ethers (PBDEs) by males and females from the Saginaw Bay walleye population so that the "hot spot" effect can be better understood. Mature males from this population are nearly three times higher in PCB and PBDE concentrations than mature females. Most of this difference between the sexes has been attributed to males spending more time in the vicinity of the contaminant "hot spot" at the mouth of Saginaw River, which is the main tributary to Saginaw Bay, and consequently feeding on prey fish considerably higher in PCB and PBDE concentrations than prey fish in Saginaw Bay (Madenjian et al. 1998, 2009, 2012c). A small portion of the sex difference in PCB and PBDE concentrations was attributable to females growing faster than males and to males expending energy at a faster rate than females (Madenjian et al. 2009; Madenjian 2011b). Much of the details of the "hot spot" effect remain unknown (Madenjian et al. 2012c). Modeling PCB and PBDE accumulation in male and female walleye from the Saginaw Bay population, using the Wisconsin walleye bioenergetics model, should provide valuable insights into this "hot spot" effect. To effectively carry out this modeling exercise, sex-specific walleye bioenergetics models, as described above, would need to be developed. The number of applications of percid bioenergetics models is mainly bounded by the imagination and ingenuity of the researchers investigating percid biology and ecology.

Acknowledgments I thank Patrick Kestemont and Konrad Dabrowski for encouraging me to prepare this chapter for their book on percid biology and culture. I also thank Jim Kitchell, Steve Hewett, and Jim Breck for sharing their knowledge of fish bioenergetics modeling with me. Patrick Hanchin, Jason Ross, and an anonymous reviewer provided helpful comments for improvement of the manuscript, and Rich Quintal prepared the figure. This article is Contribution 1868 of the U. S. Geological Survey Great Lakes Science Center.

References

Adams SM, McLean RB, Huffman MM (1982) Structuring of a predator population through temperature-mediated effects on prey availability. Can J Fish Aquat Sci 39:1175–1184

Bajer PG, Whitledge GW, Hayward RS, Zweifel RD (2003) Laboratory evaluation of two bioenergetics models applied to yellow perch: identification of a major source of systematic error. J Fish Biol 62:436–454

Bajer PG, Whitledge GW, Hayward RS (2004) Widespread consumption-dependent systematic error in fish bioenergetics models and its implications. Can J Fish Aquat Sci 61:2158–2167

Becker DS, Bigham GN (1995) Distribution of mercury in the aquatic food web of Onondaga Lake, New York. Water Air Soil Pollut 80:563–571

Boisclair D, Leggett WC (1989) The importance of activity in bioenergetics models applied to actively foraging fishes. Can J Fish Aquat Sci 46:1859–1867

Brandt SB, Constantini M, Kolesar S, Ludsin SA, Mason DM, Rae CM, Zhang H (2011) Does hypoxia reduce habitat quality for Lake Erie walleye (*Sander vitreus*)? A bioenergetics perspective. Can J Fish Aquat Sci 68:857–879

Brett JR, Groves TDD (1979) Physiological energetics. In: Hoar WS, Randall DJ, Brett JR (eds) Fish physiology, vol VIII. Academic, New York

Bunnell DB, Johnson TB, Knight CT (2005) The impact of introduced round gobies (*Neogobius melanostomus*) on phosphorus cycling in central Lake Erie. Can J Fish Aquat Sci 62:15–29

Carpenter SR, Kitchell JF, Hodgson JR (1985) Cascading trophic interactions and lake productivity. Bioscience 35:634–639

Chipps SR, Wahl DH (2008) Bioenergetics modeling in the 21st century: reviewing new insights and revisiting old constraints. Trans Am Fish Soc 137:298–313

Ciancio J, Beauchamp DA, Pascual M (2010) Marine effect of introduced salmonids: prey consumption by exotic steelhead and anadromous brown trout in the Patagonian Continental Shelf. Limnol Oceanogr 55:2181–2192

Eggers DM (1979) Comments on some recent methods for estimating food consumption by fish. J Fish Res Board Can 36:1018–1019

Elliott JM, Persson L (1978) The estimation of daily rates of food consumption for fish. J Anim Ecol 47:977–991

Forney JL (1977) Reconstruction of yellow perch (*Perca flavescens*) cohorts from examination of walleye (*Stizostedion vitreum vitreum*) stomachs. J Fish Res Board Can 34:925–932

Hansen MJ, Boisclair D, Brandt SB, Hewett SW, Kitchell JF, Lucas MC, Ney JJ (1993) Applications of bioenergetics models to fish ecology and management: where do we go from here? Trans Am Fish Soc 122:1019–1030

Hanson PC, Johnson TB, Schindler DE, Kitchell JF (1997) Fish bioenergetics 3.0. University of Wisconsin, Sea Grant Institute, WISCU-T-97-001, Madison

Hartman KJ, Margraf FJ (1992) Effects of prey and predator abundances on prey consumption and growth of walleyes in western Lake Erie. Trans Am Fish Soc 121:245–260

Hartman KJ, Margraf FJ (1993) Evidence of predatory control of yellow perch (*Perca flavescens*) recruitment in Lake Erie, U.S.A. J Fish Biol 43:109–119

Hayward RS, Margraf FJ (1987) Eutrophication effects on prey size and food available to yellow perch in Lake Erie. Trans Am Fish Soc 116:210–223

Hewett SW, Johnson BL (1987) A generalized bioenergetics model of fish growth for microcomputers. University of Wisconsin, Sea Grant Institute, WIS-SG-87-245, Madison

Hewett SW, Johnson BL (1992) Fish bioenergetics model 2. University of Wisconsin, Sea Grant Institute, WIS-SG-92-250, Madison

Hewett SW, Kraft CE, Johnson BL (1991) Consumption, growth, and allometry: a comment on Boisclair and Leggett (1989a, 1989b, 1989c, 1989d). Can J Fish Aquat Sci 48:1334–1337

Hilborn R, Walters CJ (1992) Quantitative fisheries stock assessment: choice, dynamics, and uncertainty. Chapman and Hall, New York

Hill DK, Magnuson JJ (1990) Potential effects of global climate warming on the growth and prey consumption of Great Lakes fish. Trans Am Fish Soc 119:265–275

Jobling M (1981) Mathematical models of gastric emptying and the estimation of daily rates of food consumption for fish. J Fish Biol 19:245–257

Johnson BM, Luecke C, Stewart RS, Staggs MD, Gilbert SJ, Kitchell JF (1992) Forecasting effects of harvest regulations and stocking of walleyes on prey fish communities in Lake Mendota, Wisconsin. N Am J Fish Manag 12:797–807

Jones ML, Koonce JF, O'Gorman R (1993) Sustainability of hatchery-dependent salmonine fisheries in Lake Ontario: the conflict between predator demand and prey supply. Trans Am Fish Soc 122:1002–1018

Karås P, Thoresson G (1992) An application of a bioenergetics model to Eurasian perch (*Perca fluviatilis* L.). J Fish Biol 41:217–230

Karjalainen J, Miserque D, Huuskonen H (1997) The estimation of food consumption in larval and juvenile fish: experimental evaluation of bioenergetics models. J Fish Biol 51(Suppl A):39–51

Kershner MW, Schael DM, Knight RL, Stein RA, Marschall EA (1999) Modeling sources of variation for growth and predatory demand of Lake Erie walleye (*Stizostedion vitreum*), 1986–1995. Can J Fish Aquat Sci 56:527–538

Keskinen T, Jääskeläinen J, Marjomäki TJ, Matilainen T, Karjalainen J (2008) A bioenergetics model for zander: construction, validation, and evaluation of uncertainty caused by multiple input parameters. Trans Am Fish Soc 137:1741–1755

Kitchell JF, Koonce JF, O'Neill RV, Shugart HH Jr, Magnuson JJ, Booth RS (1974) Model of fish biomass dynamics. Trans Am Fish Soc 103:786–798

Kitchell JF, Stewart DJ, Weininger D (1977) Applications of a bioenergetics model to yellow perch (*Perca flavescens*) and walleye (*Stizostedion vitreum vitreum*). J Fish Res Board Can 34:1922–1935

Kooijman SALM (1993) Dynamic energy budgets in biological systems: theory and applications in ecotoxicology. Cambridge University Press, New York

Kooijman SALM (2010) Dynamic energy budget theory for metabolic organization, 3rd edn. Cambridge University Press, Cambridge, UK

Kraft CE (1992) Estimates of phosphorus and nitrogen cycling by fish using a bioenergetics approach. Can J Fish Aquat Sci 49:2596–2604

Kraft CE (1993) Phosphorus regeneration by Lake Michigan alewives in the mid-1970s. Trans Am Fish Soc 122:749–755

Lantry BF, Rudstam LG, Forney JL, VanDeValk AJ, Mills EL, Stewart DJ, Adams JV (2008) Comparisons between consumption estimates from bioenergetics simulations and field measurements for walleyes from Oneida Lake, New York. Trans Am Fish Soc 137:1406–1421

Lyons J, Magnuson JJ (1987) Effects of walleye predation on the population dynamics of small littoral-zone fishes in a northern Wisconsin lake. Trans Am Fish Soc 116:29–39

Madenjian CP (2011a) Bioenergetics in ecosystems. In: Farrell AP (ed) Encyclopedia of fish physiology: from genome to environment. Elsevier, Oxford

Madenjian CP (2011b) Sex effect on polychlorinated biphenyl concentrations in fish: a synthesis. Fish Fish 12:451–460

Madenjian CP, Wang C (2013) Reevaluation of a walleye (*Sander vitreus*) bioenergetics model. Fish Physiol Biochem 39:749–754

Madenjian CP, Carpenter SR, Eck GW, Miller MA (1993) Accumulation of PCBs by lake trout (*Salvelinus namaycush*): an individual-based model approach. Can J Fish Aquat Sci 50:97–109

Madenjian CP, Noguchi GE, Haas RC, Schrouder KS (1998) Sexual difference in polychlorinated biphenyl accumulation rates of walleye (*Stizostedion vitreum*). Can J Fish Aquat Sci 55:1085–1092

Madenjian CP, O'Connor DV, Nortrup DA (2000) A new approach toward evaluation of fish bioenergetics models. Can J Fish Aquat Sci 57:1025–1032

Madenjian CP, O'Connor DV, Chernyak SM, Rediske RR, O'Keefe JP (2004) Evaluation of a chinook salmon (*Oncorhynchus tshawytscha*) bioenergetics model. Can J Fish Aquat Sci 61:627–635

Madenjian CP, Hanchin PA, Chernyak SM, Begnoche LJ (2009) Sexual difference in PCB concentrations of walleyes (*Sander vitreus*) from a pristine lake. Sci Total Environ 407:4526–4532

Madenjian CP, Wang C, O'Brien TP, Holuszko MJ, Ogilvie LM, Stickel RG (2010) Laboratory evaluation of a walleye (*Sander vitreus*) bioenergetics model. Fish Physiol Biochem 36:45–53

Madenjian CP, David SR, Krabbenhoft DP (2012a) Trophic transfer efficiency of methylmercury and inorganic mercury to lake trout *Salvelinus namaycush* from its prey. Arch Environ Contam Toxicol 63:262–269

Madenjian CP, David SR, Pothoven SA (2012b) Effects of activity and energy budget balancing algorithm on laboratory performance of a fish bioenergetics model. Trans Am Fish Soc 141:1328–1337

Madenjian CP, Trombka AW, Rediske RR, Jude DJ, O'Keefe JP (2012c) Sex difference in polybrominated diphenyl ether concentrations of walleyes. J Great Lakes Res 38:167–175

Madenjian CP, Pothoven SA, Kao Y-C (2013) Reevaluation of lake trout and lake whitefish bioenergetics models. J Great Lakes Res 39:358–364

Madon SP, Culver DA (1993) Bioenergetics model for larval and juvenile walleyes: an in situ approach with experimental ponds. Trans Am Fish Soc 122:797–813

Ney JJ (1993) Bioenergetics modeling today: growing pains on the cutting edge. Trans Am Fish Soc 122:736–748

Nisbet RM, Jusup M, Klanjscek T, Pecquerie L (2012) Integrating dynamic energy budget (DEB) theory with traditional bioenergetics models. J Exp Biol 215:892–902

Pauly D, Christensen V, Walters C (2000) Ecopath, Ecosim, and Ecospace as tools for evaluating ecosystem impact of fisheries. ICES J Mar Sci 57:697–706

Persson L (1979) The effects of temperature and different food organisms on the rate of gastric evacuation in perch (*Perca fluviatilis*). Freshw Biol 9:99–104

Peterson BJ, Fry B (1987) Stable isotopes in ecosystem studies. Ann Rev Ecol Syst 18:293–320

Phillipson J (1966) Ecological energetics. Edward Arnold, London

Post JR (1990) Metabolic allometry of larval and juvenile yellow perch (*Perca flavescens*): in situ estimates and bioenergetics models. Can J Fish Aquat Sci 47:554–560

Ricker WE (1975) Computation and interpretation of biological statistics of fish populations. Fish Res Board Can Bull 191

Schaeffer JS, Haas RC, Diana JS, Breck JE (1999) Field test of two energetic models for yellow perch. Trans Am Fish Soc 128:414–435

Solomon DJ, Brafield AE (1972) The energetics of feeding, metabolism and growth of perch (*Perca fluviatilis* L.). J Anim Ecol 41:699–718

Stewart DJ (1980) Salmonid predators and their forage base in Lake Michigan: a bioenergetics-modeling synthesis. Dissertation, University of Wisconsin

Stewart DJ, Kitchell JF, Crowder LB (1981) Forage fishes and their salmonid predators in Lake Michigan. Trans Am Fish Soc 110:751–763

Stewart DJ, Weininger D, Rottiers DV, Edsall TA (1983) An energetics model for lake trout, *Salvelinus namaycush*: application to the Lake Michigan population. Can J Fish Aquat Sci 40:681–698

Swenson WA, Smith LL Jr (1973) Gastric digestion, food consumption, feeding periodicity, and food conversion efficiency in walleye (*Stizostedion vitreum vitreum*). J Fish Res Board Can 30:1327–1336

Trudel M, Tremblay A, Schetagne R, Rasmussen JB (2000) Estimating food consumption rates of fish using a mercury mass balance model. Can J Fish Aquat Sci 57:414–428

Van Walleghem JLA, Blanchfield PJ, Hintelmann H (2007) Elimination of mercury by yellow perch in the wild. Environ Sci Technol 41:5895–5901

Weininger D (1978) Accumulation of PCBs by lake trout in Lake Michigan. Dissertation, University of Wisconsin

Whitledge GA, Hayward RS (1997) Laboratory evaluation of a bioenergetics model for largemouth bass at two temperatures and feeding levels. Trans Am Fish Soc 126:1030–1035

Winberg GG (1960) Rate of metabolism and food requirements of fishes. Fish Res Board Can Transl Ser 194

Worischka S, Mehner T (1998) Comparison of field-based and indirect estimates of daily food consumption in larval perch and zander. J Fish Biol 53:1050–1059

Chapter 15
Behaviour of Percid Fishes in the Wild and Its Relevance for Culture

Carin Magnhagen

Abstract The behaviour of percid fishes

The behaviour of percids, as in most other fishes, is to a great extent size dependent. As the fish grows larger there is a change in behaviour in accordance with a shift in relative costs and benefits of different alternative actions. While a diet of plankton is the most energy efficient for a small percid, with increasing size, larger, more energy rich food types become accessible and more profitable to eat. In connection with a change in diet, a change in foraging habitat also occurs. However, perch with different body shape has been found to differ in diet and occupy different habitats even within one lake. The perch and the pikeperch are, in general, social species but also group size changes with size. The young fish benefit more by being in a large shoal, as a protection against predators. With growth, the groups of fish become smaller, but still, foraging is more effective in groups than when alone. The behaviour of percids also depends on the environment in which they live. For example, risk-taking behaviour is influenced by lake- and size-specific risk of predation. Even though behaviour may be innate to a certain extent, experience is probably important to form the behaviour patterns of individual fish. For the best result in culture it is very important to have knowledge about the behaviour of percids in the wild, especially regarding feeding and social interactions.

Keywords Eurasian perch • Pikeperch • Behaviour • Social interactions • Feeding

15.1 Introduction

Perch belong to the most widespread freshwater species in the Northern hemisphere and occur in many lakes throughout their distribution areas. If you stand on a dock by a clear-water lake in the summer, you are likely to see perch in the water, of one or several size classes. Large shoals of small perch are moving here and there,

C. Magnhagen (✉)
Department of Wildlife, Fish, and Environmental Studies, Swedish University of Agricultural Sciences, SE-901 83 Umeå, Sweden
email: carin.magnhagen@slu.se

P. Kestemont et al. (eds.), *Biology and Culture of Percid Fishes*,
DOI 10.1007/978-94-017-7227-3_15

looking for food, and larger perch are swimming around, alone or in small groups, in the protective shade of the dock. The behaviour of perch changes over the course of its lifetime, and sociality, habitat choice and foraging mode vary due to the costs and benefits of different options at different sizes. These ontogenetic niche shifts are affected both by intra-specific interactions and by the structure of the whole fish community (see e.g. Persson 1988). The perch populations are size-structured, with fish of different size classes acting as either competitors with, or predators on, each other. In fact, one of the most severe threats on a young perch can be cannibalistic conspecifics (Treasurer 1989; Persson et al. 2000). Thus, depending on population density and size distribution, foraging strategy and habitat choice may be adapted in order to minimize the risk of being eaten. In this chapter, the main focus will be on the Eurasian perch (*Perca fluviatilis*) of which the author has most experience, with parallels drawn to the closely related North American species, the yellow perch (*Perca flavescens*) that has a very similar biology and behaviour. Other percid species, such as the European pikeperch (*Sander lucioperca*) and the North American walleye (*Sander vitreus*) will also be referred to.

15.2 Foraging Behaviour

15.2.1 Diet

As in many fish species, the diet of the perch changes over their life-time. This is both a result of changes in mouth size and choice of foraging habitat. The size of the mouth is determining the efficiency with which the fish can consume a prey (e.g. Hambright 1991; Schael et al. 1991). For a small foraging fish, the higher energy content of a larger prey may be offset by the longer handling time of that prey in comparison with smaller ones. In the beginning of life, the perch feeds on zooplankton. According to functional response experiments, attack rates on plankton increase up to a certain size, and then drop, and the peak for intake of small-sized plankton occurs at smaller perch sizes than that for larger plankton (Wahlström et al. 2000, Fig. 15.1a). Later on there is a shift in diet to larger macroinvertebrates. Also here the shift of diet is assumed to coincide with a change in profitability of different prey items. Young-of-the-year yellow perch increase energetic gain and decrease foraging costs for each prey type with growth (Graeb et al. 2006). According to optimal foraging theories, with only energy maximizing taken into account, the choice of prey type should change at a size when a potential growth rate (or net energy intake) gets higher on a diet of macroinvertebrates than when feeding on plankton (Persson 1988, Fig. 15.1b). In the natural habitat of perch, however, decisions on foraging behaviour must also account for other factors in the environment, such as intra- and interspecific competition (Persson 1988) and predation risk (Byström et al. 2003), in order to maximize individual fitness. For example, young-of-the-year Eurasian perch often compete with roach (*Rutilus rutilus*), a species more efficient in plankton feeding, which may decrease the size when a shift to

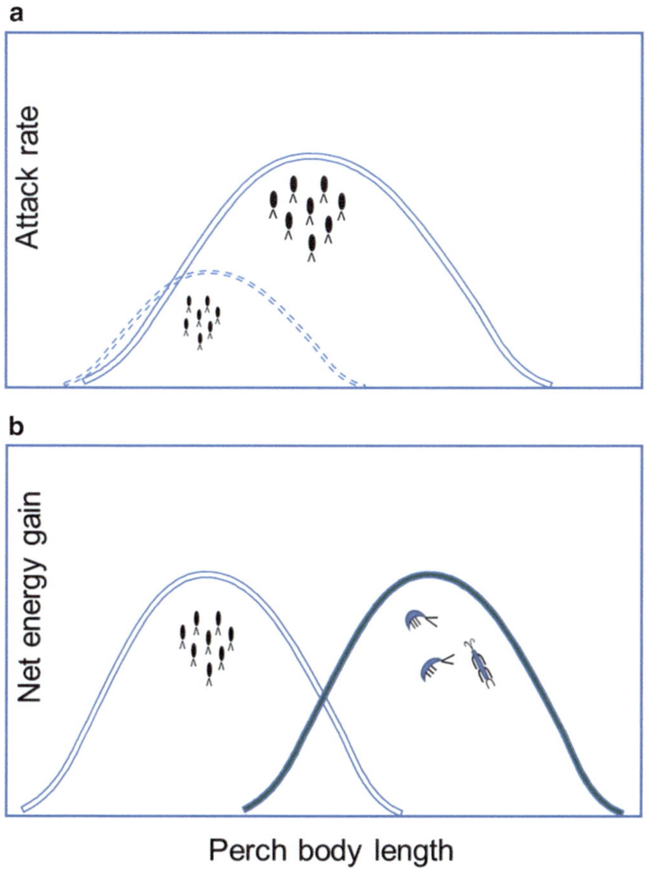

Fig. 15.1 (**a**) Attack rate on small and large zooplankton as a function of perch size (Wahlström et al. 2000). (**b**) Net energy intake (can also be shown as growth rate) for different sized perch feeding on zooplankton or macroinvertebrates (Persson 1988). Please note that the graph is a general outline and not in a correct scale

macro-invertebrates become profitable (Persson 1988). In contrast, a high density of 1-year old perch, that are better competitors on the macroinvertebrate prey, may prolong the phase of plankton feeding.

As the perch grow they become increasingly more piscivorous. Perch is considered to switch to piscivory at a size of 11–16 cm (reviewed in Mittelbach and Persson 1998), but feed on fish larvae and fry also at smaller sizes (Lundvall et al. 1999; Persson et al. 2004). Actually, even young-of-the year perch can become piscivorous if they hatch earlier than a potential prey fish, such as bream (*Abramis brama*), whose larvae are preyed upon by slightly older Eurasian perch (Borcherding et al. 2010). The perch can also be a severe predator on conspecifics, and since the size relationship between predator and prey determines foraging success of cannibalistic perch (Braband 1995; Lundvall et al. 1999; Byström et al. 2003), the risk for cannibalism depends on the size distribution and density of fish in the population.

The European pikeperch and the closely related North American walleye are more diet specialists than the perch species. With relatively bigger gape sizes, these species have a higher capacity for piscivory compared with the more omnivorous perch, and the morphology allows them to switch to piscivory earlier than perch (Persson and Brönmark 2002; Graeb et al. 2005). Walleye grew slower than yellow perch when the proportion of fish was low in the diet, but faster on a diet dominated by fish (Graeb et al. 2005). Ontogenetic changes in foraging patterns are linked to prey profitability, but relative prey densities may influence consumption, which was found to affect growth in juvenile walleye (Galarowicz et al. 2006).

15.2.2 Feeding Habitat

The ontogenetic changes in prey utilisation are partly connected with habitat shifts. In the Eurasian perch the newly hatched larvae usually occupy the pelagic zone, feeding on plankton in the water column (Byström et al. 2003). After some time, they move into the littoral zone, starting to feed on macroinvertebrates. The same pattern is found in the yellow perch (Post and McQueen 1988). The timing of the shift can vary for different populations, but also between years. The 1+ perch not only compete with the younger cohort for the macroinvertebrates, they also cannibalize on the smallest stages of the 0+ perch. The 1+ usually occupy the littoral zone, and with a high density of that age class in the population, the 0+ perch stays longer in the pelagic zone than if the density is lower. Probably the 0+ perch are better at escaping cannibalistic attacks if they obtain a larger size before shifting habitat. Larger perch occupy both the littoral and the pelagic areas. However, larger perch are not efficient as predators on the smallest stages of perch, and the 1+ perch cannibalise on these sizes more effectively than do the larger individuals (Lundvall et al. 1999; Byström et al. 2003).

Switches between habitats in connection with foraging also occur on a short-term basis. Eurasian perch, tagged with ultrasonic transmitters, showed daily movements between the limnetic and the littoral zones (Zamora and Moreno-Amich 2002). These movements were assumed to be related to feeding activity in the open waters during the day and resting in the littoral zone during the night. Yellow perch in lake Erie were found to undertake short foraging forays into hypoxic habitats, to feed on benthic prey (Roberts et al. 2012).

In many lakes, the adult perch population is divided into subpopulations with different habitat occupation and diet. Stable isotope studies showed that the trophic overlap was small between littoral and pelagic Eurasian perch from the same lake (Quevedo et al. 2009). Perch from both habitats feed on fish, but the fish caught in the pelagic zone are feeding on zooplankton to a higher proportion than on macro-invertebrates, while the opposite is true for those from the littoral zone (Svanbäck and Eklöv 2002, 2003; Quevedo et al. 2009). Furthermore, in a pond experiment, offspring from littoral parents had a higher proportion of littoral prey types in their diet than those from pelagic parents and vice versa, suggesting a genetic component in their foraging strategy (Svanbäck and Eklöv 2006). The two types of perch also differ in morphology, with the littoral perch being more deep-bodied and the pelagic one more streamlined (Svanbäck and Eklöv 2002). Aquarium studies showed that each morphotype was more efficient feeding on its usual diet than the alternative (i.e. plankton versus macroinvertebrates) (Svanbäck and Eklöv 2002). However, morphology seems to be a plastic trait that can change within a rather short time span, as a phenotypic response to food type availability (Olsson and Eklöv 2005) or predation risk (Eklöv and Jonsson 2007, see below).

15.3 Predator Avoidance

Predation risk is a strong selective force in all ecosystems, and adaptations to avoid being eaten are generally found in all species/populations subjected to predation. In perch, several morphological features are likely to have been evolved to avoid being eaten by a piscivore. For example, the spiny dorsal fins can be erected as a defence in a dangerous situation (Vainikka et al. 2005), while the black stripes may serve as a camouflage, to avoid being detected. Also body coloration varies depending on water colour (Kekalainen et al. 2010) and habitat utilisation (Svanbäck and Eklöv 2011), suggested to be partly a consequence of predator selection, but also depend on a trade-off between camouflage and intra-specific communication. The ability to develop a higher body shape in the presence of gape-limited piscivores has been explained both as a direct response to predation risk (Eklöv and Jonsson 2007), and as an indirect cause of a predator-induced change of habitat (Eklöv and Svanbäck 2006).

Also behaviour can evolve as an adaptation to the need for predator avoidance (Fig. 15.2). In fish, shoaling is considered to be a behaviour partly evolved as a protection from predators (Pitcher and Parrish 1993; Krause and Ruxton 2002). The dilution effect makes the probability of being caught in a predator attack smaller in a larger shoal, and with many individuals (and eyes) the chance of early predator detection will also increase (Pitcher and Parrish 1993). When the risk of predation increases, shoals become more cohesive, as in yellow perch exposed to alarm substances from either injured conspecifics or from other species (Mirza et al. 2003). Juvenile yellow perch also increased shelter use when exposed to adult perch that had been fed perch and spot-tail shiners (*Notropis hudsonius*), a species commonly co-occurring with yellow perch (Mirza and Chivers 2001; Mirza et al. 2003). However, the perch did not react when the piscivore had been fed an unknown species, the swordtail (*Xiphophorus helleri*). In an experiment on Eurasian perch, the fish responded to odour from pike (*Esox lucius*) by erecting their dorsal fin and decreasing swimming activity (Vainikka et al. 2005). However, they increased shoaling behaviour only when a pike was actually seen.

One way to decrease predation risk is to avoid the piscivore habitat, as mentioned earlier, with 0+ perch avoiding high densities of 1+ perch (Byström et al. 2003).

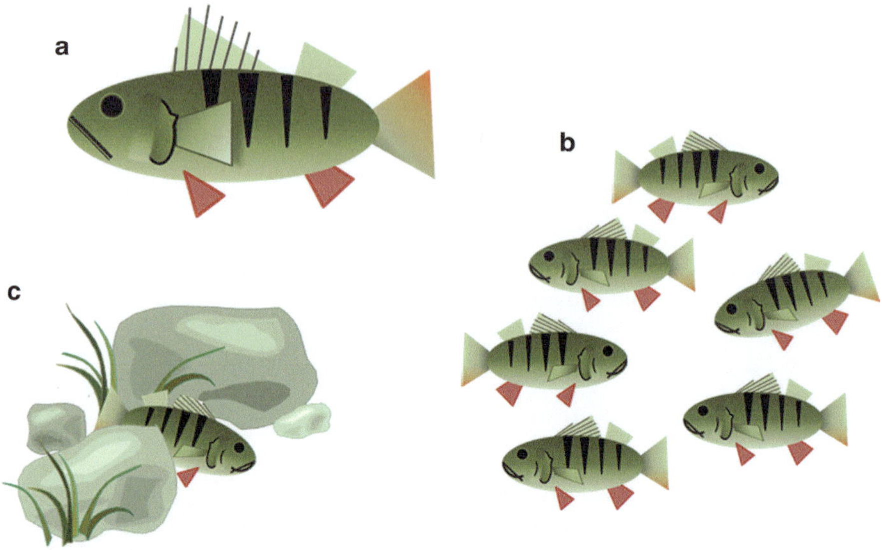

Fig. 15.2 Antipredator behaviours in perch include (**a**) erecting the spiny dorsal fin to discourage attack, (**b**) increase shoal cohesiveness to confuse attackers and to benefit from the dilution effect, (**c**) seek shelter and decrease activity to minimise risk of detection

Furthermore, daily vertical migrations were found both in 0+ Eurasian perch and in pikeperch, with lower abundances in the water column during daytime than during night (Kratochvíl et al. 2010). These migrations were considered to be governed by predation avoidance rather than by feeding. Also Eckmann and Imbrock (1996) suggested that the daily vertical migrations in perch during winter were due to predator avoidance. By staying near the bottom during the day, the fish avoid predation from piscivorous birds, like goosander (*Merganser merganser*) and cormorants (*Phalocrocorax carbo*). They may also decrease the risk of being caught by night-active bottom-dwelling burbot (*Lota lota*) by moving upwards in the water column at dusk (Eckmann and Imbrock 1996). In the yellow perch, larval and juvenile perch showed horizontal migrations, from the near shore to offshore at dusk and vice versa (Post and McQueen 1988), but predator avoidance was only one of three possible explanations (the others being prey availability and temperature changes).

15.4 Social Behaviour

The perch is a social fish, that is, it generally occurs in shoals or small groups. However, perch are not schooling in the same sense as herring (e.g. *Clupea* spp.) or mackerel and tuna (e.g. *Scomber* spp.), with synchronised movements and high cohesiveness. The shoal sizes are large in the beginning of life, but decreases with size, and large individuals can even become solitary (Bruylants et al. 1986; LeCren 1992). Still, in pond experiments, growth rates of piscivorous Eurasian perch were higher in groups of five than when they were alone (Eklöv 1992). In the laboratory, juvenile perch showed a higher growth rate in groups than when kept individually, even though food intake was higher in the singletons (Strand et al. 2007a). The results were suggested to be caused by increased stress levels in the solitary fish. This was supported in another study showing higher respiration rates in solitary perch compared to those in groups, probably due to a calming effect of the presence of conspecifics (Schleuter et al. 2007). Shoaling of young-of-the-year perch is seen to vary over the daily cycle. Eurasian perch aggregate densely in the epilimnion during daytime and disperse evenly below the surface at night (Probst et al. 2009). Yellow perch show a similar breakup of groups at twilight (Helfman 1979). The social behaviour of pikeperch is similar to that of the perch, shoaling as young fish but becoming solitary with age. In contrast, walleye live in shoals during their whole life, even as piscivores (Craig 2000).

In Lake Constance two genetically different populations of Eurasian perch occur sympatrically (Behrmann-Godel et al. 2006). Genetic studies on the pelagic perch larvae showed that shoals during early ontogeny were kin-structured. Despite females spawning in close proximity to each other, siblings stay together (Gerlach et al. 2001). This might suggest that perch possess kin preferences and kin recognition. Choice tests in a fluviarium showed preference for odours of unfamiliar kin vs unfamiliar non-kin, and the authors suggested that assortative mating is causing the divergence in lake Constance perch (Behrmann-Godel et al. 2006).

Aggressive interactions are rarely seen among perch in a shoal (Staffan et al. 2002) although it occurs occasionally (Westerberg et al. 2004). In aquarium studies on small groups of 0+ Eurasian perch, no aggression during feeding was found, although a competition for shelters provoked aggressiveness and created a social hierarchy among individuals (Mikheev et al. 2005). A size divergence in 0+ perch in pond experiments was explained by dominance structures and social interactions (Huss et al. 2008). In yellow perch, at high densities, the 0+ split into groups with faster growing littoral fish and slower growing pelagic fish, respectively, probably due to individual differences in competitive ability (Post et al. 1997). In the laboratory, when feeding young Eurasian perch, some individuals took more of the food than others, and their share of the total food provided was consistent over an experimental period of 10 days (Westerberg et al. 2004), and even over a time-span of 8 months (Staffan et al. 2005). Thus, even in the absence of overt aggression, dominance hierarchies may be created in more subtle ways, undetectable for the human eye.

15.5 Behaviour Differences Between Individuals and Populations

Competitive ability, as mentioned above, can vary among individuals in a perch shoal, hence leading to unequal feeding opportunities (e.g. Staffan et al. 2005). Also other behaviour traits can differ within a species. Studies on individual variation in behaviour have been carried out in many taxa, including numerous fish species (Gosling 2001; Dingemanse and Reale 2005). In accordance with the observed behaviour, individuals have been divided into coping style categories (e.g. Koolhaas et al. 1999; Øverli et al. 2004) or been arranged along a behaviour gradient, such as the bold/shy continuum (Wilson et al. 1993; Gosling 2001). Boldness of young Eurasian perch has been estimated in aquarium studies as the trade-off between feeding in the vicinity of a piscivorous perch and hiding in the vegetation without food (e.g. Magnhagen and Borcherding 2008, Fig. 15.3). There is a clear variation among individuals in the time spent near the predator and the latency to start feeding. Because of the social structure in perch, they have generally been tested in small groups rather than alone. However, individual fish are influenced by the behaviour of the others in the group, and behaviour differences between groups are often larger than the variation within groups (e.g. Magnhagen 2012). When tested alone, most perch showed a shyer behaviour than when in groups, but still the individual boldness scores were correlated when comparing the same fish in both treatments (Magnhagen and Bunnefeld 2009). The question is whether the specific behaviour of an individual perch depends on inherent traits, which has been found in other fish (Magurran 1990; Øverli et al. 2002; Brown et al. 2007), or if it is a phenotypic response to its natural environment. Differences in risk-taking behaviour have been found when comparing perch populations with differences in predation pressure, which seems to indicate that the behaviour is selected for (Magnhagen 2006). However, when raising perch from two lakes in a predator-free environment in a common garden experiment, the results showed that boldness mainly seemed to be shaped by experience (Hellström and Magnhagen 2011). Furthermore, a long-term

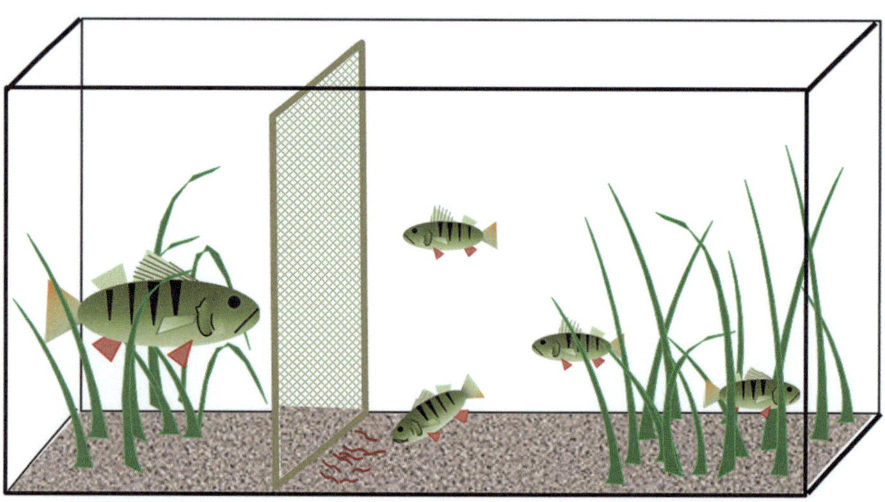

Fig. 15.3 Experimental set-up to test for variation among young perch in the behavioural trade-off between foraging (in a risky place) and hiding (with no food), in the presence of a potential predator

study showed that, even if the perch in one lake were consistently bolder than those in another lake, the magnitude of the difference varied across years, and the observed behaviour was correlated with year-specific risk of cannibalism in both lakes (Magnhagen et al. 2012). That perch behaviour has some genetic component cannot be completely ruled out, but it seems likely that divergent experience is the most important factor for the variation observed.

Between-population differences in activity have been found, using underwater video observations on 1+ yellow perch in two lakes in Quebec (Aubin-Horth et al. 1999). These differences were also connected to differences in consumption and growth rates, with slower growth in perch from the high-activity lake compared to those in the low-activity lake. From an analysis of 11 populations of yellow perch, Boisclair and Rasmussen (1996) suggested that consumption and activity rates would be connected to prey size distribution, total prey biomass, and water transparency. Thus, environmental conditions are, to a high extent, explaining behavioural differences between perch populations.

15.6 Spawning Behaviour

The spawning event of Eurasian perch was described by Treasurer (1981), who filmed the behaviour in a large aquarium, and he also refers to older studies on yellow perch, describing similar behaviour (Harrington 1947 and Hergenrader 1956 in Treasurer 1981). A female ready to spawn is followed by a group of several males, and when finding a submerged tree branch or some other suitable structure she releases her eggs in a single strand, and the males dash in and release the milt (Fig. 15.4, see also Craig 2000). According to Fabricius (1956) the female stays for a while after spawning and chase away the males. Why this would be the case is not clear, since the eggs are well protected from predation, due to a thick gelatinous coating of the egg mass. In fact, a study looking at egg predation by several fish and invertebrate species showed that the perch roe was not eaten by anyone (Newsome and Tompkins 1985). In contrast to Fabricius (1956), Wirtz and Steinmann (2006) state that perch males show three different mating tactics depending on body size, with dominant, group spawners and sneakers, respectively, but this is only referring to unpublished data. Thus, information on perch spawning behaviour is fragmentary and contradictory. There is an obvious need for more scientific studies on the spawning behaviour in perch.

Spawning habitat requirements are studied in more detail (e.g. Čech et al. 2009; Probst et al. 2009; Smith et al. 2001). Spawning habitat in perch is selected on the base of substrate, wave exposure, temperature and depth (Snickars et al. 2010). The most important factor is the substrate, with a preference for rigid and complex structures, like reeds and other aquatic plants, and dead tree branches. These structures, especially the dead branches, are beneficial for keeping the eggs oxygenated, while a higher mortality is found on egg strands located on the sediment (Smith et al. 2001).

Fig. 15.4 During spawning, a female perch is releasing all her eggs simultaneously, placing the gelatinous egg strand on vegetation or submerged dead branches. Several males can fertilize the eggs in one batch

Sheltered areas (Probst et al. 2009) and shallow depths (Snickars et al. 2010) are also preferred for spawning. A transplant experiment in Lake Zürich tested the development of perch roes at different depths (Zeh et al. 1989). There was no effect of depth on hatching success, but, as would be expected, eggs developed slower at deeper and colder areas. However, turbidity could also affect the selection of spawning depth, as in the yellow perch that spawned at deeper locations in a clear water lake than in a lake with more dissolved organic material (Huff et al. 2004). The use of the deeper spawning area, leading to longer development times, may be explained with the avoidance of high solar ultraviolet radiation that has a negative effect on perch egg survival.

The spawning behaviour of pikeperch has been reviewed by Lappalainen et al. (2003). The behaviour differs from the perch species, with the males building a nest and defending the eggs after spawning. The pikeperch spawn in pairs, and seems to be a monogamous species. After courtship, egg laying and fertilization, the male chases away the female, and then stays to guard and fan the eggs until they hatch. The eggs are glutinous and attached singly to the ground or on plant roots (Lappalainen et al. 2003). In contrast, walleye spawn randomly over suitable substrates and do not provide any parental care (Malison and Held 1996).

Spawning migrations occur in the Eurasian perch, for example, from the brackish waters of the Baltic and Bothnian sea to spawning grounds upstream into fresh water (Berglund 1978), or from reservoirs for spawning in the river habitat (Hladík and Kubečka 2003). Also in stationary lake-dwelling perch a short-ranged migration occurs in the spring, with movement from deeper waters where they spend the winter, to the shallow spawning areas. Homing behaviour has been found in yellow perch, with much movement after spawning but high returns for the next spawning season (Glover et al. 2008). Although there was a high fidelity to some areas, straying from all sites occurred, promoting gene flow. Also pikeperch show spawning migration and homing, both in freshwater and brackish populations (reviewed in Lappalainen et al. 2003). Migrations are performed in early spring from deeper waters to the spawning grounds in shallow waters, river inlets or even in rivers. In walleye, genetic differences among spawning groups in Lake Erie indicated the occurrence of natal site fidelity but also some straying (Strange and Stepien 2007).

15.7 Behaviour in Percid Aquaculture

For the best result in culture it is very important to have knowledge about the behaviour of percids in the wild, especially regarding feeding and social interactions. For example, the extensive cannibalism in the natural habitat is of high relevance in a farming situation. In contrast to the salmonids commonly farmed, the percids generally have bigger mouths and can ingest prey of large relative size, often up to almost half their own size (Fig. 15.5). Thus, size differences do not need to be very large in order for cannibalism to appear. Consequently, the degree of size heterogeneity was not found to affect amount of cannibalism in cultured larvae and juvenile perch (Kestemont et al. 2003; Mandiki et al. 2007). A higher stocking density leads to lower rate of cannibalism in perch (Baras et al. 2003). This may be because attacks are more difficult to perform successfully, but also because aggression is more costly in denser crowds (Grant 1993). Similarly, in the Arctic charr (*Salvelinus alpinus*) agonistic behaviour decreased at higher stocking densities (Brown et al. 1992), and thus lower energy expenditure lead to higher growth rates. However, no connection with stocking density was found in the high rate of cannibalism (>40 %) found in fingerling pikeperch, weaned in ponds (Molnar et al. 2004).

The occurrence of cannibalism is probably connected with social hierarchies allowing some individuals to consistently get a higher proportion of the food distributed, as found in experimental studies of perch (Staffan et al. 2002, 2005). Thereby, the most voracious fish get a higher growth rate than others and can quickly become large enough to prey on the smaller conspecifics. Even though perch are not overtly aggressive, a resource that is defendable may invoke aggression (Grant 1993). This was found in perch competing for shelters (Mikheev et al. 2005). Similarly, a feeding station is also sometimes possible to monopolise, as shown in Arctic charr and rainbow trout (*Oncorhynchus mykiss*) (Brännäs and Alanärä 1993). Thus, the distribution

Fig. 15.5 Young-of-the-year perch caught in August and kept together in a 1 m^3 tank for several months. At capture all perch were of a similar size, but some grew faster and became cannibals. Inside the mouth of the larger perch a tail fin was shown, similar in size to the one of the smaller perch (Photo: Carin Magnhagen)

of food in space and time is important to consider in fish farming, with a more unpredictable feeding scheme decreasing the heterogeneity in food intake (Kadri et al. 1996).

Social interactions within a group of fish are, of course, not only negative. Social environment seemed to affect learning in perch trained to feed on pellets, with a positive result of the presence of demonstrators, already used to the diet (Magnhagen and Staffan 2003). An effect of density on pikeperch survival rates during weaning was explained by better learning abilities at higher stocking densities (Molnar et al. 2004). In the opposite situation, when fish are farmed for subsequent stocking in open waters, social learning can be used to prepare the fish for a more natural environment (Brown and Laland 2001).

Domestication of fish may change behaviour patterns compared to wild stocks (Price 1999). Traits that are required by the fish farmer can be genetically linked to more negative characteristics. For example, selection for fast growth in salmon (*Salmo salar*) may increase aggression (Huntingford and Adams 2005). This is obviously an unwanted effect, which affects both production and welfare in the farming environment. In percids, to my knowledge, no studies have so far looked at the effect of domestication on behaviour. However, Douxfils et al. (2011) showed that the domestication of Eurasian perch may decrease stress levels, in terms of cortisol response to a confinement test. That is an encouraging finding, since perch have been found to decrease food intake and consequently growth when disturbed in rearing tanks, as compared to when not disturbed (Strand et al. 2007b). In rainbow trout, levels of stress response were connected with aggression and dominance (Øverli et al. 2004). A selection aimed at stress tolerance rather than at growth rates per se might thus get positive results for fish production. However, more research within this area is needed.

15.8 Conclusions

Within the percid family, most studies on behaviour have been done of the Eurasian perch, which has served as a model species for several questions within the field of ecology. Behaviour is studied mainly on a small scale, in aquaria with a low density of fish or in the field with different techniques. No research on behaviour in a full-scale percid farming environment have been found, but investigations of growth rates and mortality can give indirect indications of competition and cannibalism. Knowledge of behaviour in wild stocks can give some hints about which issues are important to look further into. For example, the relative importance of genetics and phenotypic plasticity in behaviour, potentially affecting stress response and aggression is one of these questions. Farming of percids has started to develop quite recently, compared with for example, salmonids, and the demand for a better understanding of behaviour in culture will increase.

References

Aubin-Horth N, Gingras J, Boisclair D (1999) Comparison of activity rates of 1+yellow perch (*Perca flavescens*) from populations of contrasting growth rates using underwater video observations. Can J Fish Aquat Sci 56(6):1122–1132. doi:10.1139/cjfas-56-6-1122

Baras E, Kestemont P, Melard C (2003) Effect of stocking density on the dynamics of cannibalism in sibling larvae of Perca fluviatilis under controlled conditions. Aquaculture 219(1–4):241–255. doi:10.1016/s0044-8486(02)00349-6

Behrmann-Godel J, Gerlach G, Eckmann R (2006) Kin and population recognition in sympatric Lake Constance perch (*Perca fluviatilis* L.): can assortative shoaling drive population divergence? Behav Ecol Sociobiol 59(4):461–468. doi:10.1007/s00265-005-0070-3

Berglund I (1978) Spawning migration of the perch, *Perca fluviatilis* L., in a subarctic Swedish coastal stream. Aquilo Ser Zool 18:43–48

Boisclair D, Rasmussen JB (1996) Empirical analysis of the influence of environmental variables associated with lake eutrophication on perch growth, consumption, and activity rates. Ann Zool Fenn 33(3–4):507–515

Borcherding J, Beeck P, DeAngelis DL, Scharf WR (2010) Match or mismatch: the influence of phenology on size-dependent life history and divergence in population structure. J Anim Ecol 79(5):1101–1112. doi:10.1111/j.1365-2656.2010.01704.x

Braband A (1995) Intracohort cannibalism among larval stages of perch (*Perca fluviatilis*). Ecol Freshw Fish 4:70–76

Brännäs E, Alanärä A (1993) Monitoring the feeding activity of individual fish with a demand feeding system. J Fish Biol 42(2):209–215. doi:10.1006/jfbi.1993.1022

Brown C, Laland K (2001) Social learning and life skills training for hatchery reared fish. J Fish Biol 59(3):471–493. doi:10.1006/jfbi.2001.1689

Brown GE, Brown JA, Srivastava RK (1992) The effect of stocking density on the behavior of Arctic charr (Salvelinus alpinus). J Fish Biol 41(6):955–963. doi:10.1111/j.1095-8649.1992.tb02722.x

Brown C, Burgess F, Braithwaite VA (2007) Heritable and experiential effects on boldness in a tropical poeciliid. Behav Ecol Sociobiol 62(2):237–243. doi:10.1007/s00265-007-0458-3

Bruylants B, Vandelannoote A, Verheyen R (1986) The movement pattern and density distribution of perch, *Perca fluviatilis* L., in a channelized lowland river. Aquac Fish Manag 17(1):49–57

Byström P, Persson L, Wahlström E, Westman E (2003) Size- and density-dependent habitat use in predators: consequences for habitat shifts in young fish. J Anim Ecol 72(1):156–168. doi:10.1046/j.1365-2656.2003.00681.x

Čech M, Peterka J, Říha M, Jůza T, Kubečka J (2009) Distribution of egg strands of perch (*Perca fluviatilis* L.) with respect to depth and spawning substrate. Hydrobiologia 630(1):105–114. doi:10.1007/s10750-009-9783-z

Craig JF (2000) Percid fishes – systematics, ecology and exploitation, Fish and aquatic resources series. Blackwell Science, Oxford

Dingemanse NJ, Reale D (2005) Natural selection and animal personality. Behaviour 142(9–10):1159–1184. doi:10.1163/156853905774539445

Douxfils J, Mandiki SNM, Marotte G, Wang N, Silvestre F, Milla S, Henrotte E, Vandecan M, Rougeot C, Melard C, Kestemont P (2011) Does domestication process affect stress response in juvenile Eurasian perch *Perca fluviatilis*? Comp Biochem Physiol A Mol Integr Physiol 159(1):92–99. doi:10.1016/j.cbpa.2011.01.021

Eckmann R, Imbrock F (1996) Distribution and diel vertical migration of Eurasian perch (*Perca fluviatilis* L) during winter. Ann Zool Fenn 33(3–4):679–686

Eklöv P (1992) Group foraging versus solitary foraging efficiency in piscivorous predators – the perch, *Perca fluviatilis* and pike, *Esox lucius*. Anim Behav 44(2):313–326. doi:10.1016/0003-3472(92)90037-a

Eklöv P, Jonsson P (2007) Pike predators induce morphological changes in young perch and roach. J Fish Biol 70(1):155–164. doi:10.1111/j.1095-8649.2006.01283.x

Eklöv P, Svanbäck R (2006) Predation risk influences adaptive morphological variation in fish populations. Am Nat 167(3):440–452. doi:10.1086/499544

Fabricius E (1956) Hur abborren leker (The spawning behaviour of the perch, *Perca fluviatilis* L.). Zoologisk Revy 18:48–55 (in Swedish)

Galarowicz TL, Adams JA, Wahl DH (2006) The influence of prey availability on ontogenetic diet shifts of a juvenile piscivore. Can J Fish Aquat Sci 63(8):1722–1733. doi:10.1139/f06-073

Gerlach G, Schardt U, Eckmann R, Meyer A (2001) Kin-structured subpopulations in Eurasian perch (*Perca fluviatilis* L.). Heredity 86(2):213–221. doi:10.1046/j.1365-2540.2001.00825.x

Glover DC, Dettmers JM, Wahl DH, Clapp DF (2008) Yellow perch (*Perca flavescens*) stock structure in Lake Michigan: an analysis using mark-recapture data. Can J Fish Aquat Sci 65(9):1919–1930. doi:10.1139/f08-100

Gosling SD (2001) From mice to men: what can we learn about personality from animal research? Psychol Bull 127(1):45–86

Graeb BDS, Galarowicz T, Wahl DH, Dettmers JM, Simpson MJ (2005) Foraging behavior, morphology, and life history variation determine the ontogeny of piscivory in two closely related predators. Can J Fish Aquat Sci 62(9):2010–2020. doi:10.1139/f05-112

Graeb BDS, Mangan MT, Jolley JC, Wahl DH, Dettmers JM (2006) Ontogenetic changes in prey preference and foraging ability of yellow perch: insights based on relative energetic return of prey. Trans Am Fish Soc 135(6):1493–1498. doi:10.1577/t05-063.1

Grant JWA (1993) Whether or not to defend – the influence of resource distribution. Mar Behav Physiol 23(1–4):137–153

Hambright KD (1991) Experimental analysis of prey selection by largemouth bass – role of predator mouth width and prey body depth. Trans Am Fish Soc 120(4):500–508. doi:10.1577/1548-8659(1991)120<0500:eaopsb>2.3.co;2

Helfman GS (1979) Twilight activities of yellow perch, *Perca flavescens*. J Fish Res Board Can 36(2):173–179

Hellström G, Magnhagen C (2011) The influence of experience on risk taking: results from a common-garden experiment on populations of Eurasian perch. Behav Ecol Sociobiol 65(10):1917–1926. doi:10.1007/s00265-011-1201-7

Hladík M, Kubečka J (2003) Fish migration between a temperate reservoir and its main tributary. Hydrobiologia 504(1–3):251–266

Huff DD, Grad G, Williamson CE (2004) Environmental constraints on spawning depth of yellow perch: the roles of low temperature and high solar ultraviolet radiation. Trans Am Fish Soc 133(3):718–726. doi:10.1577/t03-048.1

Huntingford F, Adams C (2005) Behavioural syndromes in farmed fish: implications for production and welfare. Behaviour 142(9–10):1207–1221

Huss M, Byström P, Persson L (2008) Resource heterogeneity, diet shifts and intra-cohort competition: effects on size divergence in YOY fish. Oecologia 158(2):249–257. doi:10.1007/s00442-008-1140-9

Kadri S, Huntingford FA, Metcalfe NB, Thorpe JE (1996) Social interactions and the distribution of food among one-sea-winter Atlantic salmon (*Salmo salar*) in a sea-cage. Aquaculture 139(1–2):1–10. doi:10.1016/0044-8486(95)01163-3

Kekalainen J, Huuskonen H, Kiviniemi V, Taskinen J (2010) Visual conditions and habitat shape the coloration of the Eurasian perch (*Perca fluviatilis* L.): a trade-off between camouflage and communication? Biol J Linn Soc 99(1):47–59

Kestemont P, Jourdan S, Houbart M, Melard C, Paspatis M, Fontaine P, Cuvier A, Kentouri M, Baras E (2003) Size heterogeneity, cannibalism and competition in cultured predatory fish larvae: biotic and abiotic influences. Aquaculture 227(1–4):333–356. doi:10.1016/s0044-8486(03)00513-1

Koolhaas JM, Korte SM, De Boer SF, Van Der Vegt BJ, Van Reenen CG, Hopster H, De Jong IC, Ruis MAW, Blokhuis HJ (1999) Coping styles in animals: current status in behavior and stress-physiology. Neurosci Biobehav Rev 23(7):925–935

Kratochvíl M, Čech M, Vašek M, Kubečka J, Hejzlar J, Matěna J, Peterka J, Macháček J, Seďa J (2010) Diel vertical migrations of age 0(+) percids in a shallow, well-mixed reservoir. J Limnol 69(2):305–310. doi:10.3274/jl10-69-2-12

Krause J, Ruxton GD (2002) Living in groups, Oxford series in ecology and evolution. Oxford University Press, Oxford

Lappalainen J, Dorner H, Wysujack K (2003) Reproduction biology of pikeperch (*Sander lucioperca*(L.))–a review. Ecol Freshw Fish 12(2):95–106. doi:10.1034/j.1600-0633.2003.00005.x

LeCren ED (1992) Exceptionally big individual perch (*Perca fluviatilis* L.) and their growth. J Fish Biol 40(4):599–625

Lundvall D, Svanbäck R, Persson L, Byström P (1999) Size-dependent predation in piscivores: interactions between predator foraging and prey avoidance abilities. Can J Fish Aquat Sci 56(7):1285–1292. doi:10.1139/cjfas-56-7-1285

Magnhagen C (2006) Risk-taking behaviour in foraging young-of-the-year perch varies with population size structure. Oecologia 147(4):734–743. doi:10.1007/s00442-005-0302-2

Magnhagen C (2012) Personalities in a crowd: what shapes the behaviour of Eurasian perch and other shoaling fishes? Curr Zool 58(1):35–44

Magnhagen C, Borcherding J (2008) Risk-taking behaviour in foraging perch: does predation pressure influence age-specific boldness? Anim Behav 75(2):509–517. doi:10.1016/j.anbehav.2007.06.007

Magnhagen C, Bunnefeld N (2009) Express your personality or go along with the group: what determines the behaviour of shoaling perch? Proc R Soc B Biol Sci 276(1671):3369–3375. doi:10.1098/rspb.2009.0851

Magnhagen C, Staffan F (2003) Social learning in young-of-the-year perch encountering a novel food type. J Fish Biol 63(3):824–829. doi:10.1046/j.1095-8649.2003.00189.x

Magnhagen C, Hellström G, Borcherding J, Heynen M (2012) Boldness in two perch populations – long-term differences and the effect of predation pressure. J Anim Ecol 81(6):1311–1318. doi:10.1111/j.1365-2656.2012.02007.x

Magurran AE (1990) The inheritance and development of minnow antipredator behaviour. Anim Behav 39(5):834–842

Malison JA, Held JA (1996) Reproduction and spawning in walleye (*Stizostedion vitreum*). J Appl Ichthyol 12(3-4):153–156. doi:10.1111/j.1439-0426.1996.tb00081.x

Mandiki SNM, Babiak I, Krol J, Rasolo JFR, Kestemont P (2007) How initial predator-prey ratio affects intra-cohort cannibalism and growth in Eurasian perch *Perca fluviatilis* L larvae and juveniles under controlled conditions. Aquaculture 268(1–4):149–155. doi:10.1016/j.aquaculture.2007.04.036

Mikheev VN, Pasternak AF, Tischler G, Wanzenbock J (2005) Contestable shelters provoke aggression among 0 + perch, *Perca fluviatilis*. Environ Biol Fish 73(2):227–231. doi:10.1007/s10641-005-0558-8

Mirza RS, Chivers DP (2001) Do juvenile yellow perch use diet cues to assess the level of threat posed by intraspecific predators? Behaviour 138:1249–1258. doi:10.1163/15685390152822201

Mirza RS, Fisher SA, Chivers DP (2003) Assessment of predation risk by juvenile yellow perch, *Perca flavescens*: responses to alarm cues from conspecifics and prey guild members. Environ Biol Fish 66(4):321–327. doi:10.1023/a:1023966320018

Mittelbach GG, Persson L (1998) The ontogeny of piscivory and its ecological consequences. Can J Fish Aquat Sci 55(6):1454–1465

Molnar T, Hancz C, Bodis M, Muller T, Bercsemyi M, Horn P (2004) The effect of initial stocking density on growth and survival of pike-perch fingerlings reared under intensive conditions. Aquac Int 12(2):181–189. doi:10.1023/B:AQUI.0000032079.62056.8c

Newsome GE, Tompkins J (1985) Yellow perch egg masses deter predators. Can J Zool 63(12):2882–2884

Olsson J, Eklöv P (2005) Habitat structure, feeding mode and morphological reversibility: factors influencing phenotypic plasticity in perch. Evol Ecol Res 7(8):1109–1123

Øverli Ø, Pottinger TG, Carrick TR, Øverli E, Winberg S (2002) Differences in behaviour between rainbow trout selected for high- and low-stress responsiveness. J Exp Biol 205(3):391–395

Øverli Ø, Korzan WJ, Höglund E, Winberg S, Bollig H, Watt M, Forster GL, Barton BA, Øverli E, Renner KJ, Summers CH (2004) Stress coping style predicts aggression and social dominance in rainbow trout. Horm Behav 45(4):235–241

Persson L (1988) Asymmetries in competitive and predatory interactions in fish populations. In: Ebenmann B, Persson L (eds) Size-structured populations: ecology and evolution. Springer, Berlin/Heidelberg, pp 203–218

Persson A, Brönmark C (2002) Foraging capacity and resource synchronization in an ontogenetic diet switcher, pikeperch (*Stizostedion lucioperca*). Ecology 83(11):3014–3022. doi:10.2307/3071838

Persson L, Byström P, Wahlström E (2000) Cannibalism and competition in Eurasian perch: population dynamics of an ontogenetic omnivore. Ecology 81(4):1058–1071

Persson L, Claessen D, DeRoos AM, Byström P, Sjögren S, Svanbäck R, Wahlström E, Westman E (2004) Cannibalism in a size-structured population: energy extraction and control. Ecol Monogr 74(1):135–157

Pitcher TJ, Parrish JK (1993) Functions of shoaling behaviour in teleosts. In: Pitcher TJ (ed) Behaviour of teleost fishes. Chapman & Hall, London, pp 363–439

Post JR, McQueen DJ (1988) Ontogenetic changes in the distribution of larval and juvenile yellow perch (*Perca flavescens*) – a response to prey or predators? Can J Fish Aquat Sci 45(10):1820–1826. doi:10.1139/f88-214

Post JR, Johannes MRS, McQueen DJ (1997) Evidence of density-dependent cohort splitting in age-0 yellow perch (*Perca flavescens*): potential behavioural mechanisms and population-level consequences. Can J Fish Aquat Sci 54(4):867–875. doi:10.1139/cjfas-54-4-867

Price EO (1999) Behavioral development in animals undergoing domestication. Appl Anim Behav Sci 65(3):245–271. doi:10.1016/s0168-1591(99)00087-8

Probst WN, Stoll S, Hofmann H, Fischer P, Eckmann R (2009) Spawning site selection by Eurasian perch (*Perca fluviatilis* L.) in relation to temperature and wave exposure. Ecol Freshw Fish 18(1):1–7. doi:10.1111/j.1600-0633.2008.00327.x

Quevedo M, Svanbäck R, Eklöv P (2009) Intrapopulation niche partitioning in a generalist predator limits food web connectivity. Ecology 90(8):2263–2274. doi:10.1890/07-1580.1

Roberts JJ, Grecay PA, Ludsin SA, Pothoven SA, Vanderploeg HA, Hook TO (2012) Evidence of hypoxic foraging forays by yellow perch (*Perca flavescens*) and potential consequences for prey consumption. Freshw Biol 57(5):922–937. doi:10.1111/j.1365-2427.2012.02753.x

Schael DM, Rudstam LG, Post JR (1991) Gape limitation and prey selection in larval yellow perch (*Perca flavescens*), fresh-water drum (*Aplodinotus grunniens*), and black crappie (*Pomoxis nigromaculatus*). Can J Fish Aquat Sci 48(10):1919–1925. doi:10.1139/f91-228

Schleuter D, Haertel-Borer S, Fischer P, Eckmann R (2007) Respiration rates of Eurasian perch *Perca fluviatilis* and ruffe: lower energy costs in groups. Trans Am Fish Soc 136(1):43–55. doi:10.1577/t06-123.1

Smith C, Douglas A, Jurajda P (2001) Oviposition site selection and embryo mortality in perch. J Fish Biol 58(3):880–882. doi:10.1006/jfbi.2000.1491

Snickars M, Sundblad G, Sandström A, Ljunggren L, Bergström U, Johansson G, Mattila J (2010) Habitat selectivity of substrate-spawning fish: modelling requirements for the Eurasian perch *Perca fluviatilis*. Mar Ecol Prog Ser 398:235–243. doi:10.3354/meps08313

Staffan F, Magnhagen C, Alanärä A (2002) Variation in food intake within groups of juvenile perch. J Fish Biol 60(3):771–774

Staffan F, Magnhagen C, Alanärä A (2005) Individual feeding success of juvenile perch is consistent over time in aquaria and under farming conditions. J Fish Biol 66(3):798–809

Strand A, Alanärä A, Magnhagen C (2007a) Effect of group size on feed intake, growth and feed efficiency of juvenile perch. J Fish Biol 71(2):615–619. doi:10.1111/j.1095-8649.2007.01497.x

Strand A, Magnhagen C, Alanärä A (2007b) Effects of repeated disturbances on feed intake, growth rates and energy expenditures of juvenile perch, *Perca fluviatilis*. Aquaculture 265(1–4):163–168. doi:10.1016/j.aquaculture.2007.01.030

Strange RM, Stepien CA (2007) Genetic divergence and connectivity among river and reef spawning groups of walleye (*Sander vitreus vitreus*) in Lake Erie. Can J Fish Aquat Sci 64(3):437–448. doi:10.1139/f07-022

Svanbäck R, Eklöv P (2002) Effects of habitat and food resources on morphology and ontogenetic growth trajectories in perch. Oecologia 131(1):61–70. doi:10.1007/s00442-001-0861-9

Svanbäck R, Eklöv P (2003) Morphology dependent foraging efficiency in perch: a trade-off for ecological specialization? Oikos 102(2):273–284. doi:10.1034/j.1600-0706.2003.12657.x

Svanbäck R, Eklöv P (2006) Genetic variation and phenotypic plasticity: causes of morphological and dietary variation in Eurasian perch. Evol Ecol Res 8(1):37–49

Svanbäck R, Eklöv P (2011) Catch me if you can – predation affects divergence in a polyphenic species. Evolution 65(12):3515–3526. doi:10.1111/j.1558-5646.2011.01398.x

Treasurer JW (1981) Some aspects of the reproductive biology of perch *Perca fluviatilis* L. Fecundity, maturation and spawning behavior. J Fish Biol 18(6):729–740. doi:10.1111/j.1095-8649.1981.tb03814.x

Treasurer JW (1989) Mortality and production of 0+ perch. Perca fluviatilis L. in two Scottish lakes. J Fish Biol 34(6):913–928

Vainikka A, Jokelainen T, Kortet R, Ylonen H (2005) Predation risk allocation or direct vigilance response in the predator interaction between perch (*Perca fluviatilis* L.) and pike (*Esox lucius* L.)? Ecol Freshw Fish 14(3):225–232. doi:10.1111/j.1600-0633.2005.00095.x

Wahlström E, Persson L, Diehl S, Byström P (2000) Size-dependent foraging efficiency, cannibalism and zooplankton community structure. Oecologia 123(1):138–148. doi:10.1007/s004420050999

Westerberg M, Staffan F, Magnhagen C (2004) Influence of predation risk on individual competitive ability and growth in Eurasian perch, *Perca fluviatilis*. Anim Behav 67:273–279

Wilson DS, Coleman K, Clark AB, Biederman L (1993) Shy-bold continuum in pumpkinseed sunfish (*Lepomis gibbosus*): an ecological study of a psychological trait. J Comp Psychol 107:250–260

Wirtz S, Steinmann P (2006) Sperm characteristics in perch *Perca fluviatilis* L. J Fish Biol 68(6):1896–1902. doi:10.1111/j.1095-8649.2006.01065.x

Zamora L, Moreno-Amich R (2002) Quantifying the activity and movement of perch in a temperate lake by integrating acoustic telemetry and a geographic information system. Hydrobiologia 483:209–218. doi:10.1023/a:1021396016424

Zeh M, Ritter E, Ribi G (1989) Spawning and egg development of *Perca fluviatilis* in Lake Zürich. Aquat Sci 51(2):100–107. doi:10.1007/bf00879297

Chapter 16
Culture Methods of Eurasian Perch During Ongrowing

Tomáš Policar, Azin Mohagheghi Samarin, and Charles Mélard

Abstract Three different production systems are used for perch ongrowing: (1) traditional extensive polyculture system, (2) semi-intensive culture farming and (3) intensive perch farming under RAS (Recirculating Aquaculture System(s)). Extensive and semi-intensive culture systems have many production limitations. Therefore, intensive perch farming has been developed in Europe for more continuous and predictive marketable perch production.

Marketable perch production under RAS is affected by several main factors of production system. Optimal value and condition of each factor for stable and maximal perch production under RAS are described and recommended in details in this chapter.

Overall, white, grey and black tank walls with light regime 12L:12D or 18 L:8D and light intensity 200–1,100 lx create optimal light conditions for intensive ongrowing perch culture. Freshwater or water with salinity under 4‰ with temperature 22–24 °C, oxygen saturation around 60–72 % and very low ammonia (below 0.3 mg $N\text{-}NH_3 \cdot L^{-1}$) and nitrite (below 0.5 mg $NO_2^- \cdot L^{-1}$) concentrations are optimal conditions for intensive perch production. Disturbance (cleaning of tanks, fish size-sorting etc.) must be reduced at minimum level for providing of maximal production which is the highest under optimal fish biomass from 10 to 20 $kg \cdot m^{-3}$ for 10 g perch to 60–70 $kg \cdot m^{-3}$ for 150 g perch under RAS.

Keywords *Perca fluviatilis* • Growth • Intensive rearing • Semi-intensive culture • Extensive farming

T. Policar (✉) • A.M. Samarin
Faculty of Fisheries and Protection of Waters (FFPW), South Bohemian Research Center of Aquaculture and Biodiversity of Hydrocenoses, University of South Bohemia in České Budějovice, Zátiší 728/II, 389-25 Vodňany, Czech Republic
e-mail: policar@frov.jcu.cz

C. Mélard
Aquaculture Research and Education Center (CEFRA), University of Liège, Chemin de la Justice, 10, 4500 Tihange, Belgium

P. Kestemont et al. (eds.), *Biology and Culture of Percid Fishes*,
DOI 10.1007/978-94-017-7227-3_16

16.1 Introduction

Nowadays, three different production systems for culture of Eurasian perch (*Perca fluviatilis*) during ongrowing phase (body weight ranging from 1–2 to 100–300 g) can be recognized (Kestemont et al. 2008). Ongrowing of perch has traditionally taken place in production ponds and reservoirs under extensive polyculture systems (Tamazouzt et al. 1993). Semi-intensive culture using cage farming in lakes or sea bays or the combination of pond and RAS (Recirculating Aquaculture System(s)) culture in ongrowing perch has been used in different countries such as: Switzerland (Janssens 2013, personal communication), Sweden (Òberg 2008), Germany (Schmidt and Wedekind 2008) and Czech Republic (Policar et al. 2009). However, conditions for more intensive aquaculture of perch have been investigated over last 20 years (Overton and Paulsen 2005). Therefore, intensive perch farming under RAS has been developed and used for more predictive marketable fish production mainly in Switzerland, Ireland and France (Wernicke von Siebenthal 2013; Toner 2012; Martin and Vandevorede 2008).

16.2 Extensive Perch Farming Under Pond Conditions

Perch farming in extensive polyculture system accounts for a substantial portion of marketable perch production, especially in central and eastern part of Europe (Kestemont et al. 2008) including following countries: Russia, Ukraine, Czech Republic, Romania, Latvia and Bulgaria (FAO 2013). French perch production from polyculture ponds occurs in three specific areas (Dombes, Lorraine and Brenne) of north-east France (Tamazouzt 2008).

The perch production cycle takes 3–4 years under extensive pond culture to produce a 250–400 g market size (Policar et al. 2009). Marketable perch are harvested maximum twice per year, once during autumn and once during spring harvest season when ponds are harvested (Kratochvíl 2012).

Young perch (0+; final TL around 70 mm) are produced in monoculture system with or without the presence of prey fish, when ponds are stocked at density 120,000 fish per hectare. Prey fish (e.g. roach, *Rutilus rutilus*, topmouth gudgeon, *Pseudorasbora parva*, or other small cyprinids species) up to mentioned perch size (TL = 70 mm) have not positive effect on perch growth and survival rate compared to perch culture without prey fish. Macrophytes have a positive effect on macroinvertebrate (phytomacrofauna) community that are the main food for reared perch in ponds. Their abundance seems to be more effective to increase perch production than using of prey (Bláha et al. 2013). Survival rate from larvae up to TL = 70 mm perch ranged from 12 % to 36 % with final perch density of 14–43 thousands fish per hectare. A SGR of 1.3 $\% \cdot day^{-1}$ is recorded during the rearing period from the end of April till the end of September (Bláha et al. 2013).

Ongrowing perch (1+ to 3+) are cultured in ponds of several hectares in polyculture (Policar et al. 2009). In this rearing system, the production of marketable perch represents 0.25–1 % only from the total final biomass of fish stock, which is dominated by cyprinids species such as common carp (*Cyprinus carpio*) 80–85 %, Chinese carps (5–10 %) such as grass carp (*Ctenopharyngodon idella*) and bighead carp (*Hypophthalmichthys molitrix*) (Adámek et al. 2010; Kratochvíl 2012). Perch as a predatory species plays an important role in the control and regulation of the overpopulated and less valued small cyprinids such as: roach, bleak (*Alburnus alburnus*), bream (*Abramis brama*), topmouth gudgeon and ruffe (*Gymnocephalus cernua*) in production ponds (Musil and Adámek 2003; Adámek et al. 2012). Thus, perch is an interesting supplementary fish to regulate carp production in ponds (Adámek et al. 2010).

16.3 Semi-intensive Perch Farming Using Cages and Ponds

Commercial culture of perch in cages was used by Perlac company, Switzerland under ambient water temperature in lake Neuchâtel near to Chez–le–Bart during 1999–2005 (Janssens 2013, personal communication). The Swedish company, Stannafisk AB, followed the same procedure using cages and tarpaulin tanks for perch production in the sea bay near to Östergötland between 2006 and 2011 (Öberg 2008, 2012). Very low fish growth and survival mainly during winter period, poor efficiency of culture system (Öberg 2012), fatal bacterial infection caused by *Aeromonas sobria*, as well as social and political problems with operating fish farms in Swiss lake (Janssens 2013, personal communication) resulted in the failure of this production system for marketable perch production in the mentioned countries.

The greatest drawback of this perch farming system within Europe is suboptimal temperature for growth during whole year. It requires two successive growing seasons (May–October, Fig. 16.1) to obtain marketable-size fish (80–100 g minimum) when water temperature ranges between 14 and 20 °C (Tamazouzt et al. 1993). Temperature during ongrowing season highly affected perch production ranging from 15 to 160 $g \cdot m^{-3} \cdot day^{-1}$ when fish were reared under relatively high final density (30 $kg \cdot m^{-3}$) (Kestemont and Mélard 2000).

Fontaine et al. (1996) recommended using the pond – cage combination system for perch farming. In this system, 1-year old perch of 10–20 g body weight were harvested from ponds during spring and subsequently cultured in cages for two summer seasons. However, the perch growth recorded in this cage culture system was very low and a few fish reached a marketable size (80–100 g). When Tamazouzt et al. (1996) reared perch with initial body weight 25 g in floating cages from July to September survival rate was between 70 % and 79 % and body weight perch, ranging from 48 to 49 g, far less than minimal market size. Perch cultured in cages had a higher protein and a lower lipid and energy content compared to perch reared in recirculating aquaculture system(s) (RAS) under 22 °C (Tamazouzt et al. 1996). Nowadays, this farming system is not widely used for commercial perch production

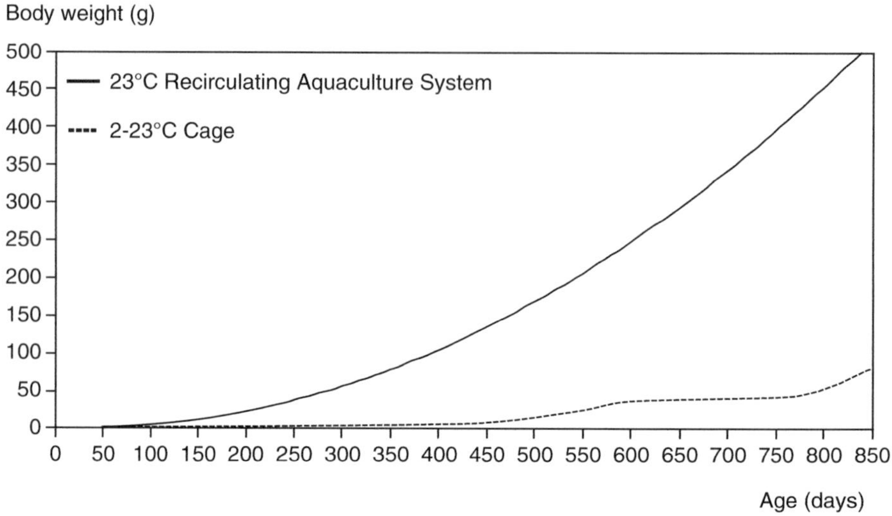

Fig. 16.1 Growth curve of perch in cage under natural fluctuated temperature (2–23 °C) or at stable temperature (23 °C) in RAS (Mélard 2008)

in Europe because it does not provide stable, high-quality and profitable production of marketable fish (Kestemont and Mélard 2000).

16.4 Semi-intensive Perch Farming Using the Combination of Pond and RAS Culture

Initial larval and juvenile culture under pond conditions up to perch of 1–3.2 g body weight has been combined with intensive ongrowing of fish to a commercial size in RAS using artificial food (Schmidt and Wedekind 2008; Stejskal et al. 2009a, 2010). This production system using advantages of both pond and RAS culture systems is very efficient.

Survival rate during habituation of the pond-reared fingerlings was 95 % after one week adaptation and weaning period when perch were fed frozen *Chironomus* or *Chaoborus* larvae with artificial food. The advantages of the pond culture system are: natural food, rapid fish growth, high-quality of produced juveniles, low production cost, elimination of abnormal fish, environmentally friendly and sustainable fish production without any fish deformity. RAS provides controlled and stable production conditions throughout year resulting in rapid growth rate and thus shorter production cycle of marketable fish (Stejskal et al. 2010; Policar et al. 2013). Currently, the greatest limitation of this system for perch rearing is only one batch production per year, when initial pond culture of this system is possible to carry out during later spring or summer period. The second limitation is the risk to introduce diseases in the RAS when stocking with juveniles fish coming from ponds (Schmidt and Wedekind 2008; Stejskal et al. 2010). This system is successfully used mainly in countries of Central Europe (Czech Republic, Germany, Hungary etc.) where large pond area is available (Policar et al. 2011).

16.5 Intensive Perch Farming Under RAS

Intensive perch culture in recirculating aquaculture system(s) provides optimal culture conditions for rapid fish growth, high survival rate, shorter production cycle, year round and predictable production, reduction of fish stress and cannibalism. In order to ensure a high productivity and reduced production costs, several rearing conditions have to be optimized (Fig. 16.1, Kestemont and Mélard 2000): colour of rearing tank walls (Staffan 2004; Mairesse et al. 2005; Strand et al. 2007a), light regime and intensity (Jourdan et al. 2000; Strand et al. 2007a; Stejskal et al. 2009a, 2010; Jacquemond 2004), water temperature (Karas and Thoresson 1992; Kestemont and Mélard 2000; Mélard et al. 1995, 1996a; Overton and Paulsen 2005; Strand et al. 2007b; Mélard 2008), water quality including mainly oxygen (Zakes and Demska-Zakes 2005; Mélard 2008; Stejskal 2009b, 2012), salinity (Overton et al. 2008), ammonia and nitrite levels (Mélard 2008; Vandecan et al. 2008; Kroupová et al. 2013), disturbance during tank cleaning, fish size-sorting (Mélard et al. 1995, 1996a; Kestemont and Baras 2001; Strand et al. 2007b; Mélard 2008) and fish density and biomass (Mélard et al. 1996a, b; Mélard 2008).

16.6 Factors Affecting Growth, Survival Rates and Productivity of Perch Under Intensive Culture

16.6.1 Colour of Tank Walls, Light Intensity and Regime

The interaction between colour of tank walls and light intensity creates specific light conditions within rearing tanks which significantly affects feed detection and feeding success of farmed perch under intensive conditions, thus influencing perch growth (Strand et al. 2007a).

In general, larvae of percid species are strongly phototactic, but older individuals may be more sensitive to high light intensities (Craig 2000; Kestemont and Mélard 2000; Kestemont et al. 2003) such as 2,200 lx (Staffan 2004). However, Strand et al. (2007a) showed that different light intensities (at least when 200 and 1,100 lx were used) did not affect feed intake, growth rate and growth efficiency. Instead, the mentioned study showed that food intake and consequently growth rate were significantly higher in white or grey tanks compared to black ones under low light intensity (200 lx). The reason was reported to be the increased feed visibility, probably due to the feed's higher contrast against the background colour of the tank walls. When the authors used higher light intensity (1,100 lx) for different wall colour tanks they didn't find any significant differences in feed intake and growth rates between different colours of tank wall. The explanation of this observation is that higher light intensity increased the ability of fish to detect feed in dark tanks (Strand et al. 2007a) and the effect of colour of tank walls on the feed intake has been reduced. Careful consideration of tank colour therefore was reported to be of greater importance at low light intensities than at high light intensities (Strand et al. 2007a). Any effect of tank colour and light intensity on the energy expenditure was not evident during mentioned experiment. Probably, stress was not induced by any tested environmental factors or their combination for the cultured perch. Staffan (2004) tested three different light intensities (16; 200 and 2,200 lx) during light regime 12L:12D and found that perch were more active during daytime in the highest light intensity compared to two lower intensities. These results can indicate increased stress at higher light intensity that is not recommended for perch aquaculture. Staffan (2004) also studied preference of perch for three different colours of tank walls (white, grey and black), when perch could move freely among tank colours. No general preferences were found for any specific colour. As growth did not differ among the three different tank colours, the study indicated that white, grey and black tank wall colours are equally suitable for farming of perch. A clear difference in body colour was interestingly noted for perch kept in black and white tanks. Almost, all perch coming from the black tanks were dark and perch coming from the white tanks were light grey (Mairesse et al. 2005; Strand et al. 2007a). This phenomenon indicated that the capacity of perch to change body colour in accordance with its background could reduce conspicuousness and thus reduce this potential source of stress in cultured fish (Strand et al. 2007a).

Light regime 12L:12D with an intensity of 105–250 lx at water surface (Stejskal 2009a, 2010; Jacquemond 2004) or 16L:8hD (Strand et al. 2007a) were used during perch ongrowing phase under controlled and intensive conditions. When Jourdan et al. (2000) increased light regime from 12L:12D to 18L:6D and even 24L:0D, specific growth rates of perch significantly increased but without any differences between both light regimes 18L:6D and 24L:0D.

16.6.2 Water Temperature

Eurasian perch is a thermophilic species and optimum temperature for rapid growth ranges from 22 to 24 °C (Mélard et al. 1996a). Intensive ongrowing of perch under this range of temperature gives the highest productivity level (Kestemont and Mélard 2000). Temperatures of 22–24 °C maintained during the whole ongrowing phase results in market size perch (130–150 g) obtained in about 14 months including larval rearing period (Mélard et al. 1996a). When perch juveniles (0.5 g initial body weight) were cultured under RAS at 23 °C the minimal commercial marketable size (100 g) could be obtained after 9 months (Fig. 16.2, Mélard 2008).

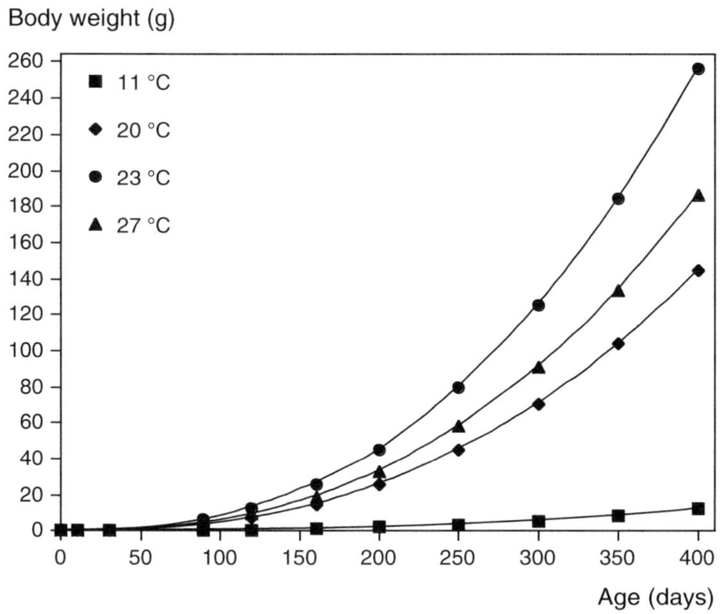

Fig. 16.2 Effect of water temperature on growth of mixed sex perch in intensive rearing conditions (Mélard 2008)

Maximum daily growth rates (0.06–1.80 g·fish^{-1}) for 3–300 g fish were also observed at 23 °C. This constant elevated temperature inhibited the sexual maturation in female. This phenomenon supported female higher investment in somatic growth rate. However, males presented a normal gonadal development in the same temperature conditions (Mélard et al. 1996a; Overton and Paulsen 2005).

Rearing at higher (27 °C) or lower temperatures (11–20 °C) reduces growth rate, i.e. the growth of 15 g fish at 27 °C was 12 % lower than at 23 °C. Similarly, 100 g perch reared under 20 °C had a 20 % lower growth compared to 23 °C (Mélard et al. 1996a). A relative low growth rate was observed at 11 °C which is 29 % of that found at 23 °C. This low temperature is not suitable for intensive perch culture and it has been used within extensive (pond) or semi-intensive (cage) perch culture during spring and autumn (Kestemont and Mélard 2000). Nevertheless, Mélard et al. (1995) found higher SGR (1.86 %·day^{-1}) in 1.9 g perch cultured at 26.5 °C compared to 22.9 °C (1.76 %·day^{-1}). These results were not significantly different because both temperatures were within optimal range of temperature for perch growth. Strand et al. (2007b) indicated the lack of temperature effect on energy expenditures of Eurasian perch and concluded that this fish is equally well adapted for growth at both 17 and 23 °C. Also they found that perch does not seem to be energetically more sensitive to disturbance at 23 than at 17 °C. However, the overall effect of temperature was significant for feed intake and growth rate, with higher feed intake and growth rate at 23 than at 17 °C. Karas and Thoresson (1992) showed a maximum food consumption of perch between 23 and 28 °C.

Despite this, Overton and Paulsen (2005) recommended temperature around 20 °C for perch rearing because this temperature is better for easier control of infections and problems occurred by *Saprolegnia* spp. Mélard et al. (1995) found the higher infection level of cultured ongrowing perch by *Heteropolaria* spp. at 26.5 °C. The survival of ongrowing perch (83.2 %) was negatively affected by this infection at 26.5 °C compared to lower temperature 22.9 °C without any infection and with survival rate 91.5 % (Mélard et al. 1995). Effect of bacteria (*Aeromonas* sp., *A. veronii*, *A. hydrophila*, *Streptococcus* sp., *Staphylococcus* sp., *Vibrio fluviatilis* and *Enterobacter agglomerans*) and protozoa (*Ichtyobodo necator*, *Trichodina* sp., *Ambiphyra* sp. and *Heteropolaria* sp.) species on the perch health inducing mass mortality of fish under stable and optimal temperature in intensive perch culture was well reviewed by Grignard et al. (1996).

16.6.3 Water Quality Including Oxygen, Salinity and Ammonia Level

Generally, water quality directly affects physiological condition of cultured perch and indirectly their feed intake and growth rate (Wedemeyer 1996). The oxygen level should be maintained above 5 mg $O_2 \cdot L^{-1}$ (60 % oxygen saturation) and higher oxygen concentrations do not significantly increase the growth rate of perch under intensive ongrowing conditions at 23 °C (Mélard 2008). However, Kestemont et al. (2008) recommended optimal oxygen level up to 6 mg $O_2 \cdot L^{-1}$, i.e. 68–72 % oxygen saturation for intensive perch juvenile culture under 20–23 °C. Stejskal et al. (2009b) used 6.8 ± 1.2 mg $O_2 \cdot L^{-1}$ oxygen level, i.e. 80 % oxygen saturation, for ongrowing perch under intensive culture. Effect of three different oxygen levels (hypoxia: 50–60 % oxygen saturation; normoxia: 90–100 % and hyperoxia: 140–150 %) on the feed intake and conversion and growth rate of intensively cultured pikeperch (*Sander lucioperca*) was examined by Stejskal et al. (2012). These authors found that higher oxygen saturation provides higher feed intake, better feed utilization (lower FCR) and growth rate. Perch has very similar environmental requirements as pikeperch (Craig 2000), therefore found results could be also generalized to intensive perch culture. Optimization of oxygen level during intensive perch culture needs additional studies mainly for obtaining more precise data regarding to the growth, feed conversion under high oxygen level and the impact on production cost in intensive perch farms. Biological oxygen consumption (OC in mg $O_2 \cdot kg^{-1} \cdot h^{-1}$) in six size groups (18.4–82.3 g) of intensively farmed perch was determined at 23 °C by Zakes and Demska-Zakes (2005). OC decreased from 336.2 to 185.0 mg $O_2 \cdot kg^{-1} \cdot h^{-1}$ in all tested size groups. An increase of body weight by 1 g led to average decrease in OC by a mean of 2.53 $O_2 \cdot kg^{-1} \cdot h^{-1}$. Stejskal et al. (2009b) observed diurnal course of OC in two groups (fed and feed-deprived) of intensively cultured perch. Fed perch with body weight from 44.8 to 279.4 g had average OC 288.3–180.6 mg $O_2 \cdot kg^{-1} \cdot h^{-1}$ with significant peak observed 6 h after the onset of feeding and relatively stable values of OC up to the end of feeding during day. Significantly lower average OC values (181.1–110.5 mg $O_2 \cdot kg^{-1} \cdot h^{-1}$) were found in feed-deprived perch with body weight 57.9–336.2 g without any significant peak of OC during day.

Oxygen levels under 5 mg $\cdot L^{-1}$ may induce significant stress leading to pathology and mortality (Stejskal et al. 2009b; Mélard 2008). Mortality of perch mainly appears when oxygen deficit is combined with intensive feeding resulting in additional oxygen consumption (Stejskal et al. 2009b).

Overton et al. (2008) found that growth rate and condition factor of perch from freshwater population (1.6 g) can be negatively affected by salinity exceeding 10‰ during 126 days rearing at 20 °C. These authors showed significantly higher growth rate in freshwater and at salinity 4‰, compared to salinity 10‰. Moreover, salinity 10‰ caused higher FCR (2.48) compared to freshwater or salinity 4–8‰ (1.80). Condition factor was reduced at salinity 10‰ compared to perch reared in freshwater with salinity 0‰. These results show that a freshwater strain of perch presents the best growth performance in fresh water or low saline conditions (4‰). Regarding to the fact that some perch strains are found in brackish waters, Overton et al. (2008) recommended choosing salinity adapted strains for brackish water production as found in the lower Baltic region.

Overton et al. (2008) found negative effect of higher salinity 13‰ and 18‰ on perch survival which was highly affected by water temperature (12, 15, 20 and 25 °C). No mortality of perch was found when fish were moved into brackish water of 13‰ at 12 and 15 °C. However, a dramatic increase in mortality rate was seen when perch were transferred into brackish water of 13‰ at 20 or 25 °C. A total of 50 % mortality was reached at 109 and 62 h for the two mentioned temperatures. Perch transfer to brackish water at 18‰ resulted in high mortality. Again, the mortality rate was positively related to temperature and 50 % mortality was reached at 178, 119, 69 and 39 h at 12, 15, 20 and 25 °C, respectively (Overton et al. 2008).

In intensive perch culture under RAS, high concentrations of ammonia and nitrite in water may take place when biological filters do not work properly. Ammonia (0.3 mg N-$NH_3 \cdot L^{-1}$) and nitrite (0.5–0.7 mg $NO_2^- \cdot L^{-1}$) concentrations induce physiological changes in perch such as reduced excretion of ammonia, conversion of haemoglobin to methaemoglobin resulting in limited oxygen transport (Jensen 2003; Svobodová 2005; Mélard 2008; Kroupová et al. 2013). These physiological changes negatively affect growth performances then induce mortality (Svobodová et al. 2005; Mélard 2008; Vandecan et al. 2008). A chronic exposure of 140–150 g perch at concentrations of 0.3–0.4 mg N-$NH_3 \cdot L^{-1}$ induces a 50 % decrease of growth rate compared to fish in control groups under temperature 23.1 °C, oxygen saturation 100 % and pH 8.1 (Fig. 16.3) (Vandecan et al. 2008; Mélard 2008). Higher ammonia concentration (0.8 mg N-$NH_3 \cdot L^{-1}$) has lethal effect for 50 % of the population after 96 h exposure at 23 °C (96hLC50) (Vandecan et al. 2008; Mélard 2008).

Acute toxicity of nitrite for 10.8 g perch was found by Kroupová et al. (2013). These authors established lethal concentration of nitrite (48hLC50) at 11 mg $NO_2^- \cdot L^{-1}$. Concentration of 3.8 mg $NO_2^- \cdot L^{-1}$ was found safe after 48 h exposure. However,

Fig. 16.3 Effect of long time exposure to ammonia on growth of 150 g perch at 23 °C (Mélard 2008)

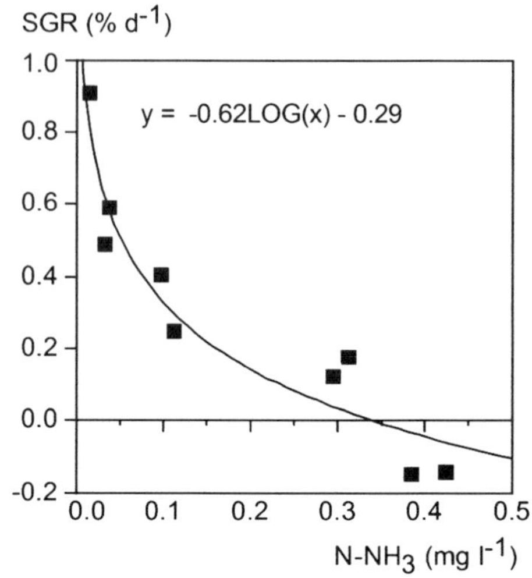

these results are obtained from laboratory experiments and according to our field experience, lethal concentration of nitrite is even significantly lower (1–3 mg $NO_2^- \cdot L^{-1}$) during intensive perch culture under RAS.

16.6.4 Disturbance

Disturbance from cultivation procedure like cleaning of tanks, service disturbance, fish size-sorting and other stressful events, reduce feed intake and increase energy expenditure (Strand et al. 2007b) and indirectly decrease growth in perch (Jentoft et al. 2005; Strand et al. 2007b). All mentioned operation must be carefully done for maintenance of good zoohygienic conditions, sorting of perch, elimination of growth heterogeneity and cannibalism with minimal negative effect on production (Mélard et al. 1996a; Acerete et al. 2004; Jentoft et al. 2005; Strand et al. 2007b). Proper culture management may reduce the response of fish to disturbance (Strand et al. 2007b).

Higher frequency of disturbance, included cleaning of tanks and other service disturbance, decreased feed intake and growth rate under lower (17 °C) and optimal temperature (23 °C). Severely disturbed perch at 17 °C had higher energy expenditures than undisturbed fish. The energetic growth efficiencies for disturbed fish were reduced by 19–38 % compared to undisturbed fish causing a reduction in weight increase of 24–56 %. No significant effect of higher frequency of disturbance on energy expenditures was observed at 23 °C (Strand et al. 2007b). The timing of this disturbance in relation to time of feeding is also very important factor affected feed intake in perch. Kestemont and Baras (2001) indicated that feed intake for perch disturbed by human activity prior to their first meal was reduced by 60 %, but if the fish were disturbed later in the day, after being fed, the daily feed intake was little affected. In contrast, Strand et al. (2007b) executed disturbance after the morning meal found 7 and 28 % reduction of feed intake at 17 °C, and 31 and 22 % at 23 °C for moderately and severely disturbed perch.

16.6.5 Growth Heterogeneity

Fish size-sorting is a very important operation within intensive perch culture for the reduction of growth heterogeneity and cannibalism with minimal negative effect on production. However this operation is stressful for fish and, paradoxically, may induce new social interactions within perch population after a grading. Generally, frequency and process of fish size-sorting must be optimized according to current situation in perch farming (Mélard et al. 1995, 1996a, b; Kestemont and Mélard 2000; Mélard 2008). Fish size-sorting to reduce growth heterogeneity results in the emergence of fast growing perch in each sorted group and strongly reduces cannibalism. However, this technique induces a new increase of heterogeneity in future culture phase (Mélard et al. 1996b; Mélard 2008; Kestemont et al. 2000). Paradoxically, Kestemont et al. (2000) found that the higher initial heterogeneity caused the lower final heterogeneity (Table 16.1). However, high initial heterogeneity promoted cannibalism exerted upon small fish and induced a decrease of heterogeneity and survival rate under its initial value.

Heterogeneity in perch growth is high. Body weight can range from 7 to 89 g for 7-month-old perch averaging 25.9 g (Mélard et al. 1996a) and from 21 to 452 g for 1-year-old perch (Fig. 16.4). In general, size heterogeneity (coefficient of variation of body weight) tends to stabilize around 40–45 % (Kestemont et al. 2000). The origin of the huge heterogeneity in perch is not only due to sexual dimorphism, when female grows around 20–30 % faster than males (Fontaine et al. 1997; Rougeot and Mélard 2008), but also to genetic traits (Mandiki et al. 2004) and social behaviour (Mélard et al. 1995). Therefore, the sorting process does not guarantee the improvement of the global productivity of perch culture. For instance, non-sorted

Table 16.1 Size model heterogeneity for perch: effect of initial size heterogeneity, body weight, tank volume and biomass (Kestemont et al. 2000)

Dependent variable: SHR (‰ day^{-1}); F-value = 100.4; R^2 = 0.839; P < 0.0001; df = 82				
Independent variable	Coefficient	S.E.	F-to-remove	P
Intercept	20.759	1.675	153.6	<0.0001
Log [CV_i (%)]	−13.404	1.092	150.8	<0.0001
W_i (g)	−0.248	0.041	37.3	<0.0001
Log [tank volume (m^3)]	1.264	0.420	9.1	0.0035
Log [Stocking biomass (kg m^{-3})]	−0.437	0.150	8.5	0.0046

Where SHR is specific heterogeneity variation rate = 100 [Ln (CV_2) – Ln (CV_1)] $[t_2\text{-}t_1]^{-1}$
Where CV_i is the initial coefficient of variation of body weight

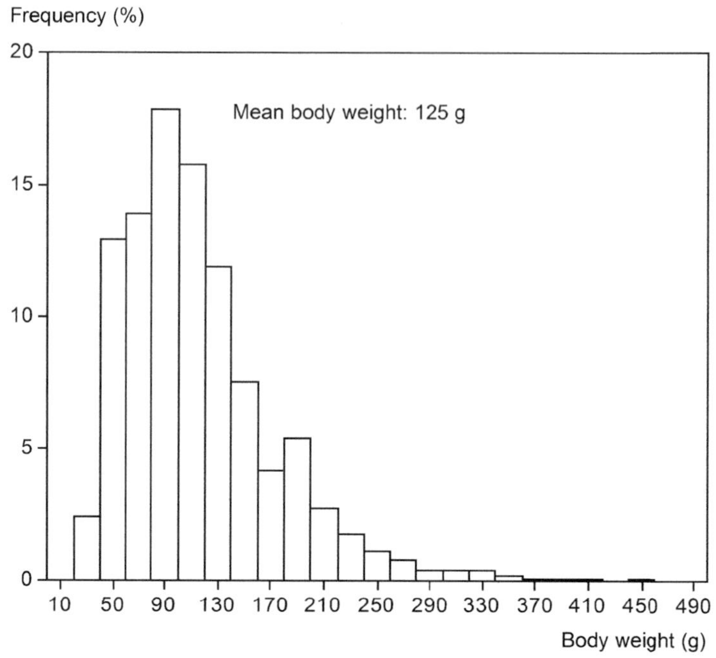

Fig. 16.4 Body weight heterogeneity of 1 year old progeny of mixed sex perch reared in RAS at 23 °C

populations had a growth rate 5–6 % higher than in size-graded populations of same origin and body weight (Mélard et al. 1995, 1996a; Mélard 2008). Generally, intensive perch farming needs to reduce frequency of fish size-sorting. Perch intensive culture with lower needs for fish size-sorting can be provide by the production of monosex female populations with higher growth rate, shortening production cycle and lowering size heterogeneity (Mélard et al. 1996a; Rougeot and Mélard 2008).

Generally, mortality resulting from direct and indirect effect of handling is only a small part of overall mortality, which is mainly due to parasites and bacteria (Mélard et al. 1996a).

16.6.6 Body Weight

Increasing body weight of intensively cultured perch under 23 °C decreased SGR and production capacity. Perch of 5 g body weight reared at an optimal biomass of 35 $kg \cdot m^{-3}$ gave production around 600 $g \cdot m^{-3} \cdot day^{-1}$. At an optimal biomass of 75 $kg \cdot m^{-3}$, the production of 150 g body weight fish was 320 $g \cdot m^{-3} \cdot day^{-1}$, only (Fig. 16.6, Mélard 2008). Overton et al. (2008) obtained the same results related to growth rate for perch reared during 4 months. Decreasing specific growth rate was found from 6 % · day^{-1} for perch reared 1–14 days after the start of the experiment up to growth rate 1 % for older perch reared 85–126 days of the experiment.

Continuous mortality may occur throughout intensive perch culture within the 1–200 g body weight range and result in an overall 50 % survival rate after 14 months (Mélard et al. 1996a). In RAS, where rearing conditions are optimal, Mélard (2008) reported that mortality takes place during first months of intensive culture of ongrowing perch from 0.3 to 200 g body weight and overall survival rate is fluctuating between 60 % and 70 % (Fig. 16.5).

16.6.7 Fish Density and Biomass

In general, higher fish density significantly increases growth rate and decreases growth heterogeneity in initial phase of perch intensive ongrowing (0.5–15 g) (Mélard et al. 1996a). Increased stocking density from 400 to 10,000 fish · m^{-3} resulted in a 67 % increased growth rate of 1 g perch juveniles cultured under 23 °C during 74 days. Growth rate was 0.12 and 0.20 g · fish^{-1} · day^{-1} at densities of 400 and 10,000 fish · m^{-3}, respectively (Mélard et al. 1996a). Mélard et al. (1996b) examined the effect of stocking density on 45-days perch juveniles and showed that higher densities (1,430 and 2,380 fish · m^{-2}) significantly increased specific growth rate of perch in comparison to low density (95 fish · m^{-2}). After 74-days rearing, juveniles with initial body weight 0.86 g reached a final body weight of 9.75 g (SGR = 3.44 % · day^{-1}) and 15.83 g (SGR = 3.87 % · day^{-1}) under low (95 fish · m^{-2}) and high fish density (2,380 fish · m^{-2}), respectively.

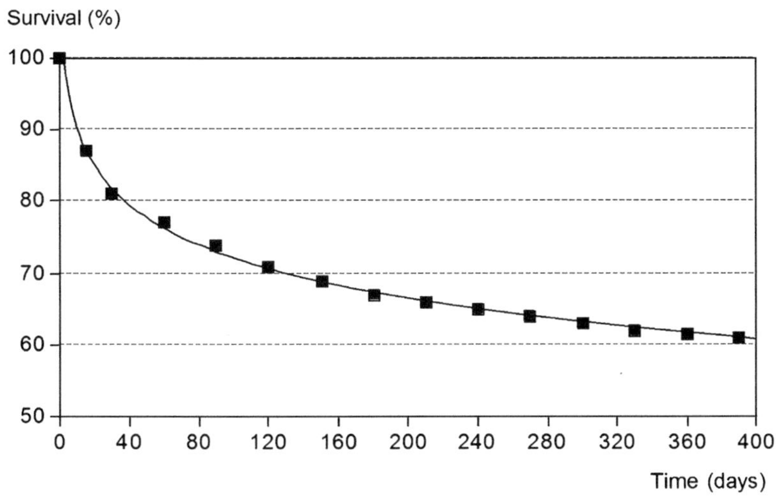

Fig. 16.5 Evolution of perch survival in RAS starting from 0.3 g weaned juveniles to 1 year 200 g fish (Mélard 2008)

However, this positive relationship between perch biomass and growth rate was valid until fish reached 10–16 g body weight (Mélard et al. 1996a). Growth rate of ongrowing perch with initial body weight up to 16 g decreased with increasing stocking biomass up to 20–60 kg · m^{-3} (Mélard 2008).

For 5–10 g perch increasing density induces a decreasing of growth heterogeneity and consequently of cannibalism (Mélard et al. 1996b). Coefficient of variation of fish body weight was decreased from 98.4 % to 57.9 % with increasing stocking density from 400 to 10,000 fish · m^{-3} (Mélard et al. 1996a). Negative effect of higher density on cannibalism rate was found by Mélard et al. (1996b). Density up to 1,430 fish · m^{-2} resulted in lower numbers of cannibals. Emergence of cannibalism depending on perch density was modelled as a second order polynomial equation ($R^2=0.995$, df = 9, $P<0.01$): $NCa = 0.157 + 6.394 \times 10^{-2}$ $ISD - 1.779 \times 10^{-5} (ISD)^2$, where NCa is numbers of cannibalism (fish · m^{-2}) and ISD is initial stocking perch density (fish · m^{-2}) (Mélard et al. 1996b). Cannibalism caused overall mortality rates of 3.0, 4.5, 6.7, 11.4 and 7.1 % at densities 95, 240, 480, 1,430 and 2,380 fish · m^{-2}, respectively. However, these authors found that cannibalism is not the main reason of overall mortality for ongrowing perch, when survival rate varied between 75.4 % and 92.2 % at perch density 1,430 and 95 fish · m^{-3}, respectively. Excluding cannibalism, the main reason of perch mortality during ongrowing phase is parasites, bacteria and stress that are indirectly affected by different perch densities and biomass (Mélard et al. 1996a, b).

16.6.8 Relationship Between Body Weight, Optimal Biomass and Production

In intensive farming, the optimal biomass of perch giving the maximal production depends of body weight. The higher is perch body weight, the higher optimal biomass is. Optimal biomass ranges from 35 kg · m^{-3} for 5 g perch to 80 kg · m^{-3} for 150 g fish. Relationship between optimal biomass (y) and body weight (x) is expressed as follow: $y = 24 \times^{0.23}$ (Fig. 16.6, Mélard 2008). Production (y) is decreasing when body weight increases (x) as shown by the following relationship: $y = 824 \times^{0.18}$ (Fig. 16.6, Mélard 2008). Maximal daily production in intensive culture ranges from 0.6 kg m^{-3} for 5 g fish to 0.35 kg m^{-3} for 150 g perch. The combination of the two models gives the maximal potential of production of a fish farm in relation to fish body weight and the total volume of rearing tanks available in the facility. Thus, the maximal fish farm productivity is obtained when perch are reared around optimal biomass.

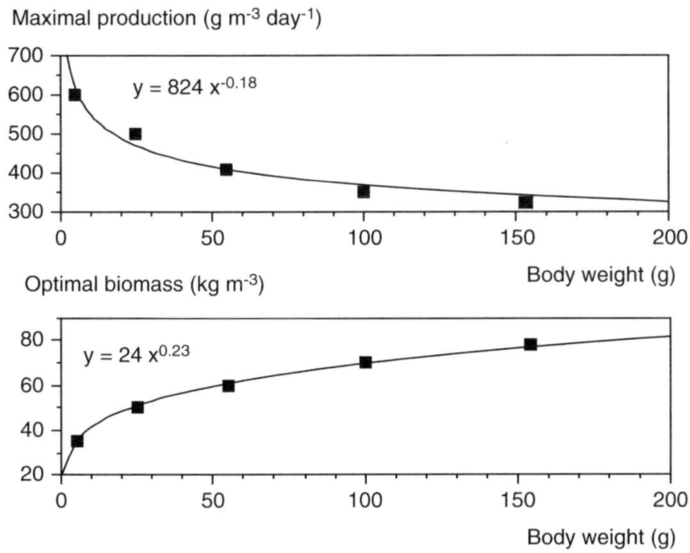

Fig. 16.6 Relationships between body weight, biomass and production level of perch reared at 23 °C (Mélard 2008)

16.7 Conclusion

Nowadays, three different production systems for perch culture during ongrowing phase are described: (1) traditional extensive polyculture system in earthen ponds, (2) semi-intensive culture farming using cage or tarpaulin tanks in lakes or sea bays or the combination of pond and RAS (Recirculating Aquaculture System(s)) culture and (3) intensive perch farming under RAS. Extensive and semi-intensive culture systems have many production limitations such as: suboptimal water temperature, acute risks of diseases, long production cycle, unpredictable production, etc. Therefore, intensive perch farming has been developed in Europe (mainly in Switzerland, Ireland and France) for more continuous and predictive marketable perch production.

Intensive ongrowing perch culture has to provide stable and optimal culture conditions for rapid fish growth, shorter production cycle, year round and predictable production, fish stress and cannibalism reduction, resulting in higher survival rate. Currently perch ongrowing under intensive culture cannot be considered as a bottleneck for stable and high-quality perch production when rearing facilities and all husbandry conditions are carefully optimized in order to ensure significant production profitability. Several factors such as: colour of tank walls, light regime and intensity, water temperature, water quality (including mainly oxygen, salinity, ammonia and nitrite levels), disturbance during tank cleaning, fish size-sorting, density and biomass affect marketable perch production under intensive conditions.

Different tank wall colours with the combination of 12L:12D or 18L:6D light regime and light intensity between 200 and 1,100 lx are equally suitable for farming of ongrowing perch. Optimum temperature for rapid growth ranges from 22 to 24 °C. Significantly higher growth and survival rates of freshwater perch strains are recorded under freshwater or water with salinity under 4‰. Optimum oxygen level and saturation for perch ongrowing should be up to 5–6 mg $O_2 \cdot L^{-1}$ and 60–72 % oxygen saturation. Very low ammonia (0.3 mg $N\text{-}NH_3 \cdot L^{-1}$) and nitrite (0.5–0.7 mg $NO_2^- \cdot L^{-1}$) concentrations may induce physiological changes and decrease growth in perch. Disturbance resulting from cultivation procedure like cleaning of tanks, service disturbance, fish size-sorting and other stressful events cause directly reduction of feed intake and an increase of energy expenditure and indirectly decreasing growth rate. The frequency of grading must be optimized by proper culture management or using of monosex female population to reduce growth heterogeneity. Mortality of perch resulting from direct and indirect effect of handling is only a very small part of overall mortality, which is mainly caused by parasite and bacteria disease. Low and high stocking densities result in slower growth rates. Optimal stocking biomass for production ranges from 10–20 $kg \cdot m^{-3}$ for 10 g perch to 60–70 $kg \cdot m^{-3}$ for 150 g fish.

Acknowledgments This chapter was prepared thanks to following projects: the Ministry of Education, Youth and Sports of the Czech Republic – CENAKVA (CZ.1.05/2.1.00/01.0024) and CENAKVA II (LO1205 under the NPU I program), NAZV (QI 101C033) and POSTDOC JU (CZ.1.07/2.3.00/30.0006).

References

Acerete L, Balasch JC, Espinnosa E, Josa A, Tort L (2004) Physical responses in Eurasian perch (*Perca fluvi*atilis, L.) subjected to stress by transport and handling. Aquaculture 237:167–178

Adámek Z, Helešic J, Maršálek B, Rulík M (2010) Applied hydrobiology. Second edition FFPW USB, 350 p, in Czech only

Adámek Z, Linhart O, Kratochvíl M, Flajšhans M, Randák T, Policar T, Masojídek J, Kozák P (2012) Aquaculture the Czech Republic in 2012: modern European prosperous sector based on thousand-year history of pond culture. Aquac Eur 37:5–14

Bláha M, Šetlíková I, Musil J, Policar T (2013) No reason for keeping 0+ perch (*Perca fluviatilis* L.ú with prey fish. Aquac Int 2:883–896

Craig JF (2000) Age, growth and mortality In: Craig JF (ed) Percids fishes systematics, ecology and exploitation. Fish and aquatic resources series, vol 3. Blackwell Sciences, Oxford, UK, pp 39–54

FAO (2013) Fisheries and aquaculture information and statistics service – 12/09/2013. http://www.fao.org/fishery/statistics/global-aquaculture-production/query/en

Fontaine P, Tamazouzt L, Capdeville B (1996) Growth of the Eurasian perch (*Perca fluviatilis*) reared in floating cages and in water recirculated system: first results. J Appl Ichthyol 12:181–184

Fontaine P, Gardeur JN, Kestemont P, Georges A (1997) Influence of feeding level on growth, intraspecific weight variability and sexual growth dimorphism of Eurasian perch *Perca fluviatilis* L. reared in a recirculation system. Aquaculture 157:1–9

Grignard JC, Mélard C, Kestemont P (1996) A preliminary study of parasites and diseases of perch in an intensive culture system. J Appl Ichthyol 12:195–199

Jacquemond F (2004) Separated breeding of perch fingerlings (*Perca fluviatilis* L.) with and without initial inflated swim bladder: comparison of swim bladder development, skeleton conformation and growth performance. Aquaculture 239:261–273

Jensen FB (2003) Nitrite disrupts multiple physiological functions in aquatic animals. Comp Biochem Physiol A Mol Integr Physiol 135:9–24

Jentoft S, Aastveit AH, Torjesen PA, Andersen Ø (2005) Effects of stress on growth, cortisol and glucose levels in non-domesticated Eurasian perch (*Perca fluviatilis*) and domesticated rainbow trout (Oncorhynchus mykiss). Comp Biochem Physiol A Mol Integr Physiol 141:353–358

Jourdan S, Fontaine P, Boujard T, Vandeloise E, Gardeur JA, Anthouard M, Kestemont P (2000) Influence of daylength on growth, heterogeneity, gonad development, sexual steroid and thyroid levels, and N and P budgets in *Perca fluviatilis*. Aquaculture 186:253–265

Karas P, Thoresson G (1992) An application of a bioenergetics model to Eurasian perch (*Perca fluviatilis* L.). J Fish Biol 41:217–230

Kestemont P, Baras E (2001) Environmental factors and feed intake: mechanisms and interactions. In: Houlihan D, Boujard T, Jobling M (eds) Food intake in fish. MPG Books, Cornwall, pp 88–156

Kestemont P, Mélard Ch (2000) Aquaculture. In: Craig JF (ed) Percids fishes systematics, ecology and exploitation. Fish and aquatic resources series, vol 3. Blackwell Sciences, Oxford, UK, pp 191–224

Kestemont P, Cuvier A, Houbart M, Mélard Ch, Baras E, Fontaine P, Jourdan S, Gardeur JN, Anthouard M, Boujard T, Kentouri M, Paspatis M, Ducarme C (2000) An integrated study of interindividual competition and its relationships with feeding physiology and behaviour in culture predatory fish. FAIR CT96-1572, final report, Facultes Universitaires Notre-Dame de la Paix, Namur, Belgium, p 118

Kestemont P, Jourdan S, Houbart M, Mélard C, Paspatis M, Fontaine P, Cuvier A, Kentouri M, Baras E (2003) Size heterogeneity, cannibalism and competition in cultured predatory fish larvae: biotic and abiotic influences. Aquaculture 227:333–356

Kestemont P, Rougeot C, Musil J, Toner D (2008) Larval and juvenile production. In: Rougeot C, Torner D (eds) Farming of Eurasian Perch, special publication BIM no 24, Bord Iascaigh Mhara, Dublin, Ireland, pp 30–41

Kratochvíl M (2012) Harvesting of marketable fish in members of Czech fish farmers association in 2011 and using of fish production in the Czech Republic during 1990–2011. Report of Czech Fish Farmers Association, 31 p, in Czech only

Kroupová H, Stejskal V, Kouřil J, Máchová J, Piačková V, Zusková E (2013) A wide difference in susceptibility to nitrite between Eurasian perch (*Perca fluviatilis* L.) and largemouth bass (*Micropterus salmoides* Lac.). Aquac Int 21:961–967

Mairesse G, Thomas M, Gardeur JN, Brun-Bellut J (2005) Appearance and technological characteristics in wild and reared Eurasian perch, *Perca fluviatilis* (L.). Aquaculture 246:295–311

Mandiki SNM, Blanchard G, Mélard C, Koskela J, Kuchaczyk D, Fontaine P, Kestemont P (2004) Effect of geographic origin on growth and food intake in Eurasian perch (*Perca fluviatilis* L.) juveniles under intensive culture conditions. Aquaculture 229:117–128

Martin B, Vandevorede D (2008) Lucas Perches: production of Eurasian perch in recirculating system. In: Fontaine P, Kestemont P, Teletchea F, Wang N (eds) Proceeding of Percid Fish Culture From Research to Production, Presses Universitaires de Namur, p 66

Mélard Ch (2008) Growth and husbandry effects in percids. In: Fontaine P, Kestemont P, Teletchea F, Wang N (eds) Proceeding of Percid Fish Culture From Research to Production, Presses Universitaires de Namur, pp 53–58

Mélard C, Baras E, Kestemont P (1995) Preliminary results of European perch (*Perca fluviatilis*) intensive rearing trials: effect of temperature and size grading on growth. Bull Fr Pêch Pisc 336:19–27

Mélard C, Kestemont P, Grignard JC (1996a) Intensive culture of juvenile and adult Eurasian perch (*P. fluuiatilis*): effect of major biotic and abiotic factors on growth. J Appl Ichthyol 12:175–180

Mélard C, Baras E, Mary L, Kestemont P (1996b) Relationships between stocking density, growth, cannibalism and survival rate in intensively cultured larvae and juveniles of perch (*Perca fluviatilis*). Ann Zool Fenn 33:643–651

Musil J, Adámek Z (2003) Predation pressure of European perch (*Perca fluviatilis*) towards topmouth gudgeon (*Pseudorasbora parva*) under the model pond condition. Bull VÚRH Vodňany 39:75–81

Öberg O (2008) Perch farming, Swedish experience. In: Fontaine P, Kestemont P, Teletchea F, Wang N (eds) Proceeding of Percid Fish Culture From Research to Production, Press Universitaires de Namur, pp 71–74

Öberg O (2012) ABBORÖS Slutrapport. KTH Architecture and the built environment, Trita-LWR report 3038: 89 p, in Swedish only

Overton JL, Paulsen H (2005) Ongrowing of perch (*Perca fluviatilis*) juveniles: (Videreopdraet af aborreyngel). Danmarks Fiskeriundersøgelser, DFU-rapport, no 151-05, p 36

Overton JL, Bayley M, Paulsen H, Wang T (2008) Salinity tolerance of cultured Eurasian perch, *Perca fluviatilis* L: effect on growth and on survival as a function of temperature. Aquaculture 277:282–286

Policar T, Stejskal V, Bláha M, Alavi SMH, Kouřil J (2009) Technology of intensive Eurasian perch (*Perca fluviatilis* L.) farming. Edition of practical handbooks, USB FFPW 89: 51 p, in Czech only

Policar T, Bláha M, Křišťan J, Stejskal V (2011). High quality and stable production of advanced juveniles in pikeperch (*Sander lucioperca*) under ponds. Edition of practical handbook, USB FFWP 110: 33 p, in Czech only

Policar T, Stejskal V, Křišťan J, Podhorec P, Švinger V, Bláha M (2013) The effect of fish size and density on the weaning success in pond-cultured pikeperch (*Sander lucioperca* L.) juveniles. Aquac Int 21:869–882

Rougeot C, Mélard Ch (2008) Genetic improvement of growth. In: Rougeot C, Torner D (eds) Farming of Eurasian perch, special publication BIM no 24, Bord Iascaigh Mhara, Dublin, Ireland, pp 42–51

Schmidt G, Wedekind H (2008) Dry feed adaptation and rearing of one-year old Eurasian perch (*Perca fluviatilis*) fed with dry feed with varying oil supplementation. In: Fontaine P, Kestemont P, Teletchea F, Wang N (eds) Proceeding of percid fish culture from research to production, Presses Universitaires de Namur, pp 130–131

Staffan F (2004) Food competition and its relation to aquaculture in juveniles *Perca fluviatilis*. PhD thesis, Acta Universitatis Agriculturae Sueciae, Silvestria 329: 24 p

Stejskal V, Kouřil J, Musil J, Hamackova J, Policar T (2009a) The growth pattern of all-female perch (*Perca fluviatilis* L.) juveniles – is monosex perch stock beneficial? J Appl Ichthyol 25:432–437

Stejskal V, Kouřil J, Valentová O, Hamáčková J, Policar T (2009b) Size-related oxygen consumption and ammonia excretion of Eurasian perch (*Perca fluviatilis* L.) cultured in a recirculating system. Aquac Res 41:135–142

Stejskal V, Policar T, Bláha M, Křišťan J (2010) Marketable Eurasian perch (*Perca fluviatilis*) production with the combination of pond and RAS. Edition of practical handbooks, USB FFPW 105: 34 p

Stejskal V, Matoušek J, Drozd B, Bláha M, Policar T, Kouřil J (2012) The effect of long-term hyperoxia and hypoxia on growth in pikeperch (*Sander lucioperca*). In: Abstract book from AQUA 2012 conference, USB of Abstracts, Prague, p 1058

Strand Å, Alanärä A, Staffan F, Magnhagen C (2007a) Effects of tank colour and light intensity on feed intake, growth rate and energy expenditure of juvenile Eurasian perch, *Perca fluviatilis* L. Aquaculture 272:312–318

Strand Å, Magnhagen C, Alanärä A (2007b) Effects of repeated disturbances on feed intake, growth rates and energy expenditures of juvenile perch, *Perca fluviatilis*. Aquaculture 265:163–168

Svobodová Z, Máchová J, Poleszczuk G, Huda J, Hamáčková J, Kroupová H (2005) Nitrite poisoning of fish in aquaculture facilities with water-recirculating systems. Acta Vet Brno 74:129–137

Tamazouzt L (2008) The French restocking market for percids. In: Fontaine P, Kestemont P, Teletchea F, Wang N (eds) Proceeding of percid fish culture from research to production, Presses Universitaires de Namur, pp 17–18

Tamazouzt L, Dubois JP, Fontaine P (1993) Enquête sur la production et les marches européens de la perche. La Pisculture Française 114:4–8

Tamazouzt L, Dubois JP, Fontaine P, Capdeville B, Terver D (1996) Zootechnical performance and body composition of *Perca fluviatilis* pelleted diet in a floating cage: effect of daily ration. Ann Zool Fenn 33:635–641

Toner D (2012) Culture of European perch Perca fluviatilis in Ireland. Recent advances and developments towards a sustainable production model. In: Abstract book from AQUA 2012 conference, USB of abstracts, Prague, p 1100

Vandecan M, Gbamou P, Mélard Ch (2008) Ammonia toxicity in percids. In: Fontaine P, Kestemont P, Teletchea F, Wang N (eds) Proceeding of percid fish culture from research to production, Presses Universitaires de Namur, pp 138–139

Wedemeyer GA (1996) Physiology of fish in intensive culture systems. Chapman and Hall, New York, 232 p

Wernicke von Siebenthal E (2013) How to protect a new RAS from disease introduction on an existing site with several health issues. In: Abstract book from Aquaculture Europe 2013, USB of abstracts, Trondheim, pp 459–460

Zakes Z, Demska-Zakes K (2005) Oxygen consumption and ammonia excretion by Eurasian perch (*Perca fluvi*atilis L.) reared under optimal thermal conditions. Archiwum Rybactwa Polskiego 13:251–258

Chapter 17
Intensive Culture Methods of Walleye During Ongrowing

J. Alan Johnson and Robert C. Summerfelt

Abstract Cultural technology is described for on-growing Walleye (*Sander vitreus*) on formulated feeds from pond- and tank-reared fingerlings to sub-adult. Current literature is reviewed and a description given of the regimen used by the Iowa Department of Natural Resources, Rathbun Fish Hatchery (RFH) for production of Walleye fingerlings for enhancement stocking. At that site, the production interval is 155 days from hatch to fall fingerling with three phases: (I) pond-culture, (II) habituation to formulated feed indoors, (III) on growing outdoors to 250 mm (140 g). For many years, the critical limitation was during phase II, but academic and applied research on feed composition, tank environment, and fingerling size now allows 85–90 % survival during habituation of pond fingerlings to formulated feeds. In phase III, use of grading fish before stocking has reduced cannibalism and a practical protocol for treatment of common diseases has increased survival. Considerations for post-stocking survival have included improvement in methods for harvest and transport. Suggestions are given for future research to further improve growth and survival, and strategies are for discussed for application of the RFH protocol to produce fish for the food-fish market.

Keywords *Sander vitreus* • Husbandry • Feeding • Cannibalism • Growth

17.1 Introduction

In 2004, more than 1000 million (1.0×10^{12}) Walleyes (*Sander vitreus*) were stocked in North America by governmental natural resource agencies for fisheries enhancement in lakes and rivers (Halverson 2008). The most recent of many surveys

J.A. Johnson (✉)
Rathbun Fish Culture Research Facility, Iowa Department of Natural Resources, 15053 Hatchery Place, 52571 Moravia, IA, USA
e-mail: Alan.johnson@dnr.iowa.gov

R.C. Summerfelt
Department of Natural Resource Ecology and Management, Iowa State University of Science and Technology, Ames, IA 50011-3221, USA

P. Kestemont et al. (eds.), *Biology and Culture of Percid Fishes*,
DOI 10.1007/978-94-017-7227-3_17

indicates that fry (pro-larvae less than 4 days posthatch) comprised 95.61 % of the 994.96 million fry, fingerling, and advanced fingerling Walleyes stocked in North America in 2006 (Kerr 2011). Post-stocking survival and economic effectiveness of stocked Walleyes is related to size and condition of the fish, predatory fish that are present, and ecological characteristics of the environment where fish are stocked. For example, cannibalism of stocked fingerlings by older Walleye already present in the lake is factor in first year survival of stocked fingerlings (Chevalier 1973; Beyerle 1978; Schneider 1983). Thus, there is substantially greater stocking success for introductions and rehabilitation events than for supplemental stocking in lakes with reproducing Walleye populations (Kerr 2011).

Annual variation in natural recruitment of Walleye in lakes of the Midwest is large and survival of stocked fry has been unpredictable, however, most agencies continue the practice because the costs for fry is low (<US $0.01), and fry stocking may occasionally produce a good year class (Mitzner 1992). Even so, research findings support a growing momentum for stocking of fingerlings, especially the largest fingerlings that can be produced by the end of the first growing season. Heidinger et al. (1987) reported relative survival of walleye fingerlings was 62 times greater than fry in one lake and 16 times greater in another lake. There is also some evidence showing higher survival of yearling fish stocked in the spring (Mitzner 2002; Harder et al. 2013). Stocking large Walleye fingerlings (186–216 mm) had highest return on investment compared to fry, small (mean total length = 48–61 mm), and medium (132–145 mm) fingerlings (Santucci and Wahl 1993).

Walleyes ≥203 mm (70 g) produced at Rathbun Fish Hatchery (RFH) of the Iowa Department of Natural Resources and stocked into some Iowa reservoirs had mean survival rate of 45 % (Mitzner 1995), and overwinter survival of these large fingerlings was twice that of fingerlings from extensive nursery lakes that averaged 125 mm. Lower and more variable survival of 4.9–31.7 % was reported for 150–178 mm average size Walleyes produced at Spirit Lake Fish Hatchery (Larscheid 1995).

Given the findings of these and other studies, the management objective for Walleye stocking of Iowa lakes with established fish populations containing several predatory species is to stock a fall fingerling of an average size ≥203 mm. The RFH has been able to produce 229 mm (108 g), range 203–254 mm (70–140 g), phase III fingerlings by mid-October, a total interval from hatch to fall harvest of about 150 days (Fig. 17.1).

The culture methods discussed in this chapter are based on research findings of the authors and others, and field-tested and refined in the production facilities at the RFH where advanced fingerling culture began in 1985. Large fingerlings are produced at RFH in a tandem pond and tank culture method consisting of three phases: Phase I, pond culture of fry (30–35 days); Phase II, raceway habituation of Phase I fingerlings to formulated diets (28–45 days); and Phase III, fingerlings on-growing on formulated diets in outdoor circular flow-through tanks (90 days). The average growth rate in the 155-day interval from pond stocking of fry (9 mm) to harvest of a fall fingerling (229 mm; 108 g) is 1.42 mm per day (0.72 g per day).

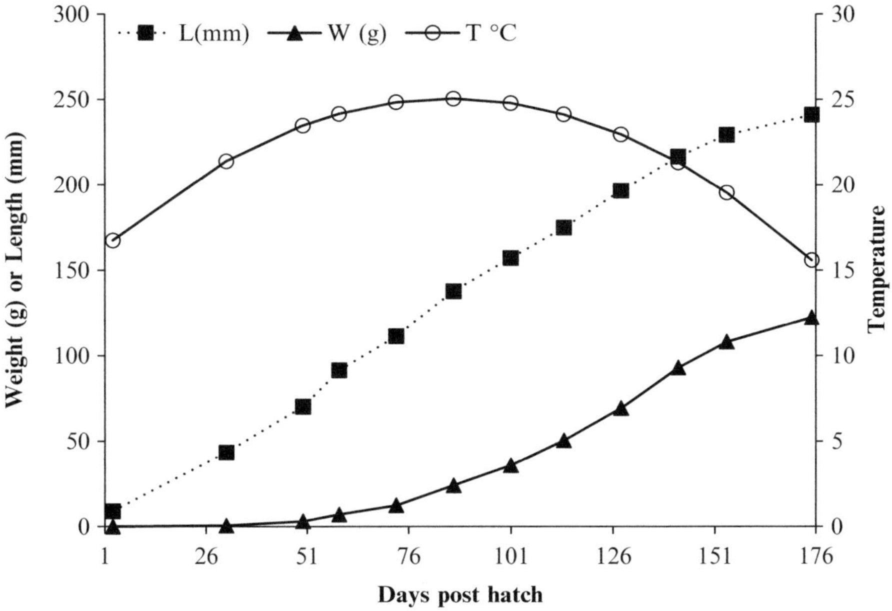

Fig. 17.1 Representative length (L) and weight (W) of Walleye at age in relation to temperature (T) at Rathbun Fish Hatchery. Three-day posthatch (dph) Walleye are stocked in fertilized ponds lined with a synthetic geomembrane and reared to 31 dph (phase I), at which time the ponds are drained and the fish (43.4 mm) harvested for transfer to indoor tanks for habituation to manufactured feed (phase II) for 28 days. At 59 dph, the 91.5 mm feed-trained fingerlings are transferred to outdoor grow-out tanks and cultured another 95–116 days (154–175 dph; 229–241 mm) until autumn temperatures reduced feeding and growth, then the juveniles are harvested for transfer to lake sites for enhancement of recreational fisheries. The temperature curve is the polynomial trend of observed temperatures in static water ponds and tanks supplied with ambient temperature surface water from a reservoir

17.2 Culture Methods for Early Juvenile Stages

17.2.1 Pond-Reared Fingerling Habituation

Habituation to formulated diets occurs during phase II culture. The protocol used for Phase II culture at RFH developed gradually from the research by the authors and from that of many others, however, a key element of the success obtained that the RFH has been the close-coordination and integration of research findings with the production team. Details of the RFH regimen are given because before critical issues were resolved, survival was highly variable and generally low; e.g., survival ranged from 26.0 % to 67.3 % between 2001 and 2006 until improved cultural technology was employed (Summerfelt et al. 2011). After improved methods were developed, survival ranged from 84.0 % to 91.8 %. The key factors responsible for improved survival were initial fingerling size, habituation environment, and diet. Ninety percent survival rates on a production-scale were obtained at RFH when each factor was implemented together.

17.2.1.1 Size of Fingerlings

Survival during phase II culture is influenced by size, size variation, and health of the fish at the time of pond-harvest as well as stress of harvest, transport and stocking in tanks used for phase II culture. Size of the pond-fingerling influences feed acceptance and size variation contributes to incidence of cannibalism. During phase II culture, some fingerlings may fail to accept a formulated diet and die. Disease and starvation rather than cannibalism may be the primary cause of mortality during the habituation process and mortality typically peaks by day 10–12 after fingerlings are transferred from the ponds. Starvation is a failure to transfer from live to manufactured feed. As well as feed, poor survival has been related to initial fish size, body condition, stressors, environmental conditions, or disease (Malison and Held 1996). Light is the dominant environmental factor affecting survival when fingerlings adjust from life in pond to confinement in tank environment. Obtaining high survival during phase II culture has been the bottleneck in production of larger Walleye on formulated diets.

Preference for habituating larger Walleye fingerlings has been noted by Cheshire and Steele (1972), Flowers (1996), and Malison and Held (1996) as larger, more robust fish survive at a higher rate than small fish or fish with poor body condition. At RFH, larger initial size of fingerlings resulted in increased survival rates during habituation (Johnson and Rudacille 2010). Pond fingerlings with an initial weight of 0.53 g had a survival rate of 69.7 %, whereas 0.32 and 0.42 g initial weight fish had survival rates of 37.1 and 54.7 %, respectively. Cannibalism rate of the large fingerlings with an initial size of 0.53 g was 4.1 %.

Malison and Held (1996) observed survival rates of 40.0 % and 68.0 % for fish habituated at 20 and 30 mm, respectively. Fish habituated in the 2005 RFH study ranged from 35 to 46 mm (0.32–0.53 g) with survival rates that ranged from 33.9 % to 69.7 %. Survival in a 28-days habituation interval using 2.2–2.6 g, 50–60 day old, pond-reared fingerlings with three cohorts ranged from 64.6 % to 85.1 % years during habituation (Kuipers and Summerfelt 1994). To produce fingerlings of that size, in pond culture, without addition of suitable forage (i.e., small minnows), would result in high cannibalism and reduced production. Thus, at RFH pond-culture is designed to produce a 0.57 g or larger fingerling in less than 40 days.

Complete scale development may be an important developmental milestone for successful handling and habituation. Scale development begins on Walleye at 24 mm and complete by 45 mm (Priegel 1964). Without scales it is likely that the skin is more susceptible to mechanical damage that allows entry of columnaris disease (*Flavobacterium columnare*; Huissain and Summerfelt 1991) and loss of electrolytes that cause stress-related mortality. Transfer in salt (NaCl) solution of 0.5 % (5 ppt) may ameliorate the problem with diuresis, which is the loss of electrolytes during stress. In spite of efforts at RFH to reduce stress, pond harvest and tank stocking requires handling in a dip net twice results in mechanical damage to many fish in the netting process, which is a problem that requires constant attention.

17.2.2 Diets

The terms habituation-, conversion-, or training-diet is used for an intermediate diet between live zooplankton and grower diets (Barrows 1996). Barrows (1996) stated that these diets should be highly palatable, low polluting, and offer extra energy to replace stores lost during habituation. Krill is often added to these diets to improve palatability and along with a high fish meal content result in protein levels of 40–60 % (Malison and Held 1996). Feeds formulated with krill and other marine-based ingredients are more expensive than feeds with lower contents of these ingredients, but they have high palatability that is required for the first 10 days of habituation, thereafter fingerlings can be transitioned to a less expensive grower diet (Barrows 1996; Bristow 1996; Flowers 1996). Most diets evaluated in previous studies are not currently available in the United States; i.e., BioKyowa (BioKyowa Inc., Chesterfield, MO) (Bristow 1996; Flowers 1996); BioMoist and BioTrainer (BioProducts Inc., Warrenton, OR; Kuipers and Summerfelt 1994; Nagel 1996). Although a soft semimoist pellet may seem to have more mouth-appeal when taken in, survival of a cohort fed a semimoist (26.5 % moisture) feed of 41 % protein was 65.4 %, but not different than the 64.6 % survival of fish fed a hard crumble (9 % moisture) with 61 % protein (Kuipers and Summerfelt 1994); however, it is not known whether the difference in survival was related to nutritive content or hardness of the feed.

In 1989, BioKyowa Fry Feed Kyowa (FFK) C-1000 became the standard Walleye habituation diet for pond fingerlings at RFH. Survival of fish fed BioKyowa FFK C diet was 63 % compared with 27 % with the W-16 diet formula (Barrows 1990). Thirty-day survival at RFH in 1989 with C-1000 ranged from 55 % to 75 % among tanks. For unknowable reasons, between 1990 and 2000, however, annual mean survival of fingerlings habituated with BioKyowa at RFH declined to 22–58 %. In 2001, bovine spongiform encephalopathy (BSE) was found in beef cattle in Japan (Johnson et al. 2008). For this reason, US importation of BioKyowa FFK formulation was stopped indefinitely.

Evaluation of other commercially available habituation diets was necessary to continue successful Walleye production at RFH. A series of feeding trials were conducted at RFH from 2001 to 2006 to compare survival of pond-reared fish fed six commercial diets during habituation (Johnson and Rudacille 2010). Of the six diets, survival of fish fed Otohime C2 (Reed Mariculture, Campbell, CA) and EPAC CW (INVE Aquaculture Inc., Ogden, Utah) was, the highest, 88.3 % and 73.5 %, respectively. The size of EPAC CW was ranged from 800 to 1200 μm and Otohime C2 was 920–1410 μm. In comparison, the habituation survival rate for Walleye fed BioKyowa FFK was 60.7 % in 2001 under similar environmental conditions (Johnson and Rudacille 2010). Given these findings, Otohime C2 has been the only feed used for habituation of pond-reared Walleye fingerlings at RFH from 2007 to 2014.

The krill content of the Japanese feeds (BioKyowa and Otohime) is regarded as the key ingredient that results in higher feed acceptance rates for Walleye fingerling habituation. Nagel (1985) fed a "krill flake diet or krill moist pelleted diet" with habituation survival rate of up to 97 %. A habituation diet developed by Barrows et al. (1992) contained krill, spray-dried egg and cod liver oil as major ingredients and >95 % survival rates during habituation were achieved (Barrows 1996). Kolkovski et al. (2000) used krill hydrolysate to enhance feed acceptance by Walleye. Krill was an ingredient in the BioKyowa FFK C-1000, and the current habituation diet, Otohime C2, and both resulted in high survival during habituation. In comparison, survival of fish fed fish meal based diets without krill had lower survival rates (Johnson and Rudacille 2010).

17.2.2.1 Feeding and Feeding Rates

At RFH, fingerlings are habituated in raceways with trough type feeders with deflectors that spread feed over most of the raceway width and distribute feed to two-thirds the length of the raceway. Feed coverage over the tank surface may be an important factor but has not been evaluated.

Tables for feeding rates of salmonids fishes are expressed as percentage body weight (BW) per day based on temperature and fish size (Piper et al. 1982). Similar tables have not been developed for Walleye. Although a rule of thumb is to feed 1 % of the fish's body weight per feeding (Wedemeyer 2001), that general rule, however, is not applicable during habituation of Walleye and keeping in mind that they are progeny of feral not domesticated fish. Feeding rates vary from 8 % to 10 % bw/day during the first weeks of habituation, 8 % has been the standard at RFH for fish larger than 0.6 g and 10 % for those smaller than 0.6 g. High feeding rates are necessary to condition fish to consume feed and to keep pace with 7.8 % bodyweight per day specific growth rates (SGR) observed during the 4 week habituation period (Johnson and Rudacille 2010). Feed rates are calculated and feeders are calibrated twice weekly to reflect growth and mortality rates. Fingerlings are transitioned from the initial Otohime feed to a 1 mm pelleted grower diet, Walleye Grower 9206, after 10 days on the habituation diet (Table 17.1).

Table 17.1 Feeding schedule for habituation of walleye fingerlings to formulated feeds at Rathbun Fish Hatchery

	Habituation diet	Grower diet	
Day post stock	0.9–1.4 mm	1.0 mm	2.0 mm
1–10	100 %		
11–15	50 %	50 %	
16–17	25 %	75 %	
18–23		100 %	
24:			100 %

During the first weeks of habituation feeding frequency has been at 5-min intervals, 18-h/day to 22h/day, which is 216 to 264 times per day (Kuipers and Summerfelt 1994; Bristow 1996; Johnson and Rudacille 2010). After 18 days the feeding interval may be expanded to 10–15 min. There is a beneficial effect of multiple feedings per day on stabilizing water quality because metabolic rates for oxygen consumption and ammonia excretion increase immediately after feeding (Yager and Summerfelt 1994).

17.2.2.2 Environment

Development of culture environment tailored to the biological requirements of the life stages of Walleye has made possible high survival rates on dry diets from fry to fingerling stages. For the first 30-day posthatch, Walleyes are attracted to direct and reflected light, but after 30-day post hatch they are repelled by light (Colby et al. 1979). A reflective layer (tapetum lucidum) develops at the back of the retina that reflects light back through the retina thereby increasing their sensitivity to light and visual acuity in low light environments (Moore 1944). This change-over in their response to light coincides with the end of phase I culture when the small fingerlings are moved from ponds to indoor tanks for habituation to manufactured feed. Thus, activities of the fish culturists during tank cleaning, feeding, and passing by tanks often cast overhead shadows that cause a flight response to escape from the overhead movements. Survival during habituation was affected by the combination of stress from pond harvest and transfer from ponds to indoor tanks where light levels were high and fish were repeated exposed to human motions.

The most effective solution to accommodate the light sensitivity and skittishness of fingerlings at the onset of habituation was the use of a tank cover or darkroom with in-tank lighting. Nagel (1976) reported a successful culture environment for habituation that included a covered raceway with in-tank lighting. The tank cover blocked the fish from seeing movement outside the tank and eliminated overhead light penetration while the in-tank light provided a low light level environment suitable for the Walleye's environmental preferences.

Fingerling Walleyes in tanks with overhead lights are easily excited by shadows caused by hand feeding, tank cleaning, and even from bumping the tank. On the contrary, fingerlings in tanks with submerged lights remained motionless with object passing over the tank as the lights eliminated shadows caused by the overhead lights; totally eliminating overhead lights in the culture room with use of submerged lights allows Walleyes to detect only the objects within the tank (Siegwarth and Summerfelt 1992). The advantage of in-tank lights is that in a dark or dimly lighted room fish can be observed without the fish actively avoiding or being adversely affected by the submerged lights.

Submerged, in-tank lighting is especially beneficial for feeding and growth of Walleyes because they are easily disturbed by shadows from overhead motion and from activities around the tank. Submerged lighting provides a well-illuminated surrounding and eliminates shadows that create a less stressful environment than bright overhead lights. Design of a simple in tank light (Fig. 17.2) is described by Johnson and Esser (2009).

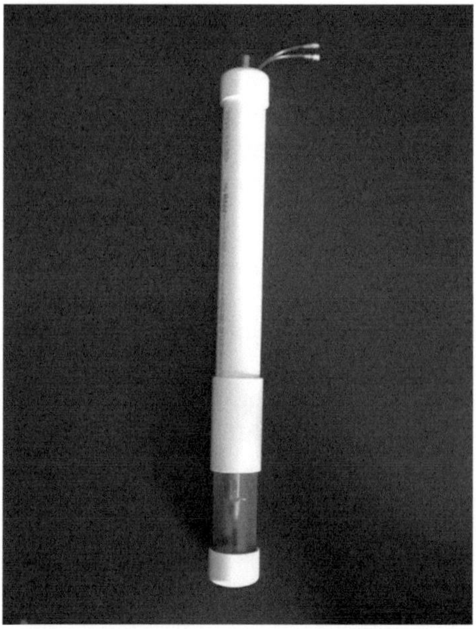

Fig. 17.2 Design of an in-tank light constructed from 50.8 mm PVC pipe and fittings and a 12-V light bulb used during phase II habituation at Rathbun Fish Hatchery (Johnson and Esser 2009)

Survival at RFH was low and inconsistent (26–50 %) (Johnson and Rudacille 2010) during habituation until the covered tank environment and in-tank lighting was used. At RFH, we first used a foam-board to cover the raceways as done by Nagel (1976), subsequently, we found that a dark room with uncovered tanks was as effective and achieved without difficulty. Johnson and Rudacille (2010) observed 60.7 % habituation survival using a covered raceway with an in-tank light environment compared to 37.3 % for tanks exposed to overhead lighting, sometimes with direct sunlight entry through overhead windows. At RFH, a portion of the raceway room traditionally used for habituation of Walleyes was converted to a dark room by installation of a black plastic fabric cover and low-voltage in-tank light fixtures were installed in each raceway. After the environment change survival during Phase II was 84–92 % from 2007 to 2010 (Johnson and Rudacille 2010).

Alternatively, light in the culture tank may be reduced by increased turbidity levels in the culture water. Turbid water (52 nephelometric turbidity units; NTU) increased habituation survival of fingerlings in an overhead light environment compared to habituation in clear water, but not greater than survival in a covered tank with in-tank lighting (Johnson and Rudacille 2010). The combination of turbid water and covered tank with in-tank lighting gave similar survival rates as the turbidity or covered tanks with in-tank light alone. Furthermore, the findings demonstrate that 52 NTU of turbidity does not reduce habituation success. An important finding given that the ambient turbidity level of RFH culture water source, Rathbun Lake, can vary from 5 to 50 NTU during habituation.

17.2.2.3 Size Grading

Size variation increases during habitation due to individual differences in feed consumption and genetic differences in growth rates, and from cannibalism of weakened and emaciated cohorts that do not consume feed. Therefore, grading to remove obvious cannibals is common practice at RFH as without cannibal removal the survival rate would decrease. The incidence (i.e., rate) of cannibalism (C) rate is determined by difference of the number stocked (S) and harvested (H) or removed as a mortality (M) in the equation $C = (S - (H + M)) / 100$.

As habituation survival rates at RFH progressively increased from 2003 to 2007 cannibalism rate during phase III growout increased to 14 % and was highly correlated to phase II habituation survival rate. Pond fingerlings begin the phase II habituation with a range in total length of 13 mm and the size difference increases to 38 mm after 27 days on feed in phase II. The high variability in the cohort length at the end of phase II was assumed to cause the increased cannibalism and mortality during phase III growout.

At RFH, a vertical-bar grader was fabricated to fit the cross-sectional area of the raceways used for phase II. Interchangeable panels of grader bars spaced at 7.5, 7.9, 9.1, and 9.9 mm were used to remove cannibals and grade fingerlings during the 27-day habituation interval (Johnson and Rudacille 2010). Fish retained by the 7.5 mm and 7.9 bar spacing, which are the larger of the fish at that time, were removed from raceways on habituation day 21 and 27 and transferred to the outdoor tanks for phase III culture. In the first year of a 2-year study to determine the benefits of grading, fish going into the outdoor tanks were of three groups: an ungraded, top grade, and pass-through groups. Ungraded and pass-through groups were 85.6 mm and 84.0 mm length, respectively, with coefficient of variation in length of 8.4 and 9.1, respectively, while the top-graded group fish were 91.5 mm and CV was 5.9 (Johnson and Rudacille 2008). Top-graded fish (i.e., the larger fish) had a significantly lower total mortality rate (6.1 %), during phase III culture compared to total mortality of the ungraded groups of fish (21.4 %), and the pass-through groups of fish (17.5 %; Johnson and Rudacille 2008). Thus, substantial benefits accrue from grading fish before they are stocked into outdoor tanks for phase III culture. In the second of the study, two grading events were used about 1 week apart during habituation. This procedure resulted in 87.9–91.3 % survival to 230 mm and only 3.4–4.1 % cannibalism among all graded groups (Johnson and Rudacille 2009a).

17.2.3 Intensive Fry to Fingerling Culture Transition

Juveniles may be produced from pond culture or intensive fry culture. Fingerlings produced in intensive systems will have consumed manufactured feed since they were 3–4 dph, which means that habituation to manufactured feed takes place at an early life stage and that survivors to 30–40 dph do not have problem of acceptance feed that is occurs with pond-reared fish at the same age. Nevertheless, post-transfer mortality rates of small juveniles (28 mm, 0.13 g) often exceeded 1 % per day during the first 10 days after transfer of the fingerlings from fry culture tanks to tanks used for habituation of fingerlings from pond-culture. The supposition if that the morality was from electrolyte loss from diuresis through mechanical injury of fish that were not fully scaled. Although research shows that size and age of pond-reared fish influences survival during habituation, the cause has not been experimentally evaluated. Poor survival is related to stress and mechanical injury during pond harvest, grading and transfer to indoor culture tanks for habituation. The problem is greater for small fish since larger and older fingerlings have a full complement of scales. Thus, at RFCRF the protocol for tank culture of fry is focused on factors that will produce larger and older fingerlings (>0.57 g and 35 dph) before they are transferred.

17.2.3.1 Diet and Feeding Rate

Juveniles may be gradually transitioned from the larval diet series to a lower cost transition diet and grower diet either just prior to removal from fry culture tanks or after stocking in a larger tank for the phase II production. The Otohime C2 diet fed at the end of fry culture can be gradually replaced with 1.0 mm pellets of the Walleye Grower 9206 over a 7-day period similar to the feeding regime used for habituation of pond fingerlings. Juveniles at 35 dph are fed 8–12 % bw/day at 21–25 °C and ration adjusted for growth rates at 3–4 day intervals to ensure adequate feed is offered.

17.2.3.2 Environment

We have assumed that the same covered tank with in-tank lighting environment proven to increase survival and growth of pond fingerlings during habituation is also advantageous for juveniles transferred from tanks used for intensive fry culture. The same reactions to light and overhead shadows apply for the fry culture juveniles as pond-reared juveniles. We have not attempted to phase the transition from a turbid water environment to a clear or low turbid environment.

17.2.4 Culture Methods for Ongrowing

17.2.4.1 Single Pass and Reuse Systems

Most advanced Walleye fingerlings reared by government agencies have been reared in single pass culture systems. The duration of the growing season in these single pass systems is often dictated by the seasonal temperature variation of the water source because heating water in a single pass system is cost prohibitive. Oneida Fish

Hatchery, Spirit Lake Hatchery, and RFH use surface water sources without temperature control, thus, the growing seasons of the Oneida (43N 15′ 6.03″) and Spirit Lake (43N 26″ 42.58″) Hatcheries are shorter than the growing season at RFH (40N 49′ 28.86″) which is 270 and 291 km more southern latitude, respectively.

Water reuse aquaculture systems (WRAS) allow the culture environment to be optimized for fish growth and designed for high biomass density. Culture of large fingerlings and food-fish size Walleye in WRAS has been successful (Summerfelt 1993; Harder et al. 2012). Walleye were reared from 5 g to food size fish (227–681 g) in a WRAS with mean temperature 23 °C in 284 days (Summerfelt 1993). Harder et al. (2012) observed 99 % survival rates during a 131-day grow-out study in a WRAS. Walleye have been reared at higher density in a WRAS 39.2 kg/m^3 for 280 mm fish (Harder et al. 2012) to 72.1 kg/m^3 for food size fish (Summerfelt 1993) compared to a flow-through system such as RFH were density seldom exceeds 16 kg/m^3. Loading rates in a WRAS of 0.4 kg/Lpm were not excessive for 254–356 mm Walleye (Summerfelt 1996), however loading rates at RFH reach 1.7 kg/Lpm at harvest from the flow-through system. Despite the advantages of optimal growing temperatures and higher density culture, the cost for new or retrofitting exciting facilities for WRAS technology has been an impediment; thus, WRAS technology not been widely adapted by government agencies for advanced Walleye fingerling culture for sport fishery enhancement at this time.

17.2.4.2 Lighting and Photoperiod

The preference of Walleye fingerling to adult stages for low light conditions has been well documented. Bulkowski and Meade (1983) described the behavioral change of Walleye becoming negatively phototaxic at 32–40 mm with larger fish preferring 2–4 lx of light. Scherer (1976) demonstrated the vertical positioning of fingerling Walleyes was inversely related to overhead light intensity (200, 20, 2 lx). In fish culture systems low (<16 lx) overhead light (Kuipers and Summerfelt 1994), tank covers (Nagel 1976; Johnson and Rudacille 2009b), and dark rooms (Johnson and Rudacille 2010) with in-tank lighting have been evaluated for pond fingerling habituation and demonstrated improved performance. In a 70-day trial Walleyes in a covered tank with in-tank lighting were 12 mm larger than fish reared in an overhead-lit raceway (Johnson and Rudacille 2009b).

At RFH, the 90-day grow-out culture period takes place outdoors in semi-round concrete tanks with no structural shade. Typical growth rates in these outdoor tanks are 1.4–1.5 mm/day during the grow-out phase. The limited amount of shade draws fish near the tank walls that may result in eye abrasions and loss of eye in about 5–10 % of fingerlings. The shallow tanks and low turbidity (5–35 NTU) also forces fish nearer the bottom of the concrete tanks resulting in a distinct pattern of erosion to the lower-lobe of the caudal fin.

Longer photoperiod is said to stimulate fish growth and result in better food conversion efficiency (Boeuf and Le Bail 1999). Harder et al. (2012) exposed Walleye fingerlings to 12/12, 18/6, and 24/0 h light/dark photoperiods and feed efficiency and growth rates were higher in the 18 and 24-h light periods than in the 12-h photoperiod. During habituation of pond fingerlings in covered tanks with in-tank light Nagel (1996) and Johnson and Rudacille (2010) used a 24-h light period and achieved good growth (1.4 mm/day) and survival (90 %) rates.

17.2.4.3 Tank Shape, Size, and Color

Walleye fingerlings have been reared in raceways or circular tanks of various sizes and colors. Nagel (1996) habituated Walleyes to formulated diets in 2.8 m^3 raceways and observed 90 % survival rates. At RFH, with fish of optimum size and in tanks with intank lighting, high survival rates (84.0–91.8 %) in phase II have been achieved in large raceways (4.5 m^3; 5.5 m length) as well as small research-scale raceways (1.4 m^3; 3.4 m length). In two separate trials at Rathbun Fish Culture Research Facility (RFCRF), fingerlings reared to 205–210 mm in circular tanks (0.99 m^3; 1.22 m diameter) were 3.3–5.5 % longer and 22–28 % heavier than fish reared in raceways (1.4 m^3, 3.4 m length; Johnson and Rudacille 2009b), survival was not affected by tank shape (Johnson and Rudacille 2010). Oneida Fish Hatchery and Spirit Lake Fish Hatchery also use raceways for Walleye fingerling habituation to formulated diets and also growout to advanced size, at Oneida raceways are green and at Spirit Lake raceways are a light aqua-green. White Lake Fish Culture Station, Ontario, Canada uses circular (2 m), cone bottom tanks for habituation of pond-reared fingerlings as well as ongrowing (Flowers 1996). Tanks are first filled to half volume and stocked with 38-mm, 0.5–1 g fingerlings for habituation, but on-growing in the same tanks with increasing water volume to harvest.

Color as well as light has been found to be influential in during phase II culture; e.g., Harder and Summerfelt (1996) obtained higher survival and growth rates by habituating pond-reared Walleyes in small (0.625 m^3) black-colored tanks than in large (1.13 m^3) blue tanks. Because Walleye prefer low light conditions and are easily disturbed by overhead shadows and tank cleaning activities, it is our recommendation to culture fingerlings to adults in a black circular tank. Black walls and bottom would reduce light reflection and disturbances from tank cleaning would be minimized with a properly designed self-cleaning circular tank.

17.2.4.4 Density

Density is expressed as the weight per volume (kg/m^3) of culture space. Generally fish are stocked at an initial density that will not exceed the maximum capacity of the unit before the end of the growth interval. Initial stocking density for pond-reared fingerlings for habituation (phase II) has varied from 1.8 to 3.3 kg/m^3 and the maximum final density of 16.3 kg/m^3 was reached with high survival rates (Table 17.4). Larger fish often show more tolerance to crowding than small fish, in phase III culture the highest density was 44.7 kg/m^3 (Yager 1991).

A systematic study to determine the maximum density that Walleye tolerate without growth reduction or compromised health has not been conducted. Data given Table 17.1 were not obtained from comprehensive survey, and for the most part the studies from which these values are derived were not designed to determine optimum density. In many hatchery situations density of fish at harvest is not the maximum density but accommodates a safety margin for unexpected delays in stock splitting or harvest keep density below the maximum for the culture system. Furthermore, the density of fish may be limited by the design of the culture system; i.e. use of ambient oxygen for the water temperature rather than supersaturated oxygen, and density may have been limited by flow rate sufficient to purge carbon dioxide and ammonia.

17.2.4.5 Temperature and Growth

Walleye are considered a coolwater species but they are eurythermal with natural distribution from the Artic Circle to the Gulf Coast. Lethal temperatures are in the range of 32–34 °C (Smith and Koenst 1975; Hokanson and Koenst 1986). Cai and Summerfelt (1992) determined a physiological optimum temperature of 25.3 °C based on measures of metabolic rate. Summerfelt and Summerfelt (1996) calculated an optimum growth temperature of 23.4 °C (range of 23.0–23.9 °C) based on a summary analysis of published growth rate and temperature data. Their data analysis also suggested that growth of Walleye was negligible below 15.6 °C. Overwinter growth rates of 184 mm fish in culture tanks for 147 days were 0.06 mm/day and 0.07 g/day at an average of 8.3 °C (Harder et al. 2013).

Growth rate is greater during phase II than during growout (phase III), and growth rate in both phases shows a positive response to mean temperature (Table 17.5, Fig. 17.1). For example, during a 27-days habituation interval in 2007 growth rate was 1.13 mm/day in the first 14 days of the interval with mean temperature of 20.9 °C, but 1.99 mm/day in the last 13 days when mean temperature was 23.1 °C (Johnson and Rudacille 2010). During growout from 102 to 211 mm in a covered tank with in-tank light environment Walleye grew 1.56 mm/day and 1.1 g/day at 24.4 °C (17.2–28.5 °C; Johnson and Rudacille 2009b). Harder et al. (2012) observed growth rates of 0.59–0.65 mm/day at a temperature of 21.9 °C for fish growth from 196 mm to 282 mm. Typical growth rates at RFH from 100 mm to 229–250 mm are 1.4–1.5 mm/day and 1.1–1.2 g/day (Summerfelt et al. 2011). Fingerlings (176–216 mm) grew at a rate of 0.55 mm/day at 25 °C (0.022 mm/d/°C; Siegwarth and Summerfelt 1992). Larger fish (285–324 mm) grew at a rate of 0.31 mm/day at 20.7 °C (0.015 mm/d/°C; Siegwarth and Summerfelt 1993).

17.2.4.6 Diet and Feeding Rate

Sinking or slow-sinking feeds and mechanical feeders are preferred over hand feeding with floating feeds because Walleye are skittish to overhead shadows and unlike Rainbow Trout (*Oncorhynchus mykiss*) they typically feed below the water surface. Held and Malison (1996) stated Walleye were difficult to feed to satiation by hand

Table 17.2 Ingredient composition (%) and proximate analysis (%) of Walleye Grower 9206 fed to walleye fingerlings produced at Rathbun Fish Hatchery (Johnson and Rudacille 2009b) and as fed by Clayton et al. (2008)

	Johnson and Rudacille (2009b)	Clayton et al. (2008)
Ingredient		
Fish meal (min. 65.0 % CP[a])	49.50	49.50
Soybean meal (47.5 % CP)	12.00	12.00
Corn gluten meal (60.0 % CP)	8.25	8.25
Blood meal (90.0 % CP)	4.75	4.75
Wheat flour	11.70	10.00
Fish oil	11.75	11.75
Vitamin premix	1.00	0.60
Choline chloride	0.45	0.45
Stabilized vitamin C	0.50	0.15
Trace mineral premix	0.10	0.05
Binder (CMC[b])		2.50
Proximate analysis (as fed)		
CP	49.21	46.60
Fat	17.33	17.88
Crude Fiber	1.46	1.03
Ash	9.63	10.13

[a]*CP* crude protein
[b]*CMC* Carboxymethyl cellulose

because their response to sinking rations was sluggish. Mechanical feeding requires presumptive feed rate calculations based on feed charts or growth estimates (Summerfelt and Summerfelt 1996). To date there have been no published feed rate tables for Walleye during habituation or ongrowing.

During phase II culture, feed conversion ratios of 0.99 to 1.17 have been observed at RFH. During phase III culture, absolute growth in mm and estimated feed conversion ratios are used by RFH staff to calculate daily feed rations for Walleye grown from 90 mm to 225–250 mm. Production tanks of fish are sampled (25–50 fish) at 2 week intervals to obtain size, growth and calculation of relative weight (Wr) with the Flammang et al. (1999) revision of the Anderson and Neumann (1996) equation.

The growth rate for total length (mm/day) is used to predict daily fish length for the next 14-day interval, then the size-appropriate relative weight equation (length <150 mm Flammang et al. 1999; >150 mm Anderson and Neumann 1996) is used to solve for weight from predicted length and *Wr*. Daily weight gain of the tank is calculated and the daily ration is calculated by an assumed feed conversion ratio of 1.2 kg feed fed per 1.0 kg of bodyweight gained. Measured daily length gain was occasionally adjusted for major changes in forecasted weather. As fish grow, feed pellet size is increased up to a 4 mm pellet (Table 17.3). This feeding calculation method has been used from 2008 to 2011 and average annual feed

Table 17.3 Relationship of walleye length and pellet size fed at Rathbun Fish Hatchery

Fish size (mm)	WG size[a]
50	1.0
75	2.0
112	3.0
175	4.0

[a]Nominal pellet size (mm) designation by manufacturer

conversion ratios of 1.3 to 1.7 have been achieved at RFH during phase III. In comparison, Harder et al. (2012) observed an FCR of 1.7 for fingerlings up to 282 mm, and 1.5 for fingerlings that grew from 197 to 251 mm in 70 day at 22.5·°C (Harder et al. 2014).

Mechanical feeders with a solenoid operated feed gate and spreader plate are used to distribute feed across 50–75 % of the tank surface at timed intervals. Typically at RFH Walleyes are fed at 15 min intervals during the first 14–28 days post stock in phase III and thereafter the interval may be increased to 30-min intervals between feeding events. Each feed event typically consists of four to five bursts of feed at 1-min intervals.

Walleye Grower 9206 diet formulation (Table 17.2) has been the standard diet for comparison with new experimental feeds however, other commercial diets have resulted in similar performance (Clayton et al. 2008). Ten alternate formulations of WG were evaluated in a series of research trials at RFCRF from 1998 to 2004 to develop a diet that provided better growth rates than the WG 9206 formulation (Johnson and Rudacille 2009b). Typical crude protein of the diet is 49 % and fat content is 17 % (Johnson and Rudacille 2009b). All tested formulations contained fish meal at 60.5–65.7 % of the diet except for one that contained only 17.4 % fish meal with white corn gluten meal comprising 55.0 % of the diet. Diet trial durations were 70–112 days and fish grew from about 50 mm to as much as 200 mm. Only one diet resulted in substantially larger fish than feeding WG 9206 and that diet had minor ingredient modifications such as wheat gluten and liver meal substituted for corn gluten meal. All other test diets resulted in fish growth equivalent or slower growth compared to fish fed WG9206.

Reduced inclusion or complete replacement of marine proteins and oil sources has been an ongoing research area in fish nutrition. Fish meal and fish oil make up over 61 % of the ingredient composition of WG. Limited research has been conducted to reduce fish meal and fish oil inclusion in Walleye diets by including higher amounts of plant oils and plant proteins. Established nutrient requirements for Walleye do not exist at this time but are needed to refine formulations for optimal growth and increase utilization of other protein and oil sources.

Nevertheless, some ingredient substitutions have been evaluated in grower diets with mixed results. A modification of the Walleye Grower diet with complete replacement of 11.75 % menhaden oil with 10 % soybean oil and a fish meal reduction to 41.6 % with inclusion of 20 % soybean meal resulted in similar growth rates to the standard WG diet (Clayton et al. 2008). A reduced fish meal (17.4 %) WG formulation with white corn gluten meal comprising 55.0 % of the diet resulted in a 0.43 mm/day reduction in daily length gain compared to the standard WG formulation. While other open formula diets have resulted in growth comparable to WG (Summerfelt and Clayton 2007; Clayton et al. 2008) commercial steelhead and salmon diets have resulted in a small (8 mm) length reduction compared to WG (Johnson and Rudacille 2009b).

17.2.5 Handling, Grading, and Transportation

17.2.5.1 Handling and Grading

Fish handling activities, grading, stock splitting, or harvest are sources of stress and mechanical damage and both may cause immediate or delayed mortality. Currently all Walleye cultured in public hatcheries in North America are derived from wild populations therefore adaptations to the culture environment, handling, and stress tolerance have not been selected or propagated to result in a stock that is more tolerant of stress compared to wild stocks. Information on performance of fingerlings from captured broodstock has not been described.

Summerfelt and Summerfelt (1996) recommended that Walleye not be handled for grading or transport above 23.9 °C because of the presumed relationship to onset of bacterial disease. Even so, fingerlings about 90 mm are often transferred from indoor tanks to outdoor tanks at RFH when water temperatures are about 23.9 °C because water temperature cannot be controlled, when indoor tank density reaches the maximum, density reduction is carried out to prevent disease. However, once fingerlings are stocked in the outdoor tanks for phase III culture they are not handled again until fall (October) when ambient water temperatures are 18.3 °C or less. Only small samples, 50 fish or less, are handled for growth monitoring during the warmer intervals of phase III culture. Delaying harvest until water temperature declines also times stocking water bodies when temperatures are lower and large predators in receiving water bodies have less appetite to consume the newly stocked fish. Because population density of preferred prey of suitable size are expected to be inadequate for fall stocked Walleyes, Harder et al. (2012, 2014) suggest retaining and cold banking Walleyes in the hatchery from fall to stocking late spring or early summer after their prey have spawned. There is also some evidence showing higher survival of yearling fish stocked in the spring (Mitzner 2002; Harder et al. 2013).

Grading during phase III has not been conducted at RFH because cannibalism loss in that interval is generally low at 3–4 % and water temperatures in excess of 25 °C growing season prohibit handling by grading equipment. Fall fingerlings of the same cohort and tank at RFH have ranged in size from 185 mm to 265 mm with an average of 227 mm. Further culture of fingerlings with this size variation may require size grading to prevent cannibalism.

17.2.5.2 Harvest and Transportation

Transportation is an important step of fish production that may greatly influence post-stocking survival and growth of Walleye when stocked for fisheries or into culture systems where fish are grown to larger size. Government agencies transport fingerlings to reach some stocking locations that may last several hours, at RFH five to 6 h of travel. The transportation process consists of three steps that can potentially stress the fish: (1) harvest and loading; (2) transportation; (3) discharge from transport tanks.

Tank harvest and loading procedures can cause mechanical damage and physiological stresses that exceeds the stress of the transport trip. During harvest, a crowder is used to confine to a smaller portion of the culture tank with fish become excited which increases oxygen demand, CO_2 production, and causes lactic acid buildup in the muscles. Fish may be netted into a smaller container to load the transport vehicle prolonging the period of confinement and stress causing cortisol levels to rise as a result. Walleye had substantially higher levels of cortisol after loading at RFH than at the end of a 6–7 h transport trip and post transport cortisol values were closer to pre-load values (Forsberg et al. 1999).

During the transport trip metabolic processes of respiration and waste excretion can reduce water quality. It is important to withhold feed for 24 h before transport to reduce excretion of ammonia and feces which impair water quality. Transport tanks must be equipped with aeration and supplemental oxygenation to maintain favorable dissolved gas concentrations during the trip. Additionally, the tanks must have ram air ventilation (RAV) to allow removal of CO_2 during the trip (Fig. 17.3; Forsberg et al. 1999) because flush mounted vents do not adequately release CO_2 from the tanks (Summerfelt et al. 1997). Tanks with RAV and mechanical aerators allow respired CO_2 to move from the water into the headspace of the tank and out of the tank via the RAV, this prevents the buildup of CO_2 in the fish and release of bicarbonate into the blood in efforts maintain blood pH homeostasis (Forsberg et al. 1999). Mechanical aeration, RAV, and continuous dissolved oxygen monitoring and control are also required to prevent oxygen super saturation of the transport water. The respiration rate of the fish is inversely related to oxygen content of the water, thus in high oxygen environments CO_2 could accumulate in the fishes blood because respiration rate is reduced. Transport tanks that lack RAV have resulted in CO_2 build up and lethargic fish stocked into a fishery.

Post-stocking survival may also be influenced by the density at which Walleye are transported. Piper et al. (1982) recommended transport densities of 66–72 g/L for 50 mm fish and up to 156 g/L for 100 mm fish for 8-h transport times. Kuipers and Summerfelt (1994) transported 50–60 day-old, 2.2–2.6 g (68–69 mm) fingerlings for 4–8 h at densities of 2.7–4.2 g/L immediately after they were seined from a production lake. After stocking these fish in laboratory tanks, most mortality over the entire 28-days phase II culture interval was during the first 8-days, which was attributed to transportation and handling stress-induced mortality from columnaris disease.

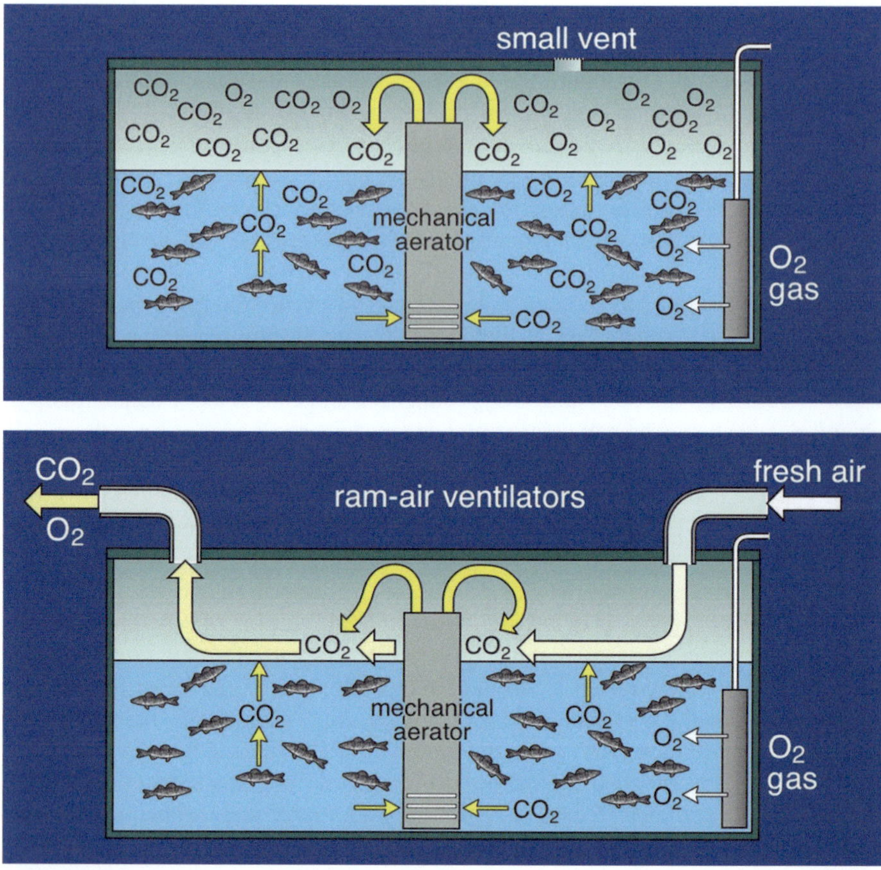

Fig. 17.3 Transport tanks with small vents (5 cm) resulted in a buildup of CO_2 gas in the headspace above the water which resulted in elevated blood CO_2 values (*top*). Ram-air ventilators forced fresh air through the transport tank while the vehicle was in motion to prevent CO_2 build up in the headspace and allowed fish to respire CO_2 and keep blood CO_2 values lower

A brief review of density for transport of fingerling Walleyes by government agencies demonstrates variety by fish size, water temperature, and agency (Table 17.5). Although water temperatures in June and July are higher than temperatures in October, summer fingerlings of 50–100 mm were transported at 30–60 g/L compared with density of 60–120 g/L for fall fingerlings of 200 mm. Walleye from RFH are transported at 30 g/L for 50 mm fish and 60 g/L for 200 mm fish (Table 17.6).

Table 17.4 Density (kg/m^3) during phase II (habituation interval) and phase III (growout) culture to fall fingerling. RFH refers to the Iowa DNR Rathbun Fish Hatchery

Initial weight (g)	Initial density (kg/m^3)	Final weight (g)	Final density (kg/m^3)	Survival (%) of number stocked	Reference
Phase II					
0.60	2.45	4.8	12.3	94	Johnson and Rudacille (2010)
0.67	3.28	5.5	16.3	91	Johnson and Rudacille (2010)
0.78	5.1	4.5	12.0	60.7	Johnson and Rudacille (2010)
2.2	3.2	6.9	8.3	83	Kuipers and Summerfelt (1994)
2.2	1.8	8.0	5.6	84	Kuipers and Summerfelt (1994)
0.5–1.0	3.0	8.1	5.6	50–60	Flowers (1996)
0.38	5.3	10.5	11.3	90	Nagel (1996)
Phase III					
61–66	10.9	189–218	39.2	99.2	Harder et al. (2012)
6.75	1.05	137	16.6	93.9	Johnson and Rudacille (2008)
9.2	3.2	87	31.5	98.6	Johnson and Rudacille (2009b)
354	37.3	425	44.7	100	Yager (1991)
235	27.9	316	36.3	100	Yager (1991)

Table 17.5 Size, temperature, and growth rates of walleye measured during phase II and phase III culture

Size (mm) initial to final	Temperature °C	Growth rate (mm/day)	Reference
45–62	20.9	1.13	Johnson and Rudacille (2010)
62–83	23.1	1.99	Johnson and Rudacille (2010)
102–211	24.4	1.56	Johnson and Rudacille (2009b)
196–282	21.9	0.59–0.65	Harder et al. (2012)
176–216	25.0	0.55	Siegwarth and Summerfelt (1993)
285–324	20.7	0.31	Siegwarth and Summerfelt (1993)
187–246	22.5	0.77	Harder et al. (2014)
197–250	22.5	0.77	Harder et. al. (2014)
313–339	23.0	0.26	Yager (1991)
258–327	23.0	0.54	Yager (1991)

Table 17.6 Hauling density (g/L) of walleye fingerlings reported by government agencies (n = 11) in the US compared to density recommendation of Piper et al. (1982)

		Survey	Piper et al.(1982)
	Temp (C)	g/L	g/L
200 mm	15–21	60–102	No recommendation
	7–13	120	
100 mm	21–29	60	156
	7–13	120	
50 mm	18–25	30–60	66–72
	15–25	60–132	

17.2.6 Diseases

17.2.6.1 Viral Diseases

Several viral diseases have been reported in Walleye (Wolf 1988), including infectious pancreatic necrosis (McAllister 1996), although isolated from Walleye, it did not cause disease. Only three viral diseases are mentioned but of these VHS requires the most attention because it is a World Organization for Animal Health (OIE) listed disease (OIE 2009) and designated as a reportable aquatic pathogen by the U.S. Department of Agriculture, and the Canadian Food Inspection Agency (Siah et al. 2014). VHS has a long history of causing severe mortality in farmed rainbow trout in Europe and it is of considerable importance for percid culture because as been implicated as the cause of fish kills in North America,

There are also four non-neoplastic virus-associated diseases that cause skin tumors in Walleye (McAllister 1996). Of these, lymphocystis and Walleye dermal sarcoma (WDS) are the more prevalent of four, both are observed on the same fish during the Walleye spawning season in Oneida Lake, New York (Bowser et al. 1999). The two other diseases are Walleye discrete epidermal hyperplasia (WEH) and diffuse epidermal sarcoma (Bowser et al. 1988). The latter disease is caused by a herpes virus and WEH is caused by a retrovirus, type 1 and 2 (Getchell et al. 2004).

Lymphocystis and WDS are not implicated in fish kills and they are not known to infect humans, however, because of their gross appearance, they are rejected by in commerce and recreational anglers. They are also important to fish propagation because skin abrasion during handling of Walleyes during spawning may increase the incidence of the disease and stocking of hatchery-reared fish may spread the disease to habitats previously free of the disease (McAllister 1996).

17.2.6.1.1 VHS

VHS is an acute to chronic disease caused by an enveloped negative-strand RNA virus belonging to the Novirhabovirus genus of the family Rhabdoviridae (Einer-Jensen et al. 2004; Batts and Winton 2014). More than 60 species of freshwater and marine fish are natural hosts of VHSV (Batts and Winton 2014).

The virus has been known as the Egtved virus (syn: VHSV) after a city in southern Denmark where the disease was first recognized (Wolf 1988). It has been regarded as the most important viral disease of farmed Rainbow Trout (*Oncorynchus mykiss*) in Europe (Olesen 1998; Amos and Hopper 1998; Einer-Jensen et al. 2004). The VHS rhabdovirus (VHSV) was not known in North America in the early 1980s (Meyer et al. 1983) but a serologically similar virus, although discriminated from the European stain by polymerase chain reaction (PCR), was isolated from returning adult coho (*Oncorynchus kisutch*) and chinook (*O. tshawytscha*) salmon at two different sites in the state of Washington in 1988 (Meyers and Winton 1995), it was not reported in freshwater fish in the United States as of 1992 (McAllister and Batts 2014). It has been isolated from frozen fish collections of Great Lakes origin that dated back to 2003 (Elsayed et al. 2006). Based on molecular methods for genetic analyses of isolates from around the world, four genotypes are recognized; genotypes I, II and III are predominantly found in Europe and Japan, while Genotype IV has to date only been isolated in North America, Japan and Korea (Einer-Jensen et al. 2004). The VHSV isolate from 2003 samples of Muskellunge (*Esox masquinogy*) in Lake St. Clair, Michigan and named the Great Lakes strain subtype VHSV-IVb (Elsayed et al. 2006).

It is "readily transmissible to all ages of fish, and survivors of infections can become longterm virus carriers" (McAllister and Batts 2014). For example, subtype IVb was associated with a large (100 metric tons) die-off Freshwater Drum (*Aplodinotus grunniens*), many Round Goby (*Neogobius melanostomus*), and few Muskellunge (*Esox masquinongy*) in the Bay of Quinte, Lake Ontario (Lumsden et al. 2007), other fish kills in wild stocks in the Great Lakes, including an epizootic in Walleye in Lake Erie in 2006, and from several inland lakes of Michigan, New York, Wisconsin, and Ohio and the upper portion of the St. Lawrence River (USDA APHIS VS 2014; Thompson et al. 2011). Bain et al. (2010) found VHSV widely dispersed but present in only 6 species from collections of 17 species of fish from 30 sites in the Great Lakes; it was not found in Walleye in that study. At least 28 freshwater fishes that are present all of the Great Lakes except Lake Superior and in several inland waters in neighboring watersheds are considerable susceptible to disease caused by VHSV IVb (USDA APHIS VS 2014; Thompson et al. 2011), but surveillance studies from 2006 to 2009 found VHSV IVb was localized to Great Lake States and neighboring watersheds (USDA APHIS VS 2009).

Because VHSV is known to cause epizootics with massive fish kills and as there is no treatment for infected fish, to prevent further spread of the virus, in 2006 the U.S. Animal and Plant Health Inspection Service (APHIS) issued a temporary order prohibiting the importation of 37 species of live fish into the US from two Canadian provinces as well as the interstate movement of these same species from the eight states bordering the Great Lakes (USDA APHIS VS 2009). That order was lifted in 2 June 2014; APHIS no longer prohibits or restricts the interstate movement of VHS-susceptible species of live fish from VHS-affected or at-risk states, and it will no longer restrict the importation of the same species of live fish from Ontario and Quebec, Canada into the United States (USDA APHIS 2014).

External signs of disease are similar to other septicemias, which means that they are not definitive for diagnosis, but they characteristically include hemorrhages in the eyes, skin, gills, brain, and at the base of pelvic and pectoral fins; internally, diffuse hemorrhage occurs in the visceral mesenteries, and abnormally large quantities of blood may cause the kidneys and liver to be swollen and discolored, and hemorrhages in skeletal muscle (Olson et al. 2013; Batts and Winton 2014). Non-symptomatic fish are believed to be carriers; it is transmissible through the water to susceptible fish of all ages.

To date, there is no commercial vaccine or treatment for VHS, therefore, biosecurity is the only preventive measure. Fish production facilities using ground (well) water are the most secure but are susceptible to contamination from wild-caught broodstock unless those sources have been tested and found to not harbor antibodies to the virus.

Human transport of infected fish is considered the greatest danger for spread of the disease. Although there is no evidence of "true" vertical transmission (Groocock et al. 2013), the virus is detectable in ovarian fluid (Kocan et al. 2001) and may adhere to the surface of the egg after spawning and then infect prolarvae (fry) at hatch. Thus, because hatcheries depend on stripping wild-caught Walleye for gametes, disinfection with 50 mg/L iodophor immediately after fertilization became a routine protocol to destroy VHSV and interrupt vertical transmission. Groocock et al. (2013), however, found that 50-mg/L dose may be insufficient because of the natural coating of the egg or because the routine of using tannic acid to reduce stickiness of the eggs may reduce the effectiveness of the iodphore treatment; thus a 100 mg/L iodophor treatment is required.

17.2.6.1.2 Lymphocystis

Lymphocystis is a cosmopolitan disease (*Lymphocystis* disease virus, LCDV) of marine and freshwater fishes caused by iridovirus of the family *Iridoviridae*. The disease is manifest by an inflammatory response with clusters of wart-like growths of hypertrophic fibroblasts on the fins, skin or gills. The infected cells are filled with virus particles that are shed into the environment when the cells burst (Yanong 2010). It is transmitted to tumor-free fish by waterborne exposure water. Although not generally acknowledged, Olson (1958) stated that handling during spawn-taking contributed to an increased incidence of the disease.

The incidence of the disease in Walleye handled during the spring spawning runs range from 1 % to 30 % (Colby et al. 1979; Ryder 1961; Margenau et al. 1988). The disease is regarded as self-limiting and although generally not considered to be a cause of mortality (Ryder 1961), however, based on a reduced rate of return of Walleyes with lymphocystis 1 year after tagging compared with Walleyes without lymphocystis, Margenau et al. (1988) stated that lymphocystis may have caused mortality in Walleyes in the St. Louis River, Wisconsin.

17.2.6.1.3 Walleye Dermal Sarcoma

The Walleye Dermal Sarcoma Virus retrovirus (WDSV) causes a benign hyperplastic and hypertrophic growth that produces grossly visible external tumors on Walleye (WDS). The lesions are similar to that of lymphocystis (Wolf 1988) and both WDSV and lymphocystis have highest prevalence in the spring, low during the summer and again high in the fall (Bowser 1988). The occurrence of WDS in Great Lakes was not noted by Meyer et al. (1983) but Walker (1969) reported it in Lake Champlain and the eastern Great Lakes. It is not listed in the Fish Health Section's Blue Book because that reference describes procedures for isolation and identification of "etiological agents responsible for epizootics of salmonids and other fishes, as well as for invertebrates" (Thoesen 1994).

The most extensive studies of WDS have been conducted on Walleye in Oneida Lake, New York, where WDS was first reported (Walker 1969). Walleyes younger than 3-years were not infected although Walleyes as young as 6 weeks were found to be highly susceptible (Bowser et al. 1997). Prevalence is seasonal with 20–30 % incidence during the cooler months and as low as 4 % during the summer (Bowser et al. 1988). It is transmitted to tumor-free fish by waterborne exposure through in as little as 5 days of cohabitation with tumor-positive fish (Bowser et al. 1999) or by an inoculum from cell-free filtrates of tumors collected in the spring (Martineau et al. 1990). Natural horizontal transmission occurs during spring spawning aggregations when tumors are present. In Walleye fingerlings experimentally infected with WDS, an inverse relationship was observed between temperature and regression of WDS tumors (Getchell et al. 2000).

17.2.6.2 Bacterial Diseases

Walleye are highly susceptible to gram-negative bacteria of the genus Flavobacterium that cause columnaris disease (*F. columnare*) and bacterial bill disease (BGD; *F. branchifilum*).

17.2.6.2.1 Columnaris Disease

Columnaris disease frequently occurs as a result of mechanical damage from handling during pond harvest and distribution to culture tanks (Hussain and Summerfelt 1991). At RFH, columnaris lesions appear on the skin from the mouth or nose to the caudal fin. The lesions appear as pale patches with the center of the lesion often with a yellow-brown appearance. Presumptive diagnosis of *F. columnare* is made by observing long flexing bacteria in colonies resembling haystacks in wet mount preparation of a skin scrape or gill filaments examined under 100–200 X. Detailed diagnostic procedures are given in the Blue Book (AFS-FHS 2014).

Columnaris disease outbreaks result in high morality rates when they are not treated early. The herbicide, diquat dibromide (Reward) is commonly used at RFH under an investigational new animal drug approval (INAD) to control mortality associated with external columnaris disease. Treatment options include 2–18 mg/L for a 1–4-h treatment or 19–28 mg/L for a 30 or 60-min treatment. At RFH, we commonly treat at 15–18 mg/L for 2 h on three consecutive days for most columnaris infections, however 28 ppm for 1 h has been applied to treat severe infections.

PEROX-AID (35 % hydrogen peroxide) is a zero-withdrawal, USFDA approved drug to control mortality of Walleye with external *F. columnarae* infection. Rach et al. (2003) determined that a 1-h bath of 50 mg/L was an effective dose but higher rates were not efficacious. Clayton and Summerfelt (1996) recommended a concentration of no more than 50 mg/L hydrogen peroxide for one hour. Our personal experience is that treatment concentrations of 75–100 mg/L causes mortality to 50–100 mm fingerlings, however, previous exposure at low concentrations (10 mg/L) were shown to increase the tolerance of Walleye to hydrogen peroxide treatment (Tripi and Bowser 2001; Tort et al. 2003).

17.2.6.2.2 Bacterial Gill Disease

High density, accumulated waste and feed, and impaired water quality may lead to outbreaks of BGD. At RFH, these conditions often occur during the Phase II period when Walleye fingerlings feed rates are highest and the water has a high concentration of dissolved organic matter and ammonia. Generally fingerlings with BGD will hang near the water surface, react slowly to overhead motion, and skin color may appear darkened. The microscopic examination of a gill filament wet mount at 200–400 X will distinguish BGD from Columnaris. Although both are gram positive, bacilli, *F. branchifilum* is longer rod compared to *F. columnarae* and only a few rods clump together into a colony.

At the RFH mortality from BGD can be controlled by a 1-h treatment with Chloramine-T in a 20 mg/L static bath treatment. In our experience one or two treatments with Chloramine-T are necessary to control mortality and to prevent recurrence of the disease the flow rate is increased or the density in the tank is reduced to improve water quality.

17.2.6.3 Protozoan Parasites

The water supply for RFH is from an artificial, flood-control lake. At the hatchery, the water is processed by filtration through 300 μm drum-screen, sand filtration, and Ultra-violet disinfection. Protozoan parasites such as *Trichodina*, *Chilodonella*, and *Ichthyoptherius multifilis* (*Ich*) pass through filtration will cause mortality in our culture system. *Trichodina* and *Chilodonella* are treated effectively with one or two consecutive day treatments with 1 % NaCl for 2 h. Both parasites are found on the gills and in high numbers impair respiration that leads to suffocation.

Infestation of Walleye with *Ich* often occurs at RFH in fish 100–225 mm during phase III culture in outdoor tanks supplied with unfiltered lake water. At the peak of infestation, water exchange is limited to 0.5 per hour, which seems to intensify the *Ich* infection. A flow-through treatment with formalin at 45 ppm for 9 h is given daily for 2–6 weeks to eradicate the infection, however, to reduce the use of formalin, the need for treatment is based on the intensity of infection, which is determined from a weekly collection of fingerlings from each tank and a count of the total number of *Ich* cells on one gill arch. Formalin treatment begins when ≥15 *Ich* cells on one gill arch are detected, and daily treatment continues until no cells are found. *Ich* outbreaks often occur multiple times during the 12 week phase III culture interval. Paradoxically, infection rates of over 250 *Ich* cells per gill arch have been observed at RFH without mortality, however, infestation of 200 or more *Ich* cells a 16-h treatment is given to speed the eradication. In a research scale experiment, where as many as 1000 *Ich* cells were counted on one gill arch, the infection was eradicated in 7 days of continuous 24-h treatment with 30 ppm formalin. Formalin treatments to eradicate *Ich* are costly and contribute over 20 % of variable cost of production for Walleye fingerlings at RFH. The monitoring and treatment protocols at RFH were implemented in 2009 and have prevented mortality related to Ich. Previous to 2009, production tanks were given daily 45 ppm 9-h treatments without monitoring infection rates.

Survival during growout is affected by the ability to control mortality caused by disease outbreaks and cannibalism. The average survival rate at RFH during phase III culture of juveniles to 225 mm has ranged from 82.8 % to 94.1 % from 2001 to 2010 (Summerfelt et al. 2011).

17.2.7 Abnormalities and Deformities

Eye, cranium, skeleton, and fin abnormalities of cultured Walleye have been documented by Clayton et al. (1998), Garcia-Abiado et al. (2004), and Johnson et al. (2008). Deformities are more often observed in fingerlings reared from larval stage with formulated diets than in fingerlings reared in culture ponds and many times are macroscopically observed in larger juveniles (>75 mm). These deformities and abnormalities reduce production performance in culture settings and impair survival and viability for sport fish enhancement stocking.

Injury to their eye from abrasion during contact with nets, tank surfaces, graders and other fish is common; e.g., eye abnormalities of 5–10 % have been observed at RFH at the end of Phase III culture, which at RFH is in outdoor concrete tanks with rough surfaces and no overhead cover. Garcia-Abiado et al. (2004) described the progression of corneal trauma in juvenile hybrid Walleye that begins with exophthalmia followed by enophthalmia then loss of globe. Secondary fungal and bacterial infection followed the mechanical damage. Although most fish showing corneal trauma died, fish that recovered had significantly reduced weight, growth, and condition factor. They stated that "eye abnormalities observed under culture conditions may jeopardize growth and condition of hybrid Walleyes reared intensively in tanks". The fish culturist should take reasonable measures to prevent ocular disease by designing culture tanks and fish handling devices to minimize abrasive surfaces to prevent the mortality and lost productivity of bacterial and fungal entry of damaged eyes.

Clayton et al. (1998) suggested that severe fin erosion of Walleye was caused by previous columnaris disease outbreaks in the fish and the presence of necrotic fin tissue. In that study, fish were reared in smooth fiberglass tanks, but were handled at 2-week intervals to measure growth for research. The severity and location of fin erosion varied between stocks and between purebred and hybrid Walleye. Eroded fins were varied from 5 % to 90 % of the length of fins for Walleye in other literature. A domesticated stock from Ohio had more severe fin erosion than a wild stock from Iowa.

17.2.8 Expectations for Growth and Survival During Growout

Growth rates (mm/day) of Walleye decrease with increasing length, for example:

- At RFH, growth rates for fingerlings between 100 and 250 mm are 1.4–1.5 mm per day when temperatures average about 24 °C,
- The rate is 0.59–0.65 mm/day for fish between 196 and 282 mm at 21.9 °C (Harder et al. 2012)
- And the rate is 0.31 mm/day from 285 to 324 mm at 20.7 °C (Siegwarth and Summerfelt 1993).

Summerfelt and Summerfelt (1996) calculated a value of 0.731 mm/day at the optimal temperature of 23.4 °C by using a regression analysis of published growth studies over a large range in length based. Walleye held in a WRAS from 55 to 368 dph and grown 87 mm to 342 mm (Fig. 17.4) displayed a steady linear decline in growth rate with age that was not characteristic of a sudden onset of sexual maturity (Summerfelt and Summerfelt 1996). Summerfelt (1993) developed a length-weight equation for 2,295 L and W values with mean length 295 mm (73–557 mm) and mean weight 250.7 g (3.1–801 g). The log-log equation was: $\log_{10}$ Weight = $4.91 + 2.929 * \log_{10}$ Length (mm), $r^2 = 0.987$.

17.2.9 Opportunities for Research and Marketing

The walleye culture techniques and success of RFH and other agency hatcheries that were reviewed in this chapter and in a previous chapter on intensive larval culture (3.4) demonstrate substantial progress in cultural technology from larval to size suitable for commercial food-fish production. Research needs are in genetics, definition of life-stage nutritional requirements and diet formulation, and food fish marketing strategy.

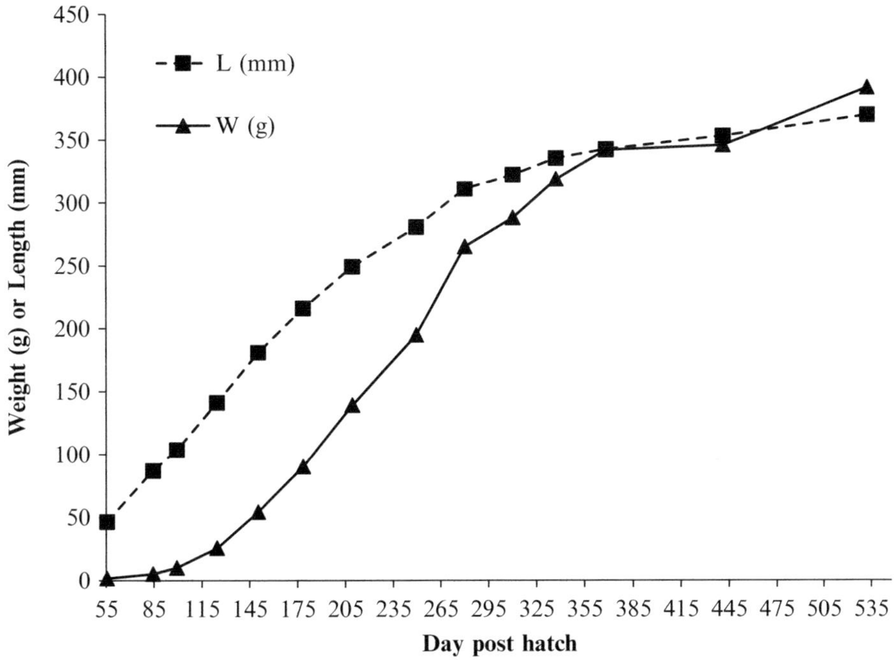

Fig. 17.4 Length and weight at age for Walleye raised in an experimental water reuse system (data from Summerfelt 1993). The temperature was initially held at 20 °C during habituation to feed then increased to an average temperature of 23 °C (maximum 28 °C) until harvest

17.2.9.1 Selective Breeding

All walleye are currently produced from wild broodstock that have not been selectively bred for performance (i.e. growth, disease resistance, etc.) in the culture setting. Walleye only spawn once per year, thus groups of captive broodstock under staggered photothermal regimes are necessary to keep a year-round supply of fry available for growout to meet food-fish market demands. Walleye are adaptable to complete lifecycles in captivity as demonstrated Ohio Department of Natural Resources (Nagel 1985) that developed a domestic stock of walleye and bred successive generations in the London State Fish Hatchery. However, the Ohio walleye broodstock program was discontinued after 20 years because the agency did not want to develop a domesticated stock for use in enhancement stocking.

17.2.9.2 Life Stage Diets

Nutrition research from fry to market size is needed to define nutrient requirements and fully exploit the genetic growth potential of walleye using economical or reduced marine ingredient formulations. Captive broodstock will require specialized diet formulations to produce high quality gametes without augmentation with live prey, a need has not been addressed because currently only wild broodstock are used.

17.2.9.3 Marketing Food Fish

A 1 kg fish has been suggested as the goal for food fish walleye production but this should be reevaluated. Yellow perch (*Perca flavescens*) are commonly marketed at a 151-g, a size that yields two 31-g fillets (Hart et al. 2006). Similar fillet sizes (35 g) were obtained from 143-g walleye in as little as 5–6 months half the time required for Yellow Perch (Summerfelt et al. 2010).

References

AFS-FHS (American Fisheries Society-Fish Health Section) (2014) FHS blue book: suggested procedures for the detection and identification of certain finfish and shellfish pathogens, 2014 edition. http://afs-fhs.org/bluebook/bluebook-index.php

Amos KJ, Hopper K (1998) A case history of adaptive management strategies for viral hemorrhagic septicemia virus (VHSV) in Washington State. J Aquat Anim Health 10:152–159

Anderson RO, Neumann RM (1996) Length, weight and associated structural indices. In: Murphy BR, Willis DW (eds) Fisheries techniques, 2nd edn. American Fisheries Society, Bethesda, pp 447–482

Bain MB, Cornwell ER, Hope KM, Eckerlin GE, Casey RN, Groocock GH, Getchell RG, Bowser PR, Winton JR, Batts WN, Cangelosi A, Casey JW (2010) Distribution of an invasive aquatic pathogen (viral hemorrhagic septicemia virus) in the great lakes and its relationship to shipping. PLoS One. doi:10.1371/journal.pone.0010156

Barrows FT (1990) Walleye research summary; 1989 season. Annual progress report: year 1. fisheries and federal aid project 81. USFWS, Denver

Barrows FT (1996) Diet and nutrition. In: Summerfelt RC (ed) Walleye culture manual, vol 101, NCRAC culture series. North Central Regional Aquaculture Center Publications Office, Iowa State University, Ames, pp 315–321

Barrows FT, Kindschi GA, Zitzow RE, Smith CE (1992) Walleye research summary; 1991 season. Annual progress report: year 3. Fisheries and federal aid project 81, 52 p

Batts WN, Winton JR (2014) Viral hemorrhagic septicemia. Section 2.2.7. In: Thoesen JN (ed) Fish health section blue book, 4th edn. American Fisheries Society, Bethesda

Beyerle GB (1978) Survival, growth, and vulnerability to angling of northern pike and walleye stocked as fingerlings in small lakes with bluegills or minnows. Am Fish Soc Spec Publ 11:135–139

Boeuf G, Le Bail PY (1999) Does light have an influence on fish growth? Aquaculture 177:129–152

Bowser PR (1988) Seasonal prevalence of skin tumors from walleye (*Stizostedion vitreum*) from Oneida Lake, New York. J Wildl Dis Assoc 24:292–298

Bowser PR, Wolfe MJ, Forney JL, Wooster GA (1988) Seasonal prevalence of skin tumors from walleye (Stizostedion vitreum) from Oneida Lake, New York. J Wildlife Dis 24:292–298

Bowser PR, Wooster GA, Earnest-Koons K (1997) Effects of fish age and challenge route in experimental transmission of walleye dermal sarcoma in walleyes by cell-free tumor filtrates. J Aquat Anim Health 9:274–278

Bowser PR, Wooster GA, Getchell RG (1999) Transmission of walleye dermal sarcoma and lymphocystis via waterborne exposure. J Aquat Anim Health 11:158–161

Bristow B (1996) Extensive-intensive production of advanced fingerling walleyes at the Spirit Lake State Fish Hatchery. In: Summerfelt RC (ed) Walleye culture manual, vol 101, NCRAC culture series. North Central Regional Aquaculture Center Publications Office, Iowa State University, Ames, pp 209–212

Bulkowski L, Meade JW (1983) Changes in phototaxis during early development of walleye. Trans Am Fish Soc 112:445–447

Cai Y, Summerfelt RC (1992) Effects of temperature and size on oxygen consumption and ammonia excretion by walleye. Aquaculture 104:127–138

Cheshire WF, Steele KL (1972) Hatchery rearing of walleye using artificial food. Prog Fish Cult 43(2):96–99

Chevalier JR (1973) Cannibalism as a factor in first year survival of walleye in Oneida Lake. Trans Am Fish Soc 102:739–744

Clayton RD, Summerfelt RC (1996) Toxicity of hydrogen peroxide to fingerling walleyes. J Appl Aquac 6:39–48

Clayton RD, Stevenson TL, Summerfelt RC (1998) Fin erosion in intensively cultured walleyes and hybrid walleyes. Prog Fish Cult 60:114–118

Clayton RD, Morris JE, Summerfelt RC (2008) Comparison of soy and fish oils in practical diets for fingerling walleyes. N Am J Aquac 70:171–174

Colby PJ, McNicol RE, Ryder RA (1979) Synopsis of biological data on the walleye *Stizostedion vitreum* (Mitchell 1818). Fisheies synopsis, vol 119. Food and Agricultural Organization of the United Nations, Rome

Einer-Jensen K, Ahrens P, Forsberg R, Lorenzen N (2004) Evolution of the fish rhabdovirus viral haemorrhagic septicaemia virus. J Gen Virol 85:1167–1179

Elsayed E, Faisal M, Thomas M, Whelan G, Batts W, Winton J (2006) Isolation of viral haemorrhagic septicaemia virus from muskellunge, *Esox masquinongy* (Mitchill), in Lake St-Clair, Michigan, USA reveals a new sublineage of the North American genotype. J Fish Dis 29:611–619

Flammang MK, Olson JA, Willis DW (1999) Application of the relative weight index to juvenile walleyes. N Am J Aquac 61:310–314

Flowers D (1996) Intensive walleye culture in Ontario: advanced fingerling production methods. In: Summerfelt RC (ed) Walleye culture manual, vol 101, NCRAC culture series. North Central Regional Aquaculture Center Publications Office, Iowa State University, Ames, pp 213–214

Forsberg JAR, Summerfelt RC, Barton BA (1999) Effects of ram-air ventilation during transportation on water quality and physiology of fingerling walleye. N Am J Aquac 61:220–229

Garcia-Abiado MA, Czesny S, Dabrowski K (2004) Tank performance of larval saugeyes (walleye & sauger) produced out-of-season and during regular season spawning. N Am J Aquac 66:48–52

Getchell RG, Wooster GA, Bowser PR (2000) Temperature-associated regression of walleye dermal sarcoma tumors. J Aquat Anim Health 12:189–195

Getchell RG, Wooster GA, Rudstam LG, Van DeValk AJ, Brooking TE, Bowser PR (2004) Prevalence of walleye discrete epidermal hyperplasia by age-class in walleyes from Oneida Lake, New York. J Aquat Anim Health 16:23–28

Groocock GH, Getchell RG, Cornwell ER (2013) Iodophor disinfection of walleye eggs exposed to viral hemorrhagic septicemia virus type IVb. North Am J Aquacult 75(1):25–33

Halverson MA (2008) Stocking trends: a quantitative review of governmental fish stocking in the United States, 1931 to 2004. Fisheries 33:69–75

Harder T, Summerfelt RC (1996) Effects of tank color and size on the success of training wall- eye fingerlings to formulated feed. In: Libey GS, Timmons MB (eds) Successes and failures in commercial recirculating aquaculture, vol 2, Northeast regional agricultural engineering service, NRAES-98. Cornell University, Ithaca, pp 631–636

Harder TM, Gotsch GG, Summerfelt RC (2012) Effect of photoperiod on growth and feed efficiency of fingerling walleye. N Am J Aquac 74:547–552

Harder TM, Gotsch GG, Summerfelt RC (2013) Survival and condition of cold banked juvenile walleyes. N Am J Aquac 75:512–516

Harder TM, Gotsch GG, Summerfelt RC (2014) Growth response of juvenile walleyes after cold banking. N Am J Aquac 76:255–260

Hart SD, Garling DL, Malison JA (eds) (2006) Yellow perch (*Perca flavescens*) culture guide, vol 103, NCRAC culture series. NCRAC, Publications Office, Iowa State University, Ames

Heidinger RC, Waddell JH, Tetzlaff BL (1987) Relative survival of walleye fry versus fingerlings in two Illinois reservoirs. Proc Annu Conf Southeast Assoc Fish Wildl Agencies 39(1985):306–311

Held JA, Malison JA (1996) Culture of walleye to food size. In: Summerfelt RC (ed) Walleye culture manual, vol 101, NCRAC culture series. North Central Regional Aquaculture Center Publications Office, Iowa State University, Ames, pp 231–232

Hokanson KEF, Koenst WM (1986) Revised estimates of growth requirements and lethal temperature limits of juvenile walleyes. Prog Fish Cult 48:90–94

Hussain M, Summerfelt RC (1991) The role of mechanical injury in an experimental transmission of *Flexibacter columnaris* to fingerling walleye. J Iowa Acad Sci 98:93–98

Johnson JA, Esser R (2009) Walleye culture—habituation to feed in the dark. Am Fish Soc Fish Cult Sect Newsl 2:15–18

Johnson JA, Rudacille JB (2008) Fish culture research. Federal aid in sport fish restoration, Annual report project F-160-R. Iowa Department of Natural Resources, Des Moines.

Johnson JA, Rudacille JB (2009a) Fish culture research. Federal aid in sport fish restoration, Annual report project F-160-R. Iowa Department of Natural Resources, Des Moines.

Johnson JA, Rudacille JB (2009b) Production scale testing of formulated diet for the intensive culture of walleye and muskellunge. Federal aid in sport fish restoration, Project F-160-R, Completion report. Iowa Department of Natural Resources, Des Moines

Johnson JA, Rudacille JB (2010) Intensive culture of walleye converted to a diet as small fingerlings. Federal aid in sport fish restoration, Project F-160-R, Completion report. Iowa Department of Natural Resources, Des Moines

Johnson JA, Summerfelt RC, Clayton RD (2008) Evaluation of replacement feeds for Fry Feed Kyowa for larval walleyes. N Am J Aquac 70:445–451

Kerr SJ (2011) Stocking and marking: lessons learned over the past century. In: Barton BA (ed) Biology, management, and culture of walleye and sauger. American Fisheries Society, Bethesda, pp 423–449

Kocan RM, Hershberger PK, Elder NE (2001) Survival of the North American strain of viral hemorrhagic septicemia virus (VHSV) in filtered seawater and seawater containing ovarian fluid, crude oil and serum-enriched culture medium. Dis Aquat Org 44:75–78

Kolkovski S, Czesny S, Dabrowski K (2000) Use of krill hydrolysate as a feed attractant for fish larvae and juveniles. J World Aquacult Soc 31:81–88

Kuipers KL, Summerfelt RC (1994) Converting pond-reared walleye fingerlings to formulated feeds: effects of diet, temperature, and stocking density. J Appl Aquac 4:31–57

Larscheid JG (1995) Contribution of stocked walleye and population dynamics of adult walleye in Spirit, East and West Okoboji Lakes. Federal aid in sport fish restoration, Project F-135-R, Completion report. Iowa Department of Natural Resources, Des Moines

Lumsden JS, Morrison B, Yason C, Russell S, Young K, Yazdanpanah A, Huber P, Al-Hussinee L, Stonhe D, Way K (2007) Mortality event in freshwater drum *Aploidinotus grunniens* from Lake Ontario, Canada, associated with viral haemorrhagic septicemia virus, type IV. Dis Aquat Org 76:99–111

Malison JA, Held JA (1996) Habituating pond-reared fingerlings to formulated feed. In: Summerfelt RC (ed) Walleye culture manual, vol 101, NCRAC culture series. North Central Regional Aquaculture Center Publications Office, Iowa State University, Ames, pp 199–204

Margenau TL, Schram ST, Blust WH (1988) Lymphocystis in a walleye population. Trans Am Fish Soc 117:308–310

Martineau D, Bowser PR, Armstrong GA, Wooster GA (1990) Experimental transmission of a dermal sarcoma in fingerling walleyes (*Stizostedion vitreum vitreum*). Vet Pathol 27:230–234

McAllister PE (1996) Viruses associated with diseases of walleye, sauger, and walleye x sauger hybrids. In: Summerfelt RC (ed) Walleye culture manual, vol 101, NCRAC culture series. North Central Regional Aquaculture Center Publications Office, Iowa State University, Ames, pp 355–367

McAllister PE, Batts WN (2014) VIII. Viral hemorrhagic septicemia. In: Thoesen JC (ed) Blue book, version 1, 4th edn, Fish health section. American Fisheries Society, Bethesda, pp 1–4
Meyer FP, Warren JW, Carey TG (1983) A guide to integrated fish health management in the Great Lakes basin, vol 83-2, Special publication. Great Lakes Fishery Commission, Ann Arbor
Meyers TR, Winton JR (1995) Viral hemorrhagic septicemia virus in North America. An Rev Fish Dis 5:3–24
Mitzner L (1992) Evaluation of walleye fingerling and fry stocking at Rathbun Lake Iowa. North Am J Fish Manag 12:321–328
Mitzner L (1995) Man-made lake investigations. Federal aid in sport fish restoration, Project F-134-R, Completion report. Iowa Department of Natural Resources, Des Moines
Mitzner L (2002) Effectiveness of walleye fry and fingerling stocking in Rathbun Lake, Iowa, 1990–1999. N Am J Fish Manag 22:1007–1013
Moore GA (1944) The retina of two North American teleosts with special reference to their tapetum lucida. J Comp Neurol 80:369–379
Nagel TO (1976) Intensive culture of fingerling walleyes on formulated feeds. Prog Fish Cult 38:90–91
Nagel T (1985) Development of a domestic walleye broodstock. Prog Fish Cult 47:121–122
Nagel T (1996) Intensive culture of fingerling walleye on formulated feeds. In: Summerfelt RC (ed) Walleye culture manual, vol 101, NCRAC culture series. North Central Regional Aquaculture Center Publications Office, Iowa State University, Ames, pp 205–207
OIE (Office International des Epizooties) (2009) Viral haemorrhagic septicemia. In: Manual of diagnostic tests for aquatic animals, 6th edn. OIE, Paris, pp 279–298
Olesen NJ (1998) Sanitation of viral haemorrhagic septicaemia (VHS). J Appl Ichthyol 14:173–177
Olson DE (1958) Statistics of a walleye sport fishery in a Minnesota lake. Trans Am Fish Soc 87(1957):52–72
Olson W, Emmenegger E, Glenn J, Winton J, Goetz F (2013) Comparative susceptibility among three stocks of yellow perch, *Perca flavescens* (Mitchill), to viral haemorrhagic septicaemia virus strain IVb from the Great Lakes. J Fish Dis 36:711–719
Piper RG, McElwain IB, Orme LE, McCraren JP, Fowler LG, Leonard JR (1982) Fish hatchery management. U.S. Fish and Wildlife Service, Washington, DC
Priegel GR (1964) Early scale development in the walleye. Trans Am Fish Soc 93:199–200
Rach JJ, Schleis SM, Gaikowski M, Johnson JA (2003) Efficacy of hydrogen peroxide in controlling mortality associated with external columnaris on walleye and channel catfish fingerlings. N Am J Aquac 65:300–305
Ryder RA (1961) Lymphocystis as a mortality factor in a walleye population. Prog Fish Cult 23:183–186
Santucci VJ Jr, Wahl DH (1993) Factors influencing survival and growth of stocked walleye (*Stizostedion vitreum*) in a centrarchid-dominated impoundment. Can J Fish Aquat Sci 50:1548–1558
Scherer E (1976) Overhead-light intensity and vertical positioning of the walleye, *Stizostedion vitreum vitreum*. J Fish Res Board Can 33:289–292
Schneider JC (1983) Experimental walleye-perch management in a small lake. Michigan Department of Natural Resources, Fisheries Research Report 1905
Siah A, Duesund H, Frisch K, Nylund A, McKenzie P, Saksida S (2014) Development of a multiplex assay to measure the effects of shipping and storage conditions on the quality of RNA used in molecular assays for detection of viral haemorrhagic septicemia virus. J Aquat Anim Health 26:173–180
Siegwarth GL, Summerfelt RC (1992) Light and temperature effects on performance of walleye and hybrid walleye fingerlings reared intensively. Prog Fish Cult 54:49–53
Siegwarth GL, Summerfelt RC (1993) Performance comparison and growth models for walleyes and walleye × sauger hybrids reared for two years in intensive culture. Prog Fish Cult 55:229–235
Smith LL Jr, Koenst WM (1975) Temperature effect on egg and fry of percid fishes. U.S. Environmental Protection Agency, Ecological research series EPA-660/3–75–017
Summerfelt ST (1993) Low-head roughing filters for enhancing recycle water treatment for aquaculture. Doctoral dissertation. Iowa State University, Ames
Summerfelt ST (1996) Aquaculture of walleye as a food fish. In: Summerfelt RC (ed) Walleye culture manual. North Central Regional Aquaculture Center Publications Office, Iowa State University, Ames, pp 215–230

Summerfelt RC, Clayton RD (2007) Evaluation of open-formula walleye grower diets for juvenile walleyes. N Am J Aquac 69:53–58

Summerfelt ST, Summerfelt RC (1996) Aquaculture of walleye as a food fish. In: Summerfelt RC (ed) Walleye culture manual, vol 101, NCRAC culture series. North Central Regional Aquaculture Center Publications Office, Iowa State University, Ames, pp 215–230

Summerfelt RC, Barton BA, Clayton RD, Pottebaum B (1997) Comparison of plasma cortisol and chloride measures of stress in fingerling walleye after handling and transportation from hatchery to stocking sites in vented and unvented transport tanks. In: Timmons MB, Losordo T (eds) Advances in aquacultural engineering: aquacultural engineering society proceedings III. Northeast Regional Agricultural Engineering service, Cooperative Extension, NRAES-105, Ithaca, pp 38–58

Summerfelt RC, Johnson JA, Clayton RD, Kinunen RE (2010) Production of walleye as potential food fish, vol 116, Fact sheet. North Central Regional Aquaculture Center, Ames

Summerfelt RC, Johnson JA, Clouse CP (2011) Culture of walleye, sauger, and hybrid walleye. In: Barton BA (ed) Biology, management, and culture of walleye and sauger. American Fisheries Society, Bethesda, pp 451–570

Thoesen JC (ed) (1994) Suggested procedures for the detection and identification of certain finfish and shellfish pathogens. Version 1, 4th edn, Fish health section. American Fisheries Society, Bethesda

Thompson TM, Batts WN, Faisal M, Bowser P, Case JW, Phillips K, Garver KA, Winton J, Kurath G (2011) Emergence of viral hemorrhagic septicemia virus in the North American Great Lakes region is associated with low viral genetic diversity. Dis Aquat Org 96:29–43

Tort MJ, Wooster GA, Bowser PR (2003) Effects of hydrogen peroxide on hematology and blood chemistry parameters of walleye *Stizostedion vitreum*. J World Aquacult Soc 34:236–242

Tripi C, Bowser PR (2001) Toxicity of hydrogen peroxide to pre-exposed young-of-the-year walleye: effects of water hardness and age of fish. J World Aquacult Soc 32:416–421

USDA APHIS (2014) APHIS to lift viral hemorrhagic septicemia federal order. Available http://www.aphis.usda.gov/stakeholders/downloads/2014/vhs_order_revocation.pdf. Oct 2014

USDA APHIS VS (United States Department of Agriculture, Animal and Plant Health Inspection Service, Veterinary Services) (2009) Viral hemorrhagic septicemia virus IVb U.S. surveillance report 2009. Available http://www.aphis.usda.gov/animal_health/animal_dis_spec/aquaculture/downloads/vhs_surv_rpt.pdf. Oct 2014

USDA APHIS VS (United States Department of Agriculture, Animal and Plant Health Inspection Service, Veterinary Services) (2014) Available http://www.aphis.usda.gov/animal_health/downloads/animal_diseases/Species_Previously_Regulated_by_APHIS_for_VHS_2014.pdf. Oct 2014

Walker R (1969) Virus associated with epidermal hyperplasia in fish. Natl Cancer Inst Monogr 31:195–207

Wedemeyer GA (2001) Fish hatchery management, 2nd edn. American Fisheries Society, Bethesda

Wolf K (1988) Fish viruses and fish viral diseases. Cornell University Press, Ithaca

Yanong RPE (2010) Lymphocystis disease in fish. Extension Publication FA181. University of Florida, Gainsville

Yager TK (1991) Effects of fish size and feeding frequency on metabolism of juvenile walleye. Master's thesis, Iowa State University, Ames

Yager TK, Summerfelt RC (1994) Effects of feeding frequency on metabolism of juvenile walleye. Aquac Eng 13:257–282

Chapter 18
Large-Scale Production of Yellow Perch, Walleye, and Hybrid Walleye in Ponds

Ruth D. Briland, Cathleen M. Doyle, and David A. Culver

Abstract Since the 1980s, angler interest in recreational fisheries has increased the demand for pond production of North American and European percids for stocking to enhance wild populations or to establish new populations. In this paper, we analyze the fish and plankton ecology in the production ponds to provide a better understanding of the ecological and biological factors involved in optimal pond production of percid fingerlings for stocking. Much of the literature uses a "black-box method" for optimizing pond management, reporting on the survival, growth, and size at harvest of fish from ponds as a function of various fertilization and pond stocking regimens. In contrast, our research since 1987 has focused on the seasonal variation in the ecological interactions among fertilizers, algae, zooplankton, benthos, and larval fish in ponds. Accordingly, in this paper we examine management of large-scale production of percid fingerlings from an ecological perspective, concentrating primarily on our research through 2012 in three Ohio state fish hatcheries, incorporating other information from the literature as appropriate. We find that despite differences in walleye, saugeye, and yellow perch growth and development, rearing ponds can be managed similarly to produce desired size and harvest yields of fingerling fish by providing adequate food resources. Management protocol for fertilization, stocking schedules, and stocking density should be site specific considering the source water quality. Further, sequential culture of ponds may boost overall hatchery production, but we show reduced springtime percid yield due to carryover effects of chemicals added during summertime catfish culture.

Keywords Percids • Pond management • Fertilization • Plankton ecology • Fry stocking

R.D. Briland (✉) • C.M. Doyle • D.A. Culver
Department of Evolution Ecology and Organismal Biology, The Ohio State University, 1315 Kinnear Road, Columbus, OH 43212, USA
e-mail: briland.1@osu.edu

P. Kestemont et al. (eds.), *Biology and Culture of Percid Fishes*,
DOI 10.1007/978-94-017-7227-3_18

18.1 Introduction

Since the 1980s, angler interest in recreational fisheries has increased the demand for stocking cultured walleye (*Sander vitreum*) and yellow perch (*Perca flavescens*), as well as saugeye (walleye ♀ X sauger *S. canadense* ♂), whether to enhance wild populations experiencing declining natural production from variable reproductive success and diminished habitat quality, or to establish new populations (Ellison and Franzin 1992; Fox 1989; Hushak et al. 1988; Mitzner 2002). Stocking fry into rivers, lakes and reservoirs can occasionally produce large year-classes, but in many instances fry stocking contributes little to recruitment (Mitzner 2002; Mathias et al. 1992; Fielder 1992; McWilliams and Larscheid 1992; Paragamian and Kingery 1992). Accordingly, managers favor stocking fingerlings (30–50 mm) or even larger fish, requiring effective methods for growing fry into stockable-sized fish. Producing really large fish (>5.0 g) typically requires feeding with manufactured feeds or providing minnows as forage, but this paper focuses on the techniques required to raise large numbers of 0.3–0.5 g percids in drainable earthen ponds, relying solely on the production of living invertebrate organisms in the ponds as prey for the fish.

Much of the literature uses a "black-box method" for optimizing pond management, reporting on the survival, growth, and size at harvest of fish from ponds as a function of various fertilization and pond stocking regimens. In contrast, our research since 1987 has focused on the seasonal variation in the ecological interactions among fertilizers, algae, zooplankton, benthos, and larval fish in ponds. Accordingly, in this paper we examine management of large-scale production of percid fingerlings from an ecological perspective, concentrating primarily on our research through 2012 in three Ohio state fish hatcheries, incorporating other information from the literature as appropriate.

During this research, we first studied existing techniques used by state fish hatchery managers for raising walleye, saugeye, and yellow perch fingerlings in a series of 0.4–2.8 ha ponds, 1–2 m in depth, with volumes of 1300–26,400 m^3. They filled the ponds from adjacent surface water sources and fertilized them with granular inorganic and organic fertilizers a month prior to stocking with percid fry. Additional fertilization continued through the production season (30–90 days), and the ponds were drained to harvest fingerlings when they stopped growing 1 mm in length per week. Because walleye and sauger spawn earlier than yellow perch (Hokanson 1977) or striped bass (*Morone saxatilis*) or white bass (*M. chrysops*), which in turn spawn before channel catfish (*Ictalurus punctatus*), pond managers often arranged to produce some combination of walleye or saugeye, hybrid striped bass and yellow perch, and channel catfish in the same ponds, sequentially filling ponds, culturing fish, and draining the ponds to harvest the fish.

However, percid fingerling production varied widely, and some ponds contained no fish upon draining, whereas others produced 12,000–20,000 percid fingerlings/ha. Accordingly, we performed a series of experiments on walleye, saugeye, and yellow perch production to decrease the variability in culture success, maximize the number of fingerlings produced at a size suitable for stocking, and minimize

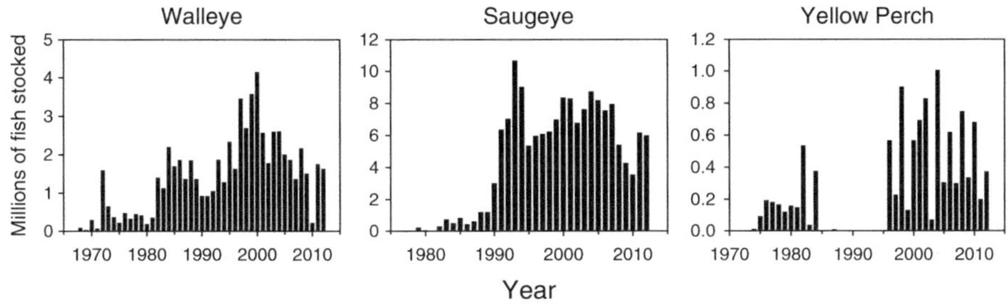

Fig. 18.1 Total walleye, saugeye, and yellow perch fingerlings produced at Ohio's St. Marys, Hebron, and Senecaville state fish hatcheries and stocked into lakes between 1970 and 2012. Variation reflects improved fertilization methods beginning in 1990 and the number of ponds placed in production for the three taxa

mortality during pond culture, while decreasing costs of labor and supplies. Our approach was to optimize planktonic production by improving the pond fertilization procedures and to adjust fry stocking density and timing to coordinate fish dietary demands with plankton dynamics in the ponds.

These studies generated data on biweekly variation in nutrients and dissolved oxygen in the ponds; seasonal variation in phytoplankton and zooplankton taxonomic composition, abundance, biomass, and productivity; how fish diets and their preferences for different zooplankton taxa varied through time; and fish growth through the production season, their survival, and numbers and biomass at harvest. New management methods for pond fertilization and fry stocking and timing based on the results of these experiments were adopted in 1990–1991, and are reflected in increases in the total number of fingerlings produced and stocked in Ohio reservoirs from the three hatcheries since that time (Fig. 18.1). In this paper, we analyze the fish and plankton ecology reflected in these increased yields to provide a better understanding of the ecological and biological factors involved in optimal pond production of percids.

18.2 Pond Management and Stocking

18.2.1 Fertilization

Managing ponds to promote zooplankton availability for planktivorous larval percids is centered on providing an abundance of edible phytoplankton. A common practice is to increase primary productivity by adding nutrients (especially nitrogen and phosphorus) to the system (Culver 1991; Soderberg et al. 1997, 2000; Tew et al. 2006; Jacob and Culver 2010; Hartleb et al. 2012). Nutrient sources can be of either inorganic or organic origin, and their relative efficacy for percid culture has been compared (see review in Tew 2003). Organic fertilizers, such as animal manures and alfalfa meal, stimulate zooplankton production by encouraging growth of heterotrophic bacteria, but they also can reduce fish production because heterotrophic intermediates consume dissolved oxygen (Qin and Culver 1992; Qin et al. 1995; Tice et al. 1996), compromising water quality. Conversely, inorganic fertilizers stimulate photosynthetic algae which produce oxygen and these fertilizers can be precisely controlled to prevent over-fertilization (Jacob and Culver 2010).

As described in Tew et al. (2006), Ohio state hatchery ponds were filled from nearby reservoirs (mesotrophic Senecaville Lake for Senecaville hatchery, eutrophic Buckeye Lake for Hebron Hatchery, and hypereutrophic Grand Lake for St. Marys hatchery (Filbrun et al. 2008), although ponds in the last are now filled from two wells) approximately 3 days before stocking fry into the ponds. Each pond has a separate filling system with a 0.5 mm mesh screen or drum filter to prevent the introduction of undesired larval fish from the lake while most phytoplankton and zooplankton taxa passed through easily.

From 1977 to 1984, ponds for percid culture were enriched to stimulate plankton growth using both inorganic and organic agricultural fertilizers, applying 6-10-4 ($N - P_2O_5 - K_2O$) granular fertilizer at the rate of 168.1 $kg{\cdot}ha^{-1}$ and alfalfa meal at the rate of 112 $kg{\cdot}ha^{-1}$ every week. This method produced an extremely low N:P ratio (1.5–1 by mass) and a weekly addition of >750 $\mu g\ PO_4\text{-}P{\cdot}L^{-1}$ which promoted the proliferation of cyanobacterial (blue-green algal) and filamentous green (Chlorophyta) phytoplankton species, and a low abundance of edible smaller algae (Helal and Culver 1991). Newly-hatched larval fish consume zooplankton for their survival and growth, and decreasing algae edible by zooplankton leads to crashes in the zooplankton population a few weeks after ponds are filled. Other water quality concerns with high abundances of cyanobacteria and filamentous green algae include low dissolved oxygen (DO) concentrations and toxic ammonia concentrations due to high pH. Identifying the appropriate amounts of nitrogen and phosphorus to apply as inorganic fertilizers is thus essential (Culver 1991; Soderberg et al. 2000; Tew et al. 2006; Jacob and Culver 2010).

Culver (1996) suggested that the percid ponds should be treated to maintain nitrogen and phosphorus at an approximate mass ratio of 20:1 by spraying sufficient liquid inorganic fertilizers (liquid 28-0-0 inorganic fertilizer (NH_4NO_3 + Urea) and phosphoric acid (H_3PO_4 or 0-54-0) both diluted with pond water) weekly to restore the pond nutrient-concentrations to 600 $\mu g\ NO_3{+}NH_4\text{-}N{\cdot}L^{-1}$ and 30 $\mu g\ PO_4\text{-}P{\cdot}L^{-1}$, based on weekly analyses of dissolved inorganic nitrogen and phosphorus in each pond. We diluted samples of fertilizer stocks by one part to one million to allow careful measurement of their phosphate, ammonia and nitrate concentrations, whereas urea content was measured by local hospital analytical labs. Not only are the actual N and P concentrations in liquid fertilizers delivered by agricultural fertilizer providers variable, but nitrogen fertilizers are often contaminated with large amounts of phosphate fertilizer. When typical amounts of 28-0-0 added are 5 l for a 5000 m^3 pond, whereas the amounts of 0-54-0 to be added are only 0.1–0.3 l, accurately measuring variation in fertilizer N and P content is required.

We estimated pond volumes for the 1990–2000 production years from area and the estimated average depth of each pond to calculate the amount of fertilizer needed to raise N and P to the targeted concentrations each week. While the ponds were empty in winter 2001, however, we performed a series of measurements on each pond using global positioning (GPS) equipment (Trimble Pathfinder Pro XR) and its depth dimensions to allow a more accurate estimation of pond volume as a function of water level measured as distance below the fill structure. We then filled the ponds and used GPS to trace the outlines of the water's edge at various water levels and determined the depths of the water column along a series of transects in each pond. The pond volume models are geometric/trigonometric constructs from the resultant areas, slopes, and depths (Details in J. Mion, Ohio Dept. Natural Resources, ms in prep.). After 2001, we used the new pond volume estimates as a function of water level to calculate the amounts of fertilizer needed each week in each pond.

This fertilization regimen improved percid survival and yield at Hebron SFH between 1991 and 1995, but in more recent years survival has both declined and become more variable, so the possibility of over-fertilization was examined (Tew et al. 2006). They found that lowering the phosphorus target concentration to 20 μg PO_4-P·L^{-1} still supported adequate prey production for percids. Importantly, however, the source water for Hebron SFH is eutrophic to hyper-eutrophic. While these methods have succeeded in Ohio, percids are reared to the fingerling stage in hatcheries throughout the northeastern US under a wide variety of source water nutrient concentrations, including those with oligotrophic water supplies. Nevertheless, Jacob and Culver (2010) recommended further lowering the phosphorus fertilization rates to 10 μg PO_4-P·L^{-1}, as this concentration still supported adequate primary production and zooplankton prey for fish growth while minimizing the potential risks of poor water quality (e.g., low DO and high pH).

Briland (2010) recently tested whether a phosphate fertilization rate as low as 10 μg PO_4-P·L^{-1}, would be effective in a less productive system, Senecaville SFH, where the water source (Seneca Lake) is mesotrophic. Similar to Jacob and Culver's (2010) results in the more eutrophic Hebron SFH, Briland found that fish yield, harvest and survival in saugeye ponds at Senecaville SFH did not differ among ponds fertilized with 30, 20, or 10 μg PO_4-P·L^{-1} (N:P ratios of 20, 30, or 60 to 1 by mass), and fish growth during the culture period was not adversely affected by reducing the phosphorus fertilization rate. In fact, no evidence of resource limitation or deteriorated water quality (from excessive nutrients) was evident from this study. Ultimately, we recommended first fertilizing ponds at Senecaville at the traditional rate, i.e., raising ponds to a target 30 μg PO_4-P·L^{-1} each week, and repeating the reduced fertilization rate experiment in future years if and when evidence of over-fertilization (i.e., low dissolved oxygen, high pH, and abundant cyanobacteria populations) occurs. This management decision was based on site-specific conditions, namely that fish sizes at Hebron SFH are consistently larger than at Senecaville and that whereas Hebron percid ponds were typically plagued with cyanobacterial dominance of the phytoplankton (Jacob and Culver 2010), Senecaville ponds predominantly produced algae types which are edible to zooplankton (Briland 2010).

18.2.2 Fry Production Methods

18.2.2.1 Walleye and Saugeye

Walleye eggs collected from nearby Ohio reservoirs were artificially inseminated with walleye sperm (for walleye fry) or sauger from the Ohio River (for saugeye fry). The fertilized eggs were treated with tannic acid to prevent clumping and Iodophor to prevent infection with Viral Hemorrhagic Septicemia (VHS), and then incubated at all three hatcheries in McDonald jars until they hatched. Fry were counted by volumetric displacement and stocked in ponds at 3–4 days posthatch in early mid-April at densities of 100,000–600,000 fry ha^{-1}.

18.2.2.2 Yellow Perch

Spawning yellow perch females extrude eggs in ribbons of a gelatinous matrix (egg skeins) and then attach them to submerged aquatic vegetation, rocks, or woody debris, such as fallen trees (Thorpe 1977; Craig 1987; Kestemont and Mélard 2000). Yellow perch females' extrusion of eggs in skeins requires different techniques for producing fry for stocking in ponds than those used for walleye and saugeye. Eggs for extensive pond production can be obtained either by natural spawning in ponds or by manual spawning in which eggs are manually stripped from ripe females, mixed with milt stripped from males, and "dry" fertilized in a plastic bowl, similar to techniques used for walleye and saugeye. Managers at Ohio's St. Marys SFH used natural spawning through 2005, and manual spawning since then, permitting us to compare the two methods' effectiveness in this paper.

For natural spawning, different densities of even numbers of males and females were stocked in ponds at either a low density (<60 parental pairs/ha), a medium density (60–120 pairs/ha), or a high density (>120 pairs/ha). Ponds were filled 1 day before stocking parents, adding suitable spawning substrates (such as dried conifer trees torched to burn off needles) before spawning begins in March through April, and the fish spawn naturally, suspending egg skeins from the branches. Unfortunately, there is little control of when the fish spawn, or the timing of hatch, and females vary in fecundity, so that there is potential for large variation in the number of fry produced per pond.

For manual spawning, spawning is synchronized by injection of females (214 fish in 2007) with human chorionic gonadotropin (hCG), 150–660 IU/kg wet weight. Spawning usually occurs a few days after injection, increasing fertility and leading to a larger percentage of fry hatching at the same time (Dabrowski et al. 1996; Hart et al. 2006). The eggs are incubated indoors on coated wire mesh in troughs with a constant water flow (Hart et al. 2006), or more recently on Heath vertical incubation trays, until they hatch, and then the fry are stocked into ponds at 740,000 ha^{-1}. Because of the gelatinous matrix surrounding the eggs, maintaining appropriate oxygen delivery to developing eggs during incubation (and removal of contaminants and waste) can be more difficult than for walleye and saugeye egg incubation. Any degradation of the matrix or excessive touching or clumping will decrease the hatch rate (Hart et al. 2006). While manual spawning is more complex than natural spawning, it is possible to assure fertilization of the eggs, control the genetic pairings of males and females, measure the numbers of fry stocked to each pond, and assure that all fry in a pond begin culture at the same size, all of which help maximize the number of fingerlings produced.

18.2.3 Timing: Pond Filling and Fry Stocking

Another ecological principle relevant to percid pond management is the match-mismatch hypothesis (Cushing 1969, 1990; Mertz and Myers 1994), which proposes that, even if both consumer (stocking rate) and producer (nutrients) controls are optimized, if fish consumptive demand and production of their prey are temporally misaligned, fish production will be reduced. In a pond aquaculture setting, managers can manipulate the relative timing between filling ponds with source water and stocking with fry, with the intention of increasing the match between prey production and fry consumptive demand. However, few studies have considered this as a management strategy (Culver et al. 1993). Under the presumption that first-feeding larval fish needed abundant zooplankton prey to survive (critical period hypothesis), percid ponds have been filled as much as 1 month prior to stocking with fish in hopes that the zooplankton forage base would build (Geiger et al. 1985). Ideally, this schedule would match the increase in zooplankton prey with the increasing consumptive demand of the growing percid juveniles. Instead of increasing zooplankton forage, zooplankton biomass likely peaked within the 4 weeks period, and zooplankton likely overgrazed the phytoplankton, and forced a low algae "clear-water phase," followed by a period of low zooplankton abundance even before fish were stocked (Munch et al. 1984; Culver 1988). Culver et al. (1993) tested the hypothesis that a short lag time (<1 week) aligns plankton dynamics with larval fish consumptive demand for a sustained period (up to 6 weeks). They found that fish predation controlled or removed large, efficient algal grazers (e.g. *Daphnia*), thus reducing grazing on algae and releasing small-bodied zooplankton (e.g., *Bosmina*) from competition, thus supplying larval fish with abundant prey. However, because the initial inocula of zooplankton from source waters likely vary with trophic status, we also tested whether a slightly longer (2 weeks) lag time for Senecaville SFH, whose source water is mesotrophic (Filbrun et al. 2008), would result in improved fish growth and production by better matching the peak in zooplankton abundance with larval percid consumption (Briland 2010).

The time-lag experiment revealed no improvement for walleye and saugeye production or growth, and both filling treatments provided similar and adequate amounts of preferred zooplankton prey (i.e., *Daphnia* spp. and adult stage copepods) during the early weeks of culture (Fig. 18.2; Briland 2010). In fact the zooplankton progression in both filling treatments followed the pattern theorized by Culver et al. (1993), whereby *Daphnia* spp. are replaced by abundant *Bosmina* due to predation. Thus, we concluded that while the normal (<7 days) filling and stock-

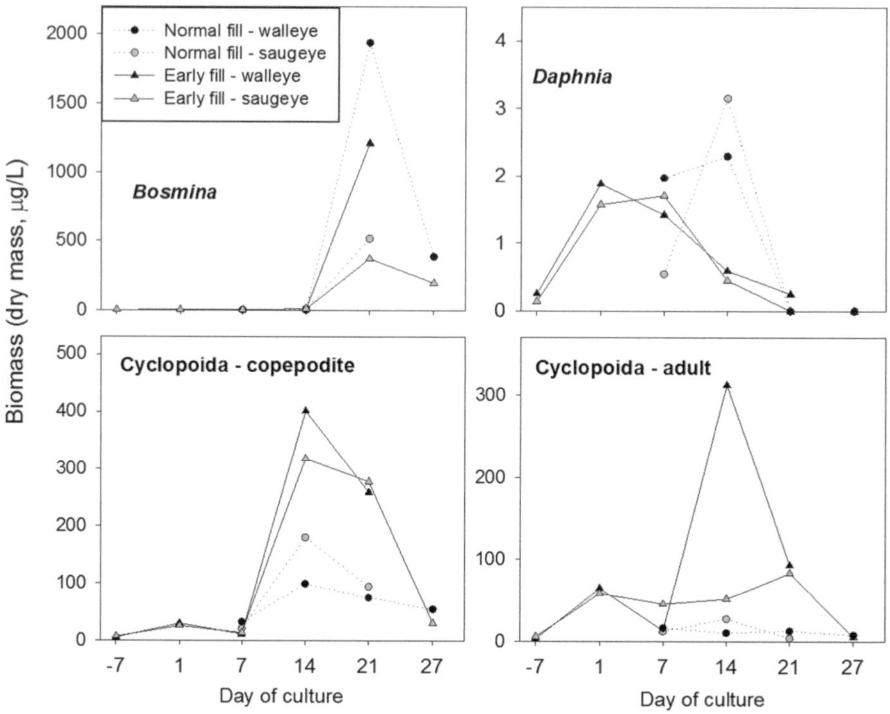

Fig. 18.2 Comparison of seasonal variation of crustacean zooplankton biomass (dry mass; µg·L^{-1}) between early and normal pond filling and stocking schedules (early fill; 2 weeks prior to stocking with percid fry; normal fill less than 1 week prior to stocking) at Senecaville State Fish Hatchery, Ohio, 2007. Values for each date represent the mean for three replicate ponds with the same fish species (saugeye or walleye)

ing schedule is appropriate for this hatchery, pond-filling schedules remain an effective pond management tool to manipulate zooplankton population dynamics, and may be particularly appropriate for ponds filled from wells or those located at higher latitudes, where management protocols call for a 2 weeks lag time before stocking with walleye fry (e.g., Flowers 1996). Further study should address the impact of initial zooplankton composition and the temperature and filling time effect on diapausing egg development to further study the role of abiotic variables in controlling the zooplankton production that is so important to planktivorous fish production.

18.2.4 Fry Stocking Densities: Trade-Off of Fingerling Size Versus Numbers at Harvest

Hatchery managers can manipulate the size and number of fish harvested by altering initial stocking density in culture ponds (Fig. 18.3). Previous studies have shown that a trade-off often exists between size of fish at harvest and the number of fish produced

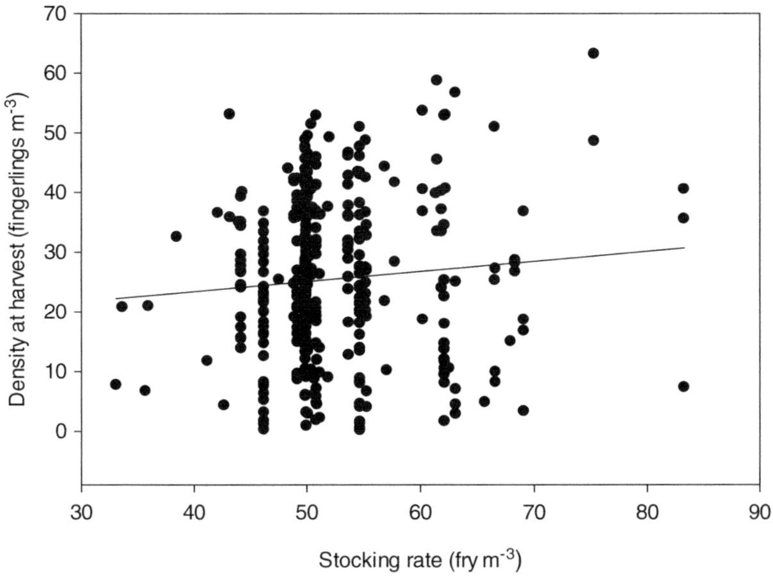

Fig. 18.3 Variation in the number of saugeye fingerlings at harvest as a function of the number of fry stocked in ponds at the Hebron State Fish Hatchery, 1998–2008

(Fig. 18.4; Fox and Flowers 1990; Qin et al. 1994). This interaction may occur due to competition for limited food resources; however, it also has been attributed to heightened activity levels of larvae stocked at high density relative to those in low density ponds (Fox and Flowers 1990). Although fish growth can be density-dependent, a threshold density appears to exist at which fish mass at harvest no longer declines (Tew 2003), such that the overall yield of fish biomass increases with increasing stocking density (Qin et al. 1994; Culver and Wu 1997). This density-dependent growth and pond yield relationship is likely mediated by the productivity and forage base of the system. Nutrient-poor environments, where prey abundance is low, appear more sensitive to predation pressure (Dettmers and Stein 1992), and competition for resources can lead to density-dependent growth (Jenkins et al. 1991). In other studies, increased fish biomass was found to increase turnover rates of zooplankton biomass (Culver et al. 1992). For example, in ponds with high fish densities, intense predation pressure on large, filter-feeding Cladocera (e.g., *Daphnia*) reduced competition with small-bodied zooplankton (e.g., *Bosmina*). This competitive release allowed an increase in the secondary productivity of *Bosmina* (DeMott 1989), ultimately resulting in abundant small-bodied prey for percid consumption (Qin et al. 1994).

Briland (2010) tested how altering the saugeye fry stocking rate from a traditional rate of about 35 fry·m^{-3} (data not shown) to a lower (20 fry·m^{-3}) or higher (50 fry·m^{-3}) rate influences the fish predation pressure on zooplankton prey resources. In her study, percid growth and size at harvest was highest at the low stocking rate (20 fry·m^{-3}), and lower at the higher stocking rates due to insufficient prey resources (Fig. 18.5). In fact, fish in ponds stocked with the highest rate of saugeye fry (50 fry·m^{-3}) reduced zoo-

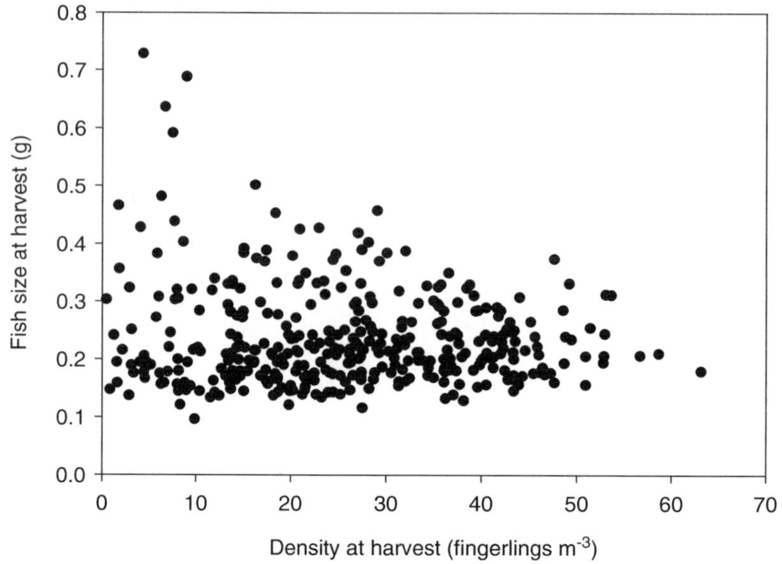

Fig. 18.4 Variation in the mean size of saugeye fingerlings at harvest as a function of the density of fish at harvest. Each data point represents saugeye fingerlings from a single pond stocked with fry at the Hebron State Fish Hatchery, Ohio, between 1998 and 2008

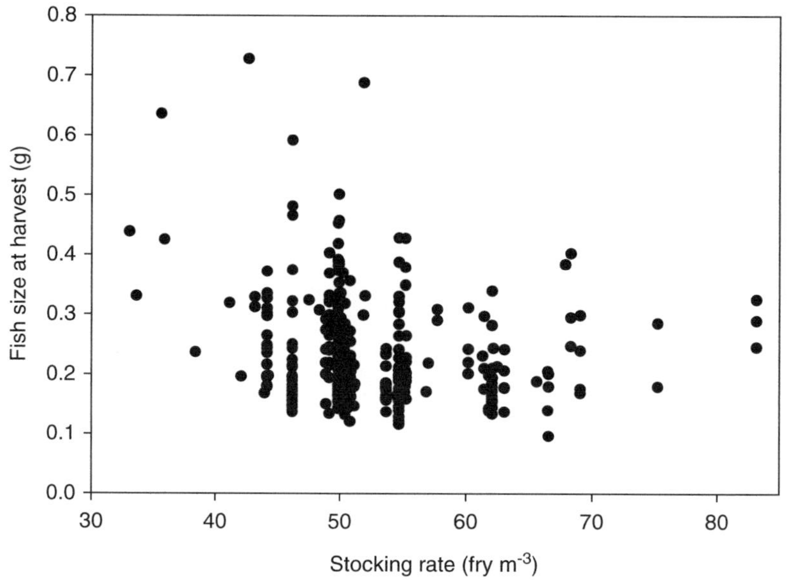

Fig. 18.5 Comparison between mean individual mass at harvest and the number of fry stocked in individual saugeye production ponds at the Hebron State Fish Hatchery, 1998–2008

plankton prey resources during the second week of culture and induced a premature diet shift to chironomid larvae prey, which were in turn exhausted, resulting in a switch back to zooplankton (primarily immature cyclopoids) during the final weeks of culture (Fig. 18.6). Further, the lower stocking density treatments allowed fish to achieve high growth toward the end of the culture period when incorporating chironomid prey in their diets; conversely, the high density fish had lower growth when they began eating chironomid prey. Likely, young fish lacked the requisite digestive enzymes to handle chironomids, or the cost of digestion (due to relative size ratio of prey to predator) was higher than for larger fish found later in the culture period, accounting for the poorer growth in the high density fish.

Between 1977–1984 and 1996–2005, it was not possible to determine yellow perch fingerling survival because allowing the adults to spawn on trees in the ponds prevented our knowing the initial number of fry in the ponds. However, between 2006 and 2009,

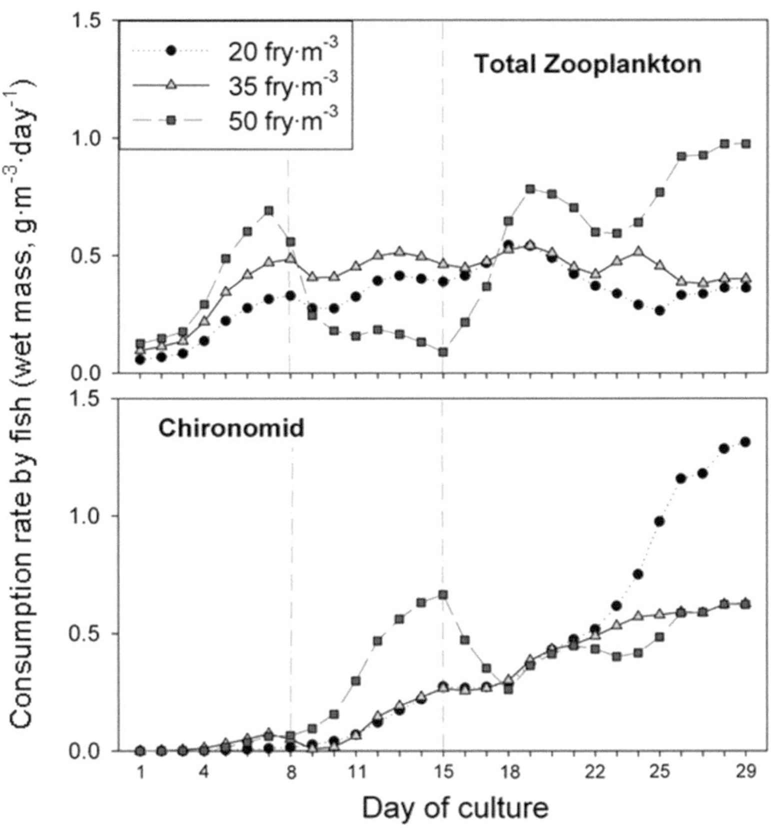

Fig. 18.6 Comparison of saugeye predation rates on zooplankton and chironomid prey across three fry stocking rates (20, 35, and 50 fry·m^{-3}) at Senecaville State Fish Hatchery during 2008, simulated with a bioenergetics model calibrated for larval walleye. Values on each date represent the mean from two replicate ponds for each treatment. *Long dash* and *dot lines* (—·—) indicate boundaries of the periods used for statistical analysis

survival was variable between ponds and between years. Individual pond survival rates ranged from 7 % to 67 %, while average yearly survival was between 20 % and 46 %. Despite this variable survival in the yellow perch ponds, we have seen a higher density of fingerlings at harvest (Figs. 18.7 and 18.8) with the current production methods, albeit at a smaller size, comparable to walleye and saugeye production.

We (Doyle and Culver, ms in prep) analyzed historic yellow perch production from St. Marys SFH to determine how the change to the new fertilization methods and the duration that fingerlings spent in the ponds affected production. As previously mentioned, between 1977 and 1985, the hatchery fertilized ponds with both inorganic and organic fertilizer and broodstock were stocked in ponds in March to spawn naturally, and the fish were harvested in September/October (duration >90 days). From 1995 to 2009, the new fertilization regimen was adopted and fish were harvested in May/June (duration <90 days). For fish reared for >90 days,

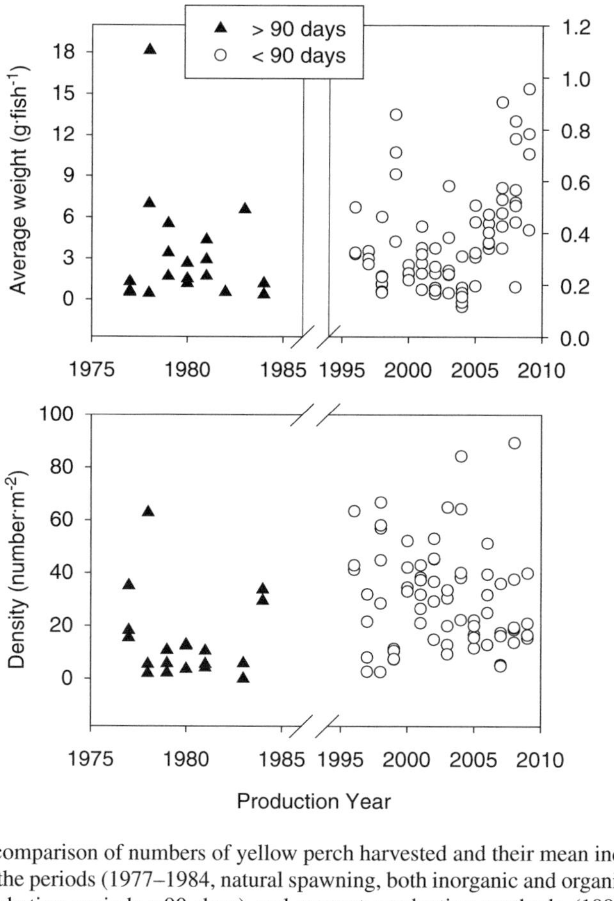

Fig. 18.7 A comparison of numbers of yellow perch harvested and their mean individual weights at harvest for the periods (1977–1984, natural spawning, both inorganic and organic pond fertilization, and incubation periods >90 days) and current production methods (1996–2009, natural spawning from 1996 to 2005 and manual spawning from 2006 to 2009, and liquid inorganic pond fertilization to target N and P concentrations, and incubation periods <90 days)

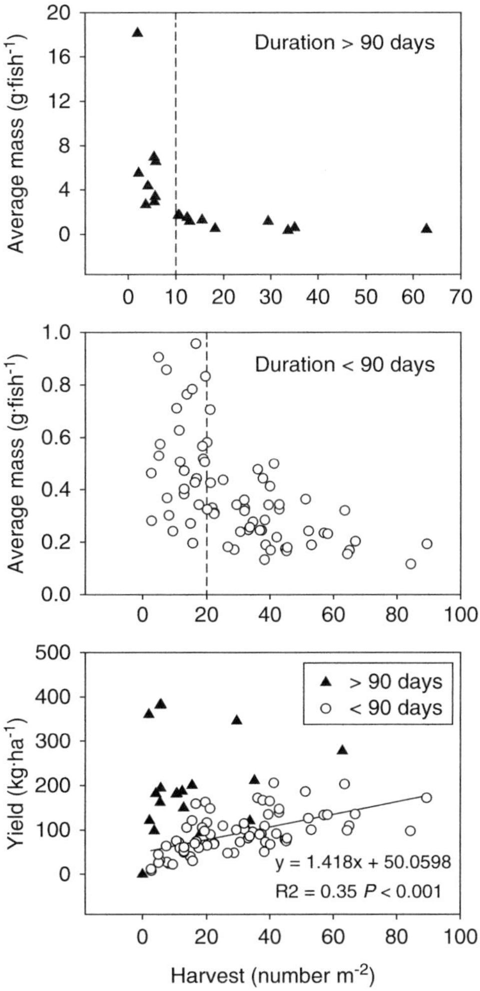

Fig. 18.8 A comparison of yellow perch yield vs. harvest density from St. Mary's hatchery for 1977–1984 (*filled triangles*; natural spawning, both inorganic and organic pond fertilization, and incubation periods >90 days) with those from 1996 to 2009 (*open circles*; natural spawning from 1996 to 2005 and manual spawning from 2006 to 2009, and liquid inorganic pond fertilization to target N and P concentrations, and incubation periods <90 days). *Dashed lines* represent the breakpoint relationship between fish harvest densities (fish·m^{-2}) and individual fish biomass

there was a significant breakpoint in the relationship between fish harvest m^{-2} and individual fish size. At harvest densities of >10 fish m^{-2}, fish were always less than 2 g (Garvey et al. 1998; 2DKS test, $D_{BKS}=0.25$, $P<0.0001$). However, for fish reared in ponds <90 days, at harvest densities >20 fish m^{-2}, fish were less than 0.4 g ($D_{BKS}=0.12$, $P<0.0001$) (Fig. 18.8). Starting in 2006, fry were produced by manual spawning and known numbers of fry were stocked into ponds. We found no differences between natural spawning vs. manual spawning from 1996 to 2009 for the number of fingerlings harvested or the density at harvest (fish m^{-2}), but stocking fry produced larger fish (average total length and average weight) and greater yield (kg ha^{-1}).

18.3 Relevant Biology and Ecology

18.3.1 Larval Through Fingerling Development

Young percids experience rapid development during approximately 4 weeks of culture in hatchery ponds. The early stages of walleye development are described by Mathias and Li (1982) and are differentiated by notable changes in body structure relative to body length and days post hatch (dph). Walleye fry total length (TL) at hatch has a relatively broad range of 4.8–9.0 mm, but this may be an artifact of inconsistent methods across studies (Summerfelt et al. 2011). Upon hatching, walleye fry are considered in the prolarval stage while the yolk sac is still present, lasting through approximately 5 dph and about 10.0 mm TL (Li and Mathias 1982). Mouth opening and even exogenous feeding can occur during the prolarval stage at around 3 dph, although the young fry is not yet nutritionally dependent on capturing prey (Summerfelt et al. 2011). Walleye and saugeye fry are typically stocked into culture ponds during the prolarval stage at 3 or 4 dph. Within a few days of pond culture young percids advance to postlarva I, after the yolk sac is absorbed, but an oil globule is still present (10–13.2 mm TL and 5–10 dph) (Li and Mathias 1982). Walleye and saugeye achieved approximately 10 mm TL after only 1 week of culture in hatchery ponds (Fig. 18.9). Diet analysis of larval saugeye captured during the postlarval I stage, after only 4–7 days in a culture pond, show active feeding on large zooplankton species including *Daphnia* (Briland 2010), although the fish are not yet dependent on external food sources. Tank-studies of walleye show the highest mortality occurs during this stage and is associated with exogenous feeding, leading the authors to suggest an optimum food density of 100 daphnids per liter during postlarva I stage (Li and Mathias 1982). However, percid culture ponds do not achieve this density ranging 1–3 daphnids per liter (density is consistent with biomass shown in Fig. 18.2), and still high survival (83–95 %) was realized (Briland 2010), indicating that access to prey items during initial feeding (i.e., critical period) does not determine survival or production success.

By the second week of culture, percids advance to postlarva II stage once the oil globule is absorbed and fish are completely dependent on exogenous feeding (Summerfelt et al. 2011). The juvenile stage follows at around 14 dph and when fish reach 16 mm TL and substantial external and internal development occurs, including fin and ray growth, scales and pigmentation, as well as gastrointestinal development (Summerfelt et al. 2011). Nunn et al. (2012) reviewed the foraging ecology of larval and juvenile fishes, and noted that improvements in vision and swimming ability, and increases in mouth gape and the digestive capability of the alimentary tract often lead to changes in diet, requiring changes in pond management.

Herein, we extend larval stages designated for walleye to saugeye, without accounting for differences in growth and/or development between the parent and hybrid species; however, we note that saugeye growth (length and weight) exceeded that of walleye after only 10 days of culture in ponds (Fig. 18.9), indicating their growth and

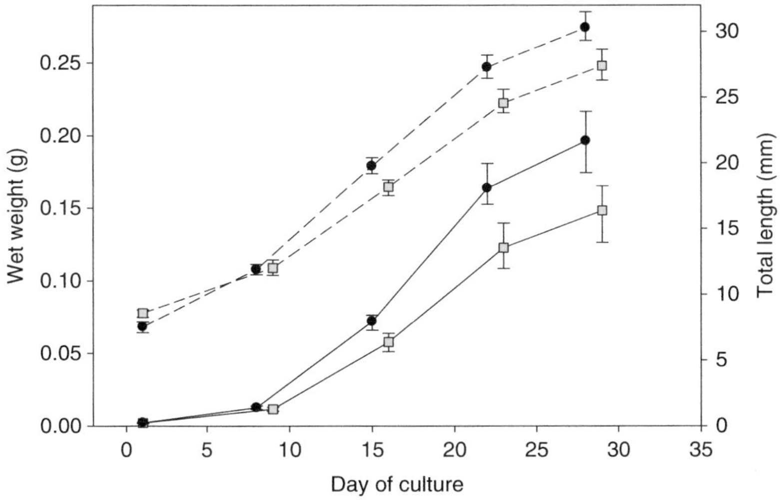

Fig. 18.9 Seasonal change in walleye (*grey squares*) and saugeye (*black circles*) weight (*solid line*) and length (*dashed line*) during culture at Senecaville SFH, 2007. *Icons* indicate mean values and *error bars* show 25th and 75th percentiles

development may differ. Compared with walleye and saugeye culture of 4–6 weeks, yellow-perch culture is extended for several months, and the highest growth rate occurs during the first week of culture (Fig. 18.10). Yellow perch require more time, 8–10 weeks, to achieve 30 mm TL (Fig. 18.10), which saugeye reach at only 4 weeks (Fig. 18.9). Due to the higher trophic status of walleye and saugeye as piscivores compared with the planktivorous-benthivorous yellow perch, yellow perch are sustained by natural prey in culture ponds and achieve positive growth for the majority of their 5 month culture period. In contrast, saugeye and walleye growth diminishes rapidly after the first 2 weeks of culture and in order to maintain positive growth abundant prey resources are required later in the culture period (Briland 2010). Thus, walleye and saugeye hatchery ponds must be managed to support fish growth for the duration of the culture period rather than being maximized at the time of fry stocking.

Our study comparing stocking rates of saugeye fry shows that at a low density of saugeye (20 fish m^{-3}), the larval fish population followed optimal consumption of prey: first and increasing consumption of zooplankton prey (days 1–8), followed by a stable rate of zooplankton consumption and supplementing it with chironomid prey from the middle to the end of culture (Fig. 18.6). However, at a high density of saugeye (50 fish m^{-3}), fish consumption exceeded zooplankton production by the second week of culture, forcing the postlarva II stage saugeye to prematurely switch to chironomid prey, and once those were depleted fish returned to a predominantly zooplankton diet (Fig. 18.6). Further, saugeye in the high density ponds did not experience higher growth generally associated with ontogenetic diet shift to larger prey, such as chironomids (Briland 2010). The lack of growth may be attributed to the underdeveloped gastrointestinal tract of the young fish which lacks digestive

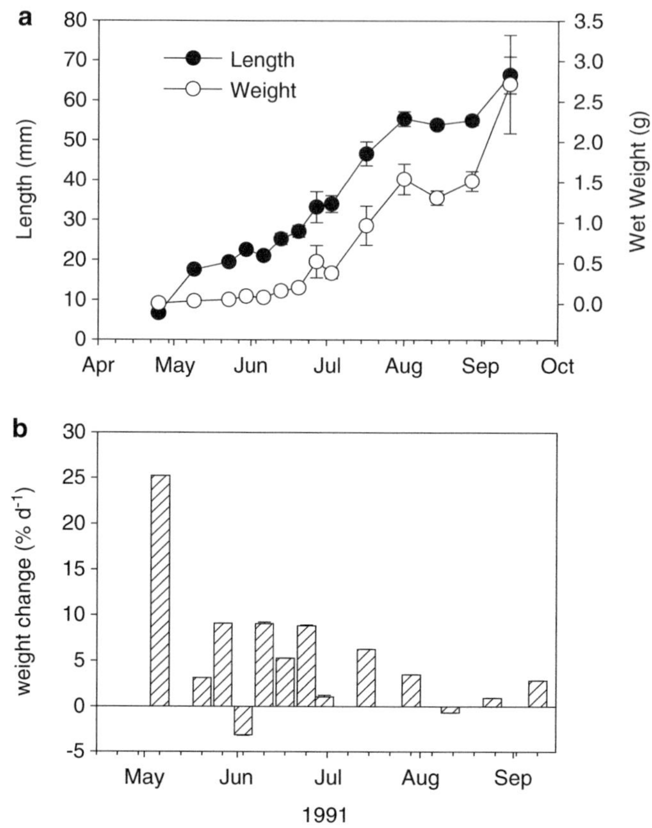

Fig. 18.10 Seasonal change in yellow perch (**a**) mean length and weight and (**b**) specific growth rate (percent change in wet weight per day) at St. Mary's SFH, 1991. (The values are means ± SE from two ponds)

enzymes required to break down chironomids (Dabrowski and Culver 1991) or to other digestive inefficiency, such as a high energetic cost considering the large prey size relative to that of the predator. Although the mode of action needs further investigation, it is clear that adequate zooplankton prey is prerequisite to desired growth and development for walleye, saugeye, and yellow perch culture.

18.3.2 Seasonal Variation in Zooplankton Species, Size, and Prey Preferences

A typical pattern of seasonal zooplankton dynamics in ponds fertilized to maintain 30 μg $PO_4 \cdot P \cdot L^{-1}$ shows a peak of cladocerans and copepods in the first 2 weeks of culture followed by their decline and an increase in rotifers, nauplii, and benthic cladocerans from then on. This pattern was observed at Hebron hatchery in 2001 in which the major zooplankton taxa present in saugeye ponds were cladocerans (*Bosmina* spp., *Chydorus* sp.), copepods (*Acanthocyclops vernalis*, *Diacyclops thomasi*, nauplii), and rotifers (*Asplanchna* sp., *Brachionus* sp., *Keratella* spp.) (Tew et al. 2006) (Fig. 18.11). The early decline in zooplankton may be the result of the early collapse of phytoplankton from high densities of filter-feeding zooplankton (Munch et al. 1984; Culver 1988). This effectively ensures large-bodied cladocerans are not as abundant as smaller cladocerans and copepods in the ponds before larval saugeye increase their consumption of prey (Qin and Culver 1992).

In 2001, we found that larval saugeye avoided rotifers and nauplii throughout the culture season (Fig. 18.12). Saugeye consumed primarily cyclopoid copepods selected during the first 2 weeks of the culture period, then their consumption declined in favor of *Bosmina* and benthic cladocerans such as chydorids and *Simocephalus* spp. in the last 2 weeks of culture. Although *Bosmina* was the most abundant zooplankter comprising the zooplankton biomass, saugeye exhibited a negative selection for them throughout the culture period (Fig. 18.12). Larval saugeye first consumed chironomids on the first date we collected fish (17 April) and they were present in their diets throughout the production season, although saugeye incorporated more benthic taxa in May (e.g., chironomid pupae, harpacticoids, ostracods, nematodes, etc.).

Tandem production of walleye and saugeye reveals that the two species exhibit similar preferences for food items (Fig. 18.13). Within only 4 weeks of culture, both percid species transition from feeding on a variety of planktonic and benthic prey items. Initially, both fish show a strong preference for large zooplankters, *Daphnia* spp. and copepods (stage V and VI copepodites). Small prey items were avoided entirely by walleye and saugeye with occurrences in fish diets of 0 % for copepod nauplii and 1 % for rotifers (Briland 2010). As the large, preferred prey decline in

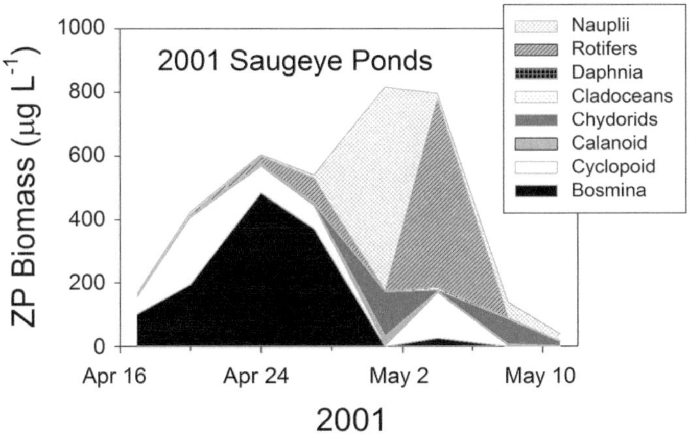

Fig. 18.11 Seasonal variation in zooplankton biomass (µg dry weight L^{-1}) for six saugeye ponds at the Hebron SFH, 2001. Cladoceran biomass is the sum of *Diaphanosoma*, *Eubosmina*, *Alona* spp., *Scapholeberis*, *Simocephalus*, *Ceriodaphnia*, and includes their egg biomass

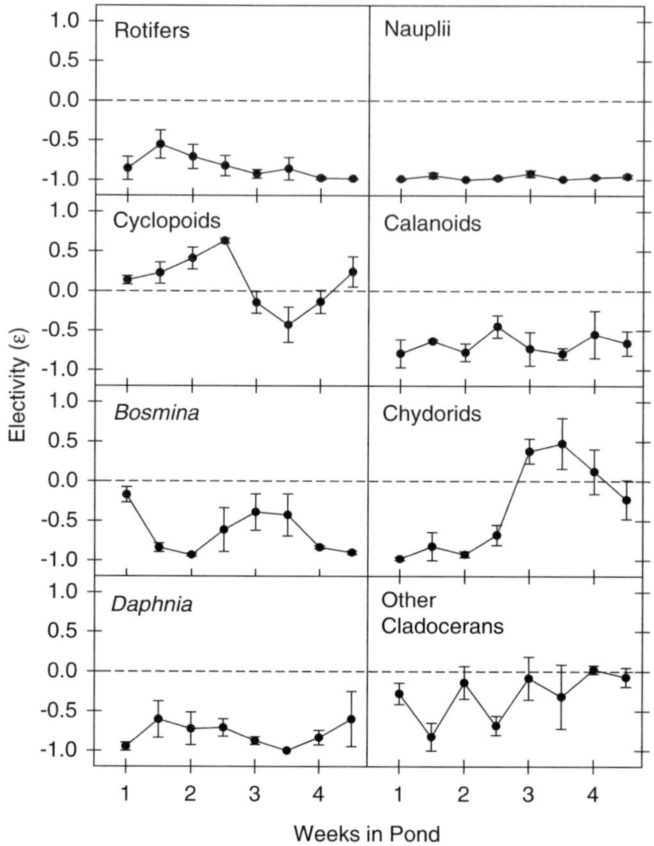

Fig. 18.12 Seasonal variation in saugeye dietary preferences for eight zooplankton taxa based on stomach analyses and net samples (Fig. 18.9) from 6 ponds at the Hebron SFH, 2001. 30 μgP/L fertilizer regimen (17 April–11 May) Rotifers were primarily *Keratella*. Cyclopoids include primarily *Acanthocyclops vernalis*, plus *Diacyclops thomasi*, *Mesocyclops edax*, *Macrocyclops albidus*, and unidentified spp. Calanoids were primarily *Skistodiaptomus oregonensis*. *Daphnia* included mostly *D. galeata mendotae* and *D. retrocurva*. Other cladocerans include *Alona* spp., *Simocephalus* spp., *Ceriodaphnia*, *Scapholeberis*, and *Diaphanosoma*. Saugeye consumed chironomids on the first date we collected fish (17 April) and were present in the diets throughout the production season

abundance during weeks 2–3, smaller zooplankters come to comprise the bulk of the diet (Briland 2010) but are not selected for disproportionate to their abundance in the pond. Near the end of the culture period (weeks 3–4), benthic cladocerans (i.e., *Chydorus* spp., *Simocephalus* spp., and *Alona* spp.) are favored, but a variety of zooplankton (*Bosmina* and copepods) and an increasing proportion of chironomids comprise walleye and saugeye diets. The strong preference by walleye and saugeye for large zooplankton prey during their initial week in the culture ponds indicates their gape size provides no limitation to the types of prey they consume.

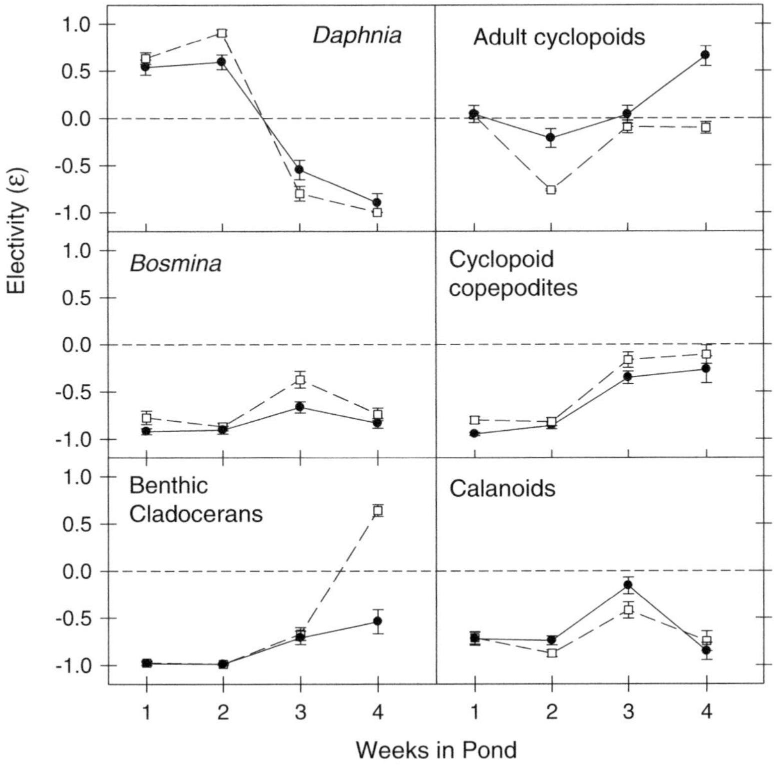

Fig. 18.13 Percid electivity of zooplankton prey items cultured at Senecaville SFH during 2007. Walleye are indicated by *open square icons*, and saugeye are indicated by *closed circles*. Data points represent mean from six ponds per fish species, and *error bars* indicate pooled standard error; approximately 10 fish per pond were analyzed on each sampling data (Total fish diets analyzed: Walleye = 200, Saugeye = 196)

The zooplankton biomass was dominated by small cladocerans from the start of production at Hebron SFH, with source water from a highly eutrophic reservoir dominated by filamentous blue-green algae (*Oscillatoria* spp.). Large cladocerans, such as *Daphnia* spp., do not thrive in systems with a high abundance of *Oscillatoria* spp. (Infante and Abella 1985). Senecaville SFH, with a mesotrophic water source, on the other hand, had *Daphnia* spp. present in the forage base in the beginning of the culture period, allowing for the strong selection for *Daphnia* spp. in the diets. Yet, saugeye selected benthic cladocerans the last 2 weeks of the culture period in both studies and also incorporated chironomids in increasing proportions. Their diets are similar, and if the source water and forage base were more similar, then the diets for both hatcheries would probably mirror one another.

It is not exclusively the fish taxonomic and size preferences which determine the dietary ontogeny of percids. For example, percid habitat use and foraging behavior are related to the structure and development of the eye (Ali and Anctil 1977). Ali et al. (1977) described how differences in photoreceptor structure and organization between *Perca* spp. and *Sander* spp. correspond to differences in their feeding habits: (1) yellow perch have relatively large and evenly distributed rods, but moderate-sized cones with smooth ellipsoids, whereas walleye and sauger have rods that are extremely small and are "distributed in groups between the processes of the pigment epithelium" and bulky cones with granular ellipsoids; (2) walleye and sauger retinomotor responses involve only the rods and pigment epithelium, but for yellow perch, the cones also contribute; and (3) yellow perch do not have a tapetum lucidum in the pigment epithelium as walleye and sauger have, but instead have a large concentration of melanin. The adaptations to the retina of walleye and sauger allow for increased scotopic sensitivity, or the ability to function in low light conditions (such as increased activity at night, feeding at dusk and dawn, and a preference for turbid waters). Braekevelt et al. (1989) determined that the tapetum lucidum did not begin to appear in the retina of walleye until 30 mm TL and it was fully developed by the time they reached 125–140 mm TL. Initially when no tapetum lucidum is present, larval walleye are positively phototactic, but as the tapetum becomes fully formed, walleye become negatively phototactic. Yellow perch, on the other hand, have a retina specialized for photopic conditions (Richmond et al. 2004), preferring clear waters and diurnal activity. In contrast, Guma'a (1982) found that larval Eurasian perch (<10 mm TL) retina consisted of pure cones and mostly of the twin type. Rods developed at 100 mm TL, when the fish became demersal. Visual acuity was dependent on the focal length of the lens, rather than the number of cones in the retina.

Loew and Wahl (1991) determined that the small, single cones at the corner of the square retinal cone mosaic are present in juvenile yellow perch up to 40 mm standard length (SL), and have a short wavelength absorbance maximum in the 400 nm (UV) region. Loew and Wahl propose these cones could be advantageous for juvenile yellow perch feeding in surface waters where the cones enhance the contrast between their planktonic prey and its background. The cones disappear when the juveniles become demersal (30–40 mm SL) and are completely absent in adults (>100 mm SL). Loew et al. (1993) determined that juvenile yellow perch obtain enough information from the near-UV receptors to identify and attack prey, whereas walleye do not have them (Wahl 1989, cited in Loew and Wahl 1991).

Wahl et al. (1993) found that juvenile yellow perch in Oneida Lake were eating smaller prey items than what was typically available in the lake and that as yellow perch grew, the average size of prey increased as well, although their preferred prey choice was not constrained by their gape. Prey detection distance increased as visual acuity increased, from <20 cm for fish <30 mm SL to 45 cm for 50 mm fish. They also found that cone morphology changed and lens diameter increased as the body length increased, corresponding to an increase in visual acuity and a movement from pelagic to demersal habitat around 24–31 mm SL, when their visual acuity is approaching adult values. Mills et al. (1984) found from field and laboratory experiments that yellow perch did not select *Daphnia* until the fish moved to the demersal habitat at about 25–30 mm TL, and that smaller prey items were preferred. Reaction distance was greater for *Daphnia* than calanoid copepods and increased as the size of the *Daphnia* increased. Miller et al. (1993) found a strong size-dependent relationship for visual acuity and that behavioral estimates underestimate the ability of fish to spot prey.

18.3.3 Gape Limitation

On the other hand, yellow perch are initially gape limited (Bremigan et al. 2003) and feed primarily on rotifers and nauplii the first few weeks in the pond (Fig. 18.14). We found that when ponds typically exhibit the "clear water" phase and the crustacean zooplankton crashes, yellow perch are in the process of incorporating larger zooplankters, such as *Acanthocyclops vernalis*, *Skistodiaptomus oregonensis*, and smaller cladocerans (*Bosmina longirostris*, *Alona* spp., *Diaphanosoma birgei*, and *Daphnia parvula*) into their diets (Fig. 18.14). Unlike walleye and saugeye, yellow perch did not incorporate benthic organisms, such as chironomids, into their diet until the end of May, and chironomids were never a major component of their diet. We also found that yellow perch exhibited slow growth during the period of low crustacean biomass in the ponds (Figs. 18.10 and 18.14) and many fish consumed primarily rotifers in the first few weeks of June. The growth pattern we observed is similar to that of yellow perch produced in Wisconsin (Hartleb 2003) and the

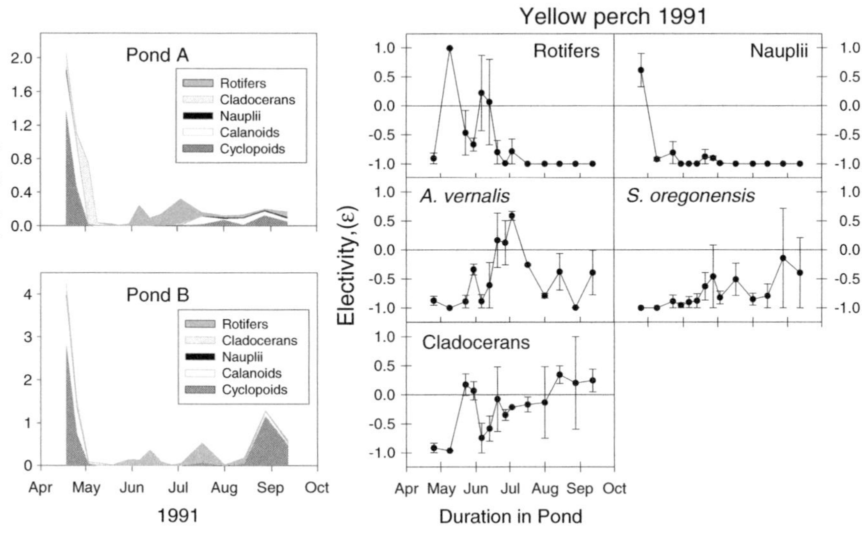

Fig. 18.14 A comparison of zooplankton biomass (dry mg L^{-1}) in two yellow perch ponds from April to September 1991 (note the different y-axis scales) and the electivity (Chesson's ε) for prey found in yellow perch diets. Crustacean biomass declined the second week of May, corresponding to a low algae "clear water" phase about 4 weeks after egg skeins were seen in the ponds. Cladocerans include *Bosmina longirostris*, *Alona* spp., *Diaphanosoma birgei*, *Daphnia parvula*, and unidentified cladocerans. Copepods included cyclopoid and calanoid nauplii, and *Acanthocyclops vernalis* and *Skistodiaptomus oregonensis* copepodites

Eurasian perch (Cuvier-Péres and Kestemont 2002). When crustacean zooplankton densities started to recover by the end of July and into August, cladocerans became the preferred prey (Fig. 18.14), which seems to correspond to increases in fish total length and weight (Fig. 18.10).

Graeb et al. (2005) found that gape width increased more quickly with size in larval walleye than larval yellow perch which allows walleye to switch to larger prey, such as other fish, sooner than yellow perch. Walleye also exhibits higher growth rates when feeding on fish compared to yellow perch (Graeb et al. 2005). Although both walleye and yellow perch captured small prey (zooplankton) with high efficiency, walleye (20 mm) were more efficient at capturing large prey (Graeb et al. 2005), further illustrating differences between these fish. Yellow perch consumed more zooplankton, while consumption of benthic organisms and fish were similar at larger sizes (Graeb et al. 2005). It would not be a selective advantage for yellow perch to feed on their conspecifics unless there were no other prey items available. We have not seen any evidence of cannibalism in the diets of yellow perch, saugeye, or walleye collected from hatchery ponds.

18.3.4 Cannibalism

Much of the current information regarding cannibalism has come from studies involving cultured fish that are used to enhance populations of sportfish species by rearing and stocking their larvae and/or fingerlings into lakes (Smith and Reay 1991; Baras and Jobling 2002; Loadman et al. 1986; van Kooten et al. 2010). Smith and Reay (1991) list seven percid species that are cannibalistic., four of which, walleye, yellow perch, pikeperch (*Sander lucioperca*), and Eurasian perch (*Perca fluviatilis*) are important species for culture as either food fish and sportfish (Malison and Held 1992; Loadman et al. 1986). Studies of cannibalism in culture typically involve observations of behaviors under intensive culture conditions in tanks for fish that are eventually habituated to formulated feed under controlled conditions in which they cannot escape predation. This intracohort cannibalism occurs during larval and juvenile stages of development (Smith and Reay 1991; Baras and Jobling 2002; Mandiki et al. 2007; Cuff 1977, 1980; Loadman et al. 1986).

Baras and Jobling (2002) suggest that Type I cannibalism (where the size of the prey is large relative to the size of the predator) occurs in early larval development. The predator ingests the prey tail first up to its head, which is usually larger than its body, and the head is eventually discarded (Cuff 1980). Type II cannibalism (where the predator is larger than the prey) occurs more often among cohorts in which a few individuals grow faster than others enabling them to completely swallow a smaller individual, and is well documented in intensive culture of walleye (Loadman et al. 1986; Li and Mathias 1982; Cuff 1977, 1980), yellow perch (Malison and Held 1992), and Eurasian perch (Mandiki et al. 2007; Kestemont et al. 2003; Baras et al. 2003). The major causes of intracohort cannibalism in culture have been linked to size at hatching (or differential/asynchronous egg hatching), low stocking densities (creating size heterogeneity), limiting food resources/or food availability, lack of refuges in order to escape predation, and limited lighting conditions above or in tanks (Smith and Reay 1991; Baras et al. 2003; Loadman et al. 1986; Cuff 1977, 1980; Colesante 1989; van Kooten et al. 2010).

Many of the studies describing observed cannibalism use a high density of fish (ranging from 1 to 100 fish L^{-1}) in relatively small glass aquaria (from 10 to 200 L) and fish are deprived of food (Loadman et al. 1986; Li and Mathias 1982; Cuff 1977, 1980; Malison and Held 1992; Kestemont et al. 2003). For example, Li and Mathias (1982) examined the effects of fish density (1, 10, and 100 fish L^{-1} in 30 L aquaria) on survival and found that the rate of cannibalistic attacks increased as fish density increased and also observed delayed feeding increased cannibalistic attacks, yet once the fish developed better avoidance capabilities, cannibalism ceased. Kestemont et al. (2003) also found that perch (*Perca fluviatilis*) post-larvae stage was impacted by cannibalism the greatest at densities of 10 fish L^{-1}, yet cannibalism was inversely proportional to stocking density in the larval stage (greater impact at 10 fish L^{-1} than 100 fish L^{-1}). We have found no incidences of cannibalism in over 5700 percid stomach analyses (4774 saugeye, 782 walleye, and 208 yellow perch). This lack of verifiable cannibalism in pond culture may be due to the low density of fish in ponds (0.001–0.006 fish L^{-1}) and the inability to observe fish behavior in the hatchery ponds.

18.3.5 Trade-Offs for Sequential Culture in Earthen Ponds

Survival of larval percids can vary from year to year and pond to pond within years for a variety of reasons such as poor water quality (low DO), inadequate forage base, fish density, or the possibility of cannibalism. An additional area of concern, however, specifically for percid culture in Ohio, is the history of pond use in the sequential culture of catfish with saugeye and walleye (Tew et al. 2010). Fertilization history may affect the accumulation in pond sediments of nitrogen and phosphorus, and detritus from decomposing algae, as well as chemicals from algaecides such as copper sulfate ($CuSO_4$) (Knud-Hansen 1992; McNevin and Boyd 2004; Tew et al. 2010).

Fertilization of ponds is a useful management strategy to optimize and manage edible algae leading to increased percid production (Qin et al. 1995). Knud-Hansen (1992) found that nutrients applied to ponds remained available for use 2 years after initial application. Phosphorus was particularly enhanced by oxygen concentration (low) and pH in the adjacent water layer. However, Boyd and Musig (1981) suggest that using a liquid fertilizer applied in small amounts more frequently would result in greater absorption by the phytoplankton community rather than a large addition of granules which would most likely be absorbed by the sediments, decreasing the ability of sediments to remove inorganic phosphorus from the water column. In addition, catfish are supplementarily fed a dry feed pellet with a high phosphorus content (TP $\approx$ 300 µg P/g dry food) (Tew 2003). According to Edwards (1993), exposing the sediments to the air mineralizes the sediments and releases the nutrients tied up in the soils, whereby they become available for use when ponds are re-filled.

Copper sulfate is an algaecide used to reduce off-flavor in channel catfish produced in ponds as well as to reduce blue-green algae and the treatment of *Icthyophthirus multifiliis* infestations during the culture period (Schlenk et al. 1998; McNevin and Boyd 2004). Copper sulfate readily absorbs into the sediment at the bottom of ponds and is concentrated in the top 6 cm of the soil, creating the potential for resuspension at the sediment-water interface (McNevin and Boyd 2004).

Tew et al. (2010) tested whether percid production was affected by the accumulation of phosphorus and copper in the sediments of ponds used for culturing catfish by comparing survival, growth and yield of percids in ponds that were single-cropped (SS ponds) with only saugeye produced in the spring for the previous two successive years with percids that were double-cropped (DD ponds) with channel catfish produced in the summer (Fig. 18.15). They found that a significant amount of copper remained in the sediment after the ponds were drained (on average 2.5–5.6 times higher) in DD ponds, while average phosphorus was consistently higher in the SS ponds for both soluble reactive phosphorus (SRP) and total phosphorus (TP), possibly due to the removal of fertile organic matter with an additional draining per year of the DD ponds. Although sediment SRP and TP concentrations were higher in SS ponds, there were no differences in total phytoplankton biomass, possibly due to the weekly addition of fertilizer as well as the growth of benthic filamentous algae and vascular plants in the catfish ponds. However, there was a significantly higher total zooplankton biomass in SS ponds than DD ponds throughout the season (Fig. 18.16). This difference was not related to the phytoplankton biomass, suggesting there could be a negative correlation with copper concentration, as other studies have shown that waterborne copper can reduce zooplankton abundance and fecundity.

These findings are consistent with the fish production results, in which there was a negative correlation between growth (TL), survival (%), and total yield (kg) in DD ponds relative to SS ponds (Tew et al. 2010). They also found that saugeye total length (TL) and wet weight were significantly higher in SS ponds throughout the 2002 growing season ($P<0.05$, one-way repeated-measures ANOVA) with an average TL at harvest of 29.7 mm in SS ponds compared with 22.7 mm in DD ponds,

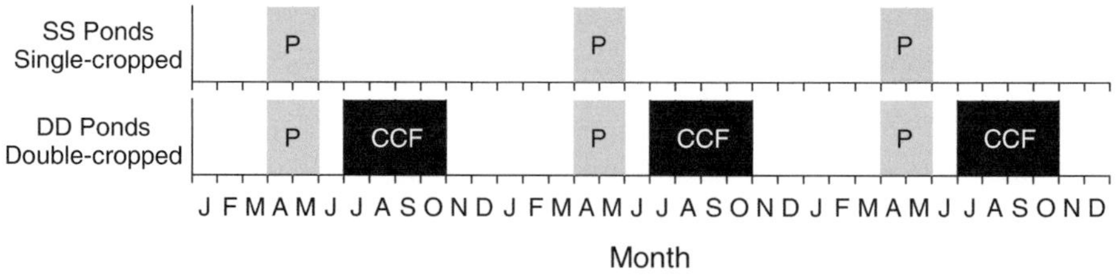

Fig. 18.15 A comparison of double cropped (DD) and single cropped (SS) ponds in a 3-year sequence at Hebron State Fish Hatchery, OH. SS ponds are used for growing percids (*P*) only, and are left dry in between growing seasons; while DD ponds are used for growing percids between April and mid-May, drained, and then refilled in mid-July for growing channel catfish (*CCF*) until late September. The open space represents the period in which the pond is empty

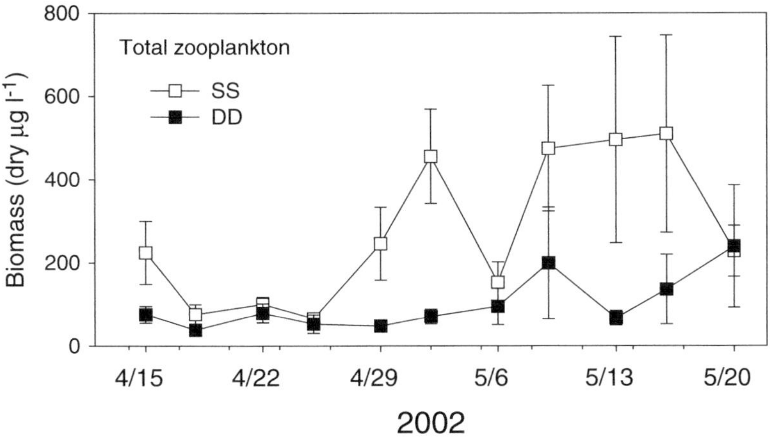

Fig. 18.16 Comparison of the seasonal dynamics of the zooplankton dry weight biomass (µg/l) in DD and SS ponds. The values for each date are the mean from six replicate ponds. *Vertical bars* refer to standard errors (± SE) associated with the mean

and average wet weights of 173.4 mg and 81.4 mg in SS and DD ponds, respectively. Analyzing 22 ponds treated with copper sulfate from 1994 to 2001 revealed a negative correlation between survival (%) and yield (kg/ha) of percids harvested in 2001 with the cumulative amount of copper applied to the ponds over the previous 7 years (Fig. 18.17).

Thus, the accumulation of sediment copper through repeated use of copper sulfate applications during catfish production has a negative effect on percid survival/production the following spring in the same pond indirectly through food web dynamics or directly on fish via chronic exposure. Using hatchery ponds for sequential culture of fish may increase the total number of species produced during the production year, but it may not maximize the production of all species, especially double-cropping percids with catfish. Single-cropped ponds may maximize percid production while mitigating the carryover effects of double-cropping.

18.4 Conclusion/Additional Considerations (Non-manageable)

Percid production in hatchery ponds is dictated by external forces beyond hatchery management such as traditional abiotic factors (i.e., temperature and weather) or land management practices that eutrophicate the hatchery's source water. Herein, we show that zooplankton composition differs between hatcheries due to the trophic status of source water, with highly eutrophic waters favoring small bodied zooplankton due to the predominance of cyanobacteria species. Further, we recommend that pond management practices be specific to the hatchery and source water's

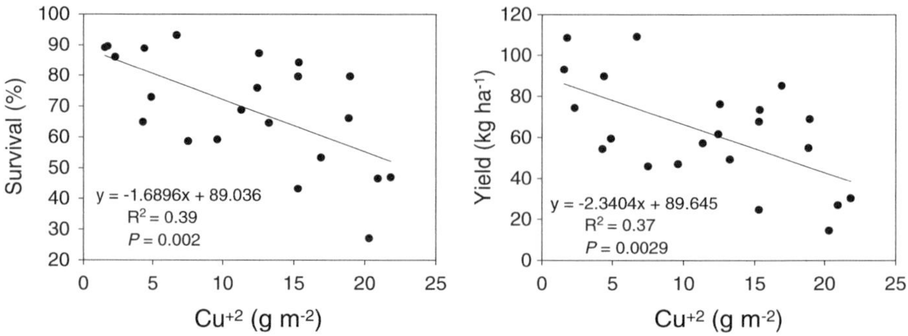

Fig. 18.17 Variation in saugeye survival and yield in 22 ponds in 2001 as a function of the cumulative amount of copper added to each pond during catfish culture from 1994 to 2000

trophic status. Specifically, a hatchery with eutrophic source water should be fertilized with the minimum necessary phosphate-phosphorus to promote adequate algal abundance (i.e., 10 μg PO_4-P·L^{-1}) while minimizing the risk of poor water quality. Alternatively, given a mesotrophic water source with phytoplankton dominated by non-cyanobacterial algae and an inoculum of low zooplankton abundance, we recommend that the hatchery ponds may benefit from a higher fertilization rate of 30 μg PO_4-P·L^{-1} without risking a negative effect from reduced water quality.

Beyond manipulatable variables, temperature and weather set the stage for good growth of cultured fish and ultimate successful fish production. During successive years of studying saugeye and walleye growth, we have seen a vast difference in temperature regime between years (Fig. 18.18) and that springtime temperatures generally fall below the optimal range for larval walleye growth in Ohio (Madon and Culver 1993). We estimated the effect of sub-optimal temperature on fish growth via a bioenergetics model for larval walleye (Madon and Culver 1993), and found that the annual temperature regime that yielded the highest growth occurred in 2007 and fish growth was reduced by 8 % when substituting temperatures from 2008, but growth was drastically reduced, by 43 %, with the cold temperature regime of 2005 (Fig. 18.18). However, temperature also reduces zooplankton production during the cooler years by increasing development time of cladoceran eggs and instar stages as well as copepod nauplii and copepodite stages. Cooler springs are generally accompanied by adverse weather such that heavy, prolonged cloud cover limits phytoplankton production, causing further limitation to zooplankton and, ultimately, fish production. Thus, despite the best efforts of hatchery managers, ponds are still subject to annual variability. However, given the ecological tools at hand (i.e., bioenergetics models, zooplankton production, and stocking records), we suggest that hatchery managers can quantify the impact of temperature relative to production goals, and allow for long-term management of the ponds and overall production while testing appropriate management action to improve our understanding of these systems and adjust management practices to maintain consistent production.

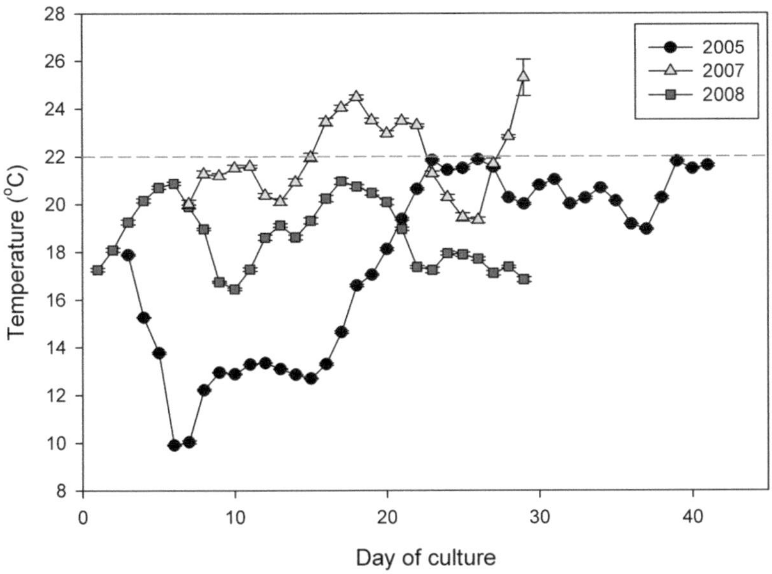

Fig. 18.18 Variation in temperature during saugeye and walleye culture at Senecaville SFH during 2005, 2007, and 2008. The *dashed line* at 22 °C indicates the lower bound of optimal temperature for larval walleye growth

References

Ali MA, Anctil M (1977) Retinal structure and function in the walleye (*Stizostedion vitreum vitreum*) and Sauger (*S. canadense*). J Fish Res Board Can 34:1467–1474

Ali MA, Ryder RA, Anctil M (1977) Photoreceptors and visual pigments as related to behavioral responses and preferred habitats of Perches (*Perca* spp.) and Pikeperches (*Stizostedion* spp.). J Fish Res Board Can 34:1475–1480

Baras E, Jobling M (2002) Dynamics of intracohort cannibalism in cultured fish. Aquac Res 33:461–479

Baras E, Kestemont P, Mélard C (2003) Effect of stocking density on the dynamics of cannibalism in sibling larvae of *Perca fluviatilis* under controlled conditions. Aquaculture 219:241–255

Boyd CE, Musig Y (1981) Orthophosphate uptake by phytoplankton and sediment. Aquaculture 22:165–173

Braekevelt CR, McIntyre DB, Ward FJ (1989) Development of the retinal tapetum lucidum of the walleye (*Stizostedion vitreum vitreum*). Histol Histopathol 4:63–70

Bremigan MT, Dettmers JM, Mahan AL (2003) Zooplankton selectivity by larval yellow perch in Green Bay, Lake Michigan. J Great Lakes Res 29:501–510

Briland RD (2010) Optimizing larval fish survival and growth through an analysis of consumer and resource interactions in percid culture ponds. M.S. thesis, The Ohio State University

Colesante RT (1989) Improved survival of walleye fry during the first 30 days of intensive rearing on brine shrimp and zooplankton. Prog Fish Cult 51:109–111

Craig JF (1987) The biology of perch and related fish. Timber Press, Portland

Cuff WR (1977) Initiation and control of cannibalism in larval walleye. Prog Fish Cult 39:29–32

Cuff WV (1980) Behavioral aspects of cannibalism in larval walleye, Stizostedion vitreum. Can J Zool 58:1504–1507

Culver DA (1988) Plankton ecology in fish hatchery ponds in Narrandera, NSW, Australia. Verh Int Ver Theor Angew Limnol 23:1085–1089
Culver DA (1991) Effects of the N:P ratio in fertilizer for fish hatchery ponds. Verh Int Ver Theor Angew Limnol 24:1503–1507
Culver DA (1996) Fertilization procedures for pond culture of walleye and saugeye. In: Summerfelt RC (ed) Walleye culture manual. North Central Regional Aquaculture Center Publications Office, Iowa State University, Ames, p. 115–122
Culver DA, Wu L (1997) Relative importance of predation and competition in the seasonal dynamics of zooplankton: results from larval fish rearing ponds. Arch Hydrobiol Spec Iss Adv Limnol 49:27–35
Culver DA, Qin J, Madon SP, Helal HA (1992) *Daphnia* production techniques for rearing fingerling walleye and saugeye. Federal aid in sport fish restoration, Project F-57-R, Final report. Ohio Department of Natural Resources, Columbus
Culver DA, Madon SP, Qin J (1993) Percid pond production techniques; timing, enrichment, and stocking density manipulation. J Appl Aquac 2:9–31
Cushing DH (1969) The regularity of the spawning season of some fishes. J Cons Int l'Explor de la Mer 33:81–92
Cushing DH (1990) Plankton production and year-class strength in fish populations: an update of the match/mismatch hypothesis. Adv Mar Biol 26:249–293
Cuvier-Péres A, Kestemont P (2002) Development of digestive enzymes in Eurasian perch larvae *Perca fluviatilis*. Fish Physiol Biochem 24:279–285
Dabrowski K, Ciereszko RE, Ciereszko A, Toth GP, Christ SA, El-Saidy D, Ottobre JS (1996) Reproductive physiology of yellow perch (*Perca flavescens*): environmental and endocrinological cues. J Appl Ichth 12:139–148
Dabrowski K, Culver DA, Brooks C, Voss A, Binkowski F, Yeo A, Balogun A (1992) Biochemical aspects of the early life history of yellow perch (Perca flavescens). In: Kaushik S, Luquet P (eds) Fish nutrition in practice, pp 531–539
DeMott WR (1989) The role of competition in zooplankton succession. In: Sommer U (ed) Plankton ecology: succession in plankton communities. Springer, New York
Dettmers JM, Stein RA (1992) Food consumption by larval gizzard shad: zooplankton effects and implications for reservoir communities. Trans Am Fish Soc 121:494–507
Edwards P (1993) Environmental issues in integrated agriculture-aquaculture wastewater-fed fish systems. In: Pullin RSV, Rosenthal H, Maclean JL (eds) Environmental and aquaculture in developing countries, vol 31, Proceedings of international center for living aquatic resources management conference. ICLARM, Manila, p 359
Ellison DG, Franzin WG (1992) Overview of the symposium on walleye stocks and stocking. N Am J Fish Manag 12:271–275
Fielder DG (1992) Evaluation of stocking walleye fry and fingerlings and factors affecting their success in lower Lake Oahe, South Dakota. N Am J Fish Manag 12:336–345
Filbrun JE, Culver DA, Briland RD, Doyle CM (2008) The quality of Ohio state fish hatcheries' water supplies, 2005–2008. Federal aid in fish restoration project F-69-P, Interim report. Ohio Department of Natural Resources, Columbus
Flowers DD (1996) Fingerling production in drainable ponds at White Lake Fish Culture Station. In: Summerfelt RC (ed) Walleye culture manual. North Central Regional Aquaculture Center Publications Office, Iowa State University, Ames, p. 123–128
Fox MG (1989) Effect of prey density and size on growth and survival of juvenile walleye (*Stizostedion vitreum vitreum*). Can J Fish Aquat Sci 46:1323–1328
Fox MG, Flowers DD (1990) Effect of fish density on growth, survival, and food consumption by juvenile walleyes in rearing ponds. Trans Am Fish Soc 119:112–121
Garvey JE, Marschall EA, Wright RA (1998) From star charts to stoneflies: detecting relationships in continuous bivariate data. Ecology 79:442–447
Geiger JG, Turner CJ, Fitzmayer K, Nicols WC (1985) Feeding habits of larval and fingerling striped bass and zooplankton dynamics in fertilized rearing ponds. Prog Fish Cult 47:213–223

Graeb BDS, Galarowicz T, Wahl DH, JM D, Simpson MJ (2005) Foraging behavior, morphology, and life history variation determine the ontogeny of piscivory in two closely related predators. Can J Fish Aquat Sci 62:2010–2020
Guma'a SA (1982) Retinal development and retinomotor responses in perch, *Perca fluviatilis* L. J Fish Biol 20:611–618
Hart SD, Garling DL, Malison JA (2006) Yellow perch (*Perca flavescens*) culture guide. Iowa State University, Ames
Hartleb CF (2003) Food chain dynamics and diets of larval and post-larval yellow perch in culture ponds. In: Barry TP, Malison JA (eds) Proceedings of Percis III: the third international percid fish symposium, University of Wisconsin Sea Grant Institute, Madison
Hartleb CF, Johnson, JA, Held, JA (2012) Walleye and yellow perch pond fertilization. In: Mischke CC (ed) Aquaculture pond fertilization: impacts of nutrient input on production. John Wiley & Sons, Ames, Iowa. doi:10.1002/9781118329443.ch11
Helal H, Culver DA (1991) N:P ratio and plankton production in fish ponds. Verh Int Ver Theor Angew Limnol 24:1508–1511
Hokanson KEF (1977) Temperature requirements of some percids and adaptations to the seasonal temperature cycle. J Fish Res Board Can 34:1524–1550
Hushak LJ, Winslow JM, Dutta N (1988) Economic value of Great Lakes sportfishing: the case of private-boat fishing in Ohio's Lake Erie. Trans Am Fish Soc 117:363–373
Infante A, Abella SEB (1985) Inhibition of *Daphnia* by *Oscillatoria* in Lake Washington. Limnol Oceanogr 30:1046–1052
Jacob AP, Culver DA (2010) Experimental evaluation of the impacts of reduced inorganic phosphorus fertilization rates on juvenile saugeye production. Aquaculture 304:22–33
Jenkins GP, Young W, Davis TLO (1991) Density-dependence of larval growth of a marine fish, the southern bluefin tuna, *Thunnus maccoyi*. Can J Fish Aquat Sci 48:1358–1363
Kestemont P, Mélard C (2000) Aquaculture. In: Craig JF (ed) Percid fishes: systematics, ecology, and exploitation. Blackwell Science, Oxford
Kestemont P, Jourdan S, Houbart M, Mélard C, Paspatis M, Fontaine P, Cuvier A, Kentouri M, Baras E (2003) Size heterogeneity, cannibalism and competition in cultured predatory fish larvae: biotic and abiotic influences. Aquaculture 227:333–356
Knud-Hansen CF (1992) Pond history as a source of error in fish culture experiments: a quantitative assessment using covariate analysis. Aquaculture 105:21–36
Li S, Mathias JA (1982) Causes of high mortality among cultured larval walleyes. Trans Am Fish Soc 111:710–721
Loadman NL, Moodie GEE, Mathias JA (1986) Significance of cannibalism in larval walleye (*Stizostedion vitreum*). Can J Fish Aquat Sci 43:613–618
Loew ER, Wahl CM (1991) A short-wavelength sensitive cone mechanism in juvenile yellow perch, *Perca flavescens*. Vis Res 31:353–360
Loew ER, McFarland WN, Mills EL, Hunter D (1993) A chromatic action spectrum for planktonic predation by juvenile yellow perch, *Perca flavescens*. Can J Zool 71:384–386
Madon SP, Culver DA (1993) Bioenergetics model for larval and juvenile walleye: an in situ approach using experimental ponds. Trans Am Fish Soc 122:797–813
Malison JA, Held JA (1992) Effects of fish size at harvest, initial stocking density and tank lighting conditions on the habituation of pond-reared yellow perch (*Perca flavescens*) to intensive culture conditions. Aquaculture 104:67–78
Mandiki SNM, Babiak I, Krol J, Rasolo JFR, Kestemont P (2007) How initial predator-prey ratio affects intra-cohort cannibalism and growth in Eurasian perch *Perca fluviatilis* L larvae and juveniles under controlled conditions. Aquaculture 268:149–155
Mathias JA, Li S (1982) Feeding habits of walleye larvae and juveniles: comparative laboratory and field studies. Trans Am Fish Soc 111:722–735
Mathias JA, Franzin WG, Craig JF, Bablauk JA, Flannagan JF (1992) Evaluation of stocking walleye fry to enhance a commercial fishery in a large, Canadian prairie lake. N Am J Fish Manag 12:299–306

McNevin AA, Boyd CB (2004) Copper concentrations in channel catfish *Ictalurus punctatus* ponds treated with copper sulfate. J World Aquac Soc 35:16–24
McWilliams RH, Larscheid JG (1992) Assessment of walleye fry and fingerling stocking in the Okoboji Lake, Iowa. N Am J Fish Manag 12:329–335
Mertz G, Myers RA (1994) Match/mismatch predictions of spawning duration versus recruitment variability. Fish Oceanogr 3:236–245
Miller TJ, Crowder LB, Rice JA (1993) Ontogenetic changes in behavioural and histological measures of visual acuity in three species of fish. Environ Biol Fish 37:1–8
Mills EL, Confer JL, Ready RC (1984) Prey selection by young yellow perch: the influence of capture success, visual acuity, and prey choice. Trans Am Fish Soc 113:579–587
Mitzner L (2002) Effectiveness of walleye fry and fingerling stocking in Rathbun Lake, Iowa, 1990–1999. N Am J Fish Manag 22:1007–1013
Munch CS, Vaga RM, Culver DA (1984) Evidence for size-selective grazing of phytoplankton species by zooplankton in fish hatchery ponds. Verh Int Ver Theor Angew Limnol 22:1640–1644
Nunn AD, Tewson LH, Cowx IG (2012) The foraging ecology of larval and juvenile fishes. Rev Fish Biol Fish 22:377–408
Paragamian VL, Kingery R (1992) A comparison of walleye fry and fingerling stockings in three rivers in Iowa. N Am J Fish Manag 12:313–320
Qin J, Culver DA (1992) The survival and growth of larval walleye, *Stizostedion vitreum*, and trophic dynamics in fertilized ponds. Aquaculture 108:257–276
Qin J, Culver DA, Yu N (1994) Comparisons of larval walleye and saugeye (walleye x sauger hybrid) growth and impacts on zooplankton in experimental ponds. Prog Fish Cult 56:91–99
Qin J, Culver DA, Yu N (1995) Effect of organic fertilizer on heterotrophs and autotrophs: implications for water quality management. Aquac Res 26:911–920
Qin J, Madon SP, Culver DA (1995) Effects of larval walleye (Stizostedion vitreum) and fertilization on the plankton community: implications for larval fish culture. Aquaculture 130:51–65
Richmond HE, Hrabik TR, Mensinger AF (2004) Light intensity, prey detection and foraging mechanisms of age 0 year yellow perch. J Fish Biol 65:195–205
Schlenk D, Gollon JL, Griffin BR (1998) Efficacy of copper sulfate for the treatment of *Ichthyophthiriasis* in channel catfish. J Aquat Anim Health 10:390–396
Smith C, Reay P (1991) Cannibalism in teleost fish. Rev Fish Biol Fish 1:41–64
Soderberg RW, Kirby JM, Lunger D, Marcinko MT (1997) Comparison of organic and inorganic fertilizers for the pond production of walleye *Stizostedion vitreum*. J Appl Aquac 7:23–30
Soderberg RW, Kirby JM, Marcinko MT (2000) Lack of response of juvenile walleyes to increased levels of fertilization or liming in soft-water ponds. N Am J Aquac 62:26–32
Summerfelt RC, Johnson JA, Clouse CP (2011) Culture of walleye, sauger, and hybrid walleye. In: Barton BA (ed) Biology, management, and culture of walleye and sauger. American Fisheries Society, Maryland
Tew KS (2003) The impacts of algal control in catfish and percid aquaculture ponds. Dissertation, The Ohio State University
Tew KS, Conroy JD, Culver DA (2006) Effect of lowered phosphorus fertilization rates on pond production of percid fingerlings. Aquaculture 255:436–446
Tew KS, Conroy JD, Culver DA (2010) The importance of phosphorus and copper carryover in pond sediments for juvenile percid survival, growth, and yield. N Am J Aquac 72:22–33
Thorpe JE (1977) Morphology, physiology, behavior, and ecology of *Perca fluviatilis* and *P flavescens* Mitchill. J Fish Res Board Can 34:1504–1514
Tice BJ, Soderberg RJ, Kirby JM, Marcinko MT (1996) Growth and survival of walleyes reared in ponds fertilized with organic or inorganic materials. Prog Fish Cult 58:135–139
van Kooten T, Andersson J, Byström P, Persson L, de Roos AM (2010) Size at hatching determines population dynamics and response to harvesting in cannibalistic fish. Can J Fish Aquac Sci 67:401–416
Wahl CM, Mills EL, McFarland WN, DeGisi JS (1993) Ontogenetic changes in prey selection and visual acuity of the yellow perch, *Perca flavescens*. Can J Fish Aquac Sci 50:743–749

Chapter 19
The Ecology of Lifetime Growth in Percid Fishes

Michael D. Rennie and Paul A. Venturelli

Abstract The factors responsible for the lifetime growth patterns of percids in natural populations can provide meaningful insights for culture operations. Here, we present a summary of a number of well-studied factors and review the current state of knowledge. We illustrate an informative approach to describe lifetime growth of percid species by applying a biphasic growth model to European perch and pikeperch populations, and discuss life-history constraints considering biphasic growth. An evaluation of proposed hypotheses for proximate mechanisms of female-biased sexual size dimorphism in percids is presented, indicating that reduced feeding in males is the most parsimonious explanation given the current evidence. Growth rates in percids are strongly temperature-dependent, and show strong evidence of countergradient growth. Percids also show significant density-dependent growth, demonstrating twofold variation in individual growth rates. Predation, food availability and prey particle size can also affect the efficiency of percid growth. Parasitism and disease in percids are not as well studied as other factors reviewed here within an ecological context, but the reported effects on percid growth vary from negative to neutral to positive. Our review of drivers of natural variation in percid growth will assist culture operators with regards to broodstock selection, husbandry and maintenance of cultured percids.

Keywords Percids • Growth • Broodstock selection • Sex dimorphism • Husbandry

M.D. Rennie (✉)
Department of Biology, Lakehead University,
955 Oliver Road, P7B 5E1 Thunder Bay, ON, Canada
e-mail: mrennie@lakeheadu.ca

P.A. Venturelli
Department of Fisheries, Wildlife and Conservation Biology, University of Minnesota Twin Cities, 135 Skok Hall, 2003 Upper Buford Circle, 55108 St. Paul, MN, USA

P. Kestemont et al. (eds.), *Biology and Culture of Percid Fishes*,
DOI 10.1007/978-94-017-7227-3_19

19.1 Introduction

As the culture of percid fishes becomes more common in both North America and Europe, it is important to keep in mind the evolutionary and environmental forces that have shaped the life histories of these fishes in the wild. Fish life histories are intimately related to individual growth rates, which are typically targeted for maximization by the culture industry. The life histories of many cultured strains of fish demonstrate incredible variability in the wild, providing the raw materials for artificial selection of desirable traits under culture conditions. For example, percids exhibit up to 3.5-fold variation in growth rate (using fork length at age 2 as a proxy for growth rate) and up to fourfold variation in asymptotic length (Table 19.1).

In this chapter, we review some of the driving forces behind this tremendous variability in growth and size. The net balance of these drivers and constraints allows for the optimization of growth within a given set of circumstances or any particular environment. Where wild populations optimize their growth given a particular set of conditions, the economics of fish culture rely on the maximization of individual growth rates to ensure the greatest rate of return on operations. Understanding the role of environment and life history in this optimization process in wild populations may assist managers in achieving their goals for cultured stocks.

We begin our review by describing how the biphasic von-Bertalanffy growth model improves our understanding of lifetime growth patterns (Sect. 19.2). Using this model as a backdrop, we then explore the primary intrinsic (Sect. 19.3) and extrinsic (Sect. 19.4) factors driving or constraining percid growth. Intrinsic factors include life history constraints or 'invariants' (Sect. 19.3.1), and sexually dimorphic growth (Sect. 19.3.2). Extrinsic factors include temperature (Sect. 19.4.1), predation (Sect. 19.4.2), density dependence (Sect. 19.4.3), resource availability and type (Sect. 19.4.4), and parasitism and disease (Sect. 19.4.5). This review focuses primarily on those species that are most familiar to the authors (walleye, *Sander vitreus*, and yellow perch, *Perca flavescens*), but also draws on relevant literature from other large, well-studied percids.

19.2 The Biphasic Growth Model

In this chapter, we introduce the biphasic growth model (BGM) as a general framework for describing and interpreting the lifetime growth pattern of percids. Lester et al. (2004) proposed the BGM as an alternative to the von Bertalanffy model (von Bertalanffy 1934, 1938), which is ubiquitous in the fish literature and routinely applied to percid species (e.g., Quist et al. 2003; Sass et al. 2004; Heibo et al. 2005; Jackson et al. 2008; Perez-Bote and Roso 2012). Although the von Bertalanffy model is based on bioenergetic principles (but see Roff 1980 and references therein, Lipinski and Roeleveld 1990) and tends to provide a good fit to lifetime growth (but see Katsanevakis and Maravelias 2008), it is criticized for failing to describe the

Table 19.1 Published range of observed asymptotic length (L_∞) and fork length at age 2 (FL_2) for common percid species. Only those species with a reasonable number of growth curves/populations listed (*N*) are included. Sex indicates the gender for which lengths apply to; either females or mixed (sexes not separated)

		Walleye (*Sander vitreus*)	Pikeperch (*Sander lucioperca*)[b]	Eurasian Perch (*Perca fluvatilis*)	Yellow perch (*Perca flavescens*)[c]	Ruffe (*Gymnocephalus cernua*)[b]
L_∞[a]		518–992	400–1413	200–700	173–318	125–353
	N	435	22	68	56	16
	Source; sex	Purchase et al. (2006); females	Fishbase.org (Froese and Pauly 2013); mixed	Heibo et al. (2005); females	Purchase et al. (2005a); females	Fishbase.org (Froese and Pauly 2013); mixed
TL_2		134–373	112–394[d]	75–215	77–169	52–159
	N	198	20	68	70	16
	Source; sex	Venturelli et al. (2010); Females	Fishbase.org (Froese and Pauly 2013); mixed	Heibo et al. (2005); females	Purchase et al. (2005a); females	Ogle (1998); mixed

[a]L_∞ (expressed as mm fork length) from the standard von Bertalanffy formula (Eq. 19.2 in text) applied to length-at-age data, to make comparisons across species and studies equivalent

[b]Lengths converted to total lengths using formulae listed for each species at www.fishbase.org (Froese and Pauly 2013). Populations for which length units were not reported or could not be standardized to *TL* were excluded from range estimates

[c]Fork lengths converted to total lengths using factor of 1.044 (Schneider 2004)

[d]TL_2 estimated from published von Bertalanffy parameters. Two outlier estimates of TL2 from a Turkish lake were removed

change in growth that occurs when a juvenile fish that is investing solely in somatic growth becomes mature and thereafter invests in both somatic growth and reproduction (Nikolskii 1969; Ricker 1975; Charnov 1993; Day and Taylor 1997; Charnov et al. 2001; Lester et al. 2004).

Unlike the von Bertalanffy growth model, the BGM explicitly accounts for the change in growth rate that occurs when a fish matures. The BGM comprises two separate length functions; a linear function that describes immature growth in length in the lead-up to maturity and an asymptotic function that describes adult growth in length thereafter (Lester et al. 2004). The qualifying phrase "in the lead-up to maturity" is important because it recognizes that growth rate over the entire immature life phase is unlikely to be constant. For example, immature percids can exhibit non-linear growth as a result of ontogeny and changes in diet or behavior (Moodie et al. 1989; Buijse and Houthuijzen 1992; Diehl and Eklov 1995; Hoxmeier et al. 2004).

In the BGM, immature length at age (L_t) growth is given by the linear function

$$L_t = h(t{-}t1), \tag{19.1}$$

where h is immature growth rate (e.g., mm/year), t is age, and t_1 – the hypothetical age at which immature length is zero – is a parameter that allows for a change in immature growth rate in the lead-up to maturity (Lester et al. 2004; Shuter et al. 2005). See Quince et al. (2008a, b) for a BGM that relaxes the assumption that immature growth is linear. Adult growth in the BGM is then described by the conventional form of the von Bertalanffy function

$$L_t = L_\infty\left(1-\mathrm{e}\left(-K\left(t-t_0\right)\right)\right), \tag{19.2}$$

where L_∞ is asymptotic length, K is the rate of deceleration of growth, and t_0 is the hypothetical age at which adult length is zero. Lester et al. (2004) showed that the von Bertalanffy parameters can be approximated by

$$L_\infty = 3h/g, \tag{19.3}$$

$$K = \ln\left(1+g/3\right), \text{ and} \tag{19.4}$$

$$t_0 = T + \ln\left(1-\mathrm{g}\left(T-t_1\right)/3\right)/\ln\left(1+g/3\right), \tag{19.5}$$

where g is annual investment in reproduction expressed as a proportion of somatic mass (a close proxy for the gonadosomatic index in female percids, Shuter et al. 2005), and T is the age at which investment in reproduction begins.

The BGM is a valuable tool because it describes growth in a way that allows for inferences about important life history traits and their constraints. For example, fitting a BGM allows one to predict length at the onset of maturity (L), investment in reproduction, and mortality rate after the first year of life (Lester et al. 2004). These inferences are simply impossible with other growth models. To date, the BGM has been used to describe the life history of two percids: walleye and yellow perch (Lester et al. 2004; Shuter et al. 2005; Quince et al. 2008a, b; Rennie et al. 2008). We applied the BGM to published data from two European percids (pikeperch, *Sander lucioperca*, and European perch, *Perca fluviatilis*) (Fig. 19.1). The model accurately describes the change in growth rate that occurs with maturity, and provides an estimate of *L*, *g* and age-1+ mortality rate for these populations. We encourage further application of the BGM to European percids, which can provide exciting new insights into growth and related life history traits.

The BGM can be fit to both males and females from a given population, but we caution that the interpretation of *g* for male fishes remains ambiguous. In female fish, *g* corresponds well to another index of reproductive investment, the gonadosomatic index (Shuter et al. 2005). However, the interpretation of *g* for male fish has not yet been mechanistically explored. Generally, when the BGM is applied separately males and females from the same population, male *g* is somewhat higher than female *g* (Rennie et al. 2008). This pattern is unexpected because male gonads are relatively small and there is little to suggest that males experience higher activity costs during

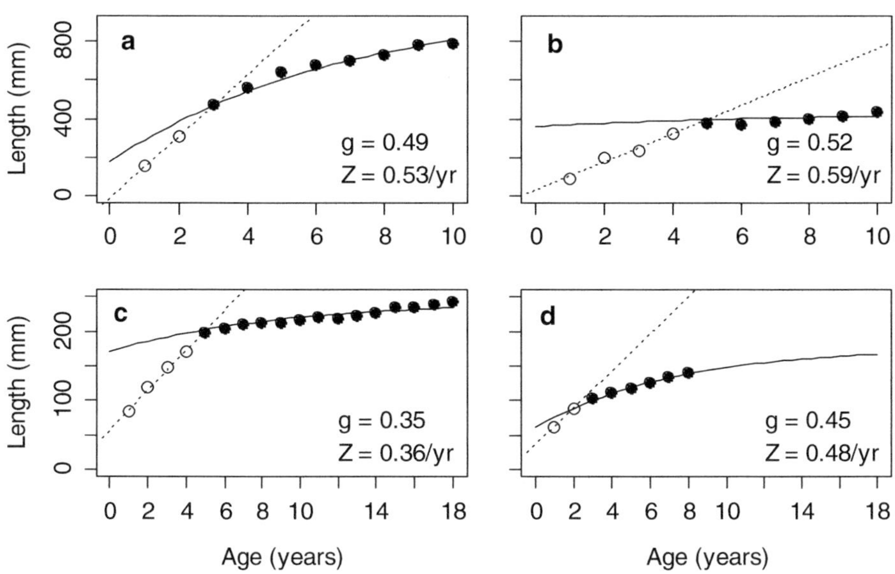

Fig. 19.1 Lifetime growth patterns of female pikeperch in (**a**) the Szczecin Lagoon, Germany/Poland (Neuhaus 1934, cited in Lehtonen et al. 1996; Lappalainen et al. 2003) and (**b**) Taivassalo Bay, Finland (Lehtonen 1987, cited in Lehtonen et al. 1996; Lappalainen et al. 2003), and female European perch in (**c**) Lake Windermere, UK (Craig 1980) and (**d**) Lake Horkkajarvi, Finland (Rask 1983). *Open* and *closed circles* are observed immature and mature length-at-age data, respectively, and *broken* and *solid lines* are predicted immature and mature growth trajectories, respectively, according to the biphasic model. *g* is investment in reproduction (expressed as a proportion of somatic mass) as predicted from the von Bertalanffy parameter *K* (Eq. 3.3 in Lester et al. 2004) and *Z* is the mean total mortality rate of age-1+ females as predicted from *g* (Eq. 4.6 in Lester et al. 2004)

reproduction (Sect. 19.3.2). Rather, higher *g* for males is likely due to lower growth efficiency in males and potential metabolic differences between genders (see Sect. 19.3.2). Gender-specific metabolic costs and/or inefficiencies are not reflected in the derivation of the BGM from first principles (Lester et al. 2004). Specifically, the mass scaling exponents around energy acquisition and losses (m_1 and m_2 in Lester et al. 2004, their Eq. 2.1) may differ between genders. Similar arguments have been made regarding a need for sex-specific parameterization of bioenergetic growth models (Rennie et al. 2008). More research is required to better understand *g* estimates from BGM as applied to male fish. Regardless, given the level of information and additional insight that can be yielded with the BGM versus traditional approaches, we strongly advocate using the BGM to describe the lifetime growth of female percids.

19.3 Intrinsic Factors Affecting Growth

19.3.1 Optimization of Growth Rates in Natural Environments and Life History Invariants

The fish culture industry seeks to bring fish to a marketable size in the most economically efficient manner. In contrast, individual growth in natural populations is optimized to a range of environmental and ecological conditions. This optimization is part of a general strategy to maximize fitness (i.e., maximize lifetime reproductive output while minimizing mortality risk) within a given set of environmental and biological constraints. These intrinsic and extrinsic forces lead to the large variation in observed growth rates both within and among wild percid populations (Table 19.1), and also explain why individuals from different populations can exhibit different growth rates in controlled environments (e.g., aquaculture).

There is a wealth of literature describing how certain life history parameters – growth among them – trade off with other life history traits in a predictable fashion. In the classic literature, the major tradeoffs that are described exist between (1) mortality rate (Z) and age at first reproduction (T_m), (2) growth rate (K) and mortality (Z) and (3) length at maturity (L_m) and asymptotic size (L_∞; Beverton and Holt 1959; Charnov 1993; Jensen 1981). The mechanism behind these relationships result from a fundamental tradeoff between survival and fecundity (Jensen 1996).

These relationships listed above form the basis for what are termed the Beverton-Holt life history invariants (Charnov 1993); that is, constants that describe the relationship between life history variables. These invariants are

$$C_1 = ZT_m, \quad (19.6)$$

$$C_2 = Z / K, \text{ and} \quad (19.7)$$

$$C_3 = L_m / L_\infty. \quad (19.8)$$

According to biphasic growth theory, C_1 can be shown analytically (Lester et al. 2014), and C_2 and C_3 reflect trade-offs associated with investment in reproduction and immature growth rate, respectively (Lester et al. 2004). Values for each of these invariants can and have been derived for fishes in general (Charnov 1993; Jensen 1996), for specific species (e.g., Beverton 1963), and for specific genders within a species (e.g., Purchase et al. 2005a, 2006).

Life history trade-offs and invariants provide useful information for broodstock selection. For instance, relationships 19.6–19.8 show that adult growth rate increases with either an increase in adult mortality rate or a reduction in age at maturity. Similarly, early reproduction has a negative impact on both reproductive potential (Jensen 1981) and asymptotic length (Lester et al. 2004). Early maturity, rapid lifetime growth and a small asymptotic length are seen in many populations subject to intense fishing mortality (Reznick et al. 1990; Law 2000), and these traits can be generated from evolutionary models that simulate the application of fishing mortality over a number of generations (Dunlop et al. 2007).

The evidence for life history invariants among percids is mixed. Although both Heibo and Vollestad (2006) and Purchase et al. (2005a) report results that are consistent with invariance, the evidence among walleye populations is limited (Beverton 1987; Purchase et al. 2006).

A potential conundrum for culturists in the traditional understanding of tradeoffs (e.g., those that assume a single lifetime growth trajectory) is the negative relationship between rapid growth and asymptotic length; the ideal fish in an aquaculture setting not only grows quickly but attains a marketable size. As discussed previously (Sect. 19.2) it is often unreasonable to describe lifetime growth as a single trajectory. Allowing for separate immature and mature growth trajectories leads to the prediction that fish can both grow rapidly and achieve large asymptotic length (Rennie et al. 2010). This prediction is based on the relationships presented in Eq. 19.3 above: if reproductive investment is invariant (i.e., does not vary among populations), then asymptotic length can increase with pre-maturation growth rate. This outcome was demonstrated in a recent study in which populations of yellow perch exposed to higher predation on immature fish demonstrated both rapid immature growth rates and larger asymptotic sizes (Rennie et al. 2010). Further, age at maturity and reproductive output in this study were invariant (similar among populations). In the scenario explored by this study, rapid growth rates allowed immature perch to escape a predator window; the consequence was a larger asymptotic length. Evolutionary models predict the same response from a fishery that targets small-bodied, immature individuals (Dunlop et al. 2007). Given that there is substantial evidence that reproductive investment in percids can vary both within (Craig 1980; Heins et al. 2004) and among populations (Heibo et al. 2005; Moles et al. 2008), variation in g in other study species or broader geographic comparisons could either dampen or enhance associations between immature growth and asymptotic length.

19.3.2 Sexual Dimorphism

Sexual dimorphism is observed throughout the natural world, and is a consequence of differential selection on sex (Darwin 1871). When sexual selection results in sexual size dimorphism (SSD) as a consequence of differences in growth rates, optimal life history strategies are also expected to differ between males and females. Proximate explanations for SSD are typically centered on differences in energy acquisition or allocation between sexes (e.g., Roff 1983; Holtby and Healey 1990; Henderson et al. 2003). Typically, the sex with the greater investment in reproduction (through either gonadal products, competition for mates, or nest defense) exhibits the larger body size (Parker 1992). In fishes, a wide range of SSD has been described, from female-biased to male-biased, though female-biased SSD appears to be most common (Breder and Rosen 1966). Among percids, female-biased SSD appears to be common in species of commercial interest, having long been documented in yellow perch (Carlander 1950), European perch (Le Cren 1958), ruffe (*Gymnocephalus cernuus*) (Lorenzoni et al. 2009 and references therein), walleye (Carlander 1945) and sauger (*Sander canadensis*) (Bozek et al. 2011). Pikeperch typically do not display sexually-dimorphic growth (Korbuly et al. 2007) and display parental care (Marshall 1977). As well, male-biased SSD has been documented in some darter species (George et al. 1996; Johnston and Haag 1996; Kelly and Alonzo 2011; Hughey et al. 2012 and references therein), some of which also exhibit paternal care (DeWoody et al. 2000; Kelly et al. 2012; Harrington et al. 2013).

Currently, there are three leading hypotheses to explain female-biased SSD in percid fishes: (1) males have smaller size as a result of decreased consumption and activity as a predator avoidance mechanism (Roff 1983), (2) males have smaller size due to increased allocation of energy to activity costs associated with spawning (Henderson et al. 2003), and (3) females have larger size due to more efficient foraging on large-particle-sized prey (Lepak et al. 2012). Although these have been described as competing hypotheses (Stacey and Lepak et al. 2012), they may not necessarily be mutually exclusive.

Direct tests of these hypotheses in the recent literature tend to support hypothesis 1 (reduced male activity and consumption). Rennie et al. (2008) compared resource allocation and consumption between male and female yellow perch and walleye populations. Specifically, they used a BGM (Sect. 19.2) to compare sex differences in reproductive investment (g from Eqs. 19.3, 19.4, and 19.5), and a combined contaminant-bioenergetic model to compare sex differences in consumption and total metabolic costs. Results showed greater g in males vs. females (1.3, 1.2 times greater on average for walleye and yellow perch, respectively). As noted previously (Sect. 19.2), g is a close approximation of gonadal investment in females (Shuter et al. 2005), but its meaning in male fish is not well characterized. If we assume that g includes reproductive costs beyond investments in gonadal tissue (e.g., spawning activity, competition for mates), then a higher g for males supports hypothesis 2 (Henderson et al. 2003). However, bioenergetic results were more consistent with hypothesis 1 (Roff

1983) in that male perch had lower consumption and total metabolic costs compared with females. Further, these differences were such that females were still afforded a greater growing efficiency (Rennie et al. 2008). While Rennie et al. (2008) highlighted the need for sex-specific bioenergetic parameters, their analytical approach (i.e., the use of mass-relative, lakes-specific energetic residuals) best reflects potential sex differences among perch in the absence of sex-specific model parameters. Another experimental study reported increased feeding activity of female European perch with increasing water clarity, whereas male feeding activity remained the same across water clarity treatments, presumably to avoid predation risk (Horppila et al. 2011). Greater SSD was observed in clear water treatments.

Sexual differences in metabolic costs among percid fishes are supported by at least two recent studies. Sexually mature male walleye collected in fall samples had 25 % greater scope for anaerobic activity (as indicated by lactate dehydrogenase enzyme assays, Kaufman et al. 2006). Another study demonstrated a higher scope for anaerobic activity in male yellow perch in spring (e.g., during spawning), as well as summer and fall (Schoenebeck and Brown 2012). However, previous work examining two independent measures of activity found no significant differences between male and female yellow perch collected from two populations in late summer (Rennie et al. 2005).

Despite these two studies reporting significant differences in putative activity rates between male and female fish, neither appears to support hypothesis 2 (higher male activity costs during spawning; Henderson et al. 2003), as both studies indicated higher male activity outside of the spawning season. If anything, Kaufman et al. (2006) and Schoenebeck and Brown 2012 support increased activity in males *generally* as a possible explanation for reduced growth efficiency and smaller size-at-age compared to female perch and walleye.

An alternative test of hypothesis 2 (and that of higher male activity in percids generally as an explanation for female-biased SSD) is an examination of sex ratios from fish captured using passive sampling. Fishes with different activity rates are predicted to have different encounter rates with passive sampling methods (Rudstam et al. 1984). Thus, if males are more active during spawning (Henderson et al. 2003), or generally (Schoenebeck and Brown 2012), we expect male-biased catches in passive sampling gear deployed during those periods. Two studies report male-biased sex ratios during spawning in yellow perch (Henderson et al. 2000) (Fig. 19.2) and European perch (Olin et al. 2012), but not at other times of the year. Outside of the spawning season, only 27 % of Ontario perch populations exhibited male-bias in sex ratio (Appendix 1 in Rennie et al. 2008, Fig. 19.3). Female-biased sex ratios for yellow perch outside of the spawning season may be the norm in the Great Lakes (Henderson et al. 2000), but may become more male-biased under high fishing effort (Lauer et al. 2008).

Spawning sex ratios in walleye appear to tell a different story. Koupal et al. (1997) compared passive (gillnetting) with active (electroshocking) fishing methods on the sex ratio of walleye collected during spawning in two Colorado reservoirs. They found that females were two to three times more common in gillnets (passive sampling) than in electroshock catches (active sampling) over the same spawning

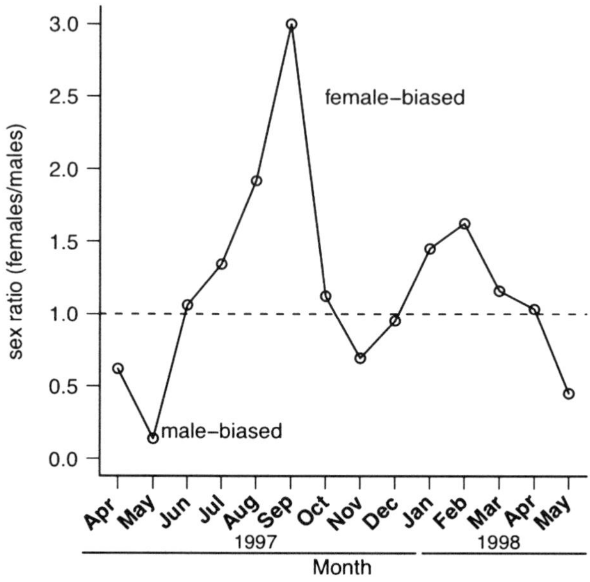

Fig. 19.2 Sex ratio (male:female) in Lake Erie index nets, April 1997–May 1998. Ratios are based on totals of age 3 and 4 perch, presented in Table 19.1 of Henderson et al. (2000)

shoals, and that males were far more common in electroshock catches. These differences in catch rates indicate that males on spawning shoals may actually be *less* active than females. During non-spawning times, 60 % of Ontario walleye populations were male-biased (Appendix 1 in Rennie et al. 2008, Fig. 19.3).

If sex ratios of percids captured in passive fishing gear reflects differential activity, and if these activity-based differences in sex ratios are linked to dimorphic growth, then SSD should be more exaggerated in populations with male-biased sex ratios. To evaluate this hypothesis, we re-evaluated previously published data by Rennie et al. (2008, Appendix 1), where fish were captured during autumn in passive fishing gear (gillnets). We detected no significant relationship between sex ratio (ratio of female:male in sample) and SSD (ratio of female:male size) in yellow perch (Fig. 19.3a). Contrary to predictions, walleye SSD *decreased* significantly as the sex ratios of populations were more male-biased (Fig. 19.3b); SSD in male-biased walleye populations (female:male ratio <1) was slightly lower than in female-biased populations (female:male ratio >1), but not significantly so (*t*-test, $t_{30.6}=1.85$, $P=0.07$; mean SSD for male-biased sex ratio= 1.07, mean SSD for female-biased sex ratio= 1.11). Neither of these results supports the hypothesis that greater male activity (as reflected in gillnet encounter rates) is the driver of female-biased sexual dimorphism in percids.

Hypothesis 3 (females are larger because they forage more efficiently on large prey) was recently proposed to explain sex-specific differences in mercury concentrations of walleye in two Colorado reservoirs (Lepak et al. 2012). Lepak et al. (2012) reported that females grew more efficiently due to their consumption of

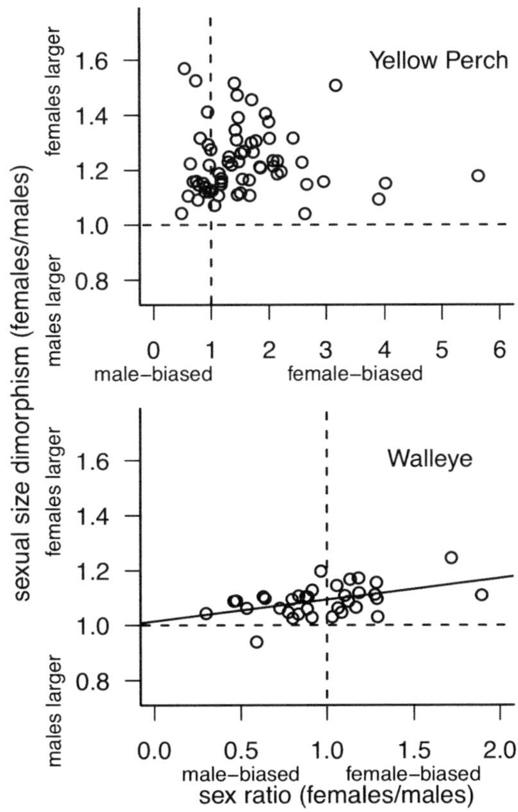

Fig. 19.3 Sexual dimorphism (expressed as the ratio of male:female mean size) as a function of sex ratio (number of male:female encountered during survey). *Top panel*, yellow perch, *bottom panel*, walleye. Regression in *bottom panel* is significant ($F_{1,33}=8.96$, $P=0.005$) (Data from Rennie et al. 2008, Appendix 1)

larger-sized prey. However, it is difficult to know whether sex-specific diets are driving SSD or if the opposite is in fact true: are females larger because they consume larger prey, or do females consume larger prey because they themselves are larger? Feeding in fishes is limited by gape, which varies with body size. Thus, larger fish would be expected to have access to larger-sized particles, if present, regardless of their sex. While Lepak et al. (2012) included body size as a covariate in their analysis of diet differences between male and female walleye, there is very little overlap between the covariate (weight) and the independent variable (sex; Lepak et al. 2012, their Fig. 19.4). A major assumption of ANCOVA and other tests that include covariates is that the value of the covariate must not differ with respect to the independent variable (Quinn and Keough 2002). A more rigorous test of this hypothesis (SSD resulting from differences in diet between sexes, leading to increased growth efficiency in females) would be to compare males and females over more similar size ranges, or use a different size covariate that permits greater overlap between sexes.

A fourth hypothesis for SSD that does not seem to have been considered in the percid literature is that of sex-specific habitat segregation. Sex-specific habitat segregation is common among other taxa that show SSD (Ruckstuhl 2007), and could

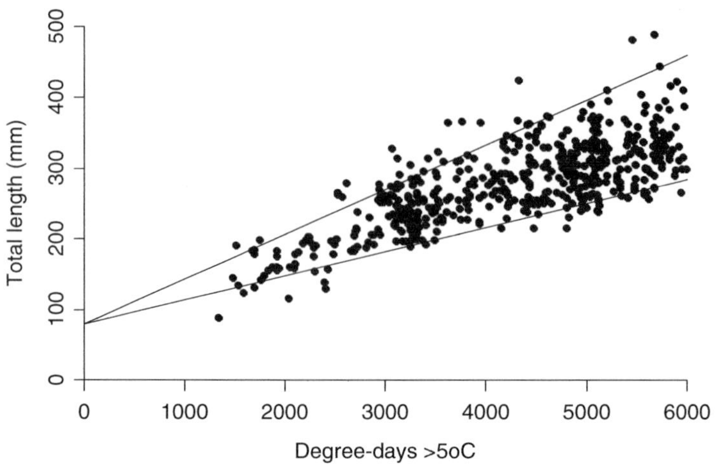

Fig. 19.4 Immature length of female walleye versus cumulative degree-days >5 °C in 85 populations. Each point ($n=551$) is an age class that was observed during a sampling year. We assumed that an age class was immature if cumulative degree-days were <6000 °C·days (Venturelli et al. 2010). *Upper* and *lower lines* are 5 % and 95 % quantile regression lines assuming a common intercept of 80 mm (Lester et al. 2014). These lines imply growth rates of 0.03 and 0.06 mm/°C·day, respectively (Walleye data are from Venturelli et al. (2010) and Colby et al. (1979) and degree-day data are from Venturelli et al. (2010) and NOAA (2013) at a base temperature of 5 °C (Chezik 2013))

be even more significant among percids and other fishes if males and females distribute themselves at different depths. Sex-specific habitat selection during the growing period could lead to sexual dimorphism in diets (sensu Lepak et al. 2012). Further, because growth is temperature-dependent, sexual segregation in stratified lakes could lead to SSD via sex-specific thermal regimes. If males and females did have different thermal preferences, then SSD should be more exaggerated in stratified lakes than in polymictic lakes. To test this hypothesis, we reexamined the data in Rennie et al. (2008, their Appendix 1) and classified lakes as either stratified or polymictic. Thermocline depth was estimated using lake area (Hanna 1990), and a lake was defined as polymictic if the estimated thermocline depth exceeded the mean lake depth. A *t*-test comparing perch SSD between lake types showed that SSD was slightly less prominent in polymictic lakes (female:male size ratio = 1.19) than in stratified lakes (1.23), but the difference was not significant ($t_{7.7}=0.9$, $P=0.2$, one-tailed test). This lack of significance may be because most lakes in this dataset were stratified, and were not selected to specifically address this question. Other studies have demonstrated that water clarity contributes to the degree of SSD in European perch via interference with feeding activity (Horppila et al. 2011), which may also play a role in reduced SSD in turbid, polymictic lakes.

Regardless of the proximate mechanism, the ultimate cause of SSD in percids very likely results from the action of sex hormones, and their influence on the behaviour and metabolism of percid fishes. Sexually dimorphic growth in percids does not appear until maturity (Rennie et al. 2008). Laboratory experiments clearly demonstrate positive effects of estrogen and negative effects of androgen hormones on growth of both yellow perch (Malison et al. 1985, 1988) and European perch (Mandiki et al. 2004, 2005). What remains unresolved in both wild and culture settings is the degree to which these responses to hormones are behavioural (influencing foraging efficacy and field rates of activity) or strictly metabolic (increased/decreased rates of basal metabolism). Evidence to date seems to suggest that a combination of both behavioural and metabolic differences are responsible.

19.4 Extrinsic Factors Affecting Growth

19.4.1 The Thermal Environment

Temperature is arguably the most important determinant of growth rates in fishes (Paloheimo and Dickie 1966; Fry 1971; Kitchell et al. 1977). Temperature affects fish growth directly through metabolic rates and processes (Chap. 15), and indirectly through the rate of food production. Direct effects of temperature on percid growth are well documented, especially early in life. For readers interested in minimum, maximum, and optimum temperatures for percid growth, we recommend existing reviews by Thorpe (1977) and Colby et al. (1979).

In this section, we use degree-days to describe the growth patterns of percid fishes. The degree-day is an agricultural index of the ambient thermal energy that is available to growth over some period of interest. Degree-day calculations include upper and lower temperature thresholds that constrain the thermal sum to temperatures that are relevant to growth. However, it appears that a precise estimate of the lower threshold is only necessary when evaluating growth across a wide thermal range (e.g., broad spatial scales, large differences in culture conditions) (Chezik 2013). We also suspect that the upper threshold is only necessary when fish experience prolonged thermal stress.

Degree-days are useful for describing general growth patterns because they (i) are more physiologically relevant than calendar time (e.g., age in years) and therefore explain more variation in growth (Neuheimer and Taggart 2007; Venturelli et al. 2010), (ii) simplify the job of estimating how other factors have contributed to growth (e.g., consumption, reproductive status, genetics) (Mooij et al. 1994, Venturelli et al. 2010), and (iii) facilitate the comparison and aggregations of growth data over broad spatial and temporal scales (Colby and Nepszy 1981).

To date, 16 studies involving four percid species have used degree-days to gain insight into growth patterns and processes (Table 19.2). The most comprehensive of these studies (Bozek et al. 2011) summarized data from 431 walleye populations spanning 1000–4630 degree-days >5 °C (~13 degrees of latitude). This analysis indicated that walleye growth over this range was remarkably consistent and characterized by (i) strong temperature-dependence in the first year of life, (ii) linear

Table 19.2 Summary of percid studies that have used degree-days to gain insight into growth patterns and processes

Species	Study	Study lake(s)	Findings
Perca flavescens	Mills et al. (1989)	Oneida Lake (New York, USA)	Age-0 growth limited by temperature in cold years, prey in warm years
	Power and Van Den Heuvel (1999)	4 lakes in Alberta and Manitoba, Canada	Temperature dominates seasonal patterns of age-0 growth
	Chong (2000)	12 lakes in Ontario, Canada	Age-0 growth per unit temperature increases with latitude
	Purchase et al. (2005b)	72 lakes in Ontario, Canada	Maximum size negatively related to length and strength of growing season
	Tardif et al. (2005)	Lake Saint-Pierre (Quebec, Canada)	Age-0 growth limited by temperature in some years, other factors in others
Perca fluviatilis	Le Cren (1958)	Lake Windermere (Westmorland, UK)	The strength of the growth-temperature relationship varies with age
	Goldspink and Goodwin (1979)	4 lakes in the UK	Lifetime growth less sensitive to temperature when food limiting
	Le Cren (1992)	Lake Windermere (Westmorland, UK)	Adult growth controlled by temperature, not density
	Houthuijzen et al. (1993)	Volkerak Lake (Noord-Brabant, The Netherlands)	Immature ages fed zooplankton and macrofauna grow quickly and large at low densities
	Mooij et al. (1994)	Tjeukemeer Lake (Friesland, The Netherlands)	Age-0 growth not food limited
	Mooij (1996)	Tjeukemeer Lake (Friesland, The Netherlands)	Temperature affects recruitment through effects of age-0 growth on survival
	Romare (2000)	Lake Dagstorpssjon (Scania, Sweden)	Age-0 growth limited by diet in mid-summer, density in late summer
	Tolonen et al. (2003)	Lake Ainijarvi (Lappi, Finland)	Age-0 growth density-dependent, even at the northern edge of the range
	Heynen et al. (2011)	Lake Speldrop (North Rhine-Westphalia, Germany)	Age-0 growth delayed but final size greater when fed fish instead of zooplankton

(continued)

Table 19.2 (continued)

Species	Study	Study lake(s)	Findings
Sander lucioperca	Mooij et al. (1994)	Tjeukemeer Lake (Friesland, The Netherlands)	Age-0 growth food limited in some years
	Mooij (1996)	Tjeukemeer Lake (Friesland, The Netherlands)	Temperature affects recruitment through effects of age-0 growth on survival
Sander vitreus	Venturelli et al. (2010)	8 lakes in Ontario and Quebec	Immature growth response to abundance changes consistent across 8 populations; adult growth sexually dimorphic
	Bozek et al. (2011)	431 lakes in Canada and USA	Age-0 growth mostly explained by temperature; growth patterns similar across species range; maximum size unaffected by length and strength of growing season

juvenile growth (~0.06 mm/dd), (iii) non-linear and sexually-dimorphic adult growth, and (iv) an asymptotic length that is unaffected by latitude. However, these data also showed twofold variation in immature growth rate that was independent of temperature (Fig. 19.4). This residual variation is likely due to among-population differences in population density (e.g., Houthuijzen et al. 1993; Tolonen et al. 2003; Venturelli et al. 2010) and food (e.g., Mills et al. 1989; Goldspink and Goodwin 1979; Mooij et al. 1994), which are factors that we review in more detail below (Sects. 19.4.3 and 19.4.4).

Analyses using degree-days also suggest that percids exhibit countergradient growth variation (e.g., Chong 2000; Purchase et al. 2005b). Countergradient growth variation is the tendency of a species to compensate for shorter growing seasons (i.e., at higher latitudes or elevations) by growing more quickly (Conover and Present 1990). We analyzed published data from yellow perch, European perch, and walleye and found positive relationships between age-0 growth rate and latitude for all three species (Fig. 19.5). The relationship for walleye was relatively weak, at least in part because of a gill net bias that selects for the fastest growing members of the youngest age classes (Walker et al. 2013). An analysis of all ages classes for which data were available show that countergradient growth in walleye persists well into adulthood (Table 19.3), but with the acknowledgement that adult growth patterns may be confounded by among-population differences in investment in reproduction. Countergradient growth variation has been observed in at least 15 non-percid fishes (Power and McKinley 1997; Conover et al. 2009; Chavarie et al. 2010; Rypel 2012a, b). Because countergradient growth is adaptive (Conover et al. 2009), latitudinal analyses to determine the genetic capacity for growth in percids (e.g., Zhao et al. 2008) have implications for both resource management and broodstock selection. However, given the potential for thermal adap-

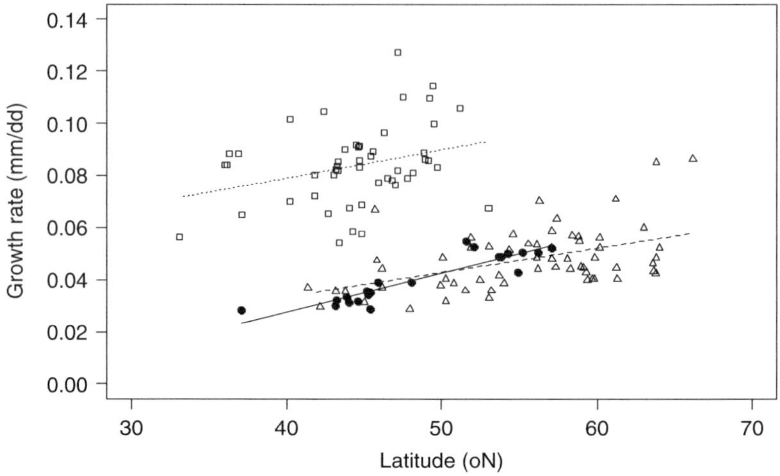

Fig. 19.5 Re-analyses of published data showing evidence for countergradient growth variation in the first year of life of yellow perch (*closed circles, solid line*) (Power and Van den Heuvel 1999), European perch (*open triangles, dashed line*) (Heibo et al. 2005), and walleye (*open squares, dotted line*) (Colby et al. 1979; Venturelli et al. 2010) (We obtained degree-day data from Venturelli et al. (2010), or calculated degree-days from online temperature data (McKenney et al. 2006; Haylock et al. 2008) using a base temperature of 5 °C for walleye and 10 °C for perch (Chezik 2013)). All linear regressions were significant (a: $F_{1,19}=87.71$, $P=1.50\cdot10^{-8}$, $r^2=0.82$; b: $F_{1,59}=19.95$, $P=3.66\cdot10^{-5}$, $r^2=0.25$; c: $F_{1,48}=5.03$, $P=2.96\cdot10^{-2}$, $r^2=0.09$)

Table 19.3 Strength of the linear relationship (r^2) between walleye growth rate (mm/°C·day) and latitude (°N) over the range of walleye ages for which data were available. Preliminary analyses found relatively weak evidence for a log-linear model at any age. Age-1 data are plotted in Fig. 19.5 (Walleye data are from Venturelli et al. (2010) and Colby et al. (1979), and degree-day data are from Venturelli et al. (2010) and NOAA (2013) at a base temperature of 5 °C (Chezik 2013))

Age (years)	*N* populations	Latitude range	Intercept	Slope	r^2	*P*
1	50	33.1–53.0	$3.50\cdot10^{-2}$	$1.10\cdot10^{-3}$	0.09	$2.96\cdot10^{-2}$
2	166	33.1–53.0	$6.21\cdot10^{-3}$	$6.48\cdot10^{-4}$	0.12	$1.53\cdot10^{-6}$
3	258	33.1–53.0	$2.33\cdot10^{-3}$	$3.84\cdot10^{-4}$	0.11	$4.27\cdot10^{-9}$
4	253	33.1–53.0	$6.09\cdot10^{-5}$	$2.85\cdot10^{-4}$	0.15	$8.17\cdot10^{-12}$
5	215	33.1–53.0	$2.34\cdot10^{-4}$	$2.06\cdot10^{-4}$	0.13	$9.63\cdot10^{-9}$
6	170	33.1–53.0	$-1.74\cdot10^{-3}$	$1.96\cdot10^{-4}$	0.24	$6.86\cdot10^{-13}$
7	123	33.1–53.0	$-1.81\cdot10^{-3}$	$1.66\cdot10^{-4}$	0.30	$1.54\cdot10^{-12}$
8	91	33.1–53.0	$-1.02\cdot10^{-3}$	$1.24\cdot10^{-4}$	0.31	$1.72\cdot10^{-10}$
9	63	36.0–53.0	$-1.16\cdot10^{-3}$	$1.09\cdot10^{-4}$	0.37	$2.52\cdot10^{-8}$
10	31	36.0–53.0	$-1.65\cdot10^{-3}$	$1.08\cdot10^{-4}$	0.51	$5.07\cdot10^{-7}$
11	20	41.8–52.3	$-1.35\cdot10^{-3}$	$8.97\cdot10^{-5}$	0.40	$5.55\cdot10^{-4}$
12	13	43.4–52.3	$-2.75\cdot10^{-3}$	$1.11\cdot10^{-4}$	0.34	$2.26\cdot10^{-2}$

tation at local or regional scales (Angilletta 2009), results of latitudinal analyses that assume a fixed growth-temperature curve should always be corroborated by common-garden experiments.

19.4.2 Predation

Mortality rate plays a major role in shaping the life histories and growth rates of fishes, including percids (Sect. 19.3.1). In nature, predation is a significant source of mortality in fishes, and can play a substantial role in shaping the life histories of prey (Abrams and Rowe 1996). For small percids and the early life stages of large percids, predation and inter-cohort cannibalism can contribute significantly to total mortality (Frankiewicz et al. 1999; Nielsen 1980; Persson et al. 2000; Post and Evans 1989; Szalai and Dick 1991).

Predation applied to immature life stages of prey may have a very different effect on prey growth and body size than when applied to adult life stages. When mortality is imposed on large adults, life-history theory predicts faster growth, earlier age/shorter length at maturity and smaller asymptotic length. However, in a stage-dependent model like the BGM, early growth is a function of reproductive investment and asymptotic length (Eq. 19.3). Thus, if early growth is rapid to escape size-dependent predation, asymptotic length is predicted to increase but age at maturity remains a function of adult mortality (Shuter et al. 2005). Both of these outcomes are supported by predictions of an evolutionary life history model that imposed mortality at both early (small) and late (large) stages of life (Dunlop et al. 2007). This evolutionary model predicted faster somatic growth rates, increased size at maturation and increased asymptotic length with an increase in mortality imposed on immature age-0 fishes.

Observations in the field appear to match these theoretical and modeled predictions closely (Rennie et al. 2010). In yellow perch, higher predator densities resulted in larger asymptotic length, later maturation and maturation at larger sizes (Fig. 19.6). Individual growth rates also appeared to increase with higher predator densities (Rennie et al. 2010).

Predator mortality can affect prey growth rates, but so too can be the threat of predator mortality. Commonly, prey adjust their behaviour to the perceived threat of predation. One strategy is to reduce activity so as to avoid detection (e.g., Peacor 2002). Reduced activity can translate into reduced foraging (Peacor 2002), though the indirect responses on prey resources and growth rates as a result of this strategy may vary (Abrams and Rowe 1996). A second strategy among prey populations facing gape-limited predators is to grow rapidly through the range of vulnerable sizes (Biro et al. 2005; Urban 2007). Increased foraging (and therefore exposure to predators) may be required to maintain rapid growth, but with the benefit of escaping a predator window (Urban 2007).

Risky behaviour so as to grow rapidly through a predator window has been referred to in the literature as "boldness" or "bold personality;" that is, an apparent disregard

Fig. 19.6 Relationships between an index of *Perca flavescens* predation {estimated as the log10-transformed $[\log 10(x+1)]$ mean catch per unit effort of *Esox lucius*, *Sander vitreus* and *Micropterus dolomieu* combined} with *P. flavescens* (**a**) asymptotic size (LF), (**b**) LF at maturity and (**c**) age at maturity for females (filled symbols, *solid lines*) and males (open symbols, *dashed lines*). Only significant relationships are shown (Reproduced with permission from Rennie et al. 2010)

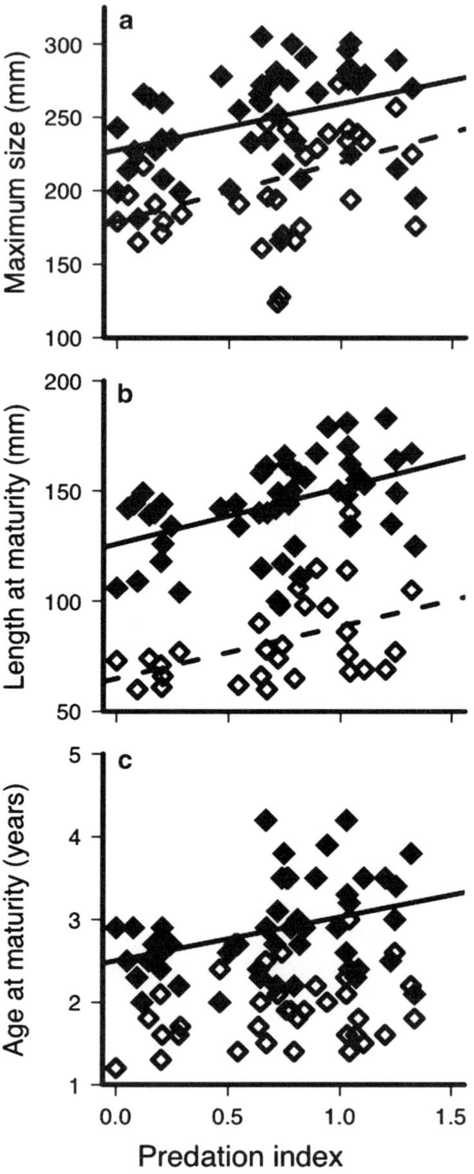

for predation risk to obtain food for growth. This behaviour has been documented in salmonid aquaculture strains (e.g., Biro and Post 2008) and among European perch populations (Sect. 15.5, Magnhagen et al. 2012 and references therein). Although boldness may seem to be a counterproductive means of avoiding predation, it can be an effective strategy for obtaining food if individual predation risk can be limited by schooling. Hellstrom et al. (2011) found that the degree of boldness observed in the presence of perceived predation risk is more common in larger groups of European perch. Further, there is evidence that risk-taking behaviour is plastic among European perch, reflecting the immediate perceived risk of predation (Magnhagen and Bocherding 2008; Magnhagen et al. 2012). By contrast, decreased swimming activity in the presence of predators seems to be the common response to predation risk among percids; European perch exposed to a perceived risk of predation reduced their swimming activity (Vainikka et al. 2005), and modelled estimates of yellow perch activity declined with increasing predator densities in lakes (Rennie et al. 2010).

It is important to note that activity also represents a metabolically costly endeavour (Weatherley 1966); given equal rates of consumption, increased activity means less energy available for growth (Johansson and Andersson 2009). Rennie et al. (2010) found that yellow perch growth rates and conversion efficiencies (rates of growth per unit food consumed) increased as the risk of predation increased, and that this was primarily due to reduced activity at higher predator densities. This relationship was present despite a reduction in consumption rates at higher predator densities. Similarly, an individual-based model found that roach (*Rutilus rutilus*) incurred higher activity costs and consumed less energy in the presence of predatory pikeperch (Holker and Mehner 2005).

Decreased activity may also result in increased growth rate by indirectly stimulating resource abundance. In a mesocosm experiment, Peacor (2002) found that the tadpoles (prey) reduced their activity in response to dragonfly larvae (predators). However, this reduction in activity released primary producers from tadpole grazing, to the point that less active tadpoles consumed more food than tadpoles that were more active. Ultimately, growth rates of low-activity tadpoles were greater in tanks with a perceived threat of predation than those with no predators present.

The growth response of prey to predators can be further modulated by the presence of a refuge. Diehl and Eklov (1995) reported that increased refuge use by European perch in the presence of predators resulted in slower growth. Similarly, Shoup et al. (2012) documented reduced growth rates of bluegill sunfish (*Lepomis macrochirus*) in refugia with no predators present. These findings imply a tradeoff between resource availability and protection from predators. However, other authors report no significant effect of refuge use on growth rates of juvenile European perch (Persson and Eklov 1995). If resources become too limiting, prey may abandon refugia and adopt more bold feeding strategies in the presence of predators (Magnhagen and Borcherding 2008). Declining resources have been found to have a positive effect on the activity rates of European perch (Olsson et al. 2007).

19.4.3 Density Dependence

Density-dependent growth is a change in individual growth rates that occurs in response to a change in population (or cohort) abundance, relative to carrying capacity. Density-dependent growth is easy to imagine in a culture setting in which fish density is allowed to vary in a fixed volume of water that receives a constant ration. Because aquaculture seeks to maximize biomass production at a given ration, the role that density plays in determining growth rates is well studied.

Density-dependent growth in nature is expected but this phenomenon can be difficult to observe and quantify in nature (Rose et al. 2001). We expect growth to change with population abundance because plasticity in growth can play an important role in determining how quickly a population recovers from a catastrophic event or to what extent a population compensates for increased mortality due to exploitation (Lester et al. 2014). However, attributing an observed change in growth to a change in population abundance relative to carrying capacity is difficult because (i) population abundance and carrying capacity are both variable and difficult to estimate; (ii) growth measurements can be biased by selectivity, both during a sampling event and via exploitation; and (iii) growth itself is influenced by numerous direct and indirect food-web linkages of varying strength. Our ability to detect density-dependent growth also depends on the extent to which data are available from both extremes of abundance. As a result of these challenges, examples with direct evidence of density-dependent growth in nature are rare (Rose et al. 2001).

The indirect evidence for density-dependent growth in walleye populations suggests that immature growth in length can approximately double in response to a large change in abundance. Much of this evidence is from heavily exploited lakes and is summarized in Bozek et al. (2011). The largest growth responses have been observed in Escanaba Lake (Kempinger and Carline 1977), Lake Erie (Spangler et al. 1977; Venturelli et al. 2010), and Red Lakes (Ostazeski and Spangler 2001). However, a 1.3-fold increase in growth in length appears to be more common, as evidenced by numerous single-lake studies (e.g., Anthony and Jorgensen 1977; Shuter and Koonce 1977; Reid and Momot 1985; Muth and Wolfert 1986; Schueller et al. 2005) and one meta-analysis (Venturelli et al. 2010). Density-dependent growth variation of this magnitude is also evident in young-of-the-year pikeperch (Table 19.4, Buijse and Houthuijzen 1992).

The indirect evidence for density-dependent growth in wild perch populations is also compelling (Table 19.4). Like walleye, both *Perca* spp. appear capable of a twofold increase in growth in length following large declines in population abundance (e.g., Linlokken and Seeland 1996; Ostazeski and Spangler 2001; Headley and Lauer 2008; Irwin et al. 2009). Large growth responses are evident in all perch life stages from YOY to juvenile to adult (Table 19.4). In perch, these growth responses may result indirectly from density-dependent habitat selection whereby crowding forces some perch to prey on sub-optimal, pelagic prey (e.g., Post et al. 1997; Svanback and Persson 2009; Bacheler et al. 2011). This mechanism has also been observed in pikeperch (Frankiewicz et al. 1999) and may be a general phenomenon among percids. Although the exact relationship between growth and population abundance in the wild is unknown, the form of this relationship is likely log-linear. LeCren (1958) found that the length increments of adult perch in Lake Windermere increased more than sevenfold as the mean number of adults per trap decreased from 610 to 10, but half of this growth response occurred at <100 adults per trap.

Table 19.4 Evidence for density-dependent growth in wild populations of yellow perch, European perch, and pikeperch (our calculations). See Bozek et al. (2011) for a summary of density-dependent growth in walleye

Species	Study	Study lake(s)	Evidence for abundance change	Evidence for growth change
P. flavescens	Beckman (1950)	2 lakes in Michigan, USA	Severe winterkill (abundance not quantified)	1.61- and 1.35-fold increases in mean growth (weighted average across age classes)
	Bardach (1951)	Lake Mendota (Wisconsin, USA)	0.33-fold decrease in gill net CPUE	1.4-fold increase in mean adult length
	El-Zarka (1959)	Saginaw Bay (Lake Huron, USA)	An estimated 19-fold increase in trap net CPUE (1929–1955)	0.75-fold decrease in age-3 length
	Ivan et al. (2011)	Saginaw Bay (Lake Huron, USA)	0.1-fold decrease in trawl CPUE (1970–2008)	1.2-fold increase in age-2 length
	Kempinger et al. (1982)	Nebish Lake (Wisconsin, USA)	6.5-fold increase in estimated biomass	0.65-fold decrease in age-3 length
	Henderson (1985)	South Bay (Lake Huron, Canada)	0.04-fold decrease in age-4+ trap CPUE	1.4-fold increase in age-3 to age-5 length
	O'Gorman and Burnett (2001)	Lake Ontario (New York, USA)	30-fold difference in YOY trawl CPUE	1.4-fold difference in YOY length
	Ostazeski and Spangler (2001)	Red Lakes (Minnesota, USA)	0.36-fold decrease in gill net CPUE	0.53-fold decrease in age 1–6 growth
	Pierce et al. (2006)	Lake Thirteen (Minnesota, USA)	Ninefold increase in electrofishing CPUE	0.65-fold decrease in age-3 length
	Headley and Lauer (2008)	Lake Michigan (Indiana, USA)	0.01-fold decrease in trawl CPUE	1.8-fold increase in immature growth rate
	Irwin et al. (2009)	Oneida Lake (New York, USA)	0.1-fold decrease in density	Twofold increase in larval growth rate

(continued)

Table 19.4 (continued)

Species	Study	Study lake(s)	Evidence for abundance change	Evidence for growth change
P. fluviatilis	LeCren (1958)	Lake Windermere (Westmorland, UK)	At least a 0.25-fold decrease in perch trap CPUE (<1941 vs. 1952)	1.4-fold increase in age-4 length
	Craig (1982)	Lake Windermere (Westmorland, UK)	0.1-fold decrease in trap CPUE of females (1955 vs. 1968)	1.2-fold increase in age-5 female length
	Linlokken and Seeland (1996)	Munksjoen (Hedmark, Norway)	0.3-fold decrease in biomass density	Twofold increase in adult instantaneous growth
	Persson et al. (2000)	Abborrtjarn (Vasterbotten, Sweden)	0.03-fold decrease in age-2+ abundance	1.6-fold increase in adult length
S. lucioperca	Buijse and Houthuijzen (1992)	Lake Ijssel (central Netherlands)	115-fold variation in YOY trawl CPUE	1.2-fold variation in YOY length

It is important to note that almost every study supporting density-dependent growth in wild percid populations also acknowledges one or more confounding variables. These include changes in temperature, nutrient status, pollution, size-selective exploitation, catchability, invasive species, and the fish community (e.g., predators and competitors). Indeed, the presence of most if not all of these confounds in the Great Lakes provide a complicated backdrop against which to interpret patterns of percid growth (e.g., El-Zarka 1959; O'Gorman and Burnett 2001; Headley and Lauer 2008). Similarly, patterns of yellow perch abundance and growth are often tied up in competitive and predatory interactions with walleye (Rose et al. 1999; Ostazeski and Spangler 2001; Pierce et al. 2006; Irwin et al. 2009; Ivan et al. 2011). Competition and predation among cohorts can also complicate growth responses. For example, although adult perch length in a Swedish lake increased 1.6-fold in response to a two-orders-of-magnitude decline in age-2+ abundance (Persson et al. 2000), this growth response may have been due to a two-orders-of-magnitude increase in YOY perch (an important food item for cannibalistic adults).

19.4.4 Food Availability and Size

Ultimately, the food energy that is consumed by a fish is a function of the quality and quantity of the food that is available to that fish. In the previous Sect. (19.4.3), we described how percid growth is affected by per-capita food availability. In this section, we describe how percid growth is affected by changes in food quality and quantity that are independent of the density of conspecifics.

Because percids are visual predators, spatially complex or turbid environments can impair foraging efficiency. For example, the growth of pikeperch was inhibited in a turbid, Finnish lake (Vinni et al. 2009). Other studies that did not measure growth point to impaired foraging efficiency as a mechanism of growth inhibition: Bartels et al. (2012) reported that European perch consumed smaller prey items as water clarity decreased. Similarly, Wellington et al. (2010) reported decreased consumption of larval and juvenile perch in the presence of phytoplanktonic turbidity. Fullerton and Lamberti (2006) reported that yellow perch in the absence of predators foraged more efficiently in open water than in macrophytes. However, other studies have shown that foraging in complex environments is adaptive when predators are present (Svanback and Eklov 2004).

Growth efficiency depends, not only on rates of energy acquisition and expenditure (e.g., consumption and activity), but also on the size spectrum of prey available for consumption (Sherwood et al. 2002). In environments in which food is limiting (arguably representative of most instances in nature), prey sizes can have significant effects on growth efficiency and growth rates. This process has been well-documented in North American percids. Walleye populations feeding on lake cisco (*Coregonus artedi*; a large-bodied prey fish) grew faster and larger than those feeding on yellow perch (a comparatively small-bodied prey fish) (Kaufman et al. 2006). After controlling for body-size, it was also shown that walleye populations feeding on cisco suffered lower activity costs, suggesting that the increased rates of growth were a direct result of more efficient foraging (i.e., less energy expended and more energy gained by searching for, capturing, and consuming larger prey).

Similarly, yellow perch activity rates increase with body size until an ontogenetic diet shift to larger prey types (e.g., zooplankton to benthos, benthos to fish) (Heath and Roff 1996; Sherwood et al. 2002; Iles and Rasmussen 2005). The authors suggest that the inability to transition to larger particle sizes represents an energetic bottleneck, and places a limit on growth and maximum size in some populations. Other studies have proposed the same mechanism (lack of large prey) in contributing to stunting in both pikeperch (van Densen et al. 1996; Vinni et al. 2009) and European perch (van Densen et al. 1996).

19.4.5 Parasitism/Disease in Natural Settings

Parasitism and disease can have significant impacts on individual growth rates of both wild and cultured populations of percids (Craig 1987; Grignard et al. 1996). In culture conditions, severe bacterial infection can prove disastrous, whereas the extent to which parasites cause concern depends on parasite load (Grignard et al. 1996). Here we briefly review the existing literature regarding the effects of parasites and diseases on percid growth. An exhaustive list of percid parasites can be found in reviews by Craig (1987) and Grignard et al. (1996).

Diseases (i.e., viral or bacterial infection) can have a significant impact on survival, but few studies have assessed impacts on individual growth rates. In one well-studied population, 98 % of the estimated European perch population in Lake Windermere were eliminated by a pathogen-related infection (Craig 1987). However, total perch biomass declined only slightly because compensatory individual growth among the surviving juveniles was so high. Simulation modeling suggests that this increase in individual growth rates was due to reduced intraspecific competition (Ohlberger et al. 2011).

Reports of the effects of parasitic infections on fishes vary, and in some cases include positive effects on growth (Arnott et al. 2000; Voutilainen et al. 2012). Table 19.5 summarizes the literature on the effects of parasites on percid growth. Of the nine parasites listed, all are involved in at least one case of a negative effect. Neutral effects were observed in four cases and positive effects were observed in three cases. Ryman et al. (2008) reported a slight but significantly higher condition factor in perch infected with *Apophallus brevis* that was unrelated to the mass of the parasites themselves. Cloutier et al. (2012) reported increased condition with parasitism by *Ichthyocotylurus spp.*, but the relationship appears to be driven by a single datapoint with high leverage (see their Fig. 3) and it is not clear whether parasite mass was accounted for. Johnson and Dick (2001) report increased reproductive allocation (GSI) under high loads of the parasite *Glugea spp.* Increased allocation of energy to reproduction vs. somatic growth in infected individuals is likely a response to maximize fitness when survival is in jeopardy.

Reduced growth rates in parasitized percids may be a result of decreased growth efficiency. Perch with heavy parasite loads were observed to have higher concentrations of MeHg (Ryman et al. 2008), which the authors suspected might be due to spatial differences in MeHg loading where fish were collected. An alternative explanation is that these fish had to consume more food (and thus retained more MeHg) in order to achieve the same body size as conspecifics with lower parasite loads. Other studies have reported a metabolic cost of carrying parasites (e.g., Seppanen et al. 2008). Additionally, Johnson et al. (2004) reported that parasites were important determinants in the nitrogen stable isotope signatures of hosts, which are also sensitive to metabolic processes and efficiencies (Ponsard and Averbusch 1999).

Additional aspects of parasitism and infection in percids remain poorly studied. Yellow perch may act as hosts to the glochidia of freshwater mussels (Jansen 1991), but effects of glochidia parasitism on percid growth rates have not been determined. Also, some evidence exists for percids provisioning eggs with anti-infection properties (e.g., yellow perch; Paxton and Willoughby 2000).

Table 19.5 Published accounts on effects of parasitism on fish growth and growth-related variables. "X" indicates no measured effect on variable of interest. Site/mode of infection listed below taxonomic name of each parasite

Parasite	Host	Source	Details	Effect	Variable measured
Apophallus brevis Metacercariae infect fish musculature	*P. flavescens*	Johnson and Dick (2001)	Infection rates > 50 per g infected tissue	–	Reduced growth
		Ryman et al. (2008)	Compared high (19 cysts/g) vs. low (1 cyst/g) rates of infection	–	Increased MeHg
		Marcogliese et al. (2005)	Effect at >10 metacercariae	–	Increased oxidative stress markers
		Marcogliese et al. (2010)	Effect only at high-pollution sites	–	Increased oxidative stress markers
		Ryman et al. (2008)	Compared high (19 cysts/g) vs. low (1 cyst/g) rates of infection	+	Higher condition
		Cloutier et al. (2012)		X	Growth
Diplostomum spp. Trematode- infects eyes, visual impairment	*P. flavescens*	Marcogliese et al. (2010)	Effect only at high-pollution sites	–	Increased oxidative stress markers
		Cloutier et al. (2012)		X	Growth
	S. vitreus	Marcogliese et al. (2001)		X	Growth
Glugea spp. Intestinal infection	*P. flavescens*	Johnson and Dick (2001)	>100 parasites per individual	–	Reduced growth
		Johnson and Dick (2001)	>100 parasites per individual	–	Reduced visceral fat
		Johnson and Dick (2001)	>100 parasites per individual	+	Increased GSI
Ichthyotylurus spp. Liver pathogen	*P. flavescens*	Marcogliese et al. (2010)		–	Increased oxidative stress markers
		Cloutier et al. (2012)	Effect only at high-pollution sites	+	Increased condition

(continued)

Table 19.5 (continued)

Parasite	Host	Source	Details	Effect	Variable measured
Ligula intestinalis Intestinal infection	*P. flavescens*	Pitt and Grundmann (1957)		–	Reduced growth by 38–55 %
Phyllodistomum superbum Urinary bladder and kidney	*P. flavescens*	Cloutier et al. (2012)		–	Reduced growth
Raphidascaris acus Liver pathogen	*P. flavescens*	Szalai and Dick (1991)	Infection rates > 50 per liver	–	Reduced condition
		Szalai and Dick (1991)	Infection rates > 50 per liver	–	Reduced growth
		Szalai and Dick (1991)	Infection rates >50 per liver	–	Increased mortality
		Marcogliese et al. (2005)	Effect only at high-pollution sites	–	Increased oxidative stress markers
		Johnson and Dick (2001)	Infections >50/g infected tissue	–	Reduced visceral fat
		Cloutier et al. (2012)		X	Growth
Triaenophorus nodulosus Liver parasite	*P. fluvatilis*	Brinker and Hamers (2007)	Infection rate 1–3 per liver	–	Reduced growth
Tylodephys scheuringi Eye pathogen, visual impairment	*P. flavescens*	Cloutier et al. (2012)		–	Reduced growth, but only in the presence of *P. superbum*

19.5 Summary

The lifetime growth of commercially-important percid fishes in the wild is the result of a complex suite of intrinsic and extrinsic factors that interact and vary in importance. Despite this complexity, our review and re-analyses have revealed a number of patterns and consistencies. In summary, percid growth in nature is:

(i) Biphasic (Sect. 19.2). The biphasic growth model allows us to accurately describe growth while gaining insight into other life history traits (e.g., investment in reproduction, mortality) and their constraints.

(ii) Optimized to maximize fitness within a given set of environmental and biological constraints (Sect. 19.3.1). Optimization results in common life history trade-offs (e.g., growth and maturity) that will allow for both rapid immature growth and large adult size.

(iii) Sexually dimorphic, with the exception of pikeperch (Sect. 19.3.2). Adult females typically grow faster and larger than adult males, which appears to result from differences in activity and consumption (as determined hormonally).
(iv) Strongly temperature-dependent (Sect. 19.4.1). Degree-days show that, on average, patterns of growth are consistent across a species' range, but that grow rates increase with latitude.
(v) Shaped by predation (Sect. 19.4.2). Percids respond to both predation mortality and perceived risk; they appear to reduce activity when faced with predation, but "bold" personalities may emerge depending on current perceived risk.
(vi) Strongly density-dependent (Sect. 19.4.3). Although growth-density relationships are confounded in nature, wild percids appear capable of up to twofold variation in individual growth in length.
(vii) A function of food availability and size (Sect. 19.4.4). Independent of population density, food quality and quantity can represent an energetic bottleneck that limits growth efficiency and leads to stunting.
(viii) Shaped by parasitism and disease (Sect. 19.4.5). Parasitism and disease are widespread but poorly studied. Although effects on growth are largely negative, some effects are neutral or even positive.

These findings are based on evidence from four, well-studied species (European perch, yellow perch, pikeperch, and walleye) from two sister sub-families (Luciopercinae and Percinae) (Sloss et al. 2004). We expect similar patterns in other members of these sub-families that are not well studied; for example sauger (a sister-species to walleye), *Gymnocephalus* spp. (which are similar to perch), and *Zindel* spp. (all of which are endangered or vulnerable). However, we recommend further study before extending these growth patterns to darters, which tend to be invertivorous stream-dwellers with very different life histories.

The insight that we have gained into percid growth rates in the wild is also relevant to percid growth in culture. Our finding that percids are capable of up to two-fold variation in growth and length under natural conditions (Sect. 19.4.3) suggests that the productivity of a facility will benefit greatly from informed design and operation. The benefits of controlling such factors as temperature, density, and food are fairly obvious. Our results also point to percids from northern populations for broodstock selection because they tend to grow faster at a given temperature and ration, than percids from southern populations (Sect. 19.4.1). Similarly, female percids may be more profitable than male percids for culturing because the former mature relatively later and larger (Sect. 19.3.1). If fish are allowed to mature, then adult females also grow relatively faster than males and reach larger sizes (Sect. 19.3.2) despite investing more in reproduction.

Our review of percid growth focused on most but not all drivers of growth in wild percid populations. Percid growth rates are optimized to a range of environmental and ecological conditions (Sect. 19.3.1), but are also shaped by selective fishing (Jorgensen et al. 2007). Theoretical models show that the selective removal of adults will lead to reductions in both age and size-at-maturity (e.g., Dunlop et al. 2007; Lester et al. 2014), although population models suggest that this response is not guaranteed (e.g., Matsumura et al. 2011). Nonetheless, the potential for fishing pressure to shape growth rates should not be ignored when selecting broodstock. Percid growth in the wild is also negatively affected by pollution and contaminants. In general, contaminants and pollutants affect growth via sub-lethal effects operating at genetic, molecular, and cellular scales. Yellow perch exposed to organic contaminants and heavy metals may have an overall impaired stress response (Hontela et al. 1992). Survival of perch <2 years old was also significantly impaired in metal-contaminated lakes (Sherwood et al. 2000). However, indirect (food-web mediated) effects of contaminants in lakes can be as significant if not more so than direct metabolic effects with respect to growth (Campbell et al. 2003; Rasmussen et al. 2008).

Finally, we note avenues of future research that are relevant to percid growth in both wild and culture settings. The degree to which positive effects of estrogen and negative effects of androgen hormones on growth (Sect. 19.2) are behavioural (influencing foraging efficacy and field rates of activity), metabolic (increased/decreased rates of basal metabolism) or some combination thereof warrants further investigation. Seemingly positive responses of fish growth and condition to parasitism, a counter-intuitive phenomenon, warrants further scrutiny and investigation (Sect. 19.4.5). Last, the mechanism behind the existence of sympatric growth forms of percids in some lakes remains unresolved. For example, dwarf and normal growth forms of walleye in Lake Winnipeg (Johnston et al. 2012) show significantly different patterns of resource allocation that do not appear to be associated with resource polymorphism (Moles et al. 2010, 2011). While increased fishing mortality on normal growth forms may facilitate the coexistence with the normal form to some degree, the driver of these sympatric forms remains unknown.

Acknowledgements Fernanda Cabrini Araujo for assistance with searching, compiling, and formatting. We acknowledge the E-OBS dataset from the EU-FP6 project ENSEMBLES (http://ensembles-eu.metoffice.com), the data providers in the ECA&D project (http://www.ecad.eu), and Dan Holloway for help with navigating OPeNDAP (http://www.opendap.org/). This manuscript was improved by the editorial assistance of two anonymous reviewers.

References

Abrams PA, Rowe L (1996) The effects of predation on the age and size of maturity of prey. Evolution 50(3):1052–1061

Angilletta MJ (2009) Thermal adaptation: a theoretical and empirical synthesis. Oxford University Press, Oxford

Anthony D, Jorgensen C (1977) Factors in the declining contributions of Walleye (*Stizostedion vitreum vitreum*) to the fishery of Lake Nipissing, Ontario, 1960–1976. J Fish Res Board Can 34:1703–1709

Arnott SA, Barber I, Huntingford FA (2000) Parasite-associated growth enhancement in a fish-cestode system. Proc R Soc B-Biol Sci 267(1444):657–663

Bacheler N, Paoli T, Schacht G (2011) Controls on abundance and distribution of yellow perch: predator, water quality, and density-dependent effects. Trans Am Fish Soc 140:989–1000. doi: 10.1080/00028487.2011.603979

Bardach JE (1951) Changes in the yellow perch population of Lake Mendota, Wisconsin, between 1916 and 1948. Ecology 32(4):719–728

Bartels P, Hirsch PE, Svanback R, Eklov P (2012) Water transparency drives intra-population divergence in Eurasian perch (*Perca fluviatilis*). Plos One 7(8):e43641. doi:10.1371/journal.pone.0043641

Beckman WC (1950) Changes in growth rates of fishes following reduction in population densities by winterkill. Trans Am Fish Soc 78(1):82–90. doi:10.1577/1548 8659(1948)78[82:CIGROF]2.0.CO;2

Beverton RJH (1963) Maturation, growth and mortality of clupeid and engraulid stocks in relation to fishing. Proc Verb Cons Explor Mer 154:44–67

Beverton RJH (1987) Longevity in fish: some ecological and evolutionary considerations. In: Woodhead AD, Thompson KH (eds) Evolution of longevity in animals: a comparative approach. Plenum Press, New York

Beverton RJH, Holt SJ (1959) A review of the lifespans and mortality rates of fish in nature, and their relation to growth and other physiological characteristics. In: Wolstenholme GEW, O'Connor M (eds) CIBA Foundation colloquia on ageing: the lifespan of animals, vol 5. J & A Churchill, London

Biro PA, Post JR (2008) Rapid depletion of genotypes with fast growth and bold personality traits from harvested fish populations. Proc Natl Acad Sci USA 105(8):2919–2922

Biro PA, Post JR, Abrahams MV (2005) Ontogeny of energy allocation reveals selective pressure promoting risk-taking behaviour in young fish cohorts. Proc R Soc B-Biol Sci 272(1571):1443–1448

Bozek MA, Baccante DA, Lester NP (2011) Walleye and sauger life history. In: Barton BA (ed) Biology, management, and culture of walleye and sauger. American Fisheries Society, Bethesda

Breder CM, Rosen DE (1966) Modes of reproduction in fishes. The Natural History Press, New York

Brinker A, Hamers R (2007) Evidence for a negative impact of plerocercoid infection of Triaenophorus nodulosus on Perca fluvatilis L. stockin in Upper Lake Constance, a water body undergoing rapid reoligotrophication. J Fish Biol 71:128–147

Buijse AD, Houthuijzen RP (1992) Piscivory, growth, and size-selective mortality of age-0 pikeperch (*Stizostedion lucioperca*). Can J Fish Aquat Sci 49:894–902

Campbell PGC, Hontela A, Rasmussen JB, Giguere A, Gravel A, Kraemer L et al (2003) Differentiating between direct (physiological) and food-chain mediated (bioenergetic) effects on fish in metal-impacted lakes. Hum Ecol Risk Assess 9(4):847–866. doi:10.1080/713610012

Carlander KD (1945) Age, growth, sexual maturity and population fluctuations of the yellow pikeperch, *Stizostedion vitreum vitreum* (Mitchill), with reference to the commercial fisheries, Lake of the Woods, Minnesota. Trans Am Fish Soc 73(1):90–107

Carlander KD (1950) Growth rate studies of saugers, *Stizostedion canadense canadense* (Smith), and yellow perch, *Perca flacescens* (Mitchill) from Lake of the Woods, Minnesota. Trans Am Fish Soc 79(1):30–42

Charnov E (1993) Life history invariants: some explorations of symmetry in evolutionary ecology. Oxford University Press, New York

Charnov EL, Turner TF, Winemiller KO (2001) Reproductive constraints and the evolution of life histories with indeterminate growth. Proc Natl Acad Sci U S A 98(16):9460–9464

Chavarie L, Dempson JB, Schwarz CJ, Reist JD, Power G, Power M (2010) Latitudinal variation in growth among Arctic charr in eastern North America: evidence for countergradient variation? Hydrobiologia 650:161–177. doi:10.1007/s10750-009-0043-z

Chezik KA (2013) Fish growth and degree-days: advice for selecting base temperatures in both within- and among-lake studies. Thesis, University of Minnesota

Chong SC (2000) The effect of latitude on recruitment variability and first year growth of yellow perch (*Perca fiavescens*). Dissertation, Queen's University

Cloutier VB, Glemet H, Ferland-Raymond B, Gendron AD, Marcogliese DJ (2012) Correlation of parasites with growth of yellow perch. J Aquat Anim Health 24(2):100–104. doi:10.1080/08997659.2012.675934

Colby PJ, Nepszy SJ (1981) Variation among stocks of walleye (*Stizostedion vitreum vitreum*) – management implications. Can J Fish Aquat Sci 38(12):1814–1831. doi:10.1139/f81-228

Colby PJ, McNicol RE, Ryder RA (1979) Synopsis of biological data on the walleye *Stizostedion v. vitreum* (Mitchill 1818), vol 119, Fisheries synopsis. Food and Agriculture Organization of the United Nations, Rome

Conover DO, Present TMC (1990) Countergradient variation in growth rate: compensation for length of the growing season among Atlantic silversides from different latitudes. Oecologia 83:316–324

Conover DO, Munch SB, Arnott SA (2009) Reversal of evolutionary downsizing caused by selective harvest of large fish. Proc R Soc B-Biol Sci 276:2015–2020. doi:10.1098/rspb.2009.0003

Craig JF (1980) Growth and production of the 1955 to 1972 cohorts of perch, *Perca fluviatilis* L., in Windermere. J Anim Ecol 49:291–315. doi:10.2307/4290

Craig JF (1982) Population dynamics of Windermere perch. Rep Freshw Biol Assoc 50:49–59

Craig JF (1987) The biology of perch and related fish. Timber Press, Portland

Darwin C (1871) The descent of man, and selection in relation to sex. D. Appleton and Company, New York

Day T, Taylor PD (1997) Von Bertalanffy's growth equation should not be used to model age and size at maturity. Am Nat 149(2):381–293, doi:0003-0147/97/4902-0009

DeWoody JA, Fletcher DE, Wilkins SD, Avise JC (2000) Parentage and nest guarding in the tessellated darter (*Etheostoma olmstedi*) assayed by microsatellite markers (Perciformes: Percidae). Copeia 3:740–747

Diehl S, Eklov P (1995) Effects of piscivore-mediated habitat use on resources, diet, and growth of perch. Ecology 76(6):1712–1726

Dunlop ES, Shuter BJ, Dieckmann U (2007) Demographic and evolutionary consequences of selective mortality: predictions from an eco-genetic model for smallmouth bass. Trans Am Fish Soc 136:749–765

El-Zarka SE-D (1959) Fluctuations in the population of yellow perch, *Perca flavescens* (Mitchill), in Saginaw Bay Lake Huron. US Fish Wildl Serv Fish Bull 59:365–415

Frankiewicz P, Dabrowski K, Martyniak A, Zalewski M (1999) Cannibalism as a regulatory force of pikeperch, *Stizostedion lucioperca* (L.), population dynamics in the lowland Sulejow reservoir (Central Poland). Hydrobiologia 408(409):47–55

Froese R, Pauly D (2013) FishBase 2013: concepts, design and data sources. ICLARM, Philippines

Fry FEJ (1971) The effect of environmental factors on the physiology of fish. In: Hoar WS, Randall DJ (eds) Fish physiology, vol 6. Academic, New York

Fullerton AH, Lamberti GA (2006) A comparison of habitat use and habitat-specific feeding efficiency by Eurasian ruffe (*Gymnocephalus cernuus*) and yellow perch (*Perca flavescens*). Ecol Freshw Fish 15(1):1–9. doi:10.1111/j.1600-0633.2005.00114.x

George SG, Slack WT, Douglas NH (1996) Demography, habitat, reproduction, and sexual dimorphism of the crystal darter, *Crystallaria asprella* (Jordan), from south-central Arkansas. Copeia 1:68–78

Goldspink CR, Goodwin D (1979) A note on the age composition, growth rate and food of perch Perca fluviatilis (L.) in four eutrophic lakes, England. J Fish Biol 14:489–505, doi:0022-1112/79/050489+17

Grignard JC, Melard C, Kestemont P (1996) A preliminary study of parasites and diseases of perch in an intensive culture system. J Appl Ichthyol 12(3–4):195–199. doi:10.1111/j.1439-0426.1996.tb00089.x

Hanna M (1990) Evaluation of models predicting mixing depth. Can J Fish Aquat Sci 47(5):940–947. doi:10.1139/f90-108

Harrington RC, Benavides E, Near TJ (2013) Phylogenetic inference of nuptial trait evolution in the context of asymmetrical introgression in North American darters (teleostei). Evolution 67(2):388–402. doi:10.1111/j.1558-5646.2012.01797.x

Hayloch MR, Hofstra N, Klein Tank AMG, Klok EJ, Jones PD, New M (2008) A European daily high-resolution gridded data set of surface temperature and precipitation for 1950–2006. J Geophys Res 113:1–12. doi:10.1029/2008JD010201

Headley HC, Lauer TE (2008) Density-dependent growth of yellow perch in southern Lake Michigan, 1984–2004. N Am J Fish Manag 28(1):57–69. doi:10.1577/m06-097.1

Heath DD, Roff DA (1996) The role of trophic bottlenecks in stunting: a field test of an allocation model of growth and reproduction in yellow perch, *Perca flavescens*. Environ Biol Fish 45(1):53–63. doi:10.1007/bf00000627

Heibo E, Vollestad LA (2006) An analysis of life-history invariants in Eurasian perch, *Perca fluviatilis*. Evol Ecol Res 8(1):51–62

Heibo E, Magnhagen C, Vollestad LA (2005) Latitudinal variation in life-history traits in Eurasian perch. Ecology 86(12):3377–3386. doi:10.1890/04-1620

Heins DC, Baker JA, Guill JM (2004) Seasonal and interannual components of intrapopulation variation in clutch size and egg size of a darter. Ecol Freshw Fish 13(4):258–265. doi:10.1111/j.1600-0633.2004.00064.x

Hellstrom G, Heynen M, Oosten J, Borcherding J, Magnhagen C (2011) The effect of group size on risk taking and social conformity in Eurasian perch. Ecol Freshw Fish 20(4):499–502. doi:10.1111/j.1600-0633.2011.00506.x

Henderson B (1985) Factors affecting growth and recruitment of yellow perch, Perca flavescens Mitchill, in South Bay, Lake Huron. J Fish Biol 26:449–458, doi:0022-1 112/85/040449

Henderson BA, Trivedi T, Collins N (2000) Annual cycle of energy allocation to growth and reproduction of yellow perch. J Fish Biol 57(1):122–133

Henderson BA, Collins N, Morgan GE, Vaillancourt A (2003) Sexual size dimorphism of walleye (*Stizostedion vitreum vitreum*). Can J Fish Aquat Sci 60(11):1345–1352

Heynen M, Heermann L, Borcherding J (2011) Does the consumption of divergent resources influence risk taking behaviour in juvenile perch (Perca fluviatilis L.)? Ecol Freshw Fish 20(1):1–4

Holker F, Mehner T (2005) Simulation of trait- and density-mediated indirect effects induced by piscivorous predators. Basic Appl Ecol 6(3):289–300. doi:10.1016/j.baae.2004.10.001

Holtby LB, Healey MC (1990) Sex-specific life-history tactics and risk-taking in coho salmon. Ecology 71(2):678–690

Hontela A, Rasmussen JB, Audet C, Chevalier G (1992) Impaired Cortisol stress response in fish from environments polluted by PAHs, PCBs, and mercury. Arch Environ Contam Toxicol 22:278–283

Horppila J, Estlander S, Olin M, Pihlajamaki J, Vinni M, Nurminen L (2011) Gender-dependent effects of water quality and conspecific density on the feeding rate of fish – factors behind sexual growth dimorphism. Oikos 120(6):855–861. doi:10.1111/j.1600-0706.2010.19056.x

Houthuijzen RP, Backx JJGM, Buuse AD (1993) Exceptionally rapid growth and early maturation of perch in a freshwater lake recently converted from an estuary. J Fish Biol 43:320–324, doi:0022-1112193/080320+05

Hoxmeier RJH, Wahl DH, Hooe ML, Pierce CL (2004) Growth and survival of larval walleyes in response to prey availability. Trans Am Fish Soc 133:45–54

Hughey MC, Heins DC, Jelks HL, Ory BA, Jordan F (2012) Variation in reproductive life history traits between two populations of blackbanded darters (*Percina nigrofasciata*). Copeia 4:714–721. doi:10.1643/ci-11-169

Iles AC, Rasmussen JB (2005) Indirect effects of metal contamination on energetics of yellow perch (*Perca flavescens*) resulting from food web simplification. Freshw Biol 50(6):976–992. doi:10.1111/j.1365-2427.2005.01380.x

Irwin BJ, Rudstam LG, Jackson JR, VanDeValk AJ, Forney JL, Fitzgerald DG (2009) Depensatory mortality, density-dependent growth, and delayed compensation: disentangling the interplay of mortality, growth, and density during early life stages of yellow perch. Trans Am Fish Soc 138(1):99–110. doi:10.1577/t07-256.1

Ivan L, Hook T, Thomas M, Fielder D (2011) Long-term and interannual dynamics of walleye and yellow perch in Saginaw Bay, Lake Huron. Trans Am Fish Soc 140:1078–1092. doi:10.1080/00028487.2011.603976

Jackson ZJ, Quist MC, Larscheid JG (2008) Growth standards for nine North American fish species. Fish Manag Ecol 15(2):107–118. doi:10.1111/j.1365-2400.2007.00591.x

Jansen WA (1991) Seasonal prevalence, intensity of infestation, and distribution of glochidia of Anodonta-grandis-simpsoniana Lea on yellow perch, *Perca flavescens*. Can J Fish Aquat Sci 69(4):964–972. doi:10.1139/z91-140

Jensen AL (1981) Population regulation in lake whitefish, *Coregonus clupeaformis* (Mitchill). J Fish Biol 19(5):557–573

Jensen AL (1996) Beverton and Holt life history invariants result from optimal trade-off of reproduction and survival. Can J Fish Aquat Sci 53:820–822

Johansson F, Andersson J (2009) Scared fish get lazy, and lazy fish get fat. J Anim Ecol 78(4):772–777. doi:10.1111/j.1365-2656.2009.01530.x

Johnson MW, Dick TA (2001) Parasite effects on the survival, growth, and reproductive potential of yellow perch (*Perca flavescens* Mitchill) in Canadian Shield lakes. Can J Zool 79(11):1980–1992. doi:10.1139/cjz-79-11-1980

Johnson MW, Hesslein RH, Dick TA (2004) Host length, age, diet, parasites and stable isotopes as predictors of yellow perch (*Perca flavescens* Mitchill), trophic status in nutrient poor Canadian Shield lakes. Environ Biol Fish 71(4):379–388. doi:10.1007/s10641-004-4189-2

Johnston CE, Haag WR (1996) Life history of the Yazoo darter (Percidae: *Etheostoma raneyi*), a species endemic to north-central Mississippi. Tulanne Stud Zool Bot 30:47–60

Johnston TA, Lysack W, Leggett WT (2012) Abundance, growth and life history characteristics of sympatric walleye (Sander vitreus) and sauger (S. canadensis) in Lake Winnipeg, Manitoba. J Great Lakes Res 38(Suppl 3):35–46

Jorgensen C, Enberg K, Dunlop ES, Arlinghaus R, Boukal DS, Brander K, Ernande B, Gardmark A, Johnston F, Matsumura S, Pardoe H, Raab K, Silva A, Vainikka A, Dieckmann U, Heino M, Rijnsdorp AD (2007) Ecology – managing evolving fish stocks. Science 318(5854):1247–1248. doi:10.1126/science.1148089

Katsanevakis S, Maravelias CD (2008) Modelling fish growth: multi-model inference as a better alternative to a priori using von Bertalanffy equation. Fish Fish 9:178–187

Kaufman SD, Gunn JM, Morgan GE, Couture P (2006) Muscle enzymes reveal walleye (*Sander vitreus*) are less active when larger prey (cisco, *Coregonus artedi*) are present. Can J Fish Aquat Sci 63:970–979. doi:10.1139/F06-004

Kelly NB, Alonzo SH (2011) The evolution of male parental care in darters (Percidae: Etheostoma). Integr Comp Biol 51:E211–E211

Kelly NB, Near TJ, Alonzo SH (2012) Diversification of egg-deposition behaviours and the evolution of male parental care in darters (Teleostei: Percidae: Etheostomatinae). J Evol Biol 25(5):836–846. doi:10.1111/j.1420-9101.2012.02473.x

Kempinger J, Carline R (1977) Dynamics of the walleye (*Stizostedion vitreum vitreum*) population in Escanaba Lake, Wisconsin 1955–72. J Fish Res Board Can 34:1800–1811

Kempinger J, Forbes A, Serns S, Snow H (1982) Population development of re-introduced smallmouth bass and yellow perch in Nebish Lake, vol 119, Research Report. Wisconsin Department of Natural Resources, Madison

Kitchell JF, Johnson MG, Minns CK, Loftus KH, Greig L, Olver CH (1977) Percid habitat: the river analogy. J Fish Res Board Can 34:1936–1940

Korbuly B, Pacala N, Grozea A, Dobrin L, Ada T (2007) Sexual morphological traits and body indices dimorphism in a Danube 2 years old pikeperch population. Lucrari Stiintifice 40(2):23–27

Koupal KD, Satterfield JR, Flickinger SA (1997) Comparative gear selectivity for male walleyes and influence of method of capture on resultant hatching success. Progress Fish Cult 59(3):218–221. doi:10.1577/1548-8640(1997)059<0218:cgsfmw>2.3.co;2

Lappalainen J, Dorner H, Wysujack K (2003) Reproduction biology of pikeperch (*Sander lucioperca*(L.))–a review. Ecol Freshw Fish 12(2):95–106. doi:10.1034/j.1600-0633.2003.00005.x

Lauer TE, Doll JC, Allen PJ, Breidert B, Palla J (2008) Changes in yellow perch length frequencies and sex ratios following closure of the commercial fishery and reduction in sport bag limits in southern Lake Michigan. Fish Manag Ecol 15(1):39–47. doi:10.1111/j.1365-2400.2007.00567.x

Law R (2000) Fishing, selection, and phenotypic evolution. ICES J Mar Sci 57(3):659–668. doi:10.1006/jmsc.2000.0731

Le Cren ED (1958) Observations on the growth of perch (*Perca fluviatilis* L.) over twenty- two years with special reference to the effects of temperature and changes in population density. J Anim Ecol 27(2):287–334. doi:10.2307/2242

Le Cren ED (1992) Exceptionally big individual perch (*Perca fluviatilis* L.) and their growth. J Fish Biol 40:599–625, doi:0022-1 1 12/92/040599+27

Lehtonen H (1987) Selection of minimum size limit for pike-perch (*Stizostedion lucioperca*) in coastal waters of Finland. Proc V Congr Eur Ichthyol Stockh 1985:351–355

Lehtonen H, Hansson S, Winkler H (1996) Biology and exploitation of pikeperch, *Stizostedion lucioperca* (L), in the Baltic Sea area. Ann Zool Fenn 33(3–4):525–535

Lepak JM, Hooten MB, Johnson BM (2012) The influence of external subsidies on diet, growth and concentrations of freshwater sport fish: implications for management and fish consumption advisories. Ecol Toxicol 21:1878–1888. doi:10.1007/s10646-012-0921-4

Lester NP, Shuter BJ, Abrams PA (2004) Interpreting the von Bertalanffy model of somatic growth in fishes: the cost of reproduction. Proc R Soc B-Biol Sci 271:1625–1631. doi:10.1098/rspb.2004.2778

Lester NP, Shuter BJ, Venturelli PA, Nadeau D (2014) Life history plasticity and sustainable exploitation: a theory of growth compensation applied to walleye management. Ecol Appl 24:38–54

Linlokken A, Seeland PAH (1996) Growth and production of perch (*Perca fluviatilis* L.) responding to biomass removal. Ann Zool Fenn 33:427–435

Lipinski MR, Roeleveld MA (1990) Minor extension of the von Bertalanffy growth theory. Fish Res 9:367–371. doi:10.1016/0165-7836(90)90054-y

Lorenzoni M, Pace R, Pedicillo G, Viali P, Carosi A (2009) Growth, catches and reproductive biology of ruffe *Gymnocephalus cernuus* in Lake Piediluco (Umbria, Italy). Folia Zool 58(4):420–435

Magnhagen C, Borcherding J (2008) Risk-taking behaviour in foraging perch: does predation pressure influence age-specific boldness? Anim Behav 75:509–517. doi:10.1016/j.anbehav.2007.06.007

Magnhagen C, Hellstrom G, Borcherding J, Heynen M (2012) Boldness in two perch populations – long-term differences and the effect of predation pressure. J Anim Ecol 81(6):1311–1318. doi:10.1111/j.1365-2656.2012.02007.x

Malison JA, Best CD, Kayes TB, Amundson CH (1985) Hormonal growth promotion and evidence for a size-related difference in response to estradiol -17 beta in yellow perch (*Perca flavescens*). Can J Fish Aquat Sci 42:1627–1633

Malison JA, Kayes TB, Wentworth BC, Amundson CH (1988) Growth and feeding responses of male versus female yellow perch (*Perca flavescens*) treated with Estradiol-17B. Can J Fish Aquat Sci 45:1942–1948

Mandiki SNM, Houbart M, Babiak I, Vandeloise E, Gardeur JN, Kestemont P (2004) Are sex steroids involved in the sexual growth dimorphism in Eurasian perch juveniles? Physiol Behav 80:603–609. doi:10.1016/j.physbeh.2003.10.016

Mandiki SNM, Babiak I, Bopopi JM, Leprieur F, Kestemont P (2005) Effects of sex steroids and their inhibitors on endocrine parameters and gender growth differences in Eurasian perch (*Perca fluviatilis*) juveniles. Steroids 70:85–94. doi:10.1016/j.steroids.2004.10.009

Marshall TR (1977) Morphological, physiological, and ethological differences between walleye (*Stizostedion vitreum vitreum*) and pikeperch (*S. lucioperca*). J Fish Res Board Can 34:1515–1523

Marcogliese DJ, Dumont P, Gendron AD, Mailhot Y, Bergeron E, McLaughlin JD (2001) Spatial and temporal variation in abundance of Diplostomum spp. in walleye (Stizostedion vitreum) and white suckers (Catostomus commersoni) from the St. Lawrence River. Can J Zool 79(3):355–369. doi:10.1139/z00-209

Marcogliese DJ, Gagnon Brambilla L, Gagné F, Genron AD (2005) Joint effects of parasitism and pollution on oxidative stress biomarkers in yellow perch Perca flavescens. Dis Aquat Org 63:77–85

Marcogliese DJ, Dautremeputis C, Gendron AD, Fournier M (2010) Interactions between parasites and pollutants in yellow perch (Perca flavescens) in the St. Lawrence River, Canada: implications for resistance and tolerance to parasites. Can J Fish Aquat Sci 88:247–258.

Matsumura S, Arlinghaus R, Dieckmann U (2011) Assessing evolutionary consequences of size-selective recreational fishing on multiple life-history traits, with an application to northern pike (*Esox lucius*). Evol Ecol 25:711–735. doi:10.1007/s10682-010-9444-8

McKenney DW, Pedlar JH, Papadopol P, Hutchinson MF (2006) The development of 1901–2000 historical monthly climate models for Canada and the United States. Agr Forest Meteorol 138:69–81. doi:10.1016/j.agrformet.2006.03.012

Mills EL, Pol MV, Sherman RE, Culver TB (1989) Interrelationships between prey body size and growth of age-0 yellow perch. Trans Am Fish Soc 118(1):1–10. doi:10.1577/1548-8659(1989)118<0001:ibpbsa>2.3.co;2

Moles MD, Johnston TA, Robinson BW, Leggett WC, Casselman JM (2008) Is gonadal investment in walleye (*Sander vitreus*) dependent on body lipid reserves? A multipopulation comparative analysis. Can J Fish Aquat Sci 65(4):600–614. doi:10.1139/f07-186

Moles MD, Robinson BW, Johnston TA, Cunjak RA, Jardine TD, Casselman JM, Leggett WC (2010) Morphological and trophic differentiation of growth morphotypes of walleye (*Sander vitreus*) from Lake Winnipeg, Canada. Can J Zool 88:950–960. doi:10.1139/Z10-062

Moles MD, Johnston TA, Robinson BW, Bernard AM, Wilson CC, Wiegand MD, Leggett WC (2011) Reproductive divergence between growth forms of Lake Winnipeg walleye (*Sander vitreus*). Ecol Freshw Fish 20:52–66. doi:10.1111/j.1600-0633.2010.00457.x

Moodie GEE, Loadman NL, Wiegand MD, Mathias JA (1989) Influence of egg characteristics on survival, growth and feeding in larval walleye (*Stizostedion vitreum*). Can J Fish Aquat Sci 46:516–521

Mooij WM (1996) Variation in abundance and survival of fish larvae in shallow eutrophic Lake Tjeukemeer. Environ Biol Fish 46:265–279

Mooij WM, Lammens E, Vandensen WLT (1994) Growth-rate of 0+ fish in relation of temperature, body-size, and food in shallow eutrophic Lake Tjeukemeer. Can J Fish Aquat Sci 51(3):516–526. doi:10.1139/f94-054

Muth K, Wolfert D (1986) Changes in growth and maturity of walleyes associated with stock rehabilitation in Western Lake Erie, 1964–1983. N Am J Fish Manag 6(2):168–175

Neuhaus E (1934) Studien uber das Stettiner Haffund seine Nebengewasser. III. Untersuchungen uber den Zander. Z Fisch 32:599–634 (in German)

Neuheimer AB, Taggart CT (2007) The growing degree-day and fish size-at-age: the overlooked metric. Can J Fish Aquat Sci 64(2):375–385. doi:10.1139/f07-003

Nielsen LA (1980) Effect on walleye (*Stizostedion vitreum vitreum*) predation on juvenile mortality and recruitment of yellow perch (*Perca flavescens*) in Oneida Lake, New York. Can J Fish Aquat Sci 37:11–19

Nikolskii GV (1969) Theory of fish population dynamics as the biological background for rational exploitation and management of fishery resources. Oliver and Boyd, Edinburg

NOAA (National Oceanic and Atmospheric Administration) (2013) NOAA National Climatic Data Center. 2011. U.S. Climate Normals 1971–2000. Available http://cdo.ncdc.noaa.gov/cgi-bin/climatenormals/climatenormals.pl. Jun 2013

O'Gorman R, Burnett J (2001) Fish community dynamics in northeastern Lake Ontario with emphasis on the growth and reproductive success of yellow perch (*Perca flavescens*) and white perch (Morone americana), 1978 to 1997. J Great Lakes Res 27(3):367–383

Ogle DH (1998) A synopsis of the biology and life history of ruffe. J Great Lakes Res 24(2):170–185

Ohlberger J, Langangen O, Edeline E, Claessen D, Winfield IJ, Stenseth NC, Vollestad LA (2011) Stage-specific biomass overcompensation by juveniles in response to increased adult mortality in a wild fish population. Ecology 92(12):2175–2182

Olin M, Jutila J, Lehtonen H, Vinni M, Ruuhijarvi J, Estlander S, Rask M, Kuparinen A, Lappalainen J (2012) Importance of maternal size on the reproductive success of perch, *Perca fluviatilis*, in small forest lakes: implications for fisheries management. Fish Manag Ecol 19(5):363–374. doi:10.1111/j.1365-2400.2012.00845.x

Olsson J, Svanback R, Eklov P (2007) Effects of resource level and habitat type on behavioral and morphological plasticity in Eurasian perch. Oecologia 152(1):48–56. doi:10.1007/s00442-006-0588-8

Ostazeski J, Spangler G (2001) Use of biochronology to examine interactions of freshwater drum, walleye and yellow perch in the Red Lakes of Minnesota. Environ Biol Fish 61:381–393

Paloheimo JE, Dickie LM (1966) Food and growth of fishes. III. Relations among food, body size and growth efficiency. J Fish Res Board Can 23(8):1209–1248

Parker GA (1992) The evolution of sexual size dimorphism in fish. J Fish Biol 41:1–20. doi:10.1111/j.1095-8649.1992.tb03864.x

Paxton CGM, Willoughby LG (2000) Resistance of perch eggs to attack by aquatic fungi. J Fish Biol 57(3):562–570. doi:10.1006/jfbi.2000.1332

Peacor SD (2002) Positive effect of predators on prey growth rate through induced modifications of prey behaviour. Ecol Lett 5(1):77–85

Pérez-Bote JL, Roso R (2012) Growth and length-weight relationships of *Sander lucioperca* (Linnaeus, 1758) in the Alcántara Reservoir, Wouth-Western Spain: comparison with other water bodies in Eurasia. J Appl Ichthyol 28:264–268. doi:10.1111/j.1439-0426.2011.01918.x

Persson L, Eklov P (1995) Prey refuges affecting interactions between piscivorous perch and juvenile perch and roach. Ecology 76(1):70–81

Persson L, Bystrom P, Wahlstrom E (2000) Cannibalism and competition in Eurasian perch: population dynamics of an ontogenetic omnivore. Ecology 81(4):1058–1071

Pierce R, Tomcko C, Negus M (2006) Interactions between stocked walleyes and native yellow perch in Lake Thirteen, Minnesota: a case history of percid community dynamics. N Am J Fish Manag 26:97–107. doi:10.1577/M05-034.1

Pitt CE, Grundmann AW (1957) A Study into the effects of parasitism on the growth of the yellow perch produced by the larvae of Ligula intestinalis (Linnaeus, 1758) Gmelin 1790. Proc Helm Soc Wash 24(2):3–10

Ponsard S, Averbuch P (1999) Should growing and adult animals fed on the same diet show different delta N-15 values? Rapid Commun Mass Spectrom 13(13):1305–1310. doi:10.1002/(sici)1097-0231(19990715)13:13<1305::aid-rcm654>3.3.co;2-4

Post JR, Evans DO (1989) Experimental evidence of size-dependent predation mortality in juvenile yellow perch. Can J Zool 67:521–523

Post JR, Johannes MRS, McQueen DJ (1997) Evidence of density-dependent cohort splitting in age-0 yellow perch (*Perca flavescens*): potential behavioural mechanisms and population-level consequences. Can J Fish Aquat Sci 54(4):867–875. doi:10.1139/cjfas-54-4-867

Power M, McKinley RS (1997) Latitudinal variation in lake sturgeon size as related to the thermal opportunity for growth. Trans Am Fish Soc 126(4):549–558

Power M, van den Heuvel MR (1999) Age-0 yellow perch growth and its relationship to temperature. Trans Am Fish Soc 128(4):687–700. doi:10.1577/1548-8659(1999)128<0687:ayp gai>2.0.co;2

Purchase CF, Collins NC, Morgan GE, Shuter BJ (2005a) Sex-specific covariation among life-history traits of yellow perch (*Perca flavescens*). Evol Ecol Res 7(4):549–566

Purchase CF, Collins NC, Morgan GE, Shuter BJ (2005b) Predicting life history traits of yellow perch from environmental characteristics of lakes. Trans Am Fish Soc 134:1369–1381

Purchase CF, Hutchings JA, Morgan GE (2006) The biological and statistical significance of life-history invariants in walleye (Sander vitreus). Evol Ecol Res 8:295–308

Quince C, Abrams PA, Shuter BJ, Lester NP (2008a) Biphasic growth in fish I: theoretical foundations. J Theor Biol 254(2):197–206. doi:10.1016/j.jtbi.2008.05.029

Quince C, Shuter BJ, Abrams PA, Lester NP (2008b) Biphasic growth in fish II: empirical assessment. J Theor Biol 254(2):207–214. doi:10.1016/j.jtbi.2008.05.030

Quinn G, Keough M (2002) Experimental design and data analysis for biologists. Cambridge University Press, Cambridge, UK

Quist MC, Guy CS, Schultz RD, Stephen JL (2003) Latitudinal comparisons of walleye growth in North America and factors influencing growth of walleyes in Kansas reservoirs. N Am J Fish Manag 23(3):677–692. doi:10.1577/m02-050

Rask M (1983) Differences in growth of perch (*Perca fluviatilis* L.) in 2 small Forest Lakes. Hydrobiologia 101:139–144. doi:10.1007/bf00008666

Rasmussen JB, Gunn JM, Sherwood GD, Iles A, Gagnon A, Campbell PGC, Hontela A (2008) Direct and indirect (foodweb mediated) effects of metal exposure on the growth of yellow perch (*Perca flavescens*): implications for ecological risk assessment. Hum Ecol Risk Assess 14(2):317–350. doi:10.1080/10807030801935017

Reid D, Momot W (1985) Evaluation of pulse fishing for the walleye, *Stizostedion vitveum vitveum*, in Henderson Lake, Ontario. J Fish Biol 27:235–251, doi:0022-1 112/85/27A235

Rennie MD, Collins NC, Shuter BJ, Rajotte JW, Couture P (2005) A comparison of methods for estimating activity costs of wild fish populations: more active fish observed to grow slower. Can J Fish Aquat Sci 62(4):767–780. doi:10.1139/f05-052

Rennie MD, Purchase CF, Lester N, Collins NC, Shuter BJ, Abrams PA (2008) Lazy males? Bioenergetic differences in energy acquisition and metabolism help to explain sexual size dimorphism in percids. J Anim Ecol 77:916–926. doi:10.1111/j.1365-2656.2008.01412.x

Rennie MD, Purchase CF, Shuter BJ, Collins NC, Abrams PA, Morgan GE (2010) Prey life-history and bioenergetic responses across a predation gradient. J Fish Biol 77:1230–1251. doi:10.1111/j.1095-8649.2010.02735.x

Reznick DA, Bryga H, Endler JA (1990) Experimentally induced life-history evolution in a natural-population. Nature 346(6282):357–359. doi:10.1038/346357a0

Ricker WE (1975) Computation and interpretation of biological statistics of fish populations, vol 191, Bulletin of the Fisheries Research Board of Canada. Department of the Environment, Fisheries and Marine Service, Ottawa

Roff DA (1980) A motion for the retirement of the von Bertalanffy function. Can J Fish Aquat Sci 37:127–129

Roff DA (1983) An allocation model of growth and reproduction in fish. Can J Fish Aquat Sci 40:1395–1404

Romare P (2000) Growth of larval and juvenile perch: the importance of diet and fish density. J Fish Biol 56(4):876–889. doi:10.1006/jfbi.1999.1208

Rose K, Rutherford E, MCDermot D, Forney J, Mills E (1999) Individual-based model of yellow perch and walleye populations in Oneida Lake. Ecol Monogr 69(2):127–154

Rose KA, Cowan JH, Winemiller KO, Myers RA, Hilborn R (2001) Compensatory density dependence in fish populations: importance, controversy, understanding and prognosis. Fish Fish 2(4):293–327

Ruckstuhl KE (2007) Sexual segregation in vertebrates: proximate and ultimate causes. Integr Comp Biol 47(2):245–257. doi:10.1093/icb/icm030

Rudstam LG, Magnuson JJ, Tonn WM (1984) Size selectivity of passive fishing gear – a correction for encounter probability applied to gill nets. Can J Fish Aquat Sci 41(8):1252–1255

Ryman JE, Van Walleghem JLA, Blanchfield PJ (2008) Methylmercury levels in a parasite (*Apophallus brevis metacercariae*) and its host, yellow perch (*Perca flavescens*). Aquat Ecol 42(3):495–501. doi:10.1007/s10452-007-9104-4

Rypel AL (2012a) Meta-analysis of growth rates for a circumpolar fish, the northern pike (*Esox lucius*), with emphasis on effects of continent, climate and latitude. Ecol Freshw Fish 21(4):521–532. doi:10.1111/j.1600-0633.2012.00570.x

Rypel AL (2012b) Concordant estimates of countergradient growth variation in striped bass (*Morone saxatilis*) using comparative life-history data. Can J Fish Aquat Sci 69:1261–1265. doi:10.1139/F2012-069

Sass GG, Hewett SW, Berd TD Jr, Fayram AH, Kitchell JF (2004) The role of density dependence in growth patterns of ceded territory walleye populations of northern Wisconsin: effects of changing management regimes. N Am J Fish Manag 24(4):1262–1278

Schneider JC (2004) Yellow perch maturity and fecundity as a function of age and growth. (Fisheries research report: 1915). University of Michigan Library, Ann Arbor, Available at http://name.umdl.umich.edu/AAG2862.1915.001

Schoenebeck C, Brown ML (2012) Does anaerobic activity differ seasonally or between sexes in yellow perch populations? Trans Am Fish Soc 141(1):199–203. doi:10.1080/00028487.2012.655119

Schueller AM, Hansen MJ, Newman SP, Edwards CJ (2005) Density dependence of walleye maturity and fecundity in Big Crooked Lake, Wisconsin, 1997–2003. N Am J Fish Manag 25(3):841–847. doi:10.1577/m04-090.1

Seppanen E, Kuukka H, Huuskonen H, Piironen J (2008) Relationship between standard metabolic rate and parasite-induced cataract of juveniles in three Atlantic salmon stocks. J Fish Biol 72(7):1659–1674. doi:10.1111/j.1095-8649.2008.01832.x

Sherwood GD, Rasmussen JB, Rowan DJ, Brodeur J, Hontela A (2000) Bioenergetic costs of heavy metal exposure in yellow perch (*Perca flavescens*): in situ estimates with a radiotracer (Cs- 137) technique. Can J Fish Aquat Sci 57(2):441–450

Sherwood GD, Pazzia I, Moeser A, Hontela A, Rasmussen JB (2002) Shifting gears: enzymatic evidence for the energetic advantage of switching diet in wild-living fish. Can J Fish Aquat Sci 59(2):229–241

Shoup DE, Nannini MA, Wahl DH (2012) The effect of vegetation density on juvenile bluegill diet and growth. J Freshwat Ecol 27(2):199–209. doi:10.1080/02705060.2011.641357

Shuter B, Koonce J (1977) A dynamic model of the western Lake Erie walleye (*Stizostedion vitreum vitreum*) population. J Fish Res Board Can 34:1972–1982

Shuter BJ, Lester NP, LaRose JJ, Purchase CF, Vascotto K, Morgan G, Collins NC, Abrams PA (2005) Optimal life histories and food web position: linkages among somatic growth, reproductive investment, and mortality. Can J Fish Aquat Sci 62(4):738–746. doi:10.1139/f05-070

Sloss BL, Billington N, Burr BM (2004) A molecular phylogeny of the Percidae (Teleostei, Perciformes) based on mitochondrial DNA sequence. Mol Phylogenet Evol 32:545–562. doi:10.1016/j.ympev.2004.01.011

Spangler G, Payne N, Winterton G (1977) Percids in the Canadian waters of Lake Huron. J Fish Res Board Can 34:1839–1848

Stacy WL, Lepak JM (2012) Relative influence of prey mercury concentration, prey energy density and predator sex on sport fish mercury concentrations. Sci Total Environ 2012(437):104–109. doi:10.1016/j.scitotenv.2012.07.064

Svanback R, Eklov P (2004) Morphology in perch affects habitat specific feeding efficiency. Funct Ecol 18(4):503–510

Svanback R, Persson L (2009) Population density fluctuations change the selection gradient in Eurasian perch. Am Nat 173(4):507–516. doi:10.1086/597223

Szalai AJ, Dick TA (1991) Role of predation and parasitism in growth and mortality of yellow perch in Dauphin Lake, Manitoba. Trans Am Fish Soc 120(6):739–751. doi:10.1577/1548-8659(1991)120<0739:ropapi>2.3.co;2

Tardif D, Glémet H, Brodeur P, Mingelbier M (2005) RNA/DNA ratio and total length of yellow perch (Perca flavescens) in managed and natural wetlands of a large fluvial lake. Can J Fish Aquat Sci 62(10):2211–2218

Thorpe J (1977) Synopsis of biological data on the perch Perca fluviatilis Linnaeus, 1758 and Perca flavescens Mitchill, 1814, vol 113, FOA fisheries synopsis. Food and Agriculture Organization of the United Nations, Rome

Tolonen A, Lappalainen J, Pulliainen E (2003) Seasonal growth and year class strength variations of perch near the northern limits of its distribution range. J Fish Biol 63:176–186. doi:10.1046/j.1095-8649.2003.00141.x

Urban MC (2007) The growth-predation risk trade-off under a growing gape-limited predation threat. Ecology 88(10):2587–2597

Vainikka A, Jokelainen T, Kortet R, Ylonen H (2005) Predation risk allocation or direct vigilance response in the predator interaction between perch (*Perca fluviatilis* L.) and pike (*Esox lucius* L.)? Ecol Freshw Fish 14(3):225–232. doi:10.1111/j.1600-0633.2005.00095.x

van Densen WLT, Ligtvoet W, Roozen RWM (1996) Intra-cohort variation in the individual size of juvenile pikeperch, *Stizostedion lucioperca*, and perch, *Perca fluvatilis*, in relation to the size spectrum of their food items. Ann Zool Fenn 33:405–506

Venturelli PA, Lester NP, Marshall TR, Shuter BJ (2010) Consistent patterns of maturity and density-dependent growth among populations of walleye (*Sander vitreus*): application of the growing degree-day metric. Can J Fish Aquat Sci 67:1057–1067. doi:10.1139/f10-041

Vinni M, Lappalainen J, Malinen T, Lehtonen H (2009) Stunted growth of pikeperch *Sander lucioperca* in Lake Sahajarvi, Finland. J Fish Biol 74(4):967–972. doi:10.1111/j.1095-8649.2009.02181.x
von Bertalanffy L (1934) Untersuchungen fiber die Gesetzlichkeit des Wachstums. I. Allgemeine Grundlagen der Theorie mathematische und physiologische Gesetzlichkeiten des Wachstums bei Wassertieren. Roux Arch Entwicklungsmech 131:613–652
von Bertalanffy L (1938) A quantitative theory of organic growth (inquiries on growth laws. II). Human Biol 10:181–213
Voutilainen A, Taskinen J, Huuskonen H (2012) Does *Diplostomum spp.* infection stimulate growth in juvenile Artic charr? Bull Eur Assoc Fish Pathol 32(4):118–126
Walker S, Addison P, Sandstrom S, Lester N (2013) Contact retention selectivity of three types of gillnet gangs. Ontario Ministry of Natural Resources. Report, 32 p.
Weatherley AH (1966) Ecology of fish growth. Nature 212(5068):1321–1324
Wellington CG, Mayer CM, Bossenbroek JM, Stroh NA (2010) Effects of turbidity and prey density on the foraging success of age 0 year yellow perch *Perca flavescens*. J Fish Biol 76(7):1729–1741. doi:10.1111/j.1095-8649.2010.02612.x
Zhao Y, Shuter BJ, Jackson DA (2008) Life history variation parallels phylogeographical patterns in North American walleye (*Sander vitreus*) populations. Can J Fish Aquat Sci 65:198–211. doi:10.1139/F07-162

Part V
Nutrition, Feeds and Feeding Practices

Chapter 20
Nutritional Requirements and Feeding of Broodstock and Early Life Stages of Eurasian Perch and Pikeperch

Patrick Kestemont and Emilie Henrotte

Abstract Up to date, the nutritional requirements of breeders and early life stages of European percid fishes, *Perca fluviatilis* and *Sander lucioperca*, have not been defined precisely, and, in fish farms, breeders are still relying on the regular supply of forage fish. It has been demonstrated in both species that the feeding conditions of broodstock largely influence the quality of gametes (especially in females) and hatched larvae. The best results, in terms of hatching rate and survival to challenge tests during the first days post-hatching, have been obtained when breeders were fed on forage fish, either as unique feed source or in combination with dry feed. Experimental diets, based on a suitable supply of phospholipids (PLs) and adequate ratio of essential long chain fatty acids (docosahexaenoic, eicosapentaenoic and arachidonic acids) have been used successfully in Eurasian perch breeders to produce high quality eggs and larvae, comparable to those obtained from perch fed forage fish. Fatty acid composition of broodstock diet significantly influenced the fatty acid composition of eggs. On the other contrary, none of the characteristics of the sperm were significantly modified by the HUFAs ratio, neither in terms of sperm volume and density, spermatozoa motility and velocity, nor in terms of seminal plasma osmolality. During the first weeks of larval rearing, *Artemia* nauplii are still currently used as starter feed for both Eurasian perch and pikeperch larvae. Enrichment of nauplii with HUFA has proved to be efficient in pikeperch but not in Eurasian perch. At the end of feeding trials with pikeperch larvae, highly significant correlations were achieved between dietary ascorbic acid content and the ascorbic acid content in larval carcass or the reduction of larval deformity. Based on commercial larval feed formulated for marine or freshwater fish, it has been shown that the freshwater feed (containing low Ca/P ratio) was more suitable for

P. Kestemont (✉)
Research Unit in Environmental and Evolutionary Biology, University of Namur,
Rue de Bruxelles, 5000 Namur, Belgium
e-mail: patrick.kestemont@unamur.be

E. Henrotte
Research Unit in Environmental and Evolutionary Biology, University of Namur,
Rue de Bruxelles, 5000 Namur, Belgium

Present Affiliation: AWE asbl – Recherche et Développement, Ciney, 5590 Belgium

P. Kestemont et al. (eds.), *Biology and Culture of Percid Fishes*,
DOI 10.1007/978-94-017-7227-3_20

pikeperch larvae and juveniles than marine fish diet, in terms of growth, survival and resistance to stress test. Recently, some experimental dry diets varying by their phospholipid content have been tested. The highest survival and growth rates of pikeperch post-larvae (first fed with *Artemia* nauplii) were obtained in groups fed 9.5 % of PLs but higher levels still need to be tested, eventually at an earlier stage of development.

Keywords Eurasian perch • Zander • Nutritional requirements • Broodstock • Larval feeding

20.1 Introduction

Reproductive strategy, including the seasonality of the reproduction cycle and the energy allocated to the production of gametes, largely influences the fecundity, success of spawning and survival of offspring. According to Wang et al. (2010), percid fishes are classified into the second functional group (among three groups including salmonids in the first one and cyprinids in the third one). In species of the second group, the three phases of the reproductive cycle (namely initiation, gametogenesis and final stages) are all driven by specific photothermal stimuli while temperature is playing a cueing role. Decreasing temperature and photoperiod are inducing the gonad recrudescence and a chilling period allows vitellogenesis (for more details, see Chap. 3), while increasing temperatures synchronize the final stages of maturation.

Energy needs are particularly high during exogenous vitellogenesis which occurs, in percid breeders, from October to late March, when exogenous food resources are low (Sulistyo et al. 1998, 2000). As shown by Blanchard et al. (2004) in Eurasian perch *Perca fluviatilis*, large amounts of nutrients are stored during summer into the perivisceral fat tissue and these lipids are progressively transferred to the ovaries during vitellogenesis. Glycogen and lipids are also stored in perch liver prior to gonad development period, in July-August. Most nutrients required for the completion of vitellogenesis in Eurasian perch and pikeperch *Sander lucioperca* are yet stored in the female body in late summer, while fish have still previtellogenic ovary or are in initiation stage, classifying these species among the "capital" breeders.

In wild Eurasian perch breeders fatty acid (FA) composition of the different tissues varies dependent on seasons. While FA composition remains rather stable throughout the year in muscle, the FA profile changed more markedly in liver and gonads, with a progressive decrease of arachidonic acid (AA) concentration in liver and its transfer to the ovaries during the oocyte maturation (Blanchard et al. 2004). n-3 polyunsaturated fatty acids (n-3 PUFA) are more abundant than n-6 PUFA In all tissues. Among the n-3 FA, docosahexaenoic acid (DHA) is by far the most abundant one in wild breeders of Eurasian perch, throughout all tissues and sampling seasons, followed by eicosapentaenoic acid (EPA). Among the n-6 FA, AA is the most abundant, representing more than 50 % of the total n-6 PUFA. These data are of particular interest and may serve as references for the determination of optimal diets for broodstock and early life stages in culture conditions.

Indeed, feeding plays a major role in the reproductive performances and gamete/larval quality of most teleosts. The influence of feeding level on the fecundity and egg quality has been investigated in several species such as brown trout *Salmo trutta fario* (Billard and de Frémont 1980), European seabass *Dicentrarchus labrax* (Cerda et al. 1994) or cod *Gadus morhua* (Karlsen et al. 1995). In most species, the fecundity is negatively correlated to feed restriction, while the diameter of the ova is influenced by factors related to fish growth rate, age, etc., although the feeding level is less important. As expected, the males are less affected by the feed restriction since the energy allocated to the gonad development is lower than in the females (Bromley et al. 2000; Imsland and Gunnarson 2011), although there are exceptions. Feed restriction also delays the onset of sexual maturation, partially due to a decrease in plasma sex steroids (testosterone and 17β-estradiol) (Cerda et al. 1994; Taranger et al. 2010).

Considering the importance of broodstock and early life stage feeding and nutrition in the whole production cycle of a given species, this chapter reviews the main results obtained so far in the determination of feeding and nutritional requirements of Eurasian perch and pikeperch breeders and early life stages.

20.2 Feeding and Nutrition of Percid Breeders

20.2.1 Feeding Level and Reproductive Performances

The effects of feeding level on the reproductive performances of percid fish have been poorly investigated, while more effort has been devoted to the influence of feed quality (live versus dry feeds, lipid and fatty acid composition). Nevertheless, using a multifactorial approach, Castets (2011) recently evaluated the effects of feed restriction in comparison with satiation feeding on several reproductive indicators such as the condition and sexual status of the breeders, the spawning rate, the fecundity, the size and mass of eggs, the fertilization and hatching rates, in respect to the period of feed restriction along the reproductive cycle of Eurasian perch that is onset of gonad recrudescence or chilling period. Breeders condition and maturation status were not significantly affected by the feed restriction while the plasma 17β-estradiol (E2) levels decreased during the chilling period in fish submitted to repeated fasting phases. These probably affected the vitellogenesis and oocyte growth. According to Castets (2011), the fecundity of the Eurasian perch females was not significantly affected by the feed restriction but, contrary to frequent observations in other species, the egg diameter was reduced. Up to date, no similar investigations have been conducted in pikeperch breeders.

20.2.2 Live and/or Artificial Feeds for Breeders

In intensive fish farming, there is a general trend to limit the use of live feed provided to the broodstock of carnivorous species, and to replace forage fish, either live or frozen by artificial feeds specially formulated for the breeders. This reduces the risk of introducing pathogens into the breeding facilities and standardizes the feeding protocol by providing specific diet of relatively stable composition.

In most salmonids, dry feed has been specifically developed to cover the feeding and nutritional requirements of breeders and natural forage are no more used. The use of live feed or a mixed diet, containing both live and/or frozen forage and dry pellets is still practiced in many species such as European seabass, gilthead seabream *Sparus aurata*, turbot *Psetta maxima* as well as other marine species. Depending on marine farms, natural forage consists of varied mixtures of trash fish, shellfish and other invertebrates, provided either together or alternating with the dry feed on a weekly basis.

Eurasian perch and pikeperch breeders in farming conditions are still largely relying on the supply of live forage, usually constituted of small by-catch either harvested from the wild or extensively produced in cyprinid nursery ponds. In some cases, slow growing fish intensively reared in the farm (including percid juveniles) are also used as food for the broodstock. The importance of providing live food complementary to dry feed has not been extensively investigated yet but existing data suggest that the reproductive performances of the breeders as well as the quality of the offsprings are greatly dependent on the diet composition and source of complimentary live food but also on the breeder stocking conditions.

A 8-month study was conducted with Eurasian perch, by Kestemont et al. (2003) in order to compare the utilization efficiencies of commercial salmonid diets enriched with vitamin E, vitamin C or highly unsaturated fatty acids (HUFAs) with the "natural" food (NF) composed of frozen chironomid larvae and live fish (small cyprinids stocked in ponds). In addition, the effects of breeders holding conditions were assessed by comparing the performance of breeders reared in earthen ponds and fed live fish with those fed very similar diet but kept in recirculating system (RAS). Temperature and photoperiod schedule in RAS simulated the natural conditions prevailing in ponds, with a chilling period in late autumn and winter and a progressive rise of temperature and day length in spring. This seasonal photothermal regime is essential to induce the gonad maturation. As the quality of the gametes may largely depend on the feeding regime of the breeders during the gonad development and oocyte growth, fish were fed the experimental diets from early fall until the next spawning season in spring. The reproductive performance was evaluated in terms of spawning time, fertilisation and hatching rates as well as biochemical composition of eggs. This included lipids, fatty acids and mineral contents, and triiodothyronine (T3) level. Both diet and culture conditions significantly affected the performance of Eurasian perch breeders, as fertilisation and hatching rates were the highest in fish reared in ponds, but differences between NF and formulated diets were also significant (Fig. 20.1). The better egg quality of fish fed NF could be

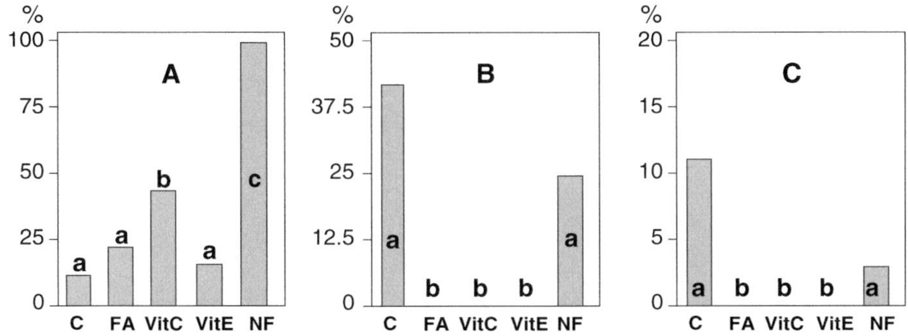

Fig. 20.1 Influence of stocking and feeding conditions on ovulation (**a**), fertilization (**b**) and hatching (**c**) rates of Eurasian perch breeders. *C* control group reared in pond and fed live preys, *FA*, *VitC* et *VitE* groups reared in recirculating system (RAS) and fed commercial salmonid diets enriched with HUFA, vitamin C and vitamin E, respectively. *NF* fish reared in recirculating system and fed « natural » diet (live fish + frozen chironomids) (From Kestemont et al. 2008)

related to an improvement in their biochemical composition as well as the T3 level and the K^+/Mg^+ and K^+/Na^+ ratios. The total lipid content and fatty acid composition of eggs did not significantly differ between treatments.

Castets (2011) demonstrated that the type of feed (live fish *vs* dry pellets) tested during either the gonad recrudescence or the chilling period significantly influenced the condition (Fulton condition factor) and maturation status (gonadosomatic index and plasma E2 level). The spawning rate and fecundity of females fed live forage significantly increased while the effects on males were rather limited.

The biochemical composition of a live fish may be substantially influenced by its own dietary regime and, as a corollary, affects the composition of eggs produced by the breeders fed those live fish as food. Actually, it has been shown that forage fish produced in intensive rearing conditions and fed dry diet can be deprived of some essential fatty acids or contain unbalanced fatty acid ratio, in comparison with fish extensively produced in ponds. In this respect, Henrotte (2009) showed that the fatty acid composition of Eurasian perch eggs was significantly affected by the fatty acid composition of the small cyprinid juveniles, such as ide *Leuciscus idus*, roach *Rutilus rutilus* and common carp *Cyprinus carpio* used as food. Fish cultivated in ponds versus fish produced in RAS fed to broodstock perch showed different n3/n6 ratio and eicosapentaenoic acid/arachidonic acid (EPA/AA ratio). These fatty acid profiles of food resulted in significant differences in perch eggs fatty acids (Table 20.1).

In pikeperch, very few studies have investigated the possibility to feed the breeders with dry diets as a replacement of live forage. Up to date, only one study (Wang et al. 2009) specifically addressed the question of the impact of breeder nutrition on reproductive performances and egg/larval quality of pikeperch. The study was rather preliminary and included comparison of the efficiency of three feeding regimes, i.e. live forage fish (common carp juveniles), commercial salmonid diet specially formulated for enhancing fish reproduction, and a mix of both forage fish

Table 20.1 Comparison of the fatty acid composition of different live preys (FF: mixture of pond reared juveniles, CC: intensively reared common carp juveniles) and Eurasian perch eggs obtained from breeders fed the corresponding preys

Fatty acid	Forage fish diet	Common carp diet	FF eggs	CC eggs
Total SFA	24.2	33.6	17.3±2.6	18.9±2.9
14:0	1.8	20.0	4.4±0.5	4.8±1.4
16:0	16.8	9.3	5.8±2.2	5.9±0.8
18:0	5.6	4.2	1.3±0.6	1.1±0.3
20:0	0.0	0.2	5.8±4.8	7.2±2.0
Total MUFA	27.1	27.3	33.3±3.6	15.2±2.0
16:1 n-7	7.4	1.6	17.9±1.8	19.6±1.1
18:1 n-9	13.3	21.4	13.3±2.3	12.4±1.2
18:1 n-7	5.3	3.0	1.5±0.6	1.6±0.2
20:1 n-9	1.2	2.4	0.6±0.8	0.3±0.2
Total PUFA	48.7	39.1	49.3±3.3	65.9±2.2
16:2 n-4	0.4	0.6	0.2±0.1	0.2±0.0
16:3 n-4	1.2	0.6	0.2±0.2	0.3±0.1
18:3 n-4	0.2	0.4	0.3±0.1	0.3±0.1
18:4 n-1	0.0	0.9	0.8±0.8	0.9±0.4
Total n-6	15.1	6.4	19.6±9.2	10.6±4.0
18:2 n-6	8.0	5.3	18.6±9.1	9.6±4.2
20:4 n-6 (AA)	7.1	1.2	1.0±0.0	1.0±0.2
Total n-3	31.8	28.8	28.3±6.1	34.1±3.4
18:3 n-3	3.9	1.7	4.1±0.2	4.2±0.8
20:4 n-3	1.0	1.0	0.5±0.1	0.5±0.1
20:5 n-3 (EPA)	7.5	11.5	3.2±1.3	5.3±1.5
22:5 n-3	2.6	2.4	1.1±0.5	1.4±0.4
22:6 n-3 (DHA)	15.9	12.0	18.6±3.7	22.5±1.8
n-3/n-6	2.1	4.5	1.7±0.9	4.4±4.9
DHA/EPA	2.1	1.0	6.4±2.1	4.6±1.4
EPA/AA	1.1	10.0	3.1±1.3	5.3±0.8

Modified from Henrotte (2009)

and salmonid feed (supplied alternatively on the basis of forage fish 3 days a week and dry diet 4 days a week). The dry feed contained 44 % proteins and 16 % lipids and was enriched in vitamin A (15,000 IE/kg). Similarly to Eurasian perch, pike-perch breeders were fed the tested regime from early September to late May, in order to cover the whole gonad maturation and breeding seasons. Fish were submitted to simulated water temperature and photoperiod profiles, allowing gravid females and spermiating males to reach maturity during the regular spawning season. Breeders were hormonally treated with hCG, and allowed to spawn naturally on nests, according to the procedure described in Chap. 4. All criteria of evaluation of reproductive performance (percentage of running males, spawning and embryo hatching rates) were significantly lower in broodstock groups fed the salmonid dry diet compared with the groups fed forage fish and mixed diet. The general

Table 20.2 Reproductive performances and larval morphology, composition and robustness in pikeperch fed with live fish (FF), dry diet (DD) or a mixture of live fish and dry diet (FD)

	Diets		
Breeders	Live fish diet	Mixed diet	Dry diet
Running males (%)	76	89	54
Spawning rate (%)	100	88	57
Broodstock mortality (%)	14	25	43
Larvae			
Body weight (mg)	0.39±0.09	0.40±0.05	0.34±0.22
Lipid content (%)	24.8±4.7	32.3±7.1	33.4±9.1
Neutral lipid (%)	84.7±8.7	81.1±5.1	82.5±2.5
Polar lipid (%)	15.3±8.7	18.9±5.1	17.5±2.5
Total MUFA (% of total lipids)	46.7±5.3	46.1±4.6	37.8±6.5
Total n-6 PUFA (% of total lipids)	6.2±0.8	5.8±0.7	8.5±1.0
Total n-3 PUFA (% of total lipids)	31.7±3.1	32.0±2.7	35.4±3.8
n-3/n-6	5.4±0.3	5.5±0.3	4.2±0.4
DHA/EPA	3.0±0.2	3.4±0.2	2.7±0.2
EPA/AA	7.1±1.1	5.4±0.9	8.1±1.3
Resistance to osmotic shock[a] (%)	53±43	50±39	42±12
Resistance to fasting[b] (h)	141±67	212±91	292±28

Modified from Wang et al. (2009)
MUFA monounsaturated fatty acids, *PUFA* polyunsaturated fatty acids
[a]Three groups of 30 larvae challenged with a 3 % saline water solution during 30 min
[b]Three groups of 100 larvae let to starve and determination of LT_{50} (time for 50 % of death)

characteristics of larvae (size at hatching, biochemical composition, resistance to osmotic shock and extended fasting) did not differ significantly between progenies from different feeding groups (Table 20.2). As suggested in regard to broodstock Eurasian perch, the fact that forage fish were also fed a dry diet could at least partially explain the relatively similar fatty acid composition of pikeperch larvae, regardless of the breeder diet.

20.2.3 Effects of Lipid and HUFA Composition on Reproductive Physiology, Spawning Performances, Gametes and Larval Quality

Aside of the amount of feed provided to the breeders, the nutritional quality of the feed is also known as an important determinant of gonad maturation and fecundity in many fish species. As shown above, low reproductive performances have been reported in Eurasian perch and pikeperch when the breeders were fed exclusively artificial diets formulated for salmonids (Kestemont et al. 2008). This suggests that the feed was not adapted to the reproductive requirement of these species. Some authors have used the biochemical composition of eggs, such as the fatty acid composition, as an indicator of egg quality, which may be manipulated via the maternal diet (Watanabe 1982; Bell et al. 1997). Most works have shown that n-3 HUFAs, especially EPA and DHA, are essential in marine fish broodstock diets, playing critical functions as the main components of phospholipids of cell membranes. DHA is particularly important during larval fish ontogeny, especially for development of neural and visual functions (Watanabe 1982; Izquierdo et al. 2001). Not only deficiency, but also excess of n-3 HUFAs negatively affect fish egg quality (Lavens et al. 1999; Furuita et al. 2000, 2002). On the other hand, as shown by Bruce et al. (1999), n-6 HUFAs play an important role in reproduction. AA (arachidonic acid 20:4 n-6) is involved in cell-mediated functions and as precursor of the eicosanoids. AA is the major precursor of the eicosanoids, including the synthesis of two-series prostaglandins (PGs) such as PGE2 and PGF2α. Those prostaglandins are endocrine factors in fish oocyte maturation and ovulation (Mercure and Van der Kraak 1996; Sorbera et al. 2001). However, EPA competitively interferes with eicosanoid production from AA because both 2- and 3-series PGs are catalysed by the same cyclooxygenase and lipoxygenase (Sargent 1995). In fish as in mammals, the eicosanoid production is influenced by the cellular ratio of EPA/AA (Sargent et al. 2002) and has an impact on the quality of reproduction.

If the effects of dietary HUFA composition and ratio provided to the breeders on egg quality and larval survival and growth have been extensively investigated in many species, little attention has been given to the effects of these HUFAs on endocrine changes. AA has been shown to be involved in steroidogenesis via its conversion into PGE2 while EPA and DHA inhibit gonadotropin-stimulated testosterone production in goldfish *Carassius auratus* and rainbow trout *Oncorhynchus mykiss* (Mercure and Van der Kraak 1996). Moreover, inadequate levels of dietary fatty acids are susceptible to influence the plasma estradiol and testosterone concentrations as well as the fatty acid composition of the liver, affecting the vitellogenin (VTG) synthesis and plasma profile in European seabass (Cerda et al. 1994; Navas et al. 1997).

The influence of dietary HUFAs ratios (namely DHA/EPA/AA) on the reproductive physiology and gametes quality of Eurasian perch males and females was recently studied by Henrotte (2009) and Henrotte et al. (2010a, b). Different experiments were conducted throughout several complete reproductive cycles with perch breeders held from early September to late April under natural photoperiod and temperature conditions, either in pond-based cages or in outdoor recirculating system (outdoor RAS) and fed formulated diets in which cod muscle meal was the unique protein source while fish and plant oils were added in such a way that the diets markedly differed by their DHA/EPA/AA ratios (diet 1: 2/1/1, diet 2: 7/4/3, diet 3: 7/3/1 and diet 4: 23/9/1) (Table 20.3). In the pond-based cages experiment, a live diet consisting in forage fish (FF) with a DHA/EPA/AA ratio similar to the one of diet 1 was also added, while in outdoor RAS, the live diet consisted in intensively reared common carp juveniles rather rich in n-3 HUFAs (DHA/EPA/AA: 10/10/1) and a commercial diet for salmonids (CDS) was used as "negative" control. Fish samplings were performed at regular intervals (October, January and March) in order to assay the plasma VTG and PGs (PGE2 and PGF2α) concentrations in

Table 20.3 Composition of a commercial salmonid diet (CDS) and different experimental diets (D1 to D4) based on fish, plant and synthetic oils and formulated to compare the effects of DHA/EPA/AA ratios on breeder performances in Eurasian perch

Ingredients (%)	Diet 1 (2/1/1)[a]	Diet 2 (7/4/3)[a]	Diet 3 (7/3/1)[a]	Diet 4 (23/9/1)[a]	CDS (14/16/1)[a]
Cod muscle meal	50.5	50.5	50.5	50.5	
Menhaden oil	0.0	7.0	0.0	16.0	
Sardine oil[b]	12.0	7.0	12.8	0.0	
Safflower oil	3.2	0.0	3.0	0.0	
Vevodar oil	0.8	2.0	0.15	0.0	
Starch	20.0	20.0	20.0	20.0	
Attractants	1.0	1.0	1.0	1.0	
Carboxymethylcellulose	2.0	2.0	2.0	2.0	
Vitamin mix	0.5	0.5	0.5	0.5	
Mineral mix[c]	2.0	2.0	2.0	2.0	
α-cellulose	6.3	6.3	6.3	6.3	
Astaxanthin	50 ppm	50 ppm	50 ppm	50 ppm	
BHA-BHT	100 ppm	100 ppm	100 ppm	100 ppm	
Proximate composition (%)					
Crude protein	45.0	45.0	45.0	45.0	44.0
Crude lipid	16.0	16.0	16.0	16.0	16.0
Carbohydrate	20.0	20.0	20.0	20.0	17.5
Fatty acid composition					
Total SFA	17.3	25.1	15.7	26.9	35.6
Total MUFA	22.7	22.5	20.3	26.2	17.0
Total PUFA	60.0	52.4	64.1	46.9	47.5
20:4 n-6 (AA)	3.0	5.5	0.7	1.1	1.0
20:5 n-3 (EPA)	3.1	7.0	2.4	10.4	15.9
22:6 n-3 (DHA)	5.1	9.1	4.8	20.3	14.4
n-3/n-6	0.2	0.7	0.1	7.0	5.6
DHA/EPA	1.6	1.3	2.0	1.9	0.9
EPA/AA	1.0	1.3	3.3	9.2	15.5

Modified from Henrotte (2009)
[a]DHA/EPA/AA ratio
[b]La sirène X21 (a mixture of sardine oil and rapeseed oil used in sea fishing)
[c](g per kg of premix): $(CaHPO_4)_2H_2O$: 727.77; $(MgSO_4)_7H_2O$: 127.5; NaCl: 60.0; KCl: 50.0;$(FeSO_4)_7H_2O$: 25.0; $(ZnSO_4)_7H_2O$: 5.5; (MnSO4)4H20: 2.54; (CuSO4)5H2O: 0.78; $(CoSO_4)_7H_2O$: 0.48; $(CaIO_3)_6H_2O$: 0.29; $(CrCl_3)_6H_2O$: 0.13

females. Just before the spawning season, semen was collected from several males per diet group (restricted to diets 1 and 4) and immediately analysed in terms of sperm volume, density, motility, velocity and osmolality, while lipid and fatty acid analyses were done on stored samples. As shown in Fig. 20.2, the fish fed similarly low dietary EPA/AA ratios (diets 1 and 2 and FF) displayed enhanced plasma PGE2 concentrations and Alkali Labile Phosphate (ALP) levels, this latter being an indicator of plasma VTG concentration in fish.

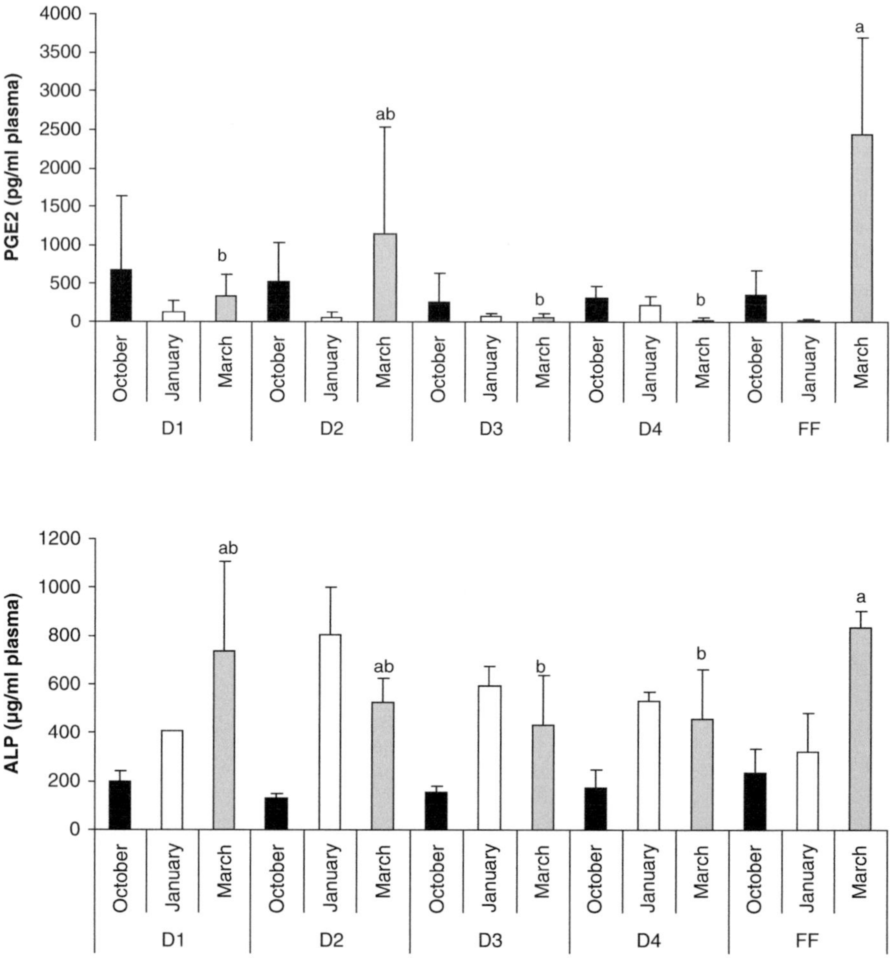

Fig. 20.2 Seasonal variations of plasma PGE2 and ALP (vitellogenin marker) levels in Eurasian perch females fed experimental diets containing different DHA/EPA/AA ratios (D1 to D4) or live forage fish (FF) (From Henrotte 2009)

In vitro investigations on Eurasian perch follicles incubated with DHA, EPA or AA confirmed the results obtained in vivo and demonstrated that only AA was able to induce the production of 17,20β-dihydroxy-4-pregnen-3-one (DHP), the hormone produced by vitellogenic follicles undergoing final meiotic maturation (ovulation), as well as the production of PGE2 and PGF2α (Henrotte et al. 2011).

The spawning performances and the quality of eggs and larvae obtained from Eurasian perch breeders held in pond-based cages and outdoor RAS (placed in outdoor conditions and thus submitted to natural photoperiod regime) were similar, while the dietary treatments significantly influenced the results. As shown in Table 20.4 where data from fish held in outdoor RAS are compared, the number of eggs per ribbon (assuming one ribbon per female) ranged between 32,100 and

Table 20.4 Fecundity, egg, larval and semen quality in Eurasian perch breeders fed experimental diets with different ratios of DHA/EPA/AA (D1, D4), live feed based on common carp juveniles (CC) or commercial salmonid diet (CDS)

	D1 (2/1/1)[a]	D4 (23/9/1)[a]	CC (10/10/1)[a]	CDS (14/16/1)[a]
Spawning quality indicators				
Number of eggs per spawn (×10³)	32.1 ± 18.0	48.5 ± 33.2	47.6 ± 27.7	38.0 ± 19.9
Fertilization rate (%)	63.5 ± 23.0a	62.1 ± 18.8a	66.1 ± 13.5a	26.1 ± 14.6b
Hatching rate (%)	63.5 ± 3.8a	44.6 ± 19.1a	40.3 ± 21.7a	14.7 ± 9.2b
Larval resistance				
Larval survival (%)[b]	39.2 ± 24.7a	4.4 ± 2.8b	10.5 ± 1.2a	0.0c
LT50 (min)[c]	87.6 ± 25.9a	35.1 ± 0.8b	35.5 ± 6.9b	18.0 ± 3.2c
Semen quality indicators				
Volume of sperm (mL)	1.44 ± 1.2	1.03 ± 0.03		
Sperm density (×10⁹spz/mL)	46.1 ± 7.5	58.9 ± 2.1		
Motility after 30s post activation (%)	74.5 ± 12.2	63.8 ± 0.8		
Velocity after 30s post activation (μm/s)	35.0 ± 2.2	29.2 ± 2.9		
Osmolality (mOsm/kg)	293.6 ± 2.9	299 ± 2.4		

Modified from Henrotte (2010a, b)
[a]DHA/EPA/AA ratio
[b]Larval survival calculated after 90 min of osmotic stress (NaCl 20 g/L)
[c]Time needed to observe 50 % of mortality within the osmotic stress

48,500 eggs without differences among dietary groups. Fertilization and hatching rates were similar between experimental and live feed diets while the values were significantly reduced in CDS fed fish. Highly significant differences were recorded in larval quality, as the larvae fed the diet D1 (2/1/1) were much more resistant to osmotic stress test (immersion in a NaCl 20 g L^{-1} saline solution during 90 min) than the other groups (Table 20.4). Lipid classes in eggs were not significantly influenced by the diets, with a predominance of neutral lipids (NL), up to 84 %, and 16 % of polar lipids (PL), reflecting the lipid class content of the diets. Phosphatidylcholine (PC) largely dominated the PL with more than 14 % of total lipid classes, regardless of diets. Fatty acid composition of eggs reflected to some extend the composition of the dietary FAs, with a higher level of PUFAs of the n-6 family (18.6 % of total FAs) and a relatively low level of EPA (2.9 % of total FAs) in fish fed diet 1 compared with eggs from breeders fed live preys (10.6 % and 5.3 %, respectively) (Henrotte et al. 2010a).

In Eurasian perch males, none of the characteristics of the sperm were significantly modified by the HUFAs ratio, neither in terms of sperm volume and density, spermatozoa motility and velocity, nor in terms of seminal plasma osmolality (Table 20.4). In contrast to observations made in females, the lipid composition of semen was influenced by the dietary lipid class composition, as observed in fatty acid profiles. Sperm contained higher levels of PL than amounts observed in eggs. Oocytes are enriched in NL in order to cover the energy required for the embryonic development and early larval stages. Neutral lipids were mainly constituted by cholesterol, while the main PLs were the phosphatidylethanolamine (PE) and the phosphatidylcholine (PC) esters. Regarding the fatty acids, Eurasian perch semen contained more saturated FA than eggs, and more EPA and DHA. While, as expected, the n-3/n-6 ratio was higher in the sperm of fish fed the diet 4 (poor in AA), a basal level of FAs of the n-3 series was maintained in the semen, mainly due to the high level of DHA in the cell membranes (Henrotte et al. 2010b).

The recent nutritional experiments conducted on Eurasian perch breeders demonstrated that a compound diet formulated with an adequate ratio of DHA/EPA/AA (2/1/1) can support spawning and gametes and larval quality were similar to those obtained with a diet composed of forage fish. The inappropriate HUFA ratio (high EPA/AA ratio) significantly impaired their reproductive performance. Egg quality appears more sensitive to the HUFA ratio than the semen quality. In pikeperch, the breeder nutritional requirements and the influence of dietary fatty acid composition on reproductive performance have not been investigated. In both species, the possible use of dry diets during the whole ongrowing stage, followed by an appropriate period of feeding with a specific reproduction-oriented "finishing" diet before the spawning season is still to be evaluated.

20.3 Feeding and Nutrition of Early Life Stages

20.3.1 Changes in Biochemical Composition

During the embryogenesis and early larval development of most species, growth and energy provision mainly depend on endogenous yolk reserves transferred by the breeders to the eggs. While salmonid species produce large eggs and, as a corollary, large fry obtaining their nutrients from the yolk during several weeks, percid fish larvae, despite their carnivorous feeding habit, are rather small at hatching (5–6 mm total length, 0.7–0.9 mg body weight) and possess a relatively small yolk sac. Yolk sac can ensure the survival of larvae during only few days if no exogenous feed is provided. Abi-ayad et al. (2000, 2004) reported that 90–95 % of Eurasian perch and pikeperch larvae died after 10 days of starvation at 15–17 °C. It is well established that biochemical composition varies during embryonic and larval development of fish (Tocher et al. 1985a, b) and nutrients are used differently and selectively by different fish species. Embryos of rainbow trout use proteins, lipids and carbohydrates to satisfy their energy requirements whereas embryos of turbot catabolize exclusively proteins and carbohydrates (Boulekbache 1981; Planas et al. 1989). Among the nutrients, lipids and fatty acids are considered as a major energy source while, as mentioned above, PUFAs are structural components during organogenesis (muscles, brain, retina, etc.) and precursors of active molecules (prostaglandins and other eicosanoids) rather than energy substrates (Bell and Tocher 1989; Sargent 1995).

When investigating the dynamics of total lipids and fatty acids during embryogenesis and larval early life of Eurasian perch, Abi-ayad et al. (2000) showed that, similarly to turbot, perch have not catabolized lipids and fatty acids as energy source during the embryonic development. The authors also indicated that between the last day of embryogenesis (8 days post-fertilization) and the first day after hatching, there is a significant increase in lipid content, possibly due to the increase of fatty acids in relation to chorion exclusion. During the early larval growth, the utilization of total lipid and fatty acids by fed larvae is intense, compared with other species such as the winter flounder *Pseudopleuronectes americanus* (Cetta and Capuzzo 1982) or the common carp (Murata and Higashi 1980). There is an apparent preference in the utilization of PUFAs and MUFAs, while saturated fatty acids are not utilized by starved Eurasian perch larvae. This observation was also reported by Dabrowski et al. (1991) in yellow perch *Perca flavescens* in which 16:0 and 18:0 did not vary during the early ontogeny. Among the MUFAs, palmitoleic (16:1n-7) and oleic (18:1n-9) acids are the main fatty acids catabolized and their contribution as energy source from total fatty acids catabolized over the first week reaches up to 37.6 % in starved fish. Among n-6 PUFAs, starved perch utilize less linoleic acid (LA, 18:2n-6) and AA. More linolenic acid (LNA, 18:3n-3) and less EPA and DHA are utilized by starved larvae, suggesting some sparing strategy for these essential HUFAs in perch is taking place (Abi-ayad et al. 2000).

Compared with lipid and fatty acid profiles of Eurasian perch, total lipid contents of pikeperch larvae are characterized by a higher level of PUFAs (55.9 % in pikeperch versus 50.6 % in Eurasian perch) (Abi-ayad et al. 2004). In contrast to Eurasian perch, pikeperch larvae utilize all, saturated and unsaturated, fatty acids, but utilization of saturated fatty acids is low compared with the one of MUFAs and PUFAs. Among MUFAs, 16:1, 18:1 and 20:1 are preferentially used as energy substrate, particularly in starved larvae. However, this assumption is highly preliminary because the difference between synthesis and catabolism was not examined. Among the n-6 PUFAs, LA is catabolized in starved larvae from the first day of food deprivation while AA concentration only decreases after 5 days, again suggesting a preservation of this essential fatty acid during the early larval development. DHA, LNA and EPA were progressively and intensely utilized by starved pikeperch larvae. According to Abi-ayad et al. (2004), the contribution of DHA for tissue renewal and energy requirements reached up to 20.3 % of the total fatty acids catabolized compared with 8.2 % in Eurasian perch larvae (Abi-ayad et al. 2000). It seems thus that DHA is more important in pikeperch during its early life stages than in Eurasian perch larvae and Senegal sole.

20.3.2 Live and/or Artificial Feeds During Early Life Stages

In its natural environment, live feed requirements of Eurasian perch and pikeperch larvae vary in type and size from the end of yolk sac absorption to the juvenile stage (Craig 2000). Rotifers and copepod nauplii are the first food for the larvae. In Lake Constance daphnids dominate the diet of Eurasian perch 2 weeks after hatching, as reported by Wang (1994). According to Wang and Eckmann (1994), the most critical period of early larval development occurs 10–15 days after hatching in a stochastically-fluctuating environment (both in respect to temperature and food density), affecting the survival of larvae and hence year-class strength in a predictable fashion.

As mentioned above, percid larvae are rather small in comparison with other freshwater fish and the mouth gape size (0.36 and 1.2 mm for Eurasian perch and walleye, respectively) is a limiting factor in the initial feeding of larvae with live prey, especially in Eurasian perch (Kestemont et al. 1996). According to the latter authors, about 60–70 % of the newly-hatched larvae of Eurasian perch ingest *Artemia* nauplii (small size strain of 420–480 µm) as a starter live feed and the remaining 30–40 % of the larvae are lost gradually during the first week of rearing. This appears to be due to exhaustion and to sibling cannibalism by the growing larvae which were able to ingest *Artemia* nauplii. Vlavonou et al. (1995) reported higher survival rates (up to 85.3 % after 14 days) with *Artemia* nauplii as the exclusive starting food. Differences in the size of larvae at hatching, which directly depend on the egg size, itself related to the size of the female, as well as in the size of newly hatched *Artemia* nauplii may explain such differences in larval performance between authors. The time of first feeding with small size *Artemia* nauplii also influences the survival and growth of Eurasian perch larvae. Better results were obtained when fish were fed from day 2 or day 3 post-hatching than from day 0 or day 1 (Kestemont et al. 1996). Lower growth rates were recorded in the two latter groups probably due to the presence of too big or too fast swimming preys which contributed to larval exhaustion. The supply of small zooplankton such as the freshwater rotifer *Brachionus calyciflorus* (mass cultivated with green algae) or sieved lake zooplankton (mainly rotifers *Polyarthra* sp. and *Conochilus* sp. and, to a lesser extent, copepod nauplii) has been tested as first feed by Awaïss et al. (1992, 1996) and Wang and Eckmann (1994) (Table 20.5). In order to assess the influence of adequate size of food on survival and growth of large populations of perch larvae reared in recirculating systems, and to reduce the early selection for "large mouth" larvae, Kestemont et al. (1996) compared the survival and growth rate of perch larvae divided into two treatments: (1) fed 100 % *Artemia* nauplii from day 2 to day 16 from mass hatching or (2) fed a mixed prey diet (rotifers + *Artemia* nauplii) during 4–5 days (all larvae being thus able to capture live prey, regardless of mouth size) and then 100 % *Artemia* nauplii. Survival and total tank production were significantly higher in the mixed live diet groups than in the 100 % *Artemia* fed groups (Table 20.5).

In contrast to Eurasian perch and yellow perch, the mouth size of pikeperch allows early ingestion of relatively large zooplankton organisms, such as copepods and cladocerans of 500–1100 µm total length (Hilge and Steffens 1996). Due to a high growth heterogeneity and subsequent cannibalism (see Chap. 10), survival of pikeperch larvae is rather low, averaging 25–30 % on day 35 post-hatching, even when secured frequent supply of *Artemia* nauplii is secured (Klein Breteler 1989; Steffens et al. 1996). Ostaszewska et al. (2005) obtained 54 % survival and 212 mg individual mass of pikeperch juveniles after 5 weeks of rearing with live *Artemia* nauplii. Ostaszewska et al. (2008) confirmed association of the highest growth rate with hyperplastic growth of white muscle fibers.

Table 20.5 Effects of live and dry diets on the survival and growth of Eurasian perch and pikeperch during their early life stages

Diet	Duration (days)	Initial mass (mg)	Final mass (mg)	Survival (%)	References
Perca fluviatilis					
Live food					
Rotifers (lake sieved)	13	0.8	4.1	41.7	Wang and Eckmann (1994)
Rotifers (mass cultured)	10	0.8	5.2	83.5	Awaïss et al. (1992)
Artemia (standard)	18	0.8	20.9	34.3	Kestemont et al. (1996)
	44	0.8	292	33	Mélard and Kestemont (1994)
	21	7.6	135	75.8	Fiogbé et al. (1995)
	29	12.6	67.7	79.7	Vlavonou et al. (1995)
Artemia (HUFA enriched)	21	7.6	100	71.7	Fiogbé et al. (1995)
Artemia (frozen)	21	7.6	55.3	45.6	
Rotifers + *Artemia* nauplii	18	0.8	17.9	50.2	Kestemont et al. (1996)
Copepodites, daphnids	12	4.1	23.9	65	Wang and Eckmann (1994)
Dry feed					
Larval shrimp feed	7	0.8	0.9	4	Awaïss et al. (1992)
Salmonid feed	15	0.8	2.6	25	Tamazouzt (1995)
NS[a]	21	8.7	90.2	12	Fiogbé et al. (1995)
Co-feeding (live + dry feed)					
Rotifers + dry feed	10	0.8	4.2	42.5	Awaïss et al. (1992)
Artemia + dry feed 75:25	44	0.8	300	20.2	Mélard and Kestemont (1994)
Artemia + dry feed 50:50	44	0.8	329	19.8	
Sander lucioperca					
Live food					
Zooplankton (lake sieved)	19–28	0.3–0.5	70–80	27–44	Schlumpberger and Schmidt (1979)
Artemia (standard)	14	3.4	105.8	64.5	Mamcarz et al. (1997)
	35	0.8[2]	212	54.4	Ostaszewska et al. (2005)
	24	3.1	363	24.8	Kestemont et al. (2007)
	18	8.1	301	71.4	
Artemia (HUFA + vitC enriched)	18	8.1	373	70.9	
Artemia + zooplankton	35	0.5	120–150	26.7	Klein Breteler (1989)

(continued)

Table 20.5 (continued)

Diet	Duration (days)	Initial mass (mg)	Final mass (mg)	Survival (%)	References
Light-attracted zooplankton	7	16	90	77	Jäger et al. (1984)
Dry feed					
NS[a]	8–18	0.5	3–4	Very low	Ruuhijärvi et al. (1991)
NS[a]	23	0.3–0.5	120–150	<1	Schlumberger and Proteau (1991)
Marine larval feed	35	0.8[b]	190	52.4	Ostaszewska et al. (2005)
Casein-based diet[c]	35	0.8[b]	53.8	28.4	
Casein hydrolysates-based diet[d]	35	0.8[b]	55	21.6	
Freshwater larval feed	18	8.1	231	77.4	Kestemont et al. (2007)
Marine larval feed	18	8.1	144	63.9	
PL1.4-based diet[e]	25	2.5	160	34.1	Hamza et al. (2008)
PL4.7-based diet[e]	25	2.5	190	36.2	
PL9.5-based diet[e]	25	2.5	238	33	

[a]*NS* not specified
[b]Day 5 post-hatching
[c]Experimental feed based on casein
[d]Experimental feed based on casein + casein hydrolysates
[e]Experimental feed containing 1.4, 4.7 or 9.5 % phospholipids from soybean lecithin (dry matter basis)

The development of formulated larval feeds for percid fishes brings several advantages over the use of live feeds, including the savings in labour costs associated with the collection or culture of live organisms and a reduction in the risk of introduction of disease and contaminants. In addition, the energetic value of a dry feed is usually higher than the one of live prey, resulting in higher growth rate when all nutrients are provided in sufficient and adequate proportions. As mentioned by Barrows et al. (1988) in respect to walleye, the cost of formulated larval feeds will probably be less than the actual costs of *Artemia* cysts. However, major constraints to feeding dry diets include poor acceptance of feed particles resulting in starvation, cannibalism and reduced water quality (uneaten feed) with associated diseases. While feeding north-American percid fishes (mainly yellow perch and walleye) with formulated diets have been investigated since the 1970s (Beyerle 1975; Nickum 1978; Best 1981; Hinshaw 1985; Heidinger and Kayes 1986; Colesante et al. 1986), similar feeding experiments on Eurasian perch and pikeperch are more recent, starting during the 1990s. The first attempts conducted by Awaïss et al. (1992) and Tamazouzt (1995) indicated that Eurasian perch larvae ingested dry diet from hatching, but the best survival rate reached was only 26 % after 15 days. The growth was very low, from 0.8 to 2.6 mg (individual mass), while fish fed *Artemia* nauplii grew

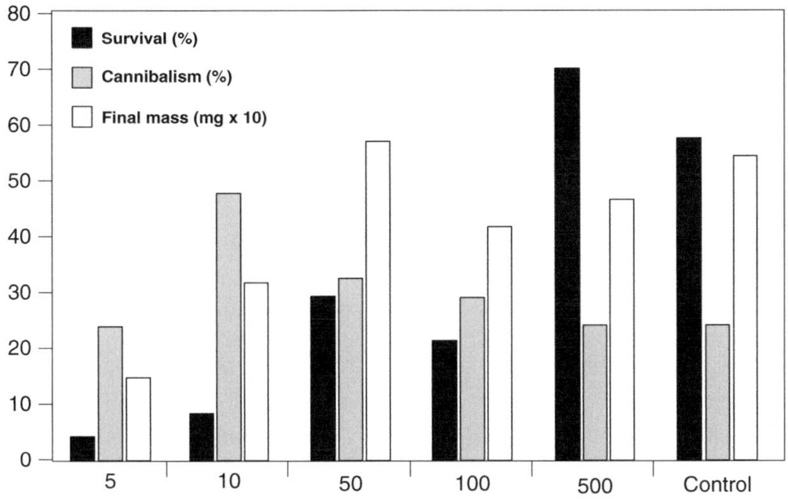

Fig. 20.3 Effects of weaning body mass and age on survival, cannibalism and growth in Eurasian perch larvae. Fish were initially fed *Artemia* nauplii and then progressively conditioned at different body mass to accept dry feed in 5 days. Control fish were fed only *Artemia* nauplii (T = 20 °C, duration = 50 days) (Modified from Kestemont and Mélard 2000, with authorization from Blackwell)

up to 33 mg during the same period (Table 20.5). Considering the poor growth obtained with a dry diet from hatching onwards, it appeared more appropriate to feed the newly hatched larvae with live feed or mixed diets (live + dry feeds) up to a stage allowing successful weaning to solely formulated feed. Mélard and Kestemont (1994) compared the growth of Eurasian perch fed with various proportions of *Artemia* nauplii and dry food (*Artemia*:dry food 100:0, 75:25 or 50:50) from hatching up to 44 days old (Table 20.5). Fish were then transitioned to dry diet exclusively, from day 45 to day 132. During the pre-weaning period, survival was the highest in the groups fed solely on *Artemia* nauplii and growth rate was similar in all treatments, with a mean weight ranging from 292 to 329 mg. However, during the weaning period, habituation to dry feed and growth of fish first-fed solely with live food (final weight = 2.47 g) was significantly lower than those of fish first-fed with mixed diets (final weight = 16.2–17.1 g). Kestemont et al. (1996) demonstrated that the time (and thus size) of weaning had a profound effect on the survival and growth rates of Eurasian perch larvae, and recommended a progressive transfer of larvae to dry feed from 50 mg body mass, when fish are about 21 days old (Fig. 20.3).

Similar investigations were conducted with pikeperch larvae by Kestemont et al. (2007) in order to determine the optimal weaning period, considering not only the survival and growth rates, but also the cannibalism, deformity and stress resistance of larvae/juveniles. Pikeperch larvae were fed *Artemia* nauplii from day 3 to day 11, 18 or 25 post-hatching and then gradually habituated within 4 consecutive days to a high quality dry diet. Three weaning ages were compared: day

12, 19 and 26 post-hatching. At day 35, the highest survival (24.8 %) was obtained in control fish fed only *Artemia* nauplii while survival in groups fed a dry diet ranged between 13.3 and 17.7 %. There were no significant differences between groups. The highest gain of body mass and specific growth rate were achieved in both control groups (mean weight of 360.2 mg and growth rate of 18.3 % day^{-1}) and fish weaned at day 19 (380.8 mg and 18.5 % day^{-1}, respectively), while the growth rate was lower in fish weaned at days 12 or 26 (Table 20.5). The highest rate of morphological deformities, mainly mandibular and maxillary bones, were found in fish fed *Artemia* only while the lowest rate of deformities was recorded in fish weaned at day 19. The authors recommended habituating the pikeperch larvae to dry feed after 19 days post-hatching. When investigating the effects of weaning age and diets on the ontogeny of digestive enzymes activities and structure of pikeperch larvae intestine, Hamza et al. (2007a) recommended a transfer to dry feed around the same period (namely at day 15 post-hatching) while earlier weaning time impaired the onset of digestive system maturation (see more details in Chap. 8). Ostaszewska et al. (2005) evaluated the performances of different commercial (AgloNorse and Biokyowa) and experimental (based on casein or casein plus casein hydrolysates) diets as starter feed, directly after mouth opening and compared their results with a control group fed *Artemia* nauplii. According to these authors, survival and growth rate of fish fed *Artemia* nauplii or commercial diets were similar, averaging 51–54 % and about 210 mg, respectively, while fish fed the experimental diets displayed significantly lower survival and growth rates (22–28 % and about 54 mg, respectively). Based on histological observations performed periodically during the first 35 days post-hatching, Ostaszewska et al. (2005) reported that the digestive development of pikeperch larvae and early juveniles was very similar between fish fed *Artemia* nauplii and those fed commercial dry diets, suggesting that pikeperch larvae can be fed directly with dry starter feeds (see also Chap. 8).

New generations of high quality dry feed, principally formulated for marine fish larvae but also suitable for percid fish larvae, are now available in the market and can be used successfully as starter diets for both Eurasian perch and pikeperch. However, the co-feeding strategy, combining the supply of *Artemia* nauplii during few days from mouth opening, followed by mixed diets (live + dry feeds) before providing dry feed exclusively from day 10 to 15 post-hatching, is still a common practice in commercial Eurasian perch and pikeperch farms.

Not only the nutritional composition of the dry feed is of importance, but also the size of the particles as well as the husbandry conditions applied during the first feeding. Food particles smaller than 125 μm are recommended during the first 10 days post-hatching. In Eurasian perch as in yellow perch, dry food detection appears to be improved in tanks with bright wall (white or light grey tank) under relatively high light intensity (800 lux, Tamazouzt 1995). On the other hand, pikeperch larvae are usually kept under lower light intensity (30 lux). Both species are fed during daytime and long day length is recommended (more details in Chap. 9).

20.3.3 Feeding and Nutritional Requirements During Early Life Stages

Compared with salmonids or other major species cultured worldwide, knowledge of percid larval nutrition is still in its infancy, however significant progresses have been made recently in the determination of their nutritional requirements. Ontogeny of the digestive system has been described by different authors (Kestemont et al. 1996; Cuvier-Péres and Kestemont 2002; Ostaszewska et al. 2005, 2008; Hamza et al. 2007) since the mid 1990s and the more detailed information is provided in Chap. 8. Some data are presented below regarding their quantitative feeding and nutritional requirements.

20.3.3.1 Quantitative Requirements

Based on the technique proposed by Kamler et al. (1986) (in which changes in prey numbers are measured in rearing tanks with and without fish), Awaïss et al. (1992) estimated the maximal daily consumption of rotifers by Eurasian perch at 20 °C as 800 rotifers larva^{-1} day^{-1} at day 3 (end of yolk sac resorption) and up to 2200 rotifers larva^{-1} day^{-1} at day 10. Densities of wild zooplankton (0–12,000 rotifers L^{-1}) were tested by Wang and Eckmann (1994) who reported, at 20 °C, the best survival and growth of Eurasian perch larvae at 6000 rotifers L^{-1}. At a larger larval size (12–24 days after hatching), a density of 400–1000 copepodites L^{-1} supported the highest growth rate. As percid larvae are able to ingest *Artemia* nauplii at the inception of the first feeding, relationships between daily feeding level (expressed as % body mass) and specific growth rate (SGR, % day^{-1}) were determined in Eurasian perch for a range of initial body mass from 8–16 to 118 mg (Kestemont and Mélard 2000). Recommended feeding levels strongly varied with the rearing temperature (tested from 14 to 26 °C). At the optimal temperature of 23 °C, a maximum SGR of 22.0 % d^{-1} was achieved when larvae were fed with *Artemia* nauplii at a daily ration of 35 % body mass (on a dry matter basis). This daily ration is reduced gradually to 10 % larval body mass on day 21 post-hatching. In pikeperch, Kestemont et al. (2007) reported a requirement of 500–600 *Artemia* nauplii fish^{-1} day^{-1} for larvae of 3.1 mg individual mass.

20.3.3.2 Nutritional Requirements

Under appropriate conditions in fertilized pond with abundance of zooplankton of appropriate size and composition, the percentage of most indispensable amino acids remained stable in the Eurasian perch from embryo (hatching) up to 2 g body mass (Fiogbé 1996) (Table 20.6). It has been suggested that such stable amino acids profiles can be used as a possible method of predicting essential dietary amino acids in perch larvae (Brown et al. 1996; Hart et al. 2010).

Table 20.6 Indispensable amino acid composition (% total amino acids) in ponds-reared perch, from egg to different larval development stages

Amino acids (%)	Eggs	1 mg body mass	75 mg body mass	340 mg body mass
Arginine		7.1	6.5	6.3
Histidine		2.9	3.3	3.1
Isoleucine		5.2	5.1	5.0
leucine		7.8	7.5	7.3
Lysine		8.6	8.0	8.3
Methionine		2.5	3.0	3.2
Phenylalanine		4.3	4.5	4.6
Tyrosine[a]		4.1	4.1	4.0
Threonine		5.5	5.3	5.3
Valine		5.8	5.7	6.0

Modified from Fiogbé (1996)
[a]Non indispensable amino acid (sparing effect of phenylalanine)

Fatty acid composition of lipids in ovaries and eggs from wild and cultured Eurasian perch (Abi-ayad et al. 1997), but also from yellow perch (Brown et al. 1996), suggested that cultured percid larvae and juveniles may require supplements of fatty acids from the n-3 and n-6 families. In this respect, requirements of Eurasian perch and pikeperch larvae seem different in such a way that no improvements of survival and growth rates were achieved by feeding Eurasian perch larvae with HUFA and vitamin C enriched *Artemia* metanauplii compared with standard *Artemia* (Fiogbé et al. 1995). A similar feeding experiment conducted with pike-perch larvae of similar age and body mass demonstrated the efficiency of *Artemia* enrichment (Kestemont et al. 2007). Enrichment consisted to incubate *Artemia* nauplii during 24 h with a Superselco solution (INVE, Dendermonde, Belgium) in order to increase the dietary content in HUFA and vitamin C. *Artemia* enrichment supported the highest growth rate of pikeperch larvae, significantly reduced the morphological deformities and increased the survival of larvae challenged with saline stress test (30 min at 3 % salinity). Similar positive effects of HUFA and vitamin C enrichment were also reported by Czesny et al. (1998) in walleye larvae although excess of PUFA can be detrimental to fish.

The composition of formulated dry diets strongly affects the performance and composition of percid larvae. A few data have been obtained so far with Eurasian perch larvae while a series of recent experiments have provided interesting results on the nutrition of pikeperch early life stages. When comparing different dry diets formulated for either marine or freshwater fish larvae from day 19 post-hatching (larvae were first-fed *Artemia* nauplii), Kestemont et al. (2007) reported higher survival and growth rate in groups fed freshwater larval diets. The best results were obtained with a diet containing 55 % of crude protein (compared with 45 %). Differences in Ca/P ratio (0.54–0.64 in freshwater fish diet vs 1.2–1.7 in marine fish diets) may explain, at least partially, these growth differences. Phospholipids (PL) have been demonstrated to significantly affect the survival, growth, deformities and resistance of stress in several fish and crustacean species (Kanazawa et al. 1985; Geurden et al. 1998; Koven et al. 1998; Cahu et al. 2003; Gisbert et al. 2005). Hamza et al. (2008) investigated the influence of dietary phospholipid levels (1.4, 4.7 and 9.5 % of total fatty acids of the phospholipid and neutral lipid fractions of the diets) on performances, enzyme activities and fatty acid composition of pikeperch larvae from day 10 to 34 post-hatching. The replacement of fish oil with soybean lecithin led to the increase in dietary PL from 1.4 % to 9.5 %. A 50 % increase in final body mass was observed, suggesting that high PL levels are needed during larval stages of pikeperch growth, but no effects on survival and deformities were evident. The specific activity of brush border membrane enzymes (aminopeptidase N and alkaline phosphatase) increased with dietary PL levels, indicating earlier or more efficient morphological changes of digestive structures. Using the ratio of alkaline phosphatase activity to leucine-alanine peptidase (a cytosolic enzyme) as indicator of gut development, these authors showed that 1.4 % of dietary PL was not sufficient to support normal enterocyte proliferation. Due to PL incorporation, the fatty acid composition differed between the diets, with a decrease of n-3 HUFA at high PL incorporation. The dietary PL contents also induced some significant changes in the liver proteome of fish sampled at day 34 post-hatching (Hamza et al. 2010). The differentially expressed proteins were involved in several cellular processes, most of them being related to intermediary metabolism. The down-regulation of several proteins of the glycolysis and gluconeogenesis pathways at high PL contents (4.7 and 9.5 %) was observed while several stress proteins (glutathione S-transferase M, glucose regulated protein 75) were down-regulated in the liver of fish fed low dietary PL (1.4 %). Hamza et al. (2010) also showed that peroxiredoxin-1 was upregulated at high dietary PL content. More recently, Hamza et al. (2012) evaluated the effects of dietary PL sources (fish gonad or soybean lecithin) and levels (50 and 90 g kg^{-1} dry matter) on the performances and fatty acid composition of pikeperch fed isoproteic and isolipidic microdiets from day 10 to day 34 post-hatching. This study confirmed the previous findings that pikeperch larvae require high dietary PL levels, regardless of their origin, suggesting that PL from plant are as efficient as those from marine fish. Based on dietary FA profiles, growth, skeletal development and fatty acid composition in the PL and NL fractions of the juveniles, optimal level of EPA + DHA for pikeperch larvae appeared to be around 12 g kg^{-1} (dry matter) associated with a PL level around 90 g kg^{-1} (Hamza et al. 2012).

20.4 Conclusions

Although feeding of Eurasian perch and pikeperch breeders in farm conditions still rely on the regular supply of forage fish, in combination or alternately with dry feed, experimental diets, based on appropriate presence of PL and ratio of n-3 and n-6 HUFA (and specially DHA/EPA/AA) have been used successfully to produce high quality eggs and larvae in Eurasian perch. In pikeperch, more research is still needed before completely replacing forage fish by compound diets. A partial replacement provides satisfactory results and allows a significant reduction of forage fish supply.

Even if *Artemia* nauplii are still used as starter feed for both Eurasian perch and pikeperch larvae, nutritional requirements of these species during their early life stage have been investigated quite thoroughly during the last decade and significant progress has been achieved regarding the efficiency of dry diets enriched in phospholipids. New commercial feeds formulated for marine fish larvae and juveniles seem promising and should be able to be used as full substitutes of live preys in a near future.

All in all, feeding and nutrition of percid fish breeder and early life stages have been significantly improved, at least at the experimental level. It is now time for a transfer of knowledge into practise, making the production of percid juveniles a reliable step towards the profitable culture of these species under fully controlled conditions.

References

Abi-ayad S-ME-A, Mélard C, Kestemont P (1997) Effects of n-3 fatty acids in Eurasian perch broodstock diet on egg fatty acid composition and larvae stress resistance. Aquac Int 5:161–168

Abi-ayad S-ME-A, Kestemont P, Mélard C (2000) Dynamics of total lipids and fatty acids during embryogenesis and larval development of Eurasian perch (*Perca fluviatilis*). Fish Physiol Biochem 23:233–243

Abi-ayad S-ME-A, Boutiba Z, Mélard C, Kestemont P (2004) Dynamics of total body fatty acids during early ontogeny of pikeperch (*Sander lucioperca*) larvae. Fish Physiol Biochem 30:129–136

Awaïss A, Kestemont P, Micha JC (1992) Nutritional suitability of the rotifer *Brachionus calyciflorus* Pallas for rearing freshwater fish larvae. J Appl Ichthyol 8:263–270

Awaïss A, Kestemont P, Micha JC (1996) Fatty acid profiles of two freshwater fish larvae (gudgeon and perch) reared with *Brachionus calyciflorus* Pallas (Rotifer). Aquacult Res 27:651–658

Barrows FT, Lellis WA, Nickum JG (1988) Intensive culture of larval walleyes with dry or formulated feed: note on swim bladder inflation. Prog Fish Cult 50:160–166

Bell MV, Tocher DR (1989) Molecular species composition of the major phospholipids in the brain and retina from rainbow trout (*Salmo gairdneri*). J Biochem 264:909–915

Bell JG, Farndale BM, Bruce MP, Navas JM, Carillo M (1997) Effects of broodstock dietary lipid on fatty acid compositions of eggs from sea bass (*Dicentrarchus labrax*). Aquaculture 149:107–119

Best CD (1981) Initiation of artificial feeding and the control of sex differentiation in yellow perch, *Perca flavescens*. M.Sc. thesis, University of Wisconsin-Madison, Madison

Beyerle GB (1975) Summary of attempts to raise walleye fry and fingerlings on artificial diets, with suggestions on needed research and procedures to be used in future tests. Prog Fish Cult 37:103–105

Billard R, de Frémont M (1980) Taux d'alimentation pendant la gamétogenèse et performance de reproduction chez la truite fario. Bull Fr Peche Piscic 279:49–56

Blanchard G, Druart X, Kestemont P (2004) Lipid content and fatty acid composition of target tissues in wild *Perca fluviatilis* females in relation to hepatic status and gonad maturation. J Fish Biol 65:1–13

Boulekbache H (1981) Energy metabolism in fish development. Am Zool 12:377–389
Bromley PJ, Ravier C, Witthames PR (2000) The influence of feeding regime on sexual maturation, fecundity and atresia in first-time spawning turbot. J Fish Biol 56:264–278
Brown PB, Dabrowski K, Garling DL (1996) Nutrition and feeding of yellow perch (*Perca flavescens*). J Appl Ichthyol 12:171–174
Bruce M, Oyen F, Bell G, Asturiano JF, Farndale B, Carillo M, Zanuy S, Ramos J, Bromage NR (1999) Development of broodstock diets for the European sea bass (*Dicentrarchus labrax*) with special emphasis on the importance of n-3 and n-6 highly unsaturated fatty acid to reproductive performance. Aquaculture 177:85–97
Cahu CL, Zambonino-Infante JL, Barbosa V (2003) Effects of dietary phospholipid level and phospholipid: neutral lipid value on the development of sea bass (*Dicentrarchus labrax*) larvae fed a compound diet. Br J Nutr 90:21–28
Castets M-D (2011) Fonction de reproduction et régulation de la qualité chez la perche commune *Perca fluviatilis*. Ph.D. thesis Nancy University – INPL, Nancy, 299 p.
Cerda J, Carillo M, Zanuy S, Ramos J (1994) Effect of food ration on estrogen and vitellogenin plasma levels, fecundity and larval survival in captive seabass *Dicentrarchus labrax*: preliminary observations. Aquat Living Resour 7:255–266
Cetta CM, Capuzzo JM (1982) Physiological and biochemical aspects of embryonic and larval development of the winter flounder (*Pseudopleuronectes americanus*). Mar Biol 71:327–337
Colesante RT, Youmans NB, Ziolkoski B (1986) Intensive culture of walleye fry with live food and formulated diets. Prog Fish Cult 48:126–127
Craig JF (ed) (2000) Percid fishes. Systematics, ecology and exploitation. Blackwell Science, Oxford
Cuvier-Péres A, Kestemont P (2002) Development of some digestive enzymes in Eurasian perch larvae *Perca fluviatilis*. Fish Physiol Biochem 24:279–285
Czesny S, Kolkovski S, Yackey C, Dabrowski K (1998) The effects of (n-3) HUFA enriched *Artemia* nauplii on growth, survival, and quality of walleye *Stizostedion vitreum* fry. Aquaculture'98, Book of abstracts, WAS symposium Las Vegas, 15–19 Feb 1998, Nevada
Dabrowski K, Culver DA, Brooks CL, Voss AC (1991) Biochemical aspects of the early life history of yellow perch (*Perca flavescens*). In: Kaushik SJ, Luquet P (eds) Fish nutrition in practice. INRA Edition, Paris, pp 531–539
Fiogbé ED, Kestemont P, Micha JC, Mélard C (1995) Comparative growth of *Perca fluviatilis* larvae fed with enriched and standard *Artemia metanauplii* (Instar II), frozen *Artemia nauplii* and dry food. In: Lavens P, Jaspers E, Roelants I (eds) Larvi'95 – fish & shellfish larviculture symposium. European aquaculture society, special publication 24, Gent, pp 166–169
Fiogbé ED (1996) Contribution à l'étude des besoins nutritionnels de la perche *Perca fluviatilis* L. PhD thesis, Presses Universitaires de Namur
Furuita H, Tanaka H, Yamamoto T, Shiraishi M, Takeuchi T (2000) Effects of n-3 HUFA levels in broodstock diet on the reproductive performance and egg and larval quality of the Japanese flounder, *Paralichthys olivaceus*. Aquaculture 187:387–398
Furuita H, Tanaka H, Yamamoto T, Suzuki N, Takeuchi T (2002) Effects of high levels of n-3 HUFA in broodstock diet on egg quality and egg fatty acid composition of Japanese flounder, *Paralichthys olivaceus*. Aquaculture 210:323–333
Geurden I, Marion D, Charlon N, Coutteau P, Bergot P (1998) Comparison of different soybean phospholipidic fractions as dietary supplements for common carp, *Cyprinus carpio* larvae. Aquaculture 161:225–235
Gisbert E, Villeneuve L, Zambonino-Infante JL, Quazuguel P, Cahu C (2005) Dietary phospholipids are more efficient than neutral lipids for long chain polyunsaturated fatty acid supply in European sea bass *Dicentrarchus labrax* development. Lipids 40:1–10
Hamza N, Mhetli M, Kestemont P (2007a) Effects of weaning age and diets on ontogeny of digestive activities and structures of pikeperch (*Sander lucioperca*) larvae. Fish Physiol Biochem 33:121–133

Hamza N, Mhetli M, Ben Khemis I, Cahu C, Kestemont P (2007b) Effect of dietary phospholipid levels on performance, enzyme activities and fatty acid composition of pikeperch (*Sander lucioperca*) larvae. Aquaculture 275:274–282

Hamza N, Mhetli M, Ben Khemis I, Cahu C, Kestemont P (2008) Effect of dietary phospholipid levels on performance, enzyme activities and fatty acid composition of pikeperch (*Sander lucioperca*) larvae. Aquaculture 275:274–282

Hamza N, Silvestre F, Mhetli M, Ben Khemis I, Dieu M, Raes M, Cahu C, Kestemont P (2010) Differential protein expression profile in the liver of pikeperch (*Sander lucioperca*) larvae fed with increasing levels of phospholipids. Comp Biochem Physiol 5D:130–137

Hamza N, Kestemont P, Khemis IB, Cahu C (2012) Effect of different sources and levels of dietary phospholipids on performances and fatty acid composition of pikeperch (*Sander lucioperca*) larvae. Aquacult Nutr 18:249–257

Hart SD, Brown BJ, Gould NL, Robar ML, Witt EM, Brown PB (2010) Predicting the optimal dietary essential amino acid profile for growth of juvenile yellow perch with whole body amino acid concentrations. Aquacult Nutr 16:248–253

Heidinger RC, Kayes TB (1986) Yellow perch. In: Stickney RR (ed) Culture of non-salmonid freshwater fishes. Iowa State University Press, Ames, pp 103–113

Henrotte E (2009) Effects of dietary n-6 and n-6 polyunsaturated fatty acids on the lipid metabolism and on reproductive physiology and performances of Eurasian perch breeders (*Perca fluviatilis*). Ph.D. thesis, Presses Universitaires de Namur, Namur, 178 p

Henrotte E, Mandiki SNM, Agbohessi P, Vandecan M, Mélard C, Kestemont P (2010a) Egg and larval quality, and egg fatty acid composition of Eurasian perch breeders (*Perca fluviatilis*) fed different dietary DHA/EPA/AA ratios. Aquacult Res 41:e53–e61

Henrotte E, Kaspar V, Rodina M, Psenicka M, Linhart O, Kestemont P (2010b) Dietary n-3/n-6 ratio affects the biochemical composition of Eurasian perch (*Perca fluviatilis*) semen but not indicators of sperm quality. Aquacult Res 41:e31–e38

Henrotte E, Milla S, Mandiki SNM, Kestemont P (2011) Arachidonic acid induces production of 17,20β-dihydroxy-4-pregnen-3-one (DHP) via a putative PGE2 receptor in fish follicles from the Eurasian perch. Lipids 46:179–187

Hilge V, Steffens W (1996) Aquaculture of fry and fingerling of pikeperch (*Stizostedion lucioperca* L.) – a short review. J Appl Ichthyol 12:167–170

Hinshaw M (1985) Effects of illumination and prey contrast on survival and growth of larval yellow perch (*Perca flavescens*). Trans Am Fish Soc 114:540–545

Imsland AK, Gunnarsson S (2011) Growth and maturation in Arctic charr (*Salvelinus alpinus*) in response to different feed rations. Aquaculture 318:407–411

Izquierdo MS, Fernadez-Palacios H, Tacon AGJ (2001) Effects of broodstock nutrition on reproductive performance of fish. Aquaculture 197:25–42

Jäger T, Nellen W, Soll H (1984) BeleuchteteNetzgehegeanlagen zur Aufzucht von Fischbrut bis zur Setzlingröβe. Eine Bauanleitung und Aufzuchbeschreibung. Ber Inst Meereskd Christian-Albrechts-Univ Kiel, n 126, 72p

Kamler E, Lewkowicz M, Lewkowicz S, Uchmanski J, Urban-Juvierska E (1986) Gravimetric techniques for measuring consumption of live feeds and artificial diets by fish larvae. Aquaculture 54:109–122

Kanazawa A, Teshima SI, Sakamoto M (1985) Effects of dietary lipids, fatty acids, and phospholipids on growth and survival of prawn (*Penaeus japonicus*) larvae. Aquaculture 50:39–49

Karlsen O, Holm JC, Kjesbu OS (1995) Effects of periodic starvation on reproductive investment in first-time spawning Atlantic cod (*Gadus mor*hua L.). Aquaculture 133:159–170

Kestemont P, Mélard C (2000) Chapter 11 – aquaculture. In: Craig JF (ed) Percid fishes. Systematics, ecology and exploitation. Blackwell Science, Oxford, pp 191–224

Kestemont P, Mélard C, Fiogbé ED, Vlavonou R, Masson G (1996) Nutritional and animal husbandry aspects of rearing early life stages of Eurasian perch *Perca fluviatilis*. J Appl Ichthyol 12:157–165

Kestemont P, Xu X, Blanchard G, Mélard C, Gielen M, Brun-Bellut J, Fontaine P (2003) Feeding and nutrition in European percid fishes – a review. In: Barry TP, Malison JA (eds) Proceedings of Percis III, the third international percid fish symposium, University of Wisconsin, Madison, pp 39–40

Kestemont P, Xu X, Hamza N, Maboudou J, Imorou Toko I (2007) Effects of weaning age and diet on pikeperch larviculture. Aquaculture 264:197–204

Kestemont P, Henrotte E, Wang N, Hamza N, Paulsen H, Overton JL (2008) Feeding and nutrition of European percid broodstock and early life stages. In: Fontaine P, Kestemont P, Teletchea F, Wang N (eds) Percid fish culture – from research to production. Presses Universitaires de Namur, Namur, pp 28–34

Klein Breteler JGP (1989) Intensive culture of pikeperch fry with live food. In: De Pauw N et al (eds) Aquaculture – a biotechnology in progress, vol 1. European Aquaculture Society, Bredene, pp 203–207

Koven WM, Parra G, Kolkovski S, Tandler A (1998) The effects of dietary phosphatidylcholine and its constituent fatty acids on microdiet ingestion and fatty acid absorption rate in gilthead sea bream, *Sparus auratus*, larvae. Aquacult Nutr 4:39–45

Lavens P, Lebegue E, Jaunet H, Brunel A, Dhert P, Sorgeloos P (1999) Effect of dietary essential fatty acids and vitamins on egg quality in turbot broodstocks. Aquac Int 7:225–240

Mamcarz A, Kucharczyck S, Kujawa R, Skrzypczak A (1997) Influence of fish density on survival, growth and cannibalism development of pikeperch (*Stizostedion lucioperca* L.) larvae. In: Harache Y, Creswell L (eds) Island aquaculture and tropical aquaculture. Second international workshop on aquaculture of percid fish. European Aquaculture Society, special publication, Oostende, pp 343–346

Mélard C, Kestemont P (1994) Diversification de la pisciculture en Région wallonne. Mise au point de l'élevage intensif de nouvelles espèces de consommation et de repeuplement – cyprinidés et percidés. Rapport de recherches à la région wallonne, Université de Liège, Facultés Universitaires N.D. de la Paix à Namur, mars 1994, 68 pp

Mercure F, Van der Kraak G (1996) Mechanisms of action of free arachidonic acid on ovarian steroid production in the goldfish. Gen Comp Endocrinol 102:130–140

Murata H, Higashi T (1980) Selective utilization of fatty acids as energy source in carp. Bull Jpn Soc Sci Fish 46:1333–1338

Navas JM, Bruce M, Thrush M, Farndale BM, Bromage NR, Zanuy S, Carillo M, Bell JG, Ramos J (1997) The impact of seasonal alteration in the lipid composition of broodstock diets on egg quality in the European sea bass. J Fish Biol 51:760–773

Nickum JG (1978) Intensive culture of walleyes: the state of the art. In: Kendall RL (ed) Selected Coolwater Fishes of North America, Special Publication 11. American Fisheries Society, Washington, DC

Ostaszewska T, Dabrowski K, Czuminska K, Olech W, Olejniczak M (2005) Rearing of pike-perch larvae using formulated diets – first success with starter feeds. Aquacult Res 36:1167–1176

Ostaszewska T, Dabrowski K, Hliwa P, Gomółka P, Kwasek K (2008) Nutritional regulation of intestine morphology in larval cyprinid fish, silver bream (*Vimba vimba*). Aquac Res 39:1268–1278

Ostaszewska T, Dabrowski K, Wegner A, Krawiec M (2008) The effects of feeding on muscle growth dynamics and the proliferation of myogenic progenitor cells during pike perch development (*Sander lucioperca*). J World Aqua Soc 39:184–195

Planas M, Ferreiro MJ, Fernandez-Reiriz MJ, Labarta U (1989) Evolution de la composicion bioquimica y actividades enzimaticas en huevos de rodaballo (*Scophthalmus maximus* L.) durante la embriognénesis. In: Yufera M (ed) Acuicultura Intermareal. Andalucia, Cadiz, pp 215–227

Ruuhijärvi J, Virtanen E, Salminen M, Myunda M (1991) The growth and survival of pike-perch *Stizostedion lucioperca* L. larvae fed on formulated feeds. In: Lavens P, Sorgeloos P, Jaspers E, Ollevier F (eds) Larvi'91 – fish and crustacean larviculture symposium. European aquaculture society, special publication n 15, Gent, pp 154–156

Sargent JR (1995) Origins and functions of egg lipids: nutritional implications. In: Bromage NR, Roberts RJ (eds) Broodstock management and egg and larval quality. Blackwell Science, Oxford, pp 353–372

Sargent JR, Henderson RJ, Tocher DR (2002) The lipids. In: Halver JE, Hardy RW (eds) Fish nutrition. Academic, Seattle, pp 153–218

Schlumpberger W, Schmidt K (1979) Untersuchungen zur Entwicklung eines industrie-mäBigen Verfahrens für die Produktion von vorgestreckten Zandern (*Stizostedion lucioperca* L.). Diss Humbold-Univ, Berlin

Schlumberger O, Proteau JP (1991) Production de juvéniles de sandre (*Stizostedion lucioperca*). Aqua Revue 36:25–28

Sorbera LA, Asturiano JF, Carillo M, Zanuy S (2001) Effects of polyunsaturated fatty acids and prostaglandins on oocyte maturation in a marine teleost, the European sea bass (*Dicentrarchus labrax*). Biol Reprod 64:382–389

Steffens W, Geldhauser F, Gerstner P, Hilge V (1996) German experiences in the propagation and rearing of fingerling pikeperch (*Stizostedion lucioperca*). Ann Zool Fenn 33:627–634

Sulistyo I, Rinchard J, Fontaine P, Gardeur JN, Capdeville B, Kestemont P (1998) Reproductive cycle and plasma levels of sex steroids in female Eurasian perch *Perca fluviatilis*. Aquat Living Resour 11(2):101–110

Sulistyo I, Fontaine P, Rinchard J, Gardeur JN, Capdeville B, Kestemont P (2000) Reproductive cycle and plasma sex steroid profiles in male Eurasian perch (*Perca fluviatilis*). Aquat Living Resour 13(2):99–106

Tamazouzt L (1995) L'alimentation artificielle de la perche *Perca fluviatilis* en milieu confinés (eau recycle, cage flottante). Thèse de doctorat, Université H. Poincaré, 128p

Tarranger GL, Carillo M, Schultz RW, Fontaine P, Zanuy S, Felip A, Weltzien FA, Dufour S, Karlsen O, Norberg B, Andersson E, Hansen T (2010) Control of puberty in farmed fish. Gen Comp Endocrinol 165:483–515

Tocher DR, Fraser AJ, Sargent JR, Gamble JC (1985a) Fatty acid composition of phospholipids and neutral lipids during embryonic and early larval development in Atlantic herring (*Clupea harengus* L.). Lipids 20:69–74

Tocher DR, Fraser AJ, Sargent JR, Gamble JC (1985b) Lipid class composition during embryonic and early larval development in Atlantic herring (*Clupea harengus* L.). Lipids 20:84–89

Vlavonou R, Masson G, Moreteau JC (1995) Use of *Artemia* as unique starting food for cultured perch *Perca fluviatilis* larvae. Percis II, second international percid fish symposium, FGRFI, Helsinki, p 79

Wang N (1994) Food and feeding of young perch (*Perca fluviatilis* L.) in Lake Constance. Verh Int Ver Limnol 25:2148–2152

Wang N, Eckmann R (1994) Effects of temperature and food density on egg development, larval survival and growth of perch (*Perca fluviatilis* L.). Aquaculture 122:323–333

Wang N, Mandiki SNM, Henrotte E, Bouyahia AG, Mairesse G, Rougeot C, Mélard C, Kestemont P (2009) Effect of partial or total replacement of forage fish by a dry diet on the quality of reproduction in pikeperch, *Sander lucioperca*. Aquacult Res 40:376–383

Wang N, Teletchea F, Kestemont P, Milla S, Fontaine P (2010) Photothermal control of the reproductive cycle in temperate fishes. Rev Aquac 2:209–222

Watanabe T (1982) Lipid nutrition in fish. Comp Biochem Physiol 73B:3–15

Chapter 21
Effects of Dietary Levels of PUFA Fed to Adult Yellow Perch on the Fatty Acid Composition of Eggs and Larvae Characteristics: New Research Directions

Konrad Dabrowski, Jacques Rinchard, Sergiusz Czesny, and Malgorzata Korzeniowska

Abstract We reviewed the current state of knowledge regarding the function of lipids and fatty acids in teleost reproduction with an emphasis on freshwater fish. Along with this review, we also provide readers with a specific case study on the effect of dietary lipids characterized by different fatty acid profiles on the reproduction of yellow perch. The unique nature of this case study stems from the use of a natural diet (fish) to feed the yellow perch broodstock over a period of two spawning seasons while monitoring reproductive efficiency and offspring viability. Furthermore, we were able to characterize the fatty acid profiles of the neutral and phospholipid fraction of egg lipid. We demonstrate significant decreases in the polyunsaturated fatty acids in gonad lipids of both, the neutral and the phospholipid fractions, an effect that was exacerbated during the second spawning season. The yellow perch however, demonstrated considerable resilience in protecting essential fatty acid concentrations in the gonads and no significant effects on offspring viability

K. Dabrowski (✉)
School of Environment and Natural Resources, Ohio State University,
2021 Coffey Rd, 43210 Columbus, OH, USA
e-mail: dabrowski.1@osu.edu

J. Rinchard
Department of Environmental Science and Biology, The College at Brockport State University of New York, 350 New Campus Drive, 14420 Brockport, NY, USA

S. Czesny
Lake Michigan Biological Station, Illinois Natural History Survey, University of Illinois, 400 17th Street, 60099 Zion, IL, USA

M. Korzeniowska
Department of Animal Products Technology and Quality Management, Wroclaw University of Environmental and Life Sciences, Wroclaw, Poland

P. Kestemont et al. (eds.), *Biology and Culture of Percid Fishes*,
DOI 10.1007/978-94-017-7227-3_21

were found. As a result of this research we propose a unique pathway for the synthesis, deposition and mobilization of wax esters in yellow perch ovaries and eggs/embryos that requires further study and provides a new approach to diet formulation for larval stages of this species.

Keywords Yellow perch • Lipids • Fatty acids • Diet formulation • Larval stages

21.1 Introduction

Fatty acids play important roles in the reproduction of teleost fish. Fatty acids mobilized from the neutral lipid reserves of female adipose tissue during gonadogenesis are transferred via vitellogenin (VTG) to developing eggs in the ovary (Wiegand 1996; Bell et al. 1997). Up to 60 % of the free saturated and monounsaturated fatty acids mobilized are catabolized to provide metabolic energy for the biosynthesis of egg lipoprotein, whereas n-3 highly unsaturated fatty acids (HUFAs), especially docosahexaenoic acid (DHA), are incorporated into the VTG (Sargent 1995). The n-3 fatty acid composition of VTG has been reported in numerous species (Norberg and Haux 1985) and can be influenced by the nutritional status of the fish as reported in rainbow trout (*Oncorhynchus mykiss*) and in sea bass (*Dicentrarchus labrax*) (Navas et al. 1998). In walleye (*Sander vitreus*) small egg size corresponded to low level of HUFA (Moodie et al. 1992). The authors sieved smaller eggs from the numerous females, so it was not certain if lower survival of larvae from low HUFA eggs was the result of small eggs as a component of all females, the eggs from smaller females or simply eggs that were immature (did not harden appropriately). In addition, these studies were carried out without replication of individual batches (large or small eggs). In conclusion, the cause of this low survival of low HUFA eggs/larvae was not determined. In comparison, Czesny et al. (2005), who examined 77 females of walleye from the same lake during an extended spawning season, found that the egg size was unrelated to lipid content. These authors were not able to correlate the size of eggs to their fatty acid profile (DHA or Eicosapentaenoic acid (EPA) concentrations) or survival (90 ± 8.7 %) during embryonic development.

Fatty acids are responsible for membrane fluidity, which is critical for cell growth during tissue differentiation of the embryo. These fatty acids also act as a major energy source in larvae (Izquierdo et al. 2001). As reported for several cultivated fish species, such as rainbow trout (Vassallo-Agius et al. 2001), gilthead sea bream *Sparus aurata* (Fernandez-Palacios et al. 1995; Rodriguez et al. 1998), Japanese flounder *Paralichthys olivaceus* (Furuita et al. 2002) and sea bass (Bell et al. 1997), the fatty acid composition of egg lipids can be altered by the nutritional status of broodstock. In rainbow trout broodstock, a diet deficient in n-3 fatty acids decreased the number and size of eggs, increased the incidence of early embryonic mortality, and caused physiological dysfunction of the developing fish (Watanabe et al. 1984). Reduced hatching success and poor larval viability have been associated with low n-3 HUFA levels in rainbow trout diets (Leray et al. 1985). A deficiency of DHA has been shown to be responsible for the non-inflation of the swimbladder in marine fish larvae (Tandler et al. 1995) and to impair vision in juvenile herring *Clupea harengus* (Bell et al. 1995), sea bass larvae (Navarro et al. 1997), and pike-perch (*Sander lucioperca*) (Lund et al. 2012). Moodie et al. (1989) also reported that the smallest eggs of freshwater, *Stizostedion vitreum* were deficient in the n-3 HUFAs. Moreover, the larvae hatched from those eggs had a high frequency of body deformation and a mortality rate of 100 % by day 10. However, high levels of dietary n-3 HUFAs reduced the amount of eggs produced, caused eggs yolk sac hypertrophy in gilthead sea bream larvae and decreased larval survival (Fernandez-Palacios et al. 1995). This was also confirmed in walleye (Czesny et al. 1999). Therefore, the ratio of DHA to EPA and n-3/n-6 HUFAs more than any single fatty acid should be considered to ensure proper larval development (Kjørsvik et al. 1990). Although arachidonic acid (ARA) is present in small amounts in fish eggs, it is essential for embryonic and larval development. This is because ARA is the major precursor of eicosanoids in salmonid eggs and testis (Sargent et al. 1995). Thus, HUFAs and eicosanoids are intimately involved in embryogenesis, hatching, early larval performance and other physiological functions (Bruce et al. 1999).

The studies mentioned previously provide evidence for the importance of dietary fatty acids in the modulation of the reproductive function in fish, since the relative levels of ARA, DHA and EPA in fish tissues are regulated by dietary intake (Rodriguez et al. 1998) (see also Fig. 21.1). Polyunsaturated fatty acids from the n-3 family (e.g., DHA and EPA) can consequently decrease the levels of synthesized steroids. Thus, the low levels of plasma estradiol-17 (E2) observed during vitellogenesis in captive female sea bass could decrease oocyte growth rate, delay production of eggs, reduced egg size and decrease egg viability as suggested by Navas et al. (1998). Dabrowski et al. (1995) also observed that a decrease of plasma E2 was correlated with a decrease of fecundity in rainbow trout.

Studies on fatty acid metabolism in coolwater percid fish, walleye and yellow perch has been conducted in our laboratory. Eggs from three distinct populations of walleye, one domesticated (London State Fish Hatchery) and two wild (Lake Erie and Salt Fork Reservoir), were compared in terms of total lipid and fatty acid content (phospholipids and neutral lipids) (Czesny and Dabrowski 1998). Domesticated walleye eggs contained significantly less total lipids and phospholipids than those from the wild populations. The concentrations of ARA in both neutral and phospholipid fractions were significantly lower in domesticated fish eggs compared to either of the wild stocks. Additionally, n-3/n-6 ratios were two to threefold higher in the eggs of wild fish versus those of cultured fish raised using formulated feeds. These data suggest that fatty acid composition of domesticated walleye eggs was altered and this effect may be a consequence of the commercial diet, which was used to feed the broodstock at London State Fish Hatchery. Thus, we hypothesize that the enrichment of the broodstock diet with HUFAs such as ARA, EPA or DHA would increase these fatty acids in the eggs. Czesny and Dabrowski (1998) also demonstrated that survival rates of embryos were positively correlated with the concentration

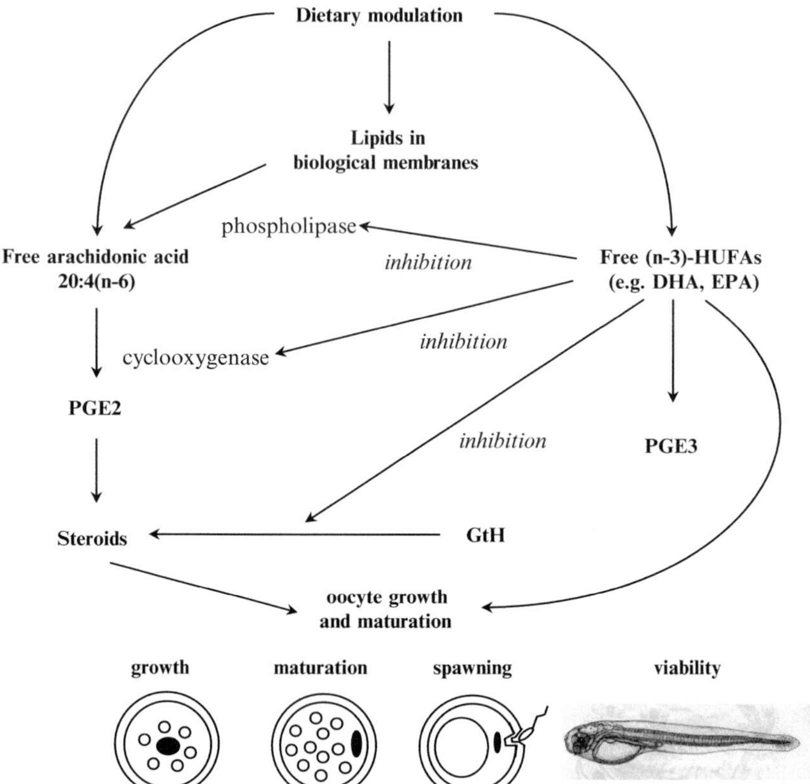

Nutritional status			
- AA and DHA	perturbation of steroid production and oocyte growth	no spawning	none
+ AA	increase of steroid production and oocyte growth rate	advanced spawning	low
+ DHA	decrease of steroid production and oocyte growth rate	delayed spawning	moderate
+ AA and DHA	normal steroid production and oocyte growth rate	normal spawning	high

Fig. 21.1 Modulation of the reproductive cycle by polyunsaturated fatty acids and expected results. *HUFAs* highly unsaturated fatty acids, *PGE* prostaglandin E, *GTH* gonadotropin, *DHA* docosahexaenoic acid, *EPA* eicosapentaenoic acid, *AA* arachidonic acid

of EPA and DHA in egg neutral lipids in walleye. Thus, increasing fatty acid concentrations have the potential to improve the quality of eggs, larval vitality and the survival of percid fish.

Arachidonic acid (ARA) is the main precursor of eicosanoids such as prostaglandins (PGE2, Fig. 21.1) in fish (Sargent et al. 1995) and these chemicals are involved in fish ovulation (Goetz et al. 1989), embryonic and larval development (Martins et al. 2013). However because AA and EPA are precursors of PGE2 and PGE3, respectively, and EPA competitively inhibits the production of PGE2 prostanoids, it is critically important to maintain the 20:4n6/20:5n3 ratio around 5:1 in marine fish diets, although much less is known about freshwater fish which, in principal, can elongate and desaturate C18 precursors. For instance, stressed Atlantic salmon (*Salmo salar*) smolts suffered heavy mortality (cardiac myopathy) when fish were fed n3/n6 ratio in the diet equal to 1:5 (Bell et al. 1991). Only recently Martins et al. (2013) reported that the whole body cortisol in Senegalese sole (*Solea senegalensis*) larvae increased from 75 to 370 ng/g as fish were fed with 1.7 % of AA in the diet in comparison to unenriched control. Prostaglandins are implicated in synthesis of cortisol, therefore AA can be linked to many physiological functions in larval fish and their viability. However, no studies, to our knowledge, have reported positive relationship between the amount of ARA in broodstock diets and subsequent egg quality measured as survival of embryos and larvae.

Ideal larval diets are those that match the nutrient profiles and concentration of the yolk sac and lipid sources in the natural food organisms. Fish larvae have a limited ability to biosynthesize phospholipids de novo but can exchange fatty acids within and between phospholipids (PL) and triglycerides (TG). Dietary PL can, in principal, be utilized (assimilated) unchanged (Izquierdo et al. 2000; Cahu et al. 2009) by larval fish. However, waxes have not previously been considered in larval fish nutrition despite the fact that they are found in a wide range of marine copepods and are transferred through the food chain to fish, accumulated in eggs, and become energy/nutrient sources in several species of fish at the larval stage. In some fish, waxes constitute up to 70 % of egg dry matter (or 91–97 % of neutral lipid content). Previous reviews of the role of lipids in larval fish nutrition did not mention their presence, function or importance. It remains unknown why the deposition of wax esters is favored over triglycerides in the body (skin, muscle, swim bladder) and consequently the eggs (Finn et al. 1996) in some fish (mullet, *Mugil*; striped bass, *Morone*; perch, *Perca*; burbot, *Lota lota*; gourami, *Trichogaster*; coelacanth, *Latimeria*; capelin, *Mallotus*; turbot *Scophthalmus*) (Eldridge et al. 1983; Finn et al. 1996; Kaitaranta and Ackman 1981; Lund et al. 2000). It is assumed that these lipids (with comparable degrees of unsaturation) have similar densities, caloric value, and compressibility. They differ though in their respective coefficients of thermal expansion, a predominance of extracellular versus intracellular location and rate of mobilization (Sargent et al. 1976).

Although the influence of dietary fatty acids on reproductive performance is becoming apparent after numerous experiments (Nandi et al. 2007; Lund et al. 2007; Xu and Kestemont 2002), the study that would address for the first time the possible mechanisms involved as a consequence of designed dietary manipulations of broodstock diets still needs to be conducted (Fig. 21.1). Therefore, to better understand the mechanisms by which this influence is exerted, we determined the effect of dietary fatty acids on the regulation of gametogenesis and consequently on gamete and larval yellow perch viability.

Yellow perch, *Perca flavescens*, was used in our research for multiple reasons. First, we documented that this species requires long chain HUFAs in its diet to sustain optimal growth (Brown et al. 1996; unpublished results). Therefore, as mentioned previously, yellow perch either lost their ability to synthesize their own long chain HUFAs from the C18 precursors due to the high concentrations of HUFAs in their diet or reduced synthesis of HUFAs due to the down regulation of gene transcription of enzymes and must therefore rely on their diet for these fatty acids.

For the last 20 years, research on the reproduction and nutrition of yellow perch has been conducted in our laboratory. A study on the early ontogeny revealed that larval yellow perch do not have a fully developed two compartmental stomach, until the body mass increases by two orders of magnitude (Dabrowski et al. 1993). This anatomical pre-meta morphosis stage of digestive tract will affect food evacuation rate and digestion process, including lipids. Among polyunsaturated fatty acids DHA was the predominant fatty acid in the phospholipids of yellow perch raised in ponds and was found to decrease from eggs (embryonic stage) to juveniles (Dabrowski et al. 1993). Furthermore, both DHA and EPA were higher in juvenile perch than in their major food item (*Daphnia* sp.) in the pond (Brown et al. 1996).

The acquisition and function of waxes in selected species of freshwaters fish remains a mystery, as the deposited wax esters are not of food origin like in marine fish (Falk-Petersen et al. 1999), their synthesis pathways are little studied and their utilization in larval stages is little known (Chu and Ozkizilcik 1995). It has been deduced that fatty alcohols of waxes are synthesized de novo from acetate in oocytes after vitellogenin cleavage based on the absence of waxes in the blood plasma of mature striped bass (Lund et al. 2000) and acetate incorporation in to fatty alcohols in the recrudescent ovaries of rainbow trout (Wiegand 1996).

21.2 Methods of the Experimental Studies

Yellow perch broodstock feeding studies were performed utilizing the results of experiments involving the conditioning of rainbow trout raised using experimental diets containing none or different proportions of PUFA. The results of these experiments with trout juveniles were described in detail earlier (Rinchard et al. 2007). Briefly, three experimental casein-gelatin based diets were formulated to be isonitrogenous (65 % protein) and isolipidic (14 %) but contained different sources of lipids. The first diet contained only the methyl esters of oleic acid (18:1n-9) and was used as a diet deficient in essential fatty acids (OA). The second diet was supplemented with olive and linseed oil (O+L) and designed to provide moderate amounts of linoleic (18:2n-6) and linolenic (18:3n-3) acids but no polyunsaturated fatty acids (PUFA). The third diet contained cod liver oil (CLO) and provided PUFA as well as their precursors. Swim-up rainbow trout alevins, previously fed a commercial diet (Starter diet, BioOregon, OR) for 1 week, were randomly assigned to one of the three experimental diets. Fish were fed either by hand or later using a belt feeder and restricted rations up to 90 % satiation for 5 weeks. At the end of the

Fig. 21.2 Tank with three compartments housing three treatments with different diets (**a**) and fish actively feeding on freshly thawed rainbow trout juveniles (**b**)

conditioning period, that is, after obtaining the desired fatty acid profile of trout lipids (see Rinchard et al. 2007), the juvenile trout were frozen in liquid nitrogen and preserved for later use. The trout were frozen in liquid nitrogen, stored at −80 °C and used for the yellow perch feeding experiment when needed.

Adult yellow perch were obtained from a local farmer (W. Lynch; Mill Creek Perch Farm, Marysville, OH). Fish were moved to our facility (October), acclimated for 2 weeks to tank conditions and then randomly distributed into four 450-L tanks divided into three compartments, with nine yellow perch per compartment (Fig. 21.2). Each yellow perch was individually weighed and implanted with a microtag (Biomark, ID) with a tag injector into the body cavity. Each tank was supplied with flow-through freshwater (60 L/h), and subjected to a regulated photoperiod and water temperature corresponding to seasonal changes (Ciereszko et al. 1996; Dabrowski et al. 1996). Fish were fed 1 % of their body wet weight every other day with freshly thawed rainbow trout (Fig. 21.2b) conditioned to contain either low levels of essential fatty acids (diet OA), a moderate amount of linoleic and linolenic acids (diet O+L), or PUFA and their precursors (diet CLO). Each dietary treatment was assigned to one of the compartments per tank (four replicates/dietary treatment) (Fig. 21.2).

After 7 months of feeding (April) during the first year of the experiment, females were observed for signs of ripeness by examining the genital papilla and pressing the abdomen. Those females that were ready to ovulate were stripped of their eggs. Each fish and its eggs were weighed. A subsample of eggs from individual females was taken and immediately frozen on dry ice for further lipid and fatty acid analysis. To evaluate the quality of the eggs, they were fertilized in duplicate with the milt of five males (100,000 spermatozoa/egg) using the dry method previously described by Ciereszko and Dabrowski (1993). Fertilized eggs from each female were incubated separately in a circular basket (two baskets per female) with a net bottom in "California type" incubators. Dechlorinated city water at 15–18 °C was used in a recirculated system equipped with a UV-sterilizing lamp and a 20 % water exchange rate per day. The survival rate was assessed 4 days after fertilization by counting the numbers of viable (eyed-stage) embryos and unfertilized eggs. At hatching, larvae were collected and fixed in glutaraldehyde and cacodylate fixative (Dabrowski and Bardega 1982) for further measurement of their morphological traits. Yellow perch females from the three dietary treatments were spawned during two consecutive years (2004 and 2005).

In the second part of the research effort we attempted to isolate neutral lipid fractions from domesticated yellow perch (OSU stock) eggs at the time of ovulation (Dabrowski et al. 1994). We analyzed the ovaries of yellow perch in order to obtain a more in-depth description of their neutral lipids (NL) by following the method described by Saito and Kotani (2000) and Stevens et al. (2004). We extracted total lipids by methanol-chloroform method (Folch et al. 1957), or hexane and subsequently separated the waxes (WE) from the NL class using dichloromethane:hexane (1:2), (WE-1) and then dichloromethane to isolate the free fatty acids (FFA). All extracts were subjected to methanolysis (Metcalfe and Schmitz 1961) followed by gas chromatography (GC, Varian) separation and quantification using a FAME-Wax™ capillary column (30 m × 0.32 mm × 0.25 μm) and helium as the carrier gas.

The differences among dietary treatments were tested by one-way ANOVA followed by Tukey's multiple comparison test. Differences were considered significant at the value of $p < 0.05$.

21.3 Results

We provide summary characteristics of the females that spawned in Table 21.1. Female size (weight) at spawning was not affected by their diet in either year. In 2004, GSI was numerically lower in fish fed the OA diet (deficient diet), but not significantly different from fish fed the CLO and O+A diets. GSI was also not significantly different among fish examined in 2005.

In 2004, the fertilization rate (survival at the eyed embryo stage) tended to be higher among the offspring of the fish fed the CLO diet compared to the offspring of the fish that were fed the OA diet (the mean of 86.4 % versus 52.8 %, respectively). But no significant differences were observed likely due to large individual variation among the females from each dietary treatment. In 2005, the fertilization rate (sur-

Table 21.1 Characteristics of spawned yellow perch female fed rainbow trout conditioned to contain different levels of fatty acids

	Year					
	2004			2005		
Parameters	OA	O+L	CLO	OA	O+L	CLO
Number of females spawned	7	7	10	10	8	10
Initial body mass (g)[a]	111.2 ± 18.6	127.8 ± 10.8	109.0 ± 10.4	135.9 ± 23.3	141.6 ± 16.4	129.4 ± 12.5
Body mass at spawning (g)	148.6 ± 25.2	180.9 ± 25.7	145.7 ± 14.5	171.1 ± 22.6	179.8 ± 17.5	181.2 ± 29.1
Egg mass (g)	43.2 ± 11.3	55.9 ± 15.9	47.5 ± 6.0	44.8 ± 10.2	51.6 ± 7.2	53.4 ± 13.0
GSI[b] (%)	28.8 ± 3.4	30.5 ± 5.0	32.5 ± 1.5	26.2 ± 4.7	29.0 ± 5.4	29.4 ± 4.7
Spawning period	4/5–5/5	4/28–5/29	4/24–5/11	5/6–5/20	5/8–5/21	5/14–5/19
Fertilization rate (%)	52.8 ± 41.3	79.5 ± 21.8	86.4 ± 17.6	64.3 ± 20.3	50.9 ± 29.5	66.5 ± 23.3

[a]Body mass measured in October 2003 for fish spawned in 2004 and October 2004 for fish spawned in 2005
[b]Gonadosomatic index, GSI = (egg mass × 100)/(total body mass)

Table 21.2 Lipid (mean ± SD) composition of yellow perch eggs fed rainbow trout conditioned to contain different levels of fatty acids. *TL* total lipid, *ww* wet weight). Means across rows within each year with different superscript differ at P<0.05

	Year					
	2004			2005		
Parameters	OA	O+L	CLO	OA	O+L	CLO
Lipids						
Total lipid (% of ww)	4.8 ± 0.4^{a}	4.8 ± 0.5^{a}	4.5 ± 0.9^{a}	4.5 ± 0.5^{a}	4.4 ± 0.5^{a}	4.0 ± 0.3^{b}
Phospholipid (% of TL)	77.6 ± 3.1^{a}	74.6 ± 1.1^{a}	76.0 ± 2.1^{a}	80.6 ± 1.7^{a}	80.8 ± 2.1^{a}	81.6 ± 4.1^{a}
Neutral lipid (% of TL)	22.4 ± 3.1^{a}	25.4 ± 1.1^{a}	24.0 ± 2.1^{a}	19.4 ± 1.7^{a}	19.2 ± 2.1^{a}	18.4 ± 4.1^{a}

vival at eyed stage) was not significantly different among treatments. Again, we observed large individual variation within dietary treatments. Interestingly, those yellow perch females fed the OA diet (deficient in PUFA), although they were mature and did spawn, produced eggs that did not resemble the usual ribbon-like form that characterizes yellow perch eggs. Rather, the eggs were loose and spawned in "chunks" or "clumps", indicating that dietary PUFA may be involved in the formation of the outer egg gelatinous substances that usually hold yellow perch eggs together.

Total lipids, neutral and phospholipids in yellow perch eggs were not affected by the dietary treatment (Table 21.2) in either year. The concentration of total lipids tended to be higher in 2004 than in 2005. Neutral lipids tended to be higher in younger fish in 2004 than in 2005.

In Tables 21.3 and 21.4, we summarize the composition of the fatty acid, neutral (NL) and phospholipid (PL) fractions of the yellow perch eggs obtained from females fed in the experiment. The saturated fatty acids (SAFA) of neutral lipids were not affected by the dietary regimen. Monounsaturated fatty acids (MUFA) were higher in the eggs of females fed the O+L diet compared to the other groups. This trend was consistent in both the 2004 and 2005 spawning seasons. With respect to polyunsaturated fatty acids (PUFA) in neutral lipids, we found no differences in their total proportions among the experimental groups, but they were markedly different in terms of their fatty acid composition. The eggs of females fed the OA diet contained mainly linoleic acid (18:2n-6) and EPA (20:5n-3) while the eggs of females fed the O+L diet had mostly linoleic acid, linolenic acid (18:3n-3), and DHA (22:6n-3) in PUFA of neutral lipids. Eggs of fish fed the CLO diet contained significantly less of linoleic and linolenic acids, but were richer in both EPA and DHA. The linoleate level in neutral lipids (NL) was particularly high in the second year in spawners of the OA (deficient) group and being a precursor of AA, resulted in 12-fold higher levels of this fatty acid in comparison to CLO group in that year.

Table 21.3 Fatty acid composition (mean ± SD) in neutral lipids of yellow perch eggs fed rainbow trout conditioned to contain different levels of fatty acids. Means across rows within a year with different superscript differ at $P<0.05$

	Year					
	2004			2005		
Parameters	OA	O+L	CLO	OA	O+L	CLO
Fatty acids (mass %)						
Saturated						
C14:0	0.2 ± 0.0^a	0.2 ± 0.0^a	0.4 ± 0.1^a	0.1 ± 0.0^a	0.1 ± 0.0^a	0.4 ± 0.1^a
C16:0	1.0 ± 0.1^a	1.2 ± 0.1^a	1.1 ± 0.2^a	1.4 ± 0.4^a	1.4 ± 0.2^a	1.2 ± 0.3^a
C18:0	0.1 ± 0.0^a	0.1 ± 0.0^a	0.2 ± 0.1^a	0.3 ± 0.2^a	0.1 ± 0.1^a	0.4 ± 0.1^a
∑ Saturated	1.3 ± 0.2^a	1.5 ± 0.1^a	1.6 ± 0.3^a	1.9 ± 0.4^a	1.6 ± 0.2^a	2.0 ± 0.4^a
Monounsaturated						
C16:1*	5.5 ± 0.8^a	6.1 ± 0.5^a	6.8 ± 1.1^a	3.7 ± 0.8^b	5.3 ± 0.4^a	5.5 ± 0.9^a
C18:1*	12.1 ± 1.8^b	15.3 ± 2.0^a	9.0 ± 1.0^c	10.9 ± 0.9^b	16.0 ± 0.9^a	8.5 ± 1.2^c
C20:1**	0.2 ± 0.0^a	0.3 ± 0.1^a	0.3 ± 0.1^a	1.5 ± 0.3^a	0.3 ± 0.1^b	0.1 ± 0.0^b
∑ Monounsaturated	17.8 ± 1.6^b	21.7 ± 2.3^a	16.0 ± 1.9^b	16.1 ± 1.7^b	22.4 ± 1.2^a	14.1 ± 2.1^b
Polyunsaturated						
C18:2n-6	8.9 ± 0.7^a	6.4 ± 0.4^b	3.5 ± 1.1^c	15.9 ± 1.8^a	7.0 ± 0.3^b	2.7 ± 0.4^c
C18:3n-3	0.7 ± 0.1^b	2.4 ± 0.4^a	0.7 ± 0.2^b	0.2 ± 0.1^b	3.0 ± 0.3^a	0.6 ± 0.1^b
C20:4n-6	0.5 ± 0.2^a	0.1 ± 0.1^b	0.5 ± 0.1^a	1.2 ± 0.2^a	0.3 ± 0.0^b	0.1 ± 0.0^c
C20:4n-3	0.5 ± 0.2^a	0.3 ± 0.1^a	0.1 ± 0.0^a	0.0 ± 0.0^a	0.1 ± 0.0^a	0.0 ± 0.0^a
C20:5n-3	1.2 ± 0.2^b	1.4 ± 0.3^b	3.4 ± 0.5^a	0.2 ± 0.1^c	0.6 ± 0.1^b	3.6 ± 0.5^a
C22:4n-6	0.8 ± 0.2^a	0.8 ± 0.1^a	0.9 ± 0.3^a	0.7 ± 0.4^a	0.7 ± 0.3^a	0.4 ± 0.5^a
C22:5n-6	0.6 ± 0.1^a	0.2 ± 0.0^b	0.2 ± 0.0^b	0.2 ± 0.1^a	0.2 ± 0.0^a	0.2 ± 0.1^a
C22:5n-3	0.5 ± 0.1^b	0.5 ± 0.1^b	1.1 ± 0.1^a	1.0 ± 0.1^a	0.9 ± 0.2^a	1.0 ± 0.1^a
C22:6n-3	11.5 ± 0.8^b	12.3 ± 09^b	17.8 ± 1.8^a	7.2 ± 2.2^c	10.4 ± 1.1^b	17.4 ± 2.5^a
∑ Polyunsaturated	25.3 ± 1.1^a	24.5 ± 1.4^a	28.4 ± 2.6^a	29.6 ± 4.5^a	24.5 ± 1.4^a	27.9 ± 3.6^a
∑ n-3	14.4 ± 1.1^c	16.9 ± 1.1^b	23.1 ± 2.0^a	10.8 ± 2.6^c	16.7 ± 1.3^b	24.0 ± 3.2^a
∑ n-6	10.8 ± 1.0^a	7.6 ± 0.5^b	5.1 ± 1.2^c	17.3 ± 2.0^a	7.5 ± 0.3^b	3.0 ± 0.5^c
n-3/n-6	1.3 ± 0.2^c	2.2 ± 0.2^b	4.9 ± 2.1^a	0.6 ± 0.1^c	2.2 ± 0.2^b	8.0 ± 0.7^a
DHA/EPA	9.9 ± 1.2^a	9.3 ± 1.5^a	5.3 ± 1.0^b	32.6 ± 9.8^a	16.6 ± 1.4^b	5.1 ± 0.4^c
EPA/AA	2.3 ± 0.8^c	10.5 ± 5.0^a	7.2 ± 1.1^b	0.2 ± 0.1^c	2.5 ± 0.3^b	47.5 ± 5.7^a

*Includes n-7 and n-9 and **includes n-9 and n-11

The phospholipid fraction of egg lipids was also affected by female dietary fatty acid intake (Table 21.4). While minor differences were observed in both SAFA and MUFA among experimental groups, major differences appeared in PUFA compositions. In both years, EPA and DHA were the lowest in the eggs of the females fed the OL diet and the highest in the eggs of the females fed the CLO diet. Females fed the O+L diet produced eggs with intermediate levels of these two key fatty acids. The level of AA was significantly higher in the OA fed fish than in CLO fed group.

Table 21.4 Fatty acid composition (mean ± SD) in phospholipids of yellow perch eggs fed rainbow trout conditioned to contain different levels of fatty acids. Means across rows within a year with different superscript differ at P<0.05

	Year					
	2004			2005		
Parameters	OA	O+L	CLO	OA	O+L	CLO
Fatty acids (mass %)						
Saturated						
C14:0	0.3 ± 0.0^{b}	0.3 ± 0.0^{b}	0.6 ± 0.0^{a}	0.2 ± 0.0^{b}	0.2 ± 0.0^{b}	0.7 ± 0.1^{a}
C16:0	8.7 ± 0.8^{b}	8.3 ± 1.0^{b}	9.1 ± 1.5^{a}	11.2 ± 1.6^{a}	11.1 ± 1.4^{a}	11.3 ± 1.2^{a}
C18:0	2.5 ± 0.3^{a}	2.3 ± 0.3^{a}	2.4 ± 0.4^{a}	3.1 ± 0.3^{a}	2.9 ± 0.3^{a}	3.3 ± 0.3^{a}
∑ Saturated	11.5 ± 1.1^{ab}	10.8 ± 1.2^{b}	12.1 ± 2.0^{a}	14.5 ± 1.4^{a}	14.2 ± 1.7^{a}	14.7 ± 2.5^{a}
Monounsaturated						
C16:1*	1.9 ± 0.2^{ab}	2.0 ± 0.2^{a}	1.7 ± 0.4^{b}	2.0 ± 0.3^{ab}	2.9 ± 0.3^{a}	1.8 ± 0.2^{b}
C18:1*	6.1 ± 0.7^{a}	5.9 ± 0.8^{a}	3.9 ± 0.6^{b}	5.9 ± 0.5^{b}	9.6 ± 1.1^{a}	5.7 ± 0.6^{b}
C20:1**	0.3 ± 0.1^{a}	0.1 ± 0.0^{b}	0.1 ± 0.0^{b}	0.1 ± 0.0^{b}	0.3 ± 0.0^{a}	0.1 ± 0.0^{b}
∑ Monounsaturated	8.4 ± 0.8^{a}	8.0 ± 1.0^{a}	5.6 ± 0.8^{b}	8.0 ± 0.7^{b}	12.8 ± 1.4^{a}	7.7 ± 0.8^{b}
Polyunsaturated						
C18:2n-6	2.7 ± 0.4^{a}	1.7 ± 0.1^{b}	0.6 ± 0.1^{c}	5.8 ± 0.5^{a}	2.9 ± 0.3^{b}	0.6 ± 0.1^{c}
C18:3n-3	0.1 ± 0.0^{b}	0.6 ± 0.1^{a}	0.1 ± 0.1^{b}	0.2 ± 0.0^{b}	1.1 ± 0.2^{a}	1.0 ± 0.1^{a}
C20:4n-6	2.8 ± 0.3^{a}	1.6 ± 0.3^{b}	1.8 ± 0.2^{b}	4.4 ± 0.6^{a}	1.8 ± 0.3^{b}	2.0 ± 0.2^{b}
C20:4n-3	0.1 ± 0.0^{a}	0.2 ± 0.0^{a}	0.1 ± 0.0^{a}	0.1 ± 0.0^{b}	0.3 ± 0.1^{a}	0.1 ± 0.0^{b}
C20:5n-3	2.1 ± 0.5^{c}	2.8 ± 0.5^{b}	5.0 ± 0.9^{a}	0.5 ± 0.2^{c}	2.2 ± 0.3^{b}	6.6 ± 0.5^{a}
C22:5n-6	0.9 ± 0.2^{a}	0.2 ± 0.0^{b}	0.1 ± 0.0^{b}	2.1 ± 0.3^{a}	0.3 ± 0.1^{b}	0.1 ± 0.0^{b}
C22:5n-3	0.4 ± 0.1^{b}	0.4 ± 0.1^{b}	0.6 ± 0.1^{a}	0.2 ± 0.1^{b}	0.3 ± 0.1^{b}	0.7 ± 0.1^{a}
C22:6n-3	12.2 ± 1.3^{c}	13.2 ± 1.3^{b}	14.6 ± 2.3^{a}	10.3 ± 1.8^{c}	15.6 ± 2.2^{b}	17.7 ± 1.5^{a}
∑ Polyunsaturated	21.2 ± 2.1^{a}	20.7 ± 2.1^{a}	23.0 ± 3.6^{a}	23.6 ± 2.2^{a}	24.4 ± 3.0^{a}	28.8 ± 2.4^{a}
∑ n-3	14.9 ± 1.8^{c}	17.3 ± 1.8^{b}	20.5 ± 3.3^{a}	11.3 ± 1.9^{c}	19.5 ± 2.6^{b}	26.1 ± 2.2^{a}
∑ n-6	6.3 ± 0.8^{a}	3.4 ± 0.3^{b}	2.5 ± 0.4^{c}	12.3 ± 0.7^{a}	4.9 ± 0.4^{b}	2.7 ± 0.3^{c}
n-3/n-6	2.4 ± 0.2^{c}	5.0 ± 0.2^{b}	8.1 ± 0.8^{a}	0.9 ± 0.1^{c}	3.9 ± 0.2^{b}	9.5 ± 0.6^{a}
DHA/EPA	6.0 ± 0.4^{a}	4.8 ± 0.4^{b}	2.9 ± 0.2^{c}	22.9 ± 4.4^{a}	7.0 ± 0.6^{b}	2.7 ± 0.1^{c}
EPA/AA	0.7 ± 0.2^{c}	1.8 ± 0.2^{b}	2.8 ± 0.3^{a}	0.1 ± 0.0^{c}	1.2 ± 0.1^{b}	3.3 ± 0.3^{a}

*Includes n-7 and n-9 and **includes n-9 and n-11

Although levels of individual fatty acids are important indicators of the nutritional state of an organism, the ratios of the n-3 and n-6 families of fatty acids are those that will provide a more comprehensive picture of the fatty acid balance. Both fractions of lipids isolated from yellow perch eggs spawned by experimental fish exhibited dramatic differences in their n-3/n-6, DHA/EPA, and EPA/AA ratios. Such differences in egg fatty acid balances and their potential effects on offspring success in terms of growth and survival are indicative of (1) the significance of a proper diet with respect to fatty acid content and (2) the potential susceptibility of yellow perch to physiological challenges such as hypoxia, metal toxicity, free radicals or other factors in their life cycle.

We conclude that maternal fatty acid intake had significant effects not only on the fatty acid composition of the eggs they produce, but also on the morphological characteristics of the embryos/larvae at hatching (Table 21.5). Size at hatching is a very important factor in fish ontogeny, because depending on larvae dimensions they may be more or less successful at first feeding attempts, predation avoidance or withstanding environmental adversities as a result of their size. The amount of yolk reserves may play an important role in withstanding initial periods of low food availability. We compared larval body dimensions including total length, tail length, and body height, as well as size of the eye and yolk volume and found significant differences depending on maternal fatty acid intake (Table 21.5). Larvae produced by females fed the OA diet which contained low levels of PUFA were consistently smaller compared to the two other groups. These differences were not as pronounced in 2004 as they were in 2005, after a second year of dietary PUFA deficiency. The yolk reserves were largest in larvae produced by females fed the O+L diet in 2004, while in 2005 the yolks of larvae produced by females fed the CLO diet were larger. This apparent tradeoff between size at hatch and yolk reserves, with larger larvae having less yolk and smaller larvae possessing larger yolk reserves has important consequences in nursery pond systems where the timing of hatching must coincide (overlap) with zooplankton blooms in order for larvae to survive and grow (match/mismatch hypothesis). We demonstrated that the fatty acid content of a female's diet will have a considerable effect on the morphological characteristics of her offspring and may affect their survival depending on the availability of food once the larvae are ready to feed exogenously in their indoor or outdoor environments.

We were able to confirm the observations of Kaitaranta and Ackman (1981) that yellow perch exhibit an unusual composition of ovarian lipids by isolating the wax esters of the yellow perch eggs at inception of embryonic development (Table 21.6). We postulate that the synthesis of wax esters de novo takes place in perch ovaries based on evidence demonstrated earlier in striped bass (*Morone saxatilis*), a species of fish that also deposits waxes in its eggs without evidence of transport of these waxes in the bloodstream of maturing females (Lund et al. 2000).

Regarding the fatty acid composition of yellow perch ovaries at the time of ovulation, a high proportion of EPA and DHA was notable in the neutral lipids, more specifically in the wax esters that constituted the majority of this lipid class (Table 21.7).

The results obtained thus far indicate that a unique pathway of synthesis, deposition, and mobilization of the wax esters contained in the eggs and then transferred to oil globule of yellow perch exists (Fig. 21.3), and may/is likely to involve exocytosis of microparticles into circulation (Jaroszewska and Dabrowski 2011), that would make the digestive tract intraluminal hydrolysis rate of secondary importance. However, a significant decline of wax esters that parallels the first exogenous feeding suggests that the role of waxes in the starter diet formulations needs to be explored.

Table 21.5 Morphological parameters of yellow perch larvae at hatching. Means across rows within each year with different superscript differ at P value listed in the last column

	2004					2005				
Parameters	OA	O+L	CLO	F value	P value	OA	O+L	CLO	F value	P value
Total length (mm)	$5.27 \pm 0.05^{a,b}$	5.42 ± 0.05^{a}	5.28 ± 0.03^{b}	3.7	0.02	5.54 ± 0.03^{c}	5.70 ± 0.03^{b}	5.91 ± 0.03^{a}	27.7	<0.01
Total tail length (mm)	2.53 ± 0.03^{a}	2.56 ± 0.03^{a}	2.49 ± 0.02^{a}	2.0	0.14	2.73 ± 0.02^{b}	2.84 ± 0.02^{a}	2.86 ± 0.03^{a}	11.5	<0.01
Mid body height (mm)	0.34 ± 0.00^{a}	0.34 ± 0.01^{a}	0.36 ± 0.02^{a}	0.3	0.72	0.40 ± 0.00^{a}	0.41 ± 0.02^{a}	0.40 ± 0.01^{a}	0.4	0.71
Yolk diameter (mm)	0.52 ± 0.00^{a}	0.54 ± 0.00^{a}	0.49 ± 0.01^{b}	23.5	<0.01	0.48 ± 0.00^{b}	0.47 ± 0.00^{b}	0.50 ± 0.00^{a}	18.4	<0.01
Yolk volume (mm^3)	0.07 ± 0.00^{a}	0.08 ± 0.00^{a}	0.07 ± 0.00^{b}	19.6	<0.01	0.06 ± 0.00^{b}	0.05 ± 0.00^{b}	0.07 ± 0.00^{a}	30.2	<0.01
Eye diameter (mm)	0.32 ± 0.00^{a}	0.32 ± 0.00^{a}	0.33 ± 0.00^{a}	0.4	0.66	0.33 ± 0.03^{b}	0.34 ± 0.03^{b}	0.35 ± 0.03^{a}	8.2	<0.01

Table 21.6 Proportion of neutral lipids (NL), phospholipids (PL), triglycerides (TG) and wax esters (WE) in oocytes (eggs) of fish (% total lipids) (After Wiegand 1996 unless reference given)

Lipid fraction	PL	NL	TG	WE
Eggs with no oil globule				
Cod	71.7	12.5	3.7	
Atlantic halibut	70.8	12.9	4.3	
Plaice	65.8	14.2	2.7	
Eggs with oil globule				
Turbot	40.0	29.0	23.0	
Rainbow trout[1,2]	49.7	46.8	0.2	
	7.3	62.8	56.5	0
Japanese eel[3]	19.4	80.6		
Walleye[4]	21.8	78.2	0	0
Burbot	12.6		4.1	81.8
Striped bass[5]	11.0		11.0	79.0
Whitefish (*Coregonus sp.*)[6]	31.7	64.9	1.8	
Perca fluviatilis[7]	13.5–17.3	82.4–86.5	8.1–12.4	68.0–68.9
(Eurasian perch)	14.3		1.1	83.7
P. flavescens[8] (yellow perch)	23.1		2.8	77.8

[1]Watanabe et al. (1984)
[2]Vassallo-Agius et al. (2001)
[3]Furuita et al. (2006)
[4]Wiegand et al. (2004)
[5]Lund et al. (2000)
[6]Dabrowski (1982)
[7]Henrotte et al. (2010); Kaitaranta and Ackman (1981)
[8]Present work

21.4 Discussion

Rinchard et al. (2007) indicated that feeding trout an increased dietary level of linoleate also resulted in a significantly higher percentage of arachidonate (ARA) in the rainbow trout body PL. Along the same line, we observed that in yellow perch fed the OA diet, the AA level in NL of eggs increase several folds (particularly in year 2005; Table 21.3). This was associated with lower quality of eggs and disrupted egg ribbons. It is a notable finding because ARA is a precursor of prostaglandins. The 17α,20β-dihydroxy-4-pregnen-3-one (DHP), a maturation inducing hormone (MIH) in yellow perch (Goetz et al. 1989; Rinchard et al. 2002) was identified as the/an inducer of prostagladins. The mechanism involved might lead to premature ovulation and compromised egg viability in "high ARA" fish. Goetz et al. (1989) indicated that although prostaglandin secretion did not precede ovulation it correlated with ovulation. It is warranted to continue this line of research and examine if increased level of ARA in the diet that resulted in high accumulation of ARA in gonads of Senegalese sole (Norambuena et al. 2013) would result in compromised embryo/larvae quality.

Table 21.7 Fatty acid composition (mean ± SD) in neutral and polar lipids and wax ester fraction of yellow perch eggs

Parameters	NL	PL	WAX
Fatty acids (mass %)			
Saturated			
C14:0	0.4 ± 0.0	2.4 ± 0.0	6.5 ± 0.2
C16:0	5.2 ± 0.0	29.1 ± 0.9	0.7 ± 0.1
C18:0	0.0 ± 0.0	0.0 ± 0.0	15.6 ± 0.8
C20:0	0.5 ± 0.0	0.1 ± 0.0	0.9 ± 0.0
C22:0	0.9 ± 0.0	0.2 ± 0.0	0.6 ± 0.2
∑ Saturated	6.8 ± 0.3[a]	31.6 ± 0.9[a]	24.3 ± 0.7
Monounsaturated			
C16:1*	23.5 ± 1.5	8.3 ± 0.2	9.4 ± 0.1
C18:1*	27.6 ± 1.7	17.5 ± 1.1	26.0 ± 0.7
C20:1**	0.4 ± 0.0	0.3 ± 0.0	0.4 ± 0.0
C22:1	0.4 ± 0.0	0.0 ± 0.0	0.3 ± 0.0
∑ Monounsaturated	51.5 ± 3.5	26.8 ± 2.5	36.0 ± 2.3
Polyunsaturated			
C18:2n-6	1.1 ± 0.1	2.7 ± 0.0	4.3 ± 0.2
C18:3n-3	0.3 ± 0.0	0.3 ± 0.0	0.3 ± 0.0
C20:2n-6	0.4 ± 0.0	2.8 ± 0.4	0.9 ± 0.0
C20:3n-3	0.1 ± 0.0	0.3 ± 0.0	0.2 ± 0.0
C20:4n-3	0.0 ± 0.0	1.1 ± 0.1	3.6 ± 0.3
C20:5n-3	9.2 ± 0.3	9.6 ± 0.2	8.2 ± 0.4
C22:6n-3	29.8 ± 1.0	23.8 ± 2.3	21.3 ± 0.7
∑ Polyunsaturated	41.4 ± 4.1	41.6 ± 3.9	38.8 ± 3.9
∑ n-3	39.5 ± 1.4	33.9 ± 1.3	33.0 ± 1.1
∑ n-6	1.6 ± 0.7	6.6 ± 0.8	8.9 ± 0.2
n-3/n-6	25.1 ± 2.1	5.1 ± 0.4	3.4 ± 0.2
DHA/EPA	3.2 ± 0.8	2.5 ± 0.3	2.6 ± 0.0
EPA/AA	20.2 ± 2.6	85.3 ± 5.1	8.9 ± 0.1

*Includes n-7 and n-9 and **includes n-9 and n-11

Growth of fish fed diets containing varying levels of oleic, linoleic, linolenic, docosahexaenoic and eicopentaenoic acids were compared to determine the fatty acid requirements for juvenile yellow perch (Brown et al. 1996; unpublished data). Each diet was formulated to contain 5 % lipids. The lowest growth rate was recorded in fish fed a diet containing 1 % of C18 precursor (linoleic and linolenic acids), but lacking EPA and DHA. Higher growth rates were observed in fish fed a diet containing either DHA alone (0.5 %) or EPA (0.5 %) and DHA (0.5 %) but without any C18 precursors. Fatty acid composition of liver polar lipids (phospholipids) revealed that yellow perch possess only a limited ability to elongate and desaturate linolenic acid to EPA and DHA. The rates of these conversions were not satisfactory to sustain optimal growth in juvenile fish. The polar lipid levels of DHA and EPA were elevated in fish fed a diet containing only linolenic acid compared to fish fed a diet free of linolenic, EPA and DHA. However, the growth of these fish was significantly

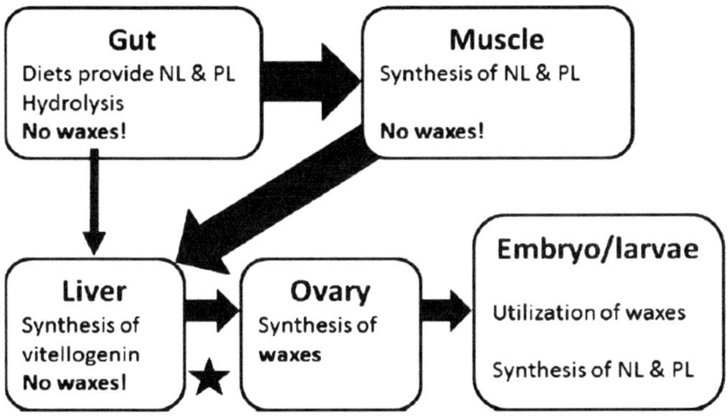

Fig. 21.3 Uptake, transport and metabolism of lipids in the life cycle of yellow perch (*NL* neutral lipids, *PL* phospholipids). Size of the *arrow* indicates the likely proportion of the nutrient transfer and deposition

inferior to that of the fish fed a diet supplemented with EPA and DHA or with DHA alone. These results suggest that juvenile yellow perch require at least 0.5 % of DHA in their diet to sustain optimal growth and these findings confirm earlier results observed for European perch (Henrotte et al. 2011b). Henrotte et al. (2011a) confirmed earlier results obtained by Kaitaranta and Ackman (1981) that wax esters constitute a major fraction of perch lipids (68 % independently of the diet source of lipids provided to broodstock fish). Furthermore, Henrotte et al. (2010) indicated that due to large variation in both fertilization and hatching rates, there was no significant difference in embryo viability among females fed diets as variable as 5.1–20.3 % in DHA level and 4.4–46.9 % in linoleate level (sardine or menhaden oils). In agreement with our results, fatty acid profiles of egg lipids of European perch have shown insignificant differences in the level of most PUFA. It demonstrates that the effect of dietary lipids on eggs lipids undergoes severe "screening" in respect to mobilization, synthesis and deposition of reserve lipids (vitellogenins) in the process of gametogenesis. In striped bass eggs, embryos (until hatching), and larvae (until 7 day after hatching, beginning of exogenous feeding) wax esters are the major lipid component (Chu and Ozkizilcik 1995). In the following period, until day 26 after hatching, waxes almost completely disappear and the level of phospholipids increases. It is interesting that fish from the genus *Perca* rely on waxes during the embryonic period and larval metamorphosis despite the fact that waxes are not obtained from food during female maturation and gametogenesis as is the case with marine fishes (Sargent et al. 1976).

The tropical fish gouramis (*Trichogaster cosby*), is a unique freshwater fish that contains in its eggs over 70 % of its lipids as wax esters (Sand and Schlenk 1969). However, the waxes of this species contain mostly oleic acid (51 %) and only 6.8 % and 1.7 % DHA and EPA, respectively. The marine fish, *Laemonema longipes*, possesses only a moderate amount of wax esters (8.2 %) in its ovarian lipids, with EPA (21.2 %) and DHA (9.0 %) comprising a large part of the PUFA (Hayashi and Kashiki 1988). While 44 % of the lipids in the ovulated eggs of turbot (*Scophthalmus maximus*) were present as wax esters (Finn et al. 1996), the DHA and EPA in the wax esters provided only a moderate proportion of the PUFA in comparison to phospholipids (17.5 % and 20 % respectively). Therefore, despite the dietary sources of wax esters in marine fish (Falk-Petersen et al. 1999), their high absorption rate in the digestive tract (Oxley et al. 2009), deposition in tissues, particularly ovaries, does not suggest any major role for these compounds in larval nutrition.

Chu and Ozkizilcik (1995) reported that wax esters constitute the major lipid component of striped bass at the time of fertilization and hatching. The mobilization of wax esters contained most of all in the oil globule of the embryo/larvae proceeds through direct trophoblastic exocytosis (as described by Jaroszewska and Dabrowski 2009) and subsequent entry into circulation. Therefore, the digestive lipolytic enzyme (wax ester hydrolase) is not of relevance in the process of the utilization of yolk sac reserves. Furthermore, timely provision of wax esters (hydrophobic lipids) to larval fish may facilitate inflation of the swim bladder by preventing inner surface adherence and collapse. Liquid waxes (with unsaturated alcohols) will diminish transepidermal water movement (loss in marine fish), prevent UV damage and microbial entrance.

It is puzzling why the synthesis/deposition of wax esters is favored over triglycerides in the ovaries of some fishes (striped bass, yellow perch, and burbot) with wax esters thus becoming the major lipid in the yolk sac of the larvae if it is assumed that these lipids (with comparable degree of fatty acid unsaturation) have similar densities, caloric value and compressibility. Sargent and Henderson (after Falk-Petersen et al. 1999) argued that the formation of fatty alcohols of wax esters removes end product inhibition of de novo fatty acid biosynthesis and results in an accelerated lipogenesis. This explanation fits marine copepods facing a short period of high food abundance as well the short period allowed for successful gametogenesis in some fish species perfectly. Wax esters differ from triglycerides with respect to their thermal expansion and their extracellular versus intracellular location. Put simply, extracellular energy/nutrient (essential fatty acids, PUFA) storage likely allows faster energy mobilization for larval fish during growth. It has been observed that when food is abundant the absorption of the oil globule is much faster than in starved larvae in striped bass (Eldridge et al. 1981).

The effect of maternal lipid composition and essential fatty acid concentration may well be extended into the early life of larval fish and their resistance/response to fatty acid profiles in their starter diets. Studies of these impacts on larval percids have just been initiated. Lund et al. (2012) were not able to demonstrate any significant effect of larval pike-perch feeding with a PUFA enriched diet on growth and survival until day 22 after hatching. However, the latter authors also reported that 17-day old pike-perch juveniles fed live *Artemia* nauplii diets enriched with DHA or ARA have showed highly increased mortality following salinity stress (19 ppt) in the case of a AA-supplemented diet.

At present we are investigating some procedures for larval rearing and transition to artificial diets (Dabrowski et al. 2014; Kwasek et al. 2013). This year, we were able to raise yellow perch larvae in tanks at high density using rotifers and live brine shrimp nauplii as food. Then fish were weaned to a dry diet. Yellow perch with non-inflated swimbladder suffered high mortality in all dietary treatments whereas survival of yellow perch with inflated swimbladder was similar across all treatments.

Fish fed live food grew faster than those fed any of the dry diets. The lack of a inflated swim bladder had a negative effect on the growth of fish. Krill hydrolysate coating of FryFeed Kyowa neither improved the growth or survival of yellow perch juveniles (Rinchard et al. 2008).

Acknowledgments The experimental studies were supported in part by the Great Lakes Fishery Commission and Great Lakes Fishery Trust Grant No. 292, and the North Central Regional Aquaculture Center, U.S.D.A., Ames, IO. We would like to thank Dr. Kornelia Dabrowska for extensive editing of the manuscript. The findings and conclusions in this article are those of the authors and do not necessarily represent the views of the US Fish and Wildlife Service.

References

Bell JG, McVicar AH, Park MT, Sargent JR (1991) Effects of high dietary linoleic acid on fatty acid compositions of individual phospholipids from tissues of Atlantic salmon (*Salmo salar*) association with a novel cardiac lesion. J Nutr 121:1163–1172

Bell MV, Batty RS, Dick JR, Fretwell K, Navarro JO, Sargent JR (1995) Dietary deficiency of docosahexaenoic acid impairs vision at low light intensity in juvenile herring (*Clupea harengus*). Lipids 30:443–449

Bell JG, Farndale BM, Bruce MP, Navas JM, Carillo M (1997) Effects of broodstock dietary lipid on fatty acid compositions of eggs from sea bass (*Dicentrarchus labrax*). Aquaculture 149:107–119

Brown PB, Dabrowski K, Garling DL (1996) Nutrition and feeding of yellow perch (*Perca flavescens*). J Appl Ichthyol 12:171–174

Bruce M, Oyen F, Bell G, Asturiano JF, Farndale B, Carrillo M, Zanuy S, Ramos J, Bromage N (1999) Development of broodstock diets for the European sea bass (*Dicentrarchus labrax*) with special emphasis on the importance of n-3 and n-6 highly unsaturated fatty acid to reproductive performance. Aquaculture 177:85–97

Cahu C, Gisbert E, Villeneuve LAN, Morais S, Hamza N, Wold P-A, Zambonino Infante JL (2009) Influence of dietary phospholipids on early ontogenesis of fish. Aquac Res 40:989–999

Chu F-LE, Ozkizilcik S (1995) Lipid and fatty acid composition of striped bass (*Morone saxatilis*) larvae during development. Comp Biochem Physiol 111B:665–674

Ciereszko A, Dabrowski K (1993) Estimation of sperm concentration of rainbow trout, whitefish and yellow perch using spectrophotometric technique. Aquaculture 109:367–373

Ciereszko RE, Dabrowski K, Ciereszko A, Ottobre JS (1996) Plasma concentrations of steroid hormones in male yellow perch, *Perca flavescens*: the effect of age and photothermal manipulation. Environ Biol Fish 51:97–105

Czesny S, Dabrowski K (1998) The effect of egg fatty acid concentrations on embryo viability in wild and domesticated walleye (*Stizostedion vitreum*). Aquat Living Resour 11:371–378

Czesny S, Kolkovski S, Dabrowski K, Culver D (1999) Growth, survival, and quality of juvenile walleye *Stizostedion vitreum* as influenced by n-3 HUFA enriched *Artemia* nauplii. Aquaculture 178:103–115

Czesny S, Rinchard J, Dabrowski K (2005) Intrapopulation variation in egg lipid and fatty acid composition and embryo viability in naturally spawning walleye population from inland reservoir. N Am J Fish Manag 25:122–129

Dabrowski K (1982) Seasonal changes in the chemical composition of fish body and nutritional value of the muscle of the pollan (*Coregonus pollan* Thompson) from Lough Neagh, Northern Ireland. Hydrobiologia 87:121–141

Dabrowski K, Bardega R (1982) The changes of fish larvae dimensions due to fixation in different preservatives. Zool Jb Anat 108:509–516

Dabrowski K, Culver DA, Brooks CL, Voss AC (1993) Biochemical aspects of the early life history of yellow perch (*Perca flavescens*). In: Kaushik SJ, Luquet P (eds) Fish nutrition in practice, vol 61, Les colloques. Institut National de la Recherche Agronomique, Paris, pp 531–540

Dabrowski K, Ciereszko RE, Ramseyer L, Culver D, Kestemont P (1994) Effects of hormonal treatment on induced spermiation and ovulation in the yellow perch (*Perca flavescens*). Aquaculture 120:171–180

Dabrowski K, Ciereszko RE, Blom JH, Ottobre JS (1995) Relationship between vitamin C and testosterone in female rainbow trout *Oncorhynchus mykiss*. Fish Physiol Biochem 14:409–414

Dabrowski K, Ciereszko RE, Ciereszko A, Toth GP, Christ SA, ElSaidy D, Ottobre JS (1996) Reproductive physiology of yellow perch (*Perca flavescens*): environmental and endocrinological cues. J Appl Ichthyol 12:139–148

Dabrowski K, Korzeniowska M, Grayson J, Kwasek K, Wojno M (2014) Effect of live food enrichment with DHA and ARA on yellow perch larvae lipids and fatty acid concentration. World Aquaculture Society Annual Meeting, Seattle

Eldridge MB, Whipple JA, Eng D, Bowers MJ, Jarvis BM (1981) Effects of food and feeding factors on laboratory-reared striped bass larvae. Trans Am Fish Soc 110:111–120

Eldridge MB, Joseph JD, Taberski KM, Seaborn GT (1983) Lipid and fatty acid composition of the endogenous energy sources of striped bass (*Morone saxatilis*) eggs. Lipids 18:510–513

Falk-Petersen SW, Sargent JR, Lonne OJ, Timofeev S (1999) Functional biodiversity of lipids in Antarctic zooplankton: *Calanoides acutus*, *Calanus propinquus*, *Thysanoessa macrura*, and *Euphausia crystallorophias*. Polar Biol 21:37–47

Fernandez-Palacios H, Izquierdo MS, Robaina L, Valencia A, Salhi M, Vergara JM (1995) Effect of n-3 HUFA level in broodstock diets on egg quality of gilthead sea bream (*Sparus aurata* L.). Aquaculture 132:325–337

Finn RN, Fyhn HJ, Henderson RJ, Evjen MS (1996) The sequence of catabolic substrates oxidation and enthalpy balance of developing embryos and yolk-sac larvae of turbot (*Scophthalmus maximus*). Comp Biochem Physiol 115A:133–151

Folch J, Lees M, Stanley GHS (1957) A simple method for the isolation and purification of total lipids from animal tissues. J Biol Chem 226:497–509

Furuita H, Tanaka H, Yamamoto T, Suzuki N, Takeuchi T (2002) Effects of high levels of n-3 HUFA in broodstock diet on egg quality and egg fatty acid composition of Japanese flounder, *Paralichthys olivaceus*. Aquaculture 210:323–333

Furuita H, Unuma T, Nomura K, Tanaka H, Okuzawa K, Sugita T, Yamamoto T (2006) Lipid and fatty acid composition of eggs producing larvae with high survival rate in the Japanese eel. J Fish Biol 69:1178–1189

Goetz FW, Duman P, Berndtson AK, Janowsky EG (1989) The role of prostaglandins in the control of ovulation in yellow perch (*Perca flavescens*). Fish Physiol Biochem 7:163–168

Hayashi K, Kashiki I (1988) Level and composition of wax esters in different tissues of deep-sea teleost fish *Lamonema longipes*. Bull Jpn Soc Sci Fish 54:135–140

Henrotte E, Kpogue D, Mandiki SNM, Wang N, Douxfils J, Dick J, Tocher D, Kestemont P (2010) Egg and larval quality, and egg fatty acid composition of Eurasian perch breeders (*Perca fluviatilis*) fed different dietary DHA/APA/AA ratios. Aquac Res 41:e53–e61

Henrotte E, Kpogue D, Mandiki SNM, Wang N, Douxfils J, Dick J, Tocher D, Kestemont P (2011a) n3 and n6 fatty acid bioconversion abilities in Eurasian perch (*Perca fluviatilis*) at two developmental stages. Aquac Nutr 17:e216–e225

Henrotte E, Milla S, Mandiki SNM, Kestemont P (2011b) Arachidonic acid induces production of 17,20β-dihydroxy-4-pregnen-3-one (DHP) via a putative PGE2 receptor in fish follicles from the Eurasian perch. Lipids 46:179–187

Izquierdo MS, Socorro J, Arantzamendi L, Hernandez-Cruz CM (2000) Recent advances in lipid nutrition in fish larvae. Fish Physiol Biochem 22:97–107

Izquierdo MS, Fernandez-Palacios H, Tacon AGJ (2001) Effect of broodstock nutrition on reproductive performance of fish. Aquaculture 197:25–42

Jaroszewska M, Dabrowski K (2009) The nature of exocytosis in the yolk trophoblastic layer of silver arowana (*Osteoglossum bicirrhosum*) juvenile, the representative of ancient teleost fishes. Anat Rec 292:1745–1755

Jaroszewska M, Dabrowski K (2011) Utilization of yolk: transition from endogenous to exogenous nutrition in fish. In: Joan Holt GJ (ed) Larval fish nutrition, 1st edn. Wiley, Ames, pp 183–218

Kaitaranta JK, Ackman RG (1981) Total lipids and lipid classes in fish roe. Comp Biochem Physiol 69:725–729

Kjørsvik E, Mangor-Jensen A, Holmefjord I (1990) Egg quality in fishes. Adv Mar Biol 6:71–113

Kwasek K, Wojno M, Niass F, Grayson J, Dabrowski K (2013) The art of raising yellow perch larvae and juveniles in captivity. World Aquaculture Society, Annual meeting, Nashville

Leray C, Monnotte C, Rouband P, Leger C (1985) Incidence of (n-3) essential fatty acid deficiency on trout reproductive processes. Reprod Nutr Dev 25:567–581

Lund ED, Sullivan CV, Place AR (2000) Annual cycle of plasma lipids in captive reared striped bass: effects of environmental conditions on reproductive cycle. Fish Physiol Biochem 22:263–275

Lund I, Steenfeld SJ, Suhr KI, Hansen BW (2007) A comparison of fatty acid composition and quality aspects of eggs and larvae from cultured and wild broodstock of common sole (*Solea solea* L.). Aquac Nutr 14:544–555

Lund I, Skov PV, Hansen BW (2012) Dietary supplementation of essential fatty acids in larval pikeperch (*Sander lucioperca*): short and long term effects on stress tolerance and metabolic physiology. Comp Biochem Physiol 162A:340–348

Martins DA, Rocha F, Castanheira F (2013) Effects of dietary arachidonic acid on cortisol production and gene expression in stress response in Senegalese sole (*Solea senegalensis*) post-larvae. Fish Physiol Biochem 39:1223–1238

Metcalfe LD, Schmitz AA (1961) The rapid preparation of fatty acid esters for gas chromatographic analysis. Anal Chem 33:363–364

Moodie GEE, Loadman NL, Wiegand MD, Mathias JA (1989) Influence of egg characteristics on survival, growth and feeding in larval walleye (*Stizostedion vitreum*). Can J Fish Aquat Sci 46:516–521

Moodie GE, Mathias JA, Loadman NL (1992) A comparison of two production-scale modules for the intensive culture of larval walleye. Aquacult Eng 11:171–182

Nandi S, Routray P, Gupta SD, Rath CS, Dasgupta S, Meher PK, Mukhopadhyay PK (2007) Reproductive performance of carp, *Catla catla* (Ham.) reared on a formulated diet with PUFA supplementation. J Appl Ichthyol 23:684–691

Navarro JC, McEvoy LA, Bell MV, Amat F, Hontoria F, Sargent JR (1997) Effect of different dietary levels of docosahexaenoic acid (DHA, 22:6w3) on the DHA composition of lipid classes in sea bass larvae eyes. Aquac Int 5:509–516

Navas JM, Mananos E, Thrush M, Ramos J, Zanuy S, Carillo M, Zohar Y, Bromage N (1998) Effect of dietary lipid composition on vitellogenin, 17-estradiol and gonadotropin plasma levels and spawning performance in captive sea bass (*Dicentrarchus labrax* L.). Aquaculture 165:65–79

Norambuena F, Morais S, Estevez A, Bell JG, Tocher DR, Navarro JC, Duncan N (2013) Dietary modulation of arachidonic acid metabolism in Senegalese sole (*Solea senegalensis*) broodstock reared in captivity. Aquaculture 372–375:80–88

Norberg B, Haux C (1985) Induction, isolation and characterization of the lipid content of plasma vitellogenin from two *Salmo* species, rainbow trout (*Salmo gairdneri*) and sea trout (*Salmo trutta*). Comp Biochem Physiol 81B:869–876

Oxley A, Bogevik AS, Henderson RJ, Waagbo R, Tocher DR, Olsen RE (2009) Digestibility of *Calanus finmarchicus* wax esters in Atlantic salmon (*Salmo salar*) freshwater presmolts and seawater postsmolots maintained at constant water temperature. Aquac Nutr 15:459–469

Rinchard J, Dabrowski K, Ottobre J (2002) Changes of plasma steroid concentrations associated with spontaneous or induced ovulation in yellow perch *Perca flavescens*. Fish Physiol Biochem 26:239–248

Rinchard J, Czesny S, Dabrowski K (2007) Influence of lipid class and fatty acid efficiency on survival, growth and fatty acid composition in rainbow trout juveniles. Aquaculture 264:363–371

Rinchard J, Becheret R, Dabrowski K (2008) Growth, survival, and body composition of yellow perch juveniles fed commercial and experimental diets. N Am J Aquac 70:74–79

Rodriguez C, Cejas JR, Martin MV, Badia P, Samper M, Lorenzo A (1998) Influence of n-3 highly unsaturated fatty acid deficiency on the lipid composition of broodstock gilthead seabream (*Sparus aurata* L.) and on egg quality. Fish Physiol Biochem 18:177–187

Saito H, Kotani Y (2000) Lipids of four boreal species of calanoid copepods: origin of monoene fats of marine animals at higher trophic levels in the grazing food chain in the subarctic ocean ecosystem. Mar Chem 71:69–82

Sand DM, Schlenk H (1969) The polyunsaturated alcohols in wax esters of fish roe. Lipids 4:303–304

Sargent JR, Lee RF, Nevenzel JC (1976) Marine waxes. In: Kolattukudy P (ed) Chemistry and biochemistry of natural waxes. Elsevier, Amsterdam

Sargent JR (1995) Origins and functions of egg lipids: nutritional implications. In: Roberts RJ, Bromage NR (eds) Broodstock management and egg and larval quality. Blackwell Science, Oxford, pp 353–372

Sargent JR, Bell JG, Bell MV, Henderson RJ, Tocher DR (1995) Requirement criteria for essential fatty acids. J Appl Ichthyol 11:183–198

Stevens CS, Deible D, Parrish CC (2004) Incorporation of bacterial fatty acids and changes in wax ester-based omnivory index during a long term incubation experiment with *Calanus glacialis*. J Exp Mar Biol Ecol 303:135–156

Tandler A, Harel M, Koven WM, Kolkovski S (1995) Broodstock and larvae nutrition in gilthead seabream *Sparus aurata* – new findings on its mode of involvement in improving growth, survival and swimbaldder inflation. Isr J Aquac Bamidgeh 47:95–111

Vassallo-Agius R, Watanabe T, Yoshizaki G (2001) Quality of eggs and spermatozoa of rainbow trout fed an n-3 essential fatty acid-deficient diet and its effects on the lipid and fatty acid components of eggs, semen and livers. Fish Sci 67:818–827

Watanabe T, Takeuchi T, Saito M, Nishimura K (1984) Effect of low protein-high calorie or essential fatty acid deficiency diet on reproduction of rainbow trout. Bull Jpn Soc Sci Fish 50:1207–1215

Wiegand M (1996) Composition, accumulation and utilization of yolk lipids in teleost fish. Rev Fish Biol Fish 6:259–286

Wiegand M, Johnston TA, Martin J, Leggett WC (2004) Variation in neutral and polar lipid compositions of ova in ten reproductively isolated populations of walleye (*Sander vitreus*). Can J Fish Aquat Sci 61:110–121

Xu X, Kestemont P (2002) Lipid metabolism and FA composition in tissues of Eurasian perch *Perca fluviatilis* as influenced by dietary fats. Lipids 37:297–304

Chapter 22
Feeding and Nutrition of Percid Fishes During Ongrowing Stages

Florian Geay and Patrick Kestemont

Abstract In the context of inland aquaculture diversification in Europe, some percid fishes, namely Eurasian perch and pikeperch, are receiving increasing attention from scientists and fish farmers. Significant improvement in the knowledge of percid fish feeding and nutrition during their ongrowing stage has been achieved during these last two decades. The relative importance of different abiotic and biotic factors on feeding activity of percid fishes was investigated. Among them, water temperature, feeding frequency, photoperiod and fish density were identified as factors of prime importance. Rearing European percid fishes at a high density in relatively warm water (22–27 °C) and fed three meals per day with a day length of up to 12 h a day significantly improve fish feeding activity, and, in the same way, the growth performances. Optimization of growth performances under artificial conditions was also investigated through the characterization of their nutritional requirements in terms of protein, lipid and carbohydrate. Depending on the fish life stage, artificial diets containing 43–50 % protein, 13–18 % lipid and 10–15 % carbohydrate cover the nutrient requirements of percid fishes and support the highest growth performance. Moreover, recent advances in the use of alternative oil sources in percid nutrition suggest a high potential of these species to biosynthesize HUFA when fish oil is replaced by plant oil rich in PUFA. In this context of fish ingredients replacement by plant sources, it could be of high interest to investigate the possibility of replacing fish meal by plant products in the future.

Keywords Eurasian perch • Pikeperch • Feeding • Nutrition • Growth • HUFA

F. Geay • P. Kestemont (✉)
Research Unit in Environmental and Evolutionary Biology, University of Namur,
61 rue de Bruxelles, 5000 Namur, Belgium
e-mail: Patrick.kestemont@unamur.be

P. Kestemont et al. (eds.), *Biology and Culture of Percid Fishes*,
DOI 10.1007/978-94-017-7227-3_22

22.1 Introduction

Eurasian perch and pikeperch are receiving increasing attention as new fish species in inland aquaculture diversification. Indeed, the Percidae family have a large geographic distribution in Europe and are appreciated by consumers for their nutritional and organoleptic qualities. Under natural conditions, percid fishes are shoaling, opportunistic predators, feeding on prey ranging from zooplankton to fish (Thorpe 1977). In aquaculture environment, growth optimization is of paramount importance for profitability. To this end, it is essential to define their behaviour and performance in control conditions. In this chapter, we will focus our attention on feeding and nutrition of percid fishes during ongrowing stages.

22.2 Feeding of Percids

In controlled environments, growth is positively correlated with food consumption. This consumption depends essentially on the feeding frequency and level, and together defines the total quantity of food ingested per day, or food ration (Brett and Grove 1979). Several indices can be calculated to characterize the effects of feeding level on fish growth. The food ration modulates the feed efficiency (FE) and is a measure of the efficiency of fish in converting the feed intake into increased body mass. On the contrary, the food conversion rate (FCR) corresponds to the ratio of feed intake to weight gain. The food ration also modulates the specific growth rate (SGR), commonly used to determine the nutritional value of a diet. The SGR is defined as the increase in fish mass per unit of time. These parameters are defined below:

Feed efficiency (FE) = (Wf-Wi)/Total Feed Intake (TFI)
Food conversion rate (FCR) = TFI/(Wf-Wi)
Specific growth rate (SGR) (% day^{-1}) = $100 \times (\mathrm{Ln}W_f - \mathrm{Ln}W_i)/\Delta t$

where Wi, Wf = the initial and final fish weight and t = time (days)

Numerous factors influence fish growth and resulting nutritional value. Experimental units and fish are influenced by abiotic and biotic factors, with high degrees of interactions, which will modulate fish behaviour and physiology. Temperature, feeding frequency, food ration, photoperiod and tank colour are among the abiotic factors of prime importance while life stage, gender, size heterogeneity and density are usually considered as the main biotic factors modulating fish growth performances.

22.2.1 Influence of Abiotic Factors on Growth Performance

22.2.1.1 Temperature

Among the parameters which modulate feeding activity, water temperature is of prime importance. Percids are temperate but quite thermophilic fish species and display the fastest growth at relatively warm temperatures (22–27 °C). The modulation of growth performances by temperature has been investigated by Mélard et al. (1995) using 1.9 g Eurasian perch fed a commercial salmonid feed and reared at 22.9 °C or 26.5 °C. During this experiment, fish were fed with automatic feeder at a feeding level of 1.7 % for 12 h per day. After 144 days, the authors concluded that there were no significant differences in growth rate between the groups maintained at 22.9 °C (SGR = 1.76 % day^{-1}) and 26.5 °C (SGR = 1.86 % day^{-1}) despite a final mean body weight 14.5 % higher in the group reared at 26.5 °C. Moreover, FCR was not significantly affected by temperature (2.95 at 22.9 °C and 2.82 at 26.5 °C). According to these authors, the absence of significant differences could be explained by the fact that both temperatures were within the optimal thermal range of perch growth, but also because of high variability of individual fish body weight observed within each group (coefficient of variation of body weight: 21.6–25.4 %). However, an infestation by a ciliate (*Heteropolaria sp.*) at 26.5 °C contributed to continuous mortality suggesting this temperature is too high for intensive rearing of Eurasian perch juveniles. As concluded by the authors, a temperature of 23 °C appeared optimal for growth of Eurasian perch juveniles fed three times a day at 23 °C (SGR = 1.76 % day^{-1}). Similar conclusions were reported by Mélard et al. (1996). Growth of 15-g fish reared at 27 °C was 12 % lower than at 23 °C. Similarly, 100-g Eurasian perch reared at 10 °C displayed a 20 % lower growth compared to fish reared at 23° (Mélard and Kestemont 1995). Thus, a rearing temperature of 23 °C can be applied during ongrowing life stages of Eurasian perch to optimize its growth.

Comparatively, pikeperch display a higher temperature preferendum. Wang et al. (2009) investigated the influence of this factor on growth performances and feed efficiency of 6.4 g pikeperch juveniles. The higher value of SGR (SGR = 2 % day^{-1}) was obtained when juveniles were fed one or three times a day at 28 °C, while the lower SGR was in juveniles reared at 20 °C (SGR: 1.2 % day^{-1}) and fed one time a day. Similarly, the feed efficiency increased from 0.73 at 20 °C to 0.97 at 28 °C. An increase of the feeding frequency from three times/day to six times/day did not affect the growth performances. These observations were reinforced by the results of Ronyai and Csengeri (2008) who investigated the growth performance of 84 pikeperch at 20 and 25 °C. In this trial, fish reared at 25 °C exhibited a higher SGR compared to fish maintained at 20 °C at the same feeding rate (Table 22.1). These results suggest that the pikeperch is one of the most thermophilic Percid species studied so far and that it could be comparable to what is found in some Cyprinids such as the tench (*Tinca tinca*) and the European catfish (*Silurus glanis*) as well as the African catfish (*Clarias gariepinus*). However, different conclusions have been raised by Frisk et al. (2012), investigating the thermal optimum for pikeperch by the use of ventilation frequency as a predicator of metabolic rate. As suggested by these latter authors, a temperature of 28 °C is very close to the critical temperature since it significantly increased the Q_{10}-values for $MO2$ consumption. Based on this variable, Frisk et al. (2012) defined an optimal temperature range between 10 and 27 °C. However, in this experiment, larger individuals were used compared to the experiments described above, and this can influence the optimal temperature of pikeperch. Indeed, as mentioned by Morita et al. (2010), lower temperatures are required to induce maximal growth rates in larger fish.

Table 22.1 Effects of temperature (T°C), feeding level and feeding frequency on growth performance and feed efficiency (FE) of pikeperch

	T°C	Duration (days)	Food frequency	Daily food ration	Photoperiod	Initial weight (g)	Final weight (g)	SGR (% day^{-1})	FE	FCR
A	22.3	42	AF	1.2 %		25.4	47.9a	1.52a	–	0.77ab
	22.3	42	AF	1.6 %		25.5	60.3b	2.05b	–	0.75a
	22.3	42	AF	2.0 %		25.6	69.4c	2.38c	–	0.79b
B	20	126	AF	0.90 %	20L:4D	84 ± 19	247 ± 4	0.81 ± 0.01a	–	–
	20	126	AF	1.22 %	20L:4D	84 ± 19	260 ± 18	0.91 ± 0.04b	–	–
	20	126	AF	1.35 %	20L:4D	84 ± 19	278 ± 9	0.97 ± 0.00c	–	–
	20	126	AF	1.70 %	20L:4D	84 ± 19	267 ± 6	0.99 ± 0.02c	–	–
	25	126	AF	0.90 %	20L:4D	84 ± 19	240 ± 4	0.80 ± 0.00a	–	–
	25	126	AF	1.22 %	20L:4D	84 ± 19	275 ± 19	0.93 ± 0.05b	–	–
	25	126	AF	1.35 %	20L:4D	84 ± 19	314 ± 0	1.03 ± 0.01d	–	–
	25	126	AF	1.70 %	20L:4D	84 ± 19	306 ± 2	1.08 ± 0.01d	–	–
C	20	56	1 meal/day	–	12L:12D	6.4 ± 0.4	12.0 ± 0.4a	1.10 ± 0.07	0.81 ± 0.05	–
	20	56	3 meals/day	–	12L:12D	6.4 ± 0.4	13.0 ± 0.4ab	1.30 ± 0.07	0.75 ± 0.05	–
	20	56	6 meals/day	–	12L:12D	6.4 ± 0.4	13.3 ± 0.4ab	1.30 ± 0.07	0.63 ± 0.05	–
	24	56	1 meal/day	–	12L:12D	6.4 ± 0.4	14.3 ± 0.4b	1.50 ± 0.07	0.92 ± 0.05	–
	24	56	3 meals/day	–	12L:12D	6.4 ± 0.4	16.9 ± 0.4cd	1.80 ± 0.07	0.97 ± 0.05	–
	24	56	6 meals/day	–	12L:12D	6.4 ± 0.4	16.3 ± 0.4c	1.60 ± 0.07	0.72 ± 0.05	–
	28	56	1 meal/day	–	12L:12D	6.4 ± 0.4	17.6 ± 0.4cd	2.00 ± 0.07	1.02 ± 0.05	–
	28	56	3 meals/day	–	12L:12D	6.4 ± 0.4	20.8 ± 0.4e	2.10 ± 0.07	1.14 ± 0.05	–
	28	56	6 meals/day	–	12L:12D	6.4 ± 0.4	18.1 ± 0.4d	1.80 ± 0.07	0.74 ± 0.05	–

Modified from Zakęś et al. (2003) (**A**), Rónyai and Csengeri (2008) (**B**), and Wang et al. (2009) (**C**)
Values are means ± SD. Values in each column with common letter are not significantly different (P > 0.05)
AF Automatic feeder

22.2.1.2 Feeding Ration and Feeding Frequency

As mentioned previously, optimal fish growth is dependent on several factors. Among them, the influence of feeding ration and feeding frequency are of prime importance to optimize fish growth performances. Fontaine et al. (1997) investigated the effects of feeding level (1, 2, 3 % BW day^{-1}) on growth performances of 10 g Eurasian perch juveniles reared at 21.4 °C for 84 days (Table 22.2). In this experiment, feed was distributed at the frequency of four meals per day. Among the three different feeding levels tested, 2 % of biomass appeared optimal since it supported a SGR of 1.7 % day^{-1} and a FCR close to one. Indeed, despite a similar FCR at a feeding level of 1 % BW day^{-1}, the growth rate was much reduced. On the contrary, an increase of the feeding level to 3 % slightly improved growth rate but resulted in a reduced feed utilization (FCR = 1.53). More recently, Fiogbé and Kestemont (2003) estimated the optimum daily ration for different sizes of Eurasian perch at the optimum temperature of 23 °C (Table 22.2). This experiment was conducted on initial body weight of 0.22, 0.73, 1.56 and 18.9 g. Fish were fed with an automatic feeder at daily feeding levels ranging from 1 % to 20 % of fish biomass, depending on the initial fish size. Authors concluded that the optimum feeding rate decreases from 7.4 % to 5.1 %, 4.5 % and 2.2 % BW day^{-1} for fish of 0.22, 0.73, 1.56 and 18.9 g initial body weight, respectively (Table 22.3). SGR values decreased with increase of fish weight at the optimal feeding rate. As described in this study, the low feeding level (2 % BW day^{-1}) significantly affected the survival of Eurasian perch, with an estimated survival of 51.5, 71 and 75 % for fish of 0.22, 0.73 and 1.56 g respectively. This mortality could be explained by a nutrient deficiency at early stages, but also by the high cannibalism rate which occurs at these fish sizes (0.73 and 1.56 g initial weight). Indeed, as reported by Kestemont et al. (2003), low food availability influences size heterogeneity in predatory species such as Percids and thus contributes to increased cannibalism. Moreover, the use of an automatic feeder at the low feeding level may contribute to increased growth heterogeneity. Indeed, food is monopolized by some dominant individuals, the predictability of food distribution in space and time with an automatic feeder contributing to reduce the accessibility of food for the dominated fish (Kadri et al. 1996).

Zakęś et al. (2003) investigated the effects of feeding levels (1.2, 1.6 and 2 % BW day^{-1}) on growth of 25 g pikeperch during 42 days (Table 22.1). In this experiment, food was distributed by automatic feeders during 19 h day^{-1}. The highest growth rate (SGR = 2.38 % day^{-1}) was obtained in the group fed a ration of 2 % BW day^{-1}. However, feed was most efficiently utilised by fish fed an intermediate ration (1.6 % BW day^{-1}). Based on its lowest SGR (1.52 % day^{-1}), the group fed 1.2 BW day^{-1} could be considered the only feed restricted group. The authors estimated that feeding rate of at least 2 % of biomass at 22 °C could be optimal for 25 g pikeperch juveniles. More recently, Bódis and Bersényi (2009) investigated the effects of three different feeding levels (2 %, 4 % and 6 %) on early 3.5 g pikeperch fed for 8 weeks in cages. In the present experiment, fish fed 2 % BW day^{-1} exhibited the best FCR (0.77) while the worst FCR was observed in the 6 % BW day^{-1} treatment. However, the maximal growth rate (SGR = 3.7 % day^{-1}) was obtained in pikeperch fingerlings

Table 22.2 Influence of the daily food ration, temperature (T°C), photoperiod, sex and initial fish size on growth performances of Eurasian perch

	Daily food ration	Duration (days)	T°C	Food frequency	Photoperiod	Initial density ($kg.m^{-3}$)	Gender	Initial weight (g)	Final weight (g)	SGR (% day^{-1})	FCR
A	1 (% biomass day^{-1})	21	23	AF	8D:16L	1.46	ND	0.73	0.9 ± 0.30a	0.83^{a}	–
	4 (% Zbiomass day^{-1})	21	23	AF	8D:16L	1.46	ND	0.73	1.2 ± 0.1ab	2.52^{a}	–
	8 (% biomass day^{-1})	21	23	AF	8D:16L	1.46	ND	0.73	1.5 ± 0.2b	3.52^{a}	–
	0.3 (% biomass day^{-1})	28	21.5	AF	8D:16L	7.2	ND	18.9	16.8 ± 1.9a	-0.42^{a}	–
	2 (% biomass day^{-1})	28	21.5	AF	8D:16L	7.2	ND	18.9	22.2 ± 1.2b	0.58^{a}	0.30 ± 0.15
	3 (% biomass day^{-1})	28	21.5	AF	8D:16L	7.2	ND	18.9	24.6 ± 1.5b	0.95^{a}	0.45 ± 0.05
B	1 (% biomass day^{-1})	84	21.4	4 meals/day	8D:16L	2.3	ND	10.4 ± 0.2	22.3 ± 0.5a	0.88 ± 0.09a	1.08 ± 0.03a
	2 (% biomass day^{-1})	84	21.4	4 meals/day	8D:16 L	2.3	ND	10.4 ± 0.3	47.9 ± 2.4b	1.77 ± 0.22b	1.01 ± 0.04a
	3 (% biomass day^{-1})	84	21.4	4 meals/day	8D:16L	2.3	ND	10.6 ± 0.2	52.8 ± 1.7c	1.86 ± 0.08c	1.53 ± 0.06b

C	1.2 pellets $fish^{-1}$ min^{-1}	100	18	AF	24L:0D	4.8[a]	M	9.8±0.8	52.7±5.6	1.68±0.01	0.88±0.05
	1.2 pellets $fish^{-1}$ min^{-1}	100	18	AF	24L:0D	4.8[a]	F	12.6±0.4	89.7±3.8	1.98±0.00	0.88±0.05
	4.2 pellets $fish^{-1}$ min^{-1}	100	18	AF	24L:0D	4.8[a]	M	12.1±0.1	63.6±8.6	1.65±0.16	1.02±0.10
	4.2 pellets $fish^{-1}$ min^{-1}	100	18	AF	24L:0D	4.8[a]	F	14.8±1.0	91.0±12.4	1.84±0.08	1.02±0.10
	14.3 pellets $fish^{-1}$ min^{-1}	100	18	AF	24L:0D	4.8[a]	M	11.9±1.2	58.3±8.8	1.57±0.02	1.04±0.11
	14.3 pellets $fish^{-1}$ min^{-1}	100	18	AF	24L:0D	4.8[a]	F	12.6±0.8	81.1±10.9	1.88±0.19	1.04±0.11

Modified from Fiogbé and Kestemont (2003) (**A**), Fontaine et al. (1997) (**B**), and Juell and Lekang (2001) (**C**)
In the same experiment, values with same letter are not significantly different among treatment conditions ($P>0.05$). Values are means ± SD
AF Automatic feeder
[a]Values estimated from data of Fiogbé and Kestemont (2003) and Juell and Lekang (2001)

Table 22.3 Maintenance, optimum and maximum daily feeding rate (% body weight day^{-1}) for different fish sizes of Eurasian perch reared at 23 °C (Fiogbé and Kestemont 2003)

Body weight (g)	Maintenance	Optimum	Maximum
0.22	1.7	7.4	17.6
0.73	2.5	5.1	6.7
1.56	1.8	4.5	6.4
18.9	0.8	2.2	3.7

fed a 6 % daily ration at 24–25 °C. The effects of temperature and feeding frequency on growth performances of larger pikeperch (150–180 g) were investigated by Rónyai and Csengeri (2008). These authors concluded that maximum feeding rates of 1.25 and 1.15 % of BW day^{-1} were optimal at 25 °C and 20 °C, respectively. Not surprisingly, different optimum feeding levels have been deduced from studies conducted on different life stages. Indeed, as it is well established, the maximal feeding rate is dependent on the initial fish size and decreases with fish weight.

The influence of feeding frequency on growth rate of pikeperch juveniles was investigated by Zakęś et al. (2006) on 4.8 and 21 g juveniles. At the end of a 6-week trial, the feeding frequency (one meal, three meals or continuous feeding per day) did not significantly affect the growth performance of 4.8 g juveniles. Similarly, FCR values were not significantly different among experimental conditions. Results obtained with 21 g juveniles reared for 8 weeks provided similar results. Indeed, SGR values (1.00–1.04 % day^{-1}) as well as FCR (0.81–0.83) were not significantly affected by feeding frequencies. These results seem to indicate that pikeperch are able to adapt to different feeding conditions. Indeed, fish that are fed less frequently can adapt to such conditions by consuming larger amounts of feed during each meal. In a long feeding schedule, this can lead to increased gut capacity and hyperphagia (Jobling 1983; Ruohonen and Grove 1996). In the study of Zakęś et al. (2006), this statement cannot be confirmed since the feeding rate did not influence the FCR. However, in addition to the feeding rate, it is also important to consider the time of each meal. For example, studies conducted on salmonid species indicated that 1 h of feeding was sufficient for fish to eat to satiation (Elliott 1975). In the study of Zakęś et al. (2006), fish were fed once a day during 3 h or three times a day during 1 h. The 3 h feeding time per day may be enough to feed fish to satiation, independently of the feeding frequency. More recently, Wang et al. (2009) evaluated the optimal daily frequency on growth of 6.4 g juveniles pikeperch reared at 20, 24 and 28 °C (Table 22.1). In this experiment, fish were hand fed one, three or six times a day for 56 days. The authors concluded that optimal feeding rates were estimated at 1.5, 1.8 and 2.0 % of the fish body weight at 20, 24 and 28 °C respectively with a frequency of three times a day. The increase of feed frequency (six times a day) did not significantly improve the growth parameters. According to Jobling and Johansen (1999), repeated feeding throughout long periods of the day can increase swimming activity of fish, resulting in energy expenditure and lower growth rate. Moreover, as concluded by Brett and Grove (1979), although any food ration between the maintenance level and the maximum one causes weight gain, the highest weight gain per unit of added ration is obtained before the maximum feeding level, at a level considered as optimum in terms of biological conversion.

22.2.1.3 Photoperiod, Intensity and Wavelengths of Light

Artificial environments that are very different from natural habitats may negatively affect fish behaviour. Among environmental factors, duration of light can profoundly affect fish. The light can be characterized by its day length (DL), intensity and quality (longer wavelengths), which modulate the feeding activity, and consequently, the growth rate capacities (Boeuf and Le Bail 1999; Brännäs et al. 2001).

Considering that Eurasian perch show a typical diurnal feeding activity when fed with self-feeders (Anthouard and Fontaine 1998), one might hypothesize that day length and intensity can influence growth, as it was already described in other perciform species such as juvenile sunfish (*Lepomis cyanellus*) (Gross et al. 1965). Indeed, in nature, the feeding success of diurnal predator fish species such as Eurasian perch is directly modulated by contrast between the prey and its background (Endler 1987; Utne-Palm 1999). For example, changing light intensity may affect the behaviour of fish and modify the activity levels, thus altering predation risk and affecting feed intake (Craig 1977; Helfman 1978, 1981).

Jourdan et al. (2000) investigated the influence of day length (DL) on growth in 5.1 g Eurasian perch for 112 days (Table 22.4). Fish were fed at optimal feeding rate as defined by Mélard et al. (1996) and maintained under lighting conditions for 12, 18 and 24 h per day. As described by the authors, specific growth rates recorded at 18 h (SGR = 1.78 % day^{-1}) and 24 h (SGR = 1.77 % day^{-1}) were significantly higher than the SGR value calculated from fish reared at 12 h of DL (SGR = 1.36 % day^{-1}). This increase of growth performance with increase of DL is in accordance with results obtained in juveniles of green sunfish *Lepomis cyanellus* (Gross et al. 1965), yellow perch (*Perca flavescens*) (Huh et al. 1976) and Sliptnose rockfish *Sebastes*

Table 22.4 Influence of photoperiod on growth of Eurasian perch juveniles (males and females) reared during 112 days in a recirculation system at 23 °C. Fish were fed on demand with self-feeder

Photoperiod	Gender	Initial weight (g)	Final weight (g)	SGR (% day^{-1})
12L:12D	M	5.2 ± 0.7	25.3 ± 4.1a	1.41[a]
12L:12D	F	5.2 ± 0.7	29.2 ± 1.6A	1.54[a]
18L:6D	M	4.9 ± 0.3	39.5 ± 3.8b	1.86[a]
18L:6D	F	4.9 ± 0.3	47.3 ± 14.7B	2.02[a]
24L:0D	M	5.4 ± 0.6	44.4 ± 2.5b	1.88[a]
24L:0D	F	5.4 ± 0.6	49.6 ± 8.3B	1.98[a]

Modified from Jourdan et al. (2000)

Values with same letter for each gender are not significantly different among photoperiod conditions ($P > 0.05$). Values are means ± SD

[a] $SGR = 100 \times \left(LnW_{final} - LnW_{initial}\right) / \Delta t\left(\%day^{-1}\right)$. Values calculated from the initial and final mean weights

diploproa (Boehlert 1981). However, increase of DL affected also the weight heterogeneity, with a maximal coefficient of variation at the continuous light (CV: 54.1 %). Indeed, continuous light accentuates fish interactions, aggressiveness and cannibalism in territorial fish species such as Percidae. As a result of cannibalism, the survival rate was significantly reduced when fish were reared under continuous light (76 %), compared to 18 h and 12 h DL (89 and 88.2 % respectively). As mentioned by the authors (Jourdan et al. 2000), the DL may also modulate gonad development. Indeed, despite few effects of DL on the final weight of males and females, the decrease in DL from 24 to 12 h induced a stimulation of gonadal development in males. This gonad development can be explained by the fact that a 12 h DL in this experiment was close to the natural day length in September, when gametogenesis begins (Sulistyo et al. 1998). In addition, it was well described that males can develop their gonads as soon as they are 150–260 days old (5–25 g). Thus, an increase of DL prevents an early maturation of males (Fontaine et al. 1997). Contrary to male gonads, the modification of DL did not affect the ovarian development at an early stage (5–40 g). Indeed, the maturity of females occur later (up to 40 g) than males, thus, they were immature at the development stage in this experiment. However, as observed in other fish species like turbot *Scophtalmus maximus* (Imsland et al. 1997), an increase of DL could delay the age at the first maturity in Percids.

Light intensity is also an important factor affecting many behavioural and biological processes in fish such as growth (Trippel and Neil 2003) and onset of sexual maturity (Porter et al. 1999). For example, pikeperch are a crepuscular predator inhabiting turbid waters (Ali et al. 1977) and it displays some retinal adaptations such as *tapetum lucidum* (Zyznar and Ali 1975) and macroreceptors (Braekevelt et al. 1989), improving vision under low ambient light. Since this species prefers turbid waters, this factor may have significant impact on rearing results. Usually, excess light intensity results in an increase of stress level associated with a decrease of feed intake and survival. According to this adaptive response to the natural environment, Luchiari et al. (2006) and Kozlowski et al. (2010) investigated the effects of light intensity on growth performance of pikeperch. Luchiari et al. (2006) reported that pikeperch prefers low intensity environments. Indeed, fish of two initial weights (6.68 ± 1.26 g and 36.90 ± 10.89 g) reared at 25–300 lx and 1–50 lx, respectively, showed a preference for the lowest light intensity. Similarly, Kozlowski et al. (2010) showed a significant preference of pikeperch juveniles of 6.1 g and 42.2 g to the lowest intensity used for each development stage (45.1 lx and 1.2 lx, respectively) (Table 22.5) when fish were fed 24 h day^{-1} with automatic band feeders. While the survival and the body weight variation coefficient were not affected by the light intensity in this experiment, the specific growth rate was significantly modulated by the light intensity. Thus, 6.68 g juveniles exhibited significant higher SGR (SGR = 3.61 % day^{-1}) at 45.1 lx compared to 385 lx (SGR = 3.37 % day^{-1}). Moreover, the FCR value decreased from 1.09 at the high intensity to 0.97 at the low one for 6.1 g juveniles. According to the experimental conditions tested in both studies, the light intensity preferendum decreased with the increase of fish weight. It is well known that fish sensitivity to light is not only species specific, but also dependent on the ontogenic stage (Blaxter 1969, 1975). More precisely, a shift from

Table 22.5 Growth of two fish size groups of pikeperch reared under different light intensities (mean value ± SD) and constant daylight during 56 days. Fish were fed continuously with an automatic feeder

Light intensity (lux)	Temperature (°C)	Initial weight (g)	Final weight (g)	SGR (% day^{-1})	FCR
385.7±36.8	24.4±0.9	6.1	40.6±1.1a	3.37±0.03a	1.09±0.01a
45.1±8.2	24.4±0.9	6.1	45.9±1.3b	3.61±0.03b	0.97±0.01b
8.0±1.2	23.6±0.5	42.2	108.6±0.6a	1.69±0.01a	0.80±0.01a
1.2±0.3	23.6±0.5	42.2	113.0±1.5b	1.76±0.02a	0.78±0.01a

Modified from Kozlowski et al. (2010)
Different letters in the same column indicated significant differences (for each experiment separately) ($P<0.05$)

positive phototaxis to negative phototaxis occurs during ontogeny. For example, Bulkowski and Meade (1983) observed that larval walleye preferred high light intensity (7800 lx) from day 1 post-hatch until the eighth week of life (TL from 9 to 33 mm), while individuals older than 8 weeks exhibited a preference for low intensity (2–4 lx). Similar results have been obtained from studies investigating the performance of Eurasian perch under artificial environments (Tamazouzt et al. 2000). Differences in light intensity preferences that occur in pikeperch during its ontogenetic development can be attributed to the changes that occur in the eye, especially in connection with the development of the reflective membrane. This reflective membrane increases retinal sensitivity by reflecting light back. The *tapetum lucidum* appears in the walleye during the first month of life. For the pikeperch, it is probable that this *tapetum lucidum* is fully developed in juveniles around 15 cm. Curiously, Strand et al. (2007) who investigated the effects of light intensity on growth performance of Eurasian perch juveniles did not confirm the results obtained on pikeperch juveniles. Indeed, no significant differences of feed intake or SGR were observed between fish reared at 200 and 1100 lx. Moreover, as mentioned by the authors, the increase of swimming activity of perch juveniles at 2200 lx compared to 200 lx, which may indicate a state of stress, did not significantly affect the energy expenditure in this experiment. The discrepancy of these results can be explained by the fish weight used in the different studies described above. In the experiment of Strand et al. (2007), large juveniles of 59.6 g were reared for only 3 weeks. As the growth rate decreases with the increase of individual weight of fish, the duration of this experiment might be not long enough to significantly influence physiological responses.

Recently, Luchiari et al. (2009) investigated the influence of the longer wavelengths of light on the growth, the feed intake and the feed efficiency of juvenile pikeperch. Fish of about 33 g were reared for 42 days under white light or specific wavelengths corresponding to the blue (λ435 nm), green (λ534 nm), yellow (λ564 nm) or red (λ610 nm) spectra. Fish were maintained individually in each tank and fed in excess twice a day. Based on the results of this study, the increase in light wavelength within the visible spectrum improves the growth rate and the feed efficiency of pikeperch juveniles despite no effect on feed intake (Table 22.6). Indeed, fish reared with the red ambient colour exhibited a significantly higher SGR (SGR=2.37 %

Table 22.6 Influence of the ambient color on the growth rate of pikeperch reared during 42 days at 22 °C under constant daylight. Fish were fed twice daily

Ambient color (λmax, nm)	Initial weight (g)	Final weight (g)	SGR (% day^{-1})
White	37.1 ± 8.5	72.1 ± 14.9	1.59 ± 0.34
Blue (435)	46.2 ± 13.8	87.0 ± 14.3	1.56 ± 0.35
Green (534)	38.5 ± 7.0	95.9 ± 23.5	2.15 ± 0.51
Yellow (564)	36.1 ± 7.2	88.2 ± 18.5	2.12 ± 0.39
Red (610)	39.8 ± 11.5	106.1 ± 24.7*	2.37 ± 0.64*

Modified from Luchiari et al. (2009)
Significant differences of the color treatments from the white colour (control) are indicated by an asterisk in each column ($P < 0.05$). Values are means ± SD

day^{-1}) compared to other conditions (SGR of 1.59–2.12 % day^{-1}). This increase in growth performance in the long-wavelength (red) environment may be related to the enhancement of visual sensitivity in the ambient where the visual pigments are able to maximize photon capture (Cohen and Forward 2002).

22.2.1.4 Tank Colour

Light can modulate the feeding activity and, thus, the growth performances of Eurasian perch and pikeperch. In connection with these factors, feed detection and feeding success of carnivorous fish species such as Percids are also under control of contrast between the prey and its background, depending on the tank colour (Endler 1987). In general, the highest growth rates of fish larvae were obtained when light conditions and tank colour optimize the contrast between the feed and the background (Barahona-Fernandes 1979; Henne and Watanabe 2003). According to Strand et al. (2007), feed intake of Eurasian perch juveniles (59.6 g) is significantly higher for fish reared in white tanks compared to black at low light intensity (220 lx). Correspondingly, fish growth rates were higher in white tanks. However, tank colour did not significantly influence the growth rate of fish at a higher light intensity (1100 lx). Indeed, feed intake was also affected by the interaction between light intensity and tank colour. Thus, as suggested by the authors, a high feed intake in lighter tanks (1100 lx) was probably a consequence of an increase of the feed visibility, due to the feed's higher contrast against the background colour of the tank. In this experiment, energy efficiency did not differ between treatments suggesting that none of the colour/light combinations were more stressful to the Eurasian perch at this fish size. This lack of light/tank colour effects on stress can be partially explained by the fact that, as most fish species, Eurasian perch are able to modify its body colour in response to tank colour (Mairesse et al. 2005). For example, Eurasian perch juveniles reared in dark tank exhibited a dark colour body while perch maintained in white tank were very pale. This morphological adaptation avoids energy expenditure to the physiological stress response (Sloman et al. 2000; Wendelaar Bonga 1997).

22.2.2 *Influence of Biotic Factors on Feeding Activity*

Among the various factors which interact in rearing conditions, biotic factors such as genetic origins, fish size and sexual dimorphism as well as stocking density may contribute to influence the growth performances of percid juveniles.

22.2.2.1 Genetic Origin

To date, juvenile Eurasian perch have been mainly obtained from wild breeders, without any genetic selection that would result in improved survival and growth performances in culture conditions. Consequently, high variations of fry quality according to the genetic quality of the breeders were observed year by year. Apart from low survival at different developmental stages, the development of intensive culture of Eurasian perch is strongly limited by slow growth compared to that of other fish species such as salmonids (Grignard et al. 1996; Kestemont and Dabrowski 1996; Kestemont et al. 1996; Tamazouzt et al. 2000). The potential of Eurasian perch domestication was investigated by Mandiki et al. (2004) by comparing under similar rearing conditions the performances of fish from wild populations from Finland, Poland, France, Belgium and Italy. In small juveniles (1.27 g), growth heterogeneity, growth rate and feed efficiency did not significantly differ between French and Belgian stocks. In contrast, growth rate and feed efficiency in large juveniles (31.8 g) were higher in French stocks than in Belgian. Growth performances of juveniles from Poland were higher than those of French and Belgian stocks. As reported for other fish species such as salmonids, this difference of growth potential between stocks may be related to a genetic variation between populations (Thodesen et al. 2001; Henryon et al. 2002). Thus, as it was achieved with carps (Vandeputte et al. 2002), it is of high interest to select wild breeders based on their growth performances to improve the domestication process of percids. This suggestion is reinforced by the results of Mairesse et al. (2007) who compared the growth performances between wild perch juveniles and domesticated ones for 103 days. As mentioned by the authors, the domestication process seems to improve the growth performance of Eurasian perch. The positive effect of domestication on growth was previously reported for more domesticated fish species such as trout or salmon (Huntingford 2004; Mambrini et al. 2004).

22.2.2.2 Life Stage

It is well established that growth rate varied highly with the fish ongrowing stage. For example, larval development is associated with high growth rate and high energy consumption for metamorphosis. At the opposite end, adult stage is characterized by low energy requirement and low growth rate. During ongrowing stage, growth rate also varied with the weight of fish. Such decrease of the maximum growth rate with fish size has been reported by Brett et al. (1969) in sockeye salmon. As suggested by the authors, growth rate is size-specific at any given temperature, but an optimum temperature coincides with optimum metabolic scope, greatest tolerance to oxygen debt and maximum sustained speed. For example, SGR values of juvenile Eurasian perch reared at 22–23 °C decrease from 1.7 % day^{-1} to 0.7 % day^{-1} for fish of 1.9 g and 48.3 g respectively. Similarly, maximum specific growth rate of Eurasian perch reared at 23 °C decreases with increasing body weight, from 5.06 % for fish of 0.22 g to 1.14 % for fish of 18.9 g (Fiogbé and Kestemont 2003).

Even if no studies have investigated so far the effect of initial size of pikeperch on maximal growth rate, comparison of results obtained from several experiments suggest a similar answer to the one of Eurasian perch. Thus, 1.36 g pikeperch reared at 23.2 °C until a final mean weight of 28.1 g displayed a SGR of 3.36 % day^{-1} (Schulz et al. 2006). Similarly, a SGR of 3.37 % day^{-1} was obtained from juvenile pikeperch of 6.1 g maintained at 24.4 °C until a final mean weight of 40.6 g (Kozłowski et al. 2010). On the contrary, larger pikeperch juveniles exhibited a SGR of only 1.59 % day^{-1} when they were reared from an initial mean weight of 37.0 g to a final mean weight of 72.0 g at 22 °C. However, it can be speculative to compare the growth capacities of juvenile pikeperch from different experiments, since, as it was described in this chapter, several factors (rearing temperature, feeding frequency, feed ration, photoperiod), can influence fish growth. For example, the relative low SGR (SGR = 1.8 % day^{-1}) obtained from pikeperch fingerlings of 6.4 g maintained at 24 °C can be due to a relative short day length period (12L:12D) compared to others studies (Wang et al. 2009).

22.2.2.3 Gender

In percids, although some environmental factors may induce size heterogeneity (Thorpe 1977), individual growth rate also depends on gender (Scott and Crossman 1973). Females grow faster than males (Malison et al. 1988). This sexual dimorphism appears at a 110 mm length in yellow perch. Indeed, an earlier sexual maturation in males compared to females favours this dimorphism (Craig 1987; Tanasichuk and MacKay 1989). Similar growth dimorphism was also observed in Eurasian perch reared in floating cages (Fontaine et al. 1995).

As suggested by Fontaine et al. (1997), sexual growth dimorphism is partially under control of feeding levels. Thus, relative low feeding level (1 and 2 %) decreases the intraspecific growth variability in Eurasian perch, in response to an inhibition of the sexual dimorphism proceeds. On the contrary, an increase of food supply to a level close to the maximal feeding level (3 %, four meals per day) allowed the females to express their growth potentialities that are superior to those of males (Fontaine et al. 1997; Mélard et al. 1995). In addition, males can develop their gonads from 150 to 260 days (5–25 g), which is relatively earlier in comparison to females (up to 40 g). As a result, energy allocation to testis development cannot be used for fish growth and reduces growth rate of males compared to females. Moreover, difference of growth rate also results from a higher feed consumption and feed conversion efficiency in females compared with males (Malison et al. 1988).

Surprisingly, an experiment focused on the effect of feed supply rate on growth of Eurasian perch juveniles seems not to confirm this information (Juell and Lekang 2001). Indeed, 10 g Eurasian perch reared in recirculated system at 18 °C and fed different ration (1.2; 4.2; 14.3 pellets fish^{-1} min^{-1}) displayed similar SGR values. Moreover, only a slight effect of the feed supply on the sexual growth dimorphism was observed in each experimental condition. The discrepancy of these results with the previous experiments can be explained by the modulation of growth rate by all the factors described before, such as the relative low rearing temperature (18 °C) and the continuous light used during this trial. Indeed, as described by Jourdan et al. (2000), a continuous light inhibited ovarian development, and potentially influence growth heterogeneity between males and females.

According to Kadri et al. (1996), the mode of food distribution is also an important factor affecting size heterogeneity between males and females. These authors suggested that, to prevent food monopolization by some dominant individuals, the food should be presented in such a manner that it is unpredictable in time and space. For example, an experiment conducted by Tamazouzt (1995) on very low density (1.4 kg.m^{-3}) of Eurasian perch suggested that restricted feed rations would induce inter-individual competition, with a clear emergence of female dominance. In such situations, the larger individuals, which are most of time females, inhibited the feeding behaviour of the smaller ones. Thus, higher growth rates and lower size heterogeneity could be achieved through the production of monosex female population (Malison et al. 1986) or sterile triploid fish (Malison et al. 1993).

22.2.2.4 Density

Growth heterogeneity is a central problem in aquaculture, especially in predatory fish species such as percids. In relation to the density, the availability and quality of food control the dominance hierarchies, individual growth and cannibalism, and thus, contribute to the success of the ongrowing process. At early stages (around 1–10 g), increased density has a positive effect on fish growth. For example, increased stocking density from 400 to 10,000 fish m^{-3} resulted in a 67 % increase in growth rate (0.2 g fish^{-1} day^{-1} at the highest density and 0.12 g fish^{-1} day^{-1} at the lowest density) for 1 g Eurasian perch reared at 23 °C during 74 days. This seems to originate from a density-dependent inhibition of territorial and agonistic tendencies that potentially limit the access to food. This hypothesis is supported by the growth heterogeneity observed at each stocking density, since coefficients of variation of fish body weight at the end of the trial decreased from 98.4 % to 57.9 % with increasing stocking density (Mélard et al. 1996).

Perch are a social species. Thus, hierarchies established at the low density could decrease access to food, whereas at higher levels such restriction is not observed (Mélard et al. 1996). Indeed, when density is low, the motivation to maintain a territory can outweigh the motivation to forage (Hecht and Uys 1997), and this might explain why growth was slowest at the lowest stocking density. Within a limited space, the available territory is limited when density is high. This decrease of heterogeneity with the increase of the density may also be due to the learning capacity of the fish reared at high density to feed on the artificial pellets more quickly than fish maintained at lower density. However, this decrease of heterogeneity with the increase of density can only be valid if the daily food ration is not a limiting factor. Indeed, as explained by Fiogbé and Kestemont (2003), a decrease of daily food ration from 20 % (% biomass) to 2 % contributes to increase heterogeneity by enhanced dominance hierarchies and competition for food. However, as already mentioned, the feeding method can contribute to reduce the heterogeneity when the daily food ration is low. In fact, one or two meals per day instead of a continuous feeding by automatic feeders help to reduce the monopolization of food by dominant individuals, and thus growth heterogeneity. Interestingly, opposite results were observed from larger juveniles (Mélard et al. 1996). Indeed, the growth of Eurasian perch reared at 2080 individuals m^{-3} was 20 % lower than for fish reared at 308 individuals m^{-3} after 103 days of trial. Similarly, the effect of density on growth heterogeneity was not significant at this development stage. Finally, the maximal production rate (0.8 kg m^{-3} day^{-1}) was achieved at a 60 kg m^{-3} stocking biomass.

Effects of density on the growth rate were also investigated in 0.91-g pikeperch (Molnár et al. 2004). In this experiment, pikeperch were reared at three initial densities; 1.25, 1.66, and 2.08 kg.m^{-3} for 4 weeks at 23 °C and fed ad libitum twice a day. According to these authors, the initial density did not significantly influence the growth parameters in this experiment. Indeed, similar SGR values (5.95–6.12 % day^{-1}) were recorded among experimental conditions. Moreover, the initial density did not affect the cannibalism and survival. These results can be explained by the low initial variability of fish weight at the beginning of the experiment. Indeed, as explained by Zakęś (2012), size variability considerably affects the success of rearing, particularly survival. If fish size varies strongly, loss due to cannibalism tends to increase. In the experiment of Zakęś (2012), fish weighing 0.7 g BW, with high heterogeneity exhibited a final survival of 50 % after 8 weeks, while in groups of uniform body weight of 1.1 g, the survival percentage recorded was up to 90 % (Zakęś, unpublished data).

22.3 Nutrition

22.3.1 Diet Composition

The majority of the gross energy of feed is contained in proteins, lipids and carbohydrates. The energetic contribution of each macronutrient can be determined from the standard values corresponding to the physiological fuel value. According to the National Research Council (NRC 2011), these values are 39.5 kJ.g^{-1} for lipids, 23.7 kJ.g^{-1} for proteins and 17.2 kJ.g^{-1} for carbohydrates. In aquaculture, artificial feeds tend to be formulated according to each species dietary requirements, with the best balance of protein, lipid and carbohydrate levels. Nevertheless, the optimum balance in nutrient not only depends of the fish species but also of its life stages. Indeed, higher dietary protein and lipids levels are required during fast-growing life stages such as larval and juvenile stages than grow-out fish. For example, crude protein and crude lipid requirements of the rainbow trout (*Oncorhynchus mykiss*) decrease respectively from 45 % to 50 % and 16–18 % for fry to 35–40 % and 14–16 % for brood-stock (Hardy 2002).

Currently, no feeds have been manufactured yet to cover specifically the pikeperch and Eurasian perch nutrient requirements. As Eurasian perch and pikeperch are freshwater carnivorous species, feed dedicated to salmonids have been frequently used in percid aquaculture. However, recent advances in the nutrition of percid fish have been completed that will define the practical diets for these species during their ongrowing stage.

22.3.2 Protein Requirements

Dietary protein is the single source of nitrogen for constructing amino acids and proteins in fish. Protein is also used as a source of energy for metabolism. The optimal dietary protein level depends on several factors including the trophic level, the life stage, the amino acid balance in diet proteins, the total energy content of the food and the feeding level. In addition, protein level incorporation in the diet depends also the digestibility of these proteins. Evaluation of protein quality in food can be estimated with the protein retention ratio (PER), as below: PER = weight gain/food protein intake. In general, the required level of dietary protein is higher at low feeding rates. However, protein sparing occurs when energy levels in the diet are sufficient to "spare" protein from being used as an energy source. In normal fish metabolism, about 14–15 % of energy is supplied by protein, but this non-productive use of protein can be reduced if a correct balance of carbohydrate (for herbivorous species) or lipid (for carnivorous species) is used in the diet.

According to fish species, life stages, protein sources and environmental conditions, protein requirements range from 30 % to 50 % (NRC 2011). In North-American walleye, a percid close to pikeperch, protein requirements for 8 g juveniles was estimated at 51 %, while for larger individuals of this species (50 g), the protein requirement decreases to 42 % (Barrows et al. 1988; Brown and Barrows 2002). The characterization of the protein requirement at early life stages of pikeperch was investigated by Schulz et al. (2007) (Table 22.7). In this experiment, 1 g fish were fed six experimental diets containing graded protein levels (26, 33, 40, 47, 54, 61 %) for 56 days. The increase of protein dietary level from 26 % to 54 % induced a significant increase in SGR (2.44–3.11 %), a significant decrease in food conversion ratio (FCR) value (2.45–1.88), but did not affect survival rate. The higher protein incorporation level (61 %) did not increase fish growth. This observation was supported by the protein efficiency ratio (PER) which decreased from 1.12 in fish fed 54 % dietary protein to 0.91 in those fed 61 % dietary protein. Thus, according to authors conclusions, protein level around 54 % (and 10 % lipids) seems optimal for growth of pikeperch juveniles. This conclusion is reinforced by the study of Schulz et al. (2008) showing that the best feed conversion ratio and specific growth rates on 1.35–1.4 g pikeperch juveniles were obtained in dietary treatments contain-

Table 22.7 Effects of dietary protein level on growth performances of juvenile pikeperch and Eurasian perch

	Duration (days)	Protein level (%)	Lipid level (%)	Initial weight (g)	Final weight (g)	SGR (% day^{-1})	FE	FCR	PER
Eurasian perch	73	20	–	2.97 ± 0.50	3.01 ± 0.60a	0.01 ± 0.42a	–	22.76 ± 10.10a	0.001 ± 0.42a
A	73	30	–	2.70 ± 0.40	4.18 ± 0.60b	0.60 ± 0.11b	–	7.58 ± 1.70b	0.23 ± 0.08b
	73	40	–	2.71 ± 0.43	5.95 ± 1.00c	1.08 ± 0.15c	–	4.62 ± 0.70c	0.38 ± 0.09c
	73	50	–	2.95 ± 0.20	6.96 ± 0.60d	1.18 ± 0.16c	–	3.91 ± 1.00d	0.42 ± 0.13c
	73	60	–	2.91 ± 0.40	7.28 ± 0.60d	1.26 ± 0.20c	–	3.81 ± 0.60d	0.43 ± 0.06c
B	123	49.2	11.9	36.8 ± 0.8	123.0 ± 2.4a	0.98 ± 0.03	0.76 ± 0.04a	–	1.37 ± 0.07a
	123	46.6	16.3	37.3 ± 1.5	140.3 ± 5.0b	1.08 ± 0.05	0.86 ± 0.09ab	–	1.65 ± 0.17b
	123	43.2	22.2	37.9 ± 0.1	130 .0 ± 7.5ab	1.00 ± 0.05	0.93 ± 0.01b	–	1.94 ± 0.03c
Pikeperch	56	26	10	1.09 ± 0.01	4.28 ± 0.18a	2.44 ± 0.07a	–	2.45 ± 0.06a	1.72 ± 0.04a
C	56	33	10	1.01 ± 0.04	5.04 ± 0.19ab	2.88 ± 0.05bc	–	2.01 ± 0.07b	1.66 ± 0.07a
	56	40	10	1.08 ± 0.08	5.01 ± 0.29ab	2.75 ± 0.13b	–	2.06 ± 0.15b	1.36 ± 0.10b
	56	47	10	1.06 ± 0.05	5.20 ± 0.32bc	2.84 ± 0.13b	–	1.97 ± 0.07b	1.18 ± 0.04bc
	56	54	10	1.05 ± 0.03	6.01 ± 0.59c	3.11 ± 0.1c	–	1.88 ± 0.18b	1.12 ± 0.12c
	56	61	10	1.00 ± 0.06	5.45 ± 0.23bc	3.03 ± 0.05bc	–	1.96 ± 0.06b	0.91 ± 0.03d
D	56	47	9	1.40 ± 0.01	7.35 ± 0.49a	2.98 ± 0.09a	–	1.94 ± 0.09a	1.08 ± 0.05ab
	56	47	17	1.37 ± 0.03	9.50 ± 0.08b	3.46 ± 0.05c	–	1.56 ± 0.04b	1.33 ± 0.04c
	56	54	9	1.40 ± 0.02	7.57 ± 0.63a	3.01 ± 0.16a	–	1.82 ± 0.07ac	1.02 ± 0.10a
	56	54	17	1.38 ± 0.03	9.43 ± 0.83b	3.43 ± 0.17cb	–	1.57 ± 0.08b	1.17 ± 0.06bc

Modified from Fiogbé et al. (1996), Mathis et al. (2003) (**B**), and Schulz et al. (2007, 2008) (**C, D**)

Values are means ± SD. Different letters in the same column and the same experiment indicated significant differences ($P < 0.05$)

ing 47–53 % dietary protein and 17 % dietary lipid (Table 22.7). Comparatively, Eurasian perch juveniles of 2.9 g exhibited also the best growth performances with relative high dietary protein levels (40–50 %) (Fiogbé et al. 1996). Moreover, an increase of the dietary protein level from 0–30 % to 40–60 % increased also the survival rate (from 17 % to 83 %). Thus, a dietary protein level of 36.8–43.6 % should be enough for 2.9 g Eurasian perch and an increase of this value up to 45 % will not significantly improve the growth.

The dietary protein requirement was also investigated in larger Eurasian perch and pikeperch. Mathis et al. (2003) evaluated the effects of three protein/energy ratios (25, 22 and 19×10^{-3} mg kJ^{-1}) on Eurasian perch of 35 g BW until a commercial size (150 g) (Table 22.7). After a 18-week period, the SGR did not significantly differ among groups while the FE and PER values increased significantly when the dietary protein level decreased. However, the decrease in protein and increase in lipid level in the diet resulted also in increased lipid deposition in liver and viscera, supported by an increased hepato-somatic index (HIS) and viscero-somatic index (VSI). This observation suggested that the protein/energy ratio 19×10^{-3} mg kJ^{-1} is too low to reach optimum protein synthesis and lipids (energy) are stored in visceral tissues. Finally, the best growth was obtained from fish fed a protein/energy ratio of 22×10^{-3} mg $kJ^{-1,}$ corresponding to 46.6 % protein and 16.3 % lipid in the diet. The similar protein level requirement was deduced from an experiment on 51.1 g pikeperch fed for 10 weeks with graded protein contents (34, 43, 50 %) (Nyina-wamwiza et al. 2005) (Table 22.7). In this experiment, the best SGR and FE values were obtained when fish were fed with 43 % and 50 % protein levels in the diet. Thus, authors concluded that the optimal protein incorporation ranged from 43 % to 50 %. In addition to the identification of the protein level requirement for Eurasian perch and pikeperch, the amino acid composition requirement was also investigated by Fiogbé et al. (1996). According to this author, the amino acid proportions remain rather similar between the different ongrowing stages, reflecting the use of these nutrients in equal proportions for tissue synthesis. Essential amino acid requirements are presented in Table 22.8 and compared to those of channel catfish and rainbow trout. The relative proportion of essential amino acid required in food for these species are globally similar. However, leucine, threonine and valine requirements are much closer to the ones of Rainbow trout which is also a carnivorous fish species. Since amino acid composition of whole body of pikeperch and Eurasian perch is very close (Fiogbé et al. 1996; Zakęś and Demska-Zakes 1998), we can suggest that amino acids requirement indicated in Table 22.8 may be used for pikeperch nutrition.

22.3.3 Lipid Requirement

Fish oil is the main source of lipids in diets formulated for finfish aquaculture. Triglycerides are the dominant constituent of fish oil but fat soluble vitamins such as vitamin A and vitamin D have also been recorded. As suggested by the relatively high amount of fatty acids in fish, it is an essential source of metabolic energy for

Table 22.8 Comparison of essential amino acid requirements (% proteins) Eurasian perch juveniles with those of catfish and rainbow trout

Essential amino acids	Eurasian perch	Catfish	Rainbow trout
Arginine	4.02	4.3	4
Histidine	2	1.5	1.8
Isoleucine	3.26	2.6	2.8
Leucine	4.99	3.5	5
Lysine	5.24	5.1	6
Methionine	1.92	–	–
(Methionine+cysteine)	2.19	2.3	3.3
Phenylalanine	2.77	–	–
(Phenylalanine+tyrosine)	5.22	5	6
Threonine	3.34	2	4.1
Tryptophan	–	0.5	0.6
Valine	3.75	3	3.6

Modified from Lovell (1991) and Fiogbé et al. (1996)

growth, reproduction and swim. This provision of metabolic energy is obtained by ß-oxidation of fatty acids to produce ATP (Sargent et al. 2002). Interestingly, different pattern of accumulation of certain dietary fatty acids are observed among fish tissues, indicating a selective retention of these fatty acids, with respect to the need and specificity of each tissue. For example, saturated (16:0) and monosaturated fatty acids (18:1n-9, 20:1n-9 and 22:1n-11) are mainly accumulated in viscera and liver of percid fish to produce energy. Fish oil is also characterized by high level of highly unsaturated fatty acids (HUFA), mainly represented by EPA (20:5n-3) and DHA (22:6n-3) of the n-3 series and ARA (20:4n-6) of the n-6 series. These HUFA can also be used to produce energy by ß-oxidation but they are preferentially incorporated as structural component in cell membranes. Indeed, incorporation of HUFA in membrane bilayers, according to their physical properties, modulates its fluidity. In addition, HUFA from cell membrane can also be used as precursor to eicosanoid synthesis. These eicosanoids are signalling molecules which exert complex control over many bodily systems, mainly in inflammation or immunity. In aquaculture experiments, HUFA have been found to be critical for maintaining high growth, survival and reproductive rates and low food conversion efficiencies.

Eurasian perch and pikeperch exhibit relative low lipid content in muscle compared to salmonids. Indeed, in percids, the triglyceride energetic source is mainly deposited in liver and viscera. The mean lipid level in whole body of wild fish ranged from 1 % to 2 % (% wet weight) and 0.64–1.88 % (% wet weight) for Eurasian perch and pikeperch respectively (Mairesse et al. 2007; Schulz et al. 2006; Zakęś and Demska-Zakes 1998).

Kestemont et al. (2001) investigated for the first time the effect of graded lipid levels (6, 12, 18 %) in isoproteic diet (40 % crude protein) on growth and nutritional status of 22.7 g Eurasian perch (Table 22.9). In addition, the importance of stabilized lipid in menhaden oil through addition of ethoxyquin as antioxidant was evaluated.

Table 22.9 Effects of dietary lipid level on growth performances of juvenile pikeperch and Eurasian perch

	Lipid level (%)	Duration (days)	Initial weight (g)	SGR (% day^{-1})	FE	FCR	PER	% lipid whole body	% lipid muscle	% lipid viscera	% lipid liver
Eurasian perch	6 + eth	56	22.3 ± 0.7	1.08 ± 0.07a	0.41 ± 0.08a	–	–	–	1.10 ± 0.21a	35.3 ± 6.2a	9.4 ± 3.3
A	12 + eth	56	22.1 ± 0.8	1.40 ± 0.09b	0.68 ± 0.09b	–	–	–	1.28 ± 0.13ab	42.6 ± 1.4ab	8.5 ± 3.2
	18 + eth	56	22.3 ± 0.5	1.42 ± 0.05b	0.60 ± 0.12b	–	–	–	1.47 ± 0.37ab	46.9 ± 6.2b	8.5 ± 0.5
	6	56	22.0 ± 0.5	0.97 ± 0.04c	0.35 ± 0.06a	–	–	–	1.52 ± 017b	31.8 ± 2.9a	9.7 ± 1.1
	12	56	22.2 ± 0.7	0.51 ± 0.06d	0.21 ± 0.03c	–	–	–	1.55 ± 0.86ab	34.0 ± 4.5a	11.1 ± 3.9
	18	56	22.3 ± 0.5	0.30 ± 0.04d	0.12 ± 0.04d	–	–	–	0.98 ± 0.15a	35.6 ± 4.9a	7.4 ± 3.7
B	12	70	35.9 ± 1.5	0.43 ± 0.01a	0.53 ± 0.05a	–	1.16 ± 0.10a	–	5.1 ± 0.14a	64.9 ± 1.8a	12.4 ± 0.1a
	15	70	33.1 ± 3.1	0.53 ± 0.02ab	0.64 ± 0.07a	–	1.45 ± 0.17a	–	5.1 ± 0.42a	78.8 ± 2.5b	17.0 ± 0.9b
	19	70	33.7 ± 2.3	0.57 ± 0.03b	0.80 ± 0.01b	–	1.93 ± 0.03b	–	5.0 ± 0.41a	71.9 ± 3.0a	23.8 ± 3.1c
Pikeperch	6	70	210 ± 17	0.26 ± 0.00a	–	3.76 ± 0.07ab	0.59 ± 0.01a	8.9 ± 0.2a	7.7 ± 0.2a	39.9 ± 0.7a	–
C	10	70	212 ± 12	0.34 ± 0.07b	–	2.93 ± 0.23a	0.80 ± 0.12b	10.7 ± 0.2b	7.6 ± 0.1a	44.6 ± 1.2b	–
	14	70	208 ± 3	0.21 ± 0.02a	–	4.65 ± 0.53b	0.48 ± 0.05a	10.3 ± 0.1b	7.7 ± 0.1a	50.5 ± 0.7c	–
D	6	42	22.1 ± 5.6	10.6 ± 1.1a	–	–	–	7.4 ± 0.5a	–	–	–
	12	42	22.1 ± 5.6	9.8 ± 2.6a	–	–	–	8.5 ± 0.8a	–	–	–
	18	42	22.1 ± 5.6	11.6 ± 1.4ab	–	–	–	7.4 ± 1.3a	–	–	–
	24	42	22.1 ± 5.6	13.4 ± 1.4b	–	–	–	11.6 ± 0.2b	–	–	–

Modified from Kestemont et al. (2001) (**A**), Xu et al. (2001) (**B**), Zakęś et al. (2004) (**C**), and Molnár et al. (2006) (**D**)
Values are means ± SD. Different letters in the same column and the same experiment indicated significant differences ($P < 0.05$)

The authors reported that the growth rate was significantly affected by both the presence of ethoxyquin and lipid levels. The best SGR value was obtained in the group fed 18 % lipid+ethoxyquin (SGR = 1.42 % day^{-1}). Moreover, higher weight gain and feed efficiency were obtained in groups fed 12 % and 18 % dietary lipids with ethoxyquin while the lowest ones were obtained in groups fed the same lipid levels, but without ethoxyquin. These results showed that juvenile Eurasian perch are highly sensitive to oxidized lipid and the necessity to add antioxidants in formulated diets. More precisely, the absence of ethoxyquin reduced the concentrations of some HUFAs as showed in the liver, muscle and viscera. Indeed, a marked decrease of EPA was detected in all analysed tissues while the tissue concentration of DHA remained stable. The relatively high level of DHA in these tissues and its stable concentration between dietary conditions suggest a more important function of DHA than EPA in fish physiology. Considering only the results of fish fed graded levels of stabilized lipids, it appears that the increase from 12 % to 18 % lipid levels did not enhance weight gain and SGR value. At the histological level, in the liver, increase of the dietary lipid level (18 %) induced an increase in glycogen storage, and then lipid droplets, both in size and abundance, which could result in "steatosis" state. Indeed, since lipids are metabolized in the liver, its histological structure can often be influenced by the quantitative and the qualitative lipid profile compositions of the diets fish consume. Thus, based on these findings, 12 % lipid level with antioxidant may be considered to an optimal level for Eurasian perch juveniles while an excessed of dietary lipid may have detrimental effects on fish health. Similar recommendations have been deduced from an experiment conducted on the same species fed with three dietary fat levels (12, 15 and 19 % lipids and 42/46 % protein) for 10 weeks (Xu et al. 2001) (Table 22.9). In this study, growth, feed efficiency and PER were correlated with dietary fat levels. In addition, the increased fat level in the diet increased the lipid deposition in viscera and liver but did not affect significantly the lipid level in the muscle. This increase in fat deposition was supported by an increase of the hepato-somatic index (HSI) and the viscera-somatic index (VSI) values. Based on the growth parameters and the nutritional state of fish among treatments, incorporation of 15 % fat in the diet may be optimal for Eurasian perch juveniles. More recently, the fatty acid profile of muscle, liver and mesenteric fat were compared between large wild and cultured Eurasian perch of 116 g and 119.3 g respectively (Jankowska et al. 2010) (Table 22.9). In this experiment, the artificial diet was composed of 16 % crude fat, close to the value recommended above, and 45 % crude protein. Similar concentrations of saturated fatty acid (SFA) and unsaturated fatty acids (UFA) in muscles, liver and mesenteric fat were found between wild and reared Eurasian perch. However, a higher proportion of MUFA were found in fish, corresponding to a higher MUFA concentration in the diet. Analogously, the higher content of PUFA in natural feed induced a higher amount of these fatty acids in wild fish. The fatty acid composition in the tissues reflecting the fatty acid proportions in the diet was previously described (Bell et al. 2002; Geay et al. 2010; Sargent et al. 1999). This observation is also supported by the difference of arachidonic acid (ARA) level between wild and reared fish. Indeed, the relatively large amount of ARA in tissues of wild Eurasian perch corresponds to a rather high level of this fatty acid in natural prey. From a nutritional point of view, this high amount of ARA suggests a relatively high importance of this fatty acid for both membrane composition and eicosanoid synthesis. However, the use of commercial diet in intensive rearing of Eurasian perch produces fish whose muscle lipids do not diverge from those of wild fish in terms of EPA and DHA contents, the most valuable to a consumer.

As for Eurasian perch, Zakęś et al. (2004) determined the effect of graded lipid content in iso-nitrogenous diets (6, 10 and 14 % lipids, 45 % protein). In this experiment, 210 g pikeperch were reared during 70 days (Table 22.9). As mentioned by the authors, the increase of dietary lipid content was correlated to an increase of this macronutrient in fish body. Indeed, the lipid content in fish body fed 10 % lipids was significantly higher than in group fed 6 % lipids. However, a further increase in dietary lipid (14 %) did not significantly increase the body lipid content. In addition, similar to the Eurasian perch, an increase of dietary lipid affects the viscera biochemical composition, with an increase from 39.9 % to 50.5 % lipids (% of wet weight), while no differences in the amount of lipids in the muscle were detected between groups (7.6–7.7 % of wet weight). Indeed, as in Eurasian perch, pikeperch accumulate excess energy mainly into viscera and liver but not into muscle (McClelland et al. 1995; Xu et al. 2001). Increased dietary energy by increasing lipid content improved the protein utilisation efficiency. As explained above, lipids are primarily used by fish to meet energy requirements, and thus allow protein utilisation as a building material (protein-sparing effect) (Cho and Kaushik 1990). This was partially confirmed in this experiment since the increase of lipid level from 6 % to 10 % resulted in an increased protein efficiency ratio (PER) from 0.59 to 0.80 and a decrease of the feed conversion ratio (FCR) from 3.76 to 2.93. As concluded by the authors, the optimal lipid incorporation in feed should be around 10 % to guarantee the fastest weight gain and the most effective feed utilisation (FCR and PER) at this size (210 g). Molnár et al. (2006) also investigated the effects of graded fat levels in the diet, but at an early life stage of pikeperch. In this experiment, 22 g pikeperch were fed three different dietary fish oil levels (6, 12 and 18 % lipids, and 44 % protein) (Table 22.9). As previously reported, increase of fat level from 6 % to 18 % in the diet induced a decrease of the FCR (from 1.2 to 1.1). Similarly, the increase of dietary fat level induced an increase of the fat deposition into the viscera but did not affect the lipid composition in the muscle. Based on these findings, a diet containing 18 % of lipid may be suggested as optimal for 22 g pikeperch juveniles.

Optimal dietary content of each nutrient depends on the balance between protein, lipid and carbohydrate incorporations. Thus, the dietary lipid level should be considered according to the protein level. Nyina-wamwiza et al. (2005) investigated the effects of the dietary balance of protein/lipid/carbohydrate (P/L/C) in diet on growth, feed efficiency and body composition of 51.1 g pikeperch during 10 weeks. In particular, several inclusion levels of lipid (10, 16 and 22 %) were tested. As described by the authors, when fish were fed diets containing 34 % of protein (low level), the lipid content in carcass increased from 5.7 % to 10.1 % (% wet weight) with the increase of dietary lipid content (from 10 % to 22 %). Similar results were also found in fish fed diets containing 43 % protein. These increments in body lipid by increasing dietary lipid level in the diet are supported by the studies described above (Zakęś et al. 2004; Molnár et al. 2006; Jankowska et al. 2010). Similar observations have been also reported in Atlantic salmon *Salmo salar* (Hillestad and Johnsen 1994), in carp *Cyprinus carpio* (Zeitler et al. 1984), and in European seabass *Dicentrarchus labrax* (Peres and Oliva-Teles 1999). Interestingly, protein sparing effect did not result in better growth with reduction in carcass lipid in fish fed high dietary lipid levels (16 and 22 %), suggesting that pikeperch might not be capable of sparing protein through the utilization of dietary lipid efficiently when dietary protein level is high. Similar results have been observed in Murray cod *Maccullochella peeli peeli* (De Silva et al. 2002), common dentex *Dentex dentex* and European seabass (Company et al. 1999). Based on this conclusion, the proximate nutrient requirements for 51.1 g juvenile pikeperch are between 10 % and 16 % lipids, with a minimum protein requirement of 43 %. More recently, the energetic balance between protein and lipid was investigated by Schulz et al. (2008) on young pikeperch of 1.35–1.40 g. During this experiment, fish were fed six experimental diets containing three dietary levels of lipid (9, 13, 17 %) combined with two levels of dietary protein (47 % and 54 %) for 56 days. As concluded by the authors, the best SGR and FCR values were obtained from fish fed the highest lipid levels (17 %). Contrary to the study of Nyina-wamwiza et al. (2005), in this experiment, the increase of dietary lipid level suggested a clear protein-sparing effect between 9 % and 17 % in pikeperch nutrition. Similarly, Mathis et al. (2003) also reported a significant protein sparing-effect up to dietary lipid level of 16.3 % in 35 g Eurasian perch. Differences of growth responses between these studies (Mathis et al. 2003; Nyina-wamwiza et al. 2005; Schulz et al. 2008) could also be influenced by fish size, since higher dietary lipid levels are beneficial for faster growing juveniles with high metabolic rates in contrast to slower growing older ones.

In conclusion, according to the life stage and the relative amount of dietary protein, incorporation of fish lipid ranging from 13 % to 18 % in the diet may improve growth of Eurasian perch and pikeperch while higher levels of lipid may have detrimental effects on fish health and nutritional value.

22.3.4 Carbohydrate Requirement

The digestibility of carbohydrates in fish is low in comparison to protein and lipid, and inappropriate level of this nutrient in aquafeeds may have negative effects on growth, metabolism and health (Wilson 1994; Erfanullah and Jafri 1998; Li et al. 2012).

The determination of optimal dietary carbohydrate level for Eurasian perch and pikeperch nutrition is poorly documented. Nyina-wamwiza et al. (2005) investigated the nutrient balance of protein, lipid and carbohydrate in nutrition of 35 g juvenile pikeperch. More precisely, three graded levels of carbohydrate (10, 15, 20 %) were fed during this experiment. Fish fed higher carbohydrate diets (15, 20 %) resulted in better growth and FE than fish fed low carbohydrate (10 %) diets at the same protein level. Finally, the best growth performances among dietary treatments were registered for fish fed diets containing 43 % protein/22 % lipid/20 % carbohydrate and 50 %

protein/16 % lipid/20 % carbohydrate. It has been reported that balance between dietary lipid and carbohydrate affects protein sparing, yielding the best growth rate when carbohydrate and lipid are supplied in equal caloric quantities. Beneficial effect of carbohydrate inclusion in diet was also deduced from an experiment on Eurasian perch (Abro et al. 2013). Based on the results of this study, incorporation of 15–20 % carbohydrate added to diets may improve growth performance. This conclusion is supported by the increase of the amylase activity in the proximal intestine until 20 % carbohydrate, but a decrease of its activity when carbohydrate inclusion is over 20 % (Abro et al. 2013).

22.3.5 Vitamin and Mineral Requirements

Most of the vitamins are not synthesized by fish or at a rate sufficient to cover the fish needs. Vitamin requirements (ascorbic acid, para-aminobenzoic acid and inositol) was investigated by Aoe and Masuda (1967) and Halver et al. (1969) on fish species of aquaculture interest such as rainbow trout and Atlantic salmon. More recently, Tacon (1992) and De Silva and Anderson (1995) characterized the effects of vitamin deficiencies or excess on fish physiology. Among the consequences identified by the authors, growth rate reduction, skin damaged and malformations are mainly observed.

Vitamin requirements for pikeperch and Eurasian perch are poorly documented. Currently, pikeperch and Eurasian perch are fed salmonid feeds, with vitamin requirements adapted to rainbow trout and Atlantic salmon nutrition. The absence of significant negative effects on percid physiology suggests that salmonids and percids have very close vitamin requirements. However, complementary experiments may be necessary to define the specific vitamin requirements for percids at the different life stages.

For fish, minerals can be assimilated from diet or from the environment, through osmotic mechanisms. The concentration of each mineral element in body tissues is closely related to its functional role. Minerals are involved as constituents in bones and teeth but also as ionic states in body fluids for osmotic balance and integration activities (nervous and endocrine systems). They play also a crucial role as enzymes and organic compounds in tissues. Mineral composition has been characterized for pikeperch juveniles (Özyurt et al. 2009) but complementary experiments should be necessary to define their optimal incorporation levels in diet.

22.4 Replacement of Marine Ingredients by Alternatives Sources

22.4.1 Economic Context

Future growth of aquaculture activity will need protein and oil sources greater than current fishmeal and fish oil production can satisfy. Thus, continuous increase of aquaculture production is fundamentally unsustainable if fish meal and fish oil remain the primary protein and oil sources used in aquafeed. In this context, the use of alternative such as plant ingredients has been investigated over the last two decades. Currently, 25–50 % of fish meal is replaced by plant meal in feeds for carnivorous fish species without reduction of growth performances, depending on species and life stage. According to the recent interest of Eurasian perch and pikeperch to the diversification of inland aquaculture, the possibility of using plant ingredients in aquafeed was investigated over the last decade. However, no experiments have been conducted on Eurasian perch and pikeperch in order to evaluate the possibility of fish meal replacement by alternative source proteins such as vegetable meal. In contrast, the effects of fish oil replacement by plant sources in food of Eurasian perch and pikeperch is well documented.

22.4.2 Effects of Fish Oil Replacement by Vegetable Oils

The relatively high level of HUFA content in muscle is maintained through the use of fish oil rich in HUFA. At the opposite, plant oil are devoid of HUFA and are rich in their C18 PUFA precursors, α-linolenic acid (ALA, C18:3n-3) and linoleic acid (LA, C18:2n-6). This absence of HUFA content in plant oils has variable consequences on fish health and fish growth, according to fish species. Indeed, it is well described that freshwater fish species such as salmonids fed a diet rich in C18 PUFA are able to maintain a certain degree of desaturase and elongase activities to produce HUFA, including EPA and DHA through the use of the "sprecher pathway". At the opposite, marine carnivorous fish species have limited endogenous capacities to biosynthesise HUFA from PUFA when fish oil is replaced by plant oil in fish farmed diet (Tocher et al. 2001, 2006; Zheng et al. 2004).

Eurasian perch and pikeperch are carnivorous freshwater species. However, at early development stages, they consume relatively small food items such as insects and zooplankton. The effects of plant oil inclusion in diets on growth performance of Eurasian perch were investigated by Xu and Kestemont (2002). During this 10-week experiment, fish were fed four experimental diets composed of 16 % cod liver oil (CLO), olive oil (OO), safflower oil (SO), or linseed oil (LO). Vegetable oils were notably characterized by different proportions of C18 precursors of the n-3 and n-6 series. LO contained high ALA level, SO contained high LA level, while OO contained low levels of both C18 fatty acids but high level of oleic acid (18:1). The lowest growth rate (30 %) and feed efficiency (0.86) were observed in fish fed the OO diet. LO and SO-fed fish increased significantly in growth response and in improved feed efficiency (1.10–1.23) compared to OO fish. However, the highest growth rate (141.7 %) and feed efficiency (1.28) were recorded in control fish fed the CLO diet. According to author's suggestion, the differences of weight between CLO and LO diets may be attributed to the better palatability of the CLO diet. In addition to an effect of oil source on fish growth performances, the total lipid content in viscera, muscle and liver were also affected by the dietary treatment. The liver lipid content of LO fish was the highest among the four dietary fat treatment groups, and the lipid content of liver in CLO fish was significantly higher than that of OO diet-fed fish, while no significant differences in liver lipid between OO and SO diet-fed fish were observed. In fish fed CLO diet, the fatty acid composition in tissues were characterized by high n-3 HUFA content and low n-6 HUFA content. This relative high n-3 HUFA concentration, especially DHA, was already described in the "Lipid requirement part" as a key fatty acid in fish physiology. Interestingly, DHA and total n-3 fatty acids in the liver of OO fish were significantly higher than those in SO fish, but they were significantly lower than those in LO and CLO fish, suggesting that n-3 desaturations were inhibited by the SO diet rich in 18:2n-6 and stimulated by the use of LO diet rich in C18:3n3. This observation is also supported by the relative high amount of desaturation and elongation products of 18:2n-6 found in fish fed SO diet, showing that desaturation enzyme also favouring 18:2n-6 substrate when dietary 18:3n-3 content was low. Thus, the use of LO rich in C18:3n-3 facilitates similar level of n-3 HUFA in fish tissues to those of CLO fish, supporting that Eurasian perch exhibited capacities to bioconvert the precursor into EPA and DHA in order to compensate a deficiency of these HUFA in the diet. Similar conclusions on the HUFA biosynthesis capacities and the importance of the n-3/n-6 ratio of C18 precursors in diet were deduced from a nutritional challenge in juvenile Eurasian perch where four diets were formulated using three different lipid sources (cod liver, safflower and linseed oils) in variable proportions (Blanchard et al. 2008). More recently, the HUFA endogenous synthesis of Eurasian perch fed with vegetable oil was confirmed by Henrotte et al. (2011) through the measure of the desaturation and elongation activities of the [1-14C]18:3n-3 and [1-14C]18:2n-6 substrates in hepatocytes. A significant increase of these activities were recorded from hepatocytes of fish fed the LO diet (with a high n-3/n-6 ratio), compare to the control diet (menhaden oil). Interestingly, the higher desaturation and elongation activities recorded with the [1-14C]18:3n-3 substrate in comparison to the [1-14C]18:2n-6 substrate reinforced the idea that these enzymes tended to favour the synthesis of n-3 HUFA than n-6 HUFA.

The effects of plant oils on the growth performance of pikeperch juveniles and its ability to biosynthesise n-3 HUFA are well documented (Kowalska et al. 2011; Molnár et al. 2006; Müller et al. 2012; Schulz et al. 2005; Zakęś et al. 2010). Schulz et al. (2005) demonstrated that 50 % replacement of fish oil (FO) by linseed oil (LO) or soybean oil (SO) did not affect the growth performance of 14.9–15.6 g pikeperch. Interestingly, triglyceride and phospholipid compositions were not affected in the same manner in response to the use of vegetable oils. Indeed, 50 % replacement of fish oil by vegetable oils (LO and SO) induced a significant decrease of the n-3 HUFA in triglycerides while the concentrations of these HUFA remained relatively stable in phospholipids. This physiological response underlines the key role of HUFA incorporated in phospholipid in the control of the membrane bilayer fluidity. High dietary α-linolenic acid content in LO diet and equal EPA and DHA contents in triglyceride and phospholipid of LO and SO diets, yielded significant increases of relative amounts of EPA and DHA in triglyceride and EPA in phospholipid of whole fish fed LO diet. This observation suggests that, as for Eurasian perch, pikeperch exhibited capacities to synthesise n-3 HUFA from their ALA precursor. However, desaturation and elongation capacities of pikeperch for ALA did not produce enough EPA and DHA amounts in either LO or SO fed fish compared with levels of FO fed fish. Furthermore, as observed in Eurasian perch, in contrast to the high dietary ALA level, the extreme relative reduction of ALA in different tissues of LO fed fish indicates that this fatty acid could be metabolized to produce energy. Finally, partial replacement of fish oil by linseed oil in diet of pikeperch seems not to reduce the growth performance and the muscle nutritional value. This conclusion is also supported by other experiments using linseed oil in replacement to fish oil (Kowalska et al. 2010, 2012).

The effect of total replacement of fish oil by plant sources was characterized by Molnár et al. (2006), Kowalska et al. (2010), and Zakęś et al. (2010). Molnár et al. (2006) demonstrated that total replacement of fish oil by linseed oil at two different incorporation levels (12.7 and 17.8 %) did not significantly affect the growth rate and the feed conversion ratio of 63 g pikeperch, compared to the control group (commercial diet). In addition, total HUFA proportion remained constant between dietary treatments. Similarly, Zakęś et al. (2010) investigated the impact of total replacement of fish oil (80 % of total lipid in the diet) by rapeseed (RO), soybean (SO) and sunflower (SFO) oils on growth rates of pikeperch. Interestingly, the improved values for feed conversion ratio (FCR), apparent net protein retention (ANPR), apparent net energy retention (ANER) and apparent lipid retention (ALR) indicated that the diets tested in this experiment were highly digestible and assimilated by pikeperch juveniles. Similar conclusions on growth performances have been deduced from an experiment on 59 g juvenile pikeperch where fish oil was totally replaced by linseed oil or peanut oil. However, in this experiment, significant reduction of n-3 HUFA contents were observed in whole body, fillet, viscera and liver of fish fed vegetable oils. The discrepancy of these results may be explained by the duration of the experiment and the fish life stage.

As described above, linseed oil provided the best results on fish growth performances and allowed maintaining relatively high n-3 HUFA contents in tissues through the bioconversion of the n-3 precursor, in comparison to other vegetable oils. However, arachidonic acid (ARA, 20:4n-6) derived from linoleic acid (18:2n-6) is also of prime importance in fish physiology since it is involved in the production of eicosanoid molecules. Thus, despite the relative low content of this n-6 HUFA in fish tissues compared to n-3 HUFA, maintaining sufficient ARA content in fish needs to be taken into account in feed formulation. In this context, the effect of a blend of vegetable oils with different n-3/n-6 ratios of C18 precursors was investigated. Kowalska et al. (2010) showed that the amount of n-6 PUFA in whole body, viscera, fillet and liver was the highest in fish fed with a blend of peanut and linseed oils (72/28). However, ARA level decreased significantly in these tissues compared to those of fish fed control diet (72 % fish oil and 28 % soy oil). In a second experiment conducted by the same author, opposite results were obtained (Kowalska et al. 2012). The authors tested two blends of vegetable oils: 70 % sunflower oil and 30 % linseed oil (70SFO/30LO) versus 30 % sunflower oil and 70 % linseed oil (30SFO/70LO) with a n-3/n-6 ratio of 0.70 and 1.35 respectively. The growth performances were similar between the dietary treatments while the fish fed 30SFO/70LO diet displayed a significantly higher DHA level in whole body. In parallel, the use of this diet seems to increase the ARA content in tissues even when significant differences were identified between dietary treatments. Thus, based on the biochemical composition of fish tissues, the diet 30SFO/70LO with n-3/n-6 dietary ratio of 1.35 was the most advantageous for feeding pikeperch juveniles with vegetable oil based diets.

In addition to growth performances and lipid composition in fish tissues, the use of vegetable oil can also influence the histological structure of internal organs. Among them, intestine and liver are the most important tissues impacted by the dietary lipids since they are directly involved in the lipid metabolism in fish. Characterization of the histological picture of liver from pikeperch juveniles fed diets containing either rapeseed (RO), soy (SO) or sunflower (SFO) oil source was investigated by Zakęś et al. (2010). Feeding pikeperch diets supplemented with vegetable oil was reflected in the morphological structure of the liver and the degree of vacuolization of the hepatocytes. In the group SO, an increase of the degree of lipid vacuolization, liver parenchyma degeneration, necrosis and congestion was observed. These changes were less pronounced from fish fed RO and SFO diets. Similar consequences on liver histology were observed by Kowalska et al. (2010, 2012). They also demonstrated that supplementing feed with vegetable oils with low contents of C14:0 and C16:0 acids and of EPA and DHA caused a significant decrease in the degree of hepatocyte vacuolization, an increase in the supranuclear zone of the enterocytes, and an accumulation of lipid droplets in the enterocytes combined with localized sloughing of the intestinal epithelium. Changes such as these can occur when the transport of hydrolysed lipid products to the circulatory system is impaired, and the lipids are stored in the liver (Ostaszewska et al. 2005). All these changes could be connected to the lipid metabolism disorders in the fish organs, underlying an imbalance of fatty acid composition in diet when fish oil is replaced by vegetable oil.

In conclusion, pikeperch and Eurasian perch display relatively high capacities to biosynthesise n-3 HUFA when fish oil is replaced by a blend of vegetable oils including linseed oil, with n-3/n-6 ratio close to 1.6. However, these capacities are still insufficient to maintain high n-3 HUFA level in fish tissues. In consequence, the use of a partial substitution of fish oil by vegetable oil can be suggested. Moreover, despite absence of effects on growth performances, excess of dietary vegetable oil induces alteration of liver structure. This might indicate limited application of these types of lipid sources in percid nutrition.

22.5 Conclusions

In the context of inland aquaculture diversification in Europe, special attention has been given to percids over the last decades. To this end, the impact of abiotic and biotic factors on feeding activity of percid fishes was investigated. Among them, water temperature, feeding frequency, photoperiod and fish density were identified as factors of prime importance. Rearing European percid fishes at a high density in relatively warm water (22–27 °C) and fed three meals per day with a day length of up to 12 h a day significantly improve fish feeding activity, and, in the same way, the growth performances. Optimization of growth performances under artificial conditions was also investigated through the characterization of their nutritional requirements in terms of protein, lipid and carbohydrate. Depending on the fish life stage, artificial diets containing 43–50 % protein, 13–18 % lipid and 10–15 % carbohydrate cover the nutrient requirements of percid fishes. Moreover, recent advances in the use of alternative oil sources in percids nutrition suggest a high potential of these species to biosynthesize HUFA when fish oil is replaced by plant oil rich in PUFA. In this context of fish ingredients replacement by plant sources, it could be also of high interest to investigate the possibility of replacing fish meal by plant products in the future.

References

Abro R, Lundh T, Lindberg J (2013) Effect of dietary starch inclusion rate on digestibility and amylase activity in arctic charr (*Salvelinus alpinus*) and Eurasian perch (*Perca fluviatilis*). J Aquac Res Dev 5:2

Ali MA, Ryder RA, Anctil M (1977) Photoreceptors and visual pigments as related to behavioral responses and preferred habitats of perches (*Perca spp.*) and pikeperches (*Stizostedion spp.*). J Fish Res Board Can 34:1475–1480

Anthouard M, Fontaine P (1998) L'auto-alimentation chez la perche (*Perca fluviatilis*): adaptation à cette modalité de nourrissage, mise en évidence du rythme nycthéméral de la prise alimentaire. Ichthyol Physiol Acta 21:1–13

Aoe H, Masuda I (1967) Water soluble vitamin requirements of carp: II. Requirements for *para-aminobenzoic* acid and inositol. Bull Jpn Soc Sci Fish 33:674–685

Barahona-Fernandes MH (1979) Some effects of light intensity and photoperiod on the sea bass larvae (*Dicentrarchus labrax* L.) reared at the Centre Oceanologique de Bretagne. Aquaculture 17:311–321

Barrows FT, Sell JL, Nickum JG (1988) Effects of dietary protein and energy levels on weight gains, body composition, and RNA: DNA ratios of fingerling walleyes. Prog Fish Cult 50:211–218

Bell JG, Henderson RJ, Tocher DR, McGhee F, Dick JR, Porter A, Smullen RP, Sargent JR (2002) Substituting fish oil with crude palm oil in the diet of Atlantic salmon (*Salmo salar*) affects muscle fatty acid composition and hepatic fatty acid metabolism. J Nutr 132:222–230

Blanchard G, Makombu JG, Kestemont P (2008) Influence of different dietary 18: 3n-3/18: 2n-6 ratio on growth performance, fatty acid composition and hepatic ultrastructure in Eurasian perch, *Perca fluviatilis*. Aquaculture 284:144–150

Blaxter JHS (1969) Visual thresholds and spectral sensitivity of flatfish larvae. J Exp Biol 51:221–230

Blaxter J (1975) The eyes of larval fish. In: Vision in fishes. Springer, New York, pp 427–443

Bódis M, Bercsényi M (2009) The effect of different daily feed rations on the growth, condition, survival and feed conversion of juvenile pikeperch (*Sander lucioperca*) reared with dry feed in net cages. Aquac Int 17:1–6

Boehlert GW (1981) The effects of photoperiod and temperature on laboratory growth of juvenile *Sebastes diploproa* and a comparison with growth in the field. Fish Bull 79:789–794

Boeuf G, Le Bail P-Y (1999) Does light have an influence on fish growth? Aquaculture 177:129–152

Braekevelt CR, Mclntyre D, Ward F (1989) Development of the retinal tapetum lucidum of the walleye (*Stizostedion vitreum vitreum*). Histol Histopathol 4:63–70

Brännäs E, Alanärä A, Magnhagen C (2001) The social behaviour of fish. In: Keeling LJ, Gonyou HW (eds) Social behaviour in farm animals. CABI, New York, pp 275–304
Brett JR, Grove TDD (1979) Physiological energetics. In: Hoar WS, Randall DJ, Brett JR (eds) Fish physiology. Academic, New York, pp 279–352
Brett J, Shelbourn JE, Shoop CT (1969) Growth rate and body composition of fingerling sockeye salmon, *Oncorhynchus nerka*, in relation to temperature and ration size. J Fish Res Board Can 26:2363–2394
Brown PB, Barrows FT (2002) Percids. In: Nutrient requirements and feeding of finfish for aquaculture. CABI, New York, pp 219–228
Bulkowski L, Meade JW (1983) Changes in phototaxis during early development of walleye. Trans Am Fish Soc 112:445–447
Cho C, Kaushik S (1990) Nutritional energetics in fish: energy and protein utilization in rainbow trout (*Salmo gairdneri*). World Rev Nutr Diet 61:132–172
Cohen JH, Forward RB (2002) Spectral sensitivity of vertically migrating marine copepods. Biol Bull 203:307–314
Company R, Calduch-Giner JA, Pérez-Sánchez J, Kaushik SJ (1999) Protein sparing effect of dietary lipids in common dentex (*Dentex dentex*): a comparative study with sea bream (*Sparus aurata*) and sea bass (*Dicentrarchus labrax*). Aquat Living Resour 12:23–30
Craig JF (1977) Seasonal changes in the day and night activity of adult perch, *Perca fluviatilis* L. J Fish Biol 11:161–166
Craig JF (1987) The biology of perch and related fish. Timber Press, Portland, p 333
De Silva SS, Anderson TA (1995) Fish nutrition in aquaculture. Springer, Berlin, p 320
De Silva SS, Gunasekera RM, Collins RA, Ingram BA (2002) Performance of juvenile Murray cod, *Maccullochella peelii peelii* (Mitchell), fed with diets of different protein to energy ratio. Aquacult Nutr 8:79–85
Elliott JM (1975) The growth rate of brown trout (*Salmo trutta*) fed on maximum rations. Br Ecol Soc 44:805–821
Endler JA (1987) Predation, light intensity and courtship behaviour in *Poecilia reticulata* (Pisces: Poeciliidae). Anim Behav 35:1376–1385
Erfanullah, Jafri AK (1998) Effect of dietary carbohydrate-to-lipid ratio on growth and body composition of walking catfish (*Clarias batrachus*). Aquaculture 161:159–168
Fiogbé E, Kestemont P (2003) Optimum daily ration for Eurasian perch *Perca fluviatilis* L. reared at its optimum growing temperature. Aquaculture 216:243–252
Fiogbé E, Kestemont P, Mélard C, Micha J (1996) The effects of dietary crude protein on growth of the Eurasian perch *Perca fluviatilis*. Aquaculture 144:239–249
Fontaine P, Tamazouzt L, Capdeville B (1995) Fitness of the common perch *Perca fluviatilis* L. for ongrowing in floating cages and in recirculated system. In: Kestemont P, Dabrowski K (eds) Short communications of the workshop on the aquaculture of percids. Université de Lorraine, Vaasa, pp 38–41
Fontaine P, Gardeur JN, Kestemont P, Georges A (1997) Influence of feeding level on growth, intraspecific weight variability and sexual growth dimorphism of Eurasian perch *Perca fluviatilis* L. reared in a recirculation system. Aquaculture 157:1–9
Frisk M, Skov PV, Steffensen JF (2012) Thermal optimum for pikeperch (*Sander lucioperca*) and the use of ventilation frequency as a predictor of metabolic rate. Aquaculture 324–325:151–157
Geay F, Santigosa I, Culi E, Corporeau C, Boudry P, Dreano Y, Corcos L, Bodin N, Vandeputte M, Zambonino-Infante JL, Mazurais D, Cahu CL (2010) Regulation of FADS2 expression and activity in European sea bass (*Dicentrarchus labrax*, L.) fed a vegetable diet. Comp Biochem Physiol B 156:237–243
Grignard J-C, Mélard C, Kestemont P (1996) A preliminary study of parasites and diseases of perch in an intensive culture system. J Appl Ichthyol 12:195–199
Gross WL, Roelofs EW, Fromm PO (1965) Influence of photoperiod on growth of green sunfish, *Lepomis cyanellus*. J Fish Res Board Can 26(22):1379–1386
Halver JE, Ashley LM, Smith RR (1969) Ascorbic acid requirements of coho salmon and rainbow trout. Trans Am Fish Soc 98:762–771

Hardy RW (2002) Utilization of plant proteins in fish diets: effects of global demand and supplies of fishmeal. Aquacult Res 41:770–776
Hecht T, Uys W (1997) Effect of density on the feeding and aggressive behaviour in juvenile African catfish, *Clarias gariepinus*. S Afr J Sci 93:537–541
Helfman G (1978) Patterns of community structure in fishes: summary and overview. Environ Biol Fish 3:129–148
Helfman GS (1981) Twilight activities and temporal structure in a freshwater fish community. Can J Fish Aquat Sci 38:1405–1420
Henne JP, Watanabe WO (2003) Effects of light intensity and salinity on growth, survival, and whole-body osmolality of larval southern flounder *Paralichthys lethostigma*. J World Aquac Soc 34:450–465
Henrotte E, Kpogue D, Mandiki SN, Wang N, Douxfils J, Dick J, Tocher D, Kestemont P (2011) n-3 and n-6 fatty acid bioconversion abilities in Eurasian perch (*Perca fluviatilis*) at two developmental stages. Aquacult Nutr 17:216–225
Henryon M, Jokumsen A, Berg P, Lund I, Pedersen PB, Olesen NJ, Slierendrecht WJ (2002) Genetic variation for growth rate, feed conversion efficiency, and disease resistance exists within a farmed population of rainbow trout. Aquaculture 209:59–76
Hillestad M, Johnsen F (1994) High-energy/low-protein diets for Atlantic salmon: effects on growth, nutrient retention and slaughter quality. Aquaculture 124:109–116
Huh HT, Calbert HE, Stuiber DA (1976) Effects of temperature and light on growth of yellow perch and walleye using formulated feed. Trans Am Fish Soc 105:254–258
Huntingford FA (2004) Implications of domestication and rearing conditions for the behaviour of cultivated fishes. J Fish Biol 65:122–142
Imsland AK, Folkvord A, Jónsdóttir ÓDB, Stefansson SO (1997) Effects of exposure to extended photoperiods during the first winter on long-term growth and age at first maturity in turbot (*Scophthalmus maximus*). Aquaculture 159:125–141
Jankowska B, Zakęś ZA, Żmijewski T, Szczepkowski MA (2010) Fatty acid profile of muscles, liver and mesenteric fat in wild and reared perch (*Perca fluviatilis* L.). Food Chem 118:764–768
Jobling M (1983) Effect of feeding frequency on food intake and growth of Arctic charr, *Salvelinus alpinus* L. J Fish Biol 23:177–185
Jobling M, Johansen SJS (1999) The lipostat, hyperphagia and catch-up growth. Aquacult Res 30:473–478
Jourdan S, Fontaine P, Boujard T, Vandeloise E, Gardeur JN, Anthouard M, Kestemont P (2000) Influence of day length on growth, heterogeneity, gonad development, sexual steroid and thyroid levels, and N and P budgets in *Perca fluviatilis*. Aquaculture 186:253–265
Juell JE, Lekang OI (2001) The effect of feed supply rate on growth of juvenile perch (*Perca fluviatilis*). Aquacult Res 32:459–464
Kadri S, Huntingford FA, Metcalfe NB, Thorpe JE (1996) Social interactions and the distribution of food among one-sea-winter Atlantic salmon (*Salmo salar*) in a sea-cage. Aquaculture 139:1–10
Kestemont P, Dabrowski D (1996) Recent advances in the aquaculture of percid fish. J Appl Ichthyol 12:137–200
Kestemont P, Mélard C, Fiogbe E, Vlavonou R, Masson G (1996) Nutritional and animal husbandry aspects of rearing early life stages of Eurasian perch *Perca fluviatilis*. J Appl Ichthyol 12:157–165
Kestemont P, Vandeloise E, Mélard C, Fontaine P, Brown PB (2001) Growth and nutritional status of Eurasian perch *Perca fluviatilis* fed graded levels of dietary lipids with or without added ethoxyquin. Aquaculture 203:85–99
Kestemont P, Jourdan S, Houbart M, Mélard C, Paspatis M, Fontaine P, Cuvier A, Kentouri M, Baras E (2003) Size heterogeneity, cannibalism and competition in cultured predatory fish larvae: biotic and abiotic influences. Aquaculture 227:333–356
Kowalska A, Zakęś ZA, Jankowska B, Siwicki A (2010) Impact of diets with vegetable oils on the growth, histological structure of internal organs, biochemical blood parameters, and proximate composition of pikeperch *Sander lucioperca* (L.). Aquaculture 301:69–77

Kowalska A, Zakęś Z, Jankowska B, Siwicki A (2011) Substituting vegetable oil for fish oil in pikeperch diets: the impact on growth, internal organ histology, blood biochemical parameters, and proximate composition. Aquacult Nutr 17:148–163

Kowalska A, Zakęś ZA, Siwicki AK, Jankowska B, Jarmołowicz S, Demska-Zakęś K (2012) Impact of diets with different proportions of linseed and sunflower oils on the growth, liver histology, immunological and chemical blood parameters, and proximate composition of pikeperch *Sander lucioperca* (L.). Fish Physiol Biochem 38:375–388

Kozłowski M, Zakęś Z, Szczepkowski M, Wunderlich K, Piotrowska I, Szczepkowska B (2010) Impact of light intensity on the results of rearing juvenile pikeperch, *Sander lucioperca* (L.), in recirculating aquaculture systems. Arch Pol Fish 18:77–84

Li X-F, Liu W-B, Lu K-L, Xu W-N, Wang Y (2012) Dietary carbohydrate/lipid ratios affect stress, oxidative status and non-specific immune responses of fingerling blunt snout bream, *Megalobrama amblycephala*. Fish Shellfish Immunol 33:316–323

Lovell RT (1991) Nutrition of aquaculture species. J Anim Sci 69:4193–4200

Luchiari AC, De Morais Freire FA, Koskela J, Pirhonen J (2006) Light intensity preference of juvenile pikeperch *Sander lucioperca* (L.). Aquacult Res 37:1572–1577

Luchiari AC, De Morais Freire FA, Pirhonen J, Koskela J (2009) Longer wavelengths of light improve the growth, intake and feed efficiency of individually reared juvenile pikeperch *Sander lucioperca* (L.). Aquacult Res 40:880–886

Mairesse G, Thomas M, Gardeur J-N, Brun-Bellut J (2005) Appearance and technological characteristics in wild and reared Eurasian perch, *Perca fluviatilis* (L.). Aquaculture 246:295–311

Mairesse G, Thomas M, Gardeur J-N, Brun-Bellut J (2007) Effects of dietary factors, stocking biomass and domestication on the nutritional and technological quality of the Eurasian perch *Perca fluviatilis*. Aquaculture 262:86–94

Malison JA, Kayes TB, Best CD, Amundson CH, Wentworth BC (1986) Sexual differentiation and use of hormones to control sex in yellow perch (*Perca flavescens*). Can J Fish Aquat Sci 43:26–35

Malison JA, Kayes TB, Wentworth BC, Amundson CH (1988) Growth and feeding responses of male versus female yellow perch (*Perca flavescens*) treated with estradiol-17ß. Can J Fish Aquat Sci 45:1942–1948

Malison JA, Procarione LS, Held JA, Kayes TB, Amundson CH (1993) The influence of triploidy and heat and hydrostatic pressure shocks on the growth and reproductive development of juvenile yellow perch (*Perca jlavescens*). Aquaculture 116:121–133

Mambrini M, Sanchez M-P, Chevassus B, Labbé L, Quillet E, Boujard T (2004) Selection for growth increases feed intake and affects feeding behavior of brown trout. Livest Prod Sci 88:85–98

Mandiki S, Blanchard G, Mélard C, Koskela J, Kucharczyk D, Fontaine P, Kestemont P (2004) Effects of geographic origin on growth and food intake in Eurasian perch (*Perca fluviatilis* L.) juveniles under intensive culture conditions. Aquaculture 229:117–128

Mathis N, Feidt C, Brun-Bellut J (2003) Influence of protein/energy ratio on carcass quality during the growing period of Eurasian perch (*Perca fluviatilis*). Aquaculture 217:453–464

McClelland G, Weber J-M, Zwingelstein G, Brichon G (1995) Lipid composition off tissue and plasma in two mediterranean fishes, the gilt-head sea bream (*Chrysophyrys auratus*) and the European seabass (*Dicentrarchus labrax*). Can J Fish Aquat Sci 52:161–170

Mélard C, Kestemont P (1995) Diversification de la pisciculture en Région Wallonne; mise au point de l'élevage intensif de cyprinidés et percidés. Rapport de recherches à la Région Wallonne. Université de Liège, Facultés Universitaires Notre Dame de la Paix, Namur, p 58

Mélard C, Kestemont P, Baras E (1995) Premiers résultats de l'élevage intensif de la perche européenne (*Perca fluviatilis*) en bassin: effet de la température et du tri sur la croissance. Bull Fr Peche Piscic 336:19–27

Mélard C, Kestemont P, Grignard JC (1996) Intensive culture of juvenile and adult Eurasian perch (*P. fluviatilis*): effect of major biotic and abiotic factors on growth. J Appl Ichthyol 12:175–180

Molnár T, Hancz C, Bódis M, Müller T, Bercsényi M, Horn P (2004) The effect of initial stocking density on growth and survival of pike-perch fingerlings reared under intensive conditions. Aquac Int 12:181–189

Molnár T, Szabo A, Szabo G, Szabó C, Hancz C (2006) Effect of different dietary fat content and fat type on the growth and body composition of intensively reared pikeperch *Sander lucioperca* (L.). Aquacult Nutr 12:173–182

Morita K, Fukuwaka M-A, Tanimata N, Yamamura O (2010) Size-dependent thermal preferences in a pelagic fish. Oikos 119:1265–1272

Müller T, Pál L, Bódis M, Kucska B, Wágner L, Bercsényi M, Molnár T (2012) Effect of dietary fat replacement on body composition of intensively reared hybrids of pikeperch and volga pikeperch. Isr J Aquacult Bamidgeh 64:7

NRC (2011) Nutrient requirements of fish and shrimp. The National Academies Press, Washington, DC, p 392

Nyina-wamwiza L, Xu XL, Blanchard G, Kestemont P (2005) Effect of dietary protein, lipid and carbohydrate ratio on growth, feed efficiency and body composition of pikeperch *Sander lucioperca* fingerlings. Aquacult Res 36:486–492

Ostaszewska T, Dabrowski K, Czumińska K, Olech W, Olejniczak M (2005) Rearing of pike-perch larvae using formulated diets-first success with starter feeds. Aquacult Res 36:1167–1176

Özyurt G, Polat A, Loker GB (2009) Vitamin and mineral content of pike perch (*Sander lucioperca*), common carp (*Cyprinus carpio*), and European catfish (*Silurus glanis*). Turk J Vet Anim Sci 33:351–356

Peres H, Oliva-Teles A (1999) Effect of dietary lipid level on growth performance and feed utilization by European sea bass juveniles (*Dicentrarchus labrax*). Aquaculture 179:325–334

Porter MJR, Duncan NJ, Mitchell D, Bromagea NR (1999) The use of cage lighting to reduce plasma melatonin in Atlantic salmon (*Salmo salar*) and its effects on the inhibition of grilsing. Aquaculture 176:237–244

Rónyai A, Csengeri I (2008) Effect of feeding regime and temperature on ongrowing results of pikeperch (*Sander lucioperca* L.). Aquacult Res 39:820–827

Ruohonen K, Grove DJ (1996) Gastrointestinal responses of rainbow trout to dry pellet and low-fat herring diets. J Fish Biol 49:501–513

Sargent J, Bell G, McEvoy L, Tocher D, Estevez A (1999) Recent developments in the essential fatty acid nutrition of fish. Aquaculture 177:191–199

Sargent JR, Tocher DR, Bell JG (2002) The lipids. In: Halver JE, Hardy RW (eds) Fish nutrition, 3rd edn. Academic, San Diego, pp 181–257

Schulz C, Knaus U, Wirth M, Rennert B (2005) Effects of varying dietary fatty acid profile on growth performance, fatty acid, body and tissue composition of juvenile pike perch (*Sander lucioperca*). Aquacult Nutr 11:403–413

Schulz C, Günther S, Wirth M, Rennert B (2006) Growth performance and body composition of pike perch (*Sander lucioperca*) fed varying formulated and natural diets. Aquac Int 14:577–586

Schulz C, Böhm M, Wirth M, Rennert B (2007) Effect of dietary protein on growth, feed conversion, body composition and survival of pike perch fingerlings (*Sander lucioperca*). Aquacult Nutr 13:373–380

Schulz C, Huber M, Ogunji J, Rennert B (2008) Effects of varying dietary protein to lipid ratios on growth performance and body composition of juvenile pike perch (*Sander lucioperca*). Aquacult Nutr 14:166–173

Scott WB, Crossman J (1973) Freshwater fishes of Canada. J Fish Res Board Can 184:755–761

Sloman K, Motherwell G, O'Connor K, Taylor A (2000) The effect of social stress on the standard metabolic rate (SMR) of brown trout, *Salmo trutta*. Fish Physiol Biochem 23:49–53

Strand Å, Alanärä A, Staffan F, Magnhagen C (2007) Effects of tank colour and light intensity on feed intake, growth rate and energy expenditure of juvenile Eurasian perch, *Perca fluviatilis* L. Aquaculture 272:312–318

Sulistyo I, Rinchard J, Fontaine P, Gardeur J-NL, Capdeville B, Kestemont P (1998) Reproductive cycle and plasma levels of sex steroids in female Eurasian perch *Perca fluviatilis*. Aquat Living Resour 11:101–110

Tacon AG (1992) Nutritional fish pathology: morphological signs of nutrient deficiency and toxicity in farmed fish. Food & Agriculture Organization, Rome

Tamazouzt L (1995) L'alimentation artificielle de la perche *Perca fluviatilis* en milieux confinés (eau recyclée, cage flottante): incidence sur la survie, la croissance et la composition corporelle. Thesis, Université Henry Poincaré, Nancy, p 126

Tamazouzt L, Chatain B, Fontaine P (2000) Tank wall colour and light level affect growth and survival of Eurasian perch larvae (*Perca fluviatilis* L.). Aquaculture 182:85–90

Tanasichuk RW, MacKay WC (1989) Quantitative and qualitative characteristics of somatic and gonadal growth of yellow perch *perca flavescens*, from lake Ste. Anne, Alberta. Can J Fish Aquat Sci 46:989–994

Thodesen Jr, Gjerde B, Grisdale-Helland B, Storebakken T (2001) Genetic variation in feed intake, growth and feed utilization in Atlantic salmon (*Salmo salar*). Aquaculture 194:273–281

Thorpe JE (1977) Daily ration of adult perch, *Perca fluviatilis* L. during summer in Loch Leven, Scotland. J Fish Biol 11:55–68

Tocher DR, Bell JG, MacGlaughlin P, McGhee F, Dick JR (2001) Hepatocyte fatty acid desaturation and polyunsaturated fatty acid composition of liver in salmonids: effects of dietary vegetable oil. Comp Biochem Physiol B 137:257–270

Tocher DR, Zheng X, Schlechtriem C, Hastings N, Dick JR, Teale AJ (2006) Highly unsaturated fatty acid synthesis in marine fish: cloning, functional characterization, and nutritional regulation of fatty acyl Δ6 desaturase of Atlantic cod (*Gadus morhua* L.). Lipids 41:1003–1016

Trippel EA, Neil SRE (2003) Effects of photoperiod and light intensity on growth and activity of juvenile haddock (*Melanogrammus aeglefinus*). Aquaculture 217:633–645

Utne-Palm AC (1999) The effect of prey mobility, prey contrast, turbidity and spectral composition on the reaction distance of *Gobiusculus flavescens* to its planktonic prey. J Fish Biol 54:1244–1258

Vandeputte M, Peignon E, Vallod D, Haffray P, Komen J, Chevassus B (2002) Comparison of growth performances of three French strains of common carp (*Cyprinus carpio*) using hemi-isogenic scaly carp as internal control. Aquaculture 205:19–36

Wang N, Xu X, Kestemont P (2009) Effect of temperature and feeding frequency on growth performances, feed efficiency and body composition of pikeperch juveniles (*Sander lucioperca*). Aquaculture 289:70–73

Wendelaar Bonga S (1997) The stress responsc in fish. Physiol Rev 77:591–625

Wilson R (1994) Utilization of dietary carbohydrate by fish. Aquaculture 124:67–80

Xu X, Kestemont P (2002) Lipid metabolism and FA composition in tissues of Eurasian perch *Perca fluviatilis* as influenced by dietary fats. Lipids 37:297–304

Xu X, Fontaine P, Mélard C, Kestemont P (2001) Effects of dietary fat levels on growth, feed efficiency and biochemical compositions of Eurasian perch *Perca fluviatilis*. Aquac Int 9:437–449

Zakęś Z (2012) The effect of body size and water temperature on the results of intensive rearing of pike-perch, *Stizostedion lucioperca* (L.) fry under controlled conditions. Arch Pol Fish 20:165–172

Zakęś Z, Demska-Zakes K (1998) Intensive rearing of juvenile *Stizostedion lucioperca* (Percidae) fed natural and artificial diets. Ital J Zool 65:507–509

Zakęś Z, Szkudlarek M, Woźniak M, Demska-Zakęś K, Czerniak S (2003) Effects of feeding regimes on growth, within-group weight variability, and chemical composition of the juvenile zander, *Sander lucioperca* (L.), body. EJPAU Fish 6:1–9

Zakęś Z, Przybyl A, Wozniak M, Szczepkowski M, Mazurkiewicz J (2004) Growth performance of juvenile pikeperch, *Sander lucioperca* (L.) fed graded levels of dietary lipids. Czech J Anim Sci 49:156–163

Zakęś Z, Kowalska A, Czerniak S, Demska-Zakęś K (2006) Effect of feeding frequency on growth and size variation in juvenile pikeperch, *Sander lucioperca* (L.). Czech J Anim Sci 51:85–91

Zakęś Z, Demska-Zakęś K, Kowalska A, Hancz C, Jarmołowicz S (2010) Impact of diets supplemented with rapeseed, soy, and sunflower oils on growth rates and the histological picture of the livers of juvenile pikeperch, *Sander lucioperca* (L.). Arch Pol Fish 18:67–75

Zeitler MH, Kirchgessner M, Schwarz FJ (1984) Effects of different protein and energy supplies on carcass composition of carp (*Cyprinus carpio* L.). Aquaculture 36:37–48

Zheng X, Tocher DR, Dickson CA, Bell JG, Teale AJ (2004) Effects of diets containing vegetable oil on expression of genes involved in polyunsaturated fatty acid biosynthesis in liver of Atlantic salmon (*Salmo salar*). Aquaculture 236:467–483

Zyznar E, Ali M (1975) An interpretative study of the organization of the visual cells and tapetum lucidum of *Stizostedion*. Can J Zool 53:180–196

Part VI
Genetic Improvement and Domestication

Chapter 23
Sex and Ploidy Manipulation in Percid Fishes

Carole Rougeot

Abstract Percid fish displayed a labile sexual development and a sexual growth dimorphism towards females that could be used to improve the productivity under controlled conditions. This chapter reviews the different methods that could be used to control the development of the phenotypic sex in order to produce all-female populations for the improvement of growth in Percid fish culture. Techniques of hormonal sex reversal treatment and chromosomes set manipulation (triploidisation and gynogenesis) are described and compared within the different Percid species.

Keywords Eurasian perch • Pikeperch • Gynogenesis • Sex reversal • Triploidisation

23.1 Introduction

As this is the case in the majority of fish species, percids display a labile sexual development that allows modifying the phenotypic sex development towards the one of interest for production (Devlin and Nagahama 2002). Percid fish did not display any morphologically differentiated sex chromosomes. Nevertheless, sex determination process is under the control of sex chromosomes, with female homogamety XX and male heterogamety XY in Eurasian perch *Perca fluviatilis* (Rougeot et al. 2002, 2005), yellow perch *Perca flavescens* (Malison et al. 1986; Malison and Garcia-Abiado 1996), walleye *Sander vitreum* (Malison and Garcia-Abiado 1996) and sander *Sander lucioperca* (Craig 2000).

Percid fish display a sexual growth dimorphism as females grow faster than males (Craig 2000), e.g. about 25 % in Eurasian perch (Mélard et al. 1996; Fontaine et al. 1998) or yellow perch. Therefore, the large-scale production of these species would be improved by rearing all-female population, as they reach commercial size long before males (Malison et al. 1986; Kestemont and Mélard 2000). These objectives could be attained using classical methods in aquaculture, from hormonal control of sex differentiation process to gynogenesis (Purdom 1986; Donaldson 1996).

C. Rougeot (✉)
CEFRA, University of Liège, Tihange, Belgium
e-mail: C.rougeot@ulg.ac.be

P. Kestemont et al. (eds.), *Biology and Culture of Percid Fishes*,
DOI 10.1007/978-94-017-7227-3_23

The use of external factors may also contribute to drive the control of gender ratio. On the other hand, sterilization of female by ploidy manipulation (triploidisation) would suppress the gonad development in female and the somatic growth rate would be improved. This method is mostly important when rearing percid fish under natural water temperature (with a chilling process) that would induce the natural development of the ovary.

23.2 Sex Control: All-Female Production

23.2.1 Hormonal Control of Sex

Hormonal sex control in fish may be achieved using two methods: the direct used of hormones or the indirect used of hormonally sex-reversed breeders. Regarding the European and USA legislations, the direct use of hormones on fish for human consumption is forbidden. Therefore, we will not extend on direct feminization of fish using estrogens. The process of hormonal sex reversal treatment is achieved within two generations: the production of hormonally sex-reversed male breeders in the first generation and the production of all-female populations by crossing XX males and females breeders in the second generation (Rougeot et al. 2002).

23.2.2 Production of XX Males

The success of hormonal sex reversal treatment depends on three main factors that are species-specific: the time of application, the duration of the treatment and the doses of hormone used. In order to be effective, hormonal treatment should be applied during the labile period of sex differentiation, before the onset of germ cell histological differentiation. In yellow perch, the onset of sex-specific gonadogenesis was described from 16 to 18 mm TL juveniles (Malison et al. 1986; Malison and Garcia-Abiado 1996). In Eurasian perch, the onset of germ cell differentiation was reported from 10 to 12 mm (Mezhnin 1978 in Craig 2000) and less than 40 mg mean body weight (Rougeot et al. 2002). In *Sander sp.*, the first sign of histological differentiation of the gonad occurs later, about 75 mm TL in walleye (Malison and Garcia-Abiado 1996) and 80 mm and 6.5 g in pikeperch (Demska-Zakes and Zakes 1995). If applied later, the treatment will be less effective or it will induce sterilization or ovotestis (reproductive organ with both ovary and testis tissues). In the same way, the increase of treatment duration would increase the percentage of female, intersex or sterile fish (Demska-Zakes and Zakes 1997; Zakes et al. 1997; Rougeot et al. 2002).

The 17α-methyltestosterone (MT) is the only synthetic steroid hormone used for sex reversal in percid fish and the production of sex reversed XX males for the female production. This synthetic steroid is administered through the food. Currently, MT is first dissolved in 95 % ethyl alcohol, added and mixed to the diet.

Table 23.1 Optimal masculinizing hormonal sex reversal treatments with 17α-methyltestosterone (MT) applied in Percid fish

Species	Doses (mg/ kg food)	Time of application	Treatment (days)	Sex-ratio (% males)	References
Perca flavescens	30	20–35 mm MBL	84	100	Malison et al. (1986), Malison and Garcia-Abiado (1996)
Perca fluviatilis	40	40 mg MBW	30	100	Rougeot et al. (2002)
Sander lucioperca	30	63 dpf–2.3 g MBW	21	88.7–96.7	Demska-Zakes and Zakes (1997), Zakes et al. (1997)
	40	70 mg MBW	30	100	Rougeot et al. (unpublished)
Sander vitreus	50	50–70 mm MBL	60	Partial sex reversal	Malison and Garcia-Abiado (1996)

MBL mean body length, *MBW* mean body weight, *dpf* days post-fertilization

Prior to the food distribution, hormonal diet was air dried for 24 h to allow the evaporation of the solvent. In Eurasian and yellow perch as well as in pikeperch, the optimal doses of MT used for masculinization is 30–40 mg/kg food (Table 23.1), applied before the onset of sexual differentiation of the gonads (Malison et al. 1986; Malison and Garcia-Abiado 1996; Demska-Zakes and Zakes 1997; Zakes et al. 1997; Rougeot et al. 2002). Using higher doses of MT in Eurasian perch (60 and 80 mg/kg food), Rougeot et al. (2002) observed on the treated progenies fish with ovotestis (20 %) or sterile fish (25 %). In this species, hormonal treatment applied after the onset of histological differentiation of the gonads (>150 mg mean body weight, MBW) leads to a decrease of sex reversal efficiency with up to 25 % of female in the progenies. In *S. lucioperca*, the administration of higher MT doses (60 mg/kg food) resulted in the production of intersex and up to 25 % of sterile fish were observed at a MT dose of 90 mg/kg food (Demska-Zakes and Zakes 1997). It seems that in walleye, only partial sex reversal was obtained with MT at 50 mg/kg food (Malison and Garcia-Abiado 1996). All the authors agreed on the fact that the most important parameter to ensure 100 % of masculinization is the timing of hormonal treatment (usually expressed in terms of days post-fertilization or fish body size).

In many cases, hormonal sex reversal treatments lead to various percentages of abnormal gonad morphology. In pikeperch, Demska-Zakes and Zakes (1997) reported 10 % of abnormal gonads with numerous small spaces and some spermatogonia after a 30 mg/kg diet MT treatment. In yellow perch and Eurasian perch, males with a single testis are considered as XX males as rudimentary paired ovaries fused during the early development to form a single ovary (Treasurer and Holliday 1981; Craig 2000). In Eurasian perch, Rougeot et al. (2002) obtained up to 20 % of males with a single twisted testis with nodules when applying a 40 mg/kg diet MT treatment on 40–70 mg MBW fish (Fig. 23.1). Similarly Malison et al. (1986)

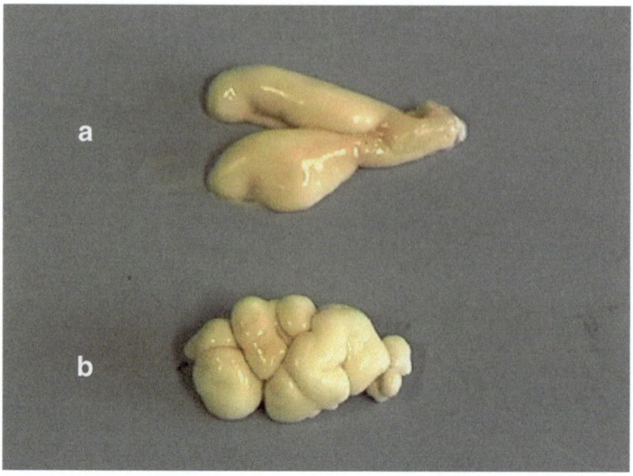

Fig. 23.1 Gross morphology of Eurasian perch gonads. (**a**) Double testis of normal XY male, (**b**) single twisted testis with nodules of hormonally (MT) sex-reversed XX males

observed up to 45 % of abnormal gonad morphology in all MT-treated yellow perch. Generally, these fish were not able to release sperm because of the abnormal morphology of the gonads or the lacks of sperm duct. Therefore, these males should be killed for intra-testicular sperm sampling for artificial reproduction. Nevertheless, Rougeot et al. (unpublished data) succeeded to obtain 40 % of XX males with a spermiduct when feeding undifferentiated Eurasian perch fry (40 mg mean body weight) with a lower dose of MT (5–10 mg/kg food) during 30 days.

Except the lack of sperm duct, hormonally sex-reversed XX males perch (both Eurasian and Yellow) are as fertile as normal males and viable progenies are obtained when theses males are artificially crossed with females (Malison et al. 1986; Rougeot et al. 2002). In E. perch, both male genotypes displayed similar sperm density, with 32.0 and 33.7×10^9 cells ml^{-1} for XY and XX males respectively (Rougeot et al. 2004). Moreover, the sperm motility of both XY and XX males, assessed with six sperm motility parameters by Computer Assisted Sperm Analysis (CASA) are comparable with, for example, 87.0 % (XY males) and 86.3 % (XX males) of motile spermatozoa (% MOT) at 15 s after activation. Finally, their sex steroid profiles (11KT, E2 and T) did not significantly differ during the spawning period (Rougeot et al. 2004). All these results confirmed that hormonal treatment induced a total sex-reversal process and allow to obtained fertile XX males.

All-female perch production was obtained by artificial crosses between normal females and hormonally sex-reversed males (Malison et al. 1986; Rougeot et al. 2002). All the progenies obtained displayed a sex-ratio of 100 % females for yellow perch and ranging from 95 % to 100 % of females for Eurasian perch. The presence of this small percentage of males in the Eurasian perch progenies is probably due to the possible action of minor genetic factors (Rougeot et al. 2002). A comparative study (Rougeot and Mélard 2008), conducted with E. perch in 0.5 m^3 tanks in a recirculating system (23 °C, O_2>6 ppm) outlined that all-female families began to

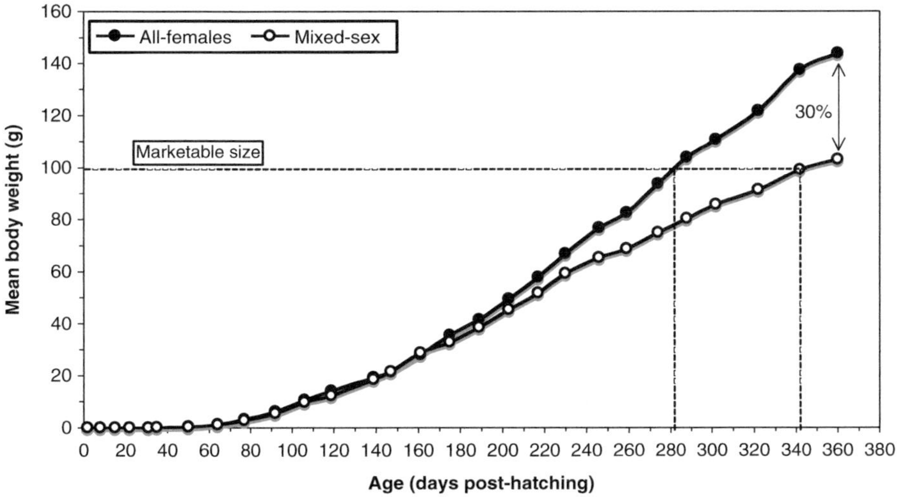

Fig. 23.2 Comparative growth curve of all-female and mixed-sex juveniles Eurasian perch reared under intensive conditions in 0.5 m^3 tank in recirculating system (23 °C) at an initial stocking density of 2000 fish.m^3 (From Rougeot and Mélard 2008)

grow faster than mixed-sex families from a mean body weight of 30 g (Fig. 23.2). After 360 days of rearing, the difference of growth performances reached 30 %. Using all-females families, the marketable size (100 g) was reached within 280 days compared to 340 days with mixed-sex families.

23.3 Ploidy Manipulation

23.3.1 Triploidisation

The use of chromosome set manipulation in aquaculture is mainly interesting for the production of triploid (to a lesser extend tetraploid) fish that are partially or totally sterile (Ihssen et al. 1990; Purdom 1993; Pandian and Koteeswaran 1998). The interest of sterility relies on the possibility to increase growth by using the energy allocated for gonadic growth to somatic growth. In triploid fish, the triplicated chromosome sets impair the meiotic division involved in germ cell formation and therefore inhibit the gonad development. In fish, triploidy is generally induced by inhibiting the second meiotic division and the extrusion of the second polar body by shocking the eggs shortly after fertilization (Purdom 1993; Pandian and Koteeswaran 1998). Many treatments are effective in inducing the polar body retention: thermal (cold or heat), chemical (colchicine or cytochalasin B) or hydrostatics pressure shocks (Ihssen et al. 1990). In percid fish, only heat shocks and hydrostatics pressure shocks are reported for the production of triploid fish (Malison et al.

1993a, b; Malison and Garcia-Abiado 1996; Rougeot et al. 2003). In Eurasian perch, Rougeot et al. (2003) obtained from 93 % to 100 % of triploids using a heat shock of 30 °C applied 5–7 min post-fertilization during 10–25 min (Table 23.2). The survival rate of embryos (6 days post-fertilization) reached 45 % resulting in a yield of triploids above 45 % among the survivals. In comparison, higher temperature (34–36 °C) and short duration (2 and 5 min) shocks allow only the production of 55 % of triploids. Therefore, the yield of triploid under these conditions was only slightly above 20 %, because of the lower survival (30 %) and triploidisation rates (55 %). Similarly, for yellow perch, Malison et al. (1993a, b) obtained a yield of triploids up to 40 % when applying a heat shock (28–30 °C) 5 min post-fertilization for a duration of 10–25 min or using a pressure shock of 9000–11,000 psi for 12 min. If both techniques are effective, these results strongly suggest using the heat shock if the percentage of triploids needed is near 100 %. On the contrary, a hydrostatic pressure shock of 8000 psi applied 4 min post-fertilization during 30 min is most effective to produce 100 % of triploids walleye with a mean survival rate at 6–7 dph above 60 % (Malison and Garcia-Abiado 1996; Malison et al. 2001).

Flow cytometry, which measures the fluorescence of the propidium iodide (a specific fluorophore) bound to nuclear DNA, is the most effective technique to date to determine rapidly and accurately the ploidy status of perch and walleye larvae (Malison et al. 1993a, b; Ewing et al. 1991; Rougeot et al. 2003). Embryos were digested in trypsine citric buffer to obtain a nuclear suspension that was stained with propidium iodide. Then the relative fluorescence of each sample was measured using a flow cytometer. The use of diploid larvae to calibrate the cytometer allows a clear separation of the two fluorescence peaks between diploid and triploid nuclei (Fig. 23.3). It is recommended to analyse a minimum of 20 embryos per batch.

23.3.2 Gynogenesis

The principle of gynogenesis is based upon the fertilization of eggs with genetically inactivated spermatozoa and the application of a thermal or physical shock to restore diploidy in the eggs (Pandian and Koteeswaran 1998). This process of genome manipulation is widely and mainly used in fish breeding programmes and genetics, but could also be used to produce all-female populations in species with female homogamety (Purdom 1993; Pandian and Koteeswaran 1998).

Different techniques of sperm irradiation and DNA inactivation are available (X-rays, gamma-irradiation or chemical), but the most commonly used protocol for Percid species is the UV-irradiation (Malison and Garcia-Abiado 1996; Rougeot et al. 2005). The main concern regarding sperm inactivation is to obtain the DNA denaturation without inhibiting the spermatozoa motility. For Percid Fish, sperm was firstly dissolved in an extender (1:10 for yellow perch, Eurasian perch and walleye), a bicine solution (Moore 1996) before UV exposition in a Petri dish. The optimal UV-irradiation duration ranged from 160 s (yellow perch and walleye) to 460 s (Eurasian perch). Up to 460 s of irradiation, Rougeot et al. (2005) observed

Table 23.2 Optimal shock treatments applied to Percid fish to induce triploidisation

Species	Type of shock	Shock amplitude	Time of application (min pf)	Treatment duration (min)	% triploids	References
Perca flavescens	Heat shock	28–30 °C	5	10–25	100	Malison et al. (1993a, b)
	Hydrostatic pressure	9000–11,000 psi	5	12	37–50	Malison et al. (1993a)
Perca fluviatilis	Heat shock	30 °C	5–7	10–25	93–100	Rougeot et al. (2003)
Sander vitreus	Hydrostatic pressure	8000 psi	4	30	100	Malison and Garcia-Abiado (1996), Malison et al. (2001)
	Heat shock	28–31 °C	1–5	25	50	Malison and Garcia-Abiado (1996), Malison et al. (2001)

pf post fertilization

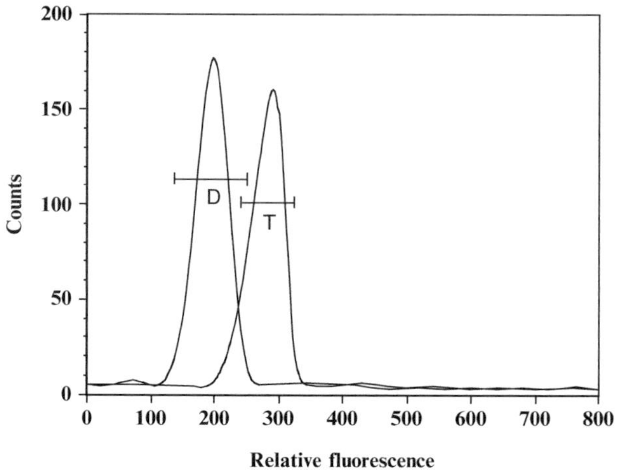

Fig. 23.3 Flow cytometry diagram showing diploid (*D*) and triploid (*T*) nuclear fluorescence intensity of 6 days post-fertilization E. perch larvae after propidium iodide staining (From Rougeot et al. 2003)

that UV-irradiated spermatozoa formed small aggregates that would impair the micropyle penetration and inhibit fertilization.

In walleye, 78 % of haploid embryos were produced using UV-irradiated semen (Malison and Garcia-Abiado 1996) and 100 % in yellow perch (Malison et al. 1993a; Malison and Garcia-Abiado 1996). No gynogen production was attempted for these two species. In Eurasian perch, meiogynogens juveniles were obtained by the fertilization of eggs with UV-irradiated semen and application of a heat shock (30 °C, 5 min post-fertilization during 25 min) to restore diploidy (Rougeot et al. 2005). From 90 % to 100 % of gynogens were obtained with a survival rate at hatching ranging from 3.4 % to 41 %. All of the four gynogens batches produced were 100 % females.

Although gynogenesis allows producing 100 % of females in one generation, this method is rarely used in percid fish production for at least two reasons (Malison and Garcia-Abiado 1996; Rougeot et al. 2005): (i) the survival rates of gynogenetics progenies is often low, probably due to expression of deleterious recessive alleles or the negative effects of heat or pressure shock on the embryos survival; (ii) the growth performances were negatively affected by the increased homozygosity. For example, after 1 year of rearing in recirculating aquaculture system(s) at 23 °C, Eurasian perch gynogens displayed a mean body weight significantly lower (108 g) than normal diploid control group (MBW: 133 g, Rougeot et al. unpublished data).

23.4 Conclusions

In Percid fish, which displayed a sexual growth dimorphism towards females, the lability of the sexual development as well as the possibility to modify the chromosomes set, could easily be used to control sex development. Using exogenous hormonal sex reversal treatment, the phenotypic sex of Percid fish could easily be changed to obtain hormonally sex-reversed males breeders, and to produce all-female population within two generations, allowing the improvement of the growth performances up to 30 %. In the same way, ploidy manipulation, mainly triploidisation, will allow sterilization of female and therefore probably induced the improvement of growth performances by the reduction of the gonad development (above 20 % during the reproductive period).

References

Craig JF (2000) Percid fishes – systematics, ecology and exploitation, vol 3, Fish and aquatic resources series. Blackwell, Oxford

Demska-Zakes K, Zakes Z (1995) Sex differentiation in pikeperch, *Stizostedion lucioperca* (L.). Arch Pol Fish 3(1):51–57

Demska-Zakes K, Zakes Z (1997) Effect of 17α-methyltestosterone on gonadal differentiation in pikeperch, *Stizostedion lucioperca* (L.). Aquacult Res 28:59–63

Devlin RH, Nagahama Y (2002) Sex determination and sex differentiation in fish: an overview of genetic, physiological, and environmental influences. Aquaculture 208:191–364

Donaldson EM (1996) Manipulation of reproduction in farmed fish. Anim Reprod Sci 42:381–392

Ewing RR, Scalet CG, Evenson DP (1991) Flow cytometric identification of larval triploid walleyes. Prog Fish Cult 53:177–180

Fontaine P, Gardeur JN, Kestemont P, Georges A (1998) Influence of feeding level on growth, intraspecific weight variability and sexual growth dimorphism of Eurasian perch *Perca fluviatilis* L. reared in a recirculating system. Aquaculture 157(1–2):1–9

Ihssen PE, McKay LR, McMillan I, Phillips RB (1990) Ploidy manipulation and gynogenesis in fishes: cytogenetic and fisheries applications. Trans Am Fish Soc 119:698–717

Kestemont P, Mélard C (2000) Chapter 11: Aquaculture. In: Craig JF (ed) Percid fishes. Systematics, ecology and exploitation. Blackwell Science, Oxford, pp 191–224

Malison JA, Garcia-Abiado AR (1996) Sex control and ploidy in yellow perch (*Perca flavescens*) and walleye (*Stizostedion vitreum*). J Appl Ichthyol 12:189–194

Malison JA, Kayes TB, Best CD, Amundson CH (1986) Sexual differentiation and use of hormones to control sex in yellow perch (*Perca flavescens*). Can J Fish Aquat Sci 43:26–35

Malison JA, Kayes TB, Held JA, Barry TP, Amundson CH (1993a) Manipulation of ploidy in yellow perch (*Perca flavescens*) by heat shock, hydrostatic pressure shock, and spermatozoa inactivation. Aquaculture 110:229–242

Malison JA, Procarione L, Held JA, Kayes TB, Amundson CH (1993b) The influence of triploidy and heat hydrostatic pressure shocks on the growth and reproductive development of juvenile yellow perch (*Perca flavescens*). Aquaculture 116:121–133

Malison JA, Held JA, Weil LS, Kayes TB, Thorgaard GH (2001) Manipulation of ploidy in walleye by heat shock and hydrostatic pressure shock. N Am J Aquac 63(1):17–24

Mélard C, Kestemont P, Grignard JC (1996) Intensive culture of juvenile and adult Eurasian perch (P. fluviatilis): effect of a major biotic and abiotic factors on growth. J Appl Ichthyol 12:175–180

Moore A (1996) Use of semen extenders for walleye. In: Summerfelt RC (ed) Walleye culture manual, vol 101, NCRAC culture series. North Central Regional Aquaculture Center Publications Office, Iowa State University, Ames, pp 51–53

Pandian TJ, Koteeswaran R (1998) Ploidy induction and sex control in fish. Hydrobiologia 384:167–243

Purdom CE (1986) Genetic techniques for control of sexuality in fish farming. Fish Physiol Biochem 2(1–4):3–8

Purdom CE (1993) Genetics and fish breeding. Fish and fisheries series, vol 8. Chapman & Hall, 2-6 Boundary Row, London SE1 8HN

Rougeot C, Mélard C (2008) Chapter 6: Genetic improvement of growth. In: Rougeot C, Toner D (eds) Farming of Eurasian perch. Volume 1: Juvenile production. Aquaculture explained, n 24. Aquaculture Development Division, Bord Iascaigh Mahra, Dublin, pp 42–51

Rougeot C, Jacobs B, Kestemont P, Mélard C (2002) Sex control and sex determinism study in Eurasian perch, *Perca fluviatilis*, by use of hormonally sex-reversed male breeders. Aquaculture 211:81–89

Rougeot C, Minet L, Prignon C, Vanderplasschen A, Detry B, Pastoret P-P, Mélard C (2003) Induce triploidy by heat shock in Eurasian perch, *Perca fluviatilis*. Aquat Living Resour 16(2):90–94

Rougeot C, Nicayenzi F, Mandiki SNM, Rurangwa E, Kestemont P, Mélard C (2004) Comparative study of the reproductive characteristics of XY male and hormonally sex-reversed XX male Eurasian perch, *Perca fluviatilis*. Theriogenology 62:790–800

Rougeot C, Virimumbalu Ngingo J, Gillet L, Vanderplasschen A, Mélard C (2005) Gynogenesis induction and sex determinism study in Eurasian perch, *Perca fluviatilis*. Aquaculture 243:411–415

Treasurer JW, Holliday FGT (1981) Some aspects of the reproductive biology of perch *Perca fluviatilis* L. A histological description of the reproductive cycle. J Fish Biol 18:359–376

Zakes Z, Demska-Zakes K, Szczerbowski A, Szkudlarek M (1997) Sex reversal in pikeperch, *Stizostedion lucioperca* (L.), by oral administration of methyltestosterone. Arch Pol Fish 5(2):325–336

Chapter 24
Performance of Hybrid Percids

James A. Held, Syaghalirwa N.M. Mandiki, Carole Rougeot, and Patrick Kestemont

Abstract Interspecific hybrids of several percids have been produced, and some of their production characteristics have been compared with purebreds. Most hybrids exhibited superior growth and survival when compared with same egg-source purebreds, however hybrids derived from parental stocks that produced smaller eggs displayed poorer survival (hybrid yellow perch) and similar growth (hybrid sauger) when compared to the purebred parental stocks that produced larger eggs (Eurasian perch and walleye, respectively). Hybrid walleye and sauger have been shown to be reproductively competent. Performance differences of walleye and sauger hybrids derived from varying geographic stocks were noted and may be a source of future production gains. The use of percid hybrids in aquaculture may be limited by regulatory concerns over the importation and introduction of alien invasive species.

Keywords Percids • Eurasian perch • Yellow perch • Sauger hybrid • Purebred

24.1 Introduction

Interspecific crossbreeding has resulted in hybrids having behavioral and growth characteristics better suited for intensive culture than those of purebred fishes. In percids several hybrids have been produced. Mandiki et al. (2004) reported the production of both reciprocal crosses of Eurasian and yellow perch. Nelson et al. (1965) produced both reciprocal crosses of walleye and sauger (*S. canadensis*). Müller et al. (2004) produced hybrids by crossing pikeperch females with Volga perch (*S. volgense*) males.

J.A. Held (✉)
University of Wisconsin-Extension Madison, Madison, WI, USA
e-mail: jaheld@wisc.edu

S.N.M. Mandiki • P. Kestemont
Research Unit in Environmental and Evolutionary Biology, University of Namur, Namur, Belgium

C. Rougeot
CEFRA, University of Liège, Tihange, Belgium

P. Kestemont et al. (eds.), *Biology and Culture of Percid Fishes*,
DOI 10.1007/978-94-017-7227-3_24

24.2 Comparison of Husbandry Performances Between Percid Hybrids and Purebreds

During the early 2000s, Kestemont and Malison exchanged fresh sperm from Eurasian perch and yellow perch diluted in extender formulated for walleye (Moore 1996), allowing the fertilization of ova and the production of hybrids (Eurasian perch females × yellow perch males, and vice-versa). These trials allowed the production characteristics of these percid hybrids to be documented. Mandiki et al. (2004) reported that the juvenile growth rate of both reciprocal crosses of Eurasian perch and yellow perch were superior to purebred Eurasian perch (by 63 % and 21 %, respectively) while they displayed a similar feed efficiency (Table 24.1). Additionally, juvenile survival of hybrids derived from Eurasian perch females (hybrid Eurasian perch) was higher than purebred Eurasian perch. During the ongrowing period (from 41 days after hatching (dah) to 179 dah) hybrids displayed significantly higher survival rates (from 84.8 % to 95.4 %) than purebred Eurasian perch (57.2–69.7 %; Rougeot et al. unpublished data). Due to their small size at hatch, fry survival of yellow perch purebreds and hybrids was relatively poor and was related to restricted feed availability, however, surviving yellow perch hybrids exhibited juvenile growth similar to that of hybrid Eurasian perch (Fig. 24.1).

In a subsequent study, Rougeot and Mélard (2008) demonstrated that hybrid Eurasian perch reached market size (100 g) 1 month sooner than purebreds. During the course of this study hybrids averaged 0.47 g/day and purebreds averaged 0.33 g/day. Rougeot et al. (unpublished data) reported that the sex-ratio of two hybrids progenies (female Eurasian perch × male yellow perch) was significantly skewed towards males (64–66 %) (Table 24.2).

To date, the reproductive capacity of hybrid perch has not been documented.

Malison et al. (1990) described juvenile survival and growth characteristics in both reciprocal crosses of walleye and sauger. This study found that hatching success of fry derived from walleye eggs (purebred and hybrid walleye) was better than hatching success of fry from sauger eggs (purebred and hybrid sauger) but hybridization had no effect on embryonic survival or hatching success within maternal egg source. In juvenile growth trials, hybrid walleye grew faster than purebred walleye and hybrid sauger which grew faster than purebred sauger. Malison et al. (1990) also commented that hybrid walleyes were not as aggressive towards each other and

Table 24.1 Comparison of growth rates and feed efficiency between Eurasian perch and both hybrids

	Eurasian perch	Hybrid Eurasian perch	Hybrid yellow perch
Weight gain (mg d^{-1})	241 ± 31^{a}	476 ± 32^{b}	397 ± 2^{b}
Relative weight gain (% EP)	1.22 ± 0.2^{a}	63	21
SGR (% d^{-1})	0.44 ± 0.0^{a}	1.5 ± 0.1^{b}	1.3 ± 0.0^{ab}
FE		0.43 ± 0.0^{a}	0.38 ± 0.1^{a}

EP Eurasian perch, *SGR* specific growth rate, *FE* feed efficiency
Hybrid Eurasian perch: Eurasian perch females, hybrid yellow perch: yellow perch females
$^{a,b}p < 0.05$

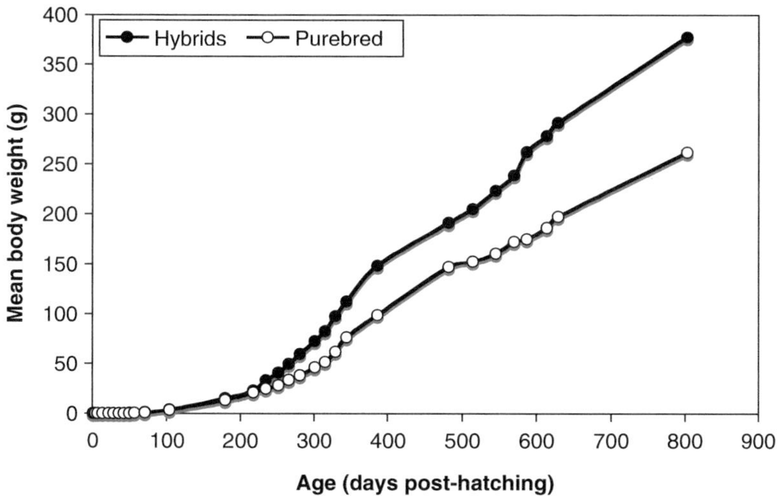

Fig. 24.1 Comparative growth of purebred Eurasian perch and hybrids female Eurasian perch *Perca fluviatilis* x male yellow perch *Perca flavescens* reared in recirculating system at 23 °C

Table 24.2 Sex-ratio of hybrids (female Eurasian perch x male yellow perch) progenies and their control at 179 days after hatching

	% females	% males	% steriles
Purebred 1	50	50	0
Hybrids 1	36	64[a]	0
Purebred 2	50	50	0
Hybrids 2	26	66[a]	8

[a]$p<0.05$

seemed to be more tolerant of routine husbandry procedures when compared to walleye, sauger and hybrid sauger, which confirmed observations by Nagel (1976) who reported that walleye were easily excited by disturbances that cause stress, interrupt their feeding and contribute to mortality. Siegwarth and Summerfelt (1992), however, found that when walleye and F_1 hybrid fingerlings (female walleye × male sauger) hybrids were longer and heavier than walleyes when reared with overhead lighting, but not when reared with submerged lighting.

Using fish produced by Malison et al. (1990), Siegwarth and Summerfelt (1990) also observed growth rates (mm/day and g/day) and survival of fingerling F_1, hybrid (female walleye x male sauger) reared in intensive culture for 73 days at both 17 and 21 °C. Held and Malison (1998) again demonstrated superior growth in hybrid walleye compared to purebred, but they also observed sexually-related dimorphic growth beginning at a size of 200 g, with the onset of gender in hybrid walleye with females growing faster than males. Table 24.3 demonstrates the differences in growth rates between hybrid and purebred walleye males and females during the

Table 24.3 Comparative growth rates of purebred and hybrid male and female walleye after the onset of sexually related dimorphic growth

Average daily weight gain (days 378–460)		
Purebred walleye	Male	0.22 g/day
	Female	0.61 g/day
Hybrid walleye	Male	0.96 g/day
	Female	1.44 g/day

Table 24.4 Growth rates of hybrid walleye produced from different geographic broodstock sources and purebred controls

	Male	Female	Growth rate (g day^{-1})
Spirit L. hybrid[a]	Mississippi R. sauger	Spirit L. walleye	0.88
Genoa hybrid	Mississippi R. sauger	Mississippi R. walleye	0.73
Rock L. hybrid	Mississippi R. sauger	Rock L. walleye	0.58
Rock L. purebred	Rock L. walleye	Rock L. walleye	0.41
Mississippi R. hybrid[a]	Mississippi R. sauger	Spirit L. walleye	0.69
Missouri R. hybrid	Missouri R. sauger	Spirit L. walleye	0.64
Ohio R. hybrid	Ohio R. sauger	Spirit L. walleye	0.64
Spirit L. purebred	Spirit L. walleye	Spirit L. walleye	0.53

[a]Same genetic cross

final stages of grow-out. As can be seen, the diminution of growth demonstrated by male purebreds can have a substantial impact on the overall average growth rate of purebreds. With this species, the culture of monosex female populations would be very beneficial for commercial food fish production.

To determine the culture interval required to reach a market size of 681 g, a growth curve model was developed by ongrowing to an age of 783 days posthatch the same stock of purebred walleye and hybrid walleye (female walleye × male sauger) described by Malison et al. (1990), and Siegwarth and Summerfelt (1990, 1992). Predictions from this model indicate that although hybrid walleye grow faster than purebred walleyes at lengths less than 325 mm, walleyes would reach the minimum market weight 31 days sooner than the hybrids (Siegwarth and Summerfelt 1993).

Held and Malison (1998) also documented differences in growth rates of hybrid walleye derived from different geographic sources of broodstock (Table 24.4, Fig. 24.2a, b).

Recently, researchers at the Northern Aquaculture Demonstration Facility (Red Cliff, WI, USA) have found that hybrid walleye can be grown from hatch to market size (200 g) in as short as 8 months using intensive fry culture methods (Summerfelt 1996) and recirculating aquaculture system(s) (Greg Fischer, personal communication). Yield for 193 g average weight hybrids was 49.6 % resulting in two scaled skin-on fillets (total weight 94.5 g) per fish. In a separate organoleptic study (Malison and Held, unpublished), consumers indicated a high preference for farm-raised purebreds and hybrids over wild caught walleyes and no preference difference

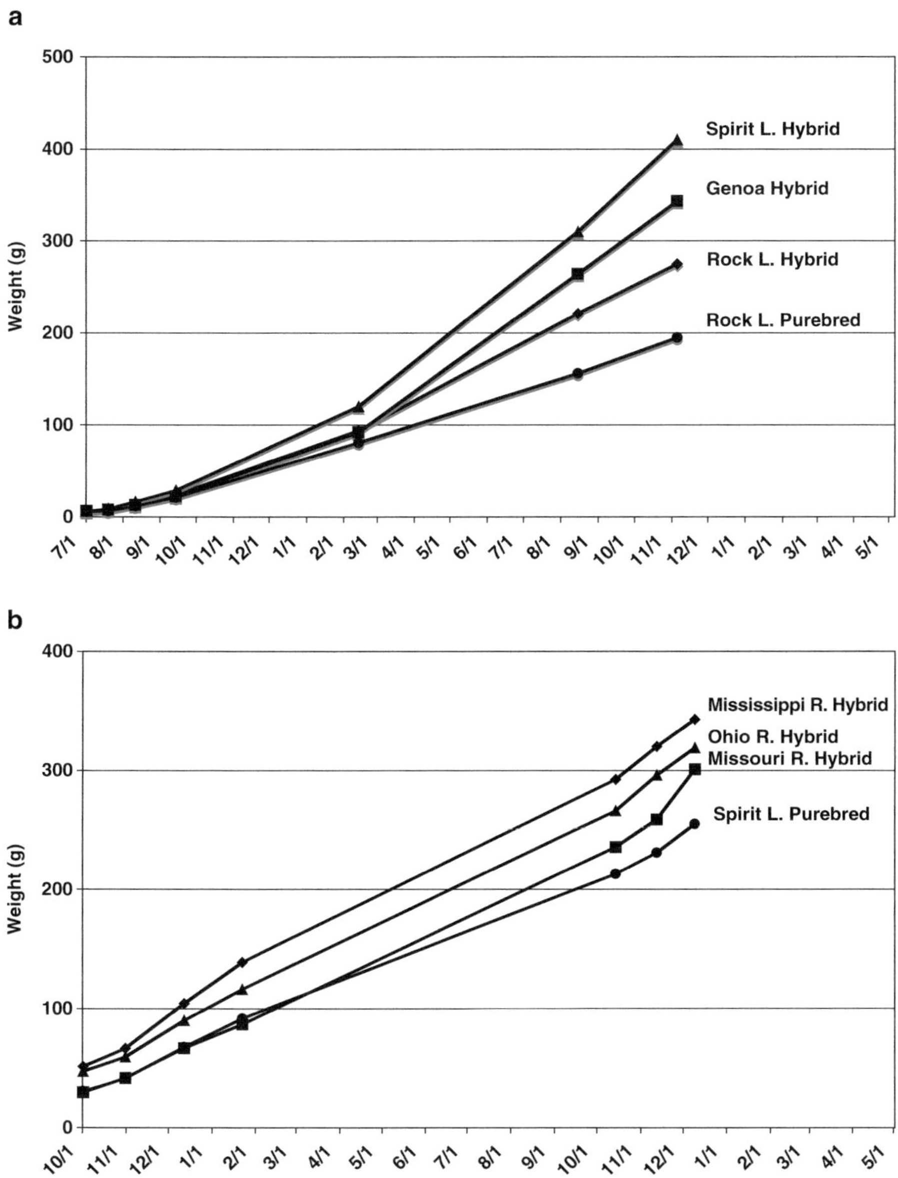

Fig. 24.2 Growth of hybrid walleye (*S. vitreum* female x *S. canadensis* male) derived from different maternal (**a**) and paternal (**b**) broodstock sources plus purebred walleye controls cultured in circular raceways at 21 °C

between farm-raised purebreds and hybrids. Although the economics of such a production protocol have not been studied, the growth results highlight the potential benefits of using hybrid walleye for food-fish production (Fig. 24.3).

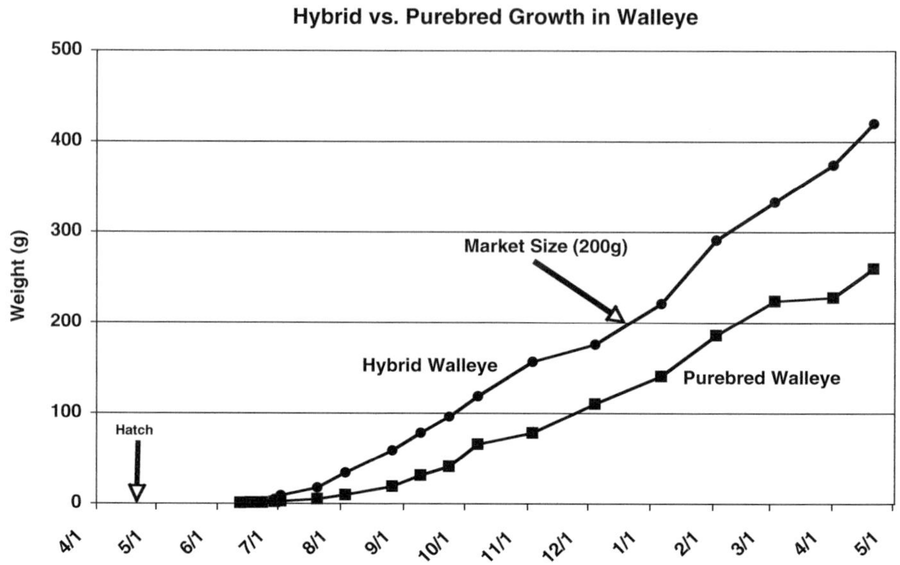

Fig. 24.3 Hybrid vs purebred growth in walleye

Barry et al. (2004) studied the stress responses of purebred and hybrid walleye, but concluded that the perceived 'docile' temperament of the hybrids described by Malison et al. (1990) was not due to physiological differences in primary and secondary stress responses between hybrids and purebreds. Hearn (1986) found that hybrid walleye were able to produce viable fry when crossed with other hybrid walleye and backcrossed with both parental species. Müller et al. (2011) reported that the juvenile growth of hybrid pikeperch was inferior to that of purebred pikeperch. To date the adult growth performance and reproductive competence of hybrid pikeperch have not been documented.

24.3 Conclusions

Performance gains displayed by interspecific hybrids have the potential to enhance aquaculture production of percids. Concerns over the importation and introduction into public waters of alien invasive species, in light of the reproductive competency displayed by some percid hybrids, may limit the practical application of hybrids in aquaculture to situations where security or sterility is ensured.

Further studies are needed to document the extent to which the development of purebred lines that result in superior hybrids and geographic strain differences can be used to further augment production performance in both *Perca* and *Sander*. The development of reliable sterilization protocols that do not impact growth performance would greatly enhance the applicability of percid hybrid aquaculture.

References

Barry TP, Ochiai M, Malison JA, Held JA (2004) Stress responses of walleye and walleye x sauger hybrids. In: Barry TP, Malison JA (eds) Proceedings of Percis III: the third international percid fish symposium. University of Wisconsin Sea Grant Institute, Madison, pp 15–16

Hearn MC (1986) Reproductive viability of sauger-walleye hybrids. Progr Fish Cult 48:149–150

Held JA, Malison JA (1998) Performance of hybrid walleye (*Stizostedion. vitreum* x *S. canadense*) derived from several geographic stocks. In: Book of abstracts aquaculture '98 –the international triennial conference and exposition of the World Aquaculture Society, Las Vegas, 15–19 Feb 1998

Malison JA, Kayes TB, Held JA, Amundson CH (1990) Comparative survival, growth and reproductive development of juvenile walleye (*Stizostedion vitreum*), sauger (*S. canadense*) and their hybrids reared under intensive culture conditions. Prog Fish Cult 52:73–82

Mandiki RSM, Malison JA, Held JA, Rougeot C, Blanchard G, Mélard C, Kestemont P (2004) Hybridization of Eurasian and yellow perch increases growth rate in offspring. In: Barry TP, Malison JA (eds) Proceedings of Percis III: the third international percid fish symposium. University of Wisconsin Sea Grant Institute, Madison, pp 47–48

Moore AA (1996) Use of semen extenders for walleye. In: Summerfelt RC (ed) Walleye culture manual, vol 101, NCRAC culture series. North Central Regional Aquaculture Center Publications Office, Iowa State University, Ames, pp 161–185

Müller T, Taller J, Nyitrai G, Kucska B, Cernák I, Bercsényi M (2004) Hybrid of pikeperch, *Sander lucioperca* L. and Volga perch, *S. volgense* (Gmelin). Aquacult Res 35:915–916

Müller T, Bódis M, Urbány B, Bercsényi M (2011) Comparison of growth in pike-perch (*Sander lucioperca*) and hybrids of pike-perch (*S. lucioperca*) x Volga pike-perch (*S. volgensis*). Bamidgeh 63(IIC):545–552

Nagel T (1976) Intensive culture of fingerling walleyes on formulated feeds. Prog Fish Cult 38:90–105

Nelson WR, Hines LH, Bergman LC (1965) Artificial propagation of saugers and hybridization with walleye. Prog Fish Cult 27:216–218

Rougeot C, Mélard C (2008) Genetic improvement of growth in perch production: domestication, sex control, hybridization and strain selection. In: Fontaine P, Kestemont P, Teletchea F, Wang N (eds) Book of abstracts: percid fish culture from research to production. Presses Universitaires de Namur, Namur, pp 35–39

Siegwarth GL, Summerfelt RC (1990) Growth comparison between fingerling walleyes and walleye × sauger hybrids reared in intensive culture. Prog Fish Cult 52:100–104

Siegwarth GL, Summerfelt RC (1992) Light and temperature effects on performance of walleye and hybrid walleye fingerlings reared intensively. Prog Fish Cult 54:49–53

Siegwarth GL, Summerfelt RC (1993) Performance comparison and growth models for walleyes and walleye × sauger hybrids reared for two years in intensive culture. Progr Fish Cult 55:229–235

Summerfelt RC (1996) Intensive culture of walleye fry. In: Summerfelt RC (ed) Walleye culture manual, vol 101, NCRAC culture series. North Central Regional Aquaculture Center Publications Office, Iowa State University, Ames, pp 161–185

Chapter 25
Comparative Genetic Diversity, Population Structure, and Adaptations of Walleye and Yellow Perch Across North America

Carol A. Stepien, Osvaldo J. Sepulveda-Villet, and Amanda E. Haponski

Abstract The yellow perch *Perca flavescens* and the walleye *Sander vitreus* are native North American percid fishes, which have considerable fishery and ecological importance across their wide geographic ranges. Over the past century, they were stocked into new habitats, often with relative disregard for conserving local genetic adaptations. This chapter focuses on their comparative population structure and genetic diversity in relationship to historical patterns, habitat connectivity, dispersal ability, distributional abundances, and reproductive behavior. Both species possess considerable genetic structure across their native ranges, exhibiting similar patterning of discontinuities among geographic regions. The two species significantly differ in levels of genetic diversity, with walleye populations possessing overall higher genetic variability than yellow perch. Genetic divergence patterns follow the opposite trend, with more pronounced differences occurring among closely spaced spawning aggregations of yellow perch than walleye. Results reveal broad-scale correspondence to isolation by geographic distance, however, their fine-scale population structures show less relationship, often with pronounced genetic

C.A. Stepien (✉) • A.E. Haponski
Great Lakes Genetics and Genomics Laboratory, Lake Erie Center and Department of Environmental Sciences, The University of Toledo, 6200 Bayshore Rd., Toledo, OH 43615, USA

Present Affiliation: Museum of Zoology and Department of Ecology and Evolutionary Biology, University of Michigan, 1109 Geddes Ave, Ann Arbor, MI 48109, USA
e-mail: carol.stepien@utoledo.edu

O.J. Sepulveda-Villet
Great Lakes Genetics and Genomics Laboratory, Lake Erie Center and Department of Environmental Sciences, The University of Toledo, 6200 Bayshore Rd, Toledo, OH 43615, USA

Present Affiliation: School of Freshwater Sciences, University of Wisconsin - Milwaukee, Milwaukee, WI, USA
e-mail: sepulveo@uwm.edu

P. Kestemont et al. (eds.), *Biology and Culture of Percid Fishes*,
DOI 10.1007/978-94-017-7227-3_25

differences among some nearby reproductive groups. Genetic composition of spawning groups is stable from year to year in walleye, according to two decades of data, and is less consistent in yellow perch. These patterns appear to reflect fundamental behavioral differences between the two species.

Keywords Yellow perch • Walleye • Genetic diversity • Adaptations • Geographical distribution

25.1 Introduction to Yellow Perch and Walleye Population Genetics

The yellow perch *Perca flavescens* and walleye *Sander vitreus* are North American percid fishes of very significant fishery value and ecological importance as top piscivores. Both genera, *Perca* and *Sander*, are exclusively native to North America and Eurasia, with their respective species significantly diverging between the two continents. In North America, *Perca* contains just a single species – the yellow perch, whereas *Sander* has two species – the walleye and the sauger *S. canadensis*. There are two native Eurasian species of *Perca*: the European perch *P. fluviatilis* and the Balkash perch *P. schrenkii*, whereas *Sander* has three: the pikeperch *S. lucioperca*, the Volga pikeperch *S. volgensis*, and the sea pikeperch *S. marinus*. Their respective phylogenies are discussed by Stepien and Haponski in Chap. 1 of this book.

Yellow perch and walleye share wide native geographic distributions across much of the northeast and north central regions of North America, with both having a few isolated relict populations in the southeast. They inhabit a wide diversity of lacustrine and fluvial habitats, ranging from large to small in geographic sizes, with their most extensive habitats and greatest abundances occurring in the Laurentian Great Lakes – especially in Lake Erie (Scott and Crossman 1973; Hubbs and Lagler 2004). Dating to the mid-1800s, walleye and yellow perch were stocked (artificially introduced to new areas), throughout much of North America to provide fishing opportunity (USFWS/GLFC 2010). These introductions often mixed nonindigenous hatchery broodstock with local genotypes, and may have influenced the overall adaptedness of some indigenous populations. Today's fishery managers increasingly recognize the importance of preserving local population variability, and it is advisable to perform any supplementation solely with native genotypes specific to that particular locale. The most prudent action is to protect the habitats of locally adapted populations and avoid negative effects of overexploitation, thereby circumventing any need to stock.

In the northern regions of the ranges of yellow perch and walleye, the habitats and basins of the Great Lakes region were formed and reshaped by the Laurentian Ice Age glaciations, leading to their present configuration about 4000–12,000 years ago (ya). During the glaciations, yellow perch, walleye, and other aquatic species migrated southward of the ice sheets, where their populations were concentrated in restricted glacial refugia areas. Three primary North American glacial refugia are recognized (marked on Fig. 25.1): the Mississippian Refugium in the central U.S., the Missourian Refugium to the west, and the Atlantic Refugium to the east (Bailey and Smith 1981; Crossman and McAllister 1986; Mandrak and Crossman 1992). Following the glacial meltwaters, aquatic taxa migrated along tributary pathways leading from the three refugia into the reformed waterbodies of the Great Lakes and other northerly habitats (see Fig. 25.1). Today's northern populations of yellow perch, walleye, and other fishes appear to retain the signatures of their genetic origins from the respective glacial refugia (summarized by Sepulveda-Villet and Stepien (2012) for yellow perch and Stepien et al. (2009) for walleye). Yellow perch and walleye, although they now are adapted to the large inland "seas" that comprise the Great Lakes, have their ecological and evolutionary origins in fluvial systems rather than large, lacustrine basins.

The genetic composition and structure of yellow perch, walleye, and other aquatic species have been shaped from the past through the present by their relative (1) dispersal abilities and migration behavior, (2) abundances and distributions across waterways, (3) habitat requirements, quality, and connectivity, and (4) reproductive behavior and spawning site fidelity. Large connected spans of suitable habitats, as found in the Great Lakes basins and tributaries, offer a variety of environmental resources for diverse populations, reflecting an interplay between migration opportunity and localized adaptation (see Lindsay et al. 2008; Vandewoestijne et al. 2008; Kunin et al. 2009). Aquatic habitats frequently are connected by narrow and relatively ephemeral connections that link populations during migration and dispersal, but whose habitats may pose distinct biological challenges. For example, small connecting channels extensively vary in size and habitat complexity, and differ in available food and shelter. These factors then influence population variability and local adaptations. In contrast, isolated relict populations with little connectivity may possess lower overall genetic diversity due to the influences of genetic drift, bottlenecks, and selection (Moran and Hopper 1983; Petit et al. 2003; Coulon et al. 2012). Such relict and "rear edge" populations (those found in fringe latitudinal portions of the native range, as in the southeastern U.S. for yellow perch and walleye) may serve as critical repositories of genetic diversity and provide possible sources for future range expansion in the face of climate change (see Hampe and Petit 2005; Diekmann and Serrão 2012). These hypotheses are explored and discussed in this chapter for yellow perch and walleye, in order to provide an understanding of their shared and different population patterns.

25.2 Reproduction of Yellow Perch and Walleye in Relation to Their Population Genetic Structure

Cued by gradual changes in water temperature and photoperiod, walleye aggregate to reproduce in early spring in slow-moving tributaries or on shallow reef complexes in lacustrine systems (Craig 1987; Kreiger et al. 1983; Colby et al. 1994).

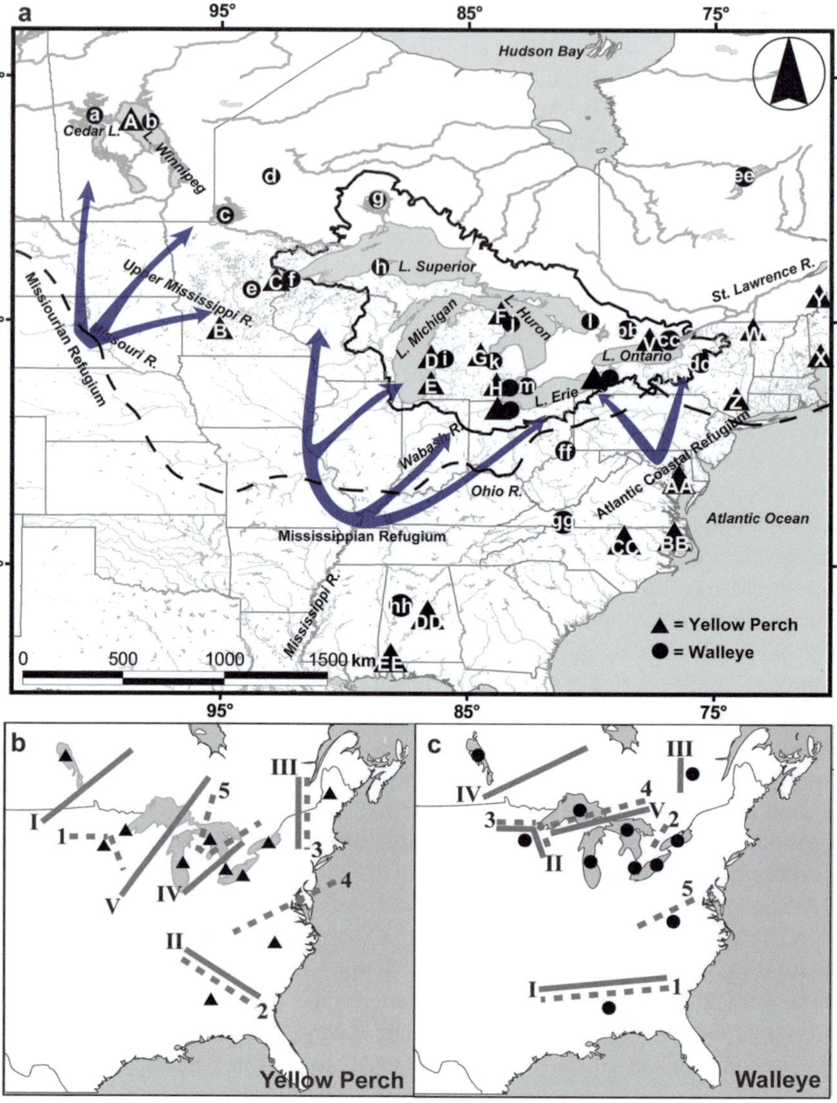

Fig. 25.1 Maps showing sampled spawning populations for (**a**) yellow perch (*triangles*) and walleye (*circles*) across North America, and genetic barriers for (**b**) yellow perch and (**c**) walleye. (**a**) *Box* shows the HEC (Huron-Erie Corridor), *thick dashed line* indicates maximum extent of the Wisconsinan glaciations, *arrows* denote likely routes of post-glacial population colonizations (adapted from Mandrak and Crossman 1992). (**b, c**) *Grey lines* (*solid* = microsatellite data, *dashed* = mtDNA control region sequences) define major barriers to gene flow calculated based on geographic coordinates (latitude and longitude) and genetic divergence (F_{ST}) from BARRIER v2.2 analysis (Manni et al. 2004; http://ecoanthropologie.mnhn.fr/software/barrier.html). These are ranked I–V for the microsatellite and 1–5 for the mtDNA control region data sets, in order of their decreasing magnitude (Barriers are modified from results of Stepien et al. (2009), Sepulveda-Villet and Stepien (2012), and Haponski and Stepien (2014a))

Yellow perch reproduce in similar habitats, about a month later at warmer temperatures (Scott and Crossman 1973; Carlander 1997; Jansen et al. 2009). Spawning of the two species occurs much earlier in the southern U.S., and much later in the Canadian north (Carlander 1997; Craig 2000).

Walleye often migrate long distances to propagate (Kreiger et al. 1983; Colby et al. 1994; Craig 2000), homing to their natal sites, as indicated by tagging studies (Ferguson and Derkson 1971; Wolfert and Van Meter 1978; Jennings et al. 1996) and genetic divergence data among their spawning groups (Stepien and Faber 1998; Strange and Stepien 2007; Stepien et al. 2009, 2010, 2012). Notably, mark-and-recapture studies have recovered most Lake Erie and Lake St. Clair walleye near their original spawning sites during the subsequent spring reproductive season(s) (Wang et al. 2007). A study by Jennings et al. (1996) tracked the spawning returns of laboratory-reared walleye after release in the field, indicating that natal homing is a genetically based response to environmental cues.

Spring spawning migrations of yellow perch are shorter and it is presumed that they also return to specific natal sites in shallow waters (Aalto and Newsome 1990; Carlander 1997; Craig 2000). Yellow perch that were captured and tagged during the reproductive season and released many km distant in the eastern basin of Lake Erie, then returned to their tagging locations (MacGregor and Witzel 1987) – implicating homing. Separate studies by Clady (1977), Rawson (1980), and the Ontario Ministry of Natural Resources (OMNR 2011) likewise found that most yellow perch tagged during spawning were recovered at or very close to their initial reproductive locations in subsequent years. For example, yellow perch spawning groups located just a few km apart (17 km) in central Lake Erie were found to diverge from one another in genetic and morphological composition (Kocovsky et al. 2013). This suggests that spawning populations likely comprise natal groups at specific locations (Kocovsky et al. 2013; Sullivan and Stepien 2015). Aalto and Newsome (1990) removed yellow perch egg masses from given spawning sites, which led to fewer fish returning to that location in subsequent years than in control sites, suggesting that they returned to the same spawning areas year after year. It is hypothesized that imprinting occurs during the early life history of walleye and yellow perch, with their highly developed olfactory systems used to detect natal sites and/or the pheromones of neighbors (see Horrall 1981; Gerlach et al. 2001).

Stepien et al. (2012) analyzed the genetic structure of walleye reproducing at three well-known spawning locations in Lake Erie, encompassing 6 years of spawning runs from 1995 to 2008. The genetic composition of each population group, based on nine nuclear microsatellite loci (see Tables 25.1 and 25.2), remained similar from year to year, among age cohorts, between the sexes, and from generation to generation (Stepien et al. 2012). In contrast, although the genetic composition of yellow perch spawning aggregations differed significantly from location to location across broad and fine geographic scales (Sepulveda-Villet and Stepien 2011), there were some significant differences at the same location from year to year (Sullivan and Stepien 2015). This suggests that although yellow perch may reproduce together with a specific population group (believed to be their natal group), specific spawning locales may vary from year to year. Thus, the site fidelity to specific reproductive locations may differ between yellow perch and walleye, which remains to be experimentally investigated.

Table 25.1 Summary of genetic variation per microsatellite locus for yellow perch (N = 892) and walleye (N = 1125) evaluated in this chapter, showing number of alleles (N_A), allelic size range (base pairs, bp), inbreeding coefficient (F_{IS}, average divergence within a spawning group), and mean genetic divergence among sampling locales (F_{ST}). These loci are used in Tables 25.2, 25.3, and 25.4, and Fig. 25.1. Results were determined using FSTAT v2.9.3.2 (Goudet 1995, 2002; http://www2.unil.ch/popgen/softwares/fstat.htm) and are summarized from Sepulveda-Villet and Stepien (2012), Stepien et al. (2009, 2010, 2012), and Haponski and Stepien (2014a)

		Yellow perch				Walleye			
Locus	Source	N_A	Size range	F_{IS}	F_{ST}	N_A	Size range	F_{IS}	F_{ST}
*Svi*2	Eldridge et al. (2002)	18	184–218	0.149	0.360	30	178–258	0.006	0.071
*Svi*7	Eldridge et al. (2002)	14	162–212	0.328	0.280	28	140–208	0.068	0.113
*Svi*4	Borer et al. (1999)	42	108–198	0.101	0.149	19	98–140	0.033	0.110
*Svi*17	Borer et al. (1999)	30	96–190	0.170	0.133	15	92–120	–0.011	0.150
*Svi*33	Borer et al. (1999)	51	76–178	0.246	0.087	26	72–128	0.020	0.067
*Svi*3	Eldridge et al. (2002)	14	112–156	0.161	0.287	–	–	–	–
*YP*13	Li et al. (2007)	23	214–280	0.289	0.348	–	–	–	–
*YP*17	Li et al. (2007)	16	191–241	0.133	0.243	–	–	–	–
*Mpf*1	Grzybowski et al. (2010)	53	171–347	0.143	0.048	–	–	–	–
*Mpf*2	Grzybowski et al. (2010)	51	203–311	0.131	0.036	–	–	–	–
*Mpf*3	Grzybowski et al. (2010)	27	103–179	0.048	0.158	–	–	–	–
*Mpf*4	Grzybowski et al. (2010)	35	171–247	0.172	0.100	–	–	–	–
*Mpf*5	Grzybowski et al. (2010)	24	127–171	0.086	0.253	–	–	–	–
*Mpf*6	Grzybowski et al. (2010)	21	100–164	0.182	0.261	–	–	–	–
*Mpf*7	Grzybowski et al. (2010)	29	128–200	0.050	0.141	–	–	–	–
*Svi*6	Borer et al. (1999)	–	–	–	–	31	126–246	0.065	0.071
*Svi*18	Borer et al. (1999)	–	–	–	–	8	114–128	0.094	0.124
*Svi*L6	Wirth et al. (1999)	–	–	–	–	22	92–140	–0.006	0.072
*Svi*L7	Wirth et al. (1999)	–	–	–	–	29	160–238	0.046	0.031
Total	–	448	–	0.150	0.175	208	–	0.034	0.079

Table 25.2 Sampling population regions tested, sample size (N), and mean genetic variability values from (a) 15 nuclear DNA microsatellite loci for yellow perch and nine loci for walleye and (b) mtDNA control region sequences. Microsatellite data include: observed (H_O) heterozygosity, inbreeding coefficient (F_{IS}), number of µsat alleles across all loci (N_A), allelic richness (A_R), proportion of private alleles (P_{PA}), and proportion of full siblings in each drainage (Sib). Values for mtDNA include number of haplotypes (N_H), haplotypic diversity (H_D), and proportion of private haplotypes (P_{PH}). Values are from GENEPOP v4.0 (Rousset 2008; http://kimura.univ-montp2.fr/~rousset/Genepop.htm), FSTAT v2.9.3.2 (Goudet 1995, 2002; http://www2.unil.ch/popgen/softwares/fstat.htm), ARLEQUIN v3.1.5.3 (Excoffier and Lischer 2010; http://cmpg.unibe.ch/software/arlequin35/) and CONVERT v1.31 (Glaubitz 2004; http://www.agriculture.purdue.edu/fnr/html/faculty/rhodes/students and staff/glaubitz/software.htm) (Data are summarized and adapted from Sepulveda-Villet and Stepien (2012), Stepien et al. (2009, 2010, 2012), and Haponski and Stepien (2014a))

a. Nuclear microsatellite DNA loci

	Yellow perch					Walleye						
Locality	N	H_O	F_{IS}	N_A	A_R	P_{PA}	N	H_O	F_{IS}	N_A	A_R	P_{PA}
Total (mean)	892	0.53	0.145	442	8.39	(0.04)	1,125	0.73	0.029	111	15.85	(0.03)
1. Lake Winnipeg	12	0.49	–0.010	68	4.53	0.02	105	0.63	0.104	51	3.64	0.00
2. Upper Mississippi R. watershed	18	0.52	0.165	112	7.47	0.04	39	0.62	0.010	38	3.22	0.05
Great Lakes region:	459	0.55	0.206	363	9.97	0.14	897	0.72	0.058	100	4.26	0.22
3. Lake Superior	25	0.64	0.080	119	7.93	0.01	114	0.72	0.059	67	4.23	0.00
4. Lake Michigan	65	0.54	0.174	298	9.93	0.05	50	0.73	0.057	57	4.20	0.00
5. Lake Huron	80	0.61	0.135	355	11.83	0.02	119	0.73	0.031	69	4.16	0.03
6. Lake St. Clair	86	0.59	0.098	225	13.22	0.03	162	0.72	0.019	77	4.09	0.03
Lake Erie:	401	0.55	0.116	313	13.26	0.09	348	0.72	0.037	85	4.14	0.01
7. Western Basin, L. Erie	189	0.55	0.100	259	12.87	0.05	211	0.70	0.035	78	4.02	0.00
8. Eastern Basin, L. Erie	212	0.54	0.122	270	13.00	0.06	137	0.74	0.034	74	4.30	0.01
9. Lake Ontario	62	0.55	0.122	213	13.79	0.04	104	0.70	0.068	76	4.13	0.00

(continued)

Table 25.2 (continued)

a. Nuclear microsatellite DNA loci

	Yellow perch					Walleye						
Locality	N	H_O	F_{IS}	N_A	A_R	P_{PA}	N	H_O	F_{IS}	N_A	A_R	P_{PA}
10. Northeastern populations	60	0.50	0.236	347	7.71	0.05	40	0.52	0.137	45	3.09	0.04
11. Southeastern populations	68	0.60	0.132	349	7.78	0.06	39	0.68	0.141	63	4.36	0.03
12. US Gulf coastal region	15	0.39	0.346	108	3.60	0.07	5	0.56	0.197	27	3.53	0.11

b. MtDNA control region sequence variation

	Yellow perch			Walleye				
Locality	N	H_D	N_H	P_{PH}	N	H_D	N_H	P_{PH}
Total (mean)	664	0.31	26	(0.12)	711	0.77	27	(0.22)
1. Lake Winnipeg	12	0.00	1	0.00	100	0.64	8	0.38
2. Upper Mississippi R. watershed	18	0.53	2	0.00	25	0.16	3	0.00
Great Lakes region (3–9):	459	0.22	14	0.07	510	0.75	22	0.59
3. Lake Superior	25	0.00	1	0.00	75	0.65	5	0.20
4. Lake Michigan	65	0.34	3	0.00	25	0.23	3	0.00
5. Lake Huron	80	0.40	4	0.15	75	0.45	5	0.00
6. Lake St. Clair	39	0.00	1	0.00	120	0.73	6	0.17
Lake Erie (7–8):	235	0.21	12	0.03	150	0.78	14	0.50
7. Western Basin, L. Erie	77	0.27	4	0.03	100	0.76	9	0.33
8. Eastern Basin, L. Erie	88	0.07	4	0.03	50	0.82	11	0.36
9. Lake Ontario	15	0.13	2	0.07	75	0.54	9	0.44
10. Northeastern populations	60	0.48	3	0.29	25	0.48	2	0.00
11. Southeastern populations	68	0.63	7	0.62	36	0.62	5	0.20
12. US Gulf coastal region	15	0.15	2	0.13	5	0.40	2	0.50

During their respective spring reproductive seasons, males move into the spawning areas first, arriving before females by a few weeks and lingering longer at the sites (Scott and Crossman 1973; Craig 2000; Simon and Wallus 2006). Single females spawn with several males. The female yellow perch lays a long gelatinous egg strand (to 2.1 m long), which contains 10,000–40,000 eggs, over submerged vegetation or other structures at night or in early morning. As the egg mass is released by the female, it is externally fertilized by a cluster of 2–25 male yellow perch, who closely follow her and often are in close proximity to other spawning clusters (Scott and Crossman 1973; Mangan 2004; Simon and Wallus 2006). In contrast, each female walleye releases thousands of separate eggs at night, which are externally fertilized by a swarm of one to six males (Kerr et al. 1997). The walleye eggs settle into crevices, where they adhere to rocks or gravel with their sticky outer coating of muco-polysaccharides (Craig 2000; Barton and Barry 2011). Female walleye tend to spawn over several days, and males of both species fertilize the eggs of multiple females over several days or weeks (Craig 2000; Simon and Wallus 2006).

Neither species provides any parental care (Trautman 1981; Zhao et al. 2009). After spawning the adults move to nearby bays and other littoral areas, and then travel offshore to summer feeding grounds (Kerr et al. 1997; Craig 2000). Eggs of both species hatch over a period ranging from 10 days to three weeks, depending on water temperature (Simon and Wallus 2006). More rapid development and better survival occurs at warmer temperatures (Scott and Crossman 1973; Kerr et al. 1997; Roseman et al. 2005).

25.3 Movements and Behavior of Larval and Juvenile Yellow Perch and Walleye, in Relation to Their Population Genetics

Larvae of yellow perch (4–7 mm) and walleye (6–9 mm) hatch with small yolk sacs (Scott and Crossman 1973; Bozek et al. 2011a). Yellow perch larvae swim in a coordinated manner at hatching (Fulford et al. 2006). In contrast, walleye are reported to swim to the surface aided by a large oil globule in the yolk, then sink back to the bottom, and by the second day are free swimming (Simon and Wallus 2006). Larvae of both species drift with currents to nearshore nursery areas in shallow vegetated areas (Craig 2000; Jones et al. 2003; Roseman et al. 2005). They begin to feed after their yolk sacs are absorbed, consuming phytoplankton and small zooplankton, and grouping with conspecifics of similar size in schools (Collette et al. 1977; Kerr et al. 1997; Craig 2000; Simon and Wallus 2006). It is hypothesized, but untested, that they may school with their natal site cohorts.

Successful location and capture of prey items of appropriate size and nutrition are critical for survival of the larvae, as well as avoiding predation (see Fulford et al. 2006; Beletsky et al. 2007). These two factors – feeding and avoiding predation – impose high selective pressures, and favor those with better-developed sensory organs and greater swimming ability (Li and Mathias 1982; Craig 2000; Simon and Wallus 2006). Water turbidity is believed to augment survival, facilitating predator avoidance

(Roseman et al. 2005; Manning et al. 2013). The larval pelagic stage lasts for about 30–40 days, with yellow perch transitioning to become increasingly benthic, but remaining in schools throughout life (summarized by Fulford et al. 2006; Simon and Wallus 2006). In contrast, most walleye remain more pelagic and become solitary, whereas other individuals are reported to school (Craig 2000; Simon and Wallus 2006).

The young-of-year move into deeper water as juveniles in late spring (Craig 2000). Larger juvenile walleye become piscivorous, eating a variety of fish species (including conspecifics), as well as zooplankton – especially when forage fish are low in abundance (Collette et al. 1977; Craig 2000; Bozek et al. 2011a). In the Great Lakes, the invasive Eurasian round goby *Neogobius melanostomus* increasingly has become important in the diets of both yellow perch and walleye (Truemper and Lauer 2005; Kornis et al. 2012).

Parker et al. (2009) found that age-1 juvenile yellow perch differed in morphology and genetically at 12 nuclear DNA microsatellite loci between populations in Lake Huron and Lake Michigan, with those from Lake Huron having deeper, longer bodies and larger dorsal fins. The researchers also discerned morphological and genetic differences between juveniles living in nearshore versus wetland habitats in Lake Michigan. Juveniles inhabiting nearshore areas from both lakes had deeper, longer bodies and larger dorsal fins than did those occupying wetlands, which might reflect an adaptive response to predators and open-water cruising. Differences between habitats across the lakes were hypothesized to reflect plasticity between phenotypic and genetic divergence (Parker et al. 2009).

25.4 Non-reproductive Movements and Behavior of Adult Yellow Perch and Walleye in Relation to Population Genetics

After the reproductive season, movements of adult yellow perch and walleye largely are determined by habitat complexity, food availability, and foraging capacity (Roseman et al. 2005; Radabaugh et al. 2010). Juvenile and adult yellow perch typically are found in schools, which likely facilitate foraging and predator avoidance (Helfman 1984; Craig 2000). In comparison, walleye are either solitary or found in smaller schools as adults (Craig 2000; Simon and Wallus 2006). Whether the schools of yellow perch are structured based on kinship has not yet been evaluated (see Sullivan and Stepien 2015). However, schools of the closely related European perch have been shown to contain large numbers of related individuals, which recognize one another via chemical and physical cues (Gerlach et al. 2001; Behrmann-Godel et al. 2006).

Walleye typically range widely to feed at non-reproductive times of the year, travelling distances from 50 to 300 km (see Colby et al. 1979; Wang et al. 2007; Bozek et al. 2011b). In contrast, a study of yellow perch tag returns by Haas et al. (1985) determined that post-spawning movements are moderate; individuals tagged at Lake Erie spawning sites did not move upstream through the Huron-Erie Corridor (HEC; see Fig. 25.1), which connects Lake Huron to Lake Erie via the St. Clair River, Lake St. Clair, and the Detroit River. Some yellow perch that were tagged in Lake St. Clair migrated to nearby tributaries (Haas et al. 1985). Likewise, Dumont

(1996) found little movement of tagged yellow perch along the St. Lawrence River. There thus appears to be much greater tendency for mixing among spawning groups of walleye than of yellow perch during non-reproductive times of the year. Note that although individuals may move among water bodies to feed, their reproductive groups determine their overall population genetic structures.

Evidence of yellow perch metapopulations inhabiting Lake Ontario embayments was discerned using otolith microchemistry (Murphy et al. 2012). Results described discrete assemblages in connected bays and impoundments (Murphy et al. 2012). This type of metapopulation structure characterized yellow perch reproductive groups along Lake Erie coastal sites (Sepulveda-Villet and Stepien (2011); these populations likely display some seasonal mixing, as described by Parker et al. (2009). In the Baltic Sea, European perch and pikeperch showed no appreciable genetic distinction among sites during the mid-summer, despite being separated by coastal features and salinity regimes (Sruoga et al. 2008). Post-spawning populations of *Perca* and *Sander* often intermingle, obscuring genetic population identities other than during reproduction.

Maturity is reported to be reached between ages 2 and 3 in yellow perch and at about age 3 in walleye (Collette et al. 1977; Barton and Barry 2011). However, fall survey results in Lake Erie showed that over half of age-1 male yellow perch and walleye are sexually mature, whereas less than half of the age-3 females had reached maturity (Dr. Patrick Kocovsky, USGS Great Lakes Science Center, personal communication, 2014). Female yellow perch and walleye mature later, grow more rapidly throughout their lives, and reach larger sizes than do the males; the females also live longer (summarized by Carlander 1997). Yellow perch attain ages of 6–21 (Craig 2000), whereas walleye reach 6–19 years (Carey and Judge 2000; Craig 2000) and live to a maximum of ~30 years (Bozek et al. 2011a). In northern habitats, both species grow more slowly and have longer life expectancies (Carlander 1997; Craig 2000). With climate warming, we predict that they likely will grow faster and may die earlier, which will influence their population structures.

25.5 Genetic Origins of Northern Yellow Perch and Walleye Populations Tracing to Glacial Refugia

Contemporary haplotypes of yellow perch appear to trace to ~6.0 million years ago (Mya; Fig. 25.2a; Sepulveda-Villet and Stepien 2012), whereas walleye date to ~10.6 Mya during the late Miocene Epoch (Fig. 25.2b). The northern haplotypes of yellow perch share a common ancestry estimated as ~0.6–4.2 Mya (Fig. 25.2a; Sepulveda-Villet and Stepien 2012), and those for walleye similarly date to ~1–2.5 Mya (Fig. 25.2b).

Northerly populations of yellow perch and walleye were subjected to drastic climate change and habitat losses during the Pleistocene glaciations 10,000 ya–2.6 Mya. As the ice sheets formed, the fish populations gradually migrated to survive in southerly areas – the Missourian, Mississippian, and Atlantic glacial refugia (Fig. 25.1; Petit et al. 2003; Hewitt 2004; Provan and Bennett 2008). Following the glacial meltwaters, migrants from these refugia surged northward, recolonizing drastically

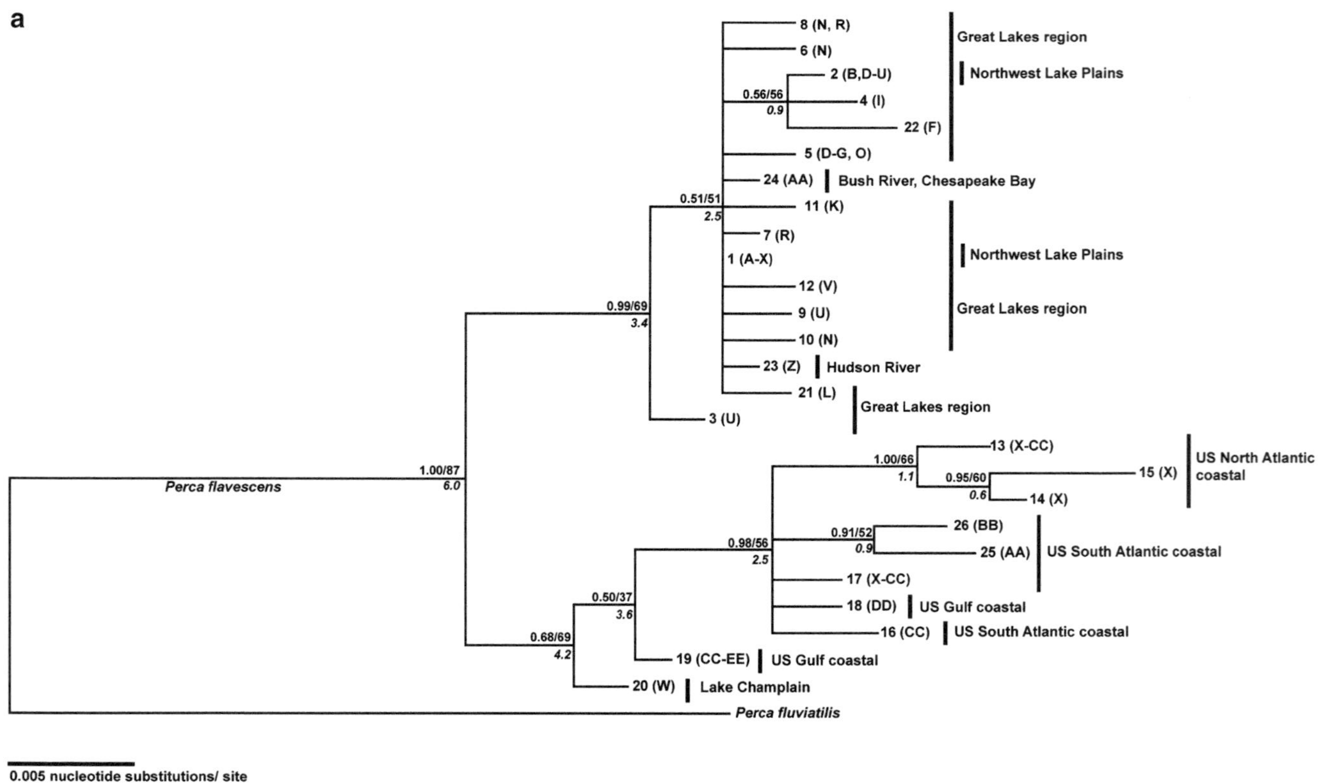

Fig. 25.2 Bayesian phylogenetic trees of mtDNA control region sequence haplotypes for (**a**) yellow perch and (**b**) walleye, calculated using MrBAYES v3.2.1 (Ronquist and Huelsenbeck 2003; http://mrbayes.sourceforge.net/). Values above nodes = Bayesian posterior probability/percentage support from 2000 bootstrap pseudo-replications in ML with PHYML v3.0 (Guindon et al. 2010; http://www.atgc-montpellier.fr/phyml/); those with ≥0.50 pp and ≥50 % bootstrap support are reported. jMODELTEST v2 (Darriba et al. 2012; https://code.google.com/p/jmodeltest2/) selected the most likely model of nucleotide substitution

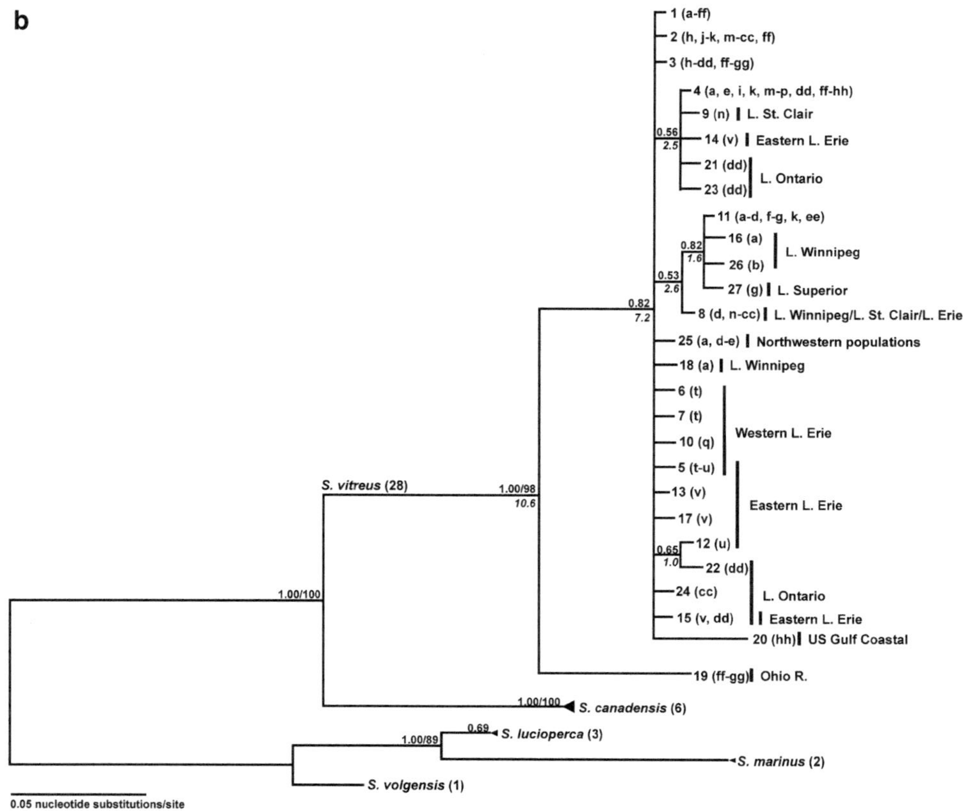

Fig. 25.2 (continued) for construction of the phylogenetic trees and divergence time estimates. Values below nodes in *italics* = estimated divergence times (given as millions of years) as determined in BEAST v1.71 (Drummond et al. 2012; http://beast2.org). Divergence times were calibrated using three fossil and four molecular calibration points following Haponski and Stepien (2013). Letters in parentheses denote sampling sites in which haplotypes were recovered (see Fig. 25.1 map). *Vertical bars* denote general geographical regions

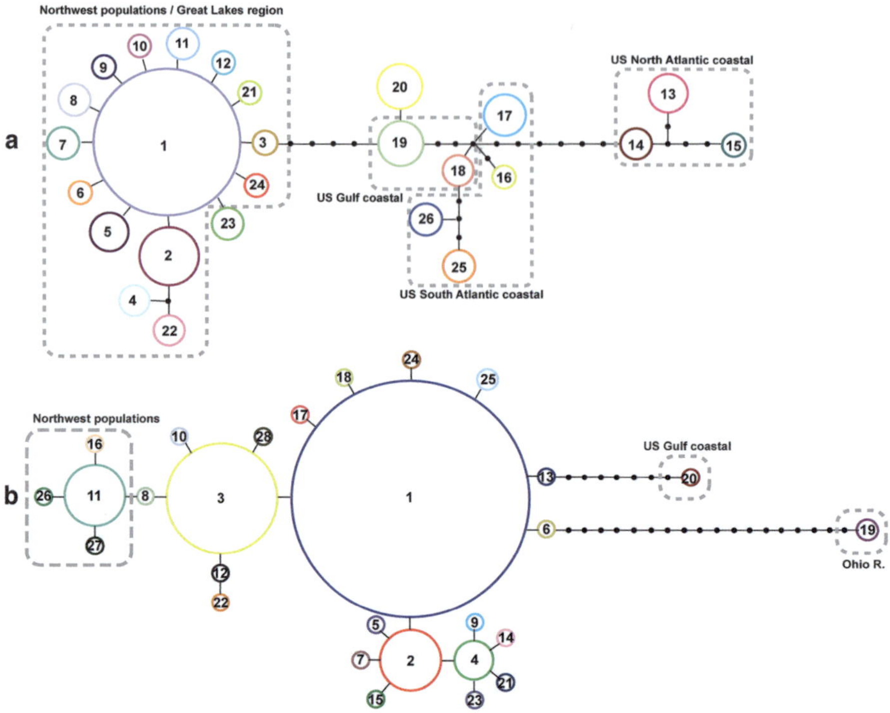

Fig. 25.3 Statistical parsimony network constructed using TCS v1.21 (Clement et al. 2000, http://darwin.uvigo.es/software/tcs.html) among mtDNA control region haplotypes of (**a**) yellow perch (Modified from Sepulveda-Villet et al. (2009) and Sepulveda-Villet and Stepien (2012)) and (**b**) walleye (Modified from Haponski and Stepien (2014a)). *Circles* are sized according to total observed frequencies of the haplotypes. *Lines* indicate a single mutational step between the haplotypes. Small, unlabelled *circles* represent hypothesized unsampled haplotypes. *Dashed lines* enclosing haplotype groups denote major regional delineations. *Circle* colours also reflect haplotype identities as portrayed in Fig. 25.4a. Note that there is no correspondence between the colors and haplotypes of yellow perch and walleye (**a** and **b** are entirely independent)

changed habitats, including the new basins of the Laurentian Great Lakes. Notably, an estimated 90 fish species migrated northward from the Mississippian glacial refugium to found modern Great Lakes populations, another 14 expanded up from the Atlantic coastal refugium, and some from each met and mixed (see Mandrak and Crossman 1992). To the west, colonists from the Missourian refugium founded yellow perch and walleye populations in the Northwest Plains, including Lake Winnipeg and the upper Mississippi River region, as well as contributing to western Lake Superior (Billington 1996; Stepien et al. 2009, 2010; Backhouse-James and Docker 2012). Patterns of genetic divergences among northern fish populations today reflect these differential

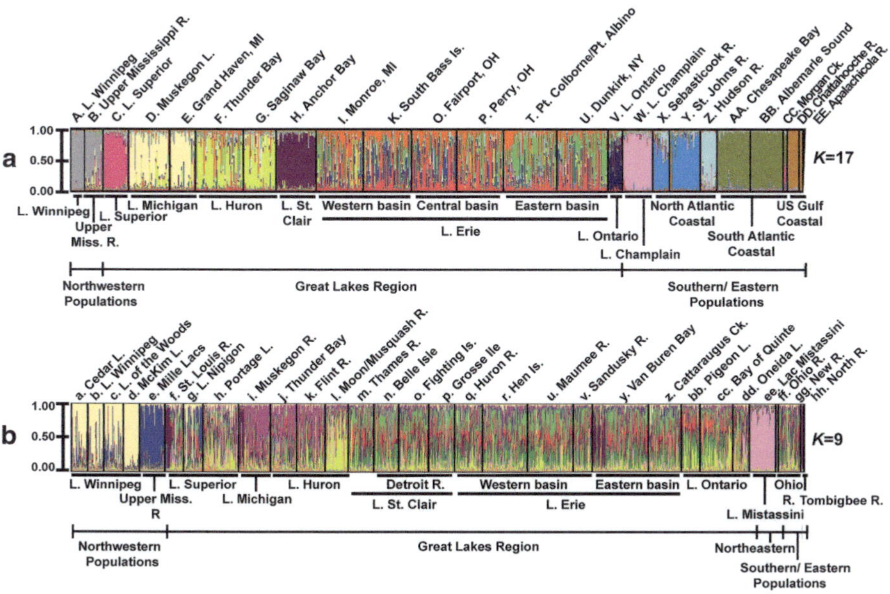

Fig. 25.4 Estimated comparative population structure from Bayesian STRUCTURE v2.3.3 analyses (Pritchard et al. 2000; Pritchard and Wen 2004; http://pritchardlab.stanford.edu/structure.html) for (**a**) 17 yellow perch groups using 15 nuclear DNA microsatellite loci (Modified from Sepulveda-Villet and Stepien 2012), and (**b**) 9 walleye groups using 9 loci (Modified from Stepien et al. 2009, 2010). Analyses were run with 100,000 burn-in and 500,000 replicates. Optimal *K* values were determined by posterior probabilities (Pritchard et al. 2000) and the Δ*K* method of Evanno et al. (2005). *Thin vertical lines* represent individuals and thicker bars separate spawning groups at given locations; these are partitioned into *K* colored segments that represent estimated population group membership. Note that there is no correspondence between the colors of yellow perch and walleye (**a** and **b** are entirely independent)

contributions originating from refugia, which were subsequently modified by drainage connections and basin isolation (Bailey and Smith 1981; Mandrak and Crossman 1992; Stepien et al. 2009, 2010).

Yellow perch and walleye in the Northwest Lake Plains region (Fig. 25.1: Lake Winnipeg and the upper Mississippi River) trace their descent to Missourian refugium colonists (Stepien et al. 2009, 2010; Sepulveda-Villet and Stepien 2012). Today, the overall differences in these populations from other regions is apparent in Figs. 25.1, 25.2, 25.3, and 25.4 (the latter depicts Bayesian STRUCTURE analyses of populations, based on nuclear DNA microsatellite loci from Tables 25.1 and 25.2), showing that the Lake Winnipeg and Upper Mississippi River populations are different from those in most of the Great Lakes. Yellow perch from western Lake Superior also are very distinctive in the nuclear microsatellite data (Fig. 25.4a; Sepulveda-Villet and Stepien 2012). Walleye reproducing in Lake Superior show a mixture of descent from the Missourian and Mississippian refugia (Fig. 25.4b; Stepien et al. 2009, 2010). Populations of lake sturgeon *Acipenser fulvescens* in the

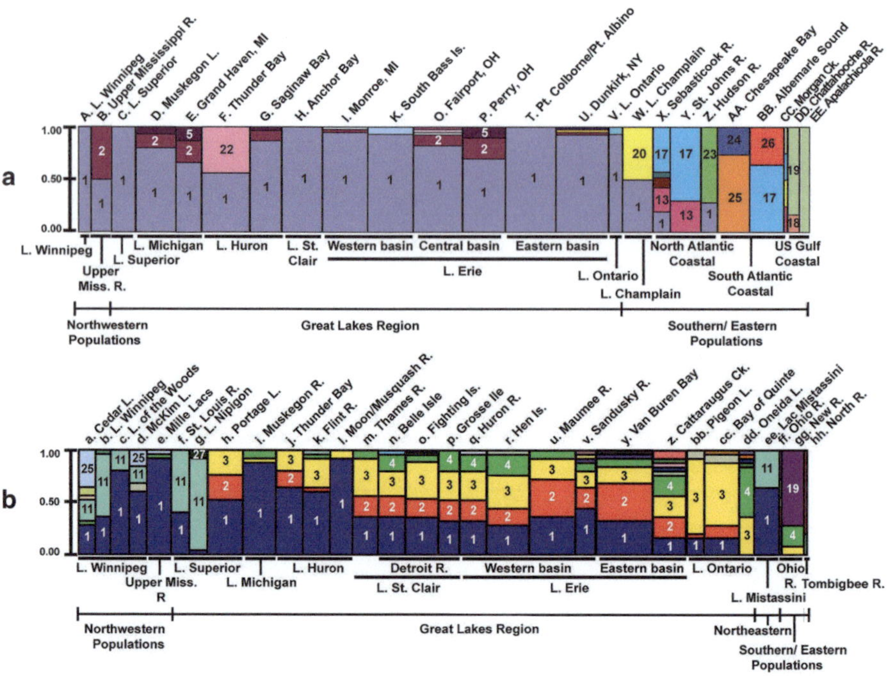

Fig. 25.5 Estimated population structure from mtDNA control region frequencies for (**a**) 26 yellow perch haplotypes (Modified from Sepulveda-Villet and Stepien 2012) and (**b**) 28 walleye haplotypes (Adapted from Haponski and Stepien 2014a) using GENEPOP v4.0 (Rousset 2008; http://kimura.univ-montp2.fr/%7Erousset/Genepop.htm), and Microsoft Excel 2008 (Redmond, VA). *Vertical black lines* separate different spawning groups (lettered). Major geographic regions are indicated in the bottom rule for each chart. Note that there is no correspondence between the colors and haplotypes of yellow perch and walleye (**a** and **b** are entirely independent)

Hudson Bay drainage (Ferguson and Duckworth 1997) and brown bullhead *Ameiurus nebulosus* in western Lake Superior (Murdoch and Hebert 1997) likewise have been attributed to Missourian refugium ancestry.

Glacial Lake Agassiz initially occupied much of the Hudson Bay watershed (including Lake Winnipeg), which probably had some southern drainage to Lake Superior (Mandrak and Crossman 1992; Rempel and Smith 1998), facilitating fish movements 8500–13,000 ya. Ice later blocked this passage (Saarnisto 1974; Teller and Mahnic 1988), isolating the yellow perch and walleye populations in the Northwest Lake Plains sites, as is shown by their high divergences from other areas (denoted by distinct colors on Figs. 25.4 and 25.5; Stepien et al. 2009, 2010; Sepulveda-Villet and Stepien 2012; Haponski and Stepien 2014a). These northwestern fish populations always have been small in size (Hoagstrom and Berry 2010). The Lake Superior region was long covered in ice, except for glacial Lake Duluth in the west until ~8500–9000 ya, thus isolating its walleye and yellow perch gene pools.

Most of the Great Lakes fauna – especially in Lakes Huron, Michigan, St. Clair, and western Lake Erie (Underhill 1986; Mandrak and Crossman 1992; Todd and Hatcher 1993) – trace their origins to the Mississippian refugium, as indicated for yellow perch (Sepulveda-Villet et al. 2009; Sepulveda-Villet and Stepien 2012), walleye (Stepien and Faber 1998; Stepien et al. 2009, 2010), smallmouth bass *Micropterus dolomieu* (Stepien et al. 2007), rainbow darter *Etheostoma caeruleum* (Haponski et al. 2009), and lake sturgeon (Ferguson and Duckworth 1997). There were apparent genetic contributions from both the Atlantic and the Mississippian refugium into the lower Great Lakes for walleye (Stepien and Faber 1998; Strange and Stepien 2007; Stepien et al. 2009, 2010), smallmouth bass (Stepien et al. 2007), brown bullhead (Murdoch and Hebert 1997), lake trout *Salvelinus namaycush* (Wilson and Hebert 1996), and ninespine stickleback *Pungitius pungitius* (Aldenhoven et al. 2010).

The closer genetic relationship between walleye reproducing in Lake Michigan and Lake Huron proper (including Saginaw Bay) likely reflects their former connection as glacial Lake Algonquin ~2000–10,600 ya, which drained west to the Mississippi River system (Bailey and Smith 1981). Lake Huron walleye diverged ~11,500 ya when Georgian Bay (the former glacial Lake Hough) was isolated from the main basin population (the former glacial Lake Stanley) (Lewis et al. 1994).

Lake Erie's formation dates to glacial Lake Maumee (~14,000 ya), which then drained west via the Ohio River to the Mississippi, changing outlets during several lake stages, to its current outlet east into Lake Ontario (~10,000 ya) (Underhill 1986; Larson and Schaetzl 2001). Lake Erie yellow perch and walleye populations today appear geographically isolated and genetically differentiated from most other Great Lakes populations. Lake Erie physically is separated from Lake Ontario by Niagara Falls and from the upper Great Lakes by the narrow and short Detroit River, which drains Lake St. Clair. Yellow perch from Lake St. Clair are separated from those spawning in Lake Erie (note the color difference between the purple-colored population from Lake St. Clair versus the mixed colors in Lake Erie in Fig. 25.4a); these appear on opposite sides of a genetic barrier IV shown in Fig. 25.1b. However, there is greater genetic exchange between walleye reproducing in Lakes St. Clair and Erie, as illustrated by their similar color on Fig. 25.4b (see Haponski and Stepien 2014b).

Yellow perch mtDNA control region haplotype 1 (Figs. 25.2a and 25.3a) likely already was widespread pre-glacially and then represented in the Mississippian and Atlantic refugia populations, but was more common in the west. Today, yellow perch haplotype 1 remains more abundant in the west (see Fig. 25.5a). Walleye possesses a similar pattern, with its most common haplotype 1 also being more common to the west, and increasing proportions of haplotypes 2, 3, and 4 to the east (Fig. 25.5b). This west-east pattern occurs in both species and presumably reflects retention of original colonization proportions from the Mississippian and Atlantic refugium to the present day.

The Atlantic coastal refugium (Fig. 25.1) formed a warm enclave of diverse habitats in coastal plains and estuaries east of the Appalachian Mountains (Schmidt 1986; Bernatchez 1997); yellow perch and walleye from that refugium migrated north to colonize the northeastern and northcentral regions after the glaciations (Russell et al. 2009). The northeastern migrating populations split to found the yellow perch populations in Maine (colored blue, sites X and Y, on Fig. 25.4a) and the

Hudson River (colored light blue, site Z); both are very divergent today (also see mtDNA haplotypes on Fig. 25.5a).

Lake Champlain (site W) drains into the St. Lawrence River and its yellow perch appear to trace to joint origins from the Atlantic and Mississippian refugia, but today have a very divergent genetic composition from other locations (see unique haplotype 20 on Figs. 25.2a and 25.3a and distinct colors for Lake Champlain yellow perch on Figs. 25.4a and 25.5a, denoting different genetic composition). Lake Champlain received meltwaters from glacial Lake St. Lawrence (~11,600 ya), and then Lake Agassiz (~8000–10,900 ya) and glacial Lake Barlow-Ojibway (~8000–9,500 ya). This produced an extensive freshwater habitat that replaced the former saline Champlain Sea, which was a temporary inlet of the Atlantic Ocean formed by the retreating glaciers (Rodrigues and Vilks 1994). Regional flooding presumably led to colonization of Lake Champlain by aquatic taxa from the Atlantic refugium, as suggested by genetic evidence from lake cisco *Coregonus artedi* (Turgeon and Bernatchez 2001); the Lake Champlain yellow perch population appears to reflect joint contribution from the Atlantic and Mississippian refugia (see Fig. 25.5a).

The overall similarities among northern populations of yellow perch, walleye, and other aquatic taxa reveal a general pattern that originated with recolonization from multiple glacial refugia, which then became modified by changes in connections and drainages, and has been maintained by reproductive site philopatry from generation through generation. Refugium origins have been shown to translate to differences in walleye growth patterns (Zhao et al. 2008); thus the influences of evolutionary history appear to persist in their physiological, life history, and genetic adaptations today.

25.6 Atlantic and Gulf Coastal Populations of Yellow Perch and Walleye

The southern genotypes of yellow perch and walleye clearly are differentiated from the northern ones, with the southern ones being older (see Fig. 25.2; Stepien and Faber 1998; Stepien et al. 2009; Sepulveda-Villet and Stepien 2012; Haponski and Stepien 2014a). The most divergent walleye identified are from the New and Ohio Rivers (Fig. 25.2b). Divergence of southern walleye haplotypes in the New/Ohio River and Gulf coastal systems was estimated at ~7.2–10.6 Mya (this chapter), whereas dates for southerly yellow perch appear to be later at ~2.5–3.6 Mya (Fig. 25.2a; Sepulveda-Villet and Stepien 2012).

The Atlantic coastal area supports high species richness and endemism today (Griffiths 2010), as discerned for its yellow perch and walleye populations (Stepien et al. 2009, 2010; Sepulveda-Villet and Stepien 2012; Haponski and Stepien 2014a). These populations have relatively high genetic diversity, possessing unique alleles (see Table 25.2). Results support the hypothesis that greater genetic diversity in southerly, unglaciated populations may be due to their long undisturbed history for evolution and local adaptation (Petit et al. 2003). The South Atlantic coastal yellow

Table 25.3 F_{ST} (Weir and Cockerham 1984) genetic divergences between pairs of sampling populations for yellow perch (below diagonal) and walleye (above diagonal), based on (a) nuclear DNA microsatellite loci and (b) mtDNA control region sequence data using FSTAT v2.9.3.2 (Goudet 1995, 2002; http://www2.unil.ch/popgen/softwares/fstat.htm) and ARLEQUIN v3.1.5.3 (Excoffier and Lischer 2010; http://cmpg.unibe.ch/software/arlequin35/) and significance from 100,000 replicates. Results are congruent to those from exact tests of differentiation comparisons (data not shown). Note that spawning populations are grouped together into regions for purpose of comparison, thus please consult the original papers to examine fine-scale patterns. **Bold** = significant with sequential Bonferroni correction (Rice 1989), *italics* = significant at 0.05 prior to Bonferroni correction. Not bold, not italics = not significant (Results modified from Stepien et al. 2009, 2010, 2012; Sepulveda-Villet and Stepien 2011, 2012; Haponski and Stepien 2014a)

a. Microsatellite divergences

Location	1.	2.	3.	4.	5.	6.	7.	8.	9.	10.	11.	12.
1. Lake Winnipeg	–	**0.097**	**0.067**	**0.121**	**0.130**	**0.137**	**0.133**	**0.124**	**0.114**	**0.146**	**0.121**	**0.250**
2. Upper Mississippi R.	**0.202**	–	**0.102**	**0.162**	**0.165**	**0.134**	**0.130**	**0.125**	**0.133**	**0.173**	**0.136**	**0.300**
3. Lake Superior	**0.116**	**0.202**	–	**0.037**	**0.035**	**0.035**	**0.035**	**0.029**	**0.038**	**0.101**	**0.033**	**0.166**
4. Lake Michigan	**0.137**	**0.228**	**0.142**	–	**0.015**	**0.037**	**0.037**	**0.037**	**0.028**	**0.145**	**0.026**	**0.143**
5. Lake Huron	**0.133**	**0.183**	**0.128**	0.030	–	**0.022**	**0.023**	**0.024**	**0.027**	**0.147**	**0.026**	**0.155**
6. Lake St. Clair	**0.217**	**0.320**	**0.226**	**0.122**	**0.117**	–	0.000	0.005	**0.023**	**0.146**	**0.015**	**0.169**
7. Lake Erie, western basin	**0.217**	**0.271**	**0.198**	**0.074**	0.055	**0.147**	–	0.005	**0.023**	**0.145**	**0.016**	**0.173**
8. Lake Erie, eastern basin	**0.218**	**0.274**	**0.202**	**0.079**	0.053	**0.145**	**0.014**	–	**0.027**	**0.127**	**0.011**	**0.153**
9. Lake Ontario	**0.244**	**0.353**	**0.213**	**0.125**	**0.111**	**0.093**	**0.118**	**0.111**	–	**0.141**	**0.021**	**0.164**
10. Northeastern populations	**0.204**	**0.246**	**0.133**	**0.166**	**0.148**	**0.222**	**0.185**	**0.188**	**0.179**	–	**0.121**	**0.293**
11. Southeastern populations	**0.250**	**0.279**	**0.169**	**0.222**	**0.203**	**0.259**	**0.273**	**0.274**	**0.231**	**0.117**	–	**0.144**
12. Gulf coastal region	**0.290**	**0.361**	**0.237**	**0.273**	**0.251**	**0.320**	**0.317**	**0.320**	**0.294**	**0.180**	**0.186**	–

b. MtDNA control region sequence divergences

Location	1.	2.	3.	4.	5.	6.	7.	8.	9.	10.	11.	12.
1. Lake Winnipeg	–	**0.176**	**0.068**	**0.162**	**0.128**	**0.160**	**0.145**	**0.168**	**0.362**	0.007	**0.335**	**0.416**
2. Upper Mississippi River	**0.552**	–	**0.359**	0.000	*0.076*	**0.257**	**0.244**	**0.314**	**0.551**	*0.226*	**0.547**	**0.793**
3. Lake Superior	**0.522**	0.000	–	**0.332**	**0.261**	**0.182**	**0.171**	**0.173**	**0.339**	*0.077*	**0.337**	**0.415**
4. Lake Michigan	**0.191**	**0.087**	**0.077**	–	0.043	**0.220**	**0.208**	**0.275**	**0.513**	**0.191**	**0.510**	**0.732**
5. Lake Huron	**0.081**	**0.107**	0.096	**0.119**	–	**0.118**	**0.117**	**0.188**	**0.369**	**0.135**	**0.433**	**0.561**
6. Lake St. Clair	**0.392**	0.000	0.000	0.027	0.045	–	0.000	**0.014**	**0.125**	**0.194**	**0.247**	**0.344**
7. Lake Erie, western basin	**0.395**	0.012	0.008	**0.037**	**0.205**	0.026	–	0.000	**0.164**	**0.179**	**0.235**	**0.327**
8. Lake Erie, eastern basin	**0.490**	0.006	0.010	**0.059**	**0.177**	0.044	0.002	–	**0.183**	**0.211**	**0.211**	**0.288**
9. Lake Ontario	**0.399**	0.050	0.036	0.054	0.069	0.029	0.011	0.004	–	**0.439**	**0.381**	**0.494**
10. Northeastern populations	**0.249**	**0.282**	**0.266**	**0.322**	**0.189**	**0.197**	**0.455**	**0.404**	**0.225**	–	**0.412**	**0.541**
11. Southeastern populations	**0.501**	**0.548**	**0.531**	**0.584**	**0.433**	**0.468**	**0.701**	**0.659**	**0.492**	**0.215**	–	**0.418**
12. Gulf coastal region	**0.895**	**0.949**	**0.943**	**0.914**	**0.537**	**0.909**	**0.951**	**0.956**	**0.915**	**0.240**	**0.436**	–

perch populations are adapted to mesohaline conditions, and likely can readily migrate from fresh to brackish waters (Grzybowski et al. 2010). In response to ongoing climate change, the unique genetic diversity of these euryhaline populations of yellow perch might provide an important genetic reservoir in the event that some inland waters become more saline.

The South Atlantic and Gulf coastal haplotypes of yellow perch are more closely related to each other than to those from the North Atlantic region (Sepulveda-Villet and Stepien 2012).

In comparison, the southern Gulf relict populations of yellow perch and walleye have relatively lower heterozygosity (Table 25.2), characteristic of their small population sizes, bottlenecks, and genetic drift. Those populations also possess high numbers and proportions of private alleles, indicative of their long-term isolation and distinctiveness (Tables 25.2 and 25.3; Figs. 25.1, 25.2, 25.3, and 25.4). The Gulf coastal walleye population is small but persistent, and is believed to represent a long-isolated unique historic strain (Boschung and Mayden 2004; Stepien et al. 2009). This also appears to be the case for yellow perch, except that there is closer relationship of its relict Gulf coastal population to the southeast Atlantic seaboard (Sepulveda-Villet et al. 2009; Sepulveda-Villet and Stepien 2012). The distinctiveness of the Gulf Coast walleye was described by Hackney and Holbrook (1978) based on life history characters, faster growth, and spawning at higher temperatures. This differentiation also was indicated by population genetic data using allozymes (Murphy 1990; Billington and Maceina 1997), mtDNA restriction haplotypes (Billington et al. 1992; Billington and Strange 1995; Billington and Maceina 1997), nuclear microsatellite loci (Stepien et al. 2009), and mtDNA sequence data (Haponski and Stepien 2014a). Boschung and Mayden (2004) noted that this unique walleye strain is in danger of potential introgression with introduced northern strains of walleye entering from the Tennessee River drainage via the Tennessee-Tombigbee Waterway.

25.7 Comparative Genetic Diversity of Yellow Perch and Walleye Populations

The overall genetic diversity of yellow perch is much lower than walleye; this is clearly evident in Tables 25.1 and 25.2, and is true for both nuclear DNA (mean heterozygosity is 0.53 for yellow perch and 0.73 for walleye) and mtDNA sequence variability (mean haplotypic diversity is 0.31 for yellow perch and 0.77 for walleye). The relatively low diversity for yellow perch likewise has been revealed by other genetic data sets, including allozymes (Leary and Booke 1982; Todd and Hatcher 1993; Moyer and Billington 2004), mtDNA restriction fragment length polymorphisms (RFLPs) (Billington 1993; Moyer and Billington 2004), mtDNA sequences (Sepulveda-Villet et al. 2009; Sepulveda-Villet and Stepien 2011, 2012), as well as nuclear microsatellites (Miller 2003;

Sepulveda-Villet and Stepien 2011, 2012). Values for yellow perch here (Table 25.2) correspond to the overall microsatellite heterozygosity average of 0.54 for 13 other freshwater fishes described by DeWoody and Avise (2000), whereas those for walleye are high.

Overall mtDNA genetic diversity of yellow perch roughly matches that of the European perch (Refseth et al. 1998; Nesbø et al. 1998, 1999), which also exhibits relatively low allozymic genetic diversity (Gyllensten et al. 1985; Bodaly et al. 1989). Relatively low genetic diversity in both mtDNA and nuclear DNA thus appears characteristic of the genus *Perca*. Moreover, mtDNA control region sequences revealed low variation in the ruffe *Gymnocephalus cernua* across Eurasia (Stepien et al. 1998, 2005), which is closely related to *Perca* (see phylogenies of Faber and Stepien 1997; Song et al. 1998; Sloss et al. 2004, and Chap. 1 of this book). This finding appears to suggest that modest genetic diversity may be characteristic of the *Perca-Gymnocephalus* lineage. Such low diversity may be a product of the life history of *Perca* species, with the yellow perch displaying considerable genetic divergence among closely-spaced reproductive groups within an aquatic system, such as large lakes (Sepulveda-Villet and Stepien 2011; Kocovsky et al. 2013; Sullivan and Stepien 2014, 2015).

Some individual yellow perch spawning groups possess a relatively high degree of kin relationship, which may result in lower diversity within samples (Sullivan and Stepien 2015). Genetically similar individuals of European perch have been found to aggregate with one another (Gerlach et al. 2001), recognizing their relatives via olfactory cues at the fry life stage and beyond (Behrmann-Godel et al. 2006). Kin recognition and olfactory cues have not yet been studied in yellow perch or walleye, but might yield important insights on their fine-scale population structure and the distribution of their respective diversities.

Across most of their North American ranges, yellow perch and walleye populations exhibit relatively consistent levels of genetic variability for the nuclear microsatellite data (Table 25.2 and Fig. 25.4). Some exceptions are that populations of both species are less variable in some of the northwest populations and the Gulf coastal region. Genetic diversities for both species are high across the Great Lakes. The southeastern populations of both species, which were never glaciated, also have high diversity.

Levels of mitochondrial DNA diversity, measured as haplotypic diversity (Table 25.2) are more subject to bottlenecks as the effective population size is one fourth that of nuclear DNA (see Avise 2004). Values of haplotypic diversity of walleye are similar to those obtained from the nuclear microsatellite data, however, those for yellow perch are much lower. This may reflect a history of bottlenecks for yellow perch. Most of the Great Lakes, as well as the northwestern populations, are dominated by a single yellow perch haplotype (haplotype 1 of Fig. 25.5). Likewise, the southern populations possess few haplotypes. In contrast, the eastern and Atlantic coastal yellow perch populations have many more haplotypes. By comparison, the diversity of walleye haplotypes is much greater across the range and is more consistent (Table 25.2 and Fig. 25.5).

Private alleles are those that are found only in a specific population or set of populations. In the Great Lakes region overall, 14 % of nuclear microsatellite alleles were private for yellow perch and 22 % for walleye (Table 25.2). For the mitochondrial DNA sequences, just 7 % of the Great Lakes' haplotypes were private in yellow perch, whereas 59 % were private in walleye. In Lake Erie, 9 % of the microsatellite alleles and 3 % of the mitochondrial haplotypes were private in yellow perch; in contrast, only 1 % of the microsatellite alleles and 50 % of the haplotypes were private for walleye. There thus appears to be a fundamental difference in the isolation and differentiation of the Lake Erie populations between the two species. In the Gulf Costal populations, 7 % of yellow perch microsatellite alleles were private, and 11 % of the walleye. Those proportions of mitochondrial haplotypes were 13 % for yellow perch and 50 % for walleye; thus, the trend was similar (Table 25.2 and Fig. 25.5).

25.8 Broadscale Genetic Divergence Patterns of Yellow Perch and Walleye Populations Across Their Native Ranges

Yellow perch and walleye congruently show greatest divergences between their populations from the upper Mississippi River and the Gulf Coast, reaching $F_{ST}=0.361$ for yellow perch and 0.300 for walleye for the microsatellite data (Table 25.3a). The greatest difference for the mitochondrial DNA data likewise occurs between the same population pairs: $F_{ST}=0.949$ for yellow perch and 0.793 for walleye (Table 25.3b). The largest divergences overall thus reflect long-term geographic separation. Most population group comparisons significantly differ, except between some adjacent systems (Table 25.3). Some of these apparent exceptions, however, actually differ at finer scales because here we grouped together different spawning groups within systems, for the purpose of comparing across large areas.

Pronounced genetic demarcations that delineate the most unique yellow perch populations identify six major geographic regions: Northwest Lake Plains, Great Lakes watershed, Lake Champlain, North Atlantic coastal, South Atlantic coastal, and Gulf coastal (Fig. 25.1b). A similar pattern is apparent for the genetic structure of walleye populations: Northwest Lake Plains (Lake Winnipeg, McKim Lake in Ontario, and the upper Mississippi River), the Great Lakes watershed (divided into six groups: Lake Superior, Lakes Michigan/Huron, Lake Huron's Georgian Bay, Lake St. Clair, Lake Erie, and Lake Ontario), North Atlantic coastal, South Atlantic coastal, and Gulf coastal groups (Stepien et al. 2009, 2010; see Fig. 25.1).

The most divergent population groups overall for yellow perch are from: the Gulf coast (mean $F_{ST}=0.275$ among 11 pairwise comparisons), the upper Mississippi River (0.257), the southeast (0.224), and then Lake Winnipeg (0.203). Three of these also are the most different with mtDNA, in order of: the Gulf Coast (mean $F_{ST}=0.786$), the southeast (0.506), Lake Winnipeg (0.424), and then Lake Erie

(0.255). According to the walleye microsatellite data, the most divergent groups likewise occur on the peripheries of the range, including: the Gulf Coast (mean $F_{ST}=0.275$), the northeast (0.153), the upper Mississippi River (0.148), and Lake Winnipeg (0.131). Two of these match those most divergent discerned by walleye mtDNA, comprising: the Gulf Coast (mean $F_{ST}=0.484$), the southeast (0.470), Lake Ontario (0.346), and the upper Mississippi River (0.322).

In the northwest, yellow perch from the Lake Winnipeg region comparatively are more different from the populations in the upper Mississippi River system and Lake Superior than are walleye; this pattern is congruent between the nuclear and mitochondrial DNA data sets (Table 25.3). However, walleye diverge more between the upper Mississippi River and Lake Superior systems than do yellow perch populations. Within the Great Lakes, the Lake Superior population samples are the most differentiated from others for both species. The greatest overall differences among the Great Lakes population groups are some of the most geographically distant ones, notably between Lakes Superior and Ontario ($F_{ST}=0.213$ for yellow perch and 0.038 for walleye microsatellites), as is predicted by the hypothesis of genetic isolation with geographic distance. This relationship between genetic and geographic distances (measured by nearest waterway) is illustrated on Fig. 25.6. Another very distant relationship occurs between Lake Superior and St. Clair yellow perch ($F_{ST}=0.226$). Walleye from Lake Superior also are very divergent from those in Lakes Michigan, Huron, St. Clair, and Erie ($F_{ST}=0.035$–0.037), and those in Lake Michigan are markedly distinct from the Lakes St. Clair and Erie populations (0.037).

A positive regression relationship of genetic versus geographic distance is supported for both species, as shown in Fig. 25.6 from the microsatellite data. Yellow perch have greater divergence per geographic distance than do walleye. Genetic BARRIER analyses, depicted in Fig. 25.1b (yellow perch) and 25.1c (walleye), denote divisions among populations that are distinguished by much greater than expected genetic differences. Several of these are congruent between the species, including those isolating the Gulf coastal region, the southeast, the Lake Winnipeg region, the upper Mississippi River, Lake Superior, and the northeastern populations. Significant demarcation of the Lake St. Clair population of yellow perch is not found in walleye, which shows more genetic exchange with nearby spawning groups (Fig. 25.1). The Bayesian STRUCTURE analyses support a greater number of distinctive population groups of yellow perch ($K=17$, Fig. 25.4a) than found for walleye ($K=9$, Fig. 25.4b). Yellow perch and walleye share many congruent population areas that are denoted by marked distinctiveness, including Lake Winnipeg, the upper Mississippi River, Lake Superior, the northeastern populations, the southeastern populations, and the Gulf Coast. Additional unique groups of yellow perch occur in Lake St. Clair (congruent with the BARRIER analysis results from Fig. 25.1b), Lake Ontario, and the Hudson River (Fig. 25.4a). Findings thus demonstrate considerable genetic divergences among most population regions for both species (Table 25.3), reflecting both broad- and fine-scale patterns of differentiation.

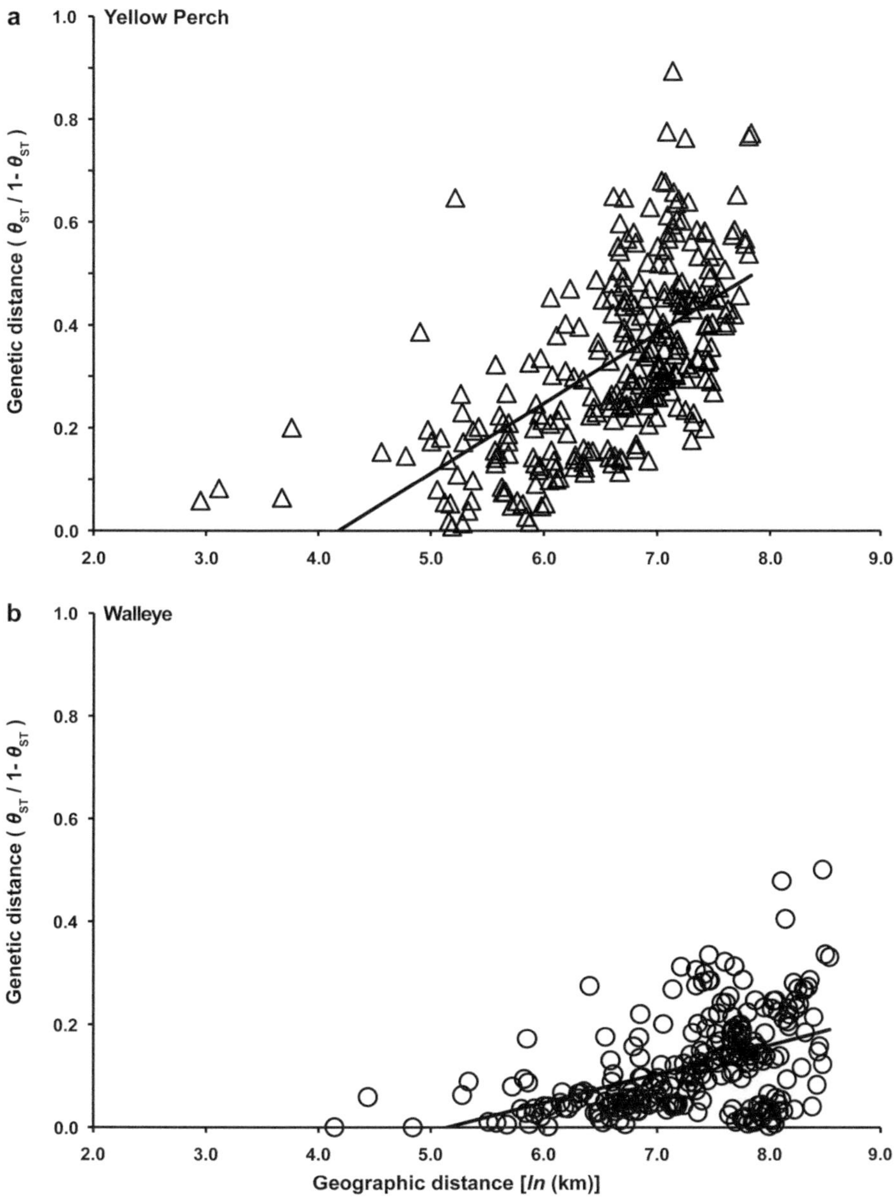

Fig. 25.6 Mantel (1967) pairwise tests for the relationship between genetic distance (θ_{ST}/1- θ_{ST}) and the natural logarithm of geographic distance (km) across the native North American range of (**a**) yellow perch and (**b**) walleye. Results determined using GENEPOP v4.0 (Rousset 2008; http://kimura.univ-montp2.fr/%7Erousset/Genepop.htm), with 10,000 permutations. Equations for: (**a**) yellow perch populations. $p<0.001$, $R^2=0.39$, $y=0.14x - 0.57$ (Modified from Sepulveda-Villet et al. 2012), and (**b**) walleye populations. $p=0.005$, $R^2=0.23$, $y=0.06x - 0.29$ (Adapted from Stepien et al. 2009; Haponski and Stepien 2014a)

25.9 Fine Scale Genetic Divergence Patterns of Yellow Perch and Walleye

Although relationships among yellow perch and walleye populations typically follow a broad-scale pattern of genetic isolation by geographic distance (Fig. 25.6), those among reproductive groups within individual lakes do not reflect geographic distance (Fig. 25.7). In both species, some closely situated spawning populations are markedly different, whereas others are more closely related. Fine-scale relationships among yellow perch and walleye reproductive groups within Lake Erie appear to be driven by spawning aggregations, natal homing behavior, and localized adaptations, rather than due to simple geographic connectivity (see Stepien et al. 2009, 2012; Sepulveda-Villet et al. 2011, 2012) (Fig. 25.8).

Sepulveda-Villet and Stepien (2011) found significant differences at 15 microsatellite loci among Lake Erie yellow perch reproductive populations (shown in Fig. 25.9), discerning no relationship between genetic distance and geographic distance between sampling locations. Kocovsky and Knight (2012) reported similar trends using morphometric data from yellow perch sampled from many of the same spawning locations used by Sepulveda-Villet and Stepien (2011). Yellow perch reproductive populations in the central basin of Lake Erie that are separated by 17–94 km were distinguished by significant genetic divergences of $F_{ST}=0.016$–0.056 using the same 15 loci, and also displayed significant morphological differences (Kocovsky et al. 2013). Similar fine-scale differentiation also was evident among yellow perch reproductive groups in the St. Lawrence River system, which comprised four distinct genetic clusters along a 310 km-long corridor (LeClerc et al. 2008). Grzybowski et al. (2010) described fine-scale genetic structure between yellow perch spawning in Lake Michigan open water versus those in Green Bay, also using microsatellite data ($F_{ST}=0.126$).

Relatively large genetic separations likewise delineated some walleye reproductive populations located in close proximity, including between the Moon and Musquash Rivers (site l) in Georgian Bay of Lake Huron ($F_{ST}=0.034$), between the Thames (m) and the Detroit Rivers (n) in Lake St. Clair ($F_{ST}=0.012$) and among spawning locations in eastern Lake Erie (sites x–aa; mean $F_{ST}=0.036$, range = 0.034–0.058; Stepien et al. 2009, 2010). Conversely, walleye spawning aggregations along the southern shore in western and central Lake Erie (q–w), which are the largest in population numbers, were linked by more connectivity and gene flow (discussed in detail by Strange and Stepien 2007; Stepien et al. 2012). This connectivity also was described by other researchers using a variety of genetic techniques (Merker and Woodruff 1996; Stepien and Faber 1998; Strange and Stepien 2007). A study by McParland et al. (1999) using mtDNA RFLPs and allozymes found no differentiation between walleye reproducing at Chickenolee Reef (site t on Fig. 25.9) and the Huron River (site q) in western Lake Erie, but this comparison differed significantly when higher-resolution microsatellites were used (Stepien et al. 2009, 2010). The latter data revealed more site-specific differentiation among walleye spawning populations, which was greater in eastern Lake Erie.

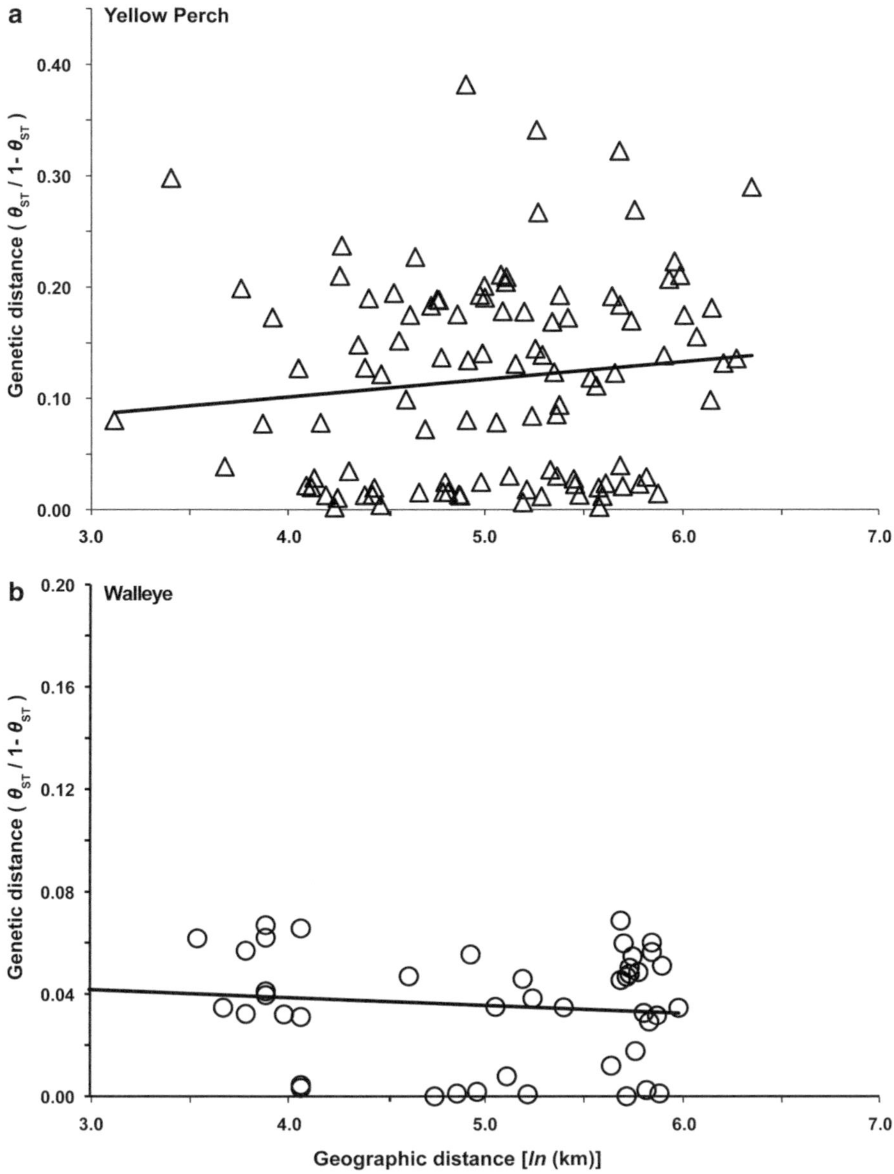

Fig. 25.7 Mantel (1967) pairwise tests in GENEPOP v4.0 (Rousset 2008; http://kimura.univ-montp2.fr/%7Erousset/Genepop.htm), with 10,000 permutations, for the relationship between genetic distance (θ_{ST}/1- θ_{ST}) and natural logarithm of geographical distance (km) across Lake Erie spawning groups of (**a**) yellow perch ($p=0.212$, $R^2=0.024$, $y=0.016x-0.038$; Sepulveda-Villet and Stepien 2011), and (**b**) walleye ($p=0.827$, $R^2=0.015$, $y=-0.003x-0.051$; Strange and Stepien 2007)

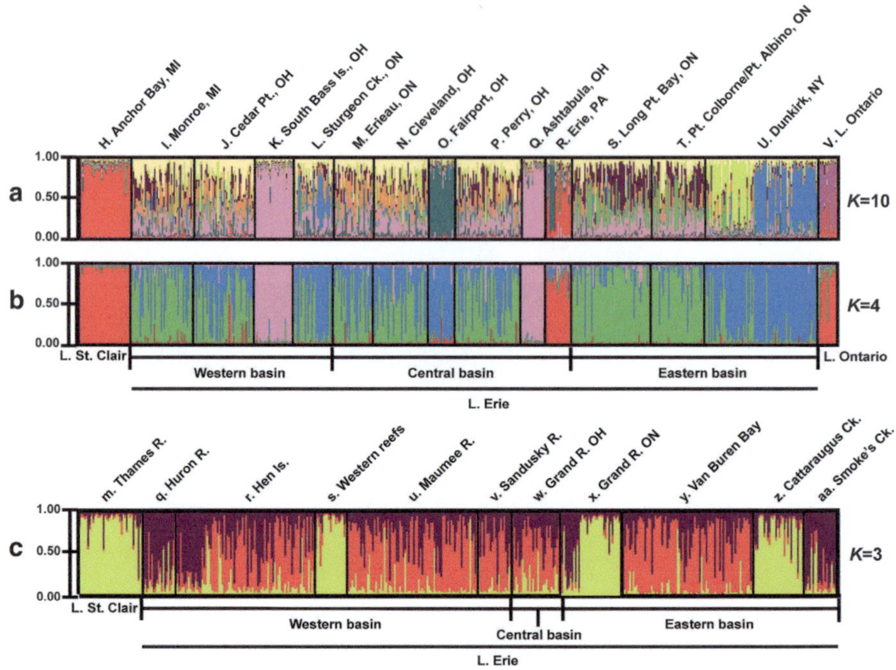

Fig. 25.8 Estimated population structure from Bayesian STRUCTURE v2.3.3 analyses (Pritchard et al. 2000; Pritchard and Wen 2004; http://pritchardlab.stanford.edu/structure.html) for Lake Erie spawning groups of yellow perch (**a**) $K=10$ and (**b**) $K=4$ four (Adapted from Sepulveda-Villet and Stepien 2011) and walleye (**c**) $K=3$ (Adapted from Strange and Stepien 2007); in reference to outlying populations from Lake St. Clair and L. Ontario (the latter for yellow perch only). Analyses were run with 100,000 burn-in and 500,000 replicates. Optimal K values were determined by posterior probabilities (Pritchard et al. 2000) and the ΔK method of Evanno et al. (2005). *Thin vertical lines*, partitioned into colored segments, represent individual fish. *Black lines* separate spawning groups from different locations. Note that there is no correspondence between the colors and haplotypes of yellow perch and walleye (**a** + **b** versus **c** are entirely independent)

Such differentiation among reproducing populations groups within a continuous system appears to result from spawning site philopatry to specific natal locations, maintained from generation to generation. European perch form long-term population groups of full and half siblings, according to microsatellite data (Bergek and Björklund 2007; Behrmann-Godel and Gerlach 2008). Reproductive success was significantly lower in non-kin groups, with reduced pre-zygotic and post-zygotic fitness manifested by lower fertilization rates and less hatching success (Behrmann-Godel and Gerlach 2008). One of the likely barriers to gene flow for European perch thus is reproductive isolation, either via kin recognition using olfactory cues (Gerlach et al. 2001) or due to reduced hybrid fitness between sympatric but divergent cohorts (Behrmann-Godel and Gerlach 2008). Likewise, it is possible that yellow perch and walleye returning to their natal locations are guided

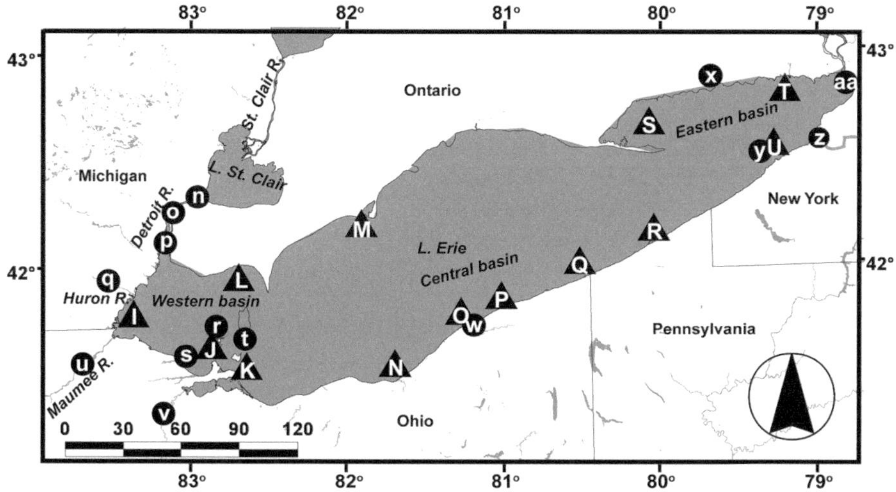

Fig. 25.9 Fine-scale map of Lake Erie showing locations of spawning groups evaluated for yellow perch (*triangles*, with capital letters) and walleye (*circles*, with lowercase letters). Yellow perch sampling sites are: I. Monroe, MI; J. Cedar Pt., OH; K. S. Bass Isl., OH; L. Sturgeon Ck., ON; M. Erieau, ON; N. Cleveland, OH; O. Fairport, OH; P. Perry, OH; Q. Ashtabula, OH; R. Erie, PA; S. Long Pt. Bay, ON; T. Pt. Colborne, ON; U. Dunkirk, NY (Modified from Sepulveda-Villet and Stepien 2011, 2012). Sites for walleye include: n. Belle Isl.; o. Fighting Isl.; p. Grosse Ile; q. Huron R.; r. Hen Isl.; s. Western Reefs; t. Chickenolee Reef; u. Maumee R.; v. Sandusky R.; w. Grand R. OH; x. Grand R. ON; y. Van Buren Bay; z. Cattaraugus Ck.; aa. Smoke's Ck (Modified from Strange and Stepien 2007; Stepien et al. 2009, 2010, 2012; Haponski and Stepien 2014a)

by olfactory information imprinted during early stages of their life history. If so, it may be the primary mechanism for maintaining divergence among spawning aggregations, but this hypothesis remains to be tested.

There is no effect of gender in the establishment of these fine-scale genetic structure trends, as both male and female yellow perch and walleye have analogous genetic patterns, and thus appear to have similar site fidelity (Stepien and Faber 1998; Stepien et al. 2009, 2012; Sepulveda-Villet et al. 2011, 2012). Eight yellow perch populations from Lake Erie locations (sites on Fig. 25.9) were all genetically distinguishable from one other (mean $F_{ST}=0.068\pm0.008$, range=0.002–0.168), but some also varied in allelic composition between two sampling time periods (2001–2004 versus 2009), at ~1/4 the magnitude of the difference among locations. Sullivan and Stepien (2015) found significant differences among yellow perch spawning groups and between sampling years at some of these sites. An example of annual variation within the yellow perch reproductive group sampled at Van Buren Bay (site y on Fig. 25.9) in eastern Lake Erie is given in Table 25.4a. A study by Demandt (2010) likewise found significant variations in microsatellite allelic frequencies of European perch among sampling years for a population in Sweden. This suggests a similar trend for annual variability at spawning sites for both European and yellow perch.

Yellow perch spawning groups varied among individual sampling years and age cohorts, with the 2003 cohort being the most distinctive of those sampled (Sepulveda-

Table 25.4 Pairwise genetic divergences for (a) spawning yellow perch from Dunkirk NY, eastern Lake Erie sampled in six different collection years: 1985, 2001, 2004, 2008, 2009, and 2010 (Modified from Sullivan and Stepien 2015) and (b) Walleye reproducing in the Maumee River sampled from five collection years (1995, 1998, 2003, 2007, and 2008; reported by Stepien et al. 2012). Calculations are from FSTAT v2.9.3.2 (Goudet 1995, 2002; http://www2.unil.ch/popgen/softwares/fstat.htm) and ARLEQUIN v3.1.5.3 (Excoffier and Lischer 2010; http://cmpg.unibe.ch/software/arlequin35/) and used 100,000 replicates to test for significance. **Bold** = significantly different following sequential Bonferroni correction (Rice 1989)

a. Yellow perch spawning at Dunkirk NY, eastern Lake Erie

Year (*N*)	1985 (34)	2001 (37)	2004 (48)	2008 (30)	2009 (30)
2001 (37)	**0.037**	–			
2004 (48)	**0.056**	**0.055**	–		
2008 (30)	**0.125**	**0.138**	**0.141**	–	
2009 (30)	**0.041**	**0.072**	**0.088**	**0.105**	–
2010 (36)	**0.011**	**0.041**	**0.057**	**0.113**	**0.014**

b. Walleye spawning in the Maumee River, western Lake Erie

Year (*N*)	1995 (53)	1998 (28)	2003 (76)	2007 (5)
1998 (28)	0.001	–		
2003 (76)	0.001	0.002	–	
2007 (43)	0.001	0.002	0.001	–
2008 (50)	0.002	0.004	0.001	0.001

Villet and Stepien 2011, 2012; Sullivan and Stepien 2015). This 2003 cohort was an especially large and successful group for both yellow perch and walleye recruitment in Lake Erie (WTG 2014, YPTG 2014). Reproductive groups of yellow perch contained high numbers of full siblings (mean = 18.5 %, ranging to 75 % for the 2001 age cohort spawning at Van Buren Bay in eastern Lake Erie; Sullivan and Stepien 2015). Temporal genetic divergence at reproductive locations was not explained by genetic isolation over time, but appeared due to yellow perch spawning in kin-related groups that varied slightly from year to year. Spatial patterns were attributed to limited migration and natal homing, whereas temporal patterns may reflect kin group structuring and differential reproductive success.

In contrast to yellow perch, walleye from three of the largest spawning populations in Lake Erie (Maumee River; site u on Fig. 25.9, Sandusky River, site v, and Van Buren Bay reefs, site y), exhibited both temporal and spatial consistency across 14 years (collected in 1995, 1998, 2003, 2007, and 2008) using nine nuclear DNA microsatellite loci (Stepien et al. 2012). There was overall year-to-year genetic consistency within walleye reproductive groups; no significant differences were found among collection dates within an annual run, between the sexes, or among age-cohorts. An example demonstrating consistency among walleye runs in the Maumee River is presented in Table 25.4b (Stepien et al. 2012). Overall, walleye spawning at the Van Buren Bay reefs were genetically divergent from those reproducing in the Maumee and Sandusky rivers, reflecting geographic distance; the latter two groups were genetically closer, with slight differences that suggested more recent divergence, higher gene flow, or both. The Van Buren Bay population group in 1995 was the most divergent sample and had the greatest genetic self-assignment (100 %); this might indicate some slight changes in its distinctive genetics over time, as it was the oldest sample examined (Stepien et al. 2012). Results illustrated the importance of sampling over several years of spawning runs to understand overall patterns of population structure, which showed remarkable genetic consistency across an open-lake system for walleye.

25.10 Color Variants and Genetics: The Extinct "Blue Pike" and Turquoise Mucus Walleye and Yellow Perch

Dark steel grey-blue colored walleye reportedly once were common in the deeper cooler eastern basin of Lake Erie and reported from western Lake Ontario. These were called "blue pike" and regarded as either a species (Hubbs 1926) or as a subspecies *S. vitreus* "*glaucus*" (Trautman 1981). Trautman (1981) documented a preponderance of morphological intergrades between the "blue pike" and common "yellow" walleye.

The "blue pike" was reported to vary from "normal" yellow-colored walleye by having a steel-grey blue color, larger eyes located higher on the head, a smaller interorbital distance, and a greater angle between the preopercle bone and branchiostegal rays (Bailey and Smith 1981; Trautman 1981; Hubbs and Lagler 2004). However, all of these characters – including color – overlapped extensively with those of the abundant and widespread "yellow" walleye, which was sympatric throughout Lakes Erie and Ontario (Scott and Crossman 1973; Trautman 1981).

The "blue pike" was reported to mostly inhabit deeper cooler waters of eastern Lake Erie, but also was caught in western Lake Erie (Trautman 1981). Spawning of the "blue pike" was reported to occur somewhat later and in deeper areas than other walleye (Stone 1948). The "blue pike" and "yellow" walleye shared a popular commercial fishery, with the "blue pike" collapsing in 1959 – attributed to exploitation, pollution, and/or habitat alteration – with walleye numbers declining concurrently (Trautman 1981). The "blue pike" had disappeared by the early 1960s (Hubbs and Lagler 2004), and officially was declared extinct in 1983 by the US Fish and Wildlife Service (Noecker 1998).

The "blue pike's" popular saga has been confused by the facts that (1) it is not a member of the pike family (i.e., is not in the family Esocidae, but is in the Percidae), it apparently had no distinctive characters, including color (Trautman 1981; Haponski and Stepien 2014a), and (3) walleye in northern waters (the Canadian Shield Lakes) frequently are bright turquoise blue in color from a protein in the mucus, termed sandercyanin (Yu et al. 2008). The turquoise-mucus colored-walleye do not match the deep steel-grey blue color or other morphological attributes of the "blue pike" (Stone 1948; Scott and Crossman 1973; Trautman 1981; Campbell 1987). The turquoise mucus walleye are reported to have a combination of sandercyanin pigment and lack of yellow xanthophores (Yu et al. 2008).

Both turquoise-mucus and yellow-colored walleye occur sympatrically in the same water bodies in Canada, with both secreting the blue mucus (Paradis and

Magnan 2005; Yu et al. 2008). Some of the turquoise blue color typically "rubs off" when the fish is collected (Yu et al. 2008, CAS, personal observation). The turquoise mucus additionally has been reported on some yellow perch in those northern waters (Trautman 1981; Haponski and Stepien 2014a). It has been hypothesized that the production of the turquoise mucus and sandercyanin in the Canadian Shield lakes is a response to ultraviolet light levels (Yu et al. 2008).

A recent study by Haponski and Stepien (2014a) collected new morphological and genetic data from the original "blue pike" specimens named and described by Hubbs (1926). They examined all morphological and meristic characters that had been alleged to be diagnostic. They also sequenced the mtDNA control region and analyzed variation at nuclear DNA microsatellite loci from the "blue pike" specimens, common "yellow" walleye from the same period, and contemporary walleye across North America from 23 spawning sites. The findings showed that the "blue pike" was indistinguishable from common yellow-colored walleye collected from that same period in Lake Erie. It is possible that the "blue pike" comprised one or more reproductive groups in eastern Lake Erie, but there is no evidence to support any greater distinctiveness than found in other walleye spawning groups.

Haponski and Stepien (2014a) also examined the genetic characters of typical "yellow" walleye and turquoise-colored mucus walleye that were collected together in McKim Lake, Ontario Canada using mtDNA control region sequence data and nine nuclear microsatellite loci. They found no genetic differences between fish of the two colors. Moreover, the McKim Lake samples showed no population genetic relationship to the extinct "blue pike" paratypes from Lake Erie (the same individuals that Carl Hubbs had named as "blue pike" Hubbs 1926). The turquoise mucus and "yellow" walleye from McKim Lake also did not show a population genetic origin shared with walleye from Lakes Erie or Ontario (see Figs. 25.4b and 25.5b). Stepien and Faber (1998) likewise analyzed several assorted turquoise mucus walleye from a variety of Canadian Shield lakes using entire mtDNA control region sequences and found no genetic distinction from the normal variation range of walleye.

Paradis and Magnan (2005) morphologically compared sympatric yellow and turquoise mucus walleye in five Canadian Shield lakes near Quebec, reporting shorter head lengths and smaller interorbital distances in the latter. Laporte et al. (2011) alleged slight genetic difference between turquoise mucus and yellow walleye populations sampled within a lake using AFLP (amplified fragment length polymorphism) markers and assignment tests, but had no diagnostic alleles and their genetic distance analyses lacked significant bootstrap support. They reported that their assignments indicated that they could respectively genetically diagnose either the yellow or turquoise mucus type within a lake, but could not assign them as a whole when more than one lake was considered. Laporte et al. (2011) also stated that turquoise mucus walleye did not warrant taxonomic recognition as a subspecies. It may be that there are some differentiated walleye variants with color differences within some lakes across the range of the Canadian Shield; many distinctions at the population level are found among walleye spawning groups within Lake Erie and other Great Lakes basins (Strange and Stepien 2007; Stepien et al. 2009, 2010, 2012; Haponski and Stepien 2014a, b).

Occasional steel-grey/blue colored walleye and yellow perch have been reported from Lake Erie (Trautman 1981; Hubbs and Lagler 2004). A recent example is that a Lake Erie commercial fisherman (Jeff Herr, personal communication, 2013) caught a dark steel-grey/blue yellow perch in summer 2013 at Fairport OH and steel-grey/blue walleye recently have been caught in the central and western basins of Lake Erie (AEH personal observation, 2013). A steel-grey/blue walleye individual sampled in the western basin near Sandusky OH had mtDNA haplotype 1 (Haponski and Stepien 2014a), which is the most common walleye haplotype (see Figs. 25.3b and 25.5b). A skin scraping showed no evidence of the turquoise mucus. A recent study by Wayne Schaeffer (personal communication, University of Wisconsin, 2013) found no turquoise mucus or sandercyanin in Lake Erie walleye. Overall, no diagnosable genetic or morphological characters have been found that distinguish historic "blue pike" from walleye, rendering its subspecies status invalid (Haponski and Stepien 2014a). In contrast, many walleye spawning groups genetically differ from one another (Stepien and Faber 1998; Stepien et al. 2009, 2010, 2012; Haponski and Stepien 2014a).

25.11 The Value of Genetic Data for Evaluating Restoration

As molecular-based population dynamics and structure analyses increasingly provide a way to better assess past and present levels of diversity in fish populations, a need for greater use of these techniques has been proposed in concert with traditional management approaches. When demographic data on exploited fisheries are collected on a larger scale than population subunits, valuable data may be lacking for management decisions to conserve local genetic and morphological diversity and adaptedness, as appears to be the case for Lake Erie yellow perch (Kocovsky et al. 2013). Likewise, a landscape genetics study by LeClerc et al. (2008) described four distinct populations of yellow perch along a 310 km stretch of the St. Lawrence River, from ten microsatellite loci. These results were in contrast with recognized fisheries management units in Quebec. Here we examine use of a combined approach linking fisheries management, conservation genetics, and historical assessment of Great Lakes' walleye populations.

Fish habitats in the Great Lakes and connecting tributaries were subject to extensive and deleterious changes in the twentieth and twenty-first centuries, with loss of wetlands, channelization of major streams, construction of dams, oxygen depletion, shoreline modification, siltation of spawning areas, nutrient enrichment, water-quality deterioration, sand and gravel extraction, and invasive species introductions (Trautman 1981; Bolsenga and Herdendorf 1993; Fielder 2002; Hoff 2002; Ryan et al. 2003). For example, Lake Huron's Saginaw Bay once housed the second-largest walleye fishery in the Great Lakes (Schneider and Leach 1977; Fielder 2002), which collapsed in the 1940s from spawning-habitat degradation and overfishing (Jude and Leach 1999). Similarly, Lake Michigan yellow perch populations underwent extensive declines, particularly in the 1980s and persisting to this day, manifested in erratic recruitment patterns and the dependence on a single year class (recently 2005) for most reproductive effort at spawning grounds in the southern lake (Redman et al. 2013). Lake Erie has the largest walleye populations, which declined throughout the 1960s and the early 1970s due to reduced habitat and water quality, coupled with high exploitation (Regier and Hartman 1973).

International management regulations reduced exploitation and improved environmental conditions, leading to the recovery of Lake Erie walleye during the 1980s and increases to historical abundance levels during the 1990s (Knight 1997). Commercial walleye fishing was banned in Lake Huron's Saginaw Bay in 1969, and a sport fishery supported by stocking came into prominence in the 1980s (Fielder 2002; USFWS/GLFC 2010). Walleye numbers in the St. Louis River of western Lake Superior also increased following river cleanup, which today comprises the largest spawning group in that lake (MacCallum and Selgeby 1987; Hoff 2002; Schram et al. 2010).

A recent study by Haponski and Stepien (2014a) found that genetic diversity of Lake Erie walleye in historic samples from the 1920s to 1940s was much less than in contemporary samples (mtDNA $H_D = 0.05 \pm 0.01$ vs. 0.79 ± 0.00; µsat $H_O = 0.47 \pm 0.06$ vs. 0.72 ± 0.01; compare to Table 25.2), suggesting population recovery after intense pollution and exploitation. The genetic composition, as shown by microsatellite alleles, underwent significant changes ($F_{ST} = 0.336$). This was attributed to Lake Erie's ecological recovery from pollution and increasing fishery regulations, with walleye abundances increasing to ~80 million fish during the 1980s and then declining to ~23 million in 2014 (see WTG 2014). Similarly, Guinand et al. (2003) found that genetic diversity of lake trout populations from Lakes Superior, Michigan, and Huron, based on five microsatellite loci, was less in 1940–1959 ($H_O = 0.47$) than in 1995–1999 (0.51). Likewise, Stott et al. (2013) found that lake whitefish *Coregonus clupeaformis* from Lakes Huron and Erie had lower diversity at seven microsatellite loci in 1927 ($H_O = 0.60$) than in 1997–2005 (0.65). These studies indicate that many fish populations recovered across the Great Lakes during the past decades.

In the past decade, Lake Erie walleye numbers have declined by about 60 % from the 1990s (Locke et al. 2005; WTG 2014). Yellow perch also have declined (YPTG 2014). Understanding and maintaining yellow perch and walleye population structure are critically important fisheries-management goals designated by the Great Lakes Fishery Commission (Ryan et al. 2003; GLFC 2011). Genomics investigations by Bélanger-Deschênes et al. (2013) and Bougas et al. (2013) examined functional polymorphisms of yellow perch from sites having an 85 year history of cadmium (Cd) and nickel (Ni) metal contamination versus clean lakes. Bélanger-Deschênes et al. (2013) discerned two non-synonymous point substitutions involving dissimilar amino acids. The authors suggest that potentially adaptive evolution selected for alleles that may increase perch fitness in polluted environments. Bougas et al. (2013) found that 475 genes had significantly different transcription levels across temperature variations, and 287 and 176 genes were differentially transcribed at different concentrations of Ni and Cd, respectively. Those metals influenced the transcription levels of genes involving iron metabolism, vitamin metabolism, blood coagulation, and calcium transport. These studies foretell the insights to be gained by investigating functional genomic adaptations of populations.

The effect of contaminants such as heavy metals have an additional, and perhaps unexpected effect on percid biology; Azizishirazi et al. (2013) discerned that yellow perch living in metal contaminated lakes had reduced olfactory acuity as compared to those living in clean lakes. Such reduction in detecting olfactory cues might negatively affect kin recognition ability described by Gerlach et al. (2001). Later improvement in water quality and the removal of metallic contaminants resulted in rapid recovery of olfactory capacities in previously impaired yellow perch populations (Azizishirazi et al. 2013), underscoring the benefits of habitat restoration beyond physical or water quality improvements.

Somatic and genetic markers were employed to evaluate the reproductive health of yellow perch populations for which fisheries monitoring revealed reduced recruitment, in urbanized and developed streams of the Chesapeake Bay watershed (Blazer et al. 2013). Results showed gonadal anomalies and changes in DNA integrity in those yellow perch population samples. These findings suggest that pollution can significantly impact reproduction and recruitment, the effects of which can be detected with molecular markers.

Genetic findings to date, as illustrated in this chapter, reveal that most yellow perch and walleye populations have appreciable genetic diversity and significantly differ from other populations, both nearby and distant, despite anthropogenic influences. These diversity and divergence patterns translate to localized adaptations, which merit preservation. Accordingly, we recommend conserving their genetic composition and differentiation patterns by maintaining and restoring spawning habitats, and continued careful management of fisheries.

25.12 Use of Genetic Data to Resolve Questions About Stocking

A literature search recovered no studies of the genetic effects of yellow perch stocking. For the most part, walleye were just occasionally stocked in the Great Lakes, showing little prevalent influence on the genetic structure and diversity of native populations. For example, Gatt et al. (2002) found that mtDNA restriction fragment length polymorphism (RFLP) diversity of walleye reproducing in Lake Huron's Georgian Bay declined across three decades, from $H_D=0.50$ in the 1960s to 0.15 in the 1990s, which they attributed to exploitation and stocking. Haponski and Stepien (2014a) also recovered similarly low mtDNA control region sequence diversity ($H_D=0.15$) for walleye spawning in the Moon/Musquash Rivers of Georgian Bay, showing that mtDNA diversity remains low there today. However, that bottleneck effect was restricted to mtDNA, since average levels of nuclear DNA variability were described for the same samples (Stepien et al. 2009, 2010; Haponski and Stepien 2014a).

Studies have revealed introgression among hatchery stocks and wild populations of walleye in inland lakes from Ontario, Canada (Cena et al. 2006; Walter et al. 2012). Despite a long history of stocking activity (>60 years) for a number of these inland lakes, the overall historical genetic signatures of native populations remain well preserved (Walter et al. 2012). The Grand River, Ontario in northeastern Lake Erie was stocked in the 1980–1990s with tagged adult walleye from the Thames River in Lake St. Clair, in effort to increase population abundance (MacDougall et al. 2007). The endeavor was regarded as unsuccessful as most adults did not remain at the site to spawn and were recaptured to the west (Timmerman 1995), which was attributed to natal homing (see Olson and Scidmore 1962; MacDougall et al. 2007). The stocking effort appeared to have little effect on the genetic composition of that native reproductive population (MacDougall et al. 2007).

Similarly, Garner et al. (2013) found that the native genetic structure of walleye reproducing in Lake Superior's Black Bay was maintained despite large releases of foreign fingerlings (from the St. Marys River) in 2004 and 2005. Although the stocked fingerlings composed 45–71 % of the individuals in their respective age classes, they appeared to have lower reproductive success and utilized different habitats from the native Black Bay walleye. The genetic composition of that spawning population, based on nine nuclear μsat loci (six were the same as used here), was not altered towards the stocked individuals (Garner et al. 2013). In contrast, walleye stocking in Escanaba Lake, Wisconsin substantially changed the genetic composition of the native population between 1952 and 2002, despite retaining consistent diversity levels (mean $H_O = 0.76$) (Franckowiak et al. 2009).

The New York State Department of Environmental Conservation (NYSDEC) stocked Cattaraugus Creek NY in eastern Lake Erie with 2.2 million fry and 44,000 fingerlings from western Lake Erie's Maumee River in 1995–2000. However, a historic group of walleye returns to spawn each spring in territorial waters of the Seneca Nation in Cattaraugus Creek. Walleye are no longer being stocked there, and the stockings were considered unsuccessful (D. Einhouse, New York Department of Environmental Conservation, personal communication 2014). Genetic analyses indicated that the stocking did not appear to genetically affect the native genotypes or diversity of walleye spawning in Cattaraugus Creek (Stepien et al. 2004; Haponski et al. 2014). There was no significant difference between samples pre- versus post-stocking, according to results from nine nuclear microsatellite loci ($F_{ST} = 0.003$) and mtDNA control region sequence data ($F_{ST} < 0.001$); the spawners also significantly differed from the Maumee River broodstock that was introduced ($F_{ST} = 0.090$). These studies indicate that it is important to retain the genetic signature of the historic reproductive groups when stocking percids and other fishes.

25.13 Genetic Patterns in the Face of Climate Change

Temperatures in temperate regimes are predicted to increase over the next 50 years, with those in the Great Lakes region predicted to increase by 5–5.5 °C to become more like today's Gulf Coast (Hayhoe et al. 2010). Warmer temperatures likely will alter growth rates and change maximum sizes and ages of yellow perch and walleye populations (see Carlander 1997). Climate change may disproportionally increase or decrease genetic variability across a taxon's range due to shifts in physical conditions or biological resources (Hewitt 1999; Petit et al. 2003; Hampe and Jump 2011), as occurred during Pleistocene glaciations (Oberdorff et al. 1997; Davis and Shaw 2001; Soltis et al. 2006) and is ongoing today (Araújo and Rahbek 2006; Harris and Taylor 2010).

Cena et al. (2006) found a positive relationship between walleye population heterozygosity and early growth rate, which was correlated with fitness and additionally may be influenced by temperature. Dupont et al. (2007) and Zhao et al. (2008) discerned that population evolutionary origin was a significant differentiating factor in biological characteristics among walleye populations. Bergek et al. (2010) suggested that environmental factors other than geographic distance distinguished among European perch spawning groups, which might have led to differences in their genetic compositions, with the most likely factor being water temperature differences among habitats during the spring reproductive season. Temperature differences may have led to genetic isolation of various spawning populations, with those in shallow waters reproducing earlier (Bergek et al. 2010). Similarly, both walleye and yellow perch were significantly affected by water level fluctuations of glacial lakes in North Dakota, with their highest recorded abundances and body weights occurring during high water periods (Dembkowski et al. 2013), underscoring potential deleterious effects of increased evaporation and water losses linked to climate change.

These findings highlight the importance of spawning habitat and localized variations among their associated reproductive groups. It appears likely that the genetic structure among spawning localities will continue to develop as a product of the interplay between ancestral lineages and environmental variation among reproductive areas, rather than isolation by distance. If this concept holds true, then we should expect effects on genetic diversity and composition from the increasing pace of climate change and higher surface water temperatures with shifting population distributions.

Evaluating diversity and divergence patterns resulting from post-glacial dispersal and adaptation in new environments, and the genetic reservoirs comprising isolated relict populations, may help us to predict the challenges faced by taxa during this era of rapid climate and habitat alterations. Populations along the lower latitudinal fringes of a species' native range likely house valuable genetic adaptations to warmer climates (Hampe and Petit 2005). In effect, global warming patterns rapidly are extending the northward post-glacial expansion trajectory of many taxa; meanwhile their southerly rear-edge groups may experience greater isolation, habitat reduction, and bottlenecks. Moreover, these southern genotypes may move northward, given connection or transport opportunity. For example, the diverse South Atlantic coastal yellow perch populations may prove especially well-adapted to tolerating salinity fluctuations and increasing water temperatures. This may facilitate their northward coastal migration, if sea levels rise to eventually connect low-lying estuaries that are currently isolated by barrier island and sandbar systems. Similarly, the large river walleye populations present in tributaries and main stem of the Mississippi River system may provide unique adaptability and resilience to warmer temperatures, and may undergo a northerly shift. These distributional changes in populations are significant in the context that they may interbreed with long-term native populations in the north. It is possible that native population adaptedness may be either positively or negatively influenced by these changes.

Warming temperatures and increases in storm events may influence fish population structure and overall productivity via biological and climate-related effects, as outlined by Newbrey and Ashworth (2004). For example, Hill and Magnuson (1990) suggested that changes in bioenergetics accompanying climate change might modify growth and prey consumption, thereby affecting food-web dynamics. Shuter and Post (1990) suggested that an increase of 4 °C may shift the distributional limit of yellow perch northward and, depending on lake morphometry and productivity, might also greatly affect survival, relative year-class strength, and ecosystem carrying capacity. To date, populations of yellow perch and walleye possess relatively consistent levels of genetic diversity and high local distinctiveness. These appear to have been maintained despite anthropogenic habitat loss, degradation, fragmentation, and exploitation, likely offset by their large population sizes and the relative abundance of habitats throughout the Great Lakes and other systems.

Genetic structure of today's yellow perch and walleye populations reflects interplay among climatic events, ephemeral waterway connections, population sizes, and likely spawning group philopatry. Delineation of the genomic adaptations that underlie the patterns of genetic diversity and diversity described here will aid predictions of likely response to changing environments, new habitat areas, and exploitation pressures (see Allendorf et al. 2010; Avise 2010). A combined fisheries management and genetics/genomic approach provides a bridge for understanding the unique challenges faced by aquatic taxa due to their constrained dispersal and gene flow via habitat connectivity. Understanding the historical and present day factors that shaped today's populations may aid their continued conservation in the face of future challenges.

Acknowledgements This is publication #2015-07 from the University of Toledo's Lake Erie Research Center. We thank Timothy Sullivan, who completed his master's degree in 2013 from our Great Lakes Genetics Laboratory under CAS for supplying his latest results. Support for our research reported here came from the National Science Foundation NSF GK-12 DGE#0742395, the USEPA #CR-83281401-0, NOAA Ohio Sea Grant R/LR-013, and USDA ARS 3655-31000-020-00D funded to CAS. We greatly appreciate samples supplied by the USGS (Bruce Manny, Patrick Kocovsky, Wendylee Stott, Edward Roseman, Jeffrey Williamson), Ohio DNR (Kevin Kayle, Cary Knight, Roger Knight, Matthew Turner, Jeffrey Tyson, Christopher Vandergoot), Michigan DNR (David Clapp, Robert Haas, Michael Thomas), New York Department of Environmental Conservation (Brian Beckwith, Donald Einhouse), Pennsylvania Fish and Boat Commission (Roger Kenyon), Ontario Ministry of Natural Resources (Christopher Wilson, Timothy Johnson), and from various colleagues including Louis Bernatchez, Neil Billington, Eric Hallerman, Wolfgang Jansen, Brandon Kulik, Christine Mayer, Ellen Marsden, Douglas Nelson, Alex Parker, Webb Pearsall, Lars Rudstam, Wayne Schaefer, Roy Stein, and Matthew White. The manuscript benefitted from careful review by Patrick Kocovsky and Edward Roseman.

References

Aalto SK, Newsome (1990) Additional evidence supporting demic behavior of a yellow perch (*Perca flavescens*) population. Can J Fish Aquat Sci 47:1959–1962

Aldenhoven JT, Miller MA, Corneli PS, Shapiro MD (2010) Phylogeography of ninespine sticklebacks. Ecology 19:4061–4076

Allendorf FW, Hohenlohe PA, Luikart G (2010) Genomics and the future of conservation genetics. Nat Rev Genet 11:697–710

Araújo MB, Rahbek C (2006) How does climate change affect biodiversity? Science 313:1396–1397

Avise JC (2004) Molecular markers, natural history, and evolution, 2nd edn. Sinauer Associates, Sunderland

Avise J (2010) Conservation genetics enters the genomics era. Conserv Genet 11:665–669

Azizishirazi A, Dew WA, Forsyth HL, Pyle GG (2013) Olfactory recovery of wild yellow perch from metal contaminated lakes. Ecotoxicol Environ Saf 88:42–47

Backhouse-James SM, Docker MF (2012) Microsatellite and mitochondrial DNA markers show no evidence of population structure in walleye (*Sander vitreus*) in Lake Winnipeg. J Great Lakes Res 38:47–57

Bailey RM, Smith GR (1981) Origin and geography of the fish fauna of the Laurentian Great Lakes basin. Can J Fish Aquat Sci 38:1539–1561

Barton BA, Barry TP (2011) Reproduction and environmental biology. In: Barton BA (ed) Biology, management, and culture of walleye and sauger. American Fisheries Society, Bethesda, pp 199–231

Behrmann–Gödel J, Gerlach G (2008) First evidence for postzygotic reproductive isolation between two populations of Eurasian perch (*Perca fluviatilis* L.) within Lake Constance. Front Zool 5:1–7

Behrmann-Godel J, Gerlach G, Eckmann R (2006) Kin and population recognition in sympatric Lake Constance perch (*Perca fluviatilis* L.): can assortative shoaling drive population divergence? Behav Ecol Sociobiol 59:461–468

Bélanger-Deschênes S, Couture P, Campbell PG, Bernatchez L (2013) Evolutionary change driven by metal exposure as revealed by coding SNP genome scan in wild yellow perch (*Perca flavescens*). Ecotoxicology 22:938–957

Beletsky D, Mason DM, Schwab DJ, Rutherford ES, Janssen J, Clapp DF, Dettmers JM (2007) Biophysical model of larval yellow perch advection and settlement in Lake Michigan. J Great Lakes Res 33:842–866

Bergek S, Björklund M (2007) Cryptic barriers to dispersal within a lake allow genetic differentiation of Eurasian perch. Evolution 61:2035–2041

Bergek S, Sundblad G, Björklund M (2010) Population differentiation in perch *Perca fluviatilis*: environmental effects on gene flow? J Fish Biol 76:1159–1172

Bernatchez L (1997) Mitochondrial DNA analysis confirms the existence of two glacial races of rainbow smelt *Osmerus mordax* and their reproductive isolation in the St Lawrence River estuary (Quebec, Canada). Mol Ecol 6:73–83

Billington N (1993) Genetic variation in Lake Erie yellow perch (*Perca flavescens*) demonstrated by mitochondrial DNA analysis. J Fish Biol 43:941–943

Billington N (1996) Geographical distribution of mitochondrial DNA (mtDNA) variation in walleye, sauger, and yellow perch. Ann Zool Fenn 33:699–706

Billington N, Maceina MJ (1997) Genetic and population characteristics of walleyes in the Mobile drainage of Alabama. Trans Am Fish Soc 126:804–814

Billington N, Strange RM (1995) Mitochondrial DNA analysis confirms the existence of a genetically divergent walleye population in northeastern Mississippi. Trans Am Fish Soc 124:770–776

Billington N, Barrette RJ, Hebert PDN (1992) Management implications of mitochondrial DNA variation in walleye stocks. N Am J Fish Manag 12:276–284

Blazer VS, Pinkney AE, Jenkins JA, Iwanowicz LR, Minkkinen S, Draugelis-Dale RO, Uphoff JH (2013) Reproductive health of yellow perch *Perca flavescens* in selected tributaries of the Chesapeake Bay. Sci Total Environ 447:198–209

Bodaly RA, Ward RD, Mills CA (1989) A genetic stock study of perch, *Perca fluviatilis* L., in Windermere. J Fish Biol 34:965–967
Bolsenga SJ, Herdendorf CE (1993) Lake Erie and Lake St. Clair handbook. Wayne State University Press, Detroit
Borer S, Miller LM, Kapuscinski AR (1999) Microsatellites in walleye *Stizostedion vitreum*. Mol Ecol 8:336–338
Boschung HT, Mayden RL (2004) Fishes of Alabama. Smithsonian Institution, Washington, DC
Bougas B, Normandeau E, Pierron F, Campbell PGC, Bernatchez L, Couture P (2013) How does exposure to nickel and cadmium affect the transcriptome of yellow perch (*Perca flavescens*) – results from a 1000 candidate-gene microarray. Aquat Toxicol 142–143C:355–364
Bozek MA, Baccante DA, Lester NP (2011a) Walleye and sauger life history. In: Barton BA (ed) Biology, management, and culture of walleye and sauger. American Fisheries Society, Bethesda, pp 233–301
Bozek MA, Haxton TJ, Raabe JK (2011b) Walleye and sauger habitat. In: Barton BA (ed) Biology, management, and culture of walleye and sauger. American Fisheries Society, Bethesda, pp 133–198
Campbell RR (1987) Status of the blue walleye, *Stizostedion vitreum glaucum*, in Canada. Can Field Nat 101:245–252
Carey JR, Judge DS (2000) Longevity records: life spans of mammals, birds, amphibians, reptiles, and fish. Odense University Press, Odense
Carlander KD (1997) Handbook of freshwater fishery biology, vol. 3: life history data on ichthyopercid and percid fishes of the United States and Canada. Iowa State University Press, Ames
Cena CJ, Morgan GE, Malette MD, Heath DD (2006) Inbreeding, outbreeding and environmental genetic diversity in 46 walleye (*Sander vitreus*) populations. Mol Ecol 15:303–320
Clady MD (1977) Distribution and relative exploitation of yellow perch tagged on spawning grounds in Oneida Lake. NY Fish Game J 24:46–52
Clement M, Posada D, Crandall KA (2000) TCS: a computer program to estimate gene genealogies. Mol Ecol 9:1657–1660. Available at http://darwin.uvigo.es/software/tcs.html
Colby PJ, McNicol RE, Ryder RA (1979) Synopsis of biological data on the walleye *Stizostedion v. vitreum*, vol 119, FAO fisheries synopsis. Food and Agriculture Organization of the United Nations, Rome
Colby PJ, Lewis CA, Eschenroder RL, Haas RC, Hushak LJ (1994) Walleye- rehabilitation guidelines for the Great Lakes area. Great Lakes Fishery Commission, technical report, Ann Arbor
Collette BB, Ali MA, Hokanson KEF, Nagiec M, Smirnov SA, Thorpe JE, Weatherly AH, Willemsen J (1977) Biology of the percids. J Fish Res Board Can 34:1891–1899
Coulon A, Fitzpatrick JW, Bowman R, Lovette IJ (2012) Mind the gap: genetic distance increases with habitat gap size in Florida scrub jays. Biol Lett 8:582–585
Craig JF (1987) The biology of perch and related fish. Croom Helm, London
Craig JF (2000) Percid fishes systematics, ecology, and exploitation. Blackwell Science, Oxford
Crossman EJ, McAllister DE (1986) Zoogeography of freshwater fishes of the Hudson Bay Drainage, Ungava Bay and the Arctic Archipelago. In: Hocutt CH, Wiley EO (eds) The zoogeography of North American freshwater fishes. Wiley, New York, pp 53–104
Darriba D, Taboada GL, Doallom R, Posada D (2012) jModelTest 2: more models, new heuristics and parallel computing. Nat Methods 9:772. Available at http://code.google.com/p/jmodeltest2/
Davis MB, Shaw RG (2001) Range shifts and adaptive responses to quaternary climate change. Science 292:673–679
Demandt MH (2010) Temporal changes in genetic diversity of isolated populations of perch and roach. Conserv Genet 11:249–255
Dembkowski DJ, Chipps SR, Blackwell BG (2013) Response of walleye and yellow perch to water-level fluctuations in glacial lakes. Fish Manag Ecol. doi:10.1111/fme.12047
DeWoody JA, Avise JC (2000) Microsatellite variation in marine, freshwater and anadromous fishes compared with other animals. J Fish Biol 56:461–473
Diekmann OE, Serrão EA (2012) Range-edge genetic diversity: locally poor extant southern patches maintain a regionally diverse hotspot in the seagrass *Zostera marina*. Mol Ecol 7:1647–1657
Drummond AJ, Suchard MA, Xie D, Rambaut A (2012) Bayesian phylogenetics with BEAUti and the BEAST 1.7. Mol Biol Evol 29:1969–1973. Available at http://beast2.org

Dumont P (1996) Comparaison de la dynamique des populations de perchaudes (*Perca flavescens*) soumises a des niveaux differents de stress anthropique. Ministere de l'Environmement et de la Faune, Service de l'amenagemetn et de l'exploitation de la faune. Report technique 06–46, Montreal

Dupont PP, Bourret V, Bernatchez L (2007) Interplay between ecological, behavioural and historical factors in shaping the genetic structure of sympatric walleye populations (*Sander vitreus*). Mol Ecol 26:937–951

Eldridge WH, Bacigalupi MD, Adelman IR, Miller LM, Kapuscinski AR (2002) Determination of relative survival of two stocked walleye populations and resident natural-origin fish by microsatellite DNA parentage assignment. Can J Fish Aquat Sci 59:282–290

Evanno G, Regnaut S, Goudet J (2005) Detecting the number of clusters of individuals using the software STRUCTURE: a simulation study. Mol Ecol 14:2611–2620

Excoffier L, Lischer HE (2010) ARLEQUIN suite ver 3.5: a new series of programs to perform population genetics analyses under Linux and Windows. Mol Ecol Resour 10:564–567. Available at http://cmpg.unibe.ch/software/arlequin35/

Faber JE, Stepien CA (1997) The utility of mitochondrial DNA control region sequences for analyzing phylogenetic relationships among populations, species, and genera of the Percidae. In: Kocher TD, Stepien CA (eds) Molecular systematics of fishes. Academic, London, pp 129–144

Ferguson RG, Derkson AJ (1971) Migrations of adult and juvenile walleyes (*Stizostedion vitreum vitreum*) in southern Lake Huron, Lake St. Clair, Lake Erie, and connecting waters. J Fish Res Board Can 8:1133–1142

Ferguson MM, Duckworth GA (1997) The status and distribution of lake sturgeon, *Ascipenser fulvescens*, in the Canadian provinces of Manitoba, Ontario and Quebec: a genetic perspective. Environ Biol Fishes 48:299–309

Fielder DG (2002) Sources of walleye recruitment in Saginaw Bay, Lake Huron. N Am J Fish Manag 22:1032–1040

Franckowiak RP, Sloss BL, Bozek MA, Newman SP (2009) Temporal effective size estimates of a managed walleye *Sander vitreus* population and implications for genetic-based management. J Fish Biol 74:1086–1103

Fulford RS, Rice JA, Miller TJ, Binkowski FP, Dettmers JM, Belonger B (2006) Foraging selectivity by larval yellow perch (*Perca flavescens*): implications for understanding recruitment in small and large lakes. Can J Fish Aquat Sci 63:28–42

Garner SR, Bobrowicz SM, Wilson CC (2013) Genetic and ecological assessment of population rehabilitation: walleye in Lake Superior. Ecol Appl 23:594–605

Gatt MH, Fraser DJ, Liskauskas AP, Ferguson MM (2002) Mitochondrial DNA variation and stock structure of walleye from eastern Lake Huron: an analysis of contemporary and historical samples. Trans Am Fish Soc 131:99–108

Gerlach G, Schardt U, Eckmann R, Meyer A (2001) Kin–structured subpopulations in Eurasian perch (*Perca fluviatilis* L.). Heredity 86:213–221

Glaubitz JC (2004) CONVERT: a user–friendly program to reformat diploid genotypic data for commonly used population genetic software packages. Mol Ecol Notes 4:309–310. http://www.agriculture.purdue.edu/fnr/html/faculty/rhodes/students%20and%20staff/glaubitz/software.htm

GLFC (Great Lakes Fishery Commission) (2011) Strategic vision of the Great Lakes Fishery Commission 2011–2020. Great Lakes Fishery Commission, special publication, Ann Arbor. Available at http://www.glfc.org/pubs/SpecialPubs/StrategicVision2012.pdf

Goudet J (1995) *F*stat version 1.2: a computer program to calculate *F*statistics. J Hered 86:485–486

Goudet J (2002) *F*stat version 2.9.3.2. Available at http://www2.unil.ch/popgen/softwares/fstat.htm

Griffiths D (2010) Pattern and process in the distribution of North American freshwater fish. Biol J Linn Soc 100:46–61

Grzybowski M, Sepulveda-Villet OJ, Stepien CA, Rosauer D, Binkowski F, Klaper R, Shepherd BS, Goetz F (2010) Genetic variation of 17 wild yellow perch populations from the Midwest and east coast analyzed via microsatellites. Trans Am Fish Soc 139:270–287

Guinand B, Scribner KT, Page KS, Burnham-Curtis MK (2003) Genetic variation over space and time: analyses of extinct and remnant lake trout populations in the Upper Great Lakes. Proc R Soc Lond B 270:425–433

Guindon S, Dufayard JF, Lefort V, Anisimova M, Hordijk W, Gascuel O (2010) New Algorithms and methods to estimate maximum–likelihood phylogenies: assessing the performance of PhyML 3.0. Syst Biol 59:307–321. Available at http://www.atgc-montpellier.fr/phyml/

Gyllensten ULF, Ryman N, Ståhl G (1985) Monomorphism of allozymes in perch (*Perca fluviatilis* L.). Hereditas 102:57–61

Haas RC, Bryant WC, Smith KD, Nuhfer AJ (1985) Movement and harvest of fish in Lake St. Clair, St. Clair River, and Detroit River. Final Report Winter Navigation Study U.S. Army Corps of Engineers

Hackney PA, Holbrook JA (1978) Sauger, walleye, and yellow perch in the southeastern United States. Am Fish Soc Spec Pub 11:74–81

Hampe A, Jump AS (2011) Climate relicts: past, present, future. Ann Rev Ecol Evol Syst 42:313–333

Hampe A, Petit R (2005) Conserving biodiversity under climate change: the rear edge matters. Ecol Lett 8:461–467

Haponski AE, Stepien CA (2013) Phylogenetic and biogeographic relationships of the *Sander* pikeperches (Perciformes: Percidae): patterns across North America and Eurasia. Biol J Linn Soc 110:156–179

Haponski AE, Stepien CA (2014a) A population genetic window in the past and future of the walleye *Sander vitreus*: relation to historic walleye and the extinct blue pike *S. v.* "*glaucus*". BMC Evol Biol 14:133

Haponski AE, Stepien CA (2014b) Genetic connectivity and diversity of walleye (*Sander vitreus*) spawning groups in the Huron–Erie corridor. J Great Lakes Res 40:89–100

Haponski AE, Bollin TL, Jedlicka MA, Stepien CA (2009) Landscape genetic patterns of the rainbow darter *Etheostoma caeruleum*: a catchment analysis of mitochondrial DNA sequences and nuclear microsatellites. J Fish Biol 75:2244–2268

Haponski AE, Dean H, Blake BE, Stepien CA (2014) Genetic history of walleye (*Sander vitreus*) spawning in Lake Erie's Cattaraugus Creek: a comparison of pre- and post-stocking. Trans Am Fish Soc. 143:1295–1307

Harris LN, Taylor EB (2010) Pleistocene glaciations and contemporary genetic diversity in a Beringian fish, the broad whitefish, *Coregonus nasus* (Pallas): inferences from microsatellite DNA variation. J Evol Biol 23:72–86

Hayhoe K, VanDorn J, Croley T, Schlegal N, Wuebbles D (2010) Regional climate change projections for Chicago and the US Great Lakes. J Great Lakes Res 36:7–21

Helfman G (1984) School fidelity in fishes: the yellow perch pattern. Anim Behav 32:663–672

Hewitt GM (1999) Post-glacial re-colonization of European biota. Biol J Linn Soc 68:87–112

Hewitt GM (2004) Genetic consequences of climatic oscillations in the Quaternary. Phil Trans R Soc B 359:183–195

Hill DK, Magnuson JJ (1990) Potential effects of global climate warming on the growth and prey consumption of Great Lakes fish. Trans Am Fish Soc 119:265–275

Hoagstrom CW, Berry CR (2010) The native range of walleyes in the Missouri River drainage. N Am J Fish Manag 30:642–654

Hoff MH (2002) A rehabilitation plan for walleye populations and habitats in Lake Superior, Great Lakes Fishery Commission miscellaneous publication 2003-01. Great Lakes Fishery Commission, Ann Arbor

Horrall RM (1981) Behavioral stock-isolating mechanisms in Great Lakes fishes with special reference to homing and site imprinting. Can J Fish Aquat Sci 38:1481–1496

Hubbs CL (1926) A check–list of the fishes of the Great Lakes and tributary waters, with nomenclatorial notes and analytical keys, vol 15, University of Michigan Museum of Zoology miscellaneous publication. University of Michigan Museum of Zoology, Ann Arbor

Hubbs CL, Lagler KF (2004) Fishes of the Great Lakes Region. (Smith GR, revised). University of Michigan, Ann Arbor

Jansen AC, Graeb BDS, Willis DW (2009) Effect of a simulated cold-front on hatching success of yellow perch eggs. J Freshw Ecol 24:651–655

Jennings MJ, Claussen JE, Philipp DP (1996) Evidence for heritable preferences for spawning habitat between two walleye populations. Trans Am Fish Soc 125:978–986

Jones ML, Netto JK, Stockwell JD, Mion JB (2003) Does the value of newly accessible spawning habitat for walleye (*Stizostedion vitreum*) depend on its location relative to nursery habitats? Can J Fish Aquat Sci 60:1527–1538

Jude DJ, Leach J (1999) Great Lakes fisheries. In: Kohler CC, Hubert WA (eds) Inland fisheries management in North America, 2nd edn. American Fisheries Society, Bethesda, pp 623–656

Kerr SJ, Corbett BW, Hutchinson NJ, Kinsman D, Leach JH, Puddister D, Stanfield L, Ward N (1997) Walleye habitat: a synthesis of current knowledge with guidelines for conservation. Percid Community Synthesis, Walleye Habitat Working Group, Ontario Ministry of Natural Resources, Peterborough

Knight RL (1997) Successful interagency rehabilitation of Lake Erie walleye. Fisheries 22:16–17

Kocovsky PM, Knight CT (2012) Morphological evidence for discrete stocks of yellow perch in Lake Erie. J Great Lakes Res 38:534–539

Kocovsky PM, Sullivan TJ, Knight CT, Stepien CA (2013) Genetic and morphometric differences demonstrate fine–scale population substructure of the yellow perch *Perca flavescens*: need for redefined management units. J Fish Biol 82:2015–2030

Kornis MS, Mercado-Silva N, Vander Zanden MJ (2012) Twenty years of invasion: a review of *Neogobius melanostomus* biology, spread, and ecological implications. J Fish Biol 80:235–285

Kreiger DA, Terrell JW, Nelson PC (1983) Habitat suitability information: yellow perch. U.S. Fish and Wildlife Service FWS/OBS-83/10.55. Washington, DC

Kunin WE, Vergeer P, Kenta T, Davey MP, Burke T, Woodward FI, Quick P, Mannarelli M-E, Watson-Haigh NS, Butlin R (2009) Variation at range margins across multiple spatial scales: environmental temperature, population genetics and metabolomic phenotype. Proc R Soc Lond B 276:1495–1506

Laporte M, Magnan P, Angers B (2011) Genetic differentiation between the blue and the yellow phenotypes of walleye (*Sander vitreus*): an example of parallel evolution. Ecoscience 18:124–129

Larson G, Schaetzl R (2001) Origin and evolution of the Great Lakes. J Great Lakes Res 27:518–546

Leary R, Booke HE (1982) Genetic stock analysis of yellow perch from Green Bay and Lake Michigan. Trans Am Fish Soc 111:52–57

LeClerc E, Mailhot Y, Mingelbier M, Bernatchez L (2008) The landscape genetics of yellow perch (*Perca flavescens*) in a large fluvial ecosystem. Mol Ecol 17:1702–1717

Lewis CFM, Moore TC, Rea DK, Dettman DL, Smith AM, Mayer LA (1994) Lakes of the Huron basin: their record of runoff from the Laurentide Ice Sheet. Quat Sci Rev 13:891–922

Li S, Mathias JA (1982) Causes of high mortality among cultured larval walleyes. Trans Am Fish Soc 111:710–721

Li L, Wang HP, Givens C, Czesny S, Brown B (2007) Isolation and characterization of microsatellites in yellow perch (*Perca flavescens*). Mol Ecol Notes 7:600–603

Lindsay DL, Barr KR, Lance RF, Tweddale SA, Hayden TJ, Leberg PL (2008) Habitat fragmentation and genetic diversity of an endangered, migratory songbird, the golden-cheeked warbler (*Dendroica chrysoparia*). Mol Ecol 17:2122–2133

Locke B, Belore M, Cook A, Einhouse D, Kenyon R, Knight R, Newman K, Ryan P, Wright E (2005) Walleye management plan. Lake Erie Committee Great Lakes Fishery Commission. Available at http://www.glfc.org/lakecom/lec/WTG_docs/other_reports_and_docs/wmp20051207.pdf

MacCallum WR, Selgeby JH (1987) Lake Superior revisited 1984. Can J Fish Aquat Sci 44:23–36

MacDougall TM, Wilson CC, Richardson LM, Lavender M, Ryan PA (2007) Walleye in the Grand River, Ontario: an overview of rehabilitation efforts, their effectiveness, and implications for eastern Lake Erie fisheries. J Great Lakes Res 33:103–117

MacGregor RB, Witzel LD (1987). A twelve year study of the fish community in the Nanticoke Region of Long Point Bay, Lake Erie. Lake Erie Fisheries Assessment Unit report 1987-3. Ontario Ministry of Natural Resources, Port Dover

Mandrak NE, Crossman EJ (1992) Postglacial dispersal of freshwater fishes into Ontario. Can J Zool 70:2247–2259

Mangan MT (2004) Yellow perch production and harvest strategies for semi- permanent wetlands in Eastern South Dakota. MSc thesis, Wildlife and Fisheries Sciences, South Dakota State University

Manni F, Guérard E, Heyer E (2004) Geographic patterns of (genetic, morphologic, linguistic) variation: how barriers can be detected by using Monmonier's algorithm. Hum Biol 76:173–190. Available at http://ecoanthropologie.mnhn.fr/software/barrier.html

Manning NF, Mayer CM, Bossenbroek JM, Tyson JT (2013) Effects of water clarity on the length and abundance of age-0 yellow perch in the Western Basin of Lake Erie. J Great Lakes Res 39:295–302

Mantel N (1967) The detection of disease clustering and a generalized regression approach. Cancer Res 27:209–220

McParland TL, Ferguson MM, Liskauskas AP (1999) Genetic population structure and mixed-stock analysis of walleyes in the Lake Erie-Lake Huron Corridor using allozymes and mitochondrial DNA markers. Trans Am Fish Soc 128:1055–1067

Merker RJ, Woodruff RC (1996) Molecular evidence for divergent breeding groups of walleye (*Stizostedion vitreum*) in tributaries to western Lake Erie. J Great Lakes Res 22:280–288

Miller LM (2003) Microsatellite DNA loci reveal genetic structure of yellow perch in Lake Michigan. Trans Am Fish Soc 132:503–513

Moran GF, Hopper SD (1983) Genetic diversity and the insular population structure of the rare granite rock species, *Eucalyptus caesia* Benth. Aust J Bot 31:161–172

Moyer GR, Billington N (2004) Stock structure among yellow perch populations throughout North America determined from allozyme and mitochondrial DNA analysis. In: Barry TP, Malison JA (eds) Proceedings of Percis III: the third international percid fish symposium, University of Wisconsin Sea Grant Institute, Madison, pp 96–97

Murdoch MH, Hebert PD (1997) Mitochondrial DNA evidence of distinct glacial refugia brown bullhead (*Ameiurus nebulosus*). Can J Fish Aquat Sci 54:1450–1460

Murphy BR (1990) Evidence for a genetically unique walleye population in the upper Tombigbee River system of northeastern Mississippi. Proc SE Fish Counc 22:14–16

Murphy S, Collins N, Doka S, Fryer B (2012) Evidence of yellow perch, largemouth bass and pumpkinseed metapopulations in coastal embayments of Lake Ontario. Environ Biol Fish 95:213–226

Nesbø CL, Magnhagen C, Jakobsen KS (1998) Genetic differentiation among stationary and anadromous perch (*Perca fluviatilis*) in the Baltic Sea. Hereditas 129:241–249

Nesbø CL, Fossheim T, Vollestad LA, Jakobsen KS (1999) Genetic divergence and phylogeographic relationships among European perch (*Perca fluviatilis*) populations reflect glacial refugia and postglacial colonization. Mol Ecol 8:1387–1404

Newbrey MG, Ashworth AC (2004) A fossil record of colonization and response of lacustrine fish populations to climate change. Can J Fish Aquat Sci 61:1807–1816

Noecker RJ (1998) Endangered species list revisions: a summary of delisting and downlisting. CRS report for congress 98-32 ENR

Oberdorff T, Hugueny B, Guégan J-F (1997) Is there an influence of historical events on contemporary fish species richness in rivers? Comparisons between western Europe and North America. J Biogeogr 24:461–467

Olson DE, Scidmore WJ (1962) Homing behavior of spawning walleyes. Trans Am Fish Soc 91:355–361

OMNR (Ontario Ministry of Natural Resources) (2011) 2006–2009 annual report. Lake Erie MU

Paradis Y, Magnan P (2005) Phenotypic variation of walleye, *Sander vitreus*, in Canadian Shield lakes: new insights on percid polymorphism. Environ Biol Fish 73:357–366

Parker AD, Stepien CA, Sepulveda-Villet OJ, Ruehl CB, Uzarski DG (2009) The interplay of morphology, habitat, resource use, and genetic relationships in young yellow perch. Trans Am Fish Soc 138:899–914

Petit RJ, Aguinaglade I, de Beaulieu J-L, Bittkau C, Brewer S, Cheddadi R, Ennos R, Fineschi S, Grivet D, Lascoux M, Mohanty A, Müller-Starck G, Demesure-Musch B, Palmé A, Martín JP, Rendell S, Vendramin GG (2003) Glacial refugia: hotspots but not melting pots of genetic diversity. Science 300:1563–1565

Pritchard JK, Wen W (2004) Documentation for STRUCTURE software: ver. 2.3.3. Stanford University. Available at http://pritchardlab.stanford.edu/software.html

Pritchard JK, Stephens M, Donnelly P (2000) Inference of population structure using multilocus genotype data. Genetics 155:945–959

Provan J, Bennett KD (2008) Phylogeographic insights into cryptic glacial refugia. Trends Ecol Evol 23:564–571

Radabaugh NB, Bauer WF, Brown ML (2010) A comparison of seasonal movement patterns of yellow perch in simple and complex lake basins. N Am J Fish Manag 30:179–190

Rawson MR (1980) Yellow perch movements. Ohio Department of Natural Resources Job Program Report, Dingell–Johnson project number F-35-R-18, study number 4, 1 Nov 1979–30 Jun 1980

Redman RA, Czesny SJ, Dettmers JM (2013) Yellow perch population assessment in Southwestern Lake Michigan. INHS technical report 2013 (25), Division of Fisheries, Illinois Department of Natural Resources, Champaign

Refseth UH, Nesbø CL, Stacy JE, Vøllestad LA, Fjeld E, Jakobsen KS (1998) Genetic evidence for different migration routes of freshwater fish into Norway revealed by analysis of current perch (*Perca fluviatilis*) populations in Scandinavia. Mol Ecol 7:1015–1027

Regier HA, Hartman WL (1973) Lake Erie's fish community: 150 years of cultural stresses. Science 180:1248–1255

Rempel LL, Smith DG (1998) Postglacial fish dispersal from the Mississippi refuge to the Mackenzie River basin. Can J Fish Aquat Sci 55:893–899

Rice RM (1989) Analyzing tables of statistical tests. Evolution 43:223–225

Rodrigues CG, Vilks G (1994) The impact of glacial lake runoff on the Goldthwait and Champlain Seas: the relationship between glacial Lake Agassiz runoff and the younger dryas. Quat Sci Rev 13:923–944

Ronquist F, Huelsenbeck JP (2003) MrBayes 3: Bayesian phylogenetic inference under mixed models. Bioinformatics 19:1572–1574. http://mrbayes.sourceforge.net/. (v3.1.2 2005)

Roseman EF, Taylor WW, Hayes DB, Tyson JT, Haas RC (2005) Spatial patterns emphasize the importance of coastal zones as nursery areas for larval walleye in western Lake Erie. J Great Lakes Res 31:28–44

Rousset F (2008) Genepop'008: a complete re-implementation of the Genepop software for Windows and Linux. Mol Ecol Resour 8:103–106. Available at http://kimura.univ-montp2.fr/%7Erousset/Genepop.htm

Russell DA, Rich FJ, Schneider V, Lynch-Stieglitz J (2009) A warm thermal enclave in the Late Pleistocene of the south-eastern United States. Biol Rev 84:173–202

Ryan PA, Knight R, MacGregor R, Towns G, Hoopes R, Culligan W (2003) Fish-community goals and objectives for Lake Erie, Great Lakes Fishery Commission special publication 03-02. Great Lakes Fishery Commission, Ann Arbor

Saarnisto M (1974) The deglaciation history of the Lake Superior region and its climatic implications. Quatern Res 4:316–339

Schmidt RE (1986) Zoogeography of the northern Appalachians. In: Hocutt CH, Wiley EO (eds) The zoogeography of North American freshwater fishes. Wiley, New York, pp 137–159

Schneider JC, Leach JH (1977) Walleye (*Stizostedion vitreum vitreum*) fluctuations in the Great Lakes and possible causes, 1800–1975. J Fish Res Board Can 34:1878–1889

Schram ST, Seider MJ, Furlong PD, Friday MJ (2010) Status of walleye in Lake Superior. In: Roseman E, Kocovsky P, Vandergoot C (eds) Status of walleye in the Great Lakes: proceedings of the 2006 symposium. Great Lakes Fishery Commission technical report 69, Ann Arbor, pp 1–13

Scott WB, Crossman EJ (1973) Freshwater fishes of Canada. J Fish Res Board Can 184:1–196

Sepulveda-Villet OJ, Stepien CA (2011) Fine-scale population genetic structure of the yellow perch *Perca flavescens* in Lake Erie. Can J Fish Aquat Sci 68:1435–1453

Sepulveda–Villet OJ, Stepien CA (2012) Waterscape genetics of the yellow perch (*Perca flavescens*): patterns across large connected ecosystems and isolated relict populations. Mol Ecol 21:5795–5826

Sepulveda-Villet OJ, Ford AM, Williams JD, Stepien CA (2009) Population genetic diversity and phylogeographic divergence patterns of the yellow perch (*Perca flavescens*). J Great Lakes Res 35:107–119

Shuter BJ, Post JR (1990) Climate, population viability, and the zoogeography of temperate fishes. Trans Am Fish Soc 119:314–336

Simon TP, Wallus R (2006) Reproductive biology and early life history of fishes in the Ohio River drainage vol. 4: Percidae – perch, pikeperch, and darters. CRC Taylor and Francis, Boca Raton

Sloss BL, Billington N, Burr BM (2004) A molecular phylogeny of the Percidae (Teleostei, Perciformes) based on mitochondrial DNA sequence. Mol Phylogenet Evol 32:545–562

Soltis DE, Morris AB, McLachlan JS, Manos PS, Soltis PS (2006) Comparative phylogeography of unglaciated eastern North America. Mol Ecol 15:4261–4293

Song CB, Near TJ, Page LM (1998) Phylogenetic relations among percid fishes as inferred from mitochondrial cytochrome *b* DNA sequence data. Mol Phylogenet Evol 10:343–353

Sruoga A, Butkauskas D, Rashal I (2008) Evaluation of genetic diversity of perch (*Perca fluviatilis*) and pikeperch (*Sander lucioperca*) populations from Curonian lagoon and inshore waters of the Baltic Sea. Acta Biol Univ Daugavpils 8:81–88

Stepien CA, Faber JE (1998) Population genetic structure, phylogeography, and spawning philopatry in Walleye (*Stizostedion vitreum*) from mtDNA control region sequences. Mol Ecol 7:1757–1769

Stepien CA, Dillon AK, Chandler MD (1998) Genetic identity, phylogeography, and systematics of ruffe *Gymnocephalus* in the North American Great Lakes and Eurasia. J Great Lakes Res 24:361–378

Stepien CA, Taylor CD, Einhouse DW (2004) An analysis of genetic risk to a native spawning stock of walleye *Sander vitreus* (*Stizostedion vitreum*) due to stocking in Cattaraugus Creek. In: Barry TP, Malison JA (eds) Proceedings of Percis III: the third international percid fish symposium, University of Wisconsin Sea Grant Institute, Madison, pp 93–94

Stepien CA, Brown JE, Neilson ME, Tumeo MA (2005) Genetic diversity of invasive species in the Great Lakes versus their Eurasian source populations: insights for risk analysis. Risk Anal 25:1043–1060

Stepien CA, Murphy DJ, Strange RM (2007) Broad- to fine-scale population genetic patterning in the smallmouth bass *Micropterus dolomieu* across the Laurentian Great Lakes and beyond: an interplay of behaviour and geography. Mol Ecol 16:1605–1624

Stepien CA, Murphy DJ, Lohner RN, Sepulveda–Villet OJ, Haponski AE (2009) Signatures of vicariance, postglacial dispersal, and spawning philopatry: population genetics and biogeography of the walleye *Sander vitreus*. Mol Ecol 18:3411–3428

Stepien CA, Murphy DJ, Lohner RN, Haponski AE, Sepulveda–Villet OJ (2010) Status and delineation of walleye (*Sander vitreus*) genetic stock structure across the Great Lakes. In: Roseman E, Kocovsky P, Vandergoot C (eds) Status of walleye in the Great Lakes: proceedings of the 2006 symposium. Great Lakes Fishery Commission technical report 69, Ann Arbor, pp 189–223

Stepien CA, Banda JA, Murphy DJ, Haponski AE (2012) Temporal and spatial genetic consistency of walleye (*Sander vitreus*) spawning groups. Trans Am Fish Soc 141:660–672

Stone FL (1948) A study of the taxonomy of the blue and yellow pikeperches (*Stizostedion*) of Lake Erie and Lake Ontario. Unpublished PhD dissertation, University of Rochester, Rochester

Stott W, Ebener MP, Mohr L, Hartman T, Johnson J, Roseman EF (2013) Spatial and temporal genetic diversity of lake whitefish (*Coregonus clupeaformis* (Mitchill)) from Lake Huron and Lake Erie. Adv Limnol 64:205–222

Strange RM, Stepien CA (2007) Genetic divergence and connectivity among river and reef spawning groups of walleye (*Sander vitreus*) in Lake Erie. Can J Fish Aquat Sci 64:437–448

Sullivan TJ, Stepien CA (2014) Genetic diversity and divergence of yellow perch spawning populations across the Huron–Erie Corridor, from Lake Huron through western Lake Erie. J Great Lakes Res 40:101–109

Sullivan TJ, Stepien CA (2015) Temporal population genetic structure of yellow perch spawning groups in the lower Great Lakes. Trans Am Fish Soc 144:211–226

Teller JT, Mahnic P (1988) History of sedimentation in the northwestern Lake Superior basin and its relation to Lake Agassiz overflow. Can J Earth Sci 25:1660–1673

Timmerman AJ (1995) Walleye assessment and enhancement projects in the middle Grand River watershed 1987–1995. Ontario Ministry of Natural Resources, Cambridge District, Guelph

Todd TN, Hatcher CO (1993) Genetic variability and glacial origins of yellow perch (*Perca flavescens*) in North America. Can J Fish Aquat Sci 50:1828–1834

Trautman MB (1981) The fishes of Ohio. Ohio State University Press, Columbus

Truemper HA, Lauer TE (2005) Gape limitation and piscine prey size-selection by yellow perch in the extreme southern area of Lake Michigan, with emphasis on two exotic prey items. J Fish Biol 66:135–149

Turgeon J, Bernatchez L (2001) Mitochondrial DNA phylogeography of lake cisco (*Coregonus artedi*): evidence supporting extensive secondary contacts between two glacial races. Mol Ecol 10:987–1001

Underhill JC (1986) The fish fauna of the Laurentian Great Lakes, the St. Lawrence lowlands, Newfoundland, and Labrador. In: Hocutt CH, Wiley EO (eds) The zoogeography of North American freshwater fishes. Wiley, New York, pp 105–136

USFWS/GLFC (United States Fish and Wildlife Service/Great Lakes Fishery Commission) (2010) Great Lakes Fish Stocking database. U.S. Fish and Wildlife Service, Region 3 Fisheries Program, and Great Lakes Fishery Commission. Available at http://www.glfc.org/fishstocking/

Vandewoestijne S, Schtickzelle N, Baguette M (2008) Positive correlation between genetic diversity and fitness in a large, well-connected metapopulation. BMC Biol 6:46

Walter RP, Cena CJ, Morgan GE, Heath DD (2012) Historical and anthropogenic factors affecting the population genetic structure of Ontario's inland lake populations of walleye (*Sander vitreus*). J Hered 103:831–841

Wang HY, Rutherford ES, Cook HA, Einhouse DW, Haas RC, Johnson TB, Kenyon R, Locke B, Turner MW (2007) Movement of walleye in Lakes Erie and St. Clair inferred from tag return and fisheries data. Trans Am Fish Soc 136:539–551

Weir BS, Cockerham CC (1984) Estimating *F*–statistics for the analysis of population structure. Evolution 38:1358–1370

Wilson CC, Hebert PD (1996) Phylogeographic origins of lake trout (*Salvelinus namaycush*) in eastern North America. Can J Fish Aquat Sci 53:2764–2775

Wirth T, Saint-Laurent R, Bernatchez L (1999) Isolation and characterization of microsatellite loci in the walleye (*Stizostedion vitreum*), and cross–species amplification within the family Percidae. Mol Ecol 8:1960–1962

Wolfert DR, Van Meter HD (1978) Movements of walleyes tagged in eastern Lake Erie. N Y Fish Game J 25:16–22

WTG (Walleye Task Group of the Lake Erie Committee, Great Lakes Fishery Commission) (2014) Report for 2013 by the Lake Erie walleye Task Group. Great Lakes Fishery Commission Ann Arbor. Available at http://www.glfc.org/lakecom/lec/WTG_docs/annual_reports/WTG_report_2014.pdf

YPTG (Yellow Perch Task Group of the Lake Erie Committee, Great Lakes Fishery Commission) (2014) Report of the Lake Erie yellow perch task group. Great Lakes Fishery Commission Ann Arbor. Available at http://www.glfc.org/lakecom/lec/YPTG_docs/annual_reports/YPTG_report_2014.pdf

Yu C, Ferraro D, Ramaswamy S, Schmitz MH, Schaefer WF, Gibson DT (2008) Purification and properties of sandercyanin, a blue protein secreted in the mucus of blue forms of walleye, *Sander vitreus*. Environ Biol Fish 82:51–58

Zhao Y, Shuter BJ, Jackson DA (2008) Life history variation parallels phylogeographical patterns in North American walleye (*Sander vitreus*) populations. Can J Fish Aquat Sci 65:198–211

Zhao Y, Jones ML, Shuter BJ, Roseman EF (2009) A biophysical model of Lake Erie walleye (*Sander vitreus*) explains interannual variations in recruitment. Can J Fish Aquat Sci 66:114–125

Chapter 26
Production of Genetically Defined Perch Broodstocks and Their Selection for Fast Growth

Frederick W. Goetz, Daniel R. Rosauer, Michael Grzybowski, Frederick P. Binkowski, and Brian S. Shepherd

Abstract Restrictions and closures of commercial freshwater fisheries, coupled with continued high consumer demand, have fueled interest in yellow perch aquaculture. However, the general slow growth of this species and the lack of commercially available broodstocks have been an impediment. A yellow perch broodstock program has been established at the University of Wisconsin Great Lakes WATER Institute in collaboration with the USDA Agricultural Research Service. The goals of this program have been to initiate genetically defined perch broodstocks from several North American geographic regions, characterize their growth, embryonic development and reproduction, and to select these stocks to obtain faster growing fish. This chapter describes the (1) genetic analysis used to define wild yellow perch populations across

F.W. Goetz (✉)
School of Freshwater Sciences, University of Wisconsin-Milwaukee, 600 E. Greenfield Avenue, 53204 Milwaukee, WI, USA

NOAA Northwest Fisheries Science Center, Manchester Research Station, 98366 Manchester, WA, USA
e-mail: rick@uwm.edu; rick.goetz@noaa.gov

D.R. Rosauer
School of Freshwater Sciences, University of Wisconsin-Milwaukee, 600 E. Greenfield Avenue, 53204 Milwaukee, WI, USA

Iowa Department of Natural Resources, Rathbun Fish Hatchery, 52571 Moravia, IA, USA

M. Grzybowski
School of Freshwater Sciences, University of Wisconsin-Milwaukee, 600 E. Greenfield Avenue, 53204 Milwaukee, WI, USA

Medical College of Wisconsin, Human Molecular Genetics Center, 53226 Milwaukee, WI, USA

F.P. Binkowski
School of Freshwater Sciences, University of Wisconsin-Milwaukee, 600 E. Greenfield Avenue, 53204 Milwaukee, WI, USA

B.S. Shepherd
USDA-ARS-Great Lakes WATER Institute, 600 E. Greenfield Avenue, 53204 Milwaukee, WI, USA

P. Kestemont et al. (eds.), *Biology and Culture of Percid Fishes*,
DOI 10.1007/978-94-017-7227-3_26

N. America from which broodstocks could potentially be derived; (2) development and characterization of perch broodstocks from three geographic regions of N. America; and (3) selection of these stocks for enhanced growth and the heritability of that growth.

Keywords Yellow perch • Breeding programme • Selection • Genetic analysis • Growth heritability

26.1 Introduction

Yellow perch are a highly valued fish in the marketplace. Restrictions and the closure of some of the commercial fisheries in the Great Lakes, coupled with continued high consumer demand, have fueled interest in yellow perch aquaculture. However, yellow perch are relatively slow growing and, compared with other aquacultured species, are smaller at market size. Genetic selection has been the basis for many of the improvements in terrestrial agriculture (Falconer 1981) and is an option for growth improvement in fish as well (Gjedrem 2000). Selection for enhanced growth has been proven successful in several species of fish including: brook trout, *Salvelinus fontinalis* (Embody and Hyford 1925), rainbow trout, *Oncorhynchus mykiss* (Kincaid et al. 1977), coho salmon, *Oncorhynchus kisutch* (Hershberger et al. 1990), Atlantic salmon, *Salmo salar* (Friars et al. 1995), red sea bream, *Pagrus major* (Murata et al. 1996), tilapia, *Oreochromis sp.* (Gall and Bakar 2002), channel catfish, *Ictalurus punctatus* (Rezk et al. 2003), and brown trout, *Salmo trutta* (Mambrini et al. 2004). However, for genetic selection to be possible, captive and genetically-defined broodstocks must be available. Commercial perch broodstocks have not been available to the industry and many growers have used stocks obtained repeatedly from the wild or continually bred within local fish farms.

A yellow perch aquaculture program is underway at the University of Wisconsin-Milwaukee's School of Freshwater Science (Milwaukee, WI). A goal of that program is to produce genetically-defined yellow perch broodstocks that are selected for enhanced growth. This goal has been approached in three major steps: (1) The genetic characterization of wild perch populations across Eastern and Midwest U.S. from which broodstocks could potentially be developed; (2) The creation of parental broodstocks from selected wild sites as determined in step 1; and (3) Crossing and selection of progeny from these parental stocks for enhanced growth. Currently there have been two rounds of genetic selection of these stocks that have resulted in significant gains in growth rates.

26.2 Genetic Characterization of Wild Perch Populations Across Eastern and Midwest U.S.

With the help of numerous State and Tribal agencies, during the 2006 spawning season we obtained fin clips for genetic analysis from (1) Perquimans River (Albemarle Sound: North Carolina); (2) Severn River (Chesapeake Bay: Maryland);

(3) Bush River (Chesapeake Bay: Maryland); (4) Nanticoke River (Chesapeake Bay: Maryland); (5) Choptank River (Chesapeake Bay: Maryland); (6) Lake Ontario; (7) Little Tail (Green Bay, Wisconsin); (8) Lake Michigan Basin (Green Can Reef: 2 year classes); (9) Lac du Flambeau (Wisconsin); (10) Bad River (Wisconsin); (11) Lake Winnebago (Wisconsin); and (12) Devils Lake (North Dakota). DNA was extracted from the fin clips and analyzed on an ABI 3730 using a total of 14 microsatellites. Six of the microsatellites were derived from published perch microsatellites (LeClerc et al. 2000) and the remaining were developed by our laboratory using oligonucleotide magnetic bead trapping (Accession #s EU153815-EU153821; EU277783-EU277832; EU281734-EU281845; EU283965-EU284009).

The genetic structure analysis showed two distinct groups, one containing East Coast yellow perch populations (Maryland and North Carolina) and the other containing Midwest populations (Wisconsin, North Dakota, and New York) (Grzybowski et al. 2010). Within these broad groups, populations could be further separated into clusters associated with drainages. The East Coast clusters included the Chesapeake Bay, MD (Severn, Bush, Choptank and Nanticoke Rivers) and the Ablemarle Sound, NC (Perquimans, Little, Pasquotank, North and Scuppernog Rivers) drainages. The Midwest clusters included the Lake Michigan (Lake Michigan, Lake Winnebago, and Little Tail Point-Green Bay), Lake Superior (Bad River), Mississippi River (Lac du Flambeau), Lake Ontario (Lake Ontario), and Red River (Devils Lake) drainages (Fig. 26.1). We found higher genetic divergence between Midwest populations than between East Coast populations that were probably a result of the potential mixing of populations within the Chesapeake Bay and the Albermarle Sound along the East Coast. Even so, there was differentiation of these East Coast populations that would allow for distinct broodstock production.

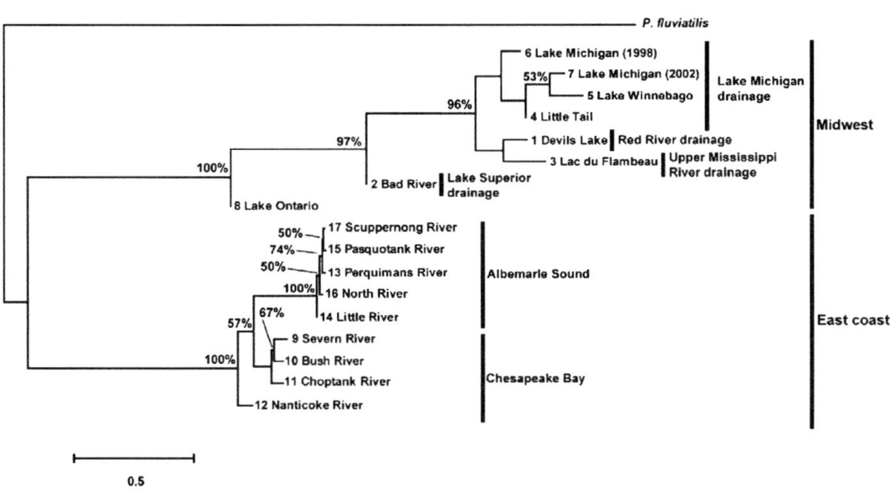

Fig. 26.1 Neighbor-joining tree of yellow perch population samples based on Nei's (1972) distance for seven microsatellite loci. Bootstrapping values indicate nodes supported by 50 % or more of 1000 resampling events. The lengths of the horizontal lines are proportional to the genetic distances (From Grzybowski et al. 2010)

The genetic analysis of wild yellow perch populations provided us with the data to select sites from which we could obtain gametes to produce the parental broodstocks from the wild.

26.3 The Creation and Analysis of Parental Yellow Perch Broodstocks

Given the population genetic study, four broodstocks were initiated. Wild populations for broodstock development were selected within separate drainage basins from each broad grouping and were based on overall genetic diversity, inbreeding coefficient, as well as the logistics of obtaining and crossing wild perch from the original sampling locations to obtain gametes for broodstock production. Based on these criteria, the four populations selected were (drainage, state): Perquimans River (Albemarle Sound, NC), Choptank River (Chesapeake Bay, MD), Lake Winnebago (Lake Michigan, WI), and Lac du Flambeau (Mississippi River, WI).

In 2007, we obtained gametes from wild fish taken from the Perquimans and Choptank Rivers and fertilized these eggs at the Tidewater Research Station (North Carolina State University) and Joseph H. Manning State Fish Hatchery (Maryland DNR), respectively. Prespawning perch adults were collected from Lac du Flambeau and Lake Winnebago and were spawned at the GLWI (University of Wisconsin-Milwaukee). All fertilized eggs were incubated, hatched and the larvae reared to adults at the GLWI under identical conditions (Rosauer et al. 2011).

From 14 to 34 single pair (one male: one female) crosses were made depending on the stock (Table 26.1) and fry from all crosses were mixed equally to obtain the resulting parental broodstocks. Of the four stocks, none of the fry from Lac du Flambeau survived and in the remaining three stocks the survival varied from 14.2 % to 38.8 %. Since all incubation and early rearing regimes were identical between stocks, it is unclear why there was no survival in Lac du Flambeau larvae. However, survival was also relatively low in Lake Winnebago fish and this might indicate that the incubation and rearing conditions used were not conducive to fish from more northern and colder climates and favored the progeny from East Coast stocks.

Table 26.1 Yellow perch stocking and survival through feed habituation from single pairwise crosses in four stocks of yellow perch. Of the surviving fry, 2400 from each stock were used in a growth trial (From Rosauer et al. 2011)

	Perquimans River	Choptank River	Lake Winnebago	Lac du Flambeau
Number of crosses	14	20	34	33
Average stocked per cross	1960	3320	2000	2003
Total stocked	27,396	66,468	67,877	66,086
Days post hatch	72	90	80	NA
Average weight	1.59±0.09 g	2.15±0.03	2.04±0.09	NA
Number surviving	7114	25,758	9691	0
Percent survival	26.0 %	38.8 %	14.2 %	0.0 %

Table 26.2 Means ± standard deviation for final weight, length, absolute growth rate, and fillet yield for three stocks of yellow perch 12 months post hatch after a growth trial. Means followed by different letters are significantly different (P<0.05) by Tukey's test (From Rosauer et al. 2011)

	Perquimans River	Choptank River	Lake Winnebago
Final weight ± Stdev (g)	138.12±6.90 g[a]	126.7±4.86 g[b]	52.08±2.48 g[c]
Final length ± Stdev (g)	208.01±2.70 mm[a]	204.10±2.70 mm[a]	162.22±1.91 mm[b]
Absolute growth rate (AGR)	0.47±0.02 g/day[a]	0.45±0.02 g/day[a]	0.18±0.01 g/day[b]
Fillet yield (%)	35.22±0.01 %	34.97±0.02 %	34.57±0.01 %

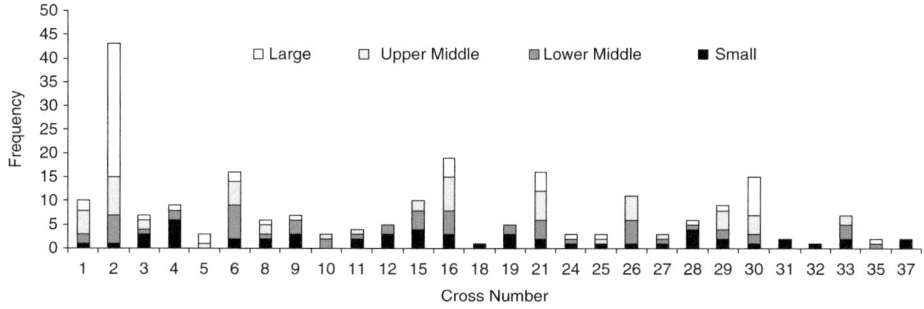

Fig. 26.2 Frequency of progeny from 37 crosses within four size classes of the Lake Winnebago yellow perch stock. Progeny (228) were divided into four size classes (Large >32 g, U. middle 32–26 g, L. middle 26–20 g, and Small <20 g) and genotyped using microsatellites (Modified from Rosauer et al. 2011)

A 12-month growth performance trial at 20 °C was conducted on progeny (F1) from the three surviving stocks (Rosauer et al. 2011). Data on length and weight were collected at 2-week intervals and fillets yield and proximate analysis was conducted on a subset of fish at the end of the trial. The results indicated that progeny from the Winnebago stock grew slower and attained a lower average weight and length as compared with the two East coast stocks (Table 26.2).

Fillet yields and most of the proximate analysis were not different between stocks except that fish from the Perquimans River had higher percentages of eicosapentaenoic and arachidonic acid (Rosauer et al. 2011). Pedigree tracking of individuals indicated that there were strong differences in the contribution of specific families to different size classes (high, medium, low) within each stock (Fig. 26.2 for example). This strongly suggested that subsequent selection of these stocks for the best performers in terms of size would result in significant gains in growth in future generations.

26.4 Selection of Broodstocks to Enhance Growth

At the conclusion (Spring 2008) of the growth performance trial on the founding broodstock (#2 above), the top 35 % of each stock in terms of weight were retained. Each of these fish was PIT tagged and genotyped using microsatellites as previously described (Gryzbowski et al. 2010; Rosauer et al. 2011). The fish were then cycled for 1 year under photoperiod and temperature regimes coincident with their geographic origin. In 2009, individuals from these retained fish were crossed within stocks to produce the first generation (F2) of growth-selected perch.

In 2009, a 12-month growth performance trial was conducted on the F2 progeny under conditions that were identical to those for the parental stocks. Data on length and weight were collected at monthly intervals. As with the F1 progeny, the results indicated that the Winnebago stock grew much slower and attained a lower average weight and length as compared with the Perquimans and Choptank stocks. However, in comparing the results of the performance trials conducted in 2007 (F1) and 2009 (F2), it was evident that growth in each of the three stocks was significantly greater in 2009 as compared to 2007. Overall, there was a 49 %, 46 % and 40 % increase in weight between the F2 and F1 Choptank, Perquimans and Winnebago juveniles, respectively. For the Choptank and Perquimans fish this would translate to a reduction of at least 2 months in the time to market.

26.5 Conclusions

We have used a stepwise approach to produce captive and genetically defined yellow perch broodstocks exhibiting enhanced growth. This process involved a genetic analysis of potential wild yellow perch populations from which broodstocks could be obtained, followed by the production of several stocks derived from the gametes of wild perch chosen on the basis of the genetic analysis. Currently there are three broodstocks; the Perquimans and Choptank River stocks from the East Coast and the Winnebago stock derived from Lake Winnebago in Wisconsin. From the analysis of growth over two generations of juveniles produced from these stocks, it is evident that the two East Coast stocks greatly outperform the Winnebago stock in terms of growth. Even though increased growth could be selected for in the Winnebago stock from the F1 to the F2 generation, the overall growth in this stock makes it undesirable for commercial production.

The basis for the differences observed in the growth between the East Coast and Winnebago stocks is unknown but was speculated to be a result of the fact that Winnebago fish did not adapt well to the culture conditions employed (Rosauer et al. 2011). In particular, the Winnebago fish exhibited behaviors that were indicative of stress and would have impacted feeding and growth. Thus, this stock does not appear to be well suited for commercial growth under the conditions used here. They might have another phenotype that would be desirable, but we have recently demonstrated that, of the three stocks, the Winnebago is also the most susceptible to contracting viral hemorrhagic septicemia (VHS) (Olson et al. 2012). Thus, further development of this stock is not promising. In contrast, the Perquimans and Choptank fish exhibited fast growth that was significantly enhanced in the F2 generation and would translate to an earlier time to market. These stocks also exhibited low susceptible to VHS (Olson et al. 2012).

References

Embody GC, Hayford CD (1925) The advantage of rearing brook trout fingerlings from selected breeders. Trans Am Fish Soc 55:135–148

Falconer DS (1981) Introduction to quantitative genetics, vol 2. Longman, New York, p 340

Friars GW, Bailey JK, O'Flynn FM (1995) Applications of selection for multiple traits in cage-reared Atlantic salmon (*Salmo salar*). Aquaculture 137:213–217

Gall G, Bakar Y (2002) Application of mixed-model techniques to fish breed improvement: analysis of breeding-value selection to increase 98-day body weight in tilapia. Aquaculture 212:93–113

Gjedrem T (2000) Genetic improvement of cold-water fish species. Aquacult Res 31:25–33

Grzybowski M, Sepulveda-Villet OJ, Stepien CA, Rosauer D, Binkowski F, Klaper R, Shepherd B, Goetz FW (2010) Genetic variation of 17 wild yellow perch populations from the Midwest and East Coast analyzed via microsatellites. Trans Am Fish Soc 139:270–287

Hershberger WK, Myers JM, Iwamoto RN, Mcauley WC, Saxton AM (1990) Genetic changes in the growth of coho salmon (*Oncorhynchus kisutch*) in marine net-pens, produced by ten years of selection. Aquaculture 85:187–197

Kincaid HL, Bridges WT, von Limbach B (1977) Three generations of selection for growth rate in fall-spawning rainbow trout. Trans Am Fish Soc 106:621–628

Leclerc D, Wirth T, Bernachez L (2000) Isolation and characterization of microsatellite loci in the yellow perch (*Perca flavescens*), and cross-species amplification within the family Percidae. Mol Ecol Prim Notes 9:955–997

Mambrini M, Medale F, Sanchez MP, Recalde B, Chevassus B, Labbe L, Quillet E, Boujard T (2004) Selection for growth in brown trout increases feed intake capacity without affecting maintenance and growth requirements. J Anim Sci 82:2865–2875

Murata O, Harada T, Miyashita S, Izumi K, Maeda S, Kato K, Kumai H (1996) Selective breeding for growth in red sea bream. Fish Sci 62:845–849

Olson W, Emmeneger E, Glenn J, Winton J, Goetz F (2012) Comparative susceptibility among three stocks of yellow perch, *Perca flavescens* (Mitchill), to viral haemorrhagic septicaemia virus (VHSV) strain IVb from the Great Lakes. J Fish Dis. doi:10.1111/jfd.12068

Rezk MA, Smitherman RO, Williams JC, Nichols A, Kucuktas H, Dunham RA (2003) Response to three generations of selection for increased body weight in channel catfish, *Ictalurus punctatus*, grown in earthen ponds. Aquaculture 228:69–79

Rosauer D, Biga P, Lindell S, Binkowski F, Shepherd B, Simchick C, Goetz FW (2011) Development of yellow perch (*Perca flavescens*) broodstocks: Initial characterization of growth and quality traits following grow-out of different stocks. Aquaculture 317:58–66

Chapter 27
Genetic Improvement of Percids

R.J.W. Blonk and J. Komen

Abstract During the past years, breeding programs for aquaculture have shown fast development. Globally, economically highly relevant species have experienced implementation of large scale breeding programs and it is impossible to imagine life today without them as they significantly improve production and profitability of enterprises. However, there are still many aquatic species cultured that rely on wild broodstock and for which there is no breeding program. The reasons for not having breeding programs are diverse: the knowledge to execute a breeding program is often not available, and more importantly, breeding programs are considered expensive. Costs for separate family rearing systems, testing environments, extensive tagging etc. are often limiting.

Farming of percids is a new sector where pioneering farmers have to develop rearing systems, reproduction methodology, fish feeds, etc., all at the same time. Especially in such cases, low-cost methods are required to get their business up and running. For this reason, many farms consider the foundation of a basic breeding program as their least concern, only to reduce costs. However, we argue that there are good reasons to start with selective breeding at the very start of an aquaculture enterprise.

In the next chapters, the principles of selective breeding programs will be described. This includes a basic description of the concept of estimating the heritable components of the phenotypic appearance of fish. Next the most commonly used selection methods and their implication for percids will be discussed. The potential traits for selection that should be relevant in percid culture are reviewed. Some insights into the optimisation of breeding programs and an overview of basic breeding program management will be presented. We present an outline of how to maintain genetic diversity within cultured stocks, with a special focus on limiting rates of inbreeding while selecting. Finally, some insights on how to manage costs and benefits of breeding programs are discussed.

Keywords Percids • Breeding programme • Genetics • Selection • Inbreeding

---

R.J.W. Blonk (✉)
Animal Breeding and Genomics Centre,
Wageningen University & Research centre, Wageningen, The Netherlands
e-mail: robbert.blonk@wur.nl

J. Komen
IMARES & Animal Breeding and Genomics Centre,
Wageningen University & Research centre, Wageningen, The Netherlands

P. Kestemont et al. (eds.), *Biology and Culture of Percid Fishes*,
DOI 10.1007/978-94-017-7227-3_27

27.1 Introduction

Selective breeding programs have two aims. First, breeding programs target to improve performance of populations on desired traits. Second, with breeding programs one can control genetic diversity, i.e. control rates of inbreeding to maintain a healthy population and to guarantee possibilities for selection in the future. The basic structure of breeding programs consists of a continuous cycle of steps from mating of the current broodstock, rearing of selection candidates, collection of data, and breeding value estimation and final selection of new broodstock to produce the next generation of selection candidates (Fig. 27.1).

During the past years, breeding programs for genetic improvement of aquaculture production populations have shown fast development. Globally, economically highly relevant species have experienced implementation of large scale breeding programs and it is impossible to imagine life today without them as they significantly improve production and profitability of enterprises. Species which nowadays are genetically improved through selective breeding are e.g. Atlantic salmon, rainbow trout, European sea bass, gilthead seabream, Nile tilapia, and several shrimp species (Chavanne et al. 2008). In 2010, the total number of breeding programs was estimated at 88, of which 62 were executed in developed countries (Neira 2010; Rye et al. 2010).

The level of technology used, both in infrastructure and in use of expertise, differs widely between breeding programs. Simple and relatively cheap programs are based on selection on own phenotype alone whereas more advanced programs combine individual selection with pedigree and sib information for multi trait selection, using separate family rearing and advanced selection methods based on statistical analysis. Genomic tools (molecular genetics) are currently only used for pedigree reconstruction and, in species with a large supporting industry such as e.g. Atlantic salmon, for marker assisted selection on specific disease resistance as e.g. IPN (Moen et al. 2009). In developing countries, the majority of breeding programs can be considered as "simple" whereas approximately 50 % of the breeding programs in developed countries can be considered as advanced.

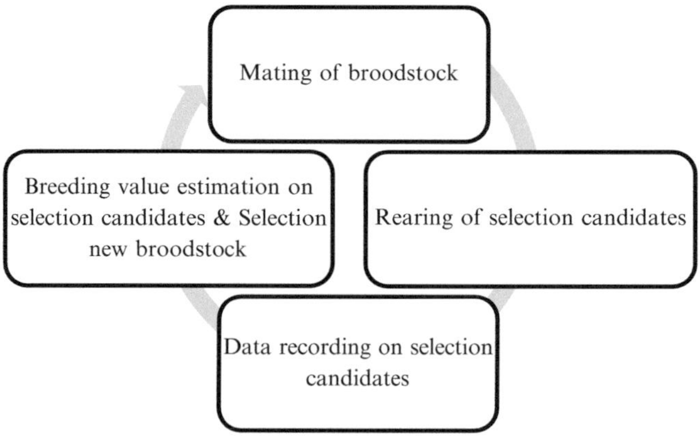

Fig. 27.1 Different steps in a breeding program with genetic selection of broodstocks

However, there are still many aquatic species currently cultured, that still rely on wild broodstock and for which there is no breeding program. The reasons for not having breeding programs are diverse: the knowledge to execute a breeding program is often not available, and more importantly, breeding programs are considered expensive. Costs for separate family rearing systems, testing environments, extensive tagging etc. are often limiting. It is therefore clear that for effective implementation of advanced breeding programs, there is a special need to lower barriers through e.g. reduction of scale and costs.

27.1.1 Breeding Programs for Percids

The availability of tailor made selective breeding programs is especially relevant for the aquaculture of percids such as eurasian perch and pikeperch, of which the production is still at a start-up phase (Meyer et al. 2012). Perch farming is a new sector where pioneering, individual farmers have to develop complete new farming systems including rearing systems, reproduction methodology, nutritional aspects, etc., all at the same time. Especially in such cases, low-cost methods are required to get their business up and running. For this reason, many farms consider the foundation of a basic breeding program as their least concern, only to reduce costs. However, there are good reasons to start with selective breeding at the very start of an aquaculture enterprise.

A first important reason is that aquaculture of new species is typically initiated with wild, undomesticated broodstock. Such animals are notoriously difficult to reproduce in captivity, if reproduction in captivity is realised at all. The use of wild broodstock also results in high variation in growth, slow growth and high (juvenile) mortality due to cannibalism, of offspring. High mortality and large variation leads to unpredictable and inefficient production which has negative impact on the economic feasibility.

A second, important reason is that farming of a new species is often initiated from the few individuals that did reproduce in captivity. This narrow genetic basis will rapidly lead to inbreeding if descendants are used as broodstock for next generations.

However, enterprises that focus on new species often try to maintain costs low by keeping few broodstock animals. Once the first hatchery reared broodstock becomes available, it is important to start using these animals in production, in order to advance domestication. However, because most farms only keep few broodstock founder animals, the genetic basis for a sound genetic variation is often too small. This means that selecting of animals cannot be done without rapidly rising rates of inbreeding. Too high rates of inbreeding directly lead to loss of genetic variation and loss of genetic potential for further genetic improvement. Inbreeding itself increases the incidence of malformations, increases disease susceptibility, and leads to reduced reproduction and suppressed growth rates (Frommen 2008; Gallardo 2004; Komen et al. 1992).

Introducing new wild broodstock, to increase the genetic diversity and to try to prevent unacceptable rates of inbreeding is the best strategy to follow as wild populations prove to be genetically diverse (Björklund et al. 2007; Nesbø et al. 1999; Säisä et al. 2010; Stepien et al. 2009), but a major drawback is that the domesticated status of the production stock will immediately be degraded, bringing back the problems mentioned earlier. Moreover, when introducing wild animals, there is also a high chance of introducing pathogens to the broodstock. Another option is to introduce unrelated domesticated stocks from a competitor, but this is not always easy or possible, especially when culturing new species. It is therefore important to start domestication and selective breeding, to reduce "wild" variation and to improve productivity at the earliest stage possible (Haffray et al. 2004; Olesen et al. 2003).

In the next chapters, the principles of selective breeding programs will be described. This includes a basic description of the concept of estimating the inheritable components of the phenotypic appearance of fish. Next the most commonly used selection methods and their implication for aquaculture will be discussed. The potential traits for selection that should be relevant in percid culture are reviewed. Some insights into the optimisation of breeding programs and an overview of basic breeding program management will be presented. In the final chapters, we present an outline of how to maintain genetic diversity within cultured stocks, with a special focus on limiting rates of inbreeding while selecting, and some insights on how to manage costs and benefits of breeding programs are discussed.

27.2 Genetic Components of Traits

Breeding programs make use of the fact that the appearance of traits is determined by both genetic and environmental factors. We can express the relation between appearance, genetic and environmental components as follows:

$$P = \mu + G + E$$

where P is the *phenotype* i.e. the observed appearance of a trait of an animal; μ is the mean of the population, G is the *genotype*, i.e. the component that is determined by the genes; E is the *environmental effect* that is determined by all other factors such as the used culture system, the feed, the management, temperatures, etc. (Falconer and Mackay 1996; Lynch and Walsh 1998). Many traits such as growth, fillet yield, (fillet) colour, shape, feed conversion rates, ability to digest plant-based diets, fecundity but also certain aspects of social behaviour such as aggressiveness are partly determined by genetic components.

In a breeding program, one may try to improve one trait or multiple traits of interest at the same time. The traits of interest that are subject to selection are collectively termed the *breeding goal*. To improve the traits of interest, individual animals are selected based on an estimate of the genotype, because genotypes cannot be observed directly on the individual. In this context, the estimated genotype of the animal is called the estimated *breeding value* (A).

27.2.1 Variance and Heritability

In order to select new parents based on their breeding values, one needs to know the genetic variation of populations. To assess genetic aspects of traits, one generally considers variation around the mean of the trait of interest in populations. For this reason, the relations between phenotypic, genotypic and environmental components are expressed in terms of variance (Falconer and Mackay 1996). Here, V_A is the genotypic variation for a trait in the population whereas V_P is the phenotypic variation. To a larger or smaller extent, phenotypes can be determined by the genotype. The magnitude to which phenotypes are set by the genotypes is called the *heritability*. The heritability (h^2) is calculated as the ratio between the genetic variance and the phenotypic variance in the population:

$$h^2 = V_A / V_P.$$

The higher the heritability, the stronger the phenotype is determined by the genotype; "favourable" phenotypes are in this case the result of "favourable" genotypes. Traits that have low heritability are less affected by the genotype and are thus less of interest for selective breeding.

Theoretically, the maximum heritability is 1. However, for most fish species and traits, heritabilities vary between approximately 0.2 and 0.5. This implies that around 20–50 % of the phenotypic variation in the population is determined by the genotype (Gjedrem 2000). For percids, heritability estimates are hardly reported. For yellow perch (*Perca flavescens*), heritability for bodyweight at 2 years was be found to be 0.14 ± 0.09 (Cao et al. 2012), but this very low value was probably due the experimental design. Heritabilities for harvest weight or growth in other cultured species that belong to the order of the *Perciformes* (sea bass, sea bream, tilapia, barramundi) are all within range of 0.2–0.5 (Antonello et al. 2009; Domingos et al. 2013; Dupont-Nivet et al. 2008; Trọng et al. 2013a).

27.2.1.1 Resemblance Within Families

The extent to which phenotypes of some traits are partly determined by genetic background is revealed by the similarity of individuals that belong to the same full sib family. Because parents transmit half of their genes to the offspring (simplified in Fig. 27.2), the genotypes of offspring are basically set by the genotype of their parents. Full sib families have the same parents in common, and therefore genotypes within families will largely resemble each other. Especially when traits are strongly heritable, genetic resemblance between members of the same family implies that there will be more phenotypic similarities and less variation within families. At the same time, there will be more phenotypic differences and more variation between families. In such cases, families with a relative favourable genotype will exhibit better performance than families with less favourable genotypes. This principle is utilised by breeding programs; selection and reproduction of animals with the best genotypes results into offspring with better genotypes.

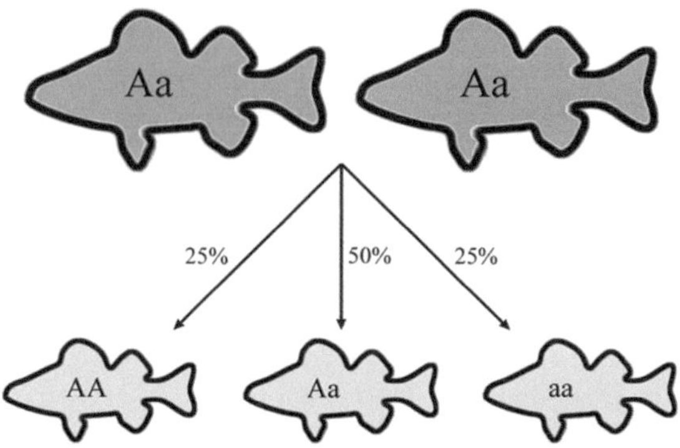

Fig. 27.2 Schematic overview of inheritance of alleles from parents to offspring. Both parents (*top*) are heterozygous for the gene with alleles "A" and "a". Produced offspring genotypes are shown (*bottom*) with probability of occurrence

Box 27.1

Although family members resemble each other for a large part, individuals within families will not be genetically identical. This is because the exact genotype of an animal is largely the result of random sampling from its parental genes. During formation of the germ cells, one copy ("allele") from each gene is transmitted to each sperm or egg cell. This process is based on chance; in an animal that has a heterozygous genotype "Aa" for a certain gene, there is a probability of 50 % that either "A" or "a" is transmitted, independently in each single produced egg or sperm cell. Having the same father and mother, full sibs will have a high chance for inheriting the same parental alleles for all genes and will thus be genetically very similar. Moreover, full sibs will often also resemble each other phenotypically, provided that they are subjected to the same environmental treatment. However, because the process of transmitting parental genes is based on chance, animals will very likely not be genetically, nor phenotypically, identical (unless when they are 100 % inbred, or clones). When crossing two heterozygous animals "Aa", possible genotypes in the offspring are either "AA" (in 25 % of the cases), "aa" (in 25 % of the cases), or "Aa" (in 50 % of the cases). This uncertainty is called the "Mendelian sampling" effect (Falconer and Mackay 1996; Lynch and Walsh 1998). Due to this "uncertainty of inheritance", the true genotype of an animal is not known (at least not without whole genome sequencing). As a consequence, because there are unknown genetic differences between sibs within families, there will also be unexplained differences in phenotypes of siblings within families (see Fig. 27.2).

The explanation in the above example was based on only one gene. However, for most traits considered, multiple to even thousands of genes play a role, of which many are still unknown. This implies that in most cases, it will be virtually impossible to know the exact genotype, i.e. all genes, of an animal. However, with the rapid development of e.g. whole genome sequencing methodologies, profiling of exact individual genotypes may come within reach, although costs for such methodology will yet be considerable.

27.2.2 Selection

Genetic selection assumes that traits are heritable and that populations will be improved by selecting and reproducing animals with the best genotypes. If done correctly, this results into offspring with favourable genotypes (and often better phenotypes). The principle that selection of parents with high breeding values leads to production of offspring with higher genotypic and phenotypic performance can also be explained using the following (simplified) equation:

$$P_{offspring} = \mu + 1/2A_{father} + 1/2\,A_{mother} + MS + E$$

where $P_{offspring}$ is the phenotype of the offspring, μ is the population mean, A_{father} is the true breeding value of the father, A_{mother} is the true breeding value of the mother, *MS* is the Mendelian sampling effect (uncertainty of inheritance) and *E* the environmental effect. It follows that the phenotypic deviation of the offspring from the population mean is partly determined by the breeding value of both parents, but also by the Mendelian sampling effect, and by the environment.

In breeding programs, animals are selected based on their breeding value. However, due to Mendelian sampling, the true breeding value of animals is not known. Therefore, breeding values of animals need to be estimated. For estimation of breeding values, there are several methods which will be discussed in the next chapter.

27.2.3 Selection Response

Figure 27.3 illustrates the effect of selection for animals with the best phenotypes for a trait with a high heritability. Consider a population (generation 0, G0) where the trait of interest is normally distributed. From this population with "selection candidates", a number of animals is selected with selection on phenotypic performance. Because heritability is high, we assume that these animals also have a better breeding value than other animals in the population. The selected group of animals is then used to produce the next generation (G1), and from this population again a group of animals is selected to produce the next generation (G2), etc. Because the selected animals have relative good breeding values, the mean of the offspring population will be improved when compared to the mean of the parental generation.

The response to selection depends on several factors. As shown before, the heritability is of importance because when a trait is not or hardly determined by genetics, genetic selection will not be effective (Falconer and Mackay 1996). The higher the heritability, the more effect can be expected from selective breeding. As explained before, the heritability can be expressed as the ratio between genetic and phenotypic variance. Rewriting the equation $h^2 = V_A/V_P$ as $V_A = h^2\,V_P$ shows that phenotypic variation is as important as heritability in determining the amount of genetic variation V_A. From Fig. 27.3 it can be seen how phenotypic variation is of importance. With more variation around the mean, it becomes easier to select the "best"

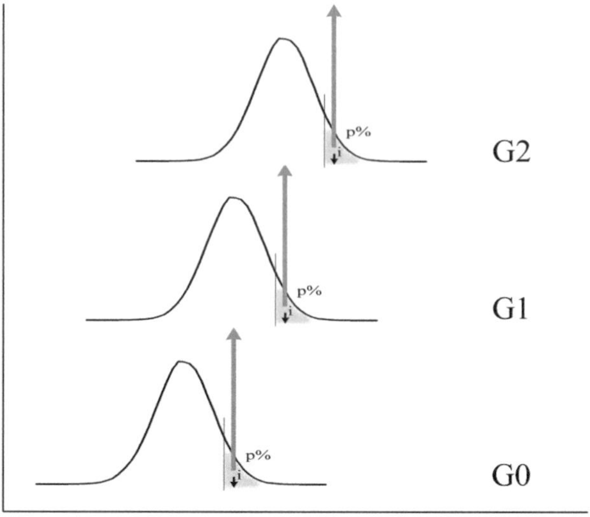

Fig. 27.3 Distribution of performance of animals for a trait with high heritability over three consecutive selected generations. Reproduction of a selection of the best performing selection candidates with a selection percentage p% = "number of selected parents"/"total number of selection candidates", (*grey arrow*) in generation 0 (G0) results in the improved mean of generation 1 (G1). Selection and reproduction of p% from generation 1 (G1) results in improvement of the mean of generation 2 (G2), etc.

animals, even when the heritability is low. With considerable heritability but very little phenotypic variation between animals, there is little difference between animals and thus selection of the "best" animals is more difficult. Under a constant rearing environment, both heritability and variation are given parameters and biological facts. This means, practically seen, that we cannot alter them, at least not in the short term, to maximise the selection response. However, one important factor that can be used to maximise selection response is the selection percentage and the selection intensity (Falconer and Mackay 1996). The selection percentage is the percentage of animals that is selected from the current generation of selection candidates and that is used as broodstock to produce the next generation of selection candidates. The selection percentage is expressed as the ratio between the number of selected parents and the number of selection candidates. The selection intensity is mathematically derived from the selection percentage. When a trait is normally distributed, and animals are selected by truncation, the selection intensity can directly be derived from the selection percentage. When the selection percentage is small, i.e. when only the few very best animals are selected from the population of selection candidates, the selection intensity is high and v.v. The exact method to calculate the selection intensity from the selection percentage is not shown here but can be found in e.g. Falconer and Mackay (1996). Some typical values for selection percentages and their selection intensities are given in Table 27.1.

Table 27.1 Typical values for selection percentages and their selection intensities

Selection (%)	Intensity i
20.00	1.400
10.00	1.755
5.00	2.063
1.00	2.665
0.50	2.892
0.10	3.367

With high selection intensities, response of selection is high. The expected approximate response of selection (R) with selection on own performance is calculated as follows:

$$R = i * h^2 * \sigma_P$$

Where R is the response to selection, i is the selection intensity, h^2 is the heritability and σ_P is the phenotypic standard deviation, a measure for the phenotypic variation of the trait of interest. The selection percentage is also an important parameter for rates of inbreeding during selective breeding. This will be explained later.

27.2.4 Correlated Response to Selection

Within breeding programs, there is a chance that selection for certain traits of interest (primary traits) has an effect on a second trait that is genetically correlated to the primary trait. When genetic correlations between primary and secondary traits are unfavourable, selection for the primary traits may have negative side effects on the secondary trait.

In some cases, growth rate and feed conversion rates show favourable correlations (Kause et al. 2006; Quinton et al. 2007). Similarly, harvest weight and fillet yield are positively correlated in Nile tilapia and rainbow trout (Rutten et al. 2004; Sae-Lim et al. 2013). This means that selection for improved harvest weight or growth rate will result in a moderate increase in these traits. Unfavourable correlations have been found in fish in several selection experiments. For example, in Atlantic salmon potentially unfavourable genetic correlations were found between body weight and muscle or intestinal fat (Rye and Gjerde 1996). In common sole, an unfavourable correlation was found between selection for fast growth and shape (Blonk et al. 2010). Although debated (Doyle and Talbot 1986), selection for growth has been associated with aggressive anti-social behaviour in the animal kingdom (Bijma et al. 2007; Ellen et al. 2008). As indicated previously, aggressiveness and cannibalism are a problem of major concern in percid culture already and if selection for growth would make the current situation worse, it should be avoided if not counteracted. Unfavourable correlations between selection for fast growth and reproductive characteristics as early puberty have been shown in some species, e.g. rainbow trout (Kause et al. 2004; Neira et al. 2006) but not in Nile tilapia (Trọng et al. 2013b). It is not clear if such an unfavourable correlation of growth rate with early puberty and maturation would also exist in perch or pikeperch. Undesired effects on secondary traits due to correlated responses can be corrected by applying negative selection against the correlated secondary trait. However, this will reduce the response to selection in the primary trait(s) in the breeding goal.

27.3 Traits for Selection

In developing sectors such as culture of most percids, there is currently no or limited targeted genetic improvement (Meyer et al. 2012). The traits for new breeding programs should therefore focus on the primary goal of (probably) most starting enterprises; enhanced growth to improve farm economics.

By selecting the first one or two generations, selection responses and effects of indirect selection for domestication will become visible. Examples of such improvements are improved growth, reduced mortality, enhanced feed uptake, reduced stress to handling and reduced variation in size at harvest (Fleming et al. 2002; Lepage et al. 2000). Gradually, with more generations of selection, the breeding goal may be fine-tuned and extended by adding more relevant traits e.g. fillet yield, production efficiency, disease resistance, product quality, digestion and performance on plant-based diets, etc. Some potentially important traits for selection in percids are outlined below.

The current aquaculture systems used, such as ponds flow-through raceways and RAS production systems, may require different breeding goals as well and the traits that one should select for may depend on the environment where production takes place. For example, when the production is done in ponds or outdoor raceways, then the breeding goal may be different in cold or temperate climates. Under cold conditions, winter survival and cold tolerance or growth in cold conditions may be important traits for selection, because in winter, growth may become depressed and mortalities may occur (Lappalainen et al. 2000). In RAS, such environmental and seasonal fluctuations will be less relevant. In recirculation systems, special attention should be paid to traits related to production efficiency because the water treatment units required for RAS heavily depend on the waste production. This means that next to growth rate, traits related to feeding efficiency (FCR, feed intake behaviour etc.) and product quality traits should be paid attention to.

Other traits of major importance may relate to the targeted market for sales. For example, one may need to select for "normal wild" shape or colour to guarantee good product appearance if fish are sold whole (i.e. not as fillets). In several fish species, it was shown that selection of shape can be of importance because consumer acceptance of the product may be negatively affected when the final product is a whole fish, as is often the case for perch and pikeperch (van Duijn et al. 2010). Shape has proved to have a unfavourable correlation with selection for fast growth, i.e. faster growing fish become more round and rotund (Blonk et al. 2010; Kause et al. 2003; Rutten et al. 2004; Trọng et al. 2013c). For fish that are sold filleted, alteration of normal shape is of less importance and may even lead to higher fillet yields (Rutten et al. 2004; Trọng et al. 2013c).

Aggression and cannibalism are a major problem in culture of percids, in particular for pikeperch (Brabrand 1995; Fessehaye et al. 2006a; Kestemont et al. 2003). It is clear that environmental conditions play an important role but large variation between individuals in a group are the main cause for many species (Baras et al. 2003). In Nile tilapia, Fessehaye et al. (2006a) showed that cannibalism is a simple function of prey size and mouth size of the predator. Reducing variation by grading and removing off cannibals is a well-established practice to counteract cannibalism. As explained before, variation is caused by environmental and genetic factors. However, recently it was shown that also genetic aspects of/in social interactions between animals were found relevant (Hecht and Pienaar 1993; Monsen et al. 2010; Smith and Reay 1991). The role of genetics in social interactions suggests that selection can be a tool to reduce losses from cannibalism and aggression. However, practical examples of responses to selection for reduced aggression are scarce. In Nile tilapia, early experiments on group selection to reduce competitive behaviour had no effect (Rutten 2005). Therefore, without direct selection to reduce aggression, it is of primary importance that competitive effects and aggression are reduced when selection candidates are challenged for fast growth, e.g. by keeping rearing conditions such that all candidates have equal access to food. One important reason for such a strategy is that selection for fast growth seems to go hand-in-hand with (indirect) selection for greedy, competitive and aggressive fish (Ruzzante and Doyle 1991). By keeping feeding conditions equal for all candidates, fast growing but less aggressive candidates will have a better chance to be selected. Additional normal culture practices such as grading should be performed to minimise effects of variation of size on aggression and cannibalism within the cultured stocks. However, it is important that cultured stocks are graded multiple times and that different size fractions are recombined continuously in order to create equal grow out conditions for all selection candidates. This is of high importance to prevent confounding of rearing environment and families and to be able to calculate unbiased breeding values (as explained in the beginning of this chapter). Methodology for these methods and the possibility to apply grading have been described by Chevassus et al. (2004) and Blonk et al. (2010).

Once the sector for a new aquaculture species is large enough, organised breeding by a (few) central breeding companies may be a feasible economic strategy. Focussing entirely on selective breeding and reproduction aspects, and providing improved fry for production to grow out farms or improved broodstocks for hatcheries, will increase cost-efficiency for a breeding program due to scaling effects. However, there are some drawbacks. In general, fish from such a centralized breeding program should be able to perform in a wide range of, possibly suboptimal production environments to preserve good performance in the different customer farm conditions, the breeding goal may need to include robustness of animals. Whether or not performance is dependent on the environment may be quantified by estimating the genetic correlation for growth (or any other trait) between the nucleus environment (i.e. the environment where the breeding program is conducted) and any production environment. When these correlations are less than 0.7–0.8 (Mulder 2007; Sae-Lim et al. 2010) genotype by environment interactions is present, and different selection lines may be required for different environments. For example, when genotype by environment interaction exists for fast growth of percids in ponds and RAS, selection for fast growth in ponds will result in sub-optimal responses for growth in RAS. Similarly fish selected for RAS conditions are likely to have lower than expected responses in ponds. Of course, the choice for having different selection lines also depends on the cost benefit ratios.

27.4 Breeding Value Estimation

27.4.1 Selection on Own Performance

The simplest method for genetic selection is the so-called selection on "own" or "phenotypic" performance, also called "mass" selection. With own performance selection, selection candidates are grown "en masse" with standardised family sizes in the same environment, being mostly one tank or pond. Under such culture conditions it is assumed that the differences between animals are solely attributable to genetic differences and that there are no major different environmental effects between animals. Animals are then selected only on their own performance and family information (e.g. from pedigrees) is not included in the analysis (Falconer and Mackay 1996). Breeding values (A) of the selection candidates can be estimated using the simple formula $A = h^2 * (P_{individual} - P_{mean})$. For own performance selection, selection candidates normally need not be genotyped nor tagged.

The absence of extensive infrastructure, techniques and expertise implies that this method is relatively simple and cheap. Good selection responses have been shown for this method for several fish species (Blonk 2010; Chevassus et al. 2004; Hulata et al. 1986; Villanueva et al. 1996). However, the major downside of selection on own performance is that animals can only be selected on one trait, and it should be possible to observe this trait on the selection candidates without killing them. Another disadvantage of this method is that there is certain chances that many selected animals originate from the same families (Bentsen and Olesen 2002; Blonk et al. 2009). Selection of too few families for production of the next generation will result in increased rates of inbreeding (see Chap. 5).

This can happen for instance when selective mortality of families due to e.g. diseases occurs during rearing of selection candidates, especially when both heritability and selection intensity are high. In such a situation, there is a larger chance of selecting families that experienced lower mortality. Although this may be a favourable method of natural selection when diseases are the reason for selective heritable mortality, this effect also increases the rate of inbreeding because the chance of selecting few families increases due to unequal family representation (Blonk et al. 2009; Fessehaye et al. 2006b). With selection on own performance, control on relationships in selected animals is limited and thus there is no room for corrective actions, such as trying to include more families in the selected group.

27.4.2 *BLUP-Selection*

A second, and more advanced method of selection is "Best Linear Unbiased Prediction" (BLUP) of breeding values. BLUP is a statistical method using mixed linear model theory to calculate breeding values of animals while accounting for fixed effects (such as different tanks or ponds, time of production, gender, etc.) and pedigree information. With BLUP of breeding values, the breeding value of an animal is estimated using not only its own performance, but also the performance of its relatives, i.e. full sibs, half sibs, parents etc. (Henderson 1984; Mrode 2005). An important advantage of this method is that the breeding value of an animal can be more accurately estimated once its relatives also have a recorded phenotype. The genetic response to selection on BLUP breeding values is therefore generally higher than with selection on own performance.

Another important advantage of this method is that many traits can be analysed, including those for which an animal needs to be killed, e.g. slaughter traits, fillet yield measured on slaughtered animals, and disease resistance traits (Gjedrem and Gjøen 1995; Rutten et al. 2005; Thodesen Da-Yong Ma et al. 2012). For example, if fillet yield is a trait of interest, as is the case for many perches, including this "lethal" trait in the breeding goal is possible. Information from slaughtered sibs then provides information on the living selection candidates.

One drawback of BLUP of breeding values is that its higher accuracy of breeding value estimation, also promotes the chance of selecting multiple members of the same family. As with selection on own performance this is likely to increase rates of inbreeding. However, because with BLUP selection family information from selection candidates is known, one can implement methods to select the optimal number of sibs/families in order to maintain genetic diversity (Blonk 2010; Hinrichs et al. 2006; Sonesson 2005) and to control levels of inbreeding. Several software packages such as GENCONT (Meuwissen 2002) are available to do this in practice.

27.4.2.1 Recording Pedigree Information

A complicating factor with regard to selection on BLUP breeding values is that the pedigree of every individual fish (selection candidate) needs to be known. For most terrestrial species in animal husbandry, such as cattle or pigs, keeping track of pedigrees is relatively easy. Parents are known and their progeny are relatively large at birth. This means they can be tagged directly with a label or transponder to be able to track them throughout their entire life. However, in fish the situation is more complicated. Parents are only known when mating is completely controlled, which is currently not the case for most percids. Furthermore, hatched fish larvae are too small to be tagged directly. This means that most fish need to be reared in tanks as separate full sib family and grown until tagging size is reached. When parents are known, fry are reared as large full sib families in one tank until tagging. One disadvantage of this procedure is that considerable facilities for separate family rearing are needed, which greatly increase costs of breeding programs. Next to that, because families are grown out in their own tank, this tank environment may bias the estimation of the breeding values of the animals in the tank. In such cases it will be impossible to disentangle genetic from environmental effects between families because both tank and breeding values of families are completely confounded, especially just after separate rearing (Blonk et al. 2010). Nevertheless, many breeding programs use this method because most of the time, selection only takes place at harvest size, where the effect of initial tank is believed to be diluted enough not to seriously affect the breeding value estimation.

When parents are not known, e.g. because fry were produced by natural mating of groups (as is often the case with pike perch) or when offspring families where mixed deliberately after egg collection from stripping (Rónyai 2007), pedigrees need to be reconstructed using genotypes of selection candidates and their potential parents (Blonk et al. 2010). With the use of genetic markers, such as microsatellites as determined for percids (Björklund et al. 2007; Kohlmann and Kersten 2008; Säisä et al. 2010; Stepien et al. 2009) or single nucleotide polymorphisms (SNP), one can determine a genotype based on a number of markers for each sampled individual animal. By comparing genotypes of offspring and parents using principles of Mendelian inheritance (Duchesne et al. 2002; Marshall et al. 1998; Taggart 2007; Vandeputte et al. 2006), pedigrees can be reconstructed and used for genetic analysis and estimation of breeding values.

An important advantage of pedigree reconstruction after genotyping is that animals (often larvae) of different families can be pooled and reared until a desired size when tagging is possible, thereby avoiding bias of BLUP due to tank effects, and thus improving genetic analyses (Saillant et al. 2007; Vandeputte et al. 2004, 2008; Wilson and McDonald 2003). However, genotyping is still rather expensive, especially for small scale industries (estimated commercial prices often start at 20,- to 30,- Euros per sample (Blonk 2010)).

Box 27.2

Genomic selection is a group of novel methods for genetic improvement which utilise extensive genome wide genotyping to associate DNA markers with traits or to provide information on (molecular) genetic relatedness (Meuwissen 2007; Sonesson and Meuwissen 2009). Using detailed genomic information of individuals, it becomes possible to perform targeted selection on traits that are determined by one or few genes, and to more accurately estimate breeding values of animals because more information on the Mendelian sampling effect becomes available (Hayes et al. 2009). The current breeding programs that perform genomic selection do so by integrating current established BLUP selection methodology with genomic data.

Until now, genomic selection methods are used for livestock species for which large industries exist, such as poultry, pig and dairy cattle. In fish first

(continued)

Box 27.2 (continued)
initiatives have been started for Atlantic salmon. At this moment, genomic selection methodology requires too high investments to be used efficiently by smaller sectors. This is mainly because large scale genomic resources are almost absent for aquatic species. However, statistical methodology to efficiently use genomic information in selective breeding programs is available. Consequently, when genomic information becomes available for more fish species, genomic selection can readily be applied. However, costs for genotyping will still be considerable and therefore only relevant for those species where investments can be matched by appropriate returns. An example is seen in salmon farming where marker assisted selection is used to select against infectious pancreatic necrosis (IPN) (Moen et al. 2009).

27.5 Design of Breeding Programs

Decisions on the most optimal number of parents and selection candidates, and the optimal method of selection of course depend on the cost benefit ratios of the breeding program. To determine the optimal design, the benefits from increased responses (improved growth rates, reduced mortality etc.) are balanced against additional costs for facilities, tagging of animals, genotyping, keeping broodstocks, efforts for testing, etc. (Gjedrem 2000; Ponzoni et al. 2007). Optimisations based on costs-benefit ratios are very enterprise specific and therefore not considered in this chapter, but in this paragraph, it is shown how chosen dimensions of breeding programs can be optimised such that too high rates of inbreeding are prevented, while responses to selection are as high as possible. Calculations in the following are modelled using SelAction software (Rutten et al. 2002).

27.5.1 Inbreeding

To ensure possibilities for selection in future generations and prevent excess inbreeding of stocks, it is crucial to collect base populations with sufficient genetic variation and low levels of inbreeding. This is mostly the case in new species because these are generally collected from the wild. However, as explained earlier, the number of broodstock animals to start with should be sufficiently large to control the *rate of inbreeding* in the subsequent generations. The rate of inbreeding is generally expressed as the increase in the proportion of homozygous animals per generation, relative to the remaining proportion of heterozygous animals:

$$\Delta F = \left(F_t - F_{t-1}\right) / \left(1 - F_{t-1}\right).$$

Where ΔF is the rate of inbreeding and F_t is the level of inbreeding in generation t and $Ft\text{-}1$ is the level of inbreeding in the previous generation. A generally accepted rate of inbreeding in breeding programs is 0.5–1 % per generation, i.e. 1 % increase in the relative number of homozygotes in the population per generation of selection.

Breeding programs should be designed in such a way that rates of inbreeding are controlled by restricting the level of relatedness in the selected population, i.e. by restricting the number of related animals (family sizes) and by maximizing the number of families (Dupont-Nivet et al. 2006). When too few parents are used to produce the next generation of selection candidates, this new generation will contain a limited number of families. A limited number of families available for the next selection step increases the chance of mating related animals, e.g. brother-sister mating or cousin mating, which directly leads to inbreeding (Falconer and Mackay 1996).

With mating of related animals, the chance increases that offspring in the population become homozygous for certain deleterious genes. This may lead to so-called inbreeding depression: increased susceptibility to diseases, malformations, depressed fertilisation rates, and depressed growth (Fessehaye et al. 2007, 2009; Frommen 2008; Gallardo 2004; Gjerde et al. 1983; Komen et al. 1992).

27.5.2 Balancing Selection and Inbreeding: Numbers of Parents and Selection Candidates

In the previous it was stated that rates of inbreeding strongly depend on the number of selected parents used to produce the next generation. This is off course true, but the rate of inbreeding also depends on the total number of selection candidates. In Sect. 27.2.3 we explained that selection response can be maximized by making a large population of selection candidates (much variation to choose from) and by selecting the best fish as future broodstock candidates. Together with the number of parents, the total number of selection candidates determines the selection percentage or selection intensity. For example, consider selecting 100 parents out of 1,000 selection candidates. In this case, the selection percentage is 10 %. Stronger selection as e.g. selection of 100 parents out of 10,000 selection candidates (selection percentage = 1 %), leads to higher responses as these fish are assumed to be the best of the best. However, rates of inbreeding will also increase because the chance that related individuals are over-represented in the selected sample becomes higher, especially when heritability is high. Furthermore, selecting 50 parents out of 5000 selection candidates or selecting 100 parents out of 10,000 selection candidates yield a selection percentage of 1 % in both cases. However, in the first situation, the rate of inbreeding will be higher because fewer parents are selected.

Figure 27.4 shows the relationship between numbers of selected broodstock and numbers of selection candidates. Seeking the right balance in these numbers, to maximise gain and restrict rates of inbreeding, is mainly important with respect to the nucleus of the actual breeding program. From a production point of view, it is not problematic to use only a few parents for production of offspring for grow out,

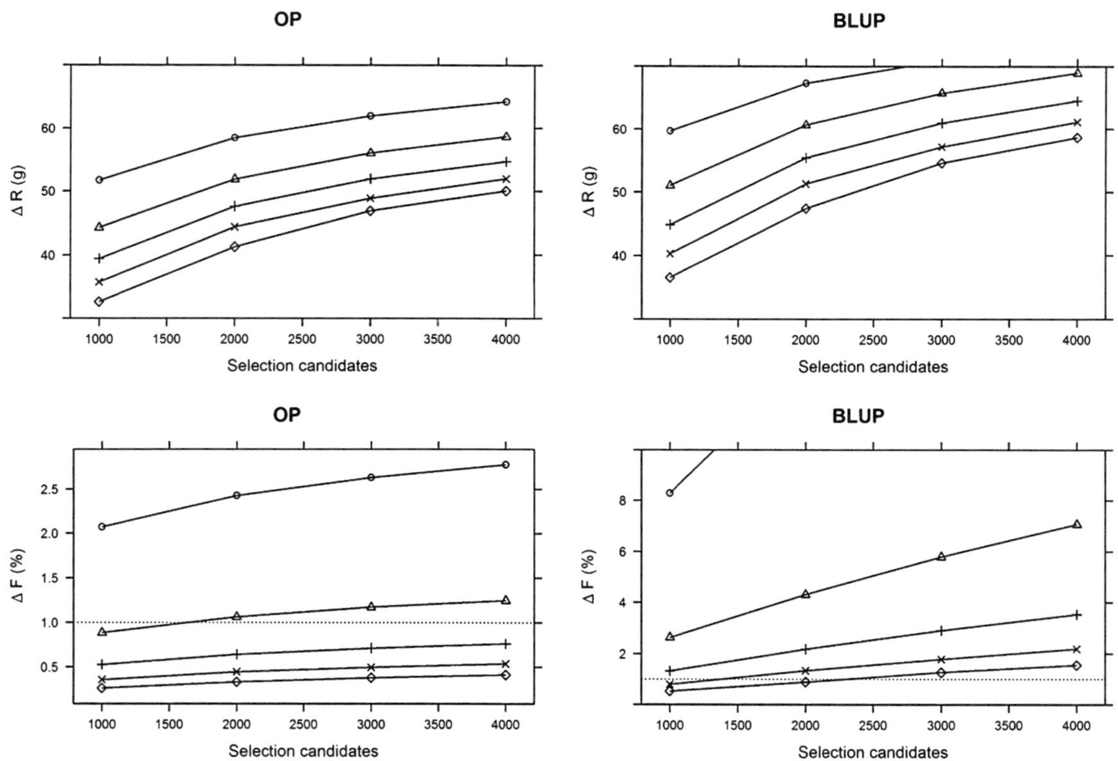

Fig. 27.4 Effect of number of selection candidates and number of selected parents on response of selection ΔR (*top*, in g per generation) and rates of inbreeding ΔF (*bottom*, in % per generation) after simulated selection on own performance (OP, *left*) and with BLUP (BLUP, *right*) of breeding values with SelAction software (Rutten et al. 2002). The 1 % rule of the maximum allowed rate of inbreeding is indicated with the *dotted line*. Mind the differences in scale for ΔF. For BLUP, the extreme values dropped off. Simulated parameters: $h^2=0.3$ and $\sigma_P=100$ g. The following numbers of parents (sex ratio 1:1) were simulated: 50 (*circle*), 100 (*triangle*), 150 (*plus*), 200 (*cross*) and 250 (*diamond*)

as the offspring is not used for selection of the next generation. In fact, using only few parents to produce the offspring for grow out even has advantages, because with few families, the variation between production animals will be lower than when many families are used. In addition, when the target of the enterprise is selling live animals such as fingerlings, providing only a few families is a form of product protection; the customer is provided with only a very limited genetic source and it will become difficult to use the acquired animals for own selection purposes because inbreeding will quickly become a problem.

27.5.3 Optimising Breeding Programs

Consider a breeding program with selection of 100 parents out of 2000 selection candidates in every generation. In this case, the selection percentage is 5 % (=100/2000). Assuming a phenotypic standard deviation of 100 g, a heritability of 0.3, and selection based on own performance, the response to selection will be approximately 51.9 g per generation whereas the rate of inbreeding will be around 1.1 % per generation. Stronger selection as e.g. selection of 50 parents out of 2000 selection candidates (selection percentage = 2.5 %), will yield 58.5 g per generation. However, in this case also the rate of inbreeding is higher: 2.4 % per generation. Performing the same scenarios with BLUP selection yields 60.7 g per generation for the response and 4.3 % per generation for the rate of inbreeding in the first example, and 67.3 g per generation for the response and 13.5 % for the rate of inbreeding in the second example. To run BLUP selection with lower rates of inbreeding, one option is to increase the selection percentage by either lowering the number of selection candidates or increasing the number of selected parents. For example, when selecting 250 parents from 2,000 selection candidates, 47.4 g per generation for the response and 0.9 % for the rate of inbreeding are achieved.

In Fig. 27.4, the effect of different numbers of parents and selection candidates on predicted response and rates of inbreeding is shown for both own performance selection (OP) and selection using BLUP of breeding values. Mind the differences in scale. It can be concluded that when using higher selection percentages, the responses will be higher, as are the rates of inbreeding. Clearly, response to selection with BLUP is higher, but rates of inbreeding are also unacceptable high for some scenarios. To maintain rates of inbreeding below the generally accepted baseline level of 1 % per generation, as a rule of thumb, broodstock sizes need to be at least 150–200 animals, while the pool of selection candidates should not exceed 3000 animals.

However, with BLUP, it is possible to optimise selection responses while restricting rates of inbreeding using optimal contribution methods (Meuwissen 2002). With BLUP, more facilities are required for e.g. tagging and/or genotyping and possibly for infrastructure as family rearing tanks. Clearly, there is need for optimisation to balance the optimal response, rate of inbreeding and requirement for facilities.

27.5.4 *Mating Designs*

It is important that all selected parents are equally contributing to the production of the next generation of selection candidates. When not all parents are contributing, the effective population size is smaller than intended, thus potentially leading to higher rates of inbreeding. Control on the reproduction and contribution of each parent thus is important. Although an occasional drop out of a parent may occur without causing any problems, any breeding program should incorporate extra backup parents, in case some animals are not reproductive or die.

It is also important that there is a correct family structure with enough "genetic links" between produced families of selection candidates. Optimal family structures are derived by using (at least partially) factorial mating designs where each parent, or at least one of both sexes, is mated with multiple partners. One reason for the use of factorial mating designs is to maximise genetic diversity during selection and thus limitation of inbreeding, because this method preserves a broad flow of genes from each generation to the next. A second reason is that this method increases response to selection (Dupont-Nivet et al. 2002, 2006; Sorensen et al. 2005). Optimal mating designs are full factorial, where each male is mated with all females (and logically, each female is mated to all males). However, such a mating design is very complex, demands extensive labour and infrastructure and may in some cases not be (biologically) feasible. For practicality, in most cases it is sufficient to create partial factorial (each male/female is mated with multiple females/males) mating designs to perform sufficiently strong genetic analyses (Dupont-Nivet et al. 2006).

It is important to realise that mating designs are only intended for production of the next generation of selection candidates and not necessarily for production of grow out animals. For the latter, one could very likely suffice with using only the best parents available.

27.6 Costs and Benefits of Breeding Programs

In practice, advanced fish breeding programs can yield improvement of growth rates with 10–15 % per generation (Gjedrem 2000). However, such conventional breeding programs are usually designed for large scale production and are mostly laborious, require a lot of infrastructure and resources, and are therefore costly to implement.

For many small scale, pioneering aquaculture enterprises under development, such large scale breeding programs may be regarded as not feasible. Breeding programs for "new" species such as percids are however important to make aquaculture of these species feasible, both from biological and economic perspectives. Small aquaculture farms can therefore implement (tailor made) low-cost breeding programs that do not require large investments or separate facilities. The basic principle of such "tailor made" breeding programs is to minimise the number of facilities and to integrate breeding activities with existent farm infrastructure as much as possible. The benefits then easily outrun the investments.

For example, by using genotyping to construct pedigrees for selection, it is no longer needed to perform separate family rearing for many families, a major source of costs in aquaculture breeding programs, and animals may be selected from production stocks directly. Post-harvest genotyping would be used to assess the molecular relatedness with other animals and even (with sufficient markers) the molecular breeding value. With these methodologies, breeding programs may also become possible or species that are (partly) dependent on natural mating in groups, such as some percids. With current insights in methodology and decreasing costs for genetic DNA profiling, it will be possible to design tailor made breeding programs that can be operated on farm at relatively low effort.

References

Antonello J, Massault C, Franch R, Haley C, Pellizzari C, Bovo G, Patarnello T, de Koning D-J, Bargelloni L (2009) Estimates of heritability and genetic correlation for body length and resistance to fish pasteurellosis in the gilthead sea bream (Sparus aurata L.). Aquaculture 298:29–35. doi:10.1016/j.aquaculture.2009.10.022

Baras E, Kestemont P, Mélard C (2003) Effect of stocking density on the dynamics of cannibalism in sibling larvae of Perca fluviatilis under controlled conditions. Aquaculture 219:241–255. doi:10.1016/S0044-8486(02)00349-6

Bentsen HB, Olesen I (2002) Designing aquaculture mass selection programs to avoid high inbreeding rates. Aquaculture 204:349–359

Bijma P, Muir WM, van Arendonk JAM (2007) Multilevel selection 1: quantitative genetics of inheritance and response to selection. Genetics 175:277–288. doi:10.1534/genetics.106.062711

Björklund M, Aho T, Larsson LC (2007) Genetic differentiation in pikeperch (Sander lucioperca): the relative importance of gene flow, drift and common history. J Fish Biol 71:264–278. doi:10.1111/j.1095-8649.2007.01609.x

Blonk RJW (2010) Selecting sole: breeding programs for natural – mating populations. PhD thesis, Wageningen University

Blonk RJW, Komen J, Kamstra A, Crooijmans RPMA, van Arendonk JAM (2009) Levels of inbreeding in group mating captive broodstock populations of common sole, (*Solea solea*), inferred from parental relatedness and contribution. Aquaculture 289:26–31

Blonk RJW, Komen H, Kamstra A, van Arendonk JAM (2010a) Estimating breeding values with molecular relatedness and reconstructed pedigrees in natural mating populations of common sole, *Solea solea*. Genetics 184:1–7

Blonk RJW, Komen H, Kamstra A, van Arendonk JAM (2010b) Effects of grading on heritability estimates under commercial conditions: a case study with common sole, Solea solea. Aquaculture 300:43–49

Blonk RJW, Komen J, Tenghe A, Kamstra A, van Arendonk JAM (2010c) Heritability of shape in common sole, Solea solea, estimated from image analysis data. Aquaculture 307:6–11

Brabrand A (1995) Intra-cohort cannibalism among larval stages of perch (Perca fluviatilis). Ecol Freshw Fish 4:70–76. doi:10.1111/j.1600-0633.1995.tb00119.x

Cao XJ, Wang HP, Yao H, O'Bryant P, Rapp D, Wang WM, MacDonald R (2012) Evaluation of 1-stage and 2-stage selection in yellow perch I: genetic and phenotypic parameters for body weight of F1 fish reared in ponds using microsatellite parentage assignment. J Anim Sci 90:27–36. doi:10.2527/jas.2011-3902

Chavanne H, Norris A, Haffray P, Sonesson AK, Vandeputte M, Chatain B, Boudry P (2008) Survey on the breeding practices in the European aquaculture industry. Reprofish Aquabreeding Workshop 2008. https://aquatrace.eu/questionnaire-portlet/reports/AquaBreedingSurvey_2009-1.pdf

Chevassus B, Quillet E, Krieg F, Hollebecq MG, Mambrini M, Faure A, Labbe L, Hiseux JP, Vandeputte M (2004) Enhanced individual selection for selecting fast growing fish: the "PROSPER" method, with application on brown trout (*Salmo trutta fario*). Genet Sel Evol 36:643–661

Domingos JA, Smith-Keune C, Robinson N, Loughnan S, Harrison P, Jerry DR (2013) Heritability of harvest growth traits and genotype–environment interactions in barramundi, Lates calcarifer (Bloch). Aquaculture 402–403:66–75. doi:10.1016/j.aquaculture.2013.03.029

Doyle RW, Talbot AJ (1986) Artificial selection on growth and correlated selection on competitive behaviour in fish. Can J Fish Aquat Sci 43:1059–1064. doi:10.1139/f86-132

Duchesne P, Godbout MH, Bernatchez L (2002) PAPA (package for the analysis of parental allocation): a computer program for simulated and real parental allocation. Mol Ecol Notes 2:191–194

Dupont-Nivet M, Vandeputte M, Chevassus B (2002) Optimization of factorial mating designs for inference on heritability in fish species. Aquaculture 204:361–370. doi:10.1016/S0044-8486(01)00839-0

Dupont-Nivet M, Vandeputte M, Haffray P, Chevassus B (2006) Effect of different mating designs on inbreeding, genetic variance and response to selection when applying individual selection in fish breeding programs. Aquaculture 252:161–170

Dupont-Nivet M, Vandeputte M, Vergnet A, Merdy O, Haffray P, Chavanne H, Chatain B (2008) Heritabilities and GxE interactions for growth in the European sea bass (Dicentrarchus labrax L.) using a marker-based pedigree. Aquaculture 275:81–87. doi:10.1016/j.aquaculture.2007.12.032

Ellen ED, Visscher J, van Arendonk JAM, Bijma P (2008) Survival of laying hens: genetic parameters for direct and associative effects in three purebred layer lines. Poult Sci 87:233–239. doi:10.3382/ps.2007-00374

Falconer DS, Mackay TFC (1996) Introduction to quantitative genetics. Pearson Prentice Hall Harlow, Essex

Fessehaye Y, Kabir A, Bovenhuis H, Komen H (2006a) Prediction of cannibalism in juvenile *Oreochromis niloticus* based on predator to prey weight ratio, and effects of age and stocking density. Aquaculture 255:314–322

Fessehaye Y, El-Bialy Z, Rezk MA, Crooijmans R, Bovenhuis H, Komen H (2006b) Mating systems and male reproductive success in Nile tilapia (*Oreochromis niloticus*) in breeding hapas: a microsatellite analysis. Aquaculture 256:148–158

Fessehaye Y, Komen H, Kezk MA, van Arendonk JAM, Bovenhuis H (2007) Effects of inbreeding on survival, body weight and fluctuating asymmetry (FA) in Nile tilapia, Oreochromis niloticus. Aquaculture 264:27–35. doi:10.1016/j.aquaculture.2006.12.038

Fessehaye Y, Bovenhuis H, Rezk MA, Crooijmans R, van Arendonk JAM, Komen H (2009) Effects of relatedness and inbreeding on reproductive success of Nile tilapia (Oreochromis niloticus). Aquaculture 294:180–186. doi:10.1016/j.aquaculture.2009.06.001

Fleming IA, Agustsson T, Finstad B, Johnsson JI, Björnsson BT (2002) Effects of domestication on growth physiology and endocrinology of Atlantic salmon (Salmo salar). Can J Fish Aquat Sci 59:1323–1330. doi:10.1139/f02-082

Frommen JG (2008) Inbreeding depression affects fertilization success and survival but not breeding coloration in threespine sticklebacks. Behaviour 145:425–441

Gallardo J (2004) Inbreeding and inbreeding depression of female reproductive traits in two populations of Coho salmon selected using BLUP predictors of breeding values. Aquaculture 234:111–122

Gjedrem T (2000) Genetic improvement of cold-water fish species. Aquacult Res 31:25–33. doi:10.1046/j.1365-2109.2000.00389.x

Gjedrem T, Gjøen HM (1995) Genetic variation in susceptibility of Atlantic salmon, Salmo salar L., to furunculosis, BKD and cold water vibriosis. Aquacult Res 26:129–134. doi:10.1111/j.1365-2109.1995.tb00892.x

Gjerde B, Gunnes K, Gjedrem T (1983) Effect of inbreeding on survival and growth in rainbow trout. Aquaculture 34:327–332. doi:10.1016/0044-8486(83)90212-0

Haffray P, Pincent C, Rault P, Coudurier B (2004) Domestication and genetic improvement of French fish farmed broodstocks in SYSAAF. Prod Anim 17:243–252

Hayes BJ, Visscher PM, Goddard ME (2009) Increased accuracy of artificial selection by using the realized relationship matrix. Genet Res 91:47–60. doi:10.1017/S0016672308009981

Hecht T, Pienaar AG (1993) A review of cannibalism and its implications in fish larviculture. J World Aquacult Soc 24:246–261. doi:10.1111/j.1749-7345.1993.tb00014.x

Henderson CR (1984) Applications of linear models in animal breeding. Guelph University Press, Guelph

Hinrichs D, Wetten M, Meuwissen THE (2006) An algorithm to compute optimal genetic contributions in selection programs with large numbers of candidates. J Anim Sci 84:3212–3218. doi:10.2527/jas.2006-145

Hulata G, Wohlfarth GW, Halevy A (1986) Mass selection for growth rate in the Nile tilapia (Oreochromis niloticus). Aquaculture 57:177–184. doi:10.1016/0044-8486(86)90195-X

Kause A, Ritola O, Paananen T, Eskelinen U, Mantysaari E (2003) Big and beautiful? Quantitative genetic parameters for appearance of large rainbow trout. J Fish Biol 62:610–622. doi:10.1046/j.0022-1112.2003.00051.x

Kause A, Ritola O, Paananen T (2004) Breeding for improved appearance of large rainbow trout in two production environments. Aquacult Res 35:924–930

Kause A, Tobin D, Houlihan DF, Martin SAM, Mantysaari EA, Ritola O, Ruohonen K (2006) Feed efficiency of rainbow trout can be improved through selection: different genetic potential on alternative diets. J Anim Sci 84:807–817

Kestemont P, Jourdan S, Houbart M, Mélard C, Paspatis M, Fontaine P, Cuvier A, Kentouri M, Baras E (2003) Size heterogeneity, cannibalism and competition in cultured predatory fish larvae: biotic and abiotic influences. Aquaculture 227:333–356. doi:10.1016/S0044-8486(03)00513-1

Kohlmann K, Kersten P (2008) Isolation and characterization of nine microsatellite loci from the pike-perch, Sander lucioperca (Linnaeus, 1758). Mol Ecol Resour 8:1085–1087. doi:10.1111/j.1755-0998.2008.02166.x

Komen J, Wiegertjes GF, van Ginneken VJT, Eding EH, Richter CJJ (1992) Gynogenesis in common carp (Cyprinus carpio L.). III. The effects of inbreeding on gonadal development of heterozygous and homozygous gynogenetic offspring. Aquaculture 104:51–66. doi:10.1016/0044-8486(92)90137-A

Lappalainen J, Erm V, Kjellman J, Lehtonen H (2000) Size-dependent winter mortality of age-0 pikeperch (Stizostedion lucioperca) in Pärnu Bay, the Baltic Sea. Can J Fish Aquat Sci 57:451–458. doi:10.1139/f99-270

Lepage O, Overli O, Petersson E, Jarvi T, Winberg S (2000) Differential stress coping in wild and domesticated sea trout. Brain Behav Evol 56:259–268

Lynch M, Walsh B (1998) Genetic and analysis of quantitative traits. Sinauer Associates, Sunderland

Marshall TC, Slate J, Kruuk LEB, Pemberton JM (1998) Statistical confidence for likelihood-based paternity inference in natural populations. Mol Ecol 7:639–655

Meuwissen THE (2002) GENCONT: an operational tool fo controlling inbreeding in selection and conservation schemes. In: Proceedings of the 7th World Congress on genetics applied to livestock production, Montpellier, 33, pp 769–770

Meuwissen T (2007) Genomic selection: marker assisted selection on a genome wide scale. J Anim Breed Gen 124:321–322. doi:10.1111/j.1439-0388.2007.00708.x

Meyer S, Teerlinck S, Policar T, Toner D (2012) European percid fish culture (EPFC) – Workshop 2012 summary

Moen T, Baranski M, Sonesson AK, Kjoglum S (2009) Confirmation and fine-mapping of a major QTL for resistance to infectious pancreatic necrosis in Atlantic salmon (Salmo salar): population-level associations between markers and trait. BMC Genomics 10:368. doi:10.1186/1471-2164-10-368

Monsen BB, Ødegård J, Arnesen KR, Toften H, Nielsen HM, Damsgård B, Bijma P, Olesen I (2010) Genetics of social interactions in Atlantic cod (Gadus morhua). 9th World Congress on genetics applied to livestock production (WCGALP), Leipzig
Mrode RA (2005) Linear models for the prediction of animal breeding values. CABI, Wallingford
Mulder HA (2007) Methods to optimize livestock breeding programs with genotype by environment interaction and genetic heterogeneity of environmental variance. PhD thesis, Wageningen University
Neira R (2010) Breeding in aquaculture species: genetic improvement programs in developing countries. 9th World Congress on genetics applied to livestock production (WCGALP), Leipzig
Neira R, Diaz NF, Gall GAE, Gallardo JA, Lhorente JP, Alert A (2006) Genetic improvement in coho salmon (Oncorhynchus kisutch). II: selection response for early spawning date. Aquaculture 257:1–8. doi:10.1016/j.aquaculture.2006.03.001
Nesbø CL, Fossheim T, Vøllestad LA, Jakobsen KS (1999) Genetic divergence and phylogeographic relationships among European perch (Perca fluviatilis) populations reflect glacial refugia and postglacial colonization. Mol Ecol 8:1387–1404. doi:10.1046/j.1365-294x.1999.00699.x
Olesen I, Gjedrem T, Bentsen HB, Gjerde B, Rye M (2003) Breeding programs for sustainable aquaculture. J Appl Aquacult 13:179–204
Ponzoni RW, Nguyen NH, Khaw HL (2007) Investment appraisal of genetic improvement programs in Nile tilapia (Oreochromis niloticus). Aquaculture 269:187–199. doi:10.1016/j.aquaculture.2007.04.054
Quinton CD, Kause A, Koskela J, Ritola O (2007) Breeding salmonids for feed efficiency in current fishmeal and future plant-based diet environments. Genet Sel Evol 39:431–446. doi:10.1051/gse:2007013
Rónyai A (2007) Induced out-of-season and seasonal tank spawning and stripping of pike perch (Sander lucioperca L.). Aquacult Res 38:1144–1151. doi:10.1111/j.1365-2109.2007.01778.x
Rutten MJM (2005) Breeding for improved production of Nile tilapia (Oreochromis niloticus L.). PhD thesis, Wageningen University
Rutten MJM, Bijma P, Woolliams JA, van Arendonk JAM (2002) SelAction: software to predict selection response and rate of inbreeding in livestock breeding programs. J Hered 93:456–458. doi:10.1093/jhered/93.6.456
Rutten MJM, Bovenhuis H, Komen H (2004) Modeling fillet traits based on body measurements in three Nile tilapia strains (Oreochromis niloticus L.). Aquaculture 231:113–122. doi:10.1016/j.aquaculture.2003.11.002
Rutten MJM, Bovenhuis H, Komen H (2005) Genetic parameters for fillet traits and body measurements in Nile tilapia (*Oreochromis niloticus L.*). Aquaculture 246:125–132
Ruzzante DE, Doyle RW (1991) Rapid behavioral-changes in Medaka (Oryzias-Latipes) caused by selection for competitive and noncompetitive growth. Evolution 45:1936–1946. doi:10.2307/2409841
Rye M, Gjerde B (1996) Phenotypic and genetic parameters of body composition traits and flesh colour in Atlantic salmon, Salmo salar L. Aquacult Res 27:121–133. doi:10.1111/j.1365-2109.1996.tb00976.x
Rye M, Gjerde B and Gjedrem T (2010) Genetic improvement programs for aquaculture species in developed countries. 9th World Congress on genetics applied to livestock production (WCGALP), Leipzig
Sae-Lim P, Komen H, Kause A (2010) Bias and precision of estimates of genotype-by-environment interaction: a simulation study. Aquaculture 310:66–73. doi:10.1016/j.aquaculture.2010.10.020
Sae-Lim P, Komen H, Kause A, Martin KE, Crooijmans R, van Arendonk JAM, Parsons JE (2013) Enhancing selective breeding for growth, slaughter traits and overall survival in rainbow trout (Oncorhynchus mykiss). Aquaculture 372:89–96. doi:10.1016/j.aquaculture.2012.10.031
Saillant E, Ma L, Wang XX, Gatlin DM, Gold JR (2007) Heritability of juvenile growth traits in red drum (Sciaenops ocellatus L.). Aquacult Res 38:781–788

Säisä M, Salminen M, Koljonen M-L, Ruuhijärvi J (2010) Coastal and freshwater pikeperch (Sander lucioperca) populations differ genetically in the Baltic Sea basin. Hereditas 147:205–214. doi:10.1111/j.1601-5223.2010.02184.x

Smith C, Reay P (1991) Cannibalism in teleost fish. Rev Fish Biol Fish 1:41–64. doi:10.1007/bf00042661

Sonesson AAK (2005) A combination of walk-back and optimum contribution selection in fish: a simulation study. Genet Sel Evol 37:587–599

Sonesson A, Meuwissen T (2009) Testing strategies for genomic selection in aquaculture breeding programs. Genet Sel Evol 41:37

Sorensen A, Berg P, Woolliams J (2005) The advantage of factorial mating under selection is uncovered by deterministically predicted rates of inbreeding. Genet Sel Evol 37:57–81

Stepien CA, Murphy DJ, Lohner RN, Sepulveda-Villet OJ, Haponski AE (2009) Signatures of vicariance, postglacial dispersal and spawning philopatry: population genetics of the walleye Sander vitreus. Mol Ecol 18:3411–3428. doi:10.1111/j.1365-294X.2009.04291.x

Taggart JB (2007) FAP: an exclusion-based parental assignment program with enhanced predictive functions. Mol Ecol Notes 7:412–415

Thodesen Da-Yong Ma J, Rye M, Wang Y-X, Bentsen HB, Gjedrem T (2012) Genetic improvement of tilapias in China: genetic parameters and selection responses in fillet traits of Nile tilapia (Oreochromis niloticus) after six generations of multi-trait selection for growth and fillet yield. Aquaculture 366–367:67–75. doi:10.1016/j.aquaculture.2012.08.028

Trọng TQ, van Arendonk JAM, Komen H (2013a) Genetic parameters for reproductive traits in female Nile tilapia (Oreochromis niloticus): I. Spawning success and time to spawn. Aquaculture 416–417:57–64. doi:10.1016/j.aquaculture.2013.08.032

Trọng TQ, van Arendonk JAM, Komen H (2013b) Genetic parameters for reproductive traits in female Nile tilapia (Oreochromis niloticus): II. Fecundity and fertility. Aquaculture 416–417:72–77. doi:10.1016/j.aquaculture.2013.08.031

Trọng TQ, Mulder HA, van Arendonk JAM, Komen H (2013c) Heritability and genotype by environment interaction estimates for harvest weight, growth rate, and shape of Nile tilapia (Oreochromis niloticus) grown in river cage and VAC in Vietnam. Aquaculture 384–387:119–127. doi:10.1016/j.aquaculture.2012.12.022

van Duijn AP, van der Mheen H, Blonk RJW, Beukers R (2010) Actieplan visteelt. Landbouw Economisch Instituut, part of Wageningen UR, The Hague

Vandeputte M, Kocour M, Mauger S, Dupont-Nivet M, De Guerry D, Rodina M, Gela D, Vallod D, Chevassus B, Linhart O (2004) Heritability estimates for growth-related traits using microsatellite parentage assignment in juvenile common carp (*Cyprinus carpio* L.). Aquaculture 235:223–236

Vandeputte M, Mauger S, Dupont-Nivet M (2006) An evaluation of allowing for mismatches as a way to manage genotyping errors in parentage assignment by exclusion. Mol Ecol Notes 6:265–267. doi:10.1111/j.1471-8286.2005.01167.x

Vandeputte M, Kocour M, Mauger S, Rodina M, Launay A, Gela D, Dupont-Nivet M, Hulak M, Linhart O (2008) Genetic variation for growth at one and two summers of age in the common carp (*Cyprinus carpio* L.): heritability estimates and response to selection. Aquaculture 277:7–13

Villanueva B, Woolliams JA, Gjerde B (1996) Optimum designs for breeding programmes under mass selection with an application in fish breeding. Anim Sci 63:563–576. doi:10.1017/S1357729800015459

Wilson AJ, McDonald (2003) Marker-assisted estimation of quantitative genetic parameters in rainbow trout, *Oncorhynchus mykiss*. Genet Res 81:145–156

Part VII
Stress, Immunology, Diseases and Health Management

Chapter 28
Corticosteroids and the Stress Response in Percid Fish

S. Milla, J. Douxfils, S.N.M. Mandiki, and M. Saroglia

Abstract While there is abundant information about the corticosteroids and the stress response in some fish families such as salmonids, there is little data in percids. Still, despite the scattered information in this fish family, accumulating evidence strongly indicates that corticosteroids are strongly regulated in percids after exposure to stressors and play essential roles in the stress response. This chapter highlights the characteristics of percids concerning the corticosteroid synthesis and receptivity, the basal blood levels, the stressors linked to husbandry conditions conducting to cortisol secretion as well as the secondary and tertiary response to stress with focus on specific biological markers. The usefulness to use cortisol as the best stress marker is discussed and attempts are made to propose other biological indicators of the stress response. The authors will also suggest other ways to prospect the stress response.

Keywords Percids • Corticosteroids • Stress • Cortisol • Biological markers

28.1 Introduction

The endocrine stress response in teleost fish has been reviewed several times (e.g. Barton and Iwama 1991; Wendelaar Bonga 1997; Tort 2011). These syntheses have focused on teleost families such as the salmonids and cyprinids, which are frequently studied as they include species of interest in the field of aquaculture, ecology or molecular biology. However, little attention has been given to the stress

S. Milla (✉)
Unité de Recherche Animal et Fonctionnalités des Produits Animaux, Université de Lorraine, 2 avenue de la Forêt de Haye, B.P. 172, 54505 Vandoeuvre-lès-Nancy, France
e-mail: sylvain.milla@univ-lorraine.fr

J. Douxfils • S.N.M. Mandiki
Research Unit in Environmental and Evolutionary Biology (URBE), University of Namur, Rue de Bruxelles 61, 5000 Namur, Belgium

M. Saroglia
Department of Biotechnology and Life Sciences, University of Insubria, Via J.H. Dunant, 3-21100 Varese, Italy

P. Kestemont et al. (eds.), *Biology and Culture of Percid Fishes*,
DOI 10.1007/978-94-017-7227-3_28

response in percid fishes. The recent development of worldwide percid aquaculture provides an incentive to summarize the current knowledge on the specializations of this fish family with regards to the stress response. While it is generally recognized that variable stressors can have deleterious impacts on biological functions, including growth, reproduction and immunity in fish, significant differences in the details of the stress response between families have only recently been addressed. Understanding the stress response in percids is essential for aquaculture development. The objective of this chapter is to highlight the characteristics of percids concerning the corticosteroid system, the stressors conducting to cortisol secretion and the secondary and tertiary response to stress with focus on specific biological markers. Potential gaps in this knowledge in percids will be also notified to delineate the progressing but limited information about this fish family.

28.2 Actors of the Corticosteroid System

28.2.1 Mammalian Corticosteroids: A Comparison with Teleosts Including Percids

Despite the divergence of tetrapod vertebrates from teleosts 450 million years ago, the range of corticosteroids present in both groups show a number of similarities. Like mammals, teleost fish including percids are able to synthesize corticosteroids, and the gonad is also a site of corticosteroid production (Theofan and Goetz 1983). While the genes encoding the enzymes involved in the corticosteroid production have not been isolated in percid fish, the presence of enzymatic activities provides evidence of common metabolic pathways with other fish families. In addition to the steroidogenic acute regulatory protein (star) and the Cholesterol side-chain cleavage enzyme (P450 scc), four key enzymes are involved in the biosynthesis of corticosteroids: 3-beta-hydroxysteroid dehydrogenase, 17-alpha-hydroxylase, 21-hydroxylase and 11-beta-hydroxylase. Among them, only 11 hydroxylase and 17 hydroxylase activities have been measured in the head kidney or gonad of Eurasian perch *Perca fluviatilis* (Kime and Hews 1978; Linderoth et al. 2006) and the relationship between enzyme activity and plasma cortisol concentrations was examined, supporting the active role of this latter enzyme in the hormonal production (Linderoth et al. 2006). But further studies would be needed to check in percids the activity of the other enzymes, in the major sites of corticosteroid production, primary the head kidney and the gonad. The role of the hypothalamus-pituitary-interrenal axis has been well established in teleosts and ACTH has been described for a long time to be one of the main secretagogue of corticosteroid production like in mammals. This regulation has been proved in vitro for cortisol in yellow perch *Perca flavescens* and interestingly the sensitivity of adreno-cortical cells to cAMP was lower in yellow perch *Perca flavescens* by comparison to rainbow trout (Lacroix and Hontela 2004). Here also, information is not abundant in percids and this preliminary result supports inter-families differences which need to be confirmed.

After corticosteroid production, these hormones are released in the blood circulation and target specific tissues. The intracellular action of these hormones is controlled by corticosteroid receptors whose characterization and functionality is still a matter of investigation in fish. The corticosteroid receptors enclose the common general structure of all nuclear receptors with five distinct regions called domains: the amino terminal (N-terminal) AB domain, the DNA binding domain (DBD, C domain), the hinge D region, and the ligand-binding domain (LBD, E domain).

Duplication of the teleost genome also concerns the corticosteroid receptors. The mineralocorticoid and the glucocorticoid receptors descend by duplication from a single ancestral corticosteroid receptor (AncCR), which existed in an ancient jawed vertebrate ~450 million years ago before the split of cartilaginous fish from bony vertebrates and the teleost radiation (Bridgham et al. 2006). Then, there is strong evidence that a second duplication event occurred in the teleost lineage which led to the presence of two glucocorticoid receptors in most teleost species while the mineralocorticoid receptor was not concerned by this duplication (Li et al. 2012). Until now, two glucocorticoid receptors (GRs), namely GR1 and GR2, have been isolated in teleosts except in the zebrafish (Terova et al. 2005; Alsop and Vijayan 2009). In Eurasian perch, two glucocorticoid receptors and one mineralocorticoid receptor have been cloned already. While the knowledge about this molecular characterization is limited to date, it seems that the present scheme in this species does not differ from the actual fish paradigm where these receptors are distinguished by their amino acid sequences and ligand specificity. In agreement with the teleost phylogeny, the Eurasian perch sequence of the corticosteroid receptors displays a high degree of conservation with the sea bass sequence (Milla et al. 2010). For instance, 100 % and 94 % of identity are observed for the C and E domains between both perciforms. However, in the other percid species, the genes of the corticosteroid receptors have not been cloned so far. The corticosteroid receptors are also ubiquitously found in Eurasian perch even if the cellular localization remains to be established (Milla et al. 2010). This ubiquitous localization is in accordance with implications in the response to different stressors such as salinity, immune and social challenges respectively investigated in other teleosts (Stolte et al. 2008; Killerich et al. 2011; Jeffrey et al. 2012). The regulation of the gene expression of these receptors following exposure to handling in Eurasian perch and common carp also indicates their involvement in the responses to stress (Stolte et al. 2008; Milla et al. 2010) even if the putative specializations of percids remain unknown. In fish, the glucocorticoid receptors have been mainly described to be involved in the osmoregulatory processes. The role of the mineralocorticoid receptor in osmoregulation appears minor while its involvement in male reproduction, behavior and stress response has been recently described (Takahashi and Sakamoto 2013). But investigations would also be needed to evaluate these implications in percids.

Teleosts differ from mammals in terms of nature of ligands for the corticosteroid receptors. While the corticosteroid receptors have been cloned in Eurasian perch, the potency of corticosteroids to activate these receptors has not been tested, and no information exists in the other percid fishes. At any case, there are now sufficient data in other teleost fishes to hypothesize that the profile of CRs ligands is common among teleosts (Sturm et al. 2005; Stolte et al. 2008; Kim et al. 2011). From these studies, it appears that the main difference with mammals is the function of 11-deoxycorticosterone (DOC) to activate the mineralocorticoid receptor even if its physiological role is still in debate. Otherwise, cortisol appears to be the main ligand of GRs, GR2 being more sensitive to cortisol than GR1 (Kim et al. 2011). Finally, 11-beta-HSD has an important role in MR function by converting cortisol into cortisone to allow the access of the mineralocorticoid hormone aldosterone to the MR. Postulating a role of DOC in fish implies some enzymatic factors such as 11-beta-HSD to protect MR from cortisol occupancy, as circulating cortisol levels in fish including percids are much higher and more variable than those of DOC. In Eurasian perch, concomitant and similar changes of MR and 11-beta-HSD2 gene expression after LPS challenge might indicate such function (Mathieu et al. 2013b).

The response to stress is therefore a complex cascade of events in which several molecules, notably belonging to the corticosteroid signaling pathways, are involved. Beside classic analytical approaches including chemical and immuno-enzymatic analyses, the use of molecular techniques to monitor stress levels using early indicators, such as molecular biomarkers, can bring considerable benefits to the enhancement of knowledge, as well as to welfare improvement of reared animals. Such approach requesting a molecular database, is facilitated in Eurasian perch aswell, following publication of an EST-based identification of genes expressed in Eurasian perch (Rossi et al. 2007). Among the genes potentially linked to the corticosteroid actions, the neurotrophin brain-derived neurotrophic factor (BDNF) is a credible candidate notably as it was recently shown to be involved in stress-induced adaptation (Tognoli et al. 2012). BDNF is a key regulator of neuronal plasticity and adaptive processes. Regulation of BDNF is complex and may reflect not only stress-specific mechanisms, but also hormonal and emotional responses. In spite a missing of information, there are reasons to assume that cortisol response as well is involved in the BDNF response cascade. For instance in sea bass, *Dicentrarchus labrax*, acute stress causes downregulation of BDNF mRNA, also inducing a significant increase in proBDNF levels and reduction in mature BDNF, suggesting that the regulation of proBDNF proteolytic processing has been altered. Here, proteolytic regulation of BDNF is simplified since the pro28kDa form, generated by the SKI-1 protease in mammals, is absent in fish. The highly predictable proBDNF/totBDNF ratio for stress in lower vertebrates indicates that BDNF processing represents a central mechanism in adapting to stress and predicts that a similar regulation of pro/mature BDNF has likely been conserved throughout evolution of vertebrates from fish to humans (Tognoli et al. 2010).

The ontogeny of the corticotroph axis may be one of the peculiarities of percids. By measuring resting and stress-induced whole-body cortisol levels in yellow perch at different development stages, pre- and post-hatching, the authors concluded that cortisol biosynthesis and maturation of the hypothalamic-pituitary-interrenal (HPI) axis is observed from 1 week after hatching onwards (Jentoft et al. 2002). Similarly, a change in basal cortisol level but also in the stress-induced cortisol rise is observed

6 days post-hatching in the perciform red drum *Sciaenops ocellatus* (Applebaum et al. 2010). The cortisol ontogeny seems to be earlier in salmonids for which a subsequent rise in cortisol prior or around hatching is observed resulting from de novo synthesis by the embryo (Feist and Schreck 2001; Fuzzen et al. 2011).

28.2.2 Circulating Levels of Corticosteroids in Percids

Commonly to what is observed in other teleosts, percid fishes possess corticosteroids including cortisol at levels substantially measured in the blood. Numerous external factors are known to modulate cortisol production and cautions must be respected to compare the data between the studies. However, the amount of data in the literature is enough to draw the main lines. Similarly to other teleosts, cortisol appears to be the main corticosteroids in percids, estimated in the range of 1–200 ng/ml in the plasma, depending on the species and the physiological situation. The baseline level of cortisol monitored during low or mild stressful situations, is measured between 2 and 40 ng/ml, a level frequently found in other fish families such as salmonids or cyprinids (Haukenes and Barton 2004; Jentoft et al. 2005; Fatemeh et al. 2008; Milla et al. 2010; Douxfils et al. 2011a). Many intrinsic factors also modulate the basal level of corticosteroids in vertebrates including fish. Among them, genetic factors, morpho-anatomical indices, the sex and the reproductive stage influence the basal plasma cortisol level (See for review Barton and Iwama 1991). But there is few information in percids. For instance, the basal cortisol levels vary during the year in yellow perch caught from the wild possibly in agreement with seasonal changes in relation with gametogenesis progress (Girard et al. 1998).

Few studies in percid fish have investigated the presence of corticosteroids other than cortisol in blood. Yet, some of them have been detected at substantial levels and these scattered data may indicate some specificity by comparison with some other fish families. The first estimates of baseline 11-deoxycorticosterone (DOC) and 11-deoxycortisol levels in immature Eurasian perch are around 2–4 ng/ml (Noaksson et al. 2005; Mathieu et al. 2013a). These data also suggest that there is low inter-individual variability for these corticosteroids and that 11-deoxycorticosteroids are present at higher levels in immature perch than in salmonids (Doyon et al. 2006; Milla et al. 2009). Indeed, on the basis of these latter studies, the plasma level of these corticosteroids in immature perch is around 5–50 fold-higher than the level measured in rainbow trout (Doyon et al. 2006; Milla et al. 2008; Kiillerich et al. 2011). However, the measurement of 11-deoxycortisol in female Eurasian Perch in April (around mating) did not reveal presence of this hormone in the plasma (Widell et al. 2011). Unfortunately, measurement of the other corticosteroids, considered today as minor has not been performed in percid plasma. Here also, it must be noted that some intrinsic factors may influence the basal levels of these hormones, for instance the reproductive stage in female Eurasian perch (Noaksson et al. 2005).

28.3 External Factors Inducing Increase of Circulating Cortisol Levels

The endocrine stress response is conserved among all vertebrates and the percid case is not excluded from this postulate even if the literature suffers from a lack of information related to some physiological processes of the cascade. Perceiving changes in external cues, the central nervous system triggers the primary stress response by activating the hypothalamic–pituitary–interrenal (HPI) axis and the autonomic nervous system (ANS), resulting in the release of stress hormones, corticosteroids and catecholamines, from the head kidney (Wendelaar Bonga 1997; Barton 2002). Importantly, fish in aquaculture may be subjected to numerous stressors: mechanical (handling, transport…), toxicological (veterinary drugs or pollutants in extensive aquaculture), physico-chemical (hypoxia, water quality…), photo-thermic and pathogen-related stressors. The aim of this part is to review the current knowledge on the effects of such stressors on cortisolemia in percids.

28.3.1 Mechanical and Social Stressors

Within the context of aquaculture, it is important to delineate the physiological consequences of mechanical stressors which constitute one of the main categories of stressors in husbandry practices. Here also, while the case of common aquaculture species has been largely documented, the promising but modest development of percid culture explains the low number of related published papers. In accordance with what has been observed in other teleost species, handling stress promotes a surge in the cortisol level in Eurasian perch (Acerete et al. 2004; Jentoft et al. 2005; Milla et al. 2010), in yellow perch (Haukenes 2001), in pikeperch (Fatemeh et al. 2008) and in walleye *Sander vitreus* (Forsberg and Summerfelt 2001). Following mechanical stress, cortisol levels usually reach a plateau in the range of 50–200 ng/ml in percids, a median concentration in comparison with the concentration measured in other fish species (30–300 ng/ml) (Barton 2002). This wide range refers to the differential sensitivity of the corticotropic axis to mechanical stressors between fish families but also within each family. These studies also show that percids are at least as sensitive as salmonids and cyprinids to mechanical stressors (Wedemeyer et al. 1990; Bau et al. 2000; Barton 2002).

Some works have investigated the responses of fish to chronic stressors in an attempt to precise maximum corticosteroid responses to continuous stressors. Following severe continuous confinement combined with handling, plasma cortisol in sturgeons, paddlefish and rainbow trout reached a plateau of 13, 60 and 160 ng/ml, respectively (Barton et al. 1980, 1998, 2000). In similar studies, the plasma cortisol concentrations reached 400–500 ng/ml in juvenile chinook salmon (Strange et al. 1978) and around 1400 ng/ml in striped bass (Noga et al. 1994), further illustrating discrepancies in the magnitude of cortisol rise between fish species.

By contrast to what is observed when fish are subjected to acute handling stressors, the exposure to constant confinement, density or repeated water emersion without handling did not induce any significant sustained cortisol elevation in Eurasian perch and yellow perch (Levesque et al. 2003; Haukenes and Barton 2004; Cairns et al. 2008; Douxfils et al. 2011a, b, 2012). This is in contrast with the effects of such stressors in other fish families, and may indicate a quicker cortisol metabolization through a rapid conversion of cortisol to cortisone. Cairns et al. (2008) reported increases in cortisol level (around 80–120 ng/ml) up to 7 days after the rainbow trout were confined into smaller experimental tanks. Other studies also showed cortisol responses to confinement and/or to crowding stressors (e.g. Barnett and Pankhurst 1998; Binuramesh et al. 2005; Saera Villa et al. 2009). In a study on European seabass, Terova et al. (2005) found that a high rearing density stressor affected GR mRNA, the abundance of which in liver showed an inverse relationship to blood cortisol levels. Nevertheless, among a fish population, a sort of "tolerance" may occur below a threshold of biomass density, GR receptors and cortisolemia being affected only when high density as 80 kg/m^3 was reached. This response may not be interpreted as an exhaustion of the corticosteroid response as the latter mounts when biomass density approaches an upper threshold.

A review from Martìnez-Porchas et al. (2009) discuss the opportunity of utilizing cortisol and glucose as stress indicators, concluding that in spite their validity to indicate acute stress, other analyses should be carried on, in particular when a chronic stress assessment is the goal. Therefore, we believe that cortisol is not an appropriate indicator of such stressor in the Eurasian perch and yellow perch even if this observation should be validated in the other percids. We do not think that this is unique to percids since similar negative responses were observed in the red porgy following chronic confinement (Fanouraki et al. 2007).

28.3.2 Water Quality

In the field of ecotoxicology, Eurasian perch and yellow perch have ever been used to reveal toxic exposure in the wild. Concerning the relationship between toxic exposure and corticosteroids, most of the studies have been related to presence of heavy metals in the framework of ecology in the Canadian lakes. Heavy metals (Cd, Cu, Fe, Zn, Pb, Ni) are suspected to interfere with the corticosteroidogenesis. In situ, exposure to heavy metals reduced cortisol production following stressor exposure, the corticotropic response to ACTH treatments, and the viability of adreno-cortical cells in yellow perch (Brodeur et al. 1997; Girard et al. 1998; Lacroix and Hontela 2004; Levesque et al. 2003). This was also confirmed in Eurasian perch in a study where the presence of heavy metals (HAP, PCB) was associated with other pollutants, although it was not possible to dissociate between the effects related to each of these pollutants (Hontela et al. 1992). It might be helpful to validate the data obtained from such field studies notably through in vivo exposure of percids to heavy metals under controlled laboratory conditions. Water-born or dietary metals have been reported to disrupt the corticosteroid axis in fish. In spite that studies related to percids are missing, care should be accounted to results obtained from other fish species, as possibly extrapolable. A recent study by Sandhu and Vijayan (2011) demonstrated that cadmium interferred with cortisol synthesis by directly disrupting melanocortin 2 receptor (MC2R) signaling. It is not excluded that other metals modulate this ACTH-independent activation of MC2R signaling possibly affecting the cortisol equilibrium.

The heavy metals are not the only substances enable to disrupt the cortisol response. Water-born selenite exposure, as well as dietary exposure to an environmentally relevant dietary concentration of Selenomethionine (Se-Met), elicited a stress response in immature female rainbow trout, with a greater concentration of blood plasma cortisol compared to control fish (Wiseman et al. 2011). Moreover, trout previously responding with enhanced cortisolemia to dietary Se-Met, showed an inability to mount a stress response to a secondary handling stressor. Although the mechanism of the attenuated cortisol response to the handling stressor is unknown, the authors suggest that it might be due to depletion of corticotrophs thereby leading to interrenal inactivation by lack of corticotropins. Alternatively, trout exposed to Se-Met might not have been able to display a cortisol response due to direct inhibition of the steroidogenic machinery by Se-Met. While further studies are needed, the utilization of organic selenium as prebiotic in fish diet should be considered carefully, to prevent fish from being enable to develop further allostatic responses.

Oxygen levels in water require to be carefully monitored in aquaculture because even short-term exposure to hypoxia can have lethal consequences in many fish species. In addition, sublethal exposures to hypoxia can cause deleterious effects on fish physiology. In Eurasian perch, cortisol level was not elevated just after a 4-h hypoxia (2 ppm) regardless of the stressor frequency; this suggests that cortisol level in Eurasian perch does not increase during hypoxia or that the primary cortisol response was of short duration (Douxfils et al. 2012). Muusze et al. (1998) exposed the cichlid *Astronotus ocellatus* to eight progressive and stepwise (4 h) reductions in oxygen levels until severely hypoxic conditions were reached (6 % air saturation, 0.5 ppm). Fish were sampled at the end of each 4-h step and no difference in cortisol level was observed throughout the experimental period. In contrast, while Pichavant et al. (2002) did not observe any cortisol elevation in turbot submitted to a 6-h period of slight or moderate hypoxia (4 and 6 ppm), an increase in cortisolemia was seen under severe hypoxia (2 ppm) throughout the 6-h period. This indicates that the severity of hypoxia might be an important contributing factor of the cortisol release. Other characteristics of the hypoxia experiments should be taken into consideration to analyze and highlight the percid response such as duration of hypoxia, the selected sampling times, the temperature level.

Disorders in the water quality, other than hypoxic conditions, may also represent a stressful situation. For instance, it is well documented that exposure to high concentrations of ammonia in water activates the HPI axis as illustrated by elevated plasma cortisol (Mommsen et al. 1999). However, in walleye, exposure to low concentrations of dissolved ammonia promoted growth rates, which was associated with reduced plasma cortisol level (Madison et al. 2009). Unfortunately, the literature suffers from a lack of information in percids to conclude on the specific response of this fish family. Salinity is another critical aspect of water quality for stenohaline species. Percid fish are considered to be freshwater species with low hypo-osmoregulatory ability. However, these fish may live or be maintained in brackish water. Eurasian perch and yellow perch are able to live in water containing up to 10 ppt salinity (Craig 2000) while pikeperch can tolerate salinities up to 12 ppt or 16 ppt (Craig 2000; Brown et al. 2001). The optimal salinity for pikeperch and walleye are probably in the range of 4–8 ppt (Craig 2000). As shown many times in euryhaline species that can tolerate a broad range of salinities, a rapid hyperosmotic challenge induces a significant rise in plasma cortisol associated with elevation of blood physiological parameters such as osmolality, ionic concentrations and glycemia. In pikeperch, plasma cortisol concentration was markedly changed by exposure to gradually increasing salinity even if challenges with moderate saline levels did not elicit clear corticotropic response (Brown et al. 2001).

28.3.3 Photothermic Factors

In temperate species, photothermal manipulations are often used to control the reproductive cycle, either by inducing its onset or by triggering the final stages (Wang et al. 2010). Changes in the level and direction of these abiotic factors allow the achievement of gametogenesis process by driving changes in the levels of sex hormones (Bromage et al. 2001). These endocrine regulations are sometimes associated with changes in the cortisol levels given the tight interaction between the Hypothalamus-Pituitary-Interrenal axis and the Hypothalamus-Pituitary-Gonad axis (Milla et al. 2009). In Eurasian perch, circulating cortisol levels were found to be significantly different depending on the tested temperatures during the inductive period (Wang et al. 2006). In pikeperch, application of different photoperiod regimes at the end of the reproductive cycle did not induce clear modifications of cortisol levels (Sarameh et al. 2012). These apparent discrepancies may be due to the target reproductive stage (initiation versus final stages), the tested species and mostly to the type of tested factors (temperature versus photoperiod).

28.3.4 Pathogens as Stressors

When fish are subjected to the presence of bacteria, virus or fungi an immune response is usually triggered in order to cope with the pathogens. Corticosteroids are secreted after a pathogen challenge acting as endocrine immunomodulators even if their roles are not yet fully understood. After lipopolysaccharide (LPS) administration, a constituent of bacterial cell walls, a huge cortisol secretion is observed in yellow perch (Haukenes and Barton 2008), reflecting the role of cortisol in fish immunomodulation (Harris and Bird 2000). Similarly, LPS administration provoked an additional cortisol secretion by comparison to handled but not injected fish, even without the vehicle (Haukenes and Barton 2004). In juvenile walleye, LPS treatment prolonged the hypercortisolemia induced by chronic density stressor (Haukenes 2001). This corticosteroid response to an inflammatory challenge appears to be a conserved trait across a wide range of vertebrate taxa (Berczi 1998) and this process is thought to reflect the adaptive response that protects the host from inflammation by reducing the likelihood of pathogen proliferation (Berczi 1998).

28.3.5 Corticosteroid Detection in Blood and in Water

Corticosteroid detection in blood is a common way to estimate hormone secretion following exposure to the stressors. But, plasma corticosteroid measurement in the blood mainly reflects the equilibrium between secretion by the interrenal and catabolism by the target tissues. The major issue of this strategy is the putative inconsistency to evaluate cortisol production because of the high difficulty to take into account corticosteroid degradation in the organs. In percids like in other fish families, radioimmunoassays (RIA) and ELISA have been routinely used to measure cortisol in the blood and the percid plasma does not seem to negatively interfere with the antibodies provided in the commercial kits. Concerning the other corticosteroids whose concentrations are barely detectable in the blood and which are thus considered as minor hormones, mass spectrometry has been used rather than the usual immunoassays. This is due to the lack of commercial immunoassays and/or specific antibodies in fish, and trust in their specificity for piscine studies (Milla et al. in preparation; Widell et al. 2011).

Measurement of water-borne steroids including corticosteroids has greatly been developed for detecting their release by the fish, notably as pheromones (Scott and Ellis 2007). This methodology presents the advantage to be non-invasive contrary to protocols involving blood sampling. This is an asset in the studies related to stress as the manipulations required in taking blood samples constitute an additional stress the effects of which are difficult to be estimated. Recent studies evaluating the effectiveness of this method to estimate cortisol release have proven a solid relationship between the increase in circulating cortisol level and the rise in water cortisol level after single or repeated handling stress in trout (Ellis et al. 2004). To date, only one study has evaluated the effectiveness of this method in percid fish by monitoring increased cortisol secretion in fish exposed to ship noise, and it is thus difficult to conclude about the specificity of the percid family by comparison to the others (Wysocki et al. 2006). This study compared the cortisol release after ship noise stressors in three fish species (Eurasian perch, *Gobio gobio* and *Cyprinus carpio*) and the data do show clear inter-species differences in the induction observed following the stressor, a higher induction being detected in the gudgeon by comparison to both other fish species (Wysocki et al. 2006).

28.4 Biological Stress Response in Relation to Activation of the HPI Axis

Although indices such as cortisol and glucose have been shown to be effective in measuring the acute and chronic stress response of a variety of small fishes, these are relatively crude measures of a complex biological response (O'Connor et al. 2011) as hypoxia is. Further biomarkers may benefit from a more refined approach to the assessment of HPI activity, beside repeated measures in individual fish by the collection of water-borne cortisol, or by the quantification of transcription factors such as hypoxia inducible factor-1α (HIF-1α) that are known to be associated with regulating the expression of physiologically relevant genes (Terova et al. 2008).

28.4.1 Secondary Stress Response, Corticosteroids and Allostatic Equilibrium

The stress response is thought to be a major adaptive process necessary to improve the probability of survival during exposure to inadequate conditions, i.e. situations capable to deeply disrupt physiological functions and internal homeostasis. To withstand the stress, the fish develops an integrated secondary stress response including anatomical, histological and physiological changes in multiple tissues with aim to provide the organism with the ability to resist external changes and invasion of foreign substances. The involvement of corticosteroids is widespread, including alterations in carbohydrate, protein, and lipid metabolism, modulation of hydromineral balance and immune functions. These modifications are clearly dependent on the type and duration of applied stressor and on the fish species. A variety of methods exist for evaluating stress responses in fish including several endpoints: whole body or organ weight measurements, histological observations, biochemical assays, immune and haematological functions, specific gene expression pattern (Harper and Wolf 2009). The present objective is not to detail them but rather to focus on putative relevant stress indicators in percids.

When fish are challenged with stressors, one of the main rapid physiological responses is the activation of metabolic functions in relation with increase of oxygen availability and release of erythrocytes in the blood. The drop in splenosomatic index in Eurasian perch (Milla et al. 2010) and elevation of hematocrit, red blood cells and mean corpuscular hemoglobin concentration ever reported in percids illustrate this secondary response (Brown et al. 2001; Acerete et al. 2004). In addition to these haematological modifications, a variety of immune responses are engaged in an attempt to counteract the threat and recover the welfare. In percids, immune alterations are also noted and among them lysozyme activity as well as complement subunits are to date the most frequently regulated parameters in Eurasian perch even if the direction of the change (increase or decrease) remains controversial (Milla et al. 2010; Douxfils et al. 2011b, 2012). Cortisol has been shown to be one of the hormonal immunoregulators in several fish species, even if its role in percids remains to be clarified (Milla et al. 2010; Mathieu et al. 2013a).

To fuel these physiological secondary stress responses, activation of the energy metabolism is necessary to provide an increase in the energy supply. In Eurasian perch, exposure to acute stressors also may lead to significant elevation of glycemia (Acerete et al. 2004; Milla et al. 2010). This rise of blood glucose is marked when fish are challenged with sharp stressors (Acerete et al. 2004; Jentoft et al. 2005; Milla et al. 2010) while the impact following long-lasting stressors is still doubtful (Douxfils et al. 2011a, b). Cortisol is involved in the regulation of the enzymes implied in energetic metabolism in fish. For example, the hyperglycemic effect of cortisol, in relation with regulation of neoglucogenetic enzymes, has been observed in several fish species following cortisol treatment (e.g. Vijayan et al. 2003). After acute stressor in percids, the parallel induction of blood cortisol and glucose levels is in agreement with the well know hyperglycemic effect of cortisol. In contrast, during chronic stress, the absence of increases in cortisol and glucose raises the question of the metabolic hormones involved in the energetic restoration. This absence of glucose regulation during chronic stress might be one peculiarity of percid fishes in comparison with the results often reported in other fish families (e.g. Cairns et al. 2008; Breves et al. 2010) supporting that the exhaustion of corticosteroid production may appear earlier in percids. Lactate which is another recognized stress markers in salmonid species seems not regulated following acute and chronic stressors in the Eurasian perch (Acerete et al. 2004; Cairns et al. 2008; Douxfils et al. 2012). Some biological markers of stress response in fish do not thus appear as very clear candidates in percid fishes. These discrepancies support the issue addressed by Barton (2002) in the interpretation of data because of apparent inconsistency among fishes in the responses of different blood chemistry characteristics.

Finally, the drop of electrolyte concentrations also belongs to the physiological changes observed post-stress, indicators of disequilibrium in the osmoregulatory processes (Forsberg and Summerfelt 2001; Barton et al. 2003). Cortisol is believed to be one mediator behind the changes in epithelial permeability and ionic exchanges observed following stressors (Segner et al. 2012). But, the relationship between plasma electrolyte levels, activation of ionic transporters and corticosteroids remains unknown in percids. Generally speaking, very few studies tried to highlight correlations between the levels of corticosteroids and the above parameters to demonstrate that such regulation is directly dependent on the corticosteroid surge. To do so, experimental studies consisting in injecting or implanting corticosteroids are useful to link the activation of the HPI axis with activation or repression of targeted metabolic pathways.

28.4.2 Tertiary Response to Stress and Exhaustion in Corticosteroid Production

When allostatic response is excessive or inefficient, the organism develops an allostatic load. the concept of "allostatic load" is often defined as the wear and tear that the body experiences due to repeated cycles of allostasis as well as the inefficient turning-on or shutting off of these responses. The stress response may represent an allostatic load as the animal quickly adjusts its physiological function to counteract changes occasioned by the external factors (Schreck et al. 2001; Segner et al. 2012; Prunet et al. 2012). Energetically, there are costs associated with allostasis and the related increased in energy consuming, may result in less energy available for other biological functions such as growth, reproduction or resistance to disease (Schreck et al. 2001; Segner et al. 2012). Further, depending on the magnitude and duration of the stressor as well as on the number of additional stressors, the energy store would be definitely consumed (allostatic overload), and the fish would not able to acclimatize any more, giving rise to pathology, disease potentially conducting to death (Barton 2002; Segner et al. 2012). Before reaching this threshold, feed intake and growth are deeply affected as shown after repeated handling stress in percids (Head and Malison 2000; Jentoft et al. 2005; Strand et al. 2007). Other tertiary responses have not been addressed in percids and more attention should be given to long term stress effects on reproductive capacity and resistance to pathogens.

The decrease of cortisol production along exposure to long-lasting stressors or multiple acute stressors is a phenomenon dubbed “habituation to stress”. As we early mentioned, there is high variability in the cortisolemia among fishes after the stressors but these inter-family differences do not necessarily preclude a similar range of habituation response. To recall, percids are known to produce quite elevated cortisol and this high induction would a priori allow detecting more easily a progressive drop of circulating hormones. When walleye fingerlings are subjected to two successive transports, the profile of cortisol following each transport is similar (Forsberg and Summerfelt 2001). In contrast, habituation to handling stress was demonstrated in Eurasian perch by comparing post-stress cortisol levels in repeatedly stressed fish (twice a week for 8 weeks) and fish exposed to the stressor only once (Jentoft et al. 2005). The observation of the cortisol response along the time course of exposure to chronic stressors is another strategy to evaluate the habituation and exhaustion to stress. In Euxrasian perch from two generation levels F1 and F4, fish confined after 30 days displayed lower levels of cortisol after acute handling than fish sampled after only 5 and 15 days even if this was only observed in F1 fish and not in the F4 generation (Douxfils et al. 2011a). Also, juvenile *Sander vitreus* exposed chronically to high concentrations of ammonia did not exhibit significant elevation after 56 days of exposure contrary to what was observed from day 28 to day 42 (Madison et al. 2009). All together, percids may also display habituation mechanisms when they are faced with chronic stress exposure or repeated stressors. The aim of this habituation would be to minimize the deleterious impact of sustained cortisol elevation on the various biological functions.

28.5 Conclusions

A gap of information exists, related to stress and corticosteroids in percid species. In spite several uncertainness on the possibility to utilise cortisol as stress indicator, it cannot be eliminated from the stress indicators list, but due to its high variability it must be complemented with other measurements, some of which requesting the molecular approach. GRs, other stress hormones, hsps, blood-cell counts, BDNF, and/or others, in order to have a more complete profile about the stress status of any fish. Cortisol may be useful only in acute stress experiments and monitored throughout time. To be used as stress indicator, the physiological status of organisms should be standardized. Non invasive methods such as measuring cortisol in water are a suitable alternative to avoid anesthetic and blood collection problems.

References

Acerete L, Balasch JC, Espinosa E, Josa A, Tort L (2004) Physiological responses in Eurasian perch (*Perca fluviatilis*, L.) subjected to stress by transport and handling. Aquaculture 237:167–178

Alsop D, Vijayan M (2009) The zebrafish stress axis: molecular fallout from the teleost-specific genome duplication event. Gen Comp Endocrinol 161:62–61

Applebaum SL, Wilson CA, Holt GJ, Nunez BS (2010) The onset of cortisol synthesis and the stress response is independent of changes in CYP11B or CYP21 mRNA levels in larval red drum (*Sciaenops ocellatus*). Gen Comp Endocrinol 165:269–276

Barnett CW, Pankhurst NW (1998) The effects of common laboratory and husbandry practices on the stress response of greenback flounder *Rhombosolea tapirina* (Günther, 1862). Aquaculture 162:313–329

Barton BA (2002) Stress in fishes: a diversity of responses with particular reference to changes in circulating corticosteroids. Integr Comp Biol 42:517–525

Barton BA, Iwama GK (1991) Physiological changes in fish from stress in aquaculture with emphasis on the response and effects of corticosteroids. Annu Rev Fish Dis 1:3–26

Barton BA, Peter RE, Paulencu CR (1980) Plasma cortisol levels of fingerling rainbow trout (*Salmo gairdneri*) at rest, and subjected to handling, confinement, transport, and stocking. Can J Fish Aquat Sci 37:805–811

Barton BA, Rahn AB, Feist G, Bollig H, Schreck CB (1998) Physiological stress responses of the freshwater chondrostean paddlefish (*Polyodon spathula*) to acute physical disturbances. Comp Biochem Physiol 120A:355–363

Barton BA, Bollig H, Hauskins BL, Jansen CR (2000) Juvenile pallid (*Scaphirhynchus albus*) and hybrid pallid 3 shovelnose (S. albus 3 platornychus) sturgeons exhibit low physiological responses to acute handling and severe confinement. Comp Biochem Physiol 126A:125–134

Barton BA, Haukenes AH, Parsons BG, Reed JR (2003) Plasma cortisol and chloride stress responses in juvenile walleyes during capture, transport, and stocking procedures. North Am J Aquacult 65(3):210–219

Bau F, Ferroni-Claverie N, Parent JP (2000) Réponses physiologiques de sept poissons lacustres à un stress de capture (filet maillant et épuisette). Bull Fr Peche Piscic 357(358):157–168

Berczi I (1998) Neurohormonal host defense in endotoxin shock. Ann N Y Acad Sci 840:787–802

Binuramesh C, Prabakaran M, Steinhagen D, Michael RD (2005) Effect of chronic confinement stress on the immune responses in different sex ration groups of *Oreochromis mossambicus* (Peters). Aquaculture 250:47–59

Breves JP, Hirano T, Grau EG (2010) Ionoregulatory and endocrine responses to disturbed salt and water balance in Mozambique tilapia exposed to confinement and handling stress. Comp Biochem Physiol A Mol Integr Physiol 155:294–300

Bridgham JT, Carroll SM, Thornton JW (2006) Evolution of hormone-receptor complexity by molecular exploitation. Science 312:97–101

Brodeur JC, Sherwood G, Rasmussen JB, Hontela A (1997) Impaired cortisol secretion in yellow perch (*Perca flavescens*) from lakes contaminated by heavy metals: in vivo and in vitro assessment. Can J Fish Aquat Sci 54:2752–2758

Bromage NR, Porter M, Randall C (2001) The environmental regulation of maturation in farmed finfish with special reference to the role of photoperiod and melatonin. Aquaculture 197:63–98
Brown JA, Moore WM, Quabius ES (2001) Physiological effects of saline waters on zander. J Fish Biol 59:1544–1555
Cairns MT, Johnson MC, Talbot AT, Pemmasani JK, McNeill RE, Houeix B, Sangrador-Vegas A, Pottinger TG (2008) A cDNA microarray assessment of gene expression in the liver of rainbow trout (*Oncorhynchus mykiss*) in response to a handling and confinement stressor. Comp Biochem Physiol Part D Genomics Proteomics 3:51–66
Craig JF (2000) Percid fishes: systematics, ecology and exploitation. Blackwell Science, Oxford
Douxfils J, Mandiki SNM, Marotte G, Wang N, Silvestre F, Milla S, Henrotte E, Vandecan M, Rougeot C, Mélard C, Kestemont P (2011a) Does domestication process affect stress response in juvenile Eurasian perch *Perca fluviatilis*? Comp Biochem Physiol A Mol Integr Physiol 159:92–99
Douxfils J, Mathieu C, Mandiki SNM, Milla S, Henrotte E, Wang N, Vandecan M, Dieu M, Dauchot N, Pigneur LM, Li X, Rougeot C, Mélard C, Silvestre F, Van Doninck K, Raes M, Kestemont P (2011b) Physiological and proteomic evidences that domestication process differentially modulates the immune status of juvenile Eurasian perch (*Perca fluviatilis*) under chronic confinement stress. Fish Shellfish Immunol 31:1113–1121
Douxfils J, Deprez M, Mandiki SNM, Milla S, Henrotte E, Mathieu C, Silvestre F, Vandecan M, Rougeot C, Mélard C, Dieu M, Raes M, Kestemont P (2012) Physiological and proteomic responses to single and repeated hypoxia in juvenile Eurasian perch under domestication process. Fish Shellfish Immunol 33:1112–1122
Doyon C, Leclair J, Trudeau VL, Moon TW (2006) Corticotropin-releasing factor and neuropeptide Y mRNA levels are modified by glucocorticoids in rainbow trout, *Oncorhynchus mykiss*. Gen Comp Endocrinol 146:126–135
Ellis T, James JD, Stewart C, Scott AP (2004) A non-invasive stress assay based upon measurement of free cortisol released into the water by rainbow trout. J Fish Biol 65:1233–1252
Fanouraki E, Divanach P, Pavlidis M (2007) Baseline values for acute and chronic stress indicators in sexually immature red porgy (*Pagrus pagrus*). Aquaculture 265:294–304
Fatemeh A, Sanaz G, Shahla J (2008) Plasma cortisol changes and body composition in *Stizostedion lucioperca* exposed to handling stress. Pak J Biol Sci 11:623–627
Feist G, Schreck CB (2001) Ontogeny of the stress response in chinook salmon Oncorhynchus tshawytscha. Fish Physiol Biochem 25:31–40
Forsberg JA, Summerfelt RC (2001) Physiological and behavioral stress responses of walleyes transported in salt and buffered-salt solutions. N Am J Aquac 63:191–200
Fuzzen MLM, Alderman SL, Bristow EN, Bernier NJ (2011) Ontogeny of the corticotropin-releasing factor system in rainbow trout and differential effects of hypoxia on the endocrine and cellular stress responses during development. Gen Comp Endocrinol 170:604–612
Girard C, Brodeur JC, Hontela A (1998) Responsiveness of the interrenal tissue of yellow perch (*Perca flavescens*) from contaminated sites to an ACTH challenge test in vivo. Can J Fish Aquat Sci 55:438–450
Harper C, Wolf JC (2009) Morphologic effects of the stress response in fish. ILAR J 50:387–396
Harris J, Bird DJ (2000) Modulation of the fish immune system by hormones. Vet Immunol Immunopathol 77(3-4):163–176
Haukenes AH (2001) Characterization of the impact of a chronic density stressor on the neuroendocrine and innate immune responses of yellow perch (*Perca flavescens*) to an acute challenge with lipopolysaccharide. PhD dissertation, University of South Dakota, Vermillion
Haukenes AH, Barton BA (2004) Characterization of the cortisol response following an acute challenge with lipopolysaccharide in yellow perch and the influence of rearing density. J Fish Biol 64:851–862
Haukenes AH, Barton BA (2008) Characterization of the cortisol response following an acute challenge with lipopolysaccharide in yellow perch and the influence of rearing density. J Fish Biol 64:851–862
Head AB, Malison JA (2000) Effects of lighting spectrum and disturbance level on the growth and stress responses of yellow perch *Perca flavescens*. J World Aquac Soc 31:73–80

Hontela A, Rasmussen JB, Audet C, Chevalier G (1992) Impaired cortisol stress response in fish from environment polluted by PAH, PCB and mercury. Arch Environ Contam Toxicol 22:278–283
Jeffrey JD, Esbaugh AJ, Vijayan MM, Gilmour KM (2012) Modulation of hypothalamic-pituitary-interrenal axis function by social status in rainbow trout. Gen Comp Endocrinol 176:201–210
Jentoft S, Held JA, Malison JA, Barry TP (2002) Ontogeny of the cortisol stress response in yellow perch (*Perca flavescens*). Fish Physiol Biochem 26:371–378
Jentoft S, Aastveit AH, Torjesen PA, Andersen O (2005) Effects of stress on growth, cortisol and glucose levels in non-domesticated Eurasian perch (*Perca fluviatilis*) and domesticated rainbow trout (*Oncorhynchus mykiss*). Comp Biochem Physiol 141:353–358
Kiilerich P, Milla S, Sturm A, Chevolleau S, Giton F, Terrien X, Fiet J, Prunet P (2011) Implication of the mineralocorticoid axis in the adaptation to salinity. J Endocrinol 209:221–235
Kim MA, Kim DS, Chang SY (2011) Characterization of two functional glucocorticoid receptors in the marine medaka *Oryzias dancena*. Gen Comp Endocrinol 171:341–349
Kime DE, Hews EA (1978) In vitro biosynthesis of 11β-hydroxy- and 11-oxotestosterone by testes of the pike (Esox lucius) and the perch (Perca fluviatilis). Gen Comp Endocrinol 36:604–608
Lacroix A, Hontela A (2004) A comparative assessment of the adrenotoxic effects of cadmium in two teleost species, rainbow trout, *Oncorhynchus mykiss*, and yellow perch, *Perca flavescens*. Aquat Toxicol 67:13–21
Levesque HM, Dorval J, Hontela A, Van Der Kraak GJ, Campbell PGC (2003) Hormonal, morphological, and physiological responses of yellow perch (*Perca flavescens*) to chronic environmental metal exposures. J Toxicol Environ Health 66:657–676
Li Y, Sturm A, Cunningham P, Bury NR (2012) Evidence for a divergence in function between two glucocorticoid receptors from a basal teleost. BMC Evol Biol 12:137
Linderoth M, Norman A, Noaksson E, Zebühr Y, Norrgren L, Balk L (2006) Steroid biosynthetic enzyme activities in leachate-exposed female perch (*Perca fluviatilis*) as biomarkers for endocrine disruption. Sci Total Environ 366:638–648
Madison BN, Hillon RSD, Tufts BL, Wang YS (2009) Exposure to low concentrations of dissolved ammonia promotes growth rate in walleye *Sander vitreus*. J Fish Biol 74:872–890
Martinez-Porchas M, Martinez-Cordova LM, Ramos-Enriquez R (2009) Cortisol and glucose: reliable indicators of fish stress? Pan-Am J Aquat Sci 4(2):158–178
Mathieu C, Milla C, Mandiki SNM, Douxfils J, Douny C, Scippo ML, De Pauw E, Kestemont P (2013a) First evidence of the possible implication of the 11-deoxycorticosterone (DOC) in immune activity of Eurasian perch (Perca fluviatilis, L.): comparison with cortisol. Comp Biochem Physiol A Mol Integr Physiol 165:149–158
Mathieu C, Milla S, Mandiki SN, Douxfils J, Kestemont P (2013b) In vivo response of some immune and endocrine variables to LPS in Eurasian perch (Perca fluviatilis, L.) and modulation of this response by two corticosteroids, cortisol and 11-deoxycorticosterone. Comp Biochem Physiol A Mol Integr Physiol 2014, 167:25–34
Milla S, Terrien X, Ibrahim F, Fiet J, Giton J, Prunet P, Le Gac F (2008) Plasma 11-deoxycorticosterone (DOC) and mineralocorticoid receptor testicular expression during rainbow trout spermiation: implication with 17, alpha, 20beta-dihydroxyprogesterone on the milt fluidity? Reprod Biol Endocrinol 6:19
Milla S, Wang N, Mandiki SNM, Kestemont P (2009) Corticosteroids: friends or foes of teleost fish reproduction? Comp Biochem Physiol A Mol Integr Physiol 153:242–251
Milla S, Wang N, Mandiki SNM, Nadzialek S, Kestemont P (2010) Spleen immune status is affected by handling stress but not regulated by cortisol in Eurasian perch, *Perca fluviatilis*. Fish Shellfish Immunol 28:931–941
Mommsen TP, Vijayan MM, Moon TW (1999) Cortisol in teleosts: dynamics, mechanisms of action, and metabolic regulation. Rev Fish Biol Fish 9:211–268
Muusze B, Marcon J, Vanden Thillart G, Almeida-Val V (1998) Hypoxia tolerance of Amazon fish. Respiratory and energy metabolism of the cichlid *Astronotus ocellatus*. Comp Biochem Physiol A 120:151–156

Noaksson E, Linderoth M, Gustavsson B, Zebühr Y, Balk L (2005) Reproductive status in female perch (*Perca fluviatilis*) outside a sewage treatment plant processing leachate from a refuse dump. Sci Total Environ 340:97–112

Noga EJ, Kerby JH, King W, Aucoin DP, Giesbrecht F (1994) Quantitative comparison of the stress response of striped bass (*Morone saxatilis*) and hybrid striped bass (*Morone saxatilis* x *Morone chrysops* and *Morone saxatilis* x *Morone americana*). Am J Vet Res 55(3):405–409

O'Connor EA, Pottinger TG, Sneddon LU (2011) The effects of acute and chronic hypoxia on cortisol, glucose and lactate concentrations in different populations of three-spined stickleback. Fish Physiol Biochem 37:461–469

Pichavant K, Maxime V, Thébault MT, Ollivier H, Garnier JP, Bousquet B, Diouris M, Bœuf G, Nonnotte G (2002) Effects of hypoxia and subsequent recovery on turbot *Scophtalmus maximus*: hormonal changes and anaerobic metabolism. Mar Ecol Prog Ser 225:275–285

Prunet P, Overli O, Douxfils J, Bernardini G, Kestemont P, Baron D (2012) Fish welfare and genomics. Fish Physiol Biochem 38(1):43–60

Rossi F, Chini V, Cattaneo AG, Bernardini G, Terova G, Saroglia M, Gornati R (2007) EST-based identification of genes expressed in perch (Perca fluviatilis, L.). Gene Expr 14:117–127

Saera-Villa A, Calduch-Giner JA, Prunet P, Pérez-Sanchez J (2009) Dynamis of liver GH/IGF axis and selected stress markers in juvenile gilthead sea bream (*Sparus aurata*) exposed to acute confinement. Differential stress response of growth hormone receptors. Comp Biochem Physiol A 154:197–203

Sandhu N, Vijayan MM (2011) Cadmium-mediated disruption of cortisol biosynthesis involved suppression of corticosteroidogenic genes in rainbow trout. Aquat Toxicol 103:92–100

Sarameh S, Falahatkar B, Azari TG, Efatpanah I (2012) Effects of different photoperiods and handling stress on spawning and reproductive performance of pikeperch, Sander lucioperca. Anim Reprod Sci 132:213–222

Schreck CB, Contreras-Sanchez W, Fitzpatrik MS (2001) Effects of stress on fish reproduction, gamete quality, and progeny. Aquaculture 197:3–24

Scott AP, Ellis T (2007) Measurement of fish steroids in water – a review. Gen Comp Endocrinol 153:392–400

Segner H, Sundh H, Buchmann K, Douxfils J, Sundell KS, Mathieu C, Ruane N, Jutfelt F, Toften H, Vaughan L (2012) Health of farmed fish: its relation to fish welfare and its utility as welfare indicator. Fish Physiol Biochem 38:85–105

Stolte EH, Nabuurs SB, Bury NR, Sturm A, Flik G, Savelkoul HFJ, Verburg-van Kemenade BML (2008) Stress and innate immunity in carp: corticosteroid receptors and pro-inflammatory cytokines. Mol Immunol 46:70–79

Strand A, Magnhagen C, Alanärä A (2007) Effects of repeated disturbances on feed intake, growth rates and energy expenditures of juvenile perch, Perca fluviatilis. Aquaculture 265:163–168

Strange RJ, Schreck CB, Ewing RD (1978) Cortisol concentrations in confined juvenile chinook salmon (*Oncorhynchus tshawytscha*). Trans Am Fish Soc 107:812–819

Sturm A, Bury N, Dengreville L, Fagart J, Flouriot G, Rafestin-Oblin ME, Prunet P (2005) 11-deoxycorticosterone is a potent agonist of the rainbow trout (*Oncorhynchus mykiss*) mineralocorticoid receptor. Endocrinology 146:47–55

Takahashi H, Sakamoto T (2013) The role of "mineralocorticoids" in teleost fish: relative importance of glucocorticoid signaling in the osmoregulation and "central" actions of mineralocorticoid receptor. Gen Comp Endocrinol 181:223–228

Terova G, Gornati R, Rimoldi S, Bernardini G, Saroglia M (2005) Quantification of a glucocorticoid receptor in sea bass (*Dicentrarchus labrax*, L.) reared at high stocking density. Gene 357:144–151

Terova G, Rimoldi S, Cora S, Bernardini G, Gornati R, Saroglia M (2008) Acute and chronic hypoxia affects HIF-1α mRNA levels in sea bass (*Dicentrarchus labrax*). Aquaculture 279:150–159

Theofan G, Goetz FW (1983) The in vitro synthesis of final maturational steroids by ovaries of brook trout (*Salvelinus fontinalis*) and yellow perch (*Perca flavescens*). Gen Comp Endocrinol 51:84–95

Tognoli C, Rossi F, Di Cola F, Baj G, Tongiorgi E, Terova G, Saroglia M, Bernardini G, Gornati R (2010) Acute stress alters transcript expression pattern and reduces processing of proBDNF to mature BDNF in *Dicentrarchus labrax*. BMC Neurosci 11:4

Tognoli C, Gornati R, Saroglia M, Terova G, Bernardini G (2012) Functional genomics of stress: molecular biomarkers for evaluating fish CNS activity. In: Saroglia M, Liu J (eds) Functional genomics in aquaculture. Wiley-Blackwell, Ames, pp 204–218

Tort L (2011) Stress and immune modulation in fish. Dev Comp Immunol 35:1366–1375

Vijayan MM, Raptis S, Sathiya R (2003) Cortisol treatment affects glucocorticoid receptor and glucocorticoid-responsive genes in the liver of rainbow trout. Gen Comp Endocrinol 132:256–263

Wang N, Gardeur JN, Henrotte E, Marie M, Kestemont P, Fontaine P (2006) Determinism of the induction of the reproductive cycle in female Eurasian perch, Perca fluviatilis: identification of environmental cues and permissive factors. Aquaculture 261(2):706–714

Wang N, Teletchea F, Kestemont P, Milla S, Fontaine P (2010) Photothermal control of the reproductive cycle in temperate fishes. Rev Aquac 2:209–222

Wedemeyer GA, Barton BA, McLeay DJ (1990) Stress and acclimation. In: Schreck CB, Moyle PB (eds) Methods for fish biology. American Fisheries Society, Bethesda, pp 451–489

Wendelaar Bonga SE (1997) The stress response in fish. Physiol Rev 77:591–625

Widell B, Noaksson E, Balk L, Zebuhr Y (2011) Simultaneous determination of several natural steroids in blood plasma from perch (Perca fluviatilis) by GC-HRMS. Int J Environ Anal Chem 91:303–318

Wiseman S, Thomas JK, McPhee L, Hursky O, Raine JC, Pietrock M, Giesy JP, Hecker M, Jana DM (2011) Attenuation of the cortisol response to stress in female rainbow trout chronically exposed to dietary selenomethionine. Aquat Toxicol 105:643–651

Wysocki LE, Dittami JP, Ladich F (2006) Ship noise and cortisol secretion in European freshwater fishes. Biol Conserv 128:501–508

Chapter 29
Domestication and Responses to Stress

J. Douxfils, S.N.M. Mandiki, C. Mathieu, S. Milla, and P. Kestemont

Abstract Domestication is a process by which animals become adapted to captive life conditions by way of natural and/or artificial selection. Genetic drift and inbreeding may also contribute to the evolution (deleterious or not) of numerous traits during domestication. In fish, domestication has been shown to influence growth, behavior (aggressiveness, dominance, alertness, feeding) and stress responsiveness. With respect to the later, it seems that stress-resistant animals may be selected during domestication as a result of an improved fitness. In percid species, some studies already investigated the effects of several husbandry stressors (chronic confinement, repeated water emersion, single and repeated hypoxia) on the physiological and immune responses along domestication process, by comparing Eurasian perch juveniles from distinct generation levels (Filial 1–5). Under chronic confinement, domestication resulted in a reduction in cellular (HSP70) and physiological (subsequent handling stress) stress response as well as the maintenance of immune status (no decrease in transferrin, complement C3 levels in Filial 4 fish). Domestication did not influence physiological and immune responses to repeated emersion stressor and repeated hypoxia. Eurasian perch has however been shown to be responsive to hypoxic conditions (hyperglycemia, spleen contractions, high transferrin level) and to potentially develop some acclimation mechanism to the repeated disturbance at the expense of some immune functions. All together, the studies demonstrated that domestication positively influenced fish tolerance to chronic confinement but not to hypoxia or water emersion and that this might be linked to the stressor severity. Moreover, F4/F5 fish groups displayed a better immune and physiological status than their F1 counterparts. A microsatellite analysis however revealed that these F4/F5 generations display a lower genetic diversity. Thus, loss of genetic diversity did not appear detrimental to the fish but will nevertheless limit the possibilities of genetic improvement in upcoming generations.

J. Douxfils (✉) • S.N.M. Mandiki • C. Mathieu • P. Kestemont
Research Unit in Environmental and Evolutionary Biology (URBE), University of Namur, 61 rue de Bruxelles, 5000 Namur, Belgium
e-mail: jessica.douxfils@unamur.be

S. Milla
Unité de Recherche Animal et Fonctionnalités des Produits Animaux, Université de Lorraine, 2 avenue de la Forêt de Haye, B.P. 172, 54505 Vandoeuvre-lès-Nancy, France

P. Kestemont et al. (eds.), *Biology and Culture of Percid Fishes*,
DOI 10.1007/978-94-017-7227-3_29

Keywords Eurasian perch • Domestication • Stress response • Hypoxia • Confinement • Emersion

29.1 Introduction

Animal farming implies a transition from nature to captive environments that usually differ from the ancestral environment. For some aspects, captivity is less challenging because of regular food supply, protection from predators and treatments for diseases. But, farm environments are also more challenging because animals are frequently disturbed by human activity and maintained under sub-optimal rearing conditions (e.g. confinement at un-naturally high densities, sometimes combined with water quality degradation), thereby increasing the risk of physiological stress and disease outbreak (Huntingford 2004). In such environments, adaptation is challenged and domestication is "*the process by which a population becomes adapted to man and to the captive environment through genetic changes occurring over generations and environmentally-induced developmental events recurring in each generation*" (Price 1999). Domestication does not mean taming, although the latter can be part of the process for numerous species (Mignon-Grasteau et al. 2005). Taming is a conditioned behavioural modification of an animal while domestication corresponds to phenotypic and genotypic changes in the captive progeny leading to a predisposition towards human association (Driscoll et al. 2009).

Domestication results from several genetic mechanisms including natural selection in captivity, relaxation of natural selection, artificial selection, genetic drift and inbreeding (Price 1999; Haffray et al. 2004; Vandeputte and Launey 2004; Mignon-Grasteau et al. 2005). Natural selection plays a substantial role in the evolution of many traits during domestication, particularly in the first generations bred in captivity. In the absence of artificial selection, natural selection provides the basic selective mechanism for genetic change in captive populations (Vandeputte and Launey 2004; Mignon-Grasteau et al. 2005). Relaxation of natural selection consists in a reduction of the selection pressure on traits that are important in the wild but not in captivity (Vandeputte and Launey 2004; Mignon-Grasteau et al. 2005). For example, traits such as competition for mating partners or the ability to hide from predators may be of less importance in captive environments where humans choose breeders and provide shelters. Consequently, relaxed traits appear to be more variable in domesticated animals (Kunzl et al. 2003; Mignon-Grasteau et al. 2005). Natural selection and relaxed natural selection in captivity are involuntary controlled by man as he determines the breeding environment. Artificial selection is a process in which humans intentionally choose the breeders. This selective mechanism is unique to animal farming and horticulture and is probably the best understood aspect of domestication (Mignon-Grasteau et al. 2005). Artificial selection may accentuate, accelerate or counteract the effects of unintentional selection during the early stages of domestication (Vandeputte and Launey 2004). Traits with direct economic impact (e.g. growth performance, flesh quality, disease resistance, age of first reproduction) are usually the first biological traits to be artificially selected in animal farming. Artificial selection on aggressiveness and/or stress response may also have indirect economic impact as they may be of great importance for ease of breeding and animal welfare (Vandeputte and Launey 2004). Inbreeding and genetic drift are processes which can result from limited population size, strong selection and/or unequal contribution of parents in natural mating groups and which can lead to random variation in the frequency of genes (Mignon-Grasteau et al. 2005). Genetic drift results in divergence that is not oriented (i.e. positive or negative consequences on productivity) while inbreeding can lead to negative effects on animal fitness and productivity (Vandeputte and Launey 2004). Domesticated animals may display behavioural, morphological, physiological and genetic divergences compared to the original wild population but also between captive populations from different farms, depending on the species considered, particular environmental conditions and human interventions in captivity (Vandeputte and Launey 2004; Mairesse et al. 2007). Domestication should thus be considered on a case-by-case basis and it is unlikely to extrapolate changes associated with one given domestication process to all domestication cases. Domestication is a continuous process. And it also seems to be, at least partly, reversible It has been demonstrated that domesticated animals (e.g. common carp *Cyprinus carpio*) can re-adapt to the wild environment after a short evolutionary period, known as feral populations (Matsuzaki et al. 2009). However, it seems that a return to the original wild status may not be expected because of the genetic changes arising over generations bred in captivity. In this respect, hybridization between feral and wild populations that could lead to outbreeding depression (i.e. introduction of maladaptive genes and/or traits causing fitness reduction and decline of the wild population) is a growing concern (Bowman et al. 2007). For instance, it has been demonstrated that hybrids between wild and escaped-farmed Atlantic salmon could cause competitive displacement of wild parr and this may result in reduced smolt production in case suitable downstream habitat is not available for these emigrant fish (McGinnity et al. 2003).

This first part of the chapter dealt with some definitions of domestication process in animals, all animal species combined. In the next sections, general information on domestication process and its effects on the physiology of fish species will first be presented, followed by a focus on the very scarce studies assessing the influence of domestication process on the stress responses of Percids. Then some data on the genetic changes and their impact on the stress sensitivity of farmed Percid species will be shown.

29.2 Domestication Process in Fish

Fish domestication is very recent and most of the farmed fish species cannot be considered domesticated at present (Teletchea and Fontaine 2014. With the exception of common carp domesticated 4000 years ago by the Chinese (Billard 1999), most of the currently-bred species are still close from their original wild population and display important genetic variability and phenotypic plasticity (Vandeputte and Prunet 2002; Haffray et al. 2004). Therefore, the adaptive potential to farming conditions remains high for numerous species.

For economic reasons, one major concern in animal farming is growth improvement and domestication is often accompanied by changes in growth performance. For instance, this has already been observed in European sea bass (*Dicentrarchus labrax*) after only one generation and in Eurasian perch *Perca fluviatilis* after less than six generations of captive rearing either under natural or artificial selection (Mairesse et al. 2007; Rougeot et al. 2007; Millot et al. 2010). Artificial selection on growth performance has been showed to lead in differences in that trait but also in other correlated morphological, behavioural and physiological characters (e.g. concordant modulation of the hormonal GH/IGF-1 growth axis and genes involved in growth regulation) (Fleming et al. 2002; Devlin et al. 2009; Tymchuk et al. 2009). Growth improvement has been associated with a decrease in stress response and disease sensitivity and might be an excellent indicator of adaptation to captivity conditions (Fontaine and Le Bail 2004).

In fish, behaviour is probably one of the first traits to be modulated in captivity (Alvarez and Nicieza 2003). Aggressiveness and dominance behaviour may be modified (e.g. in Chinook salmon *Oncorhynchus tshawytscha*) since defending territories may not be advantageous/possible for fish reared at high densities in homogenous environments (Pearsons et al. 2007). These behavioural differences may also be reinforced by intentional selection for increased growth rate in farmed strains (McGinnity et al. 2003). Moreover, shoal swimming tends to increase and inter-individual distance tends to decrease along domestication as a result of fish-to-fish proximity, suggesting a shift towards behavioural traits compatible with intensive culture conditions (Begout-Anras and Lagardere 2004). Important differences may also appear regarding feeding behaviour (Huntingford 2004). This is probably because farmed animals do not have to look for food, capture preys and/or check if it is edible (Mignon-Grasteau et al. 2005). As an example, hatchery-reared Japanese amago salmon (*Oncorhynchus rhodurus*) had emptier stomach and were more prone to eat non-food material (e.g. stones and leaves) than wild fish upon release into a natural river (Huntingford 2004). Domestication may deteriorate anti-predation behaviour, alertness and camouflage capacity (Alvarez and Nicieza 2003; Begout-Anras and Lagardere 2004; Mignon-Grasteau et al. 2005).

In fish, the influence of domestication on stress responsiveness has not been deeply studied so far but reduction of stress sensitivity has however been observed regarding both behavioral and physiological aspects. For instance, captive fish species (e.g. Salmonids, Atlantic cod *Gadus morhua*) have been characterized by lower flight distance (Begout-Anras and Lagardère 2004). In domesticated rainbow trout (*Oncorhynchus mykiss*), flight distance reduction was such that any human activity around the tanks stimulated feeding behavior rather than escape swimming, thereby suggesting a reduced stress sensitivity (Begout-Anras and Lagardère 2004). Following transfer into an unfamiliar aquarium, domesticated fighting fish (*Betta splendens*) did not display any cortisol rise conversely to wild individuals suggesting a reduction of the hypothalamo-pituitary-interrenal (HPI) axis reactivity along domestication process (Verbeek et al. 2008). Moreover, the domesticated strain showed behavioral immobility at the very beginning of a 3 h confinement period while this was not observed in wild fish. In rainbow fish (*Melanoteania duboulayi*), Zuberi et al. (2011) demonstrated that captive-reared individuals from the 15th generation displayed a substantially attenuated and late stress response after chasing in comparison with the wild population. Following a simulated predator attack, Johnsson et al. (2001) reported a lower standard heart rate and a less pronounced heart rate response in farmed Atlantic salmon after seven generations of inadvertent selection according to river origin (Namsen River, Norway) compared to wild counterparts from the same watercourse. Berejikian (1995) demonstrated that wild rainbow trout (*Oncorhynchus mykiss*) were less sensitive to predation than farmed trout if they were naive but also if they had experienced predation before. Farmed juveniles of steelhead trout have also been found to take more risks with natural predators than their wild counterparts (Johnsson and Abrahams 1991). Other studies on domesticated masu salmon (*Oncorhynchus masou*) revealed that these fish take feed closer to the surface (where they are more prone to predation) and show a shorter latency to feeding after introduction of chemical alarm signals (Reinhardt 2001; Yamamoto and Reinhardt 2003). Such behavioural deficiencies suggest a reduced sensitivity to predation stressor that may result from relaxed natural selection. Indeed, animal farming normally eliminates contact with predators and farmed fish are thus not selected for their ability to escape predator attacks (Alvarez and Nicieza 2003; Mignon-Grasteau et al. 2005). This may explain why farmed populations often experience heavier losses than wild-caught animals when they are confronted with predation (Mignon-Grasteau et al. 2005).

In captivity, management practices consistently applied to captive animals may reduce the stress responsiveness within their lifetime while both artificial and natural selection may contribute to the reduction in stress sensitivity over generations (Mignon-Grasteau et al. 2005). Indeed, stress-resistant animals may be positively selected during domestication as a result of an improved fitness (Mignon-Grasteau et al. 2005; Awata et al. 2011). Cortisol response has been shown to be moderately to highly heritable in carp ($h^2=0.60$) and rainbow trout ($h^2=0.41$–0.56) (Pottinger and Carrick 1999; Barton 2002; Vandeputte and Prunet 2002). Such heritability rates for cortisol response suggest a relatively high potential for selection on a stress sensitivity basis.

29.3 Domestication and Responses to Husbandry Stressors in Eurasian Perch

Very few studies investigated the combined effects of domestication process and husbandry stressors on the health status of Percid fish species. To date, the available data only concern Eurasian perch *Perca fluviatilis* (see Douxfils et al. 2011a, b, 2012, 2014). In these studies, captive-reared fish from different generational levels (F1 versus F4 or F5) but with a common wild origin (River Meuse in Belgium) were exposed to some potentially stressful conditions (i.e. chronic confinement, single and repeated emersion, single and repeated severe hypoxia) and stress and immune responses were evaluated. Indeed, it is well established that long-term stress may impair health, and particularly resistance to infectious diseases, since it can directly or indirectly affect the immune system. However, innate and acquired immunity are of particular importance for the host resistance to infectious diseases, especially in captivity where spatial limitation and fish to fish proximity facilitates the spreading of pathogens between individuals. In this context, it is interesting to evaluate how domestication processes, i.e. adaptation to captivity conditions, can modulate stress responsiveness and its consequences on fish immunocompetence. Detailed results of these studies are presented hereafter.

29.3.1 Chronic Confinement

Chronic confinement can be viewed as a long-term social and cognitive stressor according to the classification proposed by Tort (2011). Two experiments were conducted on the combined effects of chronic confinement and domestication in Eurasian perch (Douxfils et al. 2011a, b). Two contrasted captive generations of Eurasian perch juveniles (Filial 1 = F1 and Filial 4 = F4) that age-matched were compared in these studies. Both of them originally came from a local perch population sampled in the river Meuse in Belgium. Fish from the fourth generation (Filial 4 = F4) were used as a top-captive generation with a relatively long farming history while the F1 fish group correspond to the first generation hatched under captivity conditions and was therefore considered as a wild-close offspring (Fig. 29.1). During domestication of perch, no artificial selection was performed and only fish surviving by themselves (natural selection) under captivity conditions were reared as potential breeders for subsequent offspring. Although a high number of captive breeders (around 100 F1, F2, F3 or F4 fish) was used to constitute mating groups, only 10 F0 parents were captured from the wild to produce the F1 generation. Moreover, only higher-quality spawns (those with high fertilization rate) were collected to obtain the next generation. Such low number of wild breeders and collected spawns may have contributed to considerable reduction of genetic diversity at the very beginning of the process and unintentional selection on the basis of reproductive performances.

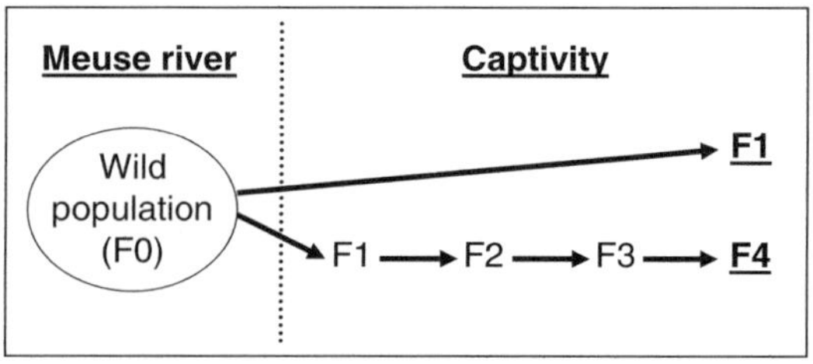

Fig. 29.1 Illustration of the way domestication was conducted for Eurasian perch by Douxfils et al. (2011a, b, 2012)

In the studies of Douxfils et al. (2011a, b), chronic confinement was managed as a reduction of the tank size in order to provide exiguous stocking area. Fish from each generation were maintained in 2000 L tanks (4 m^2) and in 100 L tanks (0.25 m^2) under similar fish density and sampled on several occasions (days 5/15/30 and days 7/55 in Douxfils et al. (2011a, b), respectively). Diverse stress (liver HSP60 and 70, cortisol, glucose, IGF-1, subsequent exposure to an acute stressor in vivo) and immune markers (lysozyme activity, immunoglobulin level, spleen macrophages respiratory burst) were measured and gel-dependent proteomic analysis was carried out to resolve serum proteins differently expressed between the two domestication levels.

During chronic confinement, no differences in cortisol and glucose levels were observed in F1 and F4 confined fish in comparison to their respective un-confined groups (Douxfils et al. 2011a, b). However the results showed that chronic confinement may indirectly affect the stress physiology of wild-close Eurasian perch (Douxfils et al. 2011a). Such impact of chronic confinement was revealed through examination of stress responses (cortisol and glucose levels) to a subsequent acute stressor (netting + anesthesia) as already reviewed by Mormede et al. (2007). Indeed, at each sampling time, some confined and un-confined fish were either directly sampled or submitted to an acute handling stressor (netting + 15 min. anaesthesia). Then, these handled fish were rapidly sampled (within 5 min). In response to the subsequent handling stressor, confined F1 individuals displayed a higher cortisol rise than un-confined F1 fish throughout the experimental period (day 5 to day 30) while confined and un-confined F4 fish were characterized by similar cortisol responses to this acute stressor (Fig. 29.2). Higher stress sensitivity at the proteic level was also observed in F1 confined fish as illustrated by an elevated HSP70 level in hepatic cells at the end of the experimental period (Fig. 29.3). Under chronic stressor exposure, cells constitutively express several stress proteins at very

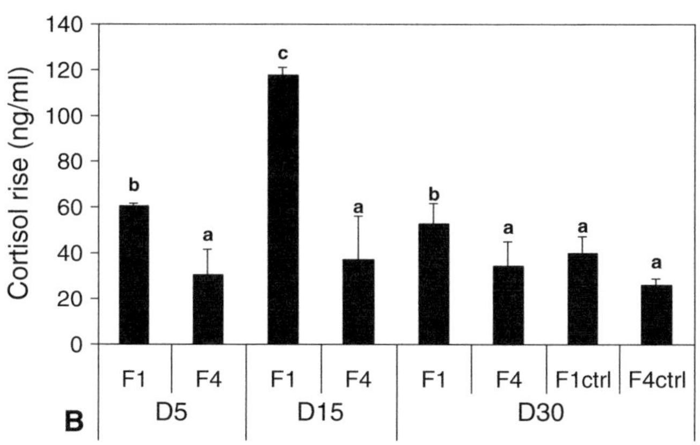

Fig. 29.2 Cortisol rise (ng ml^{-1}) related to the acute handling stress, subsequent to chronic confinement in Eurasian perch from distinct generational levels (Modified from Douxfils et al. 2011a). Results are expressed as the Mean ±SD (n = 2 as the tank is the experimental unit for statistical tests). Significant differences are indicated with different letters (*a*, *b*, and *c*). D5, D15 and D30 = day 5, 15 and 30 respectively, *Ctrl* un-confined fish

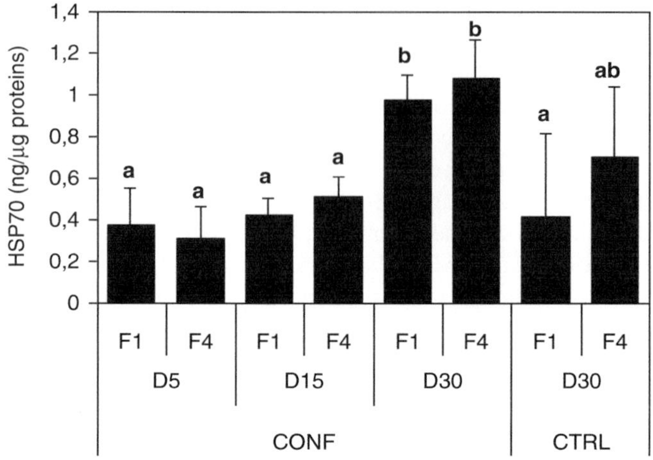

Fig. 29.3 Effect of domestication process on hepatic HSP70 level (ngEq µg prot^{-1}) during chronic confinement. Results are expressed as the Mean ±SD (n = 2 as the tank is the experimental unit for statistical tests). Significant results are indicated by different letters (a and b). *CONF* confined fish, *CTRL* "control" un-confined fish, D5 = day 5, D15 = day 15, and D30 = day 30

Fig. 29.4 Experimental scheme of the repeated emersion protocol and sampling procedure. *White stars* indicate exposure to emersion stressor. Samplings were performed on days 9, 18 and 44, 48 h after the last stressor

high level, notably HSP60 and HSP70 which can be induced during many types of stressors (Kültz 2005). These proteins make part of the cellular stress response (CRS), a defence reaction targeting a number of cellular functions including cell cycle control, protein chaperoning and repair, removal of damaged proteins and some aspects of metabolism (Kültz 2005). Moreover, it seems that HSP70, together with HSP90, helps to shape the responses to glucocorticoids (Basu et al. 2002; Grad and Picard 2007). These results indicated that domestication process reduced the sensitivity of fish to long-term confinement.

During chronic confinement, serum and spleen immune modulations were also reported in F1 and in F4 fish. In both generations, confined fish displayed increased serum lysozyme activity and respiratory burst of spleen macrophages suggesting some immune reactivity (Fig. 29.4) (Douxfils et al. 2011b). However, the consequences in terms of resistance to disease are difficult to determine since frequent inconsistencies between immune modulations and disease sensitivity have been reported in literature (Pruett 2003; Segner et al. 2012). Exposure to moderate stressors might be permissive, protective, preparatory or stimulatory with respect to immune function (Ashley 2007; Rotllant et al. 1997). But, domestication differentially modulated the immune status of fish under chronic confinement as revealed by proteomic analysis performed on day-7 samples. Transferrin abundance decreased in the serum of F1 confined fish and this was not observed in the F4 counterparts. Transferrin is an acute phase protein participating to innate immune function through sequestration of free iron which is of crucial importance for bacterial growth. Moreover, two complement C3 proteins were down-regulated in the serum of F1 confined fish. In F4 individuals, there was no decrease in abundance of these C3 molecules but rather an increase in other C3 fragments and in the abundance of apolipoprotein E. Immune activation was therefore observed in both generations subjected to chronic confinement but some immune parameters were negatively affected in F1 confined fish. The results thus highlighted the beneficial effects of domestication process on the immune status of fish exposed to long-term confinement, as far as proteomic response is concerned. However, stress-induced alterations of some immune functions (blood, internal organs, primary barriers) can occur without any deleterious consequences for the overall host resistance (Pruett 2003). Furthermore, immune modulations might be tissue specific (e.g. Milla et al. 2010). This entails the need for investigating more immune tissues and for including pathogenic challenges to validate this hypothesis.

29.3.2 Repeated Water Emersion

Impact of domestication process on the stress and immune responses of Eurasian perch has also been evaluated following repeated water emersion stressor (Douxfils et al. 2014). Emersion can be regarded as a multiple-component stressor with physico-chemical, mechanical and cognitive characteristics, and Eurasian perch has been shown to be responsive to a single emersion stressor either via netting or tank draining (Acerete et al. 2004; Milla et al. 2010), displaying short-term cortisol and glucose elevation as well as immune modulations in spleen (Milla et al. 2010). To date, the chronic effects of repeated emersion episodes (air exposure for 30 s, three times a week during 44 days) (Fig. 29.4) have been investigated in two distinct fish generations, i.e. filial 1 (F1) and 4 (F4) (as in Sect. 29.3.1), to assess the influence of domestication. Several stress indicators (cortisol, glucose, splenosomatic index) were measured at different time points (days 9, 18 and 44) but were not affected by the repeated stressor. Domestication did not influence physiological and immune responses to repeated emersion although proteomic analysis (2D-DIGE) revealed that domestication seemed to modulate, either positively or negatively, the abundance of several immune proteins (a2Ms, complement C3 component, Immunoglobulins light chains) in serum regardless of stressor exposure. However as mentioned earlier, it is difficult to establish a direct link between such immune modulations and disease resistance. Moreover, effects on the immune system have only been examined in serum but it has been shown that immune modulations were tissue-specific, notably in Percids (Milla et al. 2010).

29.3.3 Single and Repeated Hypoxia

Farmed fish are frequently exposed to low dissolved oxygen concentration, particularly following feeding and/or under rearing at high density. Hypoxia is a type of stressor with environmental and physiochemical characteristics (Tort 2011). Up to now, a study on channel catfish *Ictalurus punctatus* already observed immunosuppressive effects and increased susceptibility to high doses of bacteria following short-term sub-lethal hypoxia, suggesting that hypoxic disturbance may have a negative impact on farmed fish immunocompetence (Welker et al. 2007). In Percids, it has been reported that Eurasian perch was responsive to acute hypoxia as revealed by changes in hypoxia-inducible factor 1α (HIF-1α), a central regulating factor of the hypoxia response in mammals and fish (Rimoldi et al. 2012). An experiment investigated the physiological and immune responses of Eurasian perch to hypoxic stress and the potential influence of domestication process by comparing F1 and F5 generations (Douxfils et al. 2012). Hypoxia was progressively induced and then stabilized at 2.0 ± 0.5 mg L^{-1} during 4 h. Fish were sampled after the first hypoxic disturbance and then at the end of the experimental period, after the last hypoxic period was applied (i.e. after 12 hypoxia sessions since hypoxia was induced twice a week during 46 days) (Fig. 29.5). Several stress and immune variables were measured: serum glucose and cortisol levels, splenosomatic index, muscle lactate concentration, serum lysozyme and alternative complement pathway activities and proteomic analysis of serum by 2D-DIGE. For both generations, hyperglycemia and spleen contraction were consistently observed after a 4-h hypoxia (single or repeated), thereby showing that Eurasian perch was sensitive to the hypoxic condition tested (Figs. 29.6 and 29.7). Following repeated hypoxic disturbances, Eurasian perch displayed lower hyperglycemia, spleen contraction and higher abundance of serum transferrin level compared to fish exposed to hypoxia only once and this might indicate the development of acclimation mechanisms to this recurrent stressor. Pertaining to the immune system, repeated hypoxia resulted in a decreased

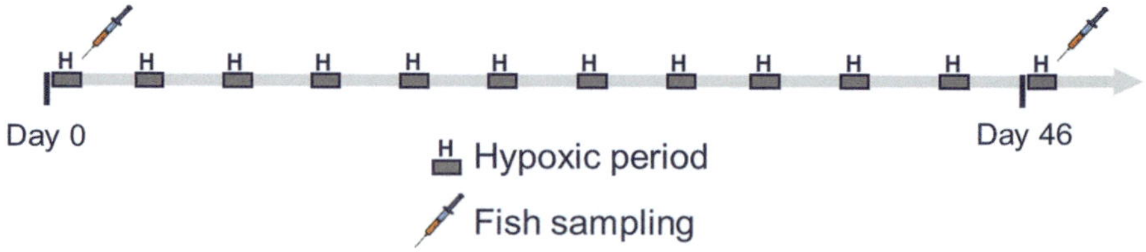

Fig. 29.5 Experimental protocol of single and repeated hypoxia. SINGLE hypoxia: the first hypoxia period was conducted on day 0 and fish (n = 5 *per* tank, three tanks *per* condition) were immediately sampled at the end of the 4-h period. REPEATED hypoxia: Fish were exposed to 4-h hypoxia, twice a week during 46 days. The last hypoxic stressor was conducted on day 46 and fish (n = 5 *per* tank, three tanks *per* condition) were immediately sampled at the end of the 4-h period

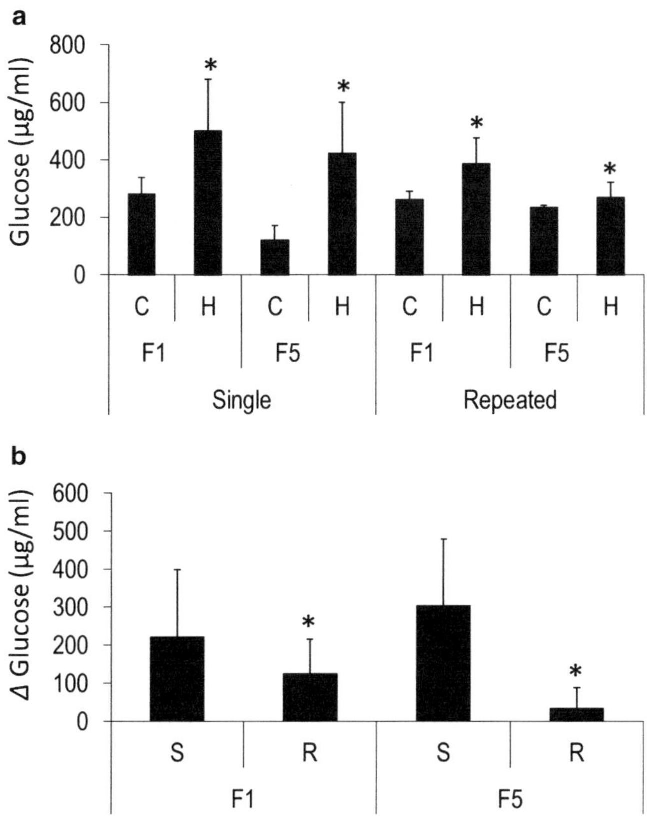

Fig. 29.6 Glucose response in the serum of Eurasian perch exposed to single or repeated hypoxia (n = 3). (**a**) total glucose level (µg/ml). *C* control fish, *H* hypoxic fish. *Asterisk* (*) indicates significant differences between C and H treatment. (**b**) Δ glucose values (µg/ml). *S* single hypoxia, *R* repeated hypoxia. *Asterisks* (*) indicates differences between S and R treatment

abundance of some complement C3 molecules and in a reduced serum lysozyme activity. Reallocation of energy and resources away from the immune system might be necessary to develop efficient physiological responses and/or long-term acclimation mechanisms to the recurrent hypoxic disturbances. Fish displayed similar responses to either single or repeated hypoxia, regardless of generational level (F1 or F5). Therefore, domestication did not seem to differentially modulate the physiology and immune system following exposure to the tested stressor. It has been hypothesized that the severe impact of hypoxic stress on homeostasis and direct survival may explain the maintenance of efficient physiological response over generations (Douxfils et al. 2012).

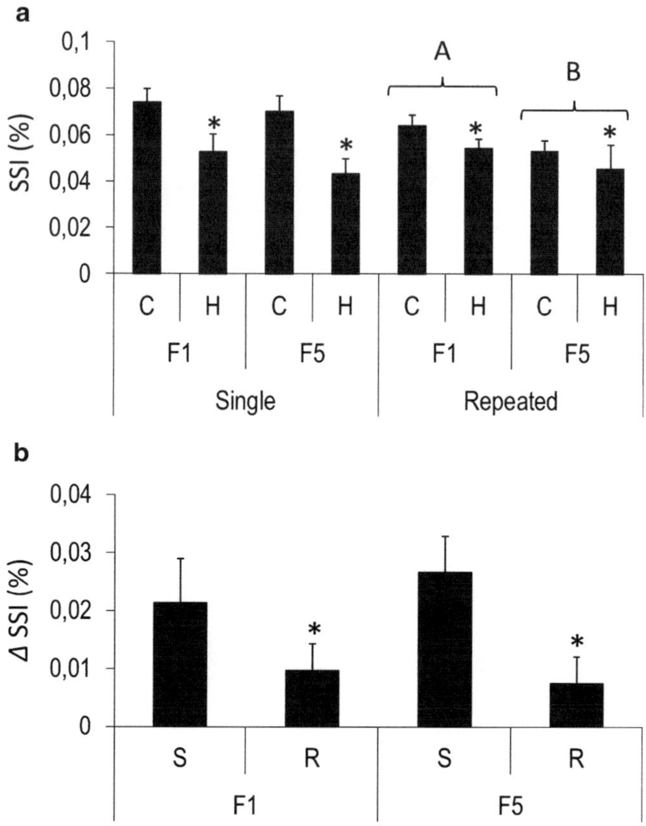

Fig. 29.7 SSI response in the serum of Eurasian perch exposed to single or repeated hypoxia (n=3). (**a**) SSI (%). *C* control fish, *H* hypoxic fish. *Asterisk* (*) indicates significant differences between C and H treatment. (**b**) Δ SSI (%). *S* single hypoxia, *R* repeated hypoxia. *Asterisks* (*) indicates differences between S and R treatment. *Capital letters* indicates differences between F1 and F5 generations

29.4 Stress Responses and Genetic Diversity in Domesticated Percid Strains

Genetic variation is fundamental for populations because it determines the potential for adaptation to changes in the environment (Pampoulie et al. 2006; Horreo et al. 2008). Genetic characterization of farmed fish stocks may thus be particularly interesting with respect to domestication processes in captivity. In the studies of Douxfils et al. (2011a, b, 2012, 2014) on Eurasian perch, a microsatellite-based genotyping of the different captive fish groups (F1, F4 and F5) as well as the initial wild population from which those groups derived revealed some reduction in genetic diversity (decreased allelic number, lower heterozygozity, increased genetic homogeneity) along generations as well as a progressive divergence from the original wild

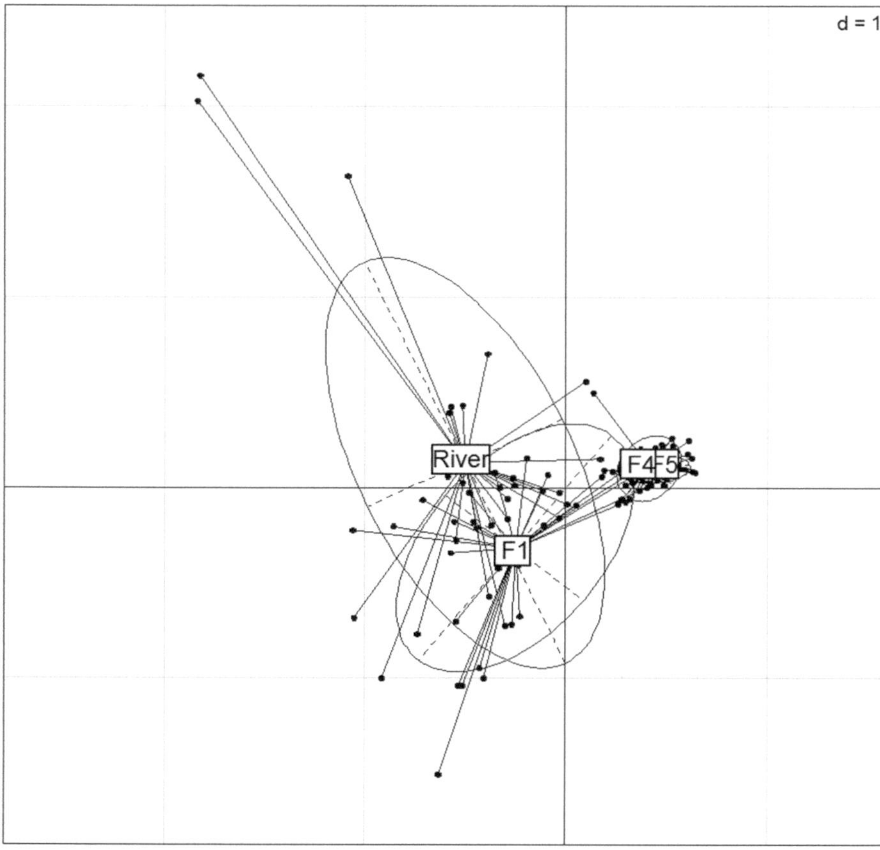

Fig. 29.8 Scatterplot graph of correspondence analysis. The Meuse River populations and captive generations are represented (n = 30 *per* population/generation). Each *dot* represents an individual (Douxfils et al. unpublished data)

population (Fig. 29.8). Decrease in genetic variability and increased homozygosity often result from inbreeding in captivity, mainly caused by limited population size and differential contribution of the parents (Blonk et al. 2009). Excessive inbreeding and the related decreasing genetic diversity can lead to inbreeding depression, i.e. a reduction in fitness resulting from the expression of rare and recessive harmful alleles which are not normally expressed and do not affect phenotype in large populations (Ditlecadet et al. 2006; Kang et al. 2006; Dixon et al. 2008). Inbred animals tend to be more sensitive to environmental stress than outbred individuals probably because adverse conditions may increase the expression of such recessive alleles or because animals are more easily pushed beyond their physiological limits (Fox and Reed 2011). Reduction of growth, viability and reproductive performances, as well as biochemical disorders, deformities and finally a decrease in the response to selection can also result from severe inbreeding (Ditlecadet et al. 2006; Dixon et al.

2008). The microsatellite analysis (using eight polymorphic loci) conducted by Douxfils et al. (2011b and unpublished data) on farmed generations of Eurasian perch indicated that the genetic homogeneity in F4 and F5 generations may be at least partly due to inbreeding although negative *Fis* values and low occurrence of linkage disequilibrium were observed. It is highly probable that skewed parental contribution affected genetic diversity in the subsequent generations since only a few spawns (those displaying high fertilization rate) were collected and used to form the next generation. However it seems that F4 individuals were able to mount similar humoral and tissular immune responses compared to F1 counterparts following chronic confinement or emersion stress despite the low genetic diversity (Douxfils et al. 2011b, 2014). Similarly, F5 fish also displayed comparable physiological and immune readjustments compared to F1 individuals following hypoxia (Douxfils et al. 2012) although genetic diversity was reduced in that fifth generation (Douxfils et al. *unpublished data*). Moreover, higher survival and growth rate have been observed in F4/F5 fish (Douxfils et al. 2011a), thereby suggesting a relatively good fitness of perch with a longer captive- life history. The relatively good immune and physiological status of F4 and F5 fish groups might be explained by the fact that inbreeding was not severe enough to cause visible and deleterious effects on the tested variables. According to Fox and Reed (2011), the effects of inbreeding on stress sensitivity are more important in highly stressful environments and may not be detected in low-stress environment. Moreover, reduced genetic diversity is not only linked to inbreeding but also to selection and drift and may not necessarily give rise to negative consequences for fish health. Only important reduction in genetic diversity is expected to lead to negative effects. Therefore, it is also probable that fish have been naturally selected for better immune function and stress tolerance. Considering the present results, it seems that the observed reduction of genetic diversity was not detrimental to farmed fish. However, physiological status was only evaluated at the level of stress and immune response and the potential impact on other biological functions are still unknown. Another point is that, although microsatellites are potent tools to reconstruct populational genetic processes, their general value as indicators of fitness and adaptive potential is unclear (Ouborg et al. 2010). Anyway, loss of genetic diversity during the first generations of breeding practices will limit the potential for future genetic improvement from artificial selection in that cultured stock (Norris et al. 1999; Pampoulie et al. 2006).

29.5 Conclusions and Perspectives

Currently, a limited number of studies examined stress responses and the influence of domestication process in captive Percid species and this field certainly needs further investigations. Some interesting conclusions and perspectives can nevertheless be expressed based on the few data obtained on Eurasian perch.

The studies mentioned earlier indicated that domestication might be a possible way to reduce stress responsiveness in some occasions. Several stressors have been tested including chronic confinement, water emersion and environmental hypoxia As above-mentioned, chronic confinement can be described as a long-term social and cognitive stressor. Water emersion may be defined as a multiple-component stressor (physico-chemical, mechanical and cognitive disturbance). Finally, sublethal hypoxia can be regarded as a strong environmental and physico-chemical stressor. The results suggest that domestication was associated with a better tolerance to chronic confinement but not to hypoxia or water emersion. The stressor capacity to severely threaten homeostasis might be a possible explanation for this observation. Since stress response is an adaptive mechanism, preservation of optimal stress responses may be necessary for immediate survival under severe conditions such as exposure to physicochemical, mechanical or thermal stressors that constitute real threat to homeostasis. On the other hand, certain stressors are not endangering physical health but are rather associated with cognitive stress (i.e. fear, anxiety and apprehension of possible danger) which can be minimized through repetition and experience by animals (Tort 2011). In this case, reduction in stress sensitivity may not be deleterious since immediate survival is not at risk. On the contrary, reduced stress responsiveness may be profitable due to lower resulting allostatic load.

However, the hypothesis that adaptive stress response to real environmental threats (e.g. hypoxia) might be preserved during domestication whereas sensitivity to cognitive stressors might be favourably reduced should be reinforced and verified by testing other stressors such as, for example, changes in water quality and temperature, physical injuries, netting and handling procedures, low water stress, shadow movements above the tanks or crowding.

The results also suggest that some reduction of genetic diversity may not cause deleterious impact on fish health regarding the tested variables. This suggests that severe inbreeding may not have already occurred and/or that natural selection process may have favoured individuals displaying suitable stress responses and immune performances.

Finally, attention must be paid on the fact that the influence of domestication may strongly depend on the particular rearing system and conditions, on the species considered as well as on the level of domestication reached (number of generations). These first results should not be necessarily generalized and further studies are needed to improve the knowledge on the field and to discriminate the global and study-specific changes associated with domestication.

References

Acerete L, Balasch JC, Espinosa E, Josa A, Tort L (2004) Physiological responses in Eurasian perch (*Perca fluviatilis*, L.) subjected to stress by transport and handling. Aquaculture 237:167–178

Alvarez D, Nicieza G (2003) Predator avoidance behaviour in wild and hatchery-reared brown trout: the role of experience and domestication. J Fish Biol 63:1565–1577

Ashley PJ (2007) Fish welfare: current issues in aquaculture. Appl Anim Behav Sci 104:199–235

Awata S, Tsuruta T, Yada T, Iguchi K (2011) Effects of suspended sediment on cortisol levels in wild and cultures strains of ayu *Plecoglossus altivelis*. Aquaculture 314:115–121
Barton BA (2002) Stress in fishes: a diversity of responses with particular reference to changes in circulating corticosteroids. Integr Comp Biol 42:517–252
Basu N, Tosgham AE, Ackerman PA, Bibeau MR, Nakano K, Schulte PM, Iwama GK (2002) Heat shock protein genes and their functional significance in fish. Gene 295:173–183
Begout-Anras ML, Lagardere JP (2004) Domestication et comportement chez les poissons téléostéens. INRA Prod Anim 17:211–215
Berejikian BA (1995) The effects of hatchery and wild ancestry and experience on the relative ability of steelhead trout fry (Oncorhynchus mykiss) to avoid a benthic predator. Can J Fish Aquat Sci 52:2476–2482
Billard R (1999) Carp: biology and culture, Praxis series in aquaculture and fisheries. Springer, London
Blonk RJW, Komen J, Kamstra A, Crooijmans RPMA, van Arendonk JAM (2009) Levels of inbreeding in group mating captive broodstock populations of common sole, (Solea solea), inferred from parental relatedness and contribution. Aquaculture 289:26–31
Bowman J, Kidd AG, Gorman RM, Schulte-Hostedde AI (2007) Assessing the potential for impacts by feral mink on wild mink in Canada. Biol Conserv 139:12–18
Devlin RH, Sakhrani D, Tymchuk WE, Rise ML, Goh B (2009) Domestication and growth hormone transgenesis cause similar changes in gene expression in coho salmon (Oncorhynchus kisutch). Proc Natl Acad Sci U S A 106:3047–3052
Ditlecadet D, Dufresne F, Le François NR, Blier PU (2006) Applying microsatellites in two commercial strains of arctic charr (*Salvelinus alpinus*): potential for a selective breeding program. Aquaculture 257:37–43
Dixon TJ, Coman GJ, Arnold SJ, Sellars MJ, Lyons RE, Dierens L, Preston NP, Li Y (2008) Shifts in genetic diversity during domestication of Black Tiger shrimp, *Penaeus monodon*, monitored using two multiplexed microsatellite systems. Aquaculture 283:1–6
Douxfils J, Mandiki SNM, Marotte G, Wang N, Silvestre F, Milla S, Henrotte E, Vandecan M, Rougeot C, Kestemont P (2011a) Does domestication process affect stress response in juvenile Eurasian perch *Perca fluviatilis*? Comp Biochem Physiol A Mol Integr physiol 159:92–99
Douxfils J, Mathieu C, Mandiki SNM, Milla S, Henrotte E, Wang N, Vandecan M, Dieu M, Dauchot N, Pigneur LM, Li X, Rougeot C, Mélard C, Silvestre F, Van Doninck K, Raes M, Kestemont P (2011b) Physiological and proteomic evidences that domestication process differentially modulates the immune status of juvenile Eurasian perch (*Perca fluviatilis*) under chronic confinement stress. Fish Shellfish Immunol 31:1113–1121
Douxfils J, Deprez M, Mandiki SNM, Milla S, Henrotte E, Mathieu C, Silvestre F, Vandecan M, Rougeot C, Mélard C, Dieu M, Raes M, Kestemont P (2012) Physiological and proteomic responses to single and repeated hypoxia in juvenile Eurasian perch under domestication – clues to physiological acclimation and humoral immune modulations. Fish Shellfish Immunol 33:1112–1122
Douxfils J, Lambert S, Mathieu C, Milla S, Mandiki SNM, Henrotte H, Wang N, Dieu M, Raes M, Rougeot C, Kestemont P (2014) Influence of domestication process on immune response to repeated emersion stressors in Eurasian perch (Perca fluviatilis, L.). Comp Biochem Physiol A Mol Integr Physiol 173:52–60
Driscoll CA, MacDonald DW, O'Brien SJ (2009) From wild animals to domestic pets, an evolutionary view of domestication. From the academy – in the light of evolution III: two centuries of Darwin Sackler colloquium. Proc Natl Acad Sci U S A 106:9971–9978
Fleming I, Agustesson T, Finstad B, Johnsson J, Bjornsson B (2002) Effects of domestication on growth physiology and endocrinology of Atlantic salmon (*Salmo salar*). Can J Fish Aquat Sci 59:1323–1330
Fontaine P, Le Bail PY (2004) Domestication et croissance chez les poissons. INRA Prod Anim 17:217–225

Fox CH, Reed DH (2011) Inbreeding depression increases with environmental stress: an experimental study and meta-analysis. Evolution 65:246–258

Grad I, Picard D (2007) The glucocorticoid responses are shaped by molecular chaperones. Mol Cell Endocrinol 275:2–12

Haffray P, Pincent C, Rault P, Coudurier B (2004) Domestication et amelioration génétique des cheptels piscicoles français dans le cadre du SYSAAF. INRA Prod Anim 17:243–252

Horreo JL, Machado-Schiaffino G, Griffiths A, Bright D, Stevens J, Garcia-Vazquez E (2008) Identification of differential broodstock contribution affecting genetic variability in hatchery stocks of Atlantic salmon (*Salmo salar*). Aquaculture 280:89–93

Huntingford FA (2004) Implications of domestication and rearing conditions for the behavior of cultivated fishes. J Fish Biol 65:122–142

Johnsson JI, Abrahams MV (1991) Interbreeding with domestic strain increases foraging under threat of predation in juvenile steelhead trout (*Oncorhynchus mykiss*): an experimental study. Can J Fish Aquat Sci 48:243–247

Johnsson J, Höjesjö J, Fleming I (2001) Behavioural and heart rate responses to predation risk in wild and domesticated Atlantic salmon. Can J Fish Aquat Sci 58:788–794

Kang JA, Noh JK, Kim JH, Lee JH, Kim HC, Kim KK, Kim BS, Lee WJ (2006) Genetic relationship between broodstocks of olive flounder, *Paralichthys olivaceus* (Temminck and Schlegel) using microsatellite markers. Aquac Res 37:701–707

Kültz D (2005) Molecular and evolutionary basis of the cellular stress response. Ann Rev Physiol 67:225–257

Kunzl C, Kaiser S, Meier E, Sachser N (2003) Is a wild mammal kept and reared in captivity still a wild animal? Horm Behav 43:187–196

Mairesse G, Thomas M, Gardeur JN, Brun-Bellut J (2007) Effects of dietary factors, stocking biomass and domestication on the nutritional and technological quality of the Eurasian perch *Perca fluviatilis*. Aquaculture 262:86–94

Matsuzaki SS, Mabuchi K, Takamura N, Nishida M, Washitani I (2009) Behavioral and morphological differences between feral and domesticated strains of common carp *Cyprinus carpio*. J Fish Biol 75:1206–1220

McGinnity P, Prodohl P, Ferguson A, Hynes R, O'Maoiléidigh N, Baker N, Cotter D, O'Hea B, Cooke D, Rogan G, Taggart J, Cross T (2003) Fitness reduction and potential extinction of wild populations of Atlantic salmon, Salmo salar, as a result of interactions with escaped farm salmon. Proc R Soc Lond 270:2443–2450

Mignon-Grasteau S, Boissy A, Bouix J, Faure JM, Fisher AD, Hinch GN, Jensen P, Le Neindre P, Mormède P, Prunet P, Vandeputte M, Beaumont C (2005) Genetics of adaptation and domestication in livestock. Livest Prod Sci 93:3–14

Milla S, Mathieu M, Wang N, Lambert S, Nadzialek S, Massart S, Henrotte E, Douxfils J, Mélard C, Mandiki SNM, Kestemont P (2010) Spleen immune status is affected after acute handling stress but not regulated by cortisol in Eurasian perch, Perca fluviatilis. Fish Shellfish Immunol 28:931–941

Millot S, Péan S, Leguay D, Vergnet A, Chatain B, Bégout ML (2010) Evaluation of behavioural changes induced by a first step of domestication or selection for growth in the European sea bass (*Dicentrarchus labrax*): a self-feeding approach under repeated acute stress. Aquaculture 306:211–217

Mormede P, Andanson S, Aupérin B, Beerda B, Guémené D, Malmkvist J, Manteca X, Manteuffel G, Prunet P, Van Reenen CG, Richard S, Veissier I (2007) Exploration of the hypothalamic-pituitary-adrenal function as a tool to evaluate fish welfare. Physiol Behav 92:317–339

Norris AT, Bradley DG, Cunningham EP (1999) Microsatellite genetic variation between and within farmed and wild Atlantic salmon (*Salmon salar*) populations. Aquaculture 180:247–264

Ouborg NJ, Angeloni F, Vergeer P (2010) An essay on the necessity and feasibility of conservation genomics. Conserv Genet 11:643–653

Pampoulie C, Jorundsdottir TD, Steinarsson A, Petursdottir G, Stefansson MO, Danielsdottir AK (2006) Genetic comparison of experimental farmed strains and wild Icelandic populations of Atlantic cod (*Gadus morhua* L.). Aquaculture 261:556–564

Pearsons TN, Fritts AL, Scott JL (2007) The effects of hatchery domestication on competitive dominance of juvenile spring Chinook salmon (*Oncorhynchus tshawytscha*). Can J Fish Aquat Sci 64:803–812

Pottinger TG, Carrick TR (1999) Modification of the plasma cortisol response to stress in rainbow trout by selective breeding. Gen Comp Endocrinol 116:122–132

Price EO (1999) Behavioral development in animals undergoing domestication. Appl Anim Behav Sci 65:245–271

Pruett BP (2003) Stress and the immune system. Pathophysiology 9:133–153

Reinhardt UG (2001) Selection for surface feeding in farmed and sea-ranched masu salmon juveniles. Trans Am Fish Soc 130:155–158

Rimoldi S, Terova G, Ceccuzzi P, Marelli S, Antonini M, Saroglia M (2012) HIF-1α mRNA levels in Eurasian perch (*Perca fluviatilis*) exposed to acute and chronic hypoxia. Mol Biol Rep 39:4009–4015

Rotllant J, Pavlidis M, Kentouri M, Abad ME, Tort L (1997) Non-specific immune responses in the red porgy *Pagrus pagrus* after crowding stress. Aquaculture 156:279–290

Rougeot C, Berviller E, Prignon C, Gustin D, Vandecan M, Mélard C (2007) Growth improvement of Eurasian perch using domesticated strains under intensive rearing conditions. Aquaculture 272:s306

Segner H, Sundh H, Buchmann K, Douxfils J, Sundell KS, Mathieu C, Ruane N, Jutfelt F, Toften H, Vaughan L (2012) Health of farmed fish: its relation to fish welfare and its utility as welfare indicator. Fish Physiol Biochem 38:85–105

Teletchea F, Fontaine P (2014) What is a domesticated fish? Implications for the sustainable future of aquaculture. Fish Fish. doi:10.1111/faf.12006

Tort L (2011) Stress and immune modulation in fish. Dev Comp Immunol 35:1366–1375

Tymchuk W, Beckman B, Devlin R (2009) Altered expression of growth hormone/insulin-like growth factor I axis hormones in domesticated fish. Endocrinology 150:1809–1816

Vandeputte M, Launey S (2004) Quelle gestion génétique de la domestication chez les poissons? INRA Prod Anim 17:237–242

Vandeputte M, Prunet P (2002) Génétique et adaptation chez les poissons: domestication, résistance au stress et adaptation aux conditions de milieu. INRA Prod Anim 15:365–371

Verbeek P, Iwamoto T, Murakami N (2008) Variable stress-responsiveness in wild and domesticated fighting fish. Physiol Behav 93:83–88

Welker TL, Mcnulty ST, Klesius PH (2007) Effect of sublethal hypoxia on the immune response and susceptibility of channel catfish, *Ictalurus punctatus*, to enteric septicemia. J World Aquac Soc 38:12–23

Yamamoto T, Reinhardt UG (2003) Dominance and predator avoidance in domesticated and wild masu salmon *Oncorhynchus masou*. Fish Sci 69:88–94

Zuberi A, Ali S, Brown C (2011) A non-invasive assay for monitoring stress responses: a comparison between wild and captive-reared rainbowfish (*Melanoteania duboulayi*). Aquaculture 321:267–272

Chapter 30
Immune Status and Immunomodulation in Percid Fish

S.N.M. Mandiki, J. Douxfils, C. Mathieu, S. Milla, E. Henrotte, H. Jansen, and P. Kestemont

Abstract Knowledge of the adaptability of the immune system is an important issue that can enable a better management of environmental conditions for increasing fish performances and welfare. Unfortunately, the immune defence of percid fish has not yet been characterized, but they should be able to accomplish innate and adaptive immunity as other fishes. Indeed, the available information about molecular characterization of some immune genes of percid fish showed similar features as that of other teleost. But some specific pathways were identified; and it could be interesting to precise their implication for the immune competence. It seems also likely that percid fish are able to adapt their immunocompetence to the seasonal changes in environmental conditions, or to respond positively to some immunomodulatory compounds as other fish species, but more studies are still needed to highlight the specific immune pathways of the relative responses.

Keywords Eurasian perch • Pikeperch • Immunostimulant • Probiotics • Prebiotics

30.1 Introduction

In comparison to other fish species, such as salmonid and cyprinid species, the immune system of percid fish has not yet been characterized, but it is obvious that the percids should be able to accomplish innate and adaptive immunity as other fishes. For instance, two anti-genetically related immunoglobulin (Ig) populations

S.N.M. Mandiki (✉) • J. Douxfils • C. Mathieu • E. Henrotte • P. Kestemont
Research Unit in Environmental and Evolutive Biology, University of Namur, Namur, Belgium
e-mail: robert.mandiki@unamur.be

S. Milla
Unité de Recherche Animal et Fonctionnalités des Produits Animaux, Université de Lorraine, 2 avenue de la Forêt de Haye, B.P. 172, 54505 Vandoeuvre-lès-Nancy, France

H. Jansen
IMARES, Wageningen, The Netherlands

P. Kestemont et al. (eds.), *Biology and Culture of Percid Fishes*,
DOI 10.1007/978-94-017-7227-3_30

have been identified in Eurasian perch serum with different chain isotypes as in other fish species (Whittington 1993), but such characterization needs to be confirmed by more sensitive techniques. As for mammals, two Ig cell forms have been identified in fish: B cell receptors, membrane-bound molecules that act as antigen receptors on the B cell surface, and antibodies secreted by blood cells, which are important molecules mediating humoral immune responses (Zhu et al. 2013). Moreover, various functional Ig groups have been characterized in different fish species, namely IgM, IgD, IgZ, and IgT. It was also demonstrated that some of these isotypes act as barrier defence, such as IgT which is a mucosal immune-related Ig that has an anti-pathogenic function only on gut, similar to IgA in warm-blooded animals and IgX in amphibians (Zhang et al. 2010; Zhu et al. 2013). Such information about the Ig pattern are still lacking for percid fish.

Major histocompatibility complex (MHC) molecules may also illustrate species differences in teleost immunity since fish MHC genes are known to be highly polymorphic and some of its alleles/haplotypes have been associated to increased disease resistance (Grimholt et al. 2003; Wynne et al. 2007). MHC molecules are a group of peptides which initiate the generation of adaptive immunity to pathogens with two class groups: MHC class I (or MHI) for initiating $CD8^+$ cytotoxic T cell-mediated cellular immunity; and MHC class II (or MHII) for activating $CD4^+$ helper T cell-mediated humoral immunity (Zhu et al. 2013). MHI molecules have been shown as heterodimer composed of an α subunit encoded by MHI gene and a β subunit encoded by another gene locus; and both α and β subunits are transmembrane proteins encoded by MHI gene clusters with polymorphism against the variable exotic antigen spectrum, while MHII can only be expressed by specialized antigen presenting cells spectrum (Zhu et al. 2013). MHI genes as well as their α and β subunits have been identified and associated with cellular responses mediated by $CD8^+$ T cells in various teleost fish, such as grass carp, Nile tilapia, common carp, half-smooth tongue sole and zebrafish (Chen et al. 2006; Zhao et al. 2008; Xu and Chen 2011; Zhu et al. 2013). MHII genes and related humoral immunity have been also characterized in several fish species (Zhang and Chen 2006; Li et al. 2011; Yu et al. 2010; Zhu et al. 2013) but differences in genomic structure were observed in some fish species. For instance, the analysis of the Atlantic cod (*Gadus morhua*) genome sequence showed a unique genomic structure lacking the genes encoding MHII loci (Star et al. 2011; Zhu et al. 2013).

A high diversity of the MHII genes was demonstrated in Eurasian perch; the occurrence of multiple MHII β loci was found to be three to eightfold higher as compared to the variability observed in salmonids for the β1 domain which possess a single MHII β locus (Kim et al. 1999; Michel et al. 2009). Moreover, the beta$_2$-microglobulin (β_2m) in salmonids which is an essential component of the MHC class I antigen signalling pathway, has been described to be similar to other known sequences encoding a protein sequence of 116 residues, with a mature peptide of 100 amino acids in length; but in the walleye *Sander vitreum*, another percid fish, this sequence possesses an additional 3 residues at the N terminus of the mature protein (Christie et al. 2007). To date, it is not known if such high diversity of the MHC genes of percid fish is an advantage for their immune defence. Such specific

MHC polymorphism may be associated to a specific immune response in percids compared to other teleost fish, since a high number of alleles or genotypes of the MHC genes was associated to a high disease resistance in Atlantic salmon (Grimholt et al. 2003; Wynne et al. 2007).

Several innate immunity genes have been also characterized in teleost fish, and the signalling pathways of some of them have been shown conserved from fish to mammals but have distinct features and show much more diversity between fish species (Michel et al. 2009; Zhou et al. 2012). Information about most of innate immunity genes of percid fish are still lacking. A non-exhaustive EST library reported (Rossi et al. 2007) indicated that some relevant innate genes are abundant in perch liver and brain as in other teleost fish, especially in the case of genes coding for apolipoproteins, transferrin, complement system, hepcidin, chemotaxin, fibrinogen and plasminogen. Their genomic organization was found almost similar to those characterized in other fish species, except for specific features in hepcidin genes. As for other fish species, the perch hepcidin genes consist of three exons and two introns, but the perch introns were found shorter than those of zebrafish or mammals, even if they were similar to those described in white bass, Atlantic salmon and Japan sea bass (Rossi et al. 2007). The study from the latter authors also showed that hepcidin expression in perch liver was up-regulated by lipopolysaccharide (LPS) challenge, indicating a potential production of immune actors to inflammatory stimuli as it is the case for hepcidin proteins of other fishes, such as striped bass, red sea bass, catfish, sea bass, zebrafish, and rainbow trout (Bao et al. 2005; Parck et al. 2005; Gerwick et al. 2007; Rossi et al. 2007).

It is also known that the distribution of fish hepcidin through lymphoid organs is not consistent across different fish species; some fish hepcidin genes are predominantly expressed in liver similar to that in mammals, while other could be detected in head kidney, spleen, and gill (Zhou et al. 2011; Zhu et al. 2013). It seems likely that hepcidin genes are expressed and regulated in liver, spleen and gill of Eurasian perch since LPS treatment induced similar increase in hepcidin mRNA in those organs (Rossi et al. 2007; Mathieu et al. 2014). LPS treatment was also reported as a modulator of both activity and transcript abundance of other immune proteins in various organs of teleosts (Paulsen et al. 2003; Acerete et al. 2007; Nayak et al. 2008; Swain et al. 2008a; Selvaraj et al. 2009; Secombes et al. 2011). However, a recent study conducted on Eurasian perch demonstrated that such LPS treatment may increase plasma lysozyme activity without any marked modulation on C-type lysozyme and apolipoprotein A1 mRNA abundance in liver, spleen or gills (Mathieu et al. 2013). This partial immune modulation by LPS may be related to the fact that primary lymphoid organs were not checked in the latter studies or perhaps the response to LPS treatment is species related and may not be fully effective in percids.

Sexually-dimorphic differences in the immune response have recently reported in yellow perch *Perca flavescens* in terms of expression of the suppressor of cytokine signalling (SOCS) genes after LPS treatment (Shepherd et al. 2012). The SOCS proteins are intracellular negative feed-back regulators which prevent the occurrence of hypersensitivity reactions and autoimmunity when immune responses are too strong or to prolonged. It has been demonstrated that they are also involved in growth and development processes (Flores-Morales et al. 2006; Shepherd et al. 2012; Zhu et al. 2013). According to the latter authors, several types of the SOCS family proteins have been identified, namely SOCS1 and SOCS3 in various fish species; and it is also known that fish SOCS molecules show similar structures and biological functions as the SOCS of mammals. However, phylogenetic analysis by Shepherd et al. (2012) showed that the yellow perch SOCS1 and SOCS3 proteins differ with other teleost SOCS1/3 molecules as separate clades, but the significance of such differences are not yet clarified. Moreover, a sex-specific expression of SOCS3 mRNA in head kidney of LPS-treated yellow perch with higher mean copy numbers occurred in males than in females. Such differences represent a specific feature of the yellow perch immune system since gender-dimorphic expression of SOCS1/3 has not been reported in a teleost fish (Zhu et al. 2013).

The aforementioned specific differences in characteristics of some percid immune genes may account for specific down or up-regulation of the related immune pathways, and thereby sustaining for specific immune responses to infection or other adverse stimuli in percid fish. Moreover, as percid fish are not yet fully domesticated compared to known aquaculture fishes, they may display high stress responsiveness under intensive rearing conditions (Jentoft et al. 2005), and such stress sensitivity may also have more specific interactions with the immune responses compared to other fish species.

30.2 Seasonality and Immune Status

30.2.1 Context

Climatic seasonal changes, mainly temperature and photoperiod synchronise various physiological events such as reproductive activities, variations in body weight and condition, food intake and locomotor activities, as well as immune response in fish as in many other organisms (Bowden et al. 2007; Morgan et al. 2008). The mechanisms underlying such synchronization are complex and are not yet clarified in fish. As for mammals, variation in photoperiod may indirectly affect immune functions of fish through its effects on hormone production pathways, such that of melatonin (Migaud et al. 2010; Esteban et al. 2013). Indeed, it is well known that pineal cell receptors intercept photoperiodic timing, and such decoding activity triggers the synthesis and secretion of melatonin during the dark period of the day cycle. This increase in melatonin may sensitize multiple receptors of the hypothalamo-pituitary axis which induces an increase in gonadotropins and thereby activates the production of sex-steroids in certain vertebrates. Increase in some steroid hormones have been reported to interfere with immune functions (Harris and Bird 2000; Bowden et al. 2007). Due to the existence of broad inter-species differences in fish, the endocrine modulation-induced by melatonin signal generation may be more complicated than in mammals (Pevet 2003; Migaud et al. 2010). Changes in photoperiod may also affect immune functions by inducing stress response, namely through the induction of glucocorticoid secretion such as cortisol, which has been pointed as immunosuppressor factor (Barton 1997; Harris and Bird 2000; Pevet 2003; Bowden et al. 2007; Milla et al. 2009a; Milla et al. 2010). In this regard, a recent study demonstrated that injections of cortisol or DOC increased the expression levels of C-type lysozyme and apolipoprotein 1 mRNA in both gills and spleen in Eurasian perch juveniles (Mathieu et al. 2013), indicating that some corticosteroids may act at short term as stimulator of the innate immune system in fish (Tort 2011).

For some fish species, the efficiency of some immune components clearly depends on the season, and it has been reported that innate pathways are more important during cold periods compared to adaptive immune system (Slater and Schreck 1998; Bowden et al. 2007). For instance, in tinch (*Tinca tinca*) leucocyte counts were significantly lower in winter and spring than those during summer and autumn, while activity of the alternative complement pathway has been found to be greater in winter (Collazos et al. 1998; Bowden et al. 2007). But, there are large differences between species for seasonal changes in the innate and adaptive immune parameters since a reduction of complement pathway, lysozyme activity and immunoglobulins (Ig) was observed in other fish species in winter, such as in gilthead sea bream *Sparus aurata* (Hernandez and Tort 2003; Tort et al. 2004). Moreover, no marked effects of seasonal changes in phagocytosis activity and blood Ig level were found in roach *Rutilus rutilus* (Kortet et al. 2003). The effect of seasonality on the immune system of percid fish has not yet described.

30.2.2 Effects of Temperature Regimes on Immune Status of Eurasian Perch Breeders

Photoperiod and temperature may be considered as the major primary components of annual seasonal cycle. Even if these components are interlinked and follow very similar cycles, variations in temperature in the same photoperiodic timing may induce specific physiological responses, especially for fish whose body temperature varies with that of its surroundings. Therefore, temperature is considered as the principal cue for anticipating seasonal events allowing immune system of fish to be adjusted accordingly (Bowden et al. 2007; Martins et al. 2011). Variation in temperature and photoperiodic timing may synergize to trigger the generation of pineal hormone melatonin since such induction has been suggested as potential mechanism which anticipates seasonal changes (Migaud et al. 2010). In this regard, it has been suggested that the level of rate limiting enzyme for melatonin synthesis, serotonin *N*-acetyl-transferase, exhibits pronounced seasonal changes and is affected by temperature (Bowden et al. 2007), but the pathways through which temperature may affect such modulation are not fully investigated as well as the immunoregulation of melatonin.

Effects of temperature variation on both innate and immune responses have been reported in many fish species. Temperatures higher than normal have been reported to enhance immune responses in fish whereas lower levels than normal adversely affect immune functions (Bly and Clem 1992; Kumari et al. 2006; Bowden et al. 2007; Martins et al. 2011). According to some authors (Bly and Clem 1992; Ibarz et al. 2010), the no observed effect level of temperature can be considered at 4 °C for salmonids, 14 °C for carp and 22 °C for channel catfish. As reported for photothermal regimes, seasonal variation in temperature has been studied in percid fish mainly with regard to reproduction, and interaction with immune responses has not yet investigated.

Therefore, a study was conducted on wild Eurasian perch breeders to compare stress and immune response to three thermal regimes in the same rearing conditions. Breeders were collected from an extensive culture pond, and transferred to an optimised recirculating aquaculture system(s). After 2 weeks of acclimatization, they were allocated in duplicate tanks to three photothermal recirculation systems (A, B, C) (Fig. 30.1). Fish were acclimatized from 10 °C to the experimental temperatures of 6 °C (system A), 12 °C (system B) and 18 °C (system C). Fish were fed earthworms and clamworms (*Nereis diversicolor*). Then, ten fish per tank were blood sampled on Days 15, 30, 43 and 62 for stress indicators, namely plasma glucose and cortisol and immune parameters, serum lysozyme activity, total immunoglobulins and hemagglutination titre. Mortality was checked from Days 0 to 62, so the fish died during the acclimatization were not considered in the comparison of mortality rate.

The results showed that average mortality rate within the experiment was 39 %, with higher values encountered at 6 °C and 18 °C compared to 12 °C regime conditions (Fig. 30.1): system A (6 °C): 42 %, system B (12 °C): 33 %, system C (18 °C): 43 %. System B conditions (12 °C) suited the temperature level often observed at the normal setting of perch spawning between the end of March and mid-May in the continental European region, and induced perhaps less allostatic charge than the one induced in presence of low or high temperature regimes. Nevertheless, mortality

Fig. 30.1 Controlled recirculation systems **a** = 6 °C, **b** = 12 °C, **c** = 18 °C

seemed mostly due to fungal infections which were frequent in all the three temperature regimes.

Regarding stress physiological response, plasma glucose levels (Fig. 30.2b) increased between D0 and D15 for the three temperature regimes but decreased afterwards for breeders submitted to 12 or 18 °C but not to those at 6 °C. In the latter fish, values were higher ($P<0.05$) than those observed in other groups between D29 to D62, or than values already reported for stressed perch by previous reports (Acerete et al. 2004; Jentoft et al. 2005). As for glucose, plasma cortisol levels (Fig. 30.2c) were high during the two first weeks of experiment as a response to the transfer from extensive pond conditions to the experimental rearing systems, but values decreased afterwards but not for breeders at 6 °C, even if no clear differences were calculated between experimental conditions. Therefore, these results indicate that low temperature conditions such as during winter may induce an enhancement of stress responsiveness in perch breeders.

In contrast to a high stress response, no marked changes were observed over the time concerning the tested immune parameters (Fig. 30.3) whatever the temperature level. However in breeders reared at 6 and 12 °C, values for lysozyme activity increased on D15 or decreased for total Ig on D29, indicating a transient modulation in the immune status, perhaps in relation to their highest stress response. Since such immune modulation was not sustained, a differential increase in parasite colonization, such as fungal infections observed during the experiment may have also contributed to the modulation in the tested immune parameters. Nevertheless, the overall results

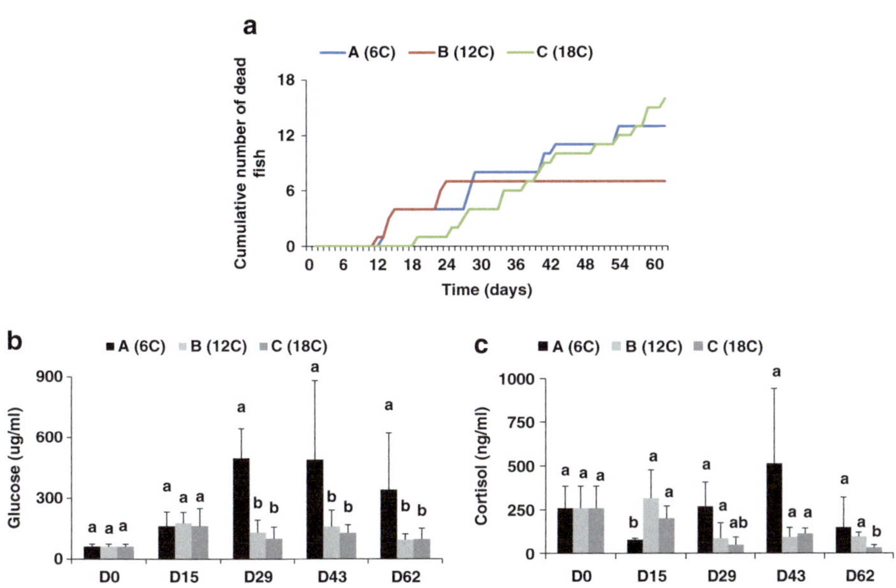

Fig. 30.2 Cumulative mortality (**a**), plasma glucose (**b**) and cortisol (**c**) levels of Eurasian perch breeders submitted to 6, 12 and 18 °C conditions of rearing during 2 months (January–February). Data are means ± SD ($n=10$). Values with *asterisk* are significantly different at $P<0.05$

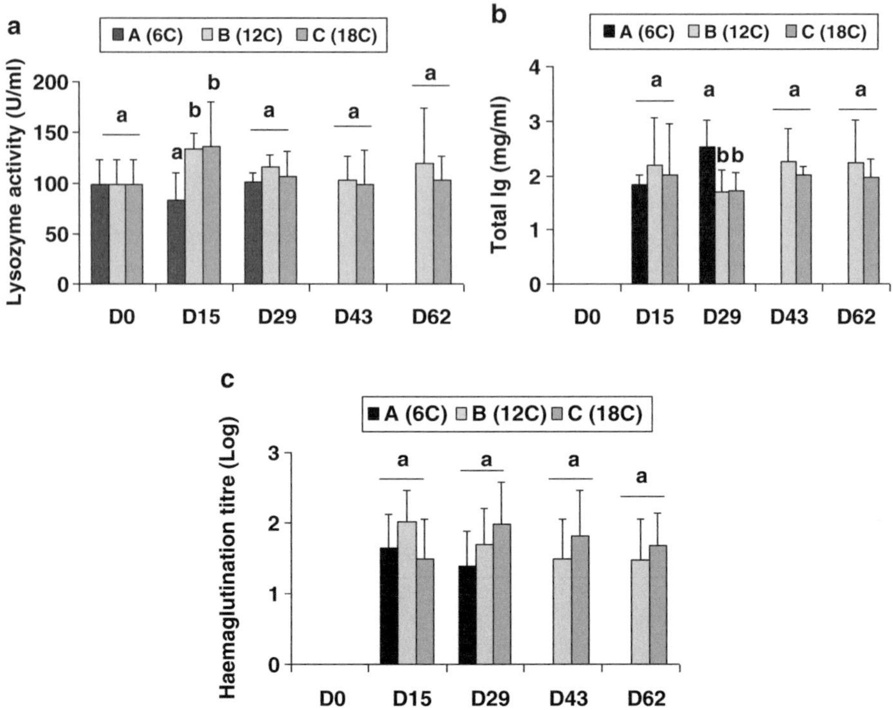

Fig. 30.3 Serum lysozyme activity (**a**), total immunoglobulins (**b**) and hemagglutination titre (**c**) levels of Eurasian perch breeders submitted to 6, 12 and 18 °C conditions of rearing during 2 months (January–February). Data are means ± SD (n = 10). Values with *asterisk* are significantly different at P < 0.05

from the current study indicate that the three tested temperature levels are not so critical for the immune status of Eurasian perch breeders, but a thermal regime of 12 °C seems to be optimal for its overall welfare. Thus, a thermal regime of 6 °C appeared as a threshold at which percid fish may progressively acclimate without a high allostatic cost susceptible to significantly depress their immune defence. It has been reported that lower temperature level than the normal thermal regime may induce immunosuppression in some fish species (Tort et al. 2004; Kumari et al. 2006). Moreover, it has been demonstrated that some non-specific immune indicators (serum complement activity, erythrocyte agglutination, lysozyme activity, lymphocyte counts and oxygen radical production by macrophages) were suppressed in gilthead sea bream sampled in winter months compared with those sampled during warmer periods of the year or after a period of recovery at higher temperature (Tort et al. 1998).

30.2.3 Breeding-Related Changes in Immune Status of Captive Eurasian Perch Breeders

Temporal changes in immune status associated with breeding process or seasonal changes have been reported in fish species (Schreck et al. 2001; Gallardo et al. 2003; Kortet et al. 2003; Tort et al. 2004; Cheng et al. 2009) but such relation is poorly studied in percid fish. Previous reports in some fish species indicate that seasonal changes in immunocompetence may be predicted by the hypothesized trade-off between reproduction and self-maintenance (Balm 1997; Schreck et al. 2001; Kortet et al. 2003; Vainikka et al. 2004). Immunological performances may be subjected to rapid temporal changes at the time of reproduction due to possible resource re-allocation between the immune system and reproduction, or through immunomodulation by reproductive hormones (Harris and Bird 2000; Vainikka et al. 2004). Other authors hypothesized that the breeding period in itself may cause physiological stress, which can per se affect the immune system (Buchanan 2000). In this way, the regulation of sex steroids considered now as immunoregulators may explain this seasonal immunomodulation along the reproductive cycle (Milla et al. 2011).

According to some authors (Wedemeyer 1997; Kent et al. 2009), susceptibility to bacterial diseases may increase in early spring because growth and invasiveness dynamics of pathogens may interact more rapidly to the rising temperature of water than that of the immune defences of fish whose physiological status had been debilitated by the overwintering environment. But, it is not clear whether the hypothesis of negative effects of temperature on immune parameters is supported in the case of percid fishes.

While photoperiodic timing of reproductive physiology of percid fish has been studied (Ciereszko et al. 1997; Migaud et al. 2004; Fontaine et al. 2006; Wang et al. 2010; Abdulfatah et al. 2013), the influence of seasonality on immune response has not yet been determined. Nevertheless, a high mortality rate during or after the spawning period is often observed under rearing conditions (Wang et al. 2003; Rodger et al. 2008). Such high mortality may be related to an immunosuppression related to reproduction effort and/or photothermal conditions. Since studies addressing this question are still lacking, we proposed to determine whether gonad maturation may be linked to stress physiology and immune status of Eurasian perch breeders maintained under natural or constant photothermal conditions. Two populations of captive breeders were compared at three sampling times during gonad maturation (January, February and March). The first broodstock was submitted to constant photothermal conditions (= CC breeders) of temperature (23 °C) and photoperiod (12 L: 12D) which is used to promote growth in Eurasian perch but to inhibit gonad development and maturation/spawning (Kestemont et al. 2003). The second broodstock was reared in outdoor tanks under wintering photothermal conditions (= NC breeders). Fish received forage fish feeding to ensure successful gonad maturation. The two groups were compared in duplicate tanks of ten fish at each sampling time. Body condition status was characterized by somato-somatic index namely factor K1, gonad somatic index and spleno-somatic index (SSI) as well as liver damage status by transaminase enzyme activities (glutamic oxalacetic transaminase and glutamic pyruvic transaminase, respectively GOT and GPT). Stress status was evaluated by plasma cortisol and glucose, and status of specific immune system was checked by leukocyte counts, lysozyme activity and spleen macrophage respiratory burst activity (RBA).

For more information concerning the stress and immune status of perch breeders during the spawning period, a second setting was undertaken using two breeder populations of two successive captive generations (F1 and F4). They were sampled during final gonad maturation (early March), during (April), or after the spawning process (May). The two breeder populations were reared in indoor greenhouse allowing natural wintering photothermal conditions to induce gonad and spawning process; and were provided forage fish feeding. At each time control, they were sampled in duplicate tanks of five fish each for blood and spleen to analyse serum cortisol, glucose, lysozyme activity and total immunoglobulins, as well as spleen macrophage RBA.

Results concerning some morpho-anatomical indexes and physiological variables as indicators of body condition status of perch breeders under the two photothermal regimes are summarized in Table 30.1. As expected, NC breeders showed an efficient gonad development while gonad maturation was suppressed in CC fish as already reported in previous studies (Migaud et al. 2002; Milla et al. 2009b). No significant difference was observed in the values for K1 factor as did those of transaminases activities, indicating no hepatic damages induced by the different photothermal regimes. No difference was observed for values of SSI between CC and CN breeders; while in other fish species a close relation was observed between the numbers of blood cells and spleen size, and it was hypothesized that a smaller spleen size may be considered as a potential indicator of a lower immune function in fish (Wester et al. 1994; Kortet et al. 2003; Vainikka et al. 2004).

Concerning the stress status, plasma cortisol (Fig. 30.4) levels did not indicate a high stress physiological response in both CN and CC groups, as compared to values reported for stressed perch by other authors (Wang et al. 2003; Acerete et al. 2004). But cortisol values showed a trend of increase in CC breeders than in NC ones, especially in February, indicating a photothermal effect relating to differences in temperature level. Results from the second setting showed no significant changes in cortisolaemia during the spawning month (April) compared to March or May in both the two captive breeder populations (Fig. 30.5). In other fish species such as rainbow trout, it has been reported that application of artificial photoperiods are not often associated to a high chronic stress physiological response detectable by blood cortisol (Biswas et al. 2006; Valenzuela et al. 2006) but may increase the susceptibility of opportunistic diseases (Valenzuela et al. 2012). In terms of plasma glucose (Fig. 30.4), stress status was also comparable between NC and CC breeders, but values were as higher as levels already reported for stressed perch fish (Haux et al. 1985; Acerete et al. 2004), especially in February for CC breeders or in March for CN fish. There was a trend of glucose increase for CN breeders, especially in March which may indicate an elevated requirement for extra energy resources for

Table 30.1 Indicators for body condition status measured between January and March, during gonad maturation for NC and CC breeders

	January		February		March	
Variable	NC	CC	NC	CC	NC	CC
K1	1.27±0.11[a]	1.56±0.16[a]	1.39±0.11[a]	1.65±0.24[a]	1.34±0.18[a]	1.46±0.22[a]
GSI (%)	11.68±1.7[a]	1.18±0.5[b]	12.40±2.1[a]	1.29±0.6[b]	15.58±3.1[a]	2.24±0.9[b]
SSI (%)	0.18±0.05[a]	0.20±0.11[a]	0.19±0.09[a]	0.13±0.07[a]	0.15±0.02[a]	0.20±0.09[a]
GOT	62.08±50.7[a]	20.98±22.6[a]	13.25±16.5[a]	20.36±19.4[a]	49.70±22.3[a]	29.36±26.7[a]
GPT	32.56±26.4[a]	17.25±18.3[a]	3.85±6.6[a]	11.06±5.7[a]	32.79±10.1[a]	28.10±13.2[a]
WBC	471±287[a]	237±91[a]	358±164[a]	135±43[b]	352±259[a]	131±31[b]

NC breeders submitted to natural photothermal conditions, *CC* breeders under constant photothermal conditions (23 °C). *K1* condition index, *GSI* gonadosomatic index, *SSI* splenosomatic index, *GOT* (SF units ml^{-1}) glutamic oxalic acid transaminase, *GPT* (SF units ml^{-1}) glutamic pyruvic acid transaminase. Significant differences are indicated with different *letters*. *WBC* (10^3 cells mm^{-3}) white blood cells

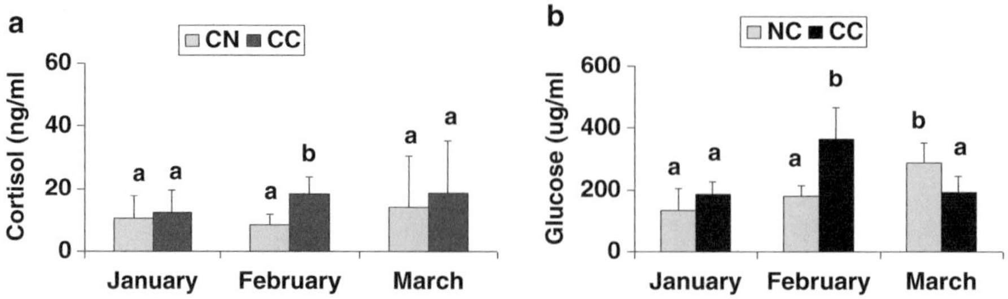

Fig. 30.4 Changes in plasma cortisol (**a**) and glucose (**b**) in Eurasian perch breeders submitted to natural (*NC*) or constant (*CC*) photothermal regimes between January and March. Data are means ± SD (n = 10). Values with *asterisk* are significantly different at P < 0.05

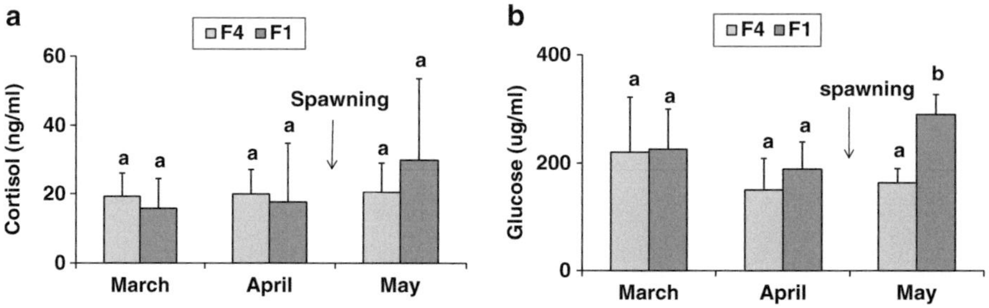

Fig. 30.5 Changes in plasma cortisol (**a**) and glucose (**b**) levels in Eurasian perch breeders undergoing final gonad maturation (early March), during (April) or after the spawning process (May). Data are means ± SD (n = 10). Values with *asterisk* are significantly different at P < 0.05

spawning processes since a decrease was observed at the same time for CC fish which did not displayed gonad maturation. But the results from the second setting showed similar profiles of glycaemia (Fig. 30.5) before, during and after the spawning process except an increase in May for one of the two tested populations. The results from previous studies demonstrated that fish acclimated to high temperature regimes have higher cortisol and glucose levels than fish acclimated to lower temperature due to a higher metabolic rate (Van Ham et al. 2003; Arnekleiv et al. 2004; Davis 2004; Jentoft et al. 2005).

Pertaining to the immune parameters, leucocyte counts were lower in CC breeders than in NC ones due to a decrease in February and March (Table 30.2). The lower WBC counts observed in CC breeders may be associated to a higher cortisol and glucose response in February compared to NC breeders, and this may indicate a kind of trade-off between stress status and immune function. Values for spleen macrophage RBA (Fig. 30.6) were higher in NC than in CC breeders in January and March, confirming that low temperature photothermal regime or spawning process may affect some immune compounds. In the second setting, spleen macrophage RBA (Fig. 30.7) seemed markedly affected by spawning process since values increased significantly in April and May compared to March. A high macrophage RBA has been reported in other fish species as an investment in non-specific defence

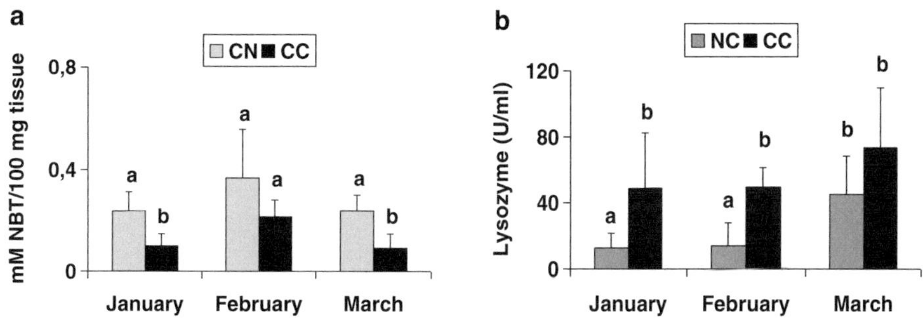

Fig. 30.6 Changes in spleen respiratory burst activity (**a**) and lysozyme activity (**b**) in Eurasian perch breeders submitted to natural (*NC*) or constant (*CC*) photothermal regimes between January and March. Data are means ± SD (n = 10). Values with *asterisk* are significantly different at P < 0.05

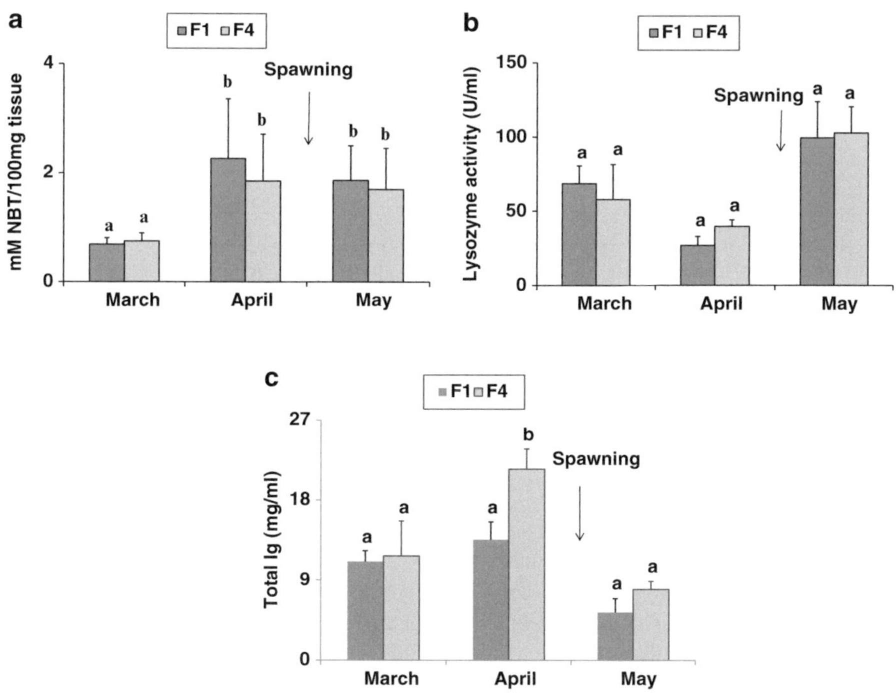

Fig. 30.7 Changes in levels of spleen respiratory burst activity (**a**) serum lysozyme activity (**b**), and serum total immunoglobulins (**c**) in Eurasian perch breeders undergoing final gonad maturation (early March), during (April) or after the spawning process (May). Data are means ± SD (n = 10). Values with *asterisk* are significantly different at P < 0.05

for an adaptation to low temperature during autumn and winter or to changes in physiological events (Secombes 1996) such reproductive process. In contrast to values observed for leukocyte numbers, plasma lysozyme activity (Fig. 30.6) was higher in CC breeders compared to NC fish in January and February, perhaps in relation to a high temperature level, but also as a compensatory process for the lower leukocyte numbers. Lysozyme values increased significantly in March for NC breeders while no marked changes were observed for CC ones, indicating an effect of temperature rise at that transient time between winter and spring for NC fish. It seems that the spawning process may depress lysozyme activity since values significantly decreased in April and increased in May after spawning in both the two breeder populations tested in the current study (Fig. 30.7). Lysozyme activity has been described as an important parameter in the specific immune defence of both invertebrates and vertebrates through hydrolysing β-[1,4] linked glycoside bonds of bacterial cell wall peptidoglycans resulting in a complete lysis action (Magnadottir 2006; Saurabh and Sahoo 2008). Chronic stress has been shown to reduce lysozyme activity in fish (Demers and Bayne 1997; Fevolden et al. 2002; Hoeger et al. 2005) but for some species, such as channel catfish (*Ictalurus punctatus*), increase in cortisol concentrations were not correlated to a decrease in lysozyme activity (Small and Bilodeau 2005) indicating that the interaction between stress-induced glucocorticoids and immune functions is complex and may be species dependent. It seems also that the spawning process may also affect the specific immune system since values for total immunoglobulins showed a trend of increase during that period and decreased afterwards in May for the two tested populations in the second setting of the current study (Fig. 30.7). The results from the current study suggest that while some parts of the Eurasian perch immune system may be changed around the spawning period such as lysozyme activity, other may remain unaffected. Also, the immune modulation overcoming during the spawning period may result both from photothermal stimuli and reproduction effort supporting the hypothesis of a somewhat trade-off between reproduction and immune system in percid fish. But more studies are still needed since this study covered only the period during gonad maturation and spawning process.

30.3 Immune Responses of Percid Fish to Immunostimulants

As for other vertebrates, both innate and acquired immune system of fish may be stimulated by various *substances. The development of effective immuno-prophylactic strategies is nowadays essential since high* occurrence of disease outbreaks is one of the major causes of mortality which hampered the sustainable aquaculture for various fish species. The use of immunostimulants may also constitute an interesting alternative to mitigate the emergence of antibiotic resistance as well as the spreading of antibiotics in the aquatic environment. According to Sakai (1999) and Bricknell and Dalmo (2005), the substances already tested as immunostimulants in fish may be divided into several groups depending on their sources (bacterial, algae-derived, nutritional and endocrine factors or their immunological responses which may depend on the binding receptor and intracellular signalling events. Although there is now clear evidence of biological effects of some immunomodulator substances, their physiological pathways are not yet fully clarified. Moreover, the application of immunostimulants has received little attention in percid fish despite high mortality rates overcoming during the early developmental stages or over the spawning period in intensive culture conditions.

30.3.1 Probiotics

Various probiotics have been tested in aquaculture to stimulate the immune system and thereby increase disease resistance (Kesarcodi-Watson et al. 2008; Newaj-Fyzul et al. 2014). Among the promising efficient probiotics for fish, several lactic acid bacteria were shown to affect the fish intestinal microbiota, with various consequences on immune defence and rearing performances, but their immunostimulation varied greatly according to their viability in the gut environment (Gatesoupe 2008). By considering the ability for adherence and penetration into fish mucus, the inhibition of pathogen growth and adhesion, and the resistance to fish bile, *Lactobacillus rhamnosus* and *Lactobacillus bulgaricus* were selected as the best candidate probiotics for rainbow trout (Nikoskelainen et al. 2001; Gatesoupe 2008). Other *Bacillus* or *Lactobacillus* genus have been also used with species related effects on husbandry parameters and on immune functions of various fish species (Kesarcodi-Watson et al. 2008; Wang et al. 2008; Al-Dohail et al. 2009; Merrified et al. 2010; Nayak 2010; Dimitroglou et al. 2011; Oke et al. 2013). Especially, dietary supplementation of *B. subtillis*, *B. licheniformis and B. pumilus* have been shown to improve either non-specific or specific immune system in various fish species (El-Haroun et al. 2006; Aly et al. 2008; Zhang et al. 2010; Merrified et al. 2010; Ai et al. 2011; Cerezuela et al. 2012; Zhao et al. 2012). Modulation of growth rate and immune defence was also observed using *Bacillus coagulans* and *Lactobacillus acidophilus* in some cyprinid fishes or catfish (Al-Dohail et al. 2009; Wang 2011). *Bacillus cereus* probiotics have been also found beneficial for enhancing the innate immune function and promoting growth of rainbow trout fingerlings (Gisbert et al. 2013).

In contrast to other fish species, information about the efficacy of probiotics as modulators of the immune system and disease resistance of percid fish is still scarce, but an in vitro study has shown that *Pseudomonas chlororaphis* are able to produce inhibitory compounds which suppressed the growth of various fish bacteria pathogens including *Aeromonas sobria*, *Yersinia rukeri and Vibrio anguillarum* (Goldschmidt-Clermont et al. 2008). Since *A. sobria* has been identified as a causative agent of ulcerative disease in Eurasian perch (Wahli et al. 2005), an in vivo study was conducted by Gobeli et al. (2009) and demonstrated that administration of *P. chlororaphis* is able to reduce mortalities caused by *A. sobria*. It was also demonstrated that administration of mixture of various probiotics may increase growth rate and decrease mortality rate when Yellow perch *Perca flavescens* larvae are subjected to stress temperature challenge (Eissa et al. 2013).

To date, few studies investigated the beneficial effects of probiotics during the early developmental stages of fish. Probiotic products are usually administered to larvae by enrichment of live food, added to dry diet or to the tank water (Panigrahi et al. 2005; Kesarcodi-Watson et al. 2008; Apun-Molina et al. 2009). Inappropriate administration mode or doses may result to detrimental effects on the balance of friendly and pathogenic microflora in the gastro-intestinal tract, and thus on health status depending on the fish species (Lategan et al. 2004; Rastall 2004; Wang et al. 2008). Since the dose-immune response to probiotics may be species-dependent in relation to the specific digestive microflora, a recent study was conducted to examine whether a commercial mixture (SANOLIFE MIC-F®, INVE Aquaculture, Belgium) composed of different *Bacillus* (*B*) species (namely *Bacillus licheniformis*, *B. subtilis* and *B. pumilus*) has beneficial actions on husbandry performances and immune system in Eurasian perch larvae (1) or juveniles (2).

30.3.1.1 Immune Responses to Probiotic Bacteria in Percid Larvae

Two settings were conducted to determine the efficacy of doses and administration mode of probiotic bacteria in Eurasian perch larvae. Experiment 1 compared two administration modes and different bacteria doses. Fourteen days after hatching (ph), larvae (IBW $=2.3\pm0.8$ mg) were submitted to four experimental treatments (D0). Control fish (B0) were fed normal size *Artemia* metanauplii without probiotics neither in live food nor in tank water. They were compared to fish which received the bacteria mixture through tank water (BW), and which were fed normal size *Artemia* metanauplii without enrichment by probiotics. Other fish received probiotic bacteria both through the tank water and were fed *Artemia* metanauplii enriched with either 2×10^{10} (BWF2) or 4×10^{10} (BWF4) CFU L^{-1} of incubation medium.

Experiment 2 emphasized the responses to bacteria administration modes combining with density level. Bacteria supplementation started on day 18 ph (D0), and seven experimental conditions were set. On the one hand, a set of three conditions was tested for which larvae (IBW $=9.5\pm4$ mg) were reared at ten fish L^{-1}. First, control groups did not receive any probiotics neither through live food nor through water (B010). In the two other groups, fish received the bacteria mixture through food (BF) composed of normal size *Artemia* metanauplii enriched with two bacteria doses: either 2×10^{10} (BF2/10) or 4×10^{10} (BF4/10) CFU L^{-1} of the incubation medium. On the other hand, a set of four treatments were tested. Fish were reared at two stocking densities (10 vs 20 larvae L^{-1}), and received the probiotics through water (BW). As in the first setting, 4.93×10^{7} CFU L^{-1} day^{-1} were mixed with the water for the two densities (BW10 vs BW20). These treatments were compared to fish reared at the same density, and which were fed *Artemia* metanauplii enriched by bacteria mixture (2×10^{10} CFU L^{-1} incubation medium) in addition to water (BF2W10 vs BF2W20). For the two settings, growth and survival parameters were recorded on days 14 and 28 (32–46 days ph); and in parallel, lysozyme activity and total immunoglobulins were measured in larval homogenates composed of 30 larvae/tank. More details on these experimental protocols have been reported in Mandiki et al. (2011).

A third larval setting tested the effects of increasing high doses of probiotic bacteria using pikeperch (*Sander lucioperca*) larvae of 34 days ph (IBW: 8 mg). The probiotic bacteria were the same commercial mixture of *Bacillus spp.* as in experiments 1 and 2 (Sanolife MIC-F = B, INVE Aquaculture, Belgium). Larvae were fed with enriched metanauplii with increasing doses of *Bacillus* (B: 4, 12, 24 × 10^{10} CFU/L incubation medium = B1, B3, B6) during 30 days. The setting was conducted in triplicate groups of 40-L tanks containing 200 larvae. Growth parameters and immune compounds (total immunoglobulins and lysozyme activity) were determined in homogenates of larvae composed of 30 larvae/tank on D15 and 30 (49–64 days ph).

Data for husbandry response of Eurasian perch larvae to bacteria supplementation are presented in Table 30.2a, b, c. The results from the two larval settings showed a significant increase (19–23 %) in growth rate when *Artemia* metanauplii were enriched with relatively high doses of bacteria (4 × 10^{10} CFU L^{-1} incubation medium), especially when probiotics were administered by live food. For the lowest stocking density (five larvae L^{-1}), growth increase was already observed on D15 (19 %, experiment 1, Table 30.2a) while for higher densities, such effect occurred only on D30 (21 %, experiment 2, Table 30.2b). The influence of bacteria supplementation varied between the two administration modes or density level. Probiotic

Table 30.2a Husbandry response (% increase/decrease vs B0) to administration mode and doses of *Bacillus* probiotics in Eurasian perch larvae reared under intensive conditions (Experiment 1)

	BW	BF2W	BF4W
SGR-D14	6	6	19*
SGR-D28	8	11	23*
Survival	−13	1	5
Type-2 cannibalism	32	−32	−35

B0 = control without probiotics. BW = water (4.93 × 10^7 CFU L^{-1} day^{-1}). BWF2 = water (4.93 × 10^7 CFU L^{-1} day^{-1}) + live food (2 × 10^{10} CFU L^{-1}). BWF4 = water (4.93 × 10^7 CFU L^{-1} day^{-1}) + live food (4 × 10^{10} CFU L^{--1}). *SGR* specific growth rate, *D14* and *D28* days 28 and 42 post-hatching. Data are means ± SD (n = 3). Values with different *letters* or *asterisk* are significantly different at $P < 0.05$

Table 30.2b Husbandry response (% increase/decrease vs B010) to administration mode of *Bacillus* probiotics and stocking density level in Eurasian perch larvae reared under intensive conditions (Experiment 2)

	BF2/10	BF4/10	BW10	BW20	BF2W10	BF4W20
SGR-D14	2	4	3	2	6	4
SGR-D28	9	21*	0	3	1	7
Survival	48*	55*	10	51*	18	48*
Type-2 cannibalism	−49*	−39*	−22*	−27*	10	−46*

B010 = control without probiotics. BF2/10 = live food (2 × 10^{10} CFU L^{-1}) (10 fish L^{-1}). BF4/10 = live food (4 × 10^{10} CFU L^{-1}) (10 fish L^{-1}). BW10 = water (4.93 × 10^7 CFU L^{-1} day^{-1}) (10 fish L^{-1}). BW20 water (4.93 × 10^7 CFU L^{-1} day^{-1}) (20 fish L^{-1}). BF2W10 = live food (2 × 10^{10} CFU L^{-1}) + water (4.93 × 10^7 CFU L^{-1} d^{-1}) (10 fish L^{-1}). BF2W20 = live food (2 × 10^{10} CFU L^{-1}) + water (4.93 × 10^7 CFU L^{-1} d^{-1}) (20 fish L^{-1}). *SGR* specific growth rate, *Can II* type II cannibalism (ingestion of whole prey), *D14* and *D28* days 32 and 46 post-hatching. Data are means ± SD (n = 3). Values with different *letters* or *asterisk* are significantly different at $P < 0.05$

Table 30.2c Husbandry response (% increase/decrease vs B0) to administration mode of *Bacillus* probiotics and stocking density level in pikeperch larvae reared under intensive conditions (Experiment 3)

	SGR-D15	SGR-D30	Survival	Type-2 cannibalism
B1	5	2	6	−9
B3	8	4	−11	−13
B6	23*	4	10	−2

B0 = control without probiotics. *SGR* specific growth rate, *B1, B3* and *B6* live food enriched with 4, 12 or 24×10^{10} CFU L^{-1}), respectively. Data are means ± SD (n = 3). Values with different *letters* or *asterisk* are significantly different at P < 0.05

administration by tank water or combined administration in live food and tank water was ineffective in improving growth performances in Eurasian perch larvae whatever the bacteria doses. When administered through water at low density (five to ten larvae L^{-1}), probiotic bacteria were not fully effective to improve fish survival, but at high density (experiment 2) an increase ranging between 48 % and 51 %) was observed. The dietary probiotic bacteria increased survival and this was associated by mortality caused by type-2 cannibalism in relation to low weight heterogeneity, except at the lowest density level (experiment 1).

There were also species related differences in the efficacy of the tested *Bacillus* mixture. Data obtained on pikeperch larvae showed poor husbandry response compared to those observed in Eurasian perch. Indeed, growth rate of pikeperch was stimulated on Day 15 (23 %, Table 30.2c) at higher bacteria doses compared to efficient doses for Eurasian perch; moreover such modulation was not sustained afterwards on D30. Furthermore, no positive effect on survival or cannibalism rates was observed whatever the bacteria doses (Table 30.2c). The overall data concerning the husbandry response showed that probiotic bacteria administered by live food may stimulate growth-related parameters in percid larvae but it seemed that pikeperch need higher doses than Eurasian perch. Using the same *Bacillus* mixture as in the present study, Decamp et al. (2006) also demonstrated an increase in growth and survival associated with low size variability in gilthead seabream and Japanese flounder. The reduction of size heterogeneity by the dietary probiotic bacteria may be an indirect effect related to the improvement of feed utilization or the increase in protein efficiency which may also induce changes in cannibalistic behaviour. Size heterogeneity-induced high cannibalistic behaviour has been reported to be reduced at high stocking density in percid larvae, thus resulting in higher survival (Baras et al. 2003; Kestemont et al. 2003). Therefore, the positive effect of the bacteria treatment on the interaction between stocking density and size heterogeneity-induced cannibalism behaviour observed in the present study needs further consideration since mortality by cannibalism is one of the imposing causes of mortality in percid larvae (Babiak et al. 2004; Mandiki et al. 2007). Testing some of the *Bacillus species* used in the present study or other bacteria species, other studies reported an increase in husbandry parameters using fingerlings or juveniles of other fish species (El-Haroun et al. 2006; Aly et al. 2008; Zhang et al. 2010; Merrified et al. 2010;

Ai et al. 2011; Zhao et al. 2012). In the latter studies, the bacteria treatments were associated with a higher feed utilization, protein efficiency and energy retention compared to control fish.

In the current study, the improvement of husbandry performances by the probiotic bacteria was associated with a positive impact on the immune system. Indeed, an increase in Ig levels was already detectable after 2 weeks of supplementation, and was sustained until D30 whatever the dietary bacteria administration for both species (Fig. 30.8). It seemed that bacteria administration by tank water was not fully effective in stimulating Ig production since marked increase was only observed for the lowest density (five larvae L^{-1}) in experiment 1 (Fig. 30.8) not for the two other density levels (10 or 20 larvae L^{-1}) in experiment 2 (Fig. 30.8). Poor lysozyme response was observed in Eurasian perch larvae whatever the administration mode. Indeed, an increase in lysozyme activity level was only observed for the higher dietary doses at low density in experiment 1 (Fig. 30.9) but such effect was not confirmed when the same type of larvae were reared at relatively higher density level in experiment 2 (Fig. 30.9). Pikeperch larvae were also responsive to the probiotic treatment. But, it seems that some immune compounds were perhaps hampered by rearing conditions since a dose related response for lysozyme activity and total Ig was observed for pikeperch larvae reared at low density in experiment 3 (Fig. 30.10). It is also worthy to note that pikeperch larvae were at a more advanced developmental stage (days 34 ph) at the beginning of the bacteria supplementation compared to perch larvae (14–18 ph). Perhaps some immune pathways which activate the lysozyme activity in percid larvae cannot enough operate before 2 weeks after hatching, thus interfering with a quick responsiveness to the bacteria supplementation. Indeed, it is well recognized in many fish species that various components of the innate immunity including lysozyme activity are activated just after egg

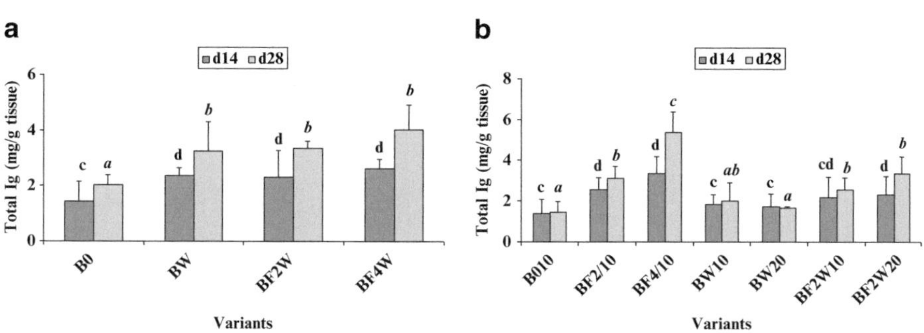

Fig. 30.8 Effects of administration mode of *Bacillus* probiotics and stocking density on total immunoglobulins (Fig. 30.8) and lysozyme activity (Fig. 30.9) in the homogenates of Eurasian perch larvae. **a**, **b** Settings 1 and 2. Data are means ± SD (n=3). Values with different letters or *asterisk* are significantly different at P<0.05. B0 = control without probiotics. BW = water (4.93×10^7 CFU L^{-1} day^{-1}). BWF2 = water (4.93×10^7 CFU L^{-1} day^{-1}) + live food (2×10^{10} CFU L^{-1}). BWF4 = water (4.93×10^7 CFU L^{-1} day^{-1}) + live food (4×10^{10} CFU L^{-1}). *D14* and *D28* days 28–32 and 42–46 ph

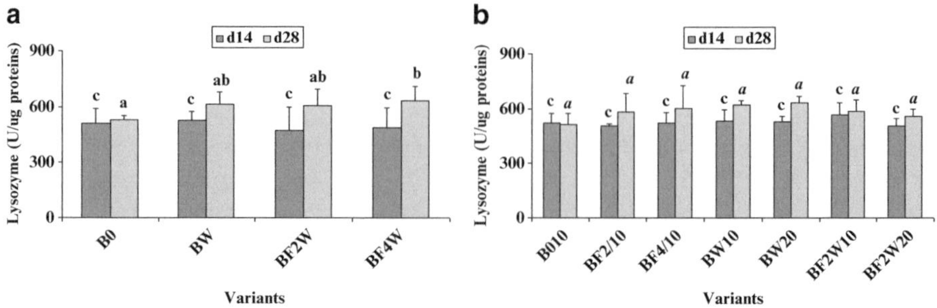

Fig. 30.9 Effects of administration mode of *Bacillus* probiotics and stocking density on total immunoglobulins (Fig. 30.8) and lysozyme activity (Fig. 30.9) in the homogenates of Eurasian perch larvae. **a**, **b** Settings 1 and 2. Data are means ± SD (n = 3). Values with different letters or *asterisk* are significantly different at P < 0.05. B0 = control without probiotics. BW = water (4.93 x 10^7 CFU L^{-1} d^{-1}). BWF2 = water (4.93 × 10^7 CFU L^{-1} d^{-1}) + live food (2 × 10^{10} CFU L^{-1}). BWF4 = water (4.93 × 10^7 CFU L^{-1} day^{-1}) + live food (4 × 10^{10} CFU L^{-1}). *D14* and *D28* days 28–32 and 42–46 ph

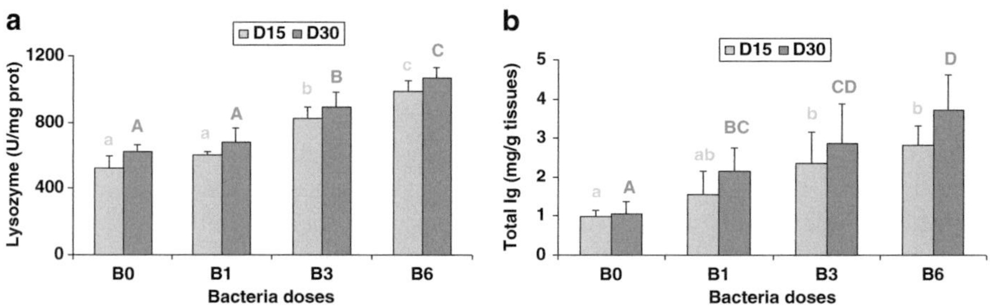

Fig. 30.10 Effects of increasing doses of *Bacillus* probiotics on lysozyme activity (**a**) and total Ig (**b**) in the homogenates of pikeperch larvae. Data are means ± SD (n = 3). Values with different letters or asterisk are significantly different at P < 0.05. B0 = control without probiotics. B1, B3 and B6: live food enriched with 4, 12 or 24 × 10^{10} CFU L^{-1}), respectively, *D15* and *D30* days 49 and 64 ph

fertilization, and become fully operational by the time of hatching (Swain and Nayak 2009; Vadstein et al. 2012). It is also well known that the adaptive immune pathways are established in the late larval developmental stage and are fully operational after the exhaustion of maternally transferred factors (Hanif et al. 2004; Swain and Nayak 2009), so the maturation of larval adaptive system may be delayed depending to the interaction with that transmitted passive immunity in some fish species. In this regard, information are still lacking in percid larvae.

The overall results from the current larval studies indicate that *Bacillus* probiotics may induce potential positive effects on some immune compounds of percid larvae, but the effective doses need to be determined depending to the administration mode or rearing conditions.

30.3.1.2 Immune Responses to Probiotic Bacteria in Eurasian Perch Juveniles

To emphasize the immune response to probiotic bacteria, Eurasian perch juveniles (IBW: 32–35 g) were used to determine whether the mixture of *Bacillus spp.* (SANOLIFE MIC-F = B) used for percid larvae has beneficial effects on immune system and disease resistance. Low ($B1 = 3.5 \times 10^9$ CFU kg^{-1} food) and high ($B5 = 16.4 \times 10^{11}$ CFU kg^{-1} food) bacteria doses were coated to a commercial feed pellets previously reground, extruded and dried. The two dose treatments were compared within disconnected recirculating water systems in triplicate groups of 100-L tanks (42 fish into each) in comparison to controls without any probiotics in food. Immune response in terms of leucocyte proportions, plasma lysozyme activity and total immunoglobulins were evaluated on Days 15 and 30 of bacteria supplementation. Growth rate was also checked on Days 30 and 40, and afterwards, fish were challenged by a virulent *Aeromonas hydrophila*, and checked during 10 days for mortality.

The results showed a significant increase in growth rate (29 %) on Days 30 and 40 for juveniles receiving high bacteria doses, while no stimulatory effect was observed at low doses. This growth improvement was associated with a significant immune modulation. Indeed, a significant increase in the proportions of macrophages and neutrophils (Fig. 30.11a) was observed with the high bacteria treatment, especially on D15. In parallel to the variations in leukocyte proportions, plasma level of total Ig increased significantly for the two bacteria treatments and values were higher for the higher doses whatever the sampling date (Fig. 30.11b). High bacteria doses also stimulated lysozyme activity on D30 (Fig. 30.12). The overall defence capacity was higher for juveniles received the higher bacteria doses as evidenced by a higher relative survival (38 % B5 vs 3.5 % B1). The immunostimulatory

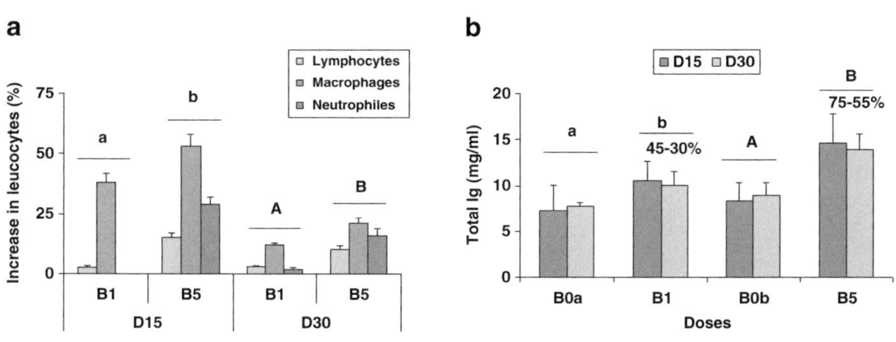

Fig. 30.11 Effects of increasing doses of *Bacillus* probiotics on leukocyte proportions (**a**), total plasma Ig (**b**), lysozyme activity (**a**) and survival rate after furunculosis (**b**) in Eurasian perch juveniles. Data are means ± SD (n=3). Values with different letters or *asterisk* are significantly different at P<0.05. Values with different letters are significantly different at *P<0.05*. B0a, b = control without probiotics. B1 or B5: dry food enriched with 3.5×10^9 or 16.4×10^{11} CFU kg^{-1} feed

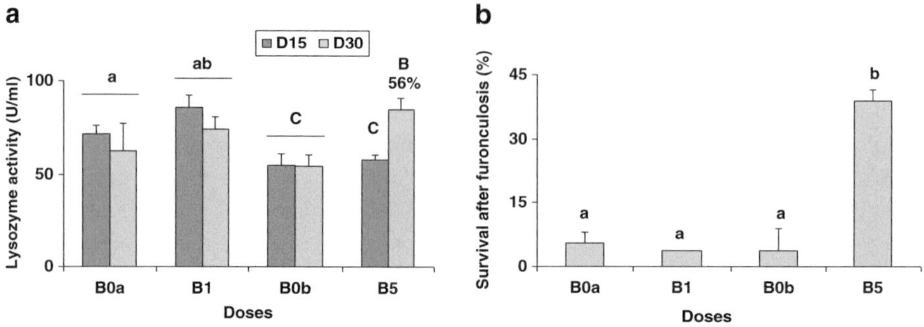

Fig. 30.12 Effects of increasing doses of *Bacillus* probiotics on leukocyte proportions (**a**), total plasma Ig (**b**), lysozyme activity (**a**) and survival rate after furunculosis (**b**) in Eurasian perch juveniles. Data are means ± SD (n = 3). Values with different letters or *asterisk* are significantly different at P < 0.05. Values with different letters are significantly different at P < *0.05*. B0a, b = control without probiotics. B1 or B5: dry food enriched with 3.5×10^9 or 16.4×10^{11} CFU kg^{-1} feed

effect of this probiotic on perch larvae is thus also observed on perch juveniles. Using *B. pumilus or L. acidophilus* as probiotics, previous studies also reported changes in immunoglobulin concentration; haematological parameters, respiratory burst activity, and improved disease resistance in other fish species such as Nile tilapia and African catfish fingerlings or juveniles (Aly et al. 2008; Al-Dohail et al. 2009). Probiotic bacteria may act on the immune defence by various ways, such as mitigating the immunosuppressive actions of some pathogens by production of antimicrobial substances (Ringo et al. 2012). Especially, it has been reported that some *Lactobacillus* or *Bacillus spp.* produce antimicrobial agents active against Gram-positive and – negative organisms (Wang et al. 2008). A competition by the probiotic bacteria for the attachment sites is also suggested as one of the action modes for enhancing immune defence and disease resistance even if little evidence has been reported in fish (Kesarcodi-Watson et al. 2008). Using the same probiotic bacteria as in the present study, Decamp et al. (2006) demonstrated an increase in survival associated with inhibition of pathogens such as *Vibrio spp.* in water and gastro-intestinal tract in gilthead seabream and Japanese flounder, but results concerning the mode of action are not yet reported. Together with our results, we hypothesize that a wide range of fish species positively respond to probiotic exposure.

All in all, the results from the larval and juveniles settings show that high dietary concentrations of probiotic bacteria have potential stimulating impact on husbandry parameters and immune compounds, at least total Ig in percid larvae and juveniles. But, further investigations are needed to determine the optimal doses for emphasizing immune defence status and disease resistance.

30.3.2 *Prebiotics and/or Other Immunostimulants*

30.3.2.1 β-1.3/1.6-Glucans

Among the numerous prebiotics already used in animal production, some non-digestible carbohydrates namely oligosaccharides are nowadays reported as effective candidates for stimulating the immune system of various fish species. Indeed, it has been demonstrated that such polysaccharides have the potential to compete with the binding process of some bacteria pathogens to the epithelial cells of the fish gut (Li and Gatlin 2004; Ringo et al. 2010). In fact, they provide a variety of sugar molecules that may attach to the same gut sites targeted by the pathogenic bacteria. Because these sugar molecules have a configuration that makes their glycosidic bonds non-digestible to the hydrolytic activity of the digestives enzymes, they pass through the gut epithelium with the pathogenic bacteria attached, then preventing colonization (Siwicki et al. 2009; Ringo et al. 2010). Despite the interest in oligosaccharides for modulating the fish immune system, only mannanoligosaccharides (MOS) have received high attention, but other polysaccharides such as inulin and fructooligosaccharides (FOS) have been also studied in some fish species (review in Ringo et al. 2010). Information concerning the application of such prebiotics is still scarce in percid fish.

Nowadays, several studies focussed on the role of MOS β-1.3/1.6-glucans, which have been reported to improve resistance to some bacteria outbreaks by enhancing various immune functions such as phagocytosis, macrophage respiratory burst activity, lysozyme activity, alternative complement pathway (Siwicki et al. 2009; Ringo et al. 2010, 2012). Only a preliminary study on the immune response to β-glucans has been reported in percid fish by Siwicki et al. (2009) using pikeperch juveniles. Feeding two doses (1 or 2 g kg^{-1} feed) of β-1.3/1.6-glucans as MacroGard (a mixture of β-glucans derived from the cell walls of *S. cerevisiae* yeast) to pikeperch juveniles during 6 weeks, the latter authors demonstrated an activation of various immune functions such as metabolic activity of spleen phagocytes, potential killing activity of spleen phagocytes, proliferation of pronephric lymphocytes, serum lysozyme activity, and serum concentration of total immunoglobulins whatever the prebiotic dose. Therefore, these results indicate that prebiotic β-glucans have potential of stimulating both the innate and acquired immune system in percid fish, but the conferred disease resistance need to be investigated.

30.3.2.2 Lipopolysaccharide

Several studies have demonstrated that lipopolysaccharide (LPS) is one of the most effective immunostimulants for modulating the fish immune system. Since LPS is a component of the cell walls of gram-negative bacteria, consisting of lipid and oligosaccharide chains; its lipid chain is recognized by the host immune system in contrast to other oligosaccharides. It has been shown that LPS can modulate both the innate and acquired immune system by enhancing the production of various immune factors such as antibodies, lysozymes, cytokines, tumor necrosis factors (TNF), and stimulating the transcription of various immune genes in different fish species (Paulsen et al. 2003; Goetz et al. 2004; Watzke et al. 2007; Swain et al. 2008b, Selvaraj et al. 2009, Bich-Hang et al. 2013). It has been also reported that LPS injections may modulate the leucocyte proportions by increasing neutrophil populations while decreasing that of lymphocytes, and may enhance disease resistance in common carp (*Cyprinus carpio*) (Selvaraj et al. 2009). The same effect on leucocyte populations was recently observed in Eurasian perch after LPS injection but disease resistance was not investigated (Mathieu et al. 2014). Moreover, injections of LPS have been used to demonstrate the bi-directional interaction between the immune system and the HPI axis in fish. In this regard, LPS has been reported to increase brain corticotropin-releasing hormone (CRH) content (Pepels et al. 2005) and plasma cortisol level (Holland et al. 2002; Haukenes and Barton 2004; Acerete et al. 2007; Haukenes et al. 2011), as well as to modulate GR gene expression in different organs (Acerete et al., 2007). There are reports that LPS injection failed to modulate cortisol response pallid sturgeon (*Scaphirhynchus albus*) (Haukenes et al. 2008) suggesting differences in interaction between the immune system and the HPI axis of fish. By elsewhere, some in vitro studies suggested that the immune responses to LPS injections, such as increase in cytokine expression could be affected by cortisol (MacKenzie et al. 2006; Castillo et al. 2009; Castro et al. 2011; Philip et al. 2012) in a more complex way than only the antagonism of LPS-induced transcriptional responses. It was also recently shown that oral administration of LPS at 20 μg/kg body weight can enhance some blood non-specific immune compounds such as hemolytic complement, lysozyme activity, bacteriolytic activity and respiratory burst in rainbow trout juveniles (Gisbert et al. 2013).

For more information concerning the immune response of LPS in fish, a recent study conducted by Mathieu et al. (2014) investigated the in vivo effects of LPS injections in Eurasian perch juveniles, and examined the corticosteroid-immune interactions. Eurasian perch juveniles were injected either with LPS (10 mg kg^{-1}) or a combination of LPS + cortisol (0.8 mg kg^{-1}) or LPS + deoxycorticosterone, a putative agonist of the mineralocorticoid receptor in fish (DOC, 0.08 mg kg^{-1}) to mimic infectious outbreaks associated or not with stress simulations as frequently observed in fish husbandry. Comparisons were made in triplicate groups of 21 fish per 100-L tank. Modulations of plasma cortisol and some immune variables (blood leucocyte populations, plasma lysozyme activity, and plasma activity of the complement alternative pathway) were evaluated 1, 3 and 7 days after injections on 12 fish per treatment. Moreover, mRNA expression levels of different immune genes or different corticosteroid receptors were determined on gills and spleen of the same fish over the same period.

Only the results concerning the effects of treatments on immune parameters are presented. The results showed that LPS injections either stimulated the immune system or co-stimulated the immune system and corticotropic axis. Indeed, a redistribution of blood leucocyte populations was suggested following the LPS injections (Fig. 30.13); the proportion of neutrophils (Fig. 30.13a) was increased while

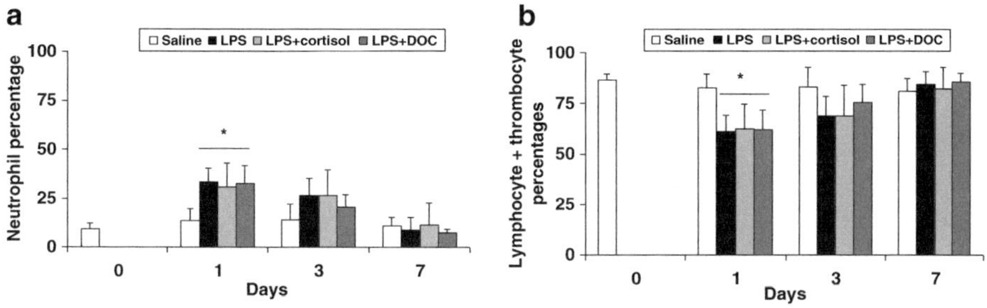

Fig. 30.13 Effect of treatments on blood proportions of lymphocytes + thrombocytes (**a**), neutrophiles (**b**). Significant differences between treatments compared with respective saline-injected are indicated by *asterisks* (*, p<0.05)

that of lymphocytes + thrombocytes (Fig.30.13b) decreased. LPS is known to regulate the proliferation of leucocytes. Indeed, it has been characterized as a mitogenic agent for B-lymphocytes in vitro for both fish (Weyts et al. 1998). In carp, an intraperitoneal injection of LPS was previously reported to increase blood neutrophil and monocyte populations without affecting that of lymphocytes (Selvaraj et al. 2004, 2006, 2009). Stimulation of neutrophils by LPS could increase the immune response of fish to opportunistic infections as these cells are the first line of defence against pathogens (Selvaraj et al. 2006). Even if it was not possible to determine the types of lymphocytes in this study, we could also hypothesize that B-lymphocytes were recruited to inflammatory sites. This recruitment of B-lymphocytes would reduce the proportion of these leucocytes in blood. In the present study, corticosteroids co-injected with LPS had no effects on the redistribution of blood leucocyte populations induced by LPS. But cortisol has been reported to increase the proportion of neutrophils and decreased that of lymphocytes in several teleost species including Eurasian perch (Weyts et al. 1998; Mathieu et al. 2013). Based on these results, we could hypothesize that 10 mg LPS kg^{-1} induced a stronger immune response than did 0.8 mg cortisol kg^{-1} in terms of modulation of blood leucocyte proportions. It could be also suggest that the leucocyte proportions resulted from LPS injections was at maximum level, thus cortisol was ineffective to modulate LPS-induced alteration of blood leucocyte proportions.

This study also showed that LPS modulated plasma lysozyme activity (Fig. 30.14a) but no significant effect was observed for C-type lysozyme mRNA abundance (Fig. 30.14b) in spleen and gills. Co-injection with corticosteroids did not further modulate such LPS effect on plasma lysozyme activity or on expression level of C-type lysozyme. In a previous study, we demonstrated that cortisol injected alone increased plasma lysozyme activity and C-type lysozyme expression in spleen of Eurasian perch (Mathieu et al. 2013). During this previous study, DOC also increased C-type lysozyme expression in gills and spleen. During the current study, similarly to leucocytes, we could suggest that the injected LPS prevented or masked the corticosteroid-induced modulation of lysozyme both in plasma and organs. The increase in plasma lysozyme activity following LPS injection has been suggested to

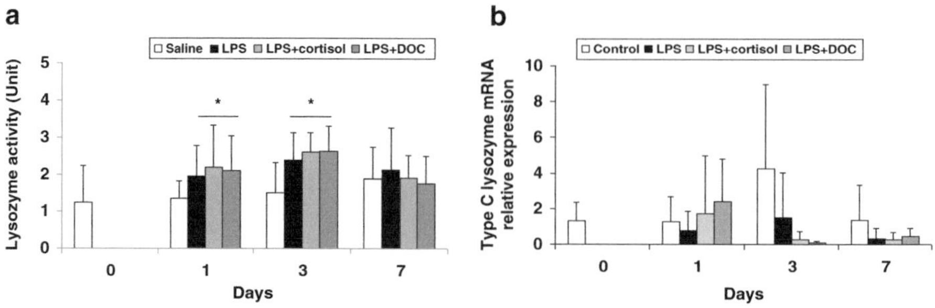

Fig. 30.14 Effect of treatments on lysozyme activity (**a**), and type c lysozyme (**b**) mRNA relative expression levels in spleen. Significant differences between treatments compared with respective saline-injected are indicated by *asterisks* (*, $p<0.05$)

result from lysozyme production by macrophages of various organs (Paulsen et al. 2003). Lysozyme is a key anti-bacterial molecule of the fish immune system, cleaving the β 1–4 glycosidic bonds from Gram-positive bacteria (Saurabh and Sahoo 2008; Ye et al. 2011). LPS is frequently described as a modulator of both plasma lysozyme activity and transcript abundance of lysozyme in various tissues of teleosts (Paulsen et al. 2003; Nayak et al. 2008). Contrary to a previous experiment performed on Atlantic salmon (*Salmo salar*; Paulsen et al. 2003), we did not observe any consistency between plasma lysozyme activity and mRNA expression level in spleen following LPS injection. This may suggest that spleen may not belong to the major organs releasing lysozyme in perch or may also indicate post-transcriptional modifications that were not noticed. Moreover, the increase in plasma lysozyme activity may be associated with the increase in blood neutrophil populations. Indeed, neutrophils are the major lysozyme production sites in fish blood (Yada et al. 2008). This relationship was also observed during a previous experiment on corticosteroid injection in Eurasian perch (Mathieu et al. 2013).

In the present study, all LPS treatments failed to induce modulation of the complement alternative pathway activity in plasma, values averaged 834 ± 269 U/ml. LPS alone did not also affect the level of C3 mRNA expression, and LPS treatment associated to cortisol or DOC induced a weak modulation of C3 mRNA (Fig. 30.15a). Since a previous study using LPS injections failed also to induce modulation of the complement alternative pathways in Eurasian perch, it can be suggested that percid fishes are not fully sensitive to LPS. Complement is an essential component of the immune system signalling the presence of pathogens and initiating or preparing for their clearance (review in Boshra et al. 2006). Contrary to lysozyme, complement targets Gram-negative bacteria. LPS is a major component of Gram-negative bacteria cell walls and as such, LPS-treatment should induce complement activity as a first immune defence mechanisms against this bacteria type (Ellis 2001). Previous studies performed on carp reported that intraperitoneal injection of LPS had no effect on activities of complement alternative and classical pathways in blood (Selvaraj et al. 2006, 2009). Rossi et al. (2007) demonstrated that LPS intraperitoneally injected to Eurasian perch at 15 mg kg^{-1} had no effect on C3 expression in liver.

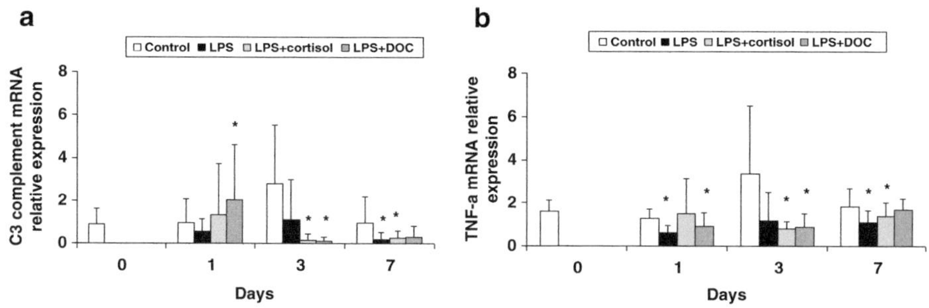

Fig. 30.15 Effect of treatments on C3 complement (**a**) and FNF-α (**b**) mRNA relative expression levels in spleen. Significant differences between treatments compared with respective saline-injected are indicated by *asterisks* (*, p<0.05)

A recent study performed in striped catfish (*Pangasianodon hypophthalmus*) revealed that a double injection of LPS at 3 or 15 mg kg^{-1}, but not at 45 mg kg^{-1}, increased the activity of plasma complement alternative pathway 14 days after the first injection (Bich Hang et al. 2013). This last study also revealed that LPS injected at 3 or 15 mg kg^{-1} increased the protein abundance of three C3 compounds in peripheral blood mononuclear cells. It is not excluded that the dose of LPS used during the present study was too high or that the time-course was too short to measure any modulation of the complement activity in plasma. LPS associated with DOC decreased C3 mRNA expression level in spleen. This may suggest that DOC could regulate C3 expression. However, a co-injection of DOC and LPS is necessary to modulate C3 expression level. Indeed, we previously demonstrated that DOC alone was ineffective to modulate plasma alternative activity or C3 expression in both spleen and gills of Eurasian perch (Mathieu et al. 2013).

Concerning the proinflammatory response to LPS treatment, TNF-α transcript level increased significantly between D1 and D7 in response to the injections (Fig. 30.15b). Moreover, values decreased for LPS and LPS + DOC but not for LPS + cortisol indicating differences in the pathway relay between the two corticosteroids. TNF-α is an important pro-inflammatory cytokine in mammals and in teleosts. LPS is well known to induce the release of some proinflammatory cytokines including TNF-α, IL-6 or IL-1β in several organs of different fish species (MacKenzie et al. 2003; Secombes et al. 2011; Teles et al. 2011). In mammals, proinflammatory cytokines produced in response to LPS, particularly IL-1β, have been identified as the effectors by which the immune system stimulates HPA (Acerete et al. 2007). In fish, there is increasing evidence that increase in the level of plasma cortisol observed following LPS treatment is a result of LPS-induced modulation of cytokine expression (Holland et al. 2002; Haukenes et al. 2008). It was also observed that proinflammatory response to LPS treatment is organ and species dependent. Indeed, following LPS injections in gilthead seabream *Sparus aurata*, an increase in TNF-α expression was observed in head kidney in contrast to a decrease in spleen at the same time (Acerete et al. 2007). It is also commonly reported that LPS induce an over-expression of TNF-α (MacKenzie et al. 2003, 2006; Iliev et al. 2008), but sometimes there is no regulation (García-Castillo et al. 2002). But in some cases, the increase in TNF-α mRNA expression may fail to induce TNF-α protein secretion suggesting that translation of the TNF-α mRNA is somehow inhibited and that the increase in the expression of a given gene is not always accompanied by the correspondent translation into protein.

At higher dose, LPS could become immunosuppressive (Nayak et al. 2008), and this may probably explain the differences reported between these different studies. Alternatively, we speculate that TNF-α regulation after LPS treatment in percids may differ from that displayed in other fish families. Contrary to DOC, cortisol seemed to neutralise the LPS-induced effect on TNF-α mRNA expression level since the fish co-injected with LPS + cortisol displayed TNF-α mRNA expression level similar to that of the saline-injected controls. A neutralising effect of cortisol on LPS-induced increase in TNF-α expression was observed in trout and carp macrophages in vitro (MacKenzie et al. 2006; Stolte et al. 2008). It was hypothesised that cortisol antagonised the immune responses induced by LPS. But, some studies indicate that cortisol-induced regulations of response to LPS are more complex than a simple antagonism (MacKenzie et al. 2006).

30.3.2.3 β-Hydroxy- β -Methylbutyrate

In mammals, it has been reported that some nutrient factors, including vitamins, amino acids and their metabolites may be considered as immunostimulants due to their positive effects on the immune status (Li et al. 2007, 2008; Zhou et al. 2012). Despite such evidence, little is yet known about the efficiency of the latter nutrients for stimulating the fish immune system. Nevertheless, there has been an interest in the potential stimulatory effects of some amino acids namely leucine, glutamine, glycine, and arginine in some fish species (Li et al. 2008). It has been reported that β-hydroxy-β-methylbutyrate (HMB, a leucine metabolite) is an effective stimulator of the immune system of pikeperch. Siwicki et al. (2005, 2012) examining the immune response of pikeperch juveniles to a 60-day HMB supplementation at a dose of 50 mg kg^{-1} feed in RAS conditions, showed an increase in the phagocytic and potential killing activities of the blood and pronephritic phagocytes, and the proliferative response of blood and pronephritic lymphocytes. Such immune response was associated to a better protection against *Y. ruckeri*; cumulative mortality rate following bacteria challenge was markedly reduced (30 %). The same dietary HMB supplementation applied in adult pikeperch during 4 weeks also increased plasma lysozyme activity and total immunoglobulins (Siwicki et al. 2006). Such response was associated to an increased protection against *A. salmonicida*; cumulative mortality rate decreased by 40 % following the bacteria challenge. But, other authors failed to demonstrate such protection of HMB against bacteria in other fish species such as tilapia, hybrid striped bass and rainbow trout (Li and Gatlin 2007; Kunttu et al. 2009; Declercq et al. 2013), indicating high species or families differences in the immune response to HMB. Such differences may be related to differences among fish species in leucine metabolism (Bystriansky et al. 2007; Li et al. 2008).

30.4 Future Perspectives

For future research, it would be first relevant to understand the molecular mechanisms behind the immunomodulation in percids. While molecular and genomic structures of most of percid immune genes are not yet well characterized, some molecular specific features have been identified for some immune genes. Studies are needed to test whether such specific differences may account for specific immune pathways, and thereby accounting for specific immune responses to infection or other adverse stimuli. For instance, concerning the interaction between the corticosteroid axis and immune system, we have recently suggested that all the corticosteroid receptors, including MR, are involved in the regulation of some immune pathways in Eurasian perch (Milla et al. 2010; Mathieu et al. 2013, 2014), while previous data (Stolte et al. 2008) did not notice such involvement of MR in fish immune response. More studies on transactivation and use of specific antagonists of perch corticosteroid receptors are still necessary to confirm the hypothesis of implication of MR in fish immune response.

The available studies on the application of immunostimulants in percid fish show positive effects on various immune compounds but information is lacking concerning the overall immunocompetence after immunomodulation since few bacteria challenges have been performed, so more investigations are recommended. Many other immunomodulatory compounds remain to be tested in order to optimize the success of the immune response and to limit the spread of pathogen outbreak in the developing aquaculture farms.

Acknowledgements The authors are grateful to the INVE Aquaculture (Belgium) for providing the mixture of *Bacillus* SANOLIFE MIC-F®. They are also grateful to André Evrard, Marie-Claire Forget and Anne Vermeylen for their technical assistance. The research was partly funded by the Belgian National Funds for Science (FNRS, contract 2.4.568.00.F).

References

Abdulfatah A, Fontaine P, Kestemont P, Milla S, Marie M (2013) Effects of the thermal threshold and the timing of temperature reduction on the initiation and course of oocyte development in cultured female of Eurasian perch *Perca fluviatilis*. Aquaculture 376–79:90–96

Acerete L, Balasch J, Espinosa E, Josa A, Tort L (2004) Physiological responses in Eurasian perch (*Perca fluviatilis*, L.) subjected to stress by transport and handling. Aquaculture 237:167–178

Acerete L, Balasch JC, Castellana B, Redruello B, Roher N, Canario AV, Planas JV, MacKenzie S, Tort L (2007) Cloning of the glucocorticoid receptor (GR) in gilthead seabream (*Sparus aurata*). Differential expression of GR and immune genes in gilthead seabream after an immune challenge. Comp Biochem Physiol B Biochem Mol Biol 148:32–43

Ai Q, Xu H, Mai K, Xu W, Wang J, Zhang W (2011) Effects of dietary supplementation of *Bacillus subtilis* and fructooligosaccharide on growth performance, survival, non-specific immune response and disease resistance of juvenile large yellow croaker, *Larimichthys crocea*. Aquaculture 317:155–161

Al-Dohail MA, Hashim R, Aliyu-Paiko A (2009) Effects of the probiotic, *Lactobacillus acidophilus*, on the growth performance, haematology parameters and immunoglobulin concentration in African catfish *Clarias gariepinus* fingerling. Aquacult Res 40:1642–1652

Aly MS, Mohamed MF, John G (2008) Effects of probiotics on the survival, growth and challenge infection in Tilapia (*Oreochromis niloticus*). Aquacult Res 39:647–656

Apun-Molina JP, Santamaria-Miranda A, Luna-Gonzalez A, Martinez-Diaz SF, Rojas-Contreras M (2009) Effect of potential probiotic bacteria on growth and survival of tilapia *Oreochromis niloticus* L., cultured in the laboratory under high density and suboptimum temperature. Aquacult Res 40:887–894

Arnekleiv JV, Urke HA, Kristensen T, Halleraker JH, Flodmarks LEW (2004) Recovery of wild, juvenile brown trout from stress of flow reduction, electrofishing, handling and transfer from river to indoor simulated stream channel. J Fish Biol 64:541–552

Babiak I, Mandiki SNM, Ratsinjomanana K, Kestemont P (2004) Initial weight and its variation in post-larval Eurasian perch affect quantitative characteristics of juvenile cohorts under controlled conditions. Aquaculture 234:263–276

Balm PHM (1997) Immune-endocrine interactions. In: Iwama GK, Sumpter AK, Schreck CB (eds) Fish stress and health aquaculture. Cambridge University Press, Cambridge, pp 195–221

Baras E, Kestemont P, Melard C (2003) Effect of stocking density on the dynamics of cannibalism in sibling larvae of Perca £uviatilis under controlled conditions. Aquaculture 219:241–255

Bao BL, Peatman E, Li P, He CB, Liu ZJ (2005) Catfish hepcidin gene is expressed in a wide range of tissues and exhibits tissue-specific up-regulation after bacterial infection. Dev Comp Immunol 29:939–950

Barton BA (1997) Stress in finfish: past, present and future – a historical perspective. In: Iwama GK, Pickering AD, Sumpter JP, Schreck CB (eds) Fish stress and health in aquaculture. Cambridge University Press, Cambridge, pp 1–33, chapter 1

Bich Hang BT, Milla S, Gillardin V, Phuong NT, Kestemont P (2013) In vivo effects of *Escherichia coli* lipopolysaccharides on regulation of immune response and protein expression in striped catfish (*Pangasianodon hypophthalmus*). Fish Shellfish Immunol 34:339–347

Biswas AK, Seoka M, Takii K, Maita M, Kumai H (2006) Stress response of red sea bream *Pagrus major* to acute handling and chronic photoperiod manipulation. Aquaculture 252:566–572

Bly J, Clem L (1992) Temperature and teleost immune functions. Fish Shellfish Immunol 2:159–171

Boshra H, Li J, Sunyer JO (2006) Recent advances on the complement system of teleost fish. Fish Shellfish Immunol 20:239–262

Bowden TJ, Thompson KD, Morgan AL, Gratacap RML, Nikoskelainen S (2007) Seasonal variation and immune response: a fish perspective. Fish Shellfish Immunol 22:695–706

Bricknell I, Dalmo RA (2005) The use of immunostimulants in fish larval aquaculture. Fish Shellfish Immunol 19:457–472

Buchanan KL (2000) Stress and the evolution of condition-dependent signals. Trends Ecol Evol 15:156–160

Bystriansky JS, Frisk NT, Ballantyne JS (2007) Intermediary metabolism of Arctic char *Salvelinus etabol* during short-term salinity exposure. J Exp Biol 210:1971–1985

Castillo J, Teles M, Mackenzie S, Tort L (2009) Stress-related hormones modulate cytokine expression in the head kidney of gilthead seabream (*Sparus aurata*). Fish Shellfish Immunol 27:493–499

Castro R, Zou J, Secombes CJ, Martin SAM (2011) Cortisol modulates the induction of inflammatory gene expression in a rainbow trout macrophage cell line. Fish Shellfish Immunol 30:215–223

Cerezuela R, Guardiola FA, González P, Meseguer J, Esteban MA (2012) Effects of dietary *Bacillus subtilis*, *Tetraselmis chuii*, and *Phaeodactylum tricornutum*, singularly or in combination, on the immune response and disease resistance of sea bream (*Sparus aurata* L.). Fish Shellfish Immunol 33:342–349

Chen SL, Zhang YX, Xu MY, Ji XS, Yu GC, Dong CF (2006) Molecular polymorphism and expression analysis of MHC class II B gene from red seabream (*Chrysophrys major*). Dev Comp Immunol 30:407–418

Cheng AC, Cheng SA, Chen YY, Chen JC (2009) Effects of temperature change on the innate cellular and humoral immune responses of orange-spotted grouper *Epinephelus coioides* and its susceptibility to *Vibrio alginolyticus*. Fish Shellfish Immunol 26:768–772

Christie D, Wei G, Fujiki K, Dixon B (2007) Cloning and characterization of a cDNA encoding walleye (Sander vitreum) β-2 microglobulin. Fish Shellfish Immunol 22:727–733

Ciereszko RE, Dabrowiski K, Ciereszko A (1997) Effects of temperature and photoperiod on reproduction of female yellow perch *Perca flavescens*: plasma concentrations of steroid hormones, spontaneous and induced ovulation, and quality of eggs. J World Aquac Soc 28:344–356

Collazos ME, Ortega E, Barriga C, Rodriguez AB (1998) Seasonal variation in haematological parameters in male and female *Tinca tinca*. Mol Cell Biochem 183:165–168

Davis KB (2004) Temperature affects physiological stress responses to acute confinement in sunshine bass (*Morone chrysops* X *Morone saxatilis*). Comp Biochem Physiol A Mol Integr Physiol 139:433–440

Decamp O, Makridis P, Qi Z, Xin N, Moriarty DJW, Sorgeloos P, Lavens P (2006) Feed ingredients: performance of selected Bacillus probiotics in marine fish culture. International Aquafeed, Magazine issue Sept–Oct, pp 16–18

Declercq AM, Haesebrouck F, Van den Broeck W, Bossier P, Decostere A (2013) Columnaris disease in fish: a review with emphasis on bacterium-host interactions. Vet Res 44:27–44

Demers NE, Bayne CJ (1997) The immediate effects of stress on hormones and plasma lysozyme in rainbow trout. Dev Comp Immunol 21:363–373

Dimitroglou A, Merrifield DL, Carnevali O, Picchietti S, Avella M, Daniels C, Güroy D, Davies SJ (2011) Microbial manipulations to improve fish health and production e A Mediterranean perspective. Fish Shellfish Immunol 30:1–16

Eissa N, Wang HP, Shen Z, Yao H, Rapp POD (2013) Physiological and molecular stress response of yellow perch subjected to handlings at different temperatures. Abstract, World Aquaculture Society Meeting, p 813

EL-Haroun ER, Goda AMAS, Chowdhury MAK (2006) Effect of a dietary probiotics Biogen® supplementation as a growth promoter on growth performance and feed utilization of Nile Tilapia *Oreochromis niloticus* (L.). Aquacult Res 37:1473–1480

Ellis A (2001) Innate host defense mechanisms of fish against viruses and bacteria. Dev Comp Immunol 25:827–839

Esteban MA, Cuesta A, Chaves-Pozo E, Meseguer J (2013) Influence of melatonin on the immune system of fish: a review. Int J Mol Sci 14:7979–7999

Fevolden SE, Røed KH, Fjalestad KT (2002) Selection response of cortisol and lysozyme in rainbow trout and correlation to growth. Aquaculture 205:61–75

Flores-Morales A, Greenhalgh CJ, Norstedt G, Rico-Bautista E (2006) Negative regulation of growth hormone receptor signaling. Mol Endocrinol 20:241–253

Fontaine P, Pereira C, Wang N, Marie M (2006) Influence of pre-inductive photoperiod variations on Eurasian perch *Perca fluviatilis* broodstock response to an inductive photothermal programme. Aquaculture 255:410–416

Gallardo AM, Sal-Rabanal M, Ibarz A, Padros F, Blasco J, Fernandez-Borras J, Sanchez J (2003) Functional alterations associated with « winter syndrome » in gilthead seabream. Aquaculture 223:15–27

García-Castillo J, Pelegrín P, Mulero V, Meseguer J (2002) Molecular cloning and expression analysis of tumor necrosis factor alpha from a marine fish reveal its constitutive expression and ubiquitous nature. Immunogenetics 54:200–207

Gisbert E, Castillo M, Skalli A, Andree KB, Badiola I (2013) Bacillus cereus var. toyoi promotes growth, affects the histological organization and microbiota of the intestinal mucosa in rainbow trout fingerlings. J Anim Sci 91:2766–2774

Gatesoupe FJ (2008) Updating the importance of lactic acid bacteria in fish farming: natural occurrence and probiotic treatments. J Mol Microbiol Biotechnol 14:107–114

Gerwick L, Corley-Smith G, Bayne CG (2007) Gene transcript changes in individual rainbow trout livers following an inflammatory stimulus. Fish Shellfish Immunol 22:157–171

Gobeli S, Goldschmidt-Clermont E, Frey J, Burr SE (2009) *Pseudomonas chlororaphis* strain JF 3835 reduces mortality of juvenile perch, *Perca fluviatilis*, caused by *Aeromonas sobria*. J Fish Dis 32:597–602

Goetz FW, Iliev DB, McCauley LAR, Liarte CQ, Tort LB, Planas JV, Mackenzie S (2004) Analysis of genes isolated from lipopolysaccharide-stimulated rainbow trout (*Oncorhynchus mykiss*) macrophages. Mol Immunol 41:1199–1210

Goldschmidt-Clermont E, Wahli T, Frey J, Burr SE (2008) Identification of bacteria from the normal flora of perch, *Perca fluviatilis*, and evaluation of their inhibitory potential towards *Aeromonas species*. J Fish Dis 31:353–359

Grimholt U, Larsen S, Nordmo R, Midtlyng P, Kjoeglum S, Storset A, Saeb S, Stet RJM (2003) MHC polymorphism and disease resistance in Atlantic salmon (*Salmo salar*); facing pathogens with single expressed major histocompatibility class I and class II loci. Immunogenetics 55:210–219

Hanif A, Bakopoulos V, Dimitriadis GJ (2004) Maternal transfer of humoral specific and non-specific immune parameters to sea bream (*Sparus aurata*) larvae. Fish Shellfish Immunol 17:411–435

Harris J, Bird DJ (2000) Modulation of the fish immune system by hormones. Vet Immunol Immunopathol 77:163–176

Haukenes Alf H, Bruce AB, Kenneth JR (2011) Plasma cortisol and hypothalamic monoamine response in yellow perch *Perca flavescens* after intraperitoneal injection of lipopolysaccharide. Fish Physiol Biochem 37:425–432

Haukenes AH, Barton BA, Bollig H (2008) Cortisol responses of pallid sturgeon and yellow perch following challenge with lipopolysaccharide. J Fish Biol 72:780–784

Haukens AH, Barton BA (2004) Characterization of the cortisol response following an acute challenge with lipopolysaccharide in yellow perch and the influence of rearing density. J Fish Biol 64:851–862

Haux C, Sjobeck ML, Larsson A (1985) Physiological stress responses in a wild fish population of perch (*Perca fluviatilis*) after capture and during subsequent recovery. Mar Environ Res 15:77–95

Hernández A, Tort L (2003) Annual variation of complement, lysozyme and haemagglutinin levels in serum of the gilthead sea bream Sparus aurata. Fish Shellfish Immunol 15:479–481

Hoeger B, Hitzfeld B, Kollner B, Dietrich DR, Van den Heuvel MR (2005) Sex and low-level sampling stress modify the impacts of sewage effluent on the rainbow trout (*Oncorhynchus mykiss*) immune system. Aquat Toxicol 73:79–90

Holland JW, Pottinger TG, Secombes CJ (2002) Recombinant interleukin-1 beta activates the hypothalamic-pituitary-interrenal axis in rainbow trout, Oncorhynchus mykiss. J Endocrinol 175:261–267

Ibarz A, Padros F, Gallardo MA, Fernandez B, Tort L (2010) Low-temperature challenges to gilthead sea bream culture: review of cold-induced alterations and winter syndrome. Rev Fish Fish 20:539–556

Iliev DB, Roach JC, Mackenzie S, Planas JV, Goetz FW (2008) Endotoxin recognition: in fish or not in fish? FEBS Lett 579:6519–6528

Jentoft S, Aastveit AH, Torjesen PA, Andersen O (2005) Effects of stress on growth, cortisol and glucose levels in non-domesticated Eurasian perch (Perca fluviatilis) and domesticated rainbow trout (*Oncorhynchus mykiss*). Comp Biochem Physiol A Mol Integr Physiol 141:353–358

Kent ML, Feist SW, Harper C, Hoogstraten-Millerd S, MacLaw J, Sanchez-Morgado JM, Tanguay RL, Sanders GE, Spitsbergen JM, Whipps CM (2009) Recommendations for control of pathogens and infectious diseases in fish research facilities. Comp Biochem Physiol C Toxicol Pharmacol 149:240–248

Kesarcodi-Watson AK, Kaspar H, Lategan M, Gibson L (2008) Probiotics in aquaculture: the need, principles and mechanisms of action and screening processes. Aquaculture 274:1–14

Kestemont P, Jourdan SJ, Houbart M, Mélard C, Paspatis M, Fontaine P, Cuvier A, Kentouri M, Baras E (2003) Size heterogeneity, cannibalism and competition in cultured predatory fish larvae: biotic and abiotic influences. Aquaculture 227:333–356

Kim TJ, Parker KM, Hedrick PW (1999) Major histocompatibility complex differentiation in Sacramento River chinook salmon. Genetics 151:1115–1122

Kortet R, Taskinen J, Sinisalo T, Jokinen I (2003) Breeding-related seasonal changes in immunocompetence, health state and condition of the cyprinid fish, *Rutilus rutilus*, L. Biol J Linn Soc 78:117–127

Kumari J, Sahoo PK, Swain T, Sahoo SK, Sahu AK, Mohanty BR (2006) Seasonal variation in the innate immune parameters of the Asian catfish *Clarias batrachus*. Aquaculture 252:121–127

Kunttu HMT, Valtonen ET, Suomalaine LR, Jokinen IE (2009) The efficacy of two immunostimulants against *Flavobacterium columnare* infection in juvenile rainbow trout (*Oncorhynchus mykiss*). Fish Shellfish Immunol 26:850–857

Lategan MJ, Torpy FR, Gibson LF (2004) Biocontrol of saprolegniosis in silver perch *Bidyanus bidyanus* (Mitchell) by *Aeromonas media* strain A199. Aquaculture 235:77–88

Li P, Galtin DM (2007) Evaluation of dietary supplementation of β-hydroxy-β-methylbutyrate for hybrid striped bass *Morone chrysops* x *Morone saxatilis*. J Appl Aquat 19:77–88

Li P, Gatlin DM (2004) Dietary brewer yeast and the prebiotic GroBiotic™ AE influence growth performance, immune responses and resistance of hybrid striped bass (*Morone chrysops×M. saxatilis*) to *Streptococcus iniae* infection. Aquaculture 231:445–456

Li P, Yin YL, Li D, Kim SW, Wu G (2007) Amino acids and immune function. Br J Nutr 98:237–252

Li P, Mai K, Trushenski J, Wu G (2008) New developments in fish amino acid nutrition: towards functional and environmentally oriented aquafeeds. Amino Acids. doi:10.1007/s00726-008-0171-1

Li H, Jiang L, Han J, Su H, Yang Q, He C (2011) Major histocompatibility complex class IIA and IIB genes of the spotted halibut *Verasper variegatus*: genomic structure, molecular polymorphism, and expression analysis. Fish Physiol Biochem 37:767–780

MacKenzie S, Planas JV, Goetz FW (2003) LPS-stimulated expression of a tumor necrosis factor-alpha mRNA in primary trout monocytes and in vitro differentiated macrophages. Dev Comp Immunol 27:393–400

MacKenzie S, Iliev D, Liarte C, Koskinen H, Planas JV, Goetz FW, Mölsä H, Krasnov A, Tort L (2006) Transcriptional analysis of LPS-stimulated activation of trout (Oncorhynchus mykiss) monocyte/macrophage cells in primary culture treated with cortisol. Mol Immunol 43:1340–1348

Magnadottir B (2006) Innate immunity of fish (review). Fish Shellfish Immunol 20:137–151

Mandiki SNM, Babiak I, Krol J, Rasolo JFR, Kestemont P (2007) How initial predator–prey ratio affects intra-cohort cannibalism and growth in Eurasian perch *Perca fluviatilis* larvae and juveniles under controlled conditions. Aquaculture 268:149–155

Mandiki SNM, Milla S, Wang N, Blanchard G, Djonkack T, Tanascaux S, Kestemont P (2011) Effects of probiotic bacteria on growth parameters and immune defence in Eurasian perch *Perca fluviatilis* larvae under intensive culture conditions. Aquacult Res 42:693–703

Martins ML, Xu DH, Shoemaker CA, Klesius PH (2011) Temperature effects on immune response and haematological of channel catfish *Ictalurus puntatus* vaccinated with live theronts of *Ichthyophthirius multifiliis*. Fish Shellfish Immunol 31:774–780

Mathieu C, Milla S, Mandiki SNM, Douxfils J, Douny C, Scippo ML, De Pauw E, Kestemont P (2013) First evidence of the possible implication of the 11-deoxycorticosterone (DOC) in immune activity of Eurasian perch (Perca fluviatilis, L.): a comparison with cortisol. Comp Biochem Physiol A Mol Integr Physiol 165:149–158

Mathieu C, Milla S, Mandiki SNM, Douxfils J, Kestemont P (2014) In vivo response of some immune and endocrine variables to LPS in Eurasian perch (*Perca fluviatilis*) and modulation of this response by two corticosteroids, cortisol and 11-deoxycorticosterone. Comp Biochem Physiol A Mol Integr Physiol 167:25–34

Merrifield DL, Bradley G, Baker RTM, Davies SJ (2010) Probiotic applications for rainbow trout. II Effects on growth performance, feed utilization, intestinal microbiota and related health criteria post-antibiotic treatment. Aquac Nutr 16:496–503

Michel C, Bernatchez L, Behrmann-Godel J (2009) Diversity and evolution of MHII beta genes in a non-model percid species – the Eurasian perch (*Perca fluviatilis* L.). Mol Immunol 46:3399–3410

Migaud H, Fontaine P, Sulistyo I, Kestemont P, Gardeur JN (2002) Induction of out-of-season spawning in Eurasian perch Perca fluviatilis: effects of rates of cooling and cooling durations on female gametogenesis and spawning. Aquaculture 205:253–267

Migaud H, Fontaine P, Kestemont P, Wang N, Brun-Bellut J (2004) Influence of photoperiod on the onset of gonadogenesis in Eurasian perch *Perca fluviatilis*. Aquaculture 241:561–574

Migaud H, Davie A, Taylor JF (2010) Current knowledge on the photoneuroendocrine regulation of reproduction in temperate fish species. J Fish Biol 76:27–68

Milla S, Mandiki SNM, Hubermont P, Rougeot C, Mélard C, Kestemont P (2009a) Ovarian steroidogenesis inhibition by constant photothermal conditions is caused by a lack of gonadotropin stimulation in Eurasian perch. Gen Comp Endocrinol 163:242–250

Milla S, Wang N, Mandiki SNM, Kestemont P (2009b) Corticosteroids: friends or foes of teleost fish reproduction? Comp Biochem Physiol A Mol Integr Physiol 153:242–251

Milla S, Mathieu C, Wang N, Lambert S, Nadzialek S, Massart S, Henrotte E, Douxfils J, Mélard C, Mandiki SNM, Kestemont P (2010) Spleen immune status is affected after acute handling stress but not regulated by cortisol in Eurasian perch, Perca fluviatilis. Fish Shellfish Immunol 28:931–941

Milla S, Depiereux S, Kestemont P (2011) The effects of estrogenic and androgenic endocrine disruptors on the immune system of fish: a review. Ecotoxicology 20:305–319

Morgan AL, Thompson KD, Auchinachie NA, Migaud H (2008) The effect of seasonality on normal haematological and innate immune parameters of rainbow trout *Oncorhynchus mykiss*. Fish Shellfish Immunol 25:791–799

Nayak SK (2010) Probiotics and immunity: a fish perspective. Fish Shellfish Immunol 29:2–14

Nayak SK, Swain P, Nanda PK, Dash S, Shukla S, Meher PK, Maiti NK (2008) Effect of endotoxin on the immunity of Indian major carp, Labeo rohita. Fish Shellfish Immunol 24:394–399

Newaj-Fyzul A, Al-Harbi AH, Austin B (2014) Review: developments in the use of probiotics for disease control in aquaculture: review. Aquaculture 431:1–11

Nikoskelainen S, Ouwehand AC, Salminen S, Bylund G (2001) Protection of rainbow trout (*Oncorhynchus mykiss*) from furunculosis by *Lactobacillus rhamnosus*. Aquaculture 198:229–236

Oke AO, Olaoye OJ, Nnali KE (2013) Recent advances in fish diseases treatment: probiotics as alternative therapy to antibiotics in aquaculture "a review". Adv Agric Sci Eng Res 3:668–676

Panigrahi A, Kiron V, Puangkaew J, Kobayashi T, Satoh S, Sugita H (2005) The variability of probiotic bacteria as a factor influencing the immune response in rainbow trout *Oncorhynchus mykiss*. Aquaculture 243:241–254

Park KC, Osborne JA, Tsoi SCM, Brown LL, Johnson SC (2005) Expressed sequence tags analysis of Atlantic halibut (*Hippoglossus hippoglossus*) liver, kidney and spleen tissues following vaccination against *Vibrio anguillarum* and *Aeromonas salmonicida*. Fish Shellfish Immunol 18:393–415

Paulsen SM, Lunde H, Engstad RE, Robertsen B (2003) In vivo effects of β-glucan and LPS on regulation of lysozyme activity and mRNA expression in Atlantic salmon (*Salmo salar* L.). Fish Shellfish Immunol 14:39–54

Pepels PPLM, Wendelaar-Bonga SE, Balm PHM (2005) Bacterial lipopolysaccharide (LPS) modulates corticotropin-releasing hormone (CRH) content and release in the brain of juvenile and adult tilapia (*Oreochromis mossambicus*; Teleostei). J Exp Biol 207:4479–4488

Pevet P (2003) Melatonin: from seasonal to circadian signal. J Neuroendocrinol 15:422–426
Philip AM, Daniel Kim S, Vijayan MM (2012) Cortisol modulates the expression of cytokines and suppressors of cytokine signaling (SOCS) in rainbow trout hepatocytes. Dev Comp Immunol 38:360–367
Rastall RA (2004) Bacteria in the gut: friends and foes and how to alter the balance. In D'Ann Finley A, Morris JG, Rogers QR (eds) Proceeding of WALTHAM international science symposium: nature, nurture, and the case of nutrition, Davis, pp 2022s–2026s, Journal of Nutrition
Ringø E, Olsen RE, Gifstad T, Dalmo RA, Amlund H, Hemre GI, Bakke AM (2010) Prebiotics in aquaculture: a review. Aquacult Nutr 16:117–136
Ringø E, Olsen RE, Vecino JLG, Wadsworth S, Song SK (2012) Use of immunostimulants and nucleotides in aquaculture: a review. J Mar Sci Res Dev 1:104–126
Rodger H, Toner D, Vandecan M (2008) Fish health. In: Toner D, Rougeot C (eds) Aquaculture explained-farming of Eurasian perch. BIM, Dun Laoghaire, Dublin, pp 52–61
Rossi F, Chini V, Cattaneo AG, Bernardini G, Terova G, Saroglia M, Gornati R (2007) EST-Based identification of genes expressed in perch (*Perca fluviatilis*). Gene Expr 14:117–127
Sakai M (1999) Current research status of fish immunostimulants. Aquaculture 172:63–92
Saurabh S, Sahoo PK (2008) Lysozyme: an important defence molecule of fish innate immune system. Aquacult Res 39:223–239
Schreck CB, Contreras-Sanchez W, Fitzpatrick MS (2001) Effects of stress on fish reproduction, gamete quality, and progeny. Aquaculture 197:3–24
Secombes C (1996) The non-specific immune system: cellular defences. In: Iwama G, Nakanishi T (eds) The fish immune system: organism, pathogen and environment. Academic Press, San Diego, pp 43–103
Secombes CJ, Wang T, Bird S (2011) The interleukins of fish. Dev Comp Immunol 35:1336–1345
Selvaraj V, Sampath K, Sekar V (2004) Extraction and characterization of lipopolysaccharide from *Aeromonas hydrophila* and its effects on survival and hematology of the carp, *Cyprinus carpio*. Asian Fish Sci 17:163–173
Selvaraj V, Samapath K, Sekar V (2006) Adjuvant and immunostimulatory effect of beta glucan administration in combination with lipopolysaccharide enhances survival and some immune parameters in carp challenged with *Aeromonas hydrophila*. Vet Immunol Immunopathol 114:15–24
Selvaraj V, Sampath K, Sekar V (2009) Administration of lipopolysaccharide increases specific and non-specific immune parameters and survival in carp (*Cyprinus carpio*) infected with Aeromonas hydrophila. Aquaculture 286:176–183
Shepherd BS, Rees CB, Binkowski FP, Goetz FW (2012) Characterization and evaluation of sex-specific expression of suppressors of cytokine signaling (SOCS)-1 and -3 in juvenile yellow perch (*Perca flavescens*) treated with lipopolysaccharide. Fish Shellfish Immunol 33:468–481
Siwicki AK, Zakes Z, Fuller JC Jr, Nissen S, Trapkowaska S, Glabski E, Kazun K, Kowalska A, Terech-Majewska E (2005) The effect of feeding the leucine metabolite β-hydroxy-β-methylbutyrate (HBM) on cell-mediated immunity and protection against *Yersinia ruckeri* in pikeperch (*Sander lucioperca*). Aquacult Res 36:16–21
Siwicki AK, Zakes Z, Fuller JC Jr, Nissen S, Trapkowaska S, Glabski E, Kowalska A, Kazun K, Terech-Majewska E (2006) Influence of β-hydroxy-β-methylbutyrate on non-specific humoral defense mechanisms and protection against furunculosis in pikeperch (*Sander lucioperca*). Aquacult Res 37:127–131
Siwicki AK, Zakes Z, Terech-Majewska E, Malaczewski J (2009) Supplementing the feed of pikeperch (*Sander lucioperca*, L.) juveniles with Macrogard and its influence on nonspecific cellular and humoral defense mechanisms. Aquacult Res 40:405–411
Siwicki AK, Lepa A, Zakes Z, Kowalska A, Kazun B, Kazun K, Glabski E (2012) Influence of dietary administration of β-hydroxy-β-methylbutyrate (HBM) in the innate immunity and resistance against bacterial infections in pikeperch (*Sander lucioperca*). J Agric Sci Technol B 2:965–970

Slater C, Schreck C (1998) Seasonal and physiological parameters modulate salmonind leucocyte androgen receptor affinity and abundance. Fish Shellfish Immunol 8:379–391

Small BC, Bilodeau AL (2005) Effects of cortisol and stress on channel catfish (*Ictalurus punctatus*) pathogen susceptibility and lysozyme activity following exposure to *Edwardsiella ictaluri*. Gen Comp Endocrinol 142:256–262

Star B, Nederbragt AJ, Gentoft S, Grimholt U, Malmstrøm M, Gregers TF, Rounge TB, Jonas Paulsen J, Solbakken MH, Sharma A, Ola F, Wetten OF et al (2011) The genome sequence of Atlantic cod reveals a unique immune system. Nature 477:207–210

Stolte EH, De Mazon AF, Leon-Koosterziel KM, Jesiak M, Bury NR, Sturm A, Savelkoul HF, van Kemenade BM, Flik G (2008) Corticosteroid receptors involved in stress regulation in common carp, *Cyprinus carpio*. J Endocrinol 198:403–417

Swain P, Nayak SK (2009) Role of maternally derived immunity in fish. Fish Shellfish Immunol 27:89–99

Swain P, Nayak SK, Nanda PK, Dash S (2008a) Biological effects of bacterial lipopolysaccharide (endotoxin) in fish: a review. Crit Care Med 25:191–201

Swain P, Nayak SK, Nanda PK, Dash S (2008b) Biological effects of bacterial lipopolysaccharide (endotoxin) in fish: a review. Fish Shellfish Immunol 25:191–201

Teles M, Mackenzie S, Boltaña S, Callol A, Tort L (2011) Gene expression and TNF-alpha secretion profile in rainbow trout macrophages following exposures to copper and bacterial lipopolysaccharide. Fish Shellfish Immunol 30:340–346

Tort L (2011) Stress and immune modulation in fish. Dev Comp Immunol 35:1366–1375

Tort L, Padros F, Rotllant J, Crespo S (1998) Winter syndrome in the gilthead sea bream *Sparus aurata*. Immunological and histopathological features. Fish Shellfish Immunol 8:37–47

Tort L, Rotllant J, Liarte C, Acerete C, Hernández A, Ceulemans C, Coutteau C, Padros F (2004) Effects of temperature decrease on feeding rates, immune indicators and histopathological changes of gilthead sea bream *Sparus aurata* fed with an experimental diet. Aquaculture 229:55–65

Vadstein O, Bergh Ø, Gatesoupe FJ, Galindo-Villegas J, Mulero V, Picchietti S, Scapigliati G, Makridis P, Olsen Y, Dierckens K, Defoirdt T, Boon N, De Schryver P, Bossier P (2012) Microbiology and immunology of fish larvae. Rev Aquac 4:1–25

Vainikka A, Jokinen EI, Kortet R, Taskinen J (2004) Gender- and season- dependent relationships between testosterone, oestradiol and immune functions in wild roach. J Fish Biol 64:227–240

Valenzuela AE, Silva VM, Klempau AE (2006) Effects of constant light on haematological parameters of cultured rainbow trout (*Oncorhynchus mykiss*) in the southern hemisphere. Fish Physiol Biochem 32:113–120

Valenzuela A, Campos V, Yanez F, Alveal K, Gutierrez P, Rivas M, Contreras N, Klempau A, Fernandez I, Oyarzun C (2012) Application of artificial photoperiod in fish: a factor that increases susceptibility to infectious diseases? Fish Physiol Biochem 38:943–950

Van Ham EH, Van Anholt RD, Kruitwagen G, Imsland AK, Foss A, Sveinsbo BO, FitzGerald R, Parpoura AC, Stefansson SO, Wendelaar Bonga SE (2003) Environment affects stress in exercised turbot. Comp Biochem Physiol A: Mol Intgr Physiol 136:525–538

Wahli T, Burr SE, Pugovkin D, Mueller O, Frey J (2005) *Aeromonas sobria*, a causative agent of disease in farmed perch, *Perca fluviatilis*. J Fish Dis 28:141–150

Wang Y (2011) Use of probiotics *Bacillus coagulans*, *Rhodopseudomonas palustris* and *Lactobacillus acidophilus* as growth promoters in grass carp (*Ctenopharyngodon idella*) fingerlings. Aquac Nutr 17:372–378

Wang N, Migaud H, Acerete L, Gardeur JN, Tort L, Fontaine P (2003) Mortality and non-specific immune response of Eurasian perch, *Perca fluviatilis*, during the spawning season. Fish Physiol Biochem 28:523–524

Wang YB, Li JR, Lin J (2008) Probiotics in aquaculture: challenge and outlook. Aquaculture 281:1–4

Wang N, Teletchea F, Kestemont P, Milla S, Fontaine P (2010) Photothermal control of the reproductive cycle in temperate fishes. Rev Aquac 2:209–222

Watzke J, Schirmer K, Scholz S (2007) Bacterial lipopolysaccharides induce genes involved in the innate immune response in embryos of the zebra fish (*Danio rerio*). Fish Shellfish Immunol 23:901–905

Wedemeyer A (1997) Effects of rearing conditions on the health and physiological quality of fish in intensive culture. In Iwama GK, Pickering AD, Sumpter, JP, Schreck CB (Ed) Fish stress and health in aquaculture. Cambridge University Press, Cambridge, UK, Chapter 2, 35–71

Wester PW, Vethaak AD, Van Muiswinkel WB (1994) Fish as biomarkers in immunotoxicology. Toxicology 86:213–232

Weyts FAA, Flik G, Rombout JHWM, Verburg-van Kemenade BML (1998) Cortisol induces apoptosis in activated B cells, not in other lymphoid cells of the common carp, *Cyprinus carpio* L. Dev Comp Immunol 22:551–562

Whittington RJ (1993) Purification and partial characterisation of serum immunoglobulin of the European perch (*Perca fluviatilis*). Fish Shellfish Immunol 3:331–343

Wynne JW, Cook MT, Nowak BF, Elliott NG (2007) Major histocompatibility polymorphism associated with resistance towards amoebic gill disease in Atlantic salmon (*Salmo salar* L.). Fish Shellfish Immunol 22:707–717

Xu TJ, Chen SL (2011) Molecular cloning, genomic structure and expression analysis of major histocompatibility complex class Ia gene of half-smooth tongue sole (*Cynoglossus semilaevis*). Fish Physiol Biochem 37:85–90

Yada T, Hyodo S, Schreck CB (2008) Effects of seawater acclimation on mRNA levels of corticosteroid receptor genes in osmoregulatory and immune systems in trout. Gen Comp Endocrinol 156:622–627

Ye JD, Wang K, Li FD, Sun YZ (2011) Single or combined effects of fructo- and mannan oligosaccharide supplements and *Bacillus clausii* on the growth, feed utilization, body composition, digestive enzyme activity, innate immune response and lipid metabolism of the Japanese flounder *Paralichthys olivaceus*. Aquac Nutr 17:902–911

Yu S, Ao J, Chen X (2010) Molecular characterization and expression analysis of MHC class II alpha and beta genes in large yellow croaker (*Pseudosciaena crocea*). Mol Biol Rep 37:1295–1307

Zhang YX, Chen SL (2006) Molecular identification, polymorphism, and expression analysis of major histocompatibility complex class IIA and B genes of turbot (*Scophthalmus maximus*). Mar Biotechnol 8:611–623

Zhang YA, Salinas I, Li J, Parra D, Bjork S, Xu Z, LaPatra SE, Bartholomew J, Sunyer JO (2010) IgT, a primitive immunoglobulin class specialized in mucosal. Nat Immunol 11:827–835

Zhao H, Hermsen T, Stet RJ, Skjodt K, Savelkoul HF (2008) Peptide-binding motif prediction by using phage display library for SasaUBA_0301, a resistance haplotype of MHC class I molecule from Atlantic Salmon (*Salmo salar*). Mol Immunol 45:1658–1664

Zhao Y, Zhang W, Xua W, Mai K, Zhang Y, Liufu Z (2012) Effects of potential probiotic *Bacillus subtilis* T13 on growth, immunity and disease resistance against Vibrio splendidus infection in juvenile sea cucumber *Apostichopus japonicas*. Fish Shellfish Immunol 32:750–755

Zhou JG, Wei JG, Xu D, Cui HC, Yan Y, Ou-Yang ZL, Huang XH, Huang YH, Qin QW (2011) Molecular cloning and characterization of two novel hepcidins from orange-spotted grouper, *Epinephelus coioides*. Fish Shellfish Immunol 30:559–568

Zhou Q, Wang L, Wang H, Xie F, Wang (2012) Effect of dietary vitamin C on the growth performance and innate immunity of juvenile cobia (*Rachycentron canadum*). Fish Shellfish Immunol 32:969–975

Zhu L-Y, Nie L, Zhu G, Xiang L-X, Shao J-Z (2013) Advances in research of fish immune-relevant genes: a comparative overview of innate and adaptive immunity in teleosts. Dev Comp Immunol 39:39–62

Chapter 31
Percid Fish Health and Disease

H.D. Rodger and N.B.D. Phelps

Abstract The farming of perch (*Perca fluviatilis* and *Perca flavescens*) is developing in both Europe and North America and the main infectious diseases associated with these species include viral diseases, such as the perch fry rhabdovirus (PFRv), bacterial diseases such as *Flavobacterium psychrophilum* and *Aeromonas sobria*, protozoan parasites including *Ichthyobodo necator* and myxosporideans such as *Myxobolus neurophilus*. Non-infectious conditions such as tail erosion, gill disease and skeletal deformities also can give rise to significant livestock challenges and welfare problems.

Keywords *Perca fluviatilis* • *Perca flavescens* • Rhabdovirus • Bacterial disease • Parasites

31.1 Introduction

The farming of perch (*Perca fluviatilis* and *Perca flavescens*) is developing in both Europe and North America and as the aquaculture of these species develops so too are some of their health challenges including infectious (viral, bacterial, parasitic and fungal) and non-infectious (genetic, congenital, environmental, nutritional) diseases (Kestemont and Mélard 2000; Rodger and Girons 2008). Some disease conditions have been associated only with wild perch, such as epizootic haematopoietic necrosis (EHN) (Whittington et al. 1999), however, it is to be expected that as aquaculture of these fish increases, some of these disease conditions from wild fish will be observed in the farmed sector. Those diseases reported, or observed by the authors, in farmed perch, and of note in wild perch, are considered in this chapter and treatments or strategies for control discussed. It should be noted that in most

H.D. Rodger (✉)
Vet-Aqua International, Oranmore, Co. Galway, Ireland
e-mail: hamishrodger@eircom.net
http://www.vetaquainter.com

N.B.D. Phelps
College of Veterinary Medicine, University of Minnesota, 55108 St. Paul, MN, USA

P. Kestemont et al. (eds.), *Biology and Culture of Percid Fishes*,
DOI 10.1007/978-94-017-7227-3_31

countries where perch aquaculture is underway there are few, if any, authorised treatments or vaccines for perch. In this communication where perch are mentioned this means *P. fluviatilis* or the Eurasian perch, unless stated otherwise.

31.2 Viral Diseases

Epizootic haematopoietic necrosis (EHN) virus causes mortalities in wild perch in Australia and has also been encountered in trout farms; however, this *Ranavirus* (in family *Iridoviridae*) has never been reported in Europe, although closely related viruses affect catfish and sheatfish in France and Germany. Clinical signs in perch include lethargy, sometimes with spiral swimming and hundreds of fish headstanding on the bottom of affected waterways (Langdon and Humphrey 1987). Studies have shown that other European species such as black bullhead (*Amerius melas*) and pike (*Esox lucius*) are susceptible to EHNV by bath exposure (OIE 2013). No vaccine is available and as with other viral diseases in fish there are no medical treatments.

Viral haemorrhagic septicaemia (VHS) caused by the rhabdovirus, viral hemorrhagic septicaemia virus, has been reported around the world in more than 80 marine and freshwater fish species, including yellow perch (Kane-Sutton et al. 2010; Phelps 2013). As the name describes, haemorrhagic lesions of the eye, skin, muscle, and internal organs characterize the disease (Fig. 31.1). Chronically infected fish may appear lethargic with abnormal swimming behaviour (Kim and Faisal 2011). However, these lesions and signs should be interpreted with caution as they are not consistent across host species or season (Kane-Sutton et al. 2010; Kim and Faisal 2010) and may be induced by other pathogens (i.e. Perch fry rhabdovirus). The virus proliferates to cause disease in cool water, often associated with host stress,

Fig. 31.1 Juvenile yellow perch affected by VHSV with characteristic haemorrhagic lesions of the skin, eyes, and base of fins (Photo courtesy of Evi Emmenegger, Western Fisheries Research Center)

such as spawning (Kane-Sutton et al. 2010). To date, VHSV not been introduced into a perch aquaculture facility and remains a significant disease of wild perch in the United States. However, given that no effective treatment currently exists for VHS, biosecurity and broodfish selection should be closely monitored.

Perch fry rhabdovirus (PFRv) has been recorded in wild and farmed perch throughout Northern Europe (Dannevig et al. 2001; Olesen et al. 2005; Henshilwood et al. 2009) and can cause high mortalities in juvenile perch. Clinical signs of the disease in farmed perch include abnormal swimming, congestion at fin bases, ascites and scale protusion, petechiae on the swim bladder and in the visceral fat (Figs. 31.2 and 31.3). Histopathology associated with this virus includes multifocal necrosis in the haematopoietic tissue in the kidney and spleen, scattered necrotic hepatocytes, necrosis in the lamina propria in the intestine, endocardial cell proliferation and congestion in the meninges (Fig. 31.4). Studies of the genetic diversity of the virus have revealed that isolates cluster together with the European lake trout rhabdovirus (Ruane et al. 2014). Careful disinfection of egg strings with buffered iodophor helps prevent apparent vertical transmission of this virus, especially if using any wild broodstock. Affected hatcheries have been able to minimise the impact of the virus by raising water temperature. When the water temperature is raised to above 17 °C the impact, in terms of mortality, is lessened and in addition the use of low levels of peroxy-type disinfectants (Virkon Aquatic®, Du Pont) (2 ppm for 3–4 days) also assists in reducing the impact from the virus in perch fry. Survivors from an outbreak should be considered as carriers of the virus and following stressful periods such fish may manifest further mortalities associated with the virus and be a source of infection to naïve fish.

Fig. 31.2 Juvenile perch affected by PFRv exhibiting congestion at the bases of the pectoral fins (*black arrow*) as well as some scale protrusion (*white arrow*)

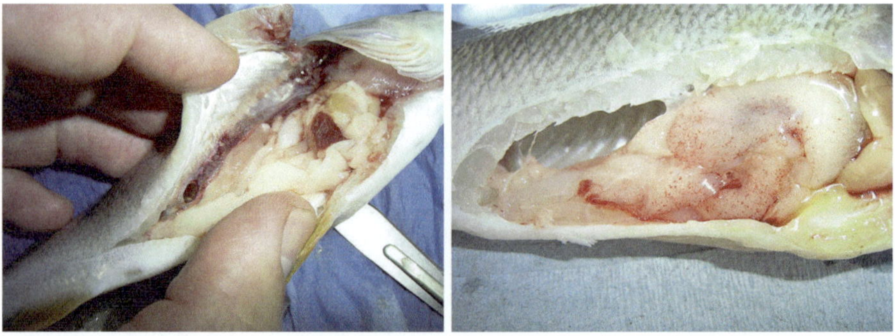

Fig. 31.3 Perch affected by PFRV presenting with blood-tinged ascites and petechiae on the swim bladder (*left*) and numerous petechiae in the visceral fat (*right*)

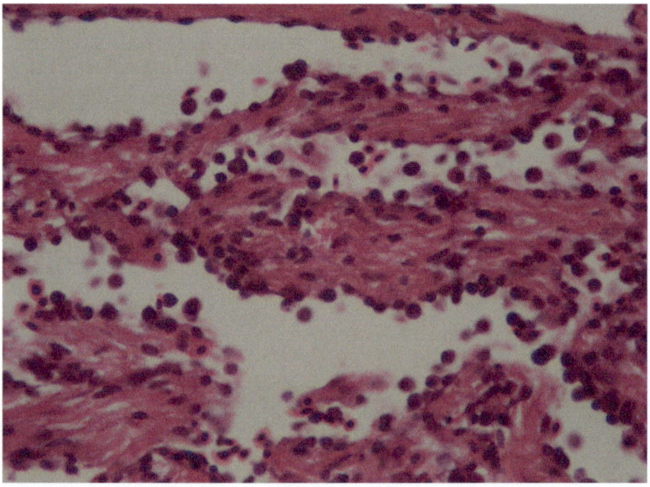

Fig. 31.4 Photomicrograph of the atrium of farmed perch showing endocardial cell proliferation due to PFRV (H & E × 200)

Lymphocystis disease, is caused by the lymphocystic disease virus (LCDV), and has been described in wild yellow perch (*P. flavescens*) in North America (van den Heuvel et al. 2000; Palmer et al. 2012). In some of these cases it caused skin and fish pathology typical of lymphocystis with white nodular lesions (Fig. 31.5), while in others there were no external signs. Asymptomatic fish have been shown to harbour an estimated viral copy number of four magnitudes lower than clinically ill fish (Palmer et al. 2012). The LCDV is a member of the genus *Lymphocystivirus* and the family *Iridoviridae* and prevalence of up to 12 % affected fish have been reported and associated with exposure to oil sands mining-associated waters in Canada (van den Heuvel et al. 2000). LCDV is not known to be a lethal disease in fish, however the unsightly lesions could affect production value in aquaculture systems.

Fig. 31.5 Lymphocystis on the body and dorsal fin of wild *Perca flavescens* demonstrating the typical white multi-nodular nature of the lesion (Photo courtesy of Dr. Laura Jane Phalen)

Walleye dermal sarcoma virus (WDSV) is characterized by raised masses on the skin (fibrosarcoma) in walleye and yellow perch (Bowser et al. 2001). Presumptive diagnosis is made based on the presence of fibrosarcoma, however confirmation of the virus remains challenging. Presumed horizontal transmission from walleye to yellow perch was observed in an aquarium, however, the aetiological agent of the tumour was not confirmed (Bowser et al. 2005). Reports of this virus remain rare in wild fish, perhaps due to the limited contact between individuals. Thus, WDSV could pose a greater threat to farmed perch held in crowded confinement, increasing the frequency of contact transmission.

31.3 Bacterial Diseases

Bacteria are a natural component to all aquatic environments, coexisting in a delicate balance with the fish. Diseases associated with bacteria infections are often associated with sudden or chronic stress factors such as poor water quality, spawning, overcrowding, etc. *Aeromonas* sp., *A. veronii*, *A. hydrophila*, *Streptococcus* sp., *Staphylococcus* sp., *Vibrio fluviatilis* and *Enterobacter agglomerans*, to name a few, have been associated with mortality in farmed perch by Grignard et al. (1996). Thus, it is imperative stress in perch aquaculture in minimized to reduce disease occurrence. Clinical lesions of bacteria infections are generally characterized by erosion or necrosis of gills, skin, and fins. In advanced infections, sepsis is observed with loss of organ function. Effective treatment is dependent on rapid and accurate diagnostic identification and antibiotic sensitivity.

Flavobacterium psychrophilum, the causal agent of rainbow trout fry syndrome (RTFS) and cold water disease in salmonids, is also encountered in perch farming where it can give rise to elevated mortalities and can present in juvenile fish as a dissolving or necrotic jaw syndrome (mouth rot) (Fig. 31.6). It can also present as fin rot or in some cases as a systemic disease (Lönnström et al. 2008). Improved environmental conditions and treatment with oral broad spectrum antibiotics are usually successful in controlling this condition, if administered when fish are still feeding. The bacterium has also been detected in otherwise healthy wild perch (Madetoja et al. 2002). Low water temperatures (7–8 °C) have been associated with this disease in farmed perch by some authors (Lönnström et al. 2008) and they recommend that culture conditions are maintained close to 20 °C for this reason. Autogenous and experimental vaccines for salmonids have been developed using *F. psychrophilum* isolated from salmon and trout and it is envisioned that perch vaccines will be developed for this pathogen using perch isolates.

Flavobacterium columnare, the causative agent of columnaris disease, is often associated with warm-water mortality events in wild and farmed fish, including Eurasian and yellow perch (Thomas-Jinu and Goodwin 2004; Morley and Lewis 2010). Columnaris clinically presents with gill necrosis, ulcerative lesions and fin erosion. In advanced infections erosion may extend into the caudal peduncle. Transmission of the bacteria occurs via contact with contaminated water, resulting in rapid dissemination throughout a farm lacking strict biosecurity.

Aeromonas species are ubiquitous in the aquatic environment and are the most frequently reported bacteria isolated from sick perch. Often, *Aeromonas* cases present as a mixed infection with ectoparasites such as *Heterpolaria* sp. or *Ichthyobodo necator*. *Aeromonas hydrophila*, which was implicated in the dramatic perch kills in Lake Windermere, England in the 1970s (Bucke et al. 1979) and *A. salmonicida*,

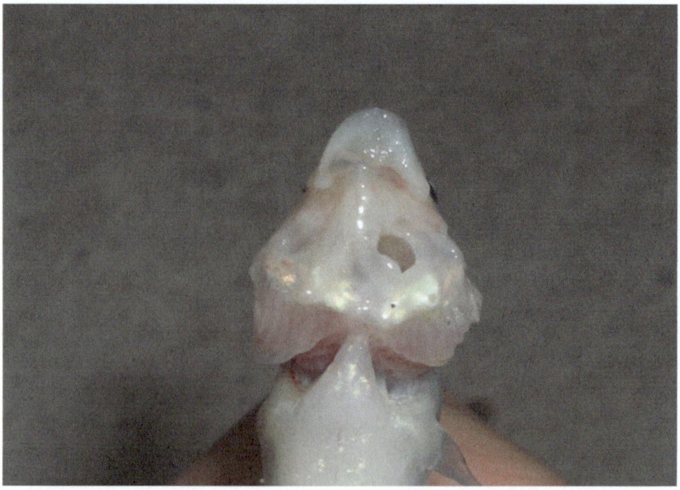

Fig. 31.6 Juvenile perch infected with *Flavobacterium psychrophilum* exhibiting necrosis of the mandible and perforation of the adjacent epithelial and dermal layers

which is more commonly associated with furunculosis of salmonids (McCarthy 1975) have been reported in farmed and wild Eurasian and yellow perch. Caudal fin necrosis is frequently observed in juvenile perch, often following repeated grading or moving of livestock, and *A. hydrophila*, *Saprolegnia* sp. and *I. necator* can often be involved. Another mixed infection of *A. hydrophila* and the parasite *Heteropolaria* sp. has been reported in farmed perch presenting as red sore disease where haemorrhagic marks on the flanks appear (Grignard et al. 1996). *A. sobria* has been reported as a significant pathogen for net pen farmed perch during winter months in Switzerland where it resulted in mortalities of 1 % total farm stock per day at the peak of the outbreaks (Wahli et al. 2005). Diseased perch were described with skin ulcers and fin rot. Burr et al. (2012) investigated the *Aeromonas* sp. infection in farmed perch further and found that the bacterial populations of wild and farmed fish were very heterogenous and that no one strain could be associated with the ulcerative disease in farmed fish. In fact, the strains found in the intestines of healthy wild fish were closely related to those in the farmed fish. The authors concluded that the presence of *Aeromonas* alone is not the sole cause of disease but, was rather a contributing factor. They recommended that improved husbandry conditions such as lower stocking densities, minimal handling or movements would help protect perch against acute infection.

Farmed perch are also susceptible to *Pseudomonas fluorescens*, which has caused elevated mortalities as a systemic disease in recirculation hatcheries (Rodger, unpublished). The infection can present with severe exophthalmos which is due to a massive retrobulbar inflammatory response (Fig. 31.7). Treatment with broad spectrum antibiotics (florfenicol, 10 mg/kg/day for 10 days) following antibiotic sensitivity testing proved effective in treating and controlling these outbreaks.

Fig. 31.7 Severe exophthalmia with panophthalmitis in juvenile farmed perch affected by systemic infection with *Pseudomonas fluorescens*

However, periods of adverse water quality due to high levels of gas supersaturation may have predisposed the fish to bacteria in these cases and again control of environmental conditions should minimise any risk of outbreaks.

Wild perch in Finland have been reported to carry *Yersinia ruckeri*, the causal agent of enteric redmouth disease, however, no clinical disease with this bacteria has been reported in farmed fish (Tellervo Valtonen et al. 1992). Epitheliocystis, caused by a chlamydia-like organism, has been observed in juvenile farmed perch in Ireland and also in farmed *P. flavescens* in Pennsylvania, USA (Fig. 31.8) and in both cases was associated with gill hyperplasia, respiratory distress and elevated mortalities (Rodger and Girons 2008). Due to the intracellular nature of this organism antibiotic treatment can be of limited benefit, however, oral oxytetracycline at 75 mg/kg body weight/day for 10 days has been used with success in some outbreaks. It should be noted that antibiotic therapy as described may not be approved or permitted in the reader's country.

In comparison to some other farmed fish species, perch appear vulnerable to skin and tail damage and if the mucus layer is removed, they are then very susceptible to secondary bacterial (and fungal) infections (Fig. 31.9). In the farming of perch there is a requirement for frequent grading to prevent cannibalism and the physical stresses this imposes on the fish mean that great care must be taken during these procedures.

Oral broad spectrum antibiotics have been used to treat various systemic bacterial infections in perch and have proven effective as follows: ampicillin (80 mg/kg), oxolinic acid (20 mg/kg), florfenicol (10 mg/kg) and oxytetracycline (75 mg/kg) for 8–10 days. It should be noted that the treatments listed may not be approved or permitted in all countries.

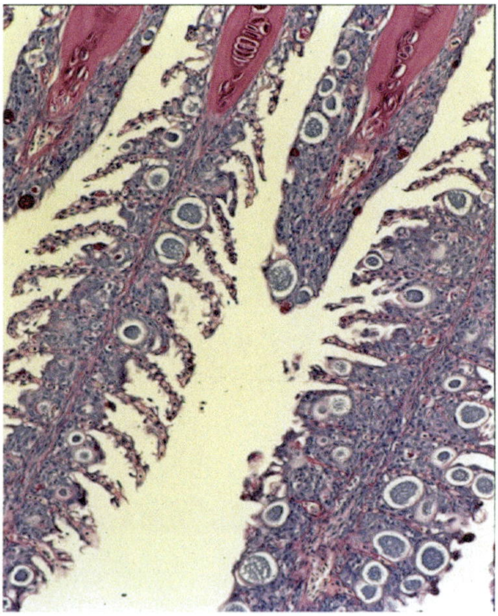

Fig. 31.8 Histopathological section of gills from *Perca flavescens* affected by epitheliocystis. Chlamydia-like organisms present as blue colonies in the gill epithelia, and there is associated hyperplasia (H & E × 400)

Fig. 31.9 Farmed *Perca fluviatilis* fry presenting with tail rot and the pathogens *Aeromonas hydrophila*, *Saprolegia* sp. and *Ichthyobodo necator* were all involved

31.4 Parasitic Diseases

Many parasites have been recorded in wild percids (Craig 2000), and some such as *Ligula intestinalis* have been associated with poor growth (Pitt and Grundman 1957), however, in the majority of reports these parasites have not been associated with disease. In this parasitic section the focus is on farmed percids and the health conditions and disease affecting them. Many of the common protozoan ectoparasitic diseases of fish are seen on juvenile farmed perch skin and gills and these include *Ichthyobodo necator* (costia), which can cause high mortalities in perch fry as well as *Trichodina* spp. (Fig. 31.10) and *Icthyophthirius multifilius* (white spot) (Grignard et al. 1996). These ectoparasites can be controlled through the use of formalin baths or flushes and/or sodium chloride, although perch fry appear more vulnerable to formalin than salmonids and the dosages used may need to be 0.3–0.5 times the standard salmonid dose levels.

Endoparasitic protozoans recorded in farmed perch include the myxosporidean, *Myxobolus neurophilus* which has been a particular problem for some yellow perch (*P. flavescens*) farms in North America (Khoo et al. 2010). This parasite invades the central nervous tissue and can cause nervous signs and high mortalities (Acland et al. 2000). Post mortem evaluation of affected fish may reveal small white nodules (1–2 mm) on the surface of the brain and Giemsa stained impression smears or histopathology sections will reveal the characteristic spores (Fig. 31.11). In the absence of any treatment for this parasite, high levels of biosecurity and parasite

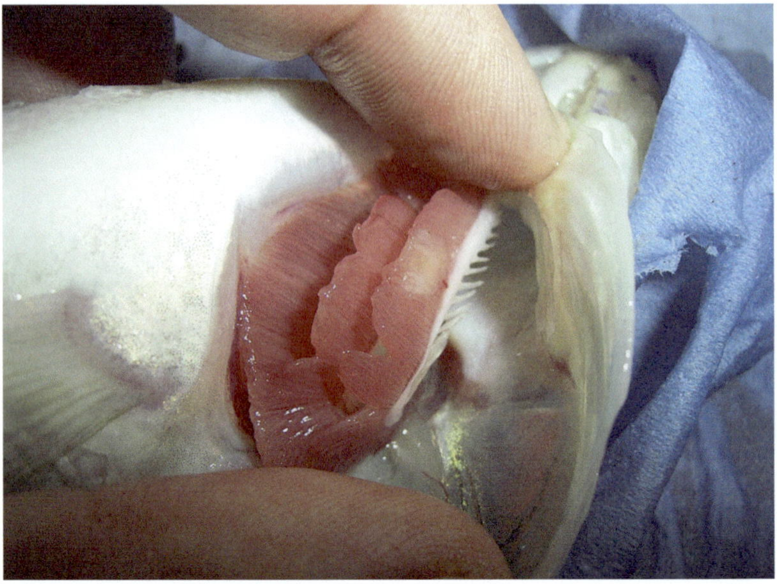

Fig. 31.10 Focal gill pathology in farmed perch appearing as a pale thickened area (hyperplasia and lamellar fusion) associated with high levels of *Trichodina* sp. protozoans

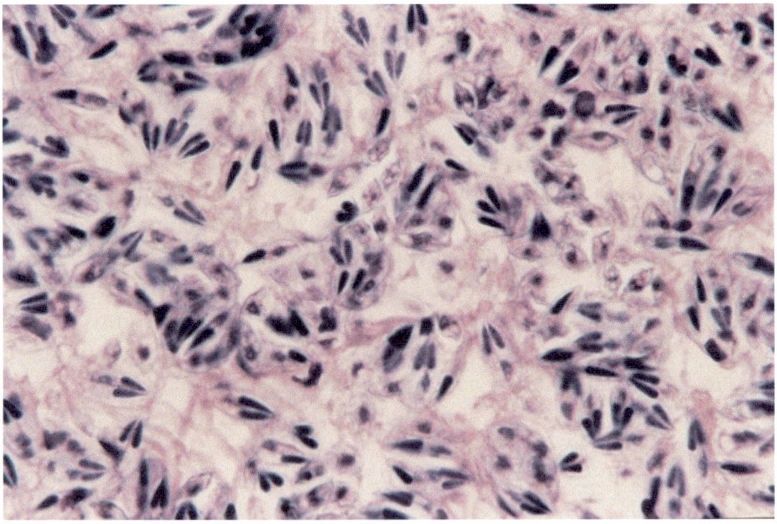

Fig. 31.11 Histopathological section of brain from *P. flavescens* displaying numerous *Myxobolus neurophilus* spores (Giemsa × 1000)

screening of any incoming livestock will reduce the risk of infection entering a farm. In wild perch there are some other myxosporidean species that have been associated with health problems and these include *Henneguya creplini*, reported to cause gill pathology in Finland (Haaparanta et al. 1994), *M. sandrae* associated with skeletal deformities in Scotland (Lom et al. 1991) and *Triangula percae* causing spinal curvatures in Australia (Langdon 1987). Coccidia such as *Eimeria* spp. have been observed in the intestines of farmed and wild perch, however, their clinical significance remains to be determined.

Heterosporosis, caused by the microsporidian, *Heterosporis sutherlandae*, is an emerging disease in the Great Lakes region of the United States and is known to infect yellow perch (Phelps et al. in review). This obligate intracellular parasite infects the muscle tissue to infected hosts, resulting in severe muscle necrosis and liquefaction (Fig. 31.12). Infected fish likely suffer from reduced performance, indirect mortality, and lowered production value with the loss of fillet quality. To date, Heterosporosis has not been reported from a fish farm, however, use of surface water and wild broodstock are a concern for introduction.

A scale-bearing amoeba (*Cochliopodium minus*) has been isolated from gills and internal organs of wild perch in the Czech Republic (Dyková et al. 1998), however, no pathology nor clinical disease was reported and it is not clear if this amoeba is pathogenic for perch or otherwise.

Although the monogeneans *Gyrodactylus* sp. and *Dactylogyrus* sp. have been observed in farmed perch, there have been no reports of these species causing serious impacts on farmed perch. The digenean, *Diplostomum spathaceum*, which causes severe cataracts in farmed trout, has been observed in perch eyes (vitreous humour and lens), however, it has not been reported as a clinical problem for farmed perch. Another digenean, *Neascus sp.*, causing black spot disease is not known to cause harm to the fish (Vaughan and Coble 1975), but would decrease the value of

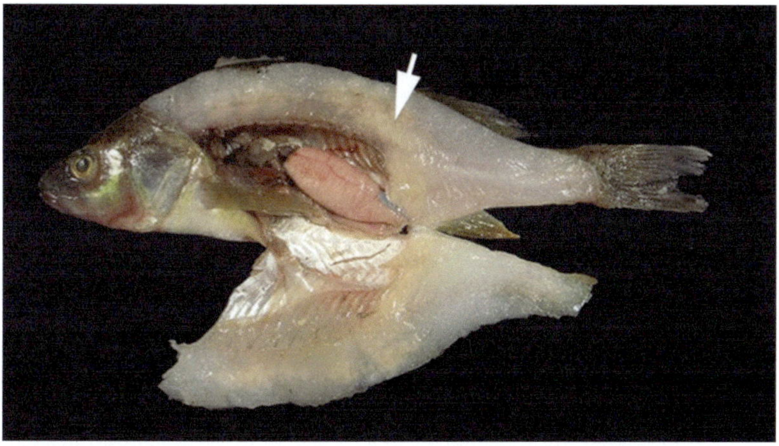

Fig. 31.12 *Perca flavescens* infected with *Heterosporis sutherlandae*, an intracellular microsporidian parasite, presenting with focal muscle necrosis and liquefaction (*arrow*)

infected filets. Crustacean parasites such as *Argulus* sp., *Ergasilus* sp. and *Achtheres* sp. (Kempter et al. 2006) have also been observed on wild fish but no clinical outbreaks in farmed fish have been recorded. The tapeworm, *Triaenophorus crassus*, has been reported in farmed perch where it presented as white nodules in the liver (Wahli et al. 2005). There are many other parasites recorded in wild perch and these are comprehensively listed with references by Kestemont and Mélard (2000).

31.5 Fungal Diseases

Saprolegniosis, caused by the ubiquitous fungus *Saprolegnia parasitica* is frequently encountered in farmed perch, usually following some physical challenge to the fish or recent movement of livestock and can affect all ages of fish as well as eggs (Grignard et al. 1996). Fungal hyphae rapidly develop in gills or skin and are often seen with concomitant ectoparasitic infections such as *Ichthyobodo* sp. (costia) or *Trichodina* sp. Repeated formalin or salt baths with frequent removal of any affected fish or mortalities has proven effective in controlling this disease.

31.6 Non-infectious Diseases

In perch fry production there are many non-infectious disease hurdles to successful aquaculture and these include failure to inflate the swim bladder, overinflation of the swim bladder and bacterial proliferation in the live feed which can give rise to bacterial enteritis and toxaemia. Use of protein skimmers, careful control of water gas saturation levels and cleaning and high levels of hygiene in live feed production can all reduce the risk of these problems appearing, however, constant vigilance is required at the crucial early life stages. Fin erosion has been recorded in wild perch and has been associated with effluent from pulp mills (Lindesjöö and Thulin 1990), however, fin erosion is also observed in farmed perch but is usually considered secondary to physical challenges which then allows in opportunistic bacterial and fungal infections. Jaw, mandibular and skeletal deformities have been observed at high prevalence (30 % of some batches) in some perch hatcheries (Fig. 31.13). Similar jaw and skeletal deformities in salmonids have been associated with nutritional deficiencies or unavailability of essential nutrients such as phosphorus (Baeverfjord et al. 1998; Hardy 2001), however, the exact aetiology of the condition in perch remains to be established. Skeletal deformities have been reported in wild perch in Scotland associated with *Myxobolus* sp. infestation (Treasurer 1992), however, in the cases examined in farmed perch there were no signs of parasitic involvement.

Recirculating aquaculture system(s) (RAS) are being increasingly used for finfish aquaculture and perch are also being reared in these systems. RAS have benefits

Fig. 31.13 Jaw deformity (*left*) and vertebral column skeletal deformity (*right*) in farmed *P. fluviatilis* which was present at a prevalence of 30 % in the affected stock

for fish health through reduced risk from pathogen entry to the farm, however, fluctuations in water quality parameters such as the accumulation of carbon dioxide and resultant alterations in pH can be an alternate challenge and parameters such as dissolved oxygen, carbon dioxide, temperature, ammonia, pH, nitrite and nitrate should be monitored continuously. Periods of high suspended solids may also lead to gill pathology and through the pathology and stress response associated with grading, handling or moving perch could give rise to increased susceptibility to pathogens (Acerete et al. 2004; Jentoft et al. 2002).

31.7 Conclusions

At the time of writing there are no authorised medicines or vaccines for the treatment or prevention of infectious disease in farmed perch destined for food. Under the veterinary cascade mechanism there is a very limited availability of medicines licensed for other food fish species, and in most countries this amounts to less than a handful of products. Non-infectious conditions or production diseases can emerge in perch farming and although some may be related to early life stage environmental conditions, others may develop as a result of nutritional imbalances and adverse environmental conditions with increasing intensification. As perch farming moves into more intensive systems with increased environmental control, such as fully enclosed recirculation systems, it is predicted that novel disease and health conditions will emerge, many which will be due to the husbandry and rearing conditions of the farms. The emphasis in all farming is towards prevention and management of the health challenges rather than "silver bullet" remedies and perch farming is no exception. Careful biosecurity, egg disinfection, improved diagnostics and screening of stocks prior to movement will reduce the risks of pathogen spread and allow this aquaculture sector to evolve in a sustainable and productive manner.

References

Acerete L, Balasch JC, Espinosa E, Josa A, Tort L (2004) Physiological responses in Eurasian perch (*Perca fluviatilis*, L.) subjected to stress by transport and handling. Aquaculture 237:167–178

Acland H, Rodger H, Rommel F (2000) Brain parasite in cultured yellow perch. Abstracts 25th Annual Eastern Fish Health Workshop, USGS, Kearneysville

Baeverfjord G, Asgård T, Shearer KD (1998) Development and detection of phosphorus deficiency in Atlantic salmon, *Salmo salar* L., parr and post-smolts. Aquacult Nutr 4:1–11

Bowser PR, Wooster GA, Getchell RG, Paul TA, Casey RN, Casey JW (2001) Experimental transmission of walleye dermal sarcoma to yellow perch. J Aquat Anim Health 13:214–219

Bowser PR, Abou-Madi N, Garner MM, Bartlett SL, Grimmett SG, Wooster GA, Paul TA, Casey RN, Casey JW (2005) Fibrosarcoma in yellow perch, *Perca flavescens* (Mitchell). J Fish Dis 28:301–305

Bucke D, Cawley GD, Craig JF, Pickering AD, Willoughby LG (1979) Further studies of an epizootic of perch, *Perca fluviatilis* L., of uncertain aetiology. J Fish Biol 2:297–311

Burr SE, Goldschmidt-Clermont E, Kuhnert P, Frey J (2012) Heterogeneity of *Aeromonas* populations in wild and farmed perch, *Perca fluviatilis* L. J Fish Dis 35:607–613

Craig JF (2000) Percid fishes: systematics, ecology & exploitation. Blackwell Science Ltd., Oxford

Dannevig BH, Olesen NJ, Jentoft S, Kvellestad A, Taksdal T, Håstein T (2001) The first isolation of a rhabdovirus from perch (*Perca fluviatilis*) in Norway. Bull Eur Assoc Fish Pathol 21:145–153

Dyková I, Lom J, Machácková B (1998) *Cochliopodium minus*, a scale-bearing amoeba isolated from organs of perch *Perca fluviatilis*. Dis Aquat Org 34:205–210

Grignard J-C, Mélard C, Kestemont P (1996) A preliminary study of parasites and diseases of perch in an intensive culture system. J Appl Ichthyol 12:195–199

Haaparanta A, Tellervo Valtonen E, Hoffmann RW (1994) Pathogenicity and seasonal occurrence of *Henneguya creplini* (Protozoa, Myxosporea) on the gills of perch *Perca fluviatilis* in central Finland. Dis Aquat Org 20:15–22

Hardy RW (2001) The nutritional pathology of teleosts. In: Roberts RJ (ed) Fish pathology, 3rd edn. W B Saunders, London

Henshilwood K, Ruane NM, McCarthy LJ, Hickey C, Rodger HD, Kerr R, Dodge M, Stone DM (2009) The emergence of rhabdovirus infections in farmed perch *Perca fluviatilis* in Ireland. In: Book of abstracts, 14th European association of fish pathologists conference, Prague, 14–19 Sept, P-021

Jentoft S, Held JA, Malison JA, Barry TP (2002) Ontogeny of the cortisol stress response in yellow perch (*Perca flavescens*). Fish Physiol Biochem 26:371–378

Kane-Sutton M, Kinter B, Dennis PM, Koonce JF (2010) Viral hemorrhagic septicemia virus infection in yellow perch, *Perca falvescens*, in Lake Erie. J Great Lakes Res 36:37–43

Kempter J, Piasecki W, Więski K, Krawczyk B (2006) Systematic position of copepods of the genus *Achtheres* (Crustacea: Copepoda: Siphonostomatoida) parasitizing perch, *Perca fluviatilis* L., and zander, *Sander lucioperca* (L.). J Fish Dis 29:103–113

Kestemont P, Mélard C (2000) Aquaculture. In: Craig JF (ed) Percid fishes: systematics, ecology & exploitation. Blackwell Science Ltd, Oxford

Khoo L, Rommel FA, Smith SA, Griffin MJ, Pote LM (2010) *Myxobolus neurophilus*: morphologic histopathologic and molecular characterisation. Dis Aquat Org 89:51–61

Kim R, Faisal M (2010) Experimental studies confirm the wide host range of the Great Lakes viral hemorrhagic septicemia virus genotype IVb. J Fish Dis 33:83–88

Kim R, Faisal M (2011) Emergence and resurgence of the viral hemorrhagic septicemia virus (*Novirhabdovirus*, *Rhabdoviridae*, *Monongavirales*). J Adv Res 2:9–23

Langdon JS (1987) Spinal curvatures and an encephalotropic myxosporean, *Triangula percae* sp. nov. (Myxozoa: Ortholineidae), enzootic in redfin perch, *Perca fluviatilis* L., in Australia. J Fish Dis 10:425–434

Langdon JS, Humphrey JD (1987) Epizootic haematopoietic necrosis, a new viral disease in redfin perch, *Perca fluviatilis* L., in Australia. J Fish Dis 10:289–297
Lindesjöö E, Thulin J (1990) Fin erosion of perch *Perca fluviatilis* and ruffe *Gymnocephalus cernua* in a pulp mill effluent area. Dis Aquat Org 8:119–126
Lom J, Pike AW, Dyková I (1991) *Myxobolus sandrae* Reuss, 1906, the agent of vertebral column deformities of perch *Perca fluviatilis* in northeast Scotland. Dis Aquat Org 12:49–53
Lönnström L-G, Hoffrén ML, Wiklund T (2008) *Flavobacterium psychrophilum* associated with mortality of farmed perch, *Perca fluviatilis* L. J Fish Dis 31:793–797
Madetoja J, Dalsgaard I, Wiklund T (2002) Occurrence of *Flavobacterium psychrophilum* in fish-farming environments. Dis Aquat Org 52:109–118
McCarthy DM (1975) Fish furunculosis caused by *Aeromonas salmonicida* var. *achromogenes*. J Wildl Dis 12:489–493
Morley NJ, Lewis JW (2010) Consequences of an outbreak of columnaris disease (*Flavobacterium columnare*) to the helminth fauna of perch (*Perca fluviatilis*) in the Queen Mary reservoir, south-east England. J Helminthol 84:186–192
OIE (2013) Chapter 2.3.1 Epizootic haematopoietic necrosis. In: Manual of diagnostic tests for aquatic animals 2013. OIE, Paris
Olesen NJ, Skall HF, Johansson T (2005) Characteristics of European perch rhabdoviruses. In: Book of abstracts, European association of fish pathologists conference, Copenhagen, 0-143
Palmer LJ, Hogan NS, van den Heuvel MR (2012) Phylogenetic analysis and molecular methods for the detection of lymphocystis disease virus from yellow perch, *Perca flavescens* (Mitchell). J Fish Dis 35:661–670
Phelps NBD (2013) Improved diagnosis and management of viral hemorrhagic septicaemia virus in fish. PhD dissertation, University of Minnesota, St. Paul
Pitt CE, Grundman AW (1957) A study into the effects of parasitism on the growth of yellow perch produced by the larvae of *Ligula intestinalis* (Linnaeus, 1758) Gmelin, 1790. Proc Helminthol Soc Wash 24:73–80
Rodger HD, Girons A (2008) Diseases and health management in perch farming. Fish Vet J 10:9–18
Ruane NM, Rodger HD, McCarthy LJ, Swords D, Dodge M, Kerr RC, Henshilwood K, Stone DM (2014) Genetic diversity and associated pathology of vesiculovirus infections in farmed and wild perch *Perca fluviatilis* in Ireland. Dis Aquat Org 112:121–130
Tellervo Valtonen E, Rintamäki P, Koskivaara M (1992) Occurrence and pathogenicity of *Yersinia ruckeri* at fish farms in northern and central Finland. J Fish Dis 15:163–171
Thomas-Jinu S, Goodwin AE (2004) Morphological and genetic characteristics of *Flavobacterium columnare* isolates: correlations with virulence in fish. J Fish Dis 27:29–35
Treasurer J (1992) Vertebral anomalies associated with *Myxobolus* sp. in perch *Perca fluviatilis* L., in a Scottish loch. Bull Eur Assoc Fish Pathol 12:61–63
van den Heuvel MR, Power M, Richards J, MacKinnon M, Dixon DG (2000) Disease and gill lesions in yellow perch (*Perca flavescens*) exposed to oil sands mining-associated waters. Ecotoxicol Environ Saf 46:334–341
Vaughan GE, Coble DW (1975) Sublethal effects of three ectoparasites on fish. J Fish Biol 7:282–294
Wahli T, Burr SE, Pugovkin D, Mueller O, Frey J (2005) *Aeromonas sobria*, a causative agent of disease in farmed perch, *Perca fluviatilis* L. J Fish Dis 28:141–150
Whittington RJ, Reddacliff LA, Marsh I, Kearns C, Zupanovic Z, Callinan RB (1999) Further observations on the epidemiology and spread of epizootic haematopoietic necrosis virus (EHNV) in farmed rainbow trout *Oncorhynchus mykiss* in southeastern Australia and a recommended sampling strategy for surveillance. Dis Aquat Org 35:125–130

Part VIII
Commercial Production, Quality, Marketing and Economics

Chapter 32
Current Status of Eurasian Percid Fishes Aquaculture

Svend Steenfeldt, Pascal Fontaine, Julia Lynne Overton, Tomáš Policar, Damien Toner, Bahram Falahatkar, Ákos Horváth, Ines Ben Khemis, Neila Hamza, and Mohammed Mhetli

Abstract This chapter presents the actual status and the perspectives of development for Eurasian perch and pikeperch in different countries, mainly in Europe (Denmark, Finland, Sweden, Ireland, The Netherlands, France, Czech Republic and Hungary) but also in Iran and Tunisia. For each country, main culture techniques are summarized and production types are specified, according to the local or international markets.

Keywords Eurasian perch • Pikeperch • Production • Culture techniques • Market

32.1 Introduction

World percid culture is limited in comparison with more prevalent species e.g. salmonids, bass and bream. In 2009, only Denmark, The Netherlands, Tunisia and Ukraine recorded an annual aquaculture production of pikeperch exceeding 100 tons

---

S. Steenfeldt (✉)
BioMar A/S, The North Sea Science Park, Willemoesvej 2, 9850 Hirtshals, Denmark
e-mail: sjs@biomar.dk

P. Fontaine
Unité de Recherche Animal et Fonctionnalités des Produits Animaux, Université de Lorraine, USC INRA 340, Vandoeuvre-lès-Nancy F-54505, France
e-mail: p.fontaine@univ-lorraine.fr

J.L. Overton
Aquapri innovation, Lergaardvej 2, 6040 Egtved, Denmark
e-mail: julia.overton@aquapri.dk

T. Policar
Faculty of Fisheries and Protection of Waters, Institute of Fish Culture and Hydrobiology, South Bohemian Research Center of Aquaculture and Biodiversity of Hydrocenoses, University of South Bohemia in České Budějovice,
Zátiší 728/II., 389 25 Vodňany, Czech Republic
e-mail: kristj01@frov.jcu.cz

P. Kestemont et al. (eds.), *Biology and Culture of Percid Fishes*,
DOI 10.1007/978-94-017-7227-3_32

Table 32.1 Pikeperch production in European countries from 2002 to 2011 (Tons) (FAO 2012)

	2002	2003	2004	2005	2006	2007	2008	2009	2010	2011
Croatia	8	7	7	8	13	16	7	6	6	6
Czech Republic	42	55	48	50	50	48	58	58	48	55
Denmark	0	6	10	49	36	47	55	106	105	105
Hungary	81	80	80	80	80	32	32	40	38	40
Netherlands	100	100	100	100	100	100	100	115	100	120
Total	231	248	245	287	279	243	252	325	297	329

Table 32.2 European perch production in European countries from 2002 to 2011 (tons) (FEAP 2011; FAO 2012)

Year	2002	2003	2004	2005	2006	2007	2008	2009	2010	2011
FAO	72	138	152	318	219	311	265	237	280	182
FEAP	24	22	14	75	90	17	26	45	45	NA

NA not available

(FAO 2011). Total pikeperch production in aquaculture in 2009 equaled 653 tons and is dwarfed by the natural fishery of 14,739 tons. EU production of pikeperch was in 2011 dominated by The Netherlands and Denmark (Table 32.1).

Culture of Eurasian perch is of the same magnitude with a maximum of 318 tons being produced in 2005 according to FAO (Table 32.2). FEAP though estimates the production of perch as lower, probably as a consequence of differing differentiation between extensive production and captive fisheries.

32.2 Denmark

Eurasian perch is an endemic species in Danish fresh- and brackish water systems. It is a popular game fish and probably the "first fish caught" by many young Danes.

D. Toner
Aquaculture Initiative EEIG, 14-15 Grays Lane, Park St, Dundalk, Co. Louth, Ireland
e-mail: toner@aquacultureinitiative.eu

B. Falahatkar
Fisheries Department, Faculty of Natural Resources, University of Guilan, Guilan, Iran
e-mail: falahatkar@guilan.ac.ir

Á. Horváth
Department of Aquaculture, Szent István University, Gödöllő, Hungary
e-mail: Horvath.Akos@mkk.szie.hu

I.B. Khemis • M. Mhetli
Institut des Sciences et Technologies de la Mer, Salambo, Tunisia
e-mail: hamza.neila@gmail.com

N. Hamza
Institut National des Sciences et Technologies de la Mer,
28 rue du 2 mars 1934, Salammbô 2025, Tunisie
e-mail: hamza.neila@gmail.com

Pikeperch on the other hand was introduced to Danish waters more than 200 years ago (Dahl 1982). Through repeated stockings with fish from German waters up until the First World War, the species was released to many Danish water systems. Success was limited though, possibly due to lack of knowledge in transport and stocking of fish. From the first to the Second World War multiple introductions of pikeperch fry from Sweden were carried out with better success, and today pikeperch are found in 70 Danish freshwater and brackish water areas. Records show that 51 of these areas are the result of direct stocking and the remaining 15 have been invaded from stocked populations in connected water systems (Dahl 1982).

Eurasian perch and pikeperch fall into the category of "new species" in Danish aquaculture. As the name implies they have not contributed significantly to Danish aquaculture as yet. Diversification has been on the private sector's agenda for decades and in recent years Eurasian perch and especially pikeperch have been identified as the most promising new fish species to Danish aquaculture. This has led to a number of governmental and EU sponsored R&D projects facilitating the development of intensive rearing methods for Eurasian perch and pikeperch in Denmark. Projects focusing on Eurasian perch fry production technology in RAS systems provided the private sector access to fry for pilot scale grow out trials in both existing pond based trout farms and RAS systems (Overton and Paulsen 2005a, b; Paulsen et al. 2005a, b). These trials were well received by the trout producers and were technically successful. They were not able though to convince the private sector of the economic sustainability of Eurasian perch culture and therefore no farmers continued to produce Eurasian perch after the end of the pilot phase. The knowledge on intensive fry production of Eurasian perch was assimilated by one governmental station that produce salmon fry for national salmon restocking programs. Today this producer is rearing Eurasian perch fry as a secondary production. The annual production is around 350,000 fry (fry size 2 g). The fry are produced twice a year with the aim to increase the production to 500,000 fry annually. Presently all fry are exported to Switzerland and Ireland.

Government and EU sponsored R&D projects on pikeperch were carried out as well. One commercial farm was in operation before the trials commenced. This farm was based on a former eel farm and produced pikeperch for ongrowing. It is not in operation at present time. This to some extent supports the fact that pikeperch culture at one point was considered a potential alternative to eel production in facilities, suffering from restrictions on elver fishery. The concept of substituting eel with pikeperch is now considered less likely to be implemented. Requirements of modern RAS systems performance makes modification of existing eel facilities to pikeperch farming resource demanding, at a scale that is outperformed by erection of new species specific facilities.

In 2005 two Danish fish farming companies ventured into commercial scale production of pikeperch. One took on fry and grow-out production of pikeperch, whereas the other focused on smaller scale production primarily for restocking. Both producers based their production solely on RAS systems. RAS systems provide optimal temperature regimes throughout the year. The considerable investment- and running costs of RAS systems is a constraint though to development of the sector.

Fig. 32.1 Pikeperch broodstock holding facility. Private sector Denmark

Leading the development in all production phases from broodstock to ongrowing adds to the burden since systems for broodstock including out of season spawning batches as well as incubators, larval rearing facilities, live feed production systems, weaning and fry holding systems, plus ongrowing systems are needed. All these facilities must be constructed and managed as well as support equipment such as fish sorting machinery, workshops, logistics and product marketing and sale.

A number of consecutive development projects in cooperation with private companies and research institutions (Figs. 32.1, 32.2, and 32.3) led to the development of a pikeperch production for one company of one million fry produced annually and a production of pikeperch consumption size fish of 250 tons (Steenfeldt et al. 2010, 2011).

Despite these considerable challenges, one dominating Danish producer of pikeperch will in 2015 increase its annual production capacity of pikeperch to 500 tons. This considerable expansion will take place in a newly erected RAS facility in Denmark, specially designed and optimized for pikeperch production.

32.3 Finland

Finland's aquaculture sector produced 11,000 tons in 2011. The main production was trout and whitefish amounting to 9900 and 1200 tons respectively.

Fig. 32.2 Upwelling system for incubation of pikeperch eggs. DTU-Aqua, Denmark

Eurasian perch is widespread in Finland's many waters. Low prices seems to have prevented Eurasian perch from being a candidate for culture or export and the recreational fishery for Eurasian perch is not demanding fish for restocking, as is the case for pikeperch.

On the contrary, pikeperch is a highly valued fish species in Finland. Capture fisheries have exceeded 3000 tons with more than half being caught in the coastal waters of the Baltic region. During the last decade catches have decreased to around 2000 tons annually. Pikeperch wild stocks experienced a decline during the 1960- and 1970s and disappeared from many lake systems. A combination of intensive restocking programs and warm summers manifested the value of restocking, and in 2003 more than nine million fry were released to Finland's lakes. Recreational fisheries contribute significantly and represent 70–90 % of the catch with the largest catch in the summer season. The commercial catch is based on gillnets during autumn, winter and spring.

Pikeperch fry is produced in most regions of Finland excluding a few regions in the northern and eastern part. The fry is almost exclusively produced for restocking purposes. Annual production of pikeperch has since 2002 been 8–11.9 million and with a production in 2011 of 8.6 million fry there is no trend towards a change in future production. The value of the pikeperch fry production in 2011 was estimated to 1.6 million euros.

Fig. 32.3 Laboratory scale intensive hatchery for development of larval rearing methodology for pikeperch at DTU-Aqua Denmark

The production methods are based on extensive pond rearing techniques which reduce the production costs to a minimum. Pikeperch thrives in warm water (Hilge and Steffens 1996; Deelder and Willemsen 1964; Frisk et al. 2012). Ongrowing is not carried out as an aquaculture production and to bridge this gap the need for recirculation technology is acknowledged. Lack of experience and high investment costs seems to limit development of this segment of aquaculture in Finland at the present.

32.4 Sweden

Both Eurasian perch and pikeperch are widespread in Sweden. Aquaculture is limited though. A concept of production of Eurasian perch in floating cages in the Swedish archipelago is being tested (Öberg 2008).

Pikeperch on the other hand have been produced in aquaculture for decades in Sweden. Three producers are producing totally approximately one million fry annually. All three producers rely on extensive methodology for production. Most fish are restocked. The majority 90 % are restocked with the purpose of bio-manipulation of lakes. The remaining 100,000 are exported for grow out in RAS systems.

At present the production is estimated to equal the market demand. Only the demand for fry for grow out is assumed to change in the future, but as intensive pikeperch fry production is developed in other countries, it is not envisaged that the present level of export is likely to expand.

32.5 Ireland

In the 1990s a renewed interest in the cultivation of perch emerged. Ireland's aquaculture industry is predominately marine based, focusing on species such as salmon, oysters, and mussels. The development of freshwater aquaculture centered largely on traditional flow through units growing trout (*Oncorhynchus mykiss*) and salmon (*Salmo salar*) smolts. Production units were mainly situated on larger river catchments and were often situated at the sites of old grain mills where artificial heads had been created using dams. It is not documented when the Eurasian perch (*Perca fluviatilis*) was first introduced into Ireland, however it is most likely that it was introduced from elsewhere in the British isles. Eurasian perch is indigenous to south-eastern England and has been widely redistributed "for their food value and ease of catch" (Maitland and Campbell 1992).

Eurasian perch is found throughout Ireland and has proliferated in certain catchments. This has inevitably led to some management issues particularly in salmonid rivers and lakes. In some areas, control of perch populations has been embarked on by Fisheries staff. Removal of Eurasian perch egg strings during spawning season and netting of stock is common in designated salmon and trout fisheries (Leggett 1966). Such control measures have by and large been unsuccessful given Eurasian perch's adaptability and suitability to the environment. In certain areas where Eurasian perch proliferated, Eurasian perch fisheries developed. Lough Corrib in Co. Galway and Lough Neagh in Northern Ireland were two such commercial fisheries. Whilst commercial fishing for Eurasian perch ended in Lough Corrib in the 1990s, it still continues in Lough Neagh where other coarse fish such as pollen *Coregonus autumnalis*, bream *Abramis brama* and European eel *Anguilla anguilla* are also fished (Wood et al. 1993).

Eurasian perch aquaculture was identified as having potential in Ireland in a review of the options for non salmonid aquaculture carried out in 1995 and subsequently in BIM's Aquaculture Explained manual, Cultivating Perch (Ashe 1997). Bord Iascaigh Mhara (BIM) is the state agency with the responsibility for development of Irelands fishing and aquaculture sectors. The agency gives producers technical and business advice and has been intrinsically involved in the development of

Irelands fledgling perch sector. As part of the review, BIM staff visited pond production systems for yellow perch *Perca flavescens* in North America. Such pond systems in Indiana and Wisconsin also produce bait fish and predominately sell their yellow perch on local markets. The pond systems are in general earthen based, utilising both surface water and groundwater sources.

The strategy at the time was to identify suitable areas in Ireland on marginal agricultural land and investigate whether such pond based systems would work for producing Eurasian perch. In 2000 PDS Irish Waters Perch established Ireland's first pond based perch farm. The farm situated in Gowna, Co. Cavan was based on marginal land in an area renowned for its waterways and angling lakes. The farm was based on 1.2 ha and utilized water from a nearby stream. The presence on the land of grey clay which has excellent water retention properties was a key factor in site selection. The farm initially began production of both Eurasian perch and Northern pike (*Esox lucius*). However in 2002 it stopped pike production to concentrate on juvenile perch production for emerging farms.

The initial production method for Eurasian perch was entirely pond based. Adult fish were allowed spawning naturally in the earthen ponds and their ribbons collected and placed in separate ponds. The resultant larvae were allowed to hatch naturally and proceeded to feed on zooplankton present. The development of zooplankton populations and the intensification of pond management were carried out with the assistance of the Institute of Vodnany in the Czech Republic. The institute had extensive experience of freshwater fish production in ponds and similar management protocols were put in place at the PDS site. Addition of natural fertilizers, timing of water fills for proper zooplankton succession and drying out of ponds for sterilization were some of the procedures which improved juvenile production at the PDS site. Whilst the farm was initially developed to grow Eurasian perch to market size, the demand for juveniles from other farms being established at the time changed the focus at PDS to one of being a hatchery.

In the mid 2000s significant interest was being expressed in Ireland in Eurasian perch aquaculture. The economy in general was prosperous and promoters were looking at new forms of aquaculture as a means of investing. In tandem with this the availability of grant aid through various EU programmes provided an incentive for farms to be developed. Keywater Fisheries Ltd (2002) based in Co. Sligo and Emlagh Fisheries Ltd (2003) based in Co. Roscommon soon followed PDS and developed Eurasian perch farms at new sites. The emphasis was still on the potential of pond production systems but interest was growing in RAS throughout Europe. In 2004, Emlagh converted much of its site to RAS through the addition of covered and insulated polytunnels, drumfilters, ultraviolet lights and a biofilter. Such additions were seen as the best way to improve growth rate, survivability and ultimately increase stocking density and production. In 2007, Clune Fisheries, Co. Tipperary established the first full RAS in Ireland for perch production and this was followed by the development of Ballybay Perch Ltd RAS in Co. Monaghan in 2008.

The development of these ongrowing farms created the impetus to improve and substantially increase juvenile production in Ireland and indeed in Europe. A European CRAFT project, PERCATECH was funded involving partners from across Europe.

The research and development on issues such as out of season spawning and improved feeding strategies allowed significant improvements to be made in hatchery production at PDS and Keywater. The importation of Eurasian perch eggs from Denmark and New Zealand from 2006 onwards enabled the hatcheries to produce the quantities of juveniles required.

The development at this time was not without serious hurdles. Inconsistent quality of juveniles became a major issue and resulted in some ongrowers developing their own hatcheries. Issues during ongrowing with disease, system failures and financial issues resulted in some farms closing with a resultant knock on effect on the sector. Such issues mirror that experienced by every other part of the aquaculture industry at some stage of their development. However the scale of the perch sector has inhibited its ability to deal with such issues. The emergence of an economic recession in Ireland from 2008 onwards has severely impacted on the sectors success in drawing in fresh investment and expanding to a fully commercial entity. Issues such as out of season spawning, juvenile quality and growth have largely been overcome, however the residual effect of these issues continues to be felt by remaining operators. Significant restructuring of Irelands perch sector is likely to take place in the short term and this should allow the companies involved to expand and grow the sector as initially envisaged.

Irelands perch sector is still at a fragile stage of development. The majority of R&D work has been carried out on semi commercial farms and predictably it has been a steep learning curve with many mistakes and failures along the way as well as significant success's. Early development of the sector was hampered by inconsistent and infrequent juvenile production. The refinement of out of season spawning strategies was a major step forward for the sector bringing with it predictable and cost efficient juvenile production. At the same time, development of skills and systems for ongrowing Eurasian perch has tended to focus on use of RAS technology. RAS is by its nature is costly to construct and implement as well as to operate. High operating costs have hampered production such that annual production has still not reached licence capacity. The development of markets, processing capability and logistics has also been a difficult path for Irish perch growers given their distance from the market. The success of the perch sector in Ireland will depend on further expansion and growth of the sector whilst at the same time, reducing production costs and further market development. Preliminary work has also begun on further development of pond systems which may be suitable for large area of cutaway peatland situated in the midlands of Ireland. Such pond production systems would tie in with Ireland's significant organic production of seafood and may lead to a premium in the environmentally conscious marketplace.

There is no market in Ireland for Eurasian perch at present and all fish produced is exported to markets in alpine countries such as Switzerland, Germany, France and Italy. In 2012, some 1500 kg of Eurasian perch was produced per week by Irish growers and this is expected to grow to 2000 kg per week by 2013. The majority of Eurasian perch produced in Ireland is in the form of skinless and skin on fillets. Whilst whole fish is sold directly to wholesalers in Europe, this invariably entails higher transport costs and curtails value addition. Whole fish are generally either filleted by the producers themselves in approved processing facilities or by third party processors on behalf of the producers. Generally speaking, Irish perch farmers produce whole fish in the 100–200 g size range. The preference in some markets for small size fillets (10–20 g) equating to whole fish of 80–100 g is generally not a market sought by growers. Harvesting fish less than 100 g is not seen at present as economical in RAS, given the high juvenile costs. Improvement of growth rate and survival may in future lead to market development of fish greater than 200 g. Whilst such a market exists in Nordic countries, it is generally small in comparison to the main market in Switzerland.

The future of percid aquaculture would seem to depend on a number of factors. Existing wild production of perch from lakes in Estonia, Poland and Russia continues to provide the large bulk of product available on the market. Aquaculture producers have tended to concentrate on the development of niche markets. Such a strategy makes sense given the price variability in wild production. Seasonality, quality, availability and market demand all feed in to market price. The success of other aquaculture sectors such as salmon, oysters, mussels and sea bass has been in having the quantity to dictate a response to such variable factors. It is likely that whilst there is still a significant wild catch of perch, aquaculture producers must concentrate their marketing efforts on building niche, high quality markets. The market ultimately dictates the success or otherwise of production strategies. Whether future expansion of the Eurasian perch sector in Ireland is based on RAS or pond systems is still very much debatable. Ultimately however SME's in any agrifood sector control their own density and must embark on a path which they believe will be successful.

32.6 Western Europe, France and The Netherlands

In Western Europe, the initial intentions to promote the production of percid fish date from the end of the 1980s and the beginning of the 1990s over last century, especially in France. Actually, percid fish were the freshwater species the most appreciated by French consumers and anglers for their flesh quality. For example, after salmonids (trout, salmon), pikeperch is the freshwater fish species the most frequent on fishmonger's stall and restaurant menu (supply only by fisheries captures). At that period, due to the stagnation and/or progressive decrease of restocking markets, pond fish farmers, especially in the North-East of France, were interested to diversify their production. Some surveys have demonstrated that important markets existed for percid fish (Tamazouzt et al. 1993). Eurasian perch was perceived as a species related to important niche markets, mainly located in Alpine areas (Switzerland, North-Italy, South-East of France) while pikeperch was considered for larger markets. For pond farmers, the development of Eurasian perch culture was welcomed due to the poor control of this production in ponds. In fact, these farmers had a high demand of large perch (>150–200 g) for restocking markets; however in ponds few individuals achieve this weight and the large majority of a cohort doesn't grow and remains at a small size (stunting). In some French regions (Sologne, Dombes), small perch were considered as undesirable fish for pond production. Consequently, the intensification of the Eurasian perch culture was not considered as competing with the production in ponds and was supported by fish farmers, whereas it was not the case with pikeperch. However, due to its high level of consumption in France, pikeperch was also considered a potential candidate species for inland aquaculture diversification and a French farm (SEPIA) in South-West of France initiated a project to promote an intensive production of pikeperch juveniles, based on the research conducted by CEMAGREF (now IRSTEA) (Schlumberger and Proteau 1991). This farm stopped its activities after few years. The first rearing trials, mainly for Eurasian perch, tested the feasibility of different rearing systems (ponds, floating cages or RAS). The on-growing of Eurasian perch was evaluated in a large eutrophic pond (Etang de Lindre, Moselle, France) and Lake reservoirs (Feronval and Plate-Taille Reservoirs, Wallonia, Belgium) (Fontaine et al. 1996; Tamazouzt et al. 1996; Kestemont et al. 1998). These comparative studies have clearly shown that floating cage or pond systems were limited by low growth rates, high mortality due to pathology (external parasites) and precocious sexual maturity, especially in males, comparing to the performances observed in RAS. Later the constraints identified with floating cages were confirmed by the results of the project Perlac developed over Neuchâtel Lake in Switzerland during the early 2000s. About 15 years ago RAS system was selected to support percid culture in Western Europe.

After this initial prospecting period, an intensive research activity was conducted during the 1990s and 2000s to develop knowledge on percid biology and promote their culture. In the framework of national or European projects, theses researches concerned initially the Eurasian perch and more recently pikeperch and were focused on the control of the reproductive cycles and gamete quality, larval rearing, nutritional requirements, on-growing, genetic improvement and flesh quality (Kestemont et al. 2000; Fontaine et al. 2001; Fontaine and Thomas 2004; Kamstra et al. 2001; Teletchea et al. 2006; Wang et al. 2008). All these researches targeted to develop intensive culture of percid in RAS.

Based on this knowledge which has been highly disseminated using scientific (Kestemont and Dabrowski 1996; Kestemont and Melard 2000) and technical documents (Kucharczyk et al. 2008; Torner and Rougeot 2008) or the organization of a percid culture workshop (Fontaine et al. 2008), completed by other activities (first rearing trials at pilot scale), several farms have been built to produce percids mainly in France (SARL Lucas Perches, SARL Asialor, EI Bresse Aquaculture), Switzerland (Percitech SA, Valperca SA), The Netherlands (Excellence fish farm B.V., HESY Aquaculture B.V., SEA FRESH B.V.). In some cases, percid production was initiated to diversify the production in farms already existing like FLAMICELL VERD S.L. in Spain. Over the last 5 years, numerous projects were studied in Western Europe (Belgium, France, North-Italy, Portugal …). Generally, these farms are compounded by building with isotherm walls for internal and thermoregulation energy economy and used drinkable water (tapwater or ground water pumping by a bore-hole). Using optimal temperature (20–22 °C for Eurasian perch and 24–26 °C for pikeperch), they produce fish at high density (on-growing phase with 60–100 kg. m^{-3}). The environmental management of the reproductive cycle in cooling chambers ensures several reproductions per year, all the year round, between 3 and 12 times per year according to SME.

In terms of markets, Eurasian perch is mainly sold as fillet (15–40 g) or sometime as whole fish (from 40 to 150 g) addressed to Swiss markets. However, in North-East of France, a local commercialization is progressively organized in order to supply restaurants; this market is promising but somewhere difficult to structure because it needs a specific logistic. The price vary between 5–8 euros/kg and 22–30 euros/kg for whole fish and filet respectively, according to season and fish or fillet size. For pikeperch, the market concerns mainly whole fish (0.8–1.2 kg) and fish are sold to wholesalers. The current prices are between 8 and 13 euros/kg. In Western Europe, percid production appears more and more as the major way to diversify and develop inland aquaculture in a sustainable way e.g. production of local species, economy of water and use of high water quality.

32.7 The Czech Republic

The Czech Republic produces both Eurasian perch and pikeperch. Production is based on commercial aquaculture and recreational fisheries (angling). According to statistics from FAO, the Czech Republic was the world's 6th biggest aquaculture producer and the 18th largest captive producer of Eurasian perch during the decade from 2000 to 2010. Regarding pikeperch the Czech aquaculture production was the 5th largest while captive production was the 11th largest (FAO 2011).

Czech annual capture production was between 14 and 37 tons of Eurasian perch and between 106 and 147 tons of pikeperch in the period 1996–2010 (Table 32.3). Aquaculture production was between 14 and 27 tons and between 31 and 58 tons in the same period respectively. The resulting total Czech production was between 32 and 61 tons of Eurasian perch and between 154 and 213 tons of pikeperch (Brožová 2005; Ženíšková and Gall 2011). Commercially this production was reported by 59 fishery companies associated to the Czech Fish Farmers Association (Kratochvíl 2012). Classical average body weight of produced marketable fish was 150–300 g for Eurasian perch and 1.3–3.5 kg for pikeperch. Cultured fish were 3, 4 or 5 years old. More than half of the Eurasian perch total production was carried out by six farms (Adámek and Kouřil 2000; Adámek et al. 2012; Policar et al. 2009, 2011b). Seventy five percent of the pikeperch were cultured by six farms.

Production of 6–12 months old juvenile pikeperch (TL = 120–180 mm) for stocking purposes has received much attention lately and one company is currently producing 100,000 fry annually for stocking only.

Generally, aquaculture production in the Czech Republic is characterized by extensive and semi-intensive fish farming in ponds. Common carp (*Cyprinus carpio* L.) is the dominant fish species produced in these systems and amounts to 87 % of total Czech fish production (Adámek et al. 2012). However, polyculture productions

Table 32.3 Czech Eurasian perch and pikeperch production by commercial aquaculture and by recreational fisheries (angling) and summarized total production during 1996–2010 periods (Brožová 2005; Ženíšková and Gall 2011)

Year	1996	1997	1998	1999	2000	2001	2002	2003	2004	2005	2006	2007	2008	2009	2010
Production by commercial aquaculture															
Perch	15	27	17	21	24	18	24	22	14	18	18	13	17	18	18
Pike-perch	32	35	40	41	40	31	42	55	48	47	47	48	58	58	48
Production by recreational fisheries (angling)															
Perch	33	34	36	37	34	34	30	30	30	26	24	20	17	18	14
Pike-perch	130	157	125	130	134	139	144	133	165	145	147	121	106	107	106
Summarized total production															
Perch	48	61	53	58	58	52	54	52	44	44	42	33	34	36	32
Pikeperch	162	192	165	171	174	170	186	188	213	192	194	169	164	165	154

are very important. Eurasian perch, pikeperch and European catfish are useful predatory species which play important roles in the control and regulation of the overpopulated and less valued small cyprinids such as: roach (*Rutilus rutilus*), bleak (*Alburnus alburnus*), bream (*Abramis brama*), topmouth gudgeon (*Pseudorasbora parva*) and Eurasian ruffe, *Gymnocephalus cernua*) (Adámek et al. 2012).

Broodstock of both Eurasian perch and pikeperch are held in production ponds with areas of 5–400 ha. Small cyprinids are used as natural food (Křišťan et al. 2012a; Policar et al. 2011b). The diet close to natural conditions results in broodstock with very well-developed gonads and high-quality gametes (Křišťan et al. 2012a). Broodstock are transported to hatcheries before the final oocyte maturation (Křišťan et al. 2012b). Out of season spawning broodstock are stocked in cooling rooms (6–9 °C) with controlled light regime (10 L:14 D, 300–400 lux) and fed small cyprinids. One spawning term in April is routinely used though (Policar et al. 2013).

Hatched larvae are stocked in small ponds with littoral macrophyte vegetation at densities of 100,000–120,000 larvae ha^{-1}. Well-timed harvesting of advanced juveniles is important to avoid cannibalism. Common average survival of advanced juveniles (TL = 30–40 mm and W = 0.35–0.8 g) is about 22 ± 8.9 % (with minimum 10.0 and maximum 33.0 %). Produced juveniles are without deformations (Policar et al. 2011b). Harvested fry are used for several purposes: (1) adaptation in RAS for intensive culture (Stejskal et al. 2010; Policar et al. 2013), (2) stocking into dams, lakes and ponds for angling and (3) stocking into production ponds such as supplementary fish within polyculture stocks.

Marketable fish have a weight of 1.3–3.5 kg which is reached within 3–5 years in pond polyculture (Policar et al. 2011b; Adámek et al. 2012).

Of the 46 % of percids marketed in the Czech Republic most are sold live. A little less i.e. 44 % are exported mainly into Germany, Austria, France and Switzerland. The price per kg is 3–5 € for Eurasian perch and 10–15 € for pikeperch at the farm gate. Wholesale prices are around 3 € for Eurasian perch and 10 € for pikeperch (Kratochvíl 2012).

Pikeperch culture is a relatively new production the Czech Republic. New methodologies are continuously being developed e.g. the use of ponds for fry production in combination with RAS for ongrowing. University of South Bohemia, Faculty Fisheries and Protection of Water has led this development.

During 2009–2011, three pilot projects were conducted at three Czech farms and supported by the European Fisheries Fund. The projects focused on: (1) optimization of hormonally induced mass spawning in pikeperch inclusive optimal methods of artificial egg fertilization and incubation, (2) optimization of mass juvenile production under pond conditions including successful weaning and (3) adaptation of juveniles to artificial food and induction of two out-of-season spawning terms in pond-cultured broodstock. One new Czech pikeperch RAS farm was established in 2011. Its planned pikeperch production is 100 tons of marketable fish in 2013 (Policar et al. 2011a, b, 2012).

32.8 Hungary

Hungary is home to eight species of the family Percidae: *Gymnocephalus cernua*, *G. baloni*, *G. schraetser*, *Zingel zingel*, *Zinger streber*, *Perca fluviatilis*, *Sander lucioperca and S. volgensis*. Of these, only *S. lucioperca* is cultured commercially at fish farms. Trials have been carried out to develop aquaculture of Eurasian perch as well as the hybrids of *S. lucioperca* and *S. volgensis*, however these remained experimental. Eurasian perch is occasionally marketed as a by-product of pond fish farming collected during fall harvest. The species is seldom stocked to ponds intentionally.

The culture of pikeperch *S. lucioperca* in Hungary is characterized by long-standing traditions of low intensity pond farming of the species. At the moment, pikeperch is farmed only in ponds, in spite of successful experimental trials to culture it in more intensive conditions.

Predator species (wels catfish, pike and pikeperch) occupy a special role in pond polyculture in Central Europe, including Hungary. They are stocked into ponds to control the populations of smaller wild fish that penetrate into the ponds during their flooding in the spring. These nuisance species would act as food competitors for the cultured cyprinids, thus, stocking of predators represents a means of biological protection against them. Each predator species has its own specific target prey species. For the pikeperch in pond aquaculture the primary prey species include the perch and the exotic topmouth gudgeon or stone moroko (*Pseudorasbora parva*). Obviously, beyond its role of controlling nuisance species, pikeperch is valued for its premium quality meat and has good market potentials. Typically the ratio of predator species (including all three) in the polyculture ponds does not exceed 3–5 %.

Generally, the culture of pikeperch in Hungary is divided into four technological steps: spawning, fry rearing, fingerling rearing and rearing of market-size fish.

Fingerling rearing starts in April-May (following fry rearing) and lasts until the end of the production season. This phase of pikeperch rearing is already done in polyculture with other fish species. Pikeperch is a fast-growing species, typically by the end of fingerling rearing they reach the individual size of 150–200 g. Typically 1000–1500 fingerlings are stocked per hectare and 20–25 % survival can be expected. At the end of the production season (October–November), the fish are harvested and stocked into wintering ponds.

Market-size rearing can last one or two additional production seasons, depending on the desired size of marketable fish. At the end of the second production season they can reach the size of 300–500 g while a further year of culture results in 1–1.5-kg fish. Survival rates are good and depend primarily on the presence of adequate prey fish. The typical yield of market-size pikeperch is 25–40 kg ha^{-1}, although, occasionally yields of 50–70 kg ha^{-1} can be achieved. As pond harvest is a seasonal activity, pikeperch is also marketed during the late fall-early winter period. Yields of pikeperch vary each year, however; generally 70–110 tons of this species are produced in Hungary each year (Table 32.4).

Table 32.4 Pond surface and volume of pikeperch produced in Hungary in the period between 2001 and 2011

Year	Pond surface (ha)	Volume of pikeperch (kg)
2001	22,462	67,093
2002	21,090	81,099
2003	22,839	67,323
2004	22,850	72,598
2005	23,078	119,821
2006	23,878	79,525
2007	24,302	82,785
2008	24,248	91,762
2009	23,967	89,098
2010	23,639	86,757
2011	24,364	101,255

Source: Statistics Department, Research Institute of Agricultural Economics, Budapest, Hungary

32.9 Iran

Among four main species of the percidae family which are highly valued for aquaculture (Kestemont and Mélard 2000), two of them are found in Iran; the Eurasian perch and the pikeperch. Between these two species, pikeperch consistently has been high in market demand. The mean weight and age caught fish from the Caspian Sea estimated to 332.7 g and 3.5 years, respectively (Abdolmaleki 2005).

However, commercial fishing of Eurasian perch is limited to some protected areas in north of Iran such as the Amirkalayeh and Anzali lagoons. The two main habitats of pikeperch in Iran i.e. the Caspian Sea and the Lake behind the Aras dam support a limited fishery of pikeperch, representing 0.2 % of the landings dominated by cyprinids and mullets. Historically though, pikeperch fishery has been considerable, representing 30 % of the landings during the period from 1927 to 1936 (Abdolmaleki and Psuty 2007).

The collapse of the pikeperch fishery from over 3000 tons in the early 1930s to less than 5 tons for the last six decades (Fig. 32.4) led to a decision in 1989 by the Iranian Fisheries Organization (Shilat) to seriously supplement natural populations by establishing a hatchery for this species. Since then the hatchery has steadily increased its production to a level where 1.6 million fry was produced in 2012 (Fig. 32.5).

Aquaculturists rearing Chinese carp in ponds are also interested in using pikeperch for biological control of fish in their ponds, with the added benefit of a pikeperch production supplementing the carps (Fig. 32.6).

There is a high demand for pikeperch in Iran. Consequently pikeperch receive 6–10 USD per kg at the farm gate which is a high price compared to other traditional fish species on the market in the northern part of Iran.

The reproductive cycle of pikeperch has not been fully investigated in Iran. The results of Rahimibashar et al. (2008) showed the highest GSI in Lake Aras (the main

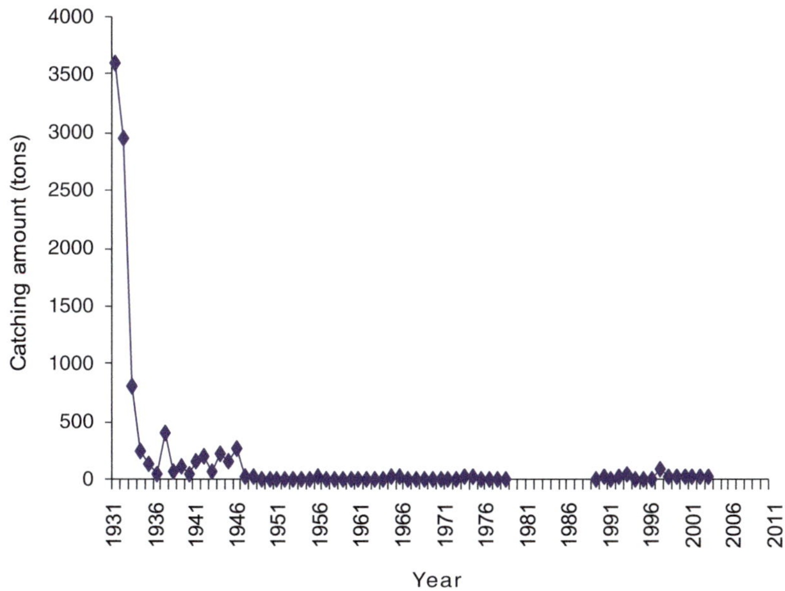

Fig. 32.4 Landing of pikeperch in Iranian coastline of the Caspian Sea

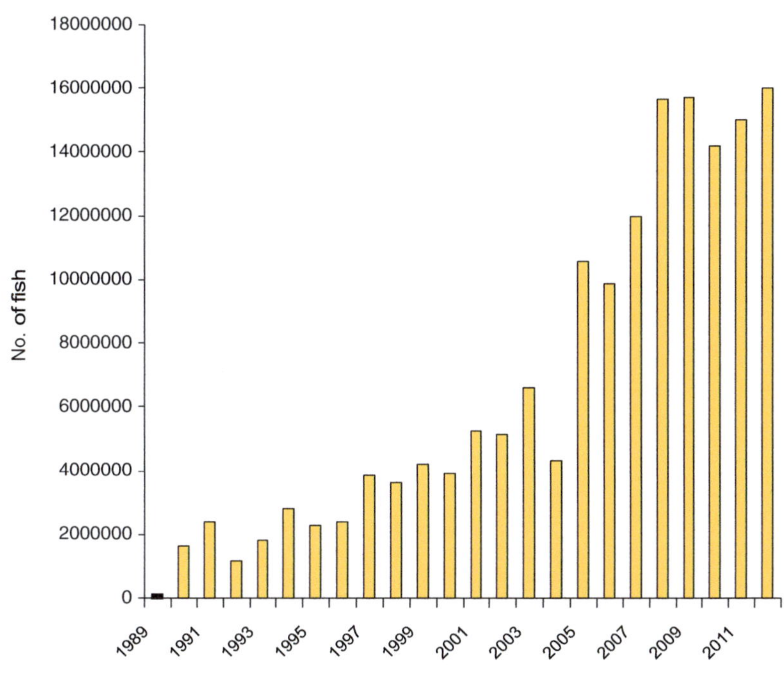

Fig. 32.5 Pikeperch juvenile production by Iranian hatcheries for stock rehabilitation

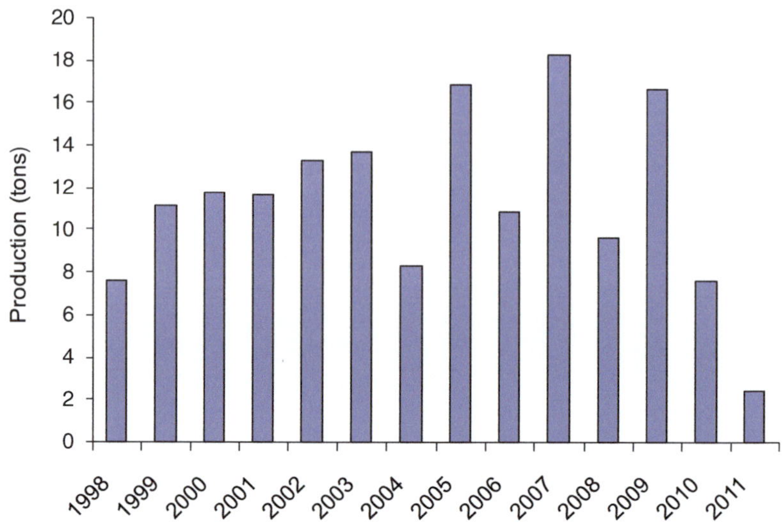

Fig. 32.6 Iranian aquaculture production of pikeperch in earthen ponds

habitat of pikeperch) during March–April. Spawning usually occurs from March to May at a water temperature of 10–16 °C. Reproductive control experiments revealed only minor effects of short term photoperiod changes. Forty days of exposure to darkness up to spawning only delayed spawning by 24 h. Changes in hematological and stress indicators were observed and were possibly related to the treatments (Pourhosein Sarameh et al. 2012, 2013).

Wild broodstock is captured by beach seine net from the lake behind the Aras dam in the northwestern part of Iran during the late fall and winter. The fish are transferred to the Dr. Yousefpour Fish Hatchery Center in Siahkal 800 km from the lake by truck . At the hatchery center fish are held in earthen ponds and fed baitfish (carp fry) till spawning. Prior to expected spawning in March, fish are introduced to large 600 m^3 concrete or small circular 1 m^3 concrete ponds, or to 2–4 ha earthen ponds. Nests made by willow branches (53×53×5 cm), are placed in the tanks or ponds (Fig. 32.7).

Recently, hormonal treatments have been implemented with success. Females are injected with 200 and 400 IU hCG kg^{-1} for the priming and resolving doses and males receive 200 IU kg^{-1}. After 28–34 h, most fish will spawn on the nests (Falahatkar et al. 2009; Golmoradizadeh et al. 2010).

Although the main propagation methods for pikeperch include the capture of wild broodstock in the autumn before spawning, reared broodstock can be used as well (Falahatkar and Poursaeid 2012).

Stripping and fertilizing of pikeperch eggs using more intensive methodology are carried out, i.e. the dry method where 2 mL of semen is added to 100 g of eggs followed by addition of a fertilization solution (4‰ Carbamide, salt and 3‰ urea) to prevent sticking of the eggs. Eggs are washed with water and transferred to upwelling jars till hatching.

Fig. 32.7 Spawning nests in the rectangular concrete ponds at Dr. Yousefpour Fish Hatchery Center in Siahkal

Larvae hatched from the upwelling systems and from the nests in the concrete systems are returned to the pond systems 4–6 days after hatching. They feed on the natural live food, mainly rotifers. Ponds are conditioned with cow (organic) and chemical (inorganic) fertilizers before and during larval rearing. Stocking densities vary from 300 to 400×10^3 larvae ha^{-1} in the earthen ponds. Yields after 50 days are 300–440 kg ha^{-1} of 1–1.1 g Survival in fertilized ponds is highly variable, but in good cases near 50 % after 5–7 weeks or rearing. High rates of cannibalism are observed and seem related to high water temperatures (up to 30 °C) during the summer months.

Currently, many carp farmers have shown interest in culture of pikeperch. One aim is to control undesirable fish species in ponds and lakes, although a production of market size pikeperch also seems to be interesting to the sector. Present limiting factors seem to be repeated failures in rearing of the species and the lack of specific formulated diets.

The average harvest size after 232 days rearing in earthen ponds together with Chinese carps was 235 g the 1 year, but growth performance and production outputs seems highly fluctuating and unpredictable (Tamazouzt et al. 1993).

Survival rates were 18 % in some water reservoirs and 41 % in earthen ponds. The main reason for the low survival rates seems related to lack of adequate adaptation with pond condition but more importantly seems the sensitivity of this species to transport (Falahatkar et al. 2012).

As mentioned previously, the main purpose of pikeperch production in Iran is wild stock rehabilitation.

32.10 Tunisia

In the geoclimatic context of Tunisia, the satisfaction of water needs for the population and the economic sectors necessitated the mobilization of renewable hydric resources by building a large number of dams and reservoirs. Even though these mobilized water resources are first destined to drinkable water, irrigation and industrial needs, they also may be object of an additional valorization through the production of freshwater fishes under extensive conditions without any alteration of water quality or perturbation of the ecosystems (Mhetli et al. 2012). Tunisian government worked in this sense since the 1960s. Knowing that native freshwater fish species are either scarce or not exploited because they are not considered economically valuable (Mhetli et al. 2011), several species have been introduced in Tunisia, including pikeperch which appeared of particular interest. Pikeperch was also introduced in Algeria in 1985 (Meddour et al. 2005). In North Africa, pikeperch populations inhabit the most southern regions of the species' geographical distribution (Mhetli et al. 2011).

This carnivorous species allow the transformation of small invaluable fishes (*Phoxinella*, *Gambusia*, young cyprinids) into exploitable biomass, of excellent nutritional quality and commercial value. Generally, fishermen are pleased with its introduction because it augments their income by increasing substantially the economic value of their catches (Lammens 2001). The species was introduced for the first time in Tunisia in 1968 by the importation of a batch of 30,000 eggs placed in incubating boxes in the lake of Nebhana dam by the National Office of Fisheries (Zaouali 1981). At the beginning of the 1990s, translocations of few adult specimens were done within the framework of a development program having as objective the utilization of the dams for the production of freshwater fish in the North of the country (Losse et al. 1991). Since 2004, a multitude of seeding has been operated based on the reproduction of wild caught pikeperch broodstocks in hatchery and the larval rearing in intensive conditions (Hamza et al. 2007, 2008, 2010, 2012). More recently, researches performed at the INSTM (Institut National des Sciences et Technologies de la Mer) allowed acclimating wild broodstocks to captivity, warranting regular collections of spawns for seeding operations (Ben Khemis et al. 2014). Nowadays, according to recent compilation of data, the species has been introduced into a total of 49 spots: 16 dams, 12 reservoirs and 21 hill lakes. Among these lakes, 18 are exploited (Table 32.5) by licensed fishermen who are mostly young people (under 30 year old) with high school graduation level (Mtimet 2010). Thus, this activity contributes to employment and creation of economic resources in the lakes surrounding regions, where opportunities are generally scarce.

Global pikeperch production increased during last decade to overpass 200 tons/year (Fig. 32.8), placing the country among the main producers with Ukraine, Netherlands and Denmark. At the national level, the production ranges between 18 % and 20 % of freshwater fish production but it contributes to more than 25 % of its value on local market. In terms of valorization, Tunisia is among the countries in which unitary value is the weakest in the world (FAO). Indeed, the prices are around

Table 32.5 Production and productivity of exploited Tunisian lakes (2004–2013)

	2004	2005	2006	2007	2008	2009	2010	2011	2012	2013
Production (mt)	58.9	89.5	192.4	196.2	198	227	190	172.4	211.8	210.9
Weighted mean productivity (kg/ha)	6.8	7.8	17.2	16.6	15.7	14.9	10.9	11.4	13.5	13.5
Minimum (kg/ha)	0.5	0.4	0.9	0.3	1.8	1.1	0.5	0.4	0.3	0.4
Maximum (kg/ha)	13.2	24.0	24.8	26.3	40.0	60.0	25.0	27.2	32.0	36.0
Number of exploited lakes	8	10	8	9	9	12	17	14	17	18
Cumulated surface (ha)	8600	11,554	11,234	11,854	12,604	14,664	17,434	15,126	15,721	15,576

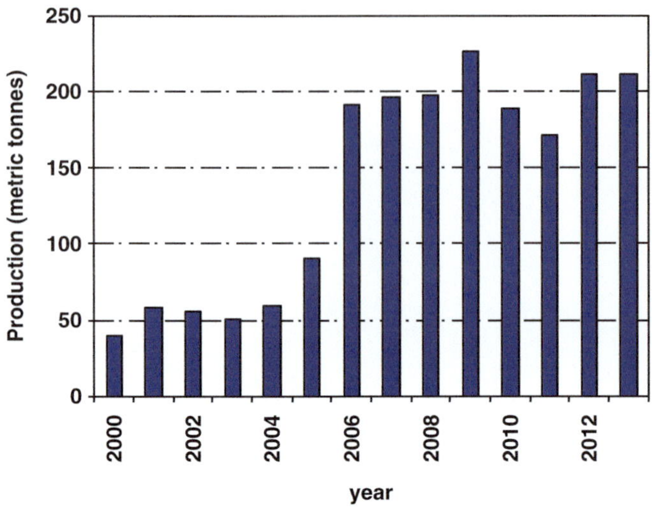

Fig. 32.8 Pikeperch production in Tunisian lakes and reservoirs from 2000 to 2013 (Statistics DGPA)

2.2 $ US/Kg. The pikeperch is exclusively produced by fishermen and license holders of reservoirs which correspond to small economical units of limited power. To allow a better valorization of the production, some of the stakeholders would like to export part of the production towards European countries where there is a strong demand for the species. The access to these markets may allow a valorization at prices six to sevenfold higher comparatively to local Tunisian market. In this way, researchers from INSTM are studying the nutritional quality and safety of pikeperch captured in Sidi Salem lake (largest dam in Tunisia) in the frame of a survey performed during 2014 (BioVecq). This survey was carried out with the aim of certification of this fish for export.

References

Abdolmaleki S (2005) Stock assessment of the Pike perch (*Stizostedion lucioperca*) in the southern coastal waters of the Caspian Sea. Iran Sci Fish J 13:95–110

Abdolmalaki S, Psuty I (2007) The effects of stock enhancement of pikeperch (Sander lucioperca) in Iranian coastal waters of the Caspian Sea. ICES J Mar Sci 64:973–980

Adámek Z, Kouřil J (2000) A long aquaculture tradition in the Czech Republic. Aquacult Eur 25:20–23

Adámek Z, Linhart O, Kratochvíl M, Flajšhans M, Randák T, Policar T, Masojídek J, Kozák P (2012) Aquaculture the Czech Republic in 2012: modern European prosperous sector based on thousand-year history of pond culture. Aquacult Eur 37:5–14

Ashe DA (1997) Cultivating perch. Aquaculture explained, Bord Iascaigh Mhara (Irish Sea fisheries Board) Ed, 20, 47 p

Ben Khemis I, Hamza N, Ben Messaoud N, Ben Rached S, Mhetli M (2014) Comparative study of pikeperch *Sander lucioperca* (Percidae; Linnaeus, 1758) eggs and larvae from wild females or from captive females fed chopped marine fish. Fish Physiol Biochem 40:375–384

Brožová M (2005) Fish – annual report of Ministry of Agriculture. Ministry of Agriculture of the Czech Republic, 40 p – in Czech only

Dahl J (1982) A century of pikeperch in Denmark. FAO EIFAC 42:344–352

Deelder CL, Willemsen J (1964) Synopsis of biological data on pike-perch *Lucioperca lucioperca* (Linneaus) 1758. Anonymous, Rome

Falahatkar B, Poursaeid S (2012) Effects of hormonal manipulation on stress responses in male and female broodstocks of pikeperch (*Sander lucioperca*). Domestication in Finfish Aquaculture, 23–25 Oct, Olsztyn

Falahatkar B, Poursaeid S, Efatpanah I, Ranaye Akhavan S, Meknatkhah B, Arzboo Z (2009) Induction of spawning pikeperch (*Sander lucioperca*) in response to various hormones. Aquaculture Europe 2009, 15–18 Aug, Trondheim

Falahatkar B, Akhavan SR, Efatpanah I, Meknatkhah B (2012) Primary and secondary responses of a teleostean, pikeperch Sander lucioperca, and a chondrostean, Persian sturgeon *Acipenser persicus* juveniles, to handling during transport. N Am J Aquacult 74:241–250

FAO (2011) Fisheries and aquaculture information and statistics service – 8/10/2012. http://www.fao.org/fishery/statistics/global-capture-production/query/en

FAO (2012) The state of world fisheries and aquaculture. Anonymous, Rome

FEAP (2011) Production and price report for fish species produced in European aquaculture. Anonymous, FEAP Secretariate, Liege, Belgium

Fontaine P, Thomas M (2004) Mise au point de protocoles de contrôle de la production de larves « hors-saison », d'optimisation de l'alimentation et de caractérisation de la qualité des produits en perciculture. Rapport final du programme d'innovation ANVAR N°A0011134LATEK / Programme franco-espagnol Eurêka ACRAPEP, 21 p. (In French)

Fontaine P, Tamazouzt L, Capdeville B (1996) Growth of the Eurasian perch *Perca fluviatilis* L. reared in floating cages and in water recirculated system: first results. J Appl Ichthyol 12(3–4):181–184

Fontaine P, Kestemont P, Mélard C, Verstraete W (2001) Nouvelles technologies microbiennes appliquées à l'élevage de la perche eurasienne (*Perca fluviatilis*) en circuit fermé. Final report FAIR CT98-9241 (1/1/99-31/12/00), 149 p

Fontaine P, Kestemont P, Teletchea F, Wang N (2008) Percid fish culture: from research to production. Abstracts and short communications, 23–24 Jan 2008, Namur, 148 p

Frisk M, Skov PV, Steffensen JF (2012) Thermal optimum for pikeperch (*Sander lucioperca*) and the use of ventilation frequency as a predictor of metabolic rate. Aquaculture 324–325:151–157

Golmoradizadeh A, Sajjadi MM, Falahatkar B, Efatpanah I (2010) Effect of human chorionic gonadotropin (HCG) and carp pituitary extract (CPE) on induced spawning of Pikeperch (*Sander lucioperca*) and fertilization parameters. J Mar Sci Technol 9:18–27

Hamza N, Mhetli M, Kestemont P (2007) Effects of weaning age and diets on ontogeny of digestive activities and structures of pikeperch (*Sander lucioperca*) larvae. Fish Physiol Biochem 33:121–133

Hamza N, Mhetli M, Ben Khemis I, Cahu C, Kestemont P (2008) Effect of dietary phospholipid levels on performance, enzyme activities and fatty acid composition of pikeperch (*Sander lucioperca*) larvae. Aquaculture 275:274–282

Hamza N, Silvestre F, Mhetli M, Ben Khemis I, Dieu M, Raes M, Cahu C, Kestemont P (2010) Differential protein expression profile in the liver of pikeperch (*Sander lucioperca*) larvae fed with increasing levels of phospholipids. Comp Biochem Physiol D 5:130–137

Hamza N, Kestemont P, Ben Khemis I, Mhetli M, Cahu C (2012) Effect of different sources and levels of dietary phospholipids on performances and fatty acid composition of pikeperch (*Sander lucioperca*) larvae. Aquacult Nutr 18:249–257

Hilge V, Steffens W (1996) Aquaculture of fry and fingerling of pike-perch (*Stizostedion lucioperca* L.) – a short review. J Appl Ichthyol 12:167–170

Kamstra A, Schram E, Span JA (2001) Trials with intensive production of pike-perch fry using dry diets. EAS Publ Ser 29:125–126
Kestemont P, Dabrowski K (1996) Recent advances in the aquaculture of percid fish. J Appl Ichthyol 12(3–4):137–200
Kestemont P, Mélard C (2000) Aquaculture (chapter 11). In: Craig JF (ed) Percid fishes – systematics, ecology and exploitation. Blackwell Science, Oxford, pp 191–224
Kestemont P, Mélard C, Etienne F, Baras E (1998) Bases biologiques de l'élevcage intenif des percidés en cage. Rapport annuel 3. Ministère de la Région Wallonne, Publication code :FP SB61/3, Namur
Kestemont P, Mélard C, Fontaine P, Anthouard M, Boujard T, Kentouri M (2000) An integrated study of interindividual competition and its relationships with feeding physiology and behaviour in cultured predatory fish. Final report FAIR CT96-1572 (1/1/97-31/12/99), 148 p
Kratochvíl M (2012) Harvesting of marketable fish in members of Czech fish farmers association in 2011 and using of fish production in the Czech Republic during 1990–2011. Report of Czech Fish Farmers Association, 31 pp – in Czech only
Křišťan J, Stejskal V, Policar T (2012a) Comparison of reproduction characteristics and broodstock mortality in farmed and wild Eurasian perch (*Perca fluviatilis* L.) females during reproduction under controlled conditions. Turk J Fish Aquat Sci 12:191–197
Křišťan J, Alavi SMH, Stejskal V, Policar T (2012b) Hormonal induction of ovulation in pikeperch (*Sander lucioperca* L.) using human chorionic gonadotropin (hCG) and mammalian GnRH analogue. Aquacult Int. doi:10.1007/s10499-012-9572-y
Kucharczyk D, Targońska K, Sosiński M, Kwiatkowski M, Szkudlarek M, Szczerbowski A (2008) Artificial reproduction of pikeperch (Sander lucioperca L.) under controlled conditions. Conference: percid fish culture from research to production, Namur, 23–24 January 2008. In: Fontaine P, Kestemont P, Teletchea F, Wang N (eds) Percid fish culture from research to production, Presses Universitaires de Namur, ISBN: 978-2-87037-582-2
Lammens EHRR (2001) Consequences of biomanipulation for fish and fisheries, FAO, Fisheries circular. No 952, 23 p
Leggett AM (1966) Some aspects of fishing management in the development of inland fisheries in Ireland. Univ Rev 3(10):34–47
Losse GF, Nau W, Winter M (1991) Le développement de la pêche en eau douce dans le nord de la Tunisie. Coopération Technique Tuniso-Allemande. Deutsche Gesellshaft für Technische Zusammenarbeit (GTZ) Eschborn, République Fédérale d'Allemagne, 418 p
Maitland PS, Campbell RN (1992) Freshwater fishes of the British Isles. HarperCollins, Somerset
Meddour A, Rouabah A, Meddour-Bouderda K, Loucif N, Remili A, Khatal Y (2005) Expérimentations sur la reproduction artificielle de *Sander lucioperca*, *Hypophthalmichthys molitrix* et *Aristichthys nobilis* en Algérie. Science et Technologie C 23:63–71
Mhetli M, Ben Khemis I, Hamza N, Turki B, Turki O (2011) Allometric growth and reproductive biology traits of pikeperch, *Sander lucioperca*, at the southern edge of its range. J Fish Biol 78:567–579
Mhetli M, Hamza N, Turki B, Ben Khemis I (2012) Valorisation des plans d'eau en Tunisie par l'introduction de populations ichthyques carnassières. In: 6ème Conférence Internationale: "Ressources en eau dans le bassin méditerranéen", 10–12 Oct 2012, Sousse, Tunisia
Mtimet N (2010) Analyse socioéconomique de la filière de la pisciculture continentale au Nord-Ouest de la Tunisie. Rapport du Groupement Interprofessionnel des produits de la Pêche en collaboration avec la coopération italienne, 150 p
Öberg O (2008) Rearing of perch in recirculating systems in Eastergötlands and Kalmars archipelago. TRITA-LWR.REPORT 3019, ISBN 978-91-7415-023-0 (in Swedish) Stockholm, Sweden
Overton JL, Paulsen H (2005a) First feeding of perch (*Perca fluviatilis*) larvae Bornholms salmon Hatchery Nexo. DFU-report 150-05. DTU-Aqua. Copenhagen, Denmark
Overton JL, Paulsen H (2005b) Ongrowing of perch (*Perca fluvia*tilis) juveniles Bornholms salmon Hatchery Nexo. DFU-report 151-05. DTU-Aqua. Copenhagen, Denmark
Paulsen H, Overton JL, Brünner L (2005a) Development of culture of perch (*Perca fluviatilis*), a potential alternative species for fresh water culture. DFU-report 149-05. ISBN 87-90968-83-2 DTU-Aqua. Copenhagen, Denmark. Hirtshals (in Danish)

Paulsen H, Overton JL, Frandsen D, Larsen MGG, Hansen KB (2005b) Evaluation of state of nutrition for perch. The North Sea Center, Hirtshals, in Danish

Policar T, Stejskal V, Bláha M, Alavi SMH, Kouřil J (2009) Technology of intensive culture of Eurasian perch (*Perca fluviatilis* L.). Practical handbook FFWP USB 89, 51 p – in Czech only

Policar T, Alavi S, Stejskal V, Křišťan J, Kouřil J (2011a) Artificial and semiartificial spawning in Eurasian perch (*Perca fluviatilis* L.) use for mass embryo production. Practical handbook FFWP USB 117, 24 p – in Czech only

Policar T, Bláha M, Křišťan J, Stejskal V (2011b) High quality and stable production of advanced juveniles in pikeperch (Sander lucioperca) under ponds. Practical handbook FFWP USB 110, 33 p – in Czech only

Policar T, Alavi S, Stejskal V, Křišťan J, Kouřil J (2012) Artificial and semiartificial spawning in Eurasian perch (*Perca fluviatilis* L.) use for mass embryo production. Edition of Methods, FFPW USB, Vodňany, No. 117, 24 pp. (in Czech)

Policar T, Stejskal V, Křišťan J, Podhorec P, Švinger V, Bláha M (2013) The effect of fish size and density on the weaning success in pond-cultured pikeperch (*Sander lucioperca* L.) juveniles. Aquacult Int 21:869–882

Pourhosein Sarameh S, Falahatkar B, Azari Takami G, Efatpanah I (2012) Effects of different photoperiods and handling stress on spawning and reproductive performance of pikeperch *Sander lucioperca*. Anim Reprod Sci 132:213–222

Pourhosein Sarameh S, Falahatkar B, Azari Takami G, Efatpanah I (2013) Physiological changes in male and female pikeperch *Sander lucioperca* (Linnaeus, 1758) subjected to different photoperiods and handling stress during the reproductive season. Fish Physiol Biochem 39:1253–1259

Rahimibashar MR, Alipoor V, Danesh M, Alinia MR (2008) Survival of biometrical characteristics, diet, gonad and liver index of (Sander lucioperca) in the lake of Aras Dam. Pajouhesh Sazandegi 79:58–65

Schlumberger O, Proteau J-P (1991) Production de juvéniles de Sandre (*Stizostedion lucioperca*). Aqua-revue 36:25–28 (in French)

Steenfeldt S, Vestergaard M, Overton JL, Paulsen H, Larsen VJ, Henriksen NH (2010) Development of intensive rearing of pikeperch in Denmark. DTU-Aqua report no. 228:2010. (in Danish)

Steenfeldt S, Lund I, Höglund E (2011) Is batch variability in hatching time related to size heterogeneity and cannibalism in pikeperch (*Sander lucioperca*)? Aquacult Res 42:727–732

Stejskal V, Policar T, Bláha M, Křišťan J (2010) Production of marketable Eurasian perch (*Perca fluviatilis*) by the combination of pond and intensive culture. Practical handbook FFWP USB 105, 34 p – in Czech only

Tamazouzt L, Dubois JP, Fontaine P (1993) Production of the perch *Perca fluviatilis* L. and its markets in Europe. La Pisciculture Française 114:4–8 (In French)

Tamazouzt L, Dubois JP, Fontaine P, Capdeville B, Terver D (1996) Zootechnical performances and body composition of *Perca fluviatilis* fed pelleted diet in a floating cage: effect of daily ration. Ann Zool Fennici 33:635–641

Teletchea F, Kestemont P, Mélard C, Schram E, Linhart O, Fontaine P (2006) Securing juveniles production of Eurasian perch by improving reproduction and larval rearing. Final report COOP-CT-2004-512629-PERCATECH, 40 p

Torner D, Rougeot C (2008) Farming of Eurasian perch. Vol. 1 – Juvenile production, Aquaculture explained, special publication. BIM, Ireland, p 24, 78 p

Wang N, Teletchea F, Schram E, Kucharczyk D, Mélard C, Fontaine P, Kestemont P (2008) Improving pikeperch larval quality and production by broodstock management and nutrition, husbandry and sex control. Final report COOP-CT-2005-017646-LUCIOPERCIMPROVE

Wood RB, Smith RV, Neagh L (1993) The ecology of a multipurpose water resource, Kluwer Academic Publ., Dordrecht, The Netherlands. ISBN 0-7923-21 12-X. 529 p

Zaouali J (1981) Problèmes d'aquaculture: eaux saumâtres et potentiel aquacole. Arch Inst Pasteur Tunis 58:93–103

Ženíšková H, Gall V (2011) Fish – annual report of Ministry of Agriculture. Ministry of Agriculture of the Czech Republic, Prague, 45 pp. (in Czech)

Chapter 33
Concept and Determinism of Quality in Percid Fishes

Marielle Thomas, Guillaume Mairesse, Jean-Noël Gardeur, and Jean Brun-Bellut

Abstract The development of the Percid fish industry calls for reflection on the concept and determinism of quality. This chapter starts with some general considerations illustrating the evolution of quality perception over time. The sense of the word 'Quality' is now polysemic; this brings together information about fish characteristics according to their origins (wild versus farmed), but also consideration on how fishes are produced. The complex picture of quality in Percid fishes is here illustrated with the study of nutritional, technological, sensory and sanitary components. We show on the basis of numerous studies that the determinism of quality is multifactorial. Quality components are thus governed by several biological (species, age, genotype, level of domestication…) and environmental (water characteristics, diet, season..) factors. However the quality objectives in Percid Fishes may vary depending on the stakeholders (fish farmer/fisherman, processor or consumer). As far as possible, the various expectations need to be addressed under the target values for the different quality components. In conclusion, we propose the adoption of multifactorial approaches to provide best information in understanding of determinism quality in Percid fishes.

Keywords Quality concept • Sensory attributes • Sanitary component • Nutritional component

M. Thomas (✉) • J.-N. Gardeur • J. Brun-Bellut
INRA – UR AFPA, Lorraine University,
2 avenue de la Forêt de Haye – B.P. 172, 54505 Vandœuvre-lès-Nancy cedex, France
e-mail: Marielle.thomas@univ-lorraine.fr

G. Mairesse
Valorex, La Messayais, 35210 Combourtille, France

P. Kestemont et al. (eds.), *Biology and Culture of Percid Fishes*,
DOI 10.1007/978-94-017-7227-3_33

33.1 General Quality Concept

While it was evident to bring up quality in Percid fishes as agri-food products in the chapter entitled "*Marketing and Economics*", it is only since the Fordist model crisis that this notion of quality has taken on its full meaning and magnitude. Generally speaking, we have switched from a mass production system to one based on developing products with intrinsic characteristics that meet user expectations. And yet, the concept of quality is complex, taking on highly heterogeneous outlines and contents, especially when it applies to food products from different sources (natural environments vs. farming systems), as is the case herein. Therefore, defining quality is a challenge for the informed consumer, wise producer, experienced processor, knowledgeable distributor and expert researcher alike! The reason is that this concept, which is common but at the same time not readily accessible, is perceived differently among the stakeholders in the sector and throughout the life cycle of the product, from its source to its use and final destination.

But let us dare to take on this challenge and thus attempt to explain what the word "Quality" means, by first referring to the International Standard Organization (ISO 8402–94), which defines quality as "*the set of characteristics of an entity that give that entity the ability to satisfy expressed and implicit needs*". A general framework is thus proposed but surely needs clarification, especially since it keeps changing. Without completely going into its background, we should remember that when pasteurisation was introduced, quality was directly connected to the bacteriological status of the food product. In the 1960s, quality referred more to the product composition (water, protein and fat content). Later on, in the 1970s and 1980s, quality focused on chemical substances and other additives, for which the terms Acceptable Daily Intake and Maximum Residue Limit (Nardone and Valfré 1999), among others, were introduced. Nowadays, we cannot overlook nutritional values and the health properties of food, which include essential amino acid and lipid content, essential fatty acid, vitamin and mineral sources, and so on. However, expectations go beyond that (Valfré and Moretti 1991) and products should also meet sensory (colour, flavour, texture, aspect) and technological (processability, preservability, etc.) requirements. Does this mean that together, all these components – hygienic, nutritional, sanitary, sensory and technological – are enough to fully define what is meant by "Quality"? The answer is a resounding "NO", as the definition takes into consideration not just the product itself, but also how it is produced and where it comes from (Hirczak 2007; Mariojouls 2000).

This complex picture of quality in its broadest terms is shown in the synoptic chart in Fig. 33.1, which illustrates the many components at different levels (product, production and/or supply systems) and for the various parties involved in the sector (fish farmer/fisherman, processor, consumer). In the area of food quality and safety, the fish supply market must meet a number of challenges, "from the net to the plate". This figure shows that while some quality components are of interest for only one party, others are pertinent to several, or even all the parties involved. Nevertheless, while the distinction between specific and common interests allows

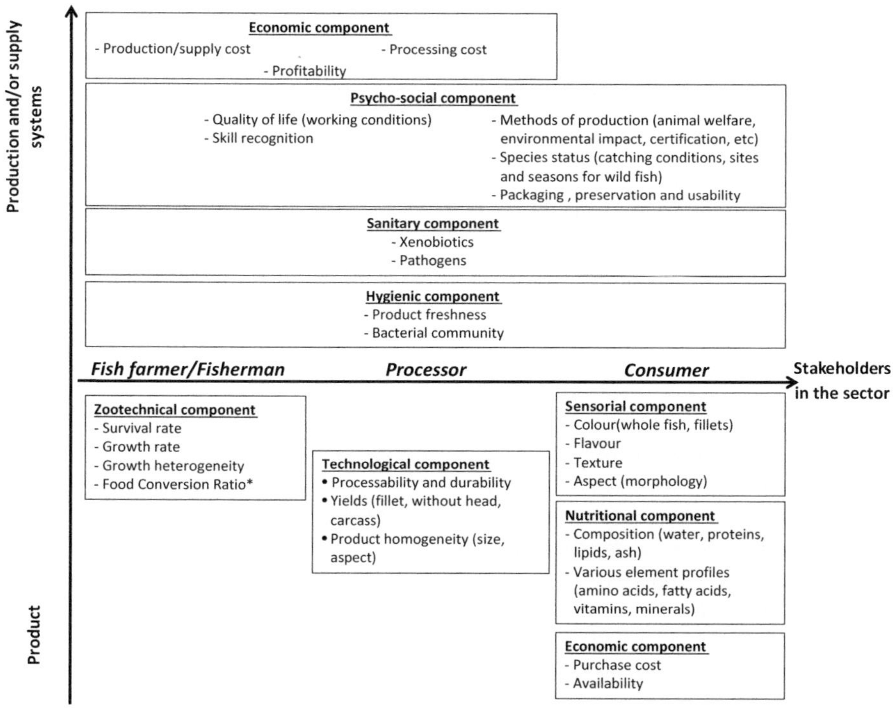

Fig. 33.1 Overview of fish quality components, at different levels (product, production and/or supply systems) and depending on various stakeholders in the sector. **Food Conversion Ratio or* $FCR = F\,x\left(w_t - w_0\right) - 1$, *where F is the total amount of feed per fish consumed during growth period* (*g*) *and* w_0 *and* w_t *are the initial and the final average fish body mass* (*g*)

for a convenient representation, it is a rather simplified view of a more complex reality. For instance, the nutritional component is essential for consumers, yet it is obviously a concern for producers, too. Likewise, we will henceforth take a look at some of these components, focusing on the different notions that overlap as well as on the exogenous and endogenous factors governing their expression.

33.2 Fish Quality Components

This section deals specifically with the nutritional, technological, sensory and sanitary components in Percid fishes. The economic component is addressed in Chap. 10.2 ("*Economics of Percid Culture*") in this book. Similarly, the reader can find, in Sects. 3 and 4, a compilation of current knowledge on the zootechnical parameters to be controlled in order to satisfy the quality criteria breeders expect (e.g., survival, growth performance, decrease in growth heterogeneity, etc.).

33.2.1 Nutritional Component

The nutritional component refers to the biochemical composition of fish. Generally speaking, the content of the main components found in fish flesh may vary widely from fish to fish and from species to species: 53–81 % water, 13–24 % proteins and 0.1–31 % lipids (Dunajski 1979). As the lipid content and fatty acid profiles are essential descriptors of this component, this is what we will deal with first in Percid fishes. Then we will give a more global overview of the other nutritional components within the same group of fish.

33.2.1.1 Fatty Acid Content and Composition

Percid fishes are lean (flesh lipid content below 2/100 g). It has been clearly established that diet plays a major part in the lipid content of fish and this relationship is tissue-dependent. An increase in the lipid content of the food does not have any significant effect on the lipid content in the muscle of Eurasian perch (Xu et al. 2001; Mathis et al. 2003). However, it affects the perivisceral tissue, a favoured storage site in Percid fishes (Jankowska et al. 2003; Kestemont et al. 2001; Xu and Kestemont 2002). In addition, while Cox and Karahadian (1998) did not find any significant differences in lipid content between wild and farmed yellow perch, other studies have shown that the muscle lipid content is usually higher in farmed fish than in wild fish (Table 33.1).

These differences in intramuscular lipid content between wild and farmed fish could be partly explained by (i) the quality of food (higher content of fat in commercial feed compared to natural food in general), (ii) the availability of food (with potential fasting periods for wild fish) and (iii) the maturation of gonads (generally prevented in farmed fish because of constant photoperiod and temperature, but occurring naturally in wild populations with a subsequent energy transfer from the muscle to the gonads).

Along with the lipid content, the fatty acid profile is a key descriptor of nutritional quality. Indeed, one of the features that distinguish fish from other meat food (pig, beef, and poultry) is their high polyunsaturated fatty acid (PUFA) content. Thus, beneficial effects of fish consumption on human health appear to be related to the high content of PUFA (15–36 % of the total fatty acids). It concerns in particular *n*-3 PUFA such as docosahexaenoic (DHA, 22:6*n*-3) and eicosapentaenoic (EPA, 20:5*n*-3) acids (Calder and Yaqoob 2009; Uauy and Valenzuela 2000).

Table 33.1 Intramuscular lipid content (% wet weight) in wild and farmed Percid fishes

Species	Intramuscular lipid content (% ww)		References
	Wild	Farmed	
Eurasian perch (*Perca fluviatilis*)	1.2	2.5	Mathis et al. (2003)
Yellow perch (*Perca flavescens*)	1.4	2.8	Gonzàlez et al. (2006)
Pikeperch (*Sander lucioperca*)	0.9	2.9	Jankowska et al. (2003)

Diet plays a major role here, with the composition of fatty acids in flesh lipids largely reflecting that of the food lipids. Thanks to Δ5 and Δ6 desaturase activity, Percid fishes, like freshwater fish in general, can use C18 precursors[1] from their diet to achieve bioconversion reactions leading to longer-chain fatty acids (Sargent et al. 2002; Xu and Kestemont 2002). Wild Percid fishes feed on prey that is mainly composed of linoleic acid, but also of α-linolenic acid and DHA (Henderson and Tocher 1987). When fish are farmed, the fatty acid profile in the flesh can be modulated by combining different sources of plant or animal dietary lipids. Fatty acid profiles have been described in different food and environmental contexts for *P. fluviatilis* (Blanchard et al. 2005; Orban et al. 2007; Mairesse et al. 2006, 2007; Stejskal et al. 2011a; Xu and Kestemont 2002; Xu et al. 2001), *P. flavescens* (Gonzàlez et al. 2006; Twibell et al. 2001), *S. lucioperca* (Celik et al. 2005; Guler et al. 2011; Jankowska et al. 2003; Kowalska et al. 2012; Molnár et al. 2006; Uysal and Aksoylar 2005) and *S. vitreum* (Czesny et al. 1999). The various studies have established that the main fatty acids found in flesh are usually palmitic acid (16:0), oleic acid (18:1*n*-9), linoleic acid (18:2*n*-6), eicosapentaenoic acid (20:5*n*-3), and docosahexaenoic acid (22:6*n*-3), to which arachidonic acid (20:4*n*-6) can be added for wild fish populations in particular, with concentrations that vary with protocol or diet.

However, lipid content and fatty acid profiles may be affected by other factors (seasons, breeding conditions, origins of the fish and so on). Indeed, it has been suggested that **rearing conditions** and **fish strain** may have an influence on the lipid content in *Perca* fish muscle. This hypothesis has been confirmed by Gardeur et al. (2007), who showed among twelve tested rearing factors that only water temperature and food lipid source (fish oil vs. fish oil and rapeseed oil) interaction had a significant effect on the lipid content in the fillet of *P. fluviatilis*. However carcass lipid content in *S. lucioperca* was no affected by water temperature ranging between 20°C and 28°C (Wang et al. 2009). Other non-food-related factors, such as the **genotype**, the **level of domestication** or the **physiological state** of the fish, have been suggested as influential factors (Mairesse et al. 2007; Gjedrem 1997). Mairesse (2005) showed differences in the fatty acid profiles for juvenile Eurasian perch of different origins, but reared from the egg stage under the same environmental and trophic conditions. In the same way, Rosauer et al. (2011) revealed differences in fillet fatty acid composition, including arachidonic acid and eicosapentaenoic acid concentrations, for three different stocks of *P. flavescens* reared over a 12-month period under similar conditions. The nutritional quality of Percid fishes thus appears to depend not only on the food factor, but on **genetic factors** as well. This relationship had already been put forth by Sargent et al. (1995) who reported that, for freshwater fish, the ability for C18 polyunsaturated fatty acids to be converted to highly unsaturated fatty acids such as DHA varies between and within species, or between different stocks. On this account, domestication has been said to be a factor of influence on saturated fatty acid content (Mairesse et al. 2007). It would also affect the polyunsaturated fatty acid profiles of *n*-3 and *n*-6 series, though just moderately

---

[1] C18 precursors such as linoleic acid [18:2*n*-6] (*n*-6 series precursor) and α-linolenic acid [18:3*n*-3] (*n*-3 series precursor) (*see* Sect. 5 "*Nutrition, feeds, and feeding practices*").

when compared to the lipid source. A change in the regulation of lipogenesis pathways for fish fed artificially over several generations could be responsible for these differences, as already mentioned by Bell and Dick (2004).

Finally, Percid fish flesh is proving to be a product of nutritional interest, with its low lipid content and high *n*-3/*n*-6 ratios. Therefore, consuming this product, rich in PUFA and even more so in DHA, makes a worthwhile contribution toward recommended EPA and DHA dietary allowances for adults (ANSES 2011). Furthermore, determinism of this quality component is mainly driven by diet, a factor that is fully controlled in fish culture.

33.2.1.2 Other Descriptors

As well as containing polyunsaturated fatty acids, which is of interest for consumers, Percid fishes are also and primarily a source of **proteins**. Indeed, proteins are the main organic compounds of fish flesh. In Percid fishes, they account for 18–21 % of wet weight (Jankowska et al. 2003; Mathis et al. 2003; Mjoun et al. 2012). The protein content in fish is usually recognized as mostly stable and not dependent on the sex and size of the fish. Moreover, while an increase in the feed protein content may have led to an increase in protein content for *Oreochromis niloticus* (Gunasekera et al. 1997), such an effect has not been demonstrated for *P. fluviatilis* (Mathis et al. 2003).

Although less documented, the **amino acid composition** and **mineral** content are other descriptors of nutritional quality. Several studies can be referred to for more information on these descriptors in Percid fishes (Gonzàlez et al. 2006; Mai et al. 1980; Mjoun et al. 2012; Tidwell et al. 1999). As an illustration, the mineral composition in *P. flavescens*, *P. fluviatilis* and *S. lucioperca* are given in Table 33.2. The interest consumers have in the mineral content in regards to essential minerals

Table 33.2 Mineral elements for three species in Percid fishes

		Mineral concentration (Mean in mg/g)					
Species	Origins	Na	K	Mg	Ca	P	References
Perca flavescens	Great Lakes (United States)	0.50	2.63	0.22	0.16	1.64	Gonzàlez et al. 2006
	Virginia Tech Aquaculture Center (United States)	0.32	3.72	0.29	0.28	2.08	
Perca fluviatilis	Bolsena lake (Italy)	0.33	3.78	0.26	0.85	2.23	Orban et al. 2007
	Bracciano lake (Italy)	0.25	3.25	0.21	0.46	2.31	
Sander lucioperca	Fish cooperative (Turkey)	0.66	3.54	0.37			Özyurt et al. 2009
	Caspian Sea	0.48	2.68	0.49	0.63	1.08	Pirestani et al. 2009

(e.g., phosphorus, calcium, magnesium, etc.) is offset by their concerns about heavy metals in the muscle (e.g., lead, cadmium, copper, etc.). This latter issue will be dealt with in Sect. "Sanitary Component" of this chapter.

33.2.2 Technological Component

Technological quality is a key quality component for fish farmers, processors and distributors, as they want products that are homogeneous in aspect and size, that are easy to process and that remain fairly stable through time. Percid fishes are consumed as fillets or provided gutted. It is thus interesting to focus on two particular descriptors of the technological component – the fillet yield and the carcass yield. A few of these yield values, which have been recorded for different Percid species, are shown in Table 33.3.

Technological quality descriptors are dependent on a large number of intrinsic and environmental factors, which shall be presented. Thus, **fish size** is a factor whose influence on the fillet yield has been shown. As noted in Table 33.3 for *P. flavescens*, Lindsay (1980) recorded higher fillet yields for the bigger farmed fish.

Table 33.3 Fillet and carcass yields (mean value ± S.E.M [min-max]) according to the fish species and different characteristics (weight, origin, etc.)

Species	Characteristics	Fillet yield (%)[a]	Carcass yield (%)[b]	References
Perca fluviatilis	Mean weight: 131±9 g	42±1 [37.3–47.3]		Mathis et al. 2003
	Wild versus farmed fish	40.1 vs.42.8		Mairesse et al. 2007
	Extensive versus intensive farming conditions	42.3 vs. 39.8		Mairesse et al. 2005
Perca flavescens	Wild versus farmed fish[c]	42.5–43.5 vs. 40–42		Lindsay 1980
	Weight range (g): 52.0±2.5 to 138.1±6.9	Between 34.57±0.01 and 35.22±0.01		Rosauer et al. 2011
Sander lucioperca	Wild versus farmed fish	51.2±0.6 vs. 48.1±0.9	88±0.7 vs. 83.8±1.3	Jankowska et al. 2003
			79–89	Bykowski and Dutkiewicz 1996
Sander vitreus	Fillet, skin-off	33.6		Summerfelt et al. 1996
	Fillet, skin-on	42.1		

[a]Versus fish weight
[b]Gutted carcass
[c]Body weight of 175–200 g (length: 20–23 cm) and 150 g (length: 18 cm) for wild and farmed fish, respectively

For *P. fluviatilis*, Mathis et al. (2003) showed no relationship between fish weight and fillet yield, whereas it generally appeared that among fish of the same age, the skeleton, fin and head share was higher for the smaller fish, which meant a decrease in fillet yield (Einen et al. 1999; Mørkøre and Austreng 2004). Furthermore, even if the impact of the fish's **sex** on fillet yields has been discussed, there did not seem to be a change in fillet yields for *P. fluviatilis* weighing less than 130 g (Mathis et al. 2003). The authors pointed out that the fish used were not yet sexually mature, a useful detail in view of the fact that technological criteria also vary with the fishing period according to the physiological status of the perch (Mairesse et al. 2005). Indeed, during the **spawning period**, gonad weight will determine fillet yield. In *P. fluviatilis* females, the ovary weight may account for up to a quarter of the body weight (Sulistyo et al. 1998; Rougeot et al. 2003), thus explaining fillet yield differences between sexes. In this species, energy mainly accumulates as perivisceral fat (Xu et al. 2001). Accordingly, it has been shown that the perivisceral fat index is at a maximum in October for perch caught in Lake Geneva and progressively decreases with gonad maturation and consequently with an increasing gonado-somatic index (Mairesse et al. 2005). So, logically, the carcass yield[2] in *P. fluviatilis* also exhibits yearly fluctuations – 93 % between May and September, and 72 % between March and April (Mathis, unpublished data) – in relation to the development of perivisceral fat reserves and the development of gonads.

The technological quality in fish also depends on **trophic factors**. Improvement of growth performance through dietary practices (highly energetic, lipid-rich feeds) leads to a significant development of fat reserves in perivisceral tissue of farmed fish. However, these fats impair their ability to be processed, with a particular decrease in yields (Borresen 1992; Fauconneau et al. 1996). Mathis et al. (2003) reported that perch which were fed with dietary energy levels of 19.6 kJ.g^{-1} had a higher fillet yield than those obtained with higher dietary energy levels (22.1 kJ. g^{-1}) – i.e., 43.1 % and 41.1 %, respectively. There is thus a threshold effect of diet: as long as it increases growth without significantly changing perivisceral fat content, fillet yield will improve. On the other hand, as soon as dietary energetic supplement is stored as perivisceral lipids, a decrease in yield will occur. Similarly, the source of dietary lipids (animal vs. plant) has also been shown to affect fillet yield and the viscero-somatic index in Eurasian perch, both of which decrease when vegetable oil is added to the feed (Mairesse et al. 2007). These authors showed that the level of **domestication** should also be taken into account when interpreting fillet yield differences.

To conclude this section, it should be noted that issues related to fish morphotypes or to gaping are not (or poorly) documented in Percid fishes. We should however keep in mind that technological quality determinism in Percid fishes is multifactorial. Yields (fillet and carcass) depend on various intrinsic, environmental and trophic factors, which can be controlled in rearing conditions. Similarly, while it is possible practically speaking to fillet fish at different times with respect to *rigor mortis*, processors most often favour *post rigor* filleting, for optimal yield but also

[2] $\text{Carcass yield} = (\text{carcass weight} / \text{body weight}) \times 100$

for a sensory quality which is said to be better. This is precisely the component which will be discussed hereafter.

33.2.3 Sensory Attributes

The sensory component encompasses all descriptors that can be perceived by the different senses: appearance (especially colour, shape and aspect of the fish), texture, odour and flavour.

33.2.3.1 Aspect: Morphology

The shape and aspect of the fish are quality descriptors, naturally perceptible by the observer, but whose assessment may sometimes be considered complex since it is based on a set of heterogeneous criteria. Indeed, some criteria may be determined objectively, such as the relationship between fish shape and fillet yield. Nevertheless, other criteria are of a more subjective nature, tied to cultural practices and consumers' preferences for example.

Beyond these considerations, studies on the phenotypic expression in Percid fishes have focused on the combined influence of the **genome** and **environment**. Turki-Missaoui et al. (2011) have recently showed the morphological differences within and between *S. lucioperca* populations from Tunisia which were nevertheless derived from a single gene pool. According to the authors, these morphological differences could be an adaptation to using different habitats, a result of dietary and reproductive behaviours that vary with the environment. In *P. fluviatilis*, Mairesse et al. (2005) similarly showed that the overall fish morphology (compact body, head and mouth relative lengths) is a factor allowing fish to be differentiated according to their origins (Lake Geneva, Rhine River). However, the latter study did not take into account the genotypic factor (Mairesse et al. 2005). The trophic environment (habitat use, resource type) would also affect morphological determinism in Percid fishes, as has already been demonstrated in *P. fluviatilis* (Svanbäck and Eklöv 2002). Environmental complexity has also been put forward as a factor to be considered for the phenotypic determinism of wild perch (Mairesse et al. 2005). In this respect, when looking at farmed Percid fishes, the environment they live in has characteristics which are less heterogeneous and, most importantly, more stable than natural ecosystems. However, rearing conditions also affect fish morphology, due to stocking densities, reduced activity, lack of flow conditions, and feed quality and amount (Sarà et al. 1999; Favaloro and Mazzola 2003). These variations in fish shape are often related to skeletal malformations and indirectly affect growth performance, mortality and acceptability of the product (Favaloro and Mazzola 2003). Moreover, rearing systems allow malformed fish to survive possible predation in the wild. These malformations, which are after all quality defects, occur in different parts of the body and have been described in *P. fluviatilis* (Jacquemond 2004). In particular fin damage was reported in intensively cultured perch (Stejskal et al. 2011b); it can affect acceptance by consumers and reduce the economic value of fish sold whole. According to these authors, water turbidity may play a role in reduction of fin damage.

33.2.3.2 Colour

Colour is one of the most important factors that encourage the consumer to buy and accept the product (Baker and Gunther 2004). This parameter applies both to the whole fish and to fish fillets. As opposed to Salmonid fishes, there are few studies on this parameter in Percid fishes, whose flesh is white. The main pigments which affect fish colour (concerning the whole fish and fillets) and which have been investigated, are:

- Carotenoids (ß-carotene commonly found in the environment, as well as astaxanthin and canthaxanthin in shellfish and molluscs) responsible for the orangey-red colour;
- Melanin, which would mainly have an effect on skin darkness (blackness).

Fish, which are unable to synthesise carotenoids *de novo*, must get them through their **diet**. Such is the case for wild perch that are yellow-red in colour when they feed on shellfish (Craig 2000). Under farming conditions, Mathis et al. (2003) further demonstrated that dietary carotenoids (4 %), together with temperature, have an effect on the colour of the fins (increasing red colour). Indeed, while ingested carotenoid pigments may be excreted unchanged in faeces, they are also absorbed, stored or converted. Fish accumulate carotenoids mainly in skin chromatophores, muscle and gonads. In the muscle, carotenoids may be either dissolved in lipid droplets between muscle fibres, or bound to other molecules such as sugar (glycosylated carotenoids) or proteins (caroteno-proteins) in the sarcoplasmic membrane of muscle fibres. Accordingly, it has already been established that fillet colour is generally affected by its **lipid content** (Baker et al. 2002) and its **composition in fatty acids**, mainly polyunsaturated ones (Sargent et al. 1989), due to significant oxidation processes (Lie 2001). Therefore, the fatty acid profile of the muscle, which is dependent on that of the feed, could at least partly explain the effect of trophic factors on fillet colour in *P. fluviatilis* (Mairesse 2005).

Fish colour is also governed by **environmental factors**. Indeed, the genus *Perca* exhibit wide variations in colour depending on where they live, with no real consensus on determinism among studies. According to Craig (2000), in shallow environments where light easily penetrates, fish are dark coloured, whereas in poorly lit environments lacking vegetation they are of brighter colours. Conversely, Mairesse et al. (2005) observed that brighter coloured fish are caught in clearer waters (i.e., from Lake Leman vs. Rhine River with water transparency of 7.5 m and 0.6 m, respectively). These authors have suggested a possible correlation between water temperature on the one hand and the colour contrast between stripes and interstripes in perch on the other hand.

A final factor noted in literature as having an effect on the colour parameter in Percid fishes is the level of **domestication**. According to Mairesse et al. (2007), this influences the colour of the tail fin, the difference in brightness between stripes and interstripes, and the saturation of the fillet *P. fluviatilis*. In this respect, Lindsay (1980) reported differences in fish colour between wild *P. flavescens* (blue-green colour) and their farmed counterparts (yellow-green colour). There also seems to be differences in fillet colour between wild and farmed fish, the latter having whiter flesh (Mathis, unpublished data). However, according to Jankowska et al. (2003), the brightness and redness of both wild and farmed pikeperch tissues were similar while the yellowness was different. Thus, wild fish are characterised by a lower intensity of tissue colour.

In brief, colour in Percid fishes is considered to be a major sensory quality attribute, which is governed by intrinsic, trophic and environmental factors.

33.2.3.3 Texture

Texture is another important attribute of fish flesh sensory quality, which covers all rheological and structural properties of a product. Depending on whether it is assessed by a sensory analysis panel or by using test instruments, texture is expressed through a wide variety of terms which include hardness, springiness, cohesiveness, gumminess and mouth feel, among others (Haard 1992; Cardello et al. 1982). In any case, its assessment requires specific conditions to be established, especially when *rigor mortis* occurs, as well as conditions for product preservation (shelf life, method) and preparation (raw vs. cooked), which may dramatically affect the various descriptors measured.

On the whole, flesh texture is influenced by many factors, including the structure of the muscle and properties of its components, in particular myofibrillar and connective tissue proteins and fat. On the other hand, growth rate has a significant impact on the content of muscle fibre, and therefore, potentially influences flesh texture (Johnston 1999). Fish muscle is known to be composed of two main types of fibres, red and white, which vary with the species. In Percid fishes, white fibres make up most of the muscle (the same holds true for most fish), whereas the red muscle appears as a single superficial layer along the horizontal septum (Lindsay 1980).

A number of comparative studies have focused on texture in Percid fishes, taking into account the origin of the fish. In *P. fluviatilis*, Mathis et al. (2000) established that hardness and mouth feel descriptors alone may allow wild fish to be distinguished from their farmed counterparts. In the same way, Gonzàlez et al. (2006) showed that farmed *P. flavescens* are not as firm as wild fish. This result could possibly be attributed to a higher fat content in farmed fish (Lie 2001) as well as higher levels of activity in wild fish, which may improve texture. On the other hand, Lindsay (1980) found no significant differences between deep-fried wild and farmed yellow perch, while Delwiche and Liggett (2004) found significant differences between skin-on, battered- and -fried farmed and wild yellow perch fillets.

Moreover, fillet texture is clearly influenced by rearing conditions in Eurasian perch (Stejskal et al. 2011a). According to these authors, the culture system has a strong effect on fish flesh texture, with lower levels of all texture parameters (hardness, springiness, cohesiveness and gumminess) in fish reared in intensive conditions (recirculation system) compared to the results recorded in fish from ponds.

33.2.3.4 Other Sensory Characteristics

Sensory quality for Percid fishes was also reported for other attributes such as odour, flavour and aftertaste (Delwiche et al. 2006; Lindsay 1980; Stejskal et al. 2011a). Although studies are scarce, a few lessons can be drawn. There are studies that suggest that differences in flavour and preference do exist between farmed and wild *P. flavescens*, with a preference toward wild-caught as reported by Delwiche and Liggett (2004). Fish caught in the wild are generally described as having a more pronounced flavour and are less bland than farmed fish. In addition, Cox and Karahadian (1998) reported differences in sweetness and oxidised flavour between butter-broiled farmed and wild *P. flavescens* fillets at different stages of storage. Usually, wild and farmed fish diverge in flavour because of differences in fatty acid profiles, oxidation processes, dietary ingredients, mineral and amino acid content (Haard 1992). On the other hand, no difference was reported between deep-fried wild and farmed *P. flavescens* (Lindsay 1980). In the same way, Gonzàlez et al. (2006) observed a similar flavour between wild and farmed yellow perch.

According to these authors, the rearing system (recirculation vs. pond) may be an important determinant with respect to off-flavour development. Furthermore, a study focusing on farmed Eurasian perch showed that the culture system (extensive polyculture in pond and natural preys vs. intensive recirculating systems and formulated feed) had no effect on sensory attributes such as odour, flavour, aftertaste and consistency (Stejskal et al. 2011a). Indeed, no aftertaste was detected in 31 % of the samples of intensively farmed fish or in 27 % of the samples of extensively farmed perch. This result must be seen in relative terms knowing that all the fish were kept in clear water for 7 days without feeding before analysis. The aftertaste reported in cultured fish flesh was found to be the result of geosmin and 2-methylisoborneol (Howgate 2004; Selli et al. 2006; Yamprayoon and Noohorm 2000). In this respect, we note that growth of micro-organisms which produced these compounds is dependent on rearing system (i.e. ponds vs. recirculating systems including the water treatment units).

Sensory quality determinism is therefore governed by several factors. While particular emphasis has been placed on *ante mortem* conditions, *peri-* and *postmortem* conditions of the product, from its origin to its consumption, not to mention storage and preparation steps, also affect the various sensory attributes. Comparisons among studies are still tricky, and the determinism of this quality component remains to be fully understood.

33.2.4 Sanitary Component

An issue often brought up by many experts in debates on the "Benefit-Risk" ratio of fish consumption is that of sanitary quality. It covers the qualitative and quantitative analysis of chemical and biological compounds (i.e., heavy metals, organic contaminants, veterinary drugs, additives, mycotoxins, phycotoxins, bacteria, viruses, etc.) that are potentially present in products. While fish is generally recognized as a healthy food providing proteins and nutrients, as well as a good source of unsaturated fatty acids, it may also be an important source of contaminants. In this respect, it is assumed that more than 90 % of human exposure to Persistent Organic Pollutants (POPs) is caused by consumption of contamined animal-based food products, with the largest contribution to this exposure being from fish (Liem et al. 2000). Similarly, fish consumption is the primary pathway to methylmercury exposure in humans in the United States (Driscoll et al. 2007). That is the reason why we have decided to focus on chemical compounds in this section.

Naturally, the sanitary aspects of fish are subject to the same regulations as other foodstuffs. Such legislation aims to strike the appropriate balance between the benefits and risks of substances intentionally used in the production chain (veterinary drugs and additives), as well as to limit contaminant levels in food (heavy metals, organic pollutants, etc.). This idea is to ultimately ensure the health of consumers. Table 33.4 lists maximum values that should not be exceeded in the muscle meat of fish – including in Percid fishes – for a set number of chemical contaminants, as defined by European regulations (Commission Regulation 2008, 2011). This table underscores concerns over metals (lead, cadmium, mercury) and organic pollutants (Polycyclic Aromatic Hydrocarbons, dioxins, Poly-Chlorinated Biphenyls).

The sanitary quality of fish is of interest for all stakeholders in the sector: fishermen, fish farmers, processors and consumers. As opposed to most animal products, the fish supply sector is unique in that it gives consumers the opportunity to choose between wild and farmed products. This is also increasingly true for Percid fishes,

Table 33.4 Maximum levels for different contaminants in muscle meat of freshwater fish, with the exception of diadromous fish species caught in freshwater (According to Commission Regulation 2008, 2011). Dioxins (Poly-Chlorinated DibenzoDioxins or PCDDs and Poly-Chlororinated DibenzoFurans or PCDFs) and Poly-Chlorinated Biphenyls or PCBs

Contaminants	Maximum levels in muscle meat of fish	
Lead	0.3 mg/kg wet weight	
Cadmium	0.05 mg/kg wet weight	
Mercury	0.5 mg/kg wet weight	
Benzo(a)pyrene	2 µg/kg wet weight	
	Sum of dioxins (WHO-PCDD/F-TEQ[a])	Sum of dioxins and dioxin-like PCBs (WHO-PCDD/F-PCB-TEQ)
Dioxins and PCBs	4 pg/g wet weight	8 pg/g wet weight

[a]System of toxic equivalents (TEQs) calculated by using the toxic equivalency factors (TEFs) of each specific compound as established by the World Health Organization (WHO)

whose culture has boomed over the past few years (see Sect. 9 '*Commercial Production*' and Session 10 '*Marketing and Economics*'). The diet and environment of farmed fish are closely monitored,[3] unlike those of their wild counterparts, which may come from potentially contaminated areas. Yet it needs to be understood that the health issue cannot be addressed without taking into account the origin of the fish. The reason for that is simple: contaminants enter organisms mainly through trophic pathways, with gills and skin being secondary pathways (Di Giulio and Hinton 2008). Consequently, apart from incidents or failings in the rearing process, chemical-related health issues mainly apply to wild fish. Nevertheless, we should also mention the particularity of fish produced in ponds. These polyculture systems are closely connected to their surrounding watersheds. As a consequence, these fish could be contaminated by substances derived from several sources (atmospheric deposition, runoff, effluent and food web). The very few studies on the matter – especially aimed to provide quantitative levels of contamination by persistent organic compounds and residues of pesticides in *P. fluviatilis* from extensive fish ponds –did not however give any cause for concern (Lazartigues et al. 2012; Thomas et al. 2012).

When contaminated food (prey) is ingested, xenobiotics are absorbed and, if excretion and metabolic rates are low,[4] they accumulate in organs (mostly in the adipose tissues of organisms for hydrophobic compounds). This process is called "bioaccumulation". Some xenobiotics are persistent, with their concentrations increasing at each trophic level (i.e., with each prey-predator transfer). Thus, carnivorous fish, with a high trophic level, may be especially vulnerable to the so-called biomagnification processes. One such example is contamination of Percid fishes by mercury (Hg), one of the most toxic heavy metals in the environment (Table 33.5). Therefore, in order to protect human health, different safety guidelines have been proposed for total mercury concentrations not to be exceeded in fish, i.e. below 0.3 mg/kg ww (wet weight) according to the U.S. Environmental Protection Agency or below 0.5 mg/kg ww as established by European Commission Regulations. The action level of the U.S. Food and Drug Administration was set at 1 mg/kg ww, the value beyond which fish is considered unfit for human consumption. Taking these values into account, published articles have reported cases which can sometimes be deemed worrisome for some Percid species – *P. flavescens*, *P. fluviatilis* and *S. vitreus* – in which safety guidelines were exceeded (Table 33.5). Conversely, mercury concentrations in *S. lucioperca* were found to be within safety guidelines.

Understanding the determinism of fish sanitary quality thus requires taking **abiotic factors** into consideration, as local conditions affect bioaccumulation processes. Indeed, it is well established that lowering the **pH** of water increases mercury accumulation in fish (Porvari 1998). Furthermore, the **level of contamination of the local environment** influences contaminant distribution in tissues. Havelková

[3] Indeed, some standards define, for example, the maximum contaminant levels in raw materials to be used in formulated feed.

[4] As such is the case for PCBs and dioxins, for example.

Table 33.5 Examples of studies that measured total mercury levels (in mg/kg wet weight) in the muscle tissue of four Percid species, according to their geographical origins

Species	Origins	Mercury levels[a]	References
Perca flavescens	The Laurentian Great Lake Region (USA)	[0.018–1.2]	Wiener et al. 2012
	Sand Lake	0.75	Dittman and Driscoll 2009
	Upper sister Lake	1.11	
	Sunday Lake	1.08	
	(Adirondack region of New York – USA)		
	Ontario lakes (Canada)	1.20	Rajotte and Couture 2002
Perca fluviatilis	Finnish reservoirs	0.65	Porvari 1998
	Russian lakes	[0.04–1.0]	Haines et al. 1995
	Lake Balkyldak (northern Kazakhstan)	0.70 ± 0.49	Ullrich et al. 2007
		[0.19–1.68]	
	Nitra River (Slovakia)	4.50 ± 1.27	Andreji et al. 2005
		[2.73–6.52]	
Sander vitreus	Supermarkets (Illinois – USA)	0.51 ± 0.13	Burger and Gochfeld 2006
	Lakes in the Abitibi area (Canada)	0.65	Abdelouahab et al. 2008
Sander lucioperca	River Irtysh (northern Kazakhstan)	0.114 ± 0.001	Ullrich et al. 2007
		[0.113–0.115]	
	Four rivers (France)	0.187 ± 0.077	Noël et al. 2012
		[0.112–0.290]	

[a]Data are mean ± standard deviation and/or [minimal – maximal values]

et al. (2008) demonstrated that in *P. fluviatilis* from contaminated locations, total mercury concentrations in the liver are significantly higher than in muscle tissue. Conversely, in fish from uncontaminated locations, mercury concentrations in muscle tissue are usually higher than in the liver. Tissue distribution also varies with **xenobiotics**. In *S. lucioperca*, the target organ of mercury deposition is the muscle, whereas the kidney is the main organ for cadmium deposition (Kenšová et al. 2010). In *P. fluviatilis*, it has also been shown that metal distribution in tissues may vary with **the seasons** (Szefer et al. 2003). This result seems to reflect differences in terms of both metal bioavailability and fish metabolism.

This leads us to look at **biological factors**, which also have an impact on sanitary quality determinism. Simoneau et al. (2005) suggested that faster-growing walleyes have lower mercury concentrations at a given length than their slower-growing counterparts. According to the authors, the **growth rate** as a biological factor outweighs all other environmental factors in accounting for differences in mercury concentrations in the walleye populations studied. This conclusion is supported by findings from other studies focusing on the influence of the fish's **sex** on contaminant levels (Gewurtz et al. 2011; Madenjian et al. 2009). They observed that male walleyes had higher mercury and total-PCB concentrations than females. They see this difference as possibly being attributed to slower growth and lower **gross growth efficiency** (amount of growth per unit of food consumption) in males than in females. PCB concentrations have been shown to be inversely proportional to gross growth efficiency (Madenjian et al. 2009). This study indicated that the release of gametes has very little effect on PCB concentrations in fish. In all the cases, these sex-specific differences in contaminant accumulation in fish in general and in *S. vitreus* in particular should be a consideration in programmes that monitor contaminant concentrations. In the same way, Dittman and Driscoll (2009) reported that total mercury concentrations increase with fish **age** in *P. flavescens*. These authors also used fish condition as an important index of the methyl status of fish. As the fish condition increases, fish can exhibit 'growth dilution' of tissue contaminants leading to lower total mercury concentrations.

In conclusion, sanitary quality determinism is also multifactorial. We decided here to focus solely on chemical contamination, for which the origin of fish (wild vs. farmed) appeared to be a determining factor. The reader should however keep in mind that other compounds may deliberately[5] or inadvertently[6] come into contact with the fish, subsequently affecting their sanitary quality.

33.3 Conclusion: Determinism of Fish Quality

The development of the Percid fish industry – encouraged by intensive research on the domestication of new species (Fontaine 2009; Teletchea et al. 2009) – calls for reflection on the concept and determinism of quality. While the complexity of this broad term has been reaffirmed through the presentation of its multiple components, its outlines and contents have yet to be completely defined, especially for Percid fishes.

First, quality determinism must be looked at differently, depending on whether the fish are wild or farmed-raised. For wild fish, quality is the outcome of an environment which is often complex, heterogeneous and within which intrinsic and extrinsic factors are little or not at all controlled. In other words, fish quality reflects the abiotic and biotic conditions of their habitat. Thereby, wild fish quality, as particularly shown for *P. fluviatilis* (Mairesse et al. 2005, 2006), may understandably vary with the seasons, geographic origins or the physiological status of individuals. Conversely, farmed fish grow in largely controlled environments, which allow quality to be orientated, provided that its determinism is controlled. This information may relevantly be considered when a new industry such as that of Percid fishes emerges. As brought up by Fauconneau (2004), the successful introduction of a new

[5] Such as veterinary drugs.

[6] As can happen with bacteria, viruses, etc.

farmed fish species on the food market depends on how it is valued and how it compares with other existing products. In Percid fishes, like in most fish, this comparison is made with wild fishery products. All in all, the latter would be seen as a quality standard, particularly with consumers who spontaneously compare what is new (farmed Percid fishes) to what they have known (wild Percid fishes). In the light of the results previously shown, the question arises whether such a comparison (farmed vs. wild Percid fishes) is relevant. Another strategy, which will be the second point of this report, may be considered.

We have previously mentioned that stakeholders in the sector cannot reach a consensus on quality indicators. For example, while fish growth heterogeneity matters little to consumers, it is a crucial quality feature for fish farmers. Therefore, quality improvements could apply just as well to the method of production (including dependence on some inputs) as they could to environmental impacts on rearing systems, their productivity, fish sensory/nutritional/sanitary/technological characteristics, animal welfare, and so on. This long but incomplete list indeed requires prioritising the various quality components so as to reach a compromise between the various expectations of the stakeholders in the sector, something which merits particular attention. The following step would then be to define target values for the different quality criteria previously selected, keeping in mind that instead of being fixed, these criteria are subject to change, such as in an evolving social context. In all cases, implementing such a strategy requires quality determinism to be clearly and concisely understood. This will be the third and final point to be considered here.

Throughout this chapter, an attempt has been made to describe the influence that some environmental and/or biological factors may have on Quality components in Percid fishes. We have made reference to numerous studies which have sought to establish the link between trophic factors (food characteristics) and fish nutritional quality (fatty acid profiles), for example. These mostly one-factor studies contribute to providing information that can be useful in understanding the determinism of quality development. Nevertheless, the recorded results can take a variety of pathways depending on the experimental context. It is thus possible to observe either excellent or poor growth in *P. fluviatilis* when given a specific food, or record divergent results in regards to nutritional characteristics, and this in relation to other environmental characteristics (Gardeur et al. 2007). These findings clearly argue the case for studies to adopt multifactorial approaches, such as those already initiated with the experimental rearing of *P. fluviatilis* (Gardeur et al. 2007; Mairesse et al. 2007). These studies allow potentially interesting combinations of factors to be formulated according to selected criteria and target values by looking at the influence of various factors and their interactions on the building of fish quality. Biotic (level of domestication) and trophic (source of lipids) factors seem to play a key role, alone or in combination, on the sensory, nutritional and technological components of fish.

However, *peri* and *post mortem* conditions must not be allowed to ruin all efforts made throughout the different stages of fish rearing. Besides, the intrinsic characteristics of the fish also appear to have a critical influence on quality determinism. Consequently, genetic and epigenetic studies should be included in future research work on fish quality determinism.

References

Abdelouahab N, Vanier C, Baldwin M, Garceau S, Lucotte M, Mergler D (2008) Ecosystems matters: fish consumption, mercury intake and exposure among fluvial lake fish-eaters. STOTEN 407:154–164

Andreji J, Stránai I, Massàyi P, Valent M (2005) Concentration of selected metals in muscle of various fish species. J Environ Sci Health 40(4):899–912

ANSES (2011) Actualisation des apports nutritionnels conseillés pour les acides gras. Rapport d'expertise collective, Mai 2011, ANSES Editions, Maisons-Alfort, France, pp. 1–323

Baker R, Gunther C (2004) The role of carotenoids in consumer choice and the likely benefits from their inclusion into products for human consumption. Trends Food Sci Technol 15(10):484–488

Baker RTM, Pfeiffer AM, Schoner FJ, Smith-Lemmon L (2002) Pigmenting efficacy of astaxanthin and canthaxanthin in fresh-water reared Atlantic salmon, *Salmo salar*. Anim Feed Sci Technol 99(1–4):97–106

Bell MV, Dick JR (2004) Changes in capacity to synthetise 22:6n-3 during early development in rainbow trout (*Oncorhynchus mykiss*). Aquaculture 235(1–4):393–409

Blanchard G, Druart X, Kestemont P (2005) Lipid content and fatty acid composition of target tissues in wild *Perca fluviatilis* females in relation to hepatic status and gonad maturation. J Fish Biol 66:73–85

Borresen T (1992) Quality aspects of wild and reared fish. In: Huss HH (ed) Quality assurance in the fish industry. Elsevier Science Publishers, Amsterdam, pp 1–17

Burger J, Gochfeld M (2006) Mercury in fish available in supermarkets in Illinois: are there regional differences. STOTEN 367:1010–1016

Bykowski P, Dutkiewicz D (1996) Freshwater fish processing and equipment in small plants. FAO, Rome, pp 1–59, N°905

Calder PC, Yaqoob P (2009) Omega-3 polyunsatured fatty acids and human health outcomes. Biofactors 35:266–272

Cardello AV, MallerO KJG, Segars RA, Sawyer FM, Murphy C, Moskowitz HR (1982) Perception of texture by trained and consumer panelists. J Food Sci 47:1186–1197

Celik M, Diler A, Kucukgulmez A (2005) A comparison of the proximate composition and fatty acid profiles of Zander (*Sander lucioperca*) from two different regions and climatic conditions. Food Chem 92:637–641

Commission Regulation (2008) Commission Regulation (EC) No 629/2008 of 2 July 2008 amending Regulation (EC) No 1881/2006 (2006) of 19 December 2006 setting maximum levels for certain contaminants in foodstuffs, L 173/6 – L 173/9

Commission Regulation (2011) Commission Regulation (EC) No1259/2011 of 2 December 2011 amending Regulation (EC) No 1881/2006 of 19 December 2006 as regards maximum levels for dioxins, dioxin-like PCBs and non dioxin-like PCBs in foodstuffs, L 320/18 – L 320/23

Cox DH, Karahadian C (1998) Evaluation of microbial counts, nucleotide degradation, and sensory attributes of cultured and wild yellow perch (Perca flavescens) during refrigerated storage. J Aquat Food Prod Technol 7(1):5–26

Craig JF (2000) Percid fishes. Systematics, ecology and exploitation. Blackwell Science, Scotland

Czesny S, Kolkovski S, Dabrowski K, Culver D (1999) Growth, survival, and quality of juvenile walleye *Stizostedion vitreum* as influenced by n-3 HUFA enriched *Artemia* nauplii. Aquaculture 178:103–115

Delwiche JF, Liggett RE (2004) Sensory preference and discrimination of wild-caught and cultured yellow perch (*Perca flavescens*). J Food Sci 69:144–147

Delwiche JF, Liggett RE, Wallat G (2006) Consumer perception of cultured yellow perch (*Perca flavescens*) and its market competitors. J Food Sci 71:579–582

Di Giulio RT, Hinton DE (2008) The toxicology of fishes. Francis and Taylor Editions, Boca Raton

Dittman JA, Driscoll CT (2009) Factors influencing changes in mercury concentrations in lake water and yellow perch (*Perca flavescens*) in Adirondack lakes. Biogeochemistry 93:179–196

Driscoll CT, Han YJ, Chen CY, Evers DC, Lambert KF, Holsen TM, Kamman NC, Munson RK (2007) Mercury contamination in forest and freshwater ecosystems in the North-eastern United States. Bioscience 57:17–28

Dunajski E (1979) Review paper: texture of fish muscle. J Texture Stud 10:301–318

Einen O, Mørkøre T, Rora AMB, Thomassen MS (1999) Feed ration prior to slaughter – a potential tool for managing product quality of Atlantic salmon (*Salmo salar*). Aquaculture 178(1–2):149–169

Fauconneau B (2004) Diversification, domestication et qualité des produits aquacoles. INRA Prod Anim 17:227–236

Fauconneau B, Faure A, Haffray P, Médale F, Vallet JL (1996) Amélioration des rendements de carcasse et maîtrise de l'état d'engraissement des poissons d'élevage. Piscic Fr Eau Vive Etang 124:25–29

Favaloro E, Mazzola A (2003) Shape change during the growth of sharpsnout seabream reared different conditions in a fish farm of southern Tyrrhenian Sea. Aquac Eng 29:57–63

Fontaine P (2009) Development of European inland fish culture and domestication of new species. Cah Agric 18(2–3):144–147

Gardeur JN, Mathis N, Kobilinsky A, Brun-Bellut J (2007) Simultaneous effects of nutritional and environmental factors on growth and flesh quality of *Perca fluviatilis* using a fractional factorial design study. Aquaculture 273:50–63

Gewurtz SB, Bhavsar SP, Fletcher R (2011) Influence of fish size and sex on mercury/PCB concentration: importance for fish consumption advisories. Environ Int 37:425–434

Gjedrem T (1997) Flesh quality improvement in fish through breeding. Aquac Int 5:197–206

Gonzàlez S, Flick GJ, O'Keefe SF, Duncan SE, McLean E, Craig SR (2006) Composition of farmed and wild yellow perch (*Perca flavescens*). J Food Compos Anal 19:720–726

Guler GO, Aktumsek A, Cakmak YS, Zengin G, Citil OB (2011) Effect of season on fatty acid composition and n-3/n-6 ratios of Zander and Carp muscle lipids in Altinapa Dam Lake. J Food Sci 76:594–597

Gunasekera RM, Shim KF, Lam TJ (1997) Influence of dietary protein content on the distribution of amino acids in oocytes, serum and muscle of Nile tilapia, *Oreochromis niloticus* (L.). Aquaculture 152(1–4):205–221

Haard NF (1992) Control of chemical composition and food quality attributes of cultured fish. Food Res Int 25:289–307

Haines TA, Komov VT, Matey VE, Jagoe CH (1995) Perch mercury content is related to acidity and color of 26 Russian lakes. Water Air Soil Pollut 85:823–828

Havelková M, Dušek L, Némethová D, Poleszczuk G, Svobodová Z (2008) Comparison of mercury distribution between liver and muscle – a biomonitoring of fish from lightly and heavily contamined localities. Sensors 8:4095–4109

Henderson RJ, Tocher DR (1987) The lipid composition and biochemistry of freshwater fish. Prog Lipid Res 26(4):281–347

Hirczak M (2007) La co-construction de la qualité agroalimentaire et environnementale dans les stratégies de développement territorial. Une analyse à partir des produits de la région Rhône-Alpes. Dissertation, Université Joseph Fourier, Grenoble 1

Howgate P (2004) Tainting of farmed fish by geosmin and 2-methyl-isoborneol: a review of sensory aspects and uptake/depuration. Aquaculture 234:155–181

Jacquemond F (2004) Separated breeding of perch fingerlings (*Perca fluviatilis* L.) with and without initial inflated swim bladder: comparison of swim bladder development, skeleton conformation and growth performances. Aquaculture 239(1–4):261–273

Jankowska B, Zakęś Z, Zmijewski T, Szczepkowski M (2003) A comparison of selected quality features of the tissue and slaughter yield of wild and cultivated pikeperch *Sander lucioperca* (L.). Eur Food Res Technol 217:401–405

Johnston IA (1999) Muscle development and growth: potential implications for flesh quality in fish. Aquaculture 177:99–115

Kenšová R, Ćelechovská O, Doubravová J, Svobodová Z (2010) Concentrations of metals in tissues of fish from the Věstonice Reservoir. Acta Vet Brno 79:335–345

Kestemont P, Vandeloise E, Mélard C, Fontaine P, Brown PB (2001) Growth and nutritional status of Eurasian perch *Perca fluviatilis* fed graded levels of dietary lipids with or without added ethoxyquin. Aquaculture 203:85–99

Kowalska A, Zakęś Z, Siwicki AK, Jankowska B, Jarmołowicz S, Demska-Zakęś K (2012) Impact of diets with different proportions of linseed and sunflower oils on the growth, liver histology, immunological and chemical blood parameters, and proximate composition of pikeperch *Sander lucioperca* (L.). Fish Physiol Biochem 38:375–388

Lazartigues A, Thomas M, Cren-Olivé C, Brun-Bellut J, Le Roux Y, Banas D, Feidt C (2012) Pesticide pressure and fish farming in barrage pond in North-eastern France. Part II: residues of 13 pesticides in water, sediments, edible fish and their relationships. Environ Sci Pollut Res 20:117–125

Lie O (2001) Flesh quality – the role of nutrition. Aquac Res 32:341–348

Liem AK, Fürst P, Rappe C (2000) Exposure of population to dioxins and related compounds. Food Addit Contam 17:241–259

Lindsay RC (1980) Comparative sensory analysis of aquacultured and wild yellow perch (*Perca flavescens*) fillets. J Food Qual 3:283–289

Madenjian CP, Hanchin PA, Chernyak SM, Begnoche LJ (2009) Sexual difference in PCB concentrations of walleyes (*Sandervitreus*) from a pristine lake. STOTEN 407:4526–4532

Mai J, Shetty JK, Kan TM, Kinsella JE (1980) Protein and amino acid composition of select freshwater fish. J Agric Food Chem 28:884–885

Mairesse G (2005) Déterminisme *ante mortem* et variabilité de la qualité nutritionnelle, technologique et organoleptique (couleur et morphologie) de la perche commune *Perca fluviatilis* (L.). Dissertation, Institut National Polytechnique de Lorraine

Mairesse G, Thomas M, Gardeur JN, Brun-Bellut J (2005) Appearance and technological characteristics in wild and reared Eurasian perch *Perca fluviatilis* (L.). Aquaculture 246:295–311

Mairesse G, Thomas M, Gardeur JN, Brun-Bellut J (2006) Effects of geographic source, rearing system, and season on the nutritional quality of wild and farmed *Perca fluviatilis*. Lipids 41(3):221–229

Mairesse G, Thomas M, Gardeur JN, Brun-Bellut J (2007) Effects of dietary factors, stocking biomass and domestication on the nutritional and technological quality of the Eurasian perch *Perca fluviatilis*. Aquaculture 262:86–94

Mariojouls C (2000) Introduction to quality. Quality concepts; quality perception by producers, clients and consumers; quality signs (geographic origin, eco-labelling, etc.); translation of quality concepts into products, procedures and services. In: Global Quality Assessment in Mediterranean Aquaculture. Zaragoza, CIHEAM, 2000, 15–21 (Cahiers Options Méditerranéennes, n. 51)

Mathis N, Feidt C, Fontaine P, Brun-Bellut J (2000) Comparative sensory and physical analysis of cultured and wild Eurasian perch. In: International Conference AQUA 2000, Nice, France, EAS Special Publication, 28, European Aquaculture Society, Ostende.

Mathis N, Feidt C, Brun-Bellut J (2003) Influence of protein/energy ratio on carcass quality during the growing period of Eurasian perch (*Perca fluviatilis*). Aquaculture 217:453–464

Mjoun K, Rosenstrater KA, Brown ML (2012) Culture performance and tissue fatty acid compositions of yellow perch (*Perca flavescens*) fed different dietary lipids. Aquaculture 360:17–24

Molnár T, Szabò A, Szabò G, Szabò C, Hancz C (2006) Effect of different dietary fat content and fat type on the growth and body composition of intensively reared pikeperch *Sander lucioperca* (L.). Aquac Nutr 12:173–182

Mørkøre T, Austreng E (2004) Temporal changes in texture, gaping, composition and copper status of Atlantic salmon (*Salmo salar*, L.) fed moist feed or extruded feed. Aquaculture 230:425–437

Nardone A, Valfré F (1999) Effects of changing production methods on quality of meat, milk and eggs. Livest Prod Sci 59(2–3):165–182

Noël L, Chekri R, Millour S, Merlo M, Leblanc JC, Guérin T (2012) Distribution and relationships of As, Cd, Pb and Hg in freshwater fish from five French fishing areas. Chemosphere 90:1900–1910

Orban E, Nevigato T, Masci M, Di Lena G, Casini I, Caproni R, Gambelli L, De Angelis P, Rampacci M (2007) Nutritional quality and safety of European perch (*Perca fluviatilis*) from three lakes of Central Italy. Food Chem 100:482–490

Özyurt G, Polat A, Loker GB (2009) Vitamin and mineral content of pike perch (*Sander lucioperca*), common carp (*Cyprinus carpio*), and European catfish (*Silurus glanis*). Turk J Vet Anim Sci 33(4):351–356

Pirestani S, Ali Sahari M, Barzegar M, Seyfabadi SJ (2009) Chemical compositions and minerals of some commercially important fish species from the South Caspian Sea. Int Food Res J 16:39–44

Porvari P (1998) Development of fish mercury concentrations in Finnish reservoirs from 1979 to 1994. STOTEN 213:279–290

Rajotte JW, Couture P (2002) Effects of environmental metal contamination on the condition, swimming performance, and tissue metabolic capacities of wild yellow perch (*Perca flavescens*). Can J Fish Aquat Sci 59:1296–1304

Rosauer DR, Biga PR, Lindell SR, Binkowski FP, Shepherd BS, Palmquist DE, Simchick CA, Goetz FW (2011) Development of yellow perch (*Perca flavescens*) broodstocks: initial characterization of growth and quality traits following grow-out of different stocks. Aquaculture 317:58–66

Rougeot C, Minet L, Prignon C, Vanderplasschen A, Detry B, Pastoret PP, Melard C (2003) Induce triploidy by heat shock in Eurasian perch, *Perca fluviatilis*. Aquat Living Resour 16(2):90–94

Sarà M, Favaloro E, Mazzola A (1999) Comparative morphometrics of sharpsnout seabream (*Diplodus puntazzo* Cetti, 1777), reared in different conditions. Aquac Eng 19:195–209

Sargent JR, Henderson RJ, Tocher DR (1989) The lipids. In: Halver JE (ed) Fish nutrition. Academic, New York

Sargent JR, Bell JG, Bell MV, Henderson RJ, Tocher DR (1995) Requirement criteria for essential fatty acids. J Appl Ichthyol 11:183–198

Sargent JR, Tocher DR, Gordon Bell J (2002) The lipids. In: Halver JE, Hardy RW (eds) Fish nutrition. Academic, Elsevier Science, New York

Selli S, Rannou C, Prost C, Robin J, Serot T (2006) Characterization of aroma-active compounds in Rainbow Trout (*Oncorhynchus mykiss*) eliciting an off-odor. J Agric Food Chem 54:9496–9502

Simoneau M, Lucotte M, Garceau S, Laliberté D (2005) Fish growth rate modulate mercury concentrations in walleye (*Sander vitreus*) from eastern Canadian lakes. Environ Res 98:73–82

Stejskal V, Vejsada P, Cepak M, Špička J, Vacha F, Kouril J, Policar T (2011a) Sensory and textural attributes and fatty acid profiles of fillets of extensively and intensively farmed Eurasian perch (*Perca fluviatilis* L.). Food Chem 129:1054–1059

Stejskal V, Policar T, Křišťan J, Kouřil J, Hamáčková J (2011b) Fin condition in intensively cultured Eurasian perch (*Perca fluviatilis*). Folia Zool 60(2):122–128

Sulistyo I, Rinchard J, Fontaine P, Gardeur JN, Capdeville B, Kestemont P (1998) Reproductive cycle and plasma levels steroids in female Eurasian perch *Perca fluviatilis*. Aquat Living Resour 11(2):101–110

Summerfelt RC, Clayton RD, Yager TK, Summerfelt ST, Kuipers KL (1996) Live weight-dressed weight relationships of walleye and hybrid walleye. In: Summerfelt RC (ed) Walleye culture manual, vol 101, NCRAC Culture Series. Iowa State University, Ames

Svanbäck R, Eklöv P (2002) Effects of habitat and food resources on morphology and ontogenetic growth trajectory in perch. Oecologia 131(1):61–70

Szefer P, Domagala-Wieloszewska M, Warzocha J, Garbacik-Wesolowska A, Ciesielski T (2003) Distribution and relationships of mercury, lead, cadmium, copper and zinc in perch (*Percafluviatilis*) from the Pomeranian Bay and Szczecin Lagoon, southern Baltic. Food Chem 81:73–83

Teletchea F, Fostier A, Kamler E, Gardeur JN, Le Bail PY, Jalabert B, Fontaine P (2009) Comparative analysis of reproductive traits in 65 freshwater fish species: application to the domestication of new fish species. Rev Fish Biol Fish 19:403–430

Thomas M, Lazartigues L, Banas D, Brun-Bellut J, Feidt C (2012) Organochlorine pesticides and polychlorinated biphenyls in sediments and fish from freshwater cultured fish ponds in different agricultural contexts in north-eastern France. Ecotoxicol Environ Saf 77:35–44

Tidwell JH, Coyle SD, Evans J, Weibel JM, Dodson K, Jones H (1999) Effects of culture temperature on growth survival and biochemical composition on yellow perch *Perca flavescens*. J World Aquac Soc 3:324–330

Turki-Missaoui O, M'Hetli M, Kraïem MM, Chriki A (2011) Morphological differentiation of introduced pikeperch (*Sander lucioperca* L., 1758) populations in Tunisian freshwaters. J Appl Ichthyol 27:1181–1189

Twibell RG, Watkins BA, Brown PB (2001) Dietary conjugated linoleic acids and lipid source alter fatty acid composition of juvenile yellow perch, *Perca flavescens*. J Nutr 131(9):2322–2328

Uauy R, Valenzuela A (2000) The health benefits of n-3 fatty acids. Nutrition 16:680–684

Ullrich SM, Ilyushchenko MA, Tanton TW, Uskov GA (2007) Mercury contamination in the vicinity of a derelict chlor-alkali plan. Part II: contamination of the aquatic and terrestrial food chain and potential risks to the local population. STOTEN 381:290–306

Uysal K, Aksoylar MY (2005) Seasonal variations in fatty acid composition and the n-6/n-3 fatty acid ratio of pikeperch (*Sander lucioperca*) muscle lipids. Ecol Food Nutr 44:23–35

Valfré F, Moretti VM (1991) Characteristics, quality and control of animal products for human consumption. In: Boyazoglu J, Renaud J (eds) EAAP Publ Proc. The round table, the livestock production sector in Eastern Europe as affected by current changes, vol 57. FAO, Rome, pp 144–148

Wang N, Xu X, Kestemont P (2009) Effect of temperature and feeding frequency on growth performances, feed efficiency and body composition of pikeperch (*Sander lucioperca*). Aquaculture 289:70–73

Wiener JG, Sandheinrich MB, Bhavsar SP, Bohr JR, Evers DC, Monson BA, Schrank CS (2012) Toxicological significance of mercury in yellow perch in the Laurentian Great Lakes region. Environ Pollut 161:350–357

Xu X, Kestemont P (2002) Lipid metabolism and FA composition in tissues of Eurasian perch *Perca fluviatilis* as influenced by dietary fats. Lipids 37:297–304

Xu X, Fontaine P, Mélard C, Kestemont P (2001) Effects of dietary levels on growth, feed efficiency and biochemical composition of Eurasian perch *Perca fluviatilis*. Aquac Int 9:437–449

Yamprayoon J, Noohorm A (2000) Geosmin and off-flavour in Nile tilapia (*Oreochromis niloticus*). J Aquat Food Prod Technol 9:29–41

Chapter 34
The Market for Eurasian Perch

Damien Toner

Abstract Whilst European perch is consumed in many countries the majority of the market revolves around Switzerland both geographically and in terms of demand and price. Whilst local fisheries such as Lac Leman supply some production, the vast majority of perch consumed in Switzerland is imported. Wild fisheries in Estonia, Poland and Russia supply the bulk of product on the marketplace whilst aquaculture production remains small. The Swiss are not large consumers of seafood but consumption is increasing and perch is likely to remain an important species. However increased consumption of perch is only likely to come on the back of product development and value added offerings as consumers tastes become more refined. The focus on sustainability is particularly important for the Swiss consumer and any production from aquaculture in the future will have to take this into consideration. It is likely in the medium term that large retail multiples will play an increasingly important role in the marketing and sale of perch products. Whilst aquaculture production will likely grow as wild catches decline, it is uncertain whether the market for perch can be extended beyond its traditional base. If market consumption is to increase, considerable resources will have to be invested in consumer education and perception.

Keywords *Perca fluviatilis* • Consumer • Market • Fisheries • Perch culture

34.1 Introduction

One of the fastest growing food industries in the world today is aquaculture. Aquaculture has been the driving force behind the growth in the seafood industry and fish is considered as a main protein source for over one billion people in developing countries and accounts for nearly 7.5 % of the world's food production (Fish Site 2008). Customers today have become more experienced, educated

D. Toner (✉)
Aquaculture Division, Bord Iascaigh Mhara, Crofton rd, Dun Laoighaire, Co. Dublin, Ireland
e-mail: toner@bim.ie

P. Kestemont et al. (eds.), *Biology and Culture of Percid Fishes*,
DOI 10.1007/978-94-017-7227-3_34

and sophisticated about their seafood selection leading to a significant change in consumer tastes and preferences.

Market information for seafood is notoriously difficult to source and percids are no exception to this rule. Seafood is a commodity traded on a global market through a complex series of distribution routes. Wild caught and farmed products sourced globally enter Europe from many different sources and the logistics of transporting chilled and frozen seafood can involve numerous traders, transporters, freight forwarders, warehousing and wholesale operatives. The last decade has however seen a significant change in the sourcing of seafood. The drive for sustainable seafood, led by consumer demand, has seen significant efforts to improve traceability across the sector. The implementation of quality standards through EU and national legislation along with partnership programmes implemented by major retail chains has seen the emergence of sustainable aquaculture and fishing practices.

In tandem with sustainable fishing, has come increased consumer demand for seafood and with an extra 40 million tons of extra seafood required globally by 2030 (Harrison 2002), there are renewed efforts to increase aquaculture production. Increased aquaculture production and a sustainable drive through implementation of vigorous quality standards should lead to more transparent and accurate trade figures in the future.

34.2 Market Supply

According to FAO estimates, European perch production in Eurasia amounted to 27,537 tons in 2010. Of this just over 318 t came from aquaculture sources (FAO, FishStatJ). The FAO data, which also includes landings of sea or brackish water perch, underline the importance of production in Eastern European countries. These countries, and in particular Estonia, Poland and the Russian Federation, are now the leading suppliers of perch to European markets. Although Finland is by far the largest perch producer in Europe the majority of this production is for the home market and relatively small quantities are exported. There is also evidence of yellow perch sourced from Canada being sold on European markets although the quantities are not verifiable. There are also reports that trade from North America has been reversed with a market now in the U.S. for European perch (Watson 2003).

34.2.1 Wild Landings

Commercial fisheries for perch exist throughout Europe. From fisheries on inland lakes such as Lac Leman in Switzerland and Lough Neagh in Ireland to Baltic Sea stocks in Finland and the Russian Federation, perch is found in a variety of territories and ecosystems. Whilst certain fisheries are in decline, others seem to be well

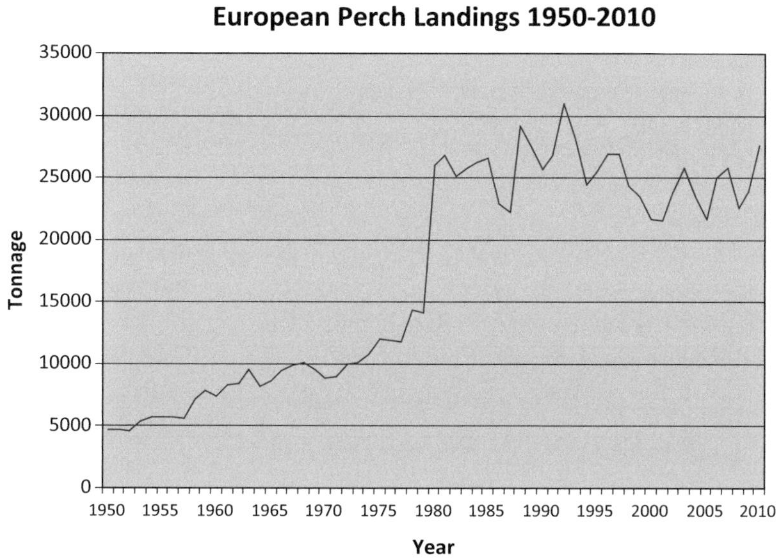

Fig. 34.1 Reported European Perch Landings in Europe 1950–2010 (FAO Fishstat J Statistical database 2013)

managed and sustainable. Overall production from wild catch has stabilised at around 24,000 tons per annum following a high of over 30,000 tons in 1993 (FAO, FishStatJ). This peak followed an exponential increase in perch landings from a low base of around 6000 t in the 1950s (Fig. 34.1). Such an increase is mirrored in other species, as Europe sought to increase food production for its citizens. Whilst other fish species have experienced sharp declines in capture fisheries, perch seems to be relatively robust although overfishing and environmental changes can always impact the fishery on a year to year basis.

Finland, Russian Federation, Estonia and Poland constitute the bulk of European perch landings. The majority of these landings come from commercial fishermen using gill nets. There is however a significant landing in Finland from recreational anglers. In 2010 recreational anglers in Finland accounted for some 7900 t of wild caught perch with a value of over €13 million (Finnish Game and Fisheries research Institute 2010). The vast majority of this fish is for home consumption and there appears to be little if any of this fish making its way to markets in Switzerland and France. This may be accounted for by the size preference of the market. Nordic countries tend to have a preference for large perch (>250 g) whilst the alpine markets source considerably smaller fish (<150 g). Whilst the above figures show a relative overall stability in perch landings, a different picture emerges when we look at the landings by country. Since 1980, Finnish landings have decreased significantly whilst at the same time landings from the Russian Federation have increased (Fig. 34.2). Notably both these fisheries operate in the Baltic Sea area.

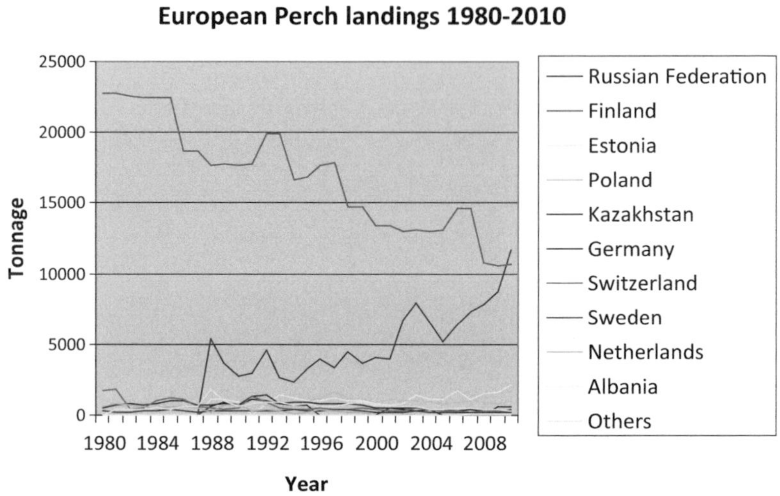

Fig. 34.2 Landings by country of European perch 1980–2010 (FAO Fishstat J Statistical database 2013)

34.2.2 Aquaculture Production

Aquaculture production of European perch still only makes up some 1 % of overall production. It is however increasing with more production units being established in a number of different countries. Notably this increased production is predominately from Recirculating Aquaculture System(s) (RAS). In recent years RAS units for the culture of perch have been established in Switzerland, Ireland, France and Denmark with more units at planning stages. Traditionally perch was produced in polyculture with other species in pond based systems. In the Czech Republic, perch is grown with a variety of coarse fish in extensive pond systems. Accurate data for such production is difficult to ascertain as statistical databases often group perch with other freshwater species. The FAO data (Fig. 34.3) does show a steady increase in farm production with Switzerland, Ireland and France being the main producers. It is likely that production will increase steadily in the coming years as existing farms fill their licensed capacity. Valperca SA in Switzerland, Lucas Perche in France and Clune Fisheries Ltd in Ireland are amongst the biggest producers of farmed perch. These farms are RAS based and capable of producing perch throughout the year on a consistent basis. Such consistency is attractive to larger multiples in target markets.

34.3 European Market Overview

The EU is increasingly dependent on imports of fish and fishery products to meet its needs. In 2009, the EU imported €15.5 billion worth of fish and fishery products, accounting for more than 60 % of it fish consumption. Europe exported €2.5 billion

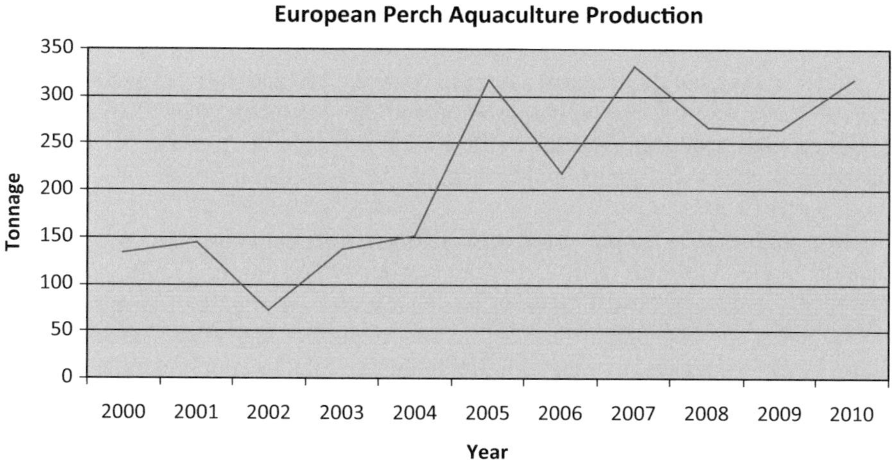

Fig. 34.3 Aquaculture production of European perch 2000–2010 (FAO Fishstat J Statistical database 2013)

worth of fisheries goods in 2009, the bulk of it to large markets like Norway, Russia and Japan. Total aquaculture production in the EU is around 1.3 million tons. (Comext, December 2010). The market for perch is small by comparison to other species such as Atlantic salmon (*Salmo salar*) and Seabass (*Dicentrarchus labrax*). Traditional markets for perch exist in some Nordic countries such as Finland and Sweden, but the majority of consumption is in the alpine regions of Switzerland, Italy, Germany and France.

34.3.1 Swiss Market

Switzerland is the most important market in Europe for perch. Market demand for perch originates from a traditional fishery for the species in Swiss lakes such as Lac Léman (Lake Geneva), Neuchâtel and Lake Constance. Within the country, Perch is regarded as a typically Swiss product, exemplified by its Alpine name of 'Egli'. The national identification with perch continues despite the fact that there are ever diminishing returns from the lakes and that the market has been import dependent for some time. During summer months the majority of lakeside restaurants advertise and sell perch fillets. Additionally in German speaking cantons there is increased demand for perch during the Christmas and Easter season. Traditionally Swiss people are meat and dairy consumers but there has been a significant trend towards seafood in recent years. In period 2006–2010, the per capita consumption of seafood increased by over 25 % from 7.2 (2006) to 9.92 kg (2010) (Genevalunch 2010). Fish consumption varies between cantons and by language population. In the German speaking areas focus is on fish fillets and fish convenience whilst in French and Italian speaking areas consumers prefer shellfish, crustaceans and fish of a

Mediterranean origin. There is a high acceptance of freshwater fish especially if it comes from Switzerland and/or is organic.

Perch landings in Switzerland averaged between 200 and 300 tons per year over the 2000–2010 period. The highest level of landings was recorded in 2003, at almost 500 tons (Swiss Federal Statistics Office 2012). Feedback from the Swiss trade, however, suggests a diminishing domestic resource and an increasing dependence on imports.

Swiss statistics do not provide specific figures on perch imports. Perch figures are included under a general freshwater fish category, which covers a range of species. For Switzerland, the most important freshwater species imported are perch, pike perch, pollen and Nile perch. However, in the category of imported fillets, the inclusion of salmon significantly inflates the total figures. Total fish imports for 2009 show a figure of 53,182 tons including some 17,116 tons of fresh fish (whole and fillets). Freshwater fish imports which include perch and salmon totalled 20,479 tons in 2009, some 38.5 % of total imports (Swiss Federal Statistics Office 2012).

According to Swiss importers, Estonia and Poland are the most important suppliers of perch to the Swiss market. Imports into Switzerland from both these countries of fish are currently running at between 1000 and 1300 tons, mostly in fillet form. It is likely that a high percentage of these imports are perch. Secondary supplier countries indicated by Swiss importers include Germany, the Netherlands and the Russian Federation. Total freshwater supplies from these countries amount to almost 1200 tons a year, again mostly in fillet form. A proportion of this figure may be accounted for by Nile perch, particularly in the case of imports from the Netherlands. It is difficult to calculate the percentage of perch in the German figures, but it is likely to be significant. Swiss importers calculate the total import of perch fillets to be in the region of 2000–3000 tons per annum.

34.3.1.1 Sources of Perch

Origin is important for Swiss perch importers. Supplying countries are ranked by importers not only on the basis of volume but also on a quality basis. Importers are agreed that Estonia is by far the most important perch supplier to the Swiss market. Perch is fished in Estonian lakes, particularly the largest lake, Chudskoye, and filleted in local processing plants for export in fresh or frozen form to Switzerland. Estonian exporters are also sourcing perch from Russia and possibly from other countries for export, together with local supplies, to Switzerland. All supplies are labelled as Estonian. Estonian prices, which are lower than other origins such as Swiss, Irish and Polish, set the rate for the market.

Poland is regarded as the second most important perch supplier by a number of Swiss importers. The Polish share is not, however, as important as it was some years ago. Over-fishing has been one factor. There is also a view that Polish perch is being exported to the U.S. One of the largest perch traders in Poland is a former Canadian company which appears to be now involved in the trade with North America. Perch fillets from Poland are exported directly to Switzerland, as well as indirectly through the Netherlands.

The main centre for German supplies of fresh perch fillets to Switzerland is the port of Friedrichshafen on Lake Constance. Lake Constance is an important source of perch within Germany. Supplies from the lakes of the former East Germany also pass through Friedrichshafen, around which there are a number of perch processors.

Netherlands exports some 500 tons of fresh-water fish to Switzerland. It is likely that a small percentage of this is perch from Dutch fisheries. Some 140 tons of perch is estimated to have been caught in Dutch fisheries in 2010 (FAO Fishstat 2010–2011). In the European freshwater fish market, the Netherlands plays essentially a trading role, importing, in the case of perch, from Scandinavia and Eastern Europe for export to Switzerland. Dutch traders also play a key role in European imports of Nile perch *Lates niloticus* and it is certain that a significant share of Dutch freshwater fish exports to Switzerland are made up of this species.

There is also a long tradition of perch exports from Lough Neagh, Northern Ireland to Switzerland. Lough Neagh is also an important fishery for eels *Anguilla anguilla* and pollan *Coregonus pollan*. Accurate statistics are not available for perch exports from Northern Ireland but however are likely to be relatively small in quantity. The volume of perch identified as Russian or Finnish on the Swiss market appears also to be relatively limited. It is likely that as already stated, perch from these countries comes through third party countries and is relabelled. Perch from Sweden and Canada also appears on the market place. Although a different species, yellow perch *Perca flavescens* is similar to European perch and does find its way on to the market place, often through distributors in Estonia. The quality of Canadian yellow perch is acceptable on the Swiss market, with just the skin considered as being thicker than that of the European perch.

Landings of perch by professional fishermen in Switzerland are from the larger lakes, particularly Lakes Geneva and Neuchâtel (Fig. 34.4). Swiss wild caught perch is now mainly sold directly by fishermen to local restaurants. There are also some sales to specialised wholesalers. At the upper end of this market, ie. the catering segment in the French speaking region, there is some differentiation between the origins of perch from within Switzerland itself. Perch from Lake Geneva is perceived as having a distinct taste, with meat firmer than perch from Neuchâtel. Top prices are paid for perch from Lake Geneva. Prices for other origins are influenced by how closely perch from these origins compare with Lake Geneva perch. Lough Neagh perch prices are traditionally high in Switzerland because it is considered to be the closest to that from Lake Geneva (Figs. 34.5 and 34.6). Perch from brackish Scandinavian waters, on the other hand, is perceived as quite different in taste to the Swiss product. It is likely, however, that in the medium and lower segments of the Swiss perch market these details are less relevant, and product is purchased solely on price and on other general quality criteria. In addition to landings of perch in Switzerland there a number of aquaculture farms including Valperca SA and Tropenhaus SA. These farms sell directly through their own restaurant (Tropenhaus) and to major retailers such as Migros and Coop. They trade heavily on the Swiss perch brand which is important to consumers in Switzerland. Clune Fisheries Ltd in Ireland also sells its fish through Swiss retail multiples whilst Lucas Perche in France distribute their fish through its wholesale operation in Geneva.

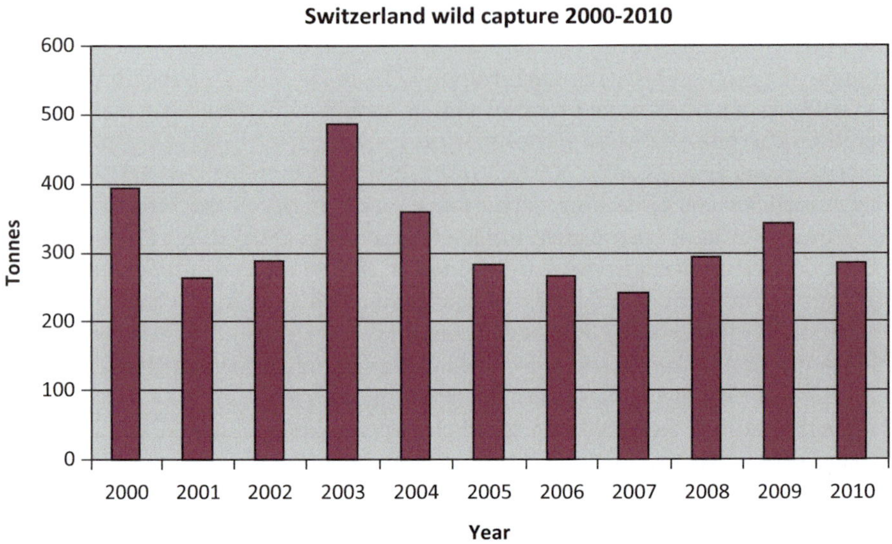

Fig. 34.4 Wild fisheries landings of European Perch in Switzerland 2000–2010 (FAO Fishstat J Statistical database 2013)

34.3.1.2 Distribution and Logistics

Importers, wholesalers and the large multiples in Switzerland generally use the same import hubs for all seafood imports. Product arrives daily into Basel, Zurich and Geneva from a variety of sources. Padborg (Denmark), Breskens (Netherlands) and Boulogne sur mer (France) are the main transport hubs. Seafood from around the globe is gathered in each of these distribution hubs before collective transport to the main Swiss cities. Transport Vooruit B.V of the Netherlands is one of the largest freight forwarders in Europe concentrating on seafood distribution and is used by all the main importers.

The Swiss market comprises a number of importers, wholesalers and retail outlets. Larger importers such as Stadel Fischimport AG have been distributing fish throughout Switzerland since 1922 and have an intricate logistics network supplying perch to various other wholesalers, retailers and restaurants. There are a large number of companies in the Geneva area involved in perch import and distribution such as Gastromer SA, Perche Import Sa, Corema Frozen Seafoods SA, Brunner &Brunner SA and Lucas Geneve SA. These importers also deliver directly to other wholesalers and retailers.

Swiss retail in general is highly concentrated. Migros (including Globus) and Coop together have a market share of over 50 %. In the seafood retail sector, their share is over 80 % followed by Manor with a little over 10 %. There are around 500 fresh fish counters in Switzerland in what is still a relatively young category. Both Migros and Coop source seafood through their agents Micarna and Bell respectively. Perch is generally sold at retail as loose fillets. In one spot market check carried out in by BIM in 2014 fresh perch fillets from Lac Leman were retailing at 68 chf (€56)/kg, Farmed Swiss and Irish perch at 68 chf (€56)/kg, Estonian perch at

Fig. 34.5 Farmed Irish perch fillets on a wet fish counter in Migros, Switzerland (Photo courtesy Simon Kaufmann)

Fig. 34.6 Skinless farmed perch fillets after trimming. Customers have strict guidelines for fillet shape and quality (Photo Damien Toner)

35.80 chf (€29)/kg and Canadian perch at 47.80 (€39.5)/kg. Such price differentiation is seen across the marketplace from the small lakeside restaurant to the large retailer. Fishermen and producers can be paid prices as low as €9/kg to as high as €30/kg for fillets depending on a whole range of factors such as source, seasonality, quality etc. Prices are generally always quoted for fillets. Fillet size and product presentation differs according to region (Table 34.1).

Some variations in demand also occur with a peak in German speaking cantons around Easter and supply shortages associated with lakes freezing over during winter in wild fisheries. Perch is also traded in frozen fillet form, particularly in the catering trade. It freezes well and some companies specialise in trade of glazed fillets.

Table 34.1 Product specifications for perch fillets in the Swiss market (Watson 2003)

French speaking cantons	
10–20 g	Skin on/(off) main size
20–30 g	Skin off
30–50 g	Skin off
50 g+	Skin off
German speaking cantons	
20–40 g	Skin on/off
40–60 g	Skin off

Sustainability of seafood is an import concept amongst the Swiss seafood buying consumer. Nearly all import stakeholders (consumer, retailer, wholesalers etc.) have mandated a NGO to care in Switzerland for seafood sustainability, WWF Switzerland. The larger retailers and wholesalers have joined the "WWF seafood group" representing over 80 % of the Swiss seafood buying power. Most prominent members are Coop (Bell), Migros (Micarna), Bianchi and Stadel. The WWF seafood group doesn't act simultaneously as one group but all of the members have signed a bilateral contract with WWF that includes an identical step-by-step process. The whole seafood assortment of the company gets a screening by WWF: product by product addressing issues such as:

(a) State of the stock
(b) Fishery methods
(c) Environmental impacts
(d) Social standards

Each screened product gets a grade from 1 (very good) to 6 (very bad); for all products. With grade 5 or 6 the company gets a 1-year-period to bring it to a better grade (e.g. buying from other suppliers or other stocks). If the new screening is still on 5 or 6 the products will be delisted. Also included in the contract is the active encouragement of MSC and organic seafood, the formulation of a 5 year buying policy, a transparent product declaration, yearly reporting and an active communication of the partnership with WWF.

Switzerland is the most important market in terms of quantity and value for European Perch. Future trends may include the development of more value added and convenience products in line with other seafood's. Price variability is significant and may pose a hurdle to new entrants particularly from aquaculture. It is important therefore for new entrants aim their products at the high niche end of the market such as sustainable and organic rather than trying to compete with the volume and scale of traditional fisheries.

34.3.2 German Market

Per capita consumption of seafood in Germany is estimated at 15.7 kg which, if higher than the Swiss level, is still well below the average for the EU (22.1 Kg). German consumers tend to be traditional and conservative in their taste for seafood, with just five species groups accounting for over 70 % of sales (Rath 2012) The balance of seafood consumption covers freshwater fish, other sea fish and shellfish. Overall, per capita consumption is higher in the north of Germany near the coastal regions, but it is likely that consumption of freshwater fish is higher in the south and east of the country near the traditional freshwater production areas.

Total perch landings from inland commercial fisheries during 2010 are estimated at over 360 tons. Of this 122 t is from inland fisheries such as Mecklenburg Vorpommern in north-east Germany, and Lake Constance in the south. Some 239 t is from brackish waters in the North of Germany (FAO Fishstat 2010–2011).

The main consumer markets for perch in Germany are close to the traditional freshwater areas, particularly around the bigger lakes, in the south and north-east of the country. As noted above, Lake Constance in the south and the state of Mecklenburg Pommern in the north-east are the main sources for the domestic perch supply. Perch is more likely to feature on menus or on distributors' product lists in Bavaria, Baden Wurtenberg or Berlin than in the northern coastal centres (Bremen, Hamburg, Kiel), which are more oriented to sea fish consumption. The German perch market, tends more towards the catering rather than the retail sector. It is also essentially a fillet market, with an emphasis on frozen rather than fresh.

There are a number of German importers specialised in freshwater fish and perch is generally included in the product range. E. Geiger, based in Meersburg beside Lake Constance, imports a range of fresh and frozen freshwater species. Geiger, like a number of other importers, is involved not only in the import and domestic distribution of perch fillets, but also in an export trade, principally with Switzerland. Other companies in this category include Ross Fisch and Fimoda.

Because of limited supplies and the regional orientation of demand, a number of the larger national fish distributors do not carry perch on their product lists. Deutsche See, the country's largest seafood wholesaler, carries fresh and frozen trout, pike and pike perch in its range, but not perch products. On the other hand, perch is sold by some, but not all, regional seafood wholesalers in both Munich and Berlin. Wholesale distribution of perch is mainly to catering outlets. The species does not generally feature with the national retail chains.

The size range for perch fillets is broader in Germany than in Switzerland. Specifications vary according to the region. Demand in Upper Bavaria and the Lake Constance area is for the 40–60 g fillet. In other parts of Germany, the preferred average fillet size is near to 80 g. In Munich, for example, Kagerer seafood and Delicatessen supplies fillets in the 40–80 g range. In general, fillets are sold mainly skin off. For certain Bavarian wholesalers, perch from Lake Constance occupies the top end of the range. Meat tenderness appears to be a key criterion, with this characteristic more likely with smaller fillets.

34.3.3 Italian Market

With a population of 58 million and an annual per capita seafood consumption of 23 kg, Italy is a major European seafood market. Total domestic consumption is estimated at over 1.3 million tons. Domestic production, while important at 800,000 tons, is insufficient to cater for the home market, which, as a result, is strongly import dependent. Italy is, in fact, the fifth largest seafood importer in the world and the third largest in Europe behind France and Spain. Since 1990, seafood consumption in Italy has fluctuated around 1.2 and 1.3 million tons per annum, while imports have averaged at over 600,000 t (Globefish 2008).

After years of stagnation and relatively low levels for a Mediterranean country, seafood consumption in Italy has started increasing significantly in the last 4 or 5 years. The current level of seafood consumption is estimated to be 23 kg per capita, with a growth rate of approximately 2 % per year. It is estimated that 7.7 kg per capita is consumed at home (21.7 kg per family), while 15.3 kg per capita is consumed in restaurants and other facilities.

Italy has about 20 000 km^2 of lakes, reservoirs and rivers. In 2003, the recorded national output of freshwater fish was 4 379 t. The number of authorized professional fishermen was about 400, located mostly in northern Italy, operating through 37 cooperatives. Most production came from the region of Lombardia, with 2 773 t in annual landings. The catch comprised cyprinid fishes (about 20 %), salmonids (10 %), pikes and bass (5 %), eels (3 %), with the rest of the catch pooled in official statistics as "other fishes" (FAO Fishstat 2010–2011). There are no separate statistics for perch landings from these lakes and it is likely to be small in volume. Perch is very popular in Milan, alpine areas and other northern cities were it is referred to as *Pesce Persico*. Fillets of perch on the Milan market tend to be sourced from abroad, mainly from Estonia (ISMEA 2011). A recent ISMEA report reflected a downturn in demand for products such as perch compared to an increase in demand for fish such as shrimps and salmon mainly due to lower import prices.

34.3.4 French Market

Commercial perch production in France is limited to several hundred tons. FAO figures estimate output in 1996 as 250 tons, but do not appear to provide estimates for subsequent years. French statistics generally classify perch production under a miscellaneous freshwater category, reflecting the limited economic value of the fishery.

France imported over 536,000 tons of fish in 2009. This figure far exceeds exports of 188,000 tons for the same period and shows the importance of the fish market in France for producers. Freshwater fish production in France in 2010 amounted to 3022 tons of which some 60 tons is estimated as perch. The establishment of new RAS units for perch by Lucas perche and Asialor will no doubt increase this figure if successful. However the majority of new farms established in France in recent years appear to be targeting the Swiss market which is situated close by.

Geographically, the main market for perch in France is in the regions close to the Swiss border such as Lyon, Annecy and Grenoble. Perch is also available in the North east where some traditional pond farms operate.

34.3.5 Market in Nordic Countries

34.3.5.1 Finland

The national fish of Finland is perch (Ahven) and as such it is very popular with recreational anglers. It is fished throughout the country in lakes and coastal waters. It is also a popular target for ice fishing and a market exists for its meat and roe which is sold in local restaurants and fish dealers.. The importance of freshwater fish in Finland has increased in recent years as fishing for species such as herring, cod and salmon fishing is regulated more strictly. The commercially important coastal and inland water species are whitefish, perch and eel. In addition to the domestic catch which is largely consumed in Finland, there are imports of freshwater fish from countries such as Canada, Russian Federation and Estonia.

As a result of the decreased supply and increasing fish processing, the price of Finnish fish is increasing constantly. According to calculations made by Statistics Finland and commissioned by *Helsingin Sanomat*, the price of fresh fish has more than doubled since the beginning of the millennium. Since 2005, the price has gone up by more than 30 %. The proportion of fresh fish in retail stores is only 30 % of all fish available, as various processed fish products have become popular among consumers (Helsinin Sanomat 2008). As noted in 8.3.2.1 inshore fishing catches have decreased considerably over the past 10 years. The main reason for the diminishing coastal fishing catches is that the number of professional fishermen has reduced. More and more fishermen have given up fishing, and there are no newcomers to the field to replace them. At the turn of the millennium, the number of professional fishermen was around 1,000, while at present the number is fewer than 700 (Finnish Game and Fisheries research Institute 2010).

34.3.5.2 Sweden

In 2000 the total consumption of fish and crustaceans for food in Sweden was estimated at about 160,000 tons. Swedes do not eat very much fish and seafood. The per capita consumption has remained fairly stable at about 18 kg per year over the past several years. There has been a long-term trend away from fresh fish towards prepared or preserved sea food products. Consumption of fresh fish, including fish frozen whole and fresh or frozen crustaceans, has stabilised at about 7 kg per person per year, while consumption of prepared or canned fish, crustacean and molluscs has increased a little in the last few years to 9.1 kg per person per year (Seafish 2010).

Sweden is not a great fishing nation but with the Baltic Sea on the east coast and the Kattegat on the west coast and with thousands of fresh water lakes and rivers, Sweden has a domestic fishing industry that nevertheless lands more fish than the market can absorb.

In addition to sea fishing, freshwater fish is professionally fished in lakes and rivers with a landed weight of around 3200 tons in 2010. Some 176 tons of this was of European perch. It is likely that the majority of this is consumed on local markets. Fillet sizes in Nordic countries tend to be bigger than for alpine regions (100 g+).

The major part of fish imported into Sweden comes from the neighbouring countries, Norway and Denmark. In 2001, by value, 92 % of the fresh fish imported into Sweden came from Norway. Norway also accounted for 68 % of Sweden's imports of frozen whole fish, 52 % of frozen fish fillets, 92 % of salted, dried or smoked fish, but only 18 % of crustaceans and 34 % of molluscs. What little of Sweden's imports has not been imported from the above countries has come from a great number of countries, mostly in Europe. The most important fish species in terms of imported fresh, chilled or frozen fish by volume are salmon and cod species, followed by rainbow trout, halibut, plaice and mackerel.

The most important fish species of which fish fillets are imported are salmon, cod, Alaska Pollock and plaice. Dried fish are mainly different cod species and smoked fish are salmon. The most important preserved and prepared fish products are frozen fish fingers and battered and bread-crumbed fish fillets, preserved herring, canned tuna, canned mackerel and sardines, in that order.

Sales of perch are difficult to assess and are likely to be generally small in volume. The market for perch in Sweden is dominated by demand for larger fish (>200 g) and is traditionally supplied by fishermen from Sweden and Finland. A market also exists for perch roe which is obviously seasonally available.

34.4 Conclusions

There is no doubting that the market for European Perch is niche and based on traditional demand in a small number of European countries. Pressure on wild fisheries and consumer demand for sustainable and traceable fish products has initiated interest in the aquaculture production of a number of niche species including perch. The majority of new species on the European market are produced in RAS and as such the sector is limited by the high cost of investment required. To date the market strategy of farmed perch has focused on existing markets and traditional products. Supermarkets see an opportunity to source farmed product when prices are high and wild supply curtailed. This is especially true during the winter months when lake supply in North Eastern Europe is restricted by frozen lakes. In order however for any aquaculture production in RAS to be economically viable, harvesting must take place year round to ensure maximum production. In order to achieve this farmed production must for the moment compete with wild production and the associated lower prices. Whilst the existing market is able to absorb some additional tonnage, the success of the farmed sector in particular will depend on its ability to develop new markets and products. The competition between wild caught and farmed perch is one which is echoed across other species such as sea bass and turbot. With both these particular species, farmed production now exceeds wild production and to a certain extent dictates market price. Such a scenario may exist with perch in the long term however at present the marketing strategy of these farmed percid companies and their ability to compete on price and quality will ultimately dictate their success or failure.

References

Comext (2010) European Commission Trade Statistics, December 2010. http://ec.europa.eu/trade/creating-opportunities/economic-sectors/fisheries/statistics/
FAO (2010–2011) Fishstat. Fisheries and Aquaculture Department, Statistics and Information Service. FishStatJ
Finnish Game and Fisheries Research Institute (2010) Recreational fishing statistics. http://www.rktl.fi/english/statistics/fishing/recreational_fishing/
Genevalunch (2010) "Swiss consumption up 25% in 3 years" News story, GenevaLunch.com, 24 Aug 2010, Geneva, Switzerland
Globefish (2008) Research programme, vol 92 – April 2008. The seafood market in Italy
Harrison P (ed.) (2002) World agriculture towards 2015/2030, an FAO Study. Rome
Helsingin Sanomat (2008) Price of Finnish fish continues to rise sharply International Consumer Edition
ISMEA (2011) Il Pesce a Tavola: Percezioni E Stili di Consumo degli italiani maggio 2011
Rath KR (2012) Germany overseas market introduction service. First activity on German seafood market. British Embassy Berlin
Seafish (2010) Seafood export profiles –Sweden. Seafish Authority, Grimsby, United Kingdom
Swiss Federal Statistics Office (2012). http://www.bfs.admin.ch/bfs/portal/en/index/infothek/onlinedb/stattab.html
The fish site (2008) World Seafood Market to Exceed US$388.9 Billion by 2010, march 2008. http://www.thefishsite.com/fishnews/6407/world-seafood-market-to-exceed-us3889-billion-by-2010
Watson L (2003) Seahorses to sea urchins, the next big splash in Irish aquaculture. BIM, Dublin, Ireland 68 pp

Chapter 35
Commercial Production: Factors for Success and Limitations in European Percid Fish Culture

Julia L. Overton, Damien Toner, Tomáš Policar, and Dariusz Kucharczyk

Abstract Percid farming is still in its infancy; however there are already a handful of commercial ventures successfully producing percids. The underlying factors for success for these companies include the introduction of technology that allows a more intensified production that is independent of season, such as RAS systems and the development of out of season spawning. General know how in these companies and research institutions is also accumulating over time. Continued investment in "learning" has resulted in some companies being able to break the barrier of pilot production and move into commercial production. There are still limitations for further upscaling of production. For established companies these include, domestication, stabilizing and streamlining production, slow growth in larger fish and nutrition. New enterprises find it difficult to find financial backing when there is a lack of general information on Percid markets; basing their sales projections on local traditional market prices. Lack of veterinary knowledge of percids and a working knowledge of RAS systems are also limiting in some countries. Finally, there are recommendations for future development necessary for improving production.

Keywords Eurasian perch • Pikeperch • Production system • Investment • Market

J.L. Overton (✉)
Aquapri Innovation, Lergårdvej 2, 6040 Egtved, Denmark
e-mail: julia.overton@aquapri.dk

D. Toner
Aquaculture division, Bord Iascaigh Mhara, Crofton rd, Dundalk, Dun Laoighaire, Ireland
e-mail: toner@bim.ie

T. Policar
Faculty of Fisheries and Protection of Waters (FFPW), Bohemian Research Centre of Aquaculture and Biodiversity of Hydrocenoses, University of South Bohemia in České Budějovice, Zátiší 728/II, 389 25 Vodňany, Czech Republic
e-mail: policar@frov.jcu.cz

D. Kucharczyk
Department of Lakes and River Fisheries, Warmia and Mazury University in Olsztyn, Al. Warszawawska 117A, 10-957 Olsztyn, Poland
e-mail: Darekk56@gmail.com

P. Kestemont et al. (eds.), *Biology and Culture of Percid Fishes*,
DOI 10.1007/978-94-017-7227-3_35

35.1 Introduction

Percid culture is still in its infancy, where the level of production development is farm specific. Limitations for further expansion and/or the criteria for success are therefore also site dependent. The challenges that affect operators in the infancy of their production (e.g. implementing husbandry protocols, reducing losses due to deformities and cannibalism, establishing controlled breeding and water quality management) are very often different to those companies that have an established production. There may still remain some bottlenecks in scaling up their production, such as market development, development of broodstock management and domestication, or financial issues in the currently depressed financial climate. The perceived success criteria and current limitations described in this section are summarised from information gathered from individuals assisting the industry or directly involved with percid farming.

35.2 Successes

35.2.1 Technology Driven Production

As described in Chap. 32, there are different forms of production, from outdoor traditional ponds mainly producing percids as bycatch, to intensive, highly recirculating indoor production. In Europe, the traditional production of percids has been extensive and in ponds. In USA percids have been largely produced for restocking, stock enhancement or replenishment of the natural fishery. Eurasian perch and pikeperch have been stocked in Europe as a byproduct or have been regarded as a method of biocontrol (feeding off smaller fish species in ponds) and secondary species to e.g. carps. What is common for all types of production is that in order to rationalize and at the same time increase commercial production, pikeperch and perch farming has been transformed from traditional extensive production in ponds to more controlled and intensified production in indoors facilities with full water treatment and recirculation; if not for all of the production cycle then for parts of it.

With a shift of focus for percids as a farmed species, traditional large earthen ponds do not lend themselves to intensifying production. Percids are highly carnivorous and require nutritional supplementation with a high protein diet if they are to be farmed in more intensive conditions. In addition to this, many locations chosen for percid culture have less than optimum ambient water temperatures required for good growth. The location of the farms is usually determined by the natural distribution of percids; within the northern hemisphere in temperate or semi temperate environments. In these instances production time can be as much as 3–5 years for both Eurasian perch and pikeperch production (Adámek and Kouřil 2000; Adámek et al. 2012; Policar et al. 2009, 2011). The use of controlled temperature conditions in indoor facilities has drastically shortened the production time (12 months for market sized Eurasian perch and 13–15 months for market sized pikeperch). Problems of "Winter kill" that are experienced by some farmers are avoided in indoor facilities.

In addition to this, production can be better rationalised with indoor recirculated systems (as the standing biomass is known continuously throughout the cycle) and harvests can be forecasted well in advance, allowing the possibility to secure a decent market price. These basic improvements in production make an indoor facility more appealing.

35.2.2 Out of Season Spawning

The spawning season for percids is controlled by water temperature and/or light profiles (as described in Chap. 3). Percids are single spawners and their natural spawning time is already well synchronized, where spawning can be completed within a matter of weeks. Traditionally in ponds there is a single spawning event per year, following the natural breeding cycle. This single spawning effect is seen as a limitation. Firstly, all production would occur at the same time resulting in a glut of fish on the market competing with the wild fishery. Secondly, development of production techniques and relevant experience are retarded. Efforts have been made to expand this window of spawning in natural systems by exposing females to hormonal intervention and/or forced temperature photoperiod changes by removing some fish from ponds and artificially heating the water earlier than is experienced in nature. This can allow for multiple spawning events a year, e.g. one in March/April and a second in August, thereby leading to efficient utilization of hatchery facilities. This protocol has been used in the Czech Republic and in Germany (e.g. Müller-Beleke and Zienert 2008). In indoor facilities, spawning time can be completely controlled and several spawnings a year can be achieved. Where several breeding periods have been achieved, there has followed acceleration in experience and development of appropriate husbandry techniques early on in the investment phase, leading to the successful production of juveniles and market sized fish.

Controlled reproduction, using hormonal intervention, is also perceived as a way of improving the synchronicity of wild breeders, and to make use of the phytoplankton community for pond reared juveniles as used in traditional systems. However, for several farms, use of hormones is not preferred, mainly due to the current legislative rules for use of hormones in farmed fish for consumption. In addition to this, the amount of handling stress for the fish under hormone treatment may affect the quality of the gametes produced. It is well known that wild breeders that are exposed to handling tend not to survive after the spawning season and cannot be reused.

35.2.3 Continued Investment

Examples of successful percid production have been based on continued investment during the "learning phase". In most cases, the successful percid farms have been established by large companies that have their main income in another service and/

or production that have branched out into percid farming as a form of diversification in production. These companies have also rationalized and accepted the long payback time on investment.

35.2.4 Knowhow

The knowledge of production has reached a critical mass where it can be the driving force to look at production of these species more seriously. In Europe, two EU CRAFT projects (Percatech and Luciopercimprove) with a main goal to secure the production of juveniles for both Eurasian perch and pikeperch succeeded in creating a network for knowledge transfer between participating companies (from Ireland, Denmark, The Netherlands, France, Poland and Czech Republic) and researchers that is still viable today. Similar networking efforts are being made by the newly established European Percid Fish Culture (EFPC) forum.

In both the US and in Germany there are practical training courses for the production of Walleye and Pikeperch respectively.

In Ireland, through support from state funded BIM, production of perch increased relatively quickly by taking the basic production knowledge created at a pilot farm and replicating it to further units that could then start producing right away. Through technology transfer such knowledge has been to the benefit of companies across Europe. It is symptomatic in the history of development of new species that it is quite often the farms created further down the road that succeed, having learnt from the mistakes of the pioneers. In Denmark, national funding supported a local university (Danish Technical University) in collaboration with a local fish farming company to improve the performance of pikeperch production in intensive recirculated systems (Steenfeldt et al. 2010).

35.3 Limitations

Potential for RAS based production is large and there is talk of many new RAS farms producing up to 500 tons. As yet, few are reality. There are still some perceived challenges in production that need to be addressed for the industry to expand and develop.

35.3.1 Domestication

Domestication of percids is at different levels depending on the individual farms level of development. However in general the level of domestication in Europe is around level four, i.e. "*where the entire life cycle is closed in captivity without wild inputs, but no selective breeding program is used*" (as defined by Teletchea and Fontaine 2012). As yet both pikeperch and Eurasian perch broodstock material has not been through a process of active selection. Moreover, the most important traits for selection have still to be identified. Up to now, a relaxed form for selection has been performed where potential candidates from the first generations of fish that have survived captivity have been selected for by the farming system itself. It is now agreed among farmers that a more proactive form for selection is required. Active selection in other fish species (such as Atlantic salmon and to a lesser degree European seabass) has shown an improvement of growth rates (around 113 % over just five generations for salmon, Gjedrem and Baranski 2009). Another issue with percids is that the potential for inbreeding is relatively high, as females are extremely fecund and there is a temptation to use few females to fill a hatchery over a short period of time. Although it is accepted that some kind of breeding program for these species could be useful, it is extremely costly to maintain a breeding station in indoor facilities, and in most cases outside the financial scope of an individual farmer. In USA, a domestication programme is underway for yellow perch (described by Rosauer et al. 2008). For restocking/stock enhancement/replenishment programs, selection is based on the best genetic representation of the standing stock of fish they are replenishing and selection is not an aim. Such a broodstock program for percids in Europe would no doubt be of benefit to the industry but requires a production scale not yet achieved. A successful broodstock program may require transnational funding and be based at an independent facility for the benefit of all commercial companies.

35.3.2 Stability of Production

As stated earlier, production of both perch and pikeperch in Europe is small and scattered, with only very few farms producing over 100 tons per year. In the low production farms, which are mainly traditional farms, production is unknown until harvest and can vary from year to year. The main bottleneck in instability is in production of juveniles. This can be due to poor environmental conditions (e.g. cold spring or late summer for outdoor systems) or bad egg/larval quality or a combination of factors. For those establishing a production there is a tendency in their management plans to budget first with growout and buy in juveniles, mainly to avoid the complexity of rearing juveniles in the startup phase. However, this results in these farms being vulnerable to both the price change for juveniles (one of the higher expenditures in production) and a lack of juveniles available on the market during the time when they need fish. Also for the juvenile producer, there is instability of sale as a result of on growers assessing the juveniles available from the local fishery (which are cheaper). This decoupling of production of juveniles and demand of fingerlings for growout needs to be solved. In many cases farms that have a larger established production have been forced to produce juveniles themselves due to a lack of juveniles available on the market. Indeed the majority of the larger successful percid farms in Switzerland, Denmark and Ireland all have vertical integration of juvenile production, on growing and processing ability. Such integration removes the reliance on external third party operators and gives the companies more control of the entire process.

35.3.2.1 Slow Growth

There is still much room for improvement with respects to growth rates (as described in Chaps. 12, 13, 14, 15, 16, 17, 18, and 19) particularly in the later stages of growth. Efforts to address this genetically are being considered, such as, all female production, genetic selection, hybridization and triploidy (as described by Rougeot and Melard 2008, and Chap. 23 of this book). As yet, the only example of this type of genetic advances is with the successful hybridization of Walleye and Sauger in the U.S. and Canada (this hybrid being known as the saugeye). This hybridization of walleye eggs with sauger sperm produces a hybrid which is a superior production fish to both pure walleye and saugere (Malison et al. 1990). Growth rates are faster and it is more robust to suboptimal water quality conditions. In general, the markets are already extremely sensitive so various possibilities for advances in production by genetic manipulation must be rationalized by the markets acceptance for these products. Percid producers can to an extent minimize the effects of slow growth by producing multiple spawnings of large quantities of larvae and selectively culling slow growers. Whilst not the most efficient method, it is none the less the most economical approach at the present time.

35.3.2.2 Nutrition

Live food is still superior for both broodstock and larvae, which is not possible to produce in indoor facilities. However, there are many commercial diets for other fish species that have most of the dietary needs, particularly for ongrowing (see Chap. 20).

Larval and juvenile production in pond systems is reliant on the natural food available. With careful control of plankton composition it may be possible to allow a longer period of potential juvenile production. This is currently still in the testing phase. It is certainly not possible to produce fingerlings during the winter months in natural systems, due to lack of available food.

Deformities in fingerlings are usually associated with larval rearing in RAS systems. Inadequate nutrition both at the broodstock and larval stages (see Chap. 20) and husbandry practice (see Chap. 9) can result in high percentages of deformity. Traditional production usually results in fewer but high quality fingerlings. A limiting factor for most percid growers is the lack of proprietary diets for pikeperch and perch. A range of diets designed for seabass, seabream, cod, catfish and turbot are used by the sector. Such diets may include most of the required constituents for acceptable percid growth but also could be limiting performance. Whilst some work has been done on assessing nutritional requirements for percids (see Chaps. 20 and 22) it is unlikely that commercial proprietary diets will be developed until the scale of the industry is commercially attractive to feed companies.

35.3.2.3 Finance

Percid species lend themselves to recirculation technology in that they are, at the moment, a high value species, which is necessary to cover the costs of current recirculation technology. However there is still some effort needed to reduce the cost of investment and maintenance of recirculation technology. RAS systems also require a substantial crop to be produced to pay off operating costs. It is suggested that a minimum production of 300 tons Eurasian perch or pikeperch (most development plans are for 500 tons pikeperch RAS systems) is necessary to cover the costs of investment in a reasonable payback time.

Establishing the production of percids in RAS is a large initial capital investment and a continued investment for operational costs until stabilization of production is in effect. Experience with small enterprises has shown early bankruptcy due to lack of financial support when it is needed at the start. Most investors are not willing to pay the "learning fees" at the start of production development.

Financial support through local, municipal and national funding schemes to support in initial building phase and through networking with other producers and local research institutes, has boosted the interest in Percid farming (e.g. Kupren et al. 2010; Turkowski et al. 2010). Production at the end of the day should not be solely dependent on funding. Where this has happened, production has only been possible for 2–3 years and then development stops.

35.3.2.4 Markets

Percids are still a niche product and the unit price is bound to the auction price for capture fisheries (Watson 2008; Dil 2008). There is a lot of tradition surrounding the products and small markets are based where the fish is traditionally caught, which in some cases can be seasonal. These markets are extremely sensitive to saturation with locally caught product, and the low prices that ensue. It is natural for these markets to demand that the product comes from the local area whether it is fished or farmed. This puts a limitation on where the fish can be produced, how much can be produced, and that the fish resembles the locally caught product. However, it has been experienced recently with other newer aquaculture species, e.g. turbot, that as more products enters the market there are new side markets being created. The market size for pikeperch in Germany is around 800–1200 g whole fish. However there is currently a growing demand for smaller fish of 500–700 g. This cuts down the production time by 3–4 months for market sized fish. In France larger pikeperch are more desired (2 kg plus) which requires a longer production time. These factors have to be included when making a business plan.

There is currently a lack of good quality data on percid catches and market prices, thus there is little data to base a business model on production needs (Watson 2008). Understanding the market needs is crucial. In most cases new farms are being rationalized around local traditional market demands. However, export markets are vital for long term survival.

Success for established percid producers has been based on a thorough knowledge of the market and an already well-established market chain for fish products that could be used to channel their farmed product on to the market already early on in the production development phase. Reliability of product delivery is important to create a stable income. It is therefore hard for a novice fish producer to have stability in their production early on so that payback on investment is within a reasonable period of time. Fish and fish processing markets in central and Eastern Europe are still underdeveloped (Turkowski and Lirski 2010).

35.3.2.5 Veterinary Control

There is still a lack of veterinary surgeons that have a working knowledge of fish diseases in general. Diseases of pikeperch are even less well known (Chap. 31). Therefore it is up to the farmer to learn what the symptoms are for percids and be able to interpret them for an effective treatment. Pikeperch (AquaPri pers. com) and walleye (Barton and Barry 2011) have a reputation for being particularly sensitive to bacterial loading and stress. Thus handling regimes during e.g. grading or transport have to be well thought out. Also, finding the right therapeutics for treatment (particularly those that are not also detrimental to the biofilter system in RAS systems) is currently empirically tested on site.

35.3.2.6 Knowhow on RAS Systems

Expertise in biological factors associated with production is well established within the university system; however, RAS still requires expertise, also with designing a new system. Carp protocols have already been successfully used for development of percid farming in Eastern Europe. However the move over to more intensive systems is currently limited by the local knowledge of large scale water treatment systems for fish culture.

35.3.3 Future Developments for Percids

Below is a list of topics requiring further development for improving the performances and profitability of percid production.

1. Suitable diets for broodstock, fry stages. Broodstock diets will influence the egg quality and therefore the larval robustness and viability. Removal of need

for live feeds would be a major step forward. Bioavailability of nutrients in fry/weaning diets. Deformities occur even when diets contain adequate levels of nutrients required for normal growth and development of larval and juvenile fish.
2. Good therapeutics that can be used in RAS systems.
3. More formalized domestication and selection programs for percids. National, regional or international (centralized) programs or a combination of these.
4. Development of all female production.
5. Triploids and hybrid productions producing sterile and/or faster growing fish. Will future markets accept these products? Reduce costs of production.
6. Understanding of recirculation technology and development for further intensifying production and introducing greener technologies (energy saving).
7. Market development and expansion local – supermarket chains.
8. Development of a market price that is not directly coupled to capture fisheries.
9. Product development to include value added products.
10. Policy and infrastructure at local and national level to allow development into fish farming.

Acknowledgements This chapter was prepared in part thanks to the following projects: CENAKVA (CZ.1.05/2.1.00/01.0024), CENAKVA II (LO1205), NAZV (QI 101C033).

References

Adámek Z, Kouřil J (2000) A long aquaculture tradition in the Czech Republic. Aquac Eur 25:20–23

Adámek Z, Linhart O, Kratochvíl M, Flajšhans M, Randák T, Policar T, Masojídek J, Kozák P (2012) Aquaculture in the Czech Republic 2012: modern European prosperous sector based on thousand-year history of pond culture. Aquac Eur 37:5–17

Barton BA, Barry TP (2011) Reproduction and environmental biology. In: Barton BA (ed) Biology, management and culture of walleye and sauger. American Fisheries Society, Bethesda, MD pp 199–223

Dil H (2008) The European market of the pikeperch for human consumption. In: Percid fish culture from research to production, Namur (Belgium), 23–24 Jan 2008. Abstracts and short communications, Presses Universitaires de Namur, Belgium, pp 15–16

Gjedrem T, Baranski M (2009) Selective breeding in aquaculture: an introduction. Springer, Dordrecht, 221 pp

Kupren K, Czarkowski TK, Turkowski A, Hakuć-Błażowska (2010) European financial instruments for the implementation of fishing and production of water resources. In: Famielec j, Kożuch (eds) The development of environmental policy in the European Union and Poland. Krakow University of Economics, Krakow, pp 297–314, in Polish

Malison JA, Kayes T, Held JA, Amundson CH (1990) Comparative survival, growth, and reproductive development of juvenile walleye and sauger and their hybrids reared under intensive culture conditions. Prog Fish Cult 52:73–82

Müller- Beleke A, Zienert S (2008) Out-of-season spawning of pike perch (*Sander lucioperca* L.) without the need for hormonal treatments. Aquac Res 39:1279–1285

Policar T, Stejskal V, Bláha M, Alavi SMH, Kouřil J (2009) Technology of intensive culture of Eurasian perch (*Perca fluviatilis* L.). Practical handbook FFWP USB 89, pp 58 (in Czech)

Policar T, Bláha M, Křišťan J, Stejska V (2011) High quality and stable production of advanced juveniles in pikeperch (*Sander leucioperca* L.) juveniles. Aquac Int. doi:10.1007/s10499-012-9563-z

Rosauer D, Grzybowski M, Binkowski F, Shepherd B, Goetz F (2008) Developing genetically defined yellow perch (*Perca flavescens*) broodstocks. In: Percid fish culture from research to production, Namur (Belgium), 23–24 Jan 2008. Abstracts and short communications, Presses Universitaires de Namur, Belgium, pp 128–129

Rougeot C, Mélard C (2008) Genetic improvement of growth in perch production: domestication, sex control, hybridization and strain selection. In: Percid fish culture from research to production, Namur (Belgium), 23–24 Jan 2008. Abstracts and short communications, Presses Universitaires de Namur, France, pp 35–39

Steenfeldt S, Vestergaard M, Overton J, Lund I, Paulsen H, Larsen V, Henriksen N (2010) Videreudvikling af intensivt opdræt af sandart i Danmark. DTU-Aqua rapport nr. 228-2010. Institut for Akvatiske Ressourcer, Danmarks Tekniske Universitet, Hirtsals, 96p. In Danish

Teletchea F, Fontaine P (2012) Levels of domestication in fish: implications for the sustainable future of aquaculture. Fish Fish. doi:10.1111/faf.12006

Turkowski K, Lirski A (2010) Utility and market value of earthen fish ponds in Poland. In: Cieśla, Lirski (eds) Multifunctionality in pond aquaculture in Poland. Current status. Warsaw University of Life Sciences. Editorial House Wieś Jutra, Warsaw, pp 26–39

Turkowski K, Hakuć-Błażowska A, Kupren K, Czarkowski K (2010) European financial instruments to support environmental functions of fisheries management in Natura 2000. In: Famielec J, Kożuch (eds) The development of environmental policy in the European Union and Poland. Krakow University of Economics, Krakow, pp 297–314, in Polish

Watson L (2008) The European market for perch (*Perca fluviatilis*). In: Percid fish culture from research to production, Namur (Belgium), 23–24 Jan 2008. Abstracts and short communications, Presses Universitaires de Namur, Belgium, pp 10–14

Index

A
Abnormalities. *See also* Deformities
  ocular
    enophthalmia, 461
    exophthalmia, 461
  skeletal
    cranial, 331
    jaw, 209, 331, 810, 811
    lordosis, spinal column, 209, 210, 287, 331
    operculum, 331
    scoliosis, spinal column, 209, 287, 331
Aeromonas
  A.hydrophila, 424, 781, 803–805, 810
  A.sobria, 419, 775, 805
  A.veronii, 424, 803
Algae, 67, 268, 269, 272, 276, 281, 286, 296, 297, 306, 325, 470, 472, 475, 487, 489, 491, 492, 494, 552, 774
Alkalinity, 268
Allostatic load, 736–737, 757
Ancestral percids, 11
Androgens, 106, 107, 125, 126, 511, 526
Anthropogenic effects, 42–46
Artemia
  A.cyst, 296, 554
  A.nauplii, 256, 257, 276–277, 281, 282, 289, 290, 325, 552–558, 560, 582

B
Bacterial diseases
  aeromonas sobria, 419, 775, 805
  flavobacterium columnare, 331, 459, 804
  flavobacterium psychrophilum, 804
BDNF. *See* Brain-derived neurotrophic factor (BDNF)
Behavior
  clinging, 281, 286–288, 322, 323, 325
  feeding, 65, 285, 286, 601, 746
  phototactic response to light, 322
  shoaling, 272, 285, 403
β-glucans, 783
β-hydroxy-β-methylbutyrate, 788
Bioenergetic models
  bias, 371, 384, 386–392
  corrections, 386, 388, 390, 392
Bioenergetics
  alewife, 382
  applications, 371, 373–383, 388, 390, 393
  consumption, 370–393
  ecosystem, 370, 376–382
  models, 356, 369–393, 479, 494, 506
  nutrients, 376, 381–382
  perch, 374–379, 381, 382, 384–391, 393
  predator-prey, 376, 380, 390
  temperature, 371–380, 384–387, 389, 391–393
  walleye, 374–381, 383, 385–393
Bioenergy budget
  commercial applications, 354, 371
  energy intake, 354, 371–372
  equations, 354, 370, 372
  feed allowance, 354, 361–362
  predictions, 384, 388, 391, 392
Biogeography
  Eurasian percids, 7, 817–838
  North American percids, 7, 240, 521, 554, 643–680
  World percids, 817

P. Kestemont et al. (eds.), *Biology and Culture of Percid Fishes*,
DOI 10.1007/978-94-017-7227-3

Biomass
optimal harvest, 471, 477, 883
optimal stocking, 432
Brain-derived neurotrophic factor (BDNF), 728, 738
Broodstock
captive, 124, 463–464
domesticated, 216
genetically-defined, 691–696
selection, 505, 513, 525
wild, 112, 463, 464, 701–702, 801, 809, 834, 836
Broodstock maturation
management, 103–117, 209, 215–217, 882
nutrition, 216, 539–560
temperature and photoperiod, 109, 111–114, 215–216, 542, 544

C

Cannibalism
type I, 282, 308, 490
type II, 282, 308, 490
Carnivorous species, 240, 251, 255, 542, 603, 836
Chaoborus, 420
Chilling process, 626
Chironomus, 420
Chorion, 153, 194, 195, 205, 212, 551
Chromosome manipulation, 629
Competition, 44, 69, 71, 72, 77, 81, 83, 283, 400, 405, 411, 475, 477, 506, 520, 522, 601, 602, 744, 782, 879
Confinement, 410, 440, 453, 730, 731, 744, 747–751, 756, 757, 803
Confinement, chronic
effects on stress physiology, 749
immune response to, 749–751
influence of domestication, 748–751, 757
Consumption, 44, 64–66, 68, 69, 79, 80, 84, 94, 103, 268, 300, 330–332, 343, 344, 370, 371, 373–393, 402, 407, 423–425, 443, 445, 475, 477, 483, 485, 490, 506–509, 511, 517, 521, 525, 557, 588, 589, 599, 600, 626, 680, 820, 827, 846, 854–856, 858, 867–869, 875–877, 883
Contaminants
PCB, 382–383, 393, 855, 858
walleye, 331, 376, 383, 393, 474–475, 506, 554, 858
Corticosteroid receptors, 727, 784, 789
Corticosteroids, 725–738, 765, 785–787, 789
Cortisol, 116, 302, 410, 453, 569, 726, 728–738, 746, 747, 749, 751, 752, 765–767, 770, 772, 774, 784–788
Culture
extensive, 423, 431, 766, 854
intensive, 272, 313–332, 421–431, 437–464, 490, 599, 635, 637, 746, 775, 827, 830
semi-intensive, 276–289, 418, 420–421, 423, 431

D

Day length (DL), 113, 114, 284–287, 362, 363, 542, 556, 595, 596, 600, 616
Deformities. *See also* Abnormalities
cause of, 331, 462
Density
at harvest in water reuse aquaculture systems, 447
in larviculture, 296, 300, 322, 330, 332
in ongrowing, 419, 429, 430, 447–449, 452–456, 588, 599, 601–602, 615
optimal, 276–280, 290, 302, 447, 483, 494, 518
Developmental abnormalities, 209–212, 331
Developmental stages, 64, 107, 205, 218, 599, 774–776, 779, 780
Digestion, 84, 246, 248, 249, 251, 252, 254, 283, 479, 570, 708
Digestive system, 201, 239–260, 283, 556, 557
Disease, kinds and treatment of
bacterial gill disease (BGD) *(Flavobacterium branchiophilum),* 331, 460
columnaris disease *(Flavobacterium columnare),* 331, 440, 459–460, 804
Great Lakes strain subtype VHSV-Ivb, 457
lymphocystis, 456, 458–459, 802, 803
protozoan
Chilodonella, 460
Ichthyoptherius multifilis (Ich), 460, 807
Trichodina, 460, 807, 808
VHS, 456–458, 474, 696, 800, 801
VHS-susceptible species, 457
walleye dermal sarcoma (WDS), 456, 459
Disease, treatment of
Chloramine-T, 331, 460
diquat dibromide, 331, 459–460
formalin, 320, 461, 807, 810
hydrogen peroxide, 320, 460
salt, 810

Disturbance, 363, 364, 423, 426, 427, 432, 448, 637, 752, 753, 757
DNA, 5, 6, 9–12, 15, 17–19, 23–25, 27, 28, 31–36, 38, 40–44, 46, 48–50, 375, 630, 646, 649, 650, 652, 654, 656–665, 668, 672, 674–676, 693, 712, 718, 727
Domesticated fish
  behaviour of, 746, 747
  growth of, 599, 746
  stress responsiveness of, 754–756
Domestication
  generalitites about, 745
  genetic mechanisms of, 744
  reversibility of, 745
Douxfils, Jessica, 410, 725–738, 743–757, 761–789

E

Early life stages, 78–80, 239–260, 265–290, 295–309, 348, 445–446, 515, 539–560, 603, 609, 810, 811
Egg
  ribbon, 124, 129, 148, 151–153, 195, 212, 272, 579
  strand, 407, 408, 651
Egg incubation
  catch tank, 319
  hatching success
    effect of temperature, 321
  incubation interval
    effect of temperature, 318
  incubation jars, 319, 321, 328
  low flow incubator, 319
  stocking density, 320, 322, 329, 330, 332
  water exchange rate, 320
Egg treatments
  formalin, 320, 810
  hydrogen peroxide, 320
  iodophor, 320, 458, 474, 801
Embryonic development
  temperature, 198, 202–203
Embryonic organogenesis, 198, 201, 203–205
Embryonic stage
  eleutheroembryo, 314
  eye-up, 318, 321
  organogenesis, 198, 201, 203–205, 314, 332, 550
  sensitive stage, 198
Embryos incubation, 232
Emersion, repeated
  effects on stress physiology, 751
  immune response to, 751, 756
  influence of domestication, 751
Energy requirement
  culture conditions, 360, 363–365
  hierarchy, 363
  metabolic rate, 355, 357, 364
  seasonal, 362–363
Eurasian perch, 62, 105, 126, 164, 195, 228, 240, 265, 309, 347, 354, 374, 400, 422, 488, 501, 539, 579, 588, 625, 635, 701, 726, 746, 762, 800, 818, 846, 865
European Percid Fish Culture (EFPC) forum, 884
Evolution, 3–50, 77, 429, 500, 505, 515, 645, 660, 676, 679, 728, 744, 745
Extender, 164, 180–187, 630, 636

F

Fertilization
  inorganic, 267–271, 470, 472, 480, 481, 835
  organic, 179, 267–268, 270, 271, 470–472, 480, 481, 835
Fingerlings, 266–275, 290, 295, 320, 322, 330, 332, 409, 420, 438–456, 459–462, 470, 471, 473, 475–484, 490, 591–594, 600, 637, 678, 716, 737, 775, 778, 782, 831, 886
  size at time of pond harvest, 440
  size for stocking, 440
Fungal diseases
  Saprolegnia parasitica, 810

G

Gape, 36, 67, 79–80, 209, 240, 269, 281, 282, 297, 308, 402, 403, 482, 486, 488–490, 509, 515, 552
Gas bladder inflation
  gallbladder, 317
  interval for, 316, 317
  pyloric sphincter, 305, 317, 326
  surfactant, 317, 326
Gastrulation, 198–203, 314, 332
Genetic
  characterization, 674, 692–694, 754
  selection, 343, 599, 692, 700, 705, 710, 886
Genetic diversity, loss of, 756
Geosmin, 854

Glucose, 170, 172, 176, 181–186, 344, 559, 731, 735, 736, 749, 751–753, 766, 767, 770, 772
Gonadogenesis, 105, 113, 566, 626
Gonadosomatic index (GSI), 106, 502, 503, 543, 573, 771, 850
Grading. *See also* Size-sorting
  benefits, 445
  frequency, 432
  vertical-bar grader, 445
Growth
  asymptotic length, 500–502, 505, 513, 515
  bioenergetic growth models, 504
  biphasic growth model (BGM), 500–504, 524
  countergradient growth, 513, 514
  density-dependent, 69, 71, 83, 285, 477, 500, 512, 517–520, 525, 601
  effect of early maturity, 71, 505
  effect of food availability, 104, 285, 376, 377, 520–521, 525
  effects of parasitism and disease, 500, 521–525
  extrinsic factors, 283, 500, 511–524
  intrinsic factors, 500, 504–511
  lifetime growth, 499–526
  sexual dimorphic growth, 427, 506–511, 599, 600
  slow growth, 65, 285, 289, 489, 599, 678, 701, 886
  temperature dependence of, 510, 511
  variability of, 500, 518, 523
  von Bertalanffy model, 500–503
Growth rate
  effect of fish length, 490
  effect of latitude, 104, 513, 525
  effect of sexual maturation, 343, 423, 541, 600
  effect of temperature, 449, 462, 510, 511, 513, 514, 525, 589, 600, 601, 678, 679, 775
  during habituation, 445, 446, 448, 449, 555
  during larviculture, 332
  during ongrowing, 420–426, 428–430, 438, 445–451, 455, 462, 589, 591, 594, 595, 597–603, 608, 611, 612, 614
GSI. *See* Gonadosomatic index (GSI)
Gymnocephalus cernuus, 26, 126, 199, 506
Gynogenesis, 630–632
Gynogens, 625, 630–632

H

Habituation
  diets, 441–445, 743
  effect of temperature, 442
  feeding frequency, 443
  feeding rates, 442–443, 591
  final density, 448–449, 455
  final size, 440–441, 445
  initial density, 448–449
  initial size, 440
Habituation to stress, 737
Handling, 46, 64, 68, 86, 87, 116, 124, 146, 164, 187, 217, 276, 297, 307, 317, 330, 357, 363, 364, 400, 432, 440, 441, 452–456, 458, 459, 462, 708, 727, 730–732, 734, 737, 749, 757, 805, 811, 883, 888. *See also* Size-sorting
Harvest
  effect of temperature
  of fish from tanks
  of pond-reared fingerlings, 440, 441
Hatching, 42, 75, 76, 78, 83, 84, 105, 124, 130, 143, 152–154, 184, 198, 199, 201, 205–206, 208–209, 212, 214, 215, 229–235, 240–242, 244–246, 249, 251, 252, 259, 272, 276, 281–286, 288, 289, 296, 306, 309, 317–319, 321, 408, 474, 482, 490, 541–545, 549–552, 554–559, 566–567, 572, 577, 578, 581, 582, 629, 632, 636, 637, 651, 670, 728, 729, 776, 777, 779, 780, 834, 835
Heritability, 703–707, 710, 714, 716, 747
Heterogeneity, 31, 107, 130, 131, 272, 282–286, 309, 409, 410, 426–430, 432, 450, 552, 588, 591, 596, 599–602, 778, 844, 845, 851, 858, 859
Homing, 409, 647, 668, 672, 678
Husbandry stressors
  contrasted effects along domestication, 747–754
Hybrid
  sauger, 636, 637
  walleye, 326, 461, 462, 469–495, 636–640
  yellow perch, 636
Hybridization, 32, 44, 636, 745, 886
Hypoxia, 66, 377, 379, 380, 424, 576, 730, 732, 735, 748, 752–754, 756, 757
Hypoxia, single or repeated
  effects on stress physiology, 752, 753
  immune response to, 752
  influence of domestication, 752

I

IGF. *See* Insulin-like growth factor (IGF)
Immune defence, 762, 768, 769, 774, 775, 782, 786

Immune gene responses, 764, 784, 789
Importers, 870, 872, 875
Inbreeding, 648, 649, 694, 699–702, 707, 710, 711, 713–717, 744, 745, 755–757, 885
Individual variation
  activity, 407
  boldness, 406
Induced maturation, 114, 115, 216, 596, 728
Insulin-like growth factor (IGF), 347, 348, 746, 749

J
Juveniles
  characteristics, 68–69, 285, 611, 615
  definition, 63, 65, 317

K
Kidney, 34, 247, 255, 458, 524, 726, 730, 763, 764, 787, 801, 857
Kin recognition, 405, 664, 670, 677
Krill
  in diets for larval fish, 441
  hydrolysate, 442, 583
  in Walleye habituation diets, 441

L
Larva, 63, 65, 68, 193–218, 241, 255, 272, 276, 281–283, 288–289, 296–308, 314–316, 322, 324–326, 328, 331, 374, 482–485, 545–550, 553, 557, 565–583, 651–652, 776–780, 835, 886
Larval development, 64, 194, 196, 198–207, 214, 232, 251, 256, 258, 490, 550–552, 558, 567, 569, 599, 780
Larval stages
  postlarva I
    exogenous feeding, 315
    gas bladder inflation (GBI), 315–317
  postlarva II, 317, 482
  prolarva
    endogenous feeding of, 314
    mouth opens, 296, 482
    sac fry, 314
    yolk sac stage, 314
Larviculture
  cultural technology
    aeration, 303
    air compressors, 329
    cannibalism, problems, 440
    clinging behavior, 286, 325
    currents and turbulence, 303
    deformities, pre-haemal lordosis, 287, 305
    disease, problems, 440
    greenwater, 276, 277, 306, 325
    light, effect of intensity, 284, 287, 306, 329, 556, 597
    light, effect of photoperiod, 342
    nitrification, 301
    non-inflation of the gas bladder (NGB), problems, 281, 322, 326
    RAS, 300
    standpipe screen, design of, 322
    standpipe screen, mesh size, 272, 299, 325
    stocking density, 320, 322, 329, 330, 332, 448
    surface skimmer, 303, 304
    surface spray, 304, 323, 326–328
    tank shape, size, and color, 448
    turbid water culture, nephelometric turbidity units (NTU), 325, 444
    water exchange, 299, 320
    weaning, 283, 289, 296, 307–308
  feeding strategy
    dietary requirements, 308
    feeding rate, 325, 330, 331
    feed, manufactured, 329, 443
    feed size, 331
  live feed
    Artemia, 296–297, 301, 302, 552, 554, 555
    bioencapsulation, 297
    emulsions, 297
    rotifers (Brachionus plicatus), 301
Light
  intensity, 284, 306, 329, 596, 597
  overhead lighting, 273, 443, 444, 447, 637
  photoperiod, 65, 113, 116, 178, 282, 342, 447–448, 595–598
  photothermal regime, 176, 178–180, 463, 542, 766, 770, 772, 773
  submerged, intank-lighting, 443
Lipopolysaccharides (LPS), 728, 733, 734, 763–764, 783–788
Live prey, 251, 254–255, 276, 281, 286, 289, 290, 301, 304, 464, 543, 544, 549, 552, 554, 560
Loading rates (kg/Lpm), 447
LPS. *See* Lipopolysaccharides (LPS)
Lux, 284, 287, 289, 306, 322–323, 329, 556, 597, 830

M
Macroinvertebrate (phytomacrofauna), 418
Market
Estonia, 866–868, 870, 871, 876, 877
Finland, 820–822, 866, 867, 869, 877, 878
France, 266, 826–828, 830, 867, 868, 871, 876–877, 884, 887
Germany, 830, 869–871, 873–874, 883, 884, 887
Italy, 825–827, 869, 876
Poland, 266, 826, 866, 867, 870, 884
Sweden, 822–823, 869, 871, 877–878
Switzerland, 418, 419, 431, 825, 826, 830, 867, 869–875, 885–886
Marketing
minimum size, 462–464
Market price, 826, 879, 883, 888, 889
Market size, 80, 266, 272, 418, 419, 422, 463, 636, 638, 692, 831, 835, 882, 883, 887
Market supply
aquaculture, 866, 868
wild landings, 866–868
Mesocosms, 276–290, 517
2-Methylisoborneol, 854
Methylmercury (MeHg), 331, 332, 855
17α-Methyltestosterone (17MT), 626, 627
Microvideography, 316
Molecular characterization, 34, 727
Monogenean diseases
Diplostomum sp., 809
Monosex, 428, 432, 601, 638
Morphotype, 403, 850
Muscle growth
genetics, 343, 344, 349
nutrition, 552
Myostatin, 347, 348
Myxospoidean diseases
Myxobolus neurophilus, 807, 808

N
Neomales (XX) males, 168, 185, 626–629
Nitrogen, 182, 268, 269, 271–272, 300, 301, 354, 373, 377, 381, 382, 471, 472, 491, 522, 570, 571, 603
Nutrients, 82, 86, 169, 187, 216, 241, 244, 245, 254, 256–259, 267–269, 297, 346, 376, 381–382, 451, 463, 471–473, 475, 477, 491, 520, 540, 550, 554, 569, 581, 582, 591, 602, 603, 605, 609, 610, 616, 675, 788, 810, 855, 889

O
Ongrowing
density at harvest, 449
density at stocking, 429, 430, 432, 448, 599, 601
diets, 438–446, 448, 451, 452, 461–464, 588, 602–616
feeders, 442, 449, 451, 570, 589, 591, 595–597, 602
feeding rates, 442–443, 446, 449–452, 589, 591, 594, 595, 603
growth rates, 420–430, 432, 438, 442, 445–451, 455, 462, 588, 589, 591, 594–604, 608, 611, 612, 614
perch, 418, 419, 424, 429–432
Walleye, 437–464, 638
Ontogenetic changes, 402
Out-of-season spawning
chill period, requirement for, 109, 114
hormonal treatments
des-Gly^{10} [D-Ala^{6}], 112
hCG, 112
LHRHa, 112
LHRH-ethylamide, 112
multifactorial regulation, 117
sexual maturity, 104–105
Ovulation, 109, 111, 112, 124, 126–128, 130, 132, 135–149, 154, 197, 215, 216, 543, 546, 548, 569, 572, 577, 579
Oxygen
ambient, 300, 449
supersaturated, 449

P
Pathogens, 45, 63, 86, 276, 288, 301, 305, 456, 522–524, 542, 702, 730, 733–734, 737, 748, 751, 762, 769, 775, 776, 782, 783, 785, 786, 789, 800, 804, 805, 807, 809, 811
Perca flavescens, 62, 126, 140, 166, 198, 203–205, 216, 217, 229, 358, 374, 400, 464, 470, 500, 501, 512, 516, 551, 570, 595, 625, 627, 631, 637, 644, 703, 726, 763, 775, 799, 803, 806, 809, 824, 846, 848, 849, 857, 871
Perca fluviatilis, 126, 165, 197, 228, 241, 276, 347, 354, 374, 400, 418, 490, 503, 540, 579, 625, 637, 726, 746, 799, 823, 846
Perch
eggs, 72, 76, 108, 134, 148–150, 152, 197, 205, 213, 216, 276, 290, 408, 543, 544, 574–577, 580, 647, 823, 825

embryos, 232, 234
Eurasian, 178–179, 393, 630
fatty acids, 543, 550–551, 565, 569, 570, 573, 575, 576
lipid metabolism, 347
neutral lipids, 197, 572, 574, 575, 577
nutrition, 281, 289, 539–560, 570, 587–616, 846–849
phospholipids, 570, 574, 576, 580
Volga, 635–636
waxes, 569, 570, 577, 580–582
Yellow, 15, 62, 105, 126, 166, 194, 229, 265, 266, 343, 358, 374, 400, 464, 470, 500, 551, 567, 595, 625, 635, 644, 692, 703, 726, 763, 800, 824, 846, 866, 885
Perch female
egg quality, 146, 149
Perch fillets, 344, 853, 869–871, 873–875
Perch production
Czech Republic, 418, 828
export, 825, 866, 870
France, 418, 431, 876–877
Ireland, 431, 824
market, 418, 419, 431, 432, 830
Netherlands, 818, 836
pond polyculture, 418
spawning out-of-season, 114, 134
Percid
culture, 456, 471, 472, 482, 491, 702, 707, 730, 817, 827, 845, 882
domestication, 103, 104, 599, 754–756, 858, 884
farming, 411, 882–885, 887, 888
fishes, 103, 124, 171, 218, 229, 240, 282, 360, 370, 500, 540, 567, 588, 625, 644, 726, 747, 761, 826, 844, 884
future development, 888–889
genetic divergence, 10–13
relationship, 5, 23, 42, 43, 659, 674
spawning groups, 75, 409
spawning season, 75, 107, 113, 116, 131, 134, 135, 146, 148, 187, 216, 409, 823
species differences, 169, 187
Percid ecological importance
geographical range, 644, 646, 648, 656, 658, 659, 665–669, 672, 679
habitats, 644, 645, 652, 653, 656, 659, 660, 673, 675–680
historical patterns, 647, 660, 663, 664, 671, 675–678, 680
Lake Erie, 644, 647, 649, 652, 653, 659, 661, 662, 665, 669–678
Percid hatcheries
walleye, 635–640
walleye–sauger hybrid, 637, 638
yellow perch, 635–637
Percid production
Europe, 500, 799, 826–828
pike-perch farming
extensive pond production, 418, 822
intensive production, 303, 827
recirculation systems, 301, 307, 811
Persistent organic pollutants (POPs), 855
Phases
I, 438, 439, 443
II, 438–440, 444–446, 448–450, 453, 455, 460
III, 438, 445, 448–452, 455, 460, 461
Phosphorus, 268, 269, 271–272, 377, 381, 382, 471–473, 491, 492, 494, 810, 848–849
Photoperiod, 65, 72–75, 105, 109, 111–116, 135–142, 144–145, 178, 215–216, 282, 342–344, 447–448, 540, 542, 544, 546, 548, 571, 588, 590, 592–593, 595, 600, 615, 645, 696, 733, 764, 765, 769, 770, 834, 846, 883
Photosynthesis, 268
Phototaxis
negative, 65, 596–597
positive, 65, 324, 596–597
Phylogenesis
diversity, 23, 31
family, 5, 6, 10
genera, 11, 17, 31
genes sequences, 5, 9–12, 31
molecular, 10, 11, 27, 29, 31, 33
nuclear DNA, 10, 11, 27, 28, 31, 32, 34
Pond, 79, 105, 124, 179, 216, 240, 266, 314, 402–403, 418, 438, 464, 470, 543, 570, 710, 766, 819, 854, 868, 883
Pond fertilization
cyanobacteria, 472, 473, 493, 494
zooplankton, 240, 268–270, 276, 470–473, 484, 486, 494, 557, 824
Pond management
percid and catfish, 470–473, 475–477, 482, 485, 487, 490–494
plankton ecology, 470, 471
stocking density, 471, 476–481, 490–491
water quality, 472, 473, 491, 494
Pond production
benthos, 470
fish harvest, 476, 481
fish survival, 470–473, 491–493

Pond production (*Cont.*)
larval fish, 470, 472, 475, 482, 483
pond ecology, 470, 471, 475, 482–494
seasonal variation, 470, 471, 476, 484–489
visual feeders, 470, 475, 482, 485, 488–491
POPs. *See* Persistent organic pollutants (POPs)
Population models, 377, 381, 526
Population structure
abundance, 644, 645, 652, 676, 678–680
behavior, 645, 651–652, 668
distribution, 644, 645, 664, 679, 680
genetic diversity
dispersal ability, 645
habitat requirement, 645
migratory behavior, 645
spawning fidelity, 645
reproduction, 645–651, 660, 670, 672, 676–679
Predation, 65, 68–69, 76, 79–81, 83–86, 274–275, 328, 344, 381, 400, 403, 404, 406, 407, 475–477, 479, 490–491, 500, 505, 507, 515–517, 520, 525, 577, 595, 651, 746, 747, 851
Predator avoidance, 231, 403–404, 506, 651, 652
Probiotic bacteria, 776–782
Production systems
extensive, 418, 431, 882
intensive, 418, 431, 882, 884, 888
ponds, 418, 431, 708
recirculated systems, 883, 884
semi-intensive, 418–420, 431
water reuse systems, 883
Progeny, 330, 428, 442, 692, 694–696, 711, 744
Proteomics
growth selection, 339, 344
protein sequencing, 259, 340, 344
Protozoan diseases
Ichthyobodo necator, 804, 805, 807
Ichthyophthirius multifiliis, 807

Q
Quality
biochemical component
amino acid, 197, 844, 848, 854
docosahexaenoic acid (DHA, 22:6n-3), 847
eicosapentaenoic (EPA), 846, 847
fatty acid profile, 846
lipid, 197, 844, 846–848, 850, 852, 859
mineral, 844, 848, 854
polyunsaturated fatty acid (PUFA), 197, 846–848
protein, 844, 846, 848, 852, 853, 855
definition, 844–845
determinism, 843–859
health component, 148–149, 194, 844, 846, 855, 856
nutritional component
food factor, 847
genetic factor, 847
sanitary safety component
cadmium, heavy metal, 855, 857
contaminants (xenobiotics), 856, 857
dioxins, oganic pollutant, 855, 856
lead, heavy metal, 855, 857
mercury, heavy metal, 855–858
methylmercury, 855
poly-chlorinated biphenyls, oganic pollutant, 855
polycyclic aromatic hydrocarbons, oganic pollutant, 855
veterinary drugs, 855
sensory component
color, 844, 851–853
flavor, 844, 851, 854
off-flavor, 854
shape, 851
texture, 844, 851, 853–854
technical component
carcass yield, 849, 850
fillet yield, 849–851
perivisceral fat, 850

R
Recirculating aquaculture systems (RAS), 103, 105, 114, 115, 215, 288–290, 300–301, 318, 418–422, 425, 426, 428, 429, 431, 542, 543, 546, 548, 632, 708, 710, 788, 810–811, 819, 820, 823–827, 830, 868, 876, 878, 884, 886–889
Recreational anglers, 69, 456, 867, 877
Reproduction
egg yolk proteins
lipovitellin, 108
phosvitins, 108
gonadotropins
follicle-stimulating hormone (FSH), 111, 125
luteinizing hormone (LH), 111, 125, 127, 134
hypothalamus-pituitary-gonad (HPG) axis, 111

oogenesis and spermatogenesis
effect of age, 104
effect of body size, 104
effect of environmental conditions, 104
ovarian steroidogenesis
17β-estradiol (E2), 108
FSH, 108
LH, 108
ovary development, stages, 107
testicle development
17,20β,21-trihydroxy-4-pregnen-3-one, 108, 125
11-ketotestosterone (11-KT), 106
maturation-inducing steroid (MIS), 108
melatonin, 111
neural hormonal regulation of spawning, 111–112
photoneuroendocrine system, 111
testicular steroidogenesis, 106–107
testosterone (T), 106, 108, 125
vitellogenesis, vitellogenin, 107
Reproduction and immune functions, 726, 774
Reproductive hormones, 769
Rotifer, 67, 240, 269–271, 276, 279, 281, 282, 296, 297, 301, 325, 484–486, 489, 551–553, 557, 582, 835
Rutilus rutilus, 400, 418, 517, 543, 765, 830

S

Salt and salinity
during ongrowing, 440
in transport, 440
Sander
S. lucioperca, 7, 9, 13, 16–21, 25, 29, 42, 44, 62, 104, 125, 142, 165, 175, 210, 213, 229, 241, 325, 342, 375, 400, 424, 490, 501, 503, 513, 520, 540, 553, 567, 625, 627, 644, 776, 831, 846–849, 851, 856, 857
S. vitreus, 17–21, 42, 44, 62, 166, 201, 202, 227, 229, 346, 374, 400, 437, 500, 501, 513, 516, 523, 566, 627, 631, 644, 673, 730, 849, 856–858
S. volgense, 9, 17–21, 134, 635, 644, 831
Sander lucioperca, 7, 9, 13, 16–21, 25, 29, 42, 44, 62, 104, 125, 142, 165, 175, 210, 213, 229, 241, 325, 342, 375, 400, 424, 490, 501, 503, 513, 520, 540, 553, 567, 625, 627, 644, 776, 831, 846–849, 851, 856, 857
Sander vitreum, 17–21, 42, 44, 62, 166, 201, 202, 227, 229, 346, 374, 400, 437, 500, 501, 513, 516, 523, 566, 627, 631, 644, 673, 730, 849, 856–858
Saprolegnia spp., 424, 805, 810
Sauger, 18, 20, 42, 44, 62, 72, 232, 313, 326, 470, 474, 488, 506, 525, 635–638, 644, 886
Saugeye, 44, 470, 471, 473–480, 482–487, 489–492, 494, 495, 886
Scale development
effect of body size and age, 85
importance, 317, 440
Seasonal variations, 362–363, 470, 471, 476, 484–489, 548, 766
Selective breeding, 129, 130, 360, 463, 700–703, 705, 707, 709, 713, 884–885
Sense organs
auditory vesicles, 234
eye, 230–233
labyrinth, 233–235
lateral line, 233–235
neuromasts, 233, 234
olfactory placodes, 228, 229
optic systems
cornea, 231, 232, 461
iris, 231, 289
lens, 231–233
pupil, 231
retina, cones, 232, 233
retina, rods, 231–233
tapetum lucidum, 232
vitreus body, 229, 232
optic vesicles, 232
otic placode, 233
otolithic organs
lagena, 234
sacculum, 233–234
semicircular canals, 233–235
utriculus, 233–234
vestibular organ, 234
taste buds, 229–231
Sensory receptors
chemoreceptors, 229, 230
electroreceptors, 228
mechanoreceptors, 227–228, 233, 234
photoreceptors, 228, 232
thermoreceptors, 227–228
Sex
chromosomes
female homogamety, XX, 625
male heterogamety, XY, 625
determination, 625
effect on growth, 601
phenotypic, 625, 632
ploidy manipulation, 625–633
ratios
female biased, 506–508

Sex (*Cont.*)
male biased, 506–508
reversal
gynogenesis, 625, 630–632
hormonal, 626–628, 632, 633
Sex control, 626–629
Sex reversal, 626–628, 632, 633
Sexual dimorphism, 131, 427, 506–511, 599, 600
Sexual size dimorphism (SSD), 506–510
SGR. *See* Specific growth rate (SGR)
Shoaling, 272, 285, 399, 403–405, 507–508, 588, 746
Size heterogeneity, 130, 272, 282–286, 309, 409, 427, 428, 490–491, 588, 591, 600, 601, 778
Size-sorting, 282, 284, 421, 426–428, 432. *See also* Grading
Social behaviour, 404–405, 427, 702, 707
Social hierarchies, 405, 409
Spawning
behaviour, 69, 70, 75–76, 107, 407–409
habitat, 70, 83, 407, 675, 679
migration, 75, 409, 647
Specific growth rate (SGR), 80, 280, 285, 289, 354, 355, 418, 422, 423, 428, 429, 442, 484, 556, 557, 588–598, 600–605, 607, 608, 610, 636, 777, 778
Spermiation, 73, 106, 107, 114, 116, 124–126, 129–135, 146, 169, 179, 544
SSD. *See* Sexual size dimorphism (SSD)
Starvation, 68, 77, 79, 80, 85, 282, 302, 315, 440, 550, 554
Steroids
estradiol, 108, 125, 541, 567
progestin, 125
testosteron, 106, 108, 125, 541, 626
Stocking
density for habituation, 266, 445, 446, 448, 455, 463, 490, 694
density for larviculture, 322, 330, 332
density for ongrowing, 289, 429, 430, 432, 448, 599, 601, 824
Stress, 45, 104, 128, 187, 214, 259, 267, 302, 317, 357, 405, 421, 440, 511, 549, 569, 596, 637, 696, 708, 725, 744, 764, 800–801, 834, 883
Subpopulations, 342, 402
Sustainable seafood, 866
Swim bladder, 23, 207, 214, 247, 282, 287, 288, 304, 569, 582, 583, 801, 802, 810

T

Tandem pond-tank, 266, 290, 329, 438
Tanks for ongrowing
color, 275, 322, 448
depth, 298
shape, 288, 307, 448
volume, 307, 322, 327, 328, 427
Temperature
lethal, 86, 449
optimum, 75, 374, 422, 432, 449, 511, 591, 600, 882
Thermophilic, 422, 588, 589
Transporting
effects of density, 453, 454
ram air ventilation (RAV), 453, 454
removal of C02, 453, 454
Triploidisation, 626, 629–630, 633
Trophic links, 376, 380–381

V

Viral diseases
haemorrhagic lesions, 800
VHSV, 86, 320, 456–458, 800, 801
Viral hemorrhagic septicemia (VHS), 86, 320, 456–458, 800, 801

W

Walleye, 18, 62, 104, 129, 164, 194, 229, 240, 269, 300, 313, 346, 374, 400, 437, 470, 500, 552, 566, 597, 625, 635, 644, 730, 762, 803, 857, 884
Water quality
ammonia, 319, 424–426, 431–432, 732
nitrite, 421, 425
oxygen, 152, 217, 268, 276, 300, 327, 328, 332, 424–426, 431–432, 443
salinity, 20, 27, 217, 424–426, 432, 733
Water reuse aquaculture systems (WRAS), 447, 462
Weaning, 209, 252–255, 260, 276, 279, 282, 283, 289, 290, 296, 305, 307–308, 409, 410, 420, 429, 555, 556, 580, 820, 830, 889

Y

Yellow perch, 15, 62, 105, 126, 166, 194, 229, 265, 266, 343, 358, 374, 400, 464, 470, 500, 551, 567, 595, 625, 635, 644, 692, 703, 726, 763, 800, 824, 846, 866, 885
  broodstocks
    Choptank River, 694, 695
    Lake Winnebago, 694, 695
    Perquimans River, 694, 695
  fillet yields, 694
  genetics of, 694
  growth performance, 694
  growth selected stocks, 696
  susceptible to VHS, 696
Yolk composition
  fatty acids, 197, 213, 577
  lipids, 77, 108, 126

MIX
Papier aus verantwortungsvollen Quellen
Paper from responsible sources
FSC® C105338

If you have any concerns about our products, you can contact us on
ProductSafety@springernature.com

In case Publisher is established outside the EU, the EU authorized representative is:
Springer Nature Customer Service Center GmbH
Europaplatz 3, 69115 Heidelberg, Germany

Printed by Libri Plureos GmbH
in Hamburg, Germany